AN INTEGRATED SYSTEM OF
CLASSIFICATION OF FLOWERING PLANTS

FRONTISPIECE. *Drimys winteri* J. R. & G. Forster, a South American member of the archaic family Winteraceae. Photograph courtesy of R. M. Schuster.

AN INTEGRATED SYSTEM OF CLASSIFICATION OF FLOWERING PLANTS

ARTHUR CRONQUIST
THE NEW YORK BOTANICAL GARDEN

COLUMBIA UNIVERSITY PRESS
NEW YORK

This book is dedicated to the many botanists,
past and present, whose studies have made it possible.

Library of Congress Cataloging-in-Publication Data

Cronquist, Arthur.
An integrated system of classification of flowering plants.

Includes bibliographical references and index.
1. Angiosperms—Classification. I. Title.
QK495.A1C76 582.13'012 80-39556
ISBN 0-231-03880-1

Columbia University Press
New York Oxford

Printed in the United States of America

c 10 9 8 7 6

CONTENTS

FOREWORD

Dr. Arthur Cronquist, who died in Provo (Utah) on March 22, 1992, at the age of seventy-three, has been for many years recognized as one of the leading evolutionary botanists. He made fundamental contributions to floristics and the phytogeography of North America—especially to the evolutionary classification of flowering plants—and influenced the growth of evolutionary botany as a whole during the past thirty years.

Arthur Cronquist was prominent among the floristicians and phylogeneticists. He was the editor and coauthor of one of the best North American floras, *Intermountain Flora: Vascular Plants of the Intermountain West, U.S.A.*, and in 1991 published a completely revised and updated edition of his and H. Gleason's *Illustrated Flora of the Northeastern United States and Adjacent Canada,* one of the most valuable regional floras of the United States. Another important book initiated by Gleason and completed by Cronquist was *The Natural Geography of Plants* (1964), the best book on North American Phytogeography.

Two of his books—*The Evolution and Classification of Flowering Plants* (1968; second edition, 1988) and *An Integrated System of Classification of Flowering Plants* (1981)—brought him worldwide fame. In the first Cronquist provided a short version of the general system of classification of flowering plants, together with an exposition of the theory underlying the system. The system of angiosperms he proposed was inspired by the works of American botanist Charles Bessey. However, as W. C. Steere, the editor of the first edition, says, "This new system reflects not only the author's own ideas and research over the years, but also his professional association and collaboration with other outstanding botanists of other countries who have turned their intellectual efforts to finding the most accurate system of higher plants." An important part of the book is an exposition of the philosophy of evolutionary classification, as well as an analysis of the evolutionary trends of characters.

Arthur Cronquist was a convinced supporter of the Darwinian philosophy of evolutionary systematics and criticized the cladistic theory developed by Willi Hennig and his followers, most notably in his brilliant article "A Botanical Critique of Cladism" (*Botanical Review,* 1987, 53: i–52) and in the second edition of his book on the evolution and classification of flowering plants. One of the main targets of his criticism was the cladistic concept of monophyly. For evolutionary taxonomists, who use the word *monophyly* in its original sense, every natural taxon is monophyletic. However, Hennig and his followers drastically changed the meaning of the word: according to cladists a monophyletic group must include all the known descendants of a common ancestor, and for groups that do not include all the descendants, they introduced a new term *paraphyletic.* This change of the traditional meaning of the word *monophyly* is unacceptable for evolutionary biologists. In agreement with Ernst Mayr's criticism of cladism, Cronquist rightly emphasizes that the cladistic concept of monophyly is destructive to our taxonomic system. According to both Mayr and Cronquist, the acceptance of this cladistic dogma would mean the destruction of many of the best-established taxa, such as prokaryotes, Reptilia, and dicotyledons. "There is something fundamentally wrong with any theory of taxonomy that does not permit the recognition of the prokaryotes as a major taxonomic group and their separation from the eukaryotes at a high taxonomic level," states Cronquist in the second edition of his *Evolution and Classification.*

The major taxonomic work of Arthur Cronquist is found in his monumental *Integrated System,* which is one of the most important books in any botanical library. In this book Cronquist used all available data from various sources, including phytochemistry and micromorphology, and it therefore justifies what he calls its "ambitious title." It is a well-documented and detailed exposition of his system of classification of flowering plants. The book contains very accurate and updated descriptions of the taxa and synoptical arrangements of the families within the orders, the orders within the subclasses, and the subclasses within the classes, as well as carefully selected references. *An Integrated System* is a magnum opus of one of the most outstanding botanists of this century.

Armen Takhtajan
June 1992

INTRODUCTION

The system of classification of flowering plants here presented is in the philosophical tradition of A. P. de Candolle, Bentham and Hooker, Hallier, and Bessey. It follows what is known as the strobilar theory of angiosperm evolution, as do most of the systems that have been presented in the past several decades.

I have tried to use all the available data, from whatever source; hence the perhaps overly ambitious title, an integrated system. At the same time, I have tried to avoid undue optimism about the present importance of data that reflect a very sparse sampling. I can hope that the classical morphological features are presented with reasonable accuracy in the following descriptions, but surprises will inevitably result from continuing expansion of the data-base for chemical and micromorphological features.

The fossil record has only recently begun to be useful to students of the evolution and classification of angiosperms. It is still not possible to put primary reliance on the fossils, but the record does begin to impose some constraints on the systems that can be seriously considered. I believe that the system here presented is fully compatible with the fossil record as presently understood, and that the fossils indeed provide significant support for some of the pillars of the system.

A detailed exposition of my views on evolutionary trends in the angiosperms was presented in my earlier book, *The Evolution and Classification of Flowering Plants*. My ideas have undergone some relatively minor changes since that time, and it is my hope and intent to present a revised version in the near future. Perhaps the most significant changes are that I am now more receptive to the view that the earliest angiosperms were shrubs rather than trees, that I now regard stipules as having originated among primitive angiosperms as petiolar flanges, that I now regard the paracytic stomatotype as probably primitive within the angiosperms, and that I now consider that perigyny can be of receptacular as well as appendicular origin. Fortunately, these changes have had only minimal impact on the system itself; indeed they promote the harmony between the principles and the system.

The system here presented is conservative in that I prefer to follow historical tradition when it is reasonably possible, and in that I am reluctant to assign the rank of family to small satellite groups if these can be comfortably accommodated within a larger group. On the other hand, there is no virtue in retaining a traditional arrangement that obstructs understanding. Thus it becomes necessary to recognize a number of small families whose members have been quietly hiding in large families where they do not belong.

Perhaps the most important thing that sets the present system apart from others is that I have consistently tried to prepare synoptical arrangements of the families within the orders, the orders within the subclasses, etc. This procedure has profoundly influenced my thinking. The pervasive parallelism within the angiosperms confounds all efforts to prepare a natural key that is consistently useful for the identification of families and higher groups, but if one cannot prepare a synoptical arrangement that will provide for most members of each group, it is time to stop and reconsider. The distinction between the Rosidae and Dilleniidae is recalcitrant to such an approach, but otherwise I have found the method very useful.

The system most nearly comparable to my own is that of Armen Takhtajan. This similarity reflects a community of interest and outlook, bolstered by frequent correspondence and personal conversation over a period of more than twenty years. We believe that the similarity reflects the requirements of the present state of knowledge and taxonomic theory.

The differences that persist between our systems have several causes. He is more likely than I to recognize small families and orders, in pursuit of sharp definition of groups. He places a little more reliance on serology, and a little less on other chemical characters, than I. He is more suspicious than I of the major significance of characters that can readily be interpreted in terms of survival value. Even with access to similar herbaria and essentially the same body of literature, no two people work with exactly the same set of data, or are exposed to exactly the same influences from their colleagues. Although we agree on the main outlines of angiosperm evolution, there remain many doubtful cases and close decisions at the present state of knowledge, and any two people will inevitably make some of these decisions differently.

The mass of literature that bears on the general system of classification of flowering plants is formidable indeed. My degree of success in encompassing and understanding it will have to be judged by my colleagues. I have examined every book and paper cited in the bibliographies, as well as others not cited. Some of the items are included merely to indicate that I have seen them, even though I disagree with the conclusions presented. The bibliographies reflect a bias against citation of unpublished theses and of works more than half a century old, the latter because they have by now had time to be incorporated into the general body of taxonomic knowledge. I have also been reluctant to cite floras and encyclopedias, although I have obtained useful information from some of them. The more significant such works are well known to botanists generally. The numerous papers directed toward a generic flora of southeastern United States are included, however, because of their comprehensive bibliographies. I regret that I do not read any Oriental language.

Daily access to the herbarium and library of the New York Botanical Garden has been a tremendous asset in the preparation of this work, and the living collection has also been of great value. Botanists and botanical institutions around the world have been uniformly helpful, and I have sought out a number of otherwise not readily available publications in the library of the Komarov Botanical Institute in Leningrad and in the library of the United States Department of Agriculture in Beltsville.

Many people have helped me in one way or another during the preparation of the manuscript and while my thoughts were developing. A complete list would probably have to include nearly all of my professional colleagues, beginning with Ray J. Davis, from whom I took my first botany course. My file of reprints continues to grow. I can not now recall to what extent some of my ideas are original, and to what extent they spring from seeds planted in my mind by others.

I owe especial thanks to my wife, Mabel, who has provided the domestic milieu conducive to my work, and who has maintained and typed the bibliography and served as a consultant on writing style; to my secretary, Lucy F. Kluska, who has skillfully typed preliminary versions and the final draft of the basic manuscript; to Armen Takhtajan, whose contribution to my thinking has already been noted; and to Margarita Baranova, who has helped me with the Russian botanical literature and with other aspects of my work in Leningrad. Most of the illustrations have been prepared by William S. Moye, a smaller number by Robin Jess, Bobbi Angell, and others.

A partial list of other botanists who have helped me follows. Many of them have reviewed some segment of the manuscript. V. Sh. Agababyan, William R. Anderson, Edward S. Ayensu, T. M. Barkley, Rupert C. Barneby, Bruce Bayer, Igor Belolipov, C. C. Berg, Barbara G. Briggs, James E. Canright, Sherwin Carlquist, Robert T. Clausen, Lincoln

Constance, David F. Cutler, Rolf Dahlgren, Thomas J. Delendick, Bernard de Winter, David Dilcher, James A. Doyle, Richard H. Eyde, Andrey Fedorov, Alvaro Fernandez Perez, Alexander Floyd, Dietrich Frohne, Eleonora Gabrielian, Alwyn H. Gentry, David E. Giannasi, Arthur C. Gibson, William T. Gillis, Peter S. Goldblatt, I. A. Grudzinskaja, Zaira Gvinianidze, Ray M. Harley, Leo J. Hickey, Noel H. Holmgren, Richard A. Howard, B. P. M. Hyland, Charles Jeffrey, Uwe Jensen, Laurie A. S. Johnson, Joseph H. Kirkbride, Tetsuo Koyama, Robert Kral, Klaus Kubitzki, Jean-F. Leroy, Richard N. Lester, Harlan Lewis, Walter H. Lewis, Eduardo Lleras, James Luteyn, Tom Mabry, Bassett Maguire, John McNeill, Rogers McVaugh, C. R. Metcalfe, Phillipe Morat, H. E. Moore, Scott A. Mori, Jan Muller, Dan H. Nicolson, Joan W. Nowicke, P. N. Ovczinnikov, Barbara F. Palser, D. Philcox, William R. Philipson, James E. Piechura, Ghillean T. Prance, Peter H. Raven, James L. Reveal, Reed C. Rollins, Irwin Rosenblum, John P. Rourke, Rolf Sattler, T. V. Shulkina, Beryl Simpson, Hermann O. Sleumer, Natalia Snigirevskaja, Thomas R. Soderstrom, Clive A. Stace, Margaret Y. Stant, W. Douglas Stevens, Benjamin C. Stone, Donald E. Stone, M. Tamura, Leonard B. Thien, Robert F. Thorne, Clifford Townsend, I. T. Vassilczenko, J. M. Veillon, Tom Wendt, Molly Ann Whalen, Del Wiens, I. J. M. Williams, Carl L. Withner, John J. Wurdack, David A. Young, Robert Faden, Enrique Forero.

OUTLINE OF CLASSIFICATION OF MAGNOLIOPHYTA

Class MAGNOLIOPSIDA

Subclass I. Magnoliidae
- Order 1. Magnoliales
  - Family 1. Winteraceae
    - 2. Degeneriaceae
    - 3. Himantandraceae
    - 4. Eupomatiaceae
    - 5. Austrobaileyaceae
    - 6. Magnoliaceae
    - 7. Lactoridaceae
    - 8. Annonaceae
    - 9. Myristacaceae
    - 10. Canellaceae
- Order 2. Laurales
  - Family 1. Amborellaceae
    - 2. Trimeniaceae
    - 3. Monimiaceae
    - 4. Gomortegaceae
    - 5. Calycanthaceae
    - 6. Idiospermaceae
    - 7. Lauraceae
    - 8. Hernandiaceae
- Order 3. Piperales
  - Family 1. Chloranthaceae
    - 2. Saururaceae
    - 3. Piperaceae
- Order 4. Aristolochiales
  - Family 1. Aristolochiaceae
- Order 5. Illiciales
  - Family 1. Illiciaceae
    - 2. Schisandraceae
- Order 6. Nymphaeales
  - Family 1. Nelumbonaceae
    - 2. Nymphaeaceae
    - 3. Barclayaceae
    - 4. Cabombaceae
    - 5. Ceratophyllaceae
- Order 7. Ranunculales
  - Family 1. Ranunculaceae
    - 2. Circaeasteraceae
    - 3. Berberidaceae
    - 4. Sargentodoxaceae
    - 5. Lardizabalaceae
    - 6. Menispermaceae
    - 7. Coriariaceae
    - 8. Sabiaceae
- Order 8. Papaverales
  - Family 1. Papaveraceae
    - 2. Fumariaceae

Subclass II. Hamamelidae
- Order 1. Trochodendrales
  - Family 1. Tetracentraceae
    - 2. Trochodendraceae
- Order 2. Hamamelidales
  - Family 1. Cercidiphyllaceae
    - 2. Eupteliaceae
    - 3. Platanaceae
    - 4. Hamamelidaceae
    - 5. Myrothamnaceae
- Order 3. Daphniphyllales
  - Family 1. Daphniphyllaceae
- Order 4. Didymelales
  - Family 1. Didymelaceae
- Order 5. Eucommiales
  - Family 1. Eucommiaceae
- Order 6. Urticales
  - Family 1. Barbeyaceae
    - 2. Ulmaceae
    - 3. Cannabaceae

4. Moraceae
5. Cecropiaceae
6. Urticaceae

Order 7. Leitneriales
Family 1. Leitneriaceae

Order 8. Juglandales
Family 1. Rhoipteleaceae
2. Juglandaceae

Order 9. Myricales
Family 1. Myricaceae

Order 10. Fagales
Family 1. Balanopaceae
2. Fagaceae
3. Betulaceae

Order 11. Casuarinales
Family 1. Casuarinaceae

Subclass III. Caryophyllidae

Order 1. Caryophyllales
Family 1. Phytolaccaceae
2. Achatocarpaceae
3. Nyctaginaceae
4. Aizoaceae
5. Didiereaceae
6. Cactaceae
7. Chenopodiaceae
8. Amaranthaceae
9. Portulacaceae
10. Basellaceae
11. Molluginaceae
12. Caryophyllaceae

Order 2. Polygonales
Family 1. Polygonaceae

Order 3. Plumbaginales
Family 1. Plumbaginaceae

Subclass IV. Dilleniidae

Order 1. Dilleniales
Family 1. Dilleniaceae
2. Paeoniaceae

Order 2. Theales
Family 1. Ochnaceae
2. Sphaerosepalaceae
3. Sarcolaenaceae
4. Dipterocarpaceae
5. Caryocaraceae
6. Theaceae
7. Actinidiaceae
8. Scytopetalaceae
9. Pentaphylacaceae
10. Tetrameristaceae
11. Pellicieraceae
12. Oncothecaceae
13. Marcgraviaceae
14. Quiinaceae
15. Elatinaceae
16. Paracryphiaceae
17. Medusagynaceae
18. Clusiaceae

Order 3. Malvales
Family 1. Elaeocarpaceae
2. Tiliaceae
3. Sterculiaceae
4. Bombacaceae
5. Malvaceae

Order 4. Lecythidales
Family 1. Lecythidaceae

Order 5. Nepenthales
Family 1. Sarraceniaceae
2. Nepenthaceae
3. Droseraceae

Order 6. Violales
Family 1. Flacourtiaceae
2. Peridiscaceae
3. Bixaceae
4. Cistaceae
5. Huaceae
6. Lacistemataceae
7. Scyphostegiaceae
8. Stachyuraceae
9. Violaceae
10. Tamaricaceae
11. Frankeniaceae
12. Dioncophyllaceae
13. Ancistrocladaceae
14. Turneraceae
15. Malesherbiaceae
16. Passifloraceae

17. Achariaceae
18. Caricaceae
19. Fouquieriaceae
20. Hoplestigmataceae
21. Cucurbitaceae
22. Datiscaceae
23. Begoniaceae
24. Loasaceae

Order 7. Salicales
Family 1. Salicaceae

Order 8. Capparales
Family 1. Tovariaceae
2. Capparaceae
3. Brassicaceae
4. Moringaceae
5. Resedaceae

Order 9. Batales
Family 1. Gyrostemonaceae
2. Bataceae

Order 10. Ericales
Family 1. Cyrillaceae
2. Clethraceae
3. Grubbiaceae
4. Empetraceae
5. Epacridaceae
6. Ericaceae
7. Pyrolaceae
8. Monotropaceae

Order 11. Diapensiales
Family 1. Diapensiaceae

Order 12. Ebenales
Family 1. Sapotaceae
2. Ebenaceae
3. Styracaceae
4. Lissocarpaceae
5. Symplocaceae

Order 13. Primulales
Family 1. Theophrastaceae
2. Myrsinaceae
3. Primulaceae

Subclass V. Rosidae

Order 1. Rosales
Family 1. Brunelliaceae
2. Connaraceae
3. Eucryphiaceae
4. Cunoniaceae
5. Davidsoniaceae
6. Dialypetalanthaceae
7. Pittosporaceae
8. Byblidaceae
9. Hydrangeaceae
10. Columelliaceae
11. Grossulariaceae
12. Greyiaceae
13. Bruniaceae
14. Anisophylleaceae
15. Alseuosmiaceae
16. Crassulaceae
17. Cephalotaceae
18. Saxifragaceae
19. Rosaceae
20. Neuradaceae
21. Crossosomataceae
22. Chrysobalanaceae
23. Surianaceae
24. Rhabdodendraceae

Order 2. Fabales
Family 1. Mimosaceae
2. Caesalpiniaceae
3. Fabaceae

Order 3. Proteales
Family 1. Elaeagnaceae
2. Proteaceae

Order 4. Podostemales
Family 1. Podostemaceae

Order 5. Haloragales
Family 1. Haloragaceae
2. Gunneraceae

Order 6. Myrtales
Family 1. Sonneratiaceae
2. Lythraceae
3. Penaeaceae
4. Crypteroniaceae
5. Thymelaeaceae
6. Trapaceae
7. Myrtaceae

8. Punicaceae
9. Onagraceae
10. Oliniaceae
11. Melastomataceae
12. Combretaceae

Order 7. Rhizophorales
Family 1. Rhizophoraceae

Order 8. Cornales
Family 1. Alangiaceae
2. Nyssaceae
3. Cornaceae
4. Garryaceae

Order 9. Santalales
Family 1. Medusandraceae
2. Dipentodontaceae
3. Olacaceae
4. Opiliaceae
5. Santalaceae
6. Misodendraceae
7. Loranthaceae
8. Viscaceae
9. Eremolepidaceae
10. Balanophoraceae

Order 10. Rafflesiales
Family 1. Hydnoraceae
2. Mitrastemonaceae
3. Rafflesiaceae

Order 11. Celastrales
Family 1. Geissolomataceae
2. Celastraceae
3. Hippocrateaceae
4. Stackhousiaceae
5. Salvadoraceae
6. Aquifoliaceae
7. Icacinaceae
8. Aextoxicaceae
9. Cardiopteridaceae
10. Corynocarpaceae
11. Dichapetalaceae

Order 12. Euphorbiales
Family 1. Buxaceae
2. Simmondsiaceae
3. Pandaceae
4. Euphorbiaceae

Order 13. Rhamnales
Family 1. Rhamnaceae
2. Leeaceae
3. Vitaceae

Order 14. Linales
Family 1. Erythroxylaceae
2. Humiriaceae
3. Ixonanthaceae
4. Hugoniaceae
5. Linaceae

Order 15. Polygalales
Family 1. Malpighiaceae
2. Vochysiaceae
3. Trigoniaceae
4. Tremandraceae
5. Polygalaceae
6. Xanthophyllaceae
7. Krameriaceae

Order 16. Sapindales
Family 1. Staphyleaceae
2. Melianthaceae
3. Bretschneideraceae
4. Akaniaceae
5. Sapindaceae
6. Hippocastanaceae
7. Aceraceae
8. Burseraceae
9. Anacardiaceae
10. Julianiaceae
11. Simaroubaceae
12. Cneoraceae
13. Meliaceae
14. Rutaceae
15. Zygophyllaceae

Order 17. Geraniales
Family 1. Oxalidaceae
2. Geraniaceae
3. Limnanthaceae
4. Tropaeolaceae
5. Balsaminaceae

Order 18. Apiales
Family 1. Araliaceae
2. Apiaceae
Subclass VI. Asteridae
Order 1. Gentianales
Family 1. Loganiaceae
2. Retziaceae
3. Gentianaceae
4. Saccifoliaceae
5. Apocynaceae
6. Asclepiadaceae
Order 2. Solanales
Family 1. Duckeodendraceae
2. Nolanaceae
3. Solanaceae
4. Convolvulaceae
5. Cuscutaceae
6. Menyanthaceae
7. Polemoniaceae
8. Hydrophyllaceae
Order 3. Lamiales
Family 1. Lennoaceae
2. Boraginaceae
3. Verbenaceae
4. Lamiaceae
Order 4. Callitrichales
Family 1. Hippuridaceae
2. Callitrichaceae
3. Hydrostachyaceae
Order 5. Plantaginales
Family 1. Plantaginaceae
Order 6. Scrophulariales
Family 1. Buddlejaceae
2. Oleaceae
3. Scrophulariaceae
4. Globulariaceae
5. Myoporaceae
6. Orobanchaceae
7. Gesneriaceae
8. Acanthaceae
9. Pedaliaceae
10. Bignoniaceae
11. Mendonciaceae
12. Lentibulariaceae
Order 7. Campanulales
Family 1. Pentaphragmataceae
2. Sphenocleaceae
3. Campanulaceae
4. Stylidiaceae
5. Donatiaceae
6. Brunoniaceae
7. Goodeniaceae
Order 8. Rubiales
Family 1. Rubiaceae
2. Theligonaceae
Order 9. Dipsacales
Family 1. Caprifoliaceae
2. Adoxaceae
3. Valerianaceae
4. Dipsacaceae
Order 10. Calycerales
Family 1. Calyceraceae
Order 11. Asterales
Family 1. Asteraceae

Class LILIOPSIDA

Subclass I. Alismatidae
Order 1. Alismatales
Family 1. Butomaceae
2. Limnocharitaceae
3. Alismataceae
Order 2. Hydrocharitales
Family 1. Hydrocharitaceae
Order 3. Najadales
Family 1. Aponogetonaceae
2. Scheuchzeriaceae
3. Juncaginaceae
4. Potamogetonaceae
5. Ruppiaceae
6. Najadaceae

7. Zannichelliaceae
8. Posidoniaceae
9. Cymodoceaceae
10. Zosteraceae

Order 4. Triuridales
Family 1. Petrosaviaceae
2. Triuridaceae

Subclass II. Arecidae
Order 1. Arecales
Family 1. Arecaceae
Order 2. Cyclanthales
Family 1. Cyclanthaceae
Order 3. Pandanales
Family 1. Pandanaceae
Order 4. Arales
Family 1. Araceae
2. Lemnaceae

Subclass III. Commelinidae
Order 1. Commelinales
Family 1. Rapateaceae
2. Xyridaceae
3. Mayacaceae
4. Commelinaceae
Order 2. Eriocaulales
Family 1. Eriocaulaceae
Order 3. Restionales
Family 1. Flagellariaceae
2. Joinvilleaceae
3. Restionaceae
4. Centrolepidaceae
Order 4. Juncales
Family 1. Juncaceae
2. Thurniaceae
Order 5. Cyperales
Family 1. Cyperaceae
2. Poaceae
Order 6. Hydatellales
Family 1. Hydatellaceae
Order 7. Typhales
Family 1. Sparganiaceae
2. Typhaceae

Subclass IV. Zingiberidae
Order 1. Bromeliales
Family 1. Bromeliaceae
Order 2. Zingiberales
Family 1. Strelitziaceae
2. Heliconiaceae
3. Musaceae
4. Lowiaceae
5. Zingiberaceae
6. Costaceae
7. Cannaceae
8. Marantaceae

Subclass V. Liliidae
Order 1. Liliales
Family 1. Philydraceae
2. Pontederiaceae
3. Haemodoraceae
4. Cyanastraceae
5. Liliaceae
6. Iridaceae
7. Velloziaceae
8. Aloeaceae
9. Agavaceae
10. Xanthorrhoeaceae
11. Hanguanaceae
12. Taccaceae
13. Stemonaceae
14. Smilacaceae
15. Dioscoreaceae
Order 2. Orchidales
Family 1. Geosiridaceae
2. Burmanniaceae
3. Corsiaceae
4. Orchidaceae

Division MAGNOLIOPHYTA
Cronquist, Takhtajan & Zimmermann 1966
the Angiosperms or Flowering Plants

Vascular plants, ordinarily with roots, stems, and leaves, the central cylinder of the stem with leaf-gaps or with scattered vascular bundles; xylem usually but not always consisting in part of vessels, at least in the roots; phloem with definite sieve-tubes and companion-cells. Sexual reproductive structures characteristically aggregated and associated with specialized leaves to form flowers; flowers, in the most typical form, with an outer set of perianth-members, the sepals, forming a calyx, an inner set of perianth-members, the petals, forming a corolla, an androecium consisting of one to many stamens (microsporophylls), and in the center a gynoecium consisting of one to many carpels (megasporophylls); carpels distinct and individually closed (rarely merely folded and unsealed) to form separate pistils, or all more or less fully connate to form a compound pistil; pistil (whether composed of one carpel or more than one) composed of an ovary, topped by one or more stigmas that are usually elevated above the ovary on a style; one or more ovules (modified megasporangia) enclosed in the ovary; each ovule containing a female gametophyte consisting of a few-nucleate (typically 8-nucleate) embryo-sac without an archegonium; stamens (microsporophylls) laminar in some archaic groups, but otherwise generally consisting of a filament and an anther, the anther most commonly with two bisporangiate pollen-sacs (thecae) joined by connective tissue; pollen-grains germinating on the stigma, each producing a pollen-tube that grows down through the pistil toward an ovule; mature male gametophyte (pollen-grain and -tube) containing 3 nuclei (the tube nucleus and two sperm nuclei); pollen-tube at maturity penetrating the embryo-sac and delivering the two naked sperms, one of which fuses with the egg to form a zygote, the other of which ordinarily fuses with two nuclei (or the product of fusion of two nuclei) of the embryo-sac to form a triple-fusion nucleus that is typically the forerunner of the endosperm in the seed; ovary (and sometimes some associated structures) maturing to form a dehiscent or indehiscent fruit containing one or more seeds. (Angiospermae)

The division Magnoliophyta consists of two classes of unequal size, the Magnoliopsida (dicotyledons) and Liliopsida (monocotyledons). The Magnoliopsida as here treated consist of 64 orders, 318 families, and about 165,000 species. The Liliopsida as here treated consist of 19 orders, 65 families, and about 50,000 species.

Angiosperms enter the fossil record in the Barremian stage of the Lower Cretaceous period, some 130 million years ago, give or take a few million years. These earliest known angiosperms had monosulcate pollen and rather small, simple leaves with a poorly organized, irregular sort of pinnate net-venation (Hickey & Doyle, 1977). Their other characters are inferred rather than established. The combination of pollen and leaf characters is compatible with only one sub-class of modern angiosperms, the Magnoliidae.

Based on the fossil pollen record, the earliest significant dichotomy in the evolutionary diversification of the angiosperms is that between the Magnoliopsida and Liliopsida (Doyle, 1973; Cronquist, 1974). The two classes are now sharply distinct, in spite of the fact that all the differences between them are subject to exception. Only one small order, the Nymphaeales, excites any present controversy over the delimitation of these two major groups.

The hypothesis of Doyle & Hickey (1976) that the early angiosperms may have been shrubs of unstable, riparian habitats provides a logical basis for considering the initial diversification of the group. This outlook is especially helpful with regard to the origin of the monocotyledons as an herbaceous, probably aquatic phylad. The Doyle & Hickey hypothesis takes as a starting point the earlier view of Stebbins (1965) that the angiosperms may have originated from the gymnosperms as a group of relatively small-leaved shrubs of arid or semi-arid habitats. Of this hypothetical transitional group, however, we have no direct evidence.

No direct connection of the angiosperms to any other group is known from the fossil record. On compelling theoretical grounds it is widely believed that the ancestry of the angiosperms must be sought among the cycadophyte group of gymnosperms, presumably among the seed ferns, sensu latissimo (Takhtajan, 1954; Cronquist, 1968). The Bennettitales and Cycadales are too specialized to be likely ancestors. The Caytoniales may stand near to the ancestry of the angiosperms, but they are probably not in the direct line of descent.

The cupule of the Caytoniales is plausibly considered to represent a possible evolutionary precursor of the ovule (not the carpel) of modern angiosperms (Gaussen, 1946; Stebbins, 1976). In this view the wall of the Caytonialean cupule is considered to be homologous with the outer integument of the angiosperm ovule, and the zigzag micropyle of some modern angiosperms is considered to reflect the asymmetrical placement of the remaining ovule in the cupule.

SELECTED REFERENCES

Arber, A. 1933. Floral anatomy and its morphological interpretation. New Phytol. 32: 231–242.

Arber, A. 1937. The interpretation of the flower: a study of some aspects of morphological thought. Biol. Rev. Cambridge Philos. Soc. 12: 157–184.

Axelrod, D. I. 1966. Origin of deciduous and evergreen habits in temperate forests. Evolution 20: 1–15.

Bailey, I. W. 1944. The development of vessels in angiosperms and its significance in morphological research. Amer. J. Bot. 31: 421–428.

Bailey, I. W. 1953. Evolution of the tracheary tissue of land plants. Amer. J. Bot. 40: 4–8.

Bailey, I. W. 1956. Nodal anatomy in retrospect. J. Arnold Arbor. 37: 269–287.

Bailey, I. W. 1957. The potentialities and limitations of wood anatomy in the study of the phylogeny and classification of angiosperms. J. Arnold Arbor. 38: 243–254.

Barroso, G. M. 1978, 1979. Sistemática de angiospermas do Brasil. Vols. 1, 2. Livras Técnicos e Científicos Editora S. A., Rio de Janeiro.

Bate-Smith, E. C. 1962, 1968. The phenolic constituents of plants and their taxonomic significance. I. Dicotyledons. II. Monocotyledons. J. Linn. Soc., Bot. 58: 95–173, 1962. 60: 325–356, 1968.

Battaglia, E. 1951. The male and female gametophytes of angiosperms—an interpretation. Phytomorphology 1: 87–116.

Baum, H. 1946. Über die postgenitale Verwachsung in Karpellen. Oesterr. Bot. Z. 95: 86–94.

Baum, H. 1948. Die Verbreitung der postgenitalen Verwachsung im Gynözeum und ihre Bedeutung für die typologische Betrachtung des coenokarpen Gynözeums. Oesterr. Bot. Z. 95: 124–128.

Behnke, H.-D. 1972. Sieve-tube plastids in relation to angiosperm systematics—An attempt towards a classification by ultrastructural analysis. Bot. Rev. 38: 155–197.

Behnke, H.-D. 1975. The bases of angiosperm phylogeny: Ultrastructure. (Preliminary considerations based on studies of sieve element plastids of some 500 species). Ann. Missouri Bot. Gard. 62: 647–663.

Behnke, H.-D. 1977. Transmission electron microscopy and systematics of flowering plants. Pl. Syst. Evol., Suppl. 1: 155–178.

Behnke, H.-D., & R. Dahlgren. 1976. The distribution of characters within an angiosperm system. 2. Sieve-element plastids. Bot. Not. 129: 287–295.

Bendz, G., & J. Santesson, eds. 1974. Chemistry in botanical classification. Nobel Symposium 25. Academic Press. New York and London.

Bessey, C. E. 1915. The phylogenetic taxonomy of flowering plants. Ann. Missouri Bot. Gard. 2: 109–164.

Bierhorst, D. W., & P. M. Zamora. 1965. Primary xylem elements and element associations of angiosperms. Amer. J. Bot. 52: 657–710.

Bocquet, G. 1959 (1960). The campylotropous ovule. Phytomorphology 9: 222–227.
Bond, G. 1976. The results of the IBP survey of root-nodule formation in non-leguminous angiosperms. *In:* P. S. Nutman, ed., Symbiotic nitrogen fixation in plants, pp. 443–474. Intern. Biol. Programme 7. Cambridge Univ. Press. London.
Bouman, F. 1971. The application of tegumentary studies to taxonomic and phylogenetic problems. Ber. Deutsch. Bot. Ges. 84: 169–177.
Brenner, G. J. 1967. Early angiosperm pollen differentiation in the Albian to Cenomanian deposits of Delaware (U.S.A.). Rev. Palaeobot. Palynol. 1: 219–227.
Brenner, G. J. 1976. Middle Cretaceous floral provinces and early migrations of angiosperms. *In:* C. B. Beck, ed., Origin and early evolution of angiosperms, pp. 23–47. Columbia Univ. Press. New York.
Brewbaker, J. 1967. The distribution and phylogenetic significance of binucleate and trinucleate pollen grains in angiosperms. Amer. J. Bot. 54: 1069–1083.
Brown, W. H. 1938. The bearing of nectaries on the phylogeny of flowering plants. Proc. Amer. Philos. Soc. 79: 549–595.
Cagnin, M. A. H., C. M. R. Gomes, O. R. Gottlieb, M. C. Marx, A. I. da Rocha, M. F. das G. F. da Silva, & J. A. Temperini. 1977. Biochemical systematics: Methods and principles. Pl. Syst. Evol., Suppl. 1: 53–76.
Carlquist, S. 1969 (1970). Toward acceptable evolutionary interpretations of floral anatomy. Phytomorphology 19: 332–362.
Carlquist, S. 1975. Ecological Strategies of Xylem Evolution. Univ. Calif. Press, Berkeley.
Carlquist, S. 1977. Ecological factors in wood evolution: A floristic approach. Amer. J. Bot. 64: 887–896.
Carr, S. G. M., & D. J. Carr. 1961. The functional significance of syncarpy. Phytomorphology 11: 249–256.
Chandler, M. E. J. 1954. Some Upper Cretaceous and Eocene fruits from Egypt. Bull. Brit. Mus. (Nat. Hist.) Geol. 2(4): 147–187.
Chenery, E. M., & K. R. Sporne. 1976. A note on the evolutionary status of aluminium-accumulators among dicotyledons. New Phytol. 76: 551–554.
Constance, L. 1964. Systematic botany—an unending synthesis. Taxon 13: 257–273.
Corner, E. J. H. 1946. Centrifugal stamens. J. Arnold Arbor. 27: 423–437.
Corner, E. J. H. 1949. The durian theory or the origin of the modern tree. Ann. Bot. (London) II. 13: 367–414.
Corner, E. J. H. 1953, 1954. The durian theory extended. I, II, III. Phytomorphology 3: 465–476, 1953. 4: 152–165, 263–274, 1954.
Crepet, W. L. 1979. Insect pollination: A paleontological perspective. BioScience 29: 102–108.
Crepet, W. L. 1979. Some aspects of the pollination biology of Middle Eocene angiosperms. Rev. Palaeobot. Palynol. 27: 213–238.
Crété, P. 1964. L'embryogénie et son rôle dans les essais de classification phylogénétique. Phytomorphology 14: 70–78.
Cronquist, A. 1963. The taxonomic significance of evolutionary parallelism. Sida 1: 109–116.
Cronquist, A. 1965. The status of the general system of classification of flowering plants. Ann. Missouri Bot. Gard. 52: 281–303.
Cronquist, A. 1969. On the relationship between taxonomy and evolution. Taxon 18: 177–187.

Cronquist, A. 1975. Some thoughts on angiosperm phylogeny and taxonomy. Ann. Missouri Bot. Gard. 62: 517–520.

Cronquist, A. 1977. On the taxonomic significance of secondary metabolites in angiosperms. Pl. Syst. Evol., Suppl. 1: 179–189.

Cronquist, A., A. Takhtajan, & W. Zimmermann. 1966. On the higher taxa of Embryobionta. Taxon 15: 129–134.

Czaja, A. T. 1978. Stärke und Stärkespeicherung bei Gefässpflanzen. Versuch einer Amylo-Taxonomie. Gustav Fischer Verlag. Stuttgart & New York.

Czaja, A. T. 1978. Structure of starch grains and classification of vascular plant families. Taxon 27: 463–470.

Dahlgren, R. 1975. A system of classification of the angiosperms to be used to demonstrate the distribution of characters. Bot. Not. 128: 119–147.

Dahlgren, R. 1975. The distribution of characters within an angiosperm system. I. Some embryological characters. Bot. Not. 128: 181–197.

Dahlgren, R. 1977. A commentary on a diagrammatic presentation of the angiosperms in relation to the distribution of character states. Pl. Syst. Evol. Suppl. 1: 253–283.

Dahlgren, R. M. T. 1980. A revised system of classification of the angiosperms. J. Linn. Soc., Bot. 80: 91–124.

Davis, G. L. 1966. Systematic embryology of the angiosperms. Wiley & Sons. New York.

Davis, P. H., & V. H. Heywood. 1963. Principles of angiosperm taxonomy. Oliver & Boyd. Edinburgh & London.

Dickison, W. C. 1975. The bases of angiosperm phylogeny: Vegetative anatomy. Ann. Missouri Bot. Gard. 62: 590–620.

Dilcher, D. L. 1971. A revision of the Eocene flora of southeastern North America. Palaeobotanist 20: 7–18.

Dilcher, D. L. 1979. Early angiosperm reproduction: An introductory report. Rev. Palaeobot. Palynol. 27: 291–328.

Douglas, G. 1944, 1957. The inferior ovary. I & II. Bot. Rev. 10: 125–186, 1944. 23: 1–46, 1957.

Doyle, J. A. 1969. Cretaceous angiosperm pollen of the Atlantic coastal plain and its evolutionary significance. J. Arnold Arbor. 50: 1–35.

Doyle, J. A. 1978. Origin of angiosperms. Ann. Rev. Ecol. Syst. 9: 365–392.

Doyle, J. A., & L. J. Hickey. 1976. Pollen and leaves from the Mid-Cretaceous Potomac Group and their bearing on early angiosperm evolution. *In:* C. B. Beck, ed., Origin and early evolution of angiosperms, pp. 139–206. Columbia Univ. Press. New York.

Doyle, J. A., M. Van Campo, & B. Lugardon. 1975. Observations on exine structure of *Eucommiidites* and Lower Cretaceous angiosperm pollen. Pollen & Spores 17: 429–486.

Dunn, B. D., G. K. Sharma, & C. C. Campbell. 1965. Structural patterns of dicotyledons and monocotyledons. Amer. Midl. Naturalist 74: 185–195.

Eames, A. J. 1931. The vascular anatomy of the flower with refutation of the theory of carpel polymorphism. Amer. J. Bot. 18: 147–188.

Eames, A. J. 1951. Again: "The new morphology." New Phytol. 50: 17–35.

Eames, A. J. 1961. Morphology of the angiosperms. McGraw-Hill. New York.

Eckardt, Th. 1957. Vergleichende Studien über die morphologischen Beziehungen zwischen Fruchtblatt, Samenanlage und Blütenachse bei einigen Angiospermen, zugleich als kritische Beleuchtung der "New Morphology." Neue Hefte Morphol. 3: 1–91.

Eckardt, Th. 1963. Zum Blütenbau der Angiospermen im Zusammenhang mit ihrer Systematik. Ber. Deutsch. Bot. Ges. 76: (38)–(49).
Eckardt, Th. 1964. Das Homologierproblem und Fälle strittiger Homologien. Phytomorphology 14: 79–92.
Eckert, G. 1966. Entwicklungsgeschichtliche und blütenanatomische Untersuchungen zum Problem der Obdiplostemonie. Bot. Jahrb. Syst. 85: 523–604.
Ehrendorfer, F. 1970. Chromosomen, Verwandtschaft und Evolution tropischer Holzpflanzen, I. Allgemeine Hinweise. Oesterr. Bot. Z. 118: 30–37.
Ehrendorfer, F. 1977. New ideas about the early differentiation of angiosperms. Pl. Syst. Evol., Suppl. 1: 227–234.
Ehrlich, P. R., & P. H. Raven. 1964. Butterflies and plants: a study in coevolution. Evolution 18: 586–608.
Erdtman, G. 1952. Pollen morphology and plant taxonomy. An introduction to palynology. I. Angiosperms. Almqvist & Wiksell. Stockholm.
Esau, K., V. I. Cheadle, & E. M. Gifford. 1953. Comparative structure and possible trends of specialization of the phloem. Amer. J. Bot. 40: 9–19.
Eyde, R. H. 1971. Evolutionary morphology: distinguishing ancestral structure from derived structure in flowering plants. Taxon 20: 63–73.
Eyde, R. H. 1975. The bases of angiosperm phylogeny: Floral anatomy. Ann. Missouri Bot. Gard. 62: 521–537.
Fahn, A. 1952. On the structure of floral nectaries. Bot. Gaz. 113: 464–470.
Fahn, A. 1953. The topography of the nectary in the flower and its phylogenetical trend. Phytomorphology 3: 424–426.
Fairbrothers, D. E. 1977. Perspectives in plant serotaxonomy. Ann. Missouri Bot. Gard. 64: 147–160.
Fraenkel, G. 1959. The raison d'être of secondary plant substances. Science 129: 1466–1470.
Fraenkel, G. 1969. Evaluation of our thoughts on secondary plant substances. Entomol. Exp. Appl. 12: 473–486.
Frohne, D., & U. Jensen. 1973. Systematik des Pflanzenreichs. Gustav Fischer Verlag. Jena. Ed. 2 in 1979.
Gaussen, H. 1946. Les gymnospermes actuelles et fossiles. Fascicule III. Chapitre V. Les autres Cycadophytes. L'origine cycadophytique des Angiospermes. pp. 1–26. [= Tom II, Vol. 1, Fasc. III of Trav. Lab. Forest. Toulouse.]
Gerassimova-Navashina, H. 1961. Fertilization and events leading up to fertilization, and their bearing on the origin of angiosperms. Phytomorphology 11: 139–146.
Gibbs, R. D. 1974. Chemotaxonomy of flowering plants. 4 volumes. McGill-Queen's Univ. Press. Montreal & London.
Gilmour, J. S. L., & S. M. Walters. 1963. Philosophy and classification. Vistas Bot. 4: 1–22.
Gornall, R. J., B. A. Bohm, & R. Dahlgren. 1979. The distribution of flavonoids in the angiosperms. Bot. Not. 132: 1–30.
Gottsberger, G. 1974. The structure and function of the primitive angiosperm flower—a discussion. Acta Bot. Neerl. 23: 461–471.
Gottsberger, G. 1977. Some aspects of beetle pollination in the evolution of flowering plants. Pl. Syst. Evol., Suppl. 1: 211–226.
Grant, V. 1950. The protection of ovules in flowering plants. Evolution 4: 179–201.
Gray, J. 1960. Temperate pollen genera in the Eocene (Claiborne) flora, Alabama. Science 132: 808–810.
Hallier, H. 1901. Über die Verwandtschaftsverhältnisse der Tubifloren und

Ebenalen, den polyphyletischen Ursprung der Sympetalen und Apetalen und die Anordnung der Angiospermen überhaupt. Abh. Naturwiss. Naturwiss. Verein Hamburg 16(2,2): 3–112.

Hallier, H. 1908. Über *Juliania*, eine Terebinthaceen-Gattung mit Cupula, und die wahren Stammeltern der Kätzchenblütler. Neue Beiträge zur Stammesgeschichte der Dicotyledonen. Beih. Bot. Centralbl. 23(2): 81–265.

Hallier, H. 1912. L'origine et le système phylétique des angiospermes exposés à l'aide de leur arbre généalogique. Arch. Néerl. Sci. Exact. Nat., Sér. 3, B. 1: 146–234.

Hamann, U. 1977. Über Konvergenzen bei embryologischen Merkmalen der Angiospermen. Ber. Deutsch. Bot. Ges. 90: 369–384.

Harborne, J. B. 1977. Flavonoids and the evolution of the angiosperms. Biochem. Syst. Ecol. 5: 7–22.

Hardin, G. 1960. The competitive exclusion principle. Science 131: 1292–1297.

Hartl, D. 1962. Die morphologische Natur und die Verbreitung des Apicalseptums. Analyse einer bisher unbekannten Gestaltungsmöglichkeit des Gynoeceums. Beitr. Biol. Pflanzen 37: 241–330.

Hegnauer, R. 1963–1973. Chemotaxonomie der Pflanzen. 2. Monocotyledoneae, 1963. 3. Dicotyledoneae: Acanthaceae-Cyrillaceae, 1964. 4. Dicotyledoneae: Daphniphyllaceae-Lythraceae, 1966. 5. Dicotyledoneae: Magnoliaceae-Quiinaceae, 1969. 6. Dicotyledoneae: Rafflesiaceae-Zygophyllaceae, 1973. Birkhouser. Basel & Stuttgart.

Hegnauer, R. 1969. Chemical evidence for the classification of some plant taxa. *In:* Harborne, J. B., & T. Swain, eds., Perspectives in Phytochemistry, pp. 121–138. Academic Press. London & New York.

Hegnauer, R. 1971. Pflanzenstoffe und Pflanzensystematik. Naturwissenschaften 58: 585–598.

Hegnauer, R. 1973. Die cyanogenen Verbindungen der Liliatae und Magnoliatae-Magnoliidae: Zur systematischen Bedeutung des Merkmals der Cyanogenese. Biochem. Syst. 1: 191–197.

Hegnauer, R. 1977. Cyanogenic compounds as systematic markers in Tracheophyta. Pl. Syst. Evol., Suppl. 1: 191–210.

Heslop-Harrison, Y., & K. R. Shivanna. 1977. The receptive surface of the angiosperm stigma. Ann. Bot. (London) II. 41: 1233–1258.

Hickey, L. J., & J. A. Doyle. 1977. Early Cretaceous fossil evidence for angiosperm evolution. Bot. Rev. 43: 3–104.

Howard, R. A. 1962. The vascular structure of the petiole as a taxonomic character. *In:* Advances in horticultural sciences and their applications, 3: 7–13. Pergamon Press. New York.

Howard, R. A. 1970. Some observations on the nodes of woody plants with special reference to the problem of the 'split-lateral' versus the 'common gap.' *In:* N. K. B. Robson, D. F. Cutler, & M. Gregory, eds., New research in plant anatomy, pp. 195–214. Suppl. 1, J. Linn. Soc., Bot. 63. Academic Press. London & New York.

Hughes, N. F. 1976. Cretaceous paleobotanic problems. *In:* C. B. Beck, ed., Origin and early evolution of angiosperms, pp. 11–22. Columbia Univ. Press. New York.

Hughes, N. F. 1977. Palaeo-succession of earliest angiosperm evolution. Bot. Rev. 43: 105–127.

Hutchinson, J. 1973. The families of flowering plants arranged according to a new system based on their probable phylogeny. 3rd ed. Clarendon Press. Oxford.

Jager, I. 1961. Vergleichend-morphologische Untersuchungen des Gefässbündelsystems peltater Nektar- und Kronblätter sowie verbildeter Staubblätter. Oesterr. Bot. Z. 108: 433–504.

Jarzen, D. M., & G. Norris. 1975. Evolutionary significance and botanical relationships of Cretaceous angiosperm pollen in the western Canadian interior. Geoscience & Man 11: 47–60.

Jeffrey, C. 1968. An introduction to plant taxonomy. J. & A. Churchill Ltd. London.

Jensen, S. R., B. J. Nielsen, & R. Dahlgren. 1975. Iridoid compounds, their occurrence and systematic importance in the angiosperms. Bot. Not. 128: 148–180.

Joshi, A. C. 1938. Systematic distribution of the *Fritillaria* type of embryo sac and the mono- and polyphyletic origin of angiosperms. Chron. Bot. 4: 507–508.

Kaussmann, B. 1941. Vergleichende Untersuchungen über die Blattnatur der Kelch-, Blumen- und Staubblätter. Bot. Arch. 42: 503–572.

Kelley, A. P. 1972. Mycotrophic nutrition. Instant Print, Henkel Press. Woodstock, Virginia.

Krassilov, V. 1973. Mesozoic plants and the problem of angiosperm ancestry. Lethaia 6: 163–178.

Krassilov, V. 1975. Dirhopalostachyaceae—A new family of proangiosperms and its bearing on the problem of angiosperm ancestry. Palaeontographica, Abt. B, 153: 100–110.

Krassilov, V. A. 1977. The origin of angiosperms. Bot. Rev. 43: 143–176.

Kubitzki, K. 1972. Probleme der Grossystematik der Blütenpflanzen. Ber. Deutsch. Bot. Ges. 85: 259–277.

Kubitzki, K. 1977. Some aspects of the classification and evolution of higher taxa. Pl. Syst. Evol., Suppl. 1: 21–31.

Kuprianova, L. A. 1967. Apertures of pollen grains and their evolution in angiosperms. Rev. Palaeobot. Palynol. 3: 73–80.

Kuprianova, L. A. 1969. On the evolutionary levels in the morphology of pollen grains and spores. Pollen & Spores 11: 333–351.

Leinfellner, W. 1950. Der Bauplan des synkarpen Gynözeums. Oesterr. Bot. Z. 97: 403–436.

Leinfellner, W. 1954. Die petaloiden Staubblätter und ihre Beziehungen zu den Kronblättern. Oesterr. Bot. Z. 101: 373–406.

Leins, P. 1964. Das zentripetale und zentrifugale Androeceum. Ber. Deutsch Bot. Ges. 77: (22)–(26).

Leins, P. 1975. Die Beziehungen zwischen multistaminaten und einfachen Androeceen. Bot. Jahrb. Syst. 96: 231–237.

Leppik, E. E. 1956. The form and function of numeral patterns in flowers. Amer. J. Bot. 43: 445–455.

Levin, D. A. 1971. Plant phenolics: an ecological perspective. Amer. Naturalist 105: 157–181.

Levin, D. A. 1973. The role of trichomes in plant defense. Quart. Rev. Biol. 48: 3–15.

Lewis, W. H. 1977. Pollen exine morphology and its adaptive significance. Sida 7: 95–102.

Li, H. L., & J. J. Willaman. 1968. Distribution of alkaloids in angiosperm phylogeny. Econ. Bot. 22: 239–252.

Lloyd, F. E. 1942. The carnivorous plants. Chronica Botanica 9. Waltham, Mass.

McNair, J. B. 1934. The evolutionary status of plant families in relation to some chemical properties. Amer. J. Bot. 21: 427–452.
Maheshwari, P. 1950. An introduction to the embryology of the angiosperms. McGraw-Hill. New York.
Maheshwari, P., & R. N. Kapil. 1966. Some Indian contributions to the embryology of angiosperms. Phytomorphology 16: 239–291.
Malloch, D. W., K. A. Pirozynski, & P. H. Raven. 1980. Ecological and evolutionary significance of mycorrhizal symbioses in vascular plants. Proc. Natl. Acad. Sci. 77: 2113–2118.
Martin, A. C. 1946. The comparative internal morphology of seeds. Amer. Midl. Naturalist 36: 513–660.
Martin, H. A. 1978. Evolution of the Australian flora and vegetation through the Tertiary: evidence from the pollen. Alcheringa 2: 181–202.
Mata Rezende, M. A., & O. R. Gottlieb. 1973. Xanthones as systematic markers. Biochem. Syst. 1: 111–118.
Melchior, H., ed. 1964. A. Engler's Syllabus der Pflanzenfamilien, Zwölfte Auflage. II Band. Angiospermen. Gebrüder Borntraeger. Berlin.
Merxmüller, H. 1967. Chemotaxonomie? Ber. Deutsch. Bot. Ges. 80: 608–620.
Merxmüller, H. 1972. Systematic botany—an unachieved synthesis. J. Linn. Soc., Biol. 4: 311–321.
Metcalfe, C. R. 1954. An anatomist's views on angiosperm classification. Kew Bull. 1954: 427–440.
Metcalfe, C. R. 1967. Distribution of latex in the plant kingdom. Econ. Bot. 21: 115–127.
Mulcahy, D. L. 1979. The rise of the angiosperms: a genecological factor. Science 206: 20–23.
Muller, J. 1970. Palynological evidence on early differentiation of angiosperms. Biol. Rev. Cambridge Philos. Soc. 45: 417–450.
Muller, J. 1979 (1980). Form and function in angiosperm pollen. Ann. Missouri Bot. Gard. 66: 593–632.
Muller, J. 1981. Fossil pollen records of extant angiosperms. Bot. Rev. 48: in press.
Nair, P. K. K. 1968. A concept on pollen evolution in the "primitive" angiosperms. J. Palyn. (Lucknow) 4: 15–20.
Němejc, F. 1956. On the problem of the origin and phylogenetic development of the angiosperms. Acta Mus. Natl. Prag. 12B: 65–144.
Novák, F. A. 1961. Vyšši rostliny. Československá Academie věd. Praha.
Nowicke, J. W., & J. J. Skvarla. 1979 (1980). Pollen morphology: The potential influence in higher order systematics. Ann. Missouri Bot. Gard. 66: 633–700.
Pacltová, B. 1977. Cretaceous angiosperms of Bohemia—Central Europe. Bot. Rev. 43: 128–142.
Page, V. M. 1967, 1968, 1970. Angiosperm wood from the Upper Cretaceous of central Calif. Parts I, II, III. Amer. J. Bot. 54: 510–514, 1967. 55: 168–172, 1968. 57: 1139–1144, 1970.
Palser, B. F. 1975. The bases of angiosperm phylogeny: Embryology. Ann. Missouri Bot. Gard. 62: 621–646.
Pant, D. D., & B. Mehra. 1964. Nodal anatomy in retrospect. Phytomorphology 14: 384–387.
Parkin, J. 1951. The protrusion of the connective beyond the anther and its bearing on the evolution of the stamen. Phytomorphology 1: 1–8.
Parkin, J. 1952. The unisexual flower—a criticism. Phytomorphology 2: 75–79.
Parkin, J. 1953. The durian theory—a criticism. Phytomorphology 3: 80–88.

Parkin, J. 1955. A plea for a simpler gynoecium. Phytomorphology 5: 46–57.
Parkin, J. 1957. The unisexual flower again—a criticism. Phytomorphology 7: 7–9.
Percival, M. S. 1961. Types of nectar in the angiosperms. New Phytol. 60: 235–281.
Pijl, L. van der. 1955. Sarcotesta, aril, pulpa and the evolution of the angiosperm fruit. I/II. Proc. K. Nederl. Akad. Wetens., Amsterdam, C, 58: 154–171; 307–312.
Pijl, L. van der. 1960, 1961. Ecological aspects of flower evolution. I. Phyletic evolution. II. Zoophilous flower classes. Evolution 14: 403–416, 1960. 15: 44–59, 1961.
Pijl, L. van der. 1966. Ecological aspects of fruit evolution. A functional study of dispersal organs. Proc. K. Nederl. Akad. Wetens. Amsterdam, C, 69: 597–640.
Plouvier, V., & J. Favre-Bonvin. 1971. Les iridoïdes et séco-iridoïdes: Répartition, structure, propriétés, biosynthèse. Phytochemistry 10: 1697–1722.
Puri, V. 1951. The role of floral anatomy in the solution of morphological problems. Bot. Rev. 17: 471–553.
Puri, V. 1952. Floral anatomy and inferior ovary. Phytomorphology 2: 122–129.
Puri, V. 1952. Placentation in angiosperms. Bot. Rev. 18: 603–651.
Puri, V. 1961. The classical concept of angiosperm carpel: a reassessment. J. Indian Bot. Soc. 40: 511–524.
Puri, V. 1970. Anther sacs and pollen grains: some aspects of their structure and function. J. Palyn. (Lucknow) 6: 1–17.
Raven, P. H. 1975. The bases of angiosperm phylogeny: Cytology. Ann. Missouri Bot. Gard. 62: 724–764.
Raven, P. H., & D. I. Axelrod. 1974. Angiosperm biogeography and past continental movements. Ann. Missouri Bot. Gard. 61: 539–673.
Rickett, H. W. 1944. The classification of inflorescences. Bot. Rev. 10: 187–231.
Rohweder, O. 1973. Angiospermen-Morphologie—Ergebnis oder Ausgangspunkt phylogenetischer Hypothesen? Bot. Jahrb. Syst. 93: 372–403.
Romeike, A. 1978. Tropane alkaloids—occurrence and systematic importance in angiosperms. Bot. Not. 131: 85–96.
Samylina, V. A. 1968. Early Cretaceous angiosperms of the Soviet Union based on leaf and fruit remains. J. Linn. Soc., Bot. 61: 207–218.
Sattler, R. 1974. A new approach to gynoecial morphology. Phytomorphology 24: 22–34.
Sattler, R. 1975. Organogenesis of flowers. A photographic text-atlas. Univ. Toronto Press.
Sattler, R. 1978. "Fusion" and "continuity" in floral morphology. Notes Roy. Bot. Gard. Edinburgh 36: 397–405.
Savile, D. B. O. 1979. Fungi as aids in higher plant classification. Bot. Rev. 45: 377–503.
Schmid, R. 1972. Floral bundle fusion and vascular conservatism. Taxon 21: 429–446.
Schoonhover, L. M. 1972. Secondary plant substances and insects. *In:* V. C. Runeckles & T. C. Tso, eds., Recent advances in phytochemistry 5: 197–224. Academic Press. New York.
Schultes, R. E. 1976. Indole alkaloids in plant hallucinogens. Pl. Med. 29: 330–342.
Schuster, R. M. 1972. Continental movements, "Wallace's Line" and Indoma-

layan-Australasian dispersal of land plants: Some eclectic concepts. Bot. Rev. 38: 3–86.
Schuster, R. M. 1976. Plate tectonics and its bearing on the geographical origin and dispersal of angiosperms. *In:* C. B. Beck, ed., Origin and early evolution of angiosperms, pp. 48–138. Columbia Univ. Press. New York.
Sell, Y. Tendances évolutives parmi les complexes inflorescentiels. Rev. Gén. Bot. 83: 247–267.
Smith, A. C. 1967. The presence of primitive angiosperms in the Amazon Basin and its significance in indicating migrational routes. Atas do Simpósio sôbre a Biota Amazônica, vol. 4 (Botânica): 37–59.
Smith, D. L. 1964. The evolution of the ovule. Biol. Rev. 39: 137–159.
Solbrig, O. T. 1966. Rol de polinización zoófila en la evolución de las angiospermas. Bol. Soc. Argent. Bot. 11: 1–18.
Sporne, K. R. 1949. A new approach to the problem of the primitive flower. New Phytol. 48: 259–276.
Sporne, K. R. 1954. A note on nuclear endosperm as a primitive character among dicotyledons. Phytomorphology 4: 275–278.
Sporne, K. R. 1954. Statistics and the evolution of dicotyledons. Evolution 8: 55–65.
Sporne, K. R. 1956. The phylogenetic classification of the angiosperms. Biol. Rev. Cambridge Philos. Soc. 31: 1–29.
Sporne, K. R. 1958. Some aspects of floral vascular systems. Proc. Linn. Soc. London 169: 75–84.
Sporne, K. R. 1969. The ovule as an indicator of evolutionary status in angiosperms. New Phytol. 68: 555–566.
Sporne, K. R. 1976. Character correlations among angiosperms and the importance of fossil evidence in assessing their significance. *In:* C. B. Beck, ed., Origin and early evolution of angiosperms, pp. 312–329. Columbia Univ. Press. New York.
Sporne, K. R. 1977. Some problems associated with character correlations. Pl. Syst. Evol., Suppl. 1: 33–51.
Stace, C. A. 1965. Cuticular studies as an aid to plant taxonomy. Bull. Brit. Mus. (Nat. Hist.), Bot. 4(1): 1–78.
Stebbins, G. L. 1951. Natural selection and differentiation of angiosperm families. Evolution 5: 299–324.
Stebbins, G. L. 1965. The probable growth habit of the earliest flowering plants. Ann. Missouri Bot. Gard. 52: 457–468.
Stebbins, G. L. 1970. Transference of function as a factor in the evolution of seeds and their accessory structures. Israel J. Bot. 19: 59–70.
Stebbins, G. L. 1973. Evolutionary trends in the inflorescence of angiosperms. Flora 162: 501–528.
Stebbins, G. L. 1974. Flowering plants: Evolution above the species level. Belknap Press of Harvard Univ. Press. Cambridge, Mass.
Stebbins, G. L. 1976. Seeds, seedlings, and the origin of angiosperms. *In:* C. B. Beck, ed., Origin and early evolution of angiosperms, pp. 300–311. Columbia Univ. Press. New York.
Swain, T. 1978. Plant-animal coevolution: a synoptic view of the Paleozoic and Mesozoic. *In:* J. B. Harborne, ed., Biochemical aspects of plant-animal coevolution, pp. 3–19. Academic Press. London.
Swamy, B. G. L., & K. Periasamy. 1964. The concept of the conduplicate carpel. Phytomorphology 14: 319–327.

Takhtajan, A. L., 1958 (1954). Origins of the angiospermous plants. Translated by Olga Gankin. Amer. Inst. Biol. Sci. Washington, 1958. (Original in Russian published by Soviet Sciences Press, 1954.)
Takhtajan, A. 1959. Die Evolution der Angiospermen. Transl. to German by W. Höppner. Gustav Fischer Verlag. Jena.
Takhtajan, A. 1968. Classification and phylogeny, with special reference to the flowering plants. Proc. Linn. Soc. London 179: 221–227.
Takhtajan, A. 1969. Flowering plants: Origin and dispersal. Transl. from the Russian by C. Jeffrey, Kew. Oliver & Boyd. Edinburgh.
Takhtajan, A. 1972. Patterns of ontogenetic alterations in the evolution of higher plants. Phytomorphology 22: 164–171.
Takhtajan, A. 1973. Evolution und Ausbreitung der Blütenpflanzen. Gustav Fischer Verlag. Jena.
Takhtajan, A. 1976. Neoteny and the origin of flowering plants. *In:* C. B. Beck, ed., Origin and early evolution of angiosperms, pp. 207–219. Columbia Univ. Press. New York.
Takhtajan, A. 1980. Outline of the classification of flowering plants (Magnoliophyta). Bot. Rev. 46: 225–359.
ter Welle, J. H. 1976. Silica grains in woody plants of the neotropics, especially Surinam. Leiden Bot. Ser. 3: 107–142.
Thorne, R. F. 1958. Some guiding principles in angiosperm phylogeny. Brittonia 10: 72–77.
Thorne, R. F. 1963. Some problems and guiding principles of angiosperm phylogeny. Amer. Naturalist 97: 287–305.
Thorne, R. F. 1968. Synopsis of a putatively phylogenetic classification of the flowering plants. Aliso 6(4): 57–66.
Thorne, R. F. 1973. Major disjunctions in the geographic ranges of seed plants. Quart. Rev. Biol. 47: 365–411.
Thorne, R. F. 1975. Angiosperm phylogeny and geography. Ann. Missouri Bot. Gard. 62: 362–367.
Thorne, R. F. 1976. A phylogenetic classification of the Angiospermae. *In:* M. K. Hecht, W. C. Steere, & B. Wallace, eds., Evolutionary biology, vol. 9, pp. 35–106. Plenum Press. New York & London.
Thorne, R. F. 1977. Some realignments in the Angiospermae. Pl. Syst. Evol., Suppl. 1: 299–319.
Thorne, R. F. 1978. Plate tectonics and angiosperm distribution. Notes Roy. Bot. Gard. Edinburgh 36: 297–315.
Tippo, O. 1946. The role of wood anatomy in phylogeny. Amer. Midl. Naturalist 36: 362–372.
Troll, W. 1932. Morphologie der schildförmigen Blätter. Planta 17: 153–314.
Van Campo, M. 1976. Patterns of pollen morphological variation within taxa. *In:* I. K. Ferguson & J. Muller, eds., The evolutionary significance of the exine, pp. 125–137. (Linn. Soc. Symp. Ser. No. 1). Academic Press. London & New York.
Vavilov, N. I. 1922. The law of homologous series in variation. J. Genetics 12: 47–89.
Vuilleumier, B. S. 1967. The origin and evolutionary development of heterostyly in the angiosperms. Evolution 21: 210–226.
Wagenitz, G. 1975. Blütenreduktion als ein zentrales Problem der Angiospermen-Systematik. Bot. Jahrb. Syst. 96: 448–470.
Walker, J. W. 1976. Evolutionary significance of the exine in the pollen of

primitive angiosperms. *In:* I. K. Ferguson & J. Muller, eds., The evolutionary significance of the exine, pp. 481–498. (Linn. Soc. Symp. Ser. No. 1). Academic Press. London & New York.

Walker, J. W., & J. A. Doyle. 1975. The bases of angiosperm phylogeny: Palynology. Ann. Missouri Bot. Gard. 62: 664–723.

Weberling, F. 1958. Die Bedeutung blattmorphologischer Untersuchungen für die Systematik (dargestellt am Beispiel der Unterblattbildungen). Bot. Jahrb. Syst. 77: 458–468.

Weberling, F. 1965. Typology of inflorescences. J. Linn. Soc., Bot. 59: 215–221.

Weberling, F. 1975. Über die Beziehungen zwischen Scheidenlappen und Stipeln. Bot. Jahrb. Syst. 96: 471–491.

Wernham, H. F. 1912. Floral evolution: with particular reference to the sympetalous dicotyledons. New Phytol. 11: 373–397.

Whitehead, D. R. 1969. Wind pollination in the angiosperms: evolutionary and environmental considerations. Evolution 23: 28–35.

Willaman, J. J., & B. G. Schubert. 1961. Alkaloid-bearing plants and their contained alkaloids. Techn. Bull. U.S.D.A. 1234: 1–287.

Willis, J. C. 1973. A dictionary of the flowering plants and ferns. 8th ed. Rev. by H. K. Airy Shaw. Univ. Press. Cambridge.

Wilson, C. L., & T. Just. 1939. The morphology of the flower. Bot. Rev. 5: 97–131.

Wunderlich, R. 1959. Zur Frage der Phylogenie der Endospermtypen bei den Angiospermen. Oesterr. Bot. Z. 106: 203–293.

Агабабян, В. Ш. 1973. Палинология и происхождение покрытосеменных. В сб.: Морфология пыльцы и спор современных растений. Труды III Международной Палинологической Конференции. СССР, Новосибирск, 1971: 7-10 изд. Наука, Ленинградское Отделение. Ленинград.

Буш, Н. А. 1935. Систематика высших растений. Издание третье. 1959. Государственное Учебно-педагогическое Изд. Министерства Просвещения Р.С.Ф.С.Р. Москва.

Вахрамеев, В. А., & В. А. Красилов. 1979. Репродуктивные органы цветковых из Альба Казахстана. Палеонтол. Ж. 1979: 121-128.

Виноградов, И. С. 1958. Сокращенное изложенне системы покрытосеменных. Пробл. Бот., Москва-Ленинград 3: 9-66.

Карташова, Н. Н. 1961. О возникновении нектарников в процессе олигомеризации частей цветка. В сб.: Морфогенез растений 2: 511-515. Иэд. Московского Унив.

Козо-Полянский, Б. М. 1922. Введение в филогенетическую систематику высших растений. Воронеж.

Кордюм, Е. Л. 1978. Эволюционная цитоэмбриология покрытосеменных растений. Киев "Наукога Думка."

Куприянова, Л. А. 1966. Апертуры пыльцевых зерен и их эволюция у покрытосеменных растений. В сб.: Значение палинологического анализа для стратиграфии и палеофлористики: 7-14 Изд. Наука. Москва.

Первухина, Н. В. 1967. Опыление первичных покрытосеменных и эволюция способов опыления. Бот. Ж. 52: 157-188.

Первухина, Н. В. 1970. Проблемы морфологии и биологии цветка. Изд. Наука, Ленинградское Отделение. Ленинград.

Поддубная-Арнольди, В. А. 1958. Значение эмбриологических исследований для построения филогенетической системы покрытосеменных растений. В сб.: Проблемы Ботаники. III. 196–247. Изд. АН СССР, Москва—Ленинград.

Поддубная-Арнольди, В. А. 1964. Общая эмбриология покрытосеменных растений. Изд. Наука. Москва.

Тахтаджян, А. Л. 1946. Об эволюционной гетерохронии признаков. Докл. АН Армянской ССР, 5: 79–86.

Тахтаджян, А. Л. 1964. Основы эволюционной морфологии покрытосеменных. Изд. Наука, Москва—Ленинград.

Тахтаджян, А. Л. 1966. Система и филогения цветковых растений. Изд. Наука, Москва—Ленинград.

Терёхин, Э. С. 1975. О способах, направлениях и эволюционном значении редукции биологических структур. Бот. Ж. 60: 1401–1412.

Трутивицки, И. В. 1965. Проблема Prephanerogamae и вопрос об эволюции семен. В сб.: Проблемы Филогении Растений. Труды Московск. общ. исп. прир. 13: 43–51.

Федоров, Ан. А., редактор. 1969. Хромосомные числа цветковых растений. Изд. Наука, Ленинградское Отделение. Ленинград.

Яковлев, М. С. 1958. Принципы выделения основных эмбриональных типов и их значение для филогении покрытосеменных. Пробл. Бот. 3: 168–195.

Яценко-Хмелевский, А. А. 1958. Происхождение покрытосеменных по данным внутренней морфологии их вегетативных органов. Бот. Ж. 43: 365–380.

CLASS MAGNOLIOPSIDA

Class MAGNOLIOPSIDA Cronquist, Takhtajan & Zimmermann, 1966, the Dicotyledons

Plants woody or herbaceous, the woody forms and many of the herbaceous ones with typical secondary growth in the stems and roots (the cambium producing xylem toward the inside and phloem toward the outside); mature root-system either primary, or adventitious, or both; vascular bundles of the stem, in herbaceous forms, open (i.e., with a cambial layer between the xylem and phloem) or less often closed, typically arranged in a ring that encloses a pith, seldom in 2 or more rings or scattered; vessels commonly well developed in the roots, stems, and leaves, but wanting in some archaic families and in some reduced ones; plastids of the sieve tubes without proteinaceous inclusions, or less often with proteinaceous inclusions that are not cuneate. LEAVES typically net-veined, most commonly with a petiole and an expanded blade, only seldom with a well defined basal sheath; blade typically developing from the distal portion of the leaf-primordium and expanding more or less simultaneously throughout its length. FLOWERS with various sorts of nectaries, but without septal nectaries, or altogether without nectaries; floral parts, when of definite number, typically borne in sets of 5, less often 4, seldom 3 or other numbers (the carpels often fewer); pollen-grains typically triaperturate or of triaperturate-derived type, except in some of the more archaic families, which are uniaperturate; derived pollen-types include biaperturate, multiaperturate, and inaperturate (or holoaperturate) forms; cotyledons 2, rarely 1, 3, or 4, or rarely (some highly mycotrophic or parasitic forms) the embryo not differentiated into parts.

FIG. 1.1 Flowering branch of *Magnolia grandiflora*, an evergreen species of the southeastern United States. U.S. Forest Service photograph by W. D. Brush.

SELECTED REFERENCES

Balfour, E. E., & W. R. Philipson. 1962. The development of the primary vascular system of certain dicots. Phytomorphology 12:110–143.

Barghoorn, E. S. 1940, 1941. The ontogenetic development and phylogenetic specialization of rays in the xylem of dicotyledons. I. The primitive ray structure. II. Modification of the multiseriate and uniseriate rays. Amer. J. Bot. 27: 918–928, 1940. 28: 273–282, 1941. III. The elimination of rays. Bull. Torrey Bot. Club 68: 317–325, 1941.

Bate-Smith, E. C., & C. R. Metcalfe. 1957. Leuco-anthocyanins. 3. The nature and systematic distribution of tannins in dicotyledonous plants. J. Linn. Soc., Bot. 55: 669–705.

Bonneman, J.-L. 1969. Le phloème interne et le phloème inclus des dicotylédones: Leur histogénèse et leur physiologie. Rev. Gén. Bot. 76: 5–36.

Carlquist, S. 1962. A theory of paedomorphosis in dicotyledonous woods. Phytomorphology 12: 30–45.

Copeland, H. F. 1957. Forecast of a system of the dicotyledons. Madroño 14: 1–9.

Corner, E. J. H. 1976. The seeds of dicotyledons. 2 vols. Cambridge Univ. Press. Cambridge, England.

Cronquist, A. 1957. Outline of a new system of families and orders of dicotyledons. Bull. Jard, Bot. État 27: 13–40.

Feldhofen, E. 1933. Beiträge zur physiologischen Anatomie der nuptialen Nektarien aus den Reihen der Dikotylen. Beih. Bot. Centralbl. 50(1): 459–634.

Frei, E. 1955. Die Innervierung der floralen Nektarien dikotyler Pflanzenfamilien. Ber. Schweiz. Bot. Ges. 65: 60–114.

Frost, F. H. 1930, 1931. Specialization in secondary xylem of dicotyledons. I. Origin of vessel. II. Evolution of end wall of vessel segment. III. Specialization of lateral wall of vessel segment. Bot. Gaz. 89: 67–94; 90: 198–212, 1930. 91: 88–96, 1931.

Gardner, R. O. 1977. Systematic distribution and ecological function of the secondary metabolites of the Rosidae-Asteridae. Biochem. Syst. Ecol. 5: 29–35.

Gottwald, H. 1977. The anatomy of secondary xylem and the classification of ancient dicotyledons. Pl. Syst. Evol., Suppl. 1: 111–121.

Gundersen, A. 1950. Families of dicotyledons. Chronica Botanica Co. Waltham, Mass.

Hess, R. W. 1950. Classification of wood parenchyma in dicotyledons. Trop. Woods 96: 1–20.

Hickey, L. J. 1973. Classification of the architecture of dicotyledonous leaves. Amer. J. Bot. 60: 17–33.

Kribs, D. A. 1935. Salient lines of structural specialization in the wood rays of dicotyledons. Bot. Gaz. 96: 547–557.

Kribs, D. A. 1937. Salient lines of structural specialization in the wood parenchyma of dicotyledons. Bull. Torrey Bot. Club 64: 177–187.

Kubitzki, K. 1969. Chemosystematische Betrachtungen zur Grossgleiderung der Dicotylen. Taxon 18: 360–368.

Lebègue, A. 1952. Recherches embryogéniques sur quelques dicotylédones dialypetales. Ann. Sci. Nat. Bot. sér. 11. 13: 1–160.

Leins, P. 1971. Das Androecium der Dikotylen. Ber. Deutsch. Bot. Ges. 84: 191–193.

Metcalfe, C. R., & L. Chalk. 1950. Anatomy of the dicotyledons; leaves, stem, and wood in relation to taxonomy, with notes on economic uses. 2 vols. Clarendon Press. Oxford.

Philipson, W. R. 1975. Evolutionary lines within the dicotyledons. New Zealand J. Bot. 13: 73–91.

Philipson, W. R. 1977. Ovular morphology and the classification of dicotyledons. Pl. Syst. Evol., Suppl. 1: 123–140.

Sporne, K. R. 1975. A note on ellagitannins as indicators of evolutionary status in dicotyledons. New Phytol. 75: 613–618.

Tiffney, B. H. 1977. Dicotyledonous angiosperm flower from the Upper Cretaceous of Martha's Vineyard, Massachusetts. Nature 265: 136–137.

Weberling, F. 1953/55. Morphologische und entwicklungsgeschichtliche Untersuchungen über die Ausbildung des Unterblattes bei dikotylen Gewächsen. Beitr. Biol. Pflanzen 32: 27–105.

Wheeler, E., R. A. Scott, & E. S. Barghoorn. 1977, 1978. Fossil dicotyledonous woods from Yellowstone National Park. J. Arnold Arbor. 58: 280–306, 1977. 59: 1–25, 1978.

Young, D. J., & L. Watson. 1970. The classification of dicotyledons: a study of the upper levels of the hierarchy. Austral. J. Bot. 18: 387–433.

SYNOPTICAL ARRANGEMENT OF THE SUBCLASSES OF MAGNOLIOPSIDA[1]

1 Plants relatively archaic, the flowers typically apocarpous, always polypetalous or apetalous (but sometimes synsepalous) and generally with an evident perianth, usually with numerous (sometimes laminar or ribbon-shaped) stamens initiated in centripetal sequence, the pollen-grains mostly binucleate and often uniaperturate or of a uniaperturate-derived type;[2] ovules bitegmic and crassinucellar; seeds very often with a tiny embryo and copious endosperm, but sometimes with a larger embryo and reduced or no endosperm; cotyledons occasionally more than 2; plants very often accumulating benzyl-isoquinoline or aporphine alkaloids, but without betalains, iridoid compounds, or mustard oils, and only seldom strongly tanniferous ....I. MAGNOLIIDAE.

1 Plants more advanced in one or more respects than the Magnoliidae; pollen-grains triaperturate or of a triaperturate-derived type; cotyledons not more than 2; stamens not laminar, generally with well defined filament and anther; plants only rarely producing benzyl-isoquinoline or aporphine alkaloids, but often with other kinds of alkaloids, or tannins, or betalains, or mustard oils, or iridoid compounds.

2 Flowers more or less strongly reduced and often unisexual, the perianth poorly developed or wanting, the flowers often borne in catkins, but never forming bisexual pseudanthia, and never with numerous seeds on parietal placentas; pollen-grains often porate and with a granular rather than columellar infratectal structure, but also often of ordinary type .... II. HAMAMELIDAE.

2 Flowers usually more or less well developed and with an evident perianth, but if not so, then usually either grouped into bisexual pseudanthia or with numerous seeds on parietal placentas, only rarely with all of the characters of the Hamamelidae as listed above; pollen-grains of various architecture, but rarely if ever both porate and with a granular infratectal structure.

[1] It should be clearly understood that the synoptical arrangements presented in this book are intended primarily as conceptual aids rather than as a means of identification. Characters that are difficult to observe are frequently used, and many of the numerous exceptions are necessarily minimized or ignored.

[2] By established convention, the term uniaperturate-derived is interpreted to exclude triaperturate pollen, even though the uniaperturate type is phyletically antecedent to the triaperturate type.

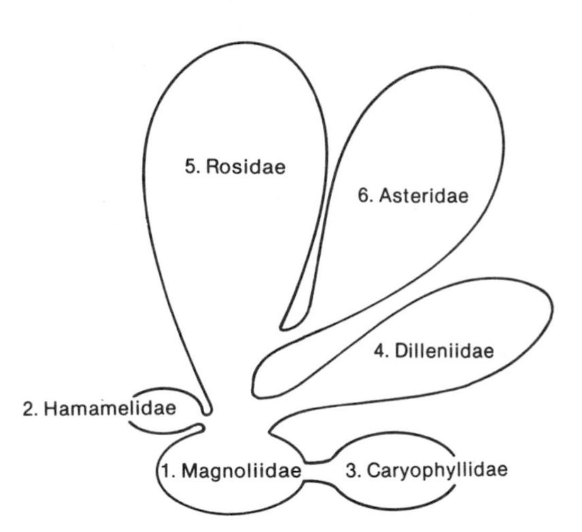

FIG. 1.2 Putative evolutionary relationships among the subclasses of dicotyledons. The size of the balloons is proportional to the number of species in each group.

3 Flowers polypetalous or less often apetalous or sympetalous, if sympetalous then usually either with more stamens than corolla-lobes, or with the stamens opposite the corolla-lobes, or with bitegmic or crassinucellar ovules; ovules only rather seldom with an integumentary tapetum; carpels 1-many, distinct or more often united to form a compound pistil; plants often tanniferous or with betalains or mustard oils.

4 Stamens, when numerous, usually initiated in centrifugal (seldom centripetal) sequence; placentation various, often parietal or free-central or basal, but also often axile; species with few stamens and axile placentation usually either bearing several or many ovules per locule, or with a sympetalous corolla, or both.

5 Plants usually either with betalains, or with free-central to basal placentation (in a compound ovary), or both, but lacking both mustard-oils and iridoid compounds, and tanniferous only in the two smaller orders; pollen-grains trinucleate or seldom binucleate; ovules bitegmic, crassinucellar, most often campylotropous or amphitropous; plants most commonly herbaceous or nearly so, the woody species usually with anomalous secondary growth or otherwise anomalous stem-structure; petals distinct or wanting, except in the Plumbaginales; in the largest order (Caryophyllales) the sieve-tubes with a unique sort of

P-type plastid, and the seeds usually with perisperm instead of endosperm .... III. CARYOPHYLLIDAE.

5 Plants without betalains, and the placentation only rarely (except in the Primulales) free-central or basal; plants often with mustard-oils or iridoid compounds or tannins; pollen-grains usually binucleate (notable exception: Brassicaceae); ovules various, but seldom campylotropous or amphitropous except in the Capparales; plants variously woody or herbaceous, many species ordinary trees; petals distinct or less often connate to form a sympetalous corolla, seldom wanting; seeds only seldom with perisperm; plastids of the sieve-tubes usually of S-type, but in any case not as in the Caryophyllales .. IV. DILLENIIDAE.

4 Stamens, when numerous, usually initiated in centripetal (seldom centrifugal) sequence; flowers seldom with parietal placentation (notable exception: many Saxifragaceae) and also seldom (except in parasitic species) with free-central or basal placentation in a unilocular, compound ovary, but very often (especially in species with few stamens) with 2-several locules that have only 1 or 2 ovules each; flowers polypetalous or less often apetalous, only rarely sympetalous; plants often tanniferous, and sometimes with iridoid compounds, but only rarely with mustard-oils, and never with betalains .... V. ROSIDAE.

3 Flowers sympetalous (rarely polypetalous or apetalous); stamens generally isomerous with the corolla-lobes or fewer, never opposite the lobes; ovules unitegmic and tenuinucellar, very often with an integumentary tapetum; carpels most commonly 2, occasionally 3-5 or more; plants only seldom tanniferous, and never with betalains or mustard-oils, but often with iridoid compounds or various other sorts of repellants .... VI. ASTERIDAE.

I. Subclass MAGNOLIIDAE Takhtajan 1966

Woody or herbaceous dicotyledons, very often producing benzyl-isoquinoline or aporphine alkaloids, less commonly other sorts of alkaloids, sometimes somewhat tanniferous, but only rarely (Nymphaeaceae, Coriariaceae) with ellagic acid, sometimes cyanogenic (based on tyrosine), but wholly without iridoid compounds; parenchymatous tissues often containing scattered spherical ethereal oil cells; vessel-segments with scalariform or simple perforations, or vessels sometimes wanting; wood-rays often with elongate ends; sieve-tube plastids usually containing protein crystalloids or filaments of one or another form, often also starch grains, less often only starch, without protein. Flowers regular or much less often irregular, hypogynous to less often perigynous or epigynous; tepals usually distinct, often not well differentiated into sepals and petals, or only the sepals present, seldom all the tepals connate, or rarely (Barclayaceae) the petals connate below to form a sympetalous corolla, or the flowers sometimes much-reduced and without perianth; stamens most commonly numerous and initiated in centripetal sequence, less often few and cyclic, not infrequently ribbon-shaped or laminar and without clearly differentiated filament and anther, or with an evidently prolonged connective, but sometimes of very ordinary form, with well defined filament and anther; pollen-grains binucleate or less often trinucleate, often uniaperturate or of uniaperturate-derived type, but in some families triaperturate or of a triaperturate-derived type; gynoecium most commonly apocarpous (or monocarpous), less often syncarpous; placentation in apocarpous and monocarpous types marginal, or sometimes laminar or laminar-lateral, or apical or basal, in syncarpous types variously parietal or axile; ovules bitegmic or much less often unitegmic, crassinucellar or much less often tenuinucellar; endosperm present or less often absent, sometimes accompanied or largely replaced by perisperm; embryo often very small, but sometimes larger and even filling the seed; cotyledons typically 2, but occasionally 3 or 4 (Degeneriaceae, Idiospermaceae), rarely 1 by suppression of one member of the pair.

The subclass Magnoliidae as here defined consists of 8 orders, 39 families, and about 12,000 species. Three of the orders (Magnoliales, Laurales, Ranunculales) make up more than two-thirds of the species.

Every order that is here referred to the Magnoliidae now seems securely placed there.

The Magnoliidae consist principally of those dicotyledons that have retained one or more features of a syndrome of primitive characters, and that are not obviously closely allied to some more advanced group. The most important items in the syndrome are uniaperturate (or uniaperturate-derived) pollen, an apocarpous (or monocarpous) gynoecium, and numerous stamens that originate in centripetal sequence. All of the dicotyledons that have uniaperturate or uniaperturate-derived pollen belong to the Magnoliidae, but about one third of the species of the subclass have triaperturate or triaperturate-derived pollen. More than nine-tenths of the species of Magnoliidae have an apocarpous or monocarpous gynoecium, but in some of the less advanced members of the Hamamelidae, Caryophyllidae, Dilleniidae, and Rosidae the gynoecium is apocarpous as in most of the Magnoliidae. These apocarpous members of other subclasses are obviously related to syncarpous members of their respective groups. Something more than half of the Magnoliidae have numerous stamens that are initiated in centripetal sequence, but in a considerable number of the Rosidae the stamens are also numerous and centripetal. With the single debatable exception of *Barclaya* (Nymphaeales), the Magnoliidae do not have a sympetalous corolla, but this feature sets them apart only from the Asteridae and more advanced Dilleniidae. The vast majority of the Magnoliidae have bitegmic, crassinucellar ovules, but these features are also widespread in other subclasses except for the Asteridae. Variously other putatively primitive features, such as vesselless wood, laminar stamens, laminar placentation, and the presence of more than 2 cotyledons, are largely or wholly restricted to the Magnoliidae, but are not standard features even within the subclass.

Finally, the Magnoliidae have their own set of chemical defenses. A great many of them have isoquinoline alkaloids and related compounds, especially benzyl-isoquinoline and aporphine types. These are rare (though not unknown) in other groups of angiosperms. The various other sorts of alkaloids found in other angiosperms are largely or wholly wanting from the Magnoliidae. Many of the Magnoliidae have characteristic volatile oils, often in specialized spherical idioblasts, but some of these same oils occur in other groups, such as for example the Apiaceae. Cyanogenic members of the Magnoliidae use a tyrosine-based pathway, so far as known. Most of the cyanogenic monocotyledons resemble the Magnoliidae in this regard. Other subclasses of dicotyledons have developed other pathways of cyanogenesis, while retaining the tyrosine-based pathway in some members. Iridoid compounds and mustard oils are unknown in the Magnoliidae. I have elsewhere (1977, in general citations) expounded the view that chemical defenses of plants tend to evolve in successive waves as old weapons become less effective. Under this concept, the chemical arsenal of the Magnoliidae may consist largely of weapons that have been superseded in more advanced groups.

Within the Magnoliidae, it is clear that the Magnoliales are the most archaic order and should come first in the linear sequence. One can reasonably suppose that if the common ancestor of all the families in any other order of the subclass were alive today, it would be referred to the Magnoliales, or to another order that can be traced back to the Magnoliales.

The position and status of the other orders of Magnoliidae in the system are to a considerable extent measured by how much they have differentiated from the Magnoliales. It seems reasonable to put the Laurales next after the Magnoliales in the sequence, since it is the Laurales that are most difficult to distinguish clearly from the Magnoliales. It is equally clear that the Papaverales are derived from the Ranunculales, and that these are the two most advanced orders of the subclass. They therefore terminate the linear sequence.

The Piperales, Aristolochiales, Illiciales, and Nymphaeales appear to be derived directly and individually from the Magnoliales. The sequence among these four is therefore more a matter of convenience than of necessity. I have chosen to arrange them in the sequence of my synoptical key. There is some further logic in this arrangement in that the Piperales, which are connected to the Magnoliales by the somewhat transitional family Chloranthaceae, come first among the group, and that the Nymphaeales, which are the most divergent from the Magnoliales, come last. A case might be made, however, for putting the

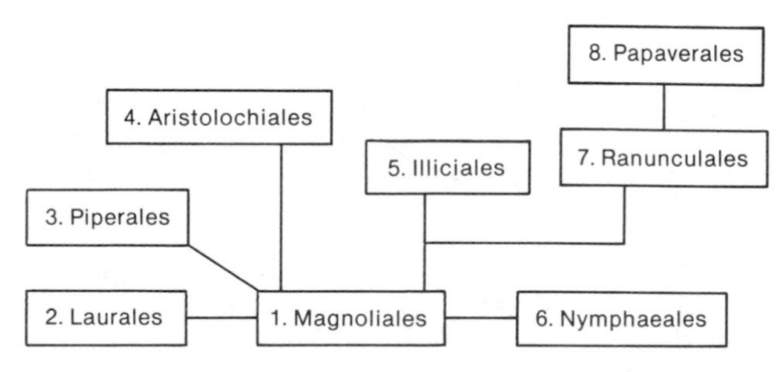

FIG. 1.3 Putative evolutionary relationships among the orders of Magnoliidae.

Illiciales immediately before the Ranunculales, inasmuch as the Ranunculales appear to take their origin in or near this group.

SELECTED REFERENCES

Bailey, I. W., & B. G. L. Swamy. 1951. The conduplicate carpel of dicotyledons and its initial trends of specialization. Amer. J. Bot. 38: 373–379.

Behnke, H.-D. 1971. Sieve-tube plastids of Magnoliidae and Ranunculidae in relation to systematics. Taxon 20: 723–730.

Benzing, D. H. 1967. Developmental patterns in stem primary xylem of woody Ranales. I. Species with unilacunar nodes. II. Species with trilacunar and multilacunar nodes. Amer. J. Bot. 54: 805–813; 813–820.

Canright, J. E. 1963. Contributions of pollen morphology to the phylogeny of some ranalean families. Grana Palynol. 4: 64–72.

Dilcher, D. L., W. L. Crepet, & C. D. Beeker. 1976. Reproductive and vegetative morphology of a Cretaceous angiosperm. Science 191: 854–856.

Ehrendorfer, F. 1976. Evolutionary significance of chromosomal differentiation patterns in gymnosperms and primitive angiosperms. *In:* C. B. Beck, ed., Origin and early evolution of angiosperms, pp. 220–240. Columbia Univ. Press. New York.

Ehrendorfer, F., F. Krendl, E. Habeler, & W. Sauer. 1968. Chromosome numbers and evolution in primitive angiosperms. Taxon 17: 337–353.

Endress, P. K. 1973. Arils and aril-like structures in woody Ranales. New Phytol. 72: 1159–1171.

Hiepko, P. 1965. Vergleichend-morphologische und entwicklungsgeschichtliche Untersuchungen über das Perianth bei den Polycarpicae. Bot. Jahrb. Syst. 84: 359–508.

Kubitzki, K., & H. Reznik. 1966. Flavonoid-Muster der Polycarpicae als systematisches Merkmal. I. Übersicht über die Familien. Beitr. Biol. Pflanzen 42: 445–470.

Lemesle, R. 1955. Contribution a l'étude de quelques familles de dicotylédones considérées comme primitives. Phytomorphology 5: 11–45.

Lemesle, R. 1956. Les éléments du xylème dans les Angiospermes à caractères primitifs. Bull. Soc. Bot. France 103: 629–677.

Leonhardt, R. 1951. Phylogenetisch-systematische Betrachtungen. II. Gedanken zur systematischen Stellung, bzw. Gliederung einiger Familien der Choripetalen. Oesterr. Bot. Z. 98: 1–43.

Okada, H. 1975. Karyomorphological studies of woody Polycarpicae. J. Sci. Hiroshima Univ. Ser. B, Div. 2, Bot. 15: 115–200.

Ozenda, P. 1949. Recherches sur les dicotylédones apocarpiques. Publ. Lab. Biol. École Norm. Supér., Paris.

Ratter, J. A., & C. Milne. 1973. Chromosome numbers of some primitive angiosperms. Notes Roy. Bot. Gard. Edinburgh 32: 423–428.

Raven, P. H., & D. W. Kyhos. 1965. New evidence concerning the original basic chromosome number of angiosperms. Evolution 19: 244–248.

Raven, P. H., D. W. Kyhos, & M. S. Cave. 1971. Chromosome numbers and relationships in Annoniflorae. Taxon 20: 479–483.

Sastri, R. L. N. 1959. The vascularization of the carpel in some Ranales. New Phytol. 58: 306–309.

Sastri, R. L. N. 1969. Comparative morphology and phylogeny of the Ranales. Biol. Rev. Cambridge Philos. Soc. 44: 291–319.
Smith, A. C. 1971 (1972). An appraisal of the orders and families of primitive extant angiosperms. J. Indian Bot. Soc. Golden Jubilee Vol. 50A: 215–226.
Thien, L. B. 1980. Patterns of pollination in primitive angiosperms. Biotropica 12: 1–13.
Thorne, R. F. 1974. A phylogenetic classification of the Annoniflorae. Aliso 8: 147–209.
Valen, F. van. 1978. Contribution to the knowledge of cyanogenesis in angiosperms. 6. Communication. Cyanogenesis in some Magnoliidae. Proc. Kon. Nederl. Akad. Wetensch. ser. C. 81: 355–362.
Walker, J. W. 1974. Aperture evolution in the pollen of primitive angiosperms. Amer. J. Bot. 61: 1112–1137.
Walker, J. W. 1974. Evolution of exine structure in the pollen of primitive angiosperms. Amer. J. Bot. 61: 891–902.
Walker, J. W. 1976. Comparative pollen morphology and phylogeny of the Ranalean complex. *In:* C. B. Beck, ed., Origin and early evolution of angiosperms, pp. 241–299. Columbia Univ. Press, New York.
West, W. C. 1969. Ontogeny of oil cells in the woody Ranales. Bull. Torrey Bot. Club 96: 329–344.
Wood, C. E. 1958. The genera of the woody Ranales in the southeastern United States. J. Arnold Arbor. 39: 296–346.

Агабабян, В. Ш. 1973. Пыльца примитивных покрытосеменных. Изд. АН Армянской ССР. Ереван, 1973.

SYNOPTICAL ARRANGEMENT OF THE ORDERS OF MAGNOLIIDAE

1 Plants ordinarily with ethereal oil cells in the parenchymatous tissues; pollen (except Illiciales) uniaperturate or derived from a uniaperturate type, not triaperturate; petals, when present, homologous with the sepals and bracts, and very often with 3 or more vascular traces.

2 Pollen variously uniaperturate, biaperturate, multiaperturate, or inaperturate, but not triaperturate.

3 Plants distinctly woody (except *Cassytha*, of the Lauraceae); flowers normally developed, usually with an evident perianth of separate tepals that may or may not be differentiated into sepals and petals (in Eupomatiaceae the tepals all connate to form a deciduous calyptra).

4 Flowers mostly hypogynous (somewhat perigynous in the Eupomatiaceae, these with numerous ovules); pollen typically uniaperturate or sometimes inaperturate (biaperturate in Eupomatiaceae and some Annonaceae); stamens often laminar, or with a prolonged and enlarged connective, but sometimes of more ordinary type; nodes most commonly trilacunar, sometimes multilacunar or unilacunar; stipules present or absent; flowers mostly solitary or in open, few-

flowered inflorescences, in most families rather large; seeds with small embryo and copious endosperm .... 1. MAGNOLIALES.

4 Flowers usually perigynous or epigynous (hypogynous in the small families Amborellaceae and Trimeniaceae, and in a few Monimiaceae); pollen mostly inaperturate or biaperturate, less often uniaperturate or multiaperturate; stamens usually of fairly ordinary type, but sometimes more or less laminar and with prolonged connective; ovule solitary (or 2, but only one maturing); nodes unilacunar; stipules wanting; flowers typically smaller and in more numerously flowered inflorescences than in most Magnoliales; seeds as in Magnoliales, or often with a large embryo and little or no endosperm ....2. LAURALES.

3 Plants otherwise, most species herbaceous or only secondarily woody.

5 Flowers with much-reduced or no perianth, often crowded into a spadix-like inflorescence, ovules mostly orthotropous; plants without aristolochic acid; seeds often with perisperm ....3. PIPERALES.

5 Flowers with a well developed perianth, often consisting of a gamosepalous, more or less corolloid calyx, not crowded into a spadix; ovules mostly anatropous; plants commonly accumulating aristolochic acid; seeds with endosperm but not perisperm .... 4. ARISTOLOCHIALES.

2 Pollen triaperturate or sexaperturate; flowers hypogynous; woody plants with unilacunar nodes and without stipules ....5. ILLICIALES.

1 Plants without ethereal oil cells; pollen (except most Nymphaeales) triaperturate or derived from a triaperturate type; petals, when present, mostly apparently staminodial in origin, very often with only a single vascular trace; plants herbaceous, less often woody, and then probably with an herbaceous ancestry.

6 Plants aquatic, lacking vessels (except in the roots of *Nelumbo*); placentation mostly laminar (apical in *Nelumbo*); pollen uniaperturate or inaperturate, or (*Nelumbo*) triaperturate ....6. NYMPHAEALES.

6 Plants terrestrial or occasionally aquatic, with vessels; placentation marginal or parietal to sometimes laminar or axile; pollen triaperturate, or derived from a triaperturate type.

7 Gynoecium mostly apocarpous or seemingly or actually unicarpellate, seldom evidently syncarpous and then with more than

2 carpels; sepals usually more than 2; plants nearly always without protopine .... 7. RANUNCULALES.

7 Gynoecium syncarpous, usually of 2 carpels (carpels more numerous in *Papaver* and some other Papaveraceae, numerous and only weakly united in *Platystemon*); sepals mostly 2, occasionally 3, seldom 4; plants commonly producing protopine (an isoquinoline alkaloid) .... 8. PAPAVERALES.

1. Order MAGNOLIALES Bromhead 1838

Trees, shrubs, or woody vines with ethereal oil cells in the parenchymatous tissues at least of the leaves, and very often producing benzylisoquinoline or aporphine alkaloids; nodes most commonly trilacunar, less often multilacunar, seldom unilacunar; xylem vesselless (Winteraceae) or with well developed vessels, these with scalariform (or reticulate) perforation plates or simply perforate. LEAVES opposite or alternate, simple, entire or seldom (*Liriodendron* in Magnoliaceae) somewhat lobed; stomates paracytic or occasionally anomocytic; stipules present or absent. FLOWERS mostly entomophilous, often cantharophilous, hypogynous, large or small, solitary (either terminal or axillary) or in open and few-flowered or more definitely cymose or racemose inflorescences; perianth well developed, with mostly free and distinct tepals, these variously cyclic (often trimerous) or spirally arranged, often not clearly differentiated into sepals and petals, sometimes in more than two series, the perianth represented or replaced in the Eupomatiaceae by a deciduous calyptra; stamens (3) 6-many, cyclic or spirally arranged, often more or less laminar or with prolonged and expanded connective, sometimes with the filaments united into a tube or column; pollen grains uniaperturate or inaperturate (many Annonaceae) or biaperturate (Eupomatiaceae and some Annonaceae), binucleate so far as known; gynoecium of 1-many separate or seldom connate (but sometimes spirally concrescent) carpels, these sometimes unsealed, with more or less decurrent stigma, varying to fully sealed and with the stigma elevated on a style; ovary superior, with 1-many laminar or marginal or nearly basal ovules, or with united carpels and distinctly parietal placentation in Canellaceae and some Annonaceae; ovules anatropous or hemitropous, bitegmic and crassinucellar so far as known. FRUITS of various types, often follicular; seeds with small embryo and copious, often ruminate endosperm, often with an aril or sarcotesta; cotyledons 2 (3–4 in Degeneriaceae). Perhaps originally $X = 7$.

The order Magnoliales as here defined consists of 10 families and nearly 3000 species. One family, the Annonaceae, has about 2300 species and makes up about three-fourths of the order. Most of the remaining fraction is made up by only three families, the Myristicaceae (300), Magnoliaceae (220), and Winteraceae (100). The other six families of the order add up to scarcely 30 species in all. Aside from the highly diversified and successful family Annonaceae, the order consists of a series of isolated end-lines. The vast majority of species of the order occur in tropical or warm-temperate regions, most of them in places with a moist, equable climate. Otherwise they occupy no particular ecological niche in contrast to other woody angiosperms, at least so far as present knowledge indicates.

Based on the comparative morphology of modern species, and on early Cretaceous fossil pollen and leaves, the Magnoliales must be considered to be the most archaic existing order of flowering plants. The fossil record clearly shows the earliest angiosperm pollen grains were monosulcate, and that simple, entire leaves with relatively unorganized pinnate venation are the ancestral archetype.

The name Annonales Lindley (1833) antedates Magnoliales, but I maintain the latter name because of its wide acceptance in recent works.

SELECTED REFERENCES

Agababian, V. Sh. 1972. Pollen morphology of the family Magnoliaceae. Grana 12: 166–176.

Armstrong, J. E., & T. K. Wilson. 1978. Floral morphology of *Horsfieldia* (Myristicaceae). Amer. J. Bot.65: 441–449.

Bailey, I. W., & C. G. Nast. 1943, 1944, 1945. The comparative morphology of the Winteraceae. I–VII. J. Arnold Arbor. 24: 340–346; 472–481, 1943. 25: 97–103; 215–221; 342–348; 454–466, 1944. 26: 37–47, 1945.

Bailey, I. W., C. G. Nast, & A. C. Smith. 1943. The family Himantandraceae. J. Arnold Arbor. 24: 190–206.

Bailey, I. W., & A. C. Smith. 1942. Degeneriaceae, a new family of flowering plants from Fiji. J. Arnold Arbor. 23: 356–365.

Bailey, I. W., & B. G. L. Swamy. 1949. The morphology and relationships of *Austrobaileya*. J. Arnold Arbor. 30: 211–226.

Baranova, M. 1972. Systematic anatomy of the leaf epidermis in the Magnoliaceae and some related families. Taxon 21: 447–469.

Bhandari, N. N. 1971. Embryology of the Magnoliales and comments on their relationships. J. Arnold Arbor. 52: 1–39; 285–304.

Bongers, J. M. 1973. Epidermal leaf characters of the Winteraceae. Blumea 21: 381–411.

Buchheim, G. 1962. Beobachtungen über den Bau der Frucht der Familie Himantandraceae. Sitzungsber. Ges. Naturf. Freunde Berlin (N. F.) 2: 78–92.

Canright, J. E. 1952, 1953, 1955, 1960. The comparative morphology and relationships of the Magnoliaceae. I. Trends of specialization in the stamens. Amer. J. Bot. 39: 484–497, 1952. II. Significance of the pollen. Phytomorphology 3: 355–365, 1953. III. Carpels. Amer. J. Bot. 47: 145–155, 1960. IV. Wood and nodal anatomy. J. Arnold Arbor. 36: 119–140, 1955.

Carlquist, S. 1964. Morphology and relationships of Lactoridaceae. Aliso 5: 421–435.

Croizat, L. 1940. Notes on the Dilleniaceae and their allies: Austrobaileyeae subfam. nov. J. Arnold Arbor. 21: 397–404.

Dahl, A. O., & J. R. Rowley. 1965. Pollen of *Degeneria vitiensis*. J. Arnold Arbor. 46: 308–323.

Dandy, J. E. 1927. The genera of Magnolieae. Kew Bull. Misc. Inform. 1927: 257–264.

Diels, L. 1932. Die Gliederung der Anonaceen und ihre Phylogenie. Sitzungsber. Preuss. Akad. Wiss.: 1932: 77–85.

Endress, P. K. 1977. Über Blütenbau und Verwandtschaft der Eupomatiaceae und Himantandraceae (Magnoliales). Ber. Deutsch. Bot. Ges. 90: 83–103.

Garratt, G. A. 1933. Bearing of wood anatomy on the relationships of the Myristicaceae. Trop. Woods 36: 20–44.

Goldblatt, P. A. 1974. A contribution to the knowledge of cytology in Magnoliales. J. Arnold Arbor. 55: 453–457.

Gottsberger, G. 1970. Beiträge zur Biologie von Annonaceen-Blüten. Oesterr. Bot. Z. 118: 237–279.

Hayashi, Y. 1965. The comparative embryology of the Magnoliaceae (s.l.) in relation to the systematic consideration of the family. Sci. Rep. Tôhoku Imp. Univ., Ser. 4, Biol. 31: 29–44.

Hiepko, P. 1966. Das Blütendiagramm von *Drimys winteri* J. R. et G. Forst. (Winteraceae). Willdenowia 4: 221–226.

Hotchkiss, A. T. 1955. Chromosome numbers and pollen tetrad size in the Winteraceae. Proc. Linn. Soc. New South Wales 80: 47–53.

Hotchkiss, A. T. 1958. Pollen and pollination in the Eupomatiaceae. Proc. Linn. Soc. New South Wales 83: 86–91.

Huynh, K.-L. 1976. L'arrangement du pollen du genre *Schisandra* (Schisandraceae) et sa signification phylogénique chez les Angiospermes. Beitr. Biol. Pflanzen 52: 227–253.

Johnson, M. A., & D. E. Fairbrothers. 1965. Comparison and interpretation of serological data in the Magnoliaceae. Bot. Gaz. 126: 260–269.

Joshi, A. C. 1946. A note on the development of pollen of *Myristica fragrans* Van Houtten and the affinities of the family Myristicaceae. J. Indian Bot. Soc. 25: 139–143.

Kapil, R. N., & N. N. Bhandari. 1964. Morphology and embryology of *Magnolia.* Proc. Natl. Inst. Sci. India 30B: 245–262.

Kavathekár, K. Y., & A. Pillai. 1976. Studies on the developmental anatomy of Ranales. II. Nodal anatomy of certain members of Annonaceae, Magnoliaceae, Menispermaceae and Ranunculaceae. Flora 165: 481–488.

Kubitzki, K., & W. Vink. 1967. Flavonoid-Muster der Polycarpicae als systematisches Merkmal. II. Untersuchungen an der Gattung *Drimys.* Bot. Jahrb. Syst. 87: 1–16.

Leinfellner, W. 1969. Über die Karpelle verschiedener Magnoliales. VIII. Überblick über alle Familien der Ordnung. Oesterr. Bot. Z. 117: 107–127.

Lemesle, R. 1936. Les vaisseaux à perforations scalariformes de l'*Eupomatia* et leur importance dans la phylogénie des Polycarpes. Compt. Rend. Hebd. Séances Acad. Sci. 203: 1538–1540.

Lemesle, R. 1938. Contribution a l'étude du genre *Eupomatia* R. Br. Rev. Gén. Bot. 50: 693–712.

Lemesle, R. 1953. Les caractères histologiques du bois secondaire des Magnoliales. Phytomorphology 3: 430–446.

Lemesle, R. 1953. Nature des éléments sclérenchymateaux de la tige du *Lactoris fernandeziana* Philippi. Rev. Gén. Bot. 60: 15–21.

Lemesle, R., & A. Duchaigne. 1955. Contribution a l'étude histologique et phylogénétique du *Degeneria vitiensis* I. W. Bailey et A. C. Sm. Rev. Gén. Bot. 62: 708–719.

Leroy, J.-F. 1977. A compound ovary with open carpels in Winteraceae (Magnoliales): Evolutionary implications. Science 196: 977–978.
Leroy, J.-F. 1978. Une sous-famille monotypique de Winteraceae endémique a Madagascar: Les Takhtajanioideae. Adansonia, Sér. 2, 17: 383–395.
Le Thomas, A., & B. Lugardon. 1976. Structure exinique chez quelques genres d'Annonacées. *In:* I. K. Ferguson and J. Muller, eds., The evolutionary significance of the exine, pp., 309–325. Linnean Society Symposium Series No. 1, 1976.
Le Thomas, A., & B. Lugardon. 1976. De la structure grenue a la structure columellaire dans le pollen des Annonacées. Adansonia, Sér. 2, 15: 543–572.
Lobreau-Callen, D. 1977. Le pollen du *Bubbia perrieri* R. Cap. Rapports palynologiques avec les autres genres de Wintéracées. Adansonia Sér. 2, 16: 445–460.
McLaughlin, R. P. 1933. Systematic anatomy of the woods of the Magnoliales. Trop. Woods 34: 3–39.
Nair, N. C., & P. N. Bahl. 1956. Vascular anatomy of the flower of *Myristica malabarica* Lamk. Phytomorphology 6: 127–134.
Occhioni, P. 1948. Contribuicao ao estudo do familia "Canellaceae." Arq. Jard. Bot. Rio de Janeiro 8: 3–20.
Padmanabhan, D. 1960. A contribution to the embryology of *Michelia champaca.* J. Madras Univ. 30B: 155–165.
Parameswaran, N. 1961. Foliar vascularization and histology in the Canellaceae. Proc. Indian Acad. Sci. 54: 306–317.
Parameswaran, N. 1961. Ruminate endosperm in the Canellaceae. Curr. Sci. 30: 344–345.
Parameswaran, N. 1962. Floral morphology and embryology in some taxa of the Canellaceae. Proc. Indian Acad. Sci. B, 55: 167–182.
Periasamy, K. 1961. Studies on seeds with ruminate endosperm. I. Morphology of ruminating tissue in *Myristica fragrans.* J. Madras Univ. 31B: 53–58.
Periasamy, K., & B. G. L. Swamy. 1959 (1960), 1961. Studies in the Annonaceae. I. Microsporogenesis in *Canaga odorata* and *Miliusa wightiana.* Phytomorphology 9: 251–263. II. The development of ovule and seed in *Canaga odorata* and *Miliusa wightiana.* J. Indian Bot. Soc. 40: 206–216, 1961.
Royen, P. van. 1962. Sertulum papuanum. 6. Himantandraceae. Nova Guinea Bot. 9: 127–135.
Rüdenberg, L. 1967. The chromosomes of *Austrobaileya.* J. Arnold Arbor. 48: 241–244.
Sampson, F. B. 1963. The floral morphology of *Pseudowintera*, the New Zealand member of the vesselless Winteraceae. Phytomorphology 13: 403–423.
Sampson, F. B., & S. C. Tucker. 1978. Placentation in *Exospermum stipitatum* (Winteraceae). Bot. Gaz. 139: 215–222.
Sastri, R. L. N. 1955. Structure and development of nutmeg seed. Curr. Sci. 24: 172.
Sastri, R. L. N. 1959. Vascularization of the carpel of *Myristica fragrans.* Bot. Gaz. 121: 92–95.
Sauer, W., & F. Ehrendorfer. 1970. Chromosomen, Verwandtschaft und Evolution tropischer Holzpflanzen. II. Himantandraceae. Oesterr. Bot. Z. 118: 38–54.
Skipworth, J. P. 1970 (1971). Development of floral vasculature in the Magnoliaceae. Phytomorphology 20: 228–236.
Smith, A. C. 1942. A nomenclatural note on the Himantandraceae. J. Arnold Arbor. 23: 366–368.

Smith, A. C. 1943. Taxonomic notes on the Old World species of Winteraceae. J. Arnold Arbor. 24: 119–164.
Smith, A. C. 1945. Geographical distribution of the Winteraceae. J. Arnold Arbor. 26:48–59.
Smith,A. C. 1949. Additional notes on *Degeneria vitiensis*. J. Arnold Arbor. 30: 1–9.
Smith, A. C., & R. P. Wodehouse. 1938. The American species of Myristicaceae. Brittonia 2: 393–510.
Srivastava, L. M. 1970. The secondary phloem of *Austrobaileya scandens*. Canad. J. Bot. 48: 341–359.
Sugiyama, M. 1976. Comparative studies of the vascular system of node-leaf continuum in woody Ranales. II. Node-leaf vascular system of *Eupomatia laurina* R. Br. J. Jap. Bot. 51: 169–174.
Swamy, B. G. L. 1949. Further contributions to the morphology of the Degeneriaceae. J. Arnold Arbor. 30: 10–38.
Swamy, B. G. L. 1952. Some aspects in the embryology of *Zygogynum bailloni* V. Tiegh. Proc. Natl. Inst. Sci. India 18: 399–406.
Thien, L. B. 1974. Floral biology of *Magnolia*. Amer. J. Bot. 61: 1037–1045.
Tiffney, B. H. 1977. Fruits and seeds of the Brandon Lignite: Magnoliaceae. J. Linn. Soc., Bot. 75: 299–323.
Tucker, S. C. 1959. Ontogeny of the inflorescence and the flower in *Drimys winteri* var. *chilensis*. Univ. Calif. Publ. Bot. 30: 257–336.
Tucker, S. C. 1960. Ontogeny of the floral apex of *Michelia fuscata*. Amer. J. Bot. 47: 266–277.
Tucker, S. C. 1961. Phyllotaxis and vascular organization of the carpels in *Michelia fuscata*. Amer. J. Bot. 48: 60–71.
Tucker, S C. 1963. Development and phyllotaxis of the vegetative axillary bud of *Michelia fuscata*. Amer. J. Bot. 50: 661–668.
Tucker, S. C. 1975. Carpellary vasculature and the ovular vascular supply in *Drimys*. Amer. J. Bot. 62: 191–197.
Tucker, S. C. 1977. Foliar sclereids in the Magnoliaceae. J. Linn. Soc., Bot. 75: 325–356.
Ueda, K. 1978. Vasculature in the carpels of *Belliolum pancheri* (Winteraceae). Acta Phytotax. Geobot. 29: 119–125.
Vander Wyk, R. W., & J. E. Canright. 1956. The anatomy and relationships of the Annonaceae. Trop. Woods 104: 1–24.
Vink, W. 1970, 1977, 1978. The Winteraceae of the Old World. I. *Pseudowintera* and *Drimys*—morphology and taxonomy. II. *Zygogynum*—morphology and taxonomy. III. Notes on the ovary of *Takhtajania*, Blumea 18: 225–354, 1970. 23: 219–250, 1977. 24: 521–525, 1978
Walker, J. W. 1971. Pollen morphology, phytogeography, and phylogeny of the Annonaceae. Contr. Gray Herb. 202: 1–131.
Walker, J. W. 1972. Chromosome numbers, phylogeny, phytogeography of the Annonaceae and their bearing on the (original) basic chromosome number of angiosperms. Taxon 21: 57–65.
Whitaker, T. W. 1933. Chromosome number and relationship in the Magnoliales. J. Arnold Arbor. 14: 376–385.
Wilson, T. K. 1960, 1965, 1964, 1966. The comparative morphology of the Canellaceae. I. Synopsis of genera and wood anatomy. Trop. Woods 112: 1–27, 1960. II. Anatomy of the young stem and node. Amer. J. Bot. 52: 369–378, 1965. III. Pollen. Bot. Gaz. 125: 192–197, 1964. IV. Floral morphology and conclusions. Amer. J. Bot. 53: 336–343, 1966.

Wilson, T. K., & L. M. Maculans. 1967. The morphology of the Myristicaceae. I. Flowers of *Myristica fragrans* and *M. malabarica*. Amer. J. Bot. 54: 214–220.

Баранова, М. А. 1962. Строение устьиц и эпидермальных клеток листа у магнолий в связи с систематикой рода *Magnolia* L. Бот. Ж. 47: 1108–1115.

Зажурило, К. К. 1940. К анатомии семенных оболочек Magnoliaceae (*Liriodendron tulipifera* L.). Бюлл. Обш. естествоисп. Воронежск. Госуд. Унив., 4: 32–40.

Скворцова, Н. Т. 1958. К анатомии цветка *Magnolia grandiflora* L. Бот. Ж. 43: 401–408.

Тахтаджян, А. Л., & Н. Р. Мейер. 1976. Некоторые дополнительные данные о морфологии пыльцы *Degeneria vitiensis* (Degeneriaceae). Бот. Ж. 61: 1531–1535.

SYNOPTICAL ARRANGEMENT OF THE FAMILIES OF MAGNOLIALES

1 Stamens distinct (rarely with the filaments connate below into a short tube in some Annonaceae).

  2 Leaves without stipules.

    3 Vessels wanting; pollen-grains uniporate, usually in tetrads; stomates more or less occluded by cutin; flowers seldom trimerous .... 1. WINTERACEAE.

    3 Vessels present in the secondary wood; pollen-grains variously monosulcate or bisulcate or inaperturate, but not uniporate, usually in monads; stomates not occluded; flowers often trimerous, at least as to the perianth.

      4 Carpel solitary, conduplicate, unsealed, the stigma running the length of the apposed margins; cotyledons 3 or 4; stamens laminar; ovules numerous; nodes pentalacunar .... 2. DEGENERIACEAE.

      4 Carpels either more than one, or sealed (with a terminal style and stigma), or often both; cotyledons 2.

        5 Stamens more or less laminar or ribbon-shaped, not clearly divided into filament and anther.

          6 Trees or shrubs with alternate leaves; perianth (or apparent perianth) calyptrate in whole or in part.

            7 Each of the 2 sepals calyptrate (the outer one enclosing the inner), but the petals distinct and not calyptrate; carpels closed, distinct, uniovulate; nodes trilacunar; pollen-grains monosulcate; endosperm not ruminate .... 3. HIMANTANDRACEAE.

            7 Perianth represented or replaced by a single calyptra; carpels individually unsealed, but all closely crowded and more or less connate, thus appearing to be deeply sunken into the broad, flat receptacle; ovules

several or many; nodes multilacunar; pollen-grains with 2 (3) parallel equatorial sulci .... 4. EUPOMATIACEAE.

6 Woody vines with opposite or subopposite leaves; perianth not at all calyptrate .... 5. AUSTROBAILEYACEAE.

5 Stamens usually with a short, thick filament and a well defined anther, the connective surpassing the anther and commonly enlarged above it, or seldom the stamens more or less laminar; perianth not calyptrate; carpels closed, each with 1-many ovules; leaves alternate .... 8. ANNONACEAE.

2 Leaves with evident, deciduous stipules.

8 Stamens numerous, spirally arranged; carpels usually more than 3, spirally arranged; ovules most commonly 2, but sometimes more numerous; nodes multilacunar; pollen-grains in monads .... 6. MAGNOLIACEAE.

8 Stamens 6, in 2 whorls of 3; carpels 3, in one whorl; ovules 4-8; nodes unilacunar; pollen-grains in tetrads .... 7. LACTORIDACEAE.

1 Stamens with the filaments united at the base into a tube or column (connate only at the base in *Brochoneura*, of the Myristicaceae).

9 Flowers unisexual; carpel solitary; ovule solitary, nearly basal; filaments united into a solid column (except as noted in *Brochoneura*) .... 9. MYRISTICACEAE.

9 Flowers perfect; carpels 2–6, united into a compound, unilocular pistil with 2–6 parietal placentas; filaments united into a tube around the ovary .... 10. CANELLACEAE.

1. Family WINTERACEAE Lindley 1830 nom. conserv., the Wintera Family

Trees or shrubs, generally glabrous except for the carpels, producing proanthocyanins and at least sometimes aporphine alkaloids, and regularly with spherical ethereal oil cells in the parenchymatous tissues, especially of the leaves; crystals seldom present; nodes trilacunar; vessels wanting; imperforate tracheary elements consisting mainly of elongate, slender tracheids but also some vascular tracheids, both types with bordered pits; woody-parenchyma scanty, commonly diffuse, sometimes in lines; wood-rays Kribs type I, heterocellular, mixed uniseriate and pluriseriate, the latter up to 10 cells wide, with elongate ends; sieve-tubes with S-type plastids. LEAVES alternate, simple, entire, finely pellucid-dotted, often pale beneath, the venation pinnate but relatively unorganized, with irregularly distributed primary lateral veins of unequal size; stomates mostly paracytic (anomocytic in *Takhtajania*), more or less occluded by the cuticle; stipules wanting. FLOWERS solitary and terminal, or often in terminal or axillary cymose inflorescences, variously entomophilous (pollinated by Coleoptera in spp. of *Drimys*, by Coleoptera and Lepidoptera in *Zygogynum*, by Thysanoptera in *Belliolum*, by Diptera in spp. of *Tasmannia*) or more or less anemophilous (as in other spp. of *Tasmannia*), or sometimes autogamous (as in *Pseudowintera*), regular or rarely somewhat irregular, perfect or sometimes (*Tasmannia*) unisexual (the plants then dioecious or polygamo-dioecious), hypogynous, the receptacle short; sepals 2–4 (–6), valvate, often distinct or connate only at the base, but initially calyptrate in *Zygogynum*, and in *Drimys* and *Tasmannia* sometimes fully connate into a deciduous calyptra; petals (2–) 5-many, commonly in 2 or more whorls, free and distinct, sometimes small and chaffy, sometimes larger and petaloid; stamens numerous, initiated in centripetal sequence, but maturing centrifugally, free and distinct, commonly more or less ribbon-shaped or laminar, tetrasporangiate, commonly with terminal or subterminal bisporangiate pollen sacs (the blade conspicuously prolonged beyond the pollen sacs only in *Belliolum*), or especially in spp. of *Tasmannia* the stamen sometimes more clearly differentiated into filament and latrorse or weakly extrorse, dithecal anther; pollen grains in tetrads or rarely in monads, binucleate, uniporate, with a distal aperture (anaulcerate), coarsely reticulate, the exine semitectate; carpels (1–) several or more or less numerous, in a single whorl, distinct or lightly connate (more especially so in fruit), conduplicate and often unsealed, with the stigma

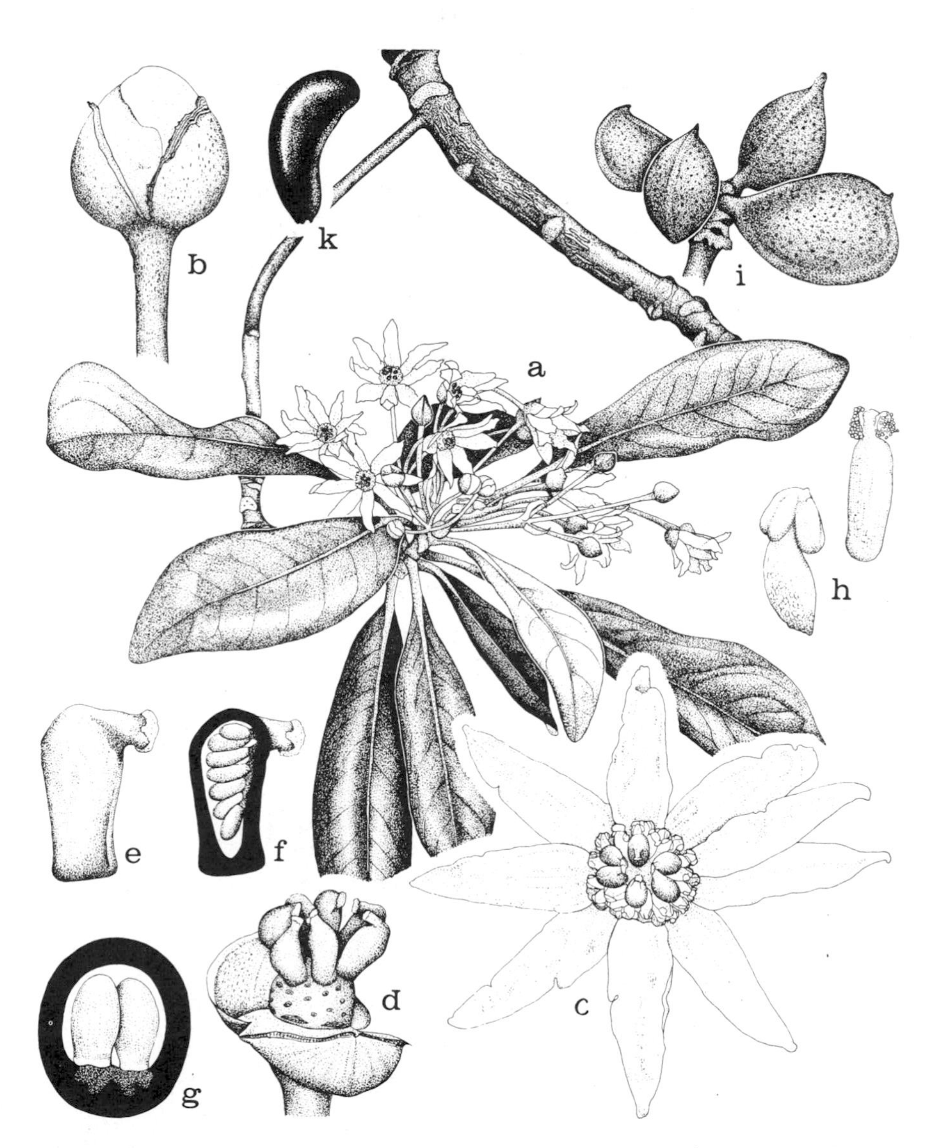

FIG. 1.4 Winteraceae. *Drimys winteri* J. R. & G. Forster. a, habit, ×½; b, flower bud, ×4; c, flower, ×2; d, flower, with stamens and petals removed, showing spirally arranged stamen-scars, ×4; e, carpel, ×8; f, carpel in long-section, ×8; g, carpel in cross-section, ×16; h, two stamens, ×8; i, fruiting carpels, ×2; k, seed, ×5.

decurrent along the apposed margins, varying to fully sealed and with a terminal style and stigma; ovules 1-several, marginal (i.e., along the ventral suture), or laminar near the margin, anatropous, bitegmic, crassinucellar; endosperm development cellular or nuclear. Carpels ripening into indehiscent, berry-like FRUITS, or into follicles, or these sometimes more or less connate or concrescent to form a multilocular capsule or a syncarp; seeds with very small, dicotyledonous embryo and abundant, oily endosperm. X = 13, 43.

The family Winteraceae consists of 9 genera and about a hundred species. *Tasmannia* (40, often considered a section of *Drimys*) and *Bubbia* (30) are the largest genera. The other genera are *Belliolum* (8), *Zygogynum* (6), *Drimys* (4), *Pseudowintera* (3), *Exospermum* (2), *Tetrathalamus* (1), and *Takhtajania* (1). The principal center of distribution lies in the islands of the southwestern Pacific (and adjacent Australia) but *Drimys* occurs from Mexico to the Straits of Magellan, and *Takhtajania* is confined to Madagascar. Winter's bark, from the South-Andean species *Drimys winteri* J. R. & G. Forst., was used in the past as a tonic and antiscorbutic, but has no present importance.

Exospermum is unusual in the family in having the ovules apparently scattered over the inner surface of the carpel, but this is an ontogenetic development. The ovules are initiated in a single row on each of the two placentas, as in other Winteraceae.

Leroy and (independently) Vink interpret the gynoecium of *Takhtajania* as a bicarpellate ovary composed of 2 connate open carpels with 2 parietal, subapical placentas and 2 sessile, more or less connate terminal stigmas. This interpretation may however be subject to challenge.

The Winteraceae are one of the most archaic families of angiosperms, and many botanists now consider that they come nearer than any other modern family to the hypothetical ancestral prototype of the division. Like the other families of the order, however, they present their own combination of primitive and more advanced characters. The pollen, in particular, does not appear to be primitive in comparison with that of other members of the Magnoliales.

The relatively unorganized venation of the leaves of Winteraceae is evidently a primitive feature, found also in many of the earliest fossil angiosperm leaves. It would be perilous, however, to identify these fossils with the Winteraceae, inasmuch as the floral structure and wood-anatomy of the plants that bore the leaves are unknown. Furthermore, a similar venation is found here and there in more advanced groups. Compare, for example, *Meryta* (Araliaceae) from New Caledonia.

The fact that the stomates of the Winteraceae are more or less occluded with cutin is probably related to the absence of vessels. As Baranova (1972) has pointed out, there must be a delicate balance in such vesselless woody plants between the need to exchange gases for photosynthesis and the need to restrict transpiration. If one takes the generally accepted view that the absence of vessels in the Winteraceae is a primitive feature, it seems reasonable to further suppose that the occlusion of the stomates evolved in relation to the increased size of the plant and the uncertainty of the water-supply to the leaves. This brings us to the Doyle & Hickey hypothesis that the angiosperms originated as relatively small (though woody) gymnosperms in tropical riparian habitats. Under such conditions small, broad-leaved plants without vessels could probably function adequately with ordinary, non-occluded stomates. Other explanations may perhaps be possible, but this one is at least plausible and internally consistent.

Pollen considered to represent the Winteraceae is known from Maestrichtian and more recent deposits. Clearly recognizable fossils of other parts of the plant date only from the early Oligocene.

2. Family DEGENERIACEAE I. W. Bailey & A. C. Smith 1942 nom. conserv., the Degeneria Family

Large trees with spherical or often axially elongate ethereal oil cells in the parenchymatous tissues; nodes pentalacunar; vessel-segments with scalariform perforations and numerous cross-bars, the side-walls scalariform-pitted; imperforate tracheary elements with bordered pits; wood-parenchyma mainly apotracheal, banded; wood-rays heterocellular, mixed uniseriate and pluriseriate, the latter up to 5 cells wide; phloem tangentially stratified into alternating fibrous and nonfibrous layers, and with wedge-shaped rays. LEAVES alternate, simple, entire, pinnately veined; stomates paracytic; petiole with a ring of vascular bundles; stipules wanting. FLOWERS rather large, solitary, pendulous on long, supra-axillary peduncles, probably cantharophilous, perfect, regular, hypogynous, the receptacle short; perianth differentiated into a calyx of 3 sepals in one whorl and a larger corolla of mostly 12–18 distinct petals in 3–5 whorls; stamens numerous, commonly 20–30 in 3–4 series, laminar, 3-nerved, the 4 microsporangia paired, embedded between the midnerve and the lateral nerves on the abaxial side; some staminodia with abortive microsporangia present between the functional stamens and the pistil; pollen-grains large, boat-shaped, monosulcate (i.e., anasulcate), nearly smooth, the exine solid and virtually structureless; gynoecium of a single conduplicate carpel that is largely unsealed at anthesis, the glandular-hairy stigmatic surfaces running along the apposed margins; ovules numerous, ca 20–32, laminar, in a single row toward each margin of the carpel, anatropous, bitegmic, crassinucellar,

FIG. 1.5 Degeneriaceae. *Degeneria vitiensis* A. C. Smith. Above, in flower, nearly 1½ times natural size; below, in fruit, about natural size. Photographs courtesy of Armen Takhtajan.

with a conspicuous funicular obturator; endosperm-development cellular. FRUIT thick and more or less fleshy but with a hard exocarp, possibly dehiscent at maturity; seeds numerous, with an orange-red sarcotesta, some of them on a slender, elongate funiculus as in the Magnoliaceae, others sessile; endosperm copious, oily, ruminate; embryo very small, with 3 or less often 4 cotyledons. 2n=24.

The family Degeneriaceae consists of a single genus and species, *Degeneria vitiensis* A. C. Smith, of Fiji. *Degeneria* has attracted much attention as an archaic angiosperm because of its laminar stamens, merely folded, unsealed carpel, laminal-lateral placentation, and 3 or sometimes 4 cotyledons. Nevertheless it is more advanced than the Winteraceae in having vessels, and more advanced than both the Winteraceae and Magnoliaceae in having only a single carpel. Like the other families of Magnoliales, it represents an isolated end-line with its own combination of primitive and more advanced characters.

3. Family HIMANTANDRACEAE Diels 1917 nom. conserv., the Himantandra Family

Large, aromatic trees with spherical ethereal oil cells in the parenchymatous tissues, and producing himandrin and related alkaloids (a special group of pyridine alkaloids) but not isoquinoline alkaloids; solitary or clustered crystals of calcium oxalate present in some of the cells; sclereids scattered in the parenchymatous tissues of the stem and to some extent also the leaves; nodes trilacunar; pith with more or less well developed sclerenchymatous diaphragms; vessel-segments elongate, with simple perforations, or some of those in young stems with scalariform perforation-plates; imperforate tracheary elements with bordered pits, considered by different authors to be tracheids or fiber-tracheids; wood-parenchyma apotracheal, crystalliferous, in concentric bands; wood-rays heterocellular, or nearly homocellular in later wood, mostly (1) 2–3 (4)-seriate; secondary phloem tangentially stratified into fibrous and nonfibrous layers, and with wedge-shaped rays. LEAVES alternate, simple, entire, pellucid-punctate, pinnately veined, the lower surface (as also the young twigs) covered with characteristic peltate trichomes; stomates paracytic, grouped in a circle of several under the shelter of each trichome; petiole with a ring of separate vascular bundles; stipules wanting. FLOWERS rather large, solitary (seldom 2–3) on axillary branches, perfect, hypogynous; sepals 2, caducous, calyptriform, one enclosing the other, the caducous outer one sometimes interpreted as being derived from 2 connate sepals, and the inner one from 4 connate petals, or sometimes both considered to be bracts (as by Endress); corolla of several (ca 7—9) spirally arranged lance-linear petals, these doubtless of staminodial origin, and interpreted as staminodes when the inner calyptrate organ is interpreted as a modified corolla; stamens numerous (ca 25–30), spirally arranged, very much like the petals in appearance, not differentiated into filament and anther, the 4 microsporangia relatively short, more or less embedded in the abaxial side of the stamen-blade below the middle, paired between the midvein and the lateral veins, each pair dehiscing by a single slit; several (ca 8–10) spirally arranged staminodes present between the functional stamens and the carpels; pollen-grains subglobose, monosulcate (i.e., anasulcate), with a more or less solid and homogeneous exine; gynoecium of ca (6) 7–10 or numerous (up to ca 28) spirally arranged, closed carpels, each clearly differentiated into an ovary and a short style with decurrent stigma, the carpels weakly connate at the base in flower, more fully connate laterally in fruit; a single (rarely 2) pendulous, laminar-lateral, anatropous ovule in each carpel. FRUIT globose, gall-like, fleshy, plurilocular by coalescence of the carpels; seeds with small, dicotyledonous embryo and abundant, oily, nonruminate endosperm. $2n = 24$.

The family Himantandraceae consists of the single genus *Galbulimima* (*Himantandra*), confined to New Guinea, the Molucca Islands, and northeastern Australia. According to one interpretation each of these 3 areas harbors its own species. According to another interpretation there is only one species, *G. belgraveana* (F. Muell.) Sprague.

The Himantandraceae are generally considered to be allied to the Magnoliaceae, Degeneriaceae, and Annonaceae. Thus they belong to the order Magnoliales even under the stricter definitions of the group. Nevertheless the Himantandraceae are aberrant in the Magnoliales (and resemble the Laurales) in usually having only a single ovule per carpel. Like the other families of the Magnoliales, they represent an isolated end-line, not really closely linked to any other family.

4. Family EUPOMATIACEAE Endlicher 1841 nom. conserv., the Eupomatia Family

Small trees or rhizomatous, slender, wiry, woody herbs, producing benzyl-isoquinoline alkaloids; parenchymatous tissues with small clus-

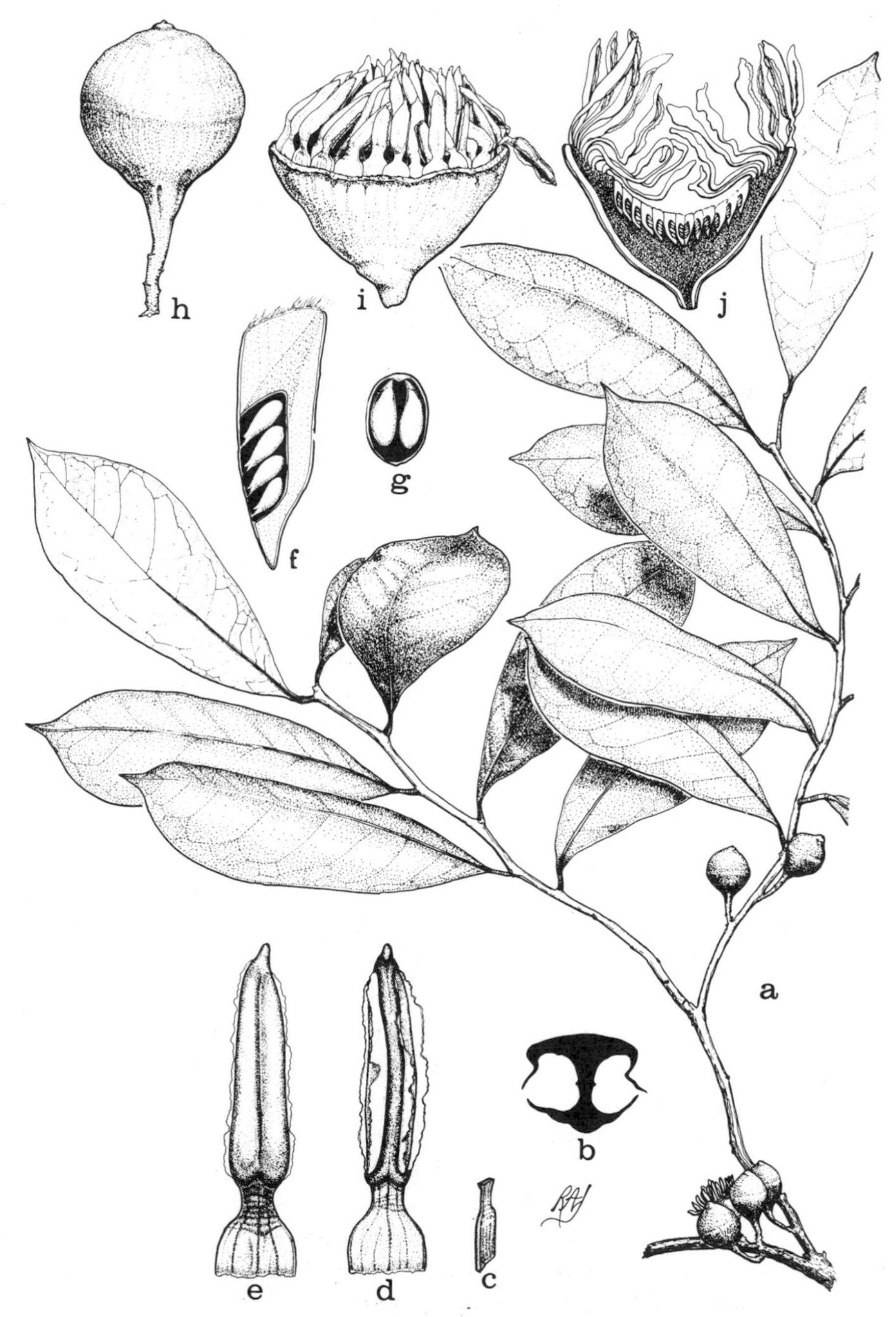

FIG. 1.6 Eupomatiaceae. *Eupomatia laurina* R. Br. a, habit, ×⅓; b, diagrammatic cross-section of anther, ×10; c, filament, ×5; d, e, two views of stamen, ×5; f, long-section of carpel, ×10; g, cross-section of carpel, ×10; h, flower bud, with calyptra intact, ×1; i, flower, after the calyptra has fallen, ×1.5; j. long-section of flower, ×1.5.

tered crystals in some of the cells, and with scattered spherical ethereal oil cells and more or less elongate idioblasts (or files of idioblasts) containing both mucilage and tannin (proanthocyanin), the tanniferous idioblasts especially evident in the pith; nodes multilacunar, with 7 (–11) traces; vessel-segments long and slender, with slanting end-walls and scalariform perforation plates that have numerous (20–150) cross-bars; imperforate tracheary elements with simple or slightly bordered pits, sometimes septate; wood-parenchyma rather sparse, diffuse or scanty-paratracheal; wood-rays heterocellular, mixed uniseriate and pluriseriate, the uniseriate ones few, the pluriseriate ones up to 8 (–15) cells wide, with elongate ends; phloem not stratified, its rays only slightly flaring; sieve-tube plastids of P-type. LEAVES alternate, simple, entire, pinnately veined; stomates paracytic; stipules wanting. FLOWERS rather large, cantharophilous, without nectar, solitary or 2 or 3 together in the axils, covered when young by a deciduous calyptra attached to the rim of an enlarged, somewhat concave, hypanthium-like receptacle; calyptra (interpreted by Endress as a bract) showing no external evidence of tepalar structure, containing masses of sclerenchymatous, *Trochodendron*-like idioblasts; stamens numerous (ca 70), attached to the receptacle near its rim, the outer ones ribbon-shaped in the manner of many Magnoliaceae, with a short, broad, laminar base and a well defined, latrorse-introrse, tetrasporangiate and dithecal anther with slightly separated pollen-sacs and prolonged, thickened connective, the inner ones sterile, somewhat fleshy, conspicuous and more or less petaloid; stamens and staminodia forming a continuous spiral; pollen grains binucleate globose-oblate, with two parallel sulci in the equatorial region, and with a solid, amorphous exine as in *Degeneria*; gynoecium of numerous (ca 13–68) carpels spirally arranged on the enlarged receptacle, all more or less connate by their margins into a sort of compound pistil, but each carpel unsealed, without a style, and with a feathery stigma across the nearly flat top; each carpel with 2-several (up to ca 11) laminar-lateral ovules, these anatropous, bitegmic, and crassinucellar. FRUIT aggregate, berry-like, subglobose but more or less truncate at the tip and bearing a rim left by abscission of the calyptra, the carpels laterally coalescent so as to appear to be sunken in a fleshy receptacle; seeds 1–2 in each carpel, with very small, dicotyledonous embryo and abundant, oily, ruminate endosperm. 2n=20.

The family consists of the single genus *Eupomatia*, with 2 species, confined to New Guinea and eastern Australia.

The Eupomatiaceae are a distinctive, highly isolated family, with their own special combination of primitive and specialized features. The unsealed carpels with laminar-lateral ovules might perhaps best be compared (in these respects) with those of the Degeneriaceae, but these two families are obviously different in a number of other features. The disulcate pollen of *Eupomatia* is nearly unique in the order, being comparable only to that of some Annonaceae. Pollen considered to represent *Eupomatia* is known from Maestrichtian (late Upper Cretaceous) and more recent deposits.

5. Family AUSTROBAILEYACEAE Croizat 1943 nom. conserv., the Austrobaileya Family

Evergreen woody vines with loosely twining main stem and straight, leafy lateral branches, with spherical ethereal oil cells in the parenchymatous tissues, at least of the cortex, and with calcium oxalate crystals in the form of crystal sand in some of the cells; alkaloids apparently wanting; nodes unilacunar, with 2 traces; pith lignified; vessel segments with scalariform perforation plates; most of the imperforate tracheary elements with evidently bordered pits and considered to be thick-walled tracheids, but some of them septate, persistently several-nucleate, and with scarcely to evidently bordered pits; wood-parenchyma paratracheal; wood-rays heterocellular, mixed uniseriate and pluriseriate, the latter up to 8 cells wide, with elongate ends; phloem of a relatively primitive type, the companion cells not all bearing a close ontogenetic relationship to the sieve elements; sieve-elements with plastids of S-type. LEAVES opposite or subopposite, glabrous, leathery, simple, entire, pinnately veined; stomates both paracytic and anomocytic; petiole short, stout, curved; stipules wanting. FLOWERS rather large, solitary in the axils of the leaves, with a putrescent odor, probably pollinated by flies, perfect, hypogynous; perianth of ca 12 light green, free and distinct members in a compact spiral, gradually larger from the outer, more sepaloid ones to the inner, more petaloid ones; stamens densely purple-dotted, 12–25, spirally arranged, the outer 6–9 fertile, the inner gradually reduced and sterile; fertile stamens light green, laminar, petaloid, not differentiated into filament and anther, tetrasporangiate, the elongate microsporangia borne in 2 juxtaposed pairs on the adaxial side; pollen grains binucleate, spherical, monosulcate (i.e., anasulcate); gynoecium of (6–) 9 (–14) free and distinct, more or less spirally arranged, stipitate carpels, each with an elongate, 2-lobed style and decurrent stigmas; ovules 8–14, laminal-lateral, anatropous, bitegmic, crassinucellar. FRUIT bright orange-yellow, ellipsoid-globose, juicy, berry-like; seeds large, with a leathery coat, the outer layer forming a sarcotesta; embryo very small, dicotyledonous; endosperm abundant, ruminate. $2n=44$.

The family Austrobaileyaceae consists of the single genus *Austrobaileya*, with a single species in northeastern Australia. Although *Austrobaileya* clearly belongs to the Magnoliidae, its position within the subclass is not so obvious. On the basis of the pollen, Agababyan (1973, cited under Magnoliidae, in Russian) thinks that *Austrobaileya* is closely allied to the subfamily Atherospermatoideae in the family Monimiaceae of the Laurales. Bailey and Swamy (1949) also consider that it is related to the Monimiaceae. Hutchinson (1973) includes the Austrobaileyaceae in the Laurales, following the Monimiaceae, "for want of a better place." Takhtajan (1966) considers the Austrobaileyaceae to be one of the most primitive families of the Laurales, which indeed must be true if they are referred to that order.

The laminar stamens, hypogynous flowers, monosulcate pollen, and pluriovulate carpels of *Austrobaileya* would all be aberrant, primitive features in the Laurales, but are perfectly compatible with the Magnoliales. On the other hand, *Austrobaileya* is aberrant in the Magnoliales in its unilacunar nodes, opposite leaves, and climbing habit. The first two of these features, at least, are fully compatible with the Laurales. Thus we have an isolated small group, not wholly compatible with the bulk of either the Laurales or Magnoliales, but not sufficiently distinctive to constitute an order of its own.

In addition to the Austrobaileyaceae, several families of this general alliance stand apart from both the clearly Magnolialean and the clearly Lauralean families, and from each other. These are the Lactoridaceae, Eupomatiaceae, and to a lesser extent the Myristicaceae and Canellaceae.

In trying to formulate a conceptually valid arrangement of the Magnoliid families I have found it useful to define the order Laurales narrowly, so as to form a fairly homogeneous group more advanced than the Magnoliales. The Magnoliales are then defined more loosely, to take in several isolated end-line families as well as the core of families that are agreed by nearly all to be fairly closely related (Winteraceae, Degeneriaceae, Himantandraceae, Magnoliaceae, Annonaceae). In such a scheme, the Austrobaileyaceae necessarily go with the Magnoliales.

6. Family MAGNOLIACEAE A. L. de Jussieu 1789 nom. conserv., the Magnolia Family

Evergreen to more often deciduous trees or shrubs, producing proanthocyanins and usually alkaloids (especially of the benzyl-isoquinoline or aporphine-type), often accumulating silica, especially in the cell walls of the epidermis of the leaves, sometimes (as in *Liriodendron*) cyanogenic through a tyrosine-based pathway, and regularly with spherical ethereal oil cells in the parenchymatous tissues, especially of the leaves, the oil cells sometimes lysing into mucilage-cavities; small crystals of calcium oxalate often present in the parenchymatous tissues; nodes multilacunar; vessels tending to be in radial groups, their segments usually with scalariform perforation-plates and few to numerous cross-bars, less often with simple perforations; imperforate tracheary elements with bordered pits and commonly with tyloses, exclusively fiber-tracheids, or some of them true tracheids; wood-parenchyma abundant, apotracheal, in terminal bands; wood-rays Kribs type IIA or IIB, heterocellular or rarely homocellular, mixed uniseriate and pluriseriate, the pluriseriate ones mostly 3–4 (7) cells wide, with short or elongate ends; phloem commonly stratified tangentially into fibrous and non-fibrous layers, and with wedge-shaped rays; sieve-tube plastids of S-type or P-type. LEAVES alternate, simple, pinnately veined, entire or seldom (*Liriodendron*) lobed, commonly with the epidermis or the hypodermis or the spongy mesophyll or the bundle-sheaths sclerified, or with idioblastic sclereids; stomates mostly paracytic (anomocytic in *Liriodendron*); stipules large, enfolding the terminal bud, often forming an ochrea, deciduous. FLOWERS large, entomophilous, often cantharophilous, terminal or seldom axillary, generally solitary, perfect or (*Kmeria*) unisexual, regular, hypogynous, very often with an elongate receptacle; perianth variously spiral to cyclic in 3 or more series, often in 3 sets of 3, the tepals 6–18, free and distinct, often more or less similar and all petaloid, less often clearly differentiated into sepals and petals; stamens numerous, free and distinct, spirally arranged, originating in centripetal sequence, tending to be more or less ribbon-shaped rather than clearly differentiated into filament and anther, the 4 microsporangia paired, often more or less embedded in the adaxial (abaxial in *Liriodendron*) surface of the narrow lamina of the microsporophyll, the connective often prolonged into a more or less distinct appendage; staminodes none; pollen grains binucleate, mostly boat-shaped, distally monosulcate (i.e.,

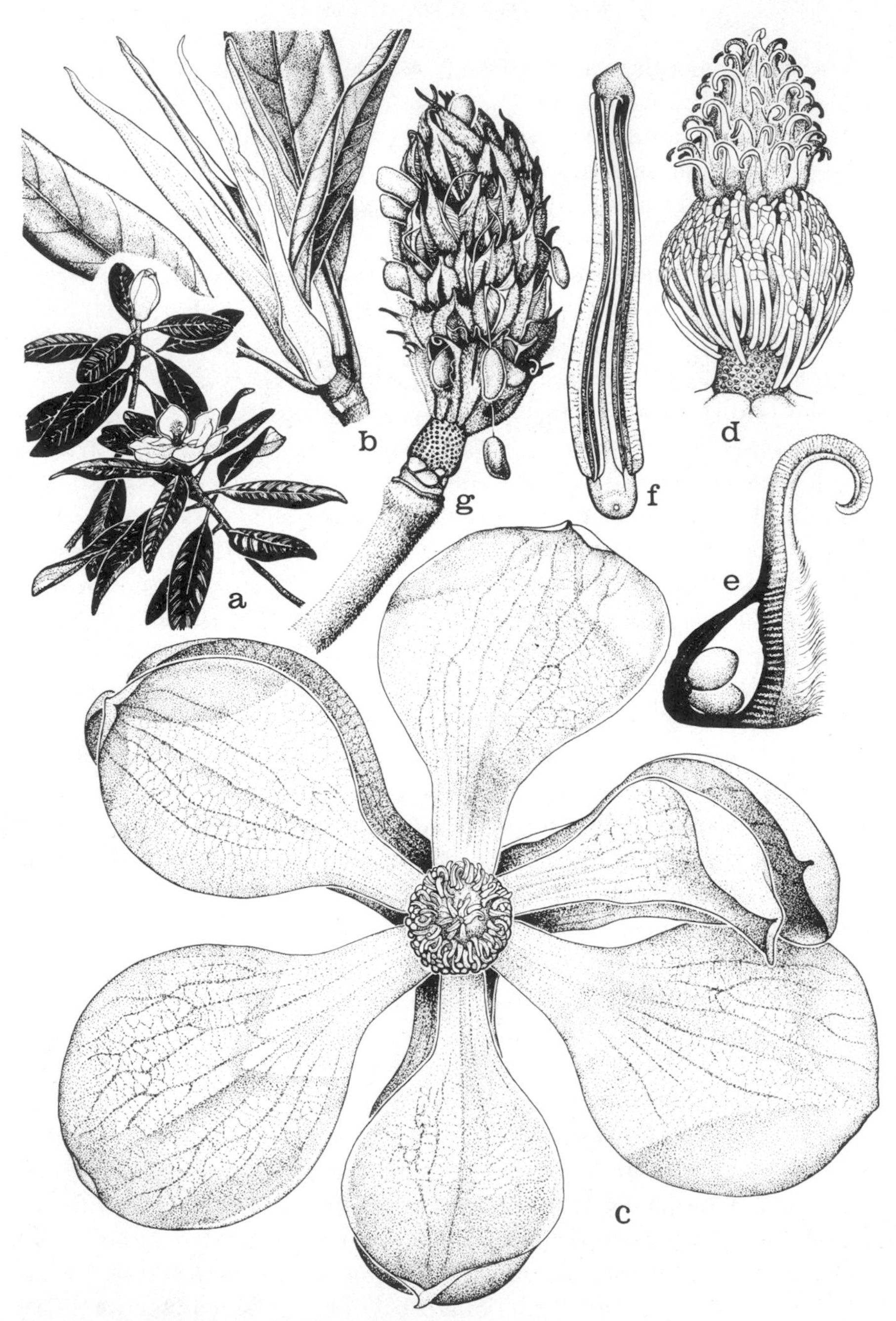

FIG. 1.7 Magnoliaceae. *Magnolia grandiflora* L. a, habit, $\times\frac{1}{16}$; b, new leaves and stipules, $\times\frac{1}{2}$; c, flower, from above, $\times\frac{1}{2}$; d, stamens and carpels of flower, showing spirally arranged stamen-scars below, $\times 1$; e, carpel, in partial long-section, $\times 4$; f, stamen, $\times 4$; g, fruiting head of carpels, $\times\frac{1}{2}$.

anasulcate), smooth to variously ornamented, often relatively large, the exine varying from solid and virtually structureless to granular or more often tectate-granular, or sometimes tectate-columellate; carpels (2) several to commonly more or less numerous, sometimes somewhat nectariferous, usually free and distinct at anthesis, conduplicate and sometimes not fully sealed, but typically sealed and with a more or less well developed style and terminal or terminal-decurrent stigma; ovules most commonly 2, but sometimes several or more or less numerous, borne on a marginal placenta, anatropous, bitegmic, crassinucellar, with a funicular obturator at least in *Liriodendron*; endosperm-development cellular. FRUIT variously follicular (often opening on the dorsal suture) or indehiscent and berry-like, or (*Liriodendron*) samaroid, the carpels often laterally more or less concrescent at maturity, sometimes forming a fleshy syncarp; seeds mostly large, commonly (except in *Liriodendron*) with a sarcotesta, with abundant, oily and proteinaceous endosperm and very small (somewhat larger and linear in *Liriodendron*), dicotyledonous embryo with a prominent suspensor, in species with dehiscent fruits the seed suspended on an elongate, slender funiculus. X = 19.

The family Magnoliaceae as currently defined (following Dandy in Hutchinson, 1964) is a coherent, well defined group of 12 genera and about 220 species. The largest genera are *Magnolia* (80), *Michelia* (40), and *Talauma* (50). *Liriodendron* (2) stands somewhat apart from the other genera in a series of features ranging from anthers and pollen grains to leaf-form, stomatal type, chemical constituents, and samaroid fruit. It is often considered to constitute a separate subfamily, but its relationship to the rest of the family is not in dispute. The bicentric distribution of *Liriodendron* (one species in China, one in eastern United States) is in conformity with a well known pattern reflecting an earlier more continuous Arcto-Tertiary distribution.

The Magnoliaceae are widespread in tropical or subtropical and warm-temperate regions of both the Old and the New World, especially in the Northern Hemisphere, but they are seldom abundant.

The Magnoliaceae are clearly one of the most archaic families of flowering plants. Like the other members of the order, however, they present their own combination of primitive and advanced features, and they cannot be considered directly ancestral to any other family.

Fossil wood compatible with the Magnoliaceae (and with several other dicotyledonous families as well) occurs in Maestrichtian (late Upper Cretaceous) deposits in California, and other fossil wood considered to represent the Magnoliaceae occurs in Eocene deposits in New York.

Seeds considered to represent the Magnoliaceae are found in late Upper Cretaceous and more recent deposits, and seeds of *Magnolia* and *Talauma* date from the Eocene epoch. Fossil pollen of *Magnolia* dates from the middle Eocene, and that of *Liriodendron* from the Oligocene. Thus in spite of its constellation of putatively primitive features, the family cannot yet be affirmed to be of any great age, on the basis of the known fossils.

7. Family LACTORIDACEAE Engler in Engler & Prantl 1888 nom. conserv., the Lactoris Family

Shrubs with spherical ethereal oil cells in the parenchymatous tissues; nodes unilacunar, with 2 traces; vessel-segments very short, simply perforate; imperforate tracheary elements with simple pits; wood-parenchyma diffuse, or partly paratracheal in older wood; wood-rays present only in the nodal regions, broad, mainly of straight, vertical cells; pith with tannin-sacs, many of them arranged in longitudinal rows. LEAVES alternate, simple, small, obovate, entire, emarginate, pellucid-punctate; stomates anomocytic; stipules large, adnate to the petiole. Plants polygamo-dioecious, the FLOWERS small, solitary in the axils or in 2–4 flowered axillary monochasia; sepals 3; petals none; stamens 6, in 2 sets of 3, short, narrowly laminar, tetrasporangiate, with abaxial, well separated, nearly marginal bisporangiate pollen sacs nearly as long as the blade, the connective only shortly prolonged; inner stamens reduced to staminodia in some flowers; pollen grains in tetrads, tectate, monosulcate (i.e., anasulcate), with a single poorly defined aperture and a granular exine; carpels 3, in a single whorl, medially more or less connate at the base, otherwise distinct; ovary narrowed above to a short style with a decurrent stigma; ovules 4–8 in each ovary on an intruded marginal placenta, anatropous, bitegmic, weakly crassinucellar, on a long funiculus; endosperm-development nuclear. FRUIT follicular, beaked; seeds with abundant, oily endosperm and very small, dicotyledonous embryo. $2n=40$ (or 42?).

The family Lactoridaceae consists of a single genus and species, *Lactoris fernandeziana* Phil., endemic to Masatierra, of the Juan Fernandez Islands, off the coast of Chile. The family is by all accounts taxonomicaly isolated. Takhtajan includes it in the order Laurales, and Carlquist leans in that direction as to its affinities. Agababyan (1973, cited under Magnoliidae, in Russian) on the other hand, maintains that on the basis of pollen structure the relationship of *Lactoris* is with *Drimys*, of the Winteraceae.

I consider it best to include the family within the rather loosely defined order Magnoliales, in which it is aberrant (and resembles the Laurales) mainly in having unilacunar nodes. In the Laurales, on the other hand, *Lactoris* would be aberrant in its stipulate leaves, hypogynous flowers, monosulcate pollen, and several (rather than only 1 or 2) ovules.

8. Family ANNONACEAE A. L. de Jussieu 1789 nom. conserv., the Custard-Apple Family

Trees, shrubs, or lianas, usually producing alkaloids, often of the benzyl-isoquinoline group, sometimes accumulating silica, especially in the cell walls, often tanniferous, at least sometimes with proanthocyanins; parenchymatous tissues with scattered spherical ethereal oil cells that occasionally lyse into mucilage cavities, also with separate or clustered crystals of calcium oxalate in some of the cells, and often with scattered sclereids; trichomes mostly simple, unicellular to uniseriate, but sometimes stellate or peltate; nodes trilacunar; pith generally septate, or sometimes sclerified more or less throughout; vessel segments with simple perforations; imperforate tracheary elements commonly with minute, bordered pits; wood-parenchyma mainly apotracheal, in numerous closely spaced fine lines mostly 1–2 cells wide, sometimes also partly paratracheal; wood-rays homocellular or slightly heterocellular, of Kribs types I and II, a few of them uniseriate, the others typically 4–10 (–14) cells wide, some even wider medullary rays commonly present; phloem commonly stratified tangentially into alternating fibrous and soft bands, and with wedge-shaped rays; sieve-tubes with P-type plastids. LEAVES simple, alternate, entire, pinnately veined, typically distichous, often pellucid-dotted with secretory cells or sclereids; stomates nearly always paracytic; petiole commonly with an arc of widely spaced vascular bundles; stipules wanting. FLOWERS solitary or in various sorts of mostly basically cymose inflorescences, mostly entomophilous, often cantharophilous, or sometimes autogamous, perfect or rarely unisexual, hypogynous, with cyclic and commonly trimerous perianth; sepals (2) 3 (4), distinct or more or less connate, valvate or imbricate; petals commonly 6 in 2 series of 3, rarely 3, 4, 8, or even 12, distinct or very rarely connate, valvate or imbricate in each series; stamens commonly numerous and spirally arranged (rarely only 3 or 6 and cyclic), generally distinct, rarely with the filaments connate below into a short tube, the filament short and stout, provided with a single vascular trace, the connective notably prolonged and often expanded above the anther

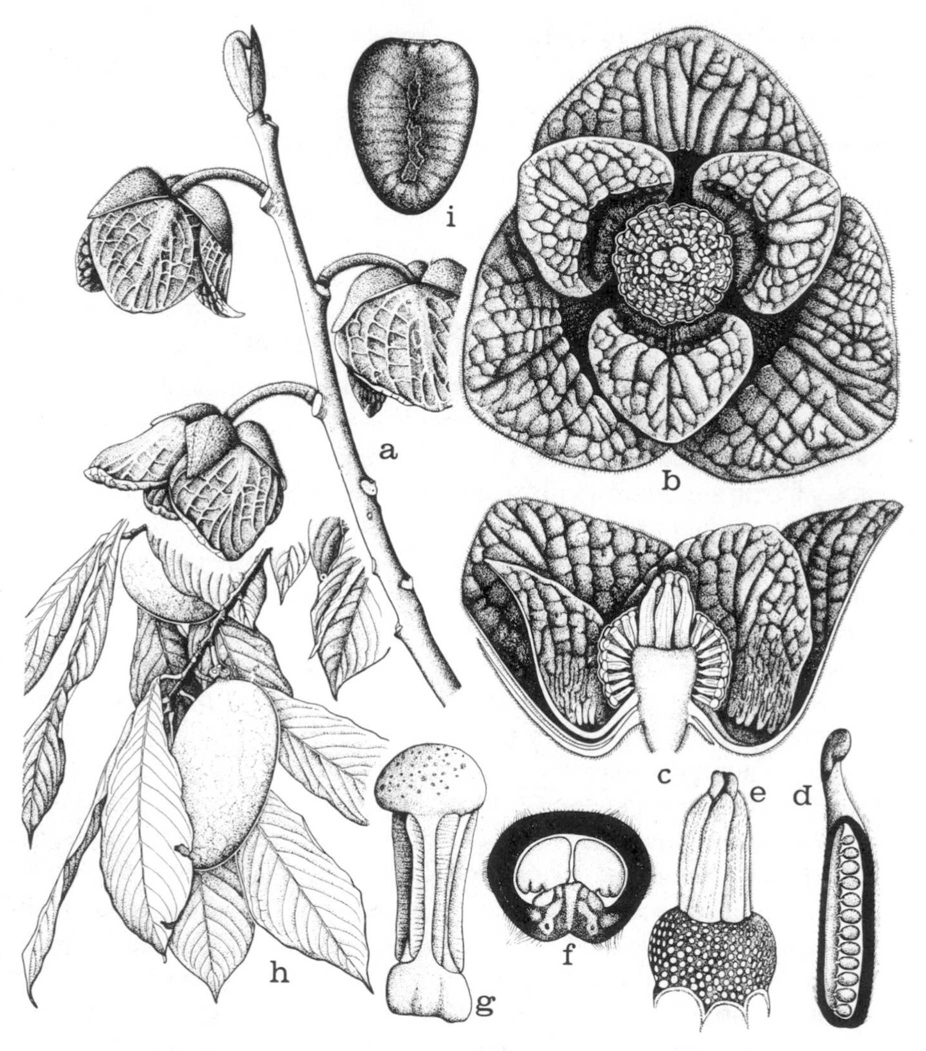

FIG. 1.8 Annonaceae. *Asimina triloba* (L.) Dunal. a, habit, ×1; b, c, flower, from above and in long-section, ×2; d, pistil, the ovary in long-section, ×8; e, receptacle, showing gynoecium and spirally arranged stamen-scars, ×4; f, cross-section of ovary, ×16; g, stamen, ×16; h, habit and fruit, ×¼; i, seed, ×1.

so that the stamen appears to be peltate, varying to rarely essentially laminar with abaxial microsporangia (*Anaxagorea*); anthers tetrasporangiate and dithecal, the pollen-sacs extrorse or latrorse, very rarely introrse, opening by longitudinal slits; staminodes wanting; pollen-grains binucleate, in monads or sometimes in tetrads or even polyads, diverse in structure, mostly uniaperturate with a distal sulcus (anasulcate), varying to inaperturate, or anaulcerate or cataulcerate, or with a proximal sulcus (catasulcate), or even bisulcate, the exine variously solid and virtually structureless, or granular-tectate, or tectate-columellate; gynoecium of (1–) more or less numerous conduplicate carpels on a flat or conic receptacle, the carpels distinct or rarely (*Monodora* and *Isolona*) connate to form a compound ovary with free styles and intruded parietal placentas; style short and thick; ovules 1-many, on marginal (rarely laminar-lateral or parietal) placentas, anatropous or sometimes campylotropous, bitegmic (seldom tritegmic), crassinucellar; endosperm development cellular. FRUIT commonly of separate, stipitate, fleshy, indehiscent, berry-like carpels, varying to sometimes dry and dehiscent (*Anaxagorea*) or indehiscent, or the carpels sometimes coalescent to form an aggregate fruit; seeds with small, basal to axile, dicotyledonous embryo and abundant, firm, ruminate, oily (sometimes also starchy) endosperm, often arillate. X=7, 8, 9.

The Annonaceae, with about 130 genera and 2300 species, are by far the largest family of Magnoliales; the other families collectively add up to only about 600 species. About a third of the species belong to only 5 genera, *Guatteria* (250), *Uvaria* (175), *Xylopia* (160), *Polyalthia* (150), and *Annona* (120). The family is well developed in both the Old and the New World, and very largely confined to tropical regions. *Asimina* (ca 10 spp.) is a notable exception, centering in southeastern United States and extending north as far as New York and Michigan. Several species of *Annona* are cultivated for their edible fruits. Notable among these is the tropical American *A. reticulata* L., the custard-apple.

The Annonaceae are generally considered to be an integral member of the group of families that is here treated as the order Magnoliales. Within that group the Myristicaceae, Canellaceae, and Eupomatiaceae are often cited as their nearest relatives. I concur.

Macrofossils considered to represent the Annonaceae are known from Eocene and later deposits, but annonaceous pollen dates only from the Oligocene.

9. Family MYRISTICACEAE R. Brown 1810 nom. conserv., the Nutmeg Family

Trees (rarely shrubs) with spherical ethereal oil cells in the parenchymatous tissues, often very aromatic, characteristically producing the psychotrophic phenolic compound myristicin, and sometimes (spp. of *Virola*) producing indole alkaloids, notably of the β-carboline and tryptamine subgroups, consistently tanniferous (at least sometimes with proanthocyanins), the tannin commonly in special sacs with yellow or red or clear contents, especially in some of the wood-rays; small, needle-like or clustered crystals of calcium oxalate commonly present in some of the parenchyma cells; nodes trilacunar; pith septate; vessel-segments usually with both simple and scalariform perforation-plates, the latter with 1–10 (–20) cross-bars, a few reticulate perforation plates sometimes also present; imperforate tracheary elements with small, inconspicuously bordered (seldom simple) pits, the elements adjoining the vessels commonly septate; wood-parenchyma paratracheal, or often apotracheal and banded or diffuse; wood-rays heterocellular, mixed uniseriate and pluriseriate, the latter 2–3 cells wide; bark commonly exuding a colored sap when slashed; phloem tangentially stratified into alternating fibrous and nonfibrous layers, and with wedge-shaped rays; sieve-tubes with S-type plastids. LEAVES alternate, simple, entire, pinnately veined, often pellucid-punctate; stomates paracytic; stipules wanting. FLOWERS in axilllary or rarely terminal, cymose to racemose inflorescences, rather small, hypogynous, unisexual, the plants commonly dioecious; calyx cupulate or campanulate, (2) 3 (–5)-lobed, the lobes valvate; petals none; staminate flowers lacking even a vestigial gynoecium; stamens 2-many, the filaments wholly united into a more or less solid column, or rarely (*Brochoneura*) connate only at the base; anthers distinct or more often laterally connate, extrorse, bisporangiate, unithecal, dehiscing by a single longitudinal slit; or, by a different interpretation, a pair of such bisporangiate pollen-sacs considered to form a tetrasporangiate, dithecal anther; pollen-grains globose to boat-shaped, monosulcate (anasulcate) to inaperturate, mostly tectate-columellate, varying to intectate; pistillate flowers without staminodes, the gynoecium of a single conduplicate, unsealed carpel with a small, simple or bilobed, sessile or subsessile stigma, or rarely (*Brochoneura*) the stigma on a long style; ovule solitary, nearly basal, more or less anatropous, bitegmic, crassinucellar; endosperm-development nuclear. FRUIT more or less leathery, generally dehiscent along two sutures; seeds with abundant, oily (sometimes also starchy) ruminate or less often non-ruminate endosperm, generally with a well developed aril that arises from the outer integument; embryo very small, dicotyledonous, the cotyledons sometimes connate at the base. X= 9, 21, 25.

The family Myristicaceae consists of some 15 genera and about 300 species, widespread in tropical regions of both the Old and the New World. The largest genera are *Myristica* (100), *Horsfieldia* (70), and

Knema (40) in the Paleotropics, and *Virola* (50) in the Neotropics. *Myristica fragrans* Houtt. is the source of nutmeg and mace. Species of *Virola* are the source of a hallucinogenic snuff (the dried and powdered exudate from the bark) used by certain Amazonian Indians. *Virola surinamensis* (Roland ex Rottb.) Warb. is a valuable timber-tree, used especially in making plywood.

It is now generally agreed that the Myristicaceae are related to the Canellaceae and Annonaceae, with which they are here associated. Armstrong & Wilson (1978) consider that the family originated in West Gondwanaland, perhaps in South America.

10. Family CANELLACEAE Martius 1832 nom. conserv., the Canella Family

Glabrous, aromatic trees (rarely shrubs) with spherical ethereal oil cells in the parenchymatous tissues (the oil consisting mainly of myrcene, a terpene), at least sometimes (*Cinnamosma*) producing benzyl-isoquinoline alkaloids, and sometimes cyanogenic; clustered crystals of calcium oxalate commonly present in some of the parenchyma-cells; nodes trilacunar; pith not septate; vessel-segments mostly very elongate, with oblique ends and mixed scalariform and reticulate perforation-plates, the scalariform plates with 10–20 or even (*Cinnamodendron*) 50–100 cross-bars; imperforate tracheary elements with evidently bordered pits, considered to be true tracheids; wood-parenchyma apotracheal or paratracheal, or both; wood-rays mostly Kribs type I or III, mixed uniseriate and pluriseriate, the former homocellular, the latter mostly 2–3 (4) cells wide and heterocellular; phloem not stratified, but the rays slightly flaring; sieve-tubes with P-type plastids. LEAVES alternate, simple, entire, pinnately veined, commonly pellucid-dotted; stomates paracytic or seldom anomocytic; stipules wanting. FLOWERS in terminal or axillary cymes or racemes or solitary in the axils, perfect, regular, hypogynous; sepals 3, thick, leathery, imbricate (these sometimes interpreted as bracts); petals (4) 5–12, in one or 2 (–4) whorls and/or spirals (the outer whorl sometimes interpreted as sepals), slender, imbricate, distinct in most genera, but connate toward the base in *Canella* and up to the middle in *Cinnamosma*; stamens mostly 6–12, or in *Cinnamodendron* up to 35 or 40, connate into a tube, to the outside of which are attached the elongate, extrorse anthers, each anther tetrasporangiate with two juxtaposed bisporangiate thecae; pollen-grains binucleate, globose to boat-shaped, tectate-columellate or tectate-perforate, monosulcate (anasulcate), or the aperture sometimes v-shaped or trichotomous in some grains; gynoecium of 2–6 carpels united to form a compound, unilocular pistil with a short, thick style and usually 2–6-lobed stigma; placentas 2–6, parietal, with a single or double row of 2-many ovules on each placenta; ovules hemitropous, crassinucellar, and bitegmic, with a zigzag micropyle. FRUIT a berry with 2 or more seeds; seeds with abundant, oily endosperm, this ruminate in *Cinnamosma* only; embryo small, dicotyledonous. 2n = 22, 26, 28.

The family Canellaceae consists of 6 genera and about 20 species, found in tropical Africa (*Warburgia*), Madagascar (*Cinnamosma*), and tropical America (*Capsicodendron* and *Cinnamodendron* in South America, *Canella* and *Pleodendron* in the Caribbean). Wilson considered the South American genera to be the most archaic.

The Canellaceae have often in the past been referred to the vicinity of the Violales, in what is here treated as the subclass Dilleniidae. It is now generally agreed, however, that they belong with the Magnoliales, in spite of the syncarpous ovary. My own description of an order Canellales in 1957 reflected an attempt to dissociate the Canellaceae from the Violales and put them alongside the Magnoliales. It now seems more useful to recognize this relationship by including the Canellaceae within the Magnoliales. It may be significant that two genera of Annonaceae resemble the Canellaceae in having a syncarpous ovary with parietal placentation.

Recent authors such as Wilson mostly see the Myristicaceae as the nearest relatives of the Canellaceae. I concur, but the relationship is not linear. A putative common ancestor of the two families would have to bear perfect flowers with several distinct carpels that have several or many ovules. Such a combination of characters would remove the plant from both families. Takhtajan (1966) thinks the Canellaceae may have arisen from some very primitive, now extinct member of the Magnoliaceae.

2. Order LAURALES Lindley 1833

Trees, shrubs, or woody vines (*Cassytha*, in the Lauraceae, a twining, parasitic herb), ordinarily with spherical ethereal oil cells in the parenchymatous tissues at least of the leaves, commonly tanniferous, producing proanthocyanins but not ellagic acid, and very often with alkaloids of the benzyl-isoquinoline or aporphine group; nodes unilacunar; xylem vesselless (Amborellaceae) or with well developed vessels, these with scalariform perforation plates or simply perforate. LEAVES opposite or alternate, simple or seldom compound, entire or rarely toothed or lobed; stomates paracytic or less often anomocytic; stipules wanting. FLOWERS mostly entomophilous, often cantharophilous, hypogynous in Amborellaceae and Trimeniaceae, otherwise mostly perigynous (the hypanthium of receptacular origin) or epigynous, often in distinctly cymose or racemose inflorescences; perianth usually well developed, with mostly free and distinct tepals, these variously cyclic (often trimerous) or spirally arranged, sometimes in more than 2 series, generally not clearly differentiated into sepals and petals, but the petals when present homologous with the sepals and bracts; stamens (3–) 5-many, cyclic or spirally arranged, ribbon-shaped to much more often organized into a more or less well defined filament and anther; anthers dehiscing by valves, or less often by longitudinal slits, rarely by pores; pollen-grains mostly inaperturate or biaperturate, seldom uniaperturate or multiaperturate, binucleate so far as known; gynoecium of 1-many carpels, fully sealed and with a definite stigma that may be borne on a distinct style or seldom (Gomortegaceae) of 2–3 united carpels, the ovary superior or much less often inferior; ovule solitary (or 2, but only one maturing), anatropous, bitegmic or seldom unitegmic, crassinucellar. FRUITS of various types, often fleshy; seeds with or without endosperm, the embryo accordingly small or large; cotyledons 2 (3–4 in Idiospermaceae).

The order Laurales as here defined consists of 8 families and about 2500 species. The Lauraceae, with about 2000 species, are by far the largest family, and the Monimiaceae, with about 450, are next. The remaining families have fewer than a hundred species in all, and three of them have only a single species each. The vast majority of the species occur in tropical or warm-temperate regions, most of them in places with a moist, equable climate. Otherwise they occupy no particular ecological niche in contrast to other woody angiosperms, at least so far as present knowledge indicates.

The Laurales are obviously allied to the Magnoliales. Several families have been shifted back and forth from one order to the other by various authors. As noted in the discussion under the Austrobaileyaceae (Magnoliales), the Laurales are here rather strictly defined so as to form a relatively homogeneous group. The doubtful or intermediate families are retained in the more heterogeneous order Magnoliales.

The Laurales are more advanced in several respects than the bulk of the Magnoliales. The mostly perigynous (or even epigynous) flowers, inaperturate or biaperturate pollen, solitary functional ovule, and usually fairly conventional stamens of the Laurales are all advanced, as compared to many or most of the Magnoliales. The unilacunar nodes of the Laurales, on the other hand, might be more primitive than the more often trilacunar or multilacunar nodes of the Magnoliales; the matter is debatable.

The status of the glandular appendages at the base of the stamens in many Lauraceae and some other families of Laurales as staminodia or mere enations has been inconclusively debated.

SELECTED REFERENCES

Bailey, I. W. 1957. Additional notes on the vesselless dicotyledon, *Amborella trichopoda* Baill. J. Arnold Arbor. 38: 374–380.

Bailey, I. W., & B. G. L. Swamy. 1948. *Amborella trichopoda* Baill., a new morphological type of vesselless dicotyledon. J. Arnold Arbor. 29: 245–254.

Blake, S. T. 1972. *Idiospermum* (Idiospermaceae), a new genus and family for *Calycanthus australiensis*. Contr. Queensland Herb. 12: 1–37.

Boyle, E. M. 1980. Vascular anatomy of the flower, seed and fruit of *Lindera Benzoin*. Bull. Torrey Bot. Club 107: 409–417.

Brizicky, G. K. 1959. Variability in the floral parts of *Gomortega* (Gomortegaceae). Willdenowia 2: 200–207.

Cheadle, V. I., & K. Esau. 1958. Secondary phloem of Calycanthaceae. Univ. Calif. Publ. Bot. 29: 397–510.

Cheng, W.-C., & S.-Y. Chang. 1964. Genus novum calycanthacearum Chinae orientalis. Acta Phytotax. Sin. 9: 137–139.

Daumann, E. 1930. Das Blütennektarium von *Magnolia* und die Futterkörper in der Blüte von *Calycanthus*. Planta 11: 108–116.

Dengler, N. G. 1972. Ontogeny of the vegetative and floral apex of *Calycanthus occidentalis*. Canad. J. Bot. 50: 1349–1356.

Endress, P. K. 1972. Zur vergleichenden Entwicklungsmorphologie, Embryologie und Systematik bei Laurales. Bot. Jahrb. Syst. 92: 331–428.

Endress, P. K. 1979. Noncarpellary pollination and 'hyperstigma' in an angiosperm (*Tambourissa religiosa*, Monimiaceae). Experientia 35: 45.

Endress, P. K. 1980. Floral structure and relationships of *Hortonia* (Monimiaceae). Pl. Syst. Evol. 133: 199–221.

Endress, P. K. 1980. Ontogeny, function and evolution of extreme floral construction in Monimiaceae. Pl. Syst. Evol. 134: 79–120.

Fahn, A., & I. W. Bailey. 1957. The nodal anatomy and the primary vascular cylinder of the Calycanthaceae. J. Arnold Arbor. 38: 107–117.

Garratt, G. A. 1934. Systematic anatomy of the woods of the Monimiaceae. Trop. Woods 39: 18–44.

Goldblatt, P. 1976. Chromosome number in *Gomortega keule*. Ann. Missouri Bot. Gard. 63: 207–208.

Gottlieb, O. R. 1972. Chemosystematics of the Lauraceae. Phytochemistry 11: 1537–1570.

Grant, V. 1950. The pollination of *Calycanthus occidentalis*. Amer. J. Bot. 37: 294–297.

Kasapligil, B. 1951. Morphological and ontogenetic studies of *Umbellularia californica* Nutt. and *Laurus nobilis* L. Univ. Calif. Publ. Bot. 25: 115–239.

Kostermans, A. J. G. H. 1957. Lauraceae. Reinwardtia 4: 193–256.

Kubitzki, K. 1969. Monographie der Hernandiaceen. Bot. Jahrb. Syst. 89: 78–209.

Leinfellner, W. 1966, 1968. Über die Karpelle verschiedener Magnoliales. II. *Xymalos*, *Hedycarya* und *Siparuna* (Monimiaceae). VI. *Gomortega keule* (Gomortegaceae). Oesterr. Bot. Z. 113: 448–458, 1966. 115: 113–119, 1968.

Lemesle, R., & Y. Pichard. 1954. Les caractères histologiques du bois des Monimiacées. Rev. Gén. Bot. 61: 69–95.

Mathur, S. N. 1968. Development of the male and female gametophytes of *Calycanthus fertilis* Walt. Proc. Natl. Inst. Sci. India, 34B: 323–329.

Money, L. L., I. W. Bailey, & B. G. L. Swamy. 1950. The morphology and relationships of the Monimiaceae. J. Arnold Arbor. 31: 372–404.

Morat, P., & H. S. MacKee. 1977. Quelques précisions sur le *Trimenia neocaledonica* Bak.f. et la famille des Triméniacées en Nouvelle-Calédonie. Adansonia sér. 2, 17: 205–213.

Nicely, K. A. 1965. A monographic study of the Calycanthaceae. Castanea 30: 38–81.

Pichon, P. 1948. Les Monimiacées, famille hétérogène. Bull. Mus. Hist. Nat. (Paris) II. 20: 383–384.

Record, S. J., & R. W. Hess. 1942. American timbers of the family Lauraceae. Trop. Woods 69: 7–35.

Rickson, F. R. 1979. Ultrastructural development of the beetle food tissue of *Calycanthus* flowers. Amer. J. Bot. 66: 80–86.

Rodenburg, W. F. 1971. A revision of the genus *Trimenia* (Trimeniaceae). Blumea 19: 3–15.

Sampson, F. B. 1969. Studies on the Monimiaceae. I. Floral morphology and gametophyte development of *Hedycarya arborea* J. R. et G. Forst. (subfamily Monimioideae). Austral. J. Bot. 17: 403–424. II. Floral morphology of *Luarelia novae-zelandiae* A. Cunn. (subfamily Atherospermoideae). New Zealand J. Bot. 7: 214–240. III. Gametophyte development of *Laurelia novae-zelandiae* A. Cunn. (subfamily Atherospermoideae). Austral. J. Bot. 17: 425–439.

Sastri, R. L. N. 1952, 1958, 1962, 1963, 1965. Studies in Lauraceae. I. Floral anatomy of *Cinnamomum iners* Reinw. and *Cassytha filiformis* Linn. II. Embryology of *Cinnamomum* and *Litsea*. J. Indian Bot. Soc. 31: 240–246, 1952. 37: 266–278, 1958. III. Embryology of *Cassytha*. Bot. Gaz. 123: 197–206, 1962. IV. Comparative embryology and phylogeny. Ann. Bot. (London) II. 27: 425–433, 1963. V. Comparative morphology of the flower. Ann. Bot. (London) II. 29: 39–44, 1965.

Schaeppi, H. 1953. Morphologische Untersuchungen an den Karpellen der Calycanthaceae. Phytomorphology 3: 112–118.
Schodde, R. 1970. Two new suprageneric taxa in the Monimiaceae alliance (Laurales). Taxon 19: 324–328.
Schroeder, C. A. 1952. Floral development, sporogenesis, and embryology in the avocado, *Persea americana*. Bot. Gaz. 113: 270–278.
Shutts, C. F. 1960. Wood anatomy of Hernandiaceae and Gyrocarpaceae. Trop. Woods 113: 85–123.
Stern, W. L. 1954. Comparative anatomy of xylem and phylogeny of Lauraceae. Trop. Woods 100: 1–72.
Stern, W. L. 1955. Xylem anatomy and relationships of Gomortegaceae. Amer. J. Bot. 42: 874–885.
Tiagi, Y. D. 1963. Vascular anatomy of the flower of certain species of the Calycanthaceae. Proc. Indian Acad. Sci. B. 58: 224–234.
Wilson, C. L. 1976. Floral anatomy of *Idiospermum australiense* (Idiospermaceae). Amer. J. Bot. 63: 987–996.

Анели, Н. А. 1956. Об анатомическом родстве растений лавровых и некоторых других семейств. Вестн. Тбилисск. Бот. Сада 63: 85-102.

SYNOPTICAL ARRANGEMENT OF THE FAMILIES OF LAURALES

1 Seeds with well developed endosperm and small to medium-sized embryo.
  2 Leaves alternate; wood without vessels; flowers unisexual; stamens numerous, opening by longitudinal slits; pollen inaperturate or obscurely uniaperturate; carpels 5–8 .... 1. AMBORELLACEAE.
  2 Leaves opposite or occasionally subopposite; wood with vessels; flowers various.
    3 Ovary superior; flowers unisexual or less often perfect.
      4 Receptacle hardly differentiated from the pedicel, the 5–25 pairs of opposite, decussate appendages passing without demarcation from the bracteoles of the pedicel to the tepals of the flower; flowers hypogynous, carpel 1 (2); pollen with 2 or 8–12 poorly marked apertures; anthers opening by longitudinal slits .... 2. TRIMENIACEAE.
      4 Receptacle well differentiated from the pedicel, the bracts not passing into the tepals; flowers mostly perigynous (varying to essentially hypogynous in *Hortonia* and *Xymalos*); carpels usually several to many (solitary in *Xymalos*); pollen inaperturate or rarely uniaperturate or biaperturate; anthers opening by slits or valves .... 3. MONIMIACEAE.
    3 Ovary inferior; flowers perfect; carpels 2–3, united; pollen inaperturate; anthers opening by valves 4. GOMORTEGACEAE.
1 Seeds without endosperm and with large embryo.
  5 Leaves opposite; pollen biaperturate; anthers opening by longitudinal slits.

6 Carpels 5–35; cotyledons 2 .... 5. CALYCANTHACEAE.
6 Carpels 1–2 (3); cotyledons 3 or 4 .... 6. IDIOSPERMACEAE.
5 Leaves alternate, very rarely opposite or whorled; pollen inaperturate; anthers nearly always opening by valves; carpel 1; cotyledons 2.
7 Ovary superior, or very rarely inferior; fruit a 1-seeded berry or a drupe, only rarely dry and indehiscent .. 7. LAURACEAE.
7 Ovary inferior; fruit dry, indehiscent .... 8. HERNANDIACEAE.

1. Family AMBORELLACEAE Pichon 1948 nom. conserv., the Amborella Family

Dioecious, evergreen, arborescent shrubs, accumulating aluminum, nearly or quite without the characteristic ethereal oil cells of the order, and also lacking mucilage cells; nodes unilacunar, with a single broad trace; pericycle of young stems containing characteristic sclereids with U-shaped thickenings; wood without vessels; tracheids elongate, with scalariform to circular bordered pits; wood-parenchyma scanty, diffuse; wood-rays Kribs type I, narrow, heterocellular, some 3–5-seriate and composed of more or less square cells; others 1–2-seriate and composed of upright cells; fibers wanting from the phloem and cortex. LEAVES alternate (spiral in early years, distichous later), simple, pinnately veined, entire or lobed; stomates paracytic to anomocytic; petiole with a single vascular arc; stipules wanting. FLOWERS in axillary, cymose inflorescences, unisexual, hypogynous or slightly perigynous; perianth of 5–8 spirally arranged tepals, these weakly united at the base; staminate flowers with more or less numerous stamens in several cycles on a slightly convex receptacle, the outer stamens basally adnate to the tepals; stamens more or less laminar, tetrasporangiate, the two pollen sacs somewhat separated, introrsely marginal just below the summit, opening by longitudinal slits; pollen-grains globose, tectate-granular, inaperturate or with an irregular distal unthickened area in the exine; pistillate flowers with 1 or 2 staminodes, and 5 or 6 separate carpels in a single whorl on a slightly concave receptacle; carpels unsealed at the tip, the stigma sessile, with 2 expanded flanges; ovule solitary, marginal, anatropous. FRUIT of separate, drupe-like carpels, these oblique, with ventro-apical stigma; seeds with abundant endosperm and a minute, basal, dicotyledonous embryo. $2n = 26$.

The family Amborellaceae consists of a single genus and species, *Amborella trichopoda* Baill., native to New Caledonia. Although it was originally included in the Monimiaceae, *Amborella* appears to be amply distinct and has been considered in most recent systems of classification to form a monotypic family. It is clearly a member of the Laurales, in which its primitively vesselless wood, alternate leaves, essentially hypogynous flowers, several carpels, abundant endosperm, and stamens dehiscent by longitudinal slits mark it as an archaic type. The virtual absence of ethereal oil cells is anomalous in the group, but it is uncertain whether this feature is primitive or secondary.

2. Family TRIMENIACEAE Gibbs 1917 nom. conserv., the Trimenia Family

Trees or scandent shrubs with ethereal oil cells in the parenchymatous tissues, and also with specialized mucilage cells, commonly accumulating aluminum, at least sometimes without alkaloids; stem without specialized

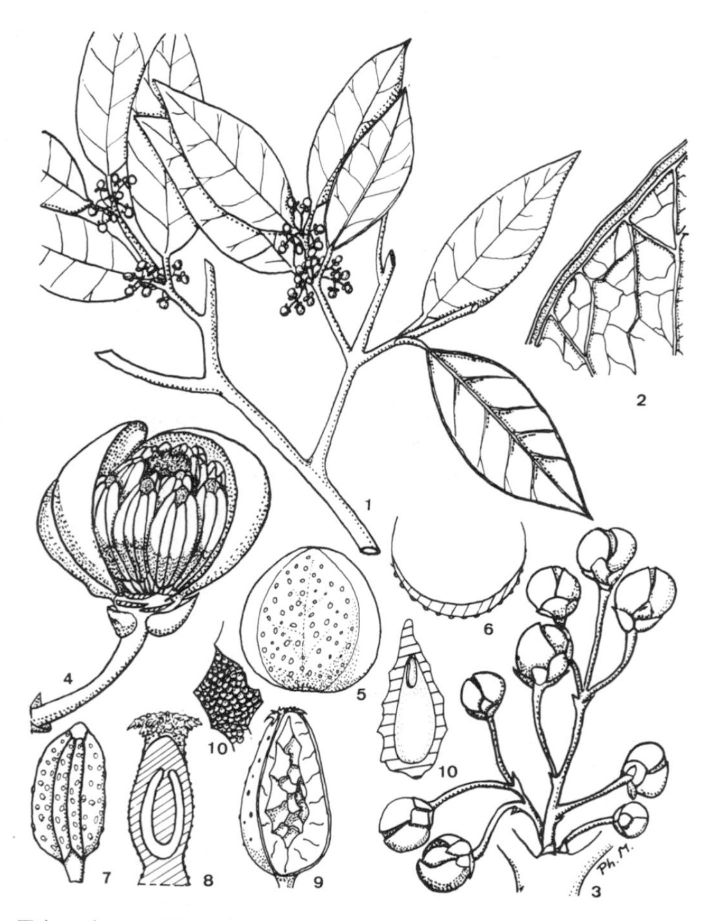

FIG. 1.9 Trimeniaceae. *Trimenia neocaledonica* Baker f. 1, habit; 2, portion of lower surface of leaf; 3, inflorescence; 4, perfect flower, with some of the tepals removed; 5, inner face of a tepal; 6, cross-section of a tepal; 7, stamen, from ventral side; 8, long-section of ovary; 9, fruit, in partial long-section; 10, at the left, detail of endocarp; 10, at the right, long-section of stone, showing embryo and endosperm. (From Quelques précisions sur le *Trimenia neocaledonica* Bak. f. et la famille des Triméniacées en Nouvelle-Calédonie. Adansonia sér. 2. 17: 205–213. 1977. Courtesy of Philippe Morat and the directors of Adansonia.)

sclereids; nodes unilacunar, with 2 traces; vessel-segments elongate, with scalariform perforation-plates that have more than 20 cross-bars; imperforate tracheary elements septate, with vestigially bordered pits; wood-parenchyma largely replaced by the septate fibers; wood-rays heterocellular, some uniseriate, others pluriseriate and up to 5 or 6 cells wide. LEAVES opposite or subopposite, simple, entire or toothed, pinnately veined; stomates paracytic; stipules wanting. FLOWERS small, in more or less pyramidal to cylindric, cymose inflorescences, hypogynous, polygamous, with perfect, unisexual, and transitional types; receptacle with flat or slightly concave tip, hardly differentiated from the pedicel, the 5–25 pairs of opposite appendages passing without clear demarcation from the bracteoles of the pedicel to the tepals of the flat or slightly convex receptacle, all deciduous at or before anthesis; stamens ca 7–23, in (1) 2–3 rows; rather broad, spirally inserted on the receptacle, without basal glands; anthers elongate, basifixed, tetrasporangiate and dithecal, latrorse or weakly extrorse, opening by longitudinal slits; connective shortly prolonged; staminodes wanting; pollen-grains globose to more or less ellipsoid, imperfectly tectate-columellate, with 2 (*Piptocalyx*) or 8–12 (*Trimenia*) slightly irregular porate apertures in the exine; gynoecium of a single carpel or seldom of 2 distinct carpels, with a broad, somewhat feathery, sessile stigma; a single anatropous ovule pendulous from the summit of the locule. FRUIT drupaceous, with rather thin, juicy, mucilaginous flesh; endosperm copious, gelatinous; embryo small, basal (next to the micropyle), short, with 2 cotyledons. 2n = 16.

The family Trimeniaceae as here defined, following Money et al. (1950), has 2 genera and 5 species, limited to New Guinea, New Caledonia, Fiji, and southeastern Australia. Botanists are agreed that the Trimeniaceae are allied to the Monimiaceae. *Xymalos*, of tropical and southern Africa, resembles the Trimeniaceae in having hypogynous flowers with only a single carpel, but Money et al. conclude that "a summation of other morphological and anatomical data indicates that the genus belongs in the florally variable subfamily Monimioideae [of the family Monimiaceae]."

3. Family MONIMIACEAE A. L. de Jussieu 1809 nom. conserv., the Monimia Family

Trees or shrubs, or occasionally woody vines, tanniferous, producing proanthocyanins, often accumulating aluminum, commonly with benzyl-isoquinoline and/or aporphine alkaloids, and with spherical ethereal oil cells in the parenchymatous tissues; raphides or small, needle-like crystals commonly present in some of the cells (often specialized sacs) of the parenchyma; pericycle of young stems commonly containing sclereids with U-shaped thickenings (except in subfamily Siparun-

FIG. 1.10 Monimiaceae. *Peumus boldus* Molina. a, habit, ×½; b, staminate flower, ×3; c, stamen and staminodes, ×9; d, pistillate flower, ×3; e, staminode from pistillate flower, ×9; f, g, gynoecium from above and from the side, ×9; h, carpel, in partial long-section, ×12; i, fruit, ×3; k, seed, ×3.

oideae); twigs commonly somewhat flattened at the nodes; nodes unilacunar, with 3-several traces or a single broad trace; vessel-segments elongate, commonly with very oblique perforation-plates that have more than 10 cross-bars (these sometimes anastomosing), varying to rarely simply perforate; imperforate tracheary elements with bordered or simple pits, ranging from tracheids through fiber-tracheids to libriform fibers, often many of them septate; wood-rays heterocellular, usually mixed uniseriate and pluriseriate, the latter up to 16 cells wide; wood-parenchyma highly variable, most commonly sparse and diffuse, sometimes wanting; sieve-tubes with P-type plastids. LEAVES opposite (rarely subopposite, alternate, or whorled), simple, with entire or toothed margins, pinnately veined, often pellucid-punctate, the cell-walls of the epidermis often silicified; stomates anomocytic; stipules wanting. FLOWERS in axillary or seldom terminal, cymose inflorescences, rarely solitary, relatively small and inconspicuous, perfect or more often unisexual, regular or rarely somewhat oblique, from essentially hypogynous (*Xymalos*, *Hortonia*) to more often definitely perigynous, with a concave to cupulate or urceolate hypanthium, the inner surface of the hypanthium often nectariferous; perianth sometimes of 2 decussate pairs of fleshy sepals and 7–20 or more petals, more often more or less reduced, often not differentiated into sepals and petals, or sometimes wholly wanting; stamens mostly numerous in 1–2 series, with short filaments, sometimes provided with a pair of basilateral, nectariferous appendages that are considered to be staminodial in origin; more ordinary staminodia sometimes present between the fertile stamens and the carpels; anthers bisporangiate or tetrasporangiate, opening by longitudinal slits or from the base upwards by valves; pollen-grains in monads or sometimes in tetrads, globose to broadly ellipsoid, binucleate, commonly inaperturate and with more or less reduced exine, but sometimes monosulcate or bisulcate and tectate; gynoecium of 1 (*Xymalos*) or more often several or many separate carpels with short or elongate style and terminal stigma, the carpels sometimes sunken in the receptacle; ovule solitary, apical and pendulous (Hortonioideae, Monimioideae) or basal and erect (Atherospermatoideae, Siparunoideae), anatropous, crassinucellar, bitegmic or (Siparunoideae) unitegmic; endosperm-development cellular. FRUIT of separate nuts or drupes, these often collectively enclosed in the hypanthium; seeds with abundant, oily endosperm and small to medium-sized, dicotyledonous embryo, in *Siparuna* with an aril. x = 18–22, 39, 43. (Atherospermataceae, Hortoniaceae, Siparunaceae)

The Monimiaceae are a somewhat heterogeneous group of some 30–35 genera and 450 species, native to tropical and subtropical regions, especially of the Southern Hemisphere, in both the Old and the New World. The largest genera are *Siparuna* (165) and *Mollinedia* (95) in the New World, and *Kibara* (40), *Steganthera* (30), *Tambourissa* (25), *Hedycarya* (20), and *Palmeria* (20) in the Old. The type genus, *Monimia*, has only 4 species and is restricted to Madagascar.

The Amborellaceae and Trimeniaceae, which were in the past included in the Monimiaceae, seem sufficiently distinct. A little further splitting would permit the segregation of each of the 4 generally recognized subfamilies of Monimiaceae (Monimioideae, Atherospermatoideae, Hortonioideae, Siparunoideae) as a separate family. That would still leave the Monimiaceae as a heterogeneous group, with more than half of the species from the more broadly defined family. It therefore seems fruitless to pursue the course of breaking up the traditional family Monimiaceae into smaller, more homogeneous, and more sharply defined families.

Although they are not the largest family of the order, the Monimiaceae are critical to an understanding of the Laurales. Search for relatives of each of the other families leads into one or another part of the Monimiaceae. Only the Calycanthaceae and Idiospermaceae form a pair collectively somewhat set off from the rest of the order.

The taxonomic identity of the several Upper Cretaceous fossil woods and leaves that have been referred to the Monimiaceae is subject to reconsideration. Pollen referrable to the family is known only from Oligocene and more recent deposits.

Endress (1979, 1980) reports that in *Tambourissa, Kibara, Hennecartia,* and *Wilkiea* the pollen germinates on a mucilaginous plug closing the small distal opening in the globular or discoid, hollow receptacle-hypanthium.

4. Family GOMORTEGACEAE Reiche 1896 nom. conserv., the Queule Family

Aromatic evergreen trees with spherical ethereal oil cells in the parenchymatous tissues of the leaves and young stems, and with small needles or prisms of calcium oxalate in some of the parenchyma-cells; nodes unilacunar, with two traces or a single broad trace; pericycle of the stem containing some sclereids with U-shaped thickenings; vessel-segments elongate, slender, with scalariform perforation-plates that commonly have 9–19 cross-bars; imperforate tracheary elements with circular bordered pits, some considered to be tracheids, others fiber-tracheids; wood-rays heterocellular, Kribs type IIB, 1–3-seriate, most of them 2-seriate; wood-parenchyma apotracheal. LEAVES opposite, simple, entire, pinnately veined; stomates paracytic; stipules wanting. FLOWERS in axillary and terminal racemes, rather small, perfect, epigynous; perianth of 5–9 (often 7) spirally arranged tepals, these progressively smaller centripetally; stamens 7–13, spirally arranged, the outer 1–3 (4) tepaloid, laminar, often with imperfectly developed anthers, the next 5–10 (often 8) differentiated into filament and anther, the filament provided with pair of short-stalked glands at the base (homologous with those of the Monimiaceae), the anther dehiscing from the base upwards by 2 valves; 1–4 (usually 3) reduced sterile stamens commonly present between the fertile stamens and the style; pollen globose, tectate, inaperturate; gynoecium of 2 or less often 3 carpels united to form a compound, inferior ovary with 2–3 locules, the short style with as many branches; each locule with a single anatropous ovule pendulous from the apex. FRUIT yellow, edible, drupaceous, commonly unilocular and single-seeded; seed with well deveoped, oily endosperm and rather large, dicotyledonous embryo. 2n = 42.

The family Gomortegaceae consists of a single species, *Gomortega keule* (Molina) I. M. Johnston (*G. nitida* Ruiz & Pavon), native to central Chile. Recent authors are agreed that *Gomortega* is allied to the Monimiaceae.

5. Family CALYCANTHACEAE Lindley 1819 nom. conserv., the Strawberry-shrub Family

Shrubs or small trees with aromatic bark and with spherical ethereal oil cells in the parenchymatous tissues, only slightly tanniferous, lacking proanthocyanins as well as ellagic acid, sometimes cyanogenic through a pathway based on tyrosine, and regularly producing the indole-related alkaloids calycanthine, calycanthidine, and folicanthine; calcium oxalate crystals present only as small prisms, or none; hairs always unicellular; young stem with 4 inverted vascular bundles in the cortex or pericycle, the xylem external; nodes unilacunar, with 2 traces; vessel-segments small, with oblique simple perforations; imperforate tracheary elements consisting of both tracheids, with bordered pits, and libriform fibers, with simple pits; wood-rays heterocellular, mixed uniseriate and pluriseriate, the latter 2–3 (4) cells wide; wood-parenchyma very scanty,

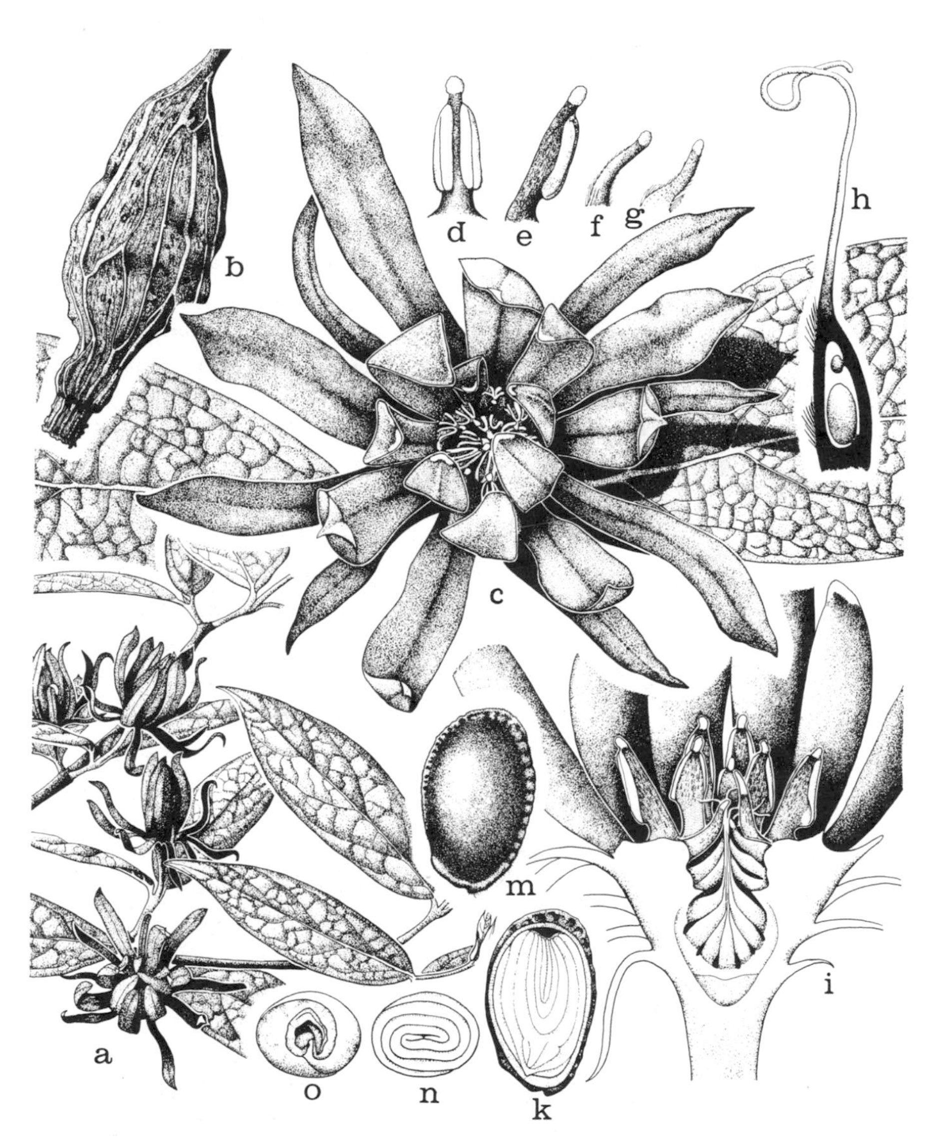

FIG. 1.11 Calycanthaceae. *Calycanthus floridus* L. a, habit, $\times\frac{1}{2}$; b, fruiting hypanthium, $\times 1$; c, flower, from above, $\times 2$; d, e, stamens, $\times 4$; f, g, staminodes, $\times 4$; h, pistil, in partial long-section, $\times 8$; i, long-section of flower, $\times 4$; k, long-section of achene, $\times 2$; m, achene, $\times 2$; n, cross-section of embryo, $\times 2$; o, top view of embryo, $\times 2$.

paratracheal; phloem soft, not fibrous; sieve-tubes with P-type plastids. LEAVES opposite, simple, entire; epidermal hairs surrounded by cells with silicified walls, stomates paracytic; stipules wanting. FLOWERS solitary at the ends of specialized short leafy branches, cantharophilous, rather large, perfect, perigynous, the hypanthium mostly urceolate, with a narrow mouth (campanulate and with a wider mouth in *Sinocalycanthus*), receptacular; perianth of 15–30 slender (broad in *Sinocalycanthus*), more or less petaloid tepals (each with 3–4 vascular traces) spirally arranged on the outside of the hypanthium; stamens 5–30, spirally arranged at the top of the hypanthium, more or less ribbon-shaped (least so in *Sinocalycanthus*), with short or no filament, abaxial, extrorse pollen-sacs that open by longitudinal slits, and broad connective that is prolonged beyond the pollen-sacs; 10–25 staminodes present internal to the fertile stamens, usually nectariferous; pollen-grains binucleate, more or less globose, tectate-columellate, bisulcate; gynoecium of 5–35 distinct, spirally arranged carpels seated within the hypanthial cup; style elongate, filiform, with a dry, decurrent stigma; ovules 2 in each carpel, the lower one anatropous, bitegmic, and crassinucellar, the upper one abortive; endosperm development cellular. FRUIT of numerous achenes enclosed in the enlarged, fleshy, oily and proteinaceous hypanthium; seeds poisonous; embryo with 2 spirally twisted cotyledons; endosperm none. X = 11, 12.

The family Calycanthaceae consists of 3 genera and 5 species, confined to China and temperate North America. *Calycanthus* (2) is confined to North America, and *Chimonanthus* (2) and *Sinocalycanthus* (1) to China. Although the family has sometimes in the past been referred to the Rosales, modern authors are agreed that its relationship is with the Laurales.

6. Family IDIOSPERMACEAE S. T. Blake 1972, the Idiospermum Family

Evergreen trees with ethereal oil cells in the parenchymatous tissues at least of the leaves; young stems with 4 inverted vascular bundles in the pericycle or cortex, the xylem external; nodes unilacunar, with a single trace; vessel-segments with scalariform perforation-plates; wood-rays heterocellular, 1–2 (3) cells wide; wood-parenchyma abundant, vasicentric and paratracheal. LEAVES opposite, simple, entire, pinnately veined; stomates paracytic; stipules wanting. FLOWERS rather large, solitary (3) on bracteate, axillary peduncles, or in few-flowered terminal infloresc-

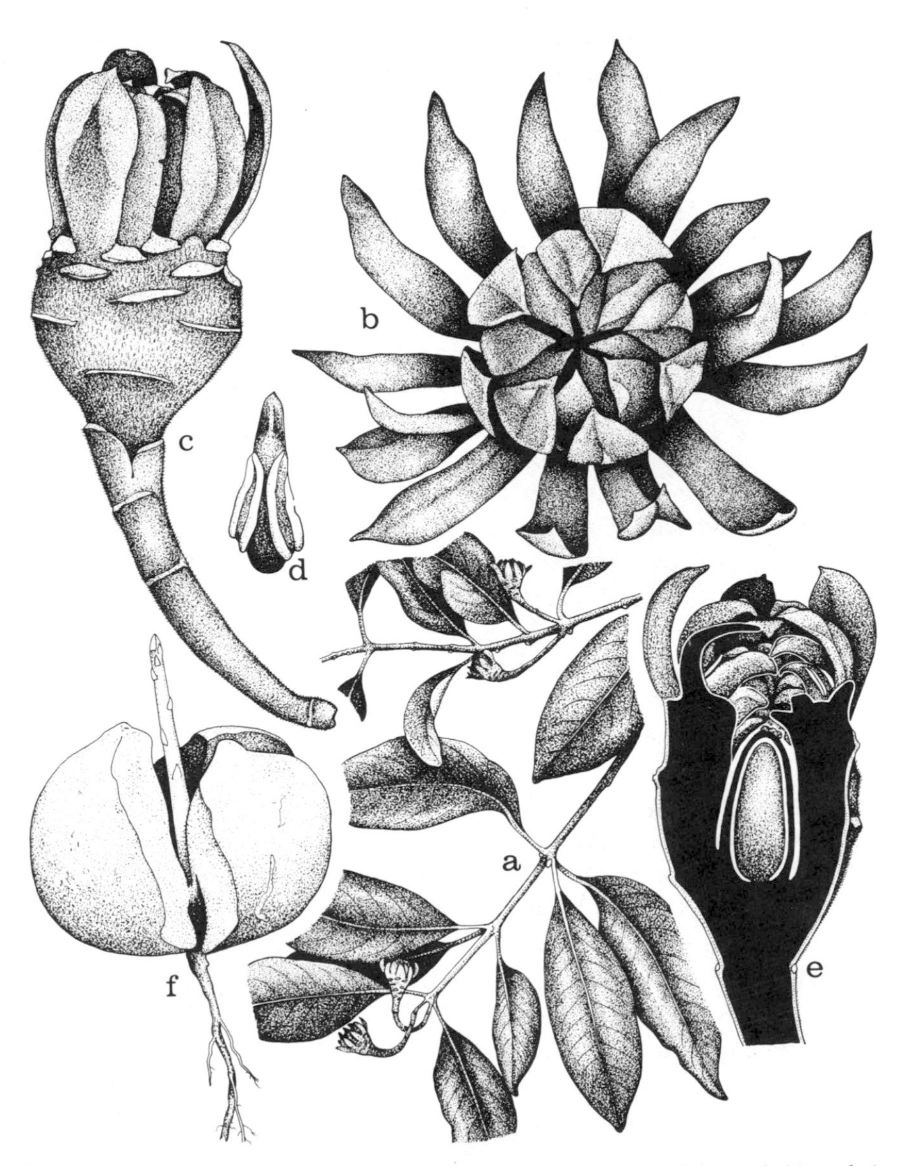

FIG. 1.12 Idiospermaceae. *Idiospermum australiense* (Diels) S. T. Blake. a, habit, $\times\frac{1}{4}$; b, flower, from above, $\times 3$; c, side view of flower after most of the tepals have fallen, $\times 3$; d, stamen, $\times 6$; e, flower in long-section, $\times 3$; f, germinating embryo, $\times\frac{1}{2}$.

ences, probably entomophilous, perfect, strongly perigynous, with a cup-shaped hypanthium; perianth of numerous (ca 30–40) spirally arranged, somewhat petaloid tepals covering the outside of the hypanthium, the outer ones the largest, the others progressively reduced, the inner ones (ca 15–18) persistent, the others deciduous; stamens ca 13–15 at the rim of the hypanthium, inflexed and covering the opening, not clearly divided into filament and anther, thick and firm, rather narrowly triangular in surface-view, tetrasporangiate, the bisporangiate pollen sacs embedded in the outer side toward the base, the inflexed tip of the stamen-blade projecting well beyond them; staminodes fleshy-thickened, more or less numerous, lining the inner surface of the hypanthium but more numerous toward the top; pollen-grains small, globose-oblate, virtually smooth, tectate, disulcate; gynoecium of one or 2 (3) distinct carpels with a subsessile, broad, fleshy, obliquely terminal stigma; ovule solitary (2), basal, anatropous, bitegmic. FRUIT large, globose or depressed-globose, indehiscent, the hypanthium becoming somewhat fleshy within a thin, hard outer layer, essentially closed at the top by the persistent stamens and upper tepals, the pericarp itself relatively thin and appressed to (but distinct from) the hypanthial wall. Seed solitary (2), poisonous, the embryo with 3 or more often 4 massive, fleshy-firm, peltate cotyledons; endosperm none; seed eventually freed from the fruit by decay of the pericarp and hypanthial wall, germinating at the surface of the ground, the hypocotyl not elongating. $2n = 22$.

The family Idiospermaceae consists of a single genus and species. *Idiospermum australiense* (Diels) S. T. Blake, native to rain-forests in northern Queensland, Australia. The species was at first referred on the basis of fragmentary material to *Calycanthus*, which indeed appears to be its nearest relative. Subsequent study of more complete specimens by S. T. Blake disclosed the numerous differences that support the establishment of a new genus and family. *Idiospermum* is reminiscent of *Degeneria* in having 3 or 4 cotyledons, but the two genera otherwise have little in common beyond the features that put them into the subclass Magnoliidae.

7. Family LAURACEAE A. L. de Jussieu 1789 nom. conserv., the Laurel Family

Aromatic evergreen trees or shrubs (*Cassytha* a twining, virtually leafless, pale green parasitic herb with little chlorophyll), sometimes storing carbohydrate as inulin, tanniferous, producing proanthocyanins and

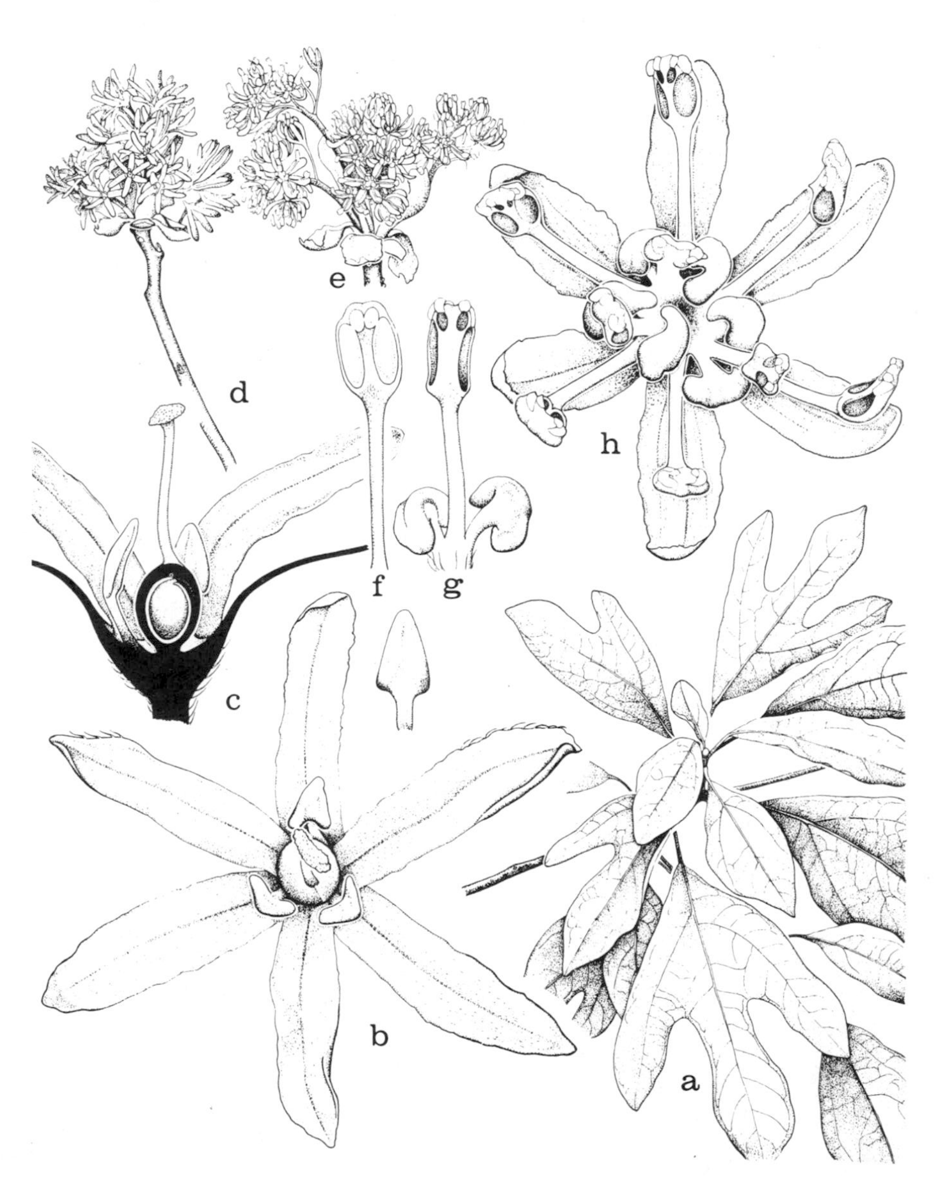

FIG. 1.13 Lauraceae. *Sassafras albidum* (Nutt.) Nees. a, habit, ×½; b, pistillate flower, ×8; c, pistillate flower in partial long-section, ×8; d, pistillate inflorescence, ×1; e, staminate inflorescence, ×1; f, outer stamen, before dehiscence, ×8; g, inner stamen, after dehiscence, with basal staminodes, ×8; h, staminate flower, ×8.

commonly benzyl-isoquinoline and/or aporphine alkaloids, generally with spherical ethereal oil cells (containing monoterpenes and/or sesquiterpenes) or mucilage cells in the parenchymatous tissues, or often with both oil cells and mucilage cells in the same tissue; small crystals of calcium oxalate sometimes present in some of the parenchyma cells; nodes unilacunar, with 2 traces; vessel-segments with simple perforations, or with scalariform perforation-plates that have up to 15 or more cross-bars, commonly both types occurring together; imperforate tracheary elements often septate, consisting mainly or wholly of libriform fibers or fiber-tracheids with simple or seldom somewhat bordered pits, but some vasicentric tracheids with clearly bordered pits sometimes also present; wood-rays heterocellular, Kribs type II B, or sometimes homocellular and type I, often containing silica-bodies or crystals of calcium oxalate, mixed uniseriate and pluriseriate, the former generally few, the latter more numerous and mostly 2–3 (–8) cells wide; wood-parenchyma sparse to abundant, largely paratracheal; sieve-tubes with S-type or P-type plastids. LEAVES alternate or rarely opposite or whorled, simple and usually entire, but sometimes (*Sassafras*) few-lobed, reduced to mere scales in *Cassytha*; venation pinnate or rarely palmate; stomates paracytic; vascular tissue of the petiole commonly forming an arc in cross-section; stipules wanting. FLOWERS in axillary, cymose to racemose inflorescences or rarely solitary, generally rather small, greenish or yellowish or white, perfect or sometimes unisexual (the plants then variously polygamous to dioecious), distinctly perigynous, with a well developed hypanthium resembling a calyx-tube, or rarely (*Hypodaphnis*) epigynous, regular, most often trimerous; tepals sepaloid, commonly 6 (4), in a single whorl or more often in 2 similar whorls, or the inner whorl sometimes (as in *Umbellularia*) slightly differentiated from the outer but scarcely petaloid, rarely the tepals 9 in 3 whorls (*Phyllostemonodaphne* and *Dicypellium*); stamens 3–12 (rarely more in *Litsea*), most often 9, the filament often with a pair of nectariferous basilateral appendages that are often considered (as by Boyle, 1980) to be staminodial in origin (each appendiculate stamen thus considered to be derived from a stamen-cluster), the innermost set of 3 stamens often reduced to staminodes; anthers primitively tetrasporangiate and tetrathecal, commonly with 2 superposed pairs of microsporangia, or sometimes with a single arc of microsporangia, or bisporangiate and dithecal by suppression of one set (at least sometimes the upper set) of microsporangia, monosporangiate and monothecal in spp. of *Potameia*, opening by (1) 2 or 4 valves from the base upwards, or rarely by valves from the outside to the inside (*Mezilaurus*) or by small pores (*Micropora*); pollen-grains globose, binucleate or rarely trinucleate, inaperturate; tectum reduced to separate spinules; gynoecium of a single carpel, or apparently so, the style topped by a capitate to disciform or lobed stigma, or the stigma sessile or decurrent on the style; ovary unilocular, superior or very rarely inferior, with a single large, pendulous, apical or subapical, anatropous, bitegmic, crassinucellar ovule; endosperm-development nuclear or (*Cassytha*) cellular. FRUIT a 1-seeded berry or a drupe, or rarely dry and indehiscent, often cupped or almost enclosed by a persistent, accrescent, fleshy or woody hypanthium; seeds with large, straight, oily (sometimes also starchy, as in *Laurus* and *Persea*), dicotyledonous embryo with a very short radicle, and without endosperm; lauric acid often a major constituent of the seed-fats. X = 12. (Cassythaceae)

The family Lauraceae consists of some 30 to 50 genera (depending on the author) and about 2000 or more species, widespread in tropical and subtropical regions throughout the world. The two greatest centers for the group are in southeast Asia and in Brasil. About two-thirds of the species belong to only 6 genera: *Ocotea* (400 +), *Litsea* (250 +), *Persea* (200), *Cinnamomum* (200), *Cryptocarya* (200), and *Beilschmiedia* (150). *Laurus nobilis L.* is the classical laurel. Cinnamon is made from the bark of *Cinnamomum zeylanicum* Nees, and camphor is extracted from the wood of *C. camphora* (L.) T. Nees & Eberm. Avocado is the fruit of *Persea americana* Miller. Oil of sassafras comes from *Sassafras albidum* (Nutt.) Nees, an eastern American species well known for variation in leaf-form.

Cassytha, with about 20 species mainly of Australia and South Africa, is very distinctive in its parasitic, twining, herbaceous, dodder-like habit. It has sometimes been considered to form a separate family Cassythaceae, but it is so similar to typical Lauraceae in other features, and its relationships are so clear, that most authors keep it in the Lauraceae. Unlike many other parasitic groups, *Cassytha* has a normal, well developed embryo.

Some Upper Cretaceous (Maestrichtian) wood from California falls well within the range of variation of the Lauraceae, and Eocene wood from Yellowstone National Park is considered to be lauraceous.

There is a long-standing disagreement as to whether the gynoecium of the Lauraceae is truly monomerous, as it appears to be, or only pseudomonomerous and derived eventually from 3 carpels. It seems reasonable to suppose that the Lauraceae originated from primitive members of the Monimiaceae that had several separate carpels, as Takhtajan has postulated. The question then becomes, did the evolutionary reduction from several separate carpels to a single carpel occur without the formation of a compound pistil at any stage along the way, or did the reduction to a single carpel occur only after a compound pistil had been evolved in some unknown intermediate ancestor? It may be significant, as Sastri (1958, 1962, 1965) has pointed out, that ontogenetically the gynoecium develops in the manner of a single conduplicate carpel, and that the appressed carpellary margins do not join until a relatively late stage in development. The occurrence of occasional flowers of *Umbellularia* with 2 distinct carpels (Kasapligil, 1951) may also be significant, although phyletic arguments from teratology are inherently dangerous. After reviewing the evidence, Endress (1972) firmly adopted the position that the gynoecium is truly monomerous, but the argument is probably not ended.

8. Family HERNANDIACEAE Blume 1826 nom. conserv., the Hernandia Family

Trees, shrubs, or woody vines, commonly producing benzyl-isoquinoline and/or aporphine alkaloids; parenchymatous tissues with scattered spherical ethereal oil cells that may lyse into mucilage cavities; proanthocyanin commonly produced in *Hernandia* and *Illigera*, but wanting from tested species of *Sparattanthelium* and *Gyrocarpus*; crystals of calcium oxalate commonly present as small needles in some of the parenchyma-cells, and cystoliths present in the leaves (and often also the stems) of *Gyrocarpus* and *Sparattanthelium*; nodes unilacunar; vessel-segments with simple perforations; imperforate tracheary elements with simple or bordered pits, not septate; wood-rays homocellular or rarely heterocellular, mixed uniseriate and pluriseriate, the latter mostly 2–3 (4) cells wide, or all pluriseriate; wood-parenchyma paratracheal; sieve tubes with P-type plastids. LEAVES alternate, simple (sometimes trilobed) or palmately compound, pinnately or palmately veined; stipules wanting. FLOWERS in cymose inflorescences, small, epigynous, regular, perfect or unisexual, in the latter case the plants polygamous or monoecious or rarely dioecious; tepals 4–8 in a single whorl or more often 3–5 in each of 2 more or less similar whorls, mostly imbricate; stamens 3–5 (–7), in a single whorl, opposite the outer tepals when the tepals are in 2 cycles, alternate with the tepals when these are in a single cycle; filament commonly with a pair of basilateral or dorsobasal, nectariferous appendages comparable to those of many Lauraceae; anthers opening longitudinally by 2 valves; pollen-grains basically similar to those of Lauraceae, globose, inaperturate, echinate, the tectum reduced to separate spinules; ovary inferior, the gynoecium of a single carpel with a short or elongate style and terminal stigma; ovule solitary, pendulous, anatropous, bitegmic, crassinucellar; endosperm-development cellular. FRUIT one-seeded, dry, often with lateral or terminal wings, sometimes more or less enclosed in an accrescent cupulate or subglobose involucre derived from 2–3 connate bracteoles; seeds without endosperm; embryo with 2 large, oily, folded or wrinkled or lobed cotyledons. X=15, 20. (Gyrocarpaceae)

The family Hernandiaceae consists of 4 genera (*Hernandia* and *Illigera* in the Hernandioideae, *Gyrocarpus* and *Sparattanthelium* in the Gyrocarpoideae) and about 60 species, widespread in tropical regions of both the Old and the New World. *Hernandia ovigera* L. is grown as a tropical street-tree. The relationships of the family are clearly with the Lauraceae and Monimiaceae.

3. Order PIPERALES Lindley 1833

Herbs, half-shrubs, or less often shrubs or trees, sometimes climbing, with spherical ethereal oil cells in the parenchymatous tissues (and sometimes elsewhere); nodes variously unilacunar to multilacunar; xylem vesselless (*Sarcandra*) or more often with well developed vessels, these with scalariform or simple perforations; sieve-tubes with S-type plastids. LEAVES opposite, alternate, or rarely whorled, simple, with or without stipules; stomates paracytic to anisocytic, laterocytic, tetracytic, or encyclocytic. FLOWERS much-reduced, nearly or quite without perianth, perfect or unisexual, borne in dense, often somewhat fleshy spikes or racemes, less often in heads or panicles; stamens 1–10, distinct or sometimes connate; pollen binucleate, tectate, variously inaperturate, uniaperturate, or less often multiporate; ovary unilocular, superior to inferior, composed of 1-several united carpels, or the gynoecium seldom of several largely distinct carpels; styles or stigmas distinct; ovules orthotropous or sometimes hemitropous, crassinucellar, bitegmic or sometimes unitegmic. FRUIT a berry, drupe, or fleshy capsule; seed with a tiny, dicotyledonous or undifferentiated embryo and abundant, more or less starchy perisperm or endosperm, the starch-grains clustered.

The order Piperales as here defined consists of 3 families and perhaps as many as 2000 species. The Chloranthaceae and Saururaceae collectively have fewer than a hundred species; the remainder belong to the Piperaceae. The close relationship of the Saururaceae to the Piperaceae is not in dispute. The position of the Chloranthaceae is more debatable and is discussed under that family.

The ancestry of the Piperales must be sought in the Magnoliales. Within the Magnoliales only the family Lactoridaceae shows some tendency toward the special features of the Piperales, having small flowers with 3 sepals, no petals, and 3 + 3 stamens. On the other hand, the wood of *Lactoris* is already much too advanced for a possible ancestor of the Piperales, having very short, simply perforate vessel-segments. Melchior (1964, general citations) has included the Lactoridaceae in the Piperales, but the well developed (though small) sepals, anatropous ovules, and follicular fruits are much more compatible with the Magnoliales. Inclusion of the Lactoridaceae in the Piperales would vitiate the characters by which the order is distinguished from the Magnoliales.

SELECTED REFERENCES

Balfour, E. 1957, 1958. The development of the vascular systems in *Macropiper excelsum* Forst. I. The embryo and seedling. II. The mature stem. Phytomorphology 7: 354–364, 1957. 8: 224–233, 1958.

Blot, J. 1960. Contribution a l'étude cytologique du genre *Peperomia.* Rev. Gén. Bot. 67: 522–535.

Dasgupta, A., & P. C. Datta. 1976. Cytotaxonomy of Piperaceae. Cytologia 41: 697–706.

Datta, P. C., & A. Dasgupta. 1977. Comparison of vegetative anatomy of Piperales. I, II. Acta Biol. Acad. Sci. Hung. 28: 81–96; 97–110.

Endress, P. K. 1971. Bau der weiblichen Blüten von *Hedyosmum mexicanum* Cordemoy (Chloranthaceae). Bot. Jahrb. Syst. 91: 39–60.

Kanta, K. 1962. Morphology and embryology of *Piper nigrum* L. Phytomorphology 12: 207–221.

Kuprianova, L. A. 1967. Palynological data for the history of the Chloranthaceae. Pollen & Spores 9: 95–100.

Murty, Y. S. 1958, 1959, 1960. Studies in the order Piperales. I. A contribution to the study of vegetative anatomy of some species of *Peperomia.* Phytomorphology 10: 50–59, 1960. II. A contribution to the study of vascular anatomy of the flower of *Peperomia.* J. Indian Bot. Soc. 37: 474–491, 1958. III. A contribution to the study of floral morphology of some species of *Peperomia.* J. Indian Bot. Soc. 38: 120–139, 1959. V. A contribution to the study of floral morphology of some species of *Piper.* VI. A contribution to the study of floral anatomy of *Pothomorphe umbellata* (L.) Miq. Proc. Indian Acad. Sci. B 49: 52–65; 82–85, 1959. VII. A contribution to the study of morphology of *Saururus cernuus* L. J. Indian Bot. Soc. 38: 195–203, 1959. VIII. A contribution to the morphology of *Houttuynia cordata* Thunb. Phytomorphology 10: 329–341, 1960.

Pant, D. D., & R. Banerji. 1965. Structure and ontogeny of stomata in some Piperaceae. J. Linn. Soc., Bot. 59: 223–228.

Quibell, C. H. 1941. Floral anatomy and morphology of *Anemopsis californica.* Bot. Gaz. 102: 749–758.

Raju, M. V. S. 1961. Morphology and anatomy of the Saururaceae. I. Floral anatomy and embryology. Ann. Missouri Bot. Gard. 48: 107–124.

Rohweder, O., & E. Treu-Koene. 1971. Bau und morphologische Bedeutung der Infloreszenz von *Houttuynia cordata* Thunb. (Saururaceae). Vierteljahrsschr. Naturf. Ges. Zürich 116(2): 195–212.

Sastrapradja, S. 1968. On the morphology of the flower in *Peperomia* (Piperaceae) species. Ann. Bogor. 4: 235–244.

Semple, K. S. 1974. Pollination in Piperaceae. Ann. Missouri Bot. Gard. 61: 868–871.

Swamy, B. G. L. 1953. *Sarcandra irvingbaileyi,* a new species of vesselless dicotyledon from South India. Proc. Natl. Inst. Sci. India 19: 301–306.

Swamy, B. G. L. 1953. The morphology and relationships of the Chloranthaceae. J. Arnold Arbor. 34: 375–408.

Swamy, B. G. L., & I. W. Bailey. 1950. *Sarcandra,* a vesselless genus of the Chloranthaceae. J. Arnold Arbor. 31: 117–129.

Tucker, S. C. 1976. Floral development in *Saururus cernuus* (Saururaceae). 2. Carpel initiation and floral vasculature. Amer. J. Bot. 63: 289–301.

Tucker, S. C. 1980. Inflorescence and flower development in the Piperaceae. I. *Peperomia*. Amer. J. Bot. 67: 686–702.
Vijayaraghavan, M. R. 1964. Morphology and embryology of a vesselless dicotyledon—*Sarcandra irvingbaileyi* Swamy, and systematic position of the Chloranthaceae. Phytomorphology 14: 429–441.
Weberling, F. 1970. Weitere Untersuchungen zur Morphologie des Unterblattes bei den Dikotylen. V. Piperales. Beitr. Biol. Pflanzen 46: 403–434.
Wood, C. E. 1971. The Saururaceae in the southeastern United States. J. Arnold Arbor. 52:479–485.
Yoshida, O. 1957, 1959, 1960, 1961. Embryologische Studien über die Ordung Piperales. I–V. J. Coll. Arts Chiba Univ. 2: 172–178, 1957; 295–303, 1959. 3: 55-60, 155–162, 1960; 311–316, 1961.
Yuncker, T. G. 1958. The Piperaceae—a family profile. Brittonia 10: 1–7.

SYNOPTICAL ARRANGEMENT OF THE FAMILIES OF PIPERALES

1 Leaves opposite, with interpetiolar stipules; ovary half to fully inferior (or nude), unicarpellate, with a single pendulous ovule; seed with copious, oily endosperm and no perisperm .... 1. CHLORANTHACEAE.

1 Leaves alternate, or rarely opposite or whorled; stipules adnate to the petiole, or wanting; ovary superior to occasionally inferior (or nude); seed with copious, starchy perisperm and no endosperm.

  2 Ovules (1) 2–10 per carpel; carpels several, distinct above the base or united to form a compound, unilocular ovary with parietal placentation .... 2. SAURURACEAE.

  2 Ovule solitary and basal or nearly so in the single locule of the compound or monomerous ovary .... 3. PIPERACEAE.

1. Family CHLORANTHACEAE R. Brown ex Lindley 1821 nom. conserv., the Chloranthus Family

Plants woody or herbaceous (sometimes even annual), not tanniferous, but with spherical ethereal oil cells in the parenchymatous tissues, and sometimes with mucilage canals as well; calcium oxalate crystals few and small, or none; nodes unilacunar, with 2 traces, or transitional to trilacunar; vessels wanting (*Sarcandra*) or with very oblique, scalariform perforation-plates that have up to about 200 cross-bars; imperforate tracheary elements with bordered pits, sometimes septate, considered to be tracheids (at least in *Sarcandra*); wood-rays heterocellular, Kribs type I, mixed uniseriate and pluriseriate, the latter up to 8 (or even as many as 22) cells wide; wood-parenchyma scanty, usually apotracheal; sieve-tube plastids of S-type. LEAVES opposite, simple, pinnately veined, toothed or entire; stomates paracytic to encyclocytic or laterocytic; petioles more or less connate at the base; stipules interpetiolar. FLOWERS in axillary or terminal compound spikes, panicles, or heads, much-reduced, perfect or more often unisexual, epigynous or hemi-epigynous, or nude, sometimes so closely crowded as to be coalescent; perianth wanting, or of a vestigial, sometimes weakly 3-toothed calyx or rim atop the ovary; stamens 1–3, in perfect flowers seeming to grow out of the ovary near midlength, usually more or less connate laterally into a lamina, the lateral ones usually with only bisporangiate half-anthers, in *Sarcandra* the solitary stamen laminar, with 2 well separated, introrsely marginal, bisporangiate pollen-sacs; anthers opening by longitudinal slits; pollen-grains medium-sized to small, globose to boat-shaped, binucleate, tectate-columellate (or semitectate), monosulcate (anasulcate) to multiaperturate; gynoecium of a single carpel, the stigma sessile or on a short style; ovule solitary, pendulous from the tip of the single locule, orthotropous, bitegmic, crassinucellar; endosperm-development cellular. FRUIT a drupe; seeds with well developed, oily and starchy endosperm (with clustered starch-grains), no perisperm, and a very small embryo with 2 tiny cotyledons, or the cotyledons scarcely differentiated. X = 8, 14, 15.

The Chloranthaceae are a taxonomically isolated family of 5 genera and about 75 species, widespread in tropical and subtropical parts of the world. The largest genus is *Hedyosmum*, with about 45 species, followed by *Chloranthus*, with about 15. *Ascarina* has about 11 species, *Sarcandra* 3, and *Ascarinopsis* only one. *Chloranthus officinalis* Blume, of the southwestern Pacific region, has been used in the past as a febrifuge, and *Sarcandra glabra* (Thunb.) Nakai, another southwestern Pacific species, is occasionally planted in mild climates as an ornamental.

It has been customary to associate the Chloranthaceae with the Piperales, but this viewpoint has not been unanimous, and recent opinion has emphasized the differences from the Piperales rather than the similarities. Takhtajan (1966) thinks the Chloranthaceae are closest to the Austrobaileyaceae and Trimeniaceae, both of which he includes in the Laurales. In the present treatment, in contrast, the Austrobaileyaceae are referred to the Magnoliales while the Trimeniaceae remain in the Laurales. The strongly reduced flowers, orthotropous ovules, and stipulate leaves of the Chloranthaceae would all be out of place in the Laurales as here defined, and the herbaceous habit of many Chloranthaceae is paralleled in the Laurales only by the anomalous genus *Cassytha* (Lauraceae). Aside from the stipulate leaves, these features would be equally out of harmony with the Magnoliales as here defined, but they are all perfectly compatible with the Piperales. Although Swamy (1953) emphasizes the differences between the Chloranthaceae on one hand and the Piperaceae and Saururaceae on the other, Thierry (cited by Metcalfe, 1950, in general citations) concludes that on anatomical grounds the Chloranthaceae are closer to the Saururaceae than either family is to the Piperaceae. Even while maintaining the Chloranthaceae in the Laurales, Takhtajan suggests that the 2 families of Piperales (in his definition) might have a common origin with the Chloranthaceae.

All told, it seems that the Chloranthaceae are not sufficiently distinctive within the Magnoliidae to constitute an order of their own, and the Piperales are the only order in which they can be accommodated without undue strain. Although the Chloranthaceae do differ in some respects from the Piperaceae and Saururaceae taken collectively, it is still possible to put all 3 families together in a distinctive and well characterized order. We should always be so lucky.

Fossil pollen that appears to be securely identified as representing the Chloranthaceae occurs in Maestrichtian and more recent deposits. Other pollen of much greater age, dating back even to the Barremian stage of the Lower Cretaceous, is rather closely comparable to that of the modern genus *Ascarina*. This ancient fossil pollen has been called *Clavatipollenites*. If it does in fact represent the Chloranthaceae, then this family antedates all other modern families in the known fossil record. On the other hand, there are several features in which the flowers of Chloranthaceae are obviously not primitive. In spite of their apparently ancient lineage, the Chloranthaceae are not among the most archaic members of the Magnoliidae. It is perfectly possible that some of the now characteristic morphological features of the Chloranthaceae did not evolve until long after the characteristic features of the pollen.

2. Family SAURURACEAE E. Meyer 1827 nom. conserv., the Lizard's-tail Family

Aromatic, perennial, rhizomatous herbs with spherical ethereal oil cells in the parenchymatous tissues, tanniferous, producing proanthocyanins but not ellagic acid, lacking alkaloids in the one species tested; cluster-crystals of calcium oxalate commonly present in some of the parenchyma-cells; vascular bundles of the stem rather widely spaced, arranged in one or sometimes 2 concentric rings; vessel-segments with scalariform perforation-plates and numerous cross-bars; sieve-tubes with S-type plastids. LEAVES alternate, simple; stipules adnate to the petiole. FLOWERS individually small, borne in dense, terminal, bracteate spikes or racemes (the lower bracts sometimes enlarged and petaloid, so that the whole inflorescence simulates a single flower), perfect, hypogynous to epigynous, without perianth; stamens 6 or 8 in 2 alternating whorls, or 3 in a single whorl, free or adnate to the base or to the whole length of the ovary; filaments slender; anthers large, tetrasporangiate and dithecal, opening by longitudinal slits; pollen-grains small to minute, globose to boat-shaped, binucleate, monosulcate (anasulcate), tectate-columellate; gynoecium of 3–5 carpels, these conduplicate and distinct above the connate base in *Saururus*, united into a unilocular compound ovary with parietal placentation in the other genera, the ovary sunken in the axis of the inflorescence in *Anemopsis*; styles distinct, not wholly closed, with decurrent stigma; ovules hemitropous to orthotropous, bitegmic, crassinucellar or tenuinucellar, (1) 2–4 and laminal-lateral in *Saururus*, 6–10 on each placenta in the other genera; embryo-sac monosporic; endosperm-development cellular or helobial. FRUIT somewhat fleshy, in *Saururus* of indehiscent, 1-seeded carpels, in the other genera an apically dehiscent capsule; seeds with minute embryo, scanty endosperm, and abundant perisperm with clustered starch-grains. X = 11, 12(?).

The family Saururaceae consists of 5 genera and only 7 species, with an interrupted distribution in eastern Asia and on both sides of North America. The genera are *Anemopsis* (1), *Circaeocarpus* (1), *Gymnotheca* (2), *Houttuynia* (1), and *Saururus* (2). Authors are agreed that the family is closely related to the Piperaceae.

FIG. 1.14 Saururaceae. *Saururus cernuus* L. a, habit, ×½; b, ventral view of mature carpel, ×16; c, mature carpel in partial long-section, ×16; d, lateral view of mature carpel, ×16; e, portion of leaf, ×2; f, lateral view of postmature flower with submature fruit, ×8; g, flower, ×8; h, stamen, ×16.

3. Family PIPERACEAE C. A. Agardh 1825 nom. conserv., the Pepper Family

Aromatic herbs or half-shrubs or less often shrubs or even small trees, sometimes vines or epiphytes, generally with spherical ethereal oil cells in the parenchymatous tissues, often producing alkaloidal amines or aporphine alkaloids, or alkaloids of the pyridine group, and sometimes accumulating aluminum, but usually not tanniferous; solitary or clustered crystals of calcium oxalate often present in some of the parenchyma cells; nodes trilacunar or multilacunar; vascular bundles characteristically in more than one ring or wholly scattered in the fashion of monocotyledons, but with an intrafascicular cambium, the outer ring of bundles often becoming continuous by cambial growth, while the

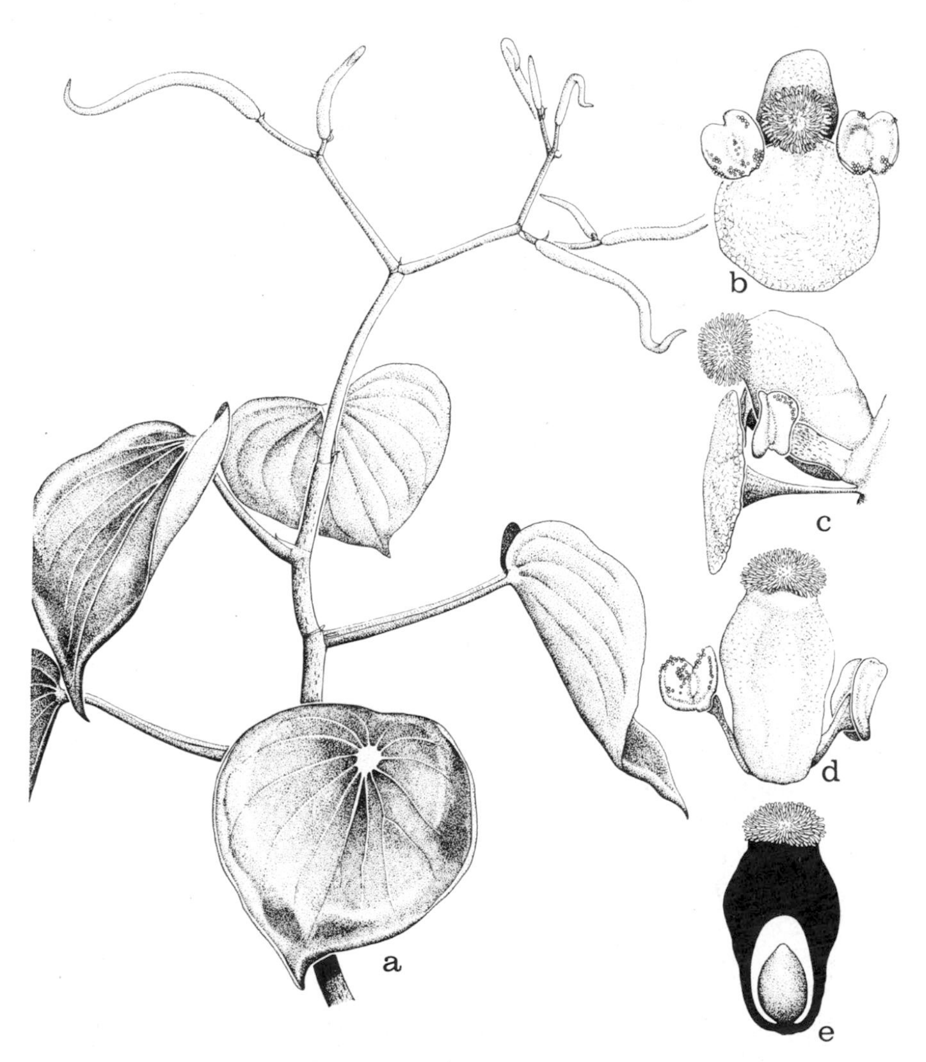

FIG. 1.15 Piperaceae. *Peperomia polybotrya* H.B.K. a, habit, $\times \frac{1}{2}$; b, c, flower with its subtending bract, front and side views, $\times 16$; d, flower, $\times 16$; e, pistil, in partial long-section, $\times 16$.

interior bundles are scattered; vessel-segments with simple perforations, or the perforation-plates sometimeswith a few cross-bars; imperforate tracheary elements with simple pits; wood-parenchyma paratracheal; sieve-tubes with S-type plastids. LEAVES simple, alternate, rarely opposite or whorled; epidermis often more or less silicified; stomates mostly tetracytic or anisocytic; stipules adnate to the petiole, or wanting. FLOWERS in dense, fleshy spikes, tiny, axillary to rather small, peltate bracts, perfect or unisexual, without perianth, at least sometimes entomophilous; stamens 1–10, often 3 + 3, the filaments commonly free; anthers bisporangiate and monothecal (*Peperomia*) or tetrasporangiate and dithecal (*Piper*), opening accordingly by one or two longitudinal slits; pollen-grains small to minute, globose, tectate, binucleate, monosulcate (anasulcate) or inaperturate, with transitional forms; ovary superior, unilocular, with 1–4 short stigmas, in *Piper* consisting of 3 or 4 carpels arising from initially distinct primordia, but in *Peperomia* consisting of a single carpel arising from a single primordium; ovule solitary, orthotropous, erect, basal or nearly so, crassinucellar, bitegmic or sometimes (*Peperomia*) unitegmic; embryo-sac tetrasporic; endosperm-development nuclear or (*Peperomia*) cellular. FRUIT baccate or drupaceous; seed solitary, with scanty endosperm and copious, starchy perisperm the starch-grains commonly clustered; embryo very small, scarcely or not at all differentiated into parts when the seed is first ripe. 2n = 8–ca 128; perhaps basically x = 11 in *Peperomia* 12 in *Piper*. (Peperomiaceae)

The family Piperaceae consists of about 10 genera and a large but highly uncertain number of species variously estimated to be from about 1400 to more than 2000. The great bulk of the species belongs to only two genera, *Piper* and *Peperomia*, but *Piper* in particular has been subjected to such an excessive and unwarranted multiplication of specific names that estimates of the true number of species are more than usually subjective. The Piperaceae are widespread in tropical regions of both the Old and the New World, often occupying mesic, shady habitats.

The Piperaceae are notable for their strong tendency toward a scattered arrangement of vascular bundles, as in monocotyledons. The bundles are open, however, rather than closed as in monocotyledons. The Piperaceae have sometimes figured in speculation about the origin of monocotyledons, but I do not find the outlook promising.

The most familiar commercial product of the Piperaceae is pepper, a condiment made from the pulverized fruits of *Piper nigrum* L. Other species of *Piper*, such as *P. betle* L., *P. methysticum* G. Forst., and *P. cubeba* L.f., are the source of other well known spices or masticatories. Some species of *Piper* have diverse local medicinal uses. Species of *Peperomia* are often cultivated as house-plants, being favored for their neat, fleshy-firm, shining, often patterned leaves and their ability to grow in filtered or indirect light.

4. Order ARISTOLOCHIALES Lindley 1833

The order consists of the single family Aristolochiaceae.

SELECTED REFERENCES

Daumann, E. 1959. Zur Kenntnis der Blütennektarien von *Aristolochia*. Preslia 31: 359–372.

Gregory, M. P. 1956. A phyletic rearrangement in the Aristolochiaceae. Amer. J. Bot. 43: 110–122.

Guédès, M. 1968. La feuille végétative et le périanthe de quelques Aristoloches. Flora B, 158: 167–179.

Hegnauer, R. 1960. Chemotaxonomische Betrachtungen. 11. Phytochemische Hinweise für die Stellung der Aristolochiaceae in System der Dicotyledonen. Die Pharmazie 15: 634–642.

Johri, B. M., & S. P. Bhatnagar. 1955. A contribution to the morphology and life history of *Aristolochia*. Phytomorphology 5: 123–137.

Nair, N. C., & K. R. Narayanan. 1962. Studies on the Aristolochiaceae. I. Nodal and floral anatomy. Proc. Natl. Inst. Sci. India 28B: 211–227.

Wyatt, R. L. 1955. An embryological study of four species of *Asarum*. J. Elisha Mitchell Sci. Soc. 71: 64–82.

1. Family ARISTOLOCHIACEAE A. L. de Jussieu 1789 nom. conserv., the Birthwort Family

Aromatic woody vines, or less often perennial herbs or small to large, erect shrubs, with spherical ethereal oil cells (containing phenylpropane derivatives, terpenes, and sesquiterpenes) in the parenchymatous tissues, commonly accumulating aristolochic acid (a series of bitter, yellow, nitrogenous compounds related to the aporphine group of isoquinoline alkaloids), or sometimes with aporphine alkaloids instead, but not tanniferous, lacking proanthocyanins as well as ellagic acid; various sorts of calcium oxalate crystals frequently present in some of the parenchyma-cells; nodes trilacunar; vascular bundles in a ring, separated by broad medullary rays, often becoming deeply and dichotomously fissured in age by the insertion of additional broad rays during the course of secondary growth; vessel-segments with simple perforations, often very large, especially in the climbing species; imperforate tracheary elements with bordered pits; wood-parenchyma usually paratracheal and rather scanty, sometimes in apotracheal bands; sieve-tube plastids of P–type, in *Asarum* with cuneate protein crystalloids as in monocots. LEAVES alternate, simple and mostly entire, but sometimes trilobed, sometimes pellucid-punctate; some of the cells of the epidermis

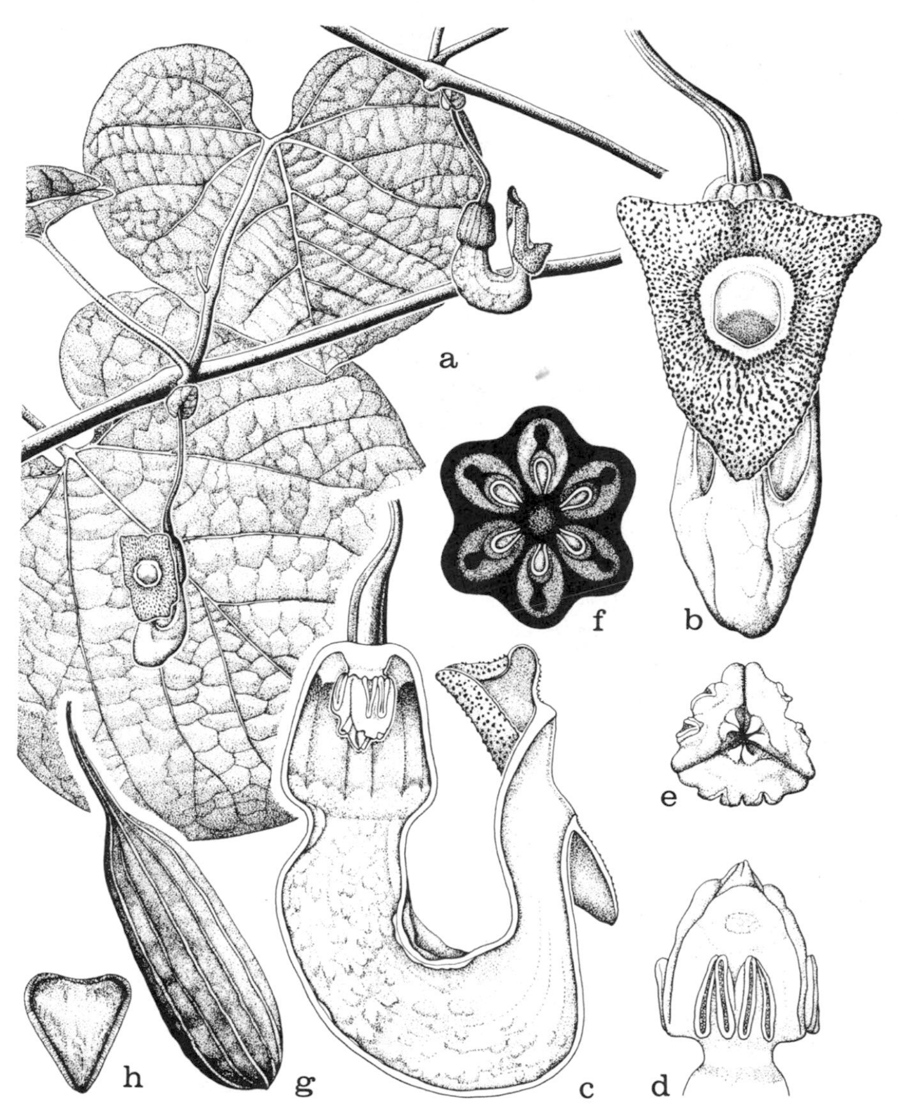

FIG. 1.16 Aristolochiaceae. *Aristolochia macrophylla* Lam. a, habit, ×½; b, flower, ×2; c, flower in partial long-section, ×2; d, e, style, with adnate anthers, from the side and from the top, ×4; f, cross-section of ovary, ×12; g, fruit, ×½; h, seed, ×2.

or even of the mesophyll often with silicified walls or containing silica bodies; stomates anomocytic; stipules ordinarily wanting. FLOWERS solitary, or in terminal or lateral racemes or cymes, perfect, regular or irregular, epigynous or sometimes only hemi-epigynous or perigynous, often smelling of carrion; calyx synsepalous, tubular at least below, regular and 3-lobed to more often distinctly irregular (often S-shaped or pipe-shaped) and then 3-lobed to 1-lobed or lobeless, often large and corolloid; petals wanting or much reduced, well developed and alternate with the 3 calyx-lobes only in the monotypic Chinese genus *Saruma*; nectaries, at least in *Aristolochia*, consisting of 2 (or more) patches of secretory hairs within the perianth-tube; stamens 4-many, most commonly 6 (in one or less often 2 cycles), or not infrequently 12, free, or with the filaments (or both filaments and anthers) joined to the style to form a gynostemium; anthers tetrasporangiate and dithecal, extrorse; pollen-grains more or less globose, large to medium-sized, binucleate, tectate, inaperturate or rarely monosulcate (*Saruma*) or even multiaperturate; gynoecium of 4–6 carpels, these in *Saruma* distinct above the base and free from the hypanthium above the middle, in *Hexastylis* joined to form a compound, superior to half-inferior ovary with separate styles, in the other genera joined to form a compound, half to more often fully inferior ovary with short, stout style and spreading stigmas; ovary 4- to 6-locular, or with 4–6 incomplete partitions, the placentation accordingly axile or parietal on intruded placentas; ovules numerous in each locule or on each placenta, bitegmic and crassinucellar, mostly anatropous; endosperm-development cellular. FRUIT usually capsular (though sometimes with a fleshy endocarp) and many-seeded, rarely follicular (*Saruma*) or indehiscent and 1-seeded; seeds with abundant, oily (sometimes also starchy) endosperm, sometimes with a thickened or winged raphe; embryo very small, basal, weakly dicotyledonous, or sometimes globular and undifferentiated. X = 4–7, 12, 13, perhaps originally 7.

The family Aristolochiaceae consists of some 8 to 10 genera and about 600 species, 500 of them belonging to the genus *Aristolochia*, and most of the rest to *Asarum*. The family is mainly tropical, but a few species occur well within temperate regions. Several species of *Aristolochia* are grown as porch-vines. Both the English name (birthwort) and the scientific name of *Aristolochia* reflect the traditional use of some species in folk-medicine. The calyx of *Aristolochia* has sometimes been interpreted on ontogenetic and teratological grounds as a false calyx, derived from the bracts beneath the flower, or from a single subtending leaf. The flower is then considered to lack a true perianth. Although this interpretation may conceivably prove to be phylogenetically correct, I see nothing to be gained by altering the formal floral description to conform to it. In its less modified forms the calyx of *Aristolochia* has every external appearance of being homologous with the admitted calyx of *Asarum*, and may well actually be so in spite of its somewhat modified ontogeny. The relationship of *Aristolochia* to the rest of the family is not in dispute, and the traditional interpretation of the floral morphology is in no way misleading about the affinities of *Aristolochia* or its family. Here, as in some other groups, a functional and topographic approach to descriptive floral morphology is more useful than a phylogenetic one, the more so because the phylogenetic interpretation is uncertain.

The affinities of the Aristolochiaceae have been much debated, but modern opinion is crystallizing in the view that the family is derived directly from the Magnoliales. The archaic genus *Saruma*, forming what Thorne has called a nonmissing link, is particularly instructive in this regard. *Saruma* has several carpels that are united only at the base, and which ripen into follicles. It has well developed petals, and monosulcate pollen. These features, together with the ethereal oil cells and numerous ovules of the family as a whole, would seem to exclude all groups but the Magnoliales as possible ancestors. Even the characteristic aristolochic acids of the family are chemically allied to the aporphine alkaloids found in many members of the Magnoliales.

5. Order ILLICIALES Hu 1950[1]

Small trees or erect to scandent shrubs or woody vines; parenchymatous tissues with scattered ethereal oil cells, usually also mucilage-cells, and very often some branching sclereids; nodes unilacunar; vessel-segments simply perforate or with scalariform or mixed scalariform and reticulate perforation-plates; sieve-tube plastids of S-type. LEAVES alternate, simple; stomates paracytic or sometimes anomocytic; stipules wanting. FLOWERS solitary or few together, most commonly axillary or supra-axillary, perfect or unisexual, regular, hypogynous; perianth of 5-many members spirally arranged in one or more often 2-several series, not clearly differentiated into sepals and petals, or the outer smaller and sepaloid and the inner larger and petaloid; stamens (4-) many, more or less spirally aranged, sometimes in more than one series, the filaments distinct or more or less connate, sometimes collectively forming a fleshy mass; anthers basifixed, tetrasporangiate and dithecal, opening by longitudinal slits; pollen-grains binucleate, semitectate-columellate, triaperturate or sexaperturate, the latter condition considered to be derived from a triaperturate one; gynoecium of (5-) 7-many distinct carpels, these spirally arranged or in a single cycle; at least the upper (stylar) portion of the carpel unsealed and with a decurrent stigma; each carpel with 1-several marginal or nearly basal, anatropous to sometimes campylotropous, bitegmic, crassinucellar ovules; endosperm-development cellular or sometimes nuclear. FRUIT of follicles or berries; seeds with copious, oily endosperm and minute, dicotyledonous embryo. 2n = 28.

The order Illiciales consists of two closely related families, the Illiciaceae and Schisandraceae, with about 90 species in all. It is generally agreed that the Illiciales are related to and derived from the Magnoliales. Indeed the only character that sets the Illiciales apart from the Magnoliales as a whole is the fundamentally triaperturate pollen, clearly an advanced feature. If they had uniaperturate pollen, the Illiciaceae and Schisandraceae would be better referred to the Magnoliales, even though their syndrome of other features would set them apart collectively from any other one family of the order.

[1] A Latin diagnosis, not given by Hu, is provided here:

Plantae ligneae, cellulis oleum aromaticum continentibus per parenchymatis texturam dispersis; folia alterna simplicia exstipulata; flores hypogyni; pollinis grana tri-vel sexaperturata; carpella (5–) 7-numerosa distincta, saltem parte supera aperta, stigmate decurrenti; embryo parvulus, cotyledonibus 2. Type: Illiciaceae A. C. Smith, Sargentia 7: 8. 1947.

SELECTED REFERENCES

Bailey, I. W., & C. G. Nast. 1948. Morphology and relationships of *Illicium, Schisandra* and *Kadsura*. I. Stem and leaf. J. Arnold Arbor. 29: 77–89.
Hayashi, Y. 1960. On the microsporogenesis and pollen morphology in the family Magnoliaceae. Sci. Rep. Tôhoku Imp. Univ. Ser. 4, Biol. 26: 45–52.
Hayashi, Y. 1963. The embryology of the family Magnoliaceae sens. lat. I. Megasporogenesis, female gametophyte and embryogeny of *Illicium anisatum* L.; II. Megasporogenesis, female gametophyte and embryogeny of *Schisandra repanda* Radlkofer and *Kadsura japonica* Dunal. Sci. Rep. Tôhoku Imp. Univ., Ser. 4, Biol. 29: 27–33; 403–411.
Jalan, S. 1962. The ontogeny of the stomata in *Schisandra grandiflora* Hook.f. & Thoms. Phytomorphology 12: 239–242.
Jalan, S. 1968. Contribution to the nodal structure of *Schisandra* Michaux. Bot. Jahrb. Syst. 88: 311–316.
Jalan, S., & R. N. Kapil. 1964. Pollen grains of *Schisandra* Michx. Grana Palyn. 5: 216–221.
Kapil, R. N., & S. Jalan. 1964. *Schisandra* Michaux—its embryology and systematic position. Bot. Not. 117: 285–306.
Keng, H. 1965. Observations on the flowers of *Illicium*. Bot. Bull. Acad. Sin. Ser. II. 6: 61–73.
Robertson, R. E., & S. C. Tucker. 1979. Floral ontogeny of *Illicium floridanum*, with emphasis on stamen and carpel development. Amer. J. Bot. 66: 605–617.
Smith, A. C. 1947. The families Illiciaceae and Schisandraceae. Sargentia 7: 1–224.
Vijayaraghavan, M. K., & U. Dhar. 1975. *Kadsura heteroclita*—Microsporangium and pollen. J. Arnold Arbor. 56: 176–182.
Yoshida, O. 1962. Embryologische Studien über *Schisandra chinensis* Baillon. J. Coll. Arts Chiba Univ. 3: 459–462.

Колбасина, Э. И. 1967. Органогенез цветка лимонника *Schizandra chinensis* (Turcz.) Baill. Бот. Ж. 52: 377-378.

SYNOPTICAL ARRANGEMENT OF THE FAMILIES OF ILLICIALES

1 Flowers perfect; trees or shrubs, not climbing; carpels in a single whorl; fruits follicular .... 1. ILLICIACEAE.
1 Flowers unisexual; scrambling or twining woody vines; carpels spirally arranged; fruits fleshy, indehiscent . 2. SCHISANDRACEAE.

1. Family ILLICIACEAE A. C. Smith 1947 nom. conserv., the Star-anise Family

Glabrous, aromatic, evergreen shrubs or small trees with scattered ethereal oil cells, mucilage cells, and branching sclereids in the parenchymatous tissues (the sclereids notably in the mesophyll), accumulating anisatin (a toxic lactone) and commonly somewhat tanniferous, producing proanthocyanins but not ellagic acid; nodes unilacunar, with a single trace; vessel-segments very elongate and slender, with mixed reticulate and scalariform perforation-plates, the latter with numerous (30–150) cross-bars; imperforate tracheary elements with numerous large, evidently bordered pits, considered to be tracheids; wood-rays heterocellular, 1–2 (3) cells wide; wood-parenchyma scanty and paratracheal or partly diffuse; sieve-tubes with S-type plastids. LEAVES alternate (sometimes so closely crowded toward the tips of the twigs as to appear almost whorled), simple, leathery, entire, pinnately veined; stomates paracytic; petiole with a single vascular arc; stipules wanting. FLOWERS small, solitary or sometimes 2 or 3 together, mostly axillary or supra-axillary, perfect, regular, hypogynous; perianth of more or less numerous (7–33) members, these often not clearly differentiated into sepals and petals, commonly arranged in several series (spiral within each series), those of the outermost series small and bract-like or sepal-like, the others commonly larger and more petaloid, or the innermost ones reduced and sometimes transitional to the stamens; stamens (4–) numerous (up to ca 50), distinct, spirally arranged in each of (1–) several series; filaments short and thick; anthers basifixed, tetrasporangiate and dithecal, opening by longitudinal slits; connective prolonged; pollen-grains medium-sized, more or less ellipsoid, binucleate, semitectate, tricolporate; gynoecium of (5–) 7–15 (–21) distinct carpels in a single cycle (ontogenetically a tight helix), these obliquely attached to the receptacle, often strongly compressed, gradually narrowed to the conduplicate, unsealed style with dry, nonpapillate, decurrent stigma, each carpel with a solitary, ventrally near-basal, anatropous, bitegmic, crassinucellar ovule; endosperm-development cellular or sometimes (fide Corner) nuclear. FRUIT of 1-seeded follicles, these often radially spreading in a stellate pattern; seed with copious, oily endosperm; embryo very small, dicotyledonous. X = 13, 14.

The family Illiciaceae consists of the single genus *Illicium*, with about 40 species. The genus is native to southeastern Asia (as far north as Korea) and some of the adjacent islands, and to southeastern United States, Cuba, Hispaniola, and Veracruz, Mexico.

FIG. 1.17 Illiciaceae. *Illicium floridanum* Ellis. a, habit, ×¼; b, portion of twig with leaf-bases, ×1; c, flower, from above, ×1½; d, flower, with perianth and some stamens removed, ×3; e, pistil, ×12; f, stamens, internal (left) and external view, ×14; g, whorl of opened fruits, ×1½; h, seed, ×5.

Some species of *Illicium*, notably *I. verum* Hook. f. (star-anise) and *I. anisatum* L. (Japanese star-anise) are traditional sources of anethole (anise oil), along with *Pimpinella anisum* L. of the Apiaceae. Anethole is used in perfumes, dentifrices, and flavoring, in the synthesis of industrial chemicals, and medically as an expectorant and carminative. Here we have a striking demonstration of the fact that the ethereal oils of the Magnoliales and their relatives are not confined to this group. Similar or sometimes identical oils are widely scattered in other groups, although seldom in spherical idioblasts like those of the Magnoliidae.

Illicium has often in the past been included in the Winteraceae, or considered to be of uncertain affinities. Since the work of Smith (1947) and of Bailey & Nast (1948), most botanists have accepted the view that *Illicium* should form a family of its own, associated with the Schisandraceae.

Pollen considered to represent *Illicium* is known from Maestrichtian and more recent deposits.

2. Family SCHISANDRACEAE Blume 1830 nom. conserv., the Schisandra Family

Glabrous, aromatic, evergreen or decidous, scrambling or twining woody vines with scattered ethereal oil cells, usually also mucilage cells, and sometimes some branching sclereids in the parenchymatous tissues, commonly somewhat tanniferous, producing proanthocyanins but not ellagic acid; nodes unilacunar, with 3 traces; vessel-segments with scalariform perforation-plates that have 1–15 cross-bars, or simply perforate; imperforate tracheary elements with evidently bordered pits, considered to be tracheids; wood-rays heterocellular, 1–3 cells wide; wood-parenchyma scanty, diffuse or paratracheal; sieve-tubes with S-type plastids. LEAVES alternate, simple, entire or more often toothed, pinnately veined, often pellucid-dotted; stomates mixed paracytic and laterocytic; petiole generally with 3 vascular bundles; stipules wanting. FLOWERS small, solitary in the axils or occasionally paired or in few-flowered axillary inflorescences, unisexual (the plants monoecious or dioecious), regular, hypogynous, with a somewhat elongate (conic to cylindric or obovoid) receptacle; perianth of 5–24 members spirally arranged in 2-several series, all more or less similar, or the outer smaller and sepaloid and the inner larger and petaloid; androecium of 4–80 more or less spirally arranged stamens; filaments short, from connate only at the base to often wholly connate in a globular, fleshy mass; anthers basifixed, tetrasporangiate and dithecal, free or partly embedded in the filament-mass, opening lengthwise, the connective somewhat expanded; pollen-grains medium-sized, more or less ellipsoid, binucleate, semitectate-columellate (the exine reticulate), typically with 6 colpi, 3 shorter ones alternating with 3 long ones that meet at one pole, seldom merely tricolpate; gynoecium of numerous (12–300) separate, spirally arranged carpels, these conduplicate, unsealed with wet, papillate, decurrent stigma; each carpel with 2–5 (–11) marginal, anatropous to campylotropous, bitegmic, crassinucellar ovules; endosperm-development cellular. FRUIT of berry-like carpels, these on a somewhat elongate axis (*Schisandra*) or in a dense head (*Kadsura*); seeds typically 2, with copious, oily and starchy endosperm; embryo minute, dicotyledonous. X = 13, 14.

The family Schisandraceae consists of 2 genera and about 50 species, found in tropical and temperate regions of eastern Asia and some of the nearby islands, and represented by a single species [*Schisandra glabra* (Brickell) Rehder] in the southeastern United States.

Pollen considered to represent the Schisandraceae is known from Maestrichtian and more recent deposits.

6. Order NYMPHAEALES J. H. Schaffner 1929

Aquatic herbs, without secondary growth and (except for the roots of Nelumbonaceae) without vessels; vascular bundles of the stem closed, scattered or less commonly in one or more rings, or (Ceratophyllaceae) the stem with only a single reduced central vascular strand; tracheids elongate, with annular, spiral, or scalariform secondary thickening, or (Ceratophyllaceae) modified into unlignified starch-containing cells; parenchymatous tissues with conspicuous schizogenous air-passages, and (except in Ceratophyllaceae) with articulated laticifers; root-hairs originating from specialized cells, as in many monocotyledons (roots wanting in *Ceratophyllum*). LEAVES either long-petiolate, with peltate to cordate or hastate, usually floating or emergent blade, or some or all of them submersed, short-petiolate or sessile, and dissected into slender segments; stomates anomocytic or (Ceratophyllaceae) wanting. FLOWERS axillary or extra-axillary, solitary, perfect or (Ceratophyllaceae) unisexual, regular, hypogynous to epigynous; perianth conspicuous and more or less differentiated into sepals and petals, or without petals and then usually small and inconspicuous; stamens variously numerous and spirally arranged, or fewer and cyclic, often more or less laminar or ribbon-shaped and not well differentiated into filament and anther, sometimes passing into the petals; pollen-grains variously uniaperturate, triaperturate, or inaperturate; carpels 1-many, distinct or united into a compound, superior to inferior ovary; placentation laminar or (Nelumbonaceae) essentially apical; ovules 1-many, anatropous or less commonly (Barclayaceae, Ceratophyllaceae) orthotropous, bitegmic or (Ceratophyllaceae) unitegmic, crassinucellar; endosperm-development variously nuclear, cellular, or helobial. FRUIT spongy, berry-like, and indehiscent or irregularly dehiscent, or of separate achenes or nuts; seeds often operculate, sometimes arillate; cotyledons 2, arising from an annular common primordium, distinct or often more or less connate below into a ring or a bilobed tube, each lobe separately vasculated; germination hypogeous.

The order Nymphaeales as here defined consists of 5 families and about 65 species. The Nymphaeaceae, with about 50 species, are by far the largest family.

Unlike most of the orders of angiosperms, the Nymphaeales have an obvious ecologic niche. They inhabit still waters, and many of their characteristics reflect adaptation to the habitat. On the other hand, they do not occupy this habitat to the exclusion of other groups. *Nymphoides*, in the Menyanthaceae, in very nymphaeaceous in aspect and habitat, and *Myriophyllum*, in the Haloragaceae, likewise recalls *Ceratophyllum*.

The plants here referred to the Nymphaeales have been variously treated in the past. At one extreme, all members except *Ceratophyllum* have been included in the Nymphaeaceae; the Nymphaeaceae and Ceratophyllaceae were then referred to a broadly defined order Ranales. This is the traditional Englerian treatment. At the other extreme, Li (1955) recognized a total of 5 families in 3 orders to provide for the traditional Nymphaeaceae.

The primitive, uniaperturate pollen of the Nymphaeaceae (as here defined) and Cabombaceae foredooms any attempt to derive these families from anything like the basically triaperturate Ranunculaceae or Berberidaceae, although such an alliance has appealed to some authors in the past. Melikyan (1964, in Russian) considers that the inaperturate pollen of *Ceratophyllum* is probably derived from a uniaperturate ancestry, and Agababyan (1973, under Magnoliidae, in Russian) would seem to concur. Thus the Ceratophyllaceae go with the Nymphaeaceae and Cabombaceae in this regard.

Another feature that distinguishes the Nymphaeaceae, Cabombaceae, Barclayaceae, and Ceratophyllaceae from the modern herbaceous groups with which they have been compared is the laminar placentation. The status of laminar placentation as a possibly primitive feature in angiosperms is debatable, but it is found only in otherwise relatively archaic groups and must surely be very ancient. If laminar placentation is not itself primitive, it must be derived from laminar-lateral placentation, which is likewise confined to archaic groups. There is no direct phyletic or morphologic connection between laminar placentation and the marginal placentation that provided the phyletic foundation for the common types of placentation in compound ovaries. Laminar-lateral placentation must provide the link.

In the light of these and other similarities among the Nymphaeaceae, Cabombaceae, Barclayaceae, and Ceratophyllaceae, it seems reasonable to suppose that they represent branches from a common stock that originated very early in the history of the angiosperms. The hypothesis of Doyle & Hickey (1976, in general citations) that the early angiosperms were riparian shrubs rather than (as commonly postulated) forest trees is highly pertinent here.

The position of the Nelumbonaceae is perhaps a little less secure. They stand apart from the other families of Nymphaeales in several features, including serological reactions, although the anomalous cross-reaction of *Nelumbo* with such a distantly related genus as *Agave* (Simon, 1971) casts some doubt on the significance of serology in this instance. More importantly, *Nelumbo* is distinctive in the Nymphaeales in having triaperturate pollen.

It might be argued that the similarities between *Nelumbo* and other Nymphaeales reflect convergence caused by the action of similar selective forces in similar habitats. This argument overlooks the fact that both groups must have originated a long time ago, probably very early in the history of the angiosperms, from things that must have been initially rather similar. Based on its morphological features, *Nelumbo*, no less than the other Nymphaeales, must be considered a highly archaic genus.

The fossil record is also compatible with and gives some support to the evolutionary interpretation here presented. Some Cretaceous fossil leaves are closely comparable to leaves of the modern genus *Brasenia* (Moseley, 1974). Peltate, *Nelumbo*-like fossil leaves occur in Lower Cretaceous (Albian, at least in part) deposits in both the United States and the Soviet Union. I have seen some of these fossils. Although I would be reluctant to assert that they must belong to the nelumbonoid evolutionary line, there is no obvious reason why they should not, even when features of venation are taken into account, in accord with modern paleobotanical practice. The generalized nymphaealean and possibly more specifically nelumbonoid affinity of some of these Lower Cretaceous fossil leaves is at least tentatively accepted even by paleobotanists who are skeptical of most identifications of Cretaceous fossils with modern families. These Albian fossil leaves, whose form and venation suggest an aquatic habitat, are coming to be called Nymphaeaphylls in the paleobotanical literature.

The most reasonable interpretation of the evidence, in my opinion, is that the modern Nymphaeales all descend from a group of primitive dicotyledons that took to an aquatic habitat and became herbaceous very early in the history of the angiosperms. There was a subsequent early dichotomy into a line leading to the modern Nelumbonaceae and a line leading eventually to the other 4 families. The modern members of the order thus represent a series of isolated end-lines, comparable on a smaller scale to the series of isolated end-lines that comprise the families of the modern orders Magnoliales and Laurales.

There has been much fruitless discussion of whether the Nymphaeales are primitively vesselless or only secondarily so. Recently, Kosakai, Moseley & Cheadle (1970) have reported very primitive, elongate, scalariform vessels in the roots of *Nelumbo*. They suggest that the Nymphaeaceae were originally vesselless, and that vessels in *Nelumbo* "have either arisen recently, or their evolution in other organs was arrested shortly after their origin in the roots, when Nymphaeaceae became aquatic." The question of course holds a certain philosophic interest, but it makes very little practical taxonomic or even phyletic difference. We have seen that the evidence suggests a very early origin of the Nymphaeales from primitive angiosperms with uniaperturate pollen. It is possible that these ancestors of the Nymphaeales were also primitively vesselless, and that the vessels in *Nelumbo* represent a later development not shared by other Nymphaeales. It is at least equally possible that the ancestors of the Nymphaeales had scalariform vessels in the secondary wood, that these vessels were lost through the evolutionary loss of cambium in association with the aquatic habit, and that the vessels in the roots of *Nelumbo* are relictual.

For whatever it may be worth, we may note that members of the archaic subfamily Erirhininae of the coleopterous family Curculionidae feed chiefly on various aquatic or semiaquatic monocots, ferns, and *Equisetum*. Two genera are terrestrial and confined to mosses, and one genus (*Argentinorhynchus*) feeds on *Nymphaea* (Kuschel, 1971, and personal communication).

SELECTED REFERENCES

Chassat, J.-F. 1962. Recherches sur la ramification chez les Nymphaeacées. Bull. Soc. Bot. France, Mém. 1962: 72–95.

Cutter, E. G. 1957, 1958, 1959 (1960). Studies of morphogenesis in the Nymphaeaceae. I. Introduction: some aspects of the morphology of *Nuphar lutea* (L.) Sm. and *Nymphaea alba* L. II. Floral development in *Nuphar* and *Nymphaea*: Bracts and calyx. III. Surgical experiments on leaf and bud formation. IV. Early floral development in species of *Nuphar*. Phytomorphology 7: 45–46; 57–73, 1957. 8: 74–95, 1958. 9: 263–275, 1959 (1960).

Cutter, E. G. 1961. The inception and distribution of flowers in the Nymphaeaceae. Proc. Linn. Soc. London 172: 93–100.

Earle, T. T. 1938. Embryology of certain Ranales. Bot. Gaz. 100: 257–275.

Esau, K., & H. Kosakai. 1975. Laticifers in *Nelumbo nucifera* Gaertn.: Distribution and structure. Ann. Bot. (London) II. 39: 713–719.

Esau, K., & H. Kosakai. 1975. Leaf arrangement in *Nelumbo nucifera:* A re-examination of a unique phyllotaxy. Phytomorphology 25: 100–112.

Fassett, N. C. 1953. A monograph of *Cabomba*. Castanea 18: 116–128.

Guttenberg, H. V., & R. Müller-Schröder. 1958. Untersuchungen über die Entwicklung des Embryos und der Keimpflanze von *Nuphar lutea*. Planta 51: 481–510.

Haines, R. W., & K. A. Lye. 1975. Seedlings of Nymphaeaceae. J. Linn. Soc., Bot. 70: 255–265.

Hartog, C. den. 1970. *Ondinea*, a new genus of Nymphaeaceae. Blumea 18: 413–416.

Jones, E. N. 1931. The morphology and biology of *Ceratophyllum demersum*. Stud. Nat. Hist. Iowa Univ. 13(3): 11–55.

Khanna, P. 1964, 1965, 1967. Morphological and embryological studies in Nymphaeaceae. I. *Euryale ferox*. Proc. Indian Acad. Sci. B 59: 237–243, 1964. II. *Brasenia schreberi* Gmel. and *Nelumbo nucifera* Gaertn. Austral. J. Bot. 13: 379–387, 1965. III. *Victoria cruziana* D'Orb. and *Nymphaea stellata* Willd. Bot. Mag. (Tokyo) 80: 305–312.

Kosakai, H., M. F. Moseley, & V. I. Cheadle. 1970. Morphological studies of the Nymphaeaceae. V. Does *Nelumbo* have vessels? Amer. J. Bot. 57: 487–494.

Leeuwen, W. A. M. van. 1963. A study of the structure of the gynoecium of *Nelumbo lutea* (Willd.) Pers. Acta Bot. Neerl. 12: 84–97.

Li, H.-L. 1955. Classification and phylogeny of Nymphaeaceae and allied families. Amer. Midl. Naturalist 54: 33–41.

Lowden, R. M. 1978. Studies on the submerged genus *Ceratophyllum* L. in the neotropics. Aquatic Bot. 4: 127–142.

Moseley, M. F. 1958, 1961, 1965, 1971 (1972). Morphological studies of the Nymphaeaceae. I. The nature of the stamens. Phytomorphology 8: 1–29, 1958. II. The flower of *Nymphaea*. Bot. Gaz. 122: 233–259, 1961. III. The floral anatomy of *Nuphar*. VI. Development of flower of *Nuphar*. Phytomorphology 15: 54–84, 1965. 21: 253–283, 1971 (1972).

Muenscher, W. C. 1940. Fruits and seedlings of *Ceratophyllum*. Amer. J. Bot. 27: 231–233.

Prance, G. T., & A. B. Anderson. 1976. Studies of the floral biology of neotropical Nymphaeaceae. 3. Acta Amazonica 6: 163–170.

Prance, G. T., & J. R. Arias. 1975. A study of the floral biology of *Victoria amazonica* (Poepp.) Sowerby (Nymphaeaceae). Acta Amazonica 5: 109–139.

Richardson, F. C. 1969. Morphological studies of the Nymphaeaceae. IV. Structure and development of the flower of *Brasenia schreberi* Gmel. Univ. Calif. Publ. Bot. 47: 1–101.

Richardson, F. C., & M. F. Moseley. 1967. The vegetative morphology and nodal structure of *Brasenia schreberi*. Amer. J. Bot. 54: 645.

Sastri, R. L. N. 1955. The embryology of *Ceratophyllum demersum* L. Proc. 42nd Indian Sci. Congr. Assoc. 3: 226.

Schneider, E. L. 1976. The floral anatomy of *Victoria* Schomb. (Nymphaeaceae). J. Linn. Soc., Bot. 72: 115–148.

Schneider, E. L. 1978. Morphological studies of the Nymphaeaceae. IX. The seed of *Barclaya longifolia* Wall. Bot. Gaz. 139:223–230.

Schneider, E. L., & J. D. Buchanan. 1980. Morphological studies of the Nymphaeaceae. XI. The floral biology of *Nelumbo pentapetala*. Amer. J. Bot. 67: 182–193.

Schneider, E. L., & E. G. Ford. 1978. Morphological studies of the Nymphaeaceae. X. The seed of *Ondinea purpurea* Den Hartog. Bull. Torrey Bot. Club 105: 192–200.

Schneider, E. L., & L. A. Moore. 1977. Morphological studies of the Nymphaeaceae. VII. The floral biology of *Nuphar lutea* subsp. *macrophylla*. Brittonia 29: 88–99.

Simon, J.-P. 1971. Comparative serology of the order Nymphaeales. II. Relationships of Nymphaeaceae and Nelumbonaceae. Aliso 7: 325–350.

Troll, W. 1933. Beiträge zur Morphologie des Gynaeceums. IV. Über das Gynaeceum der Nymphaeaceen. Planta 21: 447–485.

Van Heel, W. A. 1977. The pattern of vascular bundles in the stamens of *Nymphaea lotus* L. and its bearing on stamen morphology. Blumea 23: 345–348.
Weidlich, W. H. 1976, 1980. The organization of the vascular system in the stems of Nymphaeaceae. I. *Nymphaea* subgenera Castalia and Hydrocallis. II. *Nymphaea* subgenera Anecphya, Lotos, and Brachyceras. III. *Victoria* and *Euryale*. Amer. J. Bot. 63: 499–509, 1365–1379. 1976. 67: 790–803. 1980.
Wood, C. E. 1959. The genera of the Nymphaeaceae and Ceratophyllaceae in the southeastern United States. J. Arnold Arbor. 40: 94–112.

Вальцева, О. В., & Е. И. Савич. 1965. О развитии зародыша у *Nymphaea candida* Presl и *N. tetragona* Georgi. Бот. Ж. 50: 1323-1326.
Вахрамеев, В. А. 1952. Региональная стратиграфия СССР. Том I. Стратиграфия и ископаемая флора меловых отложений Западного Казахстана. с.181-185, *Nelumbites*. Изд. Акад. Наук С.С.С.Р. Москва.
Воронкина, Н. В. 1974. Анатомическое строение апекса корня Nymphaeales J. Schaffner. Бот. Ж. 59: 1417-1424.
Мейер, К. И. 1960. К эмбриологии *Nuphar luteum* Sm. Бюлл. Московск. Обш. испыт. прир., Отд. Биол. н. с. 65 (6): 48-58.
Мейер, Н. Р. 1964. Палинологические исследования семейства нимфейных. Бот. Ж. 49: 1421-1429.
Меликян, А. П. 1964. Гистогенез спермодермы у *Brasenia schreberi* Gmel. и *Nymphaea capensis* Thunb. Вестн. Лениградск, унив., Сер. Биол. 9(2): 121-125.
Меликян, А. П. 1964. Сравнительная анатомия спермодермы некоторых представителей семейства Nymphaeaceae. Бот. Ж. 49: 432-436.
Меликян, А. П. 1964. Сравнительная анатомия спермодермы представителей порядка Nymphaeales. Автореф. Дисс., Ленинград.
Меликян, А. П. 1967. Некоторые корреляции в строении зародыша и семенной кожуры у Нимфейных. Учен. Зап. Ереванский Госуд. Унив. Естеств. Науки. 1(105): 87-92.
Снигиревская, Н. С. 1955. К морфологии пыльцы Nymphaeales. Бот. Ж. 40: 108-115.
Снигиревская, Н. С. 1964. Материалы к морфологии и систематике рода *Nelumbo* Adans. Труды Бот. Инст. АН СССР, сер, 1, 13: 104-172.

SYNOPTICAL ARRANGEMENT OF THE FAMILIES OF NYMPHAEALES

1 Plants rooted to the substrate; flowers typically long-pedunculate and reaching or exserted from the surface of the water (seldom wholly submersed), perfect, usually with evident petals; carpels (1) 2-many; ovules bitegmic, anatropous except in Barclayaceae; leaves in most genera alternate (or apparently so), long-petiolate, with floating or emergent, cordate or peltate blade, but some or all of the leaves sometimes submersed, and in *Cabomba* these submersed leaves opposite or whorled, short-petiolate, and strongly dissected.

2 Carpels individually embedded in the enlarged, obconic receptacle; ovule 1 (2); seeds with large embryo, no perisperm, and virtually no endosperm; pollen triaperturate .... 1. NELUMBONACEAE.

2 Carpels not embedded in the receptacle; ovules 2-many; seeds with small embryo, some endosperm, and abundant perisperm; pollen uniaperturate or inaperturate.

3 Plants acaulescent, the leaves all simple and rising directly from the rhizome, usually with floating blade; petals 8-numerous (rarely none); carpels more or less firmly united to form a compound, plurilocular, superior to inferior ovary that ripens into a somewhat spongy, berry-like, irregularly (or scarcely) dehiscent fruit; ovules and seeds numerous.

4 Petals distinct (or none); stamens free from the petals; ovules anatropous; pollen uniaperturate .... 2. NYMPHAEACEAE.

4 Petals connate into a lobed tube to which the stamens are attached, the corolla-tube epigynous, arising around the top of the ovary, but the sepals hypogynous; ovules orthotropous; pollen inaperturate .... 3. BARCLAYACEAE.

3 Plants with long, slender, leafy, distally floating stems in addition to the rhizomes, with or without floating leaf-blades, often with some or all of the leaves submersed and dissected; petals (2) 3 (4); carpels distinct, ripening into leathery, indehiscent fruits; ovules and seeds 2 or 3 .... 4. CABOMBACEAE.

1 Plants rootless, free-floating, submersed; flowers inconspicuous, sessile, unisexual, apetalous; carpel 1; ovule solitary, unitegmic, orthotropous; pollen inaperturate; leaves all sessile, whorled, dissected .... 5. CERATOPHYLLACEAE.

1. Family NELUMBONACEAE Dumortier 1828 nom. conserv., the Lotus-lily Family

Aquatic, acaulescent, rhizomatous and tuberiferous herbs, producing benzyl-isoquinoline and aporphine alkaloids, somewhat tanniferous, with proanthocyanins but not ellagic acid; parenchymatous tissues with conspicuous schizogenous air-passages and with clustered crystals of calcium oxalate in some of the cells; articulated laticifers present especially in the vascular bundles and also to some extent in the parenchymatous tissues, which otherwise lack idioblasts; vessels present in the roots only, the end-walls elongate and very oblique, with scalariform perforation-plates and numerous cross-bars; vascular bundles of the stem scattered, closed, without cambium; tracheids elongate, with scalariform bordered pits; sieve-tubes with S-type plastids. LEAVES simple, long-petiolate, arising directly from the rhizome; phyllotaxy unique, the phyllomes distributed in sets of 3 (one foliage-leaf, 2 cataphylls) along the rhizome, one cataphyll on the lower side, one on the upper side and immediately subtending the foliage-leaf, the foliage-leaves thus all arising from the upper side of the rhizome; lower cataphyll sheathing, with overlapping margins, curved upwards distally, initially wrapped around the petiole and the terminal bud, but ruptured by continued growth of the rhizome; upper cataphyll wrapped around only the petiole of the subtended foliage-leaf; petiole with a 2-keeled, axillary, distally free stipule, this ochrea-like but not anatomically closed, initally forming an open sheath around the terminal bud and within the lower cataphyll; leaf-blades mostly large, circular, and centrally peltate, some floating, others distinctly raised above the water; some shorter, submerged, parallel-veined, more or less lanceolate leaves also present; stomates anomocytic. FLOWERS solitary, axillary, long-pedunculate, elevated above the water, large and showy, entomophilous (often cantharophilous), perfect, hypogynous; tepals numerous, ca 22–30, distinct, spirally arranged, the 2 outermost ones greenish and sepaloid, the others petaloid but more or less in 2 series, the 5–8 outer members smaller and less showy than the more numerous inner ones; stamens numerous (ca 200–400), spirally arranged, with slender, elongate filament and tetrasporangiate, dithecal anther, the 4 pollen-sacs introrse-latrorse on the narrowly laminar connective, which is conspicuously prolonged beyond them; staminodes none; pollen-grains tectate-columellate, tricolpate; gynoecium of numerous (ca 12–40) distinct carpels arranged in 2–4 more or less distinct cycles, individually sunken in (but

FIG. 1.18 Nelumbonaceae. *Nelumbo nucifera* Gaertn. a, habit, $\times\frac{1}{8}$; b, flower, $\times\frac{1}{4}$; c, stamens, $\times 1\frac{1}{2}$; d, fruiting receptacle, $\times\frac{1}{2}$; e, fruit, $\times 1\frac{1}{2}$.

free from) the enlarged, obconic, spongy, receptacle, only the sessile stigmas protruding; ovule solitary (2), ventral-apical, pendulous, anatropous, bitegmic, crassinucellar; endosperm-development nuclear. FRUIT of separate, hard-walled nuts loose in the cavities of the strongly accrescent receptacle, each nut with a small respiratory pore at the top; seed solitary, filled by the edible embryo, without perisperm and essentially without endosperm; cotyledons 2, originating as separately vasculated lobes of an annular (initially crescentric) primordium, becoming large and strongly thickened, connate by their margins for most of their length, forming a thick sheath around the well developed green plumule. $2n = 16$.

The family Nelumbonaceae consists of the single genus *Nelumbo*, with 2 species. *Nelumbo nucifera* Gaertn., the Oriental lotus-lily, or sacred lotus, is native to the warmer parts of Asia and Australia. *Nelumbo lutea* (Willd.) Pers., the American lotus-lily, is native to eastern United States. *Nelumbo nucifera* is famous for the longevity of its seeds, almost certainly at least 3000 years under favorable conditions of storage.

Pollen of *Nelumbo* is known from Eocene and more recent deposits.

2. Family NYMPHAEACEAE Salisbury 1805 nom. conserv., the Water-lily Family

Aquatic, rhizomatous herbs (or the rhizome shortened into an erect caudex), often with sesquiterpene alkaloids, and sometimes tanniferous, with ellagic and gallic acid and/or leuco-anthocyanins; parenchymatous tissues with conspicuous schizogenous air-passages, commonly also with articulated laticifers, and often with branching sclereids especially in the leaves; root-hairs originating from specialized cells, as in many monocotyledons; vessels wanting; vascular bundles of the stem scattered or in a single ring, closed, without cambium, often of unusual structure; tracheids elongate, with spiral or annular thickenings; sieve-tube plastids of S-type. LEAVES arising directly from the rhizome, alternate, long-petiolate, with cordate or hastate to peltate, usually floating blade (but some leaves sometimes emergent or wholly submerged); stomates anomocytic; stipules median-axillary or wanting. FLOWERS solitary, axillary or extra-axillary, sometimes (*Nymphaea*) occupying leaf-sites in the morphogenetic spiral, long-pedunculate, generally rather large, aerial and commonly entomophilous (often or usually cantharophilous), perfect, regular, hypogynous to epigynous; sepals 4–6 (–14), sometimes (*Nuphar*) more or less petaloid; petals 8-many, staminodial in origin, distinct, larger or smaller than the sepals, often passing into the stamens, rarely (*Ondinea*) absent; stamens numerous, spirally arranged, mostly laminar and 3-nerved, with elongate microsporangia more or less sunken into the adaxial side of the blade, less commonly transitional toward the familiar staminal type with well developed filament and anther, free and distinct, sometimes some of the inner or outer ones

FIG. 1.19 Nymphaeaceae. *Nymphaea odorata* Aiton. a, habit, $\times\frac{1}{2}$; b, flower, $\times 1$; c, pistil from above, showing radiating, distally distinct and incurved stigmas, $\times 2$; d, stamens of varying sizes, the largest ones outermost, $\times 2$; e, cross-section of ovary, $\times 2$; f, fruit, $\times 1$; g, seed, with aril, $\times 6$; h, seed, the aril removed, $\times 6$.

staminodial; pollen-grains binucleate (trinucleate in *Euryale*), with a granular or nearly homogeneous exine, monosulcate (sometimes zonisulcate); gynoecium of (3) 5–35 carpels more or less firmly united into a compound, plurilocular, superior to inferior ovary that is apically constricted beneath an expanded disk bearing radiate stigmatic lines (as in *Nuphar*), or the stigma a broad, radially grooved and marginally lobed disk surrounding a knob that appears to be a prolongation of the floral axis (the lobes only 3–5 in *Ondinea*, more numerous in *Nymphaea*); stigmas dry, papillate; ovules numerous, laminar (scattered on the partitions), bitegmic, crassinucellar, anatropous; endosperm-development cellular or nuclear or helobial. FRUIT spongy, at least below, berry-like, indehiscent or irregularly dehiscent; seeds small, commonly operculate, often arillate, with rather scanty endosperm and copious perisperm with clustered starch-grains; cotyledons 2, wholly distinct, or developing as separately vasculated lobes from an annular (initially crescentic) common primoridum. X = 12–29+. (Euryalaceae)

The family Nymphaeaceae as here defined consists of 5 genera (*Nymphaea, Nuphar, Euryale, Victoria,* and *Ondinea*) and about 50 species, of cosmopolitan distribution. *Nymphaea* and *Nuphar* are familiar in quiet waters in North Temperate regions, and the South American *Victoria* (royal water-lily) is often grown in conservatories.

The recently described monotypic Australian genus *Ondinea* stands apart from the other genera and is in some ways reminiscent of *Barclaya*. Although the sepals of *Ondinea* are hypogynous, the stamens are attached to the upper part of the ovary, and there are no petals. The pollen is monosulcate and similar to that of some species of *Nymphaea*, and the seeds are also much like those of *Nymphaea*.

3. Family BARCLAYACEAE Kozo-Poljansky 1922,[1] the Barclaya Family

Aquatic, rhizomatous herbs; parenchymatous tissues with conspicuous air-passages, and with articulated laticifers; vessels wanting; vascular bundles scattered, closed, without cambium. LEAVES arising directly from the short to elongate rhizome, alternate, long-petiolate, with floating, rounded-cordate to often much elongate and only basally

[1] This name was implied by Kozo-Poljansky in a diagram in his **Введение в Филогенетическую систематику вьсших растений** in 1922, but it does not appear to have been formally validated until now. It is based on and includes only the genus *Barclaya* Wall. Trans. Linnean Soc. London 15: 442-448. 1827

cordate blade; stipules wanting. FLOWERS solitary, extra-axillary, long-pedunculate but often not reaching the surface of the water, perfect, regular, at least sometimes cleistogamous; sepals 5, distinct, borne beneath the ovary; corolla-tube arising from around the top of the ovary, with 12-many lobes more or less distinctly in 3–4 series; stamens numerous, adnate to the corolla-tube in several series of about 10 each, the upper 2 series staminodial; filaments short; anthers basifixed, tetrasporangiate and dithecal, opening by longitudinal slits; pollen-grains inaperturate, with reduced exine; gynoecium of 8–14 carpels united to form a compound ovary with as many locules, capped by a short, conic style and a depressed-umblicate, obscurely radiate stigma; ovules orthotropous, bitegmic, crassinucellar, rather numerous, scattered over the surfaces of the partitions (laminar placentation); endosperm-development cellular. FRUIT globose, fleshy, crowned by the persistent corolla; seeds densely but rather softly spinulose, obscurely operculate, not arillate; embryo minute, dicotyledonous; endosperm and perisperm well developed. 2n = 34, 36.

The family Barclayaceae consists of the single genus *Barclaya*, with about 4 species native from tropical southeastern Asia to New Guinea. *Barclaya* has often been included in the Nymphaeaceae, but it differs from the other genera in so many ways as to require a separate description. It therefore seems useful to follow Kozo-Poljansky, Li, Takhtajan, and others in treating it as a distinct family. Here, as in the Magnoliales and Laurales, we are evidently dealing with an ancient group, which diversified during the Lower Cretaceous period, at the dawn of angiosperm history. The recognition of a number of presently small families may therefore be harmonious with the actual evolutionary relationships. Certainly the differences among the several families of Nymphaeales here recognized are comparable to the differences among much larger families in other groups of angiosperms.

The phyllomes at the base of the ovary of *Barclaya*, here described as sepals, are considered by some authors to be bracts. The flower is then considered to be epigynous with a prolonged hypanthium, and the outer series of petals (as here described) is treated as a set of sepals. The choice between the two interpretations is of little consequence. *Barclaya* is in either case a taxonomically isolated member of the Nymphaeales. The terms bract, sepal, and petal were developed for use in more advanced groups of angiosperms, and are as much functional and topographic as phylogenetic. I prefer the interpretation here presented because the outer phyllomes of the flower look like sepals, and because the peduncles of other members of the Nymphaeales do not have bracts.

4. Family CABOMBACEAE A. Richard 1828 nom. conserv., the Water-shield Family

Aquatic herbs with creeping rhizomes anchored in the substrate and with elongate, distally floating, leafy stems, commonly producing alkaloids (of undetermined nature) and at least sometimes gallic acid; parenchymatous tissues with conspicuous schizogenous air-passages and commonly also with articulate laticifers, but without sclereids; small crystals of calcium oxalate commonly present along the walls of some of the parenchyma-cells; vessels wanting; vascular bundles of the stem scattered, closed, without cambium; tracheids elongate, with spiral or annular thickenings; sieve-tubes with S-type plastids; young vegetative organs beset with mucilaginous hairs. LEAVES of 2 types, in *Brasenia* all alternate and long-petiolate, with floating, elongate to broadly elliptic, peltate blade, but in *Cabomba* many or all opposite or whorled, submersed, short-petiolate, and strongly dissected; stipules wanting. FLOWERS solitary, long-pedunculate from the upper nodes, rather small, but aerial and entomophilous, perfect, hypogynous, regular; sepals (2) 3 (4), free and distinct or nearly so; stamens 3–6 (*Cabomba*) or 12–18 or more (*Brasenia*) cyclic, differentiated into a slightly flattened filament and a fairly conventional, tetrasporangiate and dithecal anther; connective not prolonged; staminodes none; pollen-grains large to very large, more or less boat-shaped, tectate-granular, monosulcate (anasulcate) or sometimes trichotomosulcate; gynoecium of (1) 2–18 distinct carpels (commonly 3 in *Cabomba*, 12–18 in *Brasenia*), each tapering into a short style with a terminal (*Cabomba*) or decurrent (*Brasenia*) stigma; ovules (1) 2–3, laminar along or near the dorsal suture, anatropous, bitegmic, crassinucellar; endosperm-development helobial or cellular. FRUITS coriaceous, indehiscent, achene-like or follicle-like; seeds small, operculate, with scanty endosperm and copious perisperm; embryo small, dicotyledonous. 2n = 24, 80, 104.

The family Cabombaceae consists of 2 genera: *Cabomba*, with 7 species of tropical and warm-temperate parts of the New World; and *Brasenia*, with a single species found in tropical and warm-temperate parts of both the Old and the New World.

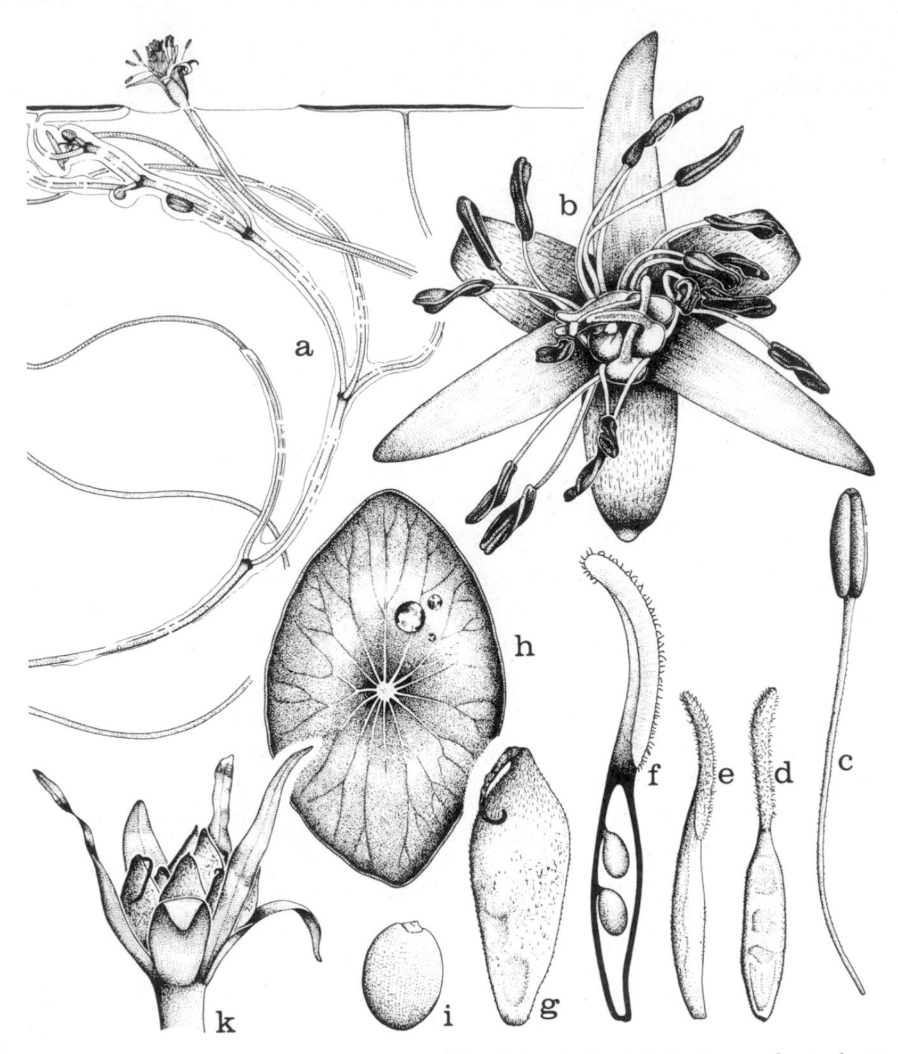

FIG. 1.20 Cabombaceae. *Brasenia schreberi* Gmelin. a, habit, ×½; b, flower, from above, ×4; c, stamen, ×4; d, e, carpel, ×4; f, carpel, the ovary in long-section, ×6; g, fruit, ×4; h, leaf, ×1; i, seed, ×4; k, achenes surrounded by persistent perianth, ×2.

5. Family CERATOPHYLLACEAE S. F. Gray 1821 nom. conserv., the Hornwort Family

Submersed, aquatic, rootless, perennial herbs with branching stems and without secondary growth; stem with a single vascular strand, this with a central air-canal surrounded by elongate, starch-bearing cells representing the modified, unlignified xylem; tanniferous secretory cells present in the stem and leaves. LEAVES whorled, sessile, somewhat cartilaginous, dichotomously dissected into slender, often serrulate segments, without stomates, exstipulate. FLOWERS small and inconspicuous, axillary, hydrogamous, hypogynous, unisexual, the male and female ordinarily solitary at different nodes on the same plant, the male often above the female; sepals (often interpreted as bracts) 8–15 in a single whorl, connate at the base; petals none; stamens (5-) 10–20 (-27), distinct, spirally arranged on the flat receptacle, not clearly differentiated into filament and anther, the "filament" short and broad, the pollen-sacs extrorse on the thickened, somewhat laminar connective, which is prolonged into a pair of prominent points; pollen-grains binucleate, globose, smooth, thin-walled, with reduced or no exine, inaperturate; gynoecium of a single carpel tapering into a terminal style with decurrent stigma; ovule solitary, pendulous, laminar-medial from near the tip of the locule, orthotropous, unitegmic, crassinucellar; endosperm-development cellular. FRUIT an achene; seed with a thin testa, lacking both perisperm and endosperm; cotyledons 2, arising from an annular common primordium but essentially distinct, thickened, more or less linear; plumule well developed, with several embryonic leaves; radicle vestigial. X mostly = 12.

The family Ceratophyllaceae consists of the single genus *Ceratophyllum*, with about 6 species, of cosmopolitan distribution. The Ceratophyllaceae are the most reduced and specialized family of the Nymphaeales, but there is no doubt about their relationship. *Cabomba*, in the Cabombaceae, provides a sort of link to the rest of the order in its elongate, leafy stems, with the submerged leaves opposite or whorled and finely dissected.

7. Order RANUNCULALES Lindley 1833

Herbs, herbaceous or woody vines, shrubs, or seldom small trees, the woody forms commonly with broad medullary rays and sometimes also with anomalous secondary growth; plants very often producing benzylisoquinoline and/or aporphine alkaloids, sometimes with various other sorts of repellents or poisons in addition or instead; nodes mostly trilacunar or multilacunar, but sometimes unilacunar; vessel-segments simply perforate, or sometimes with reticulate or scalariform perforation-plates and few cross-bars; sieve-tubes with S-type plastids, so far as known. LEAVES alternate or less commonly opposite, simple or compound; stomates most commonly anomocytic, less often paracytic or encyclocytic; stipules wanting, or sometimes present but small and inconspicuous (and then often caducous). FLOWERS borne in various sorts of inflorescences, mostly entomophilous, but sometimes ornithophilous or autogamous or more or less anemophilous, regular or less often irregular, perfect or unisexual, large or small, hypogynous, often with a large receptacle, not infrequently trimerous; perianth mostly of separate and distinct, more or less well distinguished sepals and petals, or both the sepals and the petals petaloid, or the petals wanting and the sepals then often petaloid, the sepals typically with 3 or more traces and gaps, the petals more often homologous with the stamens and with only a single trace, often nectariferous; rarely the perianth completely wanting; stamens (1-) 6–many, often spirally arranged, when few and cyclic commonly opposite the petals, generally with well differentiated filament and anther, but sometimes (notably in the Lardizabalaceae) with a laminar-expanded connective, sometimes some of them staminodial; pollen-grains binucleate or seldom trinucleate, most commonly triaperturate, less often multiaperturate or biaperturate; gynoecium of 1-many separate carpels, or less often of 2–3 (rarely more) more or less united carpels, and then sometimes with only one locule fully developed and ovuliferous; placentation marginal to laminar or axile, or sometimes essentially apical or basal; ovules 1-many, anatropous to sometimes hemitropous, campylotropous, or orthotropous, bitegmic or seldom unitegmic, crassinucellar or seldom tenuinucellar. FRUITS of various types, commonly of follicles, achenes, berries, or drupes; seeds with small embryo and copious endosperm, or less often with large embryo and little or no endosperm, sometimes arillate; cotyledons 2, seldom only one (by supression of one).

The order Ranunculales as here defined consists of 8 families and about 3200 species. More than half of the species belong to the single large family Ranunculaceae, and most of the others belong to only two families, the Berberidaceae (650) and Menispermaceae (400). Six of the families (Ranunculaceae, Circaeasteraceae, Berberidaceae, Sargentodoxaceae, Lardizabalaceae, and Menispermaceae), including the three largest ones, clearly belong to the same circle of affinity. The other two families (Coriariaceae and Sabiaceae) are less securely placed.

It is widely believed that the Ranunculales originated from the Magnoliales through something like the Illiciales, which differ from the Magnoliales most notably in having more advanced, triaperturate pollen.

Members of the Ranunculales cover a wide range of habitats, from tropical to arctic, and from strictly terrestrial to aquatic. They also embrace a wide array of growth-forms, from annual or perennial herbs to woody or herbaceous vines or erect shrubs or even small trees. Many of the Ranunculales have retained isoquinoline alkaloids from their magnolialean heritage, but other sorts of chemical defenses are also extensively exploited. One cannot even begin to characterize the order ecologically in terms that would not include a great many members of various other orders as well.

With the partial exception of the Sabiaceae (which are only doubtfully placed here), woody members of the Ranunculaceae have broad medullary rays, and the Menispermaceae also have anomalous secondary thickening. The Berberidaceae, which have many woody members, are in most respects more advanced than the more consistently herbaceous Ranunculaceae, and the genera that tend to connect these two families are herbaceous. Thus it seems likely that the order as a whole is primitively herbaceous, and that all of its woody members are only secondarily woody.

The Illiciales, on the other hand, have much more primitive xylem and appear to be primitively woody. The Sargentodoxaceae, with clearly advanced stem-anatomy, may therefore not be so closely allied to the Illiciales as has been suggested. It is plain enough that the Ranunculales must be derived from the Magnoliales through some woody group with triaperturate pollen, and the Illiciales are the only known candidates to provide such a connection. It is only the closeness of the connection that is here questioned.

My choice of the name Ranunculales rather than Berberidales for this order is purely pragmatic. Ordinal names are not now subject to priority under the Rules of Nomenclature, but they may become so in the future. Thus it is well to pay some attention to priority. As Thorne (1974, under Magnoliidae) has pointed out, the names Ranunculales and Berberidales both take their origin in Lindley's Nixus Plantarum. Lindley called his groups alliances rather than orders, and he used the spelling Ranales rather than Ranunculales, but his intent is clear enough for nomenclatural purposes. In later works the name Ranunculales (often in the form Ranales) has much more commonly been used than Berberidales, when the two are combined. Both Takhtajan (1966) and Buchheim (in the 12th edition of the Engler Syllabus, 1964) use the name Ranunculales in preference to Berberidales. It seems more useful to continue the tradition than to break with it.

The rationale for using the spelling Ranunculales instead of the more familiar form Ranales is simple. The name is based on *Ranunculus*, not on *Rana*, which is a frog.

SELECTED REFERENCES

Arnott, H. J., & S. C. Tucker. 1963. Analysis of petal variation in *Ranunculus*. I. Anastomoses in *R. repens* v. *pleniflorus*. Amer. J. Bot. 50: 821–830.

Barneby, R. C., & B. A. Krukoff. 1971. Supplementary notes on American Menispermaceae. VIII. A generic survey of the American Triclisieae and Anomospermeae. Mem. New York Bot. Gard. 22(2): 1–89.

Bhatnagar, S. P. 1965. Some observations on the embryology of *Holboellia* latifolia Wall. Curr. Sci. 34: 28–29.

Brouland, M. 1935. Recherches sur l'anatomie florale des Renonculacé Botaniste 27: 1–278.

Chapman, M. 1936. Carpel anatomy of the Berberidaceae. Amer. J. Bot 340–348.

Diels, L. 1932. *Circaeaster*, eine hochgradig reduzierte Ranunculacee. Bei Centralbl. 49 (Erg.-bd): 55–60.

Dormer, K. J. 1954. The acacian type of vascular system and so derivatives. I. Introduction, Menispermaceae, Lardizabalaceae, B ceae. New Phytol. 53: 301–311.

Ernst, W. R. 1964. The genera of Berberidaceae, Lardizabalaceae, permaceae in the southeastern United States. J. Arnold Arbor.

Ezelarab, G. E., & K. J. Dormer. 1963. The organization of the pri system in Ranunculaceae. Ann. Bot. (London) II. 27: 23–38.

Ferguson, I. K. 1975. Pollen morphology of the tribe Tr Menispermaceae in relation to its taxonomy. Kew Bull. 30:

Foster, A. S. 1961. The floral morphology and relations *uniflora*. J. Arnold Arbor. 42: 397–415.

Foster, A. S. 1963. The morphology and relationships of Arbor. 44: 299–327.

Foster, A. S., & H. J. Arnott. 1960. Morphology and di of the leaf of *Kingdonia uniflora*. Amer. J. Bot. 47: 68

Gregory, W. C. 1941. Phylogenetic and cytological studies in the Ranunculaceae. Trans. Amer. Philos. Soc. II, 31: 443–521.

Haccius, B., & E. Fischer. 1959. Embryologische und histogenetische Studien an "monokotylen Dikotylen." III. *Anemone appenina* L. Oesterr. Bot. Z. 106: 373–389.

Hammond, H. D. 1955. Systematic serological studies in Ranunculaceae. Serol. Mus. Bull. 14: 1–3.

Henderson, E. M. 1924. The stem structure of *Sargentodoxa cuneata* Rehd. et Wils. Trans. & Proc. Bot. Soc. Edinburgh 29: 57–62.

Janchen, E. 1949. Die systematische Gliederung der Ranunculaceen und Berberidaceen. Österr. Akad. Wiss., Math.-Naturwiss. Kl., Denkschr. 108 (4):1–82.

Jensen, U. 1966 (1967). Die Verwandtschaftsverhältnisse innerhalb der Ranunculaceae aus serologischer Sicht. Ber. Deutsch. Bot. Ges. 79: 407–412.

Jensen, U. 1968. Serologische Beiträge zur Systematik der Ranunculaceae. Bot. Jahrb. Syst. 88: 204–268; 269–310.

Jensen, U. 1973. The interpretation of comparative serological results. *In*: G. Bendz and J. Santesson, eds., Chemistry in botanical classification, pp. 217–227. Nobel Symposium 25. Academic Press. New York and London.

Joshi, A. C. 1939. Morphology of *Tinospora cordifolia*, with some observations on the origin of the single integument, nature of synergidae, and affinities of the Menispermaceae. Amer. J. Bot. 26: 433–439.

Junell, S. 1931. Die Entwicklungsgeschichte von *Circaeaster agrestis*. Svensk Bot. Tidskr. 25: 238–270.

Kaute, U. 1963. Beiträge zur Morphologie des Gynoeceums der Berberidaceen mit einem Anhang über die Rhizomknospe von *Plagiorhegma dubium*. Inaugural-Dissertation zur Erlangung der Doktorwürde der Mathematisch-Naturwissenschaftlichen Fakultät der Freien Universität Berlin. Dissertations-Druckstelle. Berlin.

Kordyum, E. L. 1959. Comparative embryological investigation of the family Ranunculaceae DC. Ukrajins'k. Bot. Zhurn. 16: 32–43. (In Ukrainian, with Russian and English summaries.)

Krukoff, B. A., & H. N. Moldenke. 1938. Studies of American Menispermaceae, with special reference to species used in preparations of arrow-poisons. Brittonia 3: 1–74.

Kumazawa, M. 1930. Morphology and biology of *Glaucidium palmatum* Sieb. et Zucc. with notes on affinities to the allied genera *Hydrastis, Podophyllum* and *Diphylleia*. J. Fac. Sci. Univ. Tokyo, Sect. 3, Bot. 2: 345–380.

Kumazawa, M. 1930. Structure and affinities of *Glaucidium* and its allied genera. (In Japanese.) Bot. Mag. (Tokyo) 44: 479–490.

Kumazawa, M. 1936. Pollen grain morphology in Ranunculaceae, Lardizabalaceae and Berberidaceae. Jap. J. Bot. 8: 19–46.

Kumazawa, M. 1938. On the ovular structure in the Ranunculaceae and Berberidaceae. J. Jap. Bot. 14: 10–25.

Kumazawa. M. 1938. Systematic and phylogenetic consideration of the Ranunculaceae and Berberidaceae. Bot. Mag. (Tokyo) 52: 9–15.

Leinfellner, W. 1955. Beiträge zur Kronblattmorphologie. VI. Die Nektarblätter von *Berberis*. Oesterr. Bot. Z. 102: 186–194.

Leinfellner, W. 1956. Zur Morphologie des Gynözeums von *Berberis*. Oesterr. Bot. Z. 103: 600–612.

Lemesle, R. 1943. Les trachéides à ponctuations aréolées de *Sargentodoxa cuneata* Rehd. et Wils. et leur importance dans la phylogénie des Sargentodoxacées. Bull. Soc. Bot. France 90: 104–107.

Leppik, E. E. 1964. Floral evolution in the Ranunculaceae. Iowa State Coll. J. Sci. 39: 1–101.

Lobreau-Callen, D. 1977. Nouvelle interpretation de l'"ordre" des Celastrales à l'aide de la palynologie. Compt. Rend. Hebd. Séances Acad. Sci. 284D: 915–918.

Mauritzon, J. 1936. Zur Embryologie der Berberidaceen. Acta Horti Gothob. 11: 1–18.

Metcalfe, C. R. 1936. An interpretation of the single cotyledon of *Ranunculus Ficaria* based on embryology and seedling anatomy. Ann. Bot. (London) 50: 103–120.

Pant, D. D., & B. Mehra. 1964. Ontogeny of stomata in some Ranunculaceae. Flora 155: 179–188.

Payne, W. W., & J. L. Seago. 1968. The open conduplicate carpel of *Akebia quinata* (Berberidales: Lardizabalaceae). Amer. J. Bot. 55: 575–581.

Praglowski, J. 1970. Coriariaceae. World Pollen Flora 1: 17–31.

Raju, M. V. S. 1952. Embryology of Sabiaceae. Curr. Sci. 21: 107–108.

Rohweder, O. 1967. Karpellbau und Synkarpie bei Ranunculaceen. Ber. Schweiz. Bot. Ges. 77: 376–432.

Ruijgrok, H. W. L. 1963. Chemotaxonomische Untersuchungen bei den Ranunculaceae: I. Ranunculin. Naturwissenschaften 50: 620–621. II. Über Ranunculin und verwandte Stoffe. Pl. Med. 11: 338–347.

Ruijgrok, H. W. L. 1967. Over de Verspreiding van Ranunculine en Cyanogene Verbindingen bij de Ranunculaceae. Een Bijdrage tot de Chemotaxonomie van de familie. Thesis. Leiden.

Sastri, R. L. N. 1969. Floral morphology, embryology, and relationships of the Berberidaceae. Austral. J. Bot. 17: 69–79.

Schaeppi, H., & K. Frank. 1962. Vergleichend-morphologische Untersuchungen über die Karpellgestaltung, insbesondere die Plazentation bei Anemoneen. Bot. Jahrb. Syst. 81: 337–357.

Schöffel, K. 1932. Untersuchungen über den Blütenbau der Ranunculaceen. Planta 17: 315–371.

Sharma, V. K. 1968. Floral morphology, anatomy and embryology of *Coriaria nepalensis* Wall. with a discussion on the inter-relationships of the family Coriariaceae. Phytomorphology 18: 143–153.

Shen, Y.-F. 1954. Phylogeny and wood anatomy of *Nandina*. Taiwania 5: 85–92.

Skog, L. E. 1972. The genus *Coriaria* (Coriariaceae) in the western hemisphere. Rhodora 74: 242–253.

Stapf, O. 1926. *Sargentodoxa cuneata*. Curtis's Bot. Mag. 151: pls. 9111, 9112.

Swamy, B. G. L. 1953. Some observations on the embryology of *Decaisnea insignis* Hook. et Thoms. Proc. Natl. Inst. Sci. India 19B: 307–310.

Tamura, M. 1963–68. Morphology, ecology and phylogeny of the Ranunculaceae. Parts 1–8. Sci. Rep. S. Coll. N. Coll. Osaka Univ. 11: 115–126; 12: 141–156; 13: 25–35; 14(1): 53–71; 14(2): 27–48; 15(1): 13–35; 16(2): 21–43; 17(1): 41–56.

Tamura, M. 1972. Morphology and phyletic relationship of the Glaucidiaceae. Bot. Mag. (Tokyo) 85: 29–41.

Tamura, M., & Y. Mizumoto. 1972. Stages of embryo development in ripe seeds or achenes of the Ranunculaceae. J. Jap. Bot. 47: 225–236.

Tamura, M., Y. Mizumoto, & H, Kubotia. 1977. Observations on seedlings of the Ranunculaceae. J. Jap. Bot. 52: 293–304. (In Japanese, with English summary.)

Tepfer, S. S. 1953. Floral anatomy and ontogeny in *Aquilegia formosa* var. *truncata* and *Ranunculus repens*. Univ. Calif. Publ. Bot. 25: 513–647.

Terabayashi, S. 1977, 1978. Studies in morphology and systematics of Berberidaceae. I. Floral anatomy of *Ranzania japonica*. II. Floral anatomy of *Mahonia japonica* (Thunb.) DC. and *Berberis thunbergii* DC. Acta Phytotax. Geobot. 28: 45–57, 1977. 29: 106–118, 1978.

Thanikaimoni, G. 1968. Morphologie des pollens des Ménispermacées. Inst. Franç. Pondichéry, Trav. Sect. Sci. Techn. 5(4): 1–56.

Troll, W. 1933. Beiträge zur Morphologie des Gynaeceums. III. Über das Gynaeceum von *Nigella* und einiger anderer Helleboreen. Planta 21: 266–291.

Troupin, G. 1962. Monographie des Menispermaceae africaines. Mem. Acad. Sci. Outre-Mer, Cl. Sci. Nat. Med. 13(2): 1–312.

Tucker, S. C. 1966. The gynoecial vascular supply in *Caltha*. Phytomorphology 16: 339–342.

van Beusekom, C. F. 1971. Revision of *Meliosma* (Sabiaceae), section Lorenzanea excepted, living and fossil, geography and phylogeny. Blumea 19: 355–529.

Werth, E. 1941. Die Blütennektarien der Ranunculaceen und ihre phylogenetische Bedeutung. Ber. Deutsch. Bot. Ges. 59: 246–256.

Yoshida, O., & A. Michikawa. 1973. Embryological studies of genus *Akebia* Decaisne. J. Coll. Arts Chiba Univ. B-6: 25–37.

Yoshida, O., & Y. Nakajima. 1978. Embryological study on *Stauntonia hexaphylla* Decne. J. Coll. Arts Chiba Univ. B-11: 45–57. (In Japanese; English abstract.)

Аветисян, Е. М. 1955. О морфологии микроспор рода *Adonis* L. Изв. Биол. АН Армянск ССР, 8(6): 101-104.

Архангельский Д. Б. 1973. Палинотаксономия семейства Berberidaceae. В сб.: Морфология пыльцы и спор современных растений. Труды III Международной Палинологической Конференции. СССР, Новосибирск, 1971: 18-21. Пзд Наука, Ленинградское Отделение. Ленинград.

Архарова, К. Б., & И. Г. Зубкова. 1969. Анатомическое строение черешка в семействе Berberidaceae Juss. Бот. Ж. 54: 98-103.

Благовещенский, А. В., А. В. Давыдова, & М. А. Преснякова. 1952. К биохимической характеристике семейства лютиковых. Бюлл. Главн. Бот. Сада, 14: 29-33.

Жукова, Н. А. 1958. Опыт построения системы семейства Ranunculaceae на основе анализа морфологического строя. Пробл. Бот., (Москва-Ленинград), 3: 97-107.

Кемудариа-Натадзе, Л, М. 1963. К вопросу об обьеме семейства лютиковых. Заметки по Систематике и Географии Растений, Тбилисск. Бот. Инст. 23: 44-53.

Кордюм, Е. Л., 1961. Сравнительно-эмбриологическое исследование семейства лютиковых. В сб.: Морфогенез растений 2: 473-477. Изд. Московского Унив. Москва.

Оганезова, Г. Г. 1974. Анатомическое строение листа у Berberidaceae s.l. в связи с систематикой семейства. Бот. Ж. 59: 1780-1794.

Оганезова, Г. Г. 1975. Об эволюции жизненных форм в семействе Berberidaceae s.l. Бот. Ж. 60: 1665-1675.

SYNOPTICAL ARRANGEMENT OF THE FAMILIES OF RANUNCULALES

1 Gynoecium mostly of separate carpels, or seemingly or actually of a single carpel, rarely (a few Ranunculaceae) of several more or less connate carpels; flowers often with nectariferous petals, but without an intrastaminal nectary disk; endosperm copious to

scanty or none; herbs and herbaceous or woody vines, less often shrubs or small trees.

2 Leaves mostly alternate (opposite notably in *Clematis*, of the Ranunculaceae); petals not becoming fleshy; flowers variously organized, often trimerous in part or with numerous stamens and carpels, never pentamerous throughout; plants without ellagic acid.

3 Flowers mostly perfect; herbs or less often woody plants; endosperm well developed; plants mostly of temperate or boreal regions, or of tropical mountains.

4 Leaves with dichotomous, free venation; nodes unilacunar; ovules orthotropous, unitegmic, tenuinucellar; endosperm-development cellular; flowers reduced, without petals or petaloid nectaries .... 2. CIRCAEASTERACEAE.

4 Leaves with net venation, or at least not with dichotomous, free venation; nodes mostly or always trilacunar to multilacunar; ovules anatropous to seldom hemitropous, bitegmic or seldom unitegmic; endosperm-development nuclear; flowers usually well developed, often with nectariferous petals or with petaloid sepals, but sometimes much reduced and without perianth.

5 Carpels usually 2 or more and distinct, seldom solitary or weakly united; stamens usually numerous, spirally arranged; anthers opening by longitudinal slits; herbs, rarely woody vines or low shrubs ... 1. RANUNCULACEAE.

5 Carpel solitary, or apparently so; stamens 4–18, most often 6, generally of the same number as the nectariferous petals and opposite them (opposite nectarless petals when the flowers do not have nectariferous petals); anthers usually opening by 2 uplifting valves; herbs or more often shrubs or even small trees .... 3. BERBERIDACEAE.

3 Flowers nearly always unisexual; nearly all woody plants, mostly vines; endosperm present or absent; plants mostly of tropical or warm temperate regions.

6 Leaves compound; fruit of berries or fleshy follicles; embryo small, straight; endosperm copious; flowers racemose; petals, when present, nectariferous.

7 Carpels numerous, spirally arranged; ovule solitary .... 4. SARGENTODOXACEAE.

7 Carpels 3 in a single whorl, rarely 6–15 in 2–5 whorls; ovules usually more or less numerous, seldom few ....5. LARDIZABALACEAE.

6 Leaves simple, rarely compound; fruit drupaceous or seldom of nutlets; embryo rather large, very often curved or even

coiled; endosperm copious to often scanty or none; flowers cymose or cymose-paniculate, or seldom solitary; petals, when present, not nectariferous; ovules 2, one generally abortive ....6. MENISPERMACEAE.

2 Leaves opposite or whorled; petals persistent, enlarging in fruit, becoming fleshy and more or less enveloping the achenes; flowers ordinarily wholly pentamerous; plants with ellagic acid ....7. CORIARIACEAE.

1 Gynoecium evidently of 2 (or 3) carpels united to form a compound ovary; flowers with an intrastaminal nectary-disk surrounding the base of the ovary, but without nectariferous petals; endosperm scanty or none; trees, shrubs, or woody vines ....8. SABIACEAE.

1. Family RANUNCULACEAE A. L. de Jussieu 1789, nom. conserv., the Buttercup Family

Terrestrial or sometimes aquatic herbs or seldom low shrubs, or in a few genera (notably *Clematis*) more or less woody vines, but the woody forms with broad medullary rays and probably derived from an herbaceous ancestry; plants very often with benzyl-isoquinoline or aporphine alkaloids, or alternatively with ranunculin (a lactone glycoside), and not infrequently with triterpenoid saponins, or with various other sorts of repellents, sometimes (as in *Thalictrum*) cyanogenic through a tyrosine-based pathway, but without ellagic acid and at least usually without proanthocyanins; ethereal-oil cells wanting; calcium oxalate crystals only seldom present; nodes trilacunar or more often multilacunar, only seldom unilacunar; vascular bundles usually closed (not in *Clematis*), often in concentric rings or more or less scattered, typically v-shaped in cross-section, with the point of the v inward; interfascicular cambium sometimes well developed, as in *Clematis*; vessel-segments simply perforate (scalariform with several cross-bars in *Hydrastis*); imperforate tracheary elements with simple or bordered pits; sieve-tubes with S-type plastids. LEAVES alternate or seldom (notably *Clematis*) opposite, simple or variously compound or dissected, net-veined, or at least not with dichotomous, free venation; stomates anomocytic; stipules wanting, or seldom present but vestigial. FLOWERS large to small, in more or less cymose, often racemiform or paniculiform inflorescences, or occasionally solitary, perfect or rarely unisexual, entomophilous or rarely (*Thalictrum*) anemophilous, hypogynous, generally with a more or less elongate receptacle; perianth-members spirally or cyclically arranged; sepals (3-) 5–8 or more, often petaloid, especially in apetalous genera, often caducous; petals few to rather numerous (or wanting), mostly staminodial in origin, commonly nectariferous toward the base (frequently called honey-leaves); stamens generally numerous, mostly spirally arranged and centripetal (centrifugal in *Glaucidium*), free and distinct, with slender filament and well defined anther (seldom whorled in sets of 5; 6–10 such sets in *Aquilegia*, 1–2 sets in *Xanthorhiza*); anthers tetrasporangiate and dithecal; connective not prolonged; pollen-sacs opening by longitudinal slits; pollen-grains binucleate or seldom trinucleate, tricolpate to variously multiaperturate, but without endo-apertures; carpels (1-) several or many, distinct or rarely (but notably in *Nigella*) more or less connate into a compound ovary with axile placentation, generally each carpel with an evident style, the stigma dry, often bilobed; ovules several or numerous and marginal, or solitary

FIG. 1.21 Ranunculaceae. *Ranunculus hispidus* Michx. a, habit, ×½; b, base, of petal, showing nectary-scale, ×9; c, flower, from above, ×3; d, head of achenes, ×3; e, carpel, partly in long-section, ×12; f, stamen, ×6; g, flower, in long section, ×3.

and pendulous or nearly basal, anatropous or sometimes hemitropous, bitegmic or sometimes unitegmic, crassinucellar or pseudocrassinucellar, but the nucellus commonly resorbed before fertilization; endosperm-development nuclear. FRUIT usually of follicles, achenes, or berries (dehiscent along both sutures in *Glaucidium*, capsular in *Nigella*); seeds with abundant, oily and proteinaceous (sometimes also starchy) endosperm; embryo very small, often not fully formed when the seeds first appear ripe, or sometimes linear and more elongate, the 2 cotyledons often basally connate to form a tube, or seldom one cotyledon suppressed; germination usually epigaeous. X = 6–10, 13. (Glaucidiaceae, Hydrastidaceae, Helleboraceae)

The Ranunculaceae form a large and diversified but natural family of some 50 genera and 2000 species, widespread especially in temperate and boreal regions of both the Old and the New World. Tamura recognizes 6 subfamilies, the Helleboroideae, Ranunculoideae, Isopyroideae, Thalictroideae, Hydrastidoideae, and Coptidoideae. *Paeonia*, which has often in the past been included in the Ranunculaceae, is here treated as forming a family Paeoniaceae in the order Dilleniales.

The Ranunculaceae provide many familiar garden-ornamentals, including species of *Adonis, Anemone, Aquilegia, Clematis, Consolida, Delphinium, Eranthis, Helleborus,* and *Ranunculus.*

The small genera *Glaucidium, Hydrastis,* and *Podophyllum* are more or less transitional between the Ranunculaceae and Berberidaceae. *Glaucidium* and *Hydrastis* lack the v-shaped bundles of the Ranunculaceae, and all three genera lack the specialized anthers of the Berberidaceae. These three genera, together with *Diphylleia*, are habitally much alike and have sometimes been considered to form a family lying between the Ranunculaceae and Berberidaceae. Alternatively, *Hydrastis* and *Glaucidium* have been collectively separated as a family Glaucidiaceae, or both *Glaucidium* and *Hydrastis* have been considered to form monotypic families, or only *Glaucidium* is considered to form a distinct family. I see nothing to be gained by such treatments. *Diphylleia* has valvate anthers and is reasonably at home in the Berberidaceae. Its close relative *Podophyllum* must go with it, even though it brings into the Berberidaceae the aberrant feature of anthers that open by longitudinal slits. *Glaucidum* and *Hydrastis* have distinct carpels; in this respect they would be highly anomalous in the Berberidaceae. Furthermore, *Hydrastis* is serologically more concordant with the Ranunculaceae than with the Berberidaceae (Jensen, 1968), whereas *Podophyllum* goes with *Diphylleia* and other

Berberidaceae on serological grounds. Thus it seems better to include *Hydrastis* and probably also *Glaucidium* in the variable family Ranunculaceae, within which they do not stand out so sharply as they would in the more unified family Berberidaceae. It is possible, on the other hand, that *Glaucidium* should be associated in some way with *Paeonia*, as Tamura proposes.

2. Family CIRCAEASTERACEAE Hutchinson 1926 nom. conserv., the Circaeaster Family

Herbs; nodes unilacunar; vessels simply perforate, at least in *Circaeaster*. LEAVES alternate and arising directly from the rhizome in *Kingdonia*, more or less opposite and closely clustered at the summit of an elongate, stem-like hypocotyl in *Circaeaster*, simple (*Circaeaster*) or palmately cleft (*Kingdonia*), dichotomously veined, without anastomoses; stomates anomocytic; stipules wanting. FLOWERS perfect, regular, hypogynous; sepals 2 in *Circaeaster*, 5 (–7) and petaloid in *Kingdonia*; petals none; an outer series of 8–12 apically nectariferous staminodes present (*Kingdonia*) or absent (*Circaeaster*); stamens (1) 2–6; anthers opening by longitudinal slits, bisporangiate and dithecal in *Circaeaster*, tetrasporangiate and dithecal in *Kingdonia*; pollen-grains binucleate (at least in *Circaeaster*), tricolpate or tricolporate; gynoecium of (1) 2–9 distinct carpels, each with one or two subapical ovules pendulous from the ventral margin of the carpel; ovules orthotropous, unitegmic, tenuinucellar, only one maturing; endosperm-development cellular. FRUIT of achenes; seed with abundant endosperm and a small, dicotyledonous embryo. $2n = 30$ (*Circaeaster*). (Kingdoniaceae)

The family Circaeasteraceae consists of two monotypic genera from southeastern Asia that collectively differ from the Ranunculaceae in several features as indicated in the synoptical arrangement of families. The two genera are habitally rather different, and it is not certain that they should be associated in the same family. If *Kingdonia* is kept in the Ranunculaceae, as some authors do, then there is hardly any valid reason to maintain *Circaeaster* as a separate family. It is possible that further study will call for a different disposition of these two genera than that here provided.

It may be of some interest that the veniation of the petals of some species of *Ranunculus* is openly dichotomous, as in the leaves of Circaeasteraceae.

3. Family BERBERIDACEAE A. L. de Jussieu 1789 nom. conserv., the Barberry Family

Perennial herbs, or more often evergreen or deciduous (often spiny) shrubs or even small trees, but the woody forms with broad medullary rays that suggest an herbaceous ancestry; plants mostly glabrous, only seldom somewhat tanniferous (from proanthocyanins), but commonly with various sorts of benzyl-isoquinoline and aporphine alkaloids, the tissues often colored yellow by berberine (an isoquinoline), sometimes with quinolizidine alkaloids in addition to or instead of the isoquinolines; calcium oxalate often present as single or clustered crystals in some of

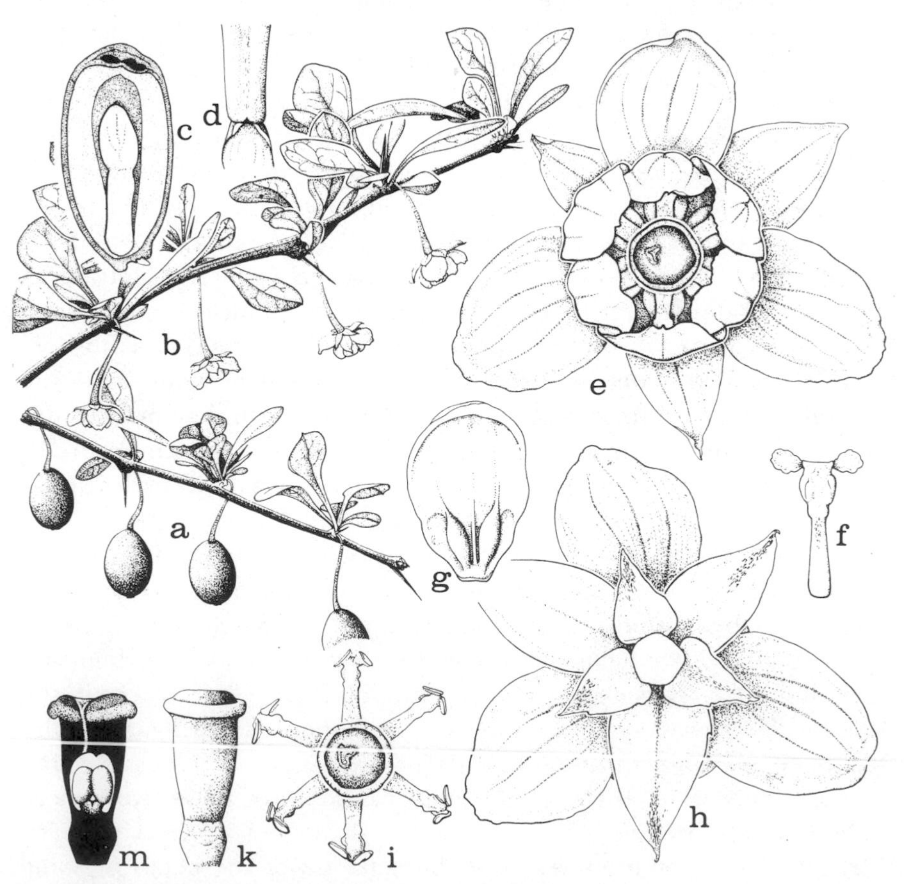

FIG. 1.22 Berberidaceae. *Berberis thunbergii* DC. a, b, habit, ×1; c, seed, in long-section, ×6; d, articulated leaf-base, ×6; e, flower, from above, ×6; f, stamen, with valves uplifted, ×6; g, nectariferous petal, ×6; h, flower, from beneath, ×6; i, stamens and pistil, ×6; k, pistil, ×6; m, pistil, in partial long-section, ×6.

the parenchyma cells; nodes trilacunar or multilacunar; vascular bundles in the herbaceous species commonly closed, often more or less scattered; vessel-segments with simple perforations, or sometimes with scalariform perforation-plates and few cross bars; imperforate tracheary elements with simple pits (bordered in *Nandina*); rays broad, homocellular (Kribs type II) or nearly so (heterocellular and mixed uniseriate and pluriseriate in *Nandina*); sieve-tubes with S-type plastids. LEAVES alternate (opposite in *Podophyllum*) or sometimes all basal, simple (but sometimes deeply cleft) to pinnately or ternately compound or decompound (simple-leaved species of *Berberis*, with an articulation at the base of the blade, may properly be unifoliolate); stomates at least sometimes (as in *Leontice*) anomocytic; petiole with 3-many vascular bundles (as many as 40 in *Diphylleia*), forming an arc or 2 rings; stipules usually wanting or vestigial, but the petiole often flared toward the base. FLOWERS commonly (not always) small, solitary or more often in racemes, spikes, typical or mixed panicles, or cymes, perfect, regular, entomophilous, hypogynous with a short receptacle, mostly trimerous, but occasionally dimerous or tetramerous; perianth (lacking in *Achlys*) usually of 6 or 7 (−9) series of distinct members, typically the outer 2 series (here called sepals, but often interpreted as bracts) relatively small and sepaloid, often caducous, the next 2 series (here called nectarless petals, but often interpreted as petaloid sepals) larger and more petaloid, but lacking nectaries, and the inner 2 (or 3) series (here called nectariferous petals, but often interpreted as staminodes) usually also petaloid but nectariferous at the base (nectaries subapical on staminodes in *Nandina*, wanting in *Podophyllum* and *Diphylleia*); stamens 4–18, most often 6, generally of the same number as the nectariferous petals and opposite them (or opposite the nectarless petals when the flowers do not have nectariferous petals) but sometimes twice as many; filaments short; anthers tetrasporangiate and dithecal, usually opening by 2 valves that lift up from the base (opening by longitudinal slits in *Podophyllum* and *Nandina*); pollen-grains binucleate, tricolpate or spiral-aperturate, or 6–12-colpate; gynoecium seemingly of a single carpel, but often interpreted as pseudomonomerous and representing 2 or 3 carpels; stigma wet or dry, sessile or on a short style, often 3-lobed; ovules anatropous or hemitropous, bitegmic, and crassinucellar, commonly numerous on a thickened marginal placenta, or the ovules only 2 or even (as in *Achlys*) solitary and basal; endosperm development nuclear. FRUIT commonly a berry, seldom dry and indehiscent or irregularly dehiscent; seeds often arillate; embryo dicotyledonous, small and basal or slender, elongate, and axile; endosperm abundant, fleshy to horny, with reserves of oil, protein, and sometimes also hemicellulose. X = 6, 7, 8, 10, 14. (Leonticaceae, Nandinaceae, Podophyllaceae)

The family Berberidaceae consists of about 13 genera and 650 species, widespread especially in temperate regions of the Northern Hemisphere. *Berberis* also extends south through the Andes of South America to the Straits of Magellan. *Berberis* is by far the largest genus, with some 500 species even if *Mahonia* (about 100 species) is not included in it. The remaining genera are mostly herbaceous and have scarcely more than 50 species in all. *Berberis* provides many ornamental shrubs; *B. vulgaris* L., the common barberry, is the alternate host of the stem rust of wheat.

Authors are agreed that the Berberidaceae are closely related to but more advanced than the Ranunculaceae. Serological and chemical studies help to support this conclusion, which was reached earlier on classical morphological grounds. The characteristic alkaloid berberine occurs in several genera of Ranunculaceae, and also in some members of the Menispermaceae and Papaveraceae, as well as in most Berberidaceae. Its occurrence in other groups is much more restricted. Some genera that link the Berberidaceae and Ranunculaceae are discussed under the latter family.

The monotypic Chinese and Japanese shrub *Nandina* stands somewhat apart from the rest of the Berberidaceae and has sometimes been treated as a separate family Nandinaceae. The relationship of *Nandina* is not in dispute; the quesion is the taxonomic rank at which to recognize the admitted differences. Recent serological studies and studies of seed-coat anatomy strengthen its connection with the Berberidaceae.

The truly monomerous or pseudomonomerous nature of the gynoecium of the Berberidaceae has been hotly but inconclusively debated. Teratological examples with two separate, simple pistils and with a single, clearly dimerous compound pistil are known. I myself see no need to interpret the pistil as anything other than the single carpel it appears to be.

4. Family SARGENTODOXACEAE Stapf ex Hutchinson 1926 nom. conserv., the Sargentodoxa Family

Twining woody vines; young stems with 4 large vascular bundles and 8 outer smaller ones, the bundles separated by well developed medullary rays; nodes trilacunar; tanniferous secretory cells present, especially in association with the vascular bundles; vessel-segments with simple perforations; imperforate tracheary elements with bordered pits, considered to be tracheids; cork arising in the innermost layer of the pericycle. LEAVES alternate, trifoliolate, exstipulate. FLOWERS small, in drooping racemes from scaly axillary buds, regular, hypogynous, unisexual, the plants dioecious; sepals 6 in 2 sets of 3, imbricate, green but petaloid in texture; petals 6, very small and scale-like, green, nectariferous; staminate flowers with 6 distinct stamens opposite the petals, and with vestigial carpels; filaments short; anthers dithecal, with a broad connective that is prolonged into a short, terminal appendage; pollen-sacs extrorse, opening by longitudinal slits; pollen-grains tricolporate, said to be similar to those of Lardizabalaceae; pistillate flowers with 6 staminodes and numerous distinct, spirally arranged carpels on an enlarged receptacle; each carpel with a solitary, pendulous, subapical, hemitropous to almost anatropous ovule. Receptacle enlarged and fleshy in FRUIT, bearing the numerous stipitate, one-seeded berries, the endocarp not hardened; seeds with minute, straight, excentric, dicotyledonous embryo; endosperm abundant, fleshy, with reserves of starch and oil.

The family Sargentodoxaceae consists of the single species *Sargentodoxa cuneata* Rehder & Wilson, native to China, Laos, and Vietnam. Authors are agreed that *Sargentodoxa* is allied to the Lardizabalaceae, to which it was originally referred, but it is so aberrant in that family that it has become customary in recent decades to accept the monotypic family Sargentodoxaceae. *Sargentodoxa* is more primitive than the Lardizabalaceae in having numerous, spirally arranged carpels (a feature that recalls the Illiciales), and more advanced in having the carpels uniovulate. The fleshy fruiting receptacle of *Sargentodoxa* is obviously a specialization, and the deep-seated cork is another distinctive feature.

5. Family LARDIZABALACEAE Decaisne 1838 nom. conserv., the Lardizabala Family

Twining woody vines, or sometimes (*Decaisnea*) straight-stemmed arborescent shrubs, commonly accumulating triterpenoid saponins, and usually strongly tanniferous, sometimes with proanthocyanins; young stems with a ring of vascular bundles separated by well developed, usually lignified medullary rays; nodes trilacunar; solitary (rarely clustered) crystals of calcium oxalate commonly present in some of the parenchyma-cells; vessel-segments with simple perforations, or seldom

FIG. 1.23 Lardizabalaceae. *Akebia trifoliata* (Thunb.) Koidz. a, habit, ×½; b, c, staminate flower, from above and from the side, ×6; e, stamens and pistillodes, ×6; f, pistillate flower, ×6; g, carpel, ×12; h, carpel, in long-section, ×12; i, carpel, in cross-section, ×12; k, staminodes from pistillate flower, ×12.

(mainly *Decaisnea*) with scalariform perforation-plates and few cross-bars; imperforate tracheary elements with bordered pits (considered to be tracheids), or rarely (*Holboellia*) with simple pits; wood-parenchyma sparse or none; cork arising superficially; sieve-tubes with S-type plastids. LEAVES alternate, palmately or seldom (*Decaisnea*) pinnately compound, or trifoliolate; stomates anomocytic; stipules usually wanting. FLOWERS small, in racemes (these often drooping) originating from scaly axillary buds, regular, hypogynous, unisexual (the plants monoecious or dioecious), or seldom (*Decaisnea*) polygamous, trimerous, with a small receptacle; sepals 6-many, or rarely (*Akebia*) 3, commonly petaloid (but small), imbricate or the outer valvate; petals 6, smaller than the sepals, nectariferous, or wanting; staminate flowers with 6 stamens, opposite the petals (when these are present); filaments distinct or more often connate; anthers tetrasporangiate and dithecal, extrorse, opening by longitudinal slits, the pollen-sacs commonly more or less embedded in a laminar-thickened connective, which is shortly prolonged into a terminal appendage; pollen-grains binucleate, tectate-columellate, (2) 3-colpate or 3-colporate; pistillodes sometimes present in the staminate flowers; pistillate flowers with 3 (or 6–15) distinct, soon divergent carpels in 1–5 whorls of 3, with or without 6 small staminodes; carpels with a terminal, often oblique, sessile or subsessile stigma, at least sometimes (*Akebia*) conduplicate and unsealed; stigma wet, not papillate; ovules numerous (seldom few) and laminar or laminar-lateral, anatropous or campylotropous to orthotropous, bitegmic, crassinucellar; endosperm-development cellular. FRUIT wholly fleshy, often edible, indehiscent, or dehiscent along the ventral suture (i.e., a berry or a fleshy follicle), the pericarp (at least in *Decaisnea*) with a laticiferous cavity-system; seeds with small, short, straight, basal-axile, dicotyledonous embryo and abundant, usually softly fleshy endosperm with reserves of oil and sometimes also starch or hemicellulose. 2n = 28, 30, 32.

The family Lardizabalaceae consists of 8 genera and about 30 species, occurring from the Himalaya Mts. to Vietnam, southeastern China, Korea, Japan, and Taiwan, and disjunct in central Chile. The most archaic genus appears to be *Decaisnea*, which has erect stems, scalariform vessel-perforations, and polygamous flowers. The two Chilean genera (*Lardizabala* and *Boquila*), on the other hand, are dioecious climbers and appear to be fairly advanced members of the family. The largest genus is *Stauntonia*, with about a dozen species. *Akebia quinata* (Thunb.) Decaisne is grown as a porch-vine in eastern United States, and has become more or less naturalized.

6. Family MENISPERMACEAE A. L. de Jussieu 1789 nom. conserv., the Moonseed Family

Woody or herbaceous twining vines or scandent shrubs, or rarely straight-stemmed shrubs or small trees or even perennial herbs, commonly containing very bitter, poisonous sesquiterpenoids (such as

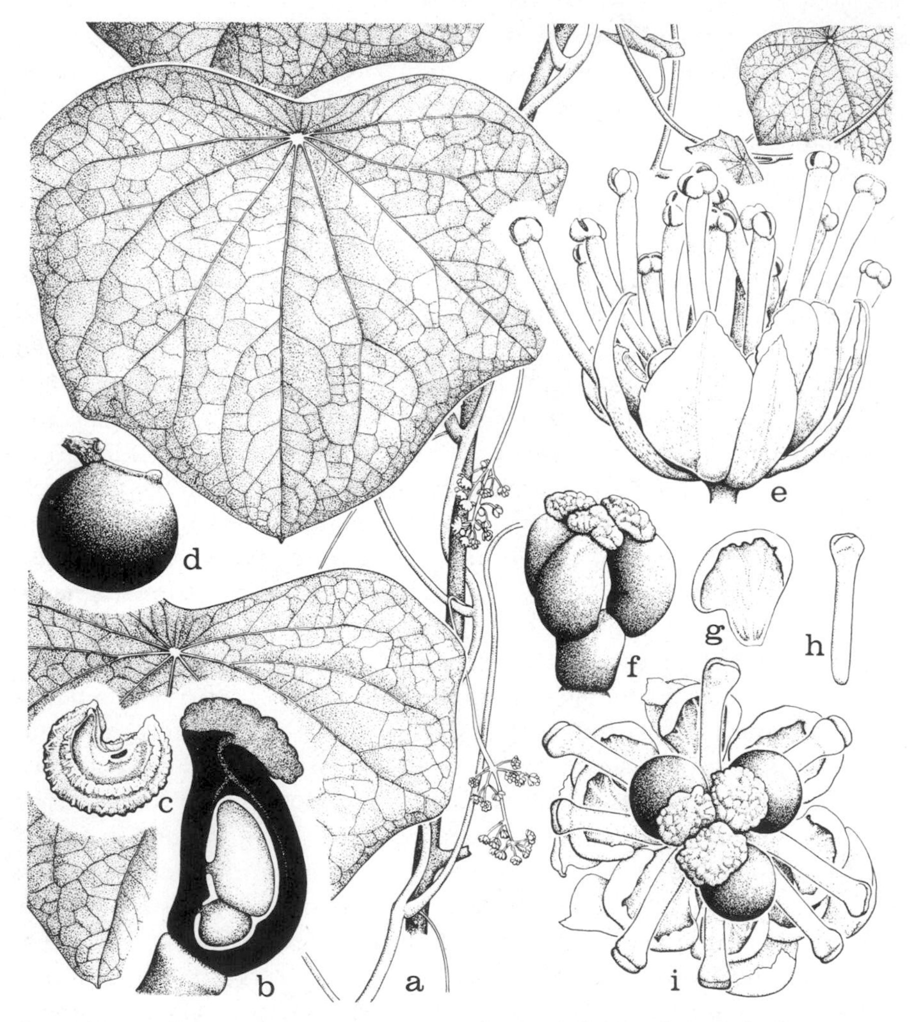

FIG. 1.24 Menispermaceae. *Menispermum canadense* L. a, habit, ×½; b, pistil, the ovary in long-section, ×12; c, seed, ×2; d, mature fruit, ×2; e, staminate flower, ×12; f, gynoecium of pistillate flower, ×12; g, petal, ×12; h, staminode of pistillate flower, ×12; i, pistillate flower, ×12.

picrotoxin) and diterpenoids, and usually with benzyl-isoquinoline alkaloids (often including berberine) and/or aporphine alkaloids; young stem with a ring of vascular bundles separated by broad medullary rays, later often with anomalous secondary thickening by one or more extrafascicular cambia that produce additional rings or arcs of bundles, the stem then commonly becoming strongly flattened through vigorous expansion at the 2 margins only; nodes trilacunar; secretory sacs with diverse sorts of contents often associated with the vascular bundles or in the parenchymatous tissues of the stem and leaves; calcium oxalate crystals of various forms (sometimes raphides) commonly present in some of the cells of the parenchymatous tissues and in the epidermis of the leaves; vessel-segments with simple perforations; imperforate tracheary elements mostly with rather indistinctly bordered pits, but some vasicentric tracheids sometimes also present; sieve-tubes with S-type plastids. LEAVES alternate, simple or rarely (as in the African genus *Syntriandrum*) trifoliolate, mostly entire, often palmately veined but only rarely lobed; inner walls of the epidermal cells often mucilaginous, or masses of mucilage present just beneath the upper epidermis; stomates of various types, most commonly paracytic or encyclocytic; stipules nearly always wanting. FLOWERS small, mostly in axillary or supra-axillary cymes or cymose panicles or pseudoracemes (the flowers fascicled at the nodes of the axis) or rarely solitary or borne on old wood, inconspicuously colored, regular or nearly so, hypogynous, ordinarily unisexual (the plants commonly dioecious), generally trimerous, commonly with 2 whorls each of sepals, petals, and stamens; sepals usually distinct (connate about half-length in *Disciphania*), imbricate or valvate, commonly 6 but sometimes only 3 or even only 2 or 1, or up to 12 or more; petals generally distinct, often biseriate, commonly 6 but sometimes more numerous or fewer or even wanting, in any case without nectaries; staminate flowers with 6 (seldom more numerous, up to about 40, or reduced to only 3 or even only 1) stamens, these commonly opposite the petals, distinct or the filaments more or less connate, sometimes forming a column; anthers tetrasporangiate and dithecal, short, opening by longitudinal or rarely transverse slits, introrse or less often extrorse, sometimes sunken into a laminar-thickened connective; pollen grains binucleate, tricolpate or tricolporate, or sometimes triporate or even tetraporate, or rarely inaperturate; pistillate flowers with (1) 3 or less often 6 (rarely up to ca 30) separate carpels in one or more whorls, often on a gynophore; stigma sessile or on a very short style; ovules in each carpel initially 2 (but one soon abortive), submarginal, pendulous, hemitropous to amphitropous (the micropyle directed upwards), bitegmic or unitegmic, crassinucellar; endosperm-development nuclear. FRUIT of drupes or sometimes nuts, in any case with a hard or bony endocarp, commonly more or less falcate-curved, the style often appearing nearly basal because of asymmetrical growth of the maturing carpel; seeds often horseshoe-shaped; embryo dicotyledonous, variously small to large, curved (sometimes even coiled) or rarely straight, with a very small radicle; endosperm ruminate or not, with reserves of oil and protein, or often scanty or none. X = 11, 12, 13, 19, 25.

The family Menispermaceae consists of some 70 genera and about 400 species, widespread in tropical and subtropical countries, with relatively few species in temperate climates. Nearly half of the species belong to the six largest genera; *Stephania* (40), *Tinospora* (35), *Abuta* (30), *Cissampelos* (30), *Tiliacora* (30), and *Cyclea* (25). *Menispermum*, the most familar genus in temperate regions, has only 2 species. Various species of several genera are used in the preparation of curare.

The Menispermaceae are generally admitted to be related to the other core families of the Ranunculales, i.e., the Lardizabalaceae, Sargentodoxaceae, Berberidaceae, and Ranunculaceae. An eventual relationship to the Illiciales is also widely assumed, but the affinity may not be so close as it at first seems. The Illiciales appear to be primitively woody, whereas the anatomy of the Menispermaceae and indeed all the woody members of the Ranunculales suggests that they are only secondarily woody. As the only clearly archaic woody Magnoliales with triaperturate pollen, the Illiciales form a sort of stepping-stone from the Magnoliales toward the Ranunculales, but there is still a gap to be crossed.

Wood resembling that of *Cocculus*, a widespread and familiar genus of Menispermaceae, occurs in Maestrichtian (late Upper Cretaceous) deposits in California. Other fossils thought to represent the Menispermaceae occur in Eocene and more recent deposits. Some endocarps from the London Clay (Eocene) are said to be unmistakably of this family.

7. Family CORIARIACEAE A. P. de Candolle 1824 nom. conserv., the Coriaria Family

Mostly shrubs, varying to suffrutescent perennial herbs or small trees, strongly tanniferous, with ellagic and often gallic acid but not proanthocyanin, containing poisonous bitter sesquiterpenoid substances resembling the picrotoxin of Menispermaceae, and commonly harboring nitrogen-fixing bacteria in nodules on the roots; leaves and branches commonly arranged so as to form flattened, frond-like sprays; twigs angular, with numerous corky lenticels; calcium oxalate crystals of various form commonly present in some of the cells of the parenchy-

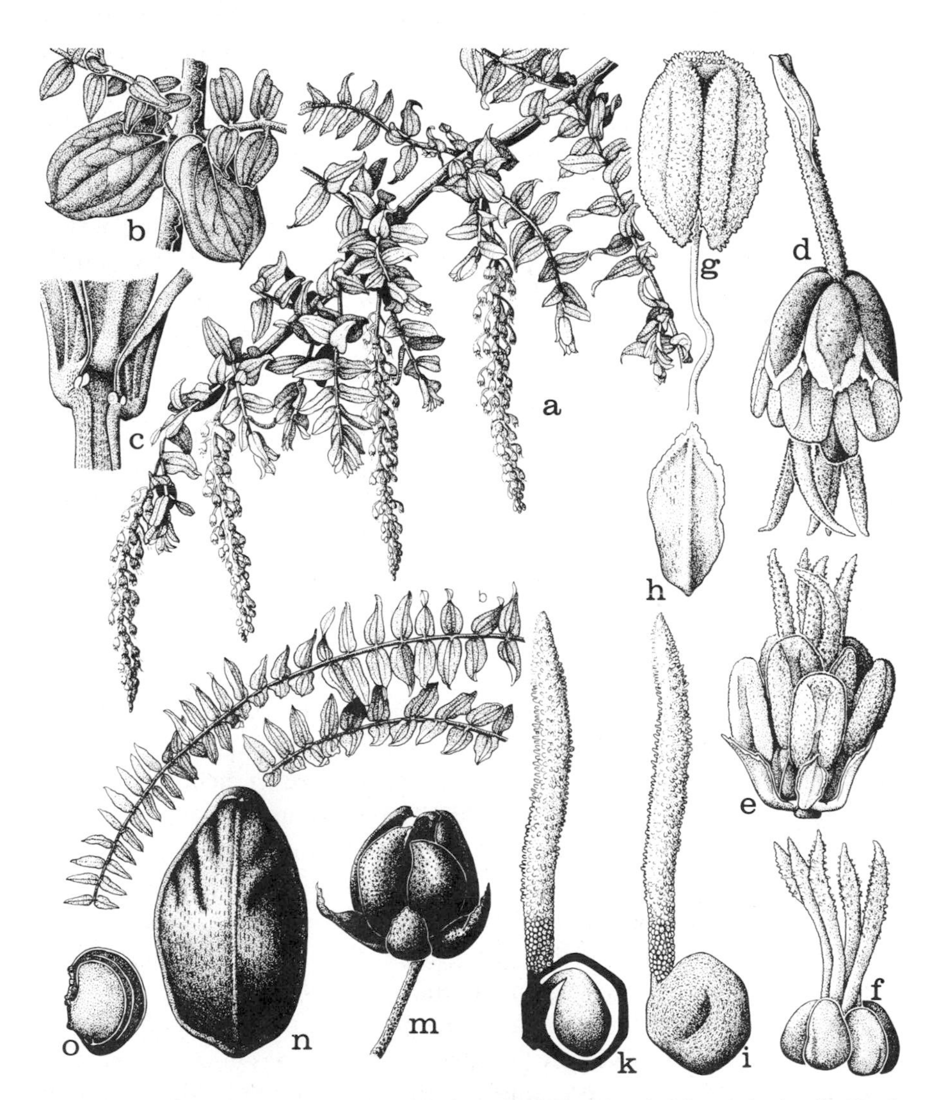

Fig. 1.25 Coriariaceae. *Coriaria thymifolia* Humb. & Bonpl. a, habit, $\times \frac{1}{2}$; b, detail of leafy twig, $\times 1$; c, detail of leaves and leaf-base, showing stipules, $\times 8$; d, flower, $\times 8$; e, flower, with the sepals removed, showing stamens in 2 series, $\times 8$; f, carpels, $\times 8$; g, stamen, $\times 16$; h, petal, $\times 16$; i, carpel, $\times 16$; k, carpel, the ovary in long-section, $\times 16$; m, flower at fruiting stage, showing persistent sepals and persistent, accrescent petals, $\times 4$; n, inner surface of petal at fruiting time, $\times 8$; o, achene, in long-section, $\times 8$.

matous tissues; young stem with large pith, a ring of vascular bundles, and lignified medullary rays; secondary wood with broad, nearly or quite homocellular rays up to 15 cells wide, without uniseriates; vessel-segments with simple perforations; imperforate tracheary elements with simple pits. LEAVES small, opposite or sometimes whorled, copiously tanniferous, simple, entire, palmately veined, with minute, often caducous stipules, stomates paracytic. FLOWERS in terminal or axillary racemes, small, regular or nearly so, hypogynous, pentamerous, perfect or sometimes some of them unisexual (the plants then polygamous); sepals 5, imbricate; petals 5, distinct, shorter than the sepals, keeled on the inner surface (in the 5-carpellate species), accrescent in fruit and becoming more or less fleshy; stamens 10, in 2 whorls, the short filaments free, or those of the antepetalous stamens adnate to the keel of the petal; anthers tetrasporangiate and dithecal, the pollen-sacs opening by longitudinal slits; pollen-grains binucleate or trinucleate, with (2) 3 (4) pores or short colpi; carpels 5 and alternate with the petals, or seldom about 10 and apparently in a single whorl, distinct or united at the base, adnate to the shortly columnar receptacle, each with a long slender, distinct style that is stigmatic all around for most of its length; stigma dry; each carpel containing a solitary, pendulous, anatropous, bitegmic, crassinucellar ovule with the micropyle directed upwards; endosperm-development nuclear. FRUIT of separate, laterally compressed, hard-walled achenes, these collectively cupped or more or less enclosed by the fleshy-accrescent petals, each achene typically subtended on each side by a half of the adjoining petal; seed somewhat compressed; embryo straight, oily, dicotyledonous; endosperm scanty or none. X probably = 10.

The family Coriariaceae consists of the single genus *Coriaria*, with 5 species forming an interrupted geographical pattern: Mexico to Chile; western Mediterranean region; Himalaya Mts. to Japan and New Guinea; New Zealand and some of the South Pacific islands.

Coriaria is a taxonomically isolated genus. It has most often been associated in one way or another with the order Sapindales as here defined, but it has no close allies there. Its wood anatomy is out of harmony with that of other families of the Sapindales (see Heimsch, 1942, under Sapindales). The pollen is also unlike that of the Sapindales, according to Praglowski (1970). The distinct carpels would also be aberrant (though not unique) in the Sapindales.

The combination of hypogynous, cyclic flowers, separate petals, bicyclic stamens, triaperturate pollen, separate carpels with the stigma running the whole length of the style, and bitegmic, crassinucellar ovules seems more compatible with the Ranunculales than with any other order. Furthermore, the anatomy of the stem, with thick pith and broad medullary rays, suggests that the plants may be secondarily woody, in accord with other woody members of the Ranunculaceae, and the small-flowered, racemose inflorescences are reminiscent of *Berberis*. The sesquiterpenoid bitter principles of *Coriaria* are similar to some of those in the Menispermaceae, and Corner (1976) considers that the structure of the seed coat resembles that of the Ranunculaceae. On the other hand, the accumulation of ellagic acid by *Coriaria* is out of harmony not only with the Ranunculales but with the Magnoliidae as a whole. A case could be made for establishing a monotypic order Coriariales, as Hutchinson has done, but the position of the order in the general system would still have to be determined. At the present state of our knowledge, I think the most reasonable position for the Coriariaceae is in the Ranunculales.

8. Family SABIACEAE Blume 1851 nom. conserv., the Sabia Family

Trees, shrubs, or woody vines, tanniferous, at least sometimes with proanthocyanins; clustered crystals of calcium oxalate commonly present in some of the parenchymatous tissues; vessel-segments with simple perforations, or with scalariform or reticulate perforation-plates; imperforate tracheary elements with simple or bordered pits, sometimes septate; wood-rays heterocellular to nearly homocellular, sometimes wholly uniseriate, more often mixed uniseriate and pluriseriate, the latter up to 3–9 cells wide; wood-parenchyma scanty and paratracheal or vasicentric, or wanting. LEAVES alternate, simple or pinnately compound; stomates anomocytic or paracytic; stipules wanting. FLOWERS in terminal or axillary panicles (or mixed panicles), small, perfect or some of them unisexual (the plants then polygamo-dioecious); sepals (3–) 5, unequal, imbricate, distinct or connate at the base; petals (4–) 5, opposite or alternate with the sepals, the 2 inner ones often distinctly smaller than the others; stamens (including staminodes) as many as and opposite the petals, all polliniferous (*Sabia*) or only the 2 inner ones (opposite the reduced petals) polliniferous and the others staminodial; pollen-grains binucleate, tricolporate; nectary-disk small, annular, surrounding the base of the ovary; gynoecium of 2 (3) carpels united to form a compound, superior ovary, the styles more or less connate; locules 2

(3), each with 1 or 2 pendulous or horizontal, axile, hemitropous, unitegmic, crassinucellar ovules; endosperm-development helobial. FRUIT unilocular or sometimes more or less dicoccous, drupaceous or dry and indehiscent; seeds with very scanty or no endosperm and a large oily embryo with a curved radicle and 2 folded or coiled cotyledons. 2n = 24 (*Sabia*), 32 (*Meliosma*). (Meliosmaceae)

The family Sabiaceae consists of 3 genera and about 60 species, occurring mainly in southeastern Asia (north to Korea and Japan) and in tropical America (north to Mexico). *Sabia* has about 30 species, *Meliosma* about 25, and *Ophiocaryon* about 7. The embryo of *Ophiocaryon paradoxum* Schomb. ex Hook. has the aspect of a coiled snake, and the seeds have been imported from South America to Europe as a novelty under the name snake-seed or snake-nut.

The position of the Sabiaceae has been inconclusively debated. Some authors associate them with the Menispermaceae, others with one or another family of the Sapindales as here defined. Pollen morphology (Erdtman, 1952) and embryology (Mauritzon, 1936) have been interpreted to favor a relationship with the Menispermaceae. Heimsch (1942, under Sapindales) considered that the wood anatomy would permit a relationship to some members of the Anacardiaceae, even though the wood is distinctly more primitive than that of the bulk of the Anacardiaceae and other members of the Sapindales. Jack Wolfe tells me that the leaf-venation of *Sabia* is highly compatible with a position near the Menispermaceae, but that *Meliosma* is more similar in this regard to some members of the Rosidae. The assignment of the Sabiaceae to the order Ranunculales in the present work must be regarded as tentative.

Leaves considered to represent the modern genus *Meliosma* have been found in late Paleocene deposits in North Dakota, and from the Eocene to the present in both Asia and North America. Some endocarps from the London Clay (Eocene) are confidently referred to the Sabiaceae.

8. Order PAPAVERALES[1]

Herbs or seldom half-shrubs, shrubs, or even arborescent (but soft-wooded) shrubs, producing diverse sorts of isoquinoline alkaloids (including benzyl-isoquinoline and aporphine types, and most notably protopine) in laticifers or elongate secretory idioblasts; plants not tanniferous, lacking both proanthocyanins and ellagic acid; vascular bundles in one or sometimes 2 or more rings, separated by broad, multiseriate medullary rays even in woody species; vessel-segments with simple perforations; imperforate tracheary elements very short, with small, simple pits; sieve-tube plastids of S-type. LEAVES alternate (sometimes all basal), or sometimes (especially the distal ones) subopposite or almost whorled, entire to more often lobed, compound, or dissected; stomates anomocytic; stipules none. FLOWERS solitary or in variously cymose, racemose, or paniculiform inflorescences, often large and showy, perfect, regular to strongly irregular, hypogynous or seldom perigynous; sepals 2 or less often 3 or even 4, commonly caducous; petals usually twice as many as the sepals and bicyclic, most commonly 4 or 6, but sometimes 8–12 or even 16, distinct or apically connate or connivent, or rarely wanting; stamens 4-many, distinct or diadelphous, when numerous originating in centripetal sequence; pollen-grains binucleate or trinucleate, triaperturate to multiaperturate, seldom biaperturate, only rarely with endo-apertures; gynoecium of 2 or less often 3 or more carpels united to form a compound, typically unilocular ovary with or without a style and with as many stigmas or stigma-lobes as carpels, or the stigma several-lobed above a bicarpellate ovary in some Fumariaceae; ovules (1) 2-many, anatropous to often amphitropous or campylotropous, bitegmic, crassinucellar, commonly on parietal placentas, but the placentas sometimes intruded as partial partitions or even meeting and joined in the center so that the ovary is fully plurilocular with axile placentation; endosperm-development nuclear. FRUIT typically capsular and longitudinally (but often incompletely) dehiscent by valves alternating with the placentas, or the openings sometimes reduced to pores, seldom opening by separation of the mature carpels, or nutlike and indehiscent or breaking transversely into 1-seeded segments; seeds with small to linear and elongate, axile, dicotyledonous embryo and abundant, oily endosperm, very often arillate.

The order Papaverales as here defined consists of 2 families of

[1] I am not sure to whom this name should be attributed, nor even whether it has yet been properly validated, in spite of its widespread recent use. It is here validated by a eference to Papaveraceae A. L. de Jussieu, Gen. Pl. 235. 1789.

somewhat unequal size, the Papaveraceae and Fumariaceae, with 600 or more species in all. Both families have often been included in a single more broadly defined family Papaveraceae. It is generally agreed that, aside from the genera *Hypecoum* and *Pteridophyllum*, the genera of the Papaverales fall into two sharply defined, readily recognizable groups.

Hypecoum and *Pteridophyllum*, although unlike in aspect, share several features in common. They differ from the remainder of the order, and are alike inter se, in having only 4 stamens (these free), and in having biaperturate pollen, but the pollen of the two genera differs in some other details. *Hypecoum* and *Pteridophyllum* resemble the Fumariaceae in lacking latex and laticifers, in having nectaries at the base of the stamens, and in having small sepals that do not enclose the developing bud. They are more like the Papaveraceae in having nearly regular flowers, the two sets of petals differing only slightly from each other. The small number of stamens is more readily compatible with the Fumariaceae, which have 6 stamens, but the further complex specialization of the androecium that marks the Fumariaceae is absent in these two genera.

Hypecoum and *Pteridophyllum* have variously been referred to the Papaveraceae (or Papaveroideae) or Fumariaceae (or Fumarioideae), or grouped together as a distinct family Hypecoaceae (or subfamily Hypecoideae), or recognized as two separate families Hypercoaceae and Pteridophyllaceae. All of these treatments can be defended.

Here as elsewhere I am reluctant to recognize small segregate families when other reasonable choices are available. *Hypecoum* and *Pteridophyllum* are aberrant in either the Papaveraceae or Fumariaceae, but in the context of the order as a whole I do not think they are so distinctive as to demand familial segregation. I choose to include them in the Fumariaceae and define and describe the families accordingly, but at the present state of our knowledge their inclusion in the Papaveraceae instead would be almost equally acceptable. Based on a comparison of modern members of the order, it seems reasonable to suppose that there was an early evolutionary dichotomy of the Papaverales into a papaveraceous and fumariaceous evolutionary line, and that *Hypecoum* and *Pteridophyllum* represent branchlets of an early side-branch from the fumariaceous line. I am well aware that such a phylogeny would be compatible with more than one taxonomic interpretation.

For many years it was customary to include the Papaverales and Capparales as here defined in a single large order. usually under the name Rhoeadales. Evidence that has accumulated in the past several decades has led most students to separate the two groups, and to associate the Papaverales with the Ranunculales. The Papaverales are rich in the benzyl-isoquinoline and aporphine-isoquinoline alkaloids that characterize the Ranunculales and other Magnoliidae, and diverse species have particular alkaloids, such as berberine, in common with members of the Ranunculales. The Capparales are poor in alkaloids, and lack isoquinoline alkaloids entirely. The Papaverales have laticifers or elongate secretory cells in which the alkaloids are borne. These are wanting from the Capparales. The Capparales, on the other hand, have scattered cells that produce myrosin, an enzyme involved in the formation of mustard oil. Myrosin cells and mustard oil are largely (though not quite entirely) confined to the Capparales, and are wholly wanting from the Papaverales. Takhtajan (1966) concludes that "biochemically they [Capparales and Papaverales] have nothing in common" (my translation). Serological studies (Jensen, 1967) also indicate a relationship of the Papaverales to the Ranunculales, and the isolation of the Capparales from both of these orders. The stamens of the Papaverales originate centripetally, when a sequence can be determined, whereas those of the Capparales originate centrifugally. The pollen of the Papaverales is considered by palynologists to be similar to and probably derived from that of the Ranunculales, and different from that of the Capparales. Significant embryological differences have also been observed. There no longer seems to be any reason to doubt that the Papaverales should be dissociated from the Capparales and associated with the Ranunculales.

SELECTED REFERENCES

Arber, A. 1931, 1932. Studies in floral morphology. III. On the Fumarioideae, with special reference to the androecium. IV. On the Hypecoideae, with special reference to the androecium. New Phytol. 30: 317–354, 1931. 31: 145–173, 1932.

Arber, A. 1938. Studies in flower structure. IV. On the gynaeceum of *Papaver* and related genera. Ann. Bot. (London) II. 2: 649–663.

Berg, R. Y. 1969. Adaptation and evolution in *Dicentra* (Fumariaceae) with special reference to seed, fruit, and dispersal mechanism. Nytt Mag. Bot. 16: 49–75.

Bersillon, G. 1955. Recherches sur les Papavéracées; contribution a l'étude du développement des dicotylédones herbacées. Ann. Sci. Nat. Bot. sér. 11. 16: 225–447.

Dickson, J. 1935. Studies in floral anatomy. II. The floral anatomy of *Glaucium flavum* with reference to other members of the Papaveraceae. J. Linn. Soc., Bot. 50: 175–224.

Ernst, W. R. 1962. The genera of Papaveraceae and Fumariaceae in the southeastern United States. J. Arnold Arbor. 43: 315–343.

Ernst, W. R. 1967. Floral morphology and systematics of *Platystemon* and its allies *Hesperomecon* and *Meconella* (Papaveraceae: Platystemonoideae). Univ. Kansas Sci. Bull. 47: 25–70.

Frohne, D. 1962. Das Verhältnis von vergleichender Serobotanik zu vergleichender Phytochemie, dargestallt an serologischen Untersuchungen im Bereich der "Rhoeadales". Pl. Med. 10: 283–297.

Günther, K.-F. 1975. Beiträge zur Morphologie und Verbreitung der Papaveraceae. Flora 164: 185–234; 393–436.

Hegnauer, R. 1961. Die Gliederung der Rhoeadales sensu Wettstein in Lichte der Inhaltstoffe. Pl. Med. 9: 37–46.

Jensen, U. 1967 (1968). Serologische Beiträge zur Frage der Verwandtschaft zwischen Ranunculaceen und Papaveraceen. Ber. Deutsch. Bot. Ges. 80: 621–624.

Norris, T. 1941. Torus anatomy and nectary characteristics as phylogenetic criteria in the Rhoeadales. Amer. J. Bot. 28: 101–113.

Röder, I. 1958. Anatomische und fluoreszenzoptische Untersuchungen an Samen von Papaveraceen. Oesterr. Bot. Z. 104: 370–381.

Ryberg, M. 1960. A morphological study of the Fumariaceae and the taxonomic significance of the characters examined. Acta Horti Berg. 19: 122–248.

Sachar, R. C., & H. Y. Mohan Ram. 1958. The embryology of *Eschscholzia californica* Cham. Phytomorphology 8: 114–124.

Saksena, H. B. 1954. Floral morphology and embryology of *Fumaria parviflora* Lamk. Phytomorphology 4: 409–417.

Souèges, R. 1943. Embryogénie des Fumariacées. La différenciation des régions fondamentales du corpe chez l'*Hypecoum procumbens* L. Compt. Rend. Hebd. Séances Acad. Sci. 216: 354–356.

Sugiura, T. 1940. Chromosome studies on Papaveraceae with special reference to the phylogeny. Cytologia 10: 558–576.

Попов, М, Г. 1957. О взаимоотношениях и истории родов *Papaver* и *Roemeria*. Бот. Ж. 42: 1389-1397.

Сардулаева, А. Л. 1959. Морфология пыльцы семейства маковых (Papaveraceae). Проблемы Ботаники 4: 11-50.

Цатурян, Т. Г. 1951. О пыльце кавказских представителей семейства маковых (Papaveraceae). Научн. Труды Ереванск. Гос. Унив., XXXIII, Биол. сер., 2: 63-75.

SYNOPTICAL ARRANGEMENT OF THE FAMILIES OF PAPAVERALES

1 Sepals fully enclosing the bud before anthesis; flowers regular; stamens distinct, numerous (only 4–6 in spp. of *Meconella*); nectaries wanting; plants producing milky or colored latex, often in articulated laticifers .... 1. PAPAVERACEAE.

1 Sepals small, not enclosing the developing bud; flowers strongly irregular, except in *Hypecoum* and *Pteridophyllum*; stamens 6 and diadelphous, or (in *Hypecoum* and *Pteridophyllum*) 4 and distinct; plants without latex or laticifers, but commonly with elongate secretory cells ....2. FUMARIACEAE.

1. Family PAPAVERACEAE A. L. de Jussieu 1789 nom. conserv., the Poppy Family

Herbs, or less commonly half-shrubs or even arborescent (but soft-wooded) shrubs, bearing diverse sorts of isoquinoline alkaloids (including benzyl-isoquinoline and aporphine types, and most notably protopine) in articulated, anastomosing or non-anastomosing laticifers or less commonly in elongate latex-cells that are solitary or arranged in longitudinal rows; plants sometimes cyanogenic through a tyrosine-based pathway, but not tanniferous, lacking both proanthocyanins and ellagic acid; nodes unilacunar or trilacunar; vascular bundles in one or sometimes 2 or more rings, separated by broad, multiseriate, heterocellular rays even in woody species; vessels with simple perforations; imperforate tracheary elements very short, with small, simple pits; sieve-tube plastids of S-type. LEAVES entire to more often lobed or dissected, basically alternate, but the distal ones sometimes subopposite or almost whorled; stomates anomocytic; stipules none. FLOWERS solitary or less often in cymose or umbelliform or seldom paniculiform inflorescences, generally rather large, perfect, regular, hypogynous or seldom (*Eschscholzia*) perigynous; sepals 2 or less often 3 or even 4, distinct or seldom more or less connate, tending to be asymmetrical, and with specialized margins, fully enclosing the bud before anthesis, commonly caducous; petals usually twice as many as the sepals and bicyclic, most commonly 4 or 6, but sometimes 8–12 or even 16, distinct, in bud imbricate and often crumpled, or rarely (*Macleaya*) wanting; stamens more or less numerous (only 4–6 in spp. of *Meconella*), originating in centripetal sequence, distinct, often in multiples of 2 (when the perianth is dimerous) or 3 (when the perianth is trimerous); anthers tetrasporangiate and dithecal; pollen-grains binucleate or trinucleate, tricolporate to pantoporate; nectaries wanting; gynoecium of 2 or more carpels united to form a superior, typically unilocular ovary with or without a style, and with the stigmas (as many as the carpels) commonly connate laterally to form a more or less discoid, often lobed structure, but in Platystemonoideae (only 5 spp.) the stigmas discrete, and in *Platystemon* (1 sp.) the carpels more or less numerous in a single whorl, conduplicate and lightly connate by their margins around a central cavity; stigmas dry, papillate; ovules usually numerous (rarely solitary), anatropous to amphitropous or more or less campylotropous, bitegmic, crassinucellar, commonly on parietal placentas, but the placentas sometimes intruded as partial partitions or even meeting and joined in the center so that the

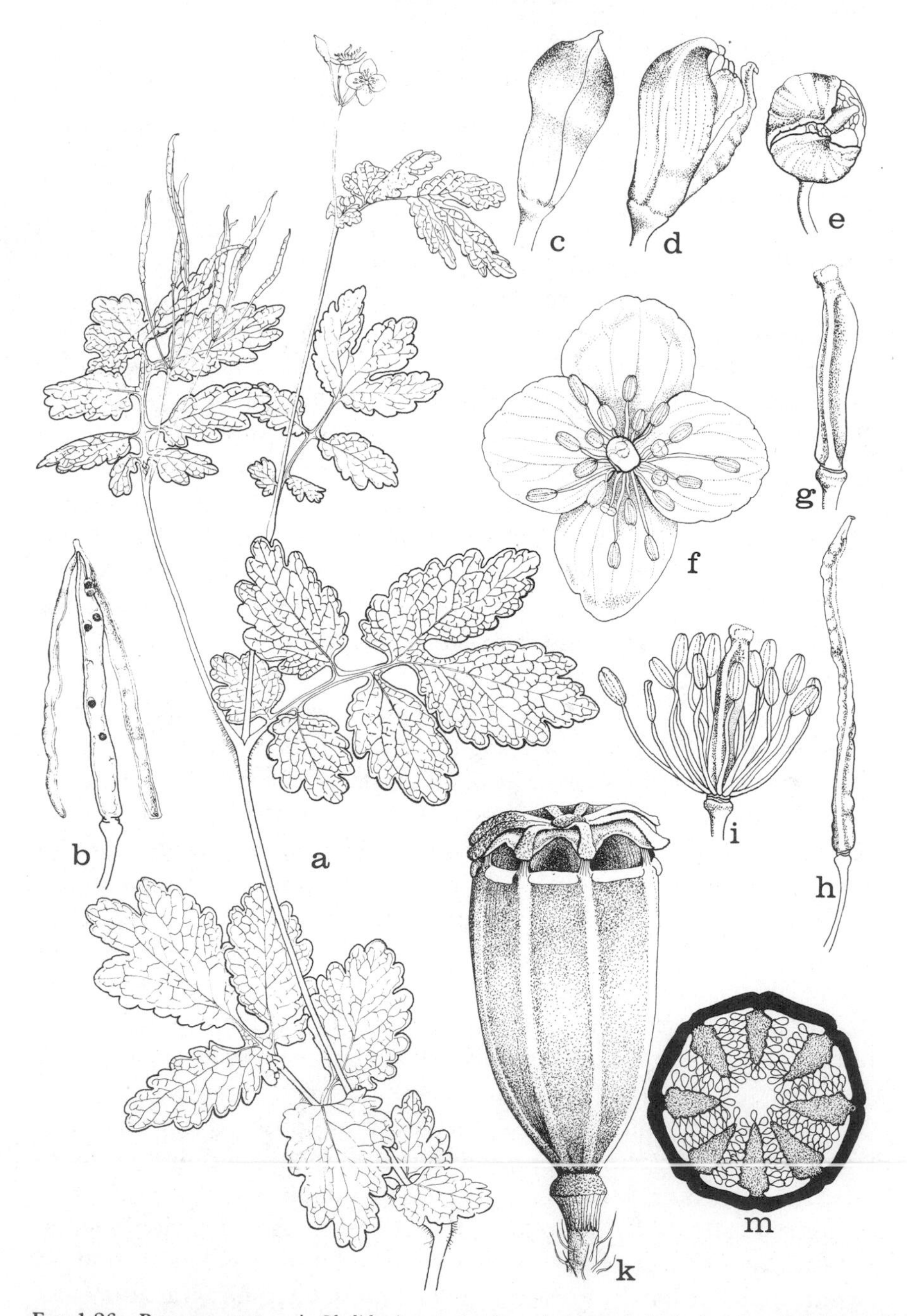

FIG. 1.26 Papaveraceae. a-i, *Chelidonium majus* L. a, habit, ×½; b, fruit, after dehiscence, showing valves and persistent replum, ×1; c, flower bud, ×4; d, e, opening flower bud, ×4; f, flower, ×4; g, pistil, ×4; h, fruit, before dehiscence, ×1; i, flower, with perianth removed, ×4; k, m, *Papaver orientale* L.; k, fruit, ×1; m, cross-section of ovary, ×4.

ovary is fully plurilocular with axile placentation; endosperm-development nuclear. FRUIT a capsule, longitudinally (but often incompletely) dehiscent by valves alternating with the placentas (which may form a replum), or the openings sometimes reduced to pores, or in the small subfamily Platystemonoideae the carpels separating at maturity, and in *Platystemon* the carpels fragmenting transversely into 1-seeded nutlets; seeds with small to linear and elongate, axile, dicotyledonous embryo and abundant, oily endosperm, sometimes arillate. X most commonly =6 or 7, less often 5, 8–11, or 19. (Chelidoniaceae, Eschscholziaceae, Platystemonaceae)

The family Papaveraceae as here defined consists of some 25 genera and about 200 species, occurring mainly in temperate and tropical parts of the Northern Hemisphere. About half of the species belong to the genus *Papaver*. The large number of species sometimes attributed to the Pacific North American genus *Eschscholzia* reflects excessive splitting rather than inherent taxonomic complexity. The family provides a number of showy garden ornamentals, especially of the genus *Papaver*, but its greatest economic importance comes from the fact that *Papaver somniferum* L. is the source of opium.

2. Family FUMARIACEAE A. P. de Candolle 1821 nom. conserv., the Fumitory Family

Perennial herbs, commonly transporting nitrogen-compounds as acetyl-ornithine, producing diverse sorts of isoquinoline alkaloids (including benzyl-isoquinoline and aporphine types, and most notably protopine) in secretory, generally elongate idioblasts, but without latex; plants not tanniferous, lacking both proanthocyanins and ellagic acid; nodes unilacunar or less often trilacunar; vessel-segments with simple perforations; imperforate tracheary elements very short, with small, simple pits; sieve tube plastids of S-type. LEAVES alternate (sometimes all basal) or sometimes subopposite, usually more or less dissected; stomates anomocytic; stipules wanting. FLOWERS (1) few-several in cymose or racemose inflorescences, perfect, hypogynous, strongly irregular or sometimes nearly regular (*Hypecoum*, *Pteridophyllum*); sepals 2, small, symmetrical, bract-like, often more or less peltate, not enclosing the developing bud, commonly caducous; petals 4, in 2 series of 2 each, in *Hypecoum* and *Pteridophyllum* nearly alike, the inner a little smaller and/or less deeply trilobed than the outer, in the other genera one or both of the outer petals with a prominent basal spur or pouch, and the inner

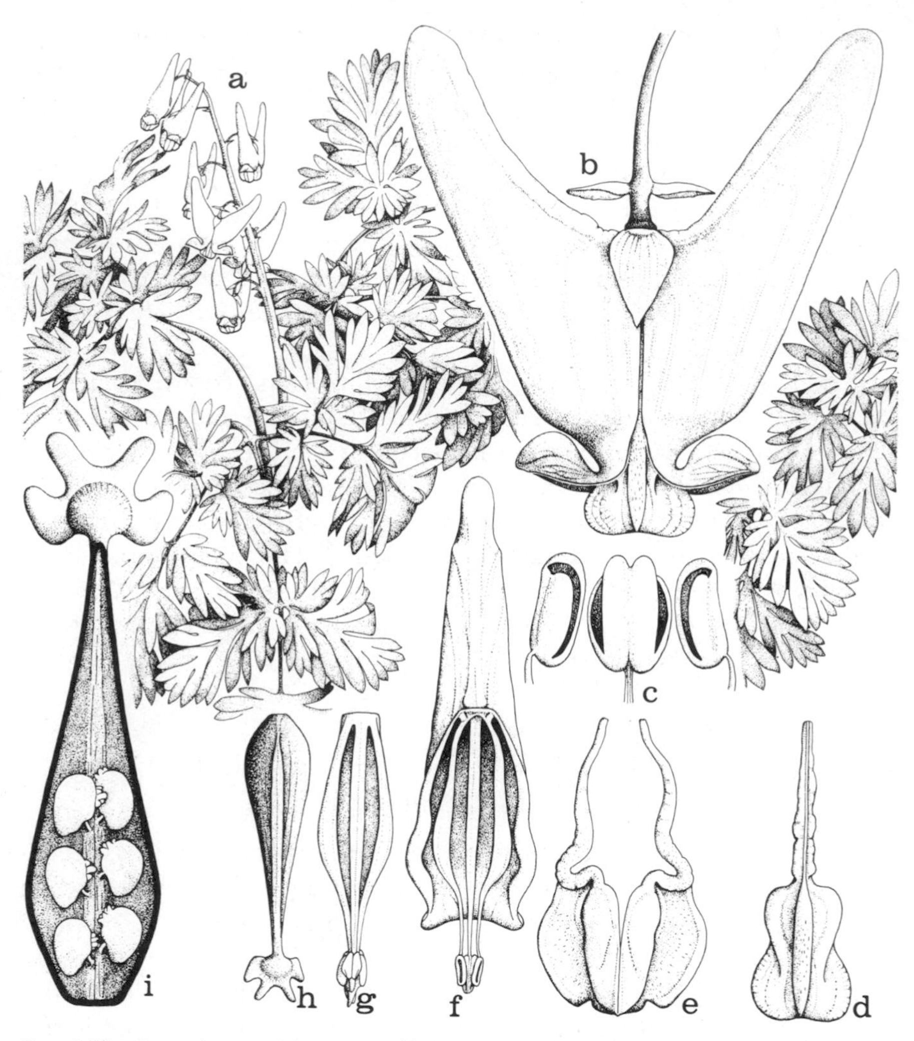

FIG. 1.27 Fumariaceae. *Dicentra cucullaria* (L.) Bernh. a, habit, ×½; b, flower, ×4; c, stamens, ×4; d, e, inner petals, ×4; f, outer petal, with 3 associated stamens, ×4; g, pistil, with connivent stamens, ×4; h, pistil, with stamens removed, ×4; i, pistil, the ovary in partial section, ×8.

petals generally more or less connate or sticky-connivent over the stigmas at the tip; stamens in most genera 6 and diadelphous into a pair of lateral phalanges opposite the outer petals, but 4 and distinct in *Hypecoum* and *Pteridophyllum*; anthers (except in *Hypecoum* and *Pteridophyllum*) dimorphic, the middle one of each phalanx fully developed, extrorse, with 4 microsporangia and 2 pollen-sacs, and the lateral ones with only 2 microsporangia and a single pollen-sac; one or more nectaries, often spur-like in form, commonly present at the base of the androecium; pollen-grains binucleate, dicolpate (in *Hypecoum* and *Pteridophyllum*) or more often tricolpate to pantoporate; gynoecium of 2 carpels united to form a compound, unilocular ovary with a usually slender (sometimes short) style and 2 stigmas or a 2- to several-lobed stigma; stigmas wet, not papillate; ovules 2-many, on parietal placentas, anatropous to campylotropous, bitegmic, crassinucellar; endosperm-development nuclear. FRUIT usually capsular, often with a replum, longitudinally (often incompletely) dehiscent, or seldom nutlike and indehiscent or breaking transversely into one-seeded joints; seeds generally arillate, with small, dicotyledonous embryo and abundant, soft, oily endosperm. X most commonly 8, seldom 6, 7, or other numbers. (Hypecoaceae, Pteridophyllaceae)

The family Fumariaceae as here defined consists of about 19 genera and 400 or more species, occurring mainly in North Temperate regions, but also in South Africa. The largest genus by far is *Corydalis*, which has nearly 300 species. The next largest genus, *Fumaria*, has about 50. Several members of the family are familiar garden ornamentals, most notably *Dicentra spectabilis* (L.) Lemaire, bleeding heart, and *D. cucullaria* (L.) Bernh., Dutchman's breeches.

II. Subclass HAMAMELIDAE[1] Takhtajan 1966

Woody or herbaceous dicotyledons, generally more or less strongly tanniferous, commonly with proanthocyanins and often with ellagic and gallic acids as well, but only seldom with alkaloids or iridoid compounds; vessel-segments with scalariform or simple perforations, or vessels sometimes wanting; sieve-tubes with S-type plastids. LEAVES simple or less often pinnately or palmately compound. FLOWERS typically anemophilous, but not infrequently (presumably only secondarily) entomophilous, mostly small and inconspicuous (considered to be reduced), commonly apetalous and often wholly without perianth, the sepals when present small and often scale-like, the petals when present distinct, small and inconspicuous except in some of the Hamamelidaceae, never very large and showy; stamens (1) 2-several or sometimes more or less numerous, often with a prolonged connective, but not laminar; pollen-grains binucleate or trinucleate, (2) 3-many-aperturate (colpate to often porate), the amentiferous orders mostly porate and often with a granular rather than columellar infratectal structure; gynoecium of 1-several distinct carpels or more often the carpels more or less firmly united to form a compound ovary, or the ovary seemingly of a single carpel but actually pseudomonomerous; placentation marginal or laminar-lateral to more often axile or apical or basal; ovules anatropous to orthotropous, more or less distinctly crassinucellar, bitegmic or less often unitegmic; fertilization porogamous, or in the amentiferous orders often chalazogamous; endosperm present or absent; embryo sometimes very small and with 2 poorly differentiated cotyledons, but more often well developed and evidently dicotyledonous.

The subclass Hamamelidae as here defined consists of 11 orders, 24 families, and about 3400 species. Nearly two-thirds of the species belong to the order Urticales, and another quarter to the Fagales. The other 9 orders have less than 300 species together. Like the Magnoliidae, the Hamamelidae have a large proportion of small, taxonomically isolated, presumably relictual families.

The Hamamelidae can be traced back through the platanoid line to near the middle of the Albian (final) stage of the Lower Cretaceous

[1] I am well aware that the pleonastic spelling Hamamelididae would be orthographically preferable.

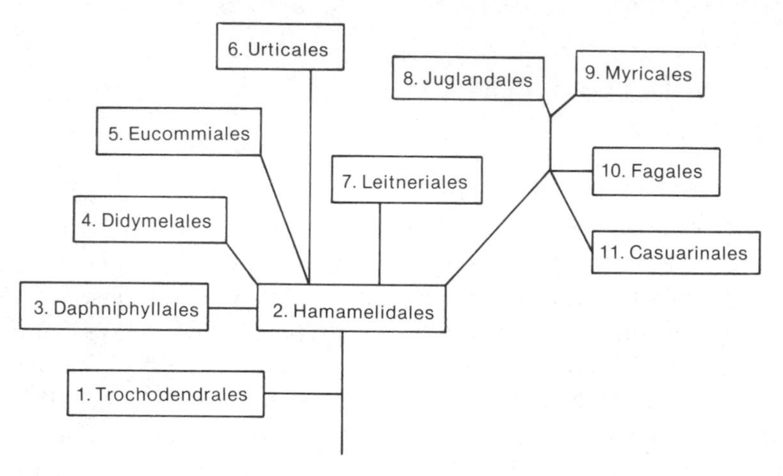

FIG. 2.1 Putative evolutionary relationships among the orders of Hamamelidae.

period, more than 100 million years ago. While the platanoids were differentiating from the ancestral magnoliids (or pre-magnoliids), a sister-group marked by pinnatifid and eventually pinnately compound leaves was differentiating as the probable ancestors of the modern Rosidae (fide Hickey & Doyle, 1977). As here interpreted, the two emerging subclasses did not become clearly distinct from each other until the appearance of the *Normapolles* pollen-complex near the middle of the Cenomanian (earliest) stage of the Upper Cretaceous.

The small size, smooth surface, and porate apertures of *Normapolles* and its allies indicate that the group was adapted to wind-pollination. Anemophily in this group was evidently an evolutionary reversion, since the pollen of the earliest angiosperms was adapted to transfer by insects rather than by wind. The immediate ancestry of the *Normapolles* group is not clearly specified. On the still scanty (or incompletely understood) fossil record it could just as well have been from pre-rosids as from distinctive pre-hamamelids.

The *Normapolles* group was evidently diversified and successful in the early part of the Upper Cretaceous. It included a wide range of subtypes, some of which do and others of which do not appear to have modern descendants. The ancestry of the modern amentiferous orders of Hamamelidae appears to be firmly fixed in the *Normapolles* group. By the Maestrichtian (latest) stage of the Upper Cretaceous, 65 to 70 million years ago, pollen had evolved that appears to be referrable to the modern orders Fagales and Juglandales.

The interpretation here presented is that the Hamamelidae originated in the Albian stage as a group characterized by wind-pollination and floral reduction, possibly in a climate with alternating wet and dry seasons. The deciduous habit that would be adaptively useful in such a climate would also be compatible with the evolution of anemophily. The combination of deciduous habit and wind-pollination might pre-adapt the early Hamamelidae to invasion of temperate regions, before these had become fully occupied by other groups of angiosperms. With the evolution of pollen more specifically adapted to transfer by wind, the group proliferated and diversified, as shown in the fossil record of the *Normapolles* complex. Later in the Upper Cretaceous the group declined, possibly as a result of increasing adaptation of insect-pollinated Rosidae (and Dilleniidae?) to similar habitats. The continuing co-evolution of insects and entomophilous angiosperms might reasonably be expected to reduce and eventually overcome any initial advantage that the anemophilous Hamamelidae had acquired in the early part of the Upper Cretaceous. Whatever the balance may be among the controlling factors, the result is that the modern Hamamelidae consist of a few still highly successful groups, plus a series of isolated end-lines.

I am receptive to current speculation that the origin and early expansion of the Hamamelidae took place in Laurasia rather than Gondwanaland.

This interpretation of the early evolutionary history of the Hamamelidae is not entirely new. It draws heavily on the work of Hickey and Doyle (e.g., 1977, in general citations), Wolfe (1973), Whitehead (1969), Axelrod (1966), Schuster (personal conversation), and others, and part of it was presented in my 1968 book. None of these botanists is to be held responsible for the use I have made of their data and ideas.

At some stage in their evolution the Hamamelidae and the Rosidae began to exploit tannins, notably hydrolyzable tannins such as ellagic acid, as an important defense against predators. Only a few of the modern Magnoliidae are significantly tanniferous, and almost none of them produces ellagic acid. In the modern Rosidae tannins are much more important among the archaic families than among the more advanced ones. One may reasonably suppose that the Hamamelidae and Rosidae took up hydrolyzable tannins as a defense when the two groups were still in the nascent stage of differentiation, and that the new weapon had something to do with their rise at the expense of their magnoliid precursors. The possible evolutionary history of the Dilleniidae cannot yet be integrated into this speculative account because of our still insufficient understanding of the fossil record.

SELECTED REFERENCES

Abbe, E. C. 1974. Flowers and inflorescences of the "Amentiferae". Bot. Rev. 40: 159–261.

Axelrod, D. I. 1966. Origin of deciduous and evergreen habits in temperate forests. Evolution 20: 1–15.

Endress, P. K. 1977. Evolutionary trends in the Hamamelidales-Fagales-group. Pl. Syst. Evol. Suppl. 1: 321–347.

Hjelmqvist, H. 1948. Studies on the floral morphology and phylogeny of the Amentiferae. Bot. Not., Suppl. 2 (No. 1): 1–171.

Janchen, E. 1950. Die Herkunft der Angiospermen-Blüte und die systematische Stellung der Apetalen. Oesterr. Bot. Z. 97: 129–167.

Mears, J. A. 1973 (1974). Chemical constituents and systematics of Amentiferae. Brittonia 25: 385–394.

Moseley, M. F. 1973 (1974). Vegetative anatomy and morphology of Amentiferae. Brittonia 25: 356–370.

Petersen, A. E. 1953. A comparison of the secondary xylem elements of certain species of the Amentiferae and Ranales. Bull. Torrey Bot. Club 80: 365–384.

Stern, W. L. 1973 (1974). Development of the amentiferous concept. Brittonia 25: 316–333.

Stone, D. E. 1973 (1974). Patterns in the evolution of amentiferous fruits. Brittonia 25: 371–384.

Thorne, R. F. 1973 (1974). The "Amentiferae" or Hamamelidae as an artificial group: a summary statement. Brittonia 25: 395–405.

Wolfe, J. 1973 (1974). Fossil forms of Amentiferae. Brittonia 25: 334–355.

Колобкова, Е. В. 1972. Сравительное изучение альбуминов и глобулинов семян сережкоцветных. В сб.: Биохимия и филогения растений, Е. В. Колобкова, Редактор.: 37-48. Изд. Наука, Москва 1972.

Куприянова, Л. А. 1962. Палинологические данные к систематике порядков Fagales и Urticales. В сб.: К Первой Международной Палинологической Конференции (Таксон, США). Докл. Совет. Палинологов.: 17-25 Изд. Академии Наук СССР. Москва.

Куприянова, Л. А. 1965. Палинология сережкоцветных. Изд. Наука. Москва-Ленинград.

Чупов, В. С. 1978. Сравительное иммуноэлектрофоретическое исследование белков пыльцы некоторых сережкоцветных. Бот. Ж. 63: 1579-1585.

SYNOPTICAL ARRANGEMENT OF THE ORDERS OF HAMAMELIDAE

1 Pistillate (or perfect) flowers either producing more than one fruit, or producing a single dehiscent fruit; ovules 1-many per carpel.

2 Vessels wanting; leaves with unique, elongate (often branched) idioblasts; flowers, or some of them, perfect; carpels 4 or more, laterally coherent, collectively ripening into a follicetum; ovules

several to many in each carpel; stipules wanting or represented only by a pair of flanges on the petiole . 1. TROCHODENDRALES.

2 Vessels present; no conspicuous, elongate idioblasts; flowers perfect to more often unisexual; carpels 2-several, distinct or more or less united to form a compound pistil, ripening into a capsule or into several separate fruits; ovules 1-many; stipules present or less commonly absent .... 2. HAMAMELIDALES.

1 Pistillate (or perfect) flowers producing a single indehiscent fruit; ovules not more than 2 per carpel.

3 Embryo minute; endosperm copious; ovary bilocular, with 2 ovules in each locule .... 3. DAPHNIPHYLLALES.

3 Embryo well developed, of normal proportions; endosperm fairly abundant to often wanting; ovary various.

4 Flowers not in aments; ovary unilocular or rarely bilocular; calyx often present; plants woody or herbaceous, endosperm well developed to wanting.

5 Staminate flowers with 2 stamens and very short, connate filaments; vessel-segments with scalariform perforations; stipules none; ovule solitary .... 4. DIDYMELALES.

5 Staminate (or perfect) flowers with distinct stamens, these usually but not always more than 2; vessel-segments with simple perforations.

6 Ovules in the single locule 2, unitegmic; stipules wanting; flowers solitary, without a perianth . 5. EUCOMMIALES.

6 Ovule solitary (or solitary in each of the 2 locules), bitegmic; stipules usually present; flowers clustered into inflorescences, usually with a vestigial perianth .... 6. URTICALES.

4 Flowers, or at least the staminate flowers, usually in aments (except some Fagaceae); ovary with 1-several locules; calyx mostly wanting or very much reduced; plants woody; endosperm scanty or none.

7 Pistil apparently of a single carpel, with a single style and locule; ovule solitary, pendulous; plants with well developed secretory canals in the pith, petioles, and leaf-veins .... 7. LEITNERIALES.

7 Pistil composed of more than one carpel, as shown by the presence of more than one style and often also more than one locule; plants without secretory canals.

8 Leaves opposite or alternate, seldom whorled, always more or less well developed; ovules each with a single embryo-sac.

9 Ovule solitary, orthotropous to less often hemitropous; aromatic plants with resinous-dotted leaves.

10 Leaves pinnately compound (unique in the subclass); ovary bilocular (seldom trilocular or seemingly quadrilocular) below but unilocular above, the partition not reaching the summit of the cavity, the ovule borne at the summit of the partial partition .... 8. JUGLANDALES.

10 Leaves simple (sometimes pinnately lobed); ovary fully unilocular, with a basal ovule 9. MYRICALES.

9 Ovules 2 or more, anatropous; plants not aromatic, or not strongly so, the leaves not resinous-dotted; ovary 2- to several-locular at least below, often unilocular above .... 10. FAGALES.

8 Leaves whorled, reduced to scales; ovules with multiple embryo-sacs .... 11. CASUARINALES.

1. Order TROCHODENDRALES Takhtajan 1966[1]

Trees with primitively vesselless wood, and with elongate, sometimes branched, secretory or nonsecretory (and then sclerified) idioblasts in the leaves and some of the parenchymatous tissues at least of the stem; wood-rays heterocellular, mixed uniseriate and pluriseriate; wood-parenchyma diffuse, the cells scattered and in short tangential lines; sieve-tubes with S-type plastids (at least in *Trochodendron*). LEAVES alternate, simple, toothed, exstipulate or with stipular flanges on the petiole; stomates laterocytic. FLOWERS anemophilous or secondarily entomophilous, in terminal inflorescences, individually small and inconspicuous, perfect or some of them unisexual; sepals 4 and distinct, or none, petals none; stamens 4 and opposite the sepals, or numerous; pollen-grains tricolpate or weakly tricolporate; gynoecium of 4 or more carpels in a single whorl, these laterally coherent but scarcely forming a compound pistil, ovules 5-many in each carpel, pendulous, anatropous, bitegmic, crassinucellar; endosperm-development cellular. FRUIT a follicetum, of laterally coherent follicles; seeds with thin testa and copious, oily and proteinaceous endosperm; embryo minute, undifferentiated or with 2 rudimentary cotyledons.

The order Trochodendrales consists of 2 monotypic families, the Trochodendraceae and Tetracentraceae. The order is the most archaic surviving group of Hamamelidae. On the basis of general floral morphology, structure of the pollen, and structure of the seed coat it is evidently linked to the Hamamelidales. On the other hand, its component families have often been associated with the "Ranalian complex," which is substantially equivalent to the Magnoliidae as here defined. If only the structure of the wood were considered, the Trochodendrales would surely be assigned to the Magnoliidae.

[1] A Latin diagnosis, not given by Takhtajan, is provided here:
Arbores ligno primitive sine vasibus, idioblastis elongatis secretoriis vel nonsecretoriis interdum ramosis in parenchymatis textura instructis; folia alterna simplicia, exstipulata vel petiolo alis stipulaceis instructo; flores parvi inconspicui, nonnulli vel omnes hermaphroditi; sepala 4 vel 0; petala 0; pollinis grana triaperturata; carpella 4 vel magis numerosiora, unicycla, lateraliter inter se cohaerentia sed vix pistillum verum componientia; ovule per carpellum pauca vel numerosa; fructus ex folliceto constans; semina albuminosa, embryone minimo simplici vel cotyledonibus 2. Type: Trochodendraceae Prantl in Engler & Prantl, Nat. Pflanzenfam. II. 2: 21. 1888.

SELECTED REFERENCES

Bailey, I. W., & C. G. Nast. 1945. Morphology and relationships of *Trochodendron* and *Tetracentron.* I. Stem, root, and leaf. J. Arnold Arbor. 26: 143–154.

Bondeson, W. 1952. Entwicklungsgeschichte und Bau der Spaltöffnungen bei den Gattungen *Trochodendron* Sieb. et Zucc., *Tetracentron* Oliv. und *Drimys* R. et G. Forst. Acta Horti Berg. 16: 169–218.

Croizat, L. 1947. *Trochodendron, Tetracentron* and their meaning in phylogeny. Bull. Torrey Bot. Club 74: 60–76.

Foster, A. S. 1945. The foliar sclereids of *Trochodendron aralioides* Sieb. & Zucc. J. Arnold Arbor. 26: 155–162.

Jørgensen, L. B., J. D. Møller, & P. Wagner. 1975. Secondary phloem of *Trochodendron aralioides.* Bot. Tidsskr. 69: 217–238.

Keng, H. 1959. Androdioecism in the flowers of *Trochodendron aralioides.* J. Arnold Arbor. 40: 158–160.

Nast, C. G., & I. W. Bailey. 1945. Morphology and relationships of *Trochodendron* and *Tetracentron.* II. Inflorescence, flower, and fruit. J. Arnold Arbor. 26: 267–276.

Praglowski, J. 1974. The pollen morphology of the Trochodendraceae, Tetracentraceae, Cercidiphyllaceae and Eupteliaceae with reference to taxonomy. Pollen & Spores 16: 449–467.

Smith, A. C. 1945. A taxonomic review of *Trochodendron* and *Tetracentron.* J. Arnold Arbor. 26: 123–142.

Valen, F. van. 1978. Contribution to the knowledge of cyanogenesis in angiosperms. 4. Communication. Cyanogenesis in *Trochodendron aralioides* Sieb. & Zucc. Proc. Kon. Nederl. Akad. Wetensch. ser. C. 81: 198–203.

Whitehead, D. R. 1969. Wind pollination in the angiosperms: evolutionary and environmental considerations. Evolution 23: 28–35.

Иоффе, М. Д. 1962. К эмбриологии *Trochodendron aralioides* Sieb. et Zucc. (Развитие пыльцы и зародышевого мешка). Труды Бот. Инст. АН СССР, сер, VII, 5: 250-259.

Иоффе, М. Д. 1965. К эмбриологии *Trochodendron aralioides* Sieb. et Zucc. (Развитие зародыша и эндосперма). В сб.: Морфология цветка и репродуктивный процесс у покрытосеменных растений, М. С. Яковлев, ред.: 177-189. Изд. Наука, Москва-Ленинград.

Первухина, Н. В. 1962. Об одной интересной особенности завязи Троходендрона, Бот. Ж. 47: 993-995.

Первухина, Н. В. 1963. К положению Троходендрона в филогенетической ситеме покрытосеменных. Бот. Ж. 48: 939-948.

Первухина, Н. В., & М. Д. Иоффе. 1962. Морфология цветка Троходендрона. Бот. Ж. 47: 1709-1730.

SYNOPTICAL ARRANGEMENT OF THE FAMILIES OF TROCHODENDRALES

1 Perianth present, of 4 sepals; stamens 4, opposite the sepals; leaves palmately veined; petiole with stipular flanges; elongate idioblasts simple or branched, secretory, not sclerified .... 1. TETRACENTRACEAE.

1 Perianth none; stamens numerous; leaves pinnately veined; stipules wanting; elongate idioblasts branched and sclerified, not secretory .... 2. TROCHODENDRACEAE.

1. Family TETRACENTRACEAE Van Tieghem 1900 (and A. C. Smith 1945) nom. conserv., the Tetracentron Family

Deciduous trees with primitively vesselless wood, trilacunar nodes, and elongate, simple or branched secretory idioblasts in the parenchymatous tissues of both the root and the shoot, somewhat tanniferous, with proanthocyanins but not ellagic acid; twigs of two types, long shoots and short shoots; ground-tissues of the xylem consisting of intermingled fiber-tracheids and vascular tracheids, both types with bordered pits; vascular tracheids arranged in longitudinal series, wider and shorter than the fiber-tracheids, vessel-like but imperforate; wood-rays heterocellular, mixed uniseriate and pluriseriate, the latter 2–4 (–7) cells wide; wood-parenchyma restricted to the late wood, where abundant and diffuse, the cells scattered and in interrupted, irregular, uniseriate lines. Short shoots alternately arranged, each bearing a single leaf each year; LEAVES simple, serrate, palmately veined, with stipular flanges at the base of the petiole; stomates laterocytic; petiole with a single arcuate or almost cylindrical vascular strand. FLOWERS anemophilous, small, perfect, hypogynous, sessile in many-flowered pendulous spikes subterminal on the short shoots; sepals 4, imbricate; petals none; stamens 4, opposite the sepals and alternate with the carpels; filaments slender; anthers basifixed, erect, tetrasporangiate and dithecal, the stout connective not prolonged; pollen-sacs latrorse, opening by longitudinal slits; pollen-grains small, more or less globose, tectate, tricolpate; gynoecium a whorl of 4 conduplicate, laterally connate but distally distinct carpels, the backs of the carpels protruding and nectariferous; styles short, ventrally terminal, with a decurrent stigma; carpels soon becoming deformed by growth on the ventral side, so that at maturity the style is reflexed and nearly basal on the abaxial side; ovules 5–6 in each carpel on a marginal placenta, pendulous, anatropous, bitegmic, crassinucellar; endosperm-development cellular. FRUIT a follicetum, consisting of laterally coherent, distally distinct follicles, seeds pendulous from near the summit of the locule, with thin testa and copious, oily and proteinaceous endosperm; embryo minute, undifferentiated, or with 2 rudimentary cotyledons. $2n = (46 \text{ or}) 48$.

The family Tetracentraceae consists of the single genus and species *Tetracentron sinense* Oliv., native to Nepal, central and southwestern China, and northern Burma.

On the basis of the structure of the seed coat, Melikyan (1973, under Hamamelidales) believes that *Tetracentron* and *Trochodendron* must be allied to *Altingia* and *Liquidambar*, genera which are here referred to the Hamamelidaceae. Likewise Corner (1976, in initial citations) considers that the seed coats of *Tetracentron*, *Trochodendron*, and *Euptelea* are "remarkably similar" and "may point to the early ancestry of the Hamamelidaceae."

According to Wolfe (1973, under Hamamelidae), *Tetracentron* can be traced back at least to the earliest Eocene, based on fossil leaves originally assigned to *Cercidiphyllum*. Some late Cretaceous leaves may also belong here, and some late Cretaceous (latest Campanian) wood may well represent either *Tetracentron* or *Trochodendron*.

2. Family TROCHODENDRACEAE Prantl in Engler & Prantl 1888 nom. conserv., the Yama-Kuruma Family

Glabrous, evergreen small trees with primitively vesselless wood, multilacunar to unilacunar nodes (depending on the size of the leaves) and elongate, branched, sclerified, nonsecretory idioblasts in the cortex, pith, and mesophyll, somewhat tanniferous, with proanthocyanins but not ellagic acid; winter-buds and inflorescences cyanogenic through a tyrosine-based pathway; twigs forming long shoots only; ground-tissue of the xylem consisting of long tracheids with scalariform-bordered pits in the early wood and circular bordered pits in the late wood; wood-rays heterocellular, many of them uniseriate, others up to 6–13 cells wide; wood-parenchyma diffuse, the cells scattered and in short, tangential, uniseriate lines in the late wood, not abundant; sieve-tubes with S-type plastids. LEAVES alternate, but closely clustered near the branch-tips in pseudowhorls, simple, serrulate, pinnately veined; stomates laterocytic; petiole with an arc of 3 vascular bundles; stipules wanting. Plants secondarily entomophilous, apparently andro-dioecious, with small, perfect or staminate, long-pedicellate FLOWERS in short, initially terminal, raceme-like cymes; perianth wanting but the flower commonly subtended by several tiny, adnate-decurrent, unvasculated scales at the top of the swollen pedicel; stamens numerous, commonly ca 70, spirally arranged, initiated in centripetal sequence; anthers basifixed, tetrasporangiate and dithecal, latrorse, opening by longitudinal slits; connective stout, not prolonged; pollen-grains more or less globose, tectate-columellate, tricolpate, with weakly developed endo-apertures; gynoecium a whorl of 4–11 (–17) conduplicate, laterally connate but distally distinct carpels, the backs of the carpels protruding and nectariferous; styles short, soon becoming outcurved, with dry, papillate, ventrally

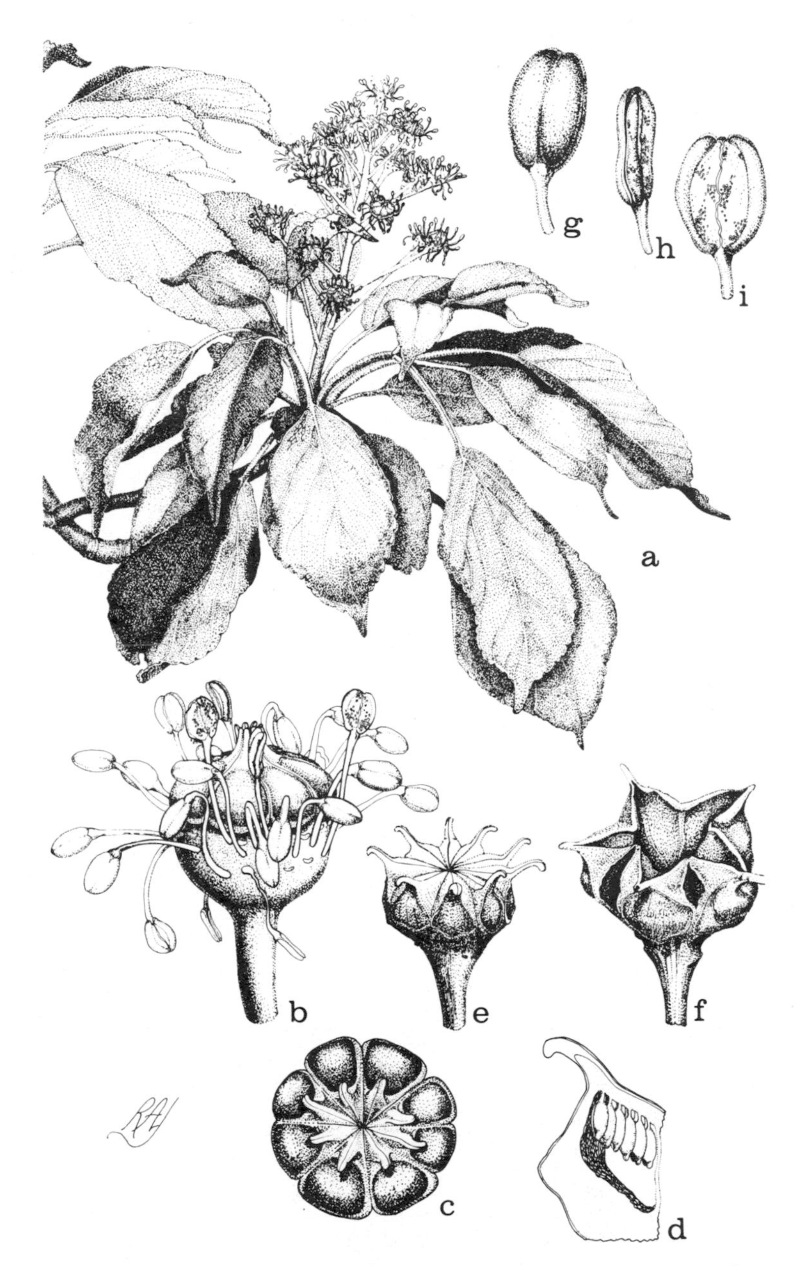

FIG. 2.2 Trochodendraceae. *Trochodendron aralioides* Sieb. & Zucc. a, habit, ×½; b, flower, ×4; c, gynoecium, from above, ×5; d, long-section of carpel, ×5; e, maturing fruit, ×3; f, mature fruit, after dehiscence, ×3; g, submature stamen, ×10; h, i, side and front views of mature stamen, ×10.

decurrent stigma; ovules 25–30 per carpel, marginal, pendulous from the broad, oblique summit of the locule (the lower part of the carpellary margins sterile), anatropous, bitegmic, crassinucellar; endosperm-development cellular. FRUIT a follicetum, consisting of laterally coherent, distally distinct and outcurved follicles with persistent, spreading, short styles; seeds with thin testa and copious, oily and proteinaceous endosperm; embryo minute, with 2 rudimentary cotyledons. 2n = 40.

The family Trochodendraceae consists of the single genus and species *Trochodendron aralioides* Sieb. & Zucc., native from Korea and Japan to Taiwan.

According to Wolfe (1973, under Hamamelidae), the leaf-venation of *Trochodendron* is the most primitive in the Hamamelidae, reminiscent of the Magnoliidae, and that of *Tetracentron* is a little more advanced. *Cercidiphyllum*, in the Hamamelidales, is still a bit more advanced in this respect.

Fossil pollen of *Trochodendron* is known from Upper Eocene and more recent deposits. Macrofossils considered to belong here date from the Paleocene.

2. Order HAMAMELIDALES Wettstein 1907

Trees, shrubs, or undershrubs; vessel-segments with oblique end-walls and scalariform (or partly reticulate) perforation-plates with few to numerous cross-bars, or seldom some of them with simple perforations; sieve-tubes with S-type plastids. LEAVES alternate or less commonly opposite, simple, entire or toothed to palmately lobed; stomates anomocytic or of various other types; stipules present or less often absent. FLOWERS anemophilous or (probably only secondarily) entomophilous, small, perfect or often unisexual, mostly in small inflorescences of often complex structure, sometimes (some Hamamelidaceae) so much reduced and crowded that what appears to be a flower may actually be a partial inflorescence; perianth present or absent, the petals when present generally small or narrow and not very showy; stamens 2-many in a single whorl, sometimes alternating with staminodes, latrorse, commonly with a more or less prolonged connective; pollen-grains triaperturate or sometimes multiaperturate; gynoecium of 1-several separate or united carpels with separate styles and very often decurrent stigmas, or (Eupteleaceae) without a style and with a marginal-decurrent stigma; ovary superior to inferior, with as many locules as carpels, and with 1-many ovules per carpel; ovules mostly anatropous (orthotropous in Platanaceae), bitegmic and crassinucellar. FRUITS dry, dehiscent or indehiscent; seeds with scanty to copious, oily and proteinaceous endosperm; embryo tiny to large, more or less evidently dicotyledonous.

The order Hamamelidales as here defined consists of 5 families. The vast majority of the species belong to the Hamamelidaceae (100+). The Platanaceae have 6 or 7 species, and the Cercidiphyllaceae, Eupteleaceae and Myrothamnaceae each have 2 species. There is a growing consensus that these 5 families are allied inter se, and that collectively they are also allied to the Tetracentraceae and Trochodendraceae. The imperfectly sealed carpels of *Platanus* and especially *Euptelea* must direct our search for the ancestry of the order back toward the Magnoliidae rather than toward anything in the modern Rosidae. The platanoid line can be traced to the middle Albian stage of the Lower Cretaceous, about 115 million years ago. It is only at this level that the ancestors of the modern Hamamelidales merge with the ancestors of the modern Rosidae and Magnoliidae.

Aside from the fact that they are all woody plants, the members of the Hamamelidales have little in common ecologically. Even the floral reduction that permeates the group has not restricted them to wind-pollination. A number of members of the Hamamelidaceae are pollinated by insects. *Hamamelis* itself appears to be adapted to insect-pollination, although it blooms at a time when few insects are about.

SELECTED REFERENCES

Baas, P. 1969. Comparative anatomy of *Platanus kerrii* Gagnep. J. Linn. Soc., Bot. 62: 413–421.

Bogle, A. L. 1970. Floral morphology and vascular anatomy of the Hamamelidaceae: The apetalous genera of Hamamelidoideae. J. Arnold Arbor. 51: 310–366.

Boothroyd, L. E. 1930. The morphology and anatomy of the inflorescence and flower of the Platanaceae. Amer. J. Bot. 17: 678–693.

Brett, D. W. 1979. Ontogeny and classification of the stomatal complex of *Platanus* L. Ann. Bot. 44: 249–251.

Carlquist, S. 1976. Wood anatomy of *Myrothamnus flabellifolia* (Myrothamnaceae) and the problem of multiperforate perforation plates. J. Arnold Arbor. 57: 119–126.

Endress, P. 1967. Systematische Studie über die verwandtschaftlichen Beziehungen zwischen den Hamamelidaceen und Betulaceen. Bot. Jahrb. Syst. 87: 431–525.

Endress, P. 1969 (1970). Gesichtspunkte zur systematischen Stellung der Eupteleaceen (Magnoliales). Untersuchungen über Bau und Entwicklung der generativen Region bei *Euptelea polyandra* (Sieb. et Zucc.). Ber. Schweiz. Bot. Ges. 79: 229–278.

Endress, P. K. 1970. Die Infloreszenzen der apetalen Hamamelidaceen, ihre grundsätzliche morphologische und systematische Bedeutung. Bot. Jahrb. Syst. 90: 1–54.

Endress, P. K. 1971. Blütenstande und morphologische Interpretation der Blüten bei apetalen Hamamelidaceen. Ber. Deutsch. Bot. Ges. 84: 183–185.

Endress, P. K. 1976. Die Androeciumanlage bei polyandrischen Hamamelidaceen und ihre systematische Bedeutung. Bot. Jahrb. Syst. 97: 436–457.

Endress, P. K. 1978. Blütenontogenese, Blütenabgrenzung und systematische Stellung der perianthlosen Hamamelidoideae. Bot. Jahrb. Syst. 100: 249–317.

Endress, P. K. 1978. Stipules in *Rhodoleia* (Hamamelidaceae). Pl. Syst. Evol. 130: 157–160.

Ernst, W. R. 1963. The genera of Hamamelidaceae and Platanaceae in the southeastern United States. J. Arnold Arbor. 44: 193–210.

Goldblatt, P., & P. K. Endress. 1977. Cytology and evolution in Hamamelidaceae. J. Arnold Arbor. 58: 67–71.

Jäger-Zürn, I. 1966. Infloreszenz- und blütenmorphologische, sowie embryologische Untersuchungen an *Myrothamnus* Welw. Beitr. Biol. Pflanzen 42: 241–271.

Jay, M. 1968. Distribution des flavonoides chez les Hamamélidacées et familles affines. Taxon. 17: 136–147.

Jha, U. N. 1978. Chemotaxonomy of the Hamamelidaceae. J. Indian Bot. Soc. 56: 44–48.

Leinfellner, W. 1969. Über die Karpelle verschiedener Magnoliales. VII. *Euptelea* (Eupteleaceae). Oesterr. Bot. Z. 116: 159–166.

Lemesle, R. 1946. Contribution a l'étude morphologique et phylogénétique des Eupteléacées, Cercidiphyllacées, Eucommiacées (ex-Trochodendracées). Ann. Sci. Nat. Bot. Sér. 11. 7: 41–52.

Leroy. J.-F. 1980. Développement et organogenèse chez le *Cercidiphyllum japonicum*; un cas semblant unique chez les Angiospermes. Compt. Rend. Hebd. Séances Acad. Sci. 10D: 679–682.

Mohana Rao, P. R. 1974. Seed anatomy in some Hamamelidaceae and phylogeny. Phytomorphology 24: 113–139.

Nast, C. G., & I. W. Bailey. 1946. Morphology of *Euptelea* and comparison with *Trochodendron*. J. Arnold Arbor. 27: 186–192.

Puff, C. 1978. Zur Biologie von *Myrothamnus flabellifolius* Welw. (Myrothamnaceae). Dinteria 14: 1–20.

Smith, A. C. 1946. A taxonomic review of *Euptelea*. J. Arnold Arbor. 27: 175–185.

Swamy, B. G. L., & I. W. Bailey. 1949. The morphology and relationships of *Cercidiphyllum*. J. Arnold Arbor. 30: 187–210.

Tong, K. 1931. Studien über die Familie der Hamamelidaceae mit besonderer Berücksichtigung der Systematik und Entwicklungsgeschichte von *Corylopsis*. Bull. Dept. Biol. Sun Yatsen Univ. 2: 1–72.

Меликян, А. П. 1973. Типы семенной кожуры Hamamelidaceae и близких семейств в связи с их систематическими взаимоотношениями. Бот. Ж. 58 : 350-359.

Скворцова, Н. Т. 1960. Строение эпидермисов у представителей сем, Hamamelidaceae. Бот. Ж. 45: 712-717.

Скворцова, Н. Т. 1960. Анатомическое строение проводящей системы черешков у листьев представителей семейств Hamamelidaceae и Altingiaceae. Докл. Акад. Наук СССР 133: 1231-1234.

Скворцова, Н. Т. 1960. О типах жилкования листьев представителей семейства Hamamelidaceae. Труды Ленингр. Химико-Фарм. Инст., 12: 75-83.

Скворцова, Н. Т. 1965. К морфологии рода *Hamamelis* L. Бот. Ж. 50: 1143-1148.

Чжан Цзинь-Тань. 1964. Морфология пыльцы семейств Hamamelidaceae & Altingiaceae. Труды Бот. Инст. АН СССР, сер, 1, 13: 173-232.

SYNOPTICAL ARRANGEMENT OF THE FAMILIES OF HAMAMELIDALES

1 Carpels distinct, or carpel solitary.
  2 Ovules numerous; fruits follicular; plants dioecious; leaves stipulate, palmately to pinnipalmately veined ....1. CERCIDIPHYLLACEAE.
  2 Ovules 1–4; fruits indehiscent; plants monoecious, or with perfect flowers; leaves various.
    3 Stamens relatively numerous, commonly 7–20; flowers mostly perfect; connective prolonged, but not apically enlarged and peltate; ovules anatropous; leaves exstipulate, pinnately veined, merely toothed; embryo tiny, with poorly differentiated cotyledons ....2. EUPTELEACEAE.

3 Stamens 3–4 (–7); flowers unisexual, the plants monoecious; connective apically enlarged and peltate; ovules orthotropous or nearly so; leaves stipulate, palmately lobed and veined; embryo of normal proportions .... 3. PLATANACEAE.

1 Carpels united (at least below) into a compound pistil.

4 Carpels 2 (3); leaves alternate (rarely opposite); pollen monadinous; flowers perfect or unisexual, often with an evident perianth; ovules 1-many, shrubs or trees 4. HAMAMELIDACEAE.

4 Carpels 3 or 4; leaves opposite; pollen tetradinous; flowers unisexual, the plants dioecious; perianth none; ovules many; low shrubs .... 5. MYROTHAMNACEAE.

1. Family CERCIDIPHYLLACEAE Engler 1909 nom. conserv., the Katsura-tree Family

Dioecious trees with long and short shoots and trilacunar nodes, tanniferous, producing both proanthocyanins and ellagic acid; some of the cells of the parenchymatous tissues containing solitary or clustered crystals of calcium oxalate; vessel-segments very small, angular, with oblique, scalariform perforation-plates that have numerous (commonly 25–50) cross-bars; imperforate tracheary elements with conspicuously bordered pits, considered to be tracheids; wood-rays heterocellular, 1–2 cells wide; wood-parenchyma scanty, apotracheal, diffuse and in irregular, uniseriate, terminal lines; sieve-tubes with S-type plastids. LEAVES deciduous, simple, palmately or pinnipalmately veined, those on long shoots elliptic or cruciform and opposite, those on short shoots cordate and alternate; stomates anomocytic; stipules deciduous. FLOWERS anemophilous, unisexual, the inflorescences terminal on the sympodial short shoots, appearing with or before the leaves; staminate flowers several in a very short, condensed raceme, the 4 lower (outer) ones each subtended by a 4-lobed bract, the inner ones apparently bractless, all without perianth; stamens 8–13, with slender, elongate filament and elongate, tetrasporangiate, dithecal, latrorse anther, the connective shortly prolonged; pollen-sacs opening by longitudinal slits; pollen-grains medium-sized, more or less globose, tectate and granular-columellate, binucleate, tricolpate, the colpi only weakly developed; pistillate inflorescences so condensed as to form pseudanthia, each pseudanthium with (2–) 4 (–6) small, sepal-like bracts, each bract subtending a naked pistillate flower consisting of a single carpel; ovary with an abaxial suture (facing the subtending bract), gradually narrowed to a slender style with a decurrent, 2-ridged stigma; ovules numerous, in 2 rows, laminar-lateral, anatropous, bitegmic, crassinucellar; endosperm-development cellular. FRUIT of separate follicles; seeds flattened, winged, with scanty, oily endosperm and large, spatulate, dicotyledonous embryo with a well developed hypocotyl; epidermis of the seed-coat tanniferous. 2n = 38.

The family Cercidiphyllaceae consists of a single genus, *Cercidiphyllum*, with two species, native to China and Japan. *Cercidiphyllum* is generally regarded as related on the one hand to the Hamamelidaceae (especially *Disanthus*) and on the other to the Trochodendrales and Magnoliales.

Megafossils of *Cercidiphyllum* can be traced with some assurance to the Paleocene. Fossil pollen goes back certainly to the Miocene. Pollen

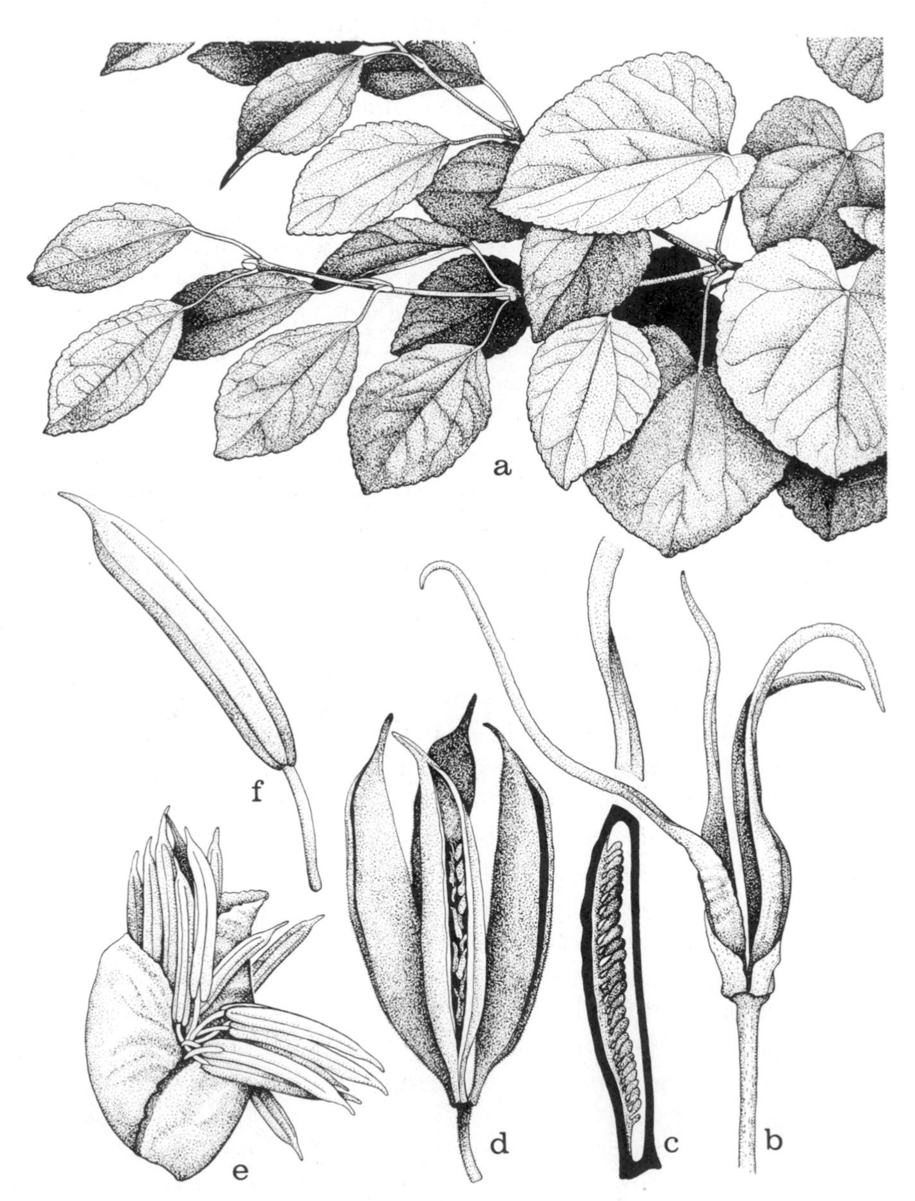

FIG. 2.3 Cercidiphyllaceae. *Cercidiphyllum japonicum* Sieb. & Zucc. a, habit, ×½; b, pistillate pseudanthium, ×4; c, carpel, the ovary in long-section, ×8; d, fruits, ×4; e, staminate inflorescence, ×4; f, stamen, ×8.

as old as the Maestrichtian or even the Campanian is probably of this alliance.

2. Family EUPTELEACEAE Wilhelm 1910 nom. conserv., the Eupteleа Family

Rather small trees or shrubs with unicellular or uniseriate hairs and modified unilacunar nodes that have 7–9 traces; secretory (tanniferous?) cells scattered in the ground-tissue of the stem and leaf; plants producing some proanthocyanin, but not ellagic acid; vessel-segments with oblique, scalariform (or in part reticulate) perforation-plates that have 20–90 cross-bars; imperforate tracheary elements with rather small, mostly bordered pits, considered to be fiber-tracheids; wood-rays heterocellular, Kribs type II (type I in twigs), mixed uniseriate and pluriseriate,

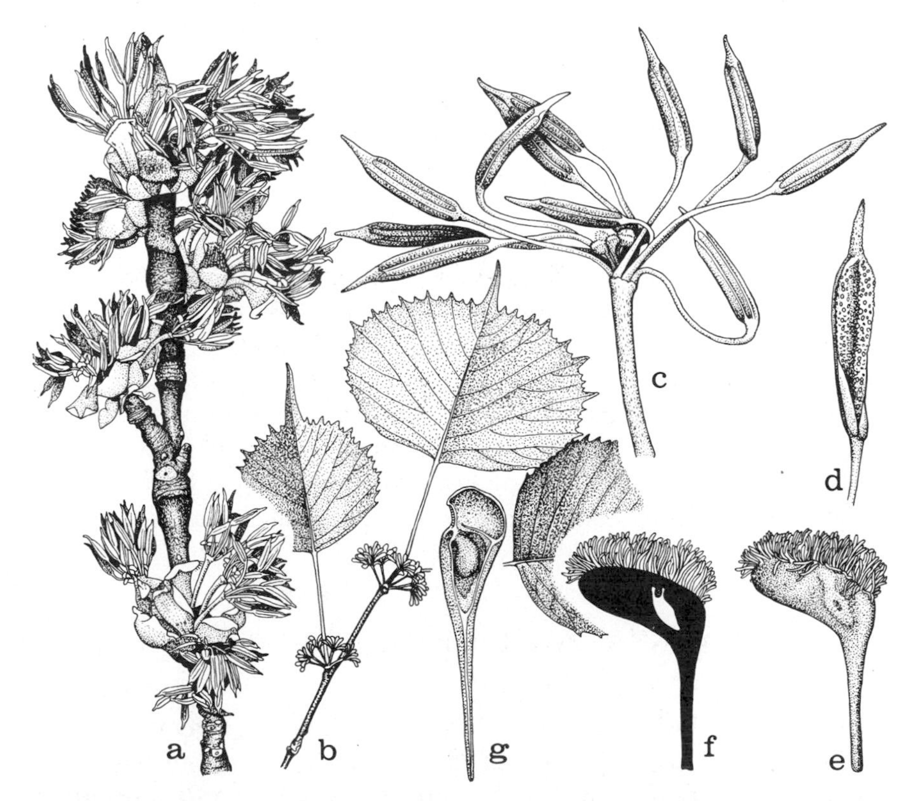

FIG. 2.4 Eupteleaceae. *Euptelea polyandra* Sieb. & Zucc. a, flowering twig, ×1; b, leafy twig with developing fruits,, ×½; c, flower, ×3; d, anther, ×6; e, carpel, ×9; f, carpel in partial long-section, ×9; g, fruiting carpel in partial long-section, ×2.

the pluriseriate ones up to 3 cells wide in young stems, up to 10 or more cells wide in older ones, with thin ends; wood-parenchyma diffuse or in short, tangentially oriented aggregates; multiseriate phloem-rays strongly sclerified; sieve-tubes with S-type plastids. LEAVES deciduous, alternate, simple, serrate or coarsely and irregularly dentate, pinnately veined; stomates anomocytic; stipules wanting. FLOWERS anemophilous or partly entomophilous, long-pedicellate, solitary in the axils of 6–12 closely crowded early bracts of vegetative shoots of the season, perfect or sometimes some of them staminate, proterandrous, without perianth; stamens more or less numerous (commonly 7–20) in a single whorl; filaments short, slender or slightly expanded; anthers elongate, red, tetrasporangiate and dithecal, latrorse, with a prolonged connective; pollen-grains tectate and granular-columellate, binucleate, tricolpate or 5–7-colpate or polycolpate; gynoecium a single cycle of 6–18 distinct, long-stipitate carpels, these conduplicate, incompletely sealed, with decurrent stigma and no style, becoming deformed by asymmetric growth after anthesis, so that the stigmatic surface does not reach the hooded summit; ovules 1–3 (4), marginal or submarginal, anatropous, bitegmic, crassinucellar, endosperm-development cellular. FRUIT of small samaras or winged nutlets with a papery pericarp; seeds with copious, oily and proteinaceous endosperm; embryo tiny, with 2 poorly differentiated cotyledons. $2n = 28$.

The family Eupteleaceae consists of the single genus *Euptelea*, with 2 species, one in Japan, the other in China and Assam. Authors have variously associated the family with the Cercidiphyllaceae, Hamamelidaceae, Platanaceae, Trochodendraceae, or Schisandraceae. The position here assigned to the family in the Hamamelidales is thus consonant with most past opinion. Wolfe (1973, under Hamamelidae) notes that the leaf-venation suggests that of *Platanus*. He considers that leaves of *Euptelea* can be traced back at least to the early Oligocene in the fossil record.

3. Family PLATANACEAE Dumortier 1829 nom. conserv., the Plane-tree Family

Trees with branching hairs and multilacunar nodes, cyanogenic (through a tyrosine-based pathway) and producing triterpenes, somewhat tanniferous, with proanthocyanins but not ellagic acid, and transporting nitrogen as allantoin; solitary or clustered crystals of calcium oxalate present in some of the parenchyma-cells; nodes multilacunar; vessel-

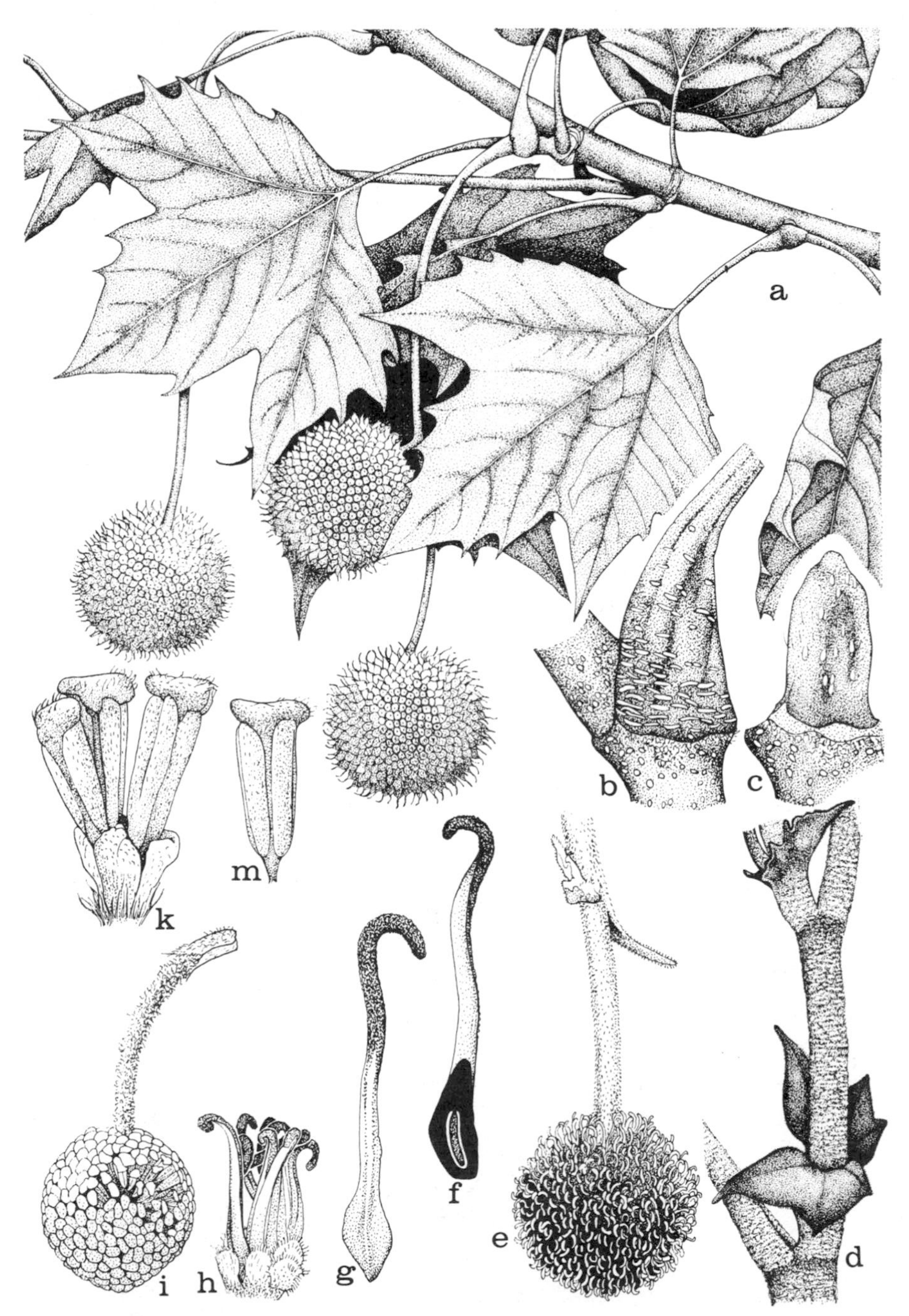

FIG. 2.5 Platanaceae. a–d, *Platanus acerifolia* (Aiton) Willd. a, habit, ×½; b, base of petiole, ×2; c, axillary bud, after removal of mitriform petiole, ×2; d, stipules, ×2. e–m, *Platanus occidentalis* L. e, pistillate inflorescence, ×2; f, pistil, the ovary in long-section, ×16; g, pistil, ×16; h, pistillate flower, ×8; i, staminate inflorescence, ×2; k, staminate flower, ×8; m, stamen, ×8.

segments with mixed simple and scalariform perforation-plates, these with relatively few (ca 12–20) cross-bars; imperforate tracheary elements with bordered pits; wood-rays homocellular, Kribs type II, all or nearly all of them pluriseriate, up to 10–20 cells wide; wood-parenchyma rather sparse, apotracheal, diffuse or in uniseriate bands; bark scaling off in large, irregular sheets, leaving a patchy, nearly smooth surface; larger phloem-rays strongly sclerified; sieve tubes with S-type plastids. LEAVES deciduous, alternate, simple, palmately lobed and veined, the petiole mitriform at the base and enclosing the axillary bud; stomates laterocytic, or some of them paracytic or anomocytic; stipules mostly large and conspicuous, united around the stem. Plants monoecious, anemophilous, the FLOWERS small, inconspicuous, hypogynous, regular, unisexual, clustered in dense, unisexual heads; perianth commonly 3–4 (–7)-merous, the sepals distinct or basally connate, not vasculated; tiny or vestigial petals alternating with the sepals in the staminate flowers, usually wanting from the pistillate flowers; stamens (in staminate flowers) as many as and opposite the sepals; filaments very short or almost obsolete; anthers tetrasporangiate and dithecal, latrorse, opening by longitudinal slits, the connective prolonged into a peltate appendage; staminate flowers sometimes with vestigial carpels; pollen-grains tricolporate, with weakly developed endo-apertures; pistillate flowers commonly with 3–4 staminodia and (3–) 5–8 (9) separate carpels in 2–3 whorls; carpels imperfectly sealed distally, the dry, papillate stigma decurrent along the linear style and encroaching on the top of the ovary; ovule solitary (2), pendulous, orthotropous or slightly hemitropous, bitegmic, crassinucellar. FRUIT of small, densely hairy achenes or nutlets in a globose head; seeds with thin testa, scanty, oily and proteinaceous endosperm, and slender, straight, dicotyledonous embryo. X = 7. Note inserted in proof: The leaves in *Platanus kerrii* Gagnep., of Laos, are pinnately veined and merely toothed.

The family Platanaceae consists of a single genus, *Platanus*, with 6 or 7 species. The genus occurs from the eastern Mediterranean region to the Himalaya Mts., and from Mexico to Canada.

The relationship of the Platanaceae to the Hamamelidaceae is generally conceded, but neither family can be derived from the other. The flowers of *Platanus* are more primitive than those of the Hamamelidaceae in having separate, imperfectly sealed carpels, but according to Tippo (1938, under Urticales) the wood of *Platanus* is somewhat more advanced than that of the Hamamelidaceae.

The platanoid phylad can be traced with reasonable certainty in the fossil record back to the middle Albian stage of the Lower Cretaceous. Leaves of very platanoid aspect and venation occur in the Albian, and platanoid wood in the latest part of the Campanian stage of the Upper Cretaceous. *Platanus* itself can be traced back at least to the Maestrichtian, on the basis of leaves and associated pollen. The Albian protoplatanoids appear to have inhabited stream-banks and flood-plains, even as modern *Platanus*.

4. Family HAMAMELIDACEAE R. Brown in Abel 1818 nom. conserv., the Witch-hazel Family

Shrubs or trees, often with stellate hairs, tanniferous, commonly with proanthocyanins and gallic acid, sometimes with ellagic acid as well, often with scattered idioblasts containing tannin and/or mucilage at least in the stem and petioles, and sometimes (*Liquidambar*) producing iridoid compounds; nodes trilacunar; secretory canals sometimes (Altingioideae) present in the bark, wood, and leaves; solitary or clustered crystals of calcium oxalate commonly present in some of the cells of the parenchymatous tissues; vessel-segments slender and often much-elongate, with slanting end-walls and scalariform (or partly reticulate) perforation-plates that have 5–50 or more cross-bars; imperforate tracheary elements large, with large, distinctly bordered pits; wood-rays heterocellular, Kribs types IIA, IIB, and III, rarely I, variously all uniseriate, or mixed uniseriate and pluriseriate, or all pluriseriate, the pluriseriate ones mostly 2–4 cells wide, often with elongate ends; wood-parenchyma apotracheal and typically diffuse, sometimes banded; sieve-tubes with S-type plastids. LEAVES alternate or rarely opposite, simple, often palmately lobed, often with sclereids in the mesophyll; stomates of various types; petiole with a trough-shaped or cylindrical vascular strand, or (Altingioideae) with 3–5 vascular bundles; stipules nearly always present. FLOWERS typically borne in a spike or system of spikes, seldom in a raceme or panicle, variously anemophilous or entomophilous, perfect or unisexual, regular or seldom irregular, usually partly or wholly epigynous, seldom hypogynous or merely perigynous, sometimes closely crowded and possibly even pseudanthial; sepals commonly 4–5 (–10), small, distinct or largely connate, sometimes much-reduced or wanting; petals commonly 4–5, usually small or

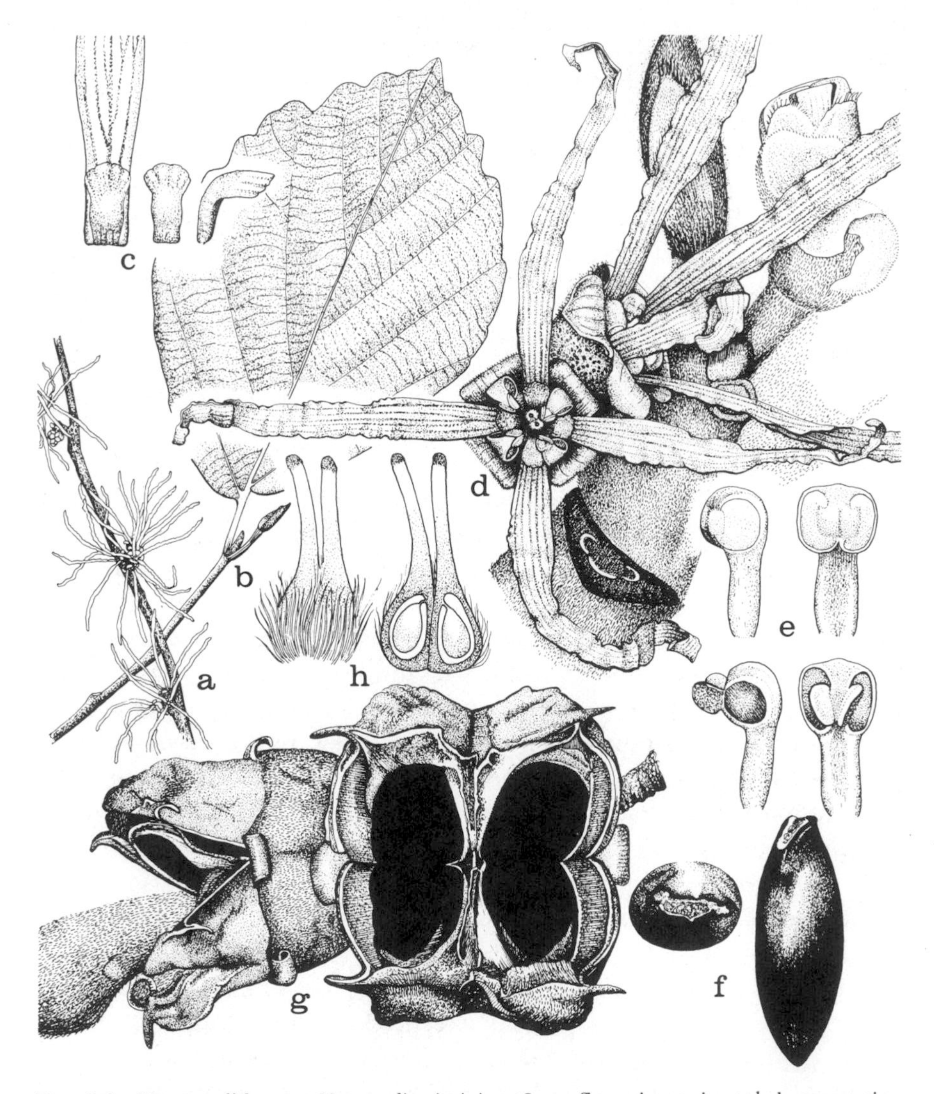

FIG. 2.6 Hamamelidaceae. *Hamamelis virginiana* L. a, flowering twig, ×½; b, vegetative twig, ×½; c, staminode, attached to a petal (left) and detached (right), ×8; d, flowers, ×4; e, stamens before (above) and after (below) dehiscence, ×8; f, seed, top and side view, ×4; g, fruits, ×4; h, pistil, external (left) and in partial long-section (right), ×12.

narrow, or wanting (in *Exbucklandia* the petals present but the sepals obsolete); stamens commonly 4–5 (–10) in a single series alternate with the petals, sometimes alternating with as many staminodia, or more numerous, up to ca 32, and then initiated in centripetal (*Matudaea*) or centrifugal (*Fothergilla*) sequence; staminodes or inner basal surface of the petals sometimes nectariferous; anthers tetrasporangiate or (*Hamamelis*) bisporangiate, dithecal, latrorse, opening by valves or longitudinal slits; connective often more or less prolonged (but not in *Hamamelis*); pollen-grains binucleate, tricolpate to sometimes (especially in Altingioideae) multiporate; gynoecium of 2 (3) carpels united at least below to form a compound ovary with axile (rarely parietal) placentation, the styles distinct, commonly persistent and indurate, the stigma dry, decurrent; each locule with one or less often 2-several more or less pendulous ovules (ovules more numerous in Altingioideae), these anatropous (orthotropous in Altingioideae), bitegmic, crassinucellar, often all but one in each locule abortive; fertilization often delayed until long after pollination; endosperm-development nuclear or (*Parrotiopsis*) cellular. FRUIT woody, capsular, septicidal, or both septicidal and loculicidal, seldom only loculicidal, seeds with thick, hard testa and thin, oily and proteinaceous endosperm; embryo large, straight, spatulate, with 2 cotyledons. X = 8, 12, 15, 16. (Altingiaceae, Rhodoleiaceae)

The family Hamamelidaceae consists of some 26 genera (13 of them monotypic) and a little more than 100 species. The family is widespread in both the Old and the New World; it is most abundant in subtropical and warm-temperate regions, especially in eastern Asia. The largest genus is *Corylopsis*, with about 30 species native to eastern Asia. *Disanthus* (a single species of China and Japan) may be the most archaic surviving genus.

Species of *Corylopsis*, *Hamamelis* (witch-hazel), and *Liquidambar* (sweet gum), are familiar as cultivated ornamentals. *Hamamelis* is the source of a well known astringent and tonic lotion, and is notable for producing some of the earliest or latest flowers of the year (according to the species). *Liquidambar* is the source of a fragrant resin; the North American species *L. styraciflua* L. has valuable wood and is often planted as a shade-tree.

The family is only very loosely knit, with many isolated, perhaps relictual genera. *Liquidambar* and *Altingia* are sometimes (and not without reason) considered to form collectively a separate family Altingiaceae, but their relationship to the rest of the group is not in dispute. Endress (1978) points out that *Rhodoleia*, usually described as lacking stipules, has well developed stipules on a few leaves of the transition region between bud scales and foliage leaves. He further argues persuasively that *Rhodoleia* properly belongs to the Hamamelidaceae.

The flowers of *Distylium*, *Distyliopsis*, and *Matudaea* have no perianth, and have been interpreted by Bogle (1970) as being pseudanthial. On the other hand, Endress (1978) considers that the flowers in these genera are euanthial, as in other Hamamelidaceae.

The Hamamelidaceae have a long fossil history. In the earliest Paleocene the family was already diversified into a number of genera (fide Wolfe 1973, under Hamamelidae), and Endress considers that *Liquidambar* can be traced back into the Upper Cretaceous. Leaves and pollen of other Hamamelidaceae also appear to be represented in late Upper Cretaceous deposits of Europe.

5. Family MYROTHAMNACEAE Niedenzu in Engler & Prantl 1891 nom. conserv., the Chanasa Family

Small, glabrous, aromatic, xerophytic shrubs, tanniferous, with both proanthocyanins and ellagic acid; clustered crystals of calcium oxalate present in some of the cells of the parenchymatous tissues; vessel-segments angular, with oblique, scalariform or reticulate-scalariform perforations and numerous cross-bars; imperforate tracheary elements with bordered pits, all considered to be tracheids; wood-rays abundant, heterocellular, uniseriate; wood-parenchyma absent; tracheids and cells of the wood-rays containing amorphous deposits. LEAVES small, opposite, cuneate-flabellate or narrowly elliptic, toothed across the broadly rounded summit, the venation palmate-flabellate, not evidently reticulate; epidermis with scattered resin- or oil-cells; stomates anomocytic; stipules more or less well developed. Plants dioecious, the FLOWERS unisexual, in erect spikes (catkins), anemophilous, regular, without perianth; stamens 4 and distinct or 3–8 and connate by their filaments; anthers dithecal, latrorse, with a small but apically prolonged connective, opening by longitudinal slits; pollen-grains binucleate, tetradinous, with poorly developed colpi; gynoecium of 3–4 carpels united to form a compound, 3–4 locular ovary with distinct, short, broad, recurved styles and ventrally decurrent stigmas; placentation axile; ovules rather numerous in each locule, anatropous, bitegmic, crassinucellar; endosperm-development nuclear. FRUIT capsular, the carpels separating above and opening ventrally; seeds numerous, with thin testa and abundant, oily endosperm; embryo small, dicotyledonous. $2n = 20$.

The family Myrothamnaceae consists of the single genus *Myrothamnus*, with 2 species, one in Africa, the other in Madagascar. Most authors are agreed that *Myrothamnus* is related to the Hamamelidaceae, and indeed some would include it in that family. Whether to lump or split in this instance is purely a matter of taste. The small number of species of

Myrothamnus, combined with the fact that there are many isolated genera in the Hamamelidaceae, would support lumping. The highly distinctive syndrome of features of *Myrothamnus*, in contrast to the Hamamelidaceae, would support splitting. It may also be noted that the Hamamelidaceae are essentially Laurasian, whereas *Myrothamnus* lives in a part of Gondwanaland. I here follow the majority of current usage, as exemplified by Takhtajan and the current Engler Syllabus, in maintaining the Myrothamnaceae as a distinct family.

The leaves of *Myrothamnus* fold up like a fan during the dry season, becoming fragile and blackish. They expand and become green again after rain.

The Myrothamnaceae are not known to be represented in the fossil record.

3. Order DAPHNIPHYLLALES Pulle 1952[1]

The order consists of the single family Daphniphyllaceae.

SELECTED REFERENCES

Satô, Y. 1972. Development of the embryo sac of *Daphniphyllum macropodum* var. *humile* (Maxim.) Rosenth. Sci. Rep. Tôhoku Univ. Biol. IV. 36: 129–133.

1. Family DAPHNIPHYLLACEAE Muell.-Arg. in A. DC. Prodr. 16 (1): 1. 1869 nom. conserv., the Daphniphyllum Family

Trees or shrubs containing a unique type of alkaloid (daphniphylline-group), sometimes with iridoid compounds, often accumulating aluminum, and regularly tanniferous, with tanniferous or mucilaginous idioblasts in the parenchymatous tissues, but without ellagic acid and probably also without proanthocyanins; clustered crystals of calcium oxalate commonly present in some of the cells of the parenchymatous tissues; vessel-segments with scalariform perforation-plates that have 20–30 to 100+ cross-bars; imperforate tracheary elements with bordered pits; wood-rays heterocellular, 1–2 cells wide; wood-parenchyma diffuse. LEAVES alternate, but sometimes closely crowded at the branch-tips and appearing almost whorled, simple, entire, pinnately veined; stomates paracytic; petiole with 3 entering vascular traces, but the small, U-shaped to cylindric lateral traces merging distally with the large, U-shaped central trace to form a single cylindrical vascular strand; stipules wanting. FLOWERS borne in axillary racemes, small and inconspicuous, unisexual (the plants dioecious), regular, hypogynous, apetalous, the pedicel subtended by a deciduous bract; sepals 2–6, more or less imbricate, or seldom wanting; stamens 5–12; filaments short, distinct; anthers basifixed, tetrasporangiate and dithecal, latrorse, opening by longitudinal slits; pollen-grains tricolporate; pistillate flowers sometimes with several staminodes; gynoecium of 2 (–4) carpels united to form a compound ovary with as many locules as carpels, the styles united only at the base, otherwise distinct, short, notably broad, divaricate to recurved or even circinate, stigmatic over the inner surface nearly to

[1] This name was proposed by Pulle in the third edition of his Compendium van de Terminologie, Nomenclatuur en Systematick der Zaadplanten, but it does not appear to have been formally validated until now.

the base; stigmas dry, papillate; ovules 2 in each locule, apical-axile, pendulous, epitropous, with ventral raphe, anatropous, bitegmic, crassinucellular; endosperm-development cellular. FRUIT a 1-seeded drupe; seeds with abundant, oily and proteinaceous endosperm and a very small, straight, apical, dicotyledonous embryo. 2n = 32.

The family Daphniphyllaceae consists of the single genus *Daphniphyllum*, with about 10 species native mostly to eastern Asia and the Malay Archipelago.

Daphniphyllum has often been included in the Euphorbiaceae, or in the Euphorbiales as a distinct family. The drupaceous fruit and scalariform vessels, and the absence of an obturator and a caruncle would all be unusual in the Euphorbiaceae, although these features can be found individually in various members of the family. More importantly, the tiny embryo is wholly out of harmony with the Euphorbiales and suggests a position nearer the base of the evolutionary tree.

The most obvious place for *Daphniphyllum*, once it is dissociated from the Euphorbiaceae, is in the Hamamelidae. Several previous authors, dating back at least to Hallier in 1912, have associated *Daphniphyllum* with the Hamamelidaceae. Only the chemistry is a little worrisome, but the alkaloid daphniphylline is unique no matter where the genus is put, and *Liquidambar* furnishes a precedent for iridoid compounds in the Hamamelidae. Although it seems reasonably at home in the Hamamelidae, *Daphniphyllum* does not fit comfortably with anything else in an established order. It appears to be another isolated end-line, comparable in that regard to *Didymeles*, *Eucommia*, *Leitneria*, and *Casuarina*.

SELECTED REFERENCES

Huang, T.-C. 1966. Monograph of Daphniphyll.

4. Order DIDYMELALES Takhtajan 1966

The order consists of the single family Didymelaceae.

SELECTED REFERENCES

Leandri, J. 1937. Sur l'aire et la position systématique du genre malgache *Didymeles* Thouars. Ann. Sci. Nat. Bot. sér. 10. 19: 309–318.

1. Family DIDYMELACEAE Leandri 1937, the Didymeles Family

Glabrous, evergreen trees; coarse, short fibers abundant in some of the non-xylem tissues; vessel-segments with scalariform perforation plates, these with 6–25 cross-bars; imperforate tracheary elements with small but evidently bordered pits; wood-rays heterocellular, mixed uniseriate and pluriseriate, the pluriseriate ones narrow, and mostly with elongate ends; wood-parenchyma replaced by lignified (thus nonparenchymatous) cells. LEAVES alternate, entire, pinnately veined; stomates encyclocytic; stipules wanting. Plants dioecious; FLOWERS small, apetalous, the staminate ones in open panicles, without perianth; stamens 2, with very short, connate filaments, the anthers almost sessile, extrorse, opening by longitudinal slits; pollen-grains tricolpate; pistillate flowers commonly paired in spike-like racemes (seldom solitary or in 3's), hypogynous, sometimes with 1–4 scale-like sepals; gynoecium seemingly or actually of a single carpel, the stigma sessile or elevated on a style; ovule solitary, hemitropous, epitropous, bitegmic, the integuments prolonged at the top into a more or less elongate collar. FRUIT drupaceous; seed without endosperm; embryo with 2 thick cotyledons.

The family Didymelaceae consists of a single taxonomically isolated genus, *Didymeles*, with 2 species, both restricted to Madagascar. Most authors have thought *Didymeles* to be related to the Hamamelidales and/ or the Leitneriaceae, but both Novak (1961, in initial citations) and Thorne (1968, 1976, in initial citations) refer it to the Euphorbiales. The primitive wood-structure argues against the latter interpretation. Wolfe (1973, under Hamamelidae) relates *Didymeles* to *Aquilaria*, a putatively archaic member of the Thymelaeaceae, but here again the wood-structure is out of harmony with such a relationship. Both the euphorbialean and the thymelaeaceous concepts of the relationship of *Didymeles* would seem to require that the ovary be pseudomonomerous rather than truly monomerous, a matter that does not appear to have been carefully investigated. The encyclocytic stomates of *Didymeles* are distinctive, but might be compared with those of certain Moraceae. It is here considered prudent to treat *Didymeles* as forming a distinct order, probably derived from the Hamamelidales, parallel in some respects to *Leitneria*.

Pollen considered to represent *Didymeles* is known from Paleocene and more recent deposits.

5. Order EUCOMMIALES F. Němejc 1956[1]

The order consists of the single family Eucommiaceae.

SELECTED REFERENCES

Eckardt, Th. 1956 (1957). Zur systematischen Stellung von *Eucommia ulmoides*. Ber. Deutsch. Bot. Ges. 69: 487–498.

Eckardt, Th. 1963. Some observations on the morphology and embryology of *Eucommia ulmoides* Oliv. J. Indian Bot. Soc. 42 A: 27–34.

Tippo, O. 1940. The comparative anatomy of the secondary xylem and the phylogeny of the Eucommiaceae. Amer. J. Bot. 27: 832–838.

Varossieau, W. W. 1942. On the taxonomic position of *Eucommia ulmoides* Oliv. (Eucommiaceae). Blumea 5: 81–92.

1. Family EUCOMMIACEAE Engler 1909 nom. conserv., the Eucommia Family

Dioecious, anemophilous trees with unilacunar nodes, storing carbohydrate as inulin, somewhat tanniferous, producing a modest amount of proanthocyanin but not ellagic acid, and also with aucubin (an iridoid compound); articulated laticifers present in the phloem and cortex, and scattered latex-cells present in some of the other tissues as well; calcium oxalate crystals wanting, but some of the cells of the bark and leaves (notably the epidermis) with more or less calcified and/or silicified walls; vessel-segments mostly with slanting ends and simple perforations; imperforate tracheary elements short, with bordered pits; wood-rays homocellular or slightly heterocellular, intermediate between Kribs type I and IIB, a few of them uniseriate, the others up to 3 (4) cells wide; sieve-tubes with S-type plastids. LEAVES deciduous, alternate, simple, toothed, pinnately veined; stomates anomocytic; stipules none. FLOWERS racemosely arranged on the proximal, bracteate part of a distally leafy shoot (as in *Euptelea*), individually solitary and short-pedicellate in the axils of the bracts, but without bracteoles, unisexual, regular, without perianth; staminate flowers consisting of 5–12 stamens with very short filament and apically prolonged connective; anthers tetrasporangiate and dithecal, opening by longitudinal slits; pollen-grains binucleate,

[1] This name was proposed but not validated by F. Němejc in Acta Mus. Nat. Prag. 12B: 99. It is here validated by a reference to the original description of its type and only genus, *Eucommia* Oliver, Hooker's Icon. Pl. 20: t. 1950. 1890, and the amplified description in the same journal 25: t. 2361. 1895.

FIG. 2.7 Eucommiaceae. *Eucommia ulmoides* Oliver. a, habit with fruit, $\times\frac{1}{2}$; b, young twig with pistillate flowers, $\times 1$; c, staminate flower, $\times 3$; d, stamen, $\times 3$; e, pistillate flower, $\times 3$; f, pistillate flower, the ovary in long-section, $\times 3$; g, ovules, $\times 12$; h, fruit, $\times 1$.

tricolporate, with solid exine as in *Degeneria,* but with well developed endexine, and with a poorly developed pore in each furrow, said to more nearly resemble the Hamamelidales than the Urticales; pistillate flowers consisting of a flattened, unilocular ovary crowned by a short style and 2 unequal stigmas, interpreted as pseudomonomerous; ovules 2 (but only one maturing), collateral, pendulous, anatropous, apotropous, unitegmic, weakly crassinucellar; endosperm-development cellular. FRUIT a samara; seed with a large, dicotyledonous embryo embedded in copious endosperm. 2n = 34.

The family Eucommiaceae consists of a single genus and species, *Eucommia ulmoides* Oliver, native to montane forests of western China. The family has usually been associated with either the Urticales or the Hamamelidales, or sometimes with the Magnoliales. The pollen has been compared to that of *Cercidiphyllum,* although it differs in detail. Morphologically *Eucommia* is perhaps most nearly at home in the Urticales, but Eckardt has pointed out a series of embryological features that make it anomalous there also. *Eucommia* is more primitive than the Urticales in having two ovules instead of only one, but more advanced in several other features (unitegmic, not fully crassinucellar ovules, absence of stipules). It seems likely that the Eucommiaceae originated from the Hamamelidales independently of the Urticales.

Because of its combination of advanced and primitive features, *Eucommia* cannot be regarded as directly transitional between the Hamamelidales and Urticales. Even so, *Eucommia* bolsters the idea of an evolutionary link between the two other orders, and I do not see why Thorne (1973, under references for Hamamelidae) finds it necessary to deny such a link.

According to Wolfe (1973, under references for Hamamelidae) *Eucommia* leaves occur in Eocene deposits, and *Eucommia* pollen in earliest Eocene, but older records are probably incorrect. Wolfe also considers that the leaf-venation of *Eucommia* has more in common with archaic groups of Rosidae than with the Hamamelidae.

6. Order URTICALES Lindley 1833

Trees, shrubs, woody or herbaceous vines, or herbs, often with characteristic cystoliths or laticifers or both, commonly tanniferous, with proanthocyanins but not (except Barbeyaceae) ellagic acid; nodes mostly trilacunar or pentalacunar (unilacunar in Barbeyaceae); vessel-segments with simple perforations only. LEAVES alternate or less often opposite, mostly simple and very often oblique at the base, sometimes very deeply pinnatifid or palmatifid, or seldom palmately compound (as in *Cannabis*), usually stipulate; stomates most commonly anomocytic, less often of other types. FLOWERS anemophilous or entomophilous, mostly in cymose inflorescences of sometimes very complex structure, rarely solitary, individually small and inconspicuous, unisexual or rarely perfect; sepals (2–) 4–5 (–12), in one or sometimes 2 cycles, commonly valvate, sometimes connate below, or rarely wanting; petals none; stamens isomerous with the sepals and opposite them, seldom fewer or more numerous (up to 15); pollen-grains 2– to many-aperturate; ovary superior or inferior, usually unilocular and with 1 or 2 styles, seldom bilocular, or (Barbeyaceae) the gynoecium of 1–3 more or less distinct, unicarpellate pistils each with a separate style; ovules solitary (or solitary in each of 2 locules), crassinucellar, bitegmic (unitegmic in Barbeyaceae); endosperm-development nuclear. FRUITS of various types, often multiple; seeds with or without endosperm; embryo with 2 cotyledons.

The order Urticales consists of 6 families and about 2200 species. The Moraceae have about a thousand species, the Urticaceae about 700, and the Cecropiaceae about 300. The Ulmaceae have about 150 species, the Cannabaceae only 3, and the Barbeyaceae only one.

There is no doubt about the close relationship among the Moraceae, Cannabaceae, Cecropiaceae, and Urticaceae. The Ulmaceae are a little farther removed, but their affinity to the other 4 families is not in dispute. The relationships of the Barbeyaceae are more debatable, and are discussed under that family. Except for the Barbeyaceae, the pollen of all families is said to be basically similar.

The Urticales have usually been associated with some of the other orders here referred to the subclass Hamamelidae, largely on the basis of floral features. Furthermore, Wolfe (1973) emphasizes the similarity in leaf-venation between several genera of the Ulmaceae and many other Hamamelidae, notably members of the Fagaceae and Betulaceae. Pollen of the Urticales is said to suggest that of the *Normapolles* fossil group, as does that of several other orders of Hamamelidae. On the other hand, there is also a long-standing school of thought, exemplified most recently by Thorne (1973, under Hamamelidae) and Berg (1977) that the Urticales are allied to the Malvales. Thorne bases his opinion at least partly on anatomical features, in contrast to Tippo (1938), who used anatomical features as important evidence favoring derivation of the Urticales from the Hamamelidales. Tippo did not discuss the possibility of an alliance with the Malvales. Sweitzer (1971) also maintains that on anatomical grounds the Urticales are probably related to the Hamamelidales and not to the Malvales.

It may be noted that if *Barbeya* is referred to the Urticales, the case for including the Urticales in the Hamamelidae is strengthened. The more or less distinct carpels of *Barbeya* are wholly compatible with the Hamamelidae but would be aberrant in a group to be associated with the Malvales. Thorne (1976, initial citations), however, considers *Barbeya* to be *incertae sedis*.

The Urticales are ecologically diversified. They all have reduced flowers that might be supposed a priori to be wind-pollinated, but many of them are in fact pollinated by insects. There is virtually nothing in the vegetative features to fit the order to any coherent set of habitats or ecological niches.

SELECTED REFERENCES

Bechtel, A. R. 1921. The floral anatomy of the Urticales. Amer. J. Bot. 8: 386–410.

Behnke, H.-D. 1973. Sieve-tube plastids of Hamamelidae—Electron microscopic investigations with special reference to Urticales. Taxon 22: 205–210.

Berg, C. C. 1977. Urticales, their differentiation and systematic position. Pl. Syst. Evol., Suppl. 1: 349–374.

Berg, C. C. 1978. Cecropiaceae, a new family of Urticales. Taxon 27: 39–44.

Corner, E. J. H. 1962. The classification of Moraceae. Gard. Bull. Singapore 19: 187–252.

Dickison, W. C., & E. M. Sweitzer. 1970. The morphology and relationships of *Barbeya oleoides*. Amer. J. Bot. 57: 468–476.

Elias, T. S. 1970. The genera of Ulmaceae in the southeastern United States. J. Arnold Arbor. 51: 18–40.

Freisleben, R. 1933. Untersuchungen über Bildung und Auflösung von Cystolithen bei den Urticales. Flora 127: 1–45.

Gangadhara, M., & J. A. Inamdar. 1977. Trichomes and stomata, and their taxonomic significance in the Urticales. Pl. Syst. Evol. 127: 121–137.

Giannasi, D. E. 1978. Generic relationships in the Ulmaceae based on flavonoid chemistry. Taxon 27: 331–344.

Johri, B. M., & R. N. Konar. 1956. The floral morphology and embryology of *Ficus religiosa* Linn. Phytomorphology 6: 97–111.

Le Coq, C. 1963. Contribution a l'étude cytotaxonomique de Moracées et des Urticacées. Rev. Gén. Bot. 70: 385–426.

Leins, P., & C. Orth. 1979. Zur Entwicklungsgeschichte männlicher Blüten von *Humulus lupulus* (Cannabaceae). Bot. Jahrb. Syst. 100: 372–378.

Macdonald, A. D. 1974. Theoretical problems of interpreting floral organogenesis of *Laportea canadensis.* Canad. J. Bot. 52: 639–644.

Miller, N. G. 1970. The genera of the Cannabaceae in the southeastern United States. J. Arnold Arbor. 51: 185–203.

Miller, N. G. 1971. The genera of the Urticaceae in the southeastern United States. J. Arnold Arbor. 52: 40–68.

Mohan Ram, H. Y., & R. Nath. 1964. The morphology and embryology of *Cannabis sativa* Linn. Phytomorphology 14: 414–429.

Rabiger, F. H. 1951. Untersuchungen an einigen Acanthaceen und Urticaceen zur Funktion der Cystolithen. Planta 40: 121–144.

Record, S. J., & R. W. Hess. 1940. American woods of the family Moraceae. Trop. Woods 61: 11–54.

Shah, A. M., & P. Kachroo. 1975. Comparative anatomy in Urticales. I. The trichomes in Moraceae. J. Indian Bot. Soc. 54: 138:153.

Singh, S. P. 1956. Floral anatomy of *Cannabis sativa* L. Agra Univ. J. Res., Sci. 5: 155–162.

Small, E. 1978. A numerical and nomenclatural analysis of morpho-geographic taxa of *Humulus.* Syst. Bot. 3: 37–76.

Small, E., & A. Cronquist. 1976. A practical and natural taxonomy for *Cannabis.* Taxon 25: 405–435.

Sweitzer, E. M. 1971. Comparative anatomy of Ulmaceae. J. Arnold Arbor. 52: 523–585.

Thurston, E. L., & N. R. Lersten. 1969 (1970). The morphology and toxicology of plant stinging hairs. Bot. Rev. 35: 393–412.

Tippo, O. 1938. Comparative anatomy of the Moraceae and their presumed allies. Bot. Gaz. 100: 1–99.

Venkataraman, K. 1972. Wood phenolics in the chemotaxonomy of the Moraceae. Phytochemistry 11: 1571–1586.

Yamada, H., & O. Yoshida. 1979. Embryological study of the family Ulmaceae I. Embryology of *Zelkova serrata* Makino. J. Coll. Arts Chiba Univ. B-12: 27–43.

Грудзинская, И. А. 1967. Ulmaceae и обоснование выделения Celtidoideae в самостоятельное семейство Celtidaceae Link. Бот. Ж. 52: 1723-1749.

Грудзинская, И. А. 1974. О гетеробластном развитии *Ulmus.* Бот. Ж. 59: 1160-1171.

Колобкова, Е. В. 1972. Эволюция белковых комплексов семян порядков Urticales и Casuarinales. В сб.: Биохимия и филогения растений, Е. В. Колобкова, ред.: 9-31. Изд. Наука, Москва 1972.

Куприянова, Л. А. 1962. Палинологические данные к систематике порядков Fagales & Urticales. В сб.: Докл. Сов. Палинологов к Первой Международной Палинолог. Конф: 17-25.

Паланджян, В. А. 1953. Строение древесины сем. ильмовых в связи с их зволюцией и систематикой. Труды Бот. Инст. АН Армянск. ССР, 9: 121-178.

Черник, В. В. 1975. Расположение и редукция частей околоцветника и андроцея у представителей Ulmaceae Mirbel и Celtidaceae Link. Бот. Ж 60: 1561-1573.

SYNOPTICAL ARRANGEMENT OF THE FAMILIES OF URTICALES

1 Leaves exstipulate, opposite; nodes unilacunar; gynoecium of 1–3 distinct or partly connate pistils, each with a separate style .... 1. BARBEYACEAE.

1 Leaves usually stipulate, variously opposite or alternate; nodes trilacunar or pentalacunar; gynoecium of a single unilocular (rarely bilocular) pistil.

2 Ovary with 2 styles (but one style sometimes more or less reduced); ovule apical, pendulous, anatropous.

3 Plants without laticifers, and without milky juice.

4 Woody plants .... 2. ULMACEAE.

4 Herbs (or herbaceous vines) ....3. CANNABACEAE.

3 Plants nearly always with laticifers and milky juice .... 4. MORACEAE.

2 Ovary with a single style, pseudomonomerous; ovule basal or nearly so, erect, more or less orthotropous.

5 Stamens more or less erect in bud, not elastically reflexed in dehiscence; woody plants; cystoliths wanting .... 5. CECROPIACEAE.

5 Stamens inflexed in bud, elastically reflexed in dehiscence; herbs or occasionally half-shrubs, or rarely small, soft-wooded trees; cystoliths generally present ....6. URTICACEAE.

1. Family BARBEYACEAE Rendle 1916 nom. conserv., the Barbeya Family

Small trees, accumulating ellagic acid, but without cystoliths, latex, or laticifers; nodes unilacunar, with a single trace; vessel-segments very short, with simple perforations; imperforate tracheary elements with simple or bordered pits; wood-rays somewhat heterocellular, mixed uniseriate and pluriseriate; wood-parenchyma very scanty, diffuse; phloem of a relatively primitive type, the sieve-tubes with very oblique, compound sieve plates. LEAVES opposite, simple, entire; stomates mainly laterocytic; stipules none. Plants dioecious, anemophilous; FLOWERS small, inconspicuous, in short, sessile or subsessile axillary cymes, regular, hypogynous, without bracts or bracteoles; sepals 3 or 4, slightly connate at the base, those of the pistillate flowers pinnately net-veined and slightly accrescent in fruit; petals none; stamens 6–9 (–12), with short filament, elongate anther, and apiculate connective; pollen-grains tricolporate, said to differ from those of other Urticales; gynoecium of 1–3 distinct or proximally more or less connate carpels, each with its own locule and elongate style with decurrent stigma; each carpel with a single pendulous, subapical, anatropous, apparently unitegmic ovule. FRUIT a short-beaked nut with associated accrescent, somewhat membranous, prominently veined sepals; seed with a straight, dicotyledonous embryo and without endosperm.

The family Barbeyaceae consists of a single species, *Barbeya oleoides* Schweinfurth, of northeastern Africa and adjacent Arabia. *Barbeya* has been variously treated as an aberrant member of the Ulmaceae, as a family Barbeyaceae in the order Urticales, as an order Barbeyales allied to the Urticales (by Takhtajan), or as a genus *incertae sedis* (by Thorne). I here follow Dickison and Sweitzer (1970) in considering that the ensemble of characters, including those of vascular anatomy, supports the recognition of the Barbeyaceae as a distinct family within the order Urticales. The more or less distinct carpels and primitive phloem of *Barbeya* indicate a position fairly low on the evolutionary tree of angiosperms, and favor the association of the Urticales with the Hamamelidae rather than with such more advanced groups as the Malvales.

2. Family ULMACEAE Mirbel 1815 nom. conserv., the Elm Family

Woody plants without latex or laticifers, somewhat tanniferous, with proanthocyanins but not ellagic acid, seldom (as in *Trema*) cyanogenic

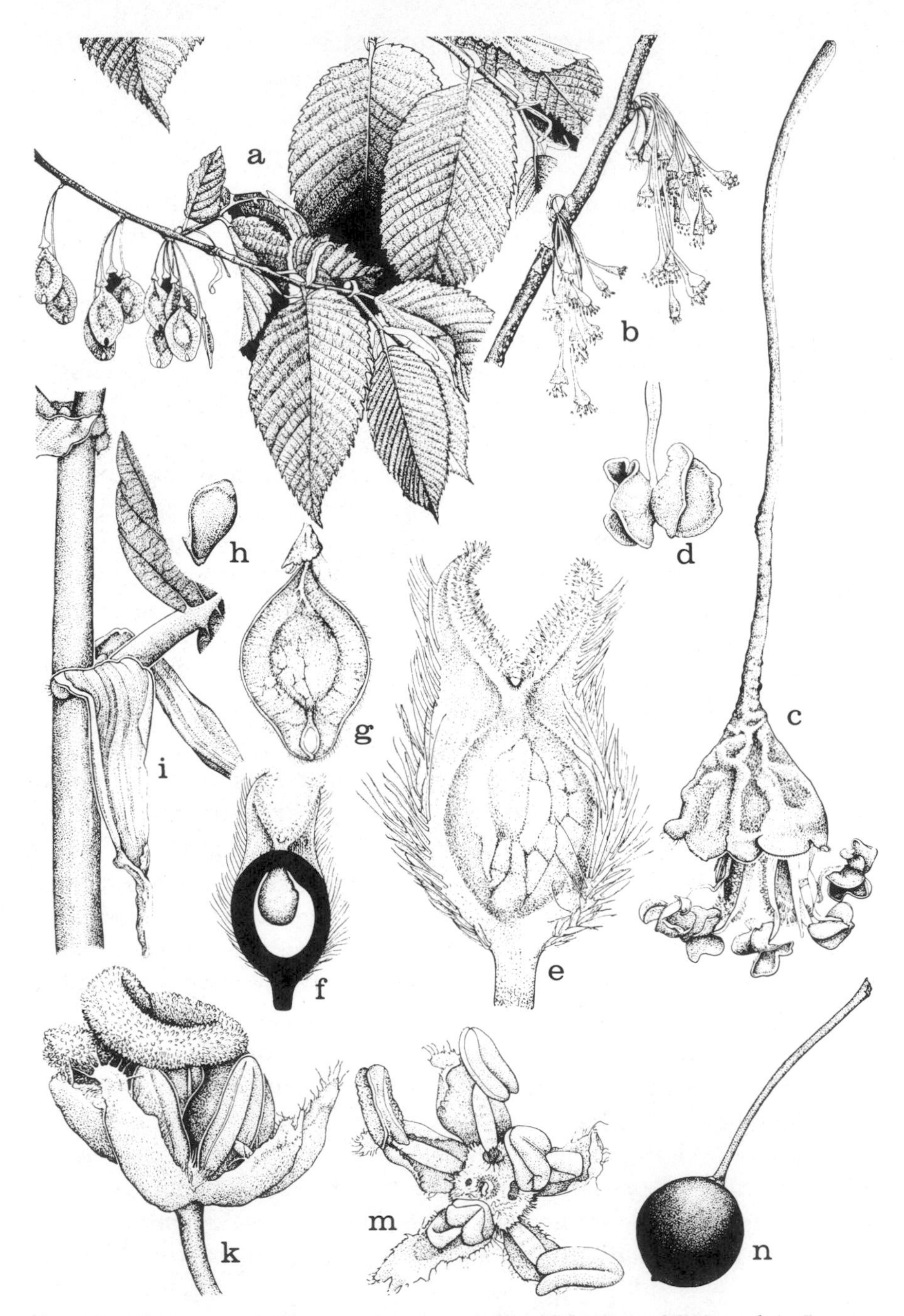

FIG. 2.8 Ulmaceae. a–i, *Ulmus americana* L. a, twig, with leaves and fruits, ×½; b, flowers, ×1; c, flower, ×12; d, anther, ×12; e, pistil, ×12; f, pistil, in partial long-section, ×6; g, fruit, ×2; h, seed, ×6; i, stipules, ×2. k–n, *Celtis laevigata* Willd. k, flower, ×6; m, flower, from above, ×6; n, fruit, ×2.

via a tyrosine-based pathway, commonly with scattered mucilage-cells or even mucilage-canals, and with a strong tendency toward mineralization of the cell-walls (especially of epidermal cells and trichomes) with calcium carbonate or silica; nodes trilacunar; solitary or clustered crystals of calcium oxalate commonly present in some of the parenchyma-cells; vessel-segments short, with simple or seldom scalariform perforations; imperforate tracheary elements with simple or slightly bordered pits, sometimes intermingled with vascular tracheids that have evidently bordered pits; wood-rays mixed uniseriate and pluriseriate, homocellular to heterocellular; wood-parenchyma commonly paratracheal; sieve tubes with P-type plastids in *Ulmus*, S-type in *Celtis* and some other genera. LEAVES alternate and commonly distichous, or rarely opposite, simple, often oblique at the base, pinnately or pinnipalmately veined; stomates anomocytic or with poorly differentiated supporting cells; cystoliths commonly present, especially in the epidermis, these with or without a cellulosic peg around which the mineral is deposited; stipules lateral or intrapetiolar, protecting the bud, deciduous. Plants anemophilous, monoecious or sometimes (as in *Ulmus*) with perfect flowers, the FLOWERS small, inconspicuous, regular or nearly so, hypogynous or perigynous, in cymose inflorescences or solitary and axillary, in some genera (e.g., *Celtis*) axillary to bracts on the proximal part of a distally leafy shoot; sepals (2–) 5 (–9), distinct or connate; petals none; stamens as many as the sepals and opposite them, or rarely twice as many or up to 15, free or with the filaments borne on the calyx-tube, usually erect in bud; pollen-grains 2–3 nucleate, 2– to many-aperturate (colpate to more often porate), with a thin exine and granular rather than columellate infratectal structure (at least in *Ulmus*); gynoecium of 2 (3) carpels united to form a compound, unilocular or rarely bilocular ovary with separate styles and decurrent stigmas; ovule solitary (or solitary in each of 2 locules), pendulous from the apex of the locule, anatropous or amphitropous, bitegmic, crassinucellar, the micropyle formed by both integuments (*Celtis*) or by only one (*Ulmus*); fertilization either porogamous (as in *Zelkova*) or chalazogamous (as in *Ulmus*); endosperm-development nuclear. FRUIT a nut, drupe, or samara; seeds with straight or curved, dicotyledonous embryo and rather scanty or no endosperm; endosperm sometimes (as in *Ulmus, Zelkova,* and *Hemiptelea*) consisting of a single layer of thick-walled cells, easily confused with the properly 3-layered seed-coat. X = 10, 11, 14.

The family Ulmaceae consists of about 18 genera and 150 or more species, widely distributed in both tropical and temperate regions, especially in the Northern Hemisphere. The largest genera are *Celtis* (70), *Trema* (30) and *Ulmus* (20). Species of *Ulmus* (elm) and *Celtis* (hackberry) are well known in cultivation in temperate regions.

According to Wolfe (1973, in references for Hamamelidae), pollen of *Ulmus*-type occurs in deposits of Maestrichtian (late Upper Cretaceous) age. Wood considered to be ulmaceous occurs in Eocene deposits in Yellowstone National Park.

It is customary to recognize two subfamilies, the Ulmoideae and Celtidoideae. Typical members of the two groups differ in an impressive list of features, as given in the accompanying table, but statements of micromorphological and chemical differences are based on a limited sampling. Grudzinskaja vigorously maintains in print and in personal conversation that the Celtidoideae should be treated as a distinct family Celtidaceae, more closely allied to the Moraceae than to the Ulmaceae. Kuprianova had earlier (1962) revived the family Celtidaceae on the basis of pollen structure, and transferred *Zelkova* to the Ulmaceae sens. strict. On the other hand, as Sweitzer has pointed out, the Ulmoideae and Celtidoideae are morphologically interconnected, and the position of individual genera could be debated. The chemical characters, insofar as they have been studied at this time, do not fully resolve the difficulties.

The problem is exemplified by *Zelkova*, which is usually included in the Celtidoideae, but which Grudzinskaja refers to the Ulmaceae proper. *Zelkova* has drupaceous fruits, as in the Celtidoideae, but strictly pinnately veined leaves, as in the Ulmoideae. In most other respects it is more nearly like the Ulmoideae. Its pollen is distinctive, but referrable to the Ulmoideae rather than the Celtidoideae. It has flavonols, as in the Ulmoideae, rather than glycoflavones, as in the Celtidoideae. It is easy enough to agree with Grudzinskaja that *Zelkova* has been misplaced and properly belongs with the Ulmoideae, but other genera present other combinations of the characters of the two subfamilies, and their placement might be subject to similar re-examination. Even Grudzinskaja considers (personal communication) that *Aphananthe* forms a link between the two groups. In such circumstances, and especially in view of the fact that we are here dealing with only about 18 genera and 150 species, I feel more comfortable with the traditional treatment in which the family Ulmaceae is broadly defined and is considered to consist of two subfamilies.

Typical members of the two subfamilies differ as follows:

ULMOIDEAE	CELTIDOIDEAE
Leaves strictly pinnately veined, craspedodromous	Leaves pinnipalmately veined, with 3 main veins from the base, brochidodromous
Fruit dry, most commonly a samara	Fruit drupaceous
Seeds flattened, with straight embryo	Seeds globose, with a curved embryo
Pollen 4–6-porate; exine rugulose	Pollen 2–3-porate; exine smooth
Wood-rays all homocellular or weakly heterocellular	Uniseriate wood-rays homocellular or nearly so, the pluriseriate ones more or less strongly heterocellular
Produce lignans, sesquiterpenes, and flavonols, but not quebrachitol and glycoflavones	Produce quebrachitol and glycoflavones, but not lignans, sesquiterpenes, and flavonols
X = 14	X = 10, 11

3. Family CANNABACEAE Endlicher 1837 nom. conserv. the Hemp Family

Erect (*Cannabis*) or twining (*Humulus*) herbs, accumulating quebrachitol and producing pyridine alkaloids, and sometimes tanniferous, with proanthocyanins; stem with secretory canals in the phloem, but without milky juice; solitary or clustered crystals of calcium oxalate commonly present in some of the parenchymatous-cells; vessel-segments with simple perforations; sieve-tubes with S-type plastids. LEAVES opposite at least below, in *Cannabis* commonly alternate above, palmately lobed (*Humulus*) or palmately compound (*Cannabis*), or the upper leaves sometimes simple and unlobed; nonglandular hairs of the epidermis intermingled with hairs bearing a multicellular, glandular cap containing aromatic or psychotrophic substances; characteristic cystoliths commonly present in the basal part of some of the nonglandular hairs of the leaves and stem; stomates anomocytic; stipules free, persistent. Plant monoecious or more often dioecious, anemophilous, with small, unisexual FLOWERS in complex, basically cymose inflorescences axillary to the upper (often reduced) leaves, the male inflorescences relatively loose, branched, and many-flowered, the female ones more compact and few-flowered; male flowers with 5 sepals, 5 stamens opposite the sepals, and

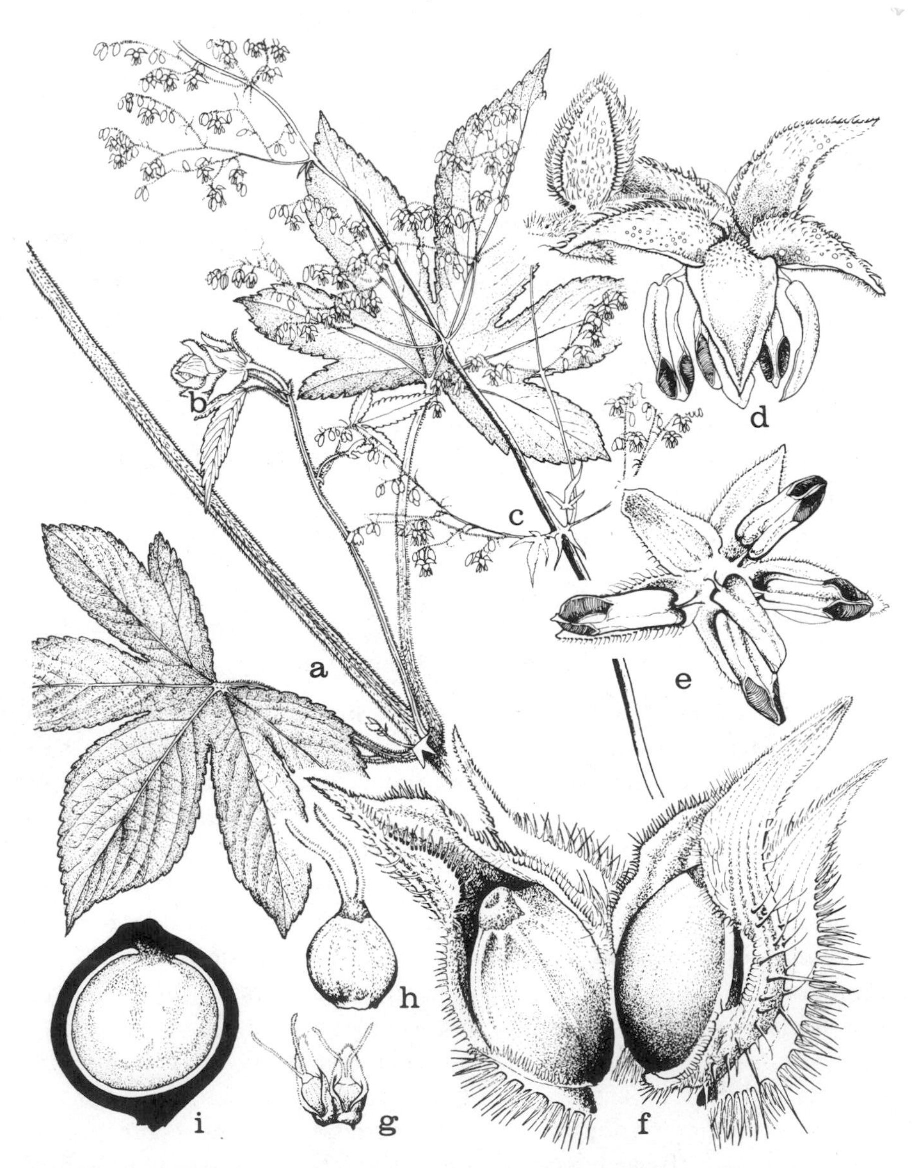

FIG. 2.9 Cannabaceae. *Humulus japonicus* Sieb. & Zucc. a, habit, $\times\frac{1}{2}$; b, pistillate inflorescence, $\times\frac{1}{2}$; c, staminate inflorescence, $\times\frac{1}{2}$; d, e, staminate flower in side and top view, $\times 6$; f, fruits with subtending bracts, $\times 6$; g, pistillate flowers, with subtending bracts, $\times 6$; h, pistillate flower, $\times 24$; i, long-section of ovary and single ovule, $\times 48$.

no gynoecium; sepals and stamens originating in a continuous spiral on a 2/5 phyllotaxy; filaments and anthers erect in bud; anthers tetrasporangiate and dithecal; pollen-grains binucleate, triporate, or seldom 2-, 4-, or 6-porate; female flowers with a short, membranous calyx-tube enclosing the ovary in *Humulus* and wild-adapted forms of *Cannabis*, or the calyx reduced to a mere ring in cultivated forms of *Cannabis*; gynoecium of 2 carpels united to form a compound, unilocular ovary capped by 2 elongate, slender, dry stigmas; ovule solitary, subapical, anatropous, bitegmic, crassinucellar, the micropyle formed by the inner integument; endosperm-development nuclear. FRUIT a small nut or an achene, in *Humulus* evidently invested by the persistent calyx, but in *Cannabis* the calyx evanescent or persistent in part as membranous portions adnate to the pericarp; seeds with curved (*Cannabis*) or coiled (*Humulus*), dicotyledonous embryo and a small amount of fleshy, oily endosperm. X = 8 (*Humulus*), 10 (*Cannabis*).

The family Cannabaceae consists of only 2 small, sharply limited genera: *Humulus* (hops), with 2 species native to North Temperate regions, and *Cannabis*, with a single highly variable species, *C. sativa* L. *C. sativa* has diversified under cultivation into a more northern subspecies *sativa*, cultivated principally for fiber (hemp), and a more tropical subspecies *indica* (Lam.) Small & Cronq., cultivated principally for psychotropic drugs (marijuana, hashish). Cultivars of subsp. *indica* contain a higher proportion of the active drug (Δ_9 tetrahydrocannabinol, abbreviated THC) than those of subsp. *sativa*. Each subspecies has a wild-adapted counterpart to the cultivated forms, but original wild material of the genus cannot be certainly identified.

The Cannabaceae are obviously related to the Moraceae, and have often been included in that family. Either treatment is defensible, but definition of the families within the order Urticales is facilitated by the removal of both the Cannabaceae and the Cecropiaceae from the Moraceae.

4. Family MORACEAE Link 1831 nom. conserv., the Mulberry Family

Trees, shrubs, lianas, or rarely herbs, often with one or another sort of unbranched, glandular hair, nearly always with milky juice borne in laticifers in the parenchymatous parts of the stem and often also in the leaves, the contents of the laticifers extraordinarily diverse in different members of the family (laticifers wanting in *Fatoua*); plants sometimes

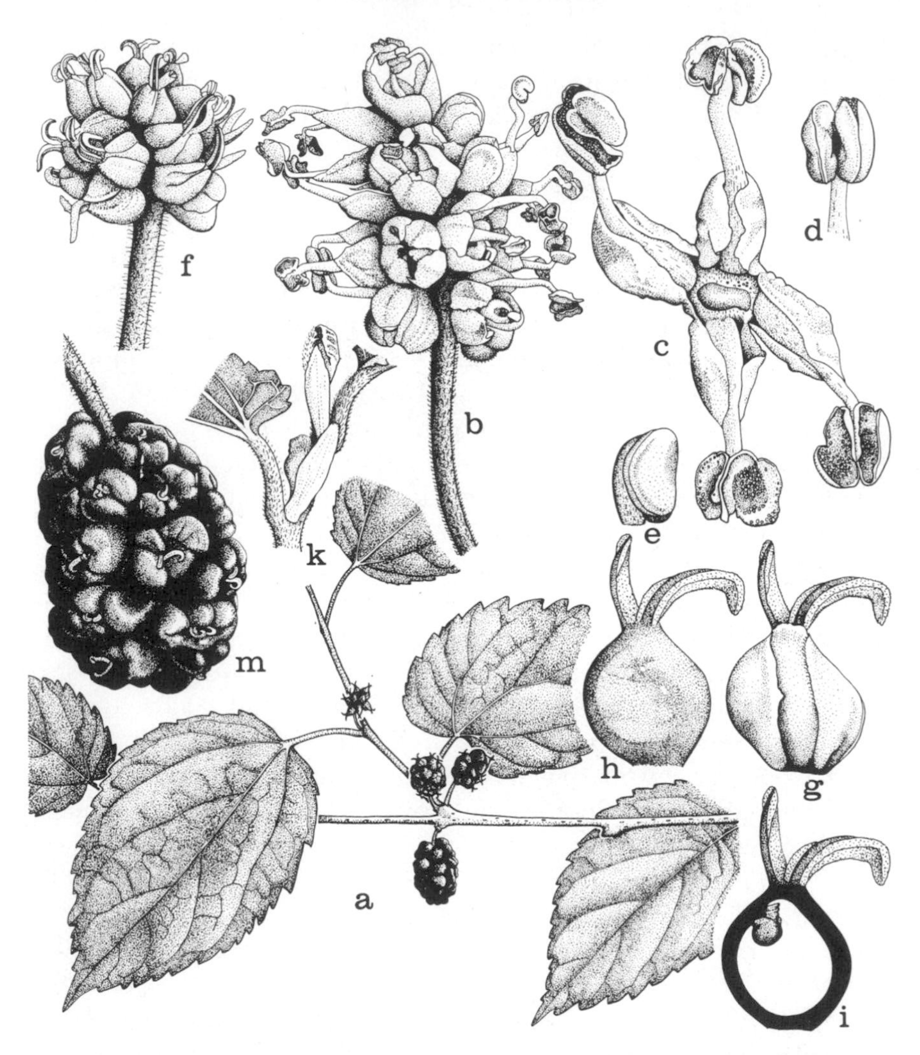

FIG. 2.10 Moraceae. *Morus alba* L. a, habit, with fruits, ×½; b, staminate inflorescence, ×4; c, staminate flower, ×8; d, anther, ×8; e, young stamen from bud, ×8; f, pistillate inflorescence, ×4; g, pistillate flower, ×8; h, pistillate flower, with sepals removed, ×8; i, pistil, the ovary in long-section, ×8; k, young twig, with stipules, ×2; m, fruit, ×2.

producing alkaloids, and often somewhat tanniferous, at least sometimes with proanthocyanins; nodes trilacunar or pentalacunar; vessel-segments with simple perforations; imperforate tracheary elements with simple or obscurely bordered pits, sometimes septate; wood-rays heterocellular or sometimes homocellular, mixed uniseriate and pluriseriate, the latter up to 5 (–12) cells wide; wood-parenchyma more or less abundant, often vasicentric or aliform; sieve-tubes with S-type plastids. LEAVES opposite or alternate, simple or rarely compound, often with cystoliths especially in the epidermis, and often with the cell walls especially of the epidermis and trichomes more or less mineralized with calcium carbonate or silica; stomates anomocytic to less often encyclocytic or anisocytic; stipules present, but sometimes much reduced, as in *Dorstenia*. Plants anemophilous or (notably in *Ficus*) entomophilous, monoecious or dioecious, with small, unisexual, apetalous FLOWERS in compact, axillary inflorescences, the axis of the inflorescence often thickened to form a head or an invaginated common receptacle; calyx of (1–) 4–5 (–8) separate or more or less connate sepals, often in 2 cycles, or sometimes wanting; stamens in the male flowers commonly as many as the sepals and opposite them, rarely only 1–3; filaments straight or inflexed in bud; anthers tetrasporangiate and dithecal (bisporangiate and monothecal in spp. of *Ficus*); pollen-grains binucleate, mostly 2–4-porate or multiporate; gynoecium in female flowers bicarpellate (rarely tricarpellate), or one carpel often more or less reduced, the superior to inferior ovary accordingly bilocular (trilocular) or much more often unilocular; styles or style-branches generally 2, sometimes one of them more or less strongly reduced; ovules solitary in each locule (or one locule empty), apical or subapical, anatropous to hemitropous or campylotropous, bitegmic, crassinucellar, the micropyle formed by the inner integument and visible only in the young ovule; endosperm-development nuclear. FRUITS mostly drupaceous but with dehiscent exocarp, varying to truly drupaceous with seed-like pyrenes, the common receptacle often ripening with the ovaries to form a fleshy syconium; seeds (pyrenes) with straight or more often curved embryo, the cotyledons often unequal, sometimes one almost wholly suppressed; endosperm fleshy and oily, or often wanting. X = 7-many.

The family Moraceae consists of about 40 genera and nearly a thousand species, widespread in tropical and subtropical regions, less common in temperate climates. The largest genus by far is *Ficus*, with more than 500 species. The next largest is *Dorstenia*, with about 100. *Ficus carica* L. (common fig), *F. elastica* Roxb. (India rubber plant), *F. benghalensis* L. (Banyan), *F. benjamina* L. (weeping fig), and *F. sycamorus* L. (sycamore fig) are well known in cultivation. *Artocarpus altilis* (Parkins.) Fosb. (breadfruit) is an important source of carbohydrate in the paleotropics. Species of *Morus* (mulberry) are well known in temperate regions, and *Maclura pomifera* (Raf.) Schneider (osage orange) is also familiar in the United States.

Shorn of the Cannabaceae and Cecropiaceae, which have often been included, the Moraceae become a well characterized and clearly limited group.

According to Wolfe (1973, in citations for Hamamelidae), fossils representing the Moraceae occur back at least to the early Eocene. Paleocene and late Cretaceous fossil leaves referred to *Artocarpus* need to be re-examined, and early fossils referred to *Ficus* are misidentified.

5. Family CECROPIACEAE Berg 1978, the Cecropia Family

Trees, shrubs, or woody vines (sometimes epiphytic), with stilt-roots and/or aerial-roots; nodes trilacunar (or pentalacunar?); laticifers more or less reduced (and restricted to the bark) or none, the plants only rarely with milky juice, but mucilage-cells and sacs often present; clustered crystals of calcium oxalate often present in some of the parenchyma-cells; vessel-segments with simple perforations; imperforate tracheary elements with simple pits, not septate; wood-rays heterocellular, mixed uniseriate and pluriseriate; wood-parenchyma mostly vasicentric, at least in *Cecropia*. LEAVES alternate, simple and entire to often so deeply lobed as to appear compound; venation pinnate or palmate; stomates anisocytic at least in *Poikilospermum*; epidermis commonly with more or less silicified cell walls; cystoliths wanting (?); stipules present. Plants entomophilous or anemophilous, monoecious, the FLOWERS small, unisexual; sepals 2–4, distinct or connate; stamens 2–4, the filaments straight in bud; anthers tetrasporangiate and dithecal; pistil pseudomonomerous, apparently of a single carpel with a single style and stigma; ovule solitary, basal or nearly so, bitegmic, crassinucellar, more or less orthotropous. FRUIT a nutlet or drupelet, or multiple as in many Moraceae; embryo straight, with 2 equal cotyledons. X = 7.

The family Cecropiaceae consists of 6 genera, all tropical: *Cecropia* (100), *Coussapoa* (75), *Pourouma* (75), *Poikilospermum* (=*Conocephalus*, 20), *Myrianthus* (5), and *Musanga* (1). *Cecropia* is a conspicuous and familiar tree of disturbed habitats in the moist American tropics. Its large leaves appear on first inspection to be palmately compound, but are instead very deeply palmatifid.

The Cecropiaceae have usually been treated as a subfamily (Conocephaloideae) of the Moraceae, or less often as a subfamily of the Urticaceae. In either of these families they form a discordant element.

Since they form a coherent, well defined group of nearly 300 species, I agree with Berg that it is better to treat them as a distinct family than to force them into either the Moraceae or the Urticaceae.

6. Family URTICACEAE A. L. de Jussieu 1789 nom. conserv., the Nettle Family

Herbs or occasionally half-shrubs or rarely small, soft-wooded trees, very rarely lianas, often provided with specialized stinging (sometimes very potent) hairs, frequently with mucilage cells, and sometimes also with nonarticulated, unbranched latex-channels, but ordinarily without milky juice, at least sometimes tanniferous, with proanthocyanins; diverse sorts of calcium carbonate cystoliths commonly present in both the stems and the leaves, often independently of the hairs; nodes trilacunar; vessel-segments with simple perforations; imperforate tracheary elements with simple pits, often septate and filled with silica; xylem-parenchyma commonly paratracheal; rays somewhat heterocellular; sieve-tubes with S-type plastids. LEAVES opposite or alternate, simple; epidermal cell-walls tending to be mineralized with silica or calcium carbonate; stomates paracytic or anomocytic; stipules generally present, but wanting in *Parietaria*. Plants mostly anemophilous, monoecious or dioecious or polygamous; FLOWERS small, individually inconspicuous, in basically cymose, often axillary inflorescences that may be reduced to a single flower, hypogynous, regular or seldom somewhat irregular; staminate flowers with (3) 4–5 (6) sepals and as many stamens opposite them, or seldom with only a single stamen; filaments inflexed in bud, elastically reflexed when the pollen is shed; anthers tetrasporangiate and dithecal, opening by longitudinal slits; pollen-grains binucleate, biporate or triporate or sometimes multiporate; gynoecium in the staminate flowers evident but vestigial and sterile; pistillate flowers with 4 (5) distinct or more or less connate sepals, or completely without perianth, sometimes with scale-like staminodes opposite the sepals; gynoecium pseudomonomerous, the ovary unilocular, with a single style and a capitate or filiform and decurrent stigma (rarely 2 stigmas, as in *Phenax*) ovule solitary, basal, orthotropous or seldom hemitropous, crassinucellar, bitegmic, the micropyle commonly formed by the inner integument; endosperm-development nuclear. FRUIT an achene or a small nut, or seldom a drupe, often enclosed in a persistent, accrescent perianth; seeds with more or less reduced or vestigial testa and a

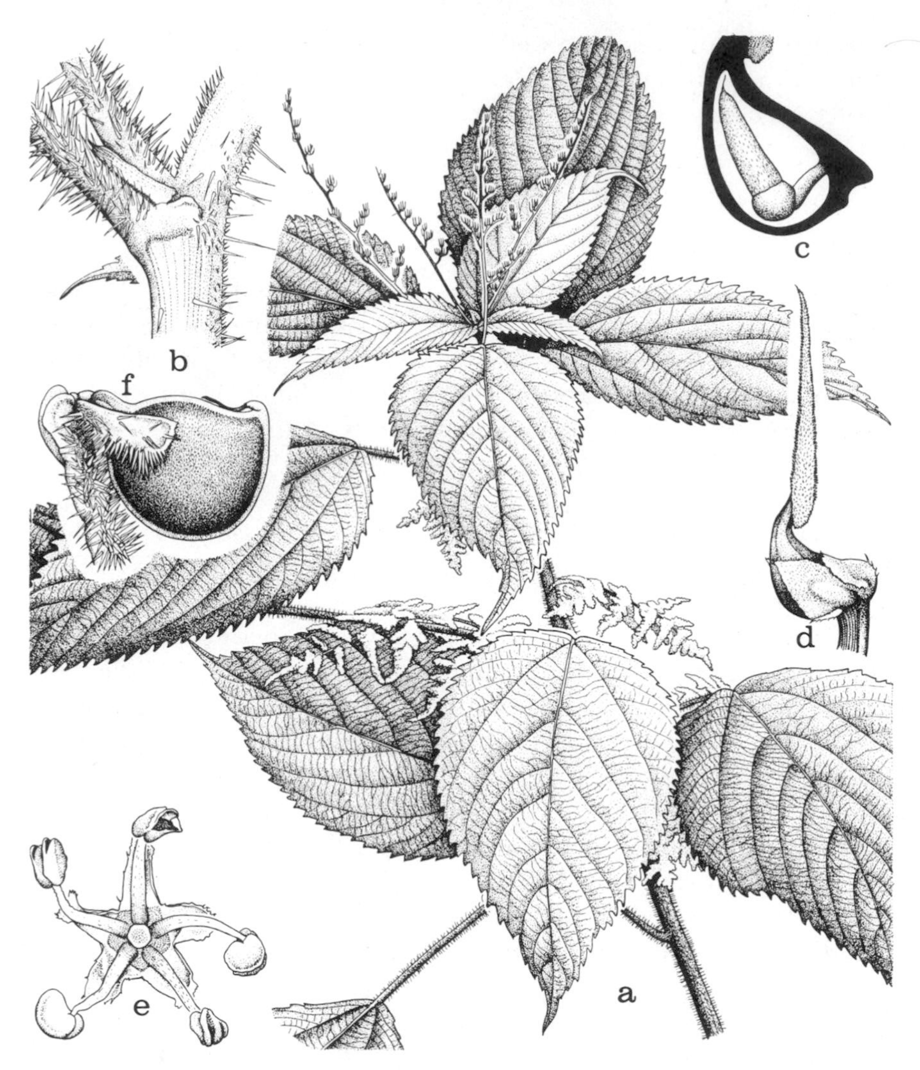

FIG. 2.11 Urticaceae. *Laportea canadensis* (L.) Willd. a, habit, ×½; b, node, with stipule abscising, ×4; c, pistil, the ovary in long-section, ×24; d, pistillate flower, ×12; e, staminate flower, ×12; f, achene, ×6.

straight, spatulate, dicotyledonous embryo, this often surrounded by a thin, oily or starchy endosperm, but sometimes (as in *Elatostema*) without endosperm. X = 6–14.

The family Urticaceae consists of about 45 genera and 700 or more species, widely distributed in tropical and subtropical regions. Relatively few species occur in the cooler parts of the world, but some of these, such as species of *Urtica* (nettle) are common and widespread. About half of the species belong to the paleotropical genus *Elatostema* (350), and another 200 to the pantropical genus *Pilea*. *Boehmeria nivea* (L.) Gaudich. is the source of the fiber ramie. Species of *Pilea* (aluminum-plant), *Helxine* (baby's tears), and *Elatostema* are sometimes grown as ornamentals.

It is still an open question whether the Urticaceae should be regarded as derived from the Moraceae or whether the two diverged from a common ancestor that would not have fit well into either family. The herbaceous or softly woody habit, pseudomonomerous gynoecium, and orthotropous, basal ovule of the Urticaceae are clearly advanced, as compared to the Moraceae. The usual absence of milky juice from the Urticaceae is another matter. Whether the poorly developed latex system of some urticaceous genera such as *Laportea* and *Urera* is vestigial or (in contrast) rudimentary remains to be determined.

Recognition of the Cannabaceae and Cecropiaceae as families distinct from the Moraceae facilitates the distinction of the Urticaceae from the Moraceae, but does not avoid all problems. The herbaceous, monotypic genus *Fatoua* is always retained in the Moraceae, but superficially it looks very urticaceous except for the inconspicuous, vestigial second style.

Macrofossils considered to represent the Urticaceae occur in early Eocene and more recent deposits.

7. Order LEITNERIALES Engler 1897

The order consists of the single family Leitneriaceae.

SELECTED REFERENCES

Abbe, E. C., & T. T. Earle. 1940. Inflorescence, floral anatomy and morphology of *Leitneria floridana*. Bull. Torrey Bot. Club 67: 173–193.
Channell, R. B., & C. E. Wood. 1962. The Leitneriaceae in the southeastern United States. J. Arnold Arbor. 43: 435–438.

1. Family LEITNERIACEAE Benth. in Benth. & Hook. 1880 nom conserv., the Corkwood Family

Mostly dioecious, anemophilous shrubs or small trees with very light and soft wood; nodes trilacunar; plants with scattered tanniferous cells, and with resin-canals in the pith, petioles, and leaf-veins; clustered crystals of calcium oxalate present in some of the parenchyma-cells; vessel-segments with simple perforations; imperforate tracheary elements thin-walled, and with very small pits; wood-rays exclusively uniseriate, or sometimes a few of them biseriate; wood-parenchyma scanty-paratracheal to vasicentric and terminal; phloem tangentially stratified into hard and soft layers; sieve-tubes with S-type plastids. LEAVES deciduous, alternate, entire, provided with uniseriate, nonglandular hairs and multicellular, clavate glands; stomates anomocytic; stipules wanting. FLOWERS small, inconspicuous, all or mostly unisexual, in erect catkins of complex structure involving numerous reduced dichasia; some perfect flowers seldom intermingled with the staminate ones; staminate flowers without perianth, strongly reduced, interpreted as occurring in 3-flowered cymules with 1–4 (5) stamens each, forming a cluster of 3–12 (–15) stamens in what might otherwise be taken as a single flower; filaments short; anthers tetrasporangiate and dithecal, opening by longitudinal slits; pollen-grains 3–6-colporate; pistillate flowers surrounded at the base by (3) 4 (–8) small, scale-like, unequal and irregularly disposed bracts or sepals (according to interpretation); gynoecium pseudomonomerous, with a linear, distally elongate-expanded style and decurrent stigma, the stigmatic groove and the placenta turned away from the axis, toward the subtending bract; teratological flowers with a compound pistil and 2 carpels, 2 locules, and 2 styles rarely occur; ovule solitary, parietal, pendulous, anatropous to amphitropous, bitegmic, crassinucellar; endosperm-development nuclear. FRUIT a dry drupe; seed with thin, starchy endosperm and straight, linear, dicotyledonous embryo. $2n = 32$.

The family Leitneriaceae consists of the single species *Leitneria floridana* Chapman, confined to the coastal plain of the southeastern United States.

Leitneria is by all accounts taxonomically isolated, and its relationships are uncertain. Some authors have compared it to *Didymeles*, but aside from the great (though surely only parallel) floral reduction the two do not have a great deal in common.

The multicellular, clavate glands of the leaves might or might not be homologous with the peltate gland-scales of the Juglandales, Myricales, and Fagales. The origin of *Leitneria* from a hamamelidalean ancestry has been postulated by several authors, and this possibility seems as likely as any. Inclusion of the Leitneriales in the subclass Hamamelidae brings no significant new characters to the larger group.

The fossil record of *Leitneria* can be traced with some assurance back to near the base of the Miocene epoch, and perhaps into the Oligocene. Earlier records, from the Eocene, are now considered to be incorrect.

8. Order JUGLANDALES Engler 1892

Aromatic, anemophilous trees or rarely shrubs, without intercellular canals; vessel-segments elongate, with simple or scalariform perforation-plates; imperforate tracheary elements with more or less clearly bordered pits. LEAVES alternate or rarely opposite, pinnately compound (or trifoliolate), provided with aromatic, resinous, peltate, basally embedded gland-scales, and often with other sorts of hairs as well; stomates anomocytic. FLOWERS individually small and inconspicuous, borne in aments, unisexual or some of them (Rhoipteleaceae) perfect; sepals (1–) 4 (5), small and scale-like, or sometimes obsolete, often adnate to the bractlets; petals none; stamens 3-many; filaments short, anthers basifixed, tetrasporangiate and dithecal, opening by longitudinal slits; pollen-grains tricolporate or triporate to multiporate; gynoecium of 2 (3) carpels united to form a compound, superior or inferior ovary with distinct styles; ovary 2 (3)-locular below (sometimes provided with additional, "false" partitions so as to seem 4–8-locular) but unilocular above, the partition not reaching the summit of the cavity; ovule solitary, attached to the partial partition, hemitropous to anatropous (Rhoipteleaceae), or orthotropous (Juglandaceae), bitegmic (Rhoipteleaceae) or unitegmic (Juglandaceae), crassinucellar; endosperm-development nuclear. FRUIT a nut (often samaroid), or drupe-like but rather dry; seeds virtually without endosperm; embryo oily, with 2 massive cotyledons.

The order Juglandales is now generally considered to include only 2 families, the Juglandaceae, with about 60 species, and the Rhoipteleaceae, with only one. In most of the respects in which the Rhoipteleaceae differ from the Juglandaceae (stipulate leaves, perfect functional flowers, superior ovary, bitegmic, hemitropous to anatropous ovule) the Rhoipteleaceae are the more archaic group. These features of the Rhoipteleaceae are all compatible with the Urticales, to which the family has sometimes been referred. On the other hand, *Rhoiptelea* has no evident close relatives in the Urticales, whereas its relationship to the Juglandaceae is now obvious to all.

There is a long-standing difference of opinion among botanists as to whether the Juglandaceae should be associated with some other families here grouped in the subclass Hamamelidae, or should be regarded as florally reduced relatives of the Anacardiaceae. The controvery still continues, with proponents of each side unable to see how those on the other side can cling to their opinion. Both Takhtajan (1959, 1966, 1969) and I (1957, 1968) have long considered the Juglandales to belong to the group now called Hamamelidae. On the other hand, Wolfe (1973, under Hamamelidae) considers that "Juglandaceae and Rhoipteleaceae have foliage that is clearly anomalous with the Hamamelidae and should be excluded from this subclass", and Thorne (1973, under Hamamelidae) professes that "it came as rather a shock to me to find out that some phylogenists still consider the family [Juglandaceae] to be derived from protohamamelid stock." Accordingly, Thorne (1968, 1976), transfers the Juglandales, Myricales, and Leitneriales to a position associated with the Rutales (=Sapindales), but he leaves the Fagales and Casuarinales in association with the Hamamelidales.

There is no doubt that some members of the Anacardiaceae are remarkably similar to *Juglans* in aspect, and both the Anacardiaceae and Juglandaceae are notably resinous. Furthermore, some of the Anacardiaceae show a trend toward floral reduction, culminating in the closely related family Julianiaceae. The Julianiaceae have often been associated with the likewise florally reduced Juglandaceae, but recent studies (Young, 1976, under Sapindales) clearly demonstrate that they (the Julianiaceae) are closely allied to the Anacardiaceae.

Thus, if there were no contra-indications, it would be reasonable to associate the Juglandaceae with the Anacardiaceae. Unfortunately for this superficially attractive conclusion, contra-indications abound. The Juglandaceae cannot be successfully extracted from the amentiferous group of Hamamelidae, and their pollen, wood, and various other features are wholly out of harmony with the Anacardiaceae

Heimsch & Wetmore (1939) have pointed out that the wood of the Juglandaceae is more primitive in several respects than that of the Anacardiaceae. Among other differences, the vessels of Anacardiaceae have almost exclusively simple perforations, whereas three genera of the Juglandaceae (*Alfaroa, Engelhardia*, and *Oreomunnea*) have scalariform perforations. In the Rhoipteleaceae, considered by Thorne as well as others to be closely allied to but more archaic than the Juglandaceae, the vessels have exclusively sclariform perforations. Thus it would seem that any phyletic link between the Juglandaceae and Anacardiaceae must go back to some remote ancestor that had wood no more specialized than that of the Rhoipteleaceae.

A series of other anatomical features that distinguish the Juglandaceae from the Anacardiaceae was long ago pointed out by Parmentier (1911) and Nagel (1914), and noted again by Takhtajan (1966). One of these features that has perhaps not been sufficiently emphasized is the form and structure of the foliar trichomes in the Juglandaceae, Rhoipteleaceae, Myricaceae, Betulaceae, and Fagaceae. These are aromatic, resinous, multicellular, peltate, and basally embedded. I would hesitate to affirm that such trichomes cannot be found outside this circle of affinity, but neither can I point to a specific example; they are certainly not common and widely distributed in other families. Even the fact that both the Anacardiaceae and the Juglandaceae are strongly resinous may be misleading. The Anacardiaceae characteristically have schizogenous resin-ducts in the bark and in the phloem of the larger veins of the leaves, often also in the flowers and fruits. These are lacking from the Juglandaceae.

The most significant new evidence on the affinities and ancestry of the Juglandales comes from the fossil and modern pollen. Stone and Broome (1975) consider that "Pollen of the Rhoipteleaceae . . . and Betulaceae . . . is very similar to the Juglandaceae at the photon and EM level of observation." The Juglandales, Myricales, Fagales, and Casuarinales have a distinctive sort of pollen that appears to take its evolutionary origin in the *Normapolles* complex of middle Cenomanian (early Upper Cretaceous) time. Van Campo (1976) considers that "in terms of their pollen they [the Juglandaceae] are incontestably Amentiferae." Wolfe (1973) maintains that the lineage of the Juglandales can be traced back to the *Normapolles* complex. He further considers that the two existing families of the order had begun to diverge by the Santonian or early Campanian stage (middle Upper Cretaceous). Some pollen-grains from the Maestrichtian (uppermost Cretaceous) are "very similar to and probably referable to *Rhoiptelea*," and by early Paleocene time pollen of distinctly juglandaceous type had evolved. Wolfe thinks that the two families differentiated from each other in tropical or subtropical climates, but he does not suggest that the dichotomy had any adaptive significance. I have no quarrel with his interpretation of the fossil history of the order, which is in my opinion wholly compatible with the taxonomic treatment here presented.

We are left with the pinnately compound leaves of the Juglandales as the only feature inharmonious with their inclusion in the Hamamelidae. The question is what taxonomic weight to attach to this feature. I simply do not believe that this one character should be allowed to overwhelm the rest of the evidence, especially inasmuch as other modern descendants of the *Normapolles* complex have simple leaves, some of them very comparable to leaves of members of the Hamamelidaceae.

Hickey and Doyle (1977, in initial citations) suggest that perhaps the whole *Normapolles*-related group should be extracted from the Hamamelidae and assigned to the Rosidae (though presumably not as close allies of the Anacardiaceae). This might be more reasonable than an attempt to divorce the Juglandales from the Fagales, but it creates other problems. As these authors admit, the Betulaceae (Fagales) "have leaves which are more similar to those of the Hamamelidaceae."

In any case, it appears that the fossil pollen record requires that any connection of the Juglandales to the Rosidae must antedate the origin of the *Normapolles* group, which first appears near the base of the Upper Cretaceous, about at the middle of the Neocomian stage. Since Hickey & Doyle consider that the Rosidae and the Hamamelidales are sister-groups, which only began to differentiate from magnoliid (or magnoliid-like) ancestors and from each other in the Albian stage (i.e., the stage just before the Neocomian), we may here be pushing the limits of the application of phylogeny to taxonomy. I have elsewhere pointed out that the determined pursuit of absolute monophylesis leads taxonomists only into absurdity. As I see the matter, the now fairly extensive fossil record would permit the assignment of the Juglandales and other *Normapolles*-related groups to either the Hamamelidae, or to an isolated position in the Rosidae, or to a subclass of their own, depending on what characters one chooses to emphasize and where he wishes to draw the lines between major groups. On the sum of the evidence I prefer the hamamelid assignment.

Recent serological studies also support a position of the Juglandaceae in the Hamamelidae. Petersen & Fairbrothers (1978, 1979) indicate a strong serological correspondence among tested genera of the Juglandaceae, Myricaceae, and Fagaceae, but very little correspondence between any of these and the two tested genera (*Rhus* and *Toxicodendron*) of the Anacardiaceae. Furthermore, Petersen informs me that still unpublished results show a similar correspondence between the Betulaceae and Juglandaceae. Studies by Chupov (1978) point in the same direction. Chupov strongly supports the consanguinity of the Betulaceae, Fagaceae, and Juglandaceae, and considers that the Rutaceae are far removed from this group. No member of the Anacardiaceae was included in his study.

SELECTED REFERENCES

Conde, L. F., & D. E. Stone. 1970. Seedling morphology in the Juglandaceae, the cotyledonary node. J. Arnold Arbor. 51: 463–477.

Crepet, W. L., D. L. Dilcher, & F. W. Potter. 1974. Eocene angiosperm flowers. Science 185: 781–782.
Crepet, W. L., D. L. Dilcher, & F. W. Potter. 1975. Investigations of angiosperms from the Eocene of North America: A catkin with Juglandaceous affinities. Amer. J. Bot. 62: 813–823.
Elias, T. S. 1972. The genera of Juglandaceae in the southeastern United States. J. Arnold Arbor. 53: 26–51.
Handel-Mazzeti, H. 1932. Rhoipteleaceae, eine neue Familie der Monochlamydeen. Repert. Spec. Nov. Regni Veg. 30: 75.
Heimsch, C., & R. H. Wetmore. 1939. The significance of wood anatomy in the taxonomy of the Juglandaceae. Amer. J. Bot. 26: 651–660.
Kribs, D. A. 1927. Comparative anatomy of the woods of the Juglandaceae. Trop. Woods 12: 16–21.
Leroy, J.-F. 1951. La théorie généralisée des carpelles-sporophylles et la fleur des Juglandales. III. Discussion et conclusions. Compt. Rend. Hebd. Séances Acad. Sci. 233: 1214–1216.
Leroy, J.-F. 1955. Étude sur les Juglandaceae. À la recherche d'une conception morphologique de la fleur femelle et du fruit. Mém. Mus. Natl. Hist. Nat., Sér. B, Bot. 6: 1–246.
Manning, W. E. 1938, 1940, 1948. The morphology of the flowers of the Juglandaceae. I. The inflorescence. II. The pistillate flowers and fruit. III. The staminate flowers. Amer. J. Bot. 25: 407–419, 1938. 27: 839–852, 1940. 35:606–621, 1948.
Manning, W. E. 1949. The genus *Alfaroa*. Bull. Torrey Bot. Club 76: 196–209.
Manning, W. E. 1959. *Alfaroa* and *Engelhardtia* in the New World. Bull. Torrey Bot. Club 86: 190–198.
Manning, W. E. 1975. An analysis of the genus *Cyclocarya* Iljinskaja (Juglandaceae). Bull. Torrey Bot. Club 102: 157–166.
Manning, W. E. 1978. The classification within the Juglandaceae. Ann. Missouri Bot. Gard. 65: 1058–1087.
Nagel, K. 1914. Studien über die Familie der Juglandaceen. Bot. Jahrb. Syst. 50: 459–530.
Parmentier, P. 1911. Recherches anatomiques et taxonomiques sur les Juglandacées. Rev. Gén. Bot. 23: 341–364.
Petersen, F., & D. Fairbrothers. 1978. A serological investigation of selected amentiferous taxa. Serol. Mus. Bull. 53: 10.
Petersen, F. P., & D. E. Fairbrothers. 1979 (1980). Serological investigation of selected amentiferous taxa. Syst. Bot. 4: 230–241.
Stachurska, A. 1961. Morphology of pollen grains of the Juglandaceae. Monogr. Bot. 12: 121–143.
Stone, D. E. 1970. Evolution of cotyledonary and nodal vasculature in the Juglandaceae. Amer. J. Bot. 57: 1219–1225.
Stone, D. E. 1972. New World Juglandaceae. III. A new perspective of the tropical members with winged fruits. Ann. Missouri Bot. Gard. 59: 297–321.
Stone, D. E., G. A. Adrouny, & R. H. Flake. 1969. New World Juglandaceae. II. Hickory nut oils, phenetic similarities, and evolutionary implications in the genus *Carya*. Amer. J. Bot. 56: 928–935.
Stone, D. E., & C. R. Broome, 1971. Pollen ultrastructure: evidence for relationship of the Juglandaceae and the Rhoipteleaceae. Pollen & Spores 13: 5–14.
Stone, D. E., & C. R. Broome, 1975. Juglandaceae A. Rich. ex Kunth. World Pollen & Spore Flora 4: 1–35.

Stone, D. E., J. Reich, & S. Whitfield, 1964. Fine structure of the walls of *Juglans* and *Carya* pollen. Pollen & Spores 6: 379–392.
Tang, Y. 1932. Timber anatomy of Rhoipteleaceae. Bull Fan. Mem. Inst. Biol. 3: 127–131.
Van Campo, M. 1976. Patterns of pollen morphological variation within taxa. *In*: I. K. Ferguson & J. Muller, eds., The evolutionary significance of the exine, pp. 125–137. Linn. Soc. Symp. Ser. No. 1. Academic Press. London & New York.
Van Heel, W. A., & F. Bouman. 1972. Note on the early development of the integument in some Juglandaceae, together with some general questions on the structure of angiosperm ovules. Blumea 20: 155–159.
Verhoog, H. 1968. A contribution towards the developmental gynoecium morphology of *Engelhardia spicata* Lechen. ex Blume (Juglandaceae). Acta Bot. Neerl. 17: 137–150.
Whitehead, D. R. 1965. Pollen morphology in the Juglandaceae, II: Survey of the family. J. Arnold Arbor. 46: 369–410.
Withner, C. L. 1941. Stem anatomy and phylogeny of the Rhoipteleaceae. Amer. J. Bot. 28: 872–878.

Ильинская, И. А. 1953. Монография рода *Pterocarya* Kunth. Труды Бот. Инст. АН СССР, сер. I, 10: 7-123.

SYNOPTICAL ARRANGEMENT OF THE FAMILIES OF JUGLANDALES

1 Flowers in triplets axillary to the bracts of the ament, the middle flower of the triplet perfect and fertile, with 4 sepals, 6 stamens, and a superior ovary; lateral flowers pistillate but sterile and more or less reduced; leaves stipulate; ovule bitegmic, hemitropous to anatropous .... 1. RHOIPTELEACEAE.

1 Flowers unisexual, solitary in the axils of the bracts of the ament, the staminate ones with (3–) 5-many stamens, the calyx adnate to the bract or wanting; pistillate flowers with an inferior ovary and 1–4 calyx-teeth, or the calyx obsolete; leaves exstipulate; ovule orthotropous, unitegmic ....2. JUGLANDACEAE.

1. Family RHOIPTELEACEAE Handel-Mazetti 1932 nom. conserv., the Rhoiptelea Family

Aromatic, anemophilous trees, without intercellular canals; vessel-segments elongate, with scalariform perforation-plates that have up to 10 (–20) cross-bars; imperforate tracheary elements mainly fiber-tracheids with small, bordered pits, some gelatinous fibers also present; wood-rays Kribs type I, heterocellular, mixed uniseriate and pluriseriate, the latter (2) 3–4 (5) cells wide; wood-parenchyma abundant, mostly vasicentric. LEAVES deciduous, alternate, pinnately compound, provided with aromatic, resinous, peltate, basally embedded gland-scales; stomates anomocytic; stipules present. FLOWERS individually small and inconspicuous, borne in aments that are clustered into large, nodding, terminal panicles; each ament with flowers in dichasial triplets axillary to the bracts, the central flower of each triplet perfect and fertile, the lateral ones pistillate but reduced and sterile (or even abortive); sepals of the perfect flowers 2 + 2, small, scarious, persistent on the fruit; petals none; disk wanting; stamens 6; filaments short; anthers basifixed, tetrasporangiate and dithecal, opening by longitudinal slits; pollen-grains tetrahedral, tricolporate or seldom tetracolporate; gynoecium of 2 carpels united to form a compound ovary with separate styles; ovary superior, bilocular below, unilocular above, the partition not reaching the top of the cavity; ovule solitary, attached to the partition in one locule, hemitropous to anatropous, bitegmic, crassinucellar; endosperm-development nuclear. FRUIT a 2-winged, samaroid nut; seed solitary, without endosperm; embryo straight, oily, with 2 thick cotyledons.

The family Rhoipteleaceae consists of a single genus and species *Rhoiptelea chiliantha* Diels & Handel-Mazzetti, native to southwestern China and northern Vietnam. Withner (1941) notes that "the flower of *Rhoiptelea* resembles closely the pre-juglandaceous flower hypothesized by Manning."

2. Family JUGLANDACEAE A. Richard ex Kunth 1824 nom. conserv., the Walnut Family

Aromatic, anemophilous trees or rarely shrubs, tanniferous, with proanthocyanins and/or ellagic acid and/or gallic acid, accumulating napthaquinones, and commonly transporting nitrogen as citrulline, often with scattered secretory cells, but without secretory canals, seldom accumulating aluminum; nodes trilacunar or sometimes pentalacunar; clustered

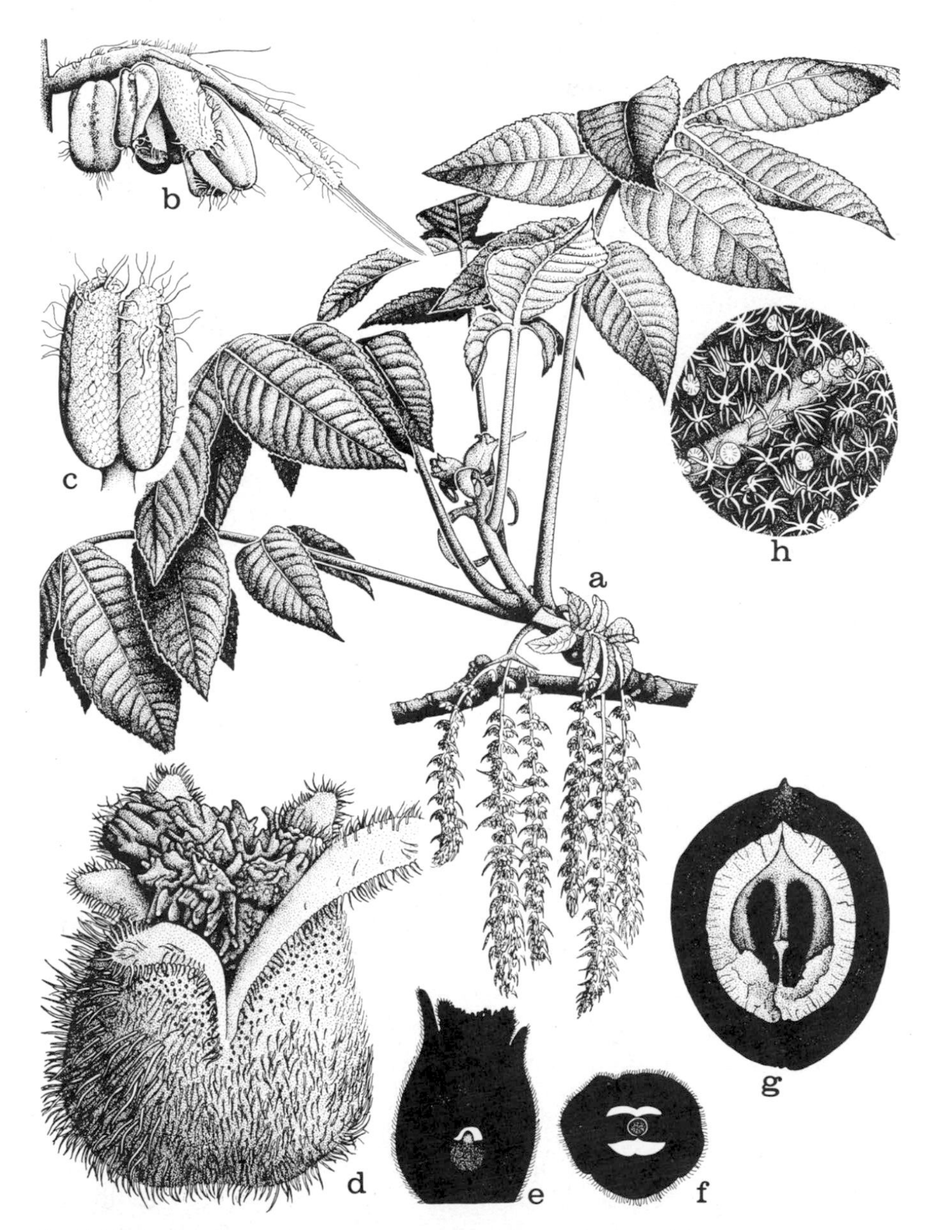

FIG. 2.12 Juglandaceae. *Carya tomentosa* (Poir.) Nutt. a, habit, with flowers and unfolding leaves, ×½; b, staminate flower, ×8; c, stamen, ×16; d, pistillate flower, ×8; e, pistillate flower in long-section, ×4; f, ovary in cross-section, ×4; g, half of nut and husk, ×1; h, lower surface of leaf, ×16.

crystals of calcium oxalate commonly present in some of the cells of the parenchymatous tissues; vessel-segments elongate, mostly with simple perforations, but in the relatively archaic genera *Alfaroa, Engelhardia,* and *Oreomunnea* some of them scalariform, with a few cross-bars; ground-mass of xylem composed mainly of fiber-tracheids with more or less distinctly bordered pits, some gelatinous fibers generally also present; wood-rays Kribs type I, or sometimes IIA and B, heterocellular to nearly homocellular, mixed uniseriate and pluriseriate, the latter 2–4 (–7) cells wide; wood-parenchyma most commonly apotracheal, occurring as scattered cells and short, uniseriate bands or continuous bands 1–3 cells wide, a little paratracheal parenchyma often present as well; sieve-tube plastids of S-type. LEAVES alternate or occasionally opposite (*Alfaroa* and *Oreomunnea*), pinnately compound (or trifoliolate), provided with aromatic, resinous, peltate, basally embedded gland-scales, and often with other sorts of glandular and eglandular hairs as well; stomates anomocytic; stipules wanting. FLOWERS individually small and inconspicuous, unisexual, apetalous, borne in erect or more often drooping aments that are sometimes grouped into terminal panicles; each ament with the flowers solitary in the axils of the bracts, the aments in archaic genera with terminal pistillate flowers and lateral staminate flowers, in more advanced genera the aments unisexual and the plants sometimes dioecious; staminate flowers with (1–) 4 (5) sepals more or less adnate to the 2 bractlets, with which they form a common structure, or the sepals obsolete, or both the sepals and the bractlets obsolete; stamens (3–) 5–40 (–100+); filaments short; anthers tetrasporangiate and dithecal, opening by longitudinal slits; pollen-grains triporate to multiporate, with a thin exine and granular rather than columellate infratectal structure; staminate flowers sometimes with a vestigial gynoecium; pistillate flowers with an inferior ovary and 4 calyx-teeth, or the teeth wholly suppressed (as in *Carya*), the bract and bracteoles subtending the pistillate flower often joined to form a cupulate involucre that ripens with the fruit to form a husk; gynoecium of 2 (3) carpels united to form a compound, inferior ovary, the styles distinct or often united at the base, or seldom the stigmas sessile; ovary 2 (3)-locular below (or seemingly 4–8-locular because of the presence of additional, "false" partitions), unilocular above, the partition not reaching the top of the cavity; ovule solitary, borne at the summit of the partition, orthotropous, unitegmic, crassinucellar; fertilization chalazogamous; endosperm-development nuclear. FRUIT a nut (sometimes samaroid), or drupaceous in form, the softer husk sometimes splitting to release the bony pericarp; seed solitary, virtually without endosperm; embryo oily, with 2 4-lobed, often massive and sculptured cotyledons. X = 16.

The family Juglandaceae consists of some 7 or 8 genera and about 60 species, widespread in temperate and subtropical parts of the Northern Hemisphere, with a few species extending into South America and into the islands of the southwestern Pacific. The two largest genera, *Juglans* (walnut) and *Carya* (hickory), include a number of familiar and economically important species.

According to Stone (1973) the fruits are primitively wind-distributed, and animal-distributed fruits have evolved in two separate lines within the family. Wind-distributed types have samaroid, relatively small and light fruits, and germination is epigeous. Animal-distributed types have larger, heavier, drupe-like fruits, and germination is hypogeous.

Based on the pollen-record, the family makes its first certain appearance in the early Paleocene, although some late Upper Cretaceous fossil pollen grains may also belong here. By the end of the Paleocence, pollen characteristic of *Carya, Juglans,* and *Pterocarya* can be clearly recognized, and *Platycarya* pollen appears in the earliest Eocene (all fide Wolfe, 1973, under Hamamelidae). Juglandaceous wood occurs in Eocene deposits in Yellowstone National Park. Some Eocene fossil catkins from southeastern United States are evidently juglandaceous, and near to the modern genus *Engelhardia.* At least in these latter fossils the evolution of the pollen appears to have progressed at about the same rate as evolution of the more gross features of the flower and inflorescence (Crepet et al., 1975).

9. Order MYRICALES Engler 1892

The order consists of the single family Myricaceae.

SELECTED REFERENCES

Baird, J. R. 1970. A taxonomic revision of the plant family Myricaceae of North America, north of Mexico. Ph.D. Thesis, Univ. North Carolina, 1968. Univ. Microfilms, 1970. Ann Arbor.
Elias, T. S. 1971. The genera of Myricaceae in the southeastern United States. J. Arnold Arbor. 52: 305–318.
Guillaumin, A. 1940 (1941). Matériaux pour la flore de la Nouvelle-Calédonie. LVII. La présence d'une Myricacée. Bull. Soc. Bot. France 87: 299–300.
Håkansson, A. 1955. Endosperm formation in *Myrica Gale* L. Bot. Not. 108: 6–16.
Halim, A. F., & R. P. Collins. 1973. Essential oil analysis of the Myricaceae of the eastern United States. Phytochemistry 12: 1077–1083.
Leroy, J.-F. 1949. De la morphologie florale et de la classification des Myricaceae. Compt. Rend. Hebd. Séances Acad. Sci. 229: 1162–1163.
Leroy, J.-F. 1957. Sur deux Amentiferes remarquables de la Flore Asiatico-Pacifique et Pacifique. Proc. Eighth Pacific Science Congress 4: 459–464.
Macdonald, A. D. 1974. Floral development of *Comptonia peregrina* (Myricaceae). Canad. J. Bot. 52: 2165–2169.
Macdonald, A. D. 1977. Myricaceae: floral hypothesis for *Gale* and *Comptonia*. Canad. J. Bot. 55: 2636–2651.
Macdonald, A. D. 1978. Organogenesis of the male inflorescence and flowers of *Myrica esculenta*. Canad. J. Bot. 56: 2415–2423.
Macdonald, A. D. 1979. Development of the female flower and gynecandrous partial inflorescence of *Myrica californica*. Canad. J. Bot. 57: 141–151.
Macdonald, A. D., & R. Sattler. 1973. Floral development of *Myrica gale* L. and the controversy over floral concepts. Canad. J. Bot. 51: 1965–1975.
Sheffy, M. V. 1972. A study of the Myricaceae from Eocene sediments of southeastern North America. Ph.D. Thesis, Indiana Univ., Bloomington. Univ. Microfilms. Ann Arbor.

Вихирева, В. В. 1957. Анатомическое строение и развитие пестичного цветка восковницы обыкновенной—*Myrica gale* L. Труды Бот. Инст. АН СССР, VII, 4: 270-287.

1. Family MYRICACEAE Blume 1829 nom. conserv., the Bayberry Family

Aromatic, anemophilous, evergreen or deciduous shrubs or small trees with two basic kinds of trichomes, one kind elongate, colorless, and unicellular, the other peltate, glandular, and usually golden-yellow, with multicellular head and multicellular, basally embedded stalk; plants accumulating triterpenes and sesquiterpenes, also tanniferous, with

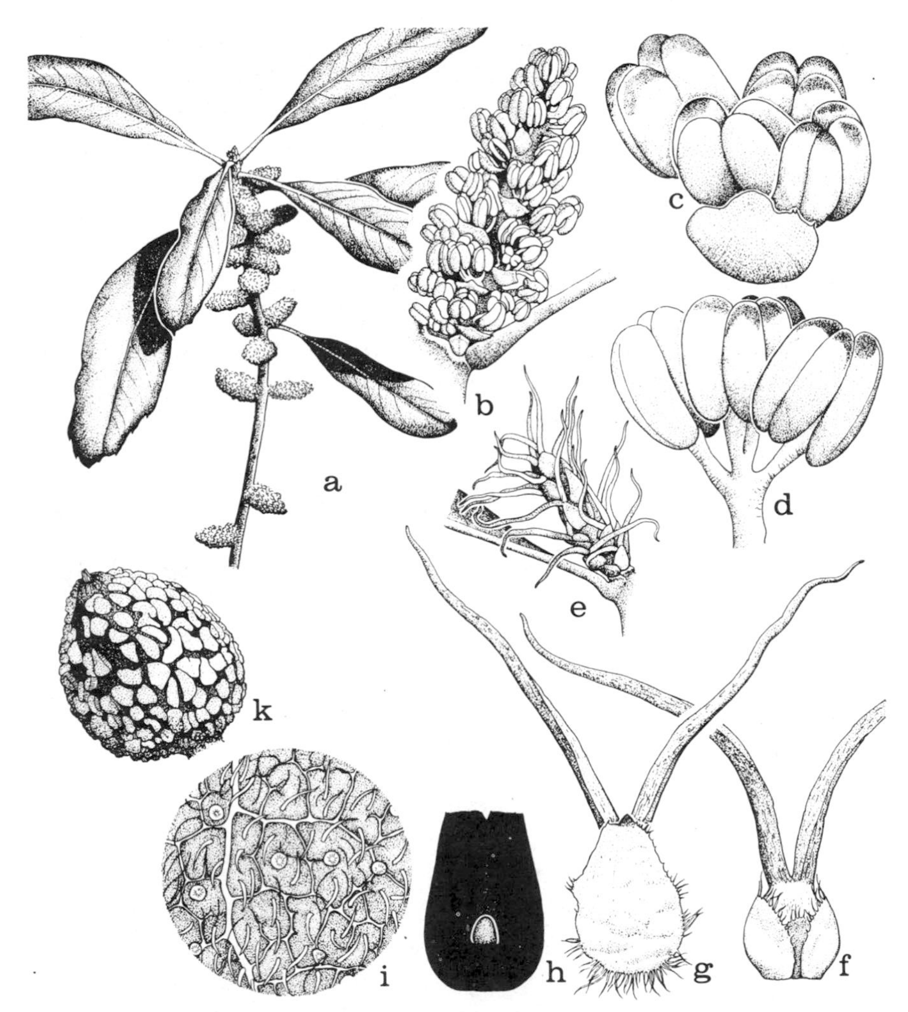

FIG. 2.13 Myricaceae. *Myrica heterophylla* Raf. a, habit, $\times \frac{1}{2}$; b, staminate inflorescence, $\times 3$; c, d, two views of staminate flower, the lower one with the bract removed, $\times 12$; e, pistillate inflorescence, $\times 3$; f, pistillate flower, the subtending bract removed, $\times 12$; g, pistillate flower, with the subtending bract in place, $\times 12$; h, long-section of ovary, $\times 24$; i, lower surface of leaf, $\times 12$; k, fruit, $\times 6$.

proanthocyanins, ellagic acid, and gallic acid; roots commonly bearing nodules that harbor nitrogen-fixing bacteria (but apparently not in *Canacomyrica*); nodes unilacunar or trilacunar; single and clustered crystals of calcium oxalate commonly present in some of the parenchyma-cells; vessel-segments with scalariform or both simple and scalariform perforations; the cross-bars mostly fewer than 15; imperforate tracheary elements with bordered or simple and bordered pits; vasicentric tracheids sometimes present; wood-rays heterocellular, all uniseriate, or mixed uniseriate and pluriseriate, the latter up to 4–8 cells wide; wood-parenchyma diffuse; sieve-tubes with S-type plastids. LEAVES alternate, simple (but pinnately lobed in *Comptonia*), pinnately veined; stomates anomocytic; stipules wanting except in *Comptonia*. FLOWERS individually small and inconspicuous, unisexual or (in *Canacomyrica*) some of them perfect, borne in aments, one flower (or pseudanthial partial inflorescence) commonly in the axil of each bract, but sometimes the bracts subtending few-flowered secondary inflorescences; perianth none; perfect flowers of *Canacomyrica* with 6 stamens atop the ovary; staminate flowers (or pseudanthia) sometimes subtended by a pair of vertically oriented bracteoles in addition to the primary bract, or the bracteoles obsolete; stamens in a single whorl, initiated simultaneously, up to 5 (aside from *Canacomyrica*), most commonly 4, progressively fewer in more acropetal flowers; anthers tetrasporangiate and dithecal, extrorse, opening by longitudinal slits; pollen (2) 3 (–6)-porate; pistillate flowers commonly subtended by 2 or more bractlets (these sometimes suggesting small sepals) as well as by a primary bract; gynoecium of 2 carpels united to form a compound, unilocular ovary, the styles distinct or united only at the base; ovule solitary, basal and erect, orthotropous (in *Canacomyrica* with an elongate, recurved micropylar tube simulating a funiculus), unitegmic, crassinucellar; fertilization (at least in *Myrica*) porogamous after slow growth of the pollen-tube; endosperm-development nuclear. FRUIT drupaceous, or almost a nutlet, sometimes enveloped by the persistent, accrescent bracteoles, which in *Comptonia* make the fruit appear bur-like, in *Canacomyrica* enveloped by an outgrowth of what appears to be a disk beneath the pistil, this disk perhaps homologous with the bracteoles of the other genera; seeds nearly or quite without endosperm; embryo straight, with 2 cotyledons. X = 8.

The family Myricaceae consists of 3 genera and some 50 species, widespread in both the Old and the New World, mostly in temperate or subtropical regions. *Comptonia*, of eastern North America, is monotypic, as is the New Caledonian genus *Canacomyrica*. The remainder of the species belong to the genus *Myrica*, which has two well marked sections that are sometimes treated as distinct genera.

Canacomyrica is so distinctive that its position as a member of the family has been questioned. Thorne (1968, 1976, in initial citations) includes it in his list of Taxa Incertae Sedis. It now appears, however, that one of its reportedly distinctive features was based on incorrect observation. Leroy (1949, 1957) has pointed out that the ovule is orthotropous and erect, as in the other genera, and that what Guillaumin originally took to be an elongate funiculus is actually a slender and reflexed prolongation of the integument.

The Myricaceae are generally considered to be related to the Juglandales, and some authors include them in that order. The conspicuous peltate gland-scales of the leaves, from which the characteristic odor emanates, are evidently homologous with similar structures in both the Juglandales and Fagales. The pollen, like that of the Juglandales, Fagales, and Casuarinales, is reminiscent of that of the *Normapolles* complex in the Cenomanian (earliest Upper Cretaceous) stage. For the purposes of a conceptual framework, it seems more useful to treat the Myricales as a distinct order than to include them in the Juglandales or any other order.

The Myricaceae can be traced with some confidence back to the Oligocene (Sheffy, 1972), and wood attributed to *Myrica* occurs in Eocene deposits in Yellowstone National Park. The known fossils do not connect the Myricaceae to any other group. Myricoid pollen dates from the Santonian, and becomes abundant in the Paleocene.

10. Order FAGALES Engler 1892

Trees or shrubs with trilacunar nodes, strongly tanniferous, commonly with gallic acid and also ellagic acid and/or proanthocyanins. LEAVES alternate, or seldom opposite or whorled, simple (but sometimes deeply lobed), pinnately veined; stomates anomocytic or laterocytic; stipules present or absent. FLOWERS anemophilous or (probably only secondarily) entomophilous, small, inconspicuous, unisexual (the plants monoecious or seldom dioecious) or seldom perfect, often (especially the staminate ones) in aments; perianth wanting, or represented only by small, scale-like sepals, but the pistillate flowers often subtended or more or less enclosed individually or in small groups by a characteristic cupule or hull; stamens (1) 2-many, distinct; anthers tetrasporangiate and dithecal, opening by longitudinal slits; pollen-grains 2- to many-aperturate; gynoecium of 2, 3, or 6 (–12) carpels united to form a compound, inferior or nude ovary, with as many locules (at least in the lower part) as carpels, but the partitions often not reaching the summit of the ovarian cavity; styles distinct or united only at the base; ovules 1–2 in each locule, anatropous, crassinucellar, unitegmic or bitegmic. FRUIT a nut, samara, or drupe, often individually or collectively subtended by a complex hull; seeds with large, dicotyledonous embryo and thin or no endosperm.

The order Fagales as here defined consists of 3 families and more than 900 species. Many of the species are common forest trees in North Temperate regions. The relationship of the Fagales to the Hamamelidales seems to be well established (cf. Endress, 1967, under Hamamelidales; Hall, 1952).

The Fagaceae and Betulaceae are by all accounts related, although some authors, such as Takhtajan, place each family in an order of its own. In my opinion nothing is to be gained by recognizing the Betulales as an order distinct from the Fagales. The purposes of taxonomy are better served by keeping the two families in the same order. I follow custom in using the name Fagales instead of the older name Betulales Bromhead (1838).

The position of the Balanopaceae is more doubtful. Certainly they stand more or less apart from the Fagaceae and Betulaceae collectively. On the basis of the less regular, less rigidly organized venation of the leaves, Wolfe (1973, under Hamamelidae) would exclude the Balanopaceae not only from the order Fagales but possibly even from the subclass Hamamelidae, although he does not suggest any other position for the group. Takhtajan (1966, in initial citations) treats the Balanopales as a distinct order, but he considers that the characteristic cupule beneath the female flowers, the structure of the wood, inflorescence, and flowers, and the details of pollen-morphology combine to indicate a common origin of the Balanopales and Fagales in the Hamamelidales. His evaluation of the relationships is reasonable and may be accepted at least until significant evidence to the contrary is produced. We have left only the question of the rank at which the groups should be recognized. In my opinion it is pointless to recognize still another small, obscure order, especially when it can be included in a larger, well known order without creating any new problems.

The pollen of the Betulaceae and at least some of the Fagaceae (*Trigonobalanus*) is so similar to that of the ancient *Normapolles* complex (although different in detail) as to suggest that the Fagales may have originated from members of the *Normapolles* group. No clearly transitional forms are known, but the Fagales can be traced back in the fossil record to the Campanian stage of the Upper Cretaceous, within the time-span of the *Normapolles* group. Wood considered to represent the tribe Coryleae of the Betulaceae has been found in uppermost Campanian deposits, and both leaves and pollen of *Alnus* appear in the Maestrichtian. Pollen attributed to *Nothofagus* also occurs in Maestrichtian deposits. *Fagopsis*, with clearly fagaceous leaves and pollen, has been thought on the basis of the reproductive structures to approach the Betulaceae. *Fagopsis* occurs from middle Eocene to Oligocene deposits, and is considered to be allied to some Paleocene fossils as well. Leaves of undoubted *Quercus* occur in late Miocene deposits in Oregon. The Balanopaceae do not appear to have a significant fossil record. This interpretation of the fossil history of the Fagales is based largely on the data presented by Wolfe (1973, under Hamamelidae).

The Fagales have a degree of ecological unity in that they are strongly tanniferous woody plants, often forest trees, most of them wind-pollinated, and with disseminules consisting of single-seeded or few-seeded fruits rather than free seeds. None of these features is especially unusual among angiosperms, however, and the whole syndrome is shared by some other groups of Hamamelidae, such as the Juglandaceae.

SELECTED REFERENCES

Abbe, E. C. 1935, 1938. Studies in the phylogeny of the Betulaceae. I. Floral and inflorescence anatomy and morphology. II. Extremes in the range of variation of floral and inflorescence morphology. Bot. Gaz. 97: 1–67, 1935. 99: 431–469, 1938.

Brett, D. W. 1964. The inflorescence of *Fagus* and *Castanea* and the evolution of the cupules of the Fagaceae. New Phytol. 63: 96–118.

Brunner, F., & D. E. Fairbrothers. 1979. Serological investigation of the Corylaceae. Bull. Torrey Bot. Club 106: 97–103.

Carlquist, S. 1980. Anatomy and systematics of Balanopaceae. Allertonia 2: 191–246.

Codaccioni, M. 1962. Recherches morphologiques et ontogénetiques sur quelques Cupulifères. Rev. Cytol. Biol. Vég. 25: 1–208.

Crepet, W. L., & C. P. Daghlian. 1980. Castaneoid inflorescences from the Middle Eocene of Tennessee and the diagnostic value of pollen (at the subfamilial level) in the Fagaceae. Amer. J. Bot. 67: 739–757.

Cutler, D. F. 1964. Anatomy of vegetative organs of *Trigonobalanus* Forman (Fagaceae). Kew Bull. 17: 401–409.

Elias, T. S. 1971. The genera of Fagaceae in the southeastern United States. J. Arnold Arbor. 52: 159–195.

Forman, L. L. 1964. *Trigonobalanus*, a new genus of Fagaceae, with notes on the classification of the family. Kew Bull. 17: 381–396.

Forman, L. L. 1966. On the evolution of cupules in the Fagaceae. Kew Bull. 18: 385–419.

Forman, L. L. 1966. Generic delimitation in the Castaneoideae. Kew Bull. 18: 421–426.

Hagerup, O. 1942. The morphology and biology of the *Corylus*-fruit. Biol. Meddel. Kongel. Danske Vidensk. Selsk. 17(6): 1–32.

Hall, J. W. 1952. The comparative anatomy and phylogeny of the Betulaceae. Bot. Gaz. 113: 235–270.

Hjelmqvist, H. 1957. Some notes on the endosperm and embryo development in Fagales and related orders. Bot. Not. 110: 173–195.

Kasapligil, B. 1964. A contribution to the histotaxonomy of *Corylus* (Betulaceae). Adansonia sér. 2, 4: 43–90.

Kuprianova, L. A. 1963. On a hitherto undescribed family belonging to the Amentiferae. Taxon 12: 12–13.

Langdon, L. M. 1939. Ontogenetic and anatomical studies of the flower and fruit of the Fagaceae and Juglandaceae. Bot. Gaz. 101: 301–327.

Langdon, L. M. 1947. The comparative morphology of the Fagaceae. I. The genus *Nothofagus*. Bot. Gaz. 108: 350–371.

Poole, A. L. 1950. Studies of New Zealand *Nothofagus*. 2. Nut and cupule development. Trans. & Proc. Roy. Soc. New Zealand 78: 502–508.

Poole, A. L. 1952. The development of *Nothofagus* seed. Trans. & Proc. Roy. Soc. New Zealand 80: 207–212.

Колобкова, Е. В. 1972. О филогенезе и эволюции белков порядка Fagales. В сб.: Биохимия и филогения растений, Е. В. Колобкова, Редактор.: 32-37. Изд. Наука, Москва 1972.

Куприянова, Л. А. 1965. Палинологические данные к систематике порядка Betulales. В сб.: Проблемы филогении растений. Труды Московск. Общ. Исп. Прир. 13: 63-70.

Палибин, И. В. (Palibin, J. W.) 1935 Sur la morphologie florale des Fagacées. Изв. АН СССР, Отд. Матем. и Естеств. Наук: 349-381. (Paper in French. Title of journal also in French: Bull. Acad. Sciences URSS. Classe des Sciences Math. et Nat.)

Туманян, С. А. 1953. Сравнительно-анатомическое исследование древесины представителей рода *Quercus* L. Труды Инст. Леса АН СССР, 9: 39–69.

SYNOPTICAL ARRANGEMENT OF THE FAMILIES OF FAGALES

1 Ovules basal or nearly so, erect, unitegmic; fruit drupaceous, with 2 or 3 stones, subtended by a multibracteate cupule; plants dioecious; stipules vestigial .... 1. BALANOPACEAE.

1 Ovules pendulous, nearly apical, bitegmic or unitegmic; fruit a nut or samara with ordinarily a single seed, with or without a subtending cupule; plants mostly monoecious, seldom dioecious or with perfect flowers; stipules more or less well developed, deciduous.

 2 Carpels (2) 3 or sometimes 6 (–12); ovules mostly bitegmic, seldom unitegmic; pistillate flowers mostly not in aments; fruits generally subtended or enclosed individually or in small groups by a characteristic multibracteate cupule .... 2. FAGACEAE.

 2 Carpels 2 (3); ovules mostly unitegmic, seldom bitegmic; pistillate flowers mostly borne in aments; fruit without a cupule, but often subtended or enclosed by a foliaceous hull derived from 2 or 3 bracts .... 3. BETULACEAE.

1. Family BALANOPACEAE Bentham in Bentham & Hooker 1880 nom. conserv., the Balonops Family

Dioecious, strongly tanniferous evergreen trees with trilacunar nodes; young shoots often beset with unicellular hairs, but soon glabrate; solitary rhomboidal crystals of calcium oxalate commonly present in some of the cells of the parenchymatous tissues; vessel-segments elongate, with scalariform perforations that have 10–30 cross-bars, or some of them with simple perforations; imperforate tracheary elements thick-walled, the pits with evident to vestigial borders; wood-rays heterocellular, mixed uniseriate and pluriseriate, the latter narrow, with linear ends; wood-parenchyma diffuse. Leaves dimorphic, basically alternate, each shoot with minute scale-leaves proximally and normal foliage-leaves distally, the latter sometimes restricted to the shoot-tip and subverticillate; foliage-leaves simple, pinnately veined, the margins recurved and toothed to subentire; stomates laterocytic; stipules vestigial, represented by a pair of minute teeth at the base of the petiole. Flowers anemophilous, small, inconspicuous, unisexual, with vestigial or no perianth; staminate flowers in small aments axillary to scale-leaves (or sometimes to foliage-leaves), each with (1–) 3–6 (–12) stamens and often a vestigial pistil; anthers almost sessile, dithecal, opening longitudinally by lateral slits; pollen-grains 3–5-colpate, the exine minutely roughened like the skin of an orange; pistillate flowers solitary and subsessile or evidently short-pedicellate in the axils of scale-leaves, the naked ovary subtended by numerous crowded, spirally arranged, deltoid bracts; gynoecium of 2 or 3 carpels united to form a compound ovary with as many locules, the styles distinct, each one bifid nearly to the base and sometimes again forked; partitions commonly not reaching the summit of the ovarian cavity; ovules 2 in each locule, nearly basal, collateral or superposed, apotropous, anatropous, unitegmic. Fruit drupaceous, with rather thin flesh, subtended at the base by a cup composed of numerous imbricate, concrescent bracts, the whole externally resembling a fleshy-walled acorn and cup; stones 2–3; seeds with a large, green, dicotyledonous embryo usually enveloped by a thin layer of endosperm; germination epigeous. N = 21.

The family Balanopaceae consists of the single genus *Balanops*, with about 9 species in the southwestern Pacific region, especially New Caledonia. Carlquist notes that the wood is relatively primitive, comparable to that of the Hamamelidaceae. The homology of the cupule with that of the Fagaceae is doubtful. In *Balanops* the cupule consists of a set of well developed bracts in a close spiral, whereas in the Fagaceae it is generally considerd to be a modified branch-system with associated bracts.

2. Family FAGACEAE Dumortier 1829 nom. conserv., the Beech Family

Trees or shrubs with trilacunar nodes, often accumulating triterpenes, strongly tanniferous, with gallic acid and also proanthocyanins and/or ellagic acid in tanniferous idioblasts scattered in the bark and often in other tissues as well, and commonly provided with unicellular or stellate hairs, sometimes also with uniseriate hairs or peltate glands; solitary or clustered crystals of calcium oxalate commonly present in some of the cells of the parenchymatous tissues; vessel-segments with simple or both simple and scalariform perforations; imperforate tracheary elements with simple or bordered pits; vasicentric tracheids often present; wood-rays homocellular or somewhat heterocellular, all uniseriate or more often mixed uniseriate and pluriseriate, the latter commonly very large, up to 20–60 or even 80 cells wide, often evidently aggregate or compound; wood-parenchyma apotracheal, diffuse or in fine lines; secondary phloem usually tangentially stratified into hard and soft layers; sieve-tubes with S-type plastids; roots characteristically forming ectotrophic mycorhizae. LEAVES alternate, or seldom opposite or whorled, simple (but sometimes deeply lobed), pinnately veined; stomates anomocytic; stipules present, deciduous. FLOWERS anemophilous or sometimes (probably only secondarily) entomophilous (as in *Castanea*), small, inconspicuous, unisexual (the plants monoecious or seldom dioecious) or some of them perfect; staminate flowers commonly in more or less reduced dichasia that are organized into aments or sometimes into small heads, or the flower-clusters distributed along a branching axis, or reduced to 1–3-flowered axillary dichasia in *Nothofagus*; sepals 4–7, most commonly 6, small, scale-like, distinct or connate below, sometimes almost obsolete; stamens (4–) 6–12 (–40); filaments slender; anthers tetrasporangiate and dithecal, opening by longitudinal slits; pollen grains binucleate, tectate-columellate (tectate-granular in *Nothofagus*), in *Nothofagus* oblate, covered with small spinules, and with 4–7 equatorial colpi, in other genera globose or prolate, very finely roughened, tricolpate or tricolporate; a vestigial pistil sometimes present; pistillate flowers 1–7 (–15) together at the base of the staminate inflorescences or from separate axils, individually or collectively sub-

FIG. 2.14 Fagaceae. *Quercus alba* L. a, habit, with flowers and unfolding leaves, ×1; b, node, showing stipules, ×3; c, staminate flower, from above, ×12; d, stamen, ×12; e, pistillate inflorescence, ×12; f, fruits, ×2.

tended by an involucre that develops into a cupule; staminodes present or absent; gynoecium of (2) 3 or sometimes 6 (7–12) carpels united to form a compound, inferior ovary with 3–7 small sepals around the summit, or these sometimes obsolete; styles distinct, as many as the carpels; stigma dry; locules as many as the carpels, but the septa sometimes not reaching the summit of the ovarian cavity; ovules axile, pendulous from near the summit of the partition, 2 in each locule, anatropous, crassinucellar, bitegmic or seldom (*Nothofagus*) unitegmic; fertilization chalazogamous, after delayed growth of the pollen-tube; endosperm-development nuclear. FRUIT generally a nut with a stony or leathery pericarp (seldom more or less samaroid) subtended at the base or more or less enclosed individually or collectively by an accrescent cupule or hull apparently composed of numerous imbricate, concrescent bracts; seed solitary, without endosperm; embryo large, straight, starchy or oily, with 2 cotyledons. X most commonly = 12, seldom 11 or 13.

The family Fagaceae consists of 6 to 8 genera, according to interpretation, and about 800 species, of cosmopolitan distribution except for tropical and southern Africa. *Quercus* (oak), *Castanea* (chestnut) and *Fagus* (beech) are well known genera in the Northern Hemisphere, and *Nothofagus* is widespread in the Southern Hemisphere. At least half of the species in the family belong to the genus *Quercus*.

The morphological nature of the cupule, especially in *Quercus*, has been much disputed. The most common view is that it represents a modified system of coalesced branches and their associated reduced leaves.

Pollen considered to represent *Nothofagus* occurs in Campanian (late Upper Cretaceous) and more recent deposits. Pollen attributed to *Castanea* and *Quercus* dates back at least to the Eocene, as does fagaceous wood. Some Eocene catkins from southeastern United States may represent several genera of Fagaceae. Pollen attributed to the Fagaceae, but not necessarily to any modern genus, occurs in deposits as old as the early Senonian.

3. Family BETULACEAE S. F. Gray 1821 nom. conserv., the Birch Family

Trees or shrubs with trilacunar nodes, strongly tanniferous, with gallic acid and commonly also proanthocyanins and ellagic acid in tanniferous idioblasts scattered in the bark and other tissues, and commonly provided with multicellular stalked glands or basally embedded gland-scales as well as unicellular or uniseriate hairs; solitary or clustered crystals of calcium oxalate commonly present in some of the cells of the parenchymatous tissues; vessel-segments with simple and/or scalariform perforations; imperforate tracheary elements of fiber-tracheids with narrowly bordered pits, and often some intermingled tracheids with

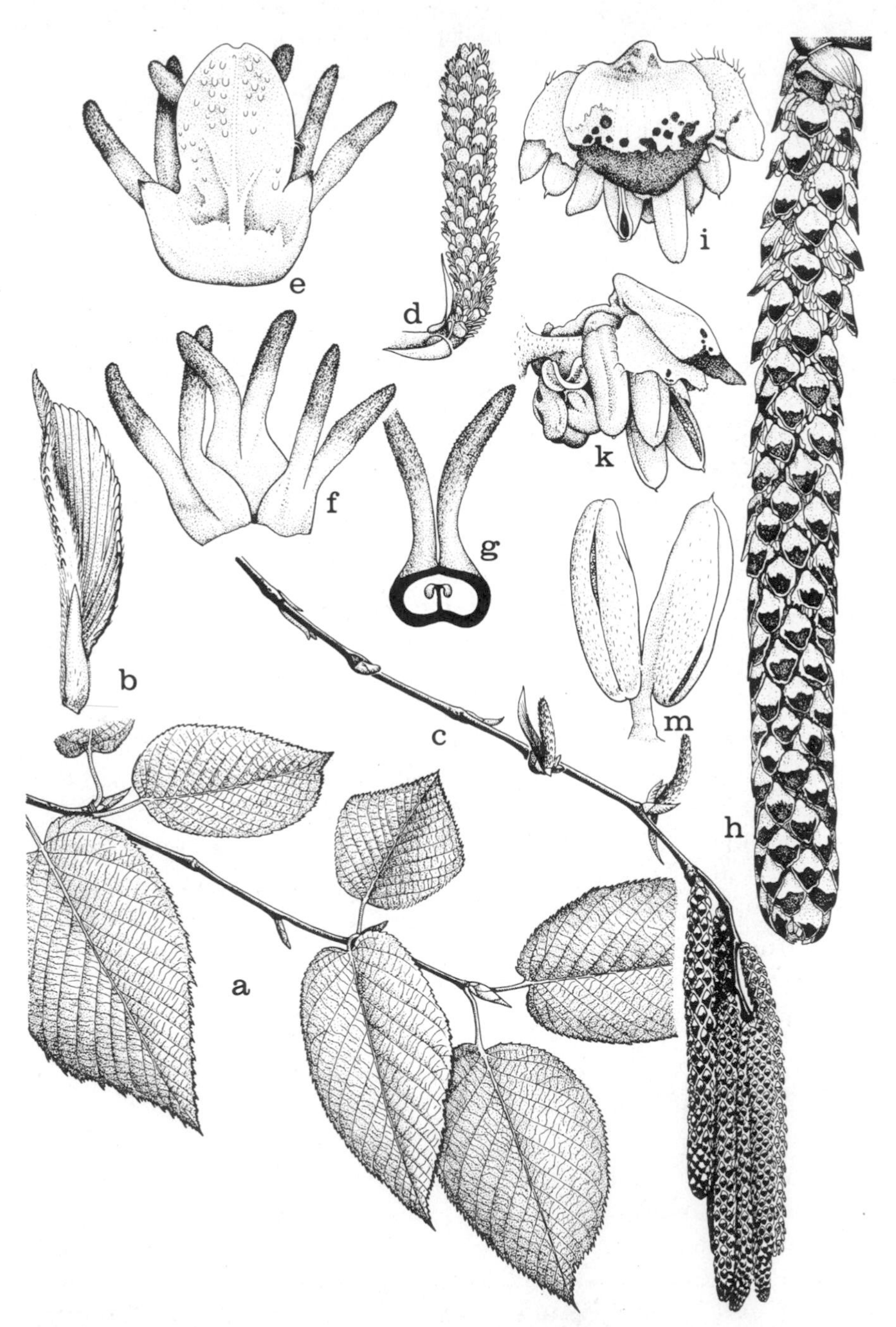

FIG. 2.15 Betulaceae. *Betula lenta* L. a, leafy twig, ×½; b, young leaf and stipule, ×2; c, flowering twig, with pistillate (erect) and staminate (drooping) inflorescences, ×½; d, pistillate inflorescence, ×2; e, bract, subtending 3 pistillate flowers, ×16; f, three pistillate flowers, without the subtending bract, ×16; g, pistil, the ovary in long-section, ×16; h, staminate inflorescence, ×2; i, k, two views of cluster of 3 staminate flowers, ×8; m, stamen, with separate pollen-sacs, ×16.

evidently bordered pits; wood-rays homocellular or heterocellular, Kribs types I, II, IIB, or III, sometimes all uniseriate, sometimes mixed uniseriate and pluriseriate, the latter up to 4 or 5 cells wide, or sometimes aggregated into much larger, compound rays; wood-parenchyma diffuse and terminal; bark often peeling in large, thin layers; secondary phloem often tangentially stratified into hard and soft layers; sieve-tubes with S-type plastids; roots characteristically forming ectotrophic mycorhizae. LEAVES alternate, simple, pinnately veined, usually toothed; epidermal cell walls very often mucilaginous; stomates anomocytic; stipules present, deciduous. FLOWERS anemophilous, small, unisexual (the plants monoecious), in unisexual aments of rather different appearance, the staminate aments pendulous and with a flexuous axis, more or less elongate, the pistillate ones pendulous or erect, short and firm, often woody; each bract of the ament commonly subtending a small, fully or incompletely developed dichasium; calyx of 1–6 small, scale-like sepals, or obsolete; petals none; staminate flowers with or without a vestigial pistil, the pistillate flowers wholly without stamens; stamens in the staminate flowers commonly as many as and opposite the sepals, or when the sepals are obsolete the stamens sometimes more numerous and up to as many as 18 in what may be phyletically a 3-flowered cymule; filaments very short, distinct or basally connate; anthers tetrasporangiate and dithecal, opening by longitudinal slits, commonly (but not in *Alnus*) deeply divided from the summit so that the pollen-sacs are more or less distinct; pollen-grains binucleate, 2–7-porate, with a thin exine and granular rather than columellate infratectal structure; pistillate flower with the gynoecium consisting of 2 (3) carpels united to form a compound, inferior or nude ovary with separate (or nearly separate) styles; stigma dry; ovary bilocular (trilocular) below, unilocular above; ovules axile, pendulous from near the summit of the partition, 1–2 in each locule, anatropous, crassinucellar, unitegmic or seldom (*Carpinus*) bitegmic; fertilization chalazogamous, after delayed growth of the pollen-tube; endosperm-development nuclear. FRUIT a nut or often a 2-winged samara, in the tribes Coryleae and Carpineae subtended or largely enclosed by a foliaceous hull derived from 2 or 3 bracts; seed ordinarily solitary, with thin, fleshy endosperm, or virtually without endosperm; embryo large, with 2 expanded, often much-thickened and oily cotyledons. X = 8, 14. (Carpinaceae, Corylaceae)

The family Betulaceae as here (and traditionally) defined, consists of 6 genera and about 120 species, mainly but not exclusively in temperate and cooler parts of the Northern Hemisphere. Abbe (1974, under Hamamelidae) considers that *Alnus* (alder) and *Betula* (birch) make up the tribe Betuleae, *Carpinus* (hornbeam) and *Ostrya* (hop-hornbeam) make up the tribe Carpineae, and *Corylus* (hazelnut, filbert) and *Ostryopsis* make up the tribe Coryleae. The three tribes have sometimes been taken as distinct, narrowly limited families, but recent serological studies emphasize the unity of the family as traditionally defined.

Fossil wood attributed to *Alnus* and *Carpinus* occurs in Eocene deposits in Yellowstone National Park, and betulaceous pollen goes back to the Senonian Stage of the Upper Creteceous.

11. Order CASUARINALES Lindley 1833

The order consists of the single family Casuarinaceae.

SELECTED REFERENCES

Barlow, B. A. 1959. Chromosome numbers in the Casuarinaceae. Austral. J. Bot. 7: 230–237.

Flores, E. M. 1977. Developmental studies in *Casuarina* III. The anatomy of the mature branchlet. Revista Biol. Trop. 25: 65–87.

Flores, E. M. 1978. The shoot apex of *Casuarina* (Casuarinaceae). Revista Biol. Trop. 26: 247–260.

Flores, E. M. 1980. Shoot vascular system and phyllotaxis of *Casuarina* (Casuarinaceae). Amer. J. Bot. 67: 131–140.

Moseley, M. F. 1948. Comparative anatomy and phylogeny of the Casuarinaceae. Bot. Gaz. 110: 231–280.

Swamy, B. G. L. 1948. A contribution to the life history of *Casuarina*. Proc. Amer. Acad. Arts 77: 1–32.

Ueno, J. 1963. On the fine structure of the pollen walls of Angiospermae. III. *Casuarina*. Grana Palynol. 4: 189–193.

1. Family CASUARINACEAE R. Brown in Flinders 1814 nom. conserv., the She-oak Family

Evergreen trees and shrubs with slender, green, equisetoid, often drooping twigs, tanniferous, producing proanthocyanins and commonly also ellagic acid; roots commonly with nodules harboring nitrogen-fixing bacteria; assimilatory branches often deciduous; nodes unilacunar; solitary or clustered crystals of calcium oxalate commonly present in some of the cells of the parenchymatous tissues; vessel-segments with simple perforations, or often some of them scalariform, with up to 30 cross-bars; imperforate tracheary elements with distinctly bordered pits, generally of tracheids, fiber-tracheids, and intermediate types, the thinner-walled tracheids commonly vasicentric; wood-rays heterocellular, Kribs type IIB, mixed uniseriate and pluriseriate, the latter up to 10 cells wide, or some of them aggregated into compound rays of indefinite width; wood-parenchyma apotracheal, diffuse and in short bands; sieve-tubes with S-type plastids. LEAVES small, scale-like, whorled in sets of 4 to about 20, more or less connate to form a toothed sheath at each node; stipules wanting. FLOWERS anemophilous, small and inconspicuous, without perianth, unisexual, the staminate ones in aments on short or elongate lateral branches, the pistillate ones in dense, head-like inflorescences on usually short lateral branches, the plants monoecious or dioecious; staminate flowers whorled at the nodes of the ament, each consisting of a single (sometimes bifurcate) stamen subtended by a single bract and 2 pairs of much-reduced bracteoles, the 2 inner bracteoles deciduous at anthesis, pushed off by the expanding stamens; anthers tetrasporangiate and dithecal; pollen-grains binucleate, (2) 3 (–5)-porate; pistillate flowers consisting of a single bicarpellate pistil subtended by a scale-like or thickened and eventually woody bract and laterally by 2 bracteoles; ovary laterally flattened, at first bilocular, but only the anterior locule fertile, the posterior one usually empty and more or less reduced or obsolete; styles 2, each with a long, decurrent stigma, united at the base into a short common style; ovules 2 in the single fertile (anterior) locule and sometimes also 2 abortive ones in the posterior locule, epitropous, hemitropous, bitegmic, crassinucellar; archesporium multicellular, embryo-sacs numerous, commonly 20 or more, but only one fully developed and fertile; fertilization chalazogamous; endosperm-development nuclear. FRUIT 1-seeded, winged and samaroid, initially enclosed by the accrescent, firm and dry, woody bracteoles, which separate at maturity, giving the appearance of an opened capsule; seed-coat adnate to the pericarp; seed pendulous; endosperm wanting; embryo large, straight, oily, with 2 cotyledons. X = 8–14, perhaps originally 9.

The family Casuarinaceae consists of the single genus *Casuarina*, with about 50 species. These are native mainly to Australia, but some occur on Pacific islands such as Java, Fiji, and New Caledonia, and some on the Asiatic mainland as far north as Burma. Various species are cultivated in tropical and subtropical dry regions elsewhere.

The Casuarinaceae were considered by Engler to be very primitive dicotyledons, but the consensus among modern workers is that the flowers are reduced rather than primitively simple. The wood is also advanced, by comparison with that of many Magnoliidae. The pollen is basically similar to that of most other amentiferous Hamamelidae. It is now generally believed that the Casuarinaceae are related to such families as the Betulaceae and Myricaceae, with which they are thought to share a common origin in or near the Hamamelidales.

Fossils attributed to *Casuarina* are known from Middle Eocene deposits in Australia, and Miocene deposits in Argentina. Pollen considered to represent *Casuarina* goes farther back, to the Paleocene and (more doubtfully) Maestrichtian.

L. A. S. Johnson proposes (unpublished) to divide *Casuarina* into 4 genera.

III. Subclass CARYOPHYLLIDAE Takhtajan 1966

Dicotyledons with trinucleate or seldom binucleate pollen, bitegmic, crassinucellar ovules and either with betalains instead of anthocyanins, or with free-central to basal placentation (in a compound ovary) or both; woody species generally with anomalous secondary growth or otherwise anomalous stem-structure; stamens, when numerous, originating in centrifugal sequence; members of the largest order (Caryophyllales) have a unique sort of P-type sieve-tube plastid and have a characteristic syndrome of embryological features; food-storage tissue of the seed typically starchy, very often with clustered starch-grains.

The subclass Caryophyllidae as here defined consists of 3 orders, 14 families, and about 11,000 species. About nine-tenths of the species belong to the single large order Caryophyllales. Thus in a sense it may be said that the subclass consists of the order Caryophyllales plus two smaller orders (Polygonales and Plumbaginales) that are customarily associated with it.

The families Bataceae, Gyrostemonaceae, and Theligonaceae, sometimes associated with the subclass Caryophyllidae or more directly with the order Caryophyllales, are here excluded from the group. The Bataceae and Gyrostemonaceae together form a small order that is somewhat doubtfully referred to the subclass Dilleniidae. The Theligonaceae are more confidently associated with the Rubiaceae in the order Rubiales (subclass Asteridae). *Rhabdodendron* and *Viviania*, sometimes included in the Caryophyllales, are here referred to the Rosidae. *Rhabdodendron* forms a distinct family in the order Rosales, and *Viviania* is included in the Geraniaceae.

The ancestry of the Caryophyllidae may lie in or near the Ranunculaceae. In the absence of known fossil connections, it may be supposed that the common ancestor of the Caryophyllidae was an herb with hypogynous flowers and separate carpels, and without well developed petals. The number of potentially ancestral groups is thus immediately limited. The possibility that the Ranunculaceae may be at least collateral ancestors is bolstered by the fact that some of them have pollen very much like that of many Caryophyllales (teste Nowicke, personal communication). The floral trimery of the Polygonaceae also has ample precedent in some of the Ranunculaceae. The evolutionary significance of the centrifugal androecium in the Caryophyllales can scarcely be evaluated until a satisfactory general interpretation of the origin of centrifugality is achieved.

The fossil record as presently interpreted carries the Caryophyllidae back only to the Maestrichtian epoch, some 70 million years ago. This relatively short fossil history, as compared to the Magnoliidae, Hamamelidae, and Rosidae, is consonant with the primitively herbaceous habit of the Caryophyllidae. Aside from the Nymphaeales, dicotyledonous herbs apparently played only a negligible role in the vegetation of the Cretaceous.

Further speculation on the ancestry of the Caryophyllidae is hampered by uncertainty about the affinity of the Polygonales and Plumbaginales to the Caryophyllales. It would be easier to see a pattern in

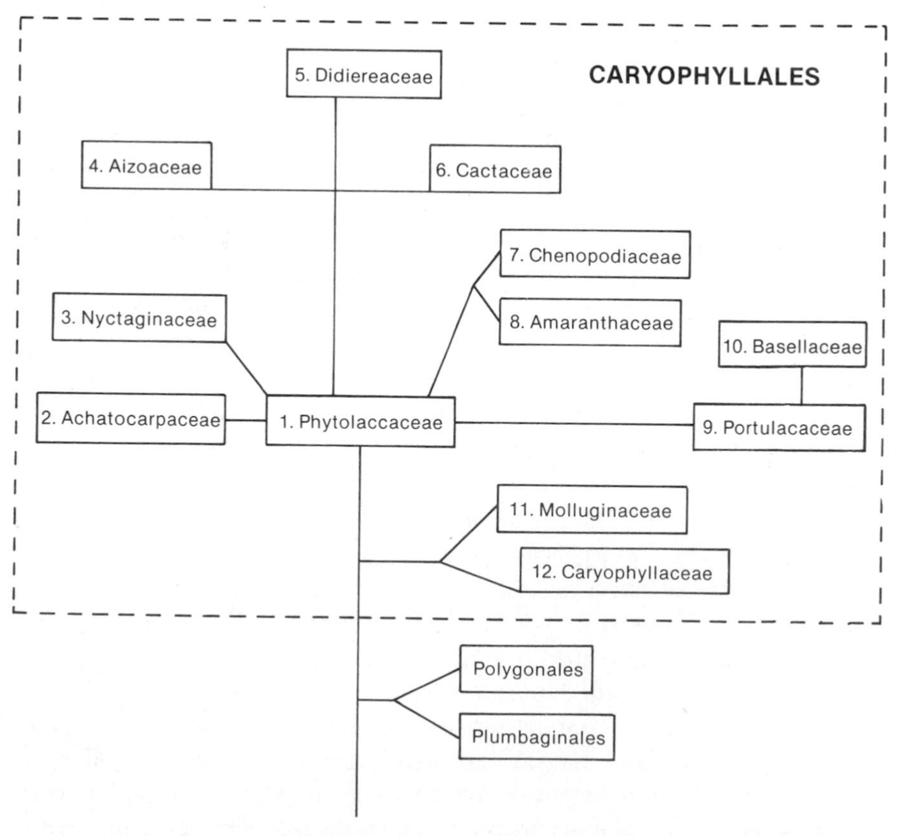

FIG. 3.1 Putative relationships among the families and orders of Caryophyllidae.

the chemical evolution if the heavily tanniferous Polygonales and Plumbaginales were divorced from the mostly nontanniferous Caryophyllales. The Plumbaginales could be accommodated in the Dilleniidae, alongside the Primulales, but the Polygonales would be out of place there. The special chemistry of the Caryophyllales may be presumed to have some evolutionary and selective significance and may indicate that the order originated after tannins had passed the peak of their effectiveness as repellents, but then what of the Polygonaceae? I leave the question unresolved.

SELECTED REFERENCES

Beck, E., H. Merxmüller, & H. Wagner. 1962. Kurze Mitteilung über die Art der Anthocyane bei Plumbaginaceen, Alsinoideen und Molluginaceen. Planta 58: 220–224.

Boulter, D. 1974. The use of amino acid sequence data in the classification of higher plants. *In*: G. Bendz and Santesson, eds., Chemistry in botanical classification, pp. 211–216. Nobel Symposium 25. Academic Press. New York and London.

Nowicke, J. W., & J. J. Skvarla. 1977. Pollen morphology and the relationship of the Plumbaginaceae, Polygonaceae, and Primulaceae to the order Centrospermae. Smithsonian Contr. Bot. 37: 1–64.

Roland, F. 1971. Données évolutives résultant de l'étude ultrastructurale des apertures de pollens appartenant au groupe Ranales Centrospermales. Rev. Gén. Bot. 78: 329–338.

SYNOPTICAL ARRANGEMENT OF THE ORDERS OF CARYOPHYLLIDAE

1 Ovules 1-many, campylotropous or amphitropous, only rarely anatropous; seeds essentially without true endosperm, the peripheral, straight to much more often curved or annular embryo generally bordering or surrounding a more or less abundant perisperm, or the perisperm sometimes scanty or wanting; sieve-tubes with a characteristic sort of P-type plastid that has a subperipheral ring of proteinaceous filaments; most families producing betalains but not anthocyanins; plants without ellagic acid, and only seldom with proanthocyanin; perianth, ovary, and placentation diverse, but the perianth unlike that of typical members of the next two orders .... 1. CARYOPHYLLALES.

1 Ovule solitary on a basal placenta, anatropous or orthotropous; seeds without perisperm, the straight or curved, peripheral or embedded embryo commonly associated with a more or less abundant endosperm, or the endosperm sometimes scanty or wanting; sieve-tube plastids of S-type; plants producing anthocyanins, but not betalains, and generally with proanthocyanins or ellagic acid or both.

2 Perianth not clearly differentiated into calyx and corolla, the 2–6 tepals either in a single whorl or more often in two more or less similar sets of (2) 3 each; carpels (2) 3 (4); stamens 2–9, rarely more, but only seldom 5, generally in 2 or 3 whorls; ovule orthotropous, only rarely anatropous, leaves often with conspicuous, sheathing stipules .... 2. POLYGONALES.

2 Perianth differentiated into a pentamerous, gamosepalous calyx and a pentamerous, generally gamopetalous corolla; carpels 5; stamens 5, opposite the petals or corolla-lobes; ovule anatropous; leaves exstipulate .... 3. PLUMBAGINALES.

1. Order CARYOPHYLLALES Bentham & Hooker 1862

Herbs, or less commonly shrubs or small (rarely rather large) trees, very often with anomalous secondary growth, producing alternating concentric rings of xylem and phloem or concentric rings of vascular bundles, often more or less succulent (as to stem or leaves or both) and with associated crassulacean acid metabolism, or with C_4 photosynthesis, often producing triterpenoid saponins, mostly not tanniferous, lacking ellagic acid and only rarely with proanthocyanin, in 10 of the 12 families producing betalains but not anthocyanins, but in the Caryophyllaceae and Molluginaceae producing anthocyanins and not betalains; vessels mostly with simple perforations; imperforate tracheary elements with simple or seldom bordered pits; sieve-tubes with a characteristic special kind of P-type plastid that has a subperipheral ring of proteinaceous filaments and often also a central protein crystalloid; phytoferritin commonly accumulated in phloem-parenchyma cells of succulent members; roots only seldom forming mycorhizae. LEAVES alternate, opposite, or seldom whorled, simple and commonly entire, sometimes reduced to spines, often showing Kranz anatomy; stipules wanting, or less often present but then usually small. FLOWERS in various sorts of inflorescences, most commonly entomophilous, perfect or less often unisexual, regular or less commonly irregular, hypogynous to perigynous or epigynous; sepals (1) 2–5 (–10), distinct or connate below to form a calyx tube, the whole calyx sometimes corolloid in appearance; petals often wanting, when present 2-many, distinct or basally connate, sometimes of staminodial origin, sometimes phyletically transformed sepals, sometimes passing into the sepals; stamens 1-many, often as many or twice as many as the sepals or petals, when numerous originating (as far as known) in centrifugal sequence; anthers mostly tetrasporangiate and dithecal, but sometimes bisporangiate and monothecal, usually opening by longitudinal slits; pollen grains trinucleate or rarely binucleate, from tricolpate to often pantoporate, the exine usually spinulose and finely perforate-punctate; gynoecium of 1-many carpels, these distinct or more often united to form a compound, superior to inferior ovary with distinct styles or with a single, usually lobed or cleft style, the locules as many as the carpels, or very often the ovary unilocular through failure of the partitions; placentation variously axile, basal, free-central, or parietal; ovules 1-many, more or less distinctly campylotropous or amphitropous to seldom hemitropous or even anatropous, crassinucellar, bitegmic or rarely unitegmic, the micropyle commonly formed by the inner integument; endosperm-development nuclear. FRUIT of various types, dry or fleshy, dehiscent or indehiscent; seeds frequently arillate; embryo large, nearly always dicotyledonous, peripheral, straight to much more often curved or annular, bordering or surrounding the more or less abundant, starchy perisperm (which commonly has clustered starch-grains), or the perisperm sometimes scanty or even wanting; true endosperm wanting or very scanty. Perhaps originally X = 9. (Centrospermae, Chenopodiales, Cactales, Opuntiales, Silenales)

The order Caryophyllales, as here defined, consists of 12 families and about 10,000 species. No one family dominates the order, but the three families Aizoaceae, Cactaceae, and Caryophyllaceae collectively contain about two-thirds of the species. On the other hand the Phytolaccaceae, usually considered to be phyletically basal in the group, have only about 120 species. The order has often been known by the descriptive name Centrospermae, in reference to the free-central or basal placentation of many members. The familiar name Caryophyllales is here retained in preference to the older names Chenopodiales and Silenales of Lindley (1833).

In addition to the traditional floral characters, the vast majority of the investigated members of the Caryophyllales have a characteristic syndrome of embryological features, which have been listed by Maheshwari (1950, p. 362, in initial citations) as follows: 1) anther tapetum glandular, and its cells two- to four-nucleate; periplasmodium absent; 2) divisions of the microspore mother cells simultaneous; 3) pollen-grains trinucleate; 4) ovules campylotropous with strongly curved and massive nucellus; 5) micropyle formed by the swollen apex of the inner integument, which protrudes and approaches the funiculus; 6) a hypodermal archesporial cell which cuts off a cell wall; 7) formation of a nucellar cap arising from periclinal divisions of the cells of the nucellar epidermis; 8) functioning of the chalazal megaspore of the tetrad; 9) formation of a monosporic eight-nucleate embryo-sac; 10) functioning of the perisperm as the chief storage region. Each of these features occurs individually in other orders, and sometimes several of them occur together, but the presence of all of them collectively is a good indication that the plant belongs to the Caryophyllales.

All investigated members of the Caryophyllales have a characteristic type of sieve-tube plastid that is unknown in other angiosperms. The plastid contains a set of bundles of proteinaceous filaments that collectively form a sub-peripheral ring. Often there is also a larger central protein crystalloid, which may be either globular or polyhedral. Sieve-tube plastids with proteinaceous inclusions also occur in some other dicotyledons (notably in some members of the Magnoliidae and Fabaceae) and are standard among the monocotyledons, but in these groups the inclusions do not form a ring of filaments. The structure of the sieve-tube plastids is beginning to take on considerable taxonomic importance because of its correlation with taxonomic groups perceived on other grounds. The functional significance of the differences in structure is still wholly obscure.

The Caryophyllales are further noteworthy in that most members of the order produce betalains and lack anthocyanins. Among the angiosperms, betalains are known only in the Caryophyllales, although they also occur in some Basidiomycetes. Betalains sometimes coexist with anthoxanthins (flavonoids of yellow to orange color), but they have never been demonstrated to coexist with anthocyanins. Betalains and flavonoids are structurally very different, but their seeming antipathy suggests the possibility of functional equivalence. They are of course similar in their frequent function as flower-pigments, but it is difficult to see how this function in common should prevent the coexistence of betalains and anthocyanins, while permitting the coexistence of betalains and anthoxanthins. Furthermore, in some other families such as the Asteraceae, flavonoid flower-pigments occur together with the chemically very different carotenoids. The preliminary report (Kimler, 1975) that betalains have fungicidal properties may provide a more useful approach to the question. It may be that the function of betalains is primarily protective, and that therefore they are better compared to proanthocyanins (also rare in the Caryophyllales) than to true anthocyanins.

Because of the widespread occurrence of betalains in the Caryophyllales and their apparent restriction to this order, their presence has taken on some taxonomic significance. Some authors have proposed to define the order (necessarily under another name) on this feature alone. Under such an interpretation the Caryophyllaceae and Molluginaceae, which have anthocyanins but not betalains, would be divorced from the other 10 families of the group. This proposal has not been well received by students of the Caryophyllales or by students of the general system of classification of flowering plants. The extensive and sometimes polemical argumentation of the opposing points of view is not reproduced here, but some of it can be found in the literature cited in the bibliography for the order. We may merely note that the most ardent proponents of the severance of the 2 anthocyanin-families from the 10 betalain-families have retreated and are now willing to include all 12 families in the same order. (See further comment under Caryophyllaceae.)

Many of the Caryophyllales, including some members of the Caryophyllaceae, have a characteristic spherical, pantoporate type of pollen-grain that has been loosely compared to a golf ball in appearance. This type of pollen is rare among other angiosperms, though not wholly unknown. It is clear that the golf ball pollen-type has evolved independently in each of the several families from a more ordinary triaperturate type. We have here a good example of the general principle that characteristics held in common by members of a taxonomic group frequently reflect independent realization of similar evolutionary potentialities, rather than direct inheritance from a common ancestor. If the members of the Caryophyllales with more primitive pollen-types had all become extinct, it would be easy to suppose that the golf ball type of pollen had originated only once in the order and was a marker of monophylesis.

Some of the evolutionary tendencies in the Caryophyllales are obviously adaptive. C_4 photosynthesis appears to be an adaptation to high temperatures and/or aridity. The succulent habit and associated crassulacean acid metabolism of many members are clearly adaptations to aridity. Anomalous secondary growth is a means of returning to a woody habit after the cambium has phyletically lost the potentiality for prolonged secondary growth. Whether anomalous secondary growth can also be interpreted as a means of protecting the phloem from desiccation, as suggested by Fahn and Shchori (1967), is more debatable. None of these adaptive features is restricted to the Caryophyllales. All can be found in various other groups of angiosperms.

Some of the other characters or evolutionary tendencies of the Caryophyllales, such as those affecting the gynoecium, the pollen, and the sieve-tube plastids, are much more difficult to interpret in Darwinian terms. We have already noted that the functional significance of the substitution of betalains for anthocyanins is still very dubious. Of course the possibility of new insight into the significance of these features cannot be denied.

The Phytolaccaceae are generally (and plausibly) regarded as the most archaic family in the Caryophyllales, possibly ancestral to all the others. The Phytolaccaceae must therefore come first in the linear sequence. The Achatocarpaceae are only marginally distinct from the Phytolaccaceae, and should follow immediately. The Nyctaginaceae, with a monomerous gynoecium, are conveniently placed before the bulk of the families that have a syncarpous gynoecium. The Aizoaceae, Didiereaceae, and Cactaceae are allied inter se and should be associated in the linear sequence. The Amaranthaceae and Chenopodiaceae must be linked together, as must also the Portulacaceae and Basellaceae. It is useful to end the sequence with the Caryophyllaceae, so as to juxtapose that family to the following order, Polygonales. It is also useful to associate the Molluginaceae with the Caryophyllaceae, since these two families collectively differ from all other members of the order in having anthocyanins and lacking betalains. The linear sequence of families here presented reflects these considerations.

SELECTED REFERENCES

Arber, A. 1939. Studies in flower structure. V. On the interpretation of the petal and "corona" in *Lychnis*. Ann. Bot. (London) II. 3: 337–346.

Bailey, I. W., and others. 1960–1968. Comparative anatomy of the leaf-bearing cacti. J. Arnold Arbor. 41: 341–356, 1960. 42: 144–156; 334–346, 1961. 43: 187–202; 234–278; 376–388, 1962. 44: 127–137; 222–231; 390–401, 1963. 45: 140–157; 374–389, 1964. 46: 74–85; 445–464, 1965. 47: 273–287, 1966. 49: 370–376, 1968.

Bailey, I. W. 1966. The significance of the reduction of vessels in the Cactaceae. J. Arnold Arbor. 47: 288–292.

Bakshi, T. S., & S. L. Chhajlani. 1954. Vascular anatomy of the flower of certain species of the Amarantaceae with a discussion of the nature of the inflorescence in the family. Phytomorphology 4: 434–446.

Balfour, E. 1965. Anomalous secondary thickening in Chenopodiaceae, Nyctaginaceae and Amaranthaceae. Phytomorphology 15: 111–122.

Behnke, H.-D. 1976. A tabulated survey of some characters of systematic importance in centrospermous families. Pl. Syst. Evol. 126: 95–98.

Behnke, H.-D. 1976. Ultrastructure of sieve-element plastids in Caryophyllales (Centrospermae), evidence for delimitation and classification of the order. Pl. Syst. Evol. 126: 31–54.

Behnke, H.-D. 1976. Die Siebelement-Plastiden der Caryophyllaceae, eine weitere specifische Form der P-Typ-Plastiden bei Centrospermen. Bot. Jahrb. Syst. 95: 327–333.

Behnke, H.-D. 1977. Regular occurring massive deposits of phytoferritin in the phloem of succulent Centrospermae. Z. Pflanzenphysiol. 85: 89–92.

Behnke, H.-D. 1978. Elektronenoptische Untersuchungen am Phloem sukkulenter Centrospermen (incl. Didiereaceen). Bot. Jahrb. Syst. 99: 341–352.

Bisalputra, T. 1960, 1961, 1962. Anatomical and morphological studies in the Chenopodiaceae. I. Inflorescence of *Atriplex* and *Bassia*. II. Vascularization of the seedlings. III. The primary vascular system and nodal anatomy. Austral. J. Bot. 8: 226–242, 1960. 9: 1–19, 1961. 10: 13–24, 1962.

Blackwell, W. H. 1977. The subfamilies of the Chenopodiaceae. Taxon 26: 395–397.

Bocquet, G. 1959 (1960). The structure of the placental column in the genus *Melandrium* (Caryophyllaceae). Phytomorphology 9: 217–221.

Bogle, A. L. 1969. The genera of Portulacaceae and Basellaceae in the southeastern United States. J. Arnold Arbor. 50: 566–598.
Bogle, A. L. 1970. The genera of Molluginaceae and Aizoaceae in the southeastern United States. J. Arnold Arbor. 51: 431–462.
Bogle, A. L. 1974. The genera of Nyctaginaceae in the southeastern United States. J. Arnold Arbor. 55: 1–37.
Bogle, A. L., et al. 1971. *Geocarpon*: Aizoaceae or Caryophyllaceae? Taxon 20: 473–477.
Boke, N. H. 1944. Histogenesis of the leaf and areole in *Opuntia cylindrica*. Amer. J. Bot. 31: 299–316.
Boke, N. H. 1954. Organogenesis of the vegetative shoot in *Pereskia*. Amer. J. Bot. 41: 619–637.
Boke, N. H. 1963. Anatomy and development of the flower and fruit of *Pereskia pititache*. Amer. J. Bot. 50: 843–858.
Boke, N. H. 1964. The cactus gynoecium: A new interpretation. Amer. J. Bot. 51: 598–610.
Boke, N. 1966. Ontogeny and structure of the flower and fruit of *Pereskia aculeata*. Amer. J. Bot. 53: 534–542.
Bortenschlager, S. 1973. Morphologie Pollinique des Phytolaccaceae. Pollen & Spores 15: 227–253.
Buell, K. M. 1952. Developmental morphology in *Dianthus*. I. Structure of the pistil and seed development. Amer. J. Bot. 39: 194–210.
Buxbaum, F. 1944. Untersuchungen zur Morphologie der Kakteenblüte. I. Das Gynoecium. Bot. Arch. 45: 190–247.
Buxbaum, F. 1949. Vorläufer des Kakteen-Habitus bei den Phytolaccaceen. Oesterr. Bot. Z. 96: 5–14.
Buxbaum, F. 1956. Das Gesetz der Verkürzung der vegetativen Phase in der Familie der Cactaceae. Oesterr. Bot. Z. 103: 353–362.
Buxbaum, F. 1958. The phylogenetic division of the subfamily Cereoideae, Cactaceae. Madroño 14: 177–206.
Buxbaum, F. 1961. Vorläufige Untersuchungen über Umfang, systematische Stellung und Gliederung der Caryophyllales (Centrospermae). Beitr. Biol. Pflanzen 36: 1–56.
Chang, C. P., & T. J. Mabry. 1973. The constitution of the order Centrospermae: $_r$RNA-DNA hybridization studies among betalain- and anthocyanin-producing families. Biochem. Syst. 1: 185–190.
Choux, P. 1934. Les Didiéréacées, xérophytes de Madagascar. Mém. Acad. Malgache XVII: 1–69.
Cranwell, L. M. 1963. The Hectorellaceae: pollen type and taxonomic speculation. Grana Palynol. 4: 195–202.
Cronquist, A. 1974. Chemical plant taxonomy: a generalist's view of a promising specialty. *In*: G. Bendz and J. Santesson, eds., Chemistry in botanical classification, pp. 29–39. Nobel Symposium 25. Academic Press. New York and London.
Devi, H. M., & T. Pullaiah. 1975. Life history of *Basella rubra* Linn and taxonomic status of the family Basellaceae. J. Indian Bot. Soc. 54: 154–166.
Dupont, S. 1960. Sur la morphologie des plantules de Mesembryanthémacées et d'Aizoacées. Bull. Soc. Hist. Nat. Toulouse 95: 356–360.
Eckardt, Th. 1954. Morphologische und systematische Auswertung der Placentation von Phytolaccaceen. Ber. Deutsch. Bot. Ges. 67: 113–128.
Eckardt, Th. 1955. Nachweis der Blattburtigkeit ("Phyllosporie") grundständiger Samenanlagen bei Centrospermen. Ber. Deutsch. Bot. Ges. 68: 167–182.

Eckardt, Th. 1967. Vergleich von *Dysphania* mit *Chenopodium* und mit Illecebraceae. Bauhinia 3: 327–344.
Eckardt, Th. 1967. Blütenbau und Blütenentwicklung von *Dysphania myriocephala* Benth. Bot. Jahrb. Syst. 86: 20–37.
Eckardt, Th. 1967 (1968). Blütenmorphologie von *Dysphania plantaginella* F. v. M. Phytomorphology 17: 165–172.
Eckardt, Th. 1974. Vom Blütenbau der Centrospermen-Gattung *Lophiocarpus* Turcz. Phyton (Horn) 16: 13–27.
Eckardt, Th. 1976. Classical morphological features of centrospermous families. Pl. Syst. Evol. 126: 5–25.
Ehrendorfer, F. 1976. Chromosome numbers and differentiation of centrospermous families. Pl. Syst. Evol. 126: 27–30.
Erdtman, G. 1948. Pollen morphology and plant taxonomy. VIII. Didiereaceae. Bull. Mus. Hist. Nat. (Paris) II. 20: 387–394.
Esau, K., & V. I. Cheadle. 1969. Secondary growth in *Bougainvillea*. Ann. Bot. (London) II. 33:807–819.
Fahn, A., & T. Arzee. 1959. Vascularization of articulated Chenopodiaceae and the nature of their fleshy cortex. Amer. J. Bot. 46: 330–338.
Fahn, A., & Y. Shchori. 1967 (1968). The organization of the secondary conducting tissues in some species of the Chenopodiaceae. Phytomorphology 17: 147–154.
Fulvio, T. E. Di. 1975. Estomatogenesis en *Halophytum ameghinoi* (Halophytaceae). Kurtziana 8: 17–29.
Gibson, A. C. 1973. Comparative anatomy of secondary xylem in Cactoideae (Cactaceae). Biotropica 5(1): 29–65.
Gibson, A. C. 1976. Vascular organization in shoots of Cactaceae. I. Development and morphology of primary vasculature in Pereskioideae and Opuntioideae. Amer. J. Bot. 63: 414–426.
Gibson, A. C. 1977. Wood anatomy of opuntias with cylindrical to globular stems. Bot. Gaz. 138: 334–351.
Gibson, A. C. 1978. Rayless secondary xylem of *Halophytum*. Bull. Torrey Bot. Club 105: 39–44.
Gibson, A. C., & K. E. Horak. 1978 (1979). Systematic anatomy and phylogeny of Mexican columnar cacti. Ann. Missouri Bot. Gard. 65: 999–1057.
Herre, H. 1971. The genera of the Mesembryanthemaceae. Tafelberg-Uitgewers Beperk. Cape Town.
Hofmann, U. 1973. Centrospermen-Studien 6. Morphologische Untersuchungen zur Umgrenzung und Gliederung der Aizoaceen. Bot. Jahrb. Syst. 93: 247–324.
Hofmann, U. 1977. Centrospermen-Studien 9. Die Stellung von *Stegnosperma* innerhalb der Centrospermen. Ber. Deutsch. Bot. Ges. 90: 39–52.
Ihlenfeldt, H.-D. 1975. Some trends in the evolution of the Mesembryanthemaceae. Boissiera 24: 249–254.
Ihlenfeldt, H.-D., & H. Straka. 1961 (1962). Über die systematische Stellung und Gliederung der Mesembryanthemen. Ber. Deutsch. Bot. Ges. 74: 485–492.
Inamdar, J. A. 1968. Epidermal structure and ontogeny of stomata in some Nyctaginaceae. Flora 158 B: 159–166.
Inamdar, J. A., M. Gangadhara, P. G. Morge, & R. M. Patel. 1977. Epidermal structure and ontogeny of stomata in some Centrospermae. Feddes Repert. 88: 465–475.

Jensen, U. 1965. Serologische Untersuchungen zur Frage der systematischen Einordnung der Didiereaceae. Bot. Jahrb. Syst. 84: 233–253.
Joshi, A. C. 1937. Some salient points in the evolution of the secondary vascular cylinder of Amarantaceae and Chenopodiaceae. Amer. J. Bot. 24: 3–9.
Joshi, A. C., & V. S. R. Rao. 1934. Vascular anatomy of flowers of four Nyctaginaceae. J. Indian Bot. Soc. 13: 169–186.
Joshi, A. C., & V. R. Rao. 1936. The embryology of *Gisekia pharnaceoides* L. Proc. Indian Acad. Sci. B. 3: 71–92.
Kajale, L. B. 1954. A contribution to the embryology of the Phytolaccaceae. II. Fertilization and the development of embryo, seed and fruit in *Rivina humilis* Linn. and *Phytolacca dioica* Linn. J. Indian Bot. Soc. 33: 206–225.
Kimler, L. M. 1975. Betanin, the red beet pigment, as an antifungal agent. Bot. Soc. Amer. Abstracts of papers, 1975: 36.
Kimler, L., J. Mears. T. J. Mabry, & H. Rösler. 1970. On the question of the mutual exclusiveness of betalains and anthocyanins. Taxon 19: 875–878.
Kowal, T. 1961. Studia nad morfologia i anatomia nasion Portulacaceae Rchb. (Subtitle: Morphology and anatomy of the seeds in Portulacaceae Rchb.) Monogr. Bot. 12: 3–47.
Leinfellner, W. 1937. Beiträge zur Kenntnis der Cactaceen-Areolen. Oesterr. Bot. Z. 86: 1–60.
Leuenberger, B. E. 1976. Die Pollenmorphologie der Cactaceae und ihre Bedeutung für die Systematik. Dissertationes Botanicae 31: 1–321.
Lewis, W. H. 1970. Extreme instability of chromosome number in *Claytonia virginica*. Taxon 19: 180–182.
Mabry, T. J. 1974. Is the order Centrospermae monophyletic? *In*: G. Bendz and J. Santesson, eds., Chemistry in botanical classification, pp. 275–285. Nobel Symposium 25. Academic Press. New York and London.
Mabry, T. J. 1977. The order Centrospermae. Ann. Missouri Bot. Gard. 64: 210–220.
Mabry, T. J., L. Kimler, & C. Chang. 1972. The betalains; structure function, and biogenesis, and the plant order Centrospermae. *In*: V. C. Runeckles and T. C. Tso, eds., Recent advances in phytochemistry 5: 105–134. Academic Press. New York.
Mabry, T. J., A. Taylor, & B. L. Turner. 1963. The betacyanins and their distribution. Phytochemistry 2: 61–64.
McNeill, J. 1974. Synopsis of a revised classification of the Portulacaceae. Taxon 23: 725–728.
McNeill, J. 1975. A generic revision of Portulacaceae tribe Montieae using techniques of numerical taxonomy. Canad. J. Bot. 53: 789–809.
Maheshwari, P., & R. N. Chopra. 1955. The structure and development of the ovule and seed of *Opuntia dillenii* Haw. Phytomorphology 5: 112–122.
Mauritzon, J. 1934. Ein Beitrag zur Embryologie der Phytolaccaceen und Cactaceen. Bot. Not. 1934: 111–135.
Nair, N. C., & V. J. Nair. 1961. Studies on the morphology of some members of the Nyctaginaceae. I. Nodal anatomy of *Boerhavia*. Proc. Indian Acad. Sci., B. 54: 281–294.
Ng Siew Yoong, W. R. Philipson, & J. R. L. Walker. 1975. Hectorellaceae—a member of the Centrospermae. New Zealand J. Bot. 13: 567–570.
Nowicke, J. W. 1975 (1976). Pollen morphology in the order Centrospermae. Grana 15: 51–77.
Paliwal, G. S. 1965. The development of stomata in *Basella rubra* Linn. Phytomorphology 15: 50–53.

Pant, D. D., & P. F. Kidwai. 1968. Structure and ontogeny of stomata in some Caryophyllaceae. J. Linn. Soc., Bot. 60: 309–314.
Pax, F. 1927. Zur Phylogenie der Caryophyllaceae. Bot Jahrb. Syst. 61: 223–241.
Payne, M. A. 1933. The morphology and anatomy of *Mollugo verticillata* L. Univ. Kansas Sci. Bull. 21: 399–419.
Philipson, W. R., & J. P. Skipworth. 1961. Hectorellaceae: A new family of dicotyledons. Trans. Roy. Soc. New Zealand, Bot. 1: 31.
Rauh, W. 1961. Weitere Untersuchungen an Didiereaceen. 1 Teil, Beitrag zur Kentniss der Wuchsformen der Didiereaceen, unter besonderer Berucksichtigung neuer Arten. Sitzungsber. Heidelberger Akad. Wiss., Math.-Naturwiss. Kl. 7: 182–300.
Rauh, W., & H. Reznik. 1961. Zur Frage der systematischen Stellung der Didiereaceen. Bot. Jahrb. Syst. 81: 94–105.
Rauh, W., & H. R. Schölch. 1964. Weitere Untersuchungen an Didiereaceen. 2 Teil: Infloreszenz-, blütenmorphologische und embryologische Untersuchungen mit Ausblick auf die systematische Stellung der Didiereaceen. Sitzungsber. Heidelberger Akad. Wiss. Math.-Naturwiss. Kl. 11: 221–434.
Reznik, H. 1955. Die Pigmente der Centrospermen als systematisches Element. Z. Bot. 43: 499–530.
Reznik, H. 1975. Betalaine. Ber. Deutsch. Bot. Ges. 88: 179–190.
Rohweder, O. 1965, 1967, 1970. Centrospermen-Studien. 1. Der Blütenbau bei *Uebelinia kiwuensis* T. C. E. Fries (Caryophyllaceae). 2. Entwicklung und morphologische Deutung des Gynöciums bei *Phytolacca*. 3. Blütenentwicklung und Blütenbau bei Silenoideen (Caryophyllaceae). 4. Morphologie und Anatomie der Blüten, Früchte und Samen bei Alsinoideen und Paronychioideen s. lat. (Caryophyllaceae). Bot. Jahrb. Syst. 83: 406–418, 1965. 84: 509–526, 1965. 86: 130–185, 1967. 90: 201–271, 1970.
Rohweder, O., & K. Huber. 1974. Centrospermen-Studien. 7. Beobachtungen und Anmerkungen zur Morphologie und Entwicklungsgeschichte einiger Nyctaginaceen. Bot. Jahrb. Syst. 94: 327–359.
Rohweder, O., & K. König. 1971. Centrospermen-Studien. 5. Bau der Blüten, Früchte und Samen von *Pteranthus dichotomus* Forsk. (Caryophyllaceae). Bot. Jahrb. Syst. 90: 447–468.
Rohweder, O., & K. Urmi-König. 1975. Centrospermen-Studien 8. Beiträge zur Morphologie, Anatomie und systematischen Stellung von *Gymnocarpos* Forsk. und *Paronychia argentea* Lam. (Caryophyllaceae). Bot. Jahrb. Syst. 96: 375–409.
Roth, I. 1961. Sobre el desarrollo del obturador en el gineceo de "*Armeria*." Acta Ci. Venez. 12: 172–174.
Roth, I. 1962. Histogenese und morphologische Deutung der basalen Plazenta von *Herniaria*. Flora 152: 179–195.
Roth, I. 1963. Histogenese und morphologische Deutung der Zentralplazenta von *Cerastium*. Bot. Jahrb. Syst. 82: 100–118.
Schaeppi, H. 1936. Zur Morphologie des Gynoeceums der Phytolaccaceen. Flora 131: 41–59.
Schölch, H.-F. 1963. Die systematische Stellung der Didiereaceen im Lichte neuer Untersuchung über ihren Blütenbereich. Ber. Deutsch. Bot. Ges. 76: (49)–(55).
Scott, A. J. 1977. Reinstatement and revision of Salicorniaceae J. Agardh (Caryophyllales). J. Linn. Soc., Bot. 75: 357–374.
Sharma, H. P. 1961. Contributions to the morphology and anatomy of *Basella rubra* Linn. Bull. Bot. Soc. Bengal. 15: 43–48.

Sharma, H. P. 1962, 1963. Contributions to the morphology of the Nyctaginaceae. I. Anatomy of the node and inflorescence of some species. II. Floral anatomy of some species. Proc. Indian Acad. Sci. B. 56: 35–50, 1962. 57: 149–163, 1963.

Sharma, H. P. 1962, 1963. Studies in the order Centrospermales. III. Vascular anatomy of the flower of some species of the family Ficoidaceae. Proc. Indian Acad. Sci. B. 56: 269–285, 1962. II. Vascular anatomy of the flower of certain species of the Molluginaceae. IV. Pollen morphology of some species of families Ficoidaceae, Molluginaceae, Nyctaginaceae, and Portulacaceae. J. Indian Bot. Soc. 42: 19–32; 637–645, 1963.

Skipworth, J. P. 1961. The taxonomic position of *Hectorella caespitosa*. Trans. Roy. Soc. New Zealand, Bot. 1: 17–30.

Skvarla, J. J., & J. W. Nowicke. 1976. Ultrastructure of pollen exine in the centrospermous families. Pl. Syst. Evol. 126: 55–78.

Straka, H. 1965. Die Pollenmorphologie der Didiereaceen. Sitzungsber. Heidelberger Akad. Wiss., Math.-Naturwiss. Kl. 11: 435–443.

Straka, H. 1975. Palynologie et différentiation systématique d'une famille endémique de Madagascar: les Didieréacées. Boissiera 24: 245–248.

Thomson, B. F. 1942. The floral morphology of the Caryophyllaceae. Amer. J. Bot. 29: 333–349.

Tiagi, Y. D. 1955. Studies in floral morphology. II. Vascular anatomy of the flower of certain species of the Cactaceae. J. Indian Bot. Soc. 34: 408–428.

Tsukada, M. 1964. Pollen morphology and identification. II. Cactaceae. Pollen & Spores 6: 45–84.

Van Campo, M. 1967. Pollen et classification. Rev. Palaeobotan. Palynol. 3: 65–71.

Wheat, D. 1977. Successive cambia in the stem of *Phytolacca dioica*. Amer. J. Bot. 64: 1209–1217.

Wohlpart, A., & T. J. Mabry. 1968. The distribution and phylogenetic significance of the betalains with respect to the Centrospermae. Taxon 17: 148–152.

Zandonella, P. 1967. Les nectaires des Alsinoideae: *Stellaria* et *Cerastium* sensu lato. Compt. Rend Hebd. Séances Acad. Sci. 264 Ser. D: 2466–2469.

Zandonella, P. 1967. Stomates des nectaires floraux chez les Centrospermales. Bull. Soc. Bot. France 114: 11–20.

Zandonella, P. 1977. Apports de l'étude comparée des nectaires floraux a la conception phylogénétique de l'ordre des Centrospermales. Ber. Deutsch. Bot. Ges. 90: 105–125.

Zandonella, P., & M. Lecocq. 1977. Morphologie pollinique et mode de pollinisation chez les Amaranthaceae. Pollen & Spores 19: 119–141.

Алешина, Л, А. 1963. Морфология пыльцевых зерен рода *Claytonia* Gronov. и близких родов. Бот. Ж. 48: 1191-1196.

Василевская, В. К. 1972. Особый тип анатомической структуры в сем. Chenopodiaceae. Бот. Ж. 57: 103-108.

Гвинианидзе, З. И. 1959.К изучению плацентации у представителей трибы Lychnideae A. Br. сем. гвоздичных. Сообщ. АН Грузинской ССР. 22: 723-728.

Гвинианидзе, З. И. 1965. Изучение Эпидермиса листа у представителей трибы Lychnideae семейства гвоздичных. Заметки по сист. и геогр. раст., Тбилисск. Бот. Инст. 24: 41-48.

Гвинианидзе, З. И. 1965. Изучение эпидермиса листа у представителей трибы Diantheae сем. гвоздичных. Заметки по сист. и геогр. раст., Тбилисск. Бот. Инст. 25: 65-68.

Замятнин, Б. Н. 1951. О нижней завязи осевого происхождения. Бот. Ж. 36: 89-92.

Кожанчиков, В. И. 1967. Морфологические признаки семян семейства Caryophyllaceae и возможные пути их эволюции Бот. Ж. 52: 1277-1286.

Моносзон, М. Х. 1952. Описание пыльцы видов семейства маревых, произрастающих на территории СССР. Труды Инст. Геогр. АН СССР, 52, Матер. по Геоморф. и Палеогеогр. СССР 7: 127-196.

Шилкина, И. А. 1953. Анатомические особенности семейства Chenopodiaceae. Бот. Ж. 38: 590-598.

SYNOPTICAL ARRANGEMENT OF THE FAMILIES OF CARYOPHYLLALES

1 Gynoecium either of 1-many distinct carpels, each with a single ovule, or of 2 or more carpels united to form a compound ovary with as many locules and ovules as carpels.

 2 Sepals distinct, or rarely connate below, but not forming a corolloid calyx-tube, generally not at all petaloid; carpels mostly 2-several, less often only one; inflorescence usually racemose or spicate or less commonly paniculate or cymose, not subtended by a conspicuous involucre; leaves alternate .... 1. PHYTOLACCACEAE.

 2 Sepals united to form a distally lobed tube that commonly simulates a sympetalous corolla and is sometimes subtended by sepaloid bracts; carpel solitary; inflorescence mostly cymose or head-like, often subtended by a conspicuous involucre; leaves opposite or rarely alternate .... 3. NYCTAGINACEAE.

1 Gynoecium otherwise, always of 2 or more carpels united to form a compound ovary, the ovary either unilocular or with as many locules as carpels, in the latter case either with more than one ovule per carpel or with some locules empty.

 3 Flowers with 10–20 stamens and a superior, unilocular ovary with 2 styles and a single basal ovule ....2. ACHATOCARPACEAE.

 3 Flowers otherwise, either with fewer stamens, or with an inferior or plurilocular ovary, or with more than one ovule, often differing in more than one of these respects.

  4 Ovary with a single fertile locule and 2 (3) empty locules; spiny, cactus-like trees or shrubs with unisexual flowers in a cymose inflorescence; seeds essentially without perisperm .... 5. DIDIEREACEAE.

  4 Ovary without empty locules; other features various, but not combined as in the Didiereaceae.

   5 Flowers epigynous or less often semi-epigynous or distinctly perigynous; tepals (and often also the stamens) usually more or less numerous, but sometimes few and in a single cycle; plants succulent.

6 Plants leaf-succulents or rarely stem-succulents, in either case nearly always unarmed; ovary superior to inferior, usually plurilocular; mainly Old-World ... 4. AIZOACEAE.

6 Plants either stem-succulents or distinctly spiny, usually both; ovary unilocular, nearly always inferior; mainly New-World .... 6. CACTACEAE.

5 Flowers hypogynous, or seldom semi-epigynous or weakly perigynous; tepals and stamens mostly few and cyclic, but sometimes more numerous in Portulacaceae; plants succulent or not.

7 Perianth evidently monochlamydeous, seldom at all petaloid, commonly small and inconspicuous, sometimes even obsolete; ovules mostly solitary (seldom 2-several) on a basal placenta; stem very commonly with anomalous secondary growth; sieve-tube plastids nearly always without a central protein crystalloid.

8 Perianth mostly green or greenish and more or less herbaceous, only seldom dry and somewhat scarious or membranous; filaments distinct, or sometimes connate at the base only .... 7. CHENOPODIACEAE.

8 Perianth generally dry and scarious or membranous; filaments often connate below (the androecial tube sometimes even simulating a small, sympetalous corolla) .... 8. AMARANTHACEAE.

7 Perianth generally dichlamydeous or seemingly so (except notably the Molluginaceae, with axile placentation and usually several or many ovules, otherwise only occasionally monochlamydeous); sieve-tube plastids with a central protein crystalloid.

9 Sepals generally 2, seldom more numerous; stamens most commonly as many as and opposite the petals, seldom more numerous or alternate with the petals; no anomalous secondary thickening; plants producing betalains but not anthocyanins; central protein crystalloid of the sieve-tube plastids globular.

10 Plants not twining or scrambling; ovules (1) 2-many; fruit capsular, or very seldom indehiscent; seldom any of the vascular bundles becoming bicollateral .... 9. PORTULACACEAE.

10 Plants twining or scrambling; ovule solitary; fruit indehiscent; larger vascular bundles becoming bicollateral (with internal phloem) at maturity .... 10. BASELLACEAE.

9 Sepals 4 or 5; stamens not at once of the same number as and opposite the petals; plants sometimes with anomalous secondary thickening, regularly producing anthocyanins but not betalains; central protein crystalloid of the sieve-tube plastids mostly polyhedral, only seldom globular.

11 Ovary with 2-several locules and axile placentation (but in the upper part of the ovary the partitions sometimes not reaching the placental column); petals small and inconspicuous or more often wanting; leaves opposite, alternate, or whorled .... 11. MOLLUGINACEAE.

11 Ovary unilocular (sometimes partitioned at the base), with central or basal placentation; petals usually more or less well developed, but sometimes wanting; leaves nearly always opposite .... 12. CARYOPHYLLACEAE.

1. FAMILY PHYTOLACCACEAE R. Brown in Tuckey 1819 nom. conserv., the Pokeweed Family

Herbs (sometimes climbing), or sometimes shrubs or even small (rarely fairly large) trees, very often glabrous, and often somewhat succulent, tending to accumulate free oxalates and nitrates, often saponiferous, but mostly not tanniferous, lacking ellagic acid and only seldom with a significant amount of proanthocyanin, consistently producing betalains, but not known to produce anthocyanins; crystals of calcium oxalate (often raphides) commonly present in some of the cells of the epidermal and parenchymatous tissues; nodes unilacunar; stem often with anomalous secondary growth, producing concentric rings of vascular bundles or alternating concentric rings of xylem and phloem; vessel-segments with simple perforations; imperforate tracheary elements with simple or seldom bordered pits; sieve-tubes with a special kind of P-type plastid

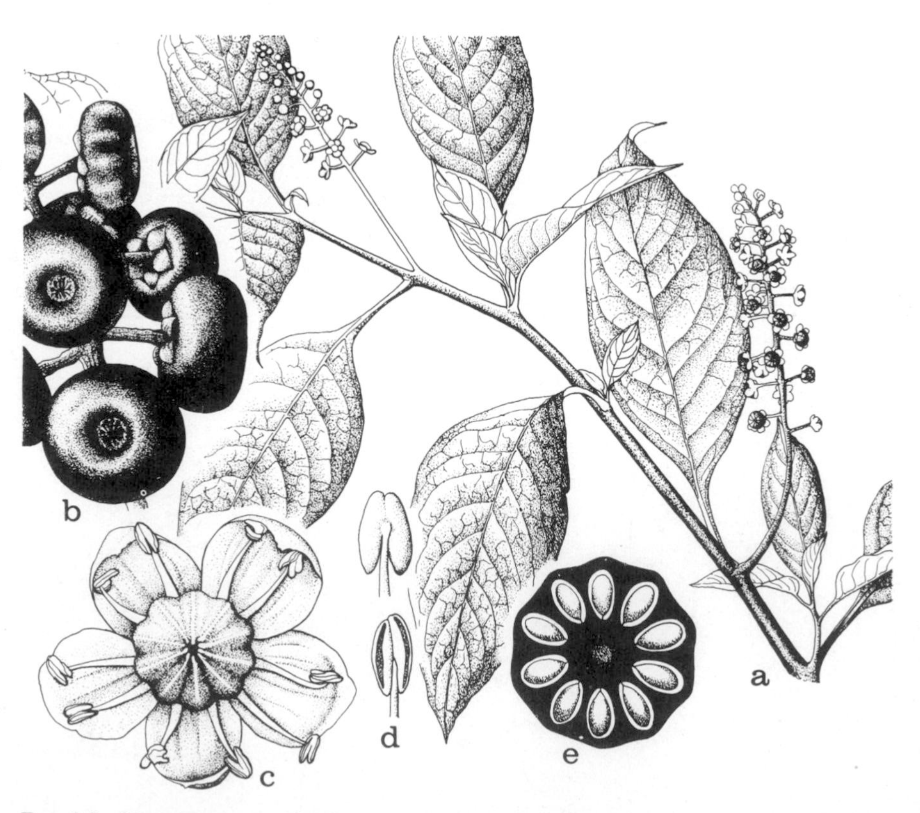

FIG. 3.2 Phytolaccaceae. *Phytolacca americana* L. a, habit, ×½; b, fruits, ×2; c, flower, from above, ×6; d, anthers, ×10; e, ovary in cross-section, ×10.

that includes a central globular or (*Stegnosperma*) polyhedral protein crystalloid as well as a subperipheral ring of proteinaceous filaments. LEAVES alternate, simple, entire; stomates paracytic or more often anomocytic; stipules wanting or vestigial. FLOWERS most commonly in axillary or leaf-opposed racemes or spikes, or sometimes in axillary panicles or mixed panicles, or (*Gisekia*) in open cymes, or solitary in the upper axils, usually regular, small, perfect or sometimes unisexual, hypogynous; sepals mostly 4 or 5 (–10), distinct or sometimes some or all of them connate below; petals wanting, or seldom (chiefly in *Stegnosperma*) present, small, distinct, and alternate with the sepals; stamens 4-many, often in 2 alternating cycles and twice as many as the sepals, when numerous originating in centrifugal sequence; filaments distinct or basally connate; anthers tetrasporangiate and dithecal, opening by longitudinal slits; pollen-grains trinucleate or reputedly (Davis) sometimes binucleate, tricolpate to pantocolpate or pantoporate; gynoecium of 1 (tribe Rivineae) or more often 2-many carpels, these distinct to more often more or less connate to form a compound pistil with distinct styles and as many locules as carpels (the partitions ephemeral in *Stegnosperma*); nectary a ring on the receptacle between the stamens and pistils, or a ring around the base of the stamens, or sometimes wanting; ovules solitary in each locule or in each simple pistil, campylotropous, bitegmic, crassinucellar, commonly basal; endosperm-development nuclear. FRUIT variously dry or fleshy, the carpels often separating at maturity; seeds with a well developed, peripheral, dicotyledonous embryo curved around a more or less abundant, hard or starchy and mealy perisperm (with clustered starch-grains), without true endosperm, sometimes (as in *Barbeuia* and *Stegnosperma*) arillate. X = 9. (Agdestidaceae, Barbeuiaceae, Gisekiaceae, Petiveriaceae, Stegnospermaceae)

The Phytolaccaceae as here broadly defined are a rather loosely knit family of some 18 genera and 125 species, widespread in tropical and subtropical regions, especially of the New World. The most familiar species in the United States is the common pokeweed, *Phytolacca americana* L.

Each of the several segregate families noted in the synonymy appears to be a natural group, but collectively they all hang together with the rest of the Phytolaccaceae. I see no reason why they cannot be accommodated at the level of tribes or subfamilies.

The most aberrant genus is *Gisekia*, which is transitional to the Molluginaceae and has often been referred to that family. It resembles the Molluginaceae in habit, inflorescence, and some details of the structure of pollen and ovules, but it is highly anomalous in that family in having separate carpels. The recent discovery that *Gisekia* has betalains rather than anthocyanins makes it seem more prudent to try to accomodate its anomalies in the Phytolaccaceae, at least until the pigments of an adequate sample of Molluginaceae are studied.

It is now widely agreed that the Phytolaccaceae are the most archaic family of the Caryophyllales. Some authors would derive all the other families of the order directly or indirectly from the Phytolaccaceae, but this may be going too far. I am not yet convinced that uniovulate carpels (characteristic of the Phytolaccaceae) are primitive in the Caryophyllales as a whole. The diagram (fig. 3.1) showing possible relationships among the families of the order should be interpreted with this reservation in mind.

Hofmann (1977) maintains that all pluricarpellate Phytolaccaceae are basically syncarpous, and that apocarpy is merely simulated in some genera by unequal growth. I am dubious.

2. Family ACHATOCARPACEAE Heimerl in Engler & Prantl 1934 nom. conserv., the Achatocarpus Family

Dioecious, sometimes spiny shrubs or small trees with normal secondary growth, somewhat tanniferous, but without ellagic acid; calcium oxalate crystals of various form (but not raphides) commonly present in some of the cells of the parenchymatous tissues; vessel-segments with simple perforations; imperforate tracheary elements with simple pits; sieve-tubes with P-type plastids like those of the Phytolaccaceae. LEAVES alternate, simple, entire; stomates anomocytic; stipules wanting. FLOWERS in axillary racemes or panicles (or the inflorescences from the bare nodes of older twigs), small, unisexual, hypogynous; sepals 4–5, small, persistent in fruit; petals none; stamens 10–20, the slender filaments distinct or connate at the base; anthers dithecal, latrorse, opening by longitudinal slits; pollen-grains 4-6-porate, coarsely granular, said to be different in detail from all other Caryophyllales; staminate flowers without even a vestigial gynoecium; gynoecium of 2 carpels united to form a compound, unilocular ovary with distinct styles; ovule solitary, basal, campylotropous. FRUIT a berry; seeds with peripheral, strongly curved, annular, dicotyledonous embryo surrounding the perisperm, without true endosperm, and lacking an aril.

The family Achatocarpaceae consists of only 2 genera (*Achatocarpus* and *Phaulothamnus*) and about 8 species, ranging from Texas and northwestern Mexico to Paraguay and Argentina. The close relationship of the Achatocarpaceae to the Phytolaccaceae is widely admitted, but the compound, unilocular ovary would be anomalous in that family. Inasmuch as the structure of the gynoecium is an important feature in the organization of the Caryophyllales into families, it seems useful to retain the Achatocarpaceae as distinct.

3. Family NYCTAGINACEAE A. L. de Jussieu 1789 nom. conserv., the Four o'clock Family

Herbs, shrubs, or trees, commonly accumulating free oxalates and at least sometimes also potassium nitrate, sometimes saponiferous, and often with secretory idioblasts in the epidermis of the leaves and parenchyma of the stems, but only seldom if ever tanniferous, lacking both proanthocyanins and ellagic acid, and producing betalains but not anthocyanins; crystals of calcium oxalate (often raphides) abundant in some of the cells of the parenchymatous tissues; nodes unilacunar; stem commonly with anomalous secondary growth (all the woody species and many of the herbaceous ones), producing concentric rings of vascular bundles or alternating concentric rings of xylem and phloem; vessel-segments nearly always with simple perforations; imperforate tracheary elements with small, simple pits; sieve-tubes with a special kind of P-type plastid that includes a central globular protein crystalloid as well as a subperipheral ring of proteinaceous filaments. LEAVES opposite or rarely alternate, simple and commonly entire, sometimes with Kranz anatomy; stomates paracytic or less often anomocytic; stipules wanting. FLOWERS perfect or seldom unisexual (sometimes gynodioecious in structure but functionally dioecious), hypogynous, commonly in cymose (sometimes head-like) inflorescences subtended by an often large and conspicuous, sometimes even corolloid involucre, the primary inflorescences sometimes reduced to a single flower, thus forming a pseudanthium with a calyx-like involucre and corolla-like calyx; calyx with a well developed, often slender and elongate tube and (3–) 5 (–8)-lobed, regular or sometimes irregular limb, valvate or plicate in bud, commonly corolloid; true corolla none; stamens commonly as many as the calyx-lobes and alternate with them, but sometimes fewer (or even only 1), or more numerous, up to 20 or 30, and then probably originating in

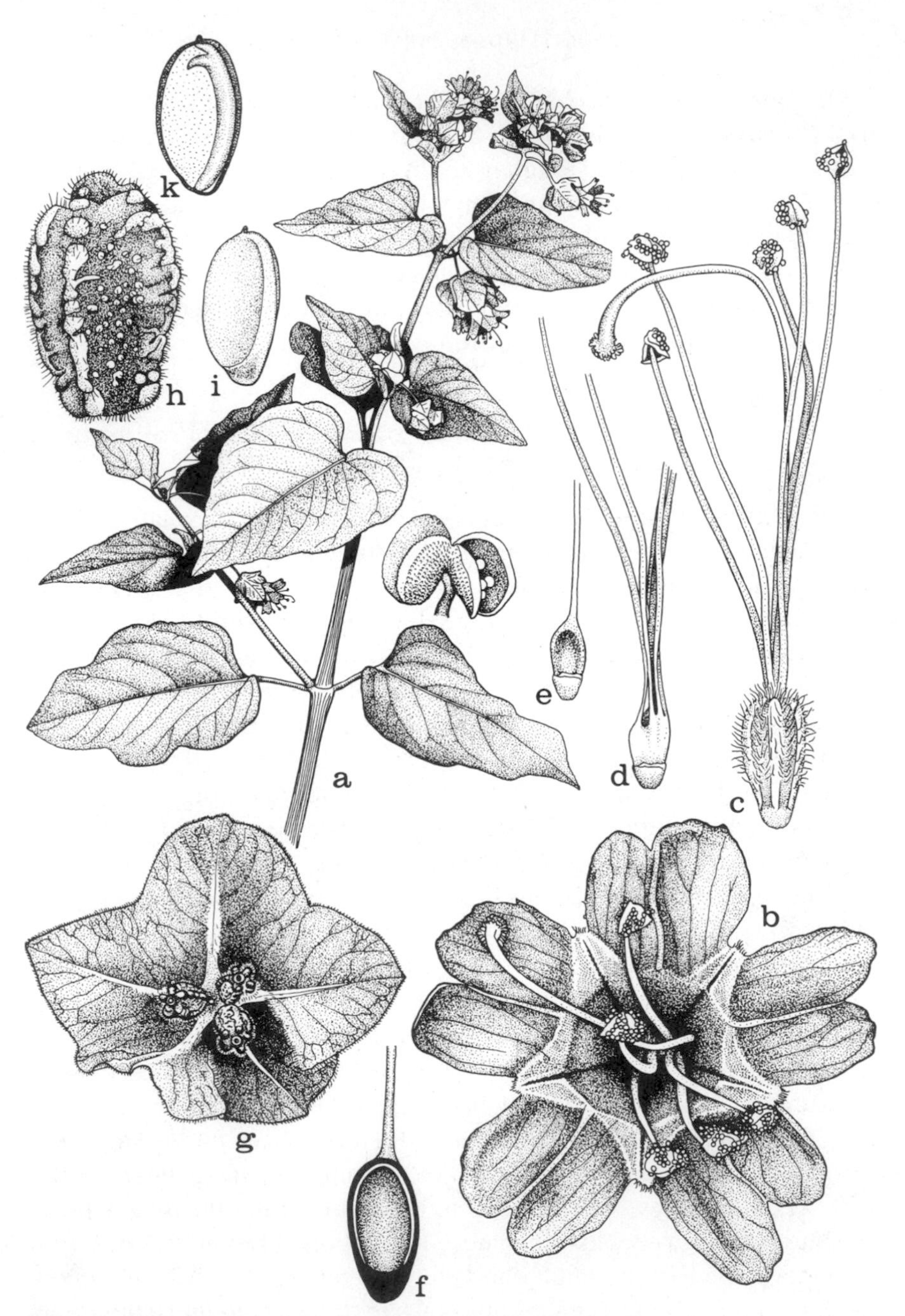

FIG. 3.3 Nyctaginaceae. *Mirabilis nyctaginea* (Michx.) MacMillan. a, habit, $\times\frac{1}{2}$; b, flower, from above, $\times 6$; c, side view of flower, with calyx-lobes removed, $\times 6$; d, base of flower, with calyx completely removed, showing basally connate filaments surrounding the ovary, $\times 6$; e, ovary and part of style, $\times 6$; f, ovary in long-section, $\times 12$; g, involucre, with 3 anthocarps, $\times 2$; h, anthocarp, $\times 6$; i, seed, $\times 6$; k, seed in long-section, showing peripheral embryo, $\times 6$.

centrifugal sequence; filaments of equal or unequal length, distinct or connate into a tube at the base; anthers tetrasporangiate and dithecal, opening by longitudinal slits; pollen-grains trinucleate or rarely binucleate, from tricolpate to pantoporate; an annular intrastaminal nectary-disk commonly present around the ovary; gynoecium of a single carpel with a long, slender style and a single basal ovule; ovule campylotropous or seldom hemitropous, crassinucellar, bitegmic or seldom unitegmic; endosperm-development nuclear. FRUIT an achene or nut, often enclosed in the persistent and indurated base of the calyx-tube, the collective structure then called an anthocarp; seeds with large, peripheral, straight or more often curved, dicotyledonous embryo and abundant or scanty, starchy (with clustered starch-grains), hard to granular or soft perisperm, without true endosperm, or the endosperm forming only a small cap over the radicle. X = 10, 13, 17, 29, 33+.

The family Nyctaginaceae consists of about 30 genera and 300 species, occurring mainly in tropical and subtropical regions of both the Old and the New World, but more numerous in the latter. Relatively few species occur in temperate climates. The most familiar members of the family are the garden four-o'clock (*Mirabilis jalapa* L., probably of Mexican origin), and species of *Bougainvillea,* widely cultivated as a porch- or arbor-vine in tropical and subtropical countries, originally South American. *Bougainvillea* well shows the corolloid involucre subtending a small cluster of flowers, as noted in the family description, whereas *Mirabilis jalapa* shows the calyx-like involucre subtending a single flower (reduced inflorescence) with a corolloid calyx. If other species of *Mirabilis* did not show the transition from a several-flowered to a single-flowered involucrate inflorescence, it would be difficult to avoid interpreting the flower of *M. jalapa* as having an ordinary sympetalous corolla and a synsepalous calyx.

The traditional placement of the Nyctaginaceae in the Caryophyllales has sometimes been challenged, as by Hutchinson (1973, in initial citations), but recent data on the floral pigments, the sieve-tube plastids, and the detailed structure of the pollen grains all support the traditional placement, as does also the structure of the nectaries (Brown, 1937, in initial citations). Furthermore, the Nyctaginaceae show the full syndrome of embryological features that have been noted by Maheshwari (1950, in initial citations) and others to be characteristic of the Caryophyllales. Most authors are now in agreement that the Nyctaginaceae are closely allied to the Phytolaccaceae and probably derived from that family.

Pollen referred to the Nyctaginaceae occurs from the base of the Eocene to the present.

4. Family AIZOACEAE Rudolphi 1830 nom. conserv., the Fig-marigold Family

Succulent herbs, or less often shrubs or subshrubs, unarmed or rarely spiny (as in *Eberlanzia*), commonly accumulating free oxalates, producing betalains but not anthocyanins, commonly with C_4 photosynthesis or crassulacean acid metabolism (but ordinary C_3 in *Tetragonia*), and often

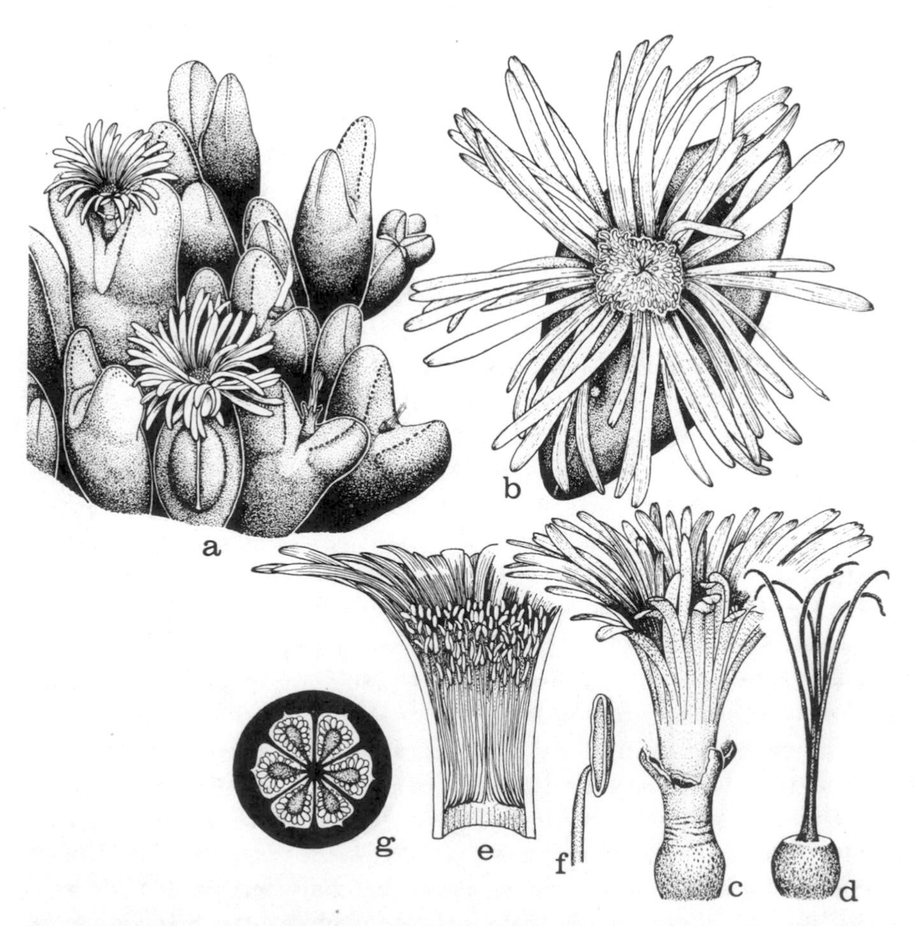

FIG. 3.4 Aizoaceae. *Conophytum* sp. a, habit, ×½; b, flower, from above, ×2; c, flower, in side view, ×2; d, pistil, ×2; e, perianth and stamens, ×2; f, stamen, ×16; g, cross-section of ovary, ×4.

with phenylalanine alkaloids, not saponiferous, and usually not tanniferous, but sometimes with scattered tanniferous cells in some of the parenchymatous tissues, in any case lacking ellagic acid; calcium oxalate crystals of various sorts (sometimes raphides or crystal sand) often present in some of the parenchyma-cells; nodes unilacunar; stem very often with anomalous secondary growth resulting in the production of concentric rings of vascular bundles or alternating concentric rings of xylem and phloem; vessel-segments generally with simple perforations; imperforate tracheary elements usually scarcely differentiated, represented by prosenchymatous elements with simple pits; sieve-tubes with a special kind of P-type plastid that includes a central globular protein crystalloid as well as a subperipheral ring of proteinaceous filaments; phytoferrin accumulating in masses in the phloem-parenchyma cells. LEAVES opposite or alternate, simple and entire or less often toothed, commonly more or less succulent and usually with centric rather than bifacial structure, very often with Kranz anatomy; epidermis of both leaf and stem commonly including many large, bladder-like cells; stomates anomocytic, or sometimes paracytic, diacytic, or anisocytic; stipules wanting, or seldom (e.g., *Trianthema*) present and interpetiolar; leaves seldom much-reduced, the stem then green and photosynthetic. FLOWERS in rather small, cymose inflorescences, or borne singly, perfect or seldom unisexual (the plants then monoecious), regular, perigynous to semi-epigynous or often epigynous; sepals (3–) 5 (–8), mostly succulent; petals (of staminodial origin) commonly numerous, in 1–6 cycles, mostly linear, attached to the hypanthium or to the summit of the ovary, distinct or sometimes connate toward the base, or sometimes wanting; stamens (1–) 4–5, or 8–10, or numerous apparently through secondary increase, and then originating in centrifugal sequence, attached to the hypanthium or to the base of the corolla or the summit of the ovary, distinct or basally connate into groups; anthers tetrasporangiate and dithecal; pollen-grains tricolpate to rarely tricolporate; nectaries typically epigynous, commonly forming a ring at the inner base of the stamens; gynoecium of 2–5 or more numerous carpels united to form a compound ovary with distinct (or distally distinct) styles, or seldom pseudomonomerous; ovary superior to half-inferior or inferior, generally with as many locules as carpels, or seldom unilocular through failure of the partitions; ovules solitary to usually more or less numerous in each locule, on axile (primitively), basal, apical, or parietal placentas, campylotropous to almost anatropous, bitegmic, crassinucellar, sometimes with a placental obturator; endosperm-development nuclear. FRUIT most commonly a loculicidal capsule, but sometimes of various other types, often included in the persistent calyx (or hypanthium); seeds with large, peripheral, dicotyledonous embryo curved around the abundant, mealy to sometimes hard, starchy (with clustered starch-grains) or somewhat oily and proteinaceous perisperm, without true endosperm, often arillate. X = 8, 9. (Ficoidaceae, Mesembryanthemaceae, Sesuviaceae, Tetragoniaceae)

The family Aizoaceae consists of some 2500 species and about a dozen genera, if the very large and highly diversified genus *Mesembryanthemum* sens. lat. is not divided into the hundred or more genera that some modern authors recognize. Perhaps 50 of these segregates from *Mesembryanthemum* can be recognized by their aspect. Some of the most familiar species of the family in cultivation belong to the segregate genus *Lampranthus*. Species of stemless genera such as *Lithops* (40, Stone-plant) and *Conophytum* (250) are often grown as curiosities.

The primary center of diversity for the Aizoaceae is in South Africa, and there is a secondary center in Australia. Relatively few species occur in other parts of the world, mostly in tropical and subtropical climates.

In contrast to the related family Cactaceae, the Aizoaceae are mostly leaf-succulents. In further contrast to the Cactaceae, they share their ecological specializations with another large family, even in the region of their greatest abundance and diversity. The family Crassulaceae (subclass Rosidae), which likewise consists of leaf-succulents with crassulacean acid metabolism, has its principal center of diversity in South Africa.

5. Family DIDIEREACEAE Drake del Castillo 1903 nom. conserv., the Didierea Family

Spiny, cactus-like, xerophytic, dioecious or seldom gynodioecious shrubs and trees with soft wood and large pith, producing betalains but not anthcyanins; long axes at first succulent, in age becoming more woody, but with broad medullary rays, the secondary growth of normal rather than concentric type; spiny short axes borne in the axils of the leaves of the long axes; cortex with abundant tanniferous cells and often large mucilage-cavities, the plants sometimes with proanthocyanins but without ellagic acid; clustered crystals of calcium oxalate commonly present in some of the cells of the parenchymatous tissues; vessel-segments with simple perforations; imperforate tracheary elements with simple pits; wood-parenchyma scanty, paratracheal and terminal; sieve tubes with a special kind of P-type plastid that includes a central globular protein crystaloid and a subperipheral ring of proteinaceous filaments; phytoferrin accumulating in masses in the phloem-parenchyma cells. LEAVES alternate, entire, small, sometimes much-reduced; stipules wanting. FLOWERS in more or less dichasial cymes terminating long or short axes, hypogynous, mostly unisexual (or perfect on some plants and pistillate on others); sepals 2, petaloid (in a phyletic sense probably modified bracteoles); petals 4, decussate in 2 cycles (in a phyletic sense probably modified sepals); staminate (and perfect) flowers with 8 (–10) stamens in 2 cycles; filaments woolly, weakly adnate at the base to the outside of an annular nectary; anthers tetrasporangiate and dithecal; pollen-grains trinucleate, (4) 5–7-zonocolpate; pistillate flowers often with staminodia; gynoecium of (2) 3 (4) carpels united to form a compound ovary with a single style and a usually irregularly (2) 3 (4)-lobed stigma; ovary with as many locules as carpels, but only the medial-adaxial one bearing an ovule, the others empty; ovule solitary, basal, erect, campylotropous, bitegmic, crassinucellar; endosperm-development nuclear. FRUIT dry, indehiscent, generally 3-angled; seeds with a small funicular aril and a large, curved or folded embryo with 2 fleshy cotyledons, nearly or quite without both endosperm and perisperm. 2n = ca 150, ca 190–200, evidently a high-polyploid number.

The family Didiereaceae consists of 4 genera (*Alluaudia, Alluaudiopsis, Decaryia,* and *Didierea*) and 11 species, confined to Madagascar. None of the species is economically important or familiar in cultivation. The family has by different botanists been assigned to several different orders, but recent authors are agreed that it belongs to the Caryophyllales (as here defined). Serological reactions, the presence of characteristic caryophylloid sieve-tube plastids, the formation of betalains instead of anthocyanins, the ornamentation of the pollen, and the successful grafting of some members of the family to some of the Cactaceae all point in this direction. The definitive importance of any one of these features might be challenged, but collectively they would seem to be conclusive.

6. Family CACTACEAE A. L. de Jussieu 1789 nom. conserv., the Cactus Family

Spiny (seldom unarmed) stem-succulents with scarcely developed leaves, or rarely (*Pereskia, Pereskiopsis*) with woody, scarcely succulent stems and well developed, succulent, alternate, simple leaves; plants with crassu-

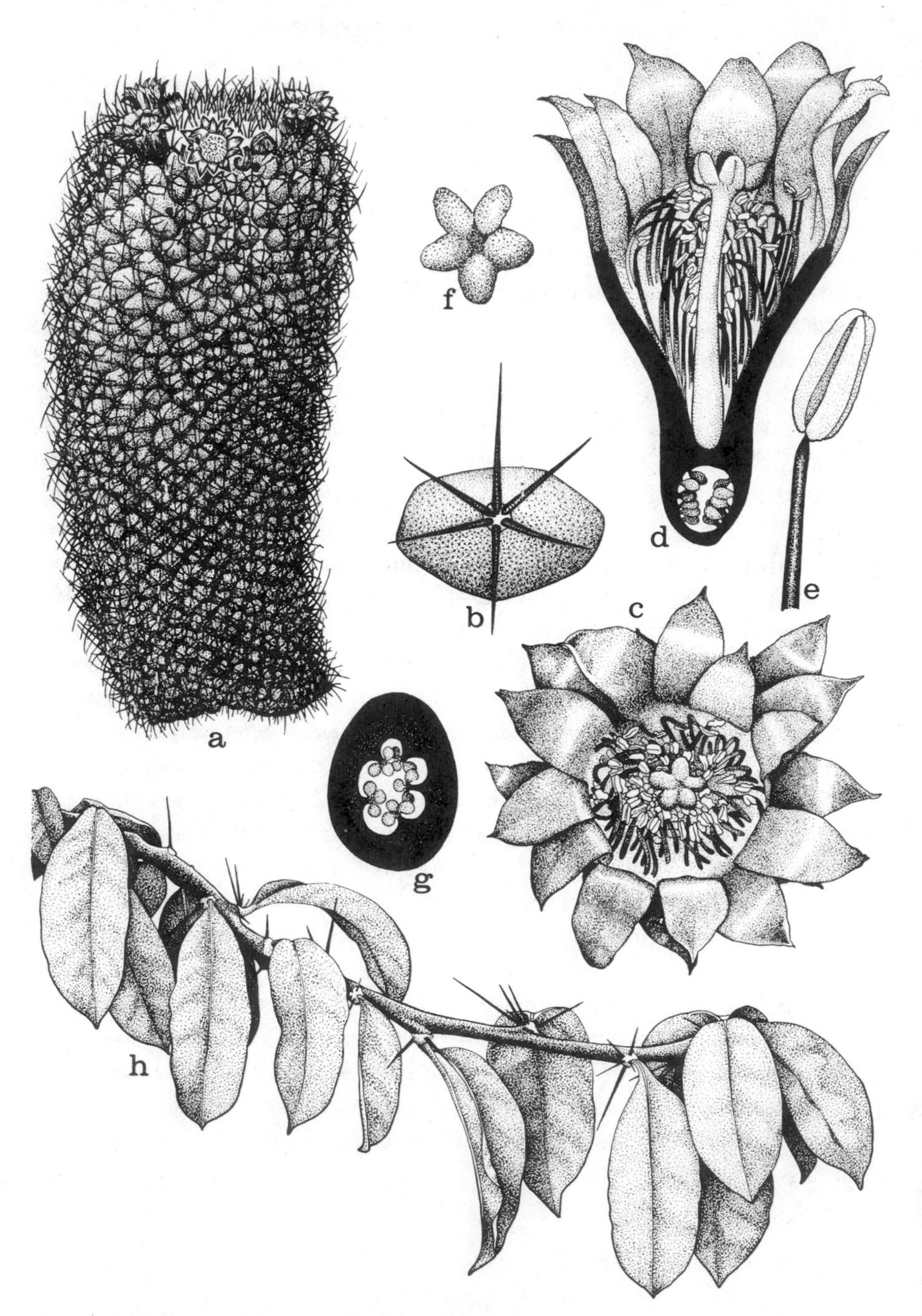

FIG. 3.5 Cactaceae. a–g, *Mammillaria polythele* Mart. a, habit, $\times\frac{1}{2}$; b, tubercle of the stem, with a terminal areole and spines, $\times 2$; c, flower, from above, $\times 4$; d, flower, in long-section, $\times 4$; e, stamen, $\times 16$; f, stigma, from above, $\times 8$; g, cross-section of ovary, $\times 6$. h, *Pereskia grandifolia* Haw., habit, $\times\frac{1}{2}$.

lacean acid metabolism, accumulating organic acids, usually producing alkaloids (often of simple isoquinoline type) or alkaloidal amines, or (alternatively) triterpenoid saponins, forming betalains but not anthocyanins, not tanniferous, lacking both proanthocyanins and ellagic acid; root-system typically very shallow and widely spreading; shoot varying from small, softly herbaceous, and with scarcely developed xylem, to arborescent and columnar or sparsely branched with normal secondary growth, commonly with scattered mucilage cells, and occasionally (as in *Coryphantha, Leuchtenbergia*, and spp. of *Mammillaria*) with anastomosing laticifers; nodes in leaf-bearing genera basically unilacunar, the basally simple trace often bifid or multifid; calcium oxalate present in some of the parenchyma-cells as crystals of one or another form, sometimes as bundles of raphides; cuticle often notably thick; stem usually with a collenchymatous hypodermis; outer cortex generally produced by an ensheathing subepidermal meristem; stomates paracytic or anomocytic; spines commonly restricted to well defined, regularly arranged (basically spiralled), very often hairy areoles on the stem, each areole representing a modified axillary bud or a short branch with its leaves or bud-scales modified into spines; vessel-segments with simple or rarely reticulate perforations, or the plants secondarily vesselless; imperforate tracheary elements commonly septate, with simple pits, or sometimes wanting, or sometimes some of them developed into vascular tracheids; wood-rays, when present, variously multiseriate or sometimes uniseriate; wood-parenchyma mostly scanty-paratracheal; sieve-tubes with a special kind of P-type plastid that has a central globular protein crystalloid and a subperipheral ring of proteinaceous filaments; phytoferrin accumulating in masses in the phloem parenchyma cells. FLOWERS typically solitary at the areoles, or sometimes (*Pereskia* and some of the unarmed genera) at the branch-tips, seldom (spp. of *Pereskia*) in terminal cymes, often large and showy, variously pollinated by bees, hummingbirds, bats, or hawkmoths, perfect or seldom unisexual, regular or sometimes irregular through curvature of the perianth-tube or unequal reflexion of the segments; tepals generally numerous and spirally arranged, all more or less showy and petaloid, or often the outer ones more sepaloid, but not sharply differentiated into two types, all united below to form a perianth-tube or a hypanthium; stamens numerous (probably only secondarily so), initiated in centrifugal sequence, arising spirally or in groups from the hypanthium, the androecium at least sometimes served by a limited number of trunk-bundles; anthers tetrasporangiate and dithecal, opening by longitudinal slits; pollen-grains trinucleate, tricolpate to 6–15-colpate or -porate; nectary a ring on the inner surface of the hypanthium; gynoecium of 3-many carpels united to form a compound, inferior ovary (in spp. of *Pereskia* the flower perigynous and the ovary superior) with a single style and as many radiating stigmas as carpels (in spp. of *Pereskia* the carpels only weakly united); ovules usually numerous, in *Pereskia* basal and the ovary partly partitioned, in the other genera on 3 or more parietal placentas in a unilocular ovary, campylotropous to rarely anatropous, bitegmic, crassinucellar; funiculi often branched and multiovulate; endosperm-development nuclear. FRUIT usually indehiscent, commonly fleshy and baccate, rarely dry or dehiscent; seeds with straight or more often curved, dicotyledonous embryo and without true endosperm, the perisperm varying from abundant, central, starchy and mealy (with clustered starch-grains) to more often scanty or wanting. X = 11.

The family Cactaceae consists of some 30 to 200 or more genera, depending on the author, and at least 1000, perhaps as many as 2000 species. They are a familiar feature of the American desert landscape, and are cultivated as pot-plants by a large cult of dedicated amateur horticulturists. This last fact may be at least partly responsible for the excessive multiplication of generic names.

Cacti are native to temperate and tropical regions of the New World, especially in warm, dry places. Some species of the epiphytic genus *Rhipsalis* have been thought to be native to parts of tropical Africa and to Madagascar and some other tropical islands in the Old World, but this is still debatable. They may be only introduced in those regions, as are various other cacti in suitable habitats in the Old World.

The Cactaceae are one of the few large families of dicotyledons with a clear ecogeographic significance. They are New-World stem-succulents, protected from grazing animals by spines. Typically they have a very widely spreading but shallow root-system, adapted to take advantage of light rains in their desert habitat. Furthermore, they have the crassulacean acid metabolic syndrome, permitting them to absorb carbon dioxide at night for use in photosynthesis during the day. The tropical spineless epiphytic cacti such as *Rhipsalis* and *Schlumbergera* (Christmas cactus, Thanksgiving cactus, and Easter cactus are species of *Schlumbergera*) are on a special ecological side-road which the xerophytic adaptations of the family permit them to follow.

The relationships of the cacti have been much debated in the past. The opinion that they should be in the same order as the Aizoaceae goes back at least as far as Bentham and Hooker, but in the Englerian system they were considered as a distinct order allied to the Parietales. Their position alongside the Aizoaceae in the Caryophyllales is strongly supported by a host of studies during the past several decades, embracing various special features as well as the classical morphological characters. They have the full syndrome of caryophylloid embryological features, they have characteristic caryophylloid sieve-tube plastids, they have betalains instead of anthocyanins, and the pollen grains show the same morphological series leading to the spherical pantoporate pollen that is seen in several other families of the order. The systematic position of the Cactaceae now seems firmly established.

The most archaic genus in the Cactaceae is obviously *Pereskia*, which has woody stems, well developed, succulent leaves, and terminal flowers that may even be arranged in a cymosely branched inflorescence. Some of the species of *Pereskia* have perigynous rather than epigynous flowers, with the ovary free from the hypanthium.

A complicated but plausible version of receptacular epigyny has been proposed by Boke (1964) to explain the gynoecial structure of the Cactaceae. He considers that the ovary is distorted by asymmetrical growth and is in effect sunken into the stem-tip. This view is certainly bolstered by the fact that at least in some species of *Pereskia* the branches of the inflorescence arise from spiny areoles on the outer surface of the ovary of the central (terminal) flower.

It is now generally believed that the Cactaceae take their origin in or near the Phytolaccaceae. Such an interpretation requires a secondary increase in the number of tepals, stamens, and ovules, and must therefore be regarded with some caution, but no other likely alternative presents itself.

7. Family CHENOPODIACEAE Ventenat 1799 nom. conserv., the Goosefoot Family

Herbs or sometimes shrubs or even (rarely) small trees, seldom climbing, the stems often succulent or jointed or both, usually showing anomalous secondary growth resulting in the formation of concentric rings of vascular bundles or sometimes alternating concentric rings of xylem and phloem; plants typically with crassulacean acid metabolism, accumulating organic acids and very often also free nitrates and/or oxalates, often with C_4 photosynthesis, producing betalains but not anthocyanins, often with one or another sort of alkaloid (variously indole, isoquinoline, quinolizidine or other alkaloids, or alkaloidal amines) and often accu-

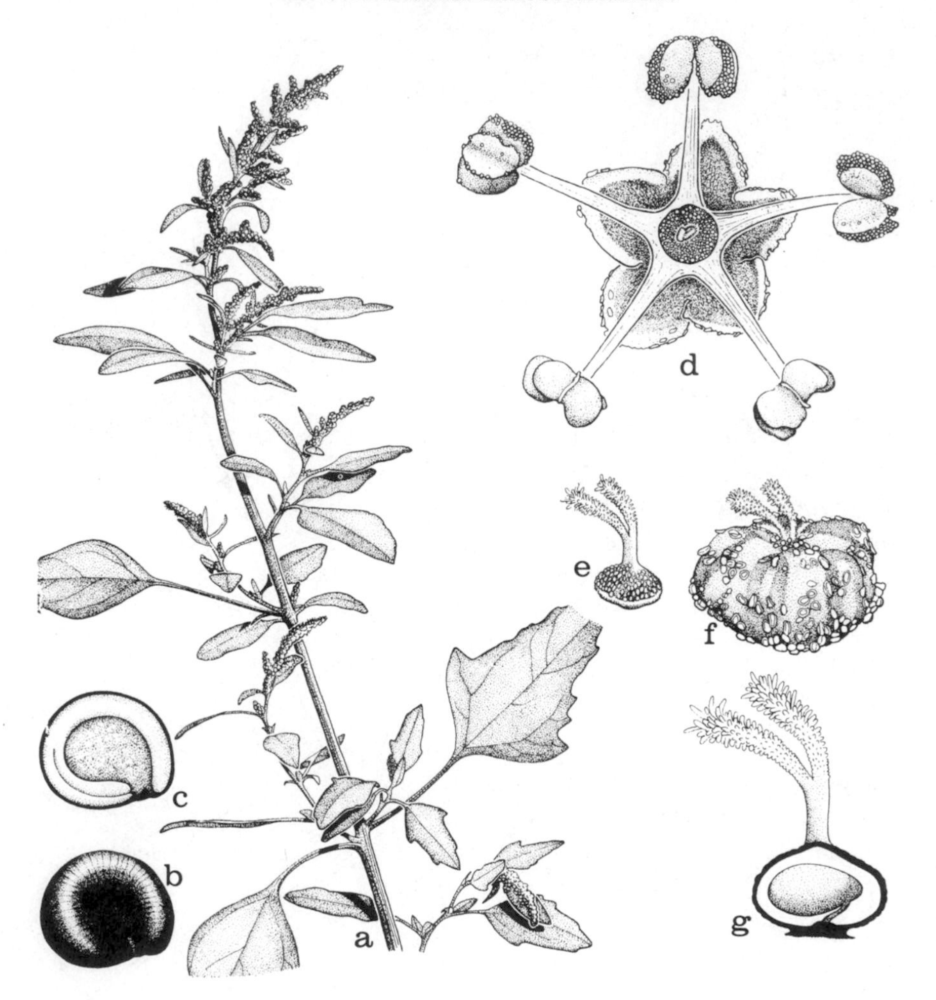

FIG. 3.6 Chenopodiaceae. *Chenopodium album* L. a, habit, ×½; b, c, seed, external and in long-section, ×18; d, flower, from above, ×18; e, pistil, ×18; f, fruit, ×18; g, pistil, the ovary in long-section, ×36.

mulating triterpenoid saponins, but not tanniferous, lacking both proanthocyanins and ellagic acid; nodes unilacunar; crystals of calcium oxalate commonly present in some of the cells of the parenchymatous tissues, often as clustered crystals or crystal sand; vessel-segments with simple perforations; imperforate tracheary elements with simple pits, often some with a persistent protoplast; wood-rays heterocellular, up to 4 cells wide; sieve-tubes with a special kind of P-type plastid that includes a subperipheral ring of proteinaceous filaments, but without a central protein crystaloid except in *Halophytum* (a genus sometimes referred to the Phytolaccaceae or treated as a separate family). LEAVES alternate or rarely opposite, simple, commonly entire but sometimes toothed or lobed, often somewhat succulent, often with Kranz anatomy, or frequently more or less strongly reduced; stomates commonly anomocytic, occasionally paracytic or anisocytic; stipules wanting. FLOWERS small, commonly green or greenish, 1-many and glomerate in the leaf-axils or in bracteate or bractless spikes, panicles, or cymes, mostly regular, perfect or less often unisexual (the plants then monoecious or dioecious); sepals (1–) 5 (6), distinct or basally connate, typically more or less herbaceous or somewhat membranous, but hardly scarious, or rarely wanting; petals wanting; stamens commonly of the same number as the sepals and opposite them, but sometimes fewer; filaments distinct, or sometimes connate at the base, hypogynous or inserted on an annular disk or adnate to the base of the calyx; anthers tetrasporangiate and dithecal, opening by longitudinal slits; pollen-grains trinucleate, 6- to multiporate (pantoporate) (cuboid with a large pore on each of the 6 faces in *Halophytum*); perfect flowers often with a nectary-ring at the inner base of the filaments, or with separate nectary-glands alternating with the filaments, but unisexual flowers with more or less reduced or no nectary; gynoecium of 2–3 (–5) carpels united to form a compound, unilocular ovary with distinct or more or less connate styles; stigmas dry; ovary superior or seldom (as in *Beta*) half-inferior; ovule solitary, basal, amphitropus to more often campylotropous, bitegmic, crassinucellar; endosperm-development nuclear. FRUIT a utricle or a small nut, indehiscent or seldom with irregular or circumscissile dehiscence, often subtended by the persistent calyx or by persistent bracteoles, sometimes several fruits ripening together with their calyces to form a multiple fruit (as in *Beta*); seeds mostly lenticular, with annular or spirally twisted or only slightly curved (*Dysphania*), dicotyledonous embryo peripheral to and usually more or less surrounding the starchy, usually hard perisperm (with clustered starch-grains), or sometimes without perisperm; true endosperm nearly or quite wanting. X = 6–9, most often 9. (Dysphaniaceae, Halophytaceae, Salicorniaceae)

The family Chenopodiaceae consists of about a hundred genera and 1500 species, of cosmopolitan distribution, but especially abundant in desert and semidesert regions. Many are halophytes; some others are weeds of waste places and cultivated fields. Some of the species have Kranz anatomy and carry on C_4 photosynthesis, but this feature varies even within the same genus, as in *Atriplex*. The largest genera are

Chenopodium (200+), *Atriplex* (150), *Obione* (100+), *Salsola* (100+), and *Suaeda* (100+). *Beta* (beet) and *Spinacia* (spinach) are familiar food-plants, and species of *Chenopodium* (goosefoot, pigweed) and *Salsola* (Russian thistle) are common weeds. Species of *Atriplex* (shadscale), *Eurotia* (winter fat) and *Sarcobatus* (greasewood) are important constituents of the desert flora in western United States, especially in alkaline soil. The most arborescent genus is *Haloxylon*, of the Irano-Turanian and eastern Mediterranean region; *H. persicum* Bunge reaches a height of 6 m, with a trunk up to 2 dm thick.

The small genera *Lophiocarpus* (5) and *Microtea* (10), often referred to the Phytolaccaceae, may perhaps better be included in the Chenopodiaceae.

Pollen referred to the Chenopodiaceae or Amaranthaceae dates from the Maestrichtian, providing the oldest known fossils in the Caryophyllidae.

The proper organization of the Chenopodiaceae into subfamilies and tribes has been a source of controversy. Blackwell (1977) recognizes two subfamilies: the Chenopodioideae, with annular embryo and usually with perisperm; and the Salsoloideae, with spirally coiled embryo and usually without perisperm. Some of the more familiar genera in the Chenopodioideae are *Allenrolfea, Atriplex, Beta, Blitum, Chenopodium, Eurotia, Salicornia,* and *Spinacia*; in the Salsoloideae, *Halogeton, Haloxylon, Salsola, Sarcobatus,* and *Suaeda.*

8. Family AMARANTHACEAE A. L. de Jussieu 1789 nom. conserv., the Amaranth Family

Mostly herbs, seldom climbers, subshrubs, shrubs, or even (rarely) small trees, commonly with anomalous secondary growth resulting in the formation of concentric rings of vascular bundles; plants often accumulating free oxalates, potassium nitrate and saponins, producing betalains but not anthocyanins, not tanniferous, lacking both proanthocyanins and ellagic acid; nodes unilacunar; crystals of calcium oxalate commonly present in some of the cells of the parenchymatous tissues, often as clustered crystals or crystal sand; vessel-segments with simple perforations; imperforate tracheary elements with simple pits; sieve-tubes with a special kind of P-type plastid that includes a subperipheral ring of proteinaceous filaments, but without a central protein crystalloid. LEAVES alternate or opposite, simple and commonly entire or nearly so, often with Kranz anatomy, the vascular bundles surrounded by large, cubical parenchyma cells; stomates variously anomocytic, anisocytic,

FIG. 3.7 Amaranthaceae. *Amaranthus retroflexus* L. a, habit, $\times \frac{1}{2}$; b, stamen and tepal from a staminate flower, $\times 12$; c, staminate flower, with associated bracts, $\times 12$; d, f, fruit, $\times 12$; e, post-mature pistillate flower, with associated bracts, $\times 12$; g, h, seed, external and in long-section, $\times 12$; i, fruit, in long-section, $\times 12$.

diacytic, or paracytic. FLOWERS small, solitary or more often in cymose or variously compound (modified cymose) inflorescences, often subtended by scarious or membranous (sometimes conspicuously pigmented) bracts or bracteoles, variously entomophilous or anemophilous, generally regular, perfect or less often unisexual, hypogynous or nearly so, apetalous; sterile flowers, forming hooks or bristles, often present, subtending normal flowers; sepals mostly 3–5, seldom only 1 or 2 or even wanting, generally dry and scarious or membranous, distinct or more or less connate at the base; stamens generally as many as the sepals and opposite them, seldom fewer; filaments distinct or more often connate at the base into a tube, the tube often produced into teeth or lobes ("pseudostaminodia") alternating with the anthers, in extreme forms (e.g., *Froelichia*) the androecium simulating a small, sympetalous corolla with the filaments attached at the sinuses; anthers tetrasporangiate and dithecal (Amaranthoideae) or less often bisporangiate and unithecal (Gomphrenoideae), opening by longitudinal slits; pollen-grains pantoporate, usually trinucleate; a nectary-ring often present at the inner base of the filament-tube; gynoecium of 2–3 (4) carpels united to form a compound, unilocular ovary with a single, often evidently lobed style; ovule usually solitary and basal (rarely apical and pendulous), or in a few genera (e.g., *Celosia*) the ovules several on a basal or short, free-central placenta; ovules more or less distinctly campylotropous, bitegmic, crassinucellar; endosperm-development nuclear. FRUIT an achene or a small nut, or a circumscissile, 1-seeded capsule (pyxis), rarely a berry, often subtended or more or less enclosed by the persistent calyx; seeds with a peripheral, annular, dicotyledonous embryo, surrounding the abundant, starchy, hard to granular perisperm (with clustered starch-grains); true endosperm nearly or quite wanting; X = 6–13, 17+.

The family Amaranthaceae consists of about 65 genera and 900 species, widespread in tropical and subtropical regions, with relatively few species in cooler countries. The largest genera are *Amaranthus* (50), *Celosia* (60), and *Ptilotus* (100) in the subfamily Amaranthoideae, and *Alternanthera* (170), *Gomphrena* (100+), and *Iresine* (70) in the subfamily Gomphrenoideae. Several members of the family, notably species of *Celosia* (cockscomb), *Amaranthus*, and *Gomphrena*, are familiar garden ornamentals, and some species of *Amaranthus* are cultivated in primitve societies for their edible seeds (grain amaranths).

The Amaranthaceae are obviously allied to the Chenopodiaceae, and the two families stand side by side in virtually all systems of classification. Together they constitute the irreducible, archtypical core of their order.

The Amaranthaceae are more advanced than the Chenopodiaceae in most of the respects in which the two families differ. However, as Takhtajan (1966) has pointed out, the Amaranthaceae cannot be regarded as directly derived from the Chenopodiaceae. The tribe Celosieae of the Amaranthaceae is more primitive than all the Chenopodiaceae in having 2-several ovules instead of a solitary ovule. The

Amaranthaceae and Chenopodiaceae must be regarded as having a common ancestry near the Phytolaccaceae.

9. Family PORTULACACEAE A. L. de Jussieu 1789 nom. conserv., the Purslane Family

Herbs, or rarely shrubs or half-shrubs, often somewhat succulent, generally with abundant mucilage cells in both stem and leaves, without anomalous secondary growth, commonly accumulating nitrates and oxalic acid, usually not tanniferous, lacking ellagic acid and usually

FIG. 3.8 Portulacaceae. *Claytonia caroliniana* Michx. a, habit, ×1; b, flower, from above, ×4; c, flower bud, ×4; d, petal, with attached stamen, ×4; e, pistil, ×8; f, ovary in long-section, ×16; g, h, fruit, ×4; i, k, seed, external and in section, ×8.

lacking proanthocyanins, and producing betalains (though not in *Lyallia*) but not anthocyanins; nodes unilacunar; calcium oxalate abundant as solitary or clustered crystals in some of the cells of the parenchymatous tissues; vessel-segments with simple perforations; imperforate tracheary elements with simple pits; vascular bundles seldom with internal phloem; sieve-tubes with a special kind of P-type plastid that has a subperipheral ring of proteinaceous filaments and a central globular protein crystalloid; phytoferrin accumulating in masses in the phloem-parenchyma cells of the more succulent species. LEAVES opposite or alternate (sometimes all basal), simple, entire, sometimes with Kranz anatomy and carrying on C_4 photosynthesis, as in *Portulaca*; stomates paracytic or diacytic or anomocytic; stipules scarious, or modified into tufts of hairs, or wanting. FLOWERS solitary or more often in various sorts of cymose or racemose to head-like inflorescences, perfect or rarely unisexual, regular or seldom (spp. of *Montia*) slightly irregular, hypogynous or (*Portulaca*) semi-epigynous; sepals (phyletically, bracteoles) 2, often a little unequal, or seldom (spp. of *Lewisia*) more numerous and up to 9, persistent or seldom (*Talinum*) deciduous; petals (phyletically, petaloid sepals) mostly imbricate, (2–) 4–6 or seldom (spp. of *Lewisia*) more numerous and up to about 18, distinct or sometimes basally connate, often ephemeral; stamens most commonly as many as and opposite the petals (alternate in the monotypic genera *Hectorella* and *Lyallia*), seldom fewer or even solitary, or secondarily numerous and then sometimes grouped into bundles; filaments free or sometimes basally adnate to the petals or to the very short corolla-tube; anthers tetrasporangiate and dithecal, opening by longitudinal slits; pollen-grains trinucleate or seldom binucleate, from tricolpate to pantocolpate or pantoporate; individual nectaries or a nectary-ring borne at the outer base of the filaments, sometimes adnate to the base of the corolla; gynoecium of 2–3 (–9) carpels united to form a compound ovary with distinct styles or seldom with a single (then usually lobed or cleft) style; ovary with as many locules as carpels at early stages of development, but soon becoming unilocular by disappearance of the partitions, with a free-central placenta bearing 2-many ovules or seldom (tribe Portulacarieae) only a single ovule, the ovules bitegmic, crassinucellar, campylotropous or amphitropous or sometimes anatropous; endosperm-development nuclear. FRUIT a loculicidal or circumscissile capsule, or seldom (tribe Portulacarieae) indehiscent; seeds commonly lenticular, often smooth and shining, but sometimes tuberculate or otherwise roughened, often strophiolate; cotyledons 2 (4 in *Anacampseros lanceolata* Sweet); embryo peripheral and more or less curved around the abundant, starchy, hard or seldom soft perisperm (with clustered starch-grains); true endosperm wanting. X = 4–42+. (Hectorellaceae)

The family Portulacaceae consists of about 20 genera and 500 species, of cosmopolitan distribution, but best developed in western North American and the Andes. More than half of the species belong to only 3 genera, *Calandrinia* (150), *Portulaca* (100+), and *Talinum* (50+). The most familiar species are rose-moss, *Portulaca grandiflora* Hook., a garden ornamental, and the common purslane, *P. oleracea* L., a cosmopolitan weed sometimes used as a pot-herb. The family is notable for instability of chromosome numbers, especially in the genus *Claytonia*. In *C. virginica* L. the somatic number ranges from 12 to nearly 200 and sometimes varies in different parts of the same individual.

The Portulacaceae are somewhat unusual in the Caryophyllales in lacking anomalous secondary growth, but otherwise they are perfectly at home in their order. Their relationships are not in dispute.

It now seems well established that the apparent sepals of the Portulacaceae are, in an evolutionary sense, modified bracteoles, and that the apparent petals are modified sepals. From a practical descriptive standpoint, however, it is useful to continue the traditional practice of naming the floral parts by their present structure, position, and function rather than by their evolutionary history. If the taxonomic system is to be useful to those who are not specialists in phylogeny, there must be a strong pragmatic component in the application of such terms as sepal and petal.

10. Family BASELLACEAE Moquin-Tandon 1840 nom. conserv., the Basella Family

Perennial herbs with slender, annual, twining or scrambling stems from fleshy-thickened or tuber-bearing rhizomes, generally with abundant mucilage-cells in both stem and leaves, without anomalous secondary growth, at least sometimes accumulating potassium nitrate, not tanniferous, lacking both proanthocyanins and ellagic acid, and producing betalains but not anthocyanins; vascular bundles in a single ring, unequal in size, the larger ones becoming bicollateral (with internal phloem) at maturity; vessel-segments with simple perforations; sieve-tubes with a special kind of P-type plastid that has a subperipheral ring of proteinaceous filaments and a central globular protein crystalloid. LEAVES opposite or alternate (sometimes all basal), simple, entire, somewhat succulent; stomates paracytic, or some of them anisocytic; stipules wanting. FLOWERS in terminal or axillary panicles, racemes, or spikes, small, regular, perfect or occasionally at least functionally unisexual; sepals (phyletically, bracteoles, as in the Portulacaceae) 2, distinct or connate at the base, sometimes adnate to the base of the corolla; petals (phytelically, sepals) 5, white or greenish to purplish or reddish, imbricate, connate below to form a 5-lobed tube, or nearly distinct, persistent in fruit; stamens 5, opposite the petals, the filaments adnate to the base of the petals or to the corolla-tube; anthers tetrasporangiate and dithecal, opening by longitudinal slits or by terminal slits or pores; pollen-grains trinucleate, pantocolpate, or cuboid with a colpus on each of the 6 faces; an annular nectary present at the outer base of the androecium; gynoecium of 3 carpels united to form a compound ovary with 3 separate styles or a single, often 3-lobed style; ovary superior, trilocular at early stages of development, but soon becoming unilocular by disappearance of the partitions; ovule solitary, basal, anatropous to campylotropous, bitegmic, crassinucellar; endosperm-development nuclear. FRUIT a utricle, surrounded by the persistent, often fleshy corolla or by wing-like persistent sepals; seeds nearly spherical, with membranous testa; embryo with 2 cotyledons, annular and surrounding the abundant, starchy perisperm (the grains clustered), or spirally twisted and the perisperm scanty or none; true endosperm wanting. X = 11, 12.

The family Basellaceae consists of 4 genera (*Anredera, Basella, Tournonia,* and *Ullucus*) and 15 or 20 species, all tropical or subtropical, most of them in the New World. *Basella alba* L., Malabar spinach, is occasionally used as a potherb, and the tubers of *Ullucus tuberosus* Caldas, ulluco, are an ancient and important crop in parts of the Andes. *Anredera* (*Boussingaultia*) *cordifolia* (Tenore) Steenis, Madeira-vine, is cultivated as an ornamental in warm countries.

The Basellaceae are generally considered to be related to but more advanced than the Portulacaceae, a view that is here adopted.

11. Family MOLLUGINACEAE Hutchinson 1926, nom. conserv., the Carpet-weed Family

Herbs, or rarely half-shrubs or shrubs, often with anomalous secondary thickening resulting in the formation of concentric rings of vascular bundles or alternating concentric rings of xylem and phloem; plants sometimes saponiferous, and so far as known producing anthocyanins

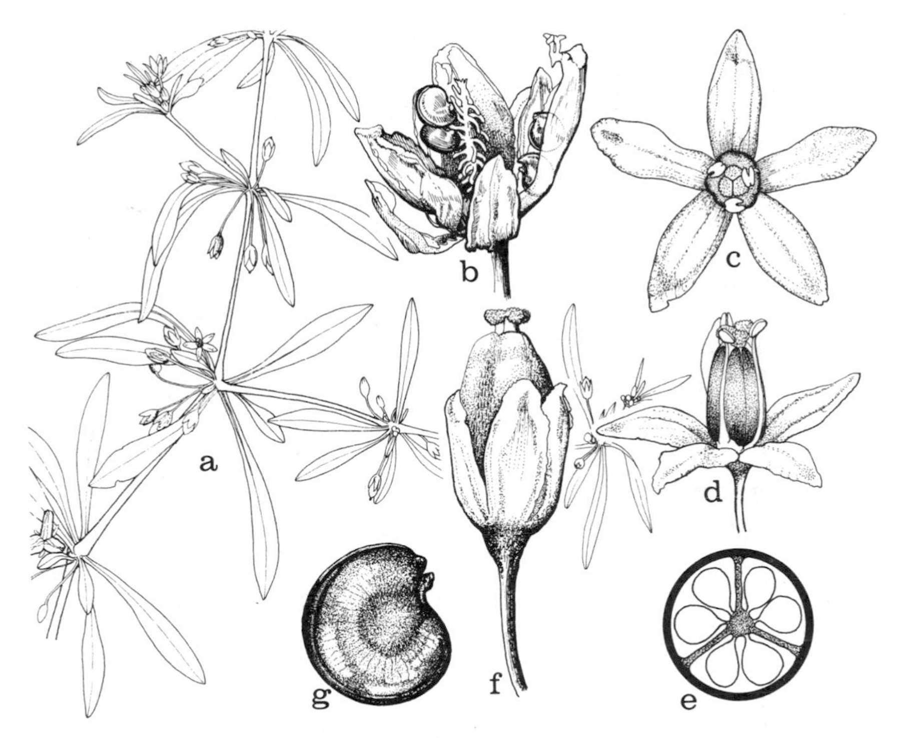

FIG. 3.9 Molluginaceae. *Mollugo verticillata* L. a, habit, ×1; b, mature fruit, after dehiscence, ×8; c, d, flower, ×8; e, cross-section of ovary, ×20; f, mature fruit, before dehiscence, ×8; g, seed, ×32.

but not betalains; crystals of calcium oxalate, sometimes in the form of raphides, commonly present in some of the cells of the parenchymatous tissues; vessel-segments with simple perforations; sieve-tubes with a special kind of P-type plastid that has a subperipheral ring of proteinaceous filaments and a central globular or polyhedral protein crystalloid. LEAVES opposite to alternate or whorled, simple, entire, bifacial, not at all or only slightly succulent, at least sometimes with Kranz anatomy; stomates anomocytic; stipules variously wanting, or small and deciduous, or conspicuous and multifid (as in *Pharnaceum*). FLOWERS commonly small and inconspicuous, borne in cymose, often loose and open inflorescences, or solitary in the axils, perfect or rarely unisexual, regular, hypogynous, with or without an annular nectary-disk around the ovary; sepals 5 or rarely (*Polpoda*) 4, persistent, distinct or rarely (*Coelanthum*) connate at the base; petals (of staminodial origin) small or more often wanting, seldom (as in *Orygia* and *Corbichonia*) connate below into a tube; androecium primitively diplostemonous, but the inner cycle of stamens generally partly or wholly suppressed, the stamens thus (2–) 5–10, or sometimes more numerous through secondary increase; filaments distinct or more or less connate at the base; anthers tetrasporangiate and dithecal, opening by longitudinal slits; pollen-grains trinucleate, tricolpate or tricolporate to pantocolpate; nectary a ring between the stamens and the ovary, or on the lower part of the ovary, or on the inner side of the filament-tube; gynoecium ordinarily of 2–5 (seldom more numerous) carpels united to form a compound, superior ovary with usually distinct styles (style solitary in *Glinus*), but in *Adenogramma* the gynoecium of a single uniovulate carpel; ovary (except in *Adenogramma*) plurilocular at least below, but in the upper part the partitions often not reaching the central axis; placentation axile (except *Adenogramma*); ovules 1 to more often numerous in each locule, campylotropous to rarely almost anatropous, bitegmic, crassinucellar; endosperm-development nuclear. FRUIT dry, opening loculicidally or by transverse slits, or rarely indehiscent, commonly surrounded by the persistent calyx; seeds sometimes arillate; embryo dicotyledonous, curved around the hard, starchy perisperm; true endosperm wanting. X = 9.

The family Molluginaceae consists of about 13 genera and nearly 100 species, found mainly in tropical and subtropical regions, especially in Africa. The largest genera are *Pharnaceum* (25), *Limeum* (20), *Mollugo* (15), and *Glinus* (10). The most familiar species in temperate regions is *Mollugo verticillata* L., the common carpet-weed.

The Molluginaceae were formerly included in the Aizoaceae, from which they differ in being mostly nonsucculent, and in having hypogynous flowers with distinct sepals. Furthermore, those few members of the Molluginaceae that have been tested have anthocyanins and lack betalains, the reverse of the situation in the tested members of the Aizoaceae. The anomalous genus *Gisekia*, which produces betalains, is here referred to the Phytolaccaceae and is discussed under that family. It will be interesting to learn whether or not a more intensive future sampling of the Molluginaceae discloses any betalain-producing members.

12. Family CARYOPHYLLACEAE A. L. de Jussieu 1789 nom. conserv., the Pink Family

Herbs, or rarely subshrubs or shrubs, the stems commonly swollen at the nodes, sometimes with anomalous secondary thickening resulting in the formation of concentric rings of xylem and phloem; plants producing anthocyanins but not betalains, not tanniferous, lacking both

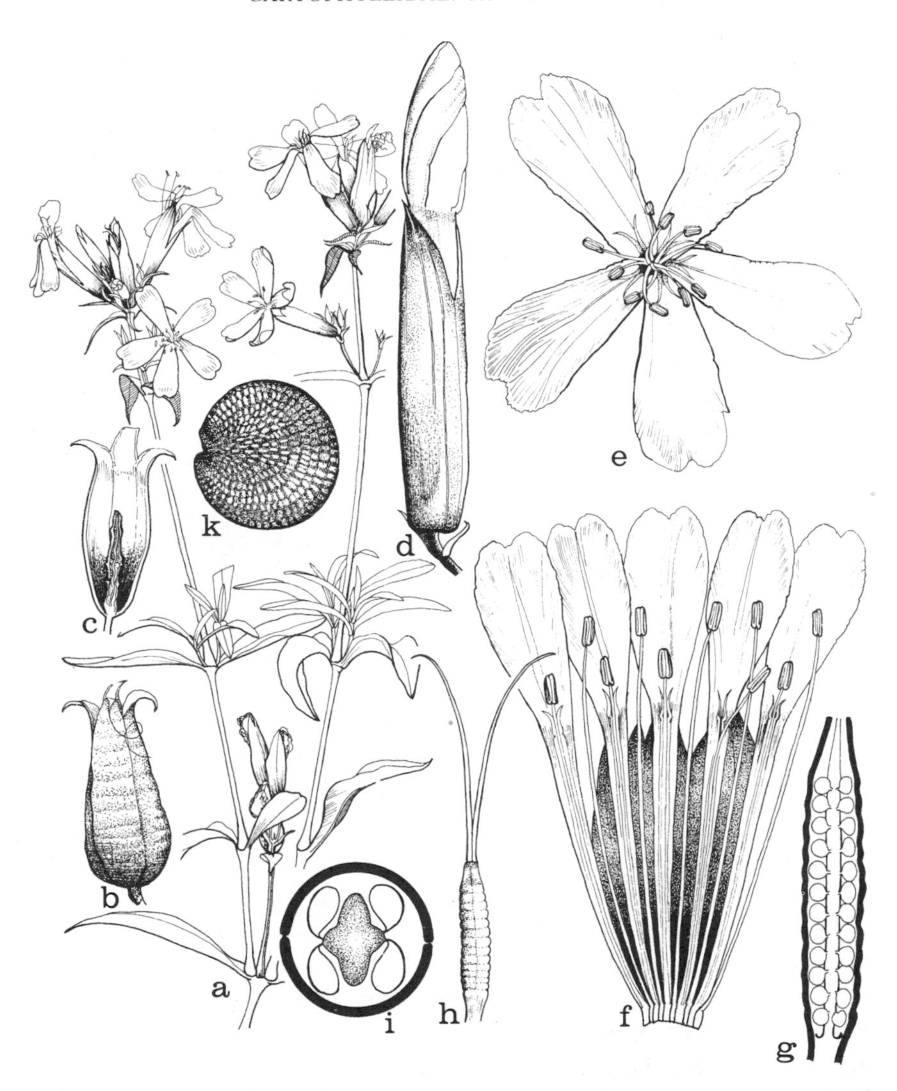

FIG. 3.10 Caryophyllaceae. *Saponaria officinalis* L. a, habit, ×½; b, mature fruit, after dehiscence, ×2; c, long-section of mature fruit, after dehiscence, showing free-central placenta, ×2; d, flower, ready to open, ×2; e, flower, from above, ×2; f, flower, opened to show perianth and stamens, with the pistil removed, ×2; g, long-section of ovary, ×4; h, pistil, ×2; i, cross-section of ovary, ×10; k, seed, ×10.

proanthocyanins and ellagic acid, commonly accumulating pinitol and often also triterpenoid saponins, and often storing carbohydrate as the trisaccharide lychnose, apparently restricted to this family; nodes unilacunar; vessel-segments with simple perforations; imperforate tracheary elements with simple or bordered pits; sieve-tubes with a special kind of P-type plastid that includes a central polyhedral protein crystalloid as well as a subperipheral ring of proteinaceous filaments. LEAVES opposite or rarely (as in *Corrigiola*) alternate, simple, entire, usually rather narrow, often connected by a transverse line at the base, sometimes with Kranz anatomy, but more often not; stomates most commonly diacytic, less often anisocytic or anomocytic; stipules present in many members of the subfamily Paronychioideae, otherwise wanting. FLOWERS commonly in dichasial cymes, or sometimes solitary, usually regular, perfect or seldom unisexual, hypogynous or seldom (as in *Scleranthus*) perigynous, 5-merous or seldom 4-merous; sepals distinct or nearly so (subfamilies Alsinoideae and Paronychioideae), or connate into an evident tube that typically has an internal nectary-ring (subfamily Silenoideae); petals variously present or absent, in the Paronychioideae often absent, in the Alsinoideae commonly present but small and often bifid, in the Silenoideae commonly well developed and differentiated into a long, basal claw and an expanded blade, and then generally appendiculate on the ventral side at the juncture of claw and blade; stamens commonly 5 to 10 in one or two cycles, or sometimes only 1–4; filaments variously hypogynous, free, and distinct, or basally adnate to the petals to form a short or more or less elongate tube that may or may not be adnate to the gynophore, or inserted at the edge of a nectary disk surrounding the ovary, or even adnate to the lower part of the calyx, the antesepalous ones often (and the others sometimes) with a dorsal nectary near the base; anthers tetrasporangiate and dithecal, opening by longitudinal slits; pollen-grains trinucleate, from tricolpate to most commonly pantoporate; gynoecium of 2–5 carpels united to form a compound ovary with distinct or less commonly more or less united styles; stigmas dry; ovary superior, often on a gynophore, unilocular above, but generally more or less distinctly partitioned toward the base at least when young, the placental column often continuous and attached to the top as well as the base of the ovary, but sometimes free at the top or reduced to a basal nubbin; ovules mostly numerous, but sometimes few or even solitary, bitegmic, crassinucellar, hemitropous to more commonly campylotropous; endosperm-development nuclear. FRUIT most commonly a capsule dehiscent by as many or twice as many valves or apical teeth as there are styles, or sometimes indehiscent, sometimes more or less utricular, and then often enclosed in the persistent, indurated calyx or hypanthium; seeds commonly finely and regularly ornamented on the surface; embryo dicotyledonous, usually peripheral and curved around the abundant, starchy (with clustered grains) hard or seldom soft perisperm, less often nearly straight but still excentric; true endosperm wanting or nearly so. X = 5–19. (Alsinaceae, Illecebraceae)

The family Caryophyllaceae consists of about 75 genera and 2000 species, widespread but mainly of temperate or warm-temperate regions in the Northern Hemisphere, best developed in the Mediterranean region and the Near East. More than half of the species belong to only 6 genera: *Silene* (400), *Dianthus* (300), *Arenaria* (250), *Gypsophila* (125), *Stellaria* (100), and *Cerastium* (100). The family includes a number of garden ornamentals, such as *Dianthus* (pink, sweet william) and *Gypsophila* (baby's breath), as well as some familiar weeds, notably species of *Cerastium* and *Stellaria* (both called chickweed). The distinctive features of the 3 subfamilies (Silenoideae, Alsinoideae, and Paronychioideae) are indicated in the description of the family.

Fossil pollen assignable to the Caryophyllaceae first appears in Oligocene deposits.

The Caryophyllaceae have traditionally been regarded as biologically typical of their order, which has often been called Centrospermae in reference to the free-central or basal placentation. The relationship of the Caryophyllaceae to the other families of the group is abundantly confirmed by both traditional macromorphological and more recently studied micromorphological characters. The latter include a syndrome of embryological features, the ornamentation of the pollen-grains, the frequent occurrence of anomalous secondary growth, and a special type of sieve-tube plastid that appears to be consistent in and restricted to the Caryophyllales as here defined. Furthermore, many members of the Caryophyllaceae accumulate triterpenoid saponins, a feature that Frohne & Jensen (1973, in initial citations) consider to be characteristic of the order as a whole.

The Caryophyllaceae and Molluginaceae do stand apart from the other families of the order in one respect: they produce anthocyanins instead of betalains. The evolutionary significance of this feature has been much debated but is not yet fully resolved. After extravagant initial claims, investigators of the pigmentation have come to agree with the remainder of the taxonomic community that the Caryophyllaceae and Molluginaceae must continue to be associated with the betalain-families.

The nature of the petals in the Caryophyllaceae has been extensively discussed in the literature. It is now generally agreed that the order Caryophyllales as here defined is primitively apetalous and that the

Phytolaccaceae are ancestral or stand near to the ancestral line of the remaining families. The most widely accepted view of the nature of the petals in the Caryophyllaceae is that they are staminodial in origin, as they appear to be in some (not all) of the other families of the order as well. This view would seem to require an ancestor with 3 cycles of stamens, a floral organization not well represented in the modern members of the Phytolaccaceae. Although the relationship of the Caryophyllaceae and Molluginaceae to the other families of the order now seems to be established beyond a reasonable doubt, it may well be that the separation of these two families from the remainder of the group was one of the earliest phyletic dichotomies in the history of the order.

Although the Caryophyllaceae do not embrace such a wide range of ecological diversity as some other large families, there is no ecological niche or adaptive zone that they can call their own. There is nothing obvious in the way they exploit the environment that is not shared by many members of other families in different orders and subclasses of dicotyledons.

2. Order POLYGONALES Lindley 1833

The order consists of the single family Polygonaceae.

SELECTED REFERENCES

Edman, G. 1929. Zur Entwicklungsgeschichte der Gattung *Oxyria* Hill, nebst zytologischen, embryologischen und systematischen Bemerkungen über einige andere Polygonaceen. Acta Horti Berg. 9: 165–291.
Emberger, L. 1939. La structure de la fleur des Polygonacées. Compt. Rend. Hebd. Séances Acad. Sci. 208: 370–372.
Galle, P. 1977. Untersuchungen zur Blütenentwicklung der Polygonaceen. Bot. Jahrb. Syst. 98: 449–489.
Graham, S. A., & C. E. Wood, 1965. The genera of Polygonaceae in the southeastern United States. J. Arnold Arbor. 46: 91–121.
Haraldson, K. 1978. Anatomy and taxonomy in Polygonaceae subfam. Polygonoideae Meisn. emend. Jaretzky. Symb. Bot. Upsal. 22(2): 1–95.
Hedberg, O. 1946. Pollen morphology in the genus *Polygonum* L. s. lat. and its taxonomical significance. Svensk Bot. Tidskr. 40: 371–404.
Laubengayer, R. A. 1937. Studies in the anatomy and morphology of the polygonaceous flower. Amer. J. Bot. 24: 329–343.
Maekawa, F. 1964. On the phylogeny in the Polygonaceae. J. Jap. Bot. 39: 366–379.
Mahony, K. L. 1935. Morphological and cytological studies on *Fagopyrum esculentum*. Amer. J. Bot. 22: 460–475.
Mitra, G. C. 1945. The origin, development and morphology of the ochrea in *Polygonum orientale* L. J. Indian Bot. Soc. 24: 191–199.
Nowicke, J. W., & J. J. Skvarla. 1977. Pollen morphology and the relationship of the Plumbaginaceae, Polygonaceae, and Primulaceae to the order Centrospermae. Smithsonian Contr. Bot. 37: 1–64.
Reveal, J. L. 1978. Distribution and phylogeny of Eriogonoideae (Polygonaceae). Great Basin Naturalist Memoirs 2: 169–190.
Roberty, G., & S. Vautier. 1964. Les genres de Polygonacées. Boissiera 10: 7–128.
Vautier, S. 1949. La vascularisation florale chez les Polygonacées. Candollea 12: 219–343.
Weberling, F. 1970. Weitere Untersuchungen zur Morphologie des Unterblattes bei den Dikotylen. VI. Polygonaceae. Beitr. Biol Pflanzen 47: 127–140.
Wodehouse, R. P. 1931. Pollen grains in the identification and classification of plants. VI. Polygonaceae. Amer. J. Bot. 18: 749–764.

1. Family POLYGONACEAE A. L. de Jussieu 1789 nom. conserv., the Buckwheat Family

Annual or perennial herbs, or shrubs, lianas, or even (in tropical areas) fairly large trees, often with one or another sort of anomalous structure; nodes multilacunar or seldom trilacunar, often evidently swollen; plants

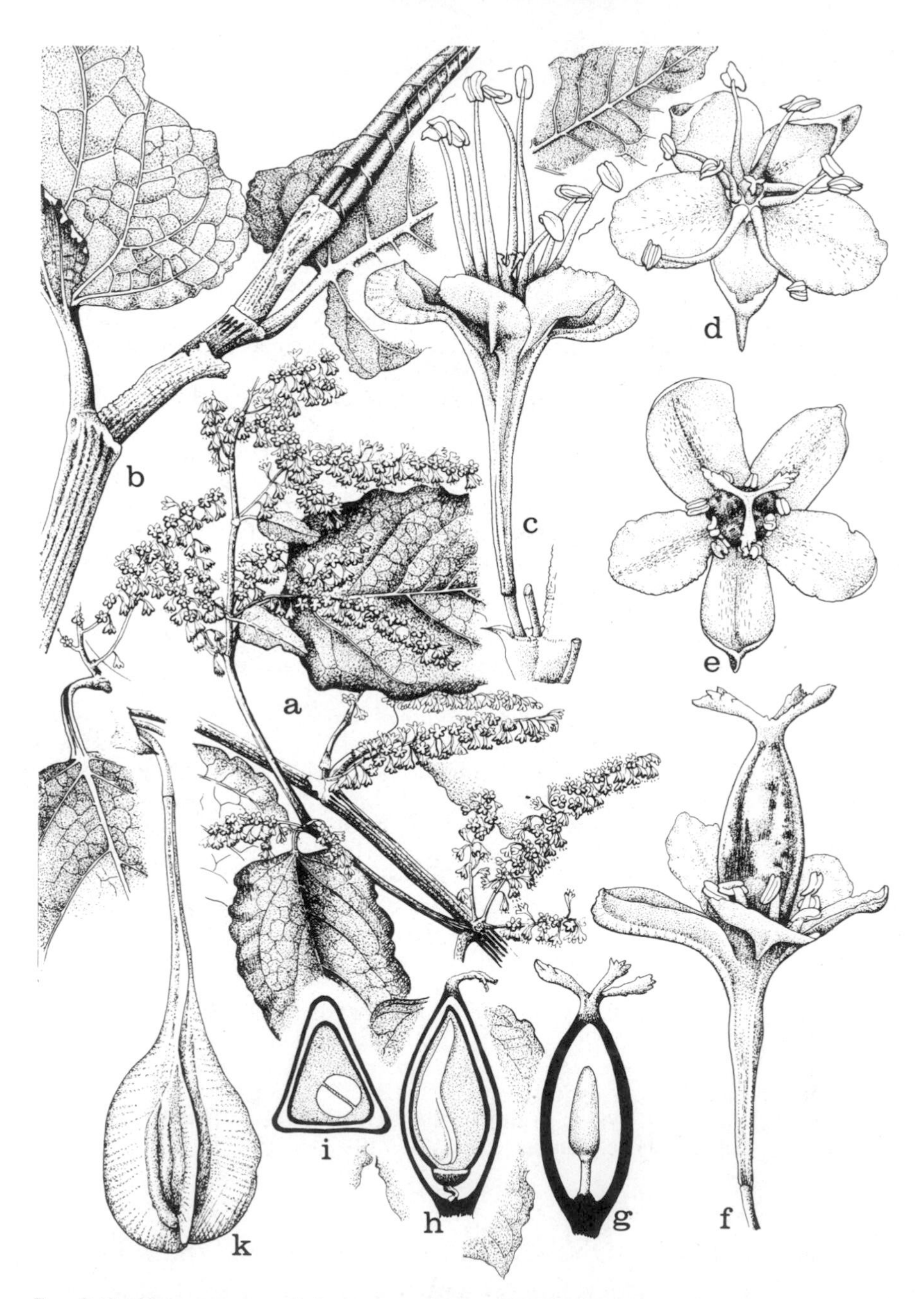

FIG. 3.11 Polygonaceae. *Polygonum cuspidatum* Sieb. & Zucc. a, habit, $\times\frac{1}{2}$; b, portion of stem and leaves, showing sheathing stipules, $\times 1$; c, d, two views of staminate flower, $\times 10$; e, f, two views of pistillate flower, $\times 10$; g, pistil, the ovary in long-section, $\times 10$; h, i, diagrammatic sections of fruit and seed, $\times 10$; k, mature fruit, $\times 5$.

producing anthocyanins but not betalains, commonly accumulating anthraquinone glycosides and often also oxalic acid, commonly tanniferous, with proanthocyanins and/or ellagic acid and sometimes also gallic acid, the parenchymatous tissues often with scattered tanniferous or mucilaginous idioblasts; calcium oxalate commonly present as clustered or less often solitary crystals in some of the cells of the parenchymatous tissues; vessel-segments with simple perforations; imperforate tracheary elements with simple pits, mostly septate; sieve-tubes with S-type plastids; roots mostly not forming mycorhizae, at least in herbaceous species. LEAVES mostly alternate, seldom opposite or whorled, simple and usually entire, seldom pinnately or palmately cleft, sometimes articulate at the base; stomates commonly anomocytic, but sometimes anisocytic, diacytic, or paracytic; stipules commonly well developed and connate into a usually scarious or hyaline (often bilobed or fringed) sheath (ocrea) around the stem, or (notably in *Eriogonum*) sometimes much-reduced or wanting. FLOWERS borne in various sorts of open or compact, simple or branched inflorescences, often in small, involucrate fascicles, individually often subtended by a persistent ocreola, evidently articulate to the pedicel, and often with a distinctly stipitate base above the articulation, always relatively small, perfect or sometimes unisexual (and then the plants generally dioecious), regular, primitively trimerous, but sometimes pentamerous (at least as to the perianth), or rarely dimerous; tepals 2–6, basally connate into a minute to evident floral tube, green and herbaceous to often colored and more or less petaloid, most commonly in 2 similar or slightly dissimilar whorls of 3, but not clearly differentiated into sepals and petals, sometimes 5 and uniseriate, originating in helical sequence even when visually in 2 whorls, commonly persistent (and sometimes accrescent) in fruit; stamens 2–9, rarely more (then usually clustered in front of the outer tepals), most commonly 6 in 2 cycles of 3, but not infrequently (notably in *Polygonum*) 8, originating in front of and in association with the tepals, 1, 2, or several per tepal (or per outer tepal only); filaments distinct or basally connate, often of two lengths, those of the inner series often dilated; anthers tetrasporangiate and dithecal, opening by longitudinal slits; pollen-grains trinucleate or rarely binucleate, highly variable in the structure of the exine, and ranging from tricolporate (most commonly) to pantoporate; an annular nectary-disk often present around the base of the ovary, or the nectaries several and placed between the bases of the stamens; gynoecium of (2) 3 (4) carpels united to form a compound, unilocular ovary (sometimes with vestigial partitions at the base), with distinct or proximally united styles; stigmas dry; ovule solitary on a basal or very shortly columnar (free–central) placenta, orthotropous or rarely anatropous, crassinucellar, bitegmic or sometimes more or less unitegmic; endosperm-development nuclear. FRUIT an achene or small nut, very often trigonous, sometimes closely subtended by the persistent, sometimes accrescent tepals, or enclosed in a fleshy hypanthium; seed with a dicotyledonous, straight or often curved, commonly excentric or peripheral (seldom centric) embryo and a well developed, starchy and oily, hard or seldom soft, sometimes ruminate endosperm with solitary starch-grains, essentially without perisperm. X = 7–13.

The family Polygonaceae consists of about 30 genera and 1000 species, chiefly of North Temperate regions. *Eriogonum* (250), *Polygonum* (200), *Rumex* (200), and *Coccoloba* (125) are the largest genera. *Polygonum* consists of several well marked sections that are often (and not without reason) treated as distinct genera. *Fagopyrum esculentum* Moench is buckwheat. Many species of *Polygonum* and *Rumex* are common weeds. *Coccoloba* and some of its allies are unusual in the family in being tropical and woody.

Pollen attributed to the Polygonaceae dates from the Paleocene epoch.

The Polygonaceae form a sharply limited and taxonomically rather isolated family. The unilocular ovary with a solitary basal ovule appears to be reduced from a basally partitioned ovary with several ovules on a free-central placenta, quite in agreement with the structure of the Caryophyllaceae and some other families of Caryophyllales. The usually peripheral embryo which may be curved around the food-storage tissue of the seed is also reminiscent of the Caryophyllales. The pollen shows a morphological series from tricolporate to pantoporate, as in several families of the Caryophyllales, although Nowicke and Skvarla (1977, p. 13) consider that the similarity of the pollen of *Polygonum* section Bistorta to that of Caryophyllales "is merely a reflection of the enormous palynological diversity in the Polygonaceae." The subfamily Paronychioideae of the Caryophyllaceae is frequently cited as a group pointing toward the Polygonaceae. For these and other reasons most authors are in agreement that the Polygonaceae are allied to the Caryophyllales. Nevertheless, the Polygonaceae differ from the Caryophyllaceae and other Caryophyllales in their mostly orthotropous ovules, in having endosperm instead of perisperm as the principal food-storage tissue of the seed, in having S-type rather than P-type sieve-tube plastids, in having primitively trimerous flowers, in being tanniferous, and in other features. Thus it seems necessary to treat the Polygonaceae as an order by themselves, rather than including them with their only evident allies in the Caryophyllales.

For whatever it may be worth, I should point out that the still very meager data on the structure of cytochrome C_3 suggest a relationship between the Polygonaceae and Caryophyllaceae (Boulter, 1974). We cannot properly place much taxonomic weight on the structure of cytochrome until many more taxa have been investigated, but at least in this instance it fits into the picture.

It might seem on first appearance that the unicyclic, pentamerous, imbricate perianth of some species of *Polygonum* represents a primitive type in the family, and that the commoner, bicyclic, trimerous type is derived. First appearances are deceiving. Pentamery in *Polygonum* has been achieved by fusion of a tepal from the outer cycle with one from the inner. This compound tepal frequently has two veins, in contrast to the solitary midvein of the other tepals.

3. Order PLUMBAGINALES Lindley 1833

The order consists of the single family Plumbaginaceae.

SELECTED REFERENCES

Baker, H. G. 1948, 1953. Dimorphism and monomorphism in the Plumbaginaceae. I. A survey of the family. II. Pollen and stigmata in the genus *Limonium*. III. Correlation of geographical distribution patterns with dimorphism and monomorphism in *Limonium*. Ann. Bot. (London) II. 12: 207–219, 1948. 17: 433–445; 615–627, 1953.

Boyes, J. W., & E. Battaglia. 1951. Embryo-sac development in the Plumbaginaceae. Caryologia 3: 305–310.

Boyes, J. W., & E. Battaglia. 1951. The tetrasporic embryo sacs of *Plumbago coccinea, P. scandens,* and *Ceratostigma willmottianum*. Bot. Gaz. 112: 485–489.

Channell, R. B., & C. E. Wood. 1959. The genera of Plumbaginaceae of the southeastern United States. J. Arnold Arbor. 40: 391–397.

Friedrich, H.-C. 1956. Studien über die natürliche Verwandtschaft der Plumbaginales und Centrospermae. Phyton (Horn) 6: 220–263.

Harborne, J. B. 1967. Comparative biochemistry of the flavonoids—IV. Correlations between chemistry, pollen morphology and systematics in the family Plumbaginaceae. Phytochemistry 6: 1415–1428.

Inamdar, J. A., & R. C. Patel. 1970. Epidermal structure and normal and abnormal stomatal development in vegetative and floral organs of *Plumbago zeylanica* Linn. Flora 159: 503–511.

Leinfellner, W. 1953. Die basiläre Plazenta von *Plumbago capensis*. Oesterr. Bot. Z. 100: 426–429.

Nowicke, J. W., & J. J. Skvarla. 1977. Pollen morphology and the relationship of the Plumbaginaceae, Polygonaceae, and Primulaceae to the order Centrospermae. Smithsonian Contr. Bot. 37: 1–64.

Roth, I. 1962. Histogenese und morphologische Deutung der basilären Plazenta von *Armeria*. Oesterr. Bot. Z. 109: 19–40.

Roth, I. 1962. Histogenese und morphologische Deutung der Kronblätter von *Armeria*. Portugaliae Acta Biol., Sér. A. 6: 211–230.

Schoute, J. C. 1935. Observations on the inflorescence in the family of the Plumbaginaceae. Recueil Trav. Bot. Néerl. 32: 406–424.

Veillet-Bartoszewska, M. 1958. Embryogénie des Plombagacées. Développement de l'embryon chez le *Plumbago europaea* L. Compt. Rend. Hebd. Séances Acad. Sci. 247: 2178–2181.

Weberling, F. 1956. Weitere Untersuchungen zur Morphologie des Unterblattes bei den Dikotylen. I. Balsaminaceae. II. Plumbaginaceae. Beitr. Biol. Pflanzen 33: 17–32.

1. Family PLUMBAGINACEAE A. L. de Jussieu 1789 nom. conserv., the Leadwort Family

Perennial (rarely annual) herbs or low shrubs or sometimes lianas, the stem often with anomalous structure, commonly with cortical and/or

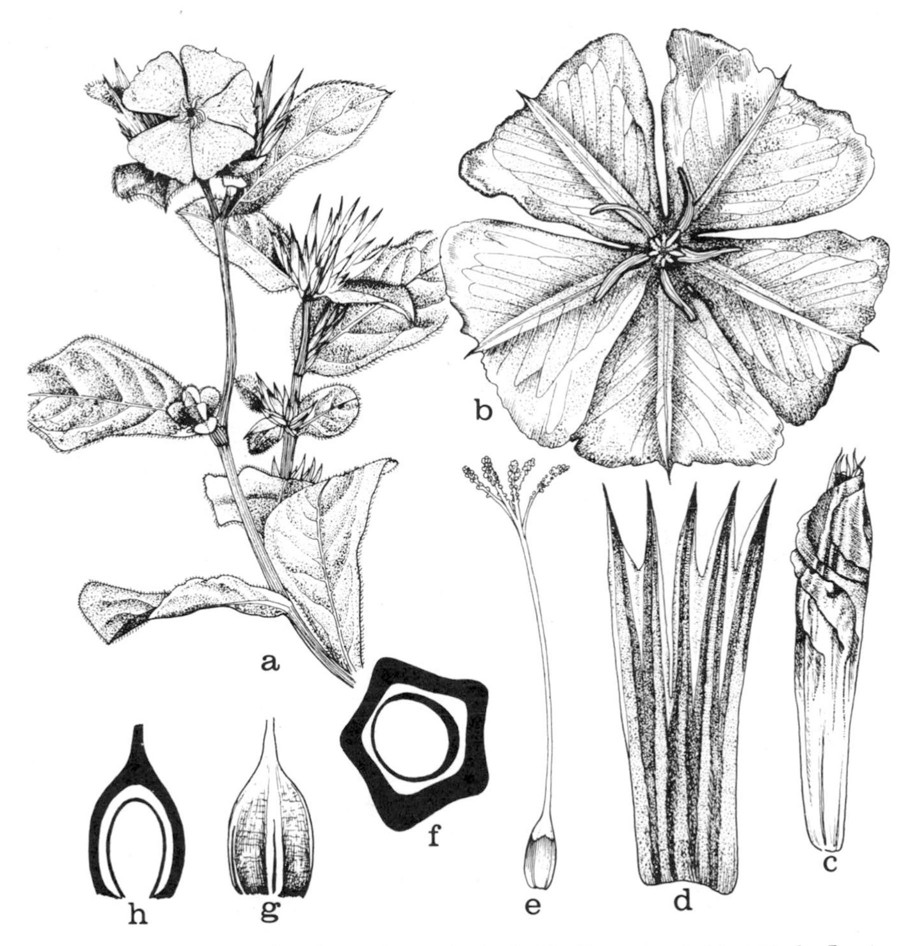

FIG. 3.12 Plumbaginaceae. *Ceratostigma plumbaginoides* Bunge. a, habit, ×1; b, flower, from above, ×4; c, corolla, before opening, ×4; d, calyx, opened out, ×4; e, pistil, ×4; f, ovary, in schematic cross-section, ×20; g, h, ovary, external and in schematic long-section, ×10.

medullary bundles, or with alternating concentric rings of secondary xylem and phloem; nodes trilacunar; plants producing anthocyanins, but not betalains, generally tanniferous, commonly with proanthocyanins, but only seldom with ellagic acid; parenchymatous tissues in the subfamily Plumbagoideae (but not in the Armerioideae), commonly with scattered secretory cells containing plumbagin, a naphthaquinone, and in the Armerioideae commonly accumulating myricetin glycosides; vessel-segments with simple perforations; imperforate tracheary elements with simple pits, often septate. LEAVES alternate, simple, entire;

stomates variously anisocytic, paracytic, encyclocytic, or anomocytic; stipules usually wanting, only seldom fairly well developed; herbage with characteristic scattered chalk-glands at (or depressed below) the surface, these exuding water and usually calcium salts, and often also with raised mucilage-glands in the axils and on the proximal part of the upper surface of the leaf. FLOWERS in panicles or cymose heads (Armerioideae), or in racemes (Plumbagoideae), perfect, regular, hypogynous, pentamerous, often heterostylic; sepals connate to form a distinct, often conspicuously 5- or 10-ribbed tube that has membranous intervals, the lobes mostly dry and membranous or scarious, often showy and somewhat petaloid; corolla sympetalous or seldom of essentially distinct, clawed petals, often persistent, the lobes or petals convolute; stamens 5, opposite the corolla-lobes (or petals), the filaments mostly free in the Plumbagoideae, more often adherent to the corolla-tube in the Armerioideae; anthers tetrasporangiate and dithecal, opening by longitudinal slits; pollen-grains trinucleate or less often binucleate, from tricolpate to seldom pantocolpate, commonly dimorphic in the Armerioideae; gynoecium of 5 carpels united to form a compound, unilocular ovary with more or less completely distinct styles (Armerioideae) or with a single, apically lobed style (Plumbagoideae); stigmas dry, papillate; ovule solitary, basal, anatropous on an elongate, slender funiculus, bitegmic, crassinucellar, the micropyle formed by the inner integument only, and penetrated by an obturator originating from the ovary-wall; embryo-sac tetrasporic, 4- to 8-nucleate, the fusion-nucleus (pro-endosperm nucleus) $3n$ or $4n$; endosperm-development nuclear. FRUIT partly or wholly enclosed by the persistent calyx, most commonly an achene, but sometimes a circumscissile capsule, or the capsule upwardly dehiscent by valves; seed with large, straight, spatulate, dicotyledonous embryo and commonly with abundant, firm, starchy endosperm with solitary starch-grains, or sometimes without endosperm, in either case without perisperm. X mostly = 6–9.

The family Plumbaginaceae consists of about a dozen genera and perhaps 400 species, the number dependent on one's taxonomic view of some of the larger genera. The family is widely distributed in diverse parts of the world, but best developed from the Mediterranean region to western and central Asia. Many of the species are xerophytic or maritime. There are two well marked subfamilies, as noted in the description. The three largest genera, *Limonium* (*Statice*), *Acantholimon*, and *Armeria*, all belong to the Armerioideae. *Plumbago*, with about 20 species, is the largest genus of the Plumbaginoideae.

The Plumbaginaceae enter the fossil record only in the Middle Miocene.

The detailed study of the Plumbaginaceae by Friedrich (1956) appeared at the time to have securely established the relationship of the family to the Caryophyllales. More recently, Corner (1976) also considered that the anatomy of the seed-coat supports an affinity of the Plumbaginaceae with the Caryophyllales.

Some other new information, however, casts doubt on the relationship. Like the Polygonales, the Plumbaginales have sieve-tubes with S-type rather than P-type plastids, and both of these orders differ from the bulk of the Caryophyllales in having anthocyanins instead of betalains. Nowicke & Skvarla (1977) consider that the pollen of the Plumbaginaceae is distinctive, unlike either the Caryophyllales or the Polygonales. The Plumbaginaceae also differ from the Caryophyllales in their sympetalous corolla, and in a series of embryological features (i.e., they do not have the full caryophylloid embryological syndrome). It therefore seems necessary to hold the Plumbaginaceae apart in their own order, rather than to include them in the Caryophyllales as proposed by Friedrich. The possibility even arises that the Plumbaginales should be transferred to the Dilleniidae, where they would have to stand alongside the Primulales, but serological studies by Frohne & John (1978, cited under Primulales) show no affinity between the Plumbaginales and Primulales. Furthermore, the lack of endo-apertures in the pollen of Plumbaginaceae militates against an assignment of the family to the Dilleniidae but is perfectly compatible with an assignment to the Caryophyllidae.

Data from fungal parasites provide some support for the treatment here presented. According to Savile (1979, in general citations) similar rusts link the Plumbaginaceae, Polygonaceae, and Caryophyllaceae, and similar smuts further link the Polygonaceae, Caryophyllaceae, and Portulacaceae.

IV. Subclass DILLENIIDAE Takhtajan 1966

Woody or herbaceous dicotyledons with various sorts of repellents, often including tannins, but without betalains and mostly poor in alkaloids; mustard oils characteristically present in the Capparales and Batales and a few Violales; iridoid compounds present only in the Ericales and some few Violales; vessel-segments with scalariform or simple perforations; wood-rays with elongate or more often short ends; sieve-tubes with S-type or seldom P-type plastids. LEAVES simple and entire or merely toothed to less often more or less dissected, only seldom compound with distinct, articulated leaflets, and then mostly palmate (as in many Capparaceae, but pinnately decompound in Moringaceae). FLOWERS polypetalous (rarely apetalous) or less often sympetalous, when sympetalous usually either with more stamens and staminodes (collectively) than corolla-lobes, or with the filaments attached directly to the receptacle, or with the stamens opposite the corolla-lobes, or with more than one of these features; stamens, when numerous, mostly initiated in centrifugal sequence (notable exceptions: Begoniaceae, many Loasaceae, at least some Ochnaceae); pollen-grains binucleate or less often trinucleate, triaperturate (typically tricolporate) or sometimes with only 2 or more than 3 apertures; gynoecium syncarpous or seldom (mainly Dilleniales) apocarpous, the ovary usually superior except in Lecythidales and many Violales; placentation various, often parietal or free-central or basal, but also often axile, the forms with axile placentation usually with numerous ovules, or if with only 1–2 ovules per locule, then usually with numerous, centrifugal stamens or sympetalous corolla or both; ovules bitegmic or unitegmic, crassinucellar or tenuinucellar, often bitegmic and tenuinucellar, in scattered orders and families with an integumentary tapetum; endosperm present or absent, but perisperm mostly wanting.

The subclass Dilleniidae as here defined consists of 13 orders, 78 families, and about 25,000 species. More than three-fourths of the species belong to only 5 orders, the Violales (5000), Capparales (4000), Ericales (4000) Theales (3500), and Malvales (3000–3500). The Primulales (1900) and Ebenales (1800) are also fairly large orders, whilst the remaining 6 orders have only about 1300 species amongst them.

The orders that make up the Dilleniidae evidently hang together as a natural group, but this group cannot be fully characterized morphologically. Like the Rosidae, the Dilleniidae are more advanced than the Magnoliidae in one or another respect, but less advanced than the Asteridae. Except for the rather small (400 spp.) order Dilleniales, the vast majority of the Dilleniidae are sharply set off from characteristic members of the Magnoliidae by being syncarpous. The Dilleniales are undoubtedly closely allied to the Theales, and they differ from at least the bulk of the Magnoliidae in having centrifugal stamens. With exceptions as noted, the species of Dilleniidae with numerous stamens have the stamens initiated in centrifugal sequence. In this respect they differ from the Rosidae, in which species with numerous stamens usually have a centripetal sequence of development. More than a third of the species of Dilleniidae have parietal placentation, in contrast to the relative rarity of this type in the Rosidae. About a third of the species (not the same third) are sympetalous, but only a very few of these (e.g., *Diapensia*) have isomerous, epipetalous stamens alternate with the corolla-lobes and also unitegmic, tenuinucellar ovules as in the Asteridae. The ovules in the Dilleniidae as a whole are bitegmic or less often unitegmic, with various transitional types, and they range from crassinucellar to tenuinucellar. Often they are bitegmic and tenuinucellar, a combination rare outside this group. Aside from the anomalous order Batales, only a single family and order (Salicales) is amentiferous. The Salicales differ from the subclass Hamamelidae in their parietal placentation and more or less numerous seeds.

It is perfectly clear that the Dilleniidae take their origin in the Magnoliidae. The apocarpous order Dilleniales, especially the family Dilleniaceae, forms the connecting link between the two subclasses. If the rest of the subclass Dilleniidae did not exist, the order Dilleniales could easily be accomodated in the Magnoliidae. Within the Magnoliidae, the family Illiciaceae in the order Illiciales may be somewhere near to the ancestry of the Dilleniaceae, but the relationship between the two families is still not particularly close.

The Theales are the central group of Dilleniidae, from which all the other orders except the Dilleniales appear to have evolved. The Malvales, Lecythidales, Violales, Capparales, and Nepenthales all appear to have arisen from a common complex in the Theales. The Salicales are an amentiferous offshoot from the Violales, and the Batales appear to be related to the Capparales. The remaining 4 orders take their origin in a different part (or parts) of the Theales. The Ericales are evidently allied to the Actinidiaceae (Theales), and the Diapensiales appear to be allied to the Ericales. The Ebenales and Primulales are somewhat more remote, but may be allied to each other and to a lesser extent to the Ericales. Only the Theales provide a reasonably likely origin for these groups. These concepts of relationships may conveniently be expressed in the accompanying phylogenetic tree.

Pollen that appears to represent the Dilleniidae dates from about the beginning of the Upper Cretaceous, but this early pollen is not clearly referrable to an order. Otherwise the fossil record as presently understood gives no clear indication of the time of origin of the group. On grounds of geographical distribution and secondary metabolites, as well as the pollen, I suggest that the Dilleniidae must have diverged from the ancestral Magnoliidae at about the same time as the Hamamelidae and Rosidae, that is, early in the Upper Cretaceous.

The adaptive significance of the characters that mark the Dilleniidae as a group is not immediately obvious. Syncarpy may have some advantage over apocarpy in relation to the access of pollen-tubes to the ovules, but it is hard to see how the advantage could be effective in the early stages of the evolution of syncarpy. The biological importance of the sequence of initiation of stamens is obscure, as is also that of the kind of placentation. The number of ovules in a locule would seem a priori to be of some possible significance in relation to seed-dispersal and the choice between quantity and quality of disseminules, but a broad-scale evaluation of that significance remains to be presented.

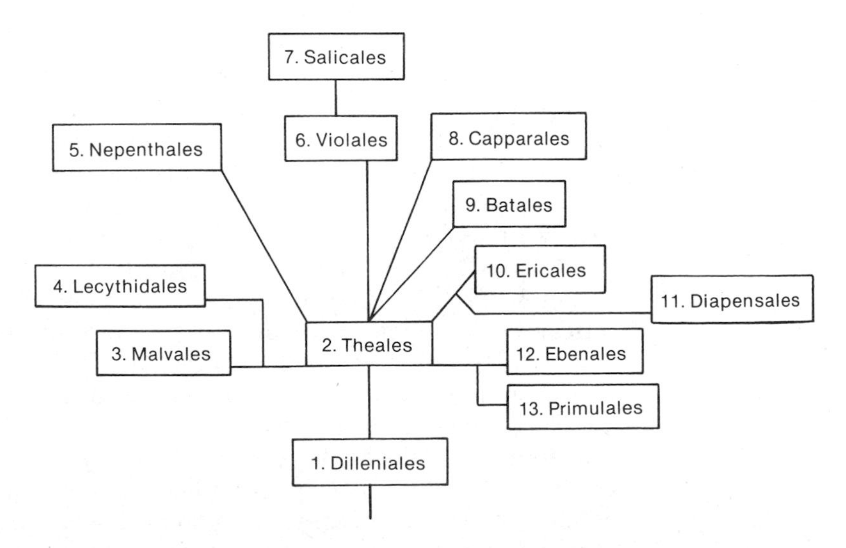

FIG. 4-1 Putative evolutionary relationships among the orders of Dilleniidae.

SELECTED REFERENCES

Lebreton, P., & M.-P. Bouchez. 1967. Recherches chimiotaxonomiques sur les plantes vasculaires. V. Distribution des composés polyphénoliques chez les Pariétales. Phytochemistry 6: 1601–1608.

SYNOPTICAL ARRANGEMENT OF THE ORDERS OF DILLENIIDAE

1 Carpels mostly distinct; stamens mostly numerous; ovules bitegmic, crassinucellar; seeds mostly arillate and with well developed endosperm .... 1. DILLENIALES.

1 Carpels almost always more or less united (at least at anthesis) to form a compound pistil; other characters various, the stamens numerous to few, the ovules bitegmic or unitegmic, crassinucellar or tenuinucellar, the seeds with or without an aril and with or without endosperm.

  2 Flowers mostly polypetalous or sometimes apetalous, only seldom (mainly the Cucurbitaceae and some other Violales) sympetalous; stamens numerous to few; ovules bitegmic or less often unitegmic, crassinucellar or often tenuinucellar; stipules present or absent.

    3 Plants insectivorous herbs or shrubs, sometimes climbing or epiphytic .... 5. NEPENTHALES.

    3 Plants not obviously insectivorous; habit various.

      4 Placentation mostly axile, seldom parietal.

        5 Flowers normally developed, with an evident perianth, perfect or less often unisexual, not borne in catkins; pollen-grains with more or less distinctly tectate-columellar (or in any case not solid) exine; plants woody or herbaceous, without mustard-oils.

          6 Flowers hypogynous or perigynous, only rarely epigynous.

            7 Sepals mostly imbricate, only seldom valvate; filaments distinct or connate into groups; plants mostly without stellate or lepidote indument, with or without mucilage, sometimes with stratified phloem, but only seldom with wedge-shaped rays; seeds not known to contain cyclopropenyl fatty acids .... 2. THEALES.

            7 Sepals valvate, only very seldom imbricate; filaments very often monadelphous, but sometimes distinct or merely connate into groups; young stems usually with stratified phloem and very wide, wedge-shaped phloem-rays; plants usually with mucilage-cells, -sacs, or-cavities; seeds commonly containing cyclopropenyl fatty acids .... 3. MALVALES.

6 Flowers epigynous or sometimes only semi-epigynous ....4. LECYTHIDALES.

5 Flowers much-reduced, unisexual, with tiny or no perianth, borne in catkins; pollen-grains with solid exine; woody plants, containing mustard-oils .... 9. BATALES.

4 Placentation mostly parietal, seldom axile, rarely basal or apical in a unilocular ovary.

8 Flowers normally developed, with an evident perianth, perfect or less often unisexual, not borne in catkins; plants woody or herbaceous.

9 Plants usually without mustard-oil and usually without myrosin-cells; fruits variously dry or fleshy and dehiscent or indehiscent, but without a replum; carpels most often 3; perianth seldom tetramerous; flowers hypogynous to often perigynous or epigynous, sometimes sympetalous .... 6. VIOLALES.

9 Plants with mustard-oil and myrosin-cells; fruit most often (but not always) a specialized type of capsule with a distinct replum; carpels most often 2; perianth very often tetramerous; flowers hypogynous to less often perigynous, not sympetalous .... 8. CAPPARALES.

8 Flowers much reduced, unisexual, without an evident perianth, borne in catkins; plants woody .... 7. SALICALES.

2 Flowers mostly sympetalous (except mainly some of the Ericales, these with unitegmic ovules and other specialized embryological features); stamens seldom more than 2 or 3 (4) times as many as the corolla-lobes; ovules tenuinucellar; stipules wanting.

10 Flowers with most or all of the features of the ericoid embryological syndrome (ovules unitegmic, tenuinucellar; endosperm-development cellular; endosperm-haustoria formed at both ends of the embryo-sac; testa a single layer of cells; or even wanting; style hollow, the cavity fluted in alignment with the locules of the ovary; anthers becoming inverted during ontogeny, so that the morphological base appears to be apical, without a fibrous layer in the wall, dehiscing by apparently terminal pores; anther-tapetum of the glandular type, with multinucleate cells; pollen-grains borne in tetrads); filaments very often attached directly to the receptacle; plants strongly mycotrophic, and often producing iridoid compounds and other special chemicals of restricted occurrence ....10. ERICALES.

10 Flowers with less than half of the features of the ericoid embryological syndrome; filaments nearly always attached

to the corolla-tube; plants not strongly mycotrophic, without iridoid compounds and mostly without the special chemicals of the Ericales.

11 Placentation mostly axile, or less often parietal; functional stamens various, very often more numerous than the corolla-lobes or alternate with them.

12 Perennial herbs or low half-shrubs, accumulating aluminum; functional stamens as many as and alternate with the corolla-lobes; ovules unitegmic; endosperm-development cellular .... 11. Diapensiales.

12 Trees or shrubs, not accumulating aluminum except in the Symplocaceae; functional stamens either more numerous than the corolla-lobes, or as many as the corolla-lobes and then usually opposite them; ovules bitegmic or unitegmic; endosperm-development nuclear or cellular .... 12. Ebenales.

11 Placentation free-central or basal in a unilocular ovary; functional stamens as many as and opposite the corolla-lobes; ovules mostly bitegmic; endosperm-development nuclear .... 13. Primulales.

1. Order DILLENIALES Hutchinson 1926

Trees, shrubs, or herbs, often tanniferous, mostly without alkaloids; vessel-segments with scalariform or simple perforations; imperforate tracheary elements with bordered pits. LEAVES alternate (seldom opposite), simple (though sometimes trilobed or pinnatifid) to ternately compound; stipules wanting or wing-like and adnate to the petiole. FLOWERS mostly perfect, regular or nearly so, hypogynous; sepals (3) 4–6 (–20), persistent, often leathery; petals (2–) 4–8 (–13), distinct; stamens typically more or less numerous, originating in centrifugal sequence, and associated with 5–15 vascular trunks, sometimes visibly clustered, sometimes not, or the androecium sometimes reduced to only one or 2 cycles or even to a single stamen; anthers tetrasporangiate and dithecal; opening by longitudinal slits or distal pores; pollen-grains binucleate, tricolpate or tricolporate, or seldom tetracolpate; gynoecium of (1–) 3–5 (–20) carpels in a single whorl or rarely in 2 whorls, the capitate, terminal stigma of each carpel elevated on a more or less definite style, or seldom sessile; carpels sometimes more or less connate below, but only rarely forming a compound, plurilocular ovary, this with distinct styles; ovules on marginal (or axile) placentas, anatropous to amphitropous or campylotropous, bitegmic, crassinucellar; endosperm-development nuclear. FRUIT mostly dry and dehiscent, but sometimes indehiscent and then either dry or fleshy; seeds provided with a more or less well developed funicular aril; embryo minute, with 2 cotyledons; endosperm copious, oily.

The order Dilleniales as here defined consists of 2 families and about 400 species. The Dilleniaceae are by far the larger family of the two. The position of the Dilleniaceae near the phyletic base of the subclass Dilleniidae seems well established.

SELECTED REFERENCES

Camp, W. H., & M. M. Hubbard, 1963. Vascular supply and structure of the ovule and aril in peony and of the aril in nutmeg. Amer. J. Bot. 50: 174–178.

Carniel, K. 1967. Über die Embryobildung in der Gattung *Paeonia*. Oesterr. Bot. Z. 114: 4–19.

Cave, M. S., H. J. Arnott, & S. A. Cook. 1961. Embryogeny in the California peonies with reference to their taxonomic position. Amer. J. Bot. 48: 397–404.

Corner, E. J. H. 1978. The inflorescence of *Dillenia*. Notes Roy. Bot. Gard. Edinburgh 36: 341–353.

Dickison, W. C. 1967–1971. Comparative morphological studies in Dilleniaceae. I. Wood anatomy. II. The pollen. III. The carpels. IV. Anatomy of the node and vascularization of the leaf. V. Leaf anatomy. VI. Stamens and young stem. VII. Additional notes on *Acrotrema*. J. Arnold Arbor. 48: 1–29; 231–240, 1967. 49: 317–329, 1968. 50: 384–400, 1969. 51:89–113; 403–418, 1970. 52: 319–333, 1971.

Eames, A. J. 1953. Floral anatomy as an aid in generic limitation. Chron. Bot. 14: 126–132.

Hiepko, P. (1964) 1965. Das zentrifugale Androeceum der Paeoniaceae. Ber. Deutsch. Bot. Ges. 77: 427–435.

Hiepko, P. 1966. Zur Morphologie, Anatomie und Funktion des Diskus der Paeoniaceae. Ber. Deutsch. Bot. Ges. 79: 233–245.

Hoogland, R. D. 1952. A revision of the genus *Dillenia*. Blumea 7: 1–145.

Keefe, J. M., & M. F. Moseley. 1978. Wood anatomy and phylogeny of *Paeonia* section Moutan. J. Arnold Arbor. 59: 274–297.

Kubitzki, K. 1968. Flavonoide und Systematik der Dilleniaceen. Ber. Deutsch. Bot. Ges. 81: 238–251.

Kumazawa, M. 1935. The structure and affinities of *Paenoia*. Bot. Mag. (Tokyo) 49: 306–315. (In Japanese; English summary).

Matthiessen, A. 1962. A contribution to the embryogeny of *Paeonia*. Acta Horti Berg. 20: 57–61.

Rao, A. N. 1957. A contribution to the embryology of Dilleniaceae. Proc. Iowa Acad. Sci. 64: 172–176.

Sastri, R. L. N. 1958. Floral morphology and embryology of some Dilleniaceae. Bot. Not. 111: 495–511.

Sawada, M. 1971. Floral vascularization of *Paeonia japonica* with some consideration on systematic position of the Paeoniaceae. Bot. Mag. (Tokyo) 84: 51–60.

Stebbins, G. L., & R. D. Hoogland, 1976. Species diversity, ecology and evolution in a primitive angiosperm genus: *Hibbertia* (Dilleniaceae). Pl. Syst. Evol. 125: 139–154.

Stern, F. C. 1946. A study of the genus *Paeonia*. Roy. Hort. Society. London.

Swamy, B. G. L., & K. Periasamy. 1955. Contributions to the embryology of *Acrotrema arnottianum*. Phytomorphology 5: 301–314.

Wilson, C. L. 1965, (1973) 1974. The floral anatomy of the Dilleniaceae. I. *Hibbertia* Andr. II. Genera other than *Hibbertia*. Phytomorphology 15: 248–274, 1965. 23: 25–42, (1973) 1974.

Wunderlich, R. 1966. Zur Deutung der eigenartigen Embryoentwicklung von *Paeonia*. Oesterr. Bot. Z. 113: 395–407.

Yakovlev, M. S., & M. D. Yoffe, 1957. On some peculiar features in the embryogeny of *Paeonia* L. Phytomorphology 7: 74–82.

Карташова, Н. Н. 1962. К вопросу о природе нектарников цветка *Paeonia*. Труды Бот. Инст. АН СССР, сер. VII, 5: 77-85.

Кемулария-Натадзе, Л. М. 1958. К вопросу о положении семейства Paeoniaceae в системе покрытосеменных растений. Заметки по сист. и геогр. раст., Тбилисский Бот. Инст. 20: 19-28. Journal title also in Latin: Notulae Systematicae ac Geographicae Instituti Botanici Thbilissiensis.

Яковлев, М. С., & М. Д. Иоффе. 1965. Эмбриология некоторых представителей рода *Paeonia* L. В сб.: Морфология цветка и репродуктивный процесс у покрытосеменных растений. М. С. Яковлев, Ред.: 140-176. Изд. Наука, Москва-Ленинград.

SYNOPTICAL ARRANGEMENT OF THE FAMILIES OF DILLENIALES

1 Flowers without a nectary-disk; leaves mostly entire, seldom trilobed or pinnatifid; plants woody (of various form) or rarely herbaceous, occurring in tropical and subtropical regions, especially Australia; raphides, crystal sand, and proanthocyanins present .... 1. DILLENIACEAE.

1 Flowers with the stamens seated beneath a well developed, perigynous nectary-disk; leaves ternately or ternate-pinnately compound or dissected; herbs or half-shrubs or soft shrubs, mostly of temperate Eurasia and western North America; raphides, crystal sand, and proanthocyanins wanting .... 2. PAEONIACEAE.

1. Family DILLENIACEAE Salisbury 1807 nom. conserv., the Dillenia Family

Evergreen trees, shrubs, lianas, or rarely half-shrubs or (as in *Acrotrema*) perennial herbs, usually containing the flavonol myricetin (which is rare in the Magnoliidae), tanniferous, commonly with both ellagic acid and proanthocyanins, but without ethereal oil cells and mostly without alkaloids, the alkaloids when present not of the benzyl-isoquinoline group; nodes variously unilacunar, trilacunar, or multilacunar; stem generally with raphides and crystal sand in the parenchymatous tissues, often in long sacs or tubes; anomalous secondary thickening of concentric type (producing interxylary phloem) known in a few species; vessel-segments more or less elongate, with scalariform perforations, or the larger vessels in some genera with simple perforations (the vessels also simply perforate in some of the xerophytic spp. of *Hibbertia*); imperforate tracheary elements with bordered pits; wood-rays mostly Kribs type I, varying to IIA, heterocellular, mixed uniseriate and pluriseriate, sometimes as much as 40 cells wide, commonly with elongate ends. LEAVES alternate, rarely opposite, simple, entire or toothed to rarely pinnatifid or trilobed, commonly with silicified epidermis, rarely much-reduced and scale-like; stomates most commonly anomocytic, but sometimes encyclocytic, anisocytic, paracytic, or of other types; stipules wanting, or wing-like and adnate to the petiole. FLOWERS solitary or in variously cymose or racemose inflorescences, yellow or white, perfect or rarely unisexual, hypogynous, regular or nearly so at least as to the perianth, often or usually without nectar; sepals (3–) 5 (–20), spirally imbricate, persistent; petals (2–) 5, imbricate, often crumpled in bud, deciduous; stamens typically more or less numerous, but sometimes reduced to few or even only one, often asymmetrically placed, associated with a limited number (mostly 5–15) of stamen-trunks (common vascular traces to several or many stamens), but not always visibly clustered in relation to the trunks, the members of each group originating in centrifugal sequence, or the whole androecium appearing to develop centrifugally; number of stamens associated with each stamen-trunk sometimes reduced to few or even only one, the number of trunks likewise sometimes reduced to one; anthers tetrasporangiate and dithecal, sometimes with an enlarged and shortly prolonged connective, or on a distally thickened filament, opening by longitudinal slits or distal pores; pollen-grains binucleate, tricolpate (seldom tetracolpate) to triporate; gynoecium of (1–) several or sometimes numerous (up to 20)

FIG. 4-2 Dilleniaceae. *Hibbertia cuneiformis* (Labill.) J. E. Smith. a, habit, ×1; b, flower, from above, ×2; c, androecium and gynoecium, from above, ×8; d, e, pistil, external and in partial long-section, ×8; f, cross-section of ovary, ×8; g, stamen-cluster, ×8; h, flower bud, ×2.

distinct carpels in a single whorl or rarely in 2 whorls, the receptacle sometimes (*Dillenia*) produced as a small cone beyond the level of attachment of the carpels; carpels more or less evidently conduplicate and sometimes not fully sealed, primitively provided with 3 vascular traces, but the traces fewer or more numerous in some more advanced forms, fully distinct (but sometimes adnate to the elongating receptacle), or rarely connate to form a compound, plurilocular ovary with distinct styles; style commonly slender and elongate, with a terminal, often capitate stigma; ovules 1-many in each carpel, apotropous (with ventral raphe), anatropous to amphitropous or campylotropous, with zigzag micropyle, bitegmic, crassinucellar; endosperm-development nuclear. FRUIT dry and dehiscent, or sometimes (*Dillenia*) indehiscent and then included in the more or less fleshy calyx; seeds with a funicular, often crested or laciniate aril closely adnate to the testa, or the aril sometimes vestigial or obsolete; embryo very small, straight, with 2 short cotyledons; endosperm copious, oily and proteinaceous. X = 4, 5, 8, 9, 10, 12, 13.

The family Dilleniaceae as now customarily defined consists of 10 genera and about 350 species, largely confined to tropical and subtropical countries, best developed in the Australian region. About one-third of the species belong to the single large genus *Hibbertia*, which shows a remarkable graded series in the androecium from numerous (ca 200) stamens not obviously grouped into bundles, to numerous stamens groups into 15 bundles, and through reduction in both the number of bundles and the number of stamens per bundle to an androecium consisting of a solitary stamen.

The Dilleniaceae are relatively primitive in many features. They commonly have more or less distinct, conduplicate carpels, and the carpels are sometimes not fully sealed. The ovules are bitegmic and crassinucellar, with a zigzag micropyle, and they mature into seeds with a tiny embryo and copious endosperm. The stamens are usually more or less numerous, and the connective is sometimes prolonged. The wood has scalariform vessels and a syndrome of other features regarded as primitive. In all of these respects the Dilleniaceae recall the Magnoliidae.

On the other hand, the Dilleniaceae do not closely resemble any one family of the Magnoliidae, and they are sharply set off from the Magnoliidae by a series of chemical features. Unlike the vast majority of the Magnoliidae, the Dilleniaceae have ellagic acid, proanthocyanins, and raphides. The ethereal oil cells so characteristic of the woody Magnoliidae are wanting from the Dilleniaceae. The Dilleniaceae are almost without alkaloids, and the few that have been found in some species have nothing to do with the characteristic benzyl-isoquinoline alkaloids of the Magnoliidae. Furthermore, there is a still unresolved controversy about the ancestry of the multistaminate, centrifugal androecium such as that found in the Dilleniaceae. If, as some authors maintain, this kind of androecium is necessarily derived from an ordinary, cyclic, paucistaminate androecium by secondary increase in the number of stamens, then the gap between the Dilleniaceae and the Magnoliidae is further widened.

If the hypothesis of chemical evolution that I presented in 1977 is correct, the Dilleniidae probably originated at about the same time in the Cretaceous as the Hamamelidae and Rosidae. The Hamamelidae and the more archaic members of the Dilleniidae and Rosidae all rely heavily on hydrolyzable tannins as defensive weapons. The more advanced members of these latter two groups have largely discarded tannins in favor of more recently evolved defenses.

The centrifugal androecium and frequently campylotropous or amphitropous ovules of the Dilleniidae recall the Caryophyllidae, but the Caryophyllidae have their own set of specialized features not found in the Dilleniaceae and other Dilleniidae. Any evolutionary relationship between the Dilleniaceae and the Caryophyllidae must be rather remote in time.

In contrast to their evident separation from the Magnoliidae and Caryophyllidae, the Dilleniaceae are obviously allied to such syncarpous families as the Actinidiaceae and Theaceae. On the other hand, Corner (1976, in initial citations) considers that the anatomy of the seed-coat is uniform within the Dilleniaceae, and unlike that of the Theaceae. For purposes of conceptual organization it is in any case useful to put the largely apocarpous family Dilleniaceae in a separate order from the largely syncarpous Theales. The Dilleniaceae appear to be the modern remnants of the group that gave rise not only to the Theales but also to the rest of the subclass Dilleniidae.

2. Family PAEONIACEAE Rudolphi 1830 nom. conserv., the Peony Family

Perennial herbs or less often half-shrubs or soft shrubs, glabrous or sometimes with unicellular hairs, lacking myricetin, proanthocyanins, ellagic acid, and raphides, but with scattered secretory cells in the parenchymatous tissues of the young stem, especially the pith; nodes trilacunar or pentalacunar; clustered crystals of calcium oxalate com-

FIG. 4-3 Paeoniaceae. *Paeonia californica* Nutt. a, habit, ×½; b, flower, from above, ×1; c, stamen, ×2; d, transitional stamen-carpel, ×2; e, f, carpel, external and in long-section, ×2; g, disk and carpels, from above, ×2; h, fruits, ×½.

monly present in some of the cells of the parenchymatous tissues; stem with supplementary cortical bundles; vessels very small and numerous, angular, with scalariform, often oblique perforations that have mostly 1–7 cross-bars or some with simple perforations; imperforate tracheary elements with bordered pits, considered to be tracheids and fiber-tracheids; wood-rays heterocellular to homocellular, mixed uniseriate and pluriseriate, the latter 2–3 (–8) cells wide, with elongate ends; wood-parenchyma diffuse. LEAVES large, alternate, ternately or ternate-pinnately mostly twice (or more times) compound or dissected; stomates anomocytic; petiole with an arc of vascular bundles; stipules wanting. FLOWERS large, terminal, solitary or few together, commonly cantharophilous, perfect, regular or with somewhat unequal sepals; receptacle slightly to evidently concave; phyllotactic spiral continuous from leaves through bracts, sepals, petals, stamen-trunks, and usually carpels; sepals (3–) 5 (–7), leathery-persistent; petals 5–8 (–13), with 3 or more vascular traces like the sepals; stamens numerous, associated with 5 vascular trunks, but not visibly clustered; the members of each set developing in centrifugal sequence; anthers tetrasporangiate and dithecal, opening by longitudinal slits; pollen-grains binucleate, tricolporate, with an incomplete tectum, similar to those of some Dilleniaceae (and some other families); gynoecium surrounded by a fleshy, lobed nectary-disk that is variously regarded as an outgrowth from the receptacle or as a modified part of the androecium; carpels (2) 3–5 (–15), distinct, coarse and thick-walled, the expanded stigma subsessile or on a short, stout style; stigma wet, papillate; ovules several or many in each carpel, marginal, anatropous, crassinucellar (but the nucellus degenerating after pollination), bitegmic, but the outer integument massive and eventually differentiating into two layers, so that the testa is 3-layered; endosperm-development nuclear; embryogenesis following a unique pattern involving unequal initial division of the zygote, degeneration of the smaller daughter-cell, and development of a coenocytic, free-nuclear stage by subsequent nuclear divisions in the larger daughter-cell. FRUIT follicular; seeds large, with a more or less well developed funicular aril; endosperm abundant, oily and proteinaceous; embryo minute, with 2 cotyledons; germination hypogaeal. X = 5.

The family Paeoniaceae consists of the single genus *Paeonia*, with about 30 species. Aside from one or two species in the western United States, the genus is wholly Eurasian, and best-developed in temperate eastern Asia.

Paeonia has traditionally been referred to the Ranunculaceae, where it is sadly misplaced. The persistent, leathery sepals, sepal-derived rather than staminodial petals, centrifugal stamens, perigynous disk, and ariliferous seeds are all out of harmony with the Ranunculaceae. The distinction is further reinforced by anatomical, cytological, palynological, and serological features. The idea that *Paeonia* is closely related to the Ranunculaceae can no longer be seriously entertained.

Beginning with Camp (in Gundersen, 1950, under dicotyledons), most recent authors have treated *Paeonia* as a separate family to be associated with the Dilleniaceae. This seems reasonable, and no other obvious possibility presents itself. Even so, there are some notable differences between the two families, and it is a matter of opinion whether taxonomic needs are best met by putting them into the same order or by keeping the Paeoniaceae in an order of their own. Either treatment is defensible. Keefe and Moseley (1978) consider that *Paeonia* is closely allied to but distinct from the Dilleniaceae, but not close to the Crossosomataceae. The latter family, often included in the Dilleniales, is here referred to the Rosales.

2. Order THEALES Lindley 1833

Trees, shrubs, lianas, or herbs, mostly poor in alkaloids, very often tanniferous, producing proanthocyanins or ellagic acid or both, and often with mucilage or resin in special cells, cavities, or channels; vessel-segments with scalariform or simple perforations; imperforate tracheary elements with simple or more often bordered pits; young stems in several families with stratified phloem, and sometimes with wedge-shaped phloem-rays. LEAVES alternate, opposite, or whorled, simple or seldom compound, stipulate or exstipulate. FLOWERS perfect or unisexual, usually regular or nearly so, hypogynous to occasionally somewhat perigynous or rarely partly epigynous; sepals (2) 3–6 (–14), imbricate or seldom valvate, distinct or often connate at the base, seldom connate to the middle; petals (2–) 4–5 (-numerous), imbricate or convolute, distinct or occasionally connate at the base, seldom forming a clearly sympetalous corolla or connate above to form a calyptra; stamens (2–) 4-many, sometimes connate in bundles, when numerous usually developing in centrifugal sequence and often served by a limited number of trunk bundles; pollen-grains binucleate or rarely trinucleate, 2- to 6-aperturate, most commonly tricolporate; gynoecium of 2-many carpels, these almost always either united to form a compound ovary or united by a common gynobasic style; ovary typically with as many locules as carpels, or sometimes unilocular by failure of the partitions to meet in the center; styles distinct and as many as the carpels, or partly or wholly united; ovules on axile placentas, or on intruded parietal placentas, rarely essentially basal, (1) 2-many per carpel (or some locules empty), anatropous to sometimes hemitropous or campylotropous, bitegmic or seldom (mainly Actinidiaceae) unitegmic, crassinucellar or more often tenuinucellar, often at once bitegmic and tenuinucellar, a combination uncommon outside the order; endosperm-development nuclear or less often cellular. FRUIT variously dry or fleshy and dehiscent or indehiscent; seeds with small to large, dicotyledonous embryo, and copious to scanty or no endosperm, sometimes arillate.

The order Theales as here defined consists of 18 families and nearly 3500 species. Three of the families, Clusiaceae (1200), Theaceae (600) and Dipterocarpaceae (600), make up fully two-thirds of the order, and most of the remaining species belong to only 2 additional families, the Ochnaceae (400) and Actinidiaceae (300). The other 13 families contain scarcely 300 species in all, and five of them (Medusagynaceae, Oncothecaceae, Paracryphiaceae, Pentaphylacaceae, and Pellicieraceae) are unispecific.

The assortment of families amongst the Theales and Violales that is presented here is conditioned partly by obvious affinities between particular pairs of families and sets of families, and partly by the rationale that (with relatively insignificant probable exceptions among the Magnoliidae) parietal placentation always has an axile phyletic antecedent. Phyletic fusion of closed carpels forms a compound, plurilocular ovary with axile placentation. Parietal placentation is derived from axile by failure of the partitions to meet and join in the center of the ovary. There is no theoretical reason why an ovary with partial partitions might not revert, phyletically, to a plurilocular one, and doubtless this has happened in some groups. The initial working hypothesis, however, must always be to the contrary. A group with parietal placentation should be derived from one with axile placentation, rather than vice versa. In this instance adherence to this rationale gives us an order in which the major families clearly belong together, and the minor ones are at least not obviously misplaced.

The rationale for distinguishing the Malvales from the Theales is discussed under the Malvales.

The characters by which the families of Theales differ among themselves are a mixed bag of obviously adaptive, possibly adaptive, and seemingly nonadaptive features. The mangrove habit of the monotypic family Pellicieraceae is obviously adaptive. The accrescent fruiting sepals of the Dipterocarpaceae appear to be associated with wind-dispersal of the fruits, although students of the family say that the mechanism is not highly effective. The highly modified floral bracts of the Marcgraviaceae are said to secrete nectar and to attract pollinators such as hummingbirds, but strictly from the standpoint of effective pollination it would seem more advantageous to have the nectar produced within the flowers, closer to the anthers and stigmas. The different sorts of fruits in the different families can reasonably be supposed to related in considerable part to dispersal-mechanisms. The raphides, sclerenchymatous idioblasts, resin or mucilage systems, tannins, and other secondary metabolites of various families are doubtless significant in the continuing evolutionary struggle between plants and their predators. One might argue that the gynobasic style and essentially separate ovary-lobes of many Ochnaceae relate in some way to dispersal mechanisms, but it is more difficult to see interim survival value in the progressive indentation of the ovary that leads to the more specialized, gynobasic structure. Someone with a lively imagination might see some advantage in reducing the number of stamens from numerous to few (or building it up from few to numerous). The prospects for finding survival-value in the following features seem to me to be dismal: opposite versus alternative leaves; presence or absence of stipules (absence is advanced!); number of carpels in the gynoecium; bitegmic versus unitegmic ovules; crassinucellar versus tenuinucellar ovules; nuclear versus cellular development of the endosperm; presence of cortical bundles in the young stem.

SELECTED REFERENCES

Bancroft, H. 1935. The wood anatomy of representative members of the Monotoideae. Amer. J. Bot. 22: 717–739.

Baretta-Kuipers, T. 1976. Comparative wood anatomy of Bonnetiaceae, Theaceae and Guttiferae. Leiden Bot. Series 3: 76–101.

Boureau, E. 1958. Contribution à l'étude anatomique des espèces actuelles de Ropalocarpaceae. Bull. Mus. Hist. Nat. (Paris) II. 30: 213–221.

Brown, E. G. S. 1935. The floral mechanism of *Saurauja subspinosa* Anth. Trans. & Proc. Bot. Soc. Edinburgh 31: 485–497.

Brundell, D. J. 1975. Flower development of the Chinese Gooseberry (*Actinidia chinensis* Planch.). I. Development of the flowering shoot. II. Development of the flower bud. New Zealand J. Bot. 13: 473–483; 485–496.

Capuron, R. 1962. Révision des Rhopalocarpacées. Adansonia sér. 2, 2: 228–267.

Capuron, R. 1970. Observations sur les Sarcolaenacées. Adansonia sér. 2, 10: 247–265.

Carlquist, S. 1964. Pollen morphology and evolution of Sarcolaenaceae (Chlaenaceae). Brittonia 16: 231–254.

Carpenter, C. S., & W. C. Dickison. 1976. The morphology and relationships of *Oncotheca balansae*. Bot. Gaz. 137: 141–153.

Cavaco, A. 1952. Chlaenacées. *In*: H. Humbert, ed., Flore de Madagascar et des Comores 126: 1–37.

Cavaco, A. 1952. Recherches sur les Chlaenacées, famille endemique de Madagascar. Mem. Inst. Sci. Madagascar, Sér. B. 4: 59–92.

Dathan, A. S. R., & D. Singh. 1971. Embryology and seed development in *Bergia* L. J. Indian Bot. Soc. 59: 362–370.

Decker, J. M. 1966. Wood anatomy and phylogeny of Luxemburgieae (Ochnaceae). Phytomorphology 16: 39–55.

Decker, J. M. 1967. Petiole vascularization of Luxembergieae (Ochnaceae). Amer. J. Bot. 54: 1175–1181.

Den Outer, R. W., & A. P. Vooren. 1980. Bark anatomy of some Sarcolaenaceae and Rhopalocarpaceae and their systematic position. Meded. Landbouwhogeschool 80(6): 1–15.

Dickison, W. C. 1972. Observations on the floral morphology of some species of *Saurauia, Actinidia* and *Clematoclethra*. J. Elisha Mitchell Sci. Soc. 88: 43–54.

Dickison, W. C., & P. Baas. 1977. The morphology and relationships of *Paracryphia* (Paracryphiaceae). Blumea 23: 417–438.

Foster, A. S. 1950. Morphology and venation of the leaf in *Quiina acutangula*. Amer. J. Bot. 37: 159–171.

Foster, A. S. 1950. Venation and histology of the leaflets in *Touroulia guianensis* Aubl., and *Froesia tricarpa* Pires. Amer. J. Bot. 37: 848–862.
Gottwald, H., & N. Parameswaran. 1967. Beiträge zur Anatomie und Systematik der Quiinaceae. Bot. Jahrb. Syst. 87: 361–381.
Graham, A. 1977. New records of *Pelliceria* (Theaceae/Pelliceriaceae) in the Tertiary of the Caribbean. Biotropica 9: 48–52.
Guédès, M., & R. Schmid. 1978. The peltate (ascidiate) carpel theory and carpel peltation in *Actinidia chinensis* (Actinidiaceae). Flora 167: 525–543.
Guillaumin, A. 1938. Observations morphologiques et anatomiques sur le genre *Oncotheca*. Rev. Gén. Bot. 50: 629–635.
Howe, M. A. 1911. A little-known mangrove of Panama. J. New York Bot. Gard. 12: 61–72.
Huard, J. 1965. Anatomie des Rhopalocarpacées. Adansonia sér. 2, 5: 103–123.
Huard, J. 1965. Palynologia Madagassica et Mascarenica. Fam. 127. Rhopalocarpaceae. Pollen & Spores 7: 303–312.
Huard, J. 1965. Remarques sur la position systématique des Rhopalocarpacées d'après leur anatomie et leur morphologie pollinique. Bull. Soc. Bot. France 112: 252–254.
Hunter, G. E. 1966. Revision of Mexican and Central American *Saurauia* (Dilleniaceae). Ann. Missouri Bot. Gard. 53: 47–89.
Kawano, S. 1965. Anatomical studies on the androecia of some members of the Guttiferae-Moronoboideae. Bot. Mag. (Tokyo) 78: 97–108.
Keng, H. 1962. Comparative morphological studies in Theaceae. Univ. Calif. Publ. Bot. 33: 269–384.
Kobuski, C. E. 1951. Studies in the Theaceae. XXIII. The genus *Pelliciera*. J. Arnold Arbor. 32: 256–262.
Kubitzki, K. 1978. The botany of the Guyana Highland—Part X. *Caraipa* and *Mahurea* (Bonnetiaceae). Mem. New York Bot. Gard. 29: 82–128.
Kubitzki, K., A. A. L. Mesquita, & O. R. Gottlieb. 1978. Chemosystematic implications of xanthones in *Bonnetia* and *Archytaea*. Biochem. Syst. Ecol. 6: 185–187.
Leins, P. 1964. Die frühe Blütenentwicklung von *Hypericum hookerianum* Wight et Arn. und *H. aegypticum* L. Ber. Deutsch. Bot. Ges. 77: 112–123.
Letouzey, R. 1961. Notes sur les Scytopétalacées (Révision des Scytopétalacées de l'herbier de Paris). Adansonia sér. 2, 1: 106–142.
Li, H.-L. 1952. A taxonomic review of the genus *Actinidia*. J. Arnold Arbor. 33: 1–61.
Maguire, B. 1972. The botany of the Guayana Highland—Part IX. Bonnetiaceae. Mem. New York Bot. Gard. 23: 131–165.
Maguire, B., et al. 1972. Botany of the Guayana Highland—Part IX. Tetrameristaceae. Mem. New York Bot. Gard. 23: 165–192.
Maguire, B., et al. 1977. Pakaraimoideae, Dipterocarpaceae of the Western Hemisphere. Taxon 26: 341–385.
Maguire, B., & P. S. Ashton. 1980. *Pakaraimaea dipterocarpacea* II. Taxon 29: 225–231.
Mauritzon, J. 1939. Über die Embryologie von *Marcgravia*. Bot. Not. 1939: 249–255.
Maury, G., J. Muller, & B. Lugardon. 1975. Notes on the morphology and fine structure of the exine of some pollen types in Dipterocarpaceae. Rev. Palaeobot. Palynol. 19: 241–289.
Muller, J. 1969. Pollen-morphological notes on Ochnaceae. Rev. Palaeobot. Palynol. 9: 149–173.

Narayana, L. L. 1975. A contribution to the floral anatomy and embryology of Ochnaceae. J. Jap. Bot. 50: 329–336.
Pauzé, F., & R. Sattler. 1979. La placentation axillaire chez *Ochna atropurpurea*. Canad. J. Bot. 57: 100–107.
Prance, G. T., & M. F. da Silva. 1973. A monograph of the Caryocaraceae. Fl. Neotropica Monograph 12: 1–75.
Raghavan, T. S., & V. K. Srinivasan. 1940. A contribution to the life-history of *Bergia capensis* Linn. J. Indian Bot. Soc. 19: 283–291.
Rao, A. N. 1957. The embryology of *Hypericum patulum* Thunb. and *H. mysorense* Heyne. Phytomorphology 7: 36–45.
Record, S. J. 1942. American woods of the family Theaceae. Trop. Woods 70: 23–33.
Schmid, R. 1978. Actinidiaceae, Davidiaceae, and Paracryphiaceae: systematic considerations. Bot. Jahrb. Syst. 100: 196–204.
Schmid, R. 1978. Reproductive anatomy of *Actinidia chinensis* (Actinidiaceae). Bot. Jahrb. Syst. 100: 149–195.
Schofield, E. K. 1968. Petiole anatomy of the Guttiferae and related families. Mem. New York Bot. Gard. 18(1): 1–55.
Soejarto, D. D. 1969. Aspects of reproduction in *Saurauia*. J. Arnold Arbor. 50: 180–196.
Soejarto, D. D. 1970. *Saurauia* species and their chromosomes. Rhodora 72: 81–93.
van Steenis, C. G. G. J. Pentaphylacaceae. *In*: Fl. Malesiana ser. 1, 5(2): 121–124. 1956.
Straka, H. 1963 (1964). Betrachtungen zur Phylogenie der Sarcolaenaceae (Chlaenaceae). Ber. Deutsch. Bot. Ges. 76: (55)–(62).
Straka, H., & F. Albers. 1978. Die Pollenmorphologie von *Diegodendron Humbertii* R. Capuron (Diegodendraceae, Ochnales bzw. Theales). Bot. Jahrb. Syst. 99: 363–369.
Swamy, B. G. L. 1948. A contribution to the embryology of the Marcgraviaceae. Amer. J. Bot. 35: 628–633.
Vestal, P. A. 1937. The significance of comparative anatomy in establishing the relationship of the Hypericaceae to the Guttiferae and their allies. Philipp. J. Sci. 64: 199–256.
Vijayaraghavan, M. R. 1965. Morphology and embryology of *Actinidia polygama*, and systematic position of the family Actinidiaceae. Phytomorphology 15: 224–235.
Vijayaraghavan, M. R., & U. Dhar. 1976. *Scytopetalum tieghemii*—embryologically unexplored taxon and affinities of the family Scytopetalaceae. Phytomorphology 26: 16–22.
Wood, C. E. 1959. The genera of Theaceae of the southeastern United States. J. Arnold Arbor. 40: 413–419.

Агабабян, В. Ш., & Э. Л. Заварян. 1971. К палиносистематике рода *Paracryphia* Bak. f. АН Армянской ССР. Биол. Ж. Армении 25: 35-40.
Колбасина, Э. И. 1969. Органогенез соцветия и цветка *Actinidia kolomikta* Maxim. Бот. Ж. 54: 1397-1400.
Первухина, Н. В. 1965. Околоцветник *Thea sinensis* L. и его происхождение. В. сб.: Морфология цветка и репродуктивный процесс у покрытосеменных растений, М. С. Яковлев, ред.: 47-60. Изд. "Наука," Москва-Ленинград.
Федоров, Ан. А. 1976. Закон гомологических рядов Н. И. Вавилова и

видообразование во влажных тропиках. Изв. АН СССР сер. Биол. 1975 (5): 705–715.

Шилкина, И. А. 1977. Сравнительно-анатомическая характеристика древесины рода *Oncotheca* Baill. (Пор. Theales). Бот. Ж. 62: 1273–1275.

SYNOPTICAL ARRANGEMENT OF THE FAMILIES OF THEALES

1 Leaves digitately 3-foliolate; embryo with enlarged, thickened hypocotyl and reduced cotyledons; tropical America .... 5. CARYOCARACEAE.

1 Leaves simple, or seldom pinnately compound; embryo usually of more ordinary proportions.

2 Leaves mostly alternate.

3 Leaves mostly stipulate.

4 Connective not prominently exserted (though sometimes laterally expanded); sepals only seldom accrescent and forming wings about the fruit; endosperm present in most spp.

5 Flowers with an intrastaminal disk, or without a disk; pollen in monads.

6 Anthers with normal connective, often opening by terminal pores; stem generally with cortical bundles; mucilage-cells and -channels absent in most genera; staminodes present (internal to the functional stamens) or not, when present either distinct or connate to form a tube or disk around the ovary, but not resembling a gynophore; flowers often with a gynobasic style; pantropical .... 1. OCHNACEAE.

6 Anthers with broad, glandular connective separating the pollen-sacs, these dehiscent by longitudinal slits; stem without cortical bundles; mucilage-cells or -channels present; flowers with a prominent, intrastaminal, gynophore-like disk in which the ovary is somewhat embedded; flowers only seldom with a gynobasic style; Madagascar .... 2. SPHAEROSEPALACEAE.

5 Flowers generally with an extrastaminal disk; pollen in tetrads; connective normal; anthers dehiscent by longitudinal slits; mucilage-cells and -channels present; cortical bundles absent; style terminal; Madagascar .... 3. SARCOLAENACEAE.

4 Connective prominently exserted; sepals, or some of them, very often accrescent and wing-like in fruit; endosperm mostly wanting; interruptedly pantropical .... 4. DIPTEROCARPACEAE.

3 Leaves exstipulate.

7 Floral bracts not highly modified, though sometimes elongate and petaloid.
 8 Stamens mostly numerous, seldom only 10, very rarely only 5; ovules (1) 2-many in each locule.
  9 Petals imbricate or convolute; styles distinct, or the style solitary and evidently bilobed or cleft, rarely only the stigmas distinct; sepals imbricate, distinct or basally connate; widespread, chiefly in tropical and subtropical regions.
   10 Anthers opening by longitudinal slits, or rarely by short, terminal, pore-like slits, not inverted; ovules bitegmic; plants with sclerenchymatous idioblasts in the parenchymatous tissues, but not raphides; trees or shrubs, only rarely climbing ....6. THEACEAE.
   10 Anthers inverted, deeply sagittate, opening by seemingly terminal pores (or the pores sometimes eventually elongating into lateral slits); ovules unitegmic; plants with raphides, but not sclerenchymatous idioblasts; trees, shrubs, or often lianas ....7. ACTINIDIACEAE.
  9 Petals valvate; style solitary, only the stigma shallowly lobed; calyx-lobes valvate, or the calyx-tube virtually entire; tropical West Africa .... 8. SCYTOPETALACEAE.
 8 Stamens 4 or 5; ovules 1–4 in each locule, or some locules empty.
  11 Corolla of distinct petals; style solitary, only the stigmas sometimes distinct.
   12 Pollen-sacs opening by a specialized terminal pore; ovules 2 in each locule; plants with neither raphides nor sclerenchymatous idioblasts; sepals without the specialized glands of the next group; se. Asia to Sumatra ....9. PENTAPHYLACACEAE.
   12 Pollen-sacs opening by a longitudinal slit; ovules solitary in each locule, or some locules empty; plants producing raphides; sepals beset toward the middle of the inner surface with tiny, embedded, flask-shaped glands.
    13 Trees or shrubs, but not mangroves; fruit fleshy, with one seed in each of the 4 or 5 locules; seeds with copious endosperm; sepals not petaloid; no sclerenchymatous idioblasts; Malaysian region, and disjunct in n. S. America .... 10. TETRAMERISTACEAE.

13 Mangroves; fruit dry, 1-seeded; seeds without endosperm; sepals petaloid; plants with sclerenchymatous idioblasts in the cortex and pith; Pacific tropical America .... 11. PELLICIERACEAE.

11 Corolla sympetalous; ovary apically shortly 5-lobed, the lobes stigmatic ventrally; ovules 2 in each locule; anthers opening by longitudinal slits; New Caledonia .... 12. ONCOTHECACEAE.

7 Floral bracts, or some of them, highly modified into pitcher-like, saccate, spurred, or hooded structures; plants mostly lianas and epiphytes, producing raphides and sclerenchymatous idioblasts; tropical America 13. MARCGRAVIACEAE.

2. Leaves opposite or verticillate (or merely subverticillate in Paracryphiaceae).

14 Leaves stipulate.

15 Trees, shrubs, or lianas; stamens 15-many; ovules 2 in each locule; fruit fleshy; tropical America ..... 14. QUIINACEAE.

15 Herbs or half-shrubs; stamens as many or twice as many as the petals, fewer than 15; ovules numerous; fruit capsular; cosmopolitan .... 15. ELATINACEAE.

14 Leaves exstipulate (stipulate in *Mahurea*, of the Clusiaceae).

16 Ovary 8–25-locular; no secretory canals or cavities; fruit capsular, opening septicidally from the base, the carpels united distally and spreading away from the summit like the ribs of an umbrella.

17 Ovules 4 in each locule; ovary 8–15-locular; stamens mostly 8, in a single cycle; vessel-segments with scalariform perforations; New Caledonia ....16. PARACRYPHIACEAE.

17 Ovules 2 in each locule; ovary 17–25-locular; stamens numerous; vessel-segments with simple perforations; Seychelles Islands ....17. MEDUSAGYNACEAE.

16. Ovary (1–) 3–5 (–20+)-locular; well developed secretory canals or cavities present in most or all organs; ovules (1) 2-many in each locule; fruit variously dehiscent or indehiscent, but not as in the previous group; cosmopolitan, mainly tropical and North Temperate .. 18. CLUSIACEAE.

1. Family OCHNACEAE A. P. de Candolle 1811 nom. conserv., the Ochna Family

Trees, shrubs, or rarely herbs, most species glabrous, often producing proanthocyanins, but lacking ellagic acid and usually lacking mucilage-sacs and channels (but these present in the cortex and pith of *Strasburgeria* and some other genera); clustered crystals of calcium oxalate commonly present in some of the cells of the parenchymatous tissues; young stem ordinarily with cortical vascular bundles, and sometimes with medullary bundles as well; nodes trilacunar or multilacunar; vessel-segments with simple or occasionally both simple and scalariform perforations; imperforate tracheary elements with simple or bordered pits, sometimes some of them septate; some vasicentric tracheids sometimes present; wood-rays of diverse types, mostly heterocellular (homocellular in *Lophira*), usually only a few of them uniseriate, the others 2–4 (–8)-seriate; wood-parenchyma sparse to abundant, variously distributed. LEAVES evergreen, alternate, mostly simple, rarely pinnately compound (*Rhytidanthera*), usually with numerous parallel lateral veins, but the veins fewer in (e.g.) *Strasburgeria*; epidermis commonly with some mucilage-cells; stomates paracytic or sometimes (as in *Lophira*) anomocytic; petiolar vascular supply usually forming a siphonostele; stipules present. FLOWERS in terminal panicles, or in various other sorts of inflorescences, perfect, mostly regular or nearly so, but in *Lophira* with 2 of the sepals notably accrescent in fruit; sepals (3–) 5 (–12), imbricate, often persisent, similar or less often dissimilar inter se; petals (4) 5 (–10), distinct, convolute or seldom imbricate; stamens distinct, often excentrically placed, sometimes borne on an elongate androgynophore, 5, 10, or more often numerous (sometimes in 3–5 whorls), when numerous commonly associated with 5 trunk bundles and (at least in *Ochna*) the members of each group originating in centripetal sequence; anthers tetrasporangiate and dithecal, opening by terminal pores or less often by longitudinal slits; pollen-grains binucleate or trinucleate, mostly tricolporate; staminodes sometimes present internal to the functional stamens, and sometimes modified into a tube around the ovary or forming a lobed intrastaminal disk; gynoecium of (1) 2–15 carpels, these united at least by their common style, the ovary sometimes partly or wholly partitioned, with an apical style and more or less distinctly axile placentation, or often so deeply lobed that the carpels appear to be distinct except for their common gynobasic style, the receptacle then commonly enlarging in fruit (in the large genus *Ouratea*

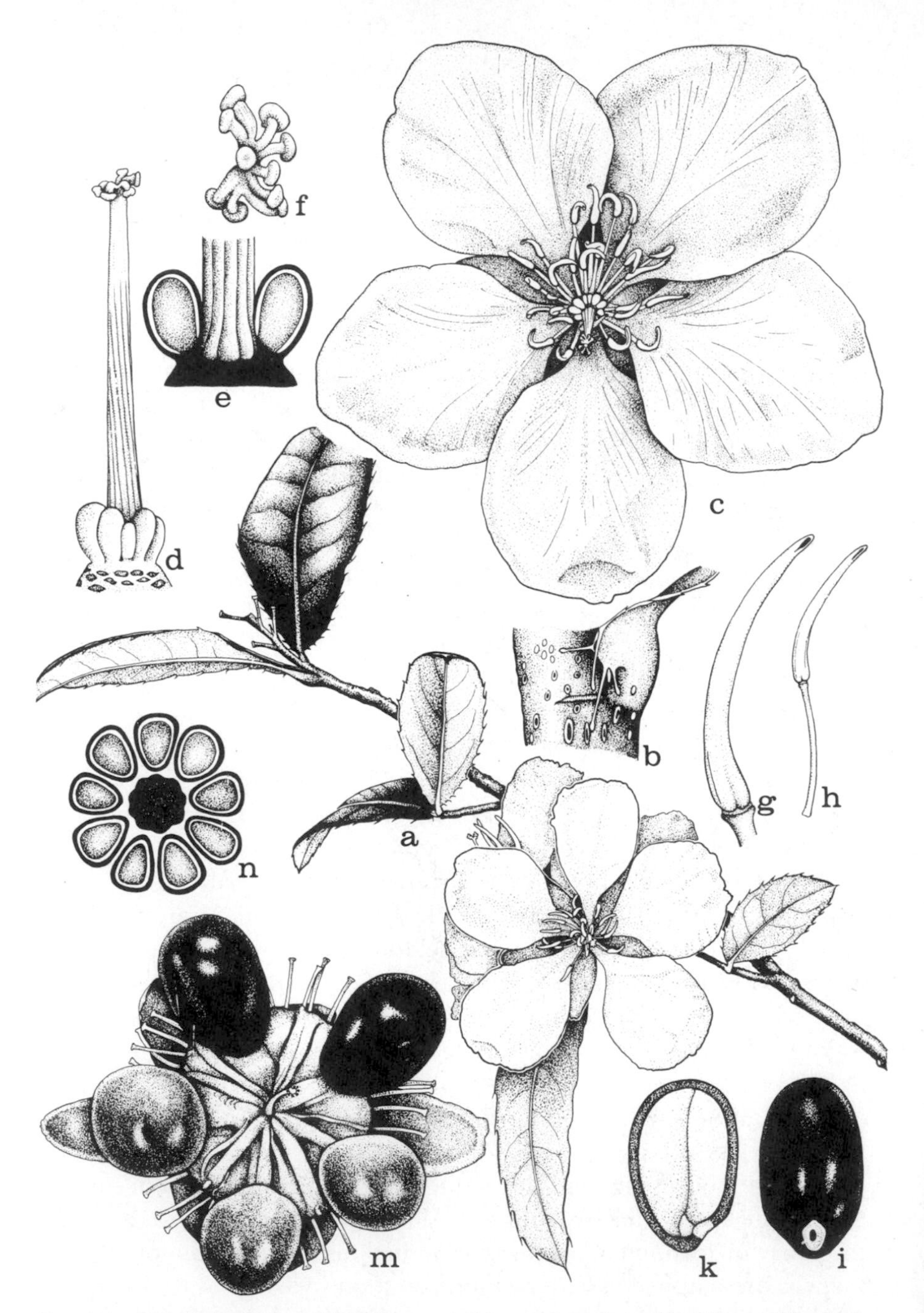

FIG. 4-4 Ochnaceae. *Ochna kirkii* Oliver. a, habit, ×1; b, node and leaf-base, with stipular scar, ×4; c, flower, from above, ×2; d, pistil, ×4; e, schematic representation of gynobasic style and distinct ovary-lobes, ×8; f, stigmas, from above, ×12; g, stamen, ×8; h, anther, ×16; i, k, mericarp, external and in long-section, ×4; m, fruit, ×4; n, schematic cross-section of gynobasic style and distinct ovary-lobes, ×8.

nearly always becoming red and swollen); ovules 1-many in each locule or carpel, anatropous or nearly campylotropous, bitegmic or seldom (*Lophira*) unitegmic, tenuinucellar; endosperm-development nuclear. FRUIT of diverse types, often of separate, 1-seeded drupes, but sometimes a septicidal capsule, or indehiscent and either woody or fleshy; seeds very often winged; embryo with 2 cotyledons; endosperm oily and proteinaceous, or absent. X = 12, 14 (or 7?). (Diegodendraceae, Lophiraceae, Luxemburgiaceae, Strasburgeriaceae, Sauvagesiaceae, Wallaceaceae)

The family Ochnaceae as here broadly defined consists of about 30 genera and 400 species, widespread in tropical regions, especially in Brazil. About half of the species belong to the single large genus *Ouratea*. Although its members are evidently related, the family is only very loosely knit. Many authors now recognize one or more small, segregate families, such as the Strasburgeriaceae (1 sp. of new Caledonia), Diegodendraceae (1 sp. of Madagascar), and Lophiraceae (2 spp. of tropical Africa). *Lophira* suggests the Dipterocarpaceae in its unequally accrescent sepals, but its pollen is that of the Ochnaceae, and the longitudinally dehiscent, latrorse anthers do not have a prolonged connective. *Strasburgeria* is somewhat like *Clusia* in aspect, but it has evident stipules and a less well developed secretory system.

Even when the more commonly recognized segregate families are removed, the Ochnaceae remain diversified. There is an especially notable difference between those members with a terminal style and usually capsular fruit, and those with a gynobasic style and fruit of separate, indehiscent cocci. Members of the former group mostly have endospermous seeds, in contrast to the nonendospermous seeds of the latter group, but the correlation is not perfect. Under such circumstances there is little to be gained by snipping off a few twigs from the divaricately branched phyletic bush, and I therefore prefer the traditional broader definition of the family.

Muller (1969) considers that the pollen morphology of the Ochnaceae supports a relationship of the family to the Actinidiaceae, Marcgraviaceae, and Theaceae.

2. Family SPHAEROSEPALACEAE Van Tieghem 1900, the Rhopalocarpus Family

Trees or arborescent shrubs with simple, unicellular hairs, tanniferous, producing ellagic and gallic acids, and more or less mucilaginous throughout, with schizogenous mucilage-channels in the wood and well developed mucilage-cells in the cortex and mesophyll; solitary or clustered crytals of calcium oxalate present in some of the cells of the parenchymatous tissues; vessel-segments with simple perforations; imperforate tracheary elements with bordered pits; wood-rays few and short, mixed uniseriate and pluriseriate, intermediate between Kribs Heterogeneous type II and Homogeneous type II, very narrow to wide; wood-parenchyma in uniseriate concentric bands; phloem rays wedge-shaped as in the Malvales, and the secondary phloem tangentially stratified into hard and soft layers as in the Malvales. LEAVES deciduous, alternate, simple, entire, pinnately veined or 3-nerved; stomates encyclocytic to anisocytic or anomocytic; petiolar anatomy diverse, the vascular supply sometimes a siphonostele with included medullary bundles; stipules large, caducous. FLOWERS in axillary and terminal subumbelliform cymules, individually subtended by a set of caducous bracts, perfect, regular or nearly so, hypogynous; sepals 2 + 2 (rarely 3 + 3), coriaceous, distinct, caducous, strongly imbricate, the innermost ones the largest; petals (3) 4 (–8), white or yellow, distinct, caducous, unequal, imbricate, densely streaked with short, resinous lines; stamens 25–100 or more, in 2–4 series, inserted below a large, thick, gynophore-like disk; filaments slender, elongate, distinct or more often shortly and irregularly connate at the base into groups, the outer filaments the shorter; anthers tetrasporangiate and dithecal, the two pollen-sacs borne on a broad, glandular connective, opening by longitudinal slits; pollen-grains 3–6-colporate; gynoecium of 2–4 (5) carpels united to form an apically indented ovary with a terminal, geniculate style and entire stigma, the locules as many as the carpels, or (*Dialyceras*) with the carpels essentially distinct except for the common gynobasic style; ovary superior, but partly sunken in the disk; ovules 2–9 in each carpel, basal-ventral, anatropous. FRUIT indehiscent, muricate, often lobed according to the number of fertile carpels, usually with a single (2) large seed in each locule, in *Dialyceras* the mericaps separating at maturity; seeds with rather small, straight, embryo and 2 thin, flat, basally cordate, sometimes ruminate cotyledons, the cotyledons bilobed; endosperm oily. (Rhopalocarpaceae)

The family Sphaerosepalaceae consists of only 2 genera and about 14 species, confined to Madagascar. *Dialyceras* is monotypic. The remaining species belong to *Rhopalocarpus*, of which *Sphaerosepalum* is a well known but junior synonym. At the familial level the name Sphaerosepalaceae (1900) is older than Rhopalocarpaceae (1925).

The Sphaerosepalaceae have variously been included in the Theales, Violales, or Malvales, or in an order Ochnales carved out of the present Theales. There is not so much actual disagreement about the relationships of the family as might at first seem, because these several orders are all closely related. There is a growing trend to ally the Sphaerosepalaceae with the Sarcolaenaceae, and to associate both of these families with the Malvales. I regard the two families as transitional between the Theales and Malvales, and I could accept them in either order. On balance, I find it more useful to retain them in the already diverse order Theales, so as to permit a sharper delimitation of the relatively homogeneous order Malvales. In the Theales, their nearest ally is the Ochnaceae.

3. Family SARCOLAENACEAE Caruel 1881 nom. conserv., the Sarcolaena Family

Trees or shrubs, usually evergreen, often with stellate hairs; mucilage cells present in the cortex, pith, and mesophyll; nodes trilacunar; clustered crystals of calcium oxalate commonly present in some of the cells of the parenchymatous tissues; vessel-segments with simple perforations; imperforate tracheary elements very short; wood-rays heterocellular, exclusively uniseriate, Kribs type II to type III; wood-parenchyma apotracheal, diffuse; secondary phloem tangentially stratified into hard and soft layers, as in the Malvales. LEAVES alternate, simple, entire; stomates surrounded by a rather large number of ordinary epidermal cells; petiole with complex anatomy, often siphonostelic with included medullary bundles; stipules present, usually caducous. FLOWERS in umbel-like or more or less paniculiform cymes, or seldom solitary, perfect, regular or nearly so, hypogynous, usually subtended individually (or in pairs) by an involucel of bractlets, these distinct or often more or less connate to form a cup, the cupulate involucel sometimes interpreted as derived from the pedicel-tip; sepals distinct, imbricate, most often 3, but sometimes 4 or 5, when 5 the outer ones smaller than the 3 inner, when 4 either one or 3 outer ones smaller than the inner; petals 5 (6), large, distinct or very slightly united at the base, convolute in bud; stamens mostly very numerous (only 5–10 in spp. of *Leptolaena*), generally seated internally to an entire or toothed, more or less cupular nectary-disk of probably staminodial origin, or the disk rarely wanting; filaments slender, distinct or weakly connate at the base into 5–10 bundles; anthers tetrasporangiate and dithecal, with normal connective, the pollen-sacs opening by longitudinal slits; pollen-grains tetradinous, 3–6-colpate; gynoecium of (1–) 3–4 (5) carpels united to form a compound, superior ovary with a terminal style and an expanded, usually lobed stigma; ovary with as many locules as carpels; ovules anatropous, (1) 2-many in each locule, on axile or basal-axile or apical-axile placentas, erect to more often pendulous. FRUIT a loculicidal capsule, or indehiscent and with only one or a few seeds, and then often only one carpel maturing; embryo with a slender, straight radicle and 2 expanded, basally cordate, thin or thickened cotyledons; endosperm copious and starchy to seldom scanty or wanting.

The family Sarcolaenaceae consists of 10 genera and about 30 species, limited to Madagascar. Together with the Sphaerosepalaceae, the Sarcolaenaceae are here regarded as lying near to the ancestry of the Malvales. None of the species is well known in cultivation or of much economic value. The largest genera are *Sarcolaena* (10) and *Leptolaena* (12).

4. Family DIPTEROCARPACEAE Blume 1825 nom. conserv., the Meranti Family

Trees or seldom shrubs, generally with fascicled, often glandular hairs or peltate scales, tanniferous, with proanthocyanins and mostly with ellagic acid, and commonly with triterpenes (notably dipterocarpol, apparently unique to this family) and sesquiterpenes, strongly resinous, and in the large subfamily Dipterocarpoideae with characteristic, branching, intercellular resin-canals in the pith, wood, and bark; mucilage canals sometimes present in the cortex, and in the two smaller subfamilies characteristically present in the pith; solitary and clustered crystals of calcium oxalate commonly present in some of the cells of the parenchymatous tissues; nodes 3- to 5-lacunar; cortical bundles present; vessel-segments short, with mostly simple perforations; imperforate tracheary elements with simple or bordered pits; wood rays heterocellular (Kribs type IIA and B) to less often homocellular (Kribs types I and III), mixed uniseriate and pluriseriate, or (Monotoideae and Pakaramoideae) nearly all uniseriate; phloem stratified into hard and soft layers, as in the Malvales, and in the Dipterocarpoideae with broad, wedge-shaped rays; roots, at least in the Dipterocarpoideae, commonly forming ectotrophic mycorhizae. LEAVES alternate, generally evergreen, leathery, simple, entire; stomates with or without clearly differentiated subsidiary cells, sometimes definitely paracytic; petiole commonly geniculate (or sometimes merely sinuate) near the tip, and with variously complex anatomical structure; stipules well developed, commonly serving to protect the terminal bud, deciduous or sometimes persistent. FLOWERS in axillary (or terminal) racemes or panicles or seldom cymes, generally perfect and regular; sepals 5, imbricate, sometimes becoming valvate in fruit, usually connate below into a calyx-tube that is sometimes adnate to the ovary; petals 5, distinct or slightly connate at the base, convolute in bud and spirally twisted, often leathery; androecium typically served by 10 trunk bundles; stamens hypogynous or nearly so (on an androgynophore in Monotoideae), (5) 10-numerous (up to about 100) in 1–3 whorls or somewhat irregularly disposed, most often 15, initiated in centrifugal sequence; filaments distinct, or sometimes connate below, short in the Dipterocarpoideae, longer in the other two subfamilies; anthers basifixed (but basally more or less cordate or sagittate and appearing versatile in the Monotoideae and Pakaraimoideae), tetrasporangiate and dithecal, opening by longitudinal slits, the connective often evidently expanded and generally conspicuously prolonged; pollen-grains binucleate, tricolpate or tricolporate (the absence of an endo-aperture thought to be secondary); gynoecium of 2–4 (5) carpels united to form a compound, plurilocular ovary with axile placentation and a terminal, apically entire or shortly lobed, basally fleshy-thickened style; ovules 2–4 in each locule, anatropous, bitegmic, crassinucellar; endosperm-development nuclear. FRUIT dry, usually with a woody pericarp, indehiscent or only very tardily dehiscent, 1-seeded; calyx persistent in fruit, the inner 2, 3, or all 5 segments very often accrescent and wing-like, serving rather ineffectively in wind-distribution; seeds without a dormant state, commonly without endosperm; cotyledons often folded and embracing the radicle. X = 6, 7, 10, 11.

The family Dipterocarpaceae consists of some 16 genera and nearly 600 species, all tropical, especially abundant in the rain forest of Malaysia. The largest genera are *Shorea* (150) and *Dipterocarpus* (75), both of which include many large trees with valuable wood. The family is generally (as here) regarded as allied to the Ochnaceae, and also to some families of the Malvales, notably the Tiliaceae and Elaeocarpaceae. Meranti is the Malay name of *Shorea*.

There are 3 well marked and geographically segregated subfamilies of very unequal size, as follows:

1 Anthers basifixed; stem with characteristic branching, intercellular resin-canals in the pith, wood, and bark; pith

without mucilage-canals; pollen tricolpate, the exine convoluted, without clear columellae; often 2 or more sepals conspicuously accrescent and wing-like; larger wood-rays pluriseriate, mostly 4–8 cells wide; ovary (2) 3-locular, each locule with 2 ovules; 13 genera and 550 species, restricted to tropical Asia and Malesia .... DIPTEROCARPOIDEAE.

1 Anthers basi-versatile; stem without resin-canals, but with mucilage-canals in the pith; pollen tricolporate, the exine tectate-columellate; sepals equally accrescent, papery, only seldom wing-like; larger wood-rays biseriate.

2 Petals longer than the sepals; ovary 2 (4)-locular, each locule with 2 ovules; flowers with an androgynophore; wood-rays homocellular or nearly so, predominantly uniseriate; 2 genera and 20 species, restricted to tropical Africa and Madagascar .... MONOTOIDEAE.

2 Petals shorter than the sepals; ovary 4–5-locular, each locule with (2) 4 ovules; flowers without an androgynophore; wood-rays heterocellular, predominantly biseriate; a single species, *Pakaraimaea dipterocarpacea* Maguire & Ashton, restricted to the Guayana highlands in tropical South America .... PAKARAIMOIDEAE.

The distribution of the Dipterocarpaceae suggests a greater age for the family than is documented by the known fossil record. The existence of 3 well marked subfamilies on 3 separate continents, together with the evidently relictual nature of the South American subfamily, calls for an origin before the pieces of Gondwanaland had drifted very far apart. Anything later than the middle of the Upper Cretaceous (or perhaps even earlier) would seem to present problems in dispersal, yet I know of no verified Cretaceous fossils of Dipterocarpaceae. According to Muller (personal communication), the family shows a differentiation of types, both pollen and macrofossil, in the Oligocene. Pollen considered to be dipterocarpaceous suddenly becomes abundant in Miocene deposits of Borneo (on the Laurasian plate). This apparent proliferation in the Miocene may mark the invasion of Laurasia after the Gondwanaland plate had drifted within range.

5. Family CARYOCARACEAE Szyszylowicz in Engler & Prantl 1893 nom. conserv., the Souari Family

Trees, or rarely shrubs or undershrubs, glabrous or with unicellular or uniseriate hairs, tanniferous and with branched, sclerenchymatous idioblasts in the mesophyll, petioles, and often also the pith; nodes multilacunar; solitary or clustered crystals of calcium oxalate present in some of the cells of the parenchymatous tissues; silica grains present in parenchyma of the wood of *Anthodiscus* but not *Caryocar*; vessel-segments mostly elongate, with slanting ends and simple perforations; imperforate tracheary elements with small, simple pits, often septate; wood-rays heterocellular, Kribs type IIA, mostly uniseriate and biseriate, or some of them up to 4 cells wide; wood-parenchyma mostly apotracheal and diffuse (*Caryocar*) or paratracheal (*Anthodiscus*); phloem not stratified. LEAVES evergreen, opposite (*Caryocar*) or alternate (*Anthodiscus*), digitately trifoliolate, at least some species poisonous to fish; leaflets toothed or seldom subentire; stomates anomocytic or seldom anisocytic or paracytic; petiole with a complex, essentially siphonostelic vascular system; stipules 2 or 4 and caducous, or none. FLOWERS in terminal, bractless, sometimes condensed racemes, large, perfect, regular, hypogynous or nearly so, in *Caryocar* pollinated by bats; sepals 5 (6) imbricate, connate at the base (*Caryocar*), or the calyx reduced and subtruncately lobed (*Anthodiscus*); petals 5 (6), imbricate, distinct or connate at the base, in *Anthodiscus* connate above to form a calyptra; stamens very numerous, shortly connate at the base into a ring or into 5 bundles alternate with the petals, the inner filaments sometimes without anthers; anthers dithecal, opening by longitudinal slits; pollen-grains (2) 3 (4)-syncolporate; gynoecium of 4–20 carpels united to form a compound ovary with distinct styles and as many locules as carpels, each carpel with a single axile, anatropous to orthotropous, bitegmic ovule. FRUIT a drupe, the flesh in some species poisonous, in others edible; stone eventually separating into 1-seeded pyrenes; seeds reniform, with thin or no endosperm; embryo edible, with much-enlarged, oily and proteinaceous, spirally twisted hypocotyl and 2 small inflexed cotyledons; germination hypogeal.

The family Caryocaraceae consists of 2 genera and about 23 species, confined to tropical America, best developed in the Amazon basin. *Caryocar* has 15 known species, and *Anthodiscus* 8. *Caryocar nuciferum* L., souari-nut, is widely cultivated in the tropics for its edible seeds.

Recent authors are agreed that the Caryocaraceae are closely allied to the Theaceae. The pollen of *Caryocar* is said to be very similar to that of *Kielmeyera*, a genus transitional between the Clusiaceae and Theaceae.

Pollen attributed to the Caryocaraceae occurs in Eocene and more recent deposits.

6. Family THEACEAE D. Don 1825 nom. conserv., the Tea Family

Trees or shrubs, rarely climbing shrubs, commonly glabrous, the hairs when present mostly simple and unicellular; plants tanniferous, commonly producing both proanthocyanins and ellagic acid, often accumulating saponins, aluminum, and fluorides, and in the Bonnetioideae often producing xanthones similar to those of the Clusiaceae, commonly with sclerenchymatous idioblasts in the parenchymatous tissues of the shoot, but only rarely (some Bonnetioideae) with resin-ducts; solitary or clustered crystals of calcium oxalate commonly present in some of the cells of the parenchymatous tissues, but not in the form of raphides;

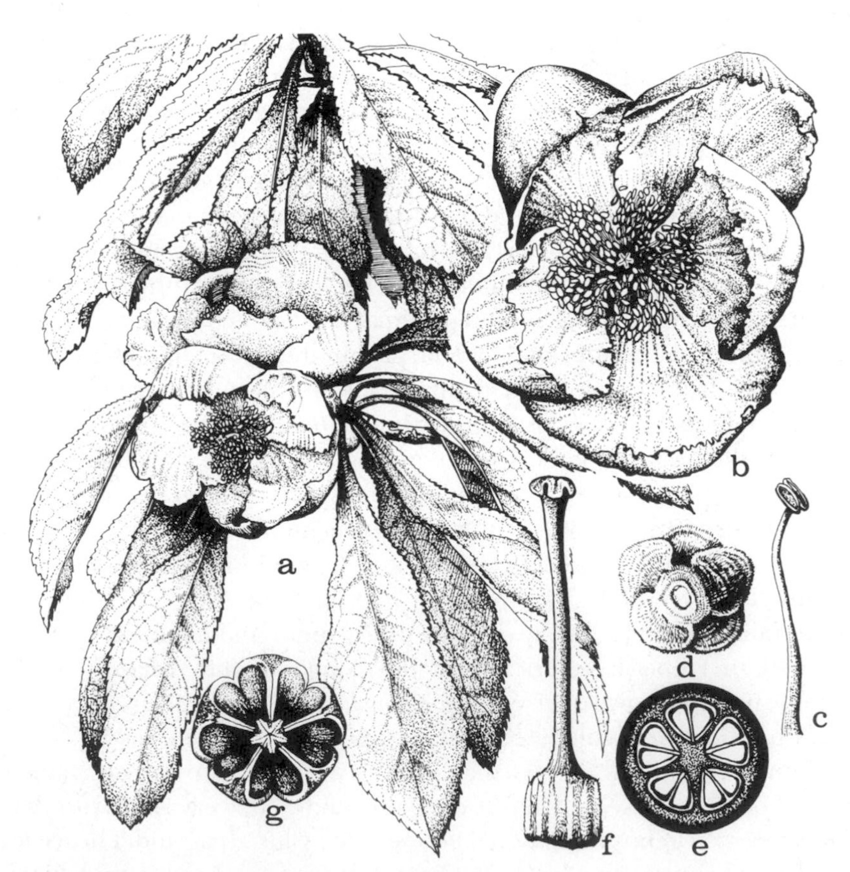

FIG. 4-5 Theaceae. *Franklinia alatamaha* Marshall. a, habit, $\times\frac{1}{2}$; b, flower, $\times 1$; c, stamen, $\times 2$; d, calyx, from beneath, $\times 1$; e, schematic cross-section of ovary, $\times 5$; f, pistil, $\times 2$; g, fruit, after dehiscence, from above, $\times 1$.

silica grains sometimes present in the xylem, especially in the Bonnetioideae; nodes unilacunar; vessel-segments with slanting end-walls and scalariform perforations, or sometimes (many Bonnetioideae, *Asteropeia*) with more or less straight ends and simple perforations; imperforate tracheary elements generally with bordered pits (except many Bonnetioideae); wood-rays heterocellular, mostly Kribs type I, (II in most Bonnetioideae) mixed uniseriate and pluriseriate, the latter 2–4 (–8) cells wide, with long or short ends, or seldom (*Asteropeia*) the rays all uniseriate and homocellular; wood-parenchyma commonly apotracheal and diffuse, varying to sometimes paratracheal. LEAVES evergreen or seldom deciduous, alternate or rarely opposite, simple, entire or toothed; epidermis often containing some mucilage-cells; stomates variously anomocytic, encyclocytic, or paracytic; petiolar vascular supply commonly trough-shaped, or sometimes a siphonostele; stipules wanting. FLOWERS commonly large and showy, mostly axillary and solitary, but sometimes (especially in *Asteropeia* and Bonnetioideae) in terminal panicles or racemes, perfect or seldom unisexual (as in the dioecious genus *Eurya*), hypogynous to somewhat perigynous or seldom partly (rarely almost wholly) epigynous; calyx often subtended by 2-several bracteoles, these rarely passing into the sepals; sepals (4) 5 (–7), imbricate, distinct or more often connate at the base, persistent or sometimes deciduous, sometimes (*Asteropeia*) accrescent and wing-like in fruit; petals (4) 5 (-many), distinct or basally connate, imbricate or seldom (Bonnetioideae) convolute, sometimes passing into the sepals; stamens usually more or less numerous and developing centrifugally, distinct or basally connate into a ring, or often connate into 5 bundles opposite (and frequently adnate to) the petals, seldom fewer, only 10, or rarely (spp. of *Eurya*) only 5; androecium commonly served by a limited number of trunk bundles; base of the filaments often nectariferous, or the ovary nectariferous at the base; anthers tetrasporangiate and dithecal, opening by longitudinal slits, or rarely (spp. of *Eurya*) by short, terminal, pore-like slits; connective sometimes shortly prolonged; pollen-grains binucleate, tricolporate; gynoecium of (2) 3–5 (–10) carpels united to form a compound ovary with as many locules as carpels (in *Piquetia* the carpels united only at the base) and with axile placentation; styles distinct, or connate below into a single, distally lobed or cleft style, or seldom only the stigmas distinct; ovules commonly more or less numerous in each locule, varying to only 2 or rarely only 1, anatropous or somewhat campylotropous, bitegmic, tenuinucellar; endosperm-development nuclear. FRUIT a septicidal capsule (Bonnetioideae) or more often a loculicidal capsule, in either case very often with a persistent central column after dehiscence, or indehiscent and either fleshy or dry; seeds 1-many in each locule, commonly with scanty or no endosperm and large, straight or curved, dicotyledonous embryo, or seldom (as in *Visnea*) with copious, oily and proteinaceous endosperm and small embryo, only rarely (*Anneslea*) arillate. X = 15, 18, 21, 25. (Asteropeiaceae, Bonnetiaceae, Camelliaceae, Sladeniaceae, Ternstroemiaceae.)

The Theaceae as here defined consist of about 40 genera and 600 species, widespread in tropical and subtropical regions, with only a few species in warm-temperate climates. The largest genera are *Ternstroemia* (130), *Camellia* (including *Thea*, 80), *Eurya* (70), and *Laplacea* (40). Tea is made from the leaves of *Camellia sinensis* (L.) Kuntze. Other species of *Camellia*, notably *C. japonica* L., are often cultivated for ornament. The Franklin tree, *Franklinia alatamaha* Marsh., originally native to Georgia (U.S.A.), is now known only in cultivation.

The Theaceae are here considered to consist of 4 subfamilies of very unequal size, the Theoideae, Ternstroemioideae, Bonnetioideae, and Asteropeioideae. The great bulk of the genera and species belong to the Theoideae and Ternstroemioideae. Although these latter two groups encompass a considerable amount of diversity, it is generally agreed that all of their members belong together in a single family. Authors who prefer small, sharply limited families elevate the Bonnetioideae and Asteropeioideae to familial status.

The Bonnetioideae are transitional toward the Clusiaceae. The group consists of 11 genera (13 if *Kielmeyera* and *Neotatea* are included) and about 100 species. Except for the Malaysian genus *Ploiarium*, all of the Bonnetioideae are neotropical. Baretta-Kuipers (1976) considers that in wood anatomy the Bonnetioideae (as Bonnetiaceae) are intermediate between the Theaceae and Clusiaceae. *Kielmeyera* and *Neotatea* have the characteristic resin-ducts of the Clusiaceae and are anatomically at home in that group, but they have alternate leaves as in the Bonnetioideae and other Theaceae.

The Asteropeioideae, consisting of a single genus with 5 species restricted to Madagascar, are an isolated end-line, not closely related to any other family.

The 4 subfamilies here recognized may be characterized as follows:

1 Petals convolute; capsule septicidal; vessel-segments mostly with simple perforations; wood-rays mixed uniseriate and

pluriseriate; anthers small, versatile; plants often accumulating xanthones .... BONNETIOIDEAE.
1 Petals imbricate; capsule loculicidal, or the fruit indehiscent; plants not known to produce xanthones.
2 Sepals strongly accrescent and wing-like in fruit; vessel-segments with simple perforations; wood-rays all uniseriate; fruit dry and indehiscent; anthers small and versatile ....ASTEROPEIOIDEAE.
2 Sepals not strongly accrescent; vessel-segments with scalariform perforations; wood-rays mixed uniseriate and pluriseriate.
3 Fruits capsular, or rarely drupaceous; anthers mostly small and versatile; embryo mostly straight .... THEOIDEAE.
3 Fruit baccate, or dry and indehiscent; anthers mostly elongate and basifixed; embryo more or less strongly curved, rarely almost straight .....TERNSTROEMIOIDEAE.

Four additional small groups that have sometimes been included in the Theaceae are here considered to form distinct families. These are the Actinidiaceae, Pellicieraceae, Pentaphylacaceae, and Tetrameristaceae. These groups are generally considered to be allied to the Theaceae, but in the interests of conceptual harmony it is here thought best to treat them as distinct.

It is generally agreed that the Theaceae are closely related to the Dilleniaceae but more advanced, differing especially in the syncarpous gynoecium and in the usually larger embryo and scanty endosperm of the seeds. Furthermore, the seeds of Theaceae generally lack an aril, in contrast to the arillate Dilleniaceae, and the Theaceae mostly lack raphides, whereas the Dilleniaceae commonly have raphides.

Pollen attributed to the Theaceae is known from Eocene and more recent deposits. Recognizable macrofossils of the Theaceae are not well represented in the fossil record. Some fossil leaves thought to be allied to *Ternstroemia* occur in Eocene deposits in North Dakota.

7. Family ACTINIDIACEAE Hutchinson 1926 nom. conserv., the Chinese Gooseberry Family

Trees, shrubs, or woody vines, tanniferous, producing proanthocyanins and iridoid compounds, and with raphides in elongate sacs or tubes in the parenchymatous tissues, but without prominent sclerenchymatous

FIG. 4-6 Actinidiaceae. *Actinidia chinensis* Planchon. a, b, habit, ×½ and ×¼; c, flower, from above, ×2; d, stamen, ×6; e, anther, ×6; f, pistil, from above, ×6; g, diagrammatic cross-section of ovary, ×6; h, diagrammatic long-section of ovary, ×9; i, fruit, ×¼.

idioblasts; nodes unilacunar or seldom trilacunar; vessel-segments with more or less oblique ends and scalariform perforations that have numerous cross-bars, or with mixed simple and scalariform perforations; imperforate tracheary elements elongate, with evidently bordered pits; wood-rays heterocellular, Kribs type I, mixed uniseriate and pluriseriate, the latter 2–6 cells wide, with elongate ends; wood-parenchyma apotracheal, often diffuse. LEAVES alternate, simple, toothed or entire; stomates anomocytic, or sometimes nearly paracytic; stipules wanting. FLOWERS in axillary or lateral, basically cymose inflorescences (these sometimes reduced to 1 or a few flowers), often on old wood, perfect or unisexual, hypogynous, often subtended by small bracteoles; sepals and petals most commonly 5 each, less often 4, seldom fewer or more numerous (up to 7); sepals generally imbricate, distinct or basally connate; petals imbricate, distinct or often shortly connate at the base; stamens mostly numerous and developing centrifugally, or sometimes as few as 10, frequently in 5 antepetalous clusters, that may be adnate to the base of the petals, the clusters at least sometimes served by trunk bundles; anthers dithecal, versatile, inverted, opening by morphologically basal but seemingly apical pores or short slits, the slits sometimes eventually elongating to run most of the length of the anther; pollen-grains tricolporate, with a continuous, complete tectum, binucleate, in monads or sometimes in tetrads; gynoecium of 3-many (up to 30 or more) carpels united to form a compound ovary with as many locules as carpels (or the partitions sometimes imperfect, hardly meeting in the center); styles distinct, or connate below or even throughout, often emerging from an apical depression in the ovary; ovules 10-many in each locule on axile or nearly axile placentas, anatropous, unitegmic, tenuinucellar, with an integumentary tapetum; endosperm-development cellular. FRUIT a berry or a more or less loculicidal capsule; seeds small, with copious, oily and proteinaceous endosperm and well developed, linear, axile, straight or slightly curved embryo with 2 short cotyledons. X = 15 (*Saurauia*). (Saurauiaceae)

The family Actinidiaceae consists of 3 genera and about 300 species, widespread in tropical and subtropical regions and in the mountains of Asia. The vast majority of the species belong to the genus *Saurauia* (including *Trematanthera*). *Actinidia* has about 35 species, and *Clematoclethra* only about a dozen. The Chinese gooseberry, or kiwi, *Actinidia chinensis* Planch., is becoming familiar as a cultivated plant in Europe and the United States. It grows vigorously as far north as New York, but produces fruit only in a milder climate.

The Actinidiaceae are widely regarded as being related to the Dilleniaceae, Theaceae, and Ericaceae, and they have by different authors been associated in one way or another with each of these groups. I consider the family to be an early and distinct branch of the Theales, not far removed from the ancestral order Dilleniales, and near to the ancestry of the Ericales.

8. Family SCYTOPETALACEAE Engler in Engler & Prantl 1897 nom. conserv., the Scytopetalum Family

Trees, shrubs, or lianas, glabrous or provided with simple, unicellular hairs, without mucilage-cells or -cavities; cortex containing scattered cristarque cells (crystal-bearing cells with U-shaped wall-thickenings); most of the vessel-segments with simple perforations, but commonly some of them with scalariform perforation-plates that have up to a dozen cross-bars; imperforate tracheary elements with simple pits; wood-rays heterocellular, Kribs type IIB, mixed uniseriate and pluriseriate, the latter up to 6 cells wide, with short ends; wood-parenchyma apotracheal; phloem of young stems tangentially stratified into hard and soft layers, but without wedge-shaped rays. LEAVES alternate, sessile or short petiolate, simple, leathery, pinnately veined; stomates anisocytic; sclerenchymatous idioblasts, connecting with the sclerenchyma of the veins, often present in the mesophyll; petiole with a large median vascular strand and 2 smaller lateral ones; stipules none. FLOWERS in terminal panicles or axillary racemes, or in clusters in old wood, perfect, regular, hypogynous; calyx shortly cupular, with an entire or valvately toothed margin; petals 3–10 (–16), distinct or often connate toward the base, linear, valvate, sometimes forming a deciduous calyptra, or rarely wanting; stamens (10–) numerous, in several (typically 3–6) series around the margin of a disk or on the disk, originating centrifugally, individually distinct or connate only at the base, but tending to be irregularly clustered; anthers tetrasporangiate and dithecal, opening by longitudinal slits or obliquely terminal pores; pollen-grains binucleate, tricolporate or triporate, smooth; gynoecium of 3–8 carpels united to form a compound ovary with a bent style and small stigma; ovary with as many locules as carpels (the partitions sometimes incomplete toward the summit), and with 2–8 apical-axile ovules in each locule; ovules pendulous, anatropous, bitegmic, tenuinucellar. FRUIT a loculicidal, often woody capsule or a single-seeded drupe; seeds with abundant, often ruminate endosperm, embryo embedded in the endosperm, with linear hypocotyl and large, thin, foliaceous cotyledons. n = 11, 18. (Rhaptopetalaceae)

The family Scytopetalaceae consists of 5 genera and about 20 species, native to tropical West Africa. None of the species is of any great economic importance.

The Scytopetalaceae have usually been included in the Malvales, where they have no close relatives and form a discordant element. Aside from the general characters of the subclass, they have only two features that suggest the Malvales: stratified phloem and a valvate calyx. Stratified phloem is also found in several families of Theales, however, and even in the Lecythidales, Bixaceae (Violales), and some of the Magnoliales. Furthermore, even the calyx of the Scytopetalaceae would be unusual in the Malvales, since it has a well developed tube and short or virtually no lobes, instead of having the sepals distinct nearly or quite to the base. Characters of the Scytopetalaceae that would be anomalous in the Malvales include the bitegmic, tenuinucellar ovules (a combination common in the Theales and some other orders of Dilleniidae, but unknown in the Malvales), the presence of cristarque cells in the cortex (similar to Ochnaceae), the scalariform vessels, simple, unicellular (not stellate or peltate) hairs, and the absence of mucilage, absence of wedge-shaped phloem rays, and absence of the characteristic sepalar nectaries of the Malvales. The cyclopropenoid fatty acids of the Malvales may or may not turn out to be present in the Scytopetalaceae; so far as I know, no one has looked for them. It might be pointed out, as a caution, that the Elaeocarpaceae also do not have the full syndrome of Malvalean characters, but this basal family in the order does not differ from typical members in so many features, and its close relationship to the Tiliaceae has been plain for all to see. In contrast to the misfit in the Malvales, the Scytopetalaceae fit comfortably into the more loosely defined and variable order Theales. In this respect I agree with Thorne (1968, 1976, in initial citations).

9. Family PENTAPHYLACACEAE Engler & Prantl 1897 nom. conserv., the Pentaphylax Family

Evergreen shrubs or small trees, accumulating aluminum, and with mucilage-cells in the cortex; single and clustered crystals of calcium oxalate present in some of the cells of the parenchymatous tissues, but not in the form of raphides; vessel-segments elongate, with scalariform perforation-plates that commonly have ca 15 (or up to 25 or more) cross-bars; imperforate tracheary elements much elongate, with evidently bordered pits; wood-rays heterocellular, mixed uniseriate and pluriseriate, the latter up to 5 cells wide, commonly with elongate ends; wood-parenchyma diffuse. LEAVES alternate, simple, entire, leathery, pinnately veined; epidermis mucilaginous; stomates paracytic; stipules wanting. FLOWERS short-pedicellate in axillary and terminal racemes that tend to have a sterile, leafy tip, small, perfect, hypogynous, pentamerous throughout, provided with a pair of persistent bracteoles appressed to the calyx; sepals 5, distinct, persistent, imbricate; petals 5, distinct, imbricate; stamens 5, distinct, alternate with the petals and equaling or shorter than them; filaments thickened and expanded, especially toward the middle, but tapering to a more slender tip; anthers dithecal, the pollen-sacs distinct, individually basifixed on the summit of the filament, short, each opening by a terminal pore formed by a small, uplifting valve; pollen-grains tricolporate; gynoecium of 5 carpels united to form a compound, 5-locular ovary tipped by a short, stout, persistent style with a shortly 5-lobed stigma; ovules 2 in each locule, pendulous on axile placentas, anatropous to somewhat campylotropous, bitegmic, crassinucellar. FRUIT a partly or fully dehiscent, rather woody capsule with a persistent central axis (as in Theaceae, subfamily Theoideae); seeds more or less winged, with scanty endosperm and horseshoe-shaped, dicotyledonous embryo.

The family Pentaphylacaceae consists of the single genus *Pentaphylax*, occurring from southern China to the Malay peninsula and Sumatra. Several species have been described, but van Steenis (Fl. Malesiana ser. 1, 5 (2): 121. 1956) regards the genus as consisting of a single species, *P. euryoides* Gardn. & Champ.

10. Family TETRAMERISTACEAE Hutchinson 1959, the Punah Family

Trees or shrubs with trilacunar nodes, producing raphides in some special cells of the parenchymatous tissues, and crystal sand in some others; vessel-segments with simple perforations, or sometimes some of them with scalariform perforations; imperforate tracheary elements with simple or bordered pits; wood rays Kribs type I, heterocellular, mixed uniseriate and pluriseriate or (*Pentamerista*) all pluriseriate, the latter up to 5 cells wide; wood-parenchyma apotracheal, diffuse to diffuse-in-aggregates. LEAVES alternate, simple, entire, pinnately veined, slightly decurrent at the base; stomates mainly or wholly anomocytic;

stipules wanting. FLOWERS in axillary, pedunculate, umbelliform or compactly corymbiform racemes, perfect, hypogynous, subtended by a pair of persistent or deciduous bracteoles, tetramerous or pentamerous throughout; sepals 4 or 5, distinct, imbricate, persistent or deciduous, with rather numerous scattered, small, glandular pits near the middle of the inner surface; petals 4 or 5, distinct, imbricate, not much if at all longer than the sepals; stamens 4 or 5, alternate with the petals; filaments shortly connate at the base; anthers tetrasporangiate and dithecal, opening by longitudinal slits; pollen-grains tricolporate; gynoecium of 4 or 5 carpels, these fully connate to form a 4- or 5-locular ovary capped by a terminal style with a punctate or minutely lobed stigma; ovules solitary in each locule, axile-basal, anatropous, bitegmic. FRUIT a 4- or 5-seeded berry; seeds relatively large, with copious endosperm surrounding a straight, basal embryo with the hypocotyl much longer than the 2 cotyledons.

The family Tetrameristaceae consists of two monotypic genera, *Tetramerista* and *Pentamerista*. The former occurs in the Malaysian region, and the latter in the Guayana highlands in southern Venezuela. *Tetramerista* has been known for more than a century, but *Pentamerista* was described only in 1972. *Tetramerista* has variously been included in the Ochnaceae, Marcgraviaceae, or Theaceae, or considered to constitute a monotypic family. The closest relationship of the Tetrameristaceae (including *Pentamerista*) is probably with the taxonomically isolated, monotypic family Pellicieraceae, as was noted by Maguire when he described *Pentamerista*. It is generally agreed that all of these taxa belong to the Theales.

11. Family PELLICIERACEAE Beauvisage 1920, the Panama Mangrove Family

Glabrous mangrove trees with raphides in special cells of the parenchymatous tissues, and with branched, sclerenchymatous idioblasts in the cortex and pith; vessel-segments with oblique, simple perforations; imperforate tracheary elements with inconspicuously bordered pits. LEAVES evergreen, leathery, alternate, simple, glandular-denticulate, sessile and shortly decurrent; mesophyll with elongate, sinuous fibers parallel to the epidermis; stomates encyclocytic; petiole with an almost cylindric vascular strand; stipules wanting. FLOWERS large, solitary, axillary, perfect, hypogynous, regular, closely subtended by 2 elongate, crimson to rose or coral-red, eventually deciduous petaloid bracts somewhat surpassing the petals; sepals 5, distinct, unequal, imbricate, petaloid, externally crimson or rosy, finely glandular-pitted near the middle of the inner surface; petals 5, distinct, imbricate, pink to white, elongate, caducous; stamens 5, alternate with the petals; filaments distinct, appressed to alternate grooves of the 10-grooved ovary; anthers elongate, appressed to the style, dithecal, opening subextrorsely by longitudinal slits, with a distinctly prolonged connective; pollen-grains globose, tricolporate, tectate-perforate and columellate; gynoecium of 2 carpels fully united to form a compound ovary capped by a thickened, smooth, elongate, gradually tapering style and punctiform stigma; ovary with 2 locules, the partition sometimes not reaching the summit of the cavity, each locule with a single campylotropous ovule on the axile placenta, or one locule ovuliferous and the other empty, or the second locule even wholly suppressed. FRUIT dry and leathery, indehiscent, ovoid-turbinate with a prominent beak, multi-ridged and -grooved, with a single locule containing a single seed; embryo with a well developed radicle pointed into the beak of the fruit, 2 broad, fleshy cotyledons, and an elongate, hooked plumule; endosperm none.

The family Pellicieraceae consists of a single genus and species, *Pelliciera rhizophora* Planchon & Triana, native to the shores of the Pacific Ocean from Costa Rica to Colombia. *Pelliciera* is generally considered to be related to the Theaceae, and is often included in that family. It is distinctive in so many ways, however, that it is more conveniently considered to form its own family. The curious small glandular pits on the inner surface of the sepals are comparable to those of the related small family Tetrameristaceae.

Pollen considered to represent *Pelliciera* occurs in Eocene and more recent deposits in several Caribbean localities, indicating that the genus was formerly widespread in the region.

12. Family ONCOTHECACEAE Airy Shaw 1965, the Oncotheca Family

Small, glabrous, evergreen tree with pentalacunar nodes, tanniferous and with clustered crystals of calcium oxalate in some of the cells of the parenchymatous tissues; vessel-segments with oblique, scalariform perforation-plates that have 10–50 cross-bars; imperforate tracheary elements with bordered pits, some with much thicker walls and narrower lumen than others; wood-rays heterocellular, Kribs type II, mixed uniseriate and pluriseriate, the pluriseriate ones 2–3 (–6) cells wide; wood-parenchyma rather scanty (fide Carpenter & Dickison) or abundant (fide Shilkina), diffuse to diffuse-in-aggregates with a tendency to form bands, and also scanty-paratracheal. LEAVES alternate, crowded toward the branch-tips, simple, pinnately but obscurely veined, leathery, minutely glandular-denticulate at least near the tip, tapering to a short petiole or petiolar base; stomates of an unusual type, nearly anomocytic; petiole proximally with 5 vascular traces in an arc, these distally confluent into a trough; stipules wanting. FLOWERS virtually sessile in terminal panicles or compound spikes, small, greenish or chloroleucous, perfect, regular, hypogynous; sepals 5, imbricate, persistent in fruit, very shortly connate and thickened at the base; corolla sympetalous, shortly campanulate, 5-lobed, the lobes imbricate; stamens 5, attached to the tube of the corolla and alternate with its lobes; filaments short; anthers basifixed, extrorse, apparently bisporangiate, dithecal, opening by longitudinal slits; anther-connectives prolonged and abruptly inflexed, forming a roof over the gynoecium; pollen-grains tricolporate; gynoecium of 5 carpels united to form a compound, superior, 5-locular, 5-grooved and -fluted, apically shortly 5-lobed ovary, each lobe ventrally stigmatic and imperfectly sealed; ovules 2 in each locule, apical-axile, pendulous on a long funiculus, epitropous, anatropous, unitegmic. FRUIT an oblate-compressed drupe with thin flesh and a very thick-walled, 5-locular stone; seeds solitary in each locule (or some locules empty); embryo centrally embedded in the copious, soft endosperm, straight, cylindric, with a long hypocotyl and 2 very short cotyledons.

The family Oncothecaceae consists of the single genus and species *Oncotheca balansae* Baillon, native to New Caledonia. Although still imperfectly known, the species is so distinctive that it can hardly be squeezed into any existing family. Its affinities are pretty clearly with the Theales and Ebenales. It is anomalous in the Ebenales in having a single set of stamens *alternate* with the corolla-lobes, and anomalous in the Theales in being sympetalous. The prolonged connective of the anthers finds some parallel in members of both the Theaceae and the Ebenaceae. Shilkina (1977) finds the wood anatomy to be more like the Theaceae than that of any other family. I here follow Carpenter & Dickison (1976) and Shilkina (1977) in referring *Oncotheca* to the Theales. *O. macrocarpa* (1981), is not here provided for.

13. Family MARCGRAVIACEAE Choisy in A. P. de Candolle 1824 nom. conserv., the Shingle Plant Family

Mostly lianas or epiphytic shrubs with clinging aerial roots, seldom erect shrubs or small trees, glabrous throughout, producing proanthocyanins but not ellagic acid, and at least sometimes accumulating inulin, but without starch; raphides and sclerenchymatous idioblasts present in the parenchymatous tissues; nodes unilacunar; vessel-segments mostly with simple perforations or with perforations having a few cross-bars, varying to sometimes with numerous cross-bars (*Norantea*); imperforate tracheary elements septate, with small, bordered pits; wood-rays heterocellular, mostly Kribs type I, mixed uniseriate and pluriseriate, the latter up to 10 cells wide or even wider, with elongate ends; wood-parenchyma variously paratracheal and vasicentric to apotracheal and diffuse. LEAVES alternate, simple, entire, leathery, often dimorphic, those on juvenile, rooting vegetative shoots sessile and oriented in a single plane, those on flowering, nonrooting shoots petiolate and spirally arranged; stomates of an unusual type called staurocytic; petiole with a trough-shaped to more often siphonestelic vascular supply; stipules wanting. FLOWERS in terminal, often pendulous racemes, spikes, or umbels, perfect, regular, hypogynous, often pollinated by hummingbirds, but sometimes self-pollinated and even cleistogamous; some of the bracts of the inflorescence (usually associated with sterile flowers) strongly modified into pitcher-like, saccate, spurred or hooded, hollow, nectariferous structures; sepals 4 or 5, distinct or basally connate, imbricate; petals 4 or 5, distinct or more or less connate at the base, and often (*Marcgravia*) distally connate to form a deciduous calyptra; stamens 3–40, the filaments distinct or basally connate, sometimes adnate to the base of the petals; anthers tetrasporangiate and dithecal, opening by longitudinal slits; pollen-grains binucleate, tricolporate; gynoecium of 2–8 carpels united to form a compound ovary topped by a simple or merely lobed, sessile stigma; ovary at first unilocular, but becoming plurilocular by intrusion and fusion of the placental partitions, so that the numerous ovules are eventually borne on axile placentas; ovules anatropous, bitegmic, tenuinucellar, with a poorly developed integumentary tapetum; endosperm-development cellular, and at least in *Marcgravia* with a micropylar haustorium. FRUIT thick and rather fleshy, indehiscent, or partly dehiscent near the base; seeds numerous, small, with very scanty or no endosperm and straight or slightly curved embryo that has a large hypocotyl and 2 small to large cotyledons.

The family Marcgraviaceae consists of 5 genera and more than a hundred species, native to tropical America. *Marcgravia* has about 50 species, *Norantea* about 35, *Souroubea* about 20, *Ruyschia* 6, and *Caracasia* only 2. None of the species is economically important or well known in cultivation.

The Marcgraviaceae are well marked by their highly modified floral bracts, mostly climbing or epiphytic habit, the presence of raphides and sclerenchymatous idioblasts, and the apparent absence of starch. Most authors are agreed that they are related to the Theaceae, but more specialized. *Norantea* is regarded as the most archaic genus.

14. Family QUIINACEAE Engler in Martius 1888 nom. conserv., the Quiina Family

Trees, shrubs, or sometimes lianas, with scattered secretory cells in the parenchymatous tissues of the young stem, and with lysigenous mucilage-channels, especially within the vascular strand of the midrib and petiole of the leaves; solitary and clustered crystals of calcium oxalate present in some of the cells of the parenchymatous tissues; nodes trilacunar; vessel-segments with simple or seldom scalariform perforations; imperforate tracheary elements with evidently bordered pits; vasicentric tracheids sometimes present; wood-rays of Kribs type I, heterocellular, mixed uniseriate and pluriseriate, the latter up to 4 (–6) cells wide, with elongate ends; wood-parenchyma apotracheal, diffuse or sometimes banded. LEAVES opposite or whorled, simple or (*Touroulia, Froesia*) pinnately compound (sometimes pinnately compound on young plants, simple in adults), with elongate, narrow areolae between the ultimate veins; veins, especially the smaller ones, sheathed by very thick-walled fibers, some of the fibers extending out into the mesophyll and sometimes lying parallel to the leaf-surface just beneath the epidermis; stomates paracytic; petiolar trace a simple or dissected siphonostele, often with medullary bundles, or sometimes trough-shaped; stipules interpetiolar, persistent, rigid or foliaceous. FLOWERS in axillary or terminal panicles or racemes, perfect or more often polygamous, hypogynous, regular or nearly so; sepals 4–5, small, unequal, imbricate; petals 4–5 (–8) imbricate; stamens 15-many; filaments distinct or connate at the base, and sometimes also adnate to the base of the petals; anthers small, basifixed, tetrasporangiate and dithecal, the pollen-sacs sharply distinct, or (*Froesia*) back to back and latrorse, opening by longitudinal slits; gynoecium of 2–13 carpels united to form a compound ovary with distinct, linear styles (3 distinct carpels in *Froesia*); stigma discoid; ovary with as many locules as carpels; ovules 2 in each locule, on basal-axile placentas, ascending, anatropous. FRUIT berry-like, fleshy but often dehiscent at full maturity, often only 1–2-locular through abortion of some of the locules, and often with only 1–4 seeds; seeds tomentose (glabrous in *Froesia*), without endosperm, the embryo straight, with short hypocotyl and 2 thick cotyledons.

The family Quiinaceae consists of 4 genera and about 40 species, native to tropical America, especially the Amazon region. *Quiina* has about 25 species, *Lacunaria* about 10, *Touroulia* about 4, and *Froesia* only 2. *Froesia*, with distinct carpels, pinnately compound leaves, and glabrous seeds, stands somewhat apart from the other genera, but it has the characteristic petiolar anatomy of the family, with a siphonostele that has included mucilage channels and medullary bundles. The position of the Quiinaceae as members of the Theales is widely accepted.

15. Family ELATINACEAE Dumortier 1829 nom. conserv., the Waterwort Family

Herbs (mostly rather small) or sometimes half-shrubs, glabrous or with unicellular or uniseriate hairs, growing in shallow water or in wet places, often creeping and rooting at the nodes, resinous more or less throughout, but without the complex secretory system of the Clusiaceae, at least sometimes with ellagic acid and proanthocyanins, but only seldom with tanniferous idioblasts; cortex often with a system of lacunae; clustered crystals of calcium oxalate present in some of the cells of the parenchymatous tissues; vessel-segments with simple perforations. LEAVES opposite or whorled, simple, entire or toothed; stomates more or less distinctly tetracytic; stipules scarious. FLOWERS small, solitary in the axils of the leaves or in small, axillary cymose inflorescences, perfect, hypogynous, regular; sepals 2–5 (6), distinct or connate for up to half their length; petals as many as the sepals, small, distinct, imbricate, persistent; stamens as many or twice as many as the petals, distinct; anthers tetrasporangiate and dithecal, opening by longitudinal slits; pollen-grains tricolporate, 2- or 3-nucleate; gynoecium of 2–5 carpels united to form a compound ovary with distinct (often short) styles and capitate stigmas; locules as many as the carpels, but the partitions not reaching the summit in spp. of *Bergia*; ovules numerous on axile placentas, anatropous, bitegmic, weakly crassinucellar, sometimes with a zigzag micropyle; endosperm-development nuclear. FRUIT a septifragal capsule; seeds without endosperm, the embryo straight or curved, with 2 relatively short cotyledons. X = 6, 9.

The family Elatinaceae consists of 2 genera and about 40 species, widespread in both tropical and temperate regions, but with more species in the tropics. The two genera, *Elatine* and *Bergia*, are of about equal size. Species of *Elatine* are sometimes grown as aquarium plants.

Widely divergent opinions about the affinities of the Elatinaceae have been expressed by different authors. Most commonly the group has been associated with one or another of the families here referred to the Theales and Violales. Takhtajan (1966) has pointed out in a series of similarities between *Elatine* and some species of *Hypericum*, and Corner (1976) considers that the seed-structure of the Elatinaceae indicates a relationship to the hypericoid group of Clusiaceae.

I believe that the Elatinaceae are indeed allied to the Hypericoideae, but a direct filial relationship is unlikely. The closest similarity of the Elatinaceae to members of the Hypericoideae is to relatively advanced, herbaceous species of *Hypericum*, yet the Elatinaceae are more primitive than the Clusiaceae as a whole in having well developed stipules. Therefore the relationship between the two families is more likely collateral, and the similarities of *Elatine* to *Hypericum* probably reflect the sort of close parallelism that often occurs in related groups.

The Elatinaceae differ from most other members of the Theales in being aquatic or semiaquatic herbs. The other features that distinguish the family from its relatives are more difficult to interpret in Darwinian terms.

16. Family PARACRYPHIACEAE Airy Shaw 1965, the Paracryphia Family

Shrub or small to middle-sized tree with unicellular hairs; nodes trilacunar; styloid crystals of calcium oxalate present in some of the cells of the parenchymatous tissues; vessel-segments slender and much elongate, with elongate, steeply slanting perforation-plates that have very numerous (commonly ca 100, sometimes as many as 200) cross-bars; imperforate tracheary elements with bordered pits; wood-rays heterocellular, Kribs type I, mixed uniseriate and pluriseriate, the pluriseriate ones 2–3 cells wide; wood-parenchyma apotracheal and scanty-paratracheal, diffuse or diffuse in aggregates. LEAVES subverticillate, simple, toothed, pinnately veined; stomates anomocytic; petiole with a complex vascular supply; stipules wanting. FLOWERS numerous, in compound spikes (the inflorescence panicle-like, but the flowers sessile), perfect or some of them staminate; perianth of 4 distinct, decussate, caducous tepals, the outermost one the largest and more or less enclosing the other 3; stamens mostly 8, in a single cycle; filaments of the staminate flowers somewhat laminar-expanded; anthers basifixed, tetrasporangiate and dithecal, opening by longitudinal slits; pollen-grains tricolporate; gynoecium of 8–15 carpels that are laterally connate and ventrally adnate to the solid core of central tissue; stigmas distinct, sessile, as many as the carpels, conduplicately folded; each locule with 4 anatropous, unitegmic, apparently crassinucellar ovules in a single row on the axile placenta; endosperm-development cellular. FRUIT capsular, the mature carpels separating from the central column except at the top, spreading out from the base and opening ventrally; seeds small, winged, with copious endosperm; embryo straight, the radicle longer than the 2 cotyledons.

The family Paracryphiaceae consists of a single species, *Paracryphia alticola* (Schlechter) van Steenis, native to New Caldeonia. The affinities and even the formal morphology of *Paracryphia* have occasioned some dispute. The interpretation here presented is based on that of Dickison and Baas (1977). Although it cannot properly be included in any other family, *Paracryphia* is in my opinion reasonably at home in the Theales. The persistent central column is a common (though by no means consistent) feature of capsular fruits in the Theales, and the manner of dehiscence in *Paracryphia* is reminiscent of *Medusagyne.*

17. Family MEDUSAGYNACEAE Engler & Gilg 1924 nom. conserv., the Medusagyne Family

Glabrous, hard-wooded shrubs or small trees, with cortical vascular bundles and with the phloem stratified into alternating concentric rings of sieve-tissue and sclerenchyma; calcium oxalate crystals clustered in some of the cells of the cortex and the leaves, borne singly in some of the cells of the pith; vessel-segments with simple perforations; xylem ring interrupted by primary rays that are mostly 5–8 cells wide, the secondary rays 2–3 cells wide. LEAVES opposite, leathery, simple, with scattered mucilaginous cells in the mesophyll; stomates anomocytic; petiole with an arc of vascular bundles; stipules wanting. FLOWERS in small, terminal mixed panicles, foetid, perfect, regular, hypogynous; sepals 5, connate at the base, persistent, imbricate; petals 5, distinct, convolute, white, becoming rosy; stamens numerous, distinct; filaments very slender, shorter than the petals; anthers basifixed, dithecal, the pollen-sacs often set at different heights, opening by longitudinal slits; pollen-grains (2) 3 (4)-porate; gynoecium of ca 17–25 carpels united to form a compound ovary with as many locules, the styles rather short, distinct, forming a circle around the shoulders of the ovary; stigmas capitate; ovules 2 in each locule, on axile placentas, one ascending, the other descending. FRUIT a capsule, opening septicidally from the base, the carpels united distally, at maturity diverging from the top of the central column like the ribs of an umbrella and opening ventrally; seeds winged all around.

The family Medusagynaceae consists of the single species *Medusagyne oppositifolia* Baker, endemic to the Seychelles Islands. *Medusagyne* is so distinctive that its taxonomic relationships are not easy to ascertain. The combination of imbricate sepals, numerous stamens, and numerous carpels united to form a compound, multilocular ovary with distinct styles would seem to restrict the possibilities to the Theales and Rosales. The cortical bundles (similar to Ochnaceae) and stratified phloem (similar to Malvales and some Theales) favor the Thealean rather than the Rosalean relationship. Clarification of the relationships of *Medusagyne* awaits more detailed studies of the genus.

18. Family CLUSIACEAE Lindley 1826 nom. conserv., the Mangosteen Family

Trees, shrubs, lianas, or herbs, glabrous or with mostly unicellular or uniseriate hairs, with resinous, often yellow or otherwise brightly colored juice in schizogenous secretory canals or cavities in most or all of the tissues, and commonly with scattered tanniferous secretory cells as well, generally producing proanthocyanins but usually not ellagic acid, and often accumulating diverse sorts of xanthones; solitary or clustered crystals of calcium oxalate present in some of the cells of the parenchymatous tissues; nodes mostly unilacunar; vessel-segments with simple (or sometimes mixed simple and scalariform) perforations; imperforate tracheary elements with simple or seldom bordered pits, often septate; vasicentric tracheids sometimes present; wood-rays heterocellular to homocellular, mixed uniseriate and pluriseriate, the latter up to 6 cells wide, with short ends; wood-parenchyma apotracheal (diffuse or in ribbons) or paratracheal, or sometimes (*Hypericum*) wanting. LEAVES opposite or whorled, simple and mostly entire, commonly with many slender lateral veins, often (including all Hypericoideae) beset with resin-cavities that make the leaf appear glandular-punctate; stomates paracytic or surrounded by 3 or more supporting cells, sometimes distinctly encyclocytic; petiolar trace trough-shaped or forming a si-

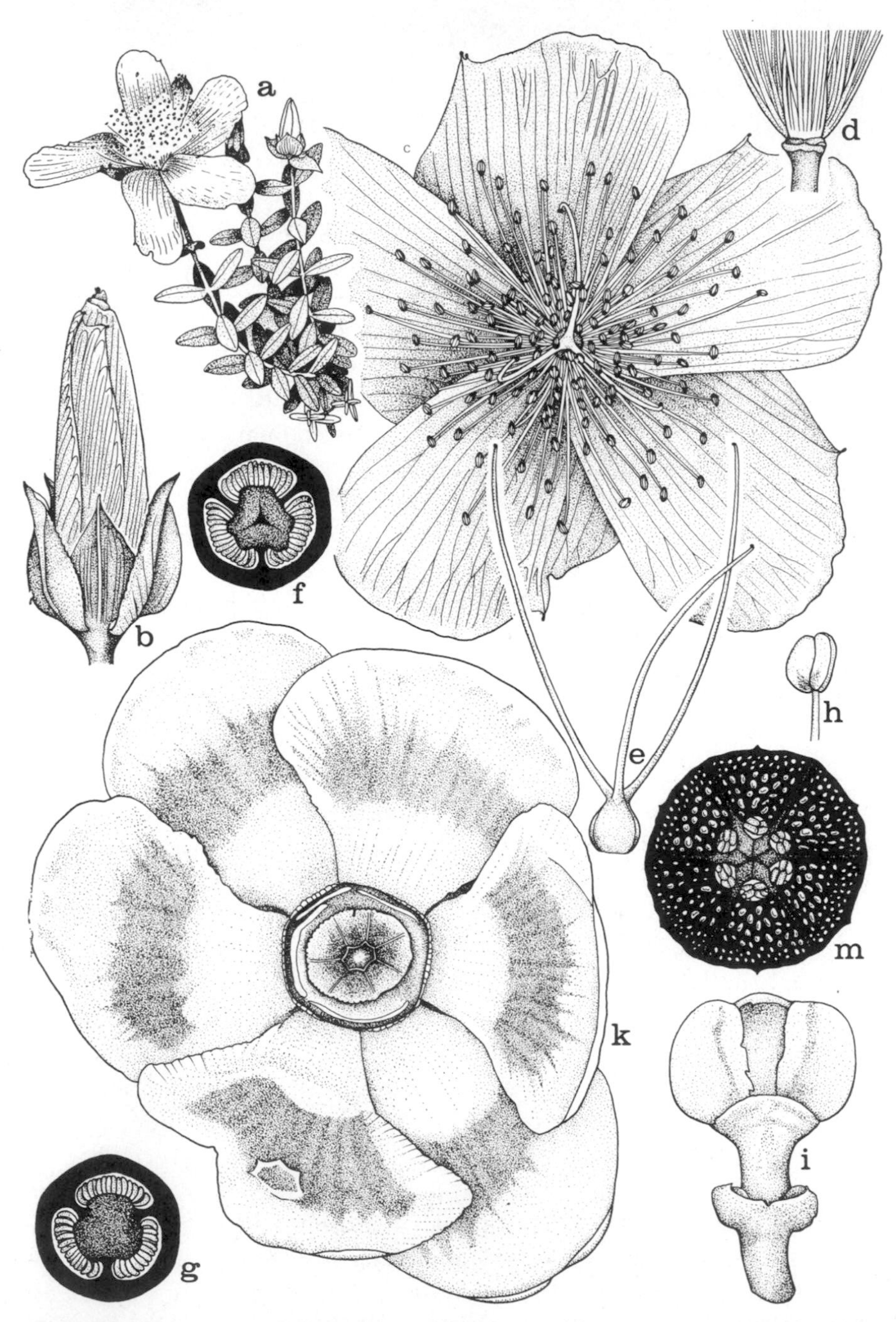

FIG. 4-7 Clusiaceae. a–h, *Hypericum olympicum* L. a, habit, ×½; b, flower bud, ×2; c, flower, from above, ×2; d, fascicled stamens of flower, ×2; e, pistil, ×2; f, schematic cross-section through upper part of ovary, ×8; g, schematic cross-section through lower part of ovary, ×8; h, anther, ×8. i–m, *Clusisa rosea* Jacq. i, flower bud, ×1; k, pistillate flower, from above, ×1; m, schematic cross-section of ovary, ×2.

phonostele; stipules wanting. FLOWERS in terminal, cymose inflorescences, or seldom solitary and terminal, perfect or unisexual, regular, hypogynous, sometimes cantharophilous; bracteoles often present just beneath the calyx, sometimes passing into the sepals; petals (2) 3–6 (–14), distinct or connate at the base, imbricate or convolute; androecium served by a limited number of trunk-bundles; stamens from numerous, distinct, and originating in centrifugal sequence to more often grouped into 2–5 centrifugal bundles opposite and often adnate to the petals, often some of them staminodial, or through reduction in the number of stamens in the bundle rarely only 3 or 5 stamens in all; anthers dithecal but at least sometimes only bisporangiate, opening by longitudinal slits; pollen-grains binucleate, (2) 3 (–5)-colporate; gynoecium of (1–) 3–5 (–20+) carpels united to form a compound ovary with as many locules as carpels, or the ovary sometimes unilocular through failure of the partitions (intruded placentas) to meet in the center; styles as many as the carpels and distinct, or connate below, or fully united to form a single style with a lobed or peltate stigma; placentation axile, or less commonly parietal on intruded placentas, or seldom basal; sometimes, as in spp. of *Clusia,* the ovules attached to the partitions even when the ovary is fully plurilocular, ovules (1) 2-many per carpel, anatropous or hemitropous, bitegmic, tenuinucellar; endosperm-development nuclear. FRUIT a berry of a drupe, or a septicidal or septifragal capsule; seeds often with a funicular or micropylar aril; embryo straight or curved, oily, sometimes with reduced cotyledons; endosperm none. X=7, 8, 9, 10. (Garciniaceae, Guttiferae, Hypericaceae)

The family Clusiaceae as here defined consists of about 50 genera and 1200 species, widespread in moist tropical regions; the subfamily Hypericoideae is also well developed in the North Temperate Zone. Well over half of the species belong to only 3 genera, *Hypericum* (350), *Clusia* (200), and *Garcinia* (200). Mangosteen is the fruit of *Garcinia mangostana* L. The mammey-apple, *Mammea americana* L., is widely cultivated in tropical America.

The Clusiaceae are here considered to consist of two well marked subfamilies, the Clusioideae and Hypericoideae, which have often been treated as distinct families. Each subfamily encompasses a considerable range of diversity, and *Hypericum* is anatomically so distinctive within its subfamily that some authors have sought its origin in the Clusioideae separate from the origin of the remainder of the genera of Hypericoideae. Admitting some exceptions, the two subfamilies may be characterized as follows:

Flowers generally unisexual; secretory system generally of canals, the leaves not appearing punctate; seeds often with an aril; plants tropical and woody .... CLUSIOIDEAE.

Flowers uniformly perfect; secretory system generally largely of shorter cavities, the leaves appearing glandular-punctate; seeds without an aril; plants from tropical and woody to herbaceous and adapted to temperate climates .... HYPERICOIDEAE.

Authors are agreed that the Clusioideae and Hypericoideae are closely related. It is only the rank at which the groups should be received that is in dispute. In these circumstances I prefer to follow the conservative and historically more widely accepted definition of the Clusiaceae (Guttiferae) to include the Hypericaceae.

In spite of the diversity among its members, the family forms a distinctive, well marked group. Only *Kielmeyera* and *Neotatea* blur the limits. These genera are anatomically at home in the Clusiaceae, but aberrant in having alternate leaves. Maguire (1972 and personal communication) believes that *Kielmeyera* and *Neotatea* are better associated with *Bonnetia* (here included in the Theaceae) than with the Clusiaceae.

Pollen referred to the Clusiaceae occurs in Eocene and more recent deposits.

3. Order MALVALES Lindley 1833

Trees, shrubs, lianas, or herbs, mostly relatively poor in alkaloids, without raphides, with or without proanthocyanins but only seldom with ellagic acid, very often with fatty acids containing a cyclopropenyl ring, especially in the seed-fats, most families with stellate or lepidote indument and producing mucilage in special cells, cavities, or canals; vessel-segments mostly with simple perforations; imperforate tracheary elements with simple or sometimes bordered pits; wood-rays commonly including some tile-cells; wood-parenchyma both apotracheal and paratracheal; phloem in young stems usually stratified tangentially into hard and soft layers, and generally with wedge-shaped rays. LEAVES alternate or rarely opposite, often palmately veined, simple to dissected or sometimes palmately compound; petiolar vascular supply usually complex, often siphonostelic or with a ring of vascular bundles, and sometimes with included medullary bundles; stipules present. FLOWERS perfect or less often unisexual, hypogynous or slightly perigynous, rarely epigynous, regular or seldom somewhat irregular, often provided with an epicalyx; nectaries typically consisting of tufts of glandular hairs, most often borne at the base of the sepals; sepals distinct or connate at the base, valvate or seldom (some Elaeocarpaceae) imbricate; petals convolute or sometimes imbricate or valvate, distinct, but sometimes basally adnate to the filament-tube, or sometimes wanting; androecium commonly served by a limited number of trunk-bundles or clusters of bundles; stamens (5–) usually more or less numerous, originating in centrifugal sequence, distinct or often with the filaments connate into a tube or into 5–10 (15) groups or phalanges that relate to the trunk-bundles or bundle-clusters, the antesepalous (or some of the other) stamens sometimes staminodial; anthers variously tetrasporangiate and dithecal to bisporangiate and monothecal, opening by longitudinal slits or less often by terminal pores; pollen-grains binucleate, (tricolpate or) tricolporate to pantoporate; gynoecium of (1) 2-many (often 5) carpels, these generally united to form a compound ovary with as many locules as carpels, the styles as many or twice as many as the carpels, often united below (carpel-bodies separate in some Sterculiaceae, only the styles more or less connate at anthesis); ovules 1-many in each locule, mostly on axile placentas (marginal in apocarpous Sterculiaceae), mostly anatropous, bitegmic, crassinucellar, commonly with a zigzag micropyle; endosperm-development nuclear. FRUIT variously capsular, or schizocarpic, or less often a berry, drupe, or even a samara; seeds sometimes arillate; embryo dicotyledonous; endosperm abundant to scanty or wanting.

The order Malvales as here defined consists of 5 families and 3000 to 3500 species, cosmopolitan in distribution, but best represented in the tropics. The Malvaceae, with a thousand to 1500 species, and the Sterculiaceae, with about a thousand, are the largest families. The Tiliaceae have about 450 species, the Elaeocarpaceae about 400, and the Bombacaceae about 200.

Most authors have agreed that all of the families here referred to the Malvales are allied. The Elaeocarpaceae stand somewhat apart from the rest of the order, but even so the relationship is so close that they have often been included in the Tiliaceae. The remaining four families (Tiliaceae, Sterculiaceae, Bombacaceae, and Malvaceae) are even more closely allied, and there has been some controversy about their definition. Individual genera have been shifted from one family to another by various authors. The whole tribe Hibisceae is referrable to the Bombacaceae if the structure of the fruit is taken as the critical character, and to the Malvaceae if (as here) the ornamentation of the pollen is stressed instead.

The small families Sarcolaenaceae (Chlaenaceae) and Sphaerosepalaceae (Rhopalocarpaceae) are often also referred to the Malvales. These families have mucilage-cells, as in the Malvales, and the Sarcolaenaceae also have stratified phloem. On the other hand, they have imbricate sepals, and they do not seem closely allied to any family of the Malvales. On balance, I find it more useful to include these two families in the Theales, partly because the Malvales are otherwise relatively homogeneous, whereas the Theales are more heterogeneous. Likewise, I find it more useful to refer the Scytopetalaceae to the Theales than to include them in the Malvales where they have no obvious relatives.

The nature of the nectaries in the Malvales is evidently correlated with the polypetalous structure of the flowers. According to Brown (1938, in initial citations):

> In this order, nectary glands are characteristic, multicellular, glandular hairs which are usually packed closely together to form cushion-like growths. In the Tiliaceae, these nectaries are found in various places, including the sepals, petals, and androgynophore. In the Malvaceae, Bombacaceae and Sterculiaceae they have become localized and are found in the sepals. . . . The honey is made available by means of openings between the overlapping bases of the petals. The lack of

gamopetalous corollas in these families is probably connected with the occurrence of nectaries in the sepals and the necessity for slits between the petals to make the nectar available.

There has been continuing controversy about whether the large number of stamens and/or carpels found in some or all members of most families of the order reflects a primitive survival or a secondary increase. The monothecal stamens of the Malvaceae and Bombacaceae are widely admitted to represent longitudinal halves of ancestrally dithecal stamens, but there the agreement stops. Polyandrous types commonly have one or more sets of 5 stamen-traces (trunk bundles), with each trace repeatedly forked so that eventually there is one vascular bundle for each stamen. Polycarpellate types sometimes show a similar branching of 5 basic carpel-traces.

Some students of floral anatomy have interpreted this patten to indicate an increase in the number of stamens and carpels, based on an originally pentamerous flower. I am more attracted to the contrary view that fairly large and indefinite numbers of stamens and carpels in the Malvales are inherited directly from a Thealean ancestry, and that consolidation in the vascular supply is merely the first step in a phyletic reduction from numerous stamens and carpels to few. The characteristic compound stamen-trunks that have been interpreted to mean a secondary increase in number are not restricted to the Malvales, but are widespread in the Dilleniidae in general, and they have also been found in the Myrtales, Alismatales, and even the Magnoliales.

At the same time, it must be admitted that when the number of parts of a particular kind is fairly large and indefinite, evolutionary increase as well as decrease in number may occur. There is no need to suppose that the largest number of dithecal stamens that can be found in any species of Malvales must be the minimum ancestral number for the order. Likewise, the arrangement of the numerous carpels of the small tribe Malopeae (Malvaceae) into 5 vertical ranks suggests the likelihood of a secondary increase in the number of carpels in this group.

The Malvales are probably derived from the less modified members of the Theales, from which they differ in the valvate rather than imbricate calyx. The vast majority of the Malvales also share a syndrome of vegetative features that are more sporadic (and less consistently associated) in the Theales. These are the stellate or lepidote indument, the mucilage-cells or -cavities, and the stratified phloem with wedge-shaped rays. Furthermore, seeds of at least the four core-families of the

Malvales have fatty acids containing a cyclopropenyl group. This type of fatty acid now appears to be restricted to the Malvales.

The only other obvious relationship of the Malvales to any large group is with the Violales. These two orders are here regarded as groups that have taken different evolutionary paths from a common ancestry in the Theales. The Elaeocarpaceae are the most archaic family of the Malvales, as judged by the general floral morphology, wood anatomy, absence of stellate hairs, and lack of the specialized type of nectary found in other families of the order. Some of the Elaeocarpaceae are very similar to some of the Flacourtiaceae, and this similarity extends even to the pollen-morphology. Any common ancestor of the Flacourtiaceae and Elaeocarpaceae would presumably have an imbricate calyx and axile placentation, and thus could be appropriately placed in the Theales.

Aside from the correlation of the type of nectary with the structure of the corolla, the ecological significance of the characters that mark the Malvales is mostly obscure. The mucilage-cells presumably have a function in discouraging predators, but this reasonable speculation is not backed by any objective evidence. If the valvate arrangement of the sepals has any importance to the plant, the fact is not obvious. If stellate hairs have any advantage over simple hairs, it has not been demonstrated. The stratified phloem with wedge-shaped rays does not convey any obvious advantage over plants with more conventional phloem-structure. The functional significance of the cyclopropenyl group in some of the fatty acids remains to be elucidated.

The ecological significance of the families that make up the order is scarcely more evident than that of the order as a whole. The Malvaceae are mostly herbaceous or nearly so, whereas the other families are mostly woody, but beyond that the problems are unresolved. The spines on the pollen of the Malvaceae, and the tendency in several families for the filaments to be connate, might relate to the kind of pollinators, or they might not. It is hard to see how the number of pollen-sacs per anther might affect the competitive ability of the plant, but here as always the possibility must be entertained that there is a hidden significance which might be discovered by a persistent and perceptive investigator.

The oldest clearly malvalean fossils are Maestrichtian pollen-grains of the Bombacaceae. Since the Bombacaceae are fairly well advanced members of the order, it may reasonably be supposed that the Malvales arose before Maestrichtian time.

SELECTED REFERENCES

Anderson, G. J. 1976. The pollination biology of *Tilia.* Amer. J. Bot. 63: 1203–1212.

Baker, H. G., & I. Baker. 1968. Chromosome numbers in the Bombacaceae. Bot. Gaz. 129: 294–296.

Bakhuizen van den Brink, R. C. 1924. Revisio Bombacacearum. Bull. Jard. Bot. Buitenzorg, III. 6: 161–240.

Bates, D. M. 1968. Generic relationships in the Malvaceae, tribe Malveae. Gentes Herb. 10: 117–135.

Bates, D. M. 1976. Chromosome numbers in the Malvales. III. Miscellaneous counts from the Byttneriaceae and Malvaceae. Gentes Herb. 11: 143–150.

Bouchet, P., & G. Deysson. 1974. Les canaux à mucilage des Angiospermes: Étude morphologique et ultrastructurale des cellules constituent les canaux à mucilage du *Sterculia bidwellii* Hook. Rev. Gén. Bot. 81: 369–402.

Brizicky, G. K. 1965. The genera of Tiliaceae and Elaeocarpaceae in the southeastern United States. J. Arnold Arbor. 46: 286–307.

Brizicky, G. K. 1966. The genera of Sterculiaceae in the southeastern United States. J. Arnold Arbor. 47: 60–74.

Burret, M. 1926. Beiträge zur Kenntnis der Tiliaceen. Notizbl. Bot. Gart. Berlin-Dahlem 9: 592–880.

Chattaway, M. 1932. The wood of Sterculiaceae. 1. Specialization of the vertical wood parenchyma within the sub-family Sterculieae. New Phytol. 31: 119–132.

Chattaway, M. M. 1933. Tile-cells in the rays of the Malvales. New Phytol. 32: 261–273.

Cuatrecasas, J. 1964. *Cacao* and its allies. A taxonomic revision of the genus *Theobroma.* Contr. U. S. Natl. Herb. 35: 379–614.

Dehay, C. 1941. L'Appareil libéro-ligneux foliaire des Sterculiacées. Ann. Sci. Nat. Bot. sér. 11. 2: 45–131.

Edlin, H. L. 1935. A critical revision of certain taxonomic groups of the Malvales. New Phytol. 34: 1–20; 122–143.

Fuchs, H. P. 1967. Pollen morphology of the family Bombacaceae. Rev. Palaeobot. Palynol. 3: 119–132.

Gazet du Chatelier, G. 1940. Recherches sur les Sterculiacées. Rev. Gén. Bot. 52: 174–191; 211–233; 257–284; 305–332.

Gazet du Chatelier, G. 1940. La structure florale des Sterculiacées. Compt. Rend. Hebd. Séances Acad. Sci. 210: 57–59.

Gottsberger, G. 1972. Blütenbiologische Beobachtungen an brasilianischen Malvaceen. II. Oesterr. Bot. Z. 120: 439–509.

Hall, J. W., & A. M. Swain. 1971. Pedunculate bracts of *Tilia* from the Tertiary of Western United States. Bull. Torrey Bot. Club 98: 95–100.

Inamdar, J. A., & A. J. Chohan. 1969. Epidermal structure and stomatal development in some Malvaceae and Bombacaceae. Ann. Bot. (London) II. 33: 865–878.

Leinfellner, W. 1960. Zur Entwicklungsgeschichte der Kronblätter der Sterculiaceae—Buettnerieae. Oesterr. Bot. Z. 107: 153–176.

Mauritzon, J. 1934. Zur Embryologie der Elaeocarpaceae. Ark. Bot. 26A(10): 1–8.

Miège, J., & H. M. Burdet. 1968. Étude du genre *Adansonia* L. I. Caryologie. Candollea 23(1): 59–66.

Record, S. J. 1939. American woods of the family Bombacaceae. Trop. Woods 59: 1–20.

Robyns, A. 1963. Essai de monographie du genre *Bombax* s. l. (Bombacaceae). Bull. Jard. Bot. État 33: 1–311.
Rohweder, O. 1972. Das Andröcium der Malvales und der "Konservatismus" des Leitgewebes. Bot. Jahrb. Syst. 92: 155–167.
Saad, S. I. 1960. The sporoderm stratification in the Malvaceae. Pollen & Spores 2: 13–41.
Sharma, B. D. 1969. Pollen morphology of Tiliaceae in relation to plant taxonomy. J. Palyn. (Lucknow) 5: 7–29.
Sharma, B. D. 1970. Contribution to the pollen morphology and plant taxonomy of the family Bombacaceae. Proc. Indian Natl. Sci. Acad. 36B: 175–191.
Shenstone, F. S., & J. R. Vickery. 1961. Occurrence of cyclo-propene acids in some plants of the order Malvales. Nature 190: 168, 169.
van Heel, W. A. 1966. Morphology of the androecium in Malvales. Blumea 13: 177–394.
van Heel, W. A. 1978. Morphology of the pistil in Malvaceae-Ureneae. Blumea 24: 123–127.
Venkata Rao, C. 1949. Floral anatomy of some Sterculiaceae with special reference to the position of stamens. J. Indian Bot. Soc. 28: 237–245.
Venkata Rao, C. 1949–1954. Contributions to the embryology of Sterculiaceae. I–V. J. Indian Bot. Soc. 28: 180–197, 1949. 29: 163–176, 1950. 30: 122–131, 1951. 31: 251–260, 1953. 32: 208–238, 1953 (1954).
Venkata Rao, C. 1950. Pollen grains of Sterculiaceae. J. Indian Bot. Soc. 29: 130–137.
Venkata Rao, C. 1952. The embryology of *Muntingia calabura* L. J. Indian Bot. Soc. 31: 87–101.
Venkata Rao, C. 1952. Floral anatomy of some Malvales and its bearing on the affinities of families included in the order. J. Indian Bot. Soc. 31: 171–203.
Venkata Rao, C. 1953. Floral anatomy and embryology of two species of *Elaeocarpus*. J. Indian Bot. Soc. 32: 21–33.
Venkata Rao, C. 1954, 1955. Embryological studies in Malvaceae. I. Development of gametophytes. II. Fertilization and seed development. Proc. Natl. Inst. Sci. India 20: 127–150, 1954. 21B: 53–67, 1955.
Venkata Rao, C., & K. V. S. Rao. 1952. A contribution to the embryology of *Triumfetta rhomboidea* Jacq. and *Corchorus acutangulus* L. J. Indian Bot. Soc. 31: 56–68.
Webber, I. E. 1934. Systematic anatomy of the woods of the Malvaceae. Trop. Woods 38: 15–36.
Weibel, R. 1945. La placentation chez les Tiliacées. Candollea 10: 155–177.
Weibel, R. 1968. Morphologie de l'embryon et de la graine des *Elaeocarpus*. Candollea 23: 101–108.

Александров, В. Г., & А. В. Добротворская. 1957. О морфологической сущности тычинок, лепестков и так называемой тычиночной трубки в цветке мальвовых. Труды Бот. Инст. Комарова АН СССР, сер. VII. 4: 83-137.
Ерамян, Е. Н. 1955. К изучению пыльцевых зерен кавказских представителей семейства мальвовых (Malvaceae). Научн. Тр. Ереванского Государск Унив., 49: сер Биол., 5: 211-228.

SYNOPTICAL ARRANGEMENT OF THE FAMILIES OF MALVALES

1 Anthers tetrasporangiate and dithecal; epicalyx only seldom present; filaments distinct or connate; mostly trees and shrubs, a few herbs.

2 Filaments distinct, or only shortly connate at the base into 5 or 10 clusters.

3 Plants without mucilage-cells, -cavities, and -canals, except that the epidermis of the leaves is often mucilaginous; phloem not stratified, and without wedge-shaped rays; indument not of stellate or peltate hairs; petals valvate or seldom imbricate .... 1. ELAEOCARPACEAE.

3 Plants with well developed mucilage-cells and often also -cavities or -canals; phloem of young stems tangentially stratified into hard and soft layers, and with wedge-shaped rays; indument often of stellate hairs or peltate scales; petals imbricate or convolute or valvate .... 2. TILIACEAE.

2 Filaments all generally connate into a tube around the ovary .... 3. STERCULIACEAE.

1 Anthers bisporangiate and monothecal (but sometimes coalescent in Bombacaceae); flowers very often with an epicalyx; filaments generally connate into a tube.

4 Pollen-grains generally smooth or merely rugose, only rarely minutely spinulose, of various shapes, triaperturate (colpate, colporate, or porate) to less often pantoporate; trees; fruit a loculicidal capsule, or seldom fleshy and indehiscent .... 4. BOMBACACEAE.

4 Pollen-grains generally minutely spiny, only rarely smooth, mostly spherical and pantoporate, but sometimes tricolporate; herbs or soft shrubs, only rarely trees, fruit variously schizocarpic, or septicidally or loculicidally dehiscent, only seldom baccate or samaraoid .... 5. MALVACEAE.

1. Family ELAEOCARPACEAE A. P. de Candolle 1824 nom. conserv., the Elaeocarpus Family

Trees or shrubs with trilacunar nodes, with or without proanthocyanins and ellagic acid, often (at least in *Elaeocarpus*) producing indolizidine alkaloids, without mucilage-cavities or -channels, but sometimes with mucilage-cells in the epidermis of the leaves; trichomes simple and unicellular, or glandular, not stellate or peltate; calcium oxalate present in some of the cells of the parenchymatous tissues, mostly as solitary crystals; vessel-segments with simple perforations, or sometimes some

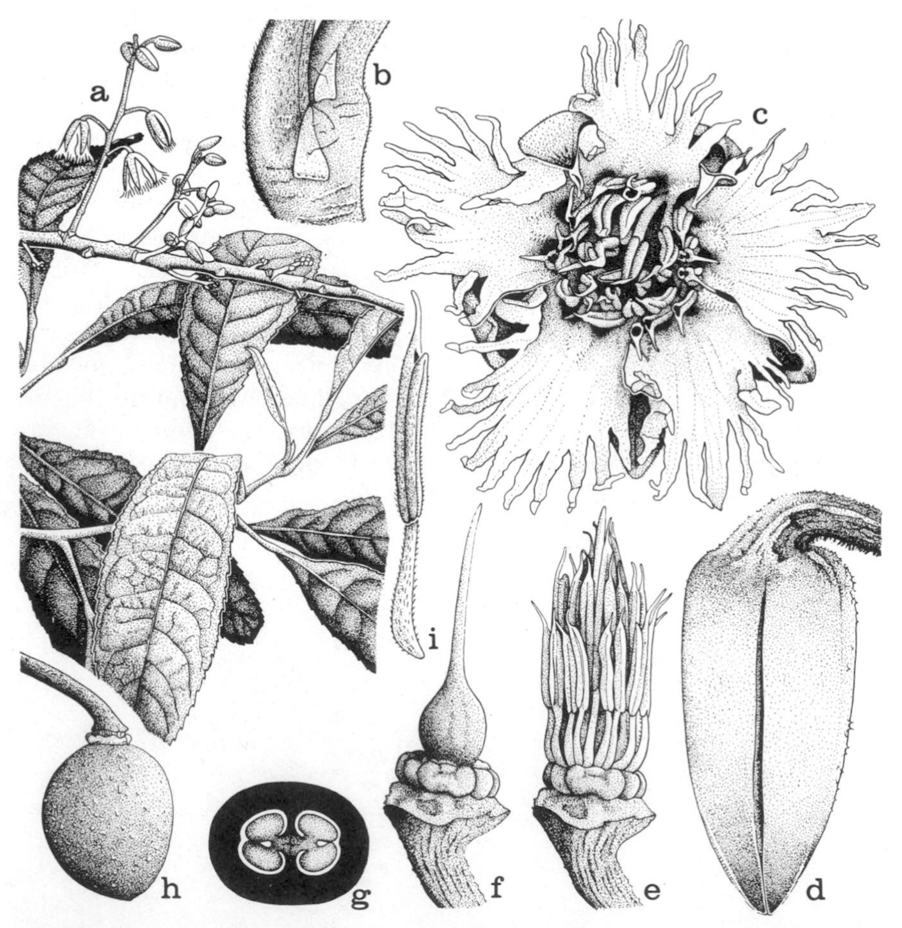

FIG. 4-8 Elaeocarpaceae. *Elaeocarpus cyaneus* Sims. a, habit, ×½; b, node, showing stipule, ×6; c, flower, from above, ×6; d, flower bud, ×6; e, flower, with perianth removed, ×6; f, flower, with perianth and stamens removed, ×6; g, schematic cross-section of ovary, ×12; h, fruit, ×2; i, stamen, ×12.

of them with a few cross-bars; imperforate tracheary elements with small, bordered pits, usually septate; wood-rays heterocellular, mixed uniseriate and pluriseriate, the latter with long or short ends; wood-parenchyma scanty, paratracheal; phloem not stratified, and without wedge-shaped rays. LEAVES alternate or sometimes opposite, simple; stomates paracytic to encyclocytic; petiolar vascular supply at least sometimes siphonostelic; stipules persistent or deciduous. FLOWERS variously in racemes, panicles, or dichasial cymes, regular, perfect or seldom unisexual, hypogynous, without an epicalyx; sepals (3) 4–5 (–11), distinct or connate at the base, valvate or sometimes imbricate; petals (3) 4–5, distinct or seldom connate at the base, often fringed at the tip, valvate or seldom imbricate, or sometimes wanting; stamens numerous, distinct, originating centrifugally, often weakly organized into 5 antesepalous groups, borne on a more or less definite disk or enlarged receptacle that is sometimes prolonged to form an androgynophore; anthers tetrasporangiate and dithecal, mostly relatively large and elongate, opening by a transverse apical slit, or by apical pores or short slits, or by rather short, longitudinal, lateral slits; connective often conspicuously prolonged; pollen-grains binucleate, small, globose, commonly tricolporate; gynoecium of (1) 2-many carpels united to form a compound ovary with a simple or shortly lobed style; ovary with as many locules as carpels, and with 2-many ovules on axile placentas in each locule; ovules anatropous, bitegmic, crassinucellar, with zigzag micropyle and an integumentary tapetum; endosperm-development nuclear. FRUIT a capsule (often spiny) or less often a drupe; seeds with copious, oily and proteinaceous endosperm, sometimes arillate; embryo dicotyledonous, straight or strongly curved (J-shaped to nearly U-shaped). (Aristoteliaceae, Maquineae.) X = 12, 14, 15.

The family Elaeocarpaceae consists of about 10 genera and 400 species, widespread in tropical and subtropical regions, but missing from continental Africa. More than half of the species belong to the single genus *Elaeocarpus* (250), and *Sloanea* accounts for another hundred.

The Elaeocarpaceae are by all accounts closely related to the Tiliaceae, and some authors even include them in that family. *Echinocarpus*, a very near ally of *Sloanea* (even included in *Sloanea* by many authors), has been cited in the literature as transitional toward *Erinocarpus*, in the Tiliaceae. At the same time, the Elaeocarpaceae are also generally regarded as closely allied to the Flacourtiaceae, in the order Violales. Members of the Flacourtiaceae are often casually confused with *Sloanea*.

The monotypic tropical American genus *Muntingia*, usually included in the Elaeocarpaceae, is here transferred to the Flacourtiaceae. Although it has a valvate calyx and plurilocular ovary, like the Elaeocarpaceae, its short, broad, longitudinally dehiscent anthers are unlike those of the Elaeocarpaceae and suggest those of such flacourtiaceous genera as *Prockia* and *Hasseltia*, which likewise have a valvate calyx and plurilocular ovary. All of these genera lie in the indefinite boundary-land between the highly diversified family Flacourtiaceae and the more narrowly limited family Elaeocarpaceae. None of them has any obvious close allies in the Elaeocarpaceae, but *Prockia* and *Hasseltia* relate to *Pleuranthodendron* and *Macrohasseltia*, which have intruded parietal placentas like those of many other Flacourtiaceae.

Fossil fruits thought to represent *Echinocarpus* occur in Eocene deposits in England, and fossil wood referred to the Elaeocarpaceae occurs in the Paleocene of Patagonia.

2. Family TILIACEAE A. L. de Jussieu 1789 nom. conserv., the Linden Family

Trees, shrubs, or rarely herbs, provided with simple or more often stellate hairs or peltate scales, tanniferous, often with proanthocyanins and sometimes with ellagic acid, and characteristically with mucilage-cells or -cavities (less often canals) in some or all of the parenchymatous tissues; solitary or clustered crystals of calcium oxalate present in some of the cells of the parenchymatous tissues; nodes trilacunar; vessel-segments with simple perforations; imperforate tracheary elements with simple pits; wood-rays heterocellular (Kribs type IIB) to almost homocellular (type I), mixed uniseriate and pluriseriate, the latter commonly up to 4–9 (–23) cells wide, with short ends, often containing tile cells; wood-parenchyma varying from apotracheal to paratracheal; phloem of young stems tangentially stratified into hard and soft layers, and provided with wedge-shaped rays; sieve-tubes with S-type plastids. LEAVES alternate or rarely opposite, simple but often toothed or lobed, usually palmately or pinnipalmately veined, commonly deciduous, often asymmetric; stomates mostly anomocytic; petiolar vascular supply diverse, ranging from a single trough-shaped strand to a siphonostele with included medullary bundles; stipules present, often deciduous. FLOWERS in various sorts of cymose inflorescences, or sometimes paired or solitary, perfect or sometimes unisexual, regular, hypogynous or rarely (*Neotessmannia*, of Peru) epigynous, sometimes with an epicalyx; sepals (3–) 5, valvate, distinct or sometimes connate at the base; petals

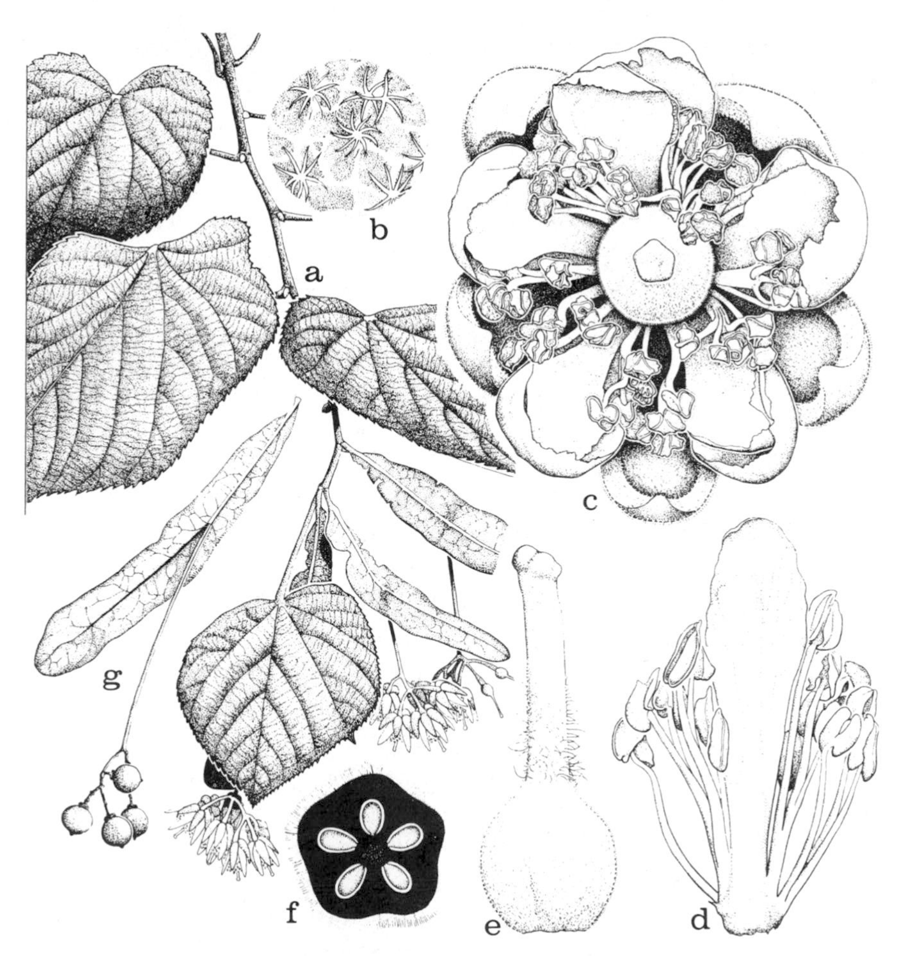

FIG. 4-9 Tiliaceae. *Tilia petiolaris* DC. a, habit, ×½; b, lower surface of leaf, ×18; c, flower, from above, ×6; d, petal and associated stamen-cluster, ×6; e, pistil, ×6; f, diagrammatic cross-section of ovary, ×9; g, fruits and subtending bract, ×½

as many as the sepals, distinct, imbricate or convolute or valvate, or sometimes wanting; nectaries commonly consisting of variously located tufts of glandular hairs; androecium served by a limited number of trunk-bundles; stamens (10–) more or less numerous, sometimes borne on a short androgynophore or internal to a nectariferous disk, not clearly differentiated into separate whorls, initiated in centrifugal sequence, distinct or the filaments sometimes basally connate into 5 or 10 groups, sometimes 5 or more of the stamens staminodial, but these not consistently alternating with stamens or stamen-clusters; anthers tetrasporangiate and dithecal, the pollen-sacs contiguous or separate at the tips of the branches of the shortly bifurcate filament, opening by longitudinal slits or sometimes by apical pores; pollen-grains smooth, variable in structure, tricolporate or sometimes triporate, binucleate; gynoecium of 2-many carpels united to form a compound ovary with a single style and a capitate or lobed, dry stigma; ovary with as many locules as carpels, or seldom (as in *Goethalsia*, *Mollia*) unilocular by failure of the partitions to meet in the center; ovules (1) 2-several on axile (or intruded parietal) placentas in each locule, anatropous to sometimes hemitropous or almost orthotropous, bitegmic, crassinucellar, with a zigzag micropyle; endosperm-development nuclear. FRUIT dry or fleshy, dehiscent or indehiscent; embryo straight or sometimes with folded cotyledons; endosperm abundant to very scanty, oily, often with fatty acids that contain cyclopropene. X = 7-41.

The family Tiliaceae as here defined consists of about 50 genera and 450 species, widespread in tropical and subtropical regions, with relatively few species in temperate climates. The largest genera are *Grewia* (150) and *Triumfetta* (70), both mainly tropical. *Tilia* (50), the linden or basswood, is the most familiar extratropical genus. Jute is obtained from the phloem fibers of species of *Corchorus*, one of the few herbaceous genera in the family.

Fossils of *Tilia*, including leaves, floral bracts, flowers, fruit, and pollen have been found in close association in deposits of late Oligocene or early Miocene age in Idaho. *Cantitha*, a fossil of British Eocene deposits, is thought to belong to the Tiliaceae, and other fossils referred to the Tiliaceae occur in Eocene deposits in India and the United States. Pollen considered to represent the modern genus *Brownlowia* occurs in Paleocene and more recent rocks. Earlier records of the family are more doubtful.

3. Family STERCULIACEAE Bartling 1830 nom. conserv., the Cacao Family

Trees or shrubs with soft to hard wood, or sometimes lianas or even herbs, with a vesture usually partly or wholly of stellate hairs or peltate scales, commonly with scattered tanniferous cells, frequently producing proanthocyanins but without ellagic acid, often accumulating methylxanthine derivatives (purine bases such as theobromine and caffeine), and also with cyclopropenoid fatty acids (i.e., fatty acids with a 3-membered—thus cyclopropene—ring) such as sterculic acid and mal-

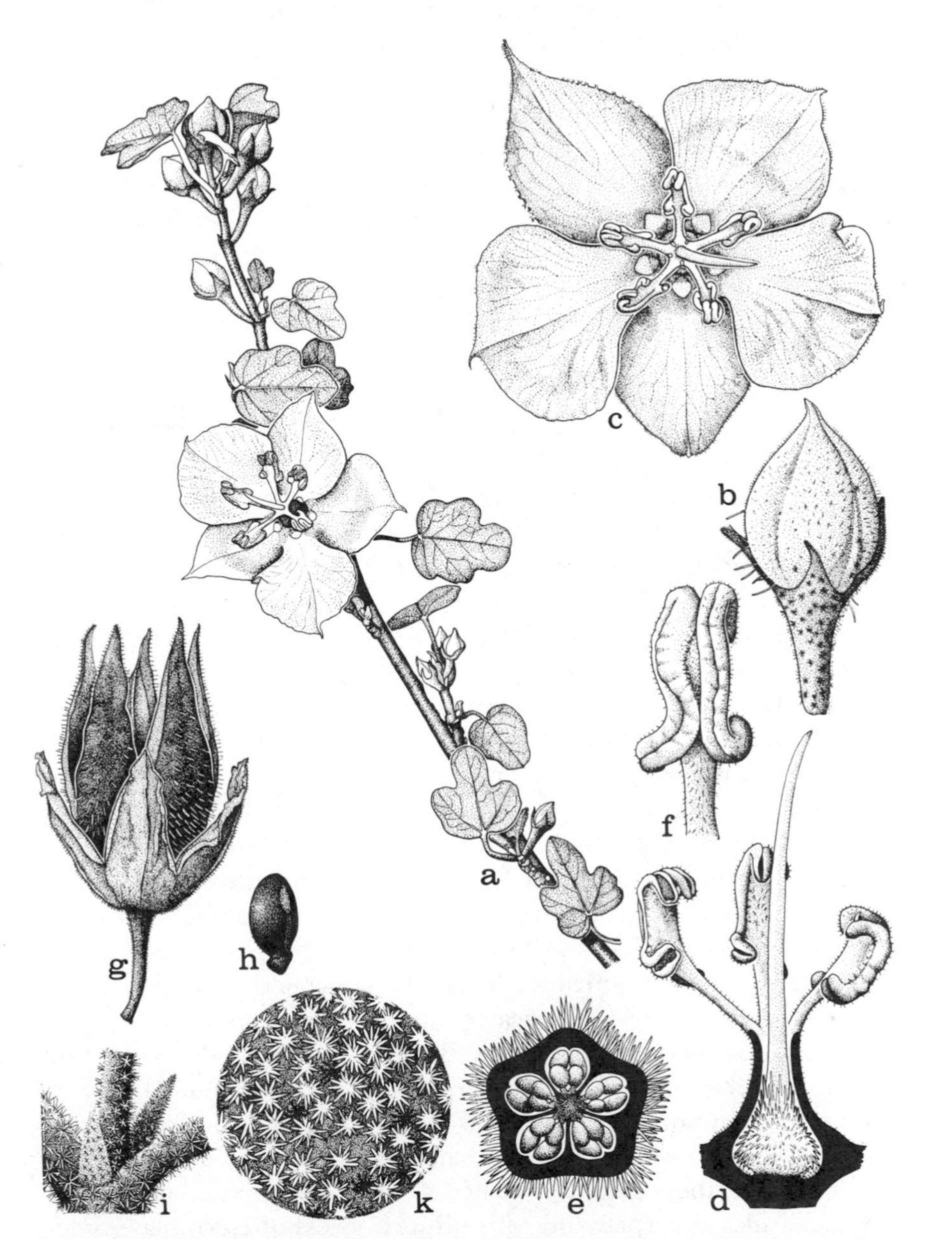

Fig. 4-10 Sterculiaceae. *Fremontia californica* Torr. a, habit, $\times\frac{1}{2}$; b, flower bud, $\times 2$; c, flower, from above, $\times 1$; d, schematic long-section showing pistil, filament-tube, and 3 stamens, $\times 2$; e, schematic cross-section of ovary, $\times 6$; f, anther, $\times 4$; g, fruit, after dehiscence, $\times 1$; h, seed, $\times 4$; i, node, with stipules, $\times 4$; k, portion of leaf-surface, $\times 16$.

valic acid, especially in the seeds; nodes trilacunar; mucilage-cells, -cavities, or -canals (either lysigenous or schizogenous) characteristically present in the parenchymatous tissues; clustered or less often solitary crystals of calcium oxalate commonly present in some of the cells of the parenchymatous tissues; vessel-segments with simple perforations; imperforate tracheary elements with simple or inconspicuously bordered pits; wood-rays heterocellular to sometimes almost homocellular (Kribs type IIA and B), mixed uniseriate and pluriseriate, the latter up to 20 or more cells wide, with short ends, and often including some tile cells; wood-parenchyma abundant, variously diffuse to vasicentric; phloem of young stems tangentially stratified into hard and soft layers, and provided with wedge-shaped rays. LEAVES alternate, simple (often palmately lobed) and pinnately to palmately veined, or often palmately compound; stomates mostly anomocytic; petiolar vascular supply of diverse complex types, often siphonostelic or with a ring of small bundles and one or more medullary bundles, or deeply trough-shaped with incurved edges; stipules present, deciduous or sometimes persistent. FLOWERS in various sorts of mostly complex inflorescences, or seldom solitary, perfect or less often unisexual, hypogynous, regular or seldom irregular, often with an epicalyx; calyx of 3–5 valvate sepals that are usually shortly connate below; nectaries consisting of tufts of glandular hairs at the base of the sepals; petals 5, distinct, convolute, sometimes adnate to the filament-tube at the base, usually clawed, sometimes also hooded, commonly small, or often wanting; stamens seeming to be basically in 2 cycles, one set of 5 antesepalous and commonly staminodial (and often petaloid) or sometimes suppressed, the other set of 5 normal, antepetalous members or often of 5 antepetalous bundles of 2–3 (–10+) members developing centrifugally from a common vascular trunk; filaments all generally connate into a tube around the ovary, often seated on an androgynophore; anthers tetrasporangiate and dithecal, the pollen-sacs parallel and adjacent, or sometimes divergent and separated, opening by longitudinal slits or seldom by apical pores; pollen-grains binucleate, from tricolporate and smooth or reticulate to pantoporate and spinulose; gynoecium of (1–) 5 (–60) carpels, these generally united to form a compound ovary with as many locules as carpels and with distinct or less often connate styles (rarely the ovary unilocular through failure of the partitions to meet in the center), but in the tribe Sterculieae the carpels united only by their styles and becoming wholly distinct at maturity, or even wholly distinct from the beginning (as in *Cola*); ovules (1) 2-many per carpel, on axile (or marginal, or deeply intruded parietal) placentas, anatropous, or hemitropous, bitegmic, crassinucellar, often with a zigzag micropyle; endosperm-development nuclear. FRUIT fleshy to leathery or even woody, dehiscent or indehiscent, often separating into mericarps; seeds sometimes arillate, embryo straight or curved, with expanded cotyledons; endosperm generally abundant, oily and/or starchy, or rarely (as in *Cola*) wanting. X = 5-many, most commonly 20. (Byttneriaceae)

The family Sterculiaceae as here (and traditionally) defined consists of about 65 genera and a thousand species, mainly confined to tropical and subtropical regions. The largest genera are *Sterculia* and *Dombeya*, with about 200 species each. *Theobroma cacao* L., *Cola nitida* (Vent.) A. Chev., and *Cola acuminata* (Beauv.) Schott & Endl. are well known economic species.

The Sterculiaceae embrace a considerable range of diversity in gross floral morphology, pollen-morphology, and the anatomy of the flowers, wood and petioles. Some authors (notably Edlin) would restrict the family to the traditional tribe Sterculieae (12 genera, including *Sterculia* and *Cola*), and refer the remaining genera to a separate family Byttneriaceae. The Sterculiaceae then become a relatively homogeneous group, marked by their unisexual, apetalous, more or less distinctly apocarpous flowers that are commonly borne in large panicles and produce smooth, tricolporate pollen. The Byttneriaceae, in contrast, have syncarpous, mostly perfect flowers, but they remain heterogeneous in other respects, including the anatomy of the wood and petioles, the nature of the inflorescence, the presence or absence of petals and an androgynophore, and the structure of the pollen. The Byttneriaceae, if recognized, stand alongside the Sterculiaceae in the taxonomic system; there is no other group to which the Byttneriaceae appear to be more closely related. Under such circumstances, I see no need to depart from the traditional broad definition of the Sterculiaceae.

Some authors have supposed that the tribe Sterculieae must be primitive within the family, because of its more or less apocarpous flowers. As Takhtajan has pointed out, however, apocarpy in this group is probably secondary, because in *Sterculia* itself the carpels are united by their styles at anthesis. A comparable evolutionary pattern is seen in the Apocynaceae and Asclepiadaceae. In the Apocynaceae there is a phyletically progressive separation of the carpels from the bottom upwards, leading to the Asclepiadaceae, in which the carpels are united only by their common stigma.

Fossils that appear to represent the Sterculiaceae occur in deposits from the latest Cretaceous onward, and leaves attributed to the modern genus *Dombeya* occur in Eocene deposits in North Dakota. Pollen attributed to the family occurs in Paleocene and more recent deposits.

4. Family BOMBACACEAE Kunth 1822 nom. conserv., the Kapok-tree Family

Trees, often very large, but commonly with soft and light wood, often with a thickened trunk containing a high proportion of parenchymatous water-storage tissue, sometimes with deciduous thorns, usually beset with stellate hairs or peltate scales, with or without proanthocyanins, without ellagic acid, and commonly producing cyclopropenoid fatty acids, especially in the seeds; mucilage-cells, -cavities, or -canals characteristically present in the parenchymatous tissues; clustered or less often solitary crystals of calcium oxalate present in some of the cells of the parenchymatous tissues; silica grains present in some of the xylem cells of some species; vessel-segments with simple perforations; imperforate tracheary elements with simple or bordered pits, sometimes septate; wood-rays heterocellular or sometimes homocellular, often including some tile cells, mixed uniseriate and pluriseriate, the latter up to about 10 (–15+) cells wide, often large and conspicuous; wood-parenchyma abundant, both apotracheal and vasicentric, sometimes forming the ground-tissue of the xylem; phloem in young stems commonly stratified tangentially into hard and soft layers, and with wedge-shaped rays. LEAVES deciduous, alternate, simple (Matisieae, Durioneae) or palmately compound (Bombaceae), when simple the venation pinnate (Durioneae) or more or less distinctly palmate (Matisieae); stomates paracytic, but the supporting cells commonly unequal, or the stomates sometimes anomocytic or anisocytic; petiolar vascular supply complex, commonly siphonostelic, often with included bundles; stipules present, deciduous. FLOWERS commonly large and showy, often pollinated by bats, solitary or in axillary or leaf-opposed cymose clusters, perfect, hypogynous or slightly perigynous, regular or rarely slightly irregular, often with an epicalyx; sepals 5, distinct or connate at the base, valvate, nectaries consisting of tufts of glandular hairs at the base of the sepals; petals 5, distinct, convolute, or seldom wanting; androecium served by a limited number of trunk-bundles; stamens 5 to very numerous, initiated in centrifugal sequence when numerous, adnate to the base of the petals, generally connate by their filaments into 5–15 phalanges or into a tube that may divide into 5–15 phalanges; often

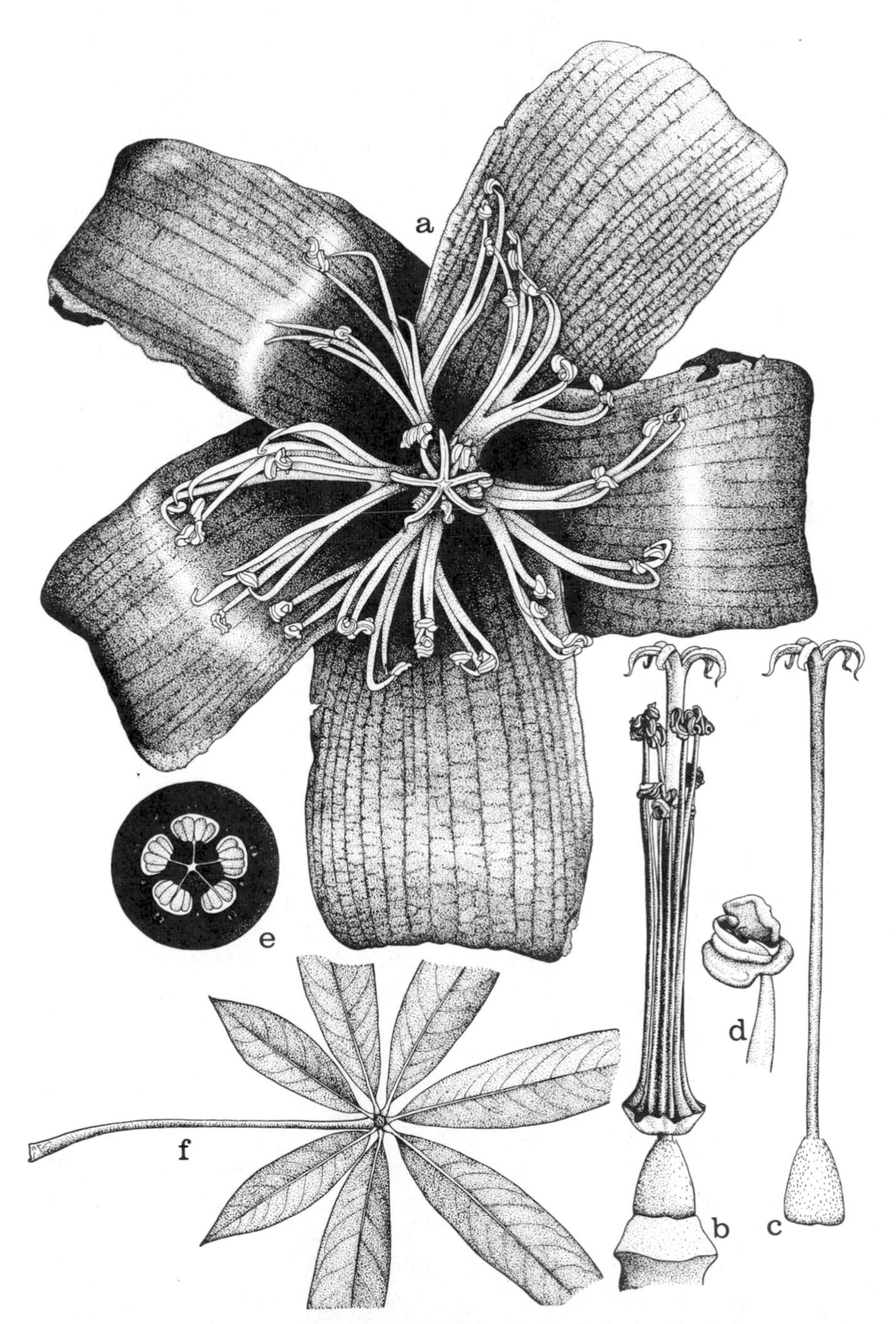

FIG. 4-11 Bombacaceae. *Bombax ceiba* L. a, flower, from above, ×1; b, pistil and stamens, ×1; c, pistil, ×1; d, anther, ×3; e, schematic cross-section of ovary, ×3; f, leaf, ×$\frac{1}{4}$.

some of the stamens staminodial; anthers opening by a longitudinal slit, monothecal, but sometimes irregularly associated on the connate filaments so that some of them may appear to have more than one theca; pollen-grains binucleate, of various shapes, most commonly oblate and angular in outline, with the apertures between the angles, triaperturate (colpate, colporate, or porate) to less often pantoporate, the surface generally smooth or reticulate-rugose, seldom minutely spinulose; gynoecium of 2–5 (–8) carpels united to form a compound ovary with a terminal, entire to deeply lobed style; ovary with as many locules as carpels, the placentation axile; ovules 2 or more in each locule, anatropous, bitegmic, crassinucellar, with zigzag micropyle; endosperm-development nuclear. FRUIT a loculicidal capsule, or seldom fleshy and indehiscent; seeds often arillate, and commonly embedded in pithy or hairy tissue derived from the inner wall of the ovary; embryo often bent and with expanded cotyledons; endosperm scanty, oily, or often wanting. X most commonly = 28, 36, 40.

The family Bombacaceae as here (and traditionally) defined, consists of some 20 to 30 genera and about 200 species, widespread in tropical countries, especially in tropical America. *Bombax*, with about 60 species, is the largest genus. Familiar members of the family include *Bombax ceiba* L., the silk-cotton tree, of tropical Asia; *Ceiba pentandra* (L.) Gaertn., the tropical American (now pantropical) kapok-tree; *Ochroma pyramidale* (Cav.) Urban, of tropical America, the source of balsa-wood; *Adansonia digitata* L., of Africa, the baobab; and *Durio zibethinus* J. Murr., of the Malay Archipelago, the durian, with a fruit famous for its delicate flavor but disagreeable odor.

The distinctive pollen of the Bombacaceae enters the fossil record in Maestrichtian deposits of southeastern United State. Progressively later entries are recorded in South America, Africa, and New Zealand.

5. Family MALVACEAE A. L. de Jussieu 1789 nom. conserv., the Mallow Family

Herbs, shrubs, or seldom small trees, generally beset with stellate hairs and sometimes also peltate scales or other sorts of hairs, with or without proanthocyanins, without ellagic acid, commonly producing cyclopropenoid fatty acids such as malvalic and sterculic acid, especially in the seeds; parenchymatous tissues characteristically with scattered mucilage-cells and often also mucilage-cavities or -canals, and frequently with

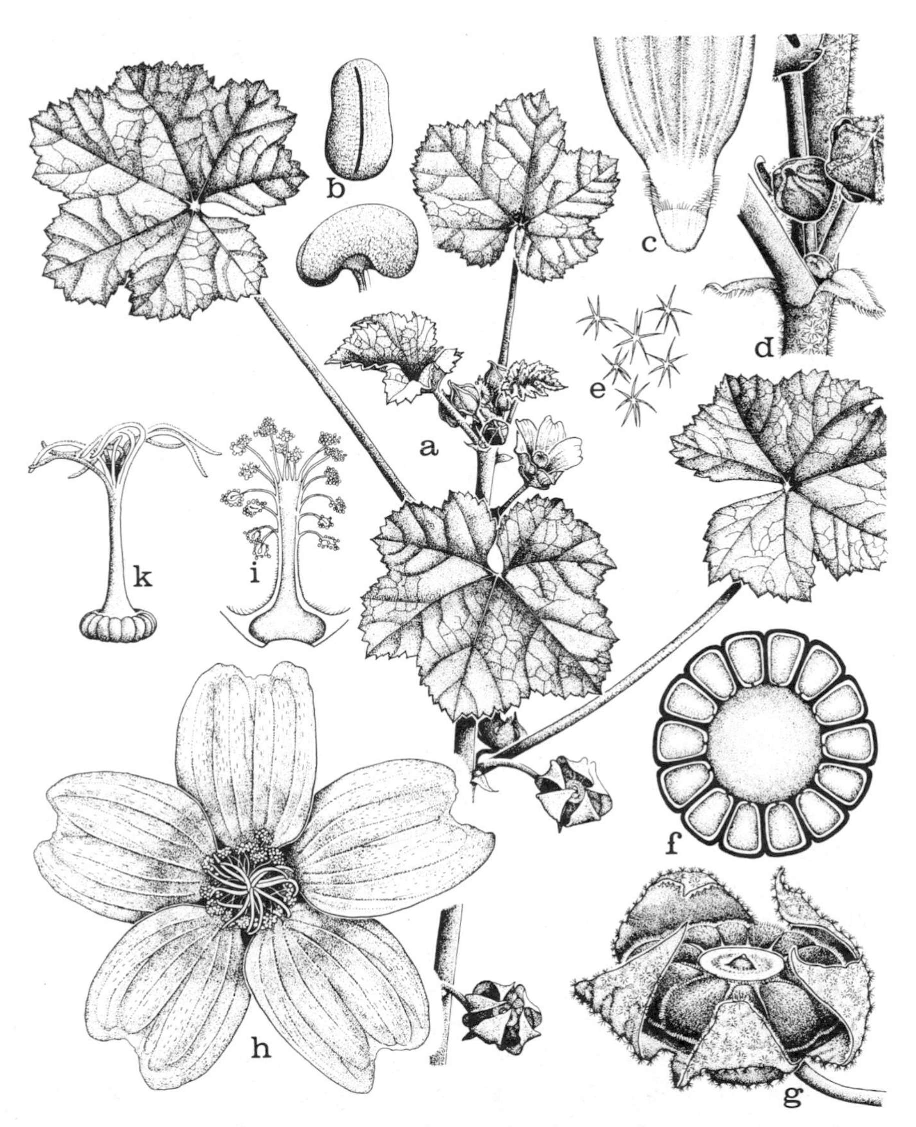

FIG. 4-12 Malvaceae. *Malva neglecta* Wallr. a, habit, × 1; b, anthers, × 16; c, base of petal, × 4; d, node, with stipules and flower buds, × 2; e, stellate hairs, × 16; f, schematic cross-section of ovary, × 16; g, fruit, with persistent calyx, × 2; h, flower, from above, × 4; i, androecium, in section, × 4; k, pistil, × 4.

other sorts of secretory cavities as well; clustered or less often solitary crystals of calcium oxalate very often present in some of the cells of the parenchymatous tissues; nodes trilacunar to multilacunar; vessel-segments with simple perforations; imperforate tracheary elements with simple or seldom bordered pits; wood-rays heterocellular to homocellular, mixed uniseriate and pluriseriate, the latter with long or short ends, often broad, often including some tile cells; wood-parenchyma variously apotracheal or paratracheal; phloem in young stems commonly stratified tangentially into hard and soft layers, and with wedge-shaped rays. LEAVES alternate, simple and entire to more or less dissected, generally palmately veined; stomates anomocytic or sometimes paracytic or anisocytic; petiolar vascular supply complex, sometimes siphonostelic, more often with a ring of separate bundles, or seldom of a single trough-shaped bundle; stipules usually present. FLOWERS solitary and axillary to more often in compound, basically cymose inflorescences, perfect or rarely unisexual, hypogynous, often with an epicalyx; sepals 5, distinct or connate at the base, valvate; nectaries consisting of tufts of glandular hairs at the base of the sepals; petals 5, distinct, often adnate to the base of the filament-tube, convolute or less often imbricate; androecium served by a limited number of trunk-bundles; stamens (5–) mostly numerous or very numerous, initiated in centrifugal sequence, the outer sometimes staminodial, the filaments all connate into a tube for most of their length; anthers bisporangiate and monothecal (thought to represent phyletically separated half-anthers), opening by a longitudinal slit; pollen-grains binucleate, from tricolporate to more often spherical and pantoporate, generally spinulose, rarely essentially smooth; gynoecium of (1) 2-many (often 5) carpels united to form a compound ovary with as many or (Ureneae) twice as many styles as carpels, the styles distinct or more often connate at least below, the stigmas dry, usually papillate, distinct, capitate or discoid to decurrent; ovary with as many locules as carpels, the placentation axile; ovules 1-many in each locule, anatropous to campylotropous, bitegmic, crassinucellar, with a zigzag micropyle; endosperm-development nuclear. FRUIT a loculicidal capsule, or often separating into mericarps, or seldom a berry or a samara; cotyledons 2, mostly folded; endosperm oily and proteinaceous, abundant to scanty, or wanting. X=6–17+, 20+.

The family Malvaceae, as here (and traditionally) defined, consists of about 75 genera and a thousand to 1500 species, essentially cosmopolitan in distribution, but best developed in the tropics. The largest genera are *Hibiscus* (200+), *Sida* (175+), *Pavonia* (150+), and *Abutilon* (100+).

The family includes a number of well known economic and ornamental plants. Cotton consists of the seed-hairs of species of *Gossypium*. Okra is the fruit of *Abelmoschus esculentus* (L.) Moench. *Alcea rosea* L. is the garden hollyhock. *Hibiscus syriacus* L., called rose-of-Sharon, is also known as *Althaea*, a name that properly belongs to another genus of the family.

The Malvaceae are obviously related to the Bombacaceae. The line between the two families is rather vague, and some genera have been shifted back and forth. The most radical proposal is that of Edlin (1935), who would restrict the Malvaceae to the genera with schizocarpic fruits, and transfer all the capsular genera (including, e.g., *Gossypium* and *Hibiscus*) to the Bombacaceae. Such a definition has the virtue of making the Malvaceae a more homogeneous and sharply defined group, but it makes the Bombacaceae more heterogeneous. Modern opinion emphasizes the sculpture of the pollen-grains as the most nearly constant distinction between the two families (spinulose in the Malvaceae, smooth or nearly so in the Bombacaceae). Reliance on pollen-structure as the fundamental criterion bolsters the traditional, though imprecise habital distinction. All the herbs and all or nearly all the shrubs go with the Malvaceae, and nearly all the trees go with the Bombacaceae.

Gottsberger (1972) proposes that the Malvaceae originated in the early Tertiary in neotropical forests as a primarily ornithophilous branch of the Tiliaceae, and that chiroptery and entomophily came later. This proposal can be properly evaluated only in the context of a more comprehensive future interpretation of the evolutionary history of the Tiliaceae, Sterculiaceae, Bombacaceae, and Malvaceae. Pollen considered to represent the Malvaceae enters the fossil record in the late Eocene, distinctly later than that of the Tiliaceae and Bombacaceae.

Van Heel (1978) reports that in the tribe Ureneae (which has a 5-locular ovary but 10 style-branches), 10 carpels are initiated in two successive whorls. The carpels of the inner set abort except for their styles.

4. Order LECYTHIDALES Cronquist 1957

The order consists of the single family Lecythidaceae.

SELECTED REFERENCES

Diehl, G. A. 1935. A study of the Lecythidaceae. Trop. Woods 43: 1–15.
Kowal, R. I., S. A. Mori, & J. A. Kallunki. 1977. Chromosome numbers of Panamanian Lecythidaceae and their use in subfamilial classification. Brittonia 29: 399–410.
Leins, P. 1972. Das zentrifugale Androeceum von *Couroupita guianensis* (Lecythidaceae). Beitr. Biol Pflanzen 48: 313–319.
Mori, S. A., J. E. Orchard, & G. T. Prance. 1980. Intrafloral pollen differentiation in the New World Lecythidaceae, subfamily Lecythidoideae. Science 209: 400–403.
Mori, S. A., G. T. Prance, & A. B. Bolten. 1978. Additional notes on the floral biology of neotropical Lecythidaceae. Brittonia 30: 113–130.
Muller, J. 1972. Pollen morphological evidence for subdivision and affinities of Lecythidaceae. Blumea 20: 350–355.
Prance, G. T. 1976. The pollination and androphore structure of some Amazonian Lecythidaceae. Biotropica 8: 235–241.
Prance, G. T., & S. A. Mori. 1979. Lecythidaceae. Part I. The actinomorphic-flowered New World Lecythidaceae (*Asteranthos*, *Gustavia*, *Grias*, *Allantoma*, & *Cariniana*). Fl. Neotropica Monograph 21.
Venkateswarlu, J. 1952. Embryological studies in Lecythidaceae. I. J. Indian Bot. Soc. 31: 103–116.
Weberling, F. 1958. Über das Vorkommen rudimentärer Stipeln bei den Lecythidaceae (s.l.) und Sonneratiaceae. Flora 145: 72–77.

1. Family LECYTHIDACEAE Poiteau 1825 nom. conserv., the Brazil-nut Family

Trees or sometimes shrubs, typically accumulating triterpenoid saponins, commonly also producing proanthocyanins and ellagic acid, often with scattered tanniferous cells, and sometimes (as in *Lecythis*) with mucilage-canals; nodes unilacunar; vessel-segments with simple perforations, or sometimes some of them with scalariform perforation-plates; wood-rays heterocellular or sometimes homocellular, mixed uniseriate and pluriseriate, the latter 2–7 (–15) cells wide, with short ends, often some of the cells containing silica bodies; wood-parenchyma commonly abundant and mainly or wholly in apotracheal bands; young stems with cortical vascular bundles and with the phloem tangentially stratified into hard and soft layers, sometimes (as in *Napoleonaea*) with wedge-shaped rays. LEAVES alternate, commonly crowded at the tips of the twigs, simple, entire or toothed; stomates commonly anisocytic; petiole

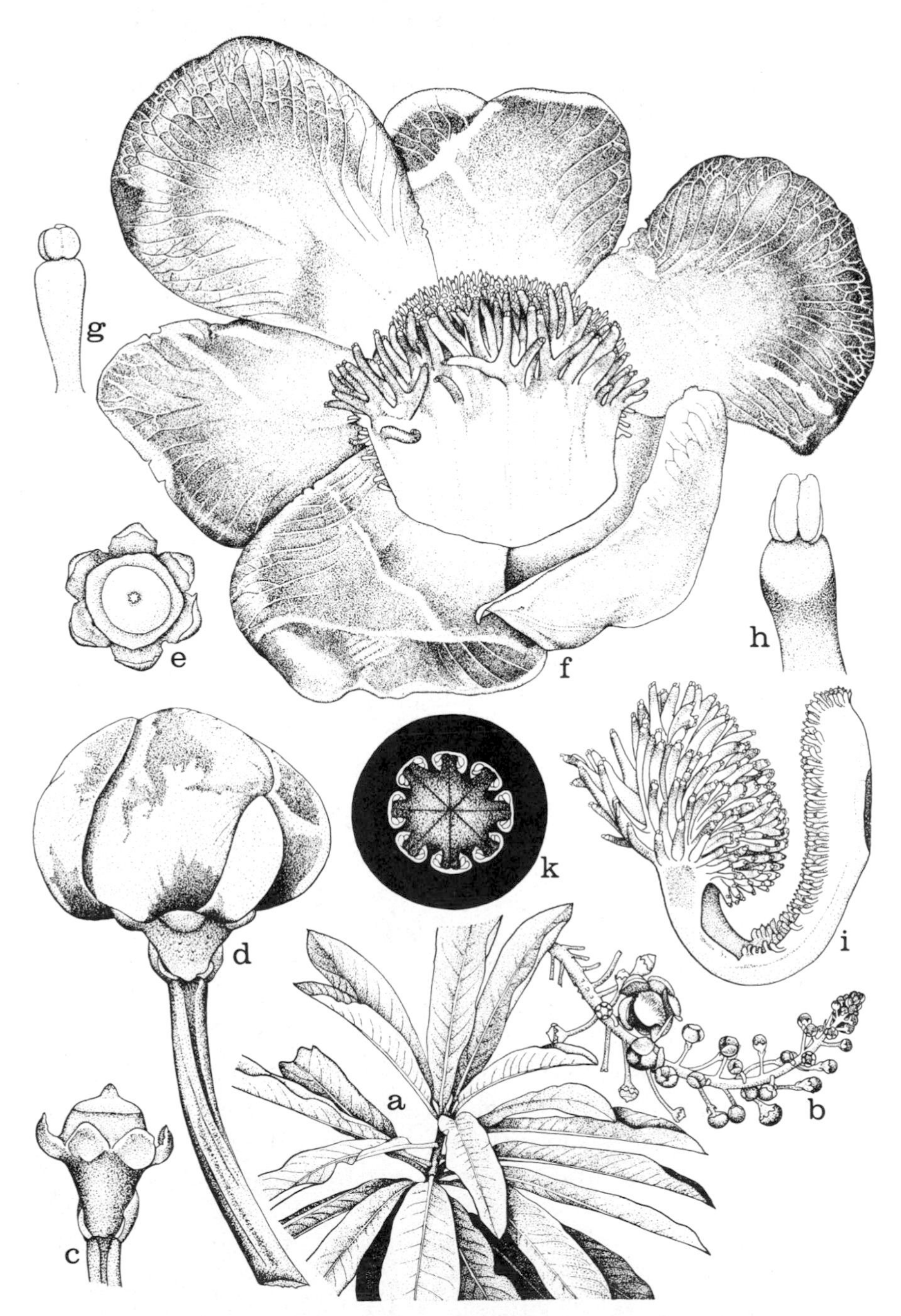

FIG. 4-13 Lecythidaceae. *Couroupita guianensis* Aubl. a, habit, $\times\frac{1}{8}$; b, inflorescence, $\times\frac{1}{8}$; c, gynoecium and calyx, side view, $\times 1$; d, flower bud, $\times 1$; e, gynoecium and calyx, from above, $\times 1$; f, flower, from above, $\times 1$; g, stamen from inner part of androecium, $\times 9$; h, stamen from distal part of androecium, $\times 9$; i, androecium, $\times 1$; k, diagrammatic cross-section of ovary, $\times 2$.

with complex vascular anatomy, typically containing numerous vascular bundles; stipules wanting, or small and caducous. FLOWERS typically ephemeral, pollinated by insects or bats, providing pollen or nectar to pollinators, solitary in the axils of leaves or bracts, or in axillary or terminal racemes, or in terminal panicles, or in fascicles or racemes from old wood, commonly rather large and showy, perfect, regular or irregular, epigynous or seldom only half-epigynous, the hypanthium sometimes prolonged beyond the ovary; calyx of (2–) 4–6 (–12) sepals, these imbricate or (in some Old World genera) valvate, or rarely fully connate and calyptrate; petals 4–6, distinct, imbricate, or sometimes wanting; stamens (10–) numerous, up to about 1200, in the more archaic genera symmetrically disposed in several centrifugally developing series, the filaments connate below and distinct above, the connate parts called the staminal ring, in the more advanced genera the staminal ring asymmetrical, produced on one side into a flat ligule that may be curved over the ovary as a hood, some or all of the ligular stamens often reduced and staminodial; in some of the apetalous genera the outer set of stamens wholly sterile and united to form an erect or spreading, toothed corona; intrastaminal nectary-disk often well developed in Old World genera, sometimes enlarged and more or less covering the top of the ovary, or scarcely developed, especially in New World genera; pollen-grains trinucleate, tricolporate or trisynocolpate; anthers tetrasporangiate and dithecal, opening by longitudinal slits or seldom by apical pores; gynoecium of 2–6 carpels united to form an inferior or seldom only half-inferior ovary with a terminal (sometimes short), simple style and a capitate or lobed stigma, or the style seldom 3–4-lobed; ovary with as many locules as carpels, the placentation axile or apical-axile or basal-axile or (*Eschweilera*) strictly basal; ovules 1-many in each locule, anatropous, bitegmic, tenuinucellar, with an integumentary tapetum; endosperm-development nuclear. FRUIT capsular with a distal operculum, often very large, or less often drupaceous or baccate; seeds commonly nutlike or samaroid, often with a funicular aril, usually without endosperm (endosperm well developed only in the monotypic genus *Asteranthos*); embryo large, oily and proteinaceous; cotyledons 2, either laminar or fleshy-thickened, or virtually suppressed, the embryo sometimes consisting mainly of the much-thickened hypocotyl; alternately arranged cataphylls often present between the cotyledons and the first ordinary leaves. X = 13 (Planchonioideae), 16 (Napoleonaeoideae), 17 (Lecythidoideae). (Asteranthaceae, Barringtoniaceae, Foetidiaceae, Napoleonaeaceae)

The family Lecythidaceae consists of some 20 genera and about 400 species, confined to tropical regions, best developed in rain forests, especially in South America. The largest genus is *Eschweilera*, with nearly a hundred species. *Gustavia* (40) is considered to be the most archaic genus. The seeds of *Bertholletia excelsa* Humboldt & Bonpland are the Brazil-nuts of commerce.

The striking differences in the androecium of various members of the Lecythidaceae have led some authors to recognize several segregate families, but these must in any case stand side by side in the system of classification. I see nothing to be gained by elevating the subfamilies to familial rank.

The complex evolutionary series of modifications of the androecium of the Lecythidaceae is considered to reflect progressive restriction of possible pollinators. The cortical vascular bundles and the bitegmic, tenuinucellar structure of the ovules are more difficult to explain in Darwinian terms. Epigyny among angiosperms in general has been claimed by some authors to function in protection of the ovules, but I remain skeptical.

Pollen considered to represent the genus *Barringtonia* is known from lower Eocene and more recent deposits.

The Lecythidaceae have traditionally been referred to the order Myrtales because of their combination of separate petals, numerous stamens, and syncarpous, inferior ovary with axile placentation. They differ from characteristic members of the Myrtales, however, in their alternate leaves, bitegmic, tenuinucellar ovules, lack of internal phloem, and a series of embryological features that have been elucidated by Mauritzon (1939). The differences are too formidable to ignore, and the Lecythidaceae must be removed from the Myrtales. No other order can accommodate the Lecythidaceae without undue strain, and it therefore becomes necessary to recognize an order Lecythidales.

Once the Lecythidales are removed from the Myrtales, their proper position in the general system can be considered de novo. The Rosidae and Dilleniidae are the only subclasses that can be seriously considered as a possible haven.

Althouth no one feature is by itself definitive, the Lecythidales are much more at home in the Dillenidae than in the Rosidae. The bitegmic, tenuinucellar ovules of the Lecythidales may be particularly significant. As Philipson (e.g., 1977) has emphasized, this type of ovule is common in the Theales, Primulales, and Ebenales (all members of the Dilleniidae), but rare and scattered elsewhere except for the Lecythidales.

Although centrifugal stamens have been demonstrated in the Myrtales, this kind of androecium is much more common in the Dilleniidae than in the Rosidae. Stratified phloem occurs in scattered families in the Magnoliidae, Hamamelidae, Rosidae, and Asteridae, but is especially common in the Dilleniidae, notably in the Malvales and to a lesser extent in the Theales. Wedge-shaped phloem-rays are likewise more common in the Malvales and Theales than in most other groups. *Barringtonia*, in the Lecythidaceae, produces 3-sambubiosides, similar to those of *Hibiscus*. Not much weight can be placed on the presence of these rare anthocyanins until their taxonomic distribution is more fully documented, but they do provide another straw in the wind.

Within the Dilleniidae, the Lecythidales differ from other polyandrous groups by the combination of epigynous flowers and axile placentation. Epigyny is obviously an advanced condition, so it is reasonable to seek the ancestry of the Lecythidales among those orders that have a superior, syncarpous ovary and numerous centrifugal stamens.

The Theales and Malvales immediately present themselves as possible relatives. The Lecythidales resemble characteristic members of the Malvales in their frequently valvate calyx, in their connate filaments, and in their stratified phloem, but they lack the stellate pubescence of the Malvales, and they have not been demonstrated to have the characteristic malvalean cyclopropenoid fatty acids. The prominent nectary-disk internal to the stamens in some genera is quite out of harmony with the Malvales, but perfectly compatible with an ancestry in the Theales. The Lecythidales resemble several families of the Theales (but not Malvales) in their bitegmic, tenuinucellar ovules, and they resemble the Ochnaceae (Theales) in having cortical vascular bundles. They resemble both the Theales and the Malvales in the complex vascular anatomy of the petiole. The enlarged hypocotyl of many Lecythidales might be compared with that of the Caryocaraceae and Marcgraviaceae in the Theales. Muller finds that the pollen structure is readily compatible with that of the Theales and Malvales, as well as with that of several other orders not under consideration here, but difficult to reconcile with that of the Myrtales.

My interpretation of this set of similarities and differences is that the Lecythidales and Malvales have undergone partly parallel and partly divergent specializations from a common ancestry in the Theales.

5. Order NEPENTHALES Lindley 1833

Insectivorous perennial herbs, half-shrubs, or shrubs, sometimes climbing or epiphytic, tanniferous, commonly producing proanthocyanins and sometimes also ellagic acid, mostly fibrous-rooted. LEAVES simple, alternate or seldom whorled, modified in one way or another to catch insects, often forming pitchers, with or without small stipules; stomates anomocytic. FLOWERS regular, hypogynous, perfect or unisexual; sepals (3) 4–5 (–8), distinct or connate below, imbricate, often persistent; petals distinct, generally as many as the sepals, or wanting; stamens 4-numerous, distinct or united by their filaments; pollen variously in monads or tetrads, binucleate or trinucleate; gynoecium of 3–5 carpels united to form a compound, plurilocular or unilocular ovary with axile, parietal, or basal placentation, the styles variously distinct (and sometimes bifid) or united, or the stigma sometimes sessile; ovules (3–) numerous, anatropous, variously bitegmic or unitegmic, crassinucellar or tenuinucellar; endosperm-development nuclear or cellular. FRUIT a loculicidal capsule, or rarely indehiscent; seeds (3–) numerous, with tiny to straight and elongate embryo surrounded by endosperm. (Sarraceniales)

The order Nepenthales as here defined consists of 3 well marked small families, scarcely 200 species in all. The Droseraceae have about a hundred species, the Nepenthaceae about 75, and the Sarraceniaceae only about 15. A fourth family, the Byblidaceae (including Roridulaceae), has often been referred to this order, but is now generally associated with the Pittosporaceae in the Rosales. The ordinal name Nepenthales (1833) is here used in preference to the later name Sarraceniales (1892).

The mutual affinity of the three families of Nepenthales has been affirmed and denied by various authors, and competent opinion is still divided. Any two of the three have been associated, to the exclusion of the third. I claim no special expertise in the group, but I am more impressed by the similarities than by the differences. In addition to the obvious exomorphic characters, the Droseraceae and Nepenthaceae have very similar pollen. (Takhtajan, on the other hand, points out the similarity of the pollen of *Sarracenia* to that of *Dendromecon,* in the Papaveraceae.) Markgraf (1955) has concluded that the insect-catching leaves in all three families are homologous, in spite of the obvious differences between the Droseraceae and the other two families. All considered, there is perhaps a little more reason to question the relationship of the Droseraceae to the Sarraceniaceae and Nepenthaceae than the relationship of the latter two families to each other.

Opinion is also divided about the affinities of the families of Nepenthales. Members of the Theales, Violales, Rosales, and Papaverales as here defined have been suggested as allies of one or more of the families. Some authors have changed positions in succeeding publications. Thus in 1966 Takhtajan assigned the Nepenthaceae and Sarraceniaceae to the order Nepenthales, following on the Papaverales in the subclass Magnoliidae, and included the Droseraceae in his Saxifragales, immediately after the Parnassiaceae. In 1973 he kept only the Sarraceniaceae, as an order Sarraceniales, in alliance with the Papaverales, and treated the Nepenthales, including Droseraceae, as an order of Rosidae allied to the Saxifragales. In 1980 he moved the Droseraceae to a position in the Saxifragales, so that the 3 families were in 3 different orders, two of them monotypic.

In my opinion the ancestry of the Nepenthales is to be sought in the Theales. Except for their insectivorous habit, the Nepenthales would fit very well into the Theales (assuming that the stamens of Sarraceniaceae turn out to be centrifugal, a point not yet fully established). Inasmuch as two of the families have axile placentation, the Violales do not seem very likely ancestors. The similarities of *Ancistrocladus* and *Dioncophyllum*, in the Violales, to members of the Nepenthales (especially *Nepenthes*) are here regarded as reflecting a common ancestry rather than a more direct relationship.

None of the three families of Nepenthales can be considered ancestral to any of the others. They represent distinct lines that have undergone more or less similar changes from a similar ancestry.

The insectivorous habit of the Nepenthales may be presumed to be an evolutionary response to their growth in habitats deficient in available nitrogen. The Sarraceniaceae and Droseraceae commonly grow in water-logged soils containing little or no soluble nitrate. The Nepenthaceae occur in wet, tropical forests, which characteristically have nutrient-poor soils. Many other groups of plants have faced similar problems, but very few have learned to meet them by trapping insects.

Aside from the insect-catching apparatus, the characters that mark the order and the individual families are of doubtful ecological significance.

SELECTED REFERENCES

Arber, A. 1941. On the morphology of the pitcher-leaves in *Heliamphora*, *Sarracenia*, *Darlingtonia*, *Cephalotus*, and *Nepenthes*. Ann. Bot. (London) II. 5: 563–578.

Basak, R. K., & K. Subramanyan. 1966. Pollen grains of some species of *Nepenthes*. Phytomorphology 16: 334–338.
Bell, C. R. 1949. A cytotaxonomic study of the Sarraceniaceae of North America. J. Elisha Mitchell Sci. Soc. 65: 137–166.
Chanda, S. 1965. The pollen morphology of Droseraceae with special reference to Taxonomy. Pollen & Spores 7: 509–528.
DeBuhr, L. E. 1975. Phylogenetic relationships of the Sarraceniaceae. Taxon 24: 297–306.
DeBuhr, L. E. 1977. Wood anatomy of the Sarraceniaceae; ecological and evolutionary implications. Pl. Syst. Evol. 128: 159–169.
Franck, D. H. 1975. Early histogenesis of the adult leaves of *Darlingtonia californica* (Sarraceniaceae) and its bearing on the nature of epiascidiate foliar appendages. Amer. J. Bot. 62: 116–132.
Kuprianova, L. A. 1973. Pollen morphology within the genus *Drosera*. Grana 13: 103–107.
McDaniel, S. 1971. The genus *Sarracenia* (Sarraceniaceae). Bull. Tall Timbers Res. Sta. #9, Sept. 1971: 1–36.
Maguire, B. 1978. Sarraceniaceae. *In:* The botany of the Guayana Highland—Part X. Mem. New York Bot. Gard. 29: 36–62.
Markgraf, F. 1955. Über Laubblat-Homologien und verwandtschaftliche Zusammenhänge bei Sarraceniales. Planta 46: 414–446.
Ragleti, H. W. J., M. Weintraub, & E. Lo. 1972. Characteristics of *Drosera* tentacles. I. Anatomical and cytological detail. Canad. J. Bot. 50: 159–168.
Roth, I. 1953. Entwicklungsgeschichtliche und histogenetische Studien an *Sarracenia*-Schlauchblättern. Planta 43: 133–162.
Roth, I. 1953. Zur Entwicklungsgeschichte und Histogenese der Schlauchblätter von *Nepenthes*. Planta 42: 177–208.
Sahashi, N., & M. Ikuse. 1973. Pollen morphology of *Aldrovanda vesiculosa* L. J. Jap. Bot. 48: 374–379.
Schmid, R. 1970. *Nepenthes*-Studien I: Homologien von Deckel (operculum, lid) und Spitzchen (calcar, spur). Bot. Jahrb. Syst. 90: 275–296.
Wood, C. E. 1960. The genera of Sarraceniaceae and Droseraceae in the southeastern United States. J. Arnold Arbor. 41: 152–163.

SYNOPTICAL ARRANGEMENT OF THE FAMILIES OF NEPENTHALES

1 Leaves, or some of them, modified to form pitchers; ovary plurilocular, with axile placentation; style solitary, or sometimes very short or none.

  2 Flowers perfect; filaments distinct; pollen-grains in monads; ovules tenuinucellar; terrestrial herbs, not climbing; New World .... 1. SARRACENIACEAE.

  2 Flowers unisexual; filaments united into a column; pollen-grains in tetrads; ovules crassinucellar; herbs or more often shrubs or half-shrubs, often climbing or epiphytic; Old World .... 2. NEPENTHACEAE.

1 Leaves not forming pitchers; ovary unilocular, with parietal or basal placentation; styles usually distinct (and often deeply bifid), only seldom united; cosmopolitan herbs .... 3. DROSERACEAE.

1. Family SARRACENIACEAE Dumortier 1829 nom. conserv., the Pitcher-plant Family

Perennial, rhizomatous, mostly acaulescent, fibrous-rooted insectivorous herbs, tanniferous, producing proanthocyanins but not ellagic acid and sometimes producing alkaloids; clustered crystals of calcium oxalate sometimes present in some of the cells of the parenchymatous tissues; rhizome with vascular bundles of varying size and shape, forming an irregular ring interrupted by medullary rays of unequal width; vessel-segments with oblique, scalariform perforations; imperforate tracheary elements with bordered pits, considered to be tracheids. LEAVES alternate, borne in a basal rosette or (species of *Heliamphora*) on an upright, sometimes even branching and somewhat scrambling stem, highly modified, the principal ones ascidiate, forming pitcher-like traps that are partly filled with digestive liquid; petiole short, with complex vascular anatomy, passing into the hollow, often more or less elongate central portion of the leaf, which bears a more or less well developed ridge or laminar wing on the ventral side and a flattened but relatively small, often hood-like blade as a prolongation on the dorsal side, the terminal or subterminal opening of the trap thus more or less distinctly ventral in orientation; epidermis of the outer side of the trap provided with nectar-glands and often also with stiff, antrorse hairs, that of the inner side also glandular (or the glands restricted to the hood) and provided with stiff, retrorse hairs, the lower part of the trap smooth within; stomates anomocytic; stipules wanting; some reduced, scale-like or sword-like leaves often produced late in the season. FLOWERS large, solitary on a scape (or in *Heliamphora* in few-flowered, sometimes axillary racemes), nodding, perfect, regular, hypogynous; sepals (3–) 5 (6), distinct, imbricate, persistent, often colored and somewhat petaloid; petals 5, distinct, imbricate, deciduous (wanting in *Heliamphora*); stamens (10–) numerous, in *Sarracenia* several arising from each of a limited number (commonly 10) of primordia, initiated in centrifugal sequence(?), in the other genera each stamen with its own vascular trace and primordium; anthers basifixed or (*Sarracenia*) versatile, tetrasporangiate and dithecal; pollen-grains borne in monads, binucleate, (3) 4– to multicolporate, nearly smooth to finely tuberculate, but not spinulose; gynoecium of 5 (in *Heliamphora* 3) carpels united to form a compound ovary with as many locules as carpels; partitions in the upper part of the ovary often not meeting or not joined, so that the placentation is axile below and intruded-parietal above; style solitary, subentire (*He-*

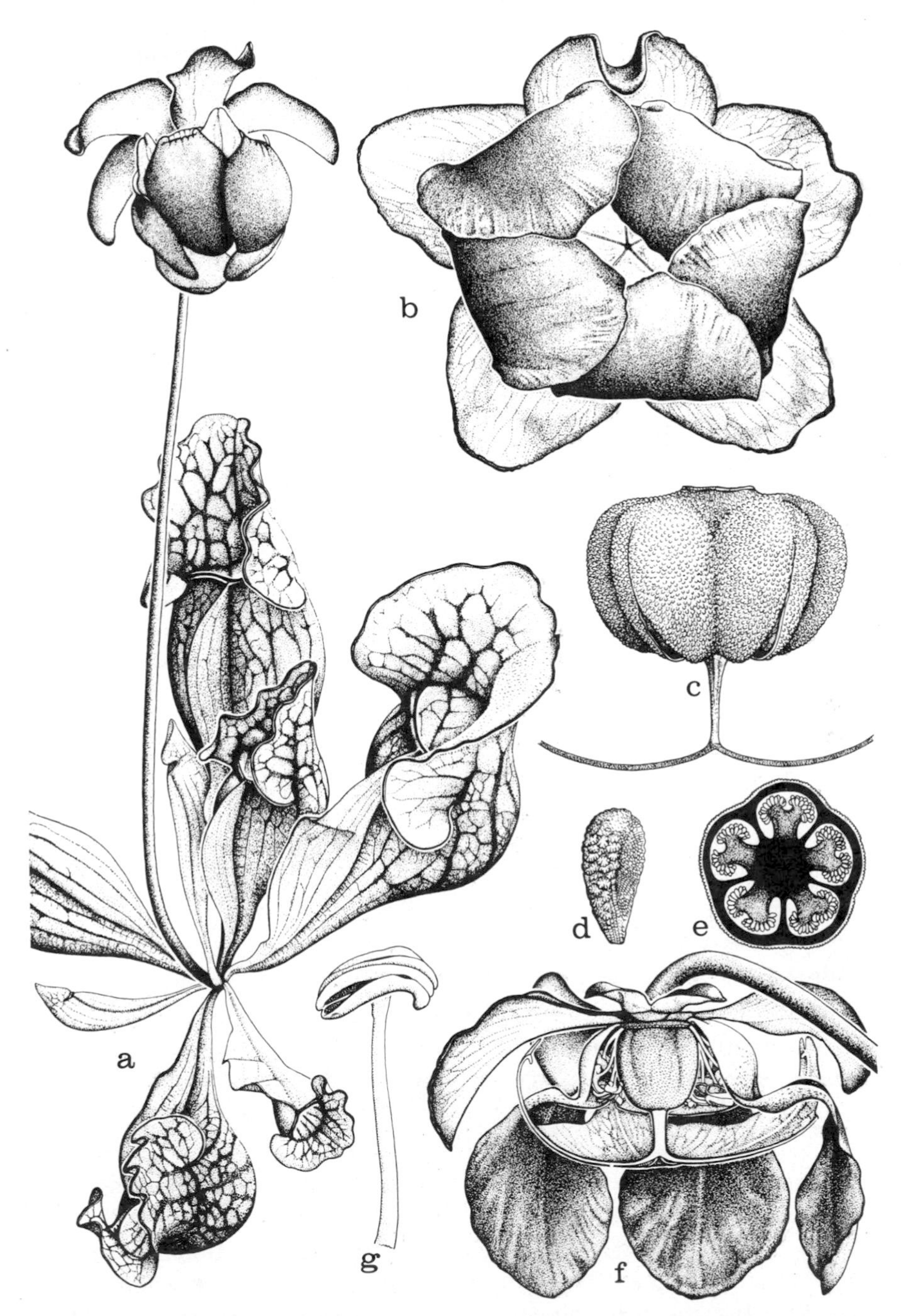

FIG. 4-14 Sarraceniaceae. *Sarracenia purpurea* L. a, habit, ×½; b, flower, ×1; c, fruit, ×2; d, seed, ×16; e, diagrammatic cross-section of ovary, ×2; f, flower, in partial section, ×1; g, stamen, ×4.

liamphora), or with 5 short branches, each with a terminal stigma (*Darlingtonia*), or expanded and peltate or umbrella-like with a small stigma under the tip of each of the 5 lobes (*Sarracenia*); ovules numerous, anatropous, unitegmic or bitegmic, tenuinucellar, with an integumentary tapetum. FRUIT a loculicidal capsule; seeds numerous, small, often with a wing-like beak; embryo minute, linear, dicotyledonous, basal, embedded in the copious, firm-fleshy, oily and proteinaceous endosperm. X = 13 (*Sarracenia*), 15 (*Darlingtonia*), 21 (*Heliamphora*).

The family Sarraceniaceae consists of only 3 genera: *Sarracenia*, with 8 species in eastern North America; *Darlingtonia*, with a single species near the coast of California and Oregon; and *Heliamphora*, with 6 species on isolated, table-top mountains in the Guayana Highlands of northern South America.

The Sarraceniaceae are the New World correlatives of the Old World family Nepenthaceae. The two families are sharply distinct. It may be significant that *Heliamphora* is the genus of Sarraceniaceae that is least unlike *Nepenthes*. The Guayana Highlands harbor a number of other relic genera that help to connect Old World and New World groups in other orders. DeBuhr (1977) considers that "the wood anatomy suggests that *Heliamphora* is growing in a habitat more similar to the original habitat for the family than *Darlingtonia* and *Sarracenia*. The wood of the Sarraceniaceae is similar to the wood of Theales."

2. Family NEPENTHACEAE Dumortier 1829 nom. conserv., the East Indian Pitcher-plant Family

Insectivorous, erect or prostrate to more often climbing shrubs or half-shrubs, often epiphytic, tanniferous, producing proanthocyanins but not ellagic acid, often with scattered mucilage-cells or tanniferous cells, and often with cortical or medullary vascular bundles in addition to the principal bundles, which are numerous and form a ring traversed by narrow medullary rays; clustered crystals of calcium oxalate sometimes present in some of the cells of the parenchymatous tissues; vessel-segments with transverse ends and simple perforations; imperforate tracheary elements with bordered pits. LEAVES alternate, highly modified, consisting, when fully developed, of a basal petiole, bearing a flattened blade, which is apically narrowed into a rather stout tendril that connects to a large, open pitcher (ascidium) with an expanded, flattened, dorso-terminal flap (operculum); complex, multicellular nectar-glands and peltate hydathodes widely distributed on the stem and

FIG. 4-15 Nepenthaceae. a–d, *Nepenthes maxima* Reinw. a, habit, ×⅛; b, pitcher, ×½; c, another view of base of pitcher, ×½; d, mouth of pitcher, ×½. e–g, *Nepenthes* hybrid. e, staminate inflorescence, ×1; f, staminal head, from above, ×6; g, staminate flower, ×3.

leaves, and digestive glands also present within the pitcher, which is partly filled with digestive liquid; stomates anomocytic; stipules wanting. FLOWERS in racemes or mixed panicles, small, regular, hypogynous, apetalous, unisexual, the plants dioecious; sepals (3) 4, imbricate, distinct or seldom basally connate, glandular and nectariferous within; stamens (4–) 8–25; filaments united into a central column; anthers crowded, tetrasporangiate and dithecal; pollen-grains borne in tetrads, spinulose, with indistinct apertures; gynoecium of (3) 4 carpels united to form a compound ovary with as many locules as carpels; style very short or none; stigma dry, papillate, discoid; ovules numerous, multiseriate on axile placentas, anatropous, bitegmic, crassinucellar. FRUIT a loculicidal capsule; seeds numerous, filiform, with a small, straight, cylindrical, dicotyledonous embryo surrounded by starchy (as well as oily and proteinaceous) endosperm, or without endosperm.

The family Nepenthaceae consists of the single genus *Nepenthes*, with about 75 species ranging from the East Indies to Madagascar, barely encroaching onto the mainland of northern Australia and southeastern Asia. The insectivorous habit of the group may reasonably be supposed to have facilitated the evolution of epiphytes, by reducing the need to obtain nitrogen and phosphorus from the soil.

3. Family DROSERACEAE Salisbury 1808 nom. conserv., the Sundew Family

Insectivorous perennial herbs or seldom (*Drosophyllum*) half shrubs, usually or always without a functional cambium, commonly acaulescent and with a rosette of leaves, mostly fibrous-rooted, tanniferous, producing proanthocyanins and often ellagic acid, commonly also accumulating naphthaquinones, and often cyanogenic. LEAVES alternate or seldom whorled, often circinate in bud, simple, the blade either modified as an active trap (*Aldrovanda*, *Dionaea*) or provided with irritable, mucilage-tipped tentacle-hairs (*Drosera*, *Drosophyllum*); stomates anomocytic; stipules often present. FLOWERS in cymose inflorescences (solitary in *Aldrovanda*), perfect, regular, hypogynous, the calyx, corolla, and stamens typically withering persistent; sepals (4) 5 (–8), more or less connate at the base, imbricate; petals as many as the sepals, distinct, convolute; stamens (4) 5 (10–20), distinct or (*Dionaea*) connate at the base; pollen in tetrads or (*Drosophyllum*) in monads, (2) 3-nucleate, spinulose, tricolpate or triporate to more often multiporate (in *Drosophyllum* pantoporate with 30–35 pores), or with indistinct apertures,

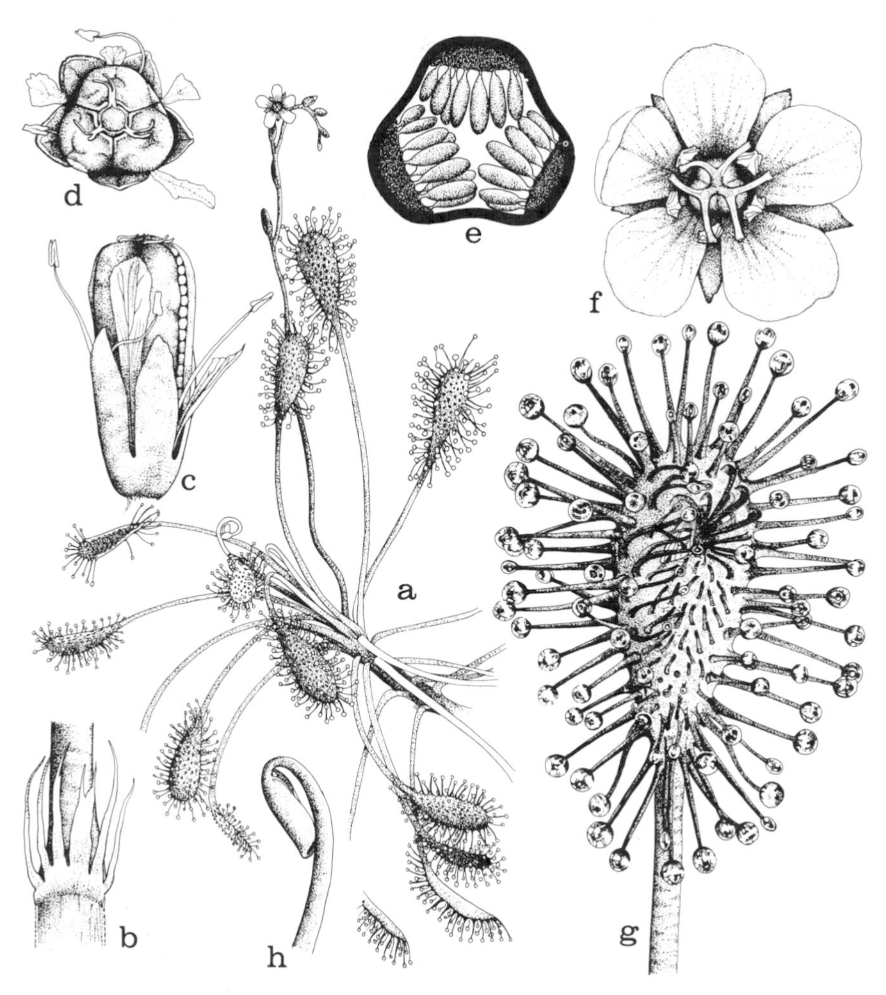

FIG. 4-16 Droseraceae. *Drosera intermedia* Hayne a, habit, ×1; b, intrapetiolar stipule, ×4; c, d, mature fruit, ×4; e, cross-section of ovary, ×20; f, flower, from above, ×4; g, leaf, ×4; h, leaf bud, ×4.

said to be intectate except in *Drosophyllum*; gynoecium of 3 (5) carpels united to form a compound, unilocular ovary; styles distinct and often also deeply bifid, or (*Dionaea*) united to form a common style; stigmas terminal, dry, not papillate; ovules (3–) numerous, on parietal placentas or on an essentially basal placenta, anatropous, bitegmic, either crassinucellar or tenuinucellar; endosperm-development nuclear. FRUIT a loculicidal capsule, or rarely indehiscent; seeds 3-many, spindle-shaped, with short, straight, basal, dicotyledonous embryo embedded in the copious, crystalline-granular, starchy (as well as oily and proteinaceous) endosperm. X=6–17+. (Dionaeaceae)

The family Droseraceae consists of 4 genera: *Aldrovanda*, *Dionaea*, *Drosera*, and *Drosophyllum*. *Drosera*, the sundew, is widespread in both temperate and tropical regions. It has about a hundred species. The other three genera are monotypic. *Dionaea*, of southeastern United States, the Venus' flytrap, has attracted a great deal of scientific and popular interest because of the active trap-mechanism of its leaves. Fossil pollen thought to represent the Droseraceae occurs in Eocene and more recent deposits. Macrofossils thought to represent *Aldrovanda* occur in Eocene deposits in England.

6. Order VIOLALES Lindley 1833

Trees, shrubs, lianas, or herbs with various sorts of chemical defenses, often cyanogenic, with or without proanthocyanins and ellagic acid, sometimes producing latex or mucilage or alkaloids or diverse other repellants, but only rarely with mustard oil, and without iridoid compounds except in the Loasaceae and Fouquieriaceae; cyclopropenoid fatty acids wanting; trichomes of various sorts, but only seldom stellate or peltate (as for example in the Cistaceae); vessel-segments with simple or less often scalariform perforations; imperforate tracheary elements with simple or bordered pits; phloem only seldom stratified or with wedge-shaped rays; sieve-tubes with S-type plastids. LEAVES alternate or opposite, simple or occasionally compound, with or without stipules. FLOWERS perfect or unisexual, regular or less often evidently irregular, hypogynous to perigynous or epigynous; sepals (3) 4–5 (–15), most commonly imbricate or convolute, seldom valvate, distinct or sometimes connate below, seldom connate to form a tube with terminal teeth, or initially closed and then splitting irregularly; petals (3) 4–5 (–15), most commonly imbricate or convolute, but sometimes valvate or induplicate-valvate, distinct or less often united to form a sympetalous corolla, or wholly wanting; stamens (1–) 5-many, sometimes connate into bundles or monadelphous, when numerous initiated (so far as known) in centrifugal sequence, except in the Begoniaceae and the subfamily Mentzelioideae of the Loasaceae; pollen-grains binucleate or seldom trinucleate, 2– to multiaperturate, most commonly tricolporate; gynoecium of (2) 3–5 (–13) but most often 3 carpels united to form a compound, typically unilocular ovary with parietal (seldom apical or basal) placentation, but the placentas often more or less deeply intruded as partial partitions, and the partitions occasionally meeting and joined in the center so that the ovary is plurilocular with axile placentas, or the ovary rarely pseudomonomerous with a single parietal placenta, or even (some Cucurbitaceae) monomerous *ab initio*; styles distinct or partly or wholly connate to form a single lobed or entire style; ovules (1–) usually more or less numerous, anatropous to sometimes hemitropous or campylotropous or orthotropous, mostly bitegmic and crassinucellar, but sometimes unitegmic (Hoplestigmataceae) or tenuinucellar (Fouquieriaceae) or both unitegmic and tenuinucellar (Loasaceae); endosperm-development nuclear or seldom cellular. FRUIT variously dry or fleshy and dehiscent or indehiscent; seeds with small to large, dicotyledonous embryo and copious to scanty, oily and proteinaceous or less often starchy endosperm, or without endosperm, sometimes arillate.

The order Violales as here defined consists of 24 families and nearly 5000 species. Fully four-fifths of the species belong to only five families, the Begoniaceae (1000), Flacourtiaceae (800+), Violaceae (800), Cucurbitaceae (700), and Passifloraceae (650). The Loasaceae, Begoniaceae, and a few of the smaller families may conceivably prove to belong somewhere else, but the bulk of the order hangs together as a natural group with the Flacourtiaceae at or near its evolutionary base.

The more archaic members of the Violales are trees with alternate, stipulate leaves, perfect, hypogynous, polypetalous flowers with numerous centrifugal stamens, a compound pistil with free styles and parietal placentation, bitegmic, crassinucellar ovules, and seeds with a well developed endosperm. Such a combination of characters immediately suggests the Flacourtiaceae, which are usually considered to be the most archaic family in the order. It may also be interesting to note that a few of the Flacourtiaceae have a plurilocular ovary with axile placentation; on formal morphological characters these would be perfectly at home in the Theales. The basis of assorting families among the Theales and Violales has been discussed under the former order.

Tendencies toward perigyny and epigyny, unisexuality, reduction in the number of stamens, fusion of filaments into groups, the development of a corona, reduction in the number of carpels, fusion of styles, and loss of endosperm from the seed can all be seen in the family Flacourtiaceae. These are some of the more prominent characters that are used in combination to define many of the families of the order.

The Violales and Capparales may be considered as more or less parallel offshoots from the Theales, each having mostly parietal instead of axile placentation. The Violales differ from the Capparales in the absence of a replum and the usual absence of mustard oils. Furthermore, the Violales have a much higher proportion of woody species, they only seldom have compound leaves, they often have perigynous to epigynous flowers, and they most commonly have 3 carpels (a rare number in the Capparales), only seldom two (the commonest number in the Capparales).

The adaptive significance of most of the characters that distinguish the Violales as a group is obscure. Placentation, number of carpels, ovular structure, and sequence of initiation of stamens are difficult to relate to survival value. The most important ecological distinction between the Violales and the Capparales is almost surely the nature of their chemical defenses. The Capparales heavily exploit mustard oils, to the virtual exclusion of most other weapons, whereas the Violales have a wide array of repellants in different families, but only rarely mustard oils.

Some of the families of the order do show some ecological correlations. The Fouquieriaceae are spiny xerophytes with small leaves that fall off as the soil dries out. The Tamaricaceae, Frankeniaceae, and many of the Cistaceae meet problems of water-stress by having small, firm, persistent leaves that can survive desiccation, and the Tamaricaceae and Frankeniaceae further have specialized salt-excreting foliar glands. The Passifloraceae and Cucurbitaceae are chiefly tendriliferous vines, and the Dioncophyllaceae and Ancistrocladaceae are woody vines that climb by means of stout hooks from the branch-tips (Ancistrocladaceae) or leaf-tips (Dioncophyllaceae). Some of the families have distinctive chemical defenses. The cucurbitacins appear to be unique to the Cucurbitaceae. The Loasaceae and Fouquieriaceae are unusual in the order in having iridoid compounds, although these are common enough in some other orders (including some in the Dilleniidae). The Flacourtiaceae, Passifloraceae, and probably also the other cyanogenic families produce cyanogens in an unusual way involving special types of fatty acids that have a cyclopentenoid ring. Aside from these features, one sees the familiar pattern of families defined by characters of little or no obvious biological importance, and some of the larger families embrace a wide range of growth-forms that occur in diverse habitats.

For those who find it useful to organize the Violales into suborders, I have indicated a scheme in the synoptical arrangement of the families.

SELECTED REFERENCES

Ayensu, E. S., & W. L. Stern. 1964. Systematic anatomy and ontogeny of the stem in Passifloraceae. Contr. U. S. Natl. Herb. 34: 45–73.

Baas, P. 1972. Anatomical contributions to plant taxonomy. II. The affinities *Hua* Pierre and *Afrostyrax* Perkins et Gilg. Blumea 20: 161–192.

Baas, P., R. Geesink, W. A. van Heel, & J. Muller. 1979. The affinities of *Plagiopteron suaveolens* Griff. (Plagiopteraceae.) Grana 18: 69–89.

Badillo, V. M. 1971. Monografia de la familia Caricaceae. Publ. Asoc. Prof., Univ. Central de Venezuela. Maracay.

Bate-Smith, E. C. 1964. Chemistry and taxonomy of *Fouquieria splendens* Engelm., a new member of the asperuloside group. Phytochemistry 3: 623–625.

Baum, B. R., I. J. Bassett, & C. W. Crompton. 1971. Pollen morphology of *Tamarix* species and its relationship to the taxonomy of the genus. Pollen & Spores 13: 495–521.

Beattie, A. J. 1974. Floral evolution in *Viola*. Ann. Missouri Bot. Gard. 61: 781–793.

Behnke, H.-D. 1976. Sieve-element plastids of *Fouquieria, Frankenia* (Tamaricales), and *Rhabdodendron* (Rutaceae), taxa sometimes allied with Centrospermae (Caryophyllales). Taxon 25: 265–268.

Beijersbergen, A. 1972. Note on the chemotaxonomy of Huaceae. Blumea 20: 160.

Brenan, J. P. M. 1954. *Soyauxia,* a second genus of Medusandraceae. Kew Bull. 1953: 507–511.

Brizicky, G. K. 1961. The genera of Turneraceae and Passifloraceae in the southeastern United States. J. Arnold Arbor. 42: 204–218.

Brizicky, G. K. 1961. The genera of Violaceae in the southeastern United States. J. Arnold Arbor. 42: 321–333.

Brizicky, G. K. 1964. The genera of Cistaceae in the southeastern United States. J. Arnold Arbor. 45: 346–357.

Bugnon, P. 1956. Valeur morphologique du complexe axillaires chez les Cucurbitacées. Ann. Sci. Nat. Bot. sér. 11. 17: 313–323.

Chakravarty, H. L. 1958. Morphology of the staminate flowers in the Cucurbitaceae with special reference to the evolution of the stamens. Lloydia 21: 49–87.

Chevalier, A. 1947. La famille des Huacaceae et ses affinités. Rev. Int. Bot. Appl. Agric. Trop. 27 (No. 291–292): 26–29.

Chopra, R. N. 1955. Some observations on endosperm development in the Cucurbitaceae. Phytomorphology 5: 219–230.

Chopra, R. N., & B. Basu. 1965. Female gametophyte and endosperm of some members of the Cucurbitaceae. Phytomorphology 15: 217–223.

Chopra, R. N., & H. Kaur. 1965. Embryology of *Bixa orellana* Linn. Phytomorphology 15: 211–214.

Crété, P. 1946. Embryogénie des Loasacées. Développement de l'embryon chez le *Loasa lateritia* Gill. Compt. Rend. Hebd. Séances Acad. Sci. 222: 920–921.

Cusset, G. 1968. Les vrilles des Passifloracées. Bull. Soc. Bot. France 115: 45–61.

Dahlgren, R., S. R. Jensen, & B. J. Nielsen. 1976. Iridoid compounds in Fouquieriaceae and notes on its possible affinities. Bot. Not. 129: 207–212.

Davidson, C. 1976. Anatomy of xylem and phloem of the Datiscaceae. Los Angeles County Mus. Contr. Sci. 280: 1–28.

Davis, W. S., & H. J. Thompson. 1967. A revision of *Petalonyx* (Loasaceae) with a consideration of affinities in subfamily Gronovioideae. Madroño 19: 1–18.

Devi, H. M., & K. C. Naidu. 1979. Embryological studies in the family Begoniaceae. Indian J. Bot. 2: 1–7.

Devi, S. 1952. Studies in the order Parietales. III. Vascular anatomy of the flower of *Carica papaya* L., with special reference to the structure of the gynaeceum. Proc. Indian Acad. Sci. B 36: 59–69.

DeWilde, W. J. J. O. 1971. The systematic position of the tribe Paropsieae, in particular the genus *Ancistrothyrsus,* and a key to the genera of Passifloraceae. Blumea 19: 99–104.

DeWilde, W. J. J. O. 1974 (1975). The genera of tribe Passifloreae (Passifloraceae) with special reference to flower morphology. Blumea 22: 37–50.

Erdtman, G. 1958. A note on the pollen morphology in the Ancistrocladaceae and Dioncophyllaceae. Veröff. Geobot. Inst. ETH Stiftung Rübel Zürich 33: 47–49.

Ernst, W. R., & H. J. Thompson. 1963. The Loasaceae in the southeastern United States. J. Arnold Arbor. 44: 138–142.

Gauthier, R. 1959. L'anatomie vasculaire et l'interprétation de la fleur pistillée de l'*Hillebrandia sandwicensis* Oliv. Phytomorphology 9: 72–87.

Gauthier, R., & J. Arros. 1963. L'anatomie de la fleur staminée de l'*Hillebrandia sandwicensis* Oliver et la vascularisation de l'étamine. Phytomorphology 13: 115–127.

Gmelin, R., & A. Kjaer. 1970. Glucosinolates in the Caricaceae. Phytochemistry 9: 591–593.

Gottwald, H., & N. Parameswaran. 1968. Das sekundäre Xylem and die systematische Stellung der Ancistrocladaceae und Dioncophyllaceae. Bot. Jahrb. Syst. 88: 49–69.

Hagerup, O. 1930. Vergleichende morphologische und systematische Studien über die Ranken und andre vegetative Organe der Cucurbitaceen und Passifloraceen. Dansk Bot. Ark. 6(8): 1–104.

Harborne, J. B. 1975. Flavonoid bisulphates and their co-occurrences with ellagic acid in the Bixaceae, Frankeniaceae and related families. Phytochemistry 14: 1331–1337.

Henrickson, J. 1972. A taxonomic revision of the Fouquieriaceae. Aliso 7: 439–537.

Henrickson, J. 1973. Fouquieriaceae DC. World Pollen & Spore Flora 1: 1–12.

Heydacker, F. 1963. Les types polliniques dans la famille des Cistaceae. Pollen & Spores 5: 41–49.

Hou, D. 1972. Germination, seedling and chromosome number of *Scyphostegia borneensis* Stapf (Scyphostegiaceae) Blumea 20: 88–92.

Humphrey, R. R. 1931. Thorn formation in *Fouquieria splendens* and *Idria columnaris*. Bull. Torrey Bot. Club 58: 263–264.

Humphrey, R. R. 1935. A study of *Idria columnaris* and *Fouquieria splendens*. Amer. J. Bot. 22: 184–207.

Inamdar, J. A., & M. Gangadhara. 1976. Structure, ontogeny and taxonomic significance of stomata in some Cucurbitaceae. Feddes Repert. 87: 293–310.

Jeffrey, C. 1962. Notes on Cucurbitaceae, including a proposed new classification of the family. Kew Bull. 15: 337–371.

Jeffrey, C. 1966. On the classification of the Cucurbitaceae. Kew Bull. 20: 417–426.

Johansen, D. A. 1936. Morphology and embryology of *Fouquieria*. Amer. J. Bot. 23: 95–99.

Johri, B. M., & D. Kak. 1954. The embryology of *Tamarix* Linn. Phytomorphology 4: 230–247.

Judson, J. E. 1929. The morphology and vascular anatomy of the pistillate flower of the cucumber. Amer. J. Bot. 16: 69–86.

Kapil, R. N., & R. Maheshwari. 1964. Embryology of *Helianthemum vulgare* Gaertn. Phytomorphology 14: 547–557.

Kaur, H. 1969. Embryological investigations on *Bixa orellana* Linn. Proc. Natl. Inst. Sci. India 35: 487–506.

Keating, R. C. 1968, 1970, 1972. Comparative morphology of Cochlospermaceae. I. Synopsis of the family and wood anatomy. Phytomorphology 18: 379–392, 1968. II. Anatomy of the young vegetative shoot. Amer. J. Bot. 57: 889–898, 1970. III. The flower and pollen. Ann. Missouri Bot. Gard. 59: 282–296, 1972.

Keating, R. C. 1973. Pollen morphology and relationships of the Flacourtiaceae. Ann. Missouri Bot. Gard. 60: 273–305.

Keating, R. C. 1976. Trends of specialization in pollen of Flacourtiaceae with comparative observations of Cochlospermaceae and Bixaceae. Grana 15: 29–49.

Keng, H. 1967. Observations on *Ancistrocladus*. Gard. Bull. Singapore 22: 113–121.

Khan, R. 1943. The ovule and embryo sac of *Fouquiera*. Proc. Natl. Inst. Sci. India 9: 253–256.

Killip, E. P. 1938. The American species of Passifloraceae. Publ. Field Mus. Nat. Hist., Bot. Ser., 19: 1–613.

Kooiman, P. 1974. Iridoid substances in the Loasaceae and the taxonomic position of the family. Acta Bot. Neerl. 23: 677–679.

Kumazawa, M. 1964. Morphological interpretations of axillary organs in the Cucurbitaceae. Phytomorphology 14: 287–298.

Leinfellner, W. 1959. Die falschen Rollblätter der Frankeniaceen, in Vergleich gesetzt mit jenen der Ericaceen. Oesterr. Bot. Z. 106: 325–351.

Leins, P., & R. Bonnery-Brachtendorf. 1977. Entwicklungsgeschichtliche Untersuchungen an Blüten von *Datisca cannabina* (Datiscaceae). Beitr. Biol. Pflanzen 53: 143–155.

Leins, P., & W. Winhard. 1973. Entwicklungsgeschichtliche Studien an Loasaceen-Blüten. Oesterr. Bot. Z. 122: 145–165.

Marburger, J. E. 1979. Glandular leaf structure of *Triphyophyllum peltatum* (Dioncophyllaceae): A "fly-paper" insect trapper. Amer. J. Bot. 66: 404–411.

Mauritzon, J. 1933. Über die Embryologie der Turneraceae und Frankeniaceae. Bot. Not. 1933: 543–554.

Mauritzon, J. 1936. Zur Embryologie einiger Parietales-Familien. Svensk Bot. Tidskr. 30: 79–113.

Melchior, H. 1925. Die phylogenetische Entwicklung der Violaceen und die natürlichen Verwandtschaftsverhältnisse ihrer Gattungen. Repert. Spec. Nov. Regni. Veg. Beih. 36: 83–125.

Merxmüller, H., & P. Leins. 1971. Zur Entwicklungsgeschichte männlicher Begonienblüten. Flora 160: 333–339.

Metcalfe, C. R. 1952. The anatomical structure of the Dioncophyllaceae in relation to the taxonomic affinities of the family. Kew Bull. 1951: 351–368.

Metcalfe, C. R. 1956. *Scyphostegia borneensis* Stapf. Anatomy of stem and leaf in relation to its taxonomic position. Reinwardtia 4: 99–104.

Metcalfe, C. R. 1962. Notes on the systematic anatomy of *Whittonia* and *Peridiscus*. Kew Bull. 15: 472–475.

Miller, R. B. 1975. Systematic anatomy of the xylem and comments in the relationships of the Flacourtiaceae. J. Arnold Arbor. 56: 20–102.

Murty, Y. S. 1954. Studies in the order Parietales. IV. Vascular anatomy of the flower of Tamaricaceae. J. Indian Bot. Soc. 33: 226–238.

Prance, G. T. & M. Freites da Silva. 1973. Caryocaraceae. Fl. Neotropica Monogr. 12: 1–77.

Presting, D. 1964 (1965). Die Systematik der Passifloraceen aus pollenmorphologischer Sicht. Ber. Deutsch. Bot. Ges. 77: (40)–(44).

Presting, D. 1965. Zur Morphologie der Pollenkörner der Passifloraceen. Pollen & Spores 7: 193–247.

Puri, V. 1947, 1948, 1954. Studies in floral anatomy. IV. Vascular anatomy of the flower of certain species of the Passifloraceae. Amer. J. Bot. 34: 562–573, 1947. V. On the structure and nature of the corona in certain species of the Passifloraceae. J. Indian Bot. Soc. 27: 130–149, 1948. VII. On placentation in the Cucurbitaceae. Phytomorphology 4: 278–299, 1954.

Raju, M. V. S. 1954. Pollination mechanism in *Passiflora foetida* Linn. Proc. Natl. Inst. Sci. India 20: 431–436.
Raju, M. V. S. 1956. Embryology of the Passifloraceae. I. Gametogenesis and seed development of *Passiflora Calcarata* Mast. J. Indian Bot. Soc. 35: 126–138.
Rao, V. S. 1949. The morphology of the calyx-tube and the origin of perigyny in Turneraceae. J. Indian Bot. Soc. 28: 198–201.
Rao, V. S. 1969. The floral anatomy of *Ancistrocladus.* Proc. Indian Acad. Sci. B 70: 215–222.
Record, S. J. 1941. American woods of the family Flacourtiaceae. Trop. Woods 68: 40–57.
Rehm, S., P. R. Enslin, A. D. J. Meeuse, & J. H. Wessels. 1957. Bitter principles of the Cucurbitaceae. VII. The distribution of bitter principles in this plant family. J. Sci. Food Agric. 12: 679–686.
Sandwith, N. Y. 1962. Contribution to the flora of tropical America LXIX. A new genus of Peridiscaceae. Kew Bull. 15: 467–471.
Schmid, R. 1964. Die systematische Stellung der Dioncophyllaceen. Bot. Jahrb. Syst. 83: 1–56.
Schnarf, K. 1931. Ein Beitrag zur Kenntnis der Samenentwicklung der Gattung *Cochlospermum.* Oesterr. Bot. Z. 80: 45–50.
Scogin, R. 1978. Leaf phenolics of the Fouquieriaceae. Biochem. Syst. Ecol. 6: 297–298.
Sensarma, P. 1955. Tendrils of the Cucurbitaceae: Their morphological nature on anatomical evidence. Proc. Natl. Inst. Sci. India 21B: 162–169.
Shaw, H. K. Airy. 1952. On the Dioncophyllaceae, a remarkable new family of flowering plants. Kew Bull. 1951: 327–347.
Singh, B. 1953. Studies on the structure and development of seeds of Cucurbitaceae. Phytomorphology 3: 224–239.
Spirlet, M.-L. 1965. Utilisation taxonomique des grains de pollen de Passifloracées. Pollen & Spores 7: 249–301.
Summerhayes, V. S. 1930. A revision of the Australian species of *Frankenia.* J. Linn. Soc., Bot. 48: 337–387.
Swamy, B. G. L. 1953. On the floral structure of *Scyphostegia.* Proc. Natl. Inst. Sci. India 19: 127–142.
Taylor, F. H. 1972. The secondary xylem of the Violaceae: a comparative study. Bot. Gaz. 133: 230–242.
van Heel, W. A. 1967. Anatomical and ontogenetic investigations on the morphology of the flowers and the fruit of *Scyphostegia borneensis* Stapf (Scyphostegiaceae). Blumea 15: 107–125.
van Heel, W. A. 1973, 1977. Flowers and fruits in Flacourtiaceae. I. *Scaphocalyx spathacea* Ridl. III. Some Oncobeae. Blumea 21: 259–279, 1973. 23: 349–369, 1977.
Venkatesh, C. S. 1956. The curious anther of *Bixa*—its structure and dehiscence. Amer. Midl. Naturalist 55: 473–476.
Vijayaraghavan, M. R., & D. Kaur. 1966. Morphology and embryology of *Turnera ulmifolia* L. and affinities of the family Turneraceae. Phytomorphology 19: 539–553.
Walia, K., & R. N. Kapil. 1965. Embryology of *Frankenia* Linn. with some comments on the systematic position of the Frankeniaceae. Bot. Not. 118: 412–429.
Woodworth, R. H. 1935. Fibriform vessel members in the Passifloraceae. Trop. Woods 41: 8–16.

Алешина, Л. А. 1964. О пыльце тыквенных. Бот. Ж. 49: 1773–1776.

Алешина, Л. А. 1966. Морфология пыльцевых зерен семейства Cucurbitaceae Juss. В сб.: Значение палинологического анализа для стратиграфии и палеофлористики: 15–22. Изд. Наука. Москва.

Алешина, Л. А. 1971. Палинологические данные к систематике и филогении семейства Cucurbitaceae Juss. В сб.: Морфология пыльцы Cucurbitaceae, Thymelaeaceae, Cornaceae. Л. А. Куприянова и М. С. Яковлев, редакторы: 3–103. Изд Наука. Ленинградское Отделение. Ленинград.

Васильченко, И. Т., & Л. И. Васильева. 1976. Пустынное ли дерево Тамарикс? Бюлл. Московск. Общ. Исп. Прир. Отд. Биол. 81 (1): 139–143.

Голышева, М. Д. 1975. Анатомия листьев *Idesia polycarpa* Maxim. и других флакуртиевых в связи с вопросом о родственных взаимоотношениях сем. Salicaceae и Flacourtiaceae. Бот. Ж. 60: 787–799.

Гуляев, В. А. 1963. Справнительная эмбриология Cucurbitaceae и ее значение для систематики семейства. Бот. Ж. 48: 80–85.

Дзевалтовский, А. К. 1963. Цитоэмбриологические исследования некоторых представителей семейства Тыквенных. Укр. Бот. Ж. 20(4): 16–29. (In Ukrainian. Russian and English summaries.)

Қамилова, Ф. Н., & Е. А. Мокеева. 1961. Природа и строение усиков тыквенных. В. сб.: Морфогенез растений, 2: 56–59. Изд. Московского Унив.

Карташова, Н. Н., & Е. Н. Немирович-Данченко. 1968. К эволюции нектарников у тыквенных (Cucurbitaceae Juss.) Бот. Ж. 53: 1219–1225.

Матиенко, Б. Т. 1957. Об анатомо-морфологической природе цветка и плода тыквенных. Труды Бот. Инст. АН СССР, сер VII, 4: 288–322.

Мокеева, Е. А. 1963. Развитие и строение узла стеблей у тыквенных. Бот. Ж. 48: 1472–1483.

Сойфер, В. Н. 1964. Анатомия семян семейства Cucurbitaceae Juss. как систематический признак. Бюлл. Московск. Общ. Исп. Прир., Отд. Биол. нов. сер 69(1): 86–101.

SYNOPTICAL ARRANGEMENT OF THE FAMILIES OF VIOLALES

1 Ovary usually superior (half-inferior in Ancistrocladaceae and some Turneraceae, inferior in a very few Flacourtiaceae); plants of various habit.

2 Flowers mostly polypetalous or apetalous, or the petals seldom shortly connate at the base.

3 Flowers without a corona, or very rarely (a monotypic genus of Flacourtiaceae) with an intrastaminal corona, mostly hypogynous, seldom evidently perigynous or even epigynous.

4 Plants mostly not climbing, in any case without hooked or twining leaf-tips or branch-tips.

5 Stamens mostly 10 or more, rarely as few as 5 or even 3 (Violineae, in part).

6 Styles distinct or united to varying degrees, the stigmas distinct; plants often cyanogenic in association with a cyclopentenoid ring system; endosperm oily, often containing cyclopentenyl (chaulmoogric) fatty acids.

7 Ovules on parietal (sometimes more or less deeply intruded) or seldom virtually basal or even axile placentas; anthers dithecal; petals variously present or absent; carpels 2–10 .... 1. FLACOURTIACEAE.

7 Ovules pendulous from the top of a unilocular ovary; anthers monothecal; petals wanting; carpels 3–4 .... 2. PERIDISCACEAE.

6 Style solitary, terminated by the simple or lobed stigma, or the stigmas seldom distinct; plants not cyanogenic; endosperm starchy or oily, not known to contain cyclopentenyl fatty acids.

8 Placentation parietal to more or less distinctly axile; stamens (3–) more or less numerous; petals imbricate or convolute; plants not with the odor of garlic.

9 Plants with an orange or red latex; anthers opening by pores or short slits; ovules anatropous; leaves alternate, from simple, entire, and palmately veined to palmately lobed or compound .... 3. BIXACEAE.

9 Plants with colorless juice; anthers opening by longitudinal slits; ovules orthotropous or rarely anatropous; leaves opposite or less often alternate, simple, variously veined, often more or less reduced and ericoid or scale-like .... 4. CISTACEAE.

8 Placentation basal; stamens (8) 10; petals induplicate-valvate; plants with a garlic-like odor 5. HUACEAE.

5 Stamens 1–8, or sometimes 10 or even more numerous in the Tamaricaceae.

10 Stamens 1–3; flowers very small, in catkin-like spikes or racemes (Violineae, in part).

11 Stamen solitary; carpels 2–3; flowers generally perfect .... 6. LACISTEMATACEAE.

11 Stamens 3, connate into a column; carpels 8–13; flowers unisexual .... 7. SCYPHOSTEGIACEAE.

10 Stamens 4 or more; flowers variously large or small.

12 Endosperm oily; leaves of normal proportions, stipulate, without salt-excreting glands; flowers regular or very often irregular (Violineae, in part).

13 Stamens 8; sepals, petals, and carpels each 4; flowers regular .... 8. STACHYURACEAE.

13 Stamens (3–) 5; sepals and petals each 5, carpels

(2) 3 (–5); flowers regular or very often irregular .... 9. VIOLACEAE.

12 Endosperm starchy or none; leaves exstipulate, often much-reduced and ericoid or scale-like, commonly with embedded, multicellular, salt-excreting glands; flowers regular (Tamaricineae).

14 Styles 2–5 (or the stigmas sessile); leaves alternate; sepals distinct or connate only near the base .... 10. TAMARICACEAE.

14 Style solitary, slender and elongate; leaves opposite; sepals connate to form a shortly lobed tube .... 11. FRANKENIACEAE.

4 Plants woody climbers with stout, hooked or twining branch-tips or leaf-tips (Ancistrocladineae).

15 Ovules numerous on the 2–5 parietal placentas, conspicuously exserted from the developing capsule well before maturity; plants climbing by hooked or twining leaf-tips; ovary superior .... 12. DIONCOPHYLLACEAE.

15 Ovule solitary, basilateral, included in the mature nut; plants climbing by hooked or twining branch-tips; ovary half-inferior .... 13. ANCISTROCLADACEAE.

3 Flowers with an extrastaminal corona (except most Turneraceae), more or less strongly perigynous; stamens 5 (-numerous) (Passiflorineae, in part).

16 Corona wanting, or seldom present; flowers without a gynophore or androgynophore; plants not climbing .... 14. TURNERACEAE.

16 Corona nearly always present; flowers usually with a gynophore or androgynophore.

17 Seeds exarillate; undershrubs or herbs, not climbing; sepals and petals valvate in bud 15. MALESHERBIACEAE.

17 Seeds arillate; herbaceous or woody vines, climbing by tendrils, or less often erect shrubs or even trees; sepals and petals imbricate in bud, or the petals seldom wanting .... 16. PASSIFLORACEAE.

2 Flowers evidently sympetalous.

18 Flowers nearly always unisexual; leaves simple (but often lobed) or compound; stamens 10 or fewer (Passiflorineae, in part).

19 Herbs or half-shrubs without a latex-system; fruit capsular; style solitary, more or less deeply cleft; stamens 3–5 .... 17. ACHARIACEAE.

19 Soft-stemmed shrubs or small trees (rarely herbs) with a well developed latex-system; fruit fleshy, melon-like; styles distinct; stamens 10 or seldom only 5 .... 18. CARICACEAE.

18 Flowers perfect, leaves simple, entire or nearly so; stamens 10 or more.

20 Spiny, xerophytic shrubs or small, fleshy trees with small, ephemeral leaves; carpels 3; fruit capsular; stamens 10–18 (–23) (Fouquieriineae) .... 19. FOUQUIERIACEAE.

20 Unarmed trees with large, persistent leaves; carpels 2; fruit drupaceous; stamens about 20–30 (Hoplestigmatineae) .... 20. HOPLESTIGMATACEAE.

1 Ovary mostly inferior, rarely only half-inferior; plants mostly herbs or herbaceous vines, a few trees and shrubs; endosperm scanty or none except in Loasaceae.

21 Flowers mostly unisexual (rarely polygamous or perfect); ovules bitegmic, crassinucellar, plants without iridoid compounds.

22 Stamens apparently 2–5, typically 3 with one monothecal and two dithecal anthers; mostly tendriliferous vines; corolla mostly sympetalous; styles 1–3; plants producing cucurbitacins; leaves exstipulate (Cucurbitineae) 21. CUCURBITACEAE.

22 Stamens 4-many, all with dithecal anthers; plants without tendrils; petals distinct or none; styles distinct or merely connate at the base; plants without cucurbitacins (Begoniineae).

23 Ovary unilocular, with parietal placentas; leaves exstipulate; sepals 4–8 .... 22. DATISCACEAE.

23 Ovary mostly plurilocular, with axile placentas, or seldom unilocular with deeply intruded parietal placentas; leaves stipulate; sepals 2–5 .... 23. BEGONIACEAE.

21 Flowers perfect; ovules unitegmic, tenuinucellar; leaves exstipulate; style solitary; plants commonly producing iridoid compounds (Loasineae) .... 24. LOASACEAE.

1. Family FLACOURTIACEAE A. P. de Candolle 1824 nom. conserv., the Flacourtia Family

Evergreen or deciduous shrubs or small to seldom rather large trees, often cyanogenic in association with a cyclopentenoid ring system, often accumulating proanthocyanins, but not ellagic acid, and sometimes producing alkaloids; flavonoid pigments, when present, commonly including flavones but not flavonols; nodes trilacunar; solitary or clustered crystals of calcium oxalate very often present in the parenchymatous tissues and in the leaf-epidermis, frequently in more or less distinctive idioblasts; vessel-segments commonly with simple perforations, but in some of the more archaic genera some or all of them with

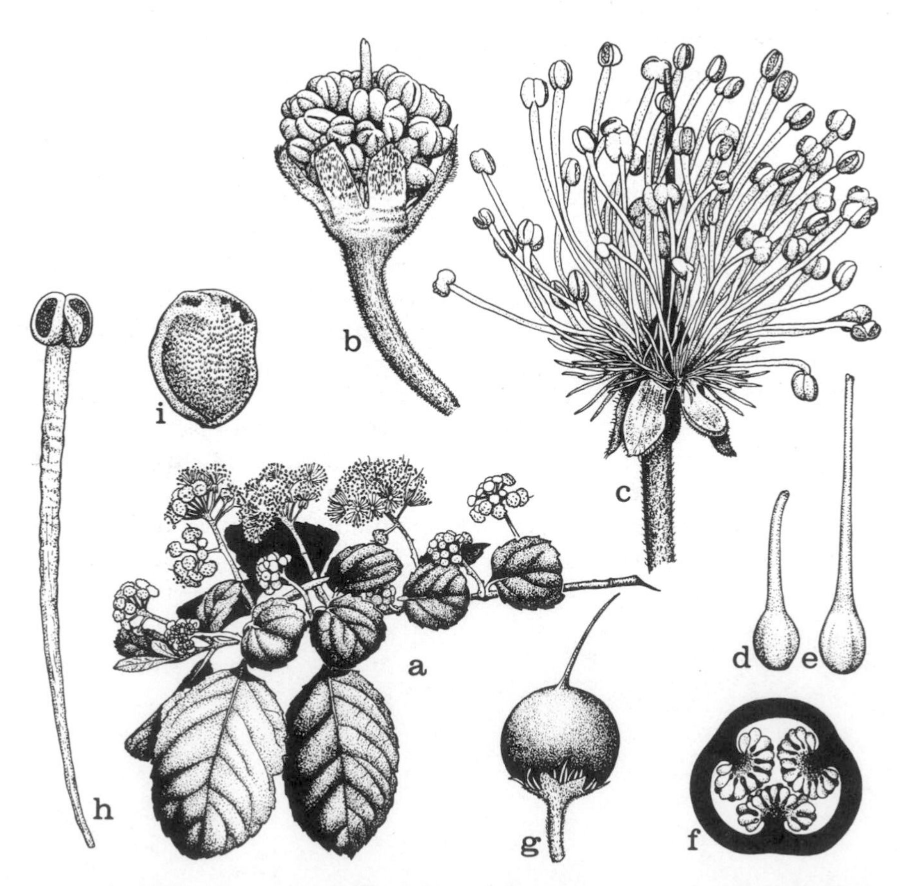

FIG. 4-17 Flacourtiaceae. *Azara serrata* Ruiz & Pavon. a, habit, ×½; b, opening flower bud, ×6; c, flower, ×6; d, e, pistil, at different stages of development, ×6; f, schematic cross-section of ovary, ×24; g, fruit, ×2; h, stamen, ×12; i, seed, ×8.

scalariform perforation-plates instead; imperforate tracheary elements with simple or bordered pits, generally septate; wood-rays very numerous and mostly narrow, mixed uniseriate and pluriseriate, the uniseriate ones mostly homocellular, and pluriseriate ones heterocellular, 2–5 (–10+) cells wide, with elongate ends, some of the ray-cells in most genera containing prismatic crystals; wood-parenchyma usually wanting or very scanty and paratracheal. LEAVES alternate or rarely opposite or whorled, simple, entire or glandular-toothed, sometimes glandular-punctate or -lineate, commonly pinnately veined; stomates paracytic or anisocytic; petiolar anatomy diverse; stipules present but often caducous; an accessory bud sometimes present in addition to the main axillary bud. FLOWERS small or seldom large, in diverse sorts of inflorescences, these usually cymose and axillary or terminal, but sometimes racemose, or epiphyllous, or the flowers solitary and axillary, or (Bembicieae, consisting of 2 monotypic genera of Madagascar) in cone-like, involucrate, pseudanthial heads, perfect or less often unisexual, regular, hypogynous or seldom perigynous, or even (Bembicieae) epigynous; sepals 3–8 (–15), distinct or sometimes shortly connate below, mostly imbricate but sometimes valvate, sometimes accrescent in fruit; petals generally present, 3–8 (–15), distinct, imbricate, alternate with the sepals, or sometimes spirally arranged and then poorly differentiated from the sepals and not regularly placed in relation to them; nectaries of various sorts, sometimes an extrastaminal or intrastaminal disk, sometimes of separate glands among or external to the stamens, sometimes a basal scale on each petal; stamens commonly numerous and centrifugal, but sometimes as few as 4, distinct or sometimes grouped into antepetalous clusters associated with trunk-bundles, sometimes some of them staminodial; anthers tetrasporangiate and dithecal, opening by longitudinal slits or seldom by terminal pores (as in *Kiggelaria*), sometimes with a prolonged connective; pollen-grains binucleate, tricolporate or seldom tricolpate or triporate or with 4 or 5 apertures; corona wanting, or rarely (monotypic African genus *Trichostephanus*) present and intrastaminal; gynoecium of 2–10 carpels (or apparently a single carpel in *Aphloia*) united to form a compound, superior or rarely (Bembicieae) inferior, ordinarily unilocular ovary with parietal (or virtually basal) placentas, or the placentas sometimes so deeply intruded that the ovary is essentially plurilocular, seldom distinctly plurilocular and with axile placentas, as in *Prockia*; styles distinct or united to varying degrees, the stigmas distinct; ovules 2-many on each placenta, anatropous to amphitropous or orthotropous, bitegmic, crassinucellar, endosperm-development nuclear. FRUIT a berry, or less often a loculicidal capsule or a drupe; seeds often arillate or woolly; embryo straight (horseshoe-shaped in *Aphloia*); cotyledons 2, generally broad, often cordate; endosperm commonly abundant, oily and proteinaceous, often containing fatty acids of the cyclopentenoid (chaulmoogric) series. X = 10–12. (Neumanniaceae, Plagiopteraceae, Soyauxiaceae)

The family Flacourtiaceae as here defined consists of about 85 genera and more than 800 species, widespread in tropical regions, with a few members extending into temperate parts of America, Asia, and Africa. *Homalium* and *Casearia*, each with more than a hundred species, are the largest genera, followed by *Xylosma*, with nearly a hundred. Seeds of species of *Hydnocarpus*, native to the East Indies, are the source of chaulmoogra oil, formerly used in the treatment of leprosy.

The Flacourtiaceae are generally considered to be the most archaic family of the Violales, ancestral or standing near to the ancestry of all the other families. The family is highly diversified, and tendencies toward the features of each of several other families of the order can be seen within it. The boundaries of the Flacourtiaceae have been variously drawn, to include one or more of the smaller families here recognized, or to include certain genera that are here referred to other families. The small tribes Paropsieae and Abatieae (7 genera in all), which are generally admitted to form a connecting link between the Flacourtiaceae and Passifloraceae, are here referred to the Passifloraceae on the basis of wood anatomy (Metcalfe & Chalk), pollen morphology (Keating, 1973), and the extrastaminal corona (this is poorly developed, however, in the 2 genera of Abatieae).

Given the basal position of the Flacourtiaceae in the Violales, it does not much matter which other families follow which in the linear sequence. It is necessary only to associate the Ancistrocladaceae with the Dioncophyllaceae, the Frankeniaceae with the Tamaricaceae, and the Passifloraceae, Malesherbiaceae, and possibly Turneraceae with each other. The sequence here presented is merely one that is convenient in relation to the synoptical arrangement of the families.

The relationship between the Flacourtiaceae and Elaeocarpaceae (Malvales) is discussed under the latter family.

The oldest fossil pollen that clearly represents the Flacourtiaceae comes from Upper Miocene deposits. Pollen that might represent the Flacourtiaceae or any of several other dilleniid families extends well back into the Cretaceous.

2. Family PERIDISCACEAE Kuhlmann 1950 nom. conserv., the Peridiscus Family

Trees, with or without crystals of calcium oxalate in some of the cells of the parenchymatous tissues; vessel-segments with scalariform perforation-plates that have up to 15 or more cross-bars; imperforate tracheary elements with large, bordered pits, sometimes some of them septate; wood-rays mixed uniseriate and pluriseriate, the uniseriate ones homocellular, the others heterocellular, with elongate ends; wood-parenchyma diffuse. LEAVES large, alternate, simple, entire, leathery, trinerved at the base, with a large pit in the axil of each of the basal lateral veins beneath, otherwise pinnately veined; stomates anomocytic; petiole with 2 cylindric vascular strands, at least distally; stipules intrapetiolar, deciduous. FLOWERS in axillary fascicles or in clusters of small racemes, small, perfect, regular, hypogynous, apetalous, with large, persistent bracteoles; sepals 4–7, imbricate; stamens numerous, seated on or around the outside of a large, fleshy, cupulate or annular, multilobate disk; filaments distinct or rather irregularly connate toward the base; anthers small, monothecal, opening by longitudinal slits; pollen-grains tricolporate; gynoecium of 3 or 4 carpels united to form a compound, unilocular ovary that in *Peridiscus* is half sunken into the disk; styles short, distinct; ovules 6–8, pendulous from the top of the ovary. FRUIT drupaceous, 1-seeded; seeds with a very small dicotyledonous embryo lying alongside the abundant, horny endosperm.

The family Peridiscaceae consists of 2 monotypic genera from tropical South America, *Peridiscus* and *Whittonia.* Recent authors are agreed that the family is related to the Flacourtiaceae, from which it differs most notably in its apical placentation and monothecal anthers.

3. Family BIXACEAE Link 1831 nom. conserv., the Lipstick-tree Family

Trees, shrubs, or sometimes rhizomatous herbs or half-shrubs, tanniferous, producing ellagic acid and at least sometimes also proanthocyanins, and bearing a red or orange juice in the abundant secretory canals or cells, sometimes also with mucilage-cells or -canals, at least sometimes (*Bixa*) producing flavonoid bisulphates, but not cyanogenic, and without alkaloids so far as known, glabrous or provided with unicellular hairs or (*Bixa*) multicellular, peltate hairs; nodes trilacunar; clustered or solitary crystals of calcium oxalate often present in some

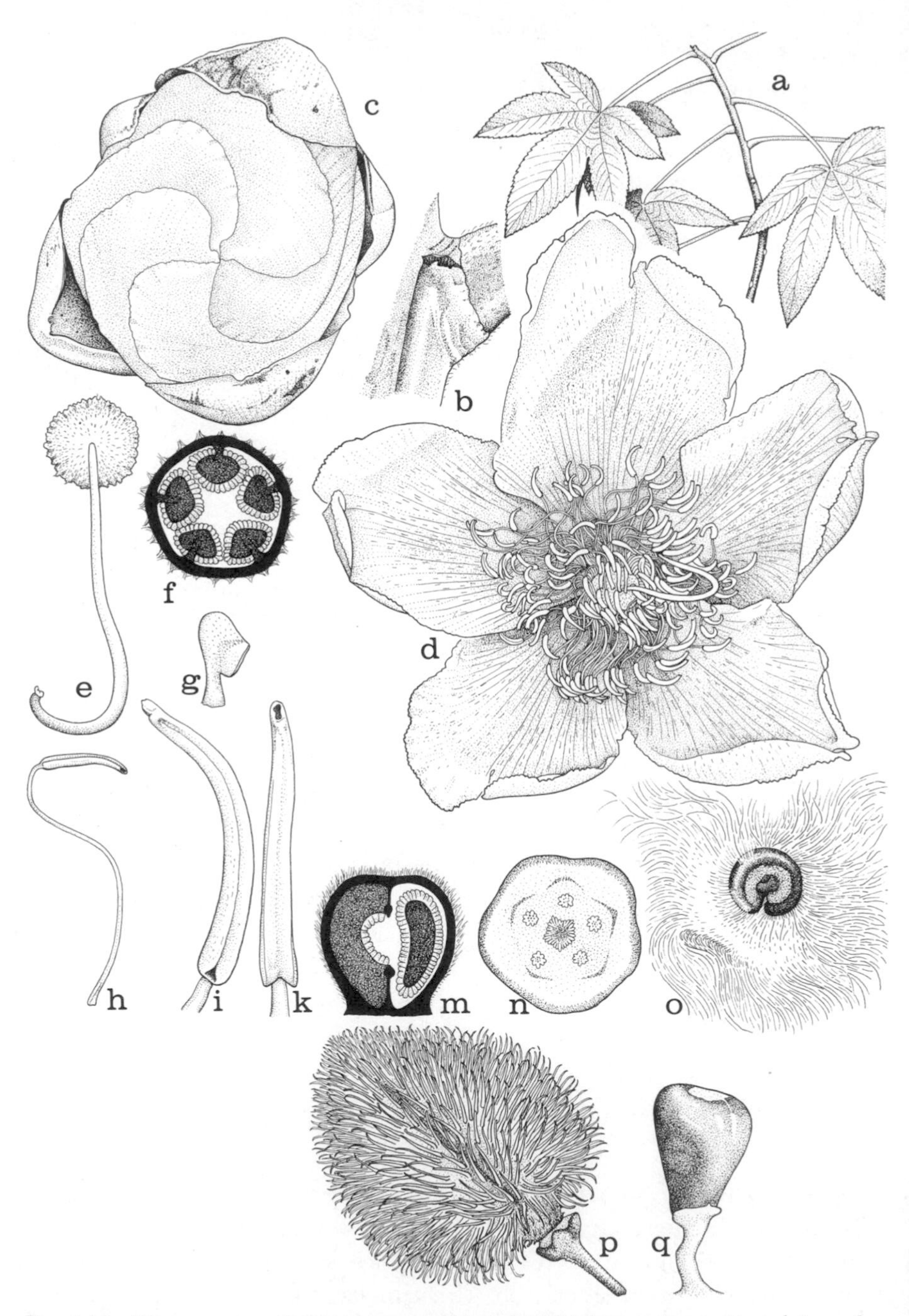

FIG. 4-18 Bixaceae. a–o, *Cochlospermum vitifolium* (Willd.) Sprengel. a, habit, ×⅛; b, node, with stipular scar, ×4; c, flower bud, from above, ×2; d, flower, from above, ×1; e, pistil, from above, ×2; f, schematic cross-section of ovary, ×4; g, ovule, ×24; h, stamen, ×2; i, k, anthers, ×6; m, schematic long-section of ovary, ×4; n, cross-section of receptacle, below the ovary, showing 5 trunk-bundles for androecium, ×2; o, seed, ×2; p, q, *Bixa orellana* L. p, fruit, ×1.5; q, seed, ×3.

of the cells of the parenchymatous tissues; vessel-segments with simple perforations; imperforate tracheary elements with small, bordered pits; wood-rays heterocellular, mixed uniseriate and pluriseriate, the latter up to 5 or 6 cells wide; wood-parenchyma apotracheal and diffuse to mainly in broad, apotracheal ribbons and also partly vasicentric; phloem in young stems tangentially stratified into hard and soft layers, and with wedge-shaped rays. LEAVES alternate, from simple, entire, and palmately veined to palmately lobed or palmately compound, often with a mucilaginous epidermis; stomates anomocytic; petiole often with complex vascular anatomy; stipules well developed, these in *Bixa* protecting the terminal bud. FLOWERS in panicles or racemes, sometimes appearing before the leaves, perfect, regular or slightly irregular, hypogynous; sepals 5, distinct, imbricate, deciduous; petals 5, distinct, imbricate or convolute; stamens numerous, initiated in centrifugal sequence, often associated with 5 (–10) trunk bundles, regularly arranged or more or less distinctly in 2–5 groups; anthers tetrasporangiate and dithecal, opening by short slits or pores, in *Bixa* interpreted to be folded so that the apparently terminal part, bearing the pore-like slits, is morphologically the middle, the top being bent down alongside the bottom, in the other genera the apertures terminal and sometimes also basilateral; pollen-grains binucleate or trinucleate, tricolporate; an intrastaminal nectary disk present around the ovary, or the stamens seated on the disk, which is nectariferous within; gynoecium of 2–5 carpels united to form a compound ovary with more or less deeply intruded partitions, these in *Cochlospermum* and *Amoreuxia* commonly meeting and partitioning the ovary toward the base and toward the summit, but not at the middle, the placentation accordingly partly axile and partly parietal, but in *Bixa* parietal (on intruded placentas) throughout; style solitary, shortly or scarcely lobed; ovules numerous, anatropous, bitegmic, crassinucellar, at least sometimes with a zigzag micropyle; endosperm-development nuclear. FRUIT a loculicidal capsule; seeds glabrous or woolly; embryo well developed, with spatulate cotyledons, embedded in the oily and proteinaceous or (*Bixa*) starchy endosperm. X = 6–8. (Cochlospermaceae)

The family Bixaceae as here defined is widespread in tropical regions and consists of only 3 genera. *Cochlospermum* has about 15 species, *Amoreuxia* about 4, and *Bixa* 3 or 4. *Bixa orellana* L., lipstick-tree, is the source of an orange dye, called annato, which is used for coloring food-products. The species is native to tropical America, but now widely introduced in the tropics of the Old World as well.

Cochlospermum and *Amoreuxia* are sometimes collectively segregated as a family Cochlospermaceae, leaving only *Bixa* in the Bixaceae. *Bixa* does stand somewhat apart from the other two, but all three genera are obviously allied, and I see nothing to be gained by the separation.

Most authors have considered the Bixaceae to stand near the Flacourtiaceae. On the other hand, Keating (1968, 1970, 1973) emphasizes a series of vegetative similarities to the Malvales, although he does not wholly reject a flacourtiaceous relationship. The stratified phloem, wedge-shaped phloem-rays, and often well developed mucilage system of the Bixaceae certainly recall the Malvales, and Keating also notes a chemical similarity in the gums produced by species of *Cochlospermum* and *Sterculia*. In spite of these similarities, the Bixaceae do not have the valvate calyx and fully septate ovary of the Malvales, and except for *Bixa* itself they do not have the stellate indument. Neither have they been reported to produce the characteristic cyclopropenoid fatty acids of the Malvales. The Malvales as here constituted are a relatively homogenous group. Inclusion of the Bixaceae in the Malvales would complicate the distinction of that order from both the Theales and the Violales.

As I have emphasized in other works, the pervasive parallelism in the angiosperms, together with the general nature of major evolutionary diversification, permits only a loose rather than a precise correlation between phylogeny and taxonomy. The Dilleniales, Theales, Malvales, and Violales together form a plexus with recognizable clusters of families, but these clusters can be delimited only arbitrarily. I think it useful to emphasize the flacourtiaceous rather than the malvalean relationship of the Bixaceae, and to note that the vegetative similarities of the Bixaceae to the Malvales probably reflect parallel developments from a similar ancestry.

4. Family CISTACEAE A. L. de Jussieu 1789 nom. conserv., the Rock-rose Family

Shrubs, half-shrubs, and herbs, generally provided with a peculiar sort of unicellular hair that appears to be double toward the base, these hairs often clustered so as to appear stellate, the plants often also with multicellular peltate or gland-tipped hairs; plants tanniferous, commonly producing proanthocyanins and ellagic acid, not cyanogenic, and without alkaloids so far as known; nodes unilacunar; clustered (and sometimes also discrete) crystals of calcium oxalate commonly present

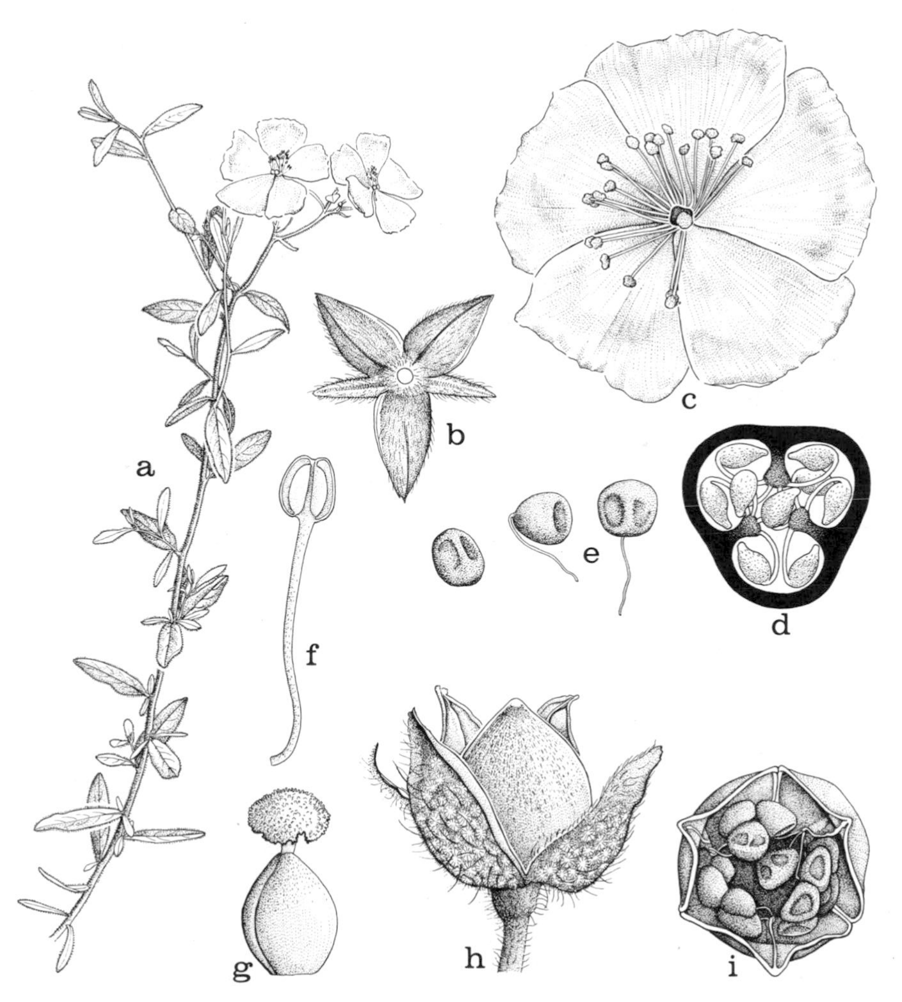

FIG. 4-19 Cistaceae. *Helianthemum canadense* (L.) Michx. a, habit, ×½; b, calyx, from beneath, ×2; c, flower, from above, ×2; d, schematic cross-section of ovary, ×20; e, seeds, ×8; f, stamen, ×8; g, pistil, ×8; h, opened fruit, with persistent calyx, ×2; i, opened fruit, from above, ×2.

in some of the cells of the parenchymatous tissues; vessel-segments with simple perforations; imperforate tracheary elements very short, commonly with bordered pits; wood-rays low and heterocellular, in most species uniseriate, seldom 2- or 3-seriate; wood-parenchyma wanting or very scanty. LEAVES opposite or less often alternate or whorled, simple, pinnately or palmately veined or univeined, often more or less reduced and ericoid or scale-like; stomates anomocytic; petiole variously with 1 or several vascular bundles; stipules present or absent. FLOWERS solitary or more often in various sorts of cymose, terminal or axillary inflorescences, perfect, regular (except as to the calyx), hypogynous; sepals 5 (the 2 outer often notably narrower than the 3 inner and sometimes adnate to them) or 3, distinct, convolute, generally persistent; petals 5 or seldom (*Lechea*) 3, convolute in the opposite direction to the sepals, or seldom imbricate, commonly ephemeral, sometimes wanting from cleistogamous flowers; stamens (3–) more or less numerous, borne on or just outside of an annular, often lobulate nectary-disk, initiated in centrifugal sequence; anthers tetrasporangiate and dithecal, opening by longitudinal slits; pollen-grains binucleate, (2) 3 (4)-colporate; gynoecium of 3 (5–10) carpels united to form a compound, typically unilocular ovary with parietal placentation, the placentas often more or less deeply intruded and sometimes (as in *Cistus*) even meeting and partitioning the ovary into discrete locules; style solitary, undivided, sometimes very short or virtually wanting; stigma minute to large and capitate or discoid and often lobed, or rarely the stigmas 3 and distinct; ovules (1–) 4-many on each placenta, orthotropous or seldom (as in *Fumana*) anatropous, bitegmic, crassinucellar; endosperm-development nuclear. FRUIT a loculicidal capsule; seeds (1–) 3-many, usually small, with starchy, often hard endosperm; embryo mostly with expanded, flat cotyledons, commonly curved or bent into a hook or ring, or folded or circinately more or less coiled, rarely almost straight. X = 5–11.

The family Cistaceae consists of 8 genera and about 200 species, half of them belonging to the single genus *Helianthemum*. The family is widely but irregularly distributed, mostly in temperate or warm-temperate regions. The principal center of diversity is in the Mediterranean region, and there is a secondary center in the eastern United States. Ladanum, a resin used in perfumery, exudes from the herbage of species of *Cistus*, such as *C. ladanifer* L.

Pollen identifiable as belonging to the Cistaceae occurs in lower Miocene and more recent deposits.

Most authors are agreed that the Cistaceae are related to the Bixaceae and Flacourtiaceae. I concur.

5. Family HUACEAE Chevalier 1947, the Garlic-tree Family

Shrubs or trees (sometimes lianoid) with a garlic-like odor, without mucilage-cells or -cavities, and lacking both proanthocyanins and ellagic acid; trichomes variously of simple or stellate hairs or peltate scales; nodes trilacunar; vessel-segments mostly with oblique, simple perforations, but some of the first-formed ones of the secondary xylem with scalariform perforation-plates; imperforate tracheary elements with minutely bordered or simple pits; wood-rays heterocellular; wood-parenchyma paratracheal, banded; phloem with wedge-shaped rays, but not stratified; cristarque cells present in various tissues, notably the petiole. LEAVES alternate, simple, entire, stomates paracytic; petiole with complex and variable anatomy, commonly with a more or less siphon-stelic vascular strand enclosing one or more vascular bundles; stipules present, caducous in *Afrostyrax*. FLOWERS small, axillary, solitary or in small clusters, regular, perfect, hypogynous; sepals 5, distinct, valvate (*Hua*), or the calyx closed in bud and opening by 3–5 irregular lobes (*Afrostyrax*); petals (4) 5, distinct, induplicate-valvate, long-clawed (*Hua*) or with a very short, broad base below an obovate blade (*Afrostyrax*); stamens (8) 10, distinct, uniseriate; anthers tetrasporangiate and dithecal; pollen-grains triporate; gynoecium of 5 carpels united to form a compound, unilocular ovary with a single terminal style and small stigma; ovule solitary (*Hua*) or ovules (4–) 6 (*Afrostyrax*), basal, erect, anatropous, bitegmic. FRUIT dry, dehiscent by 5 valves (*Hua*) or indehiscent (*Afrostyrax*); seed solitary (2); embryo straight, with 2 broad, flattened cotyledons, enveloped in the copious endosperm, which has a strong odor of garlic and lacks cyclopropenoid fatty acids.

The family Huaceae consists of *Hua*, with a single species, and *Afrostyrax*, with 2 species, all native to tropical Africa. The two genera have not always been associated taxonomically, and there have been widely divergent views about their affinities. The treatment here presented draws on the work of Baas (1972), who shows that the two genera have many anatomical as well as other features in common. I cannot follow Baas, however, in referring the Huaceae to the Malvales, where they would be a notably discordant element. The unilocular ovary, the nonstratified phloem, the absence of mucilage-cells, and the absence of cyclopropenoid fatty acids collectively form too large a set of exceptions to be tolerated in the Malvales. Within the more diversified order Violales, on the other hand, the Huaceae do not seem extraordinary. The garlic-like odor is distinctive, but does not suggest any other affinity for the family.

6. Family LACISTEMATACEAE Martius 1826 nom. conserv., the Lacistema Family

Shrubs and small trees with unbranched, multicellular hairs that commonly have elongate pits in the basal cell; parenchymatous tissues strongly tanniferous, but without special tanniferous idioblasts; clustered or solitary crystals of calcium oxalate commonly present in some of the cells of the parenchymatous tissues; vessel-segments with scalariform perforation-plates that have numerous cross-bars; imperforate tracheary elements commonly septate, with simple or obscurely bordered pits; wood-rays heterocellular, mixed uniseriate and pluriseriate, the latter 2–4 cells wide; wood-parenchyma diffuse. LEAVES alternate, simple, entire (*Lacistema*) or toothed (*Lozania*); clustered crystals present in some cells of the mesophyll; stomates more or less distinctly anomocytic; petiolar anatomy complex; stipules present, deciduous. FLOWERS very small, in slender, axillary, catkin-like spikes or racemes, generally perfect, apetalous, bibracteolate, the primary bracts of the inflorescences small and inconspicuous (*Lozania*) or conspicuous and imbricately overlapping (*Lacistema*); sepals 4–6, unequal, or none, stamen solitary, seated on or within a fleshy, sometimes cupular disk. Pollen-sacs 2, well separated on an expanded connective, or sometimes individually stipitate, opening by longitudinal slits; pollen-grains tricolporate; gynoecium of 2–3 carpels united to form a compound, unilocular ovary with parietal placentas; style solitary; stigmas distinct; ovules 1–2 on each placenta, pendulous, anatropous or hemitropous, bitegmic, crassinucellar, with a rather thick, elongate funiculus; endosperm-development nuclear. FRUIT capsular; seeds 1–2; embryo straight, with broad, foliar cotyledons, embedded in the rather copious, oily endosperm.

The family Lacistemataceae consists of 2 genera and about 20 species, native to tropical America. The affinities of the family have been much disputed. Recent opinion links it to the Flacourtiaceae, and some authors now include the Lacistemataceae in that family. Some but not all authors see *Prockia*, of the Flacourtiaceae, as the connecting link.

7. Family SCYPHOSTEGIACEAE Hutchinson 1926 nom. conserv., the Scyphostegia Family

Small, glabrous, dioecious trees with trilacunar nodes; clustered crystals of calcium oxalate present in some of the cells of the parenchymatous tissues; vessel-segments with slanting end-walls and simple perforations, or a few of them scalariform with a few bars; imperforate tracheary elements making up most of the ground-tissue of the stem, with strongly thickened walls, large, bordered pits, and numerous slender cross-partitions; wood-rays mostly uniseriate and homocellular, but some of them 2–3-seriate and heterocellular, Kribs type I; wood-parenchyma nearly or quite wanting. LEAVES alternate, simple, toothed, pinnately veined, with scattered secretory (tanniferous?) cells in the mesophyll; stomates paracytic; stipules very small, deciduous. FLOWERS unisexual, tiny, in terminal inflorescences of racemosely arranged spikes or racemes, the lower branches of the inflorescence subtended by foliage-leaves; each raceme with a series of overlapping, nestling tubular bracts subtending each a single flower; staminate flowers with a tubular, 6-lobed perianth, the lobes arranged into an outer and an inner set of 3 each; 3 large nectary-glands present on the receptacle between the perianth and the androecium, aligned with the petals (inner perianth-segments); stamens 3, fully united into a column, aligned with the nectaries and petals; anthers tetrasporangiate and dithecal, with a prolonged connective, opening by longitudinal slits; pollen-grains tricolpate; perianth of the pistillate flowers of 3 distinct sepals alternating with 3 distinct petals; gynoecium of 8–13 carpels united to form a compound ovary that is more or less definitely septate near the summit but unilocular below; ovary crowned by a thick, discoid, centrally imperfect, essentially sessile stigma with as many radiating ridges as carpels; ovules numerous, basal, erect on a prominent funiculus, anatropous with the raphe external, bitegmic, crassinucellar; endosperm-development nuclear. FRUIT a fleshy capsule opening from the top by recurving valves that separate at the carpellary midveins; seeds arillate, with a fairly large, straight, dicotyledonous embryo, rather scanty, oily endosperm, and a very thin layer of perisperm; germination epigaeal. N = 9.

The family Scyphostegiaceae consists of a single species, *Scyphostegia borneensis* Stapf, native to Borneo. The morphology of the pistillate flowers has been variously interpreted, with the result that the Monimiaceae, Moraceae, and Celastrales, in addition to the Violales, have been suggested as possible allies. I here accept the interpretation of Swamy (1953) and van Heel (1967), rather than that of Hutchinson (1926 et seq.). Under this interpretation, which is also accepted by Takhtajan and by Melchior in the current Engler Syllabus, *Scyphostegia* can most reasonably be placed in the Violales, somewhere near the Flacourtiaceae. Metcalfe (1956) likewise found that the vegetative anatomy of *Scyphostegia* supports a flacourtiaceous relationship.

8. Family STACHYURACEAE J. G. Agardh 1858 nom. conserv., the Stachyurus Family

Evergreen or deciduous shrubs or small trees, tanniferous, with both proanthocyanins and ellagic acid; nodes trilacunar; clustered crystals of

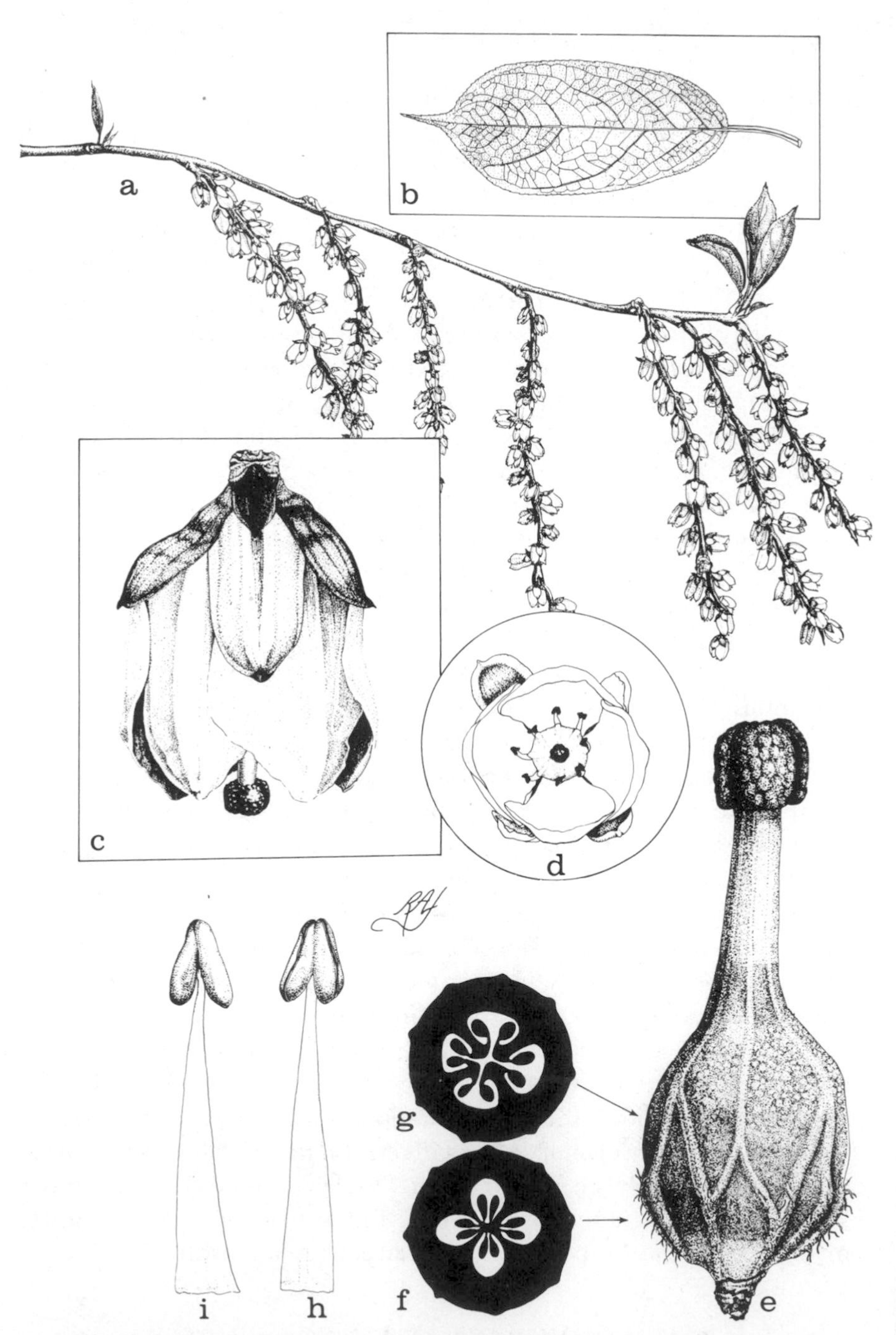

FIG. 4-20 Stachyuraceae. *Stachyurus chinensis* Franchet. a, twig with inflorescences and young leaves, $\times\frac{1}{2}$; b, mature leaf, $\times\frac{1}{2}$; c, flower, $\times 7$; d, flower, from above, $\times 5$; e, pistil, $\times 14$; f, g, schematic cross-section of ovary, near the base and just above midlength, $\times 10$; i, h, dorsal and ventral views of stamen, $\times 16$.

calcium oxalate commonly present in some of the cells of the parenchymatous tissues; vessel-segments with scalariform perforations that have numerous cross-bars; imperforate tracheary elements with large bordered pits; wood-rays heterocellular, mixed uniseriate and pluriseriate, with elongate ends; wood-parenchyma apotracheal, diffuse. LEAVES alternate, simple, toothed, pinnately veined; stomates anomocytic; stipules small, deciduous. FLOWERS in axillary racemes or spikes, small, perfect or sometimes unisexual, hypogynous, tetramerous throughout, closely subtended by 2 well developed bracteoles; sepals 2 + 2, distinct, petals 4, distinct, imbricate; stamens 8, in 2 cycles, distinct; anthers short, deeply sagittate (appearing versatile) tetrasporangiate and dithecal, opening by longitudinal slits; pollen-grains tricolporate, 2- or 3-nucleate; gynoecium of 4 carpels united to form a compound ovary with a short style and a wet, papillate, peltate-capitate, obscurely 4-lobed stigma; overy fully 4-locular (and with axile placentation) in the lower part, but in the middle and upper part not quite so, the 4 anchor-shaped partitions deeply intruded and ventrally juxtaposed but not organically united, leaving a small central cavity amongst the 4 anchor-heads, the placentation in that part of the ovary therefore parietal on deeply intruded placentas; ovules numerous, anatropous, bitegmic, crassinucellar; endosperm-development nuclear. FRUIT a fully 4-locular, rather firm and dry berry; seeds numerous, small, arillate; embryo straight, the radicle shorter than the 2 flat cotyledons; endosperm hard, oily and proteinaceous, not starchy.

The family Stachyuraceae consists of the single genus *Stachyurus*, with 5 or 6 species native from the Himalayan region to Japan.

Stachyurus has usually been thought to be allied to the Flacourtiaceae, but a relationship to the Hamamelidales has also been suggested. Metcalfe & Chalk (1950) find the wood-anatomy compatible with either relationship. The numerous and small seeds of *Stachyurus* are more compatible with the Flacourtiaceae, and Takhtajan (1966) considers that the pollen also supports a flacourtiaceous relationship. It is interesting and perhaps significant that the inflorescence of *Stachyurus* suggests that of *Populus*, in the Salicaceae. The Salicaceae are now widely regarded as being descended from ancestors similar to the Flacourtiaceae.

The Stachyuraceae have usually been described as having a 4-locular ovary with axile placentation. My own observations of fresh material of *Stachyurus chinensis* Franch. and *S. praecox* Sieb. & Zucc. indicate that over the greater part of its length the ovary at anthesis has deeply intruded parietal placentas, anchor-shaped in cross-section, that are juxtaposed in the center of the ovary but not organically united. Only at the base is the ovary fully partitioned. The mature fruit, on the other hand, appears to be fully 4-locular, as judged from herbarium specimens.

Even though I leave *Stachyurus* in its accustomed position, it should be noted that Corner (1976) considers that its seed structure is compatible with that of the Theaceae, but not the Flacourtiaceae. The boundary between the Theales and Violales as here defined is arbitrary, and it may well be that the nearest common ancestor of *Stachyurus* and other Violales such as the Flacourtiaceae was something that would have been properly referrable to the Theales.

9. Family VIOLACEAE Batsch 1802 nom. conserv., the Violet Family

Annual or more often perennial herbs or shrubs, less commonly rather small trees or even lianas, often saponiferous and often with alkaloids, and often with scattered resinous secretory cells that have yellowish contents, but not cyanogenic and not strongly tanniferous, lacking ellagic acid and with only small amounts of proanthocyanins (or none); solitary or clustered crystals of calcium oxalate often present in some of the cells of the parenchymatous tissues, especially in the mesophyll and wood-rays; nodes trilacunar; vessel-segments with scalariform perforation-plates or (especially in the herbaceous species) with simple perforations; imperforate tracheary elements septate, with simple or bordered pits; wood-rays variously heterocellular or homocellular, Kribs types I, IIA, and IIB, mixed uniseriate and pluriseriate, the latter mostly 2–7 cells wide; wood-parenchyma wanting or represented only by occasional cells about the vessels. LEAVES alternate or seldom (some spp. of *Hybanthus* and *Rinorea*) opposite, simple and entire or toothed to sometimes lobed or even dissected, but without well defined leaflets; epidermis often with mucilaginous inner walls or with scattered mucilaginous cells; stomates anisocytic or paracytic; stipules present. FLOWERS in racemes, heads, or panicles, or often solitary and axillary, perfect or rarely some or all of them unisexual, hypogynous, sometimes some of them cleistogamous, regular or slightly irregular (Rinoreae, Leonioideae), or strongly irregular (Violeae), bibracteolate; sepals 5, distinct or nearly so, imbricate, commonly persistent; petals 5, imbricate or convolute, in irregular flowers the lowermost one commonly prolonged behind into a spur; stamens 5 (3 in *Leonia triandra*), the filaments very

FIG. 4-21 Violaceae. *Viola cucullata* Aiton, a, habit, $\times\frac{1}{2}$; b, flower, $\times 4$; c, schematic cross-section of ovary, $\times 12$; d, portion of flower, in long-section, $\times 4$; e, pistil, $\times 8$; f, spurred lower stamen, $\times 8$; g, lateral stamen, $\times 8$; h, fruit, $\times 4$.

short, distinct or more or less connate, the anthers commonly connivent around the ovary, often the two anterior ones (*Viola*) or all of them (the woody genera) with a gland-like or spur-like nectary on the back; connective often prolonged into a membranous appendage; pollen-grains binucleate, 3 (–5)-colporate; gynoecium of (2) 3 (–5) carpels united to form a compound, unilocular ovary with parietal placentation;

style solitary, often distally enlarged or otherwise modified, the stigma simple or lobed; ovules 1-many on each placenta, anatropous, bitegmic, crassinucellar; endosperm-development nuclear. FRUIT a loculicidal capsule or sometimes a berry, seldom (*Leonia*) nutlike; seeds with straight embryo and flat cotyledons embedded in the abundant, softly fleshy, oily endosperm, often arillate. X = 6–13, 17, 21, 23. (Leoniaceae)

The family Violaceae consists of about 16 genera and 800 species, of cosmopolitan distribution. Nearly half of the species belong to the large, mainly herbaceous genus *Viola*, which is best developed in North Temperate regions and in tropical mountains. At the other extreme is the pantropical woody genus *Rinorea*, with nearly 300 species commonly found in the understory of rain-forests. *Rinorea* is generally considered to be archaic within the family and to indicate a relationship of the Violaceae to the Flacourtiaceae.

10. Family TAMARICACEAE Link 1821 nom. conserv., the Tamarix Family

Shrubs or rather small trees or seldom half-shrubs, with slender branches, often evergreen, mostly halophytic or xerophytic, generally (not always) tanniferous, with proanthocyanins and gallic and ellagic acids, commonly producing flavonoid bisulphates, but not cyanogenic; clustered crystals of gypsum and sometimes also calcium oxalate commonly present in some of the cells of the parenchymatous tissues; vessel-segments with simple perforations; imperforate tracheary elements mostly with few, simple pits, often with a persistent protoplast; some true tracheids sometimes also present; wood-rays heterocellular, all or nearly all pluriseriate, very broad and high, up to 10 or 15 or even 25 cells wide; wood-parenchyma scanty-paratracheal to vasicentric; sieve-tubes with S-type plastids. LEAVES alternate, small, commonly subulate or scale-like, mostly sessile, usually centric in structure, commonly with embedded, salt-excreting, multicellular external glands; stomates mostly anomocytic; stipules wanting. FLOWERS small, solitary or more often in slender, scaly-bracteate racemes, spikes, or panicles, hypogynous, without bracteoles; sepals 4–5 (6), distinct or less often connate below, imbricate, persistent; petals as many as and alternate with the sepals, distinct, sometimes persistent, seated (along with the stamens) on a fleshy nectary-disk, or the disk sometimes intrapetalar or intrastaminal or wanting; a pair of scale-like appendages present on the inner side at the base of each petal in *Reaumuria*; stamens as many or

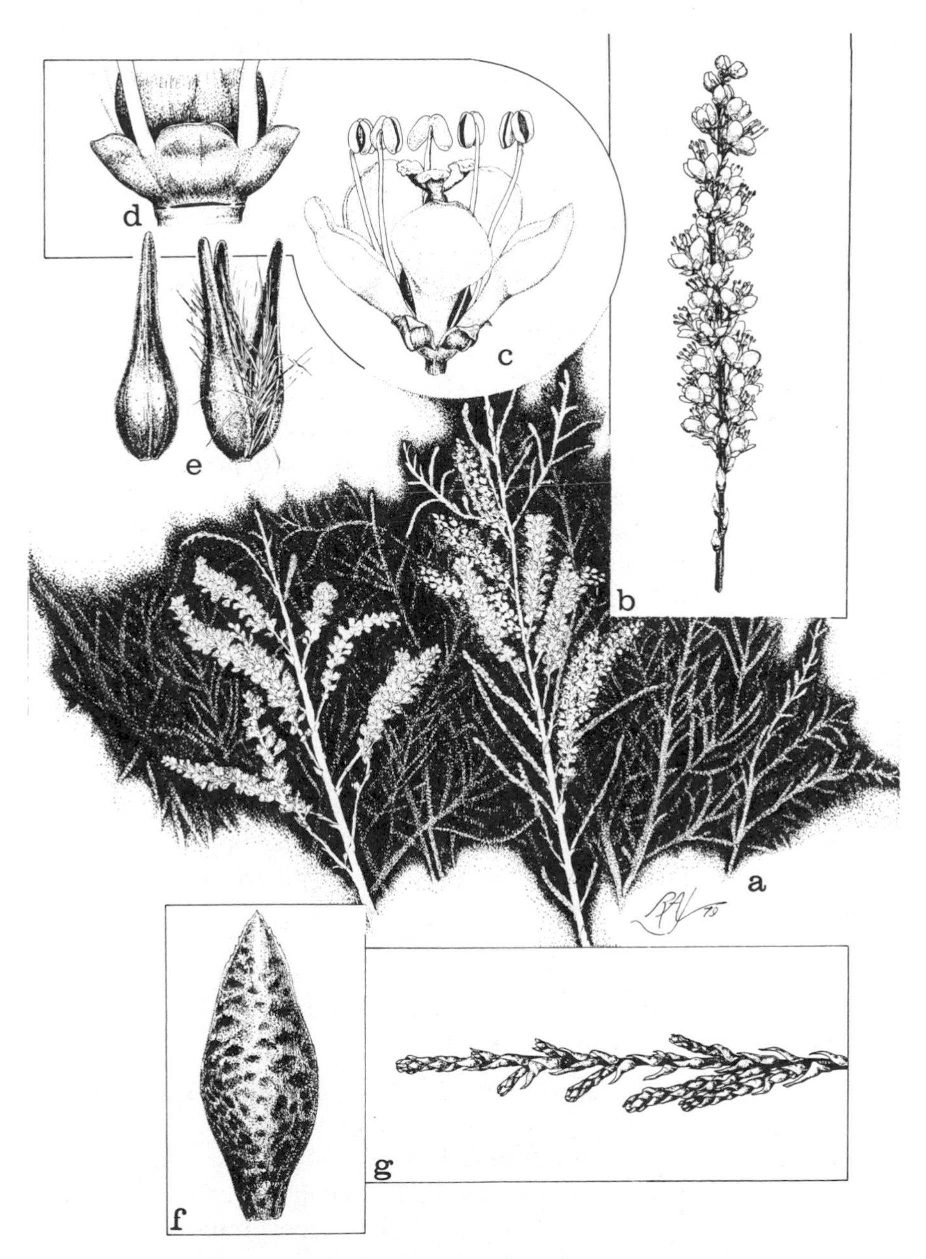

FIG. 4-22 Tamaricaceae. *Tamarix gallica* L. a, habit, ×1; b, flowering branch, ×3; c, flower, ×15; d, base of flower, with perianth removed, showing disk and lower parts of stamens and ovary, ×30; e, fruits, ×8; f, leaf, ×30, g, vegetative twig, ×4.

often twice as many as the petals, or more or less numerous and then often connate at the base into 5 bundles; anthers tetrasporangiate and dithecal, opening by longitudinal slits; pollen-grains in monads or tetrads, binucleate (2) 3 (4)-colpate; gynoecium of (2) 3–4 (5) carpels united to form a compound, unilocular ovary with parietal or (*Tamarix*) basal or parietal-basal placentation, the placental partitions sometimes so deeply intruded that the ovary appears to be plurilocular, especially at the top and bottom; styles distinct or connate at the base (in *Myricaria* the stigmas sessile); ovules 2-many on each placenta, anatropous, bitegmic, rather weakly crassinucellar; embryo-sac tetrasporic; endosperm-development nuclear or cellular, without haustoria. FRUIT a loculicidal capsule; seeds covered with long hairs, or more often the hairs forming a coma at one end; endosperm scanty and starchy, or more often wanting; a thin layer of perisperm often present; embryo dicotyledonous, straight. N = 12.

The family Tamaricaceae consists of 4 or 5 genera and about 100 species, native to Eurasia and Africa, especially in the Mediterranean region and eastward into the Irano-Turanian region of central Asia. About three-fourths of the species belong to the single genus *Tamarix*, which is cultivated in warm, dry regions for its graceful beauty. Several species of *Tamarix* are introduced and well established along dry or intermittent watercourses in the southwestern United States and northern Mexico. Vassilczenko and Vassileva (1976) consider that *Tamarix* originated in the valleys of mountainous islands of Paleogene Tethys.

The Tamaricaceae are generally considered to be allied to the Frankeniaceae. Those who prefer small orders often take these two families to form an order Tamaricales.

11. Family FRANKENIACEAE S. F. Gray 1821 nom. conserv., the Frankenia Family

Halophytic herbs or shrubs, with or without proanthocyanins and ellagic acid, at least sometimes producing flavonoid bisulphates, but not cyanogenic, sometimes with clustered crystals of calcium oxalate (but not gypsum) in some of the cells of the parenchymatous tissues; vessel-segments small, with slender lumen and simple perforations; ground-tissues of the xylem consisting of prosenchyma with simple pits; sieve-tube with S-type plastids. LEAVES opposite, simple, often small and ericoid and revolute-margined, commonly with embedded, salt-excreting, multicellular external glands; stomates anomocytic; stipules wanting. Flowers small, solitary in the axils or in axillary cymes, bibracteolate, hypogynous, regular, usually perfect; sepals 4–7, connate into a tube with short, induplicate-valvate lobes; petals 4–7, distinct, imbricate, long-clawed, with a scale-like or claw-like internal appendage at the base of the blade; stamens 4–7, or up to 24, most often 6 in 2 cycles of 3, distinct or shortly connate at the base; anthers opening lengthwise; pollen-grains binucleate or trinucleate, (2) 3 (4)-colpate; gynoecium of (2) 3 (4) carpels united to form a compound, unilocular ovary with as many parietal (sometimes intruded) or parietal-basal placentas as carpels, or sometimes with a single strictly basal placenta; style solitary, slender and elongate, with distinct stigmatic branches; ovules (1) 2–6 (–many) on each placenta, anatropous, bitegmic, pseudocrassinucellar, with a more or less elongate funiculus; embryo-sac monosporic; endosperm-development nuclear. Fruit a loculicidal capsule, enclosed in the persistent calyx; seeds with a central, straight embryo flanked on both sides of the cotyledons by the abundant, starchy endosperm. N = 10, 15.

The family Frankeniaceae consists of 3 genera and about 80 species, interruptedly cosmopolitan in distribution, but best developed in the Mediterranean region and eastward into Iran. *Hypericopsis* is monotypic, *Anthobryum* has about 3 species, and the remainder of the species belong to *Frankenia* (including *Niederleinia*). None of the species is of much economic interest. Recent authors are agreed that the Frankeniaceae are related to the Tamaricaceae.

12. Family DIONCOPHYLLACEAE Airy Shaw 1952 nom. conserv., the Dioncophyllum Family

Lianas or shrubs, climbing by hooked or cirrhose leaf-tips, provided (at least at first) on both leaves and stems with peltate hairs and with characteristic multicellular, stalked or sessile glands, recalling those of the Ancistrocladaceae and Droseraceae, which secrete a sticky, acid mucilage that traps insects; stem with anomalous secondary growth, producing successive vascular bundles without regular arrangement; vessel-segments large (except in *Habropetalum*), with simple perforations; imperforate tracheary elements with large, bordered pits, not septate; wood-rays homocellular, uniseriate. Leaves alternate, simple, pinnately veined, the midrib prolonged and forked into a pair of hooks or pigtailed tendrils; stomates actinocytic to encyclocytic; stipules wanting. Flowers in lax, axillary or supra-axillary cymes, perfect, regular; sepals

5, small, distinct or shortly connate at the base, valvate or open in bud, persistent; petals 5, distinct, convolute; stamens 10–30; anthers opening by longitudinal slits; pollen-grains 3 (4)-colporate, more or less similar to those of *Ancistrocladus*; nectary-disk wanting; gynoecium of 2 or 5 carpels united to form a compound, superior, unilocular ovary with parietal placentas; styles as many as the carpels and distinct, or somewhat connate at the base, or short and wholly connate, the stigmas variously capitate or feathery; ovules numerous, anatropous, bitegmic, crassinucellar. FRUIT a loculicidal capsule, opening well before maturity, the developing ovules individually long-exserted on an elongate, thickened funiculus; seeds large, circular-winged, with a large, discoid-obconic, dicotyledonous embryo and copious, starchy endosperm. 2n = 36 (*Triphyophyllum*). (Triphyophyllaceae)

The family Dioncophyllaceae consists of 3 closely allied monotypic genera (*Dioncophyllum*, *Triphyophyllum*, and *Habropetalum*), native to rainforests of tropical Africa. These genera were usually included in the Flacourtiaceae before Airy Shaw segregated them as a distinct family. They clearly belong to the Dilleniid subclass, in the vicinity of the Theales, Violales, and Sarraceniales. At the present state of our knowledge it seems most useful to associate the Dioncophyllaceae with the Ancistrocladaceae in a broadly defined order Violales.

13. Family ANCISTROCLADACEAE Walpers 1851 nom. conserv., the Ancistrocladus Family

Sympodially branched shrubs, climbing by hooked or twining branch-tips, sometimes producing alkaloids, and commonly with clustered or solitary crystals of calcium oxalate in some of the cells of the parenchymatous tissues; cortex with scattered thick-walled secretory cells; vessel-segments with very oblique ends and simple perforations; imperforate tracheary elements with large, evidently bordered pits; wood-rays homocellular, uniseriate or some of them biseriate; wood-parenchyma commonly in tangential bands. LEAVES alternate, simple, entire, beset with characteristic embedded, peltate, multicellular, waxy glands; stomates more or less distinctly actinocytic; stipules tiny, mostly caducous. FLOWERS small, in axillary or terminal mixed panicles or racemes or panicled spikes, perfect, epigynous or half-epigynous, regular except for the unequal sepals, sepals 5, seated on the middle or upper part of the ovary, imbricate, unequal, accrescent and wing-like in fruit; petals 5, distinct or slightly connate at the base, convolute; stamens 10, 5 somewhat larger than the others, or seldom only 5, the filaments somewhat connate at the base and adnate to the base of the petals; anthers basifixed, tetrasporangiate and dithecal, opening by longitudinal slits; pollen-grains 3 (4)-colpate, resembling those of the Dioncophyllaceae, especially *Triphyophyllum*; gynoecium of 3 carpels united to form a compound, unilocular ovary that has an inferior locular part tapering into a more or less solid, nipple-shaped, persistent free tip to which the 3 styles are articulated; ovule solitary, basilateral, hemitropous, bitegmic, on a short funiculus; endosperm-development cellular. FRUIT a nut surrounded by the corky hypanthium and crowned by the unequal sepals; seed with hard, starchy, ruminate endosperm in which the small, straight, dicotyledonous embryo is embedded.

The family Ancistrocladaceae consists of the single genus *Ancistrocladus*, with 15–20 species native to continental southeast Asia, India, and western to central tropical Africa. *Ancistrocladus* has in the past often been associated with or included in the Dipterocarpaceae, largely because of the accrescent, unequal sepals that form wings on the fruit, but it differs in so many respects that it is a highly discordant element in that family. Modern opinion tends to favor an association with the Dioncophyllaceae, but the genus is so distinctive that it must in any case be taken as a separate family.

14. Family TURNERACEAE A. P. de Candolle 1828 nom. conserv., the Turnera Family

Herbs, shrubs, or seldom trees, provided with diverse sorts of hairs in different species, sometimes producing alkaloids, often cyanogenic, lacking ellagic acid but at least sometimes with proanthocyanins, and frequently with thick-walled tanniferous cells in the pith and cortex; clustered crystals of calcium oxalate commonly present in some of the cells of the parenchymatous tissues; vessel-segments with simple or both simple and scalariform perforations; imperforate tracheary elements with small, bordered pits; wood-rays heterocellular or homocellular, mostly uniseriate or biseriate, but sometimes as much as 5 cells wide; wood-parenchyma scanty, apotracheal; cortical vascular bundles (longitudinally oriented leaf-traces) commonly present; phloem devoid of sclerenchyma. LEAVES alternate, entire or toothed to lobed, often provided with a pair of glands or extrafloral nectaries at the base of the blade, the epidermis often mucilaginous; stomates variously anomocytic, paracytic, or anisocytic; petiolar anatomy diverse; stipules small or more

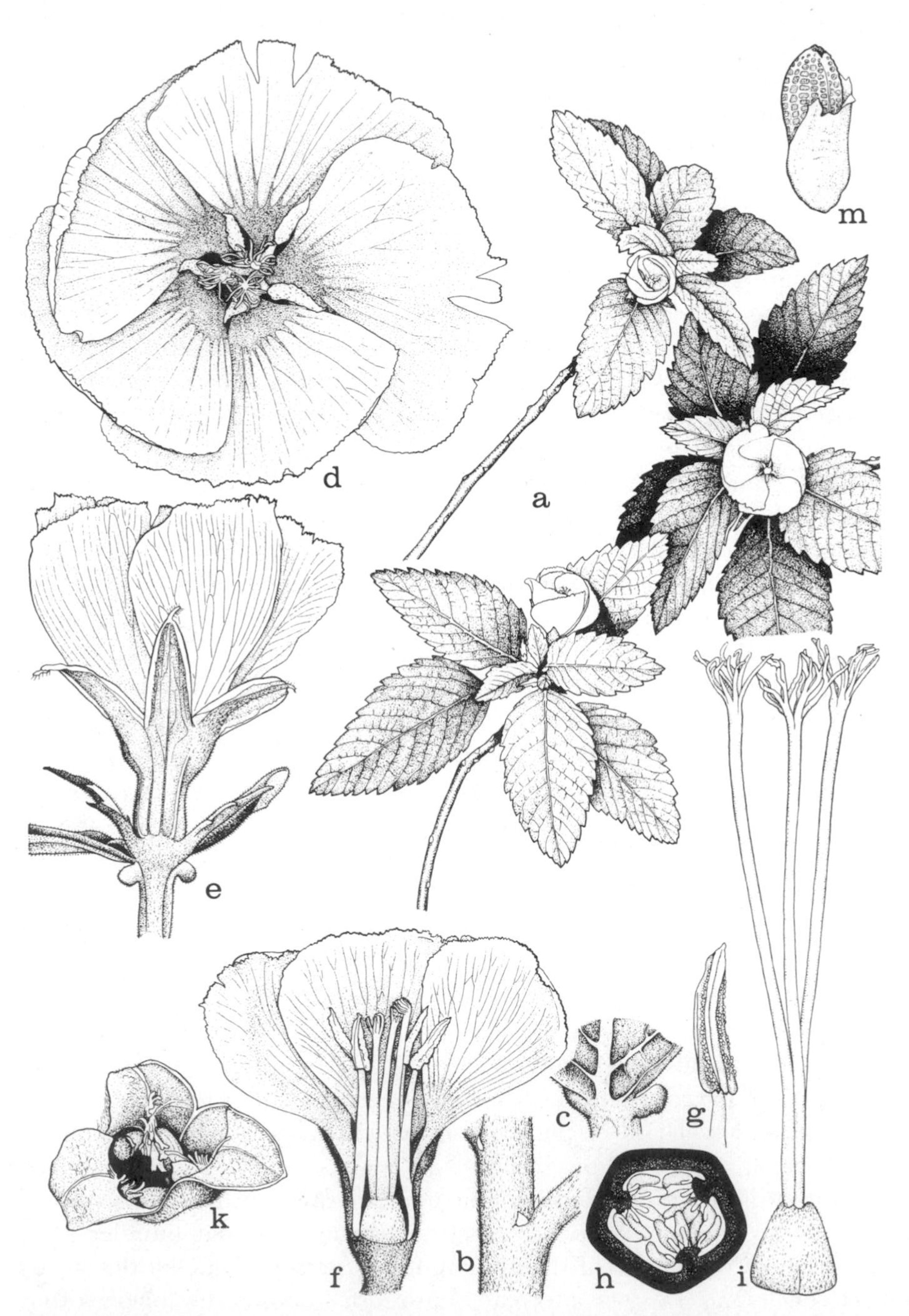

FIG. 4-23 Turneraceae. *Turnera ulmifolia* L. a, habit, $\times\frac{1}{2}$; b, portion of twig, showing 2 nodes and minute stipules, $\times 4$; c, leaf-base, showing glands, $\times 4$; d, e, flower, from above and from the side, $\times 2$; f, flower, in long-section, $\times 2$; g, anther, $\times 4$; h, schematic cross-section of ovary, $\times 8$; i, pistil, $\times 4$; k, opened capsule, from above, $\times 2$; m, seed, $\times 8$.

often wanting. FLOWERS solitary and axillary, or in various sorts of axillary or terminal inflorescences, red or yellow, perfect, regular, strongly perigynous, with a short to more often tubular hypanthium, often subtended by 2 bracteoles; sepals 5, imbricate; petals 5, clawed at the base, inserted at the summit of the hypanthium ("calyx tube"), commonly ephemeral, convolute at least in *Turnera*; hypanthium often bearing 5 glands or protuberances between the stamens and petals, or sometimes (*Piriqueta, Erblichia*) with a narrow, fringed, extrastaminal corona; stamens 5, alternate with the petals, inserted well down in the hypanthium; anthers dithecal, opening by longitudinal slits; pollen-grains tricolporate, binucleate, similar to those of the Passifloraceae; gynoecium of 3 carpels united to form a compound, superior to half-inferior, sessile, unilocular ovary with parietal placentas; styles distinct, sometimes bifid, often with a fringed or brush-like stigma; ovules anatropous, bitegmic, crassinucellar, commonly (3–) numerous on each placenta; endosperm-development nuclear. FRUIT a loculicidal (rarely septicidal) capsule; seeds (1–) 3-many, reticulate-ridged and pitted, with a membranous, unilateral aril and a straight or slightly curved, spathulate, dicotyledonous embryo embedded in the abundant, softly fleshy, oily endosperm. X = 7, 10.

The family Turneraceae consists of 8 genera and about 120 species, native to tropical and subtropical or warm-temperate parts of America and Africa, and to the Islands of Madagascar and Rodriguez. More than half of the species belong to the single genus *Turnera*, native from Texas to Argentina. One of the genera, *Erblichia*, occurs in Central America, Africa, and Madagascar.

Botanists are agreed that the Turneraceae are allied to the Passifloraceae and Malesherbiaceae. It is here suggested that these 3 families have a common origin in or near the Flacourtiaceae.

15. Family MALESHERBIACEAE D. Don 1827 nom. conserv., the Malesherbia Family

Herbs or half-shrubs, often cyanogenic, provided with unicellular hairs and long, multiseriate, frequently glandular and malodorous hairs; vessel-segments with simple perforations, or sometimes some of them with scalariform perforations; imperforate tracheary elements rather short, with very small pits; wood-rays mostly uniseriate or biseriate, seldom some of them triseriate. LEAVES alternate, simple but sometimes deeply pinnatifid, exstipulate. FLOWERS solitary or in panicles or racemes, perfect, regular, strongly perigynous, with a long, slender, straight or curved hypanthium ("calyx tube"); sepals 5, valvate; petals 5, valvate; a membranous, denticulate corona present; androecium and gynoecium borne on a villous, lobed androgynophore; stamens 5; pollen-sacs opening by longitudinal slits; pollen-grains tricolporate, similar to those of the Turneraceae; gynoecium of 3–4 carpels united to form a compound, unilocular ovary with parietal placentas; styles distinct, filamentous; ovules numerous, anatropous. FRUIT capsular, included in the persistent hypanthium; seeds exarillate, reticulate-ridged and pitted, with straight, dicotyledonous embryo and oily endosperm.

The family Malesherbiaceae consists of 1 or 2 genera and about 25 species, native to dry habitats in the Andes from Peru to Chile and Argentina. Most of the species grow in northern Chile.

Authors are agreed that the Malesherbiaceae constitute a distinctive family related to the Turneraceae and Passifloraceae.

16. Family PASSIFLORACEAE A. L. de Jussieu ex Kunth 1817 nom. conserv., the Passion-flower Family

Herbaceous or woody vines, climbing by axillary tendrils that represent modified inflorescences or parts of inflorescences, less often erect shrubs or even trees, frequently with one or another sort of anomalous secondary growth, commonly cyanogenic in association with a cyclopentenoid ring-system, often accumulating β-carboline alkaloids such as passiflorine, sometimes with scattered tanniferous cells in the parenchymatous tissues, but only seldom with ellagic acid and proanthocyanins; nodes trilacunar; solitary and clustered crystals of calcium oxalate commonly present in some of the cells of the parenchymatous tissues and in the epidermis of the leaves; vessel-segments commonly with simple perforations, but in certain genera some of them with scalariform perforations; slender, elongate fiber-tracheids often apically perforate and thus vessel-like; wood-rays of various types; wood-parenchyma commonly apotracheal. LEAVES alternate, entire or often palmately lobed, seldom (*Deidamia*) compound, very often with extrafloral nectaries on the petiole; stomates mostly anomocytic; stipules usually present, but generally small and deciduous; an accessory bud generally present in addition to the primary axillary bud, the primary bud often either abortive or developing into an inflorescence or a tendril, the vegetative branches then developing from the accessory

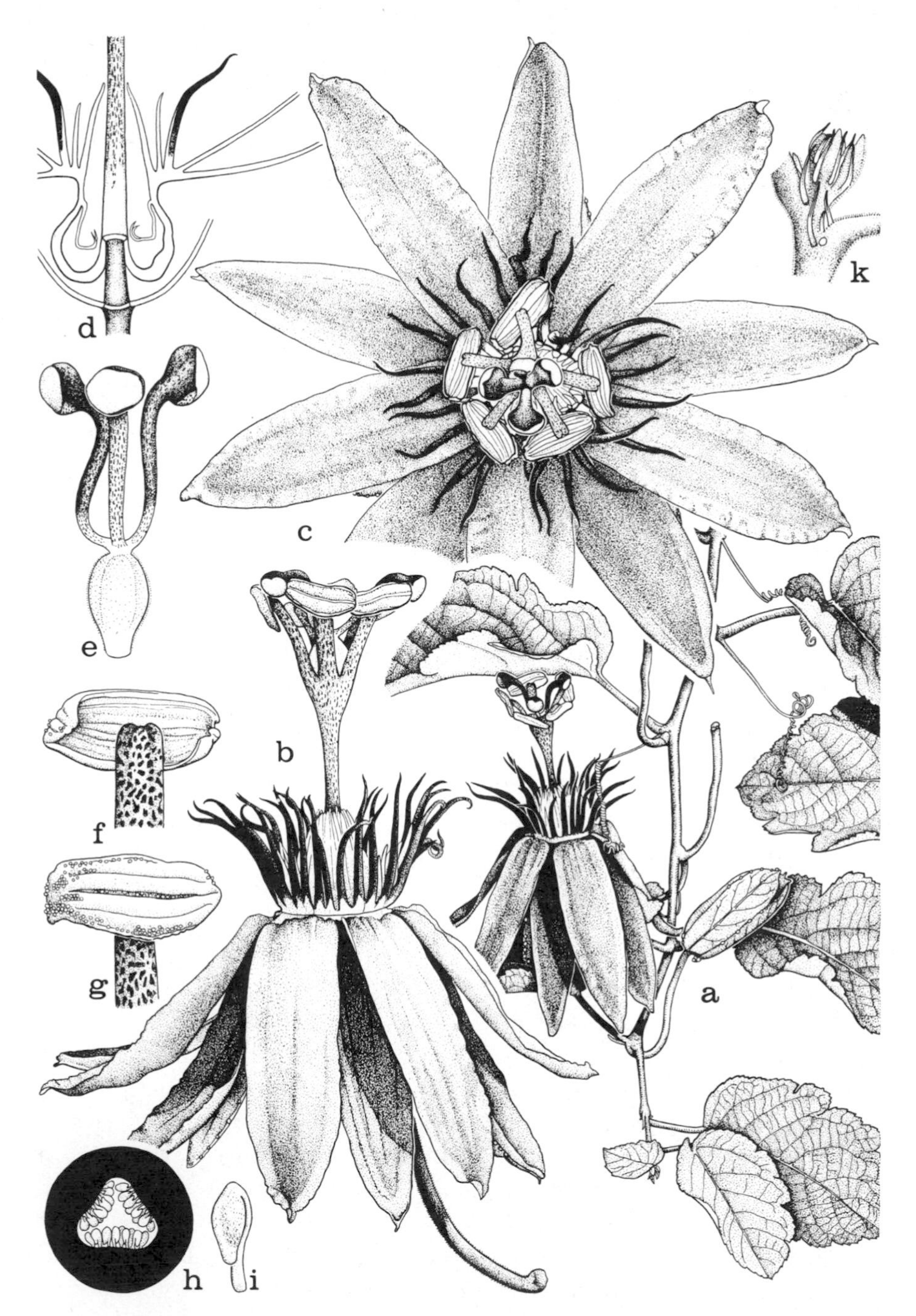

FIG. 4-24 Passifloraceae. *Passiflora coccinea* Aubl. a, habit, ×½; b, c, flower, from the side and from above, ×1; d, schematic long-section through base of flower, ×1; e, pistil, ×2; f, g, two views of anther; h, schematic cross-section of ovary, ×4; i, ovule, ×24; k, nodal region, showing stipules, ×2.

buds. FLOWERS in various sorts of cymose (seldom racemose) inflorescences, or seldom solitary, perfect or less often unisexual, regular, perigynous, with a saucer-shaped to tubular hypanthium (seldom hypogynous, as in *Paropsia*), commonly with an elongate androgynophore, less often with a mostly short gynophore, or rarely the ovary sessile; sepals (3–) 5 (–8), distinct or often connate below, imbricate, persistent; petals as many as and alternate with the sepals, distinct or shortly connate at the base, imbricate, or seldom wanting; extrastaminal corona nearly always present (poorly developed in *Adenia*), borne on the hypanthium (in perigynous flowers) within the corolla, variously developed, commonly consisting of one or more rows of filaments or scales; stamens (4) 5 (–numerous), mostly alternate with the petals, free or raised on the androgynophore, seldom (*Androsiphonia*) the filaments connate into a tube around the ovary; anthers tetrasporangiate and dithecal, versatile, opening by longitudinal slits; pollen-grains binucleate, 3– to 12-colporate; a nectary disk of staminodial origin often present around the ovary; gynoecium of (2) 3 (–5) carpels united to form a compound, unilocular ovary with parietal placentas; styles mostly distinct or connate only at the base, each with a capitate to clavate or discoid stigma (style solitary and with a single stigma in *Barteria* and *Crossostemma*); ovules more or less numerous, anatropous (rarely orthotropous), mostly on a long funiculus, bitegmic, crassinucellar; endosperm-development nuclear. FRUIT a capsule or often a berry; seeds usually much-compressed, with a bony testa, commonly pitted or ridged-reticulate, with a fleshy apical aril; embryo large, straight, spathulate, dicotyledonous, embedded in the copious, oily, softly fleshy endosperm. X = 6, 9–11.

The family Passifloraceae as here defined consists of about 16 genera and 650 species, widespread in tropical and warm-temperate regions, but best-developed in tropical America and Africa. The family is dominated by the large genus *Passiflora*, with about 400 species. The only other large genus is *Adenia* (100). The fruits of several species of *Passiflora* are edible, but the widespread production of cyanide and passiflorine by members of the family makes all species suspect until proven innocent.

The Passifloraceae are widely regarded as being related to and derived from the Flacourtiaceae. *Paropsia* and some related genera that are here included in the Passifloraceae have often been referred to the Flacourtiaceae instead, and recent authors consider that these several genera form a link between the two families. They are anomalous in the Passifloraceae because of their arborescent habit, and in the Flacourtiaceae because of their more or less well developed extrastaminal corona. Keating (1973) considers that the pollen of some of these genera is intermediate between the two families, but more in harmony with the Passifloraceae. Ayensu & Stern (1964) note the presence in *Paropsia* of elongate, slender, apically perforate fiber-tracheids (fibriform vessel-elements), which are common in the Passifloraceae but rare elsewhere and unknown in the Flacourtiaceae. Furthermore, the tribe Paropsieae is closely linked to the Passifloraceae by some transitional genera. Thus it seems useful to draw the arbitrary distinction between the Passifloraceae and Flacourtiaceae on the basis of the presence of the corona, rather than on the growth-habit.

17. Family ACHARIACEAE Harms in Engler & Prantl 1897 nom. conserv., the Acharia Family

Climbing or acaulescent herbs or half-shrubs; vessel-segments mostly with simple perforations; fibers or prosenchymatous cells with simple pits, sometimes septate. LEAVES alternate, palmately lobed, stipulate. FLOWERS solitary or racemose, hypogynous, regular, unisexual, the plants monoecious; sepals 3–5, distinct; corolla sympetalous, 3–5-lobed; stamens as many as and alternate with the corolla-lobes, attached to the corolla-tube near its base or at the throat; pollen-sacs separated on a broad connective, opening by longitudinal slits; pollen-grains tricolporate; gynoecium of 3–5 carpels united to form a compound, unilocular pistil with parietal placentas; style more or less deeply cleft; stigmas 2-lobed; ovules (2–) several or numerous on each placenta. FRUIT stipitate, capsular; seeds grooved or pitted, with small, straight, dicotyledonous embryo and abundant endosperm.

The family Achariaceae consists of 3 monotypic genera, *Acharia*, *Ceratiosicyos*, and *Guthriea*, confined to South Africa. Modern authors are agreed that the family is related to the Passifloraceae. Similarities to the Cucurbitaceae have also been noted.

18. Family CARICACEAE Dumortier 1829, nom. conserv., the Papaya Family

Soft-stemmed shrubs or small trees, commonly with an unbranched trunk and a terminal cluster of leaves in the manner of palms, or

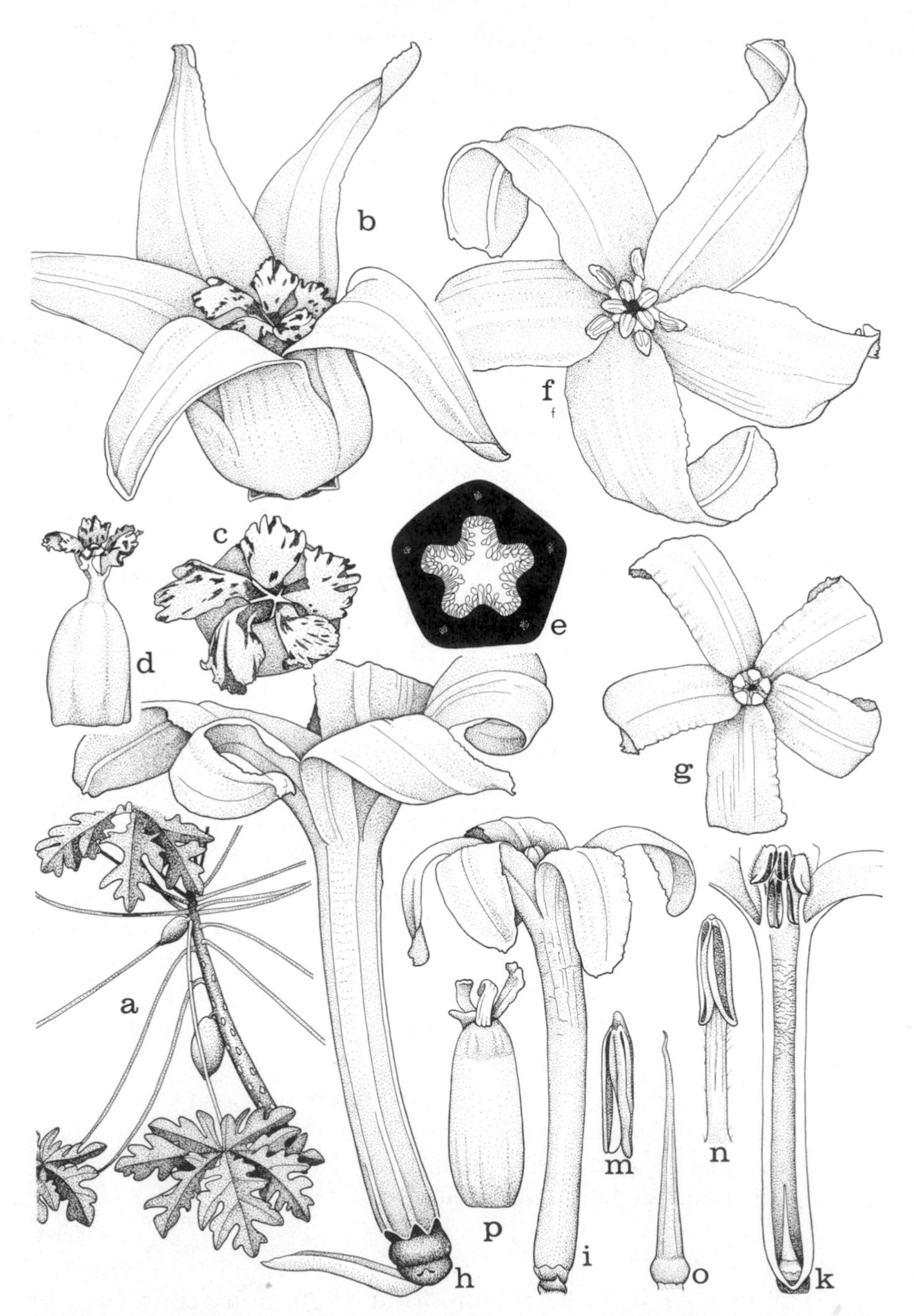

FIG. 4-25 Caricaceae. *Carica papaya* L. a, habit, $\times\frac{1}{20}$; b, pistillate flower, $\times 1$; c, pistil, from above, $\times 2$; d, pistil, $\times 1$; e, schematic cross-section of ovary; f, h, quasiperfect flower, from otherwise pistillate plant, $\times 2$; g, i, staminate flower, $\times 2$; k, staminate flower in long-section, $\times 2$; m, n, long and short stamens, $\times 6$; o, vestigial pistil from staminate flower, $\times 6$; p, pistil from quasiperfect flower, $\times 2$.

seldom (*Jarilla*) prostrate herbs, commonly producing myrosin and mustard oil, often producing carpaine, an alkaloid of unusual structure, and consistently with a well developed system of anastomosing, articulated laticifers, but lacking ellagic acid, proanthocyanins, and iridoid substances; nodes trilacunar or multilacunar; clustered crystals of calcium oxalate commonly present in some of the cells of the parenchymatous tissues; stem with a ring of narrow, radially elongated vascular bundles separated by broad, parenchymatous medullary rays; vessel-segments with simple perforations; imperforate tracheary elements wanting; secondary phloem tangentially stratified into hard and soft layers. LEAVES alternate, commonly large and long-petiolate, mostly palmately 3–13-veined and palmately lobed or palmately compound, rarely pinnately veined and entire to pinnatifid; stipules wanting, or rarely present and spine-like. FLOWERS in cymose, axillary inflorescences, or sometimes solitary in the axils, regular, hypogynous, pentamerous, unisexual (the plants dioecious or sometimes monoecious) or sometimes some of them perfect; sepals 5, small, united into a toothed or lobed calyx; petals 5, connate to form an elongate, slender corolla-tube in staminate flowers, and a short (or very short) tube in pistillate flowers, the lobes convolute or valvate; stamens 10, in 2 cycles, or sometimes 5 and then alternate with the corolla-lobes; filaments attached to the corolla-tube, distinct (*Carica*) or basally connate into a short tube; anthers tetrasporangiate and dithecal, with a shortly prolonged connective; pollen-sacs opening by longitudinal slits; pollen-grains binucleate, tricolporate; gynoecium of 5 carpels united to form a compound, unilocular ovary with more or less deeply intruded parietal placentas, or the partitions meeting and joined in the center to form a plurilocular ovary with axile placentas; styles distinct; ovules numerous, anatropous, bitegmic, crassinucellular, with a more or less enlarged funiculus; endosperm-development nuclear. FRUIT large, fleshy, melon-like; embryo straight, spatulate, with 2 broad, flat cotyledons, embedded in the endosperm, which is softly fleshy and contains reserves of both oil and protein. X = 9.

The family Caricaceae consists of 4 genera and about 30 species, occurring mainly in tropical and subtropical America from Mexico and the West Indies to northern Chile and Argentina. A single genus, *Cylicomorpha*, with 2 species, occurs in tropical Africa. The largest genus is *Carica*, with about 22 species. *Jacaratia* has about 6 species, and *Jarilla* only one. *Carica papaya* L., the papaya, produces an esteemed edible fruit. The latex of this species is the source of the enzymatic meattenderizer papain. It requires no great imagination to suppose that the latex may serve to discourage predators.

The Caricaceae are generally considered to be related to the Passifloraceae. *Adenia*, in the Passifloraceae, has been cited by some authors (e.g., Hallier, Takhtajan) as a connecting link.

19. Family FOUQUIERIACEAE A. P. de Candolle 1828 nom. conserv., the Ocotillo Family

Woody or fleshy-succulent, xerophytic, spiny shrubs or small trees with differentiated long and short shoots, often with a highly parenchymatized xylem that functions in water-storage, the plant at least sometimes producing triterpenes, triterpenoid and steroidal saponins, iridoid compounds, ellagic acid and proanthocyanins, but not cyanide; nodes unilacunar, one-trace; crystals of calcium oxalate commonly present in some of the cells of the parenchymatous tissues; vessel-segments with simple perforations; imperforate tracheary elements with bordered pits; wood-rays mostly homocellular or nearly so, up to about 8 cells wide, relatively few of them uniseriate; wood-parenchyma diffuse; sieve-tubes with S-type plastids. LEAVES alternate, simple, small, commonly produced after rain and quickly deciduous when the soil dries, those of the long shoots each surmounting a decurrent ridge of the stem, a portion of the petiole becoming indurate and persisting as a spine; short shoots axillary to the spines, producing closely clustered leaves that do not form spines; stomates anomocytic; petiolar vascular strand trough-shaped. FLOWERS in various sorts of axillary or terminal inflorescences, perfect, hypogynous, essentially regular, at least sometimes containing anthocyanin 3-galactosides, which are common also in the Ericales; sepals 5, imbricate, persistent, the 2 outer ones often somewhat smaller than the other 3; petals 5, connate for most of their length to form a tubular or salverform corolla with imbricate lobes; stamens 10–18 (–23), exserted, borne in a single whorl on the receptacle, but the antesepalous ones sometimes larger and more outwardly directed so that there appear to be two whorls; filaments commonly hairy toward the base; anthers tetrasporangiate and dithecal, opening by longitudinal slits; pollen-grains binucleate, tricolporate, pitted or ridged-reticulate; gynoecium of 3 carpels united to form a compound pistil with a single style that is branched near or above the middle, the stigmas terminal; basal, solid part of the ovary nectariferous; locular cavity of the ovary fully partitioned at the base, but at anthesis the partitions not meeting or not joined in the upper part of the ovary, so that the placentation is axile

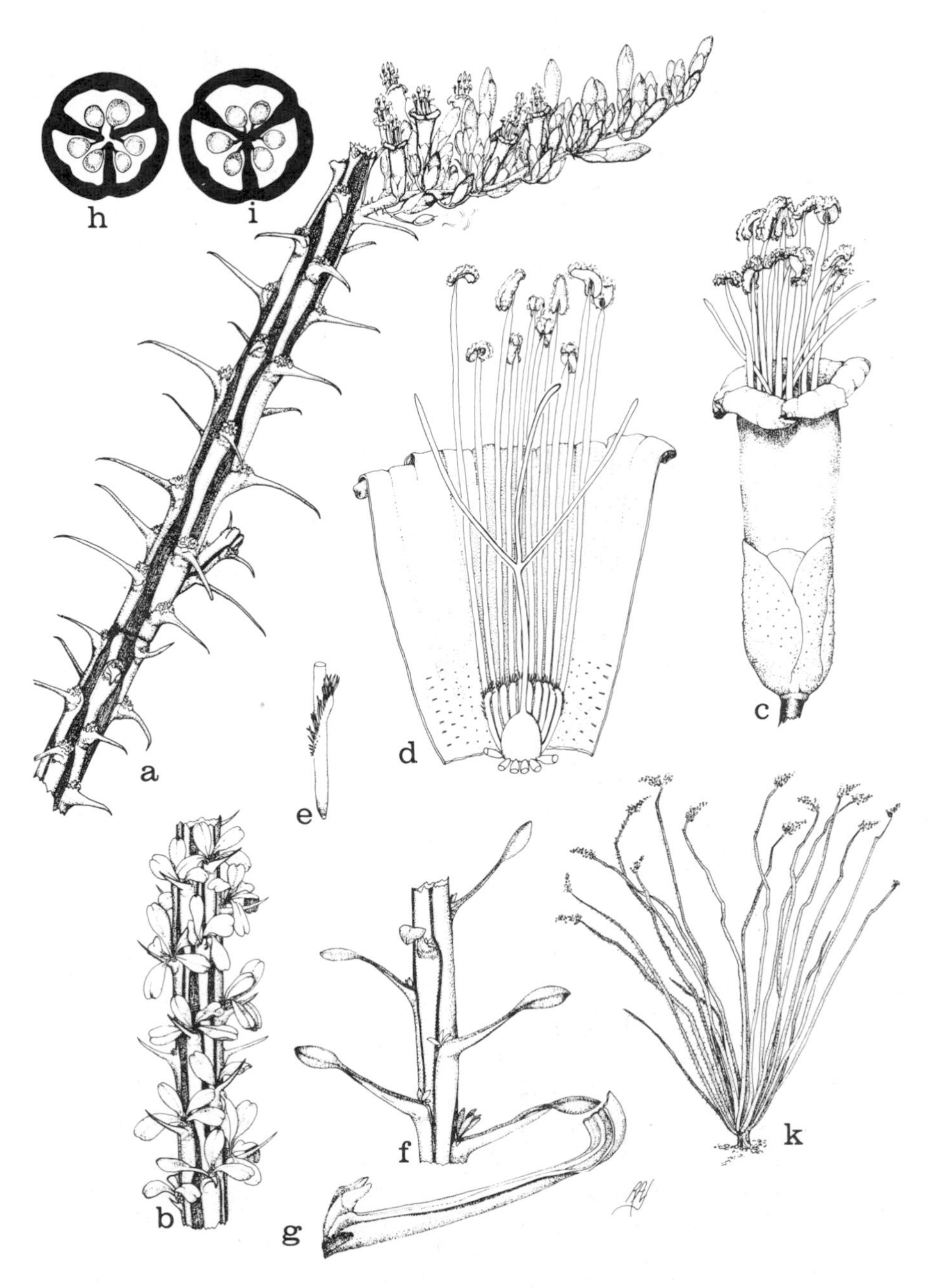

FIG. 4-26 Fouquieriaceae. *Fouquieria splendens* Engelm. a, flowering branch, $\times\frac{1}{2}$; b, leafy branch, $\times\frac{1}{2}$; c, flower, $\times 2\frac{1}{2}$; d, flower, opened up to show stamens and pistil, $\times 2\frac{1}{2}$; e, base of filament, $\times 5$; f, twig with developing primary leaves, the petioles of which will become spines, $\times\frac{1}{2}$; g, petiole of primary leaf, $\times 1\frac{1}{2}$; h, schematic cross-section through upper part of ovary, $\times 18$; i, schematic cross-section through lower part of ovary, $\times 18$; k, habit, $\times\frac{1}{50}$.

at the base but otherwise parietal with deeply intruded placentas; placentas meeting and joining as the fruit matures, eventually forming a massive central column, and the dissepiments then sometimes disintegrating, so that at maturity the placentation appears to be axile or free-central; ovules (6) 14–18 (–20) in all, anatropous, bitegmic, tenuinucellar (the nucellus evanescent) and with an integumentary tapetum; endosperm-development cellular, with a chalazal haustorium. FRUIT a loculicidal capsule; seeds winged; embryo straight, spatulate, with 2 cotyledons; endosperm thin, oily and proteinaceous, or wanting. X = 12.

The family Fouquieriaceae is here considered (following Henrickson) to consist of the single genus *Fouquieria*, with 11 species native to arid parts of Mexico and southwestern United States. *Fouquieria splendens* Engelmann, the Ocotillo, with long, slender, wand-like stems and terminal panicles of orange or red flowers, is a familiar species in the deserts of southwestern United States. *Fouquieria columnaris* (Kellogg) Kellogg (*Idria columnaris*), the bizarre Boojum tree of Baja California, has a succulent, columnar trunk sometimes 20 m high.

The relationships of the Fouquieriaceae have been much-debated. Some authors have assigned them to a position near the Polemoniaceae, in the subclass Asteridae. This is in my opinion totally unlikely, since they have bitegmic ovules and at least twice as many stamens as petals, with the stamens attached directly to the receptacle instead of to the corolla-tube. All of these features are out of harmony with the Asteridae. A position in the Ebenales, which has also been suggested by some authors, is less unreasonable, but the staminal attachment and the number of ovules are aberrant in that order and suggest a less specialized ancestry. In contrast, I see nothing at all to negate a position in the Violales. No other family of Violales is known to have bitegmic, tenuinucellar ovules, but this combination is frequent in the closely related order Theales, and its appearance in the Violales as well should cause no great concern. Outside the subclass Dilleniidae this type of ovule is rare and scattered. The xerophytic adaptations of the Fouquieriaceae make the family stand out wherever it is put, but deprived of these features it is just another group of Violales.

20. Family HOPLESTIGMATACEAE Gilg 1924 nom. conserv., the Hoplestigma Family

Trees. LEAVES large, alternate, simple, exstipulate. FLOWERS in terminal, bractless, cymose inflorescences, perfect, regular, hypogynous; calyx globose in bud, splitting irregularly into 2–4 lobes; corolla of 11–14 petals united into a short tube, the lobes rounded, imbricate in 2–4 irregular series; stamens more or less numerous, about 20–35, in about 3 irregular series, attached to the base of the corolla-tube; anthers tetrasporangiate and dithecal, dorsifixed a little above the base, opening by longitudinal slits; pollen-grains tricolporate, reported to resemble those of the Boraginaceae, especially *Ehretia*; gynoecium of 2 carpels united to form a compound, unilocular ovary with 2 intruded, forked parietal placentas; style divided nearly to the base, its elongate, slender branches curved or bent near the middle, each with an expanded terminal stigma; 2 pendulous, anatropous, unitegmic ovules borne on each placenta. FRUIT drupaceous, with a leathery exocarp and a hard endocarp; seeds with scanty endosperm; embryo rather large, nearly straight, with an elongate hypocotyl and 2 expanded cotyledons.

The family Hoplestigmataceae consists of the single genus *Hoplestigma*, with 2 species native to West Tropical Africa.

Recent authors are agreed that *Hoplestigma* should constitute a family of its own, but they are not in agreement about its relationships. The Hoplestigmataceae have variously been assigned to the Violales, to the Ebenales, or to a position near or even in the Boraginaceae. Each of these three positions has something to recommend it, as well as some problems. The inflorescence, the sympetalous structure of the corolla, the ornamentation of the pollen-grains, and the bicarpellate ovary with 4 unitegmic ovules suggest an alliance with the Boraginaceae, which is favored by Takhtajan. On the other hand, the parietal placentation is out of harmony with the Boraginaceae, and the numerous stamens, nearly free from the corolla, are aberrant not only in the Boraginaceae but in the Asteridae as a whole. The numerous lobes of the corolla suggest a secondary increase, however, and conceivably the same sort of increase might have affected the gynoecium. The sympetalous corolla, the androecium consisting of about twice as many stamens as corolla-lobes, and the small number of ovules suggest the Ebenales. Both unitegmic and bitegmic ovules occur in the Ebenales. The possible secondary increase in the number of corolla-lobes might be compared to a similar secondary increase in members of the Sapotaceae (Ebenales). On the other hand, the parietal placentation is out of harmony with the Ebenales, which otherwise seem to constitute a natural, well defined group. The parietal placentation and rather numerous stamens of *Hoplestigma* suggest a position in the Violales. The bicarpellate ovary, the small number of ovules, and the sympetalous corolla are not precisely typical of the order, but are certainly well known among its members. The only really aberrant feature of *Hoplestigma*, if it is referred to the Violales, is the unitegmic condition of the ovules. The change from bitegmic to unitegmic ovules has occurred several times within the angiosperms, and both types are included in the related orders Theales and Ebenales. I see no reason why the same change may not have occurred in the Violales. I prefer to assign *Hoplestigma* to the Violales until further studies permit its position to be more clearly established.

21. Family CUCURBITACEAE A. L. de Jussieu 1789 nom. conserv., the Cucumber or Cucurbit Family

Herbaceous or sometimes softly woody, mostly climbing or trailing, juicy plants, commonly with spirally coiled (often branched) tendrils (usually 1 at each node) that may represent modified shoots, or the tendrils rarely reduced to spines or even wanting; trichomes often with calcified walls and typically with cystoliths at the base and in nearby cells; plants using citrulline for the transport of nitrogen, often accumulating silica or calcium carbonate (but not calcium oxalate) in the cell-lumina, frequently producing pyridine (or other?) alkaloids, characteristically containing bitter, purgative, tetracyclic triterpenoid substances called cucurbitacins and usually also bitter pentacyclic triterpenoid saponins (these triterpenoid compounds localized in idioblasts or in secretory canals that sometimes approach the appearance of laticifers), but without iridoid compounds and not tanniferous, lacking both proanthocyanins and ellagic acid; nodes trilacunar; vessel-segments with simple perforations; vascular bundles of the stem very often in 2 rings, and nearly always bicollateral, the sieve-tubes sometimes also scattered in the ground tissue of the stem, or the structure of the stem otherwise anomalous. LEAVES alternate, palmately (seldom pinnately) veined and often also lobed, or sometimes palmately compound, frequently with extrafloral nectaries of one or another sort; stomates mostly anomocytic; petiole commonly with a crescent or ring of unequal vascular bundles, the larger ones bicollateral; stipules wanting. FLOWERS mostly in axillary inflorescences or solitary in the axils, unisexual (the plants monoecious or dioecious) or very rarely perfect, regular or seldom irregular, epigynous or rarely only semi-epigynous; hypanthium shortly or strongly prolonged beyond the ovary, with (3–) 5 (6) sepals or lobes, these imbricate or open; petals (3–) 5 (6), distinct or more often connate into a sympetalous, usually yellow or white corolla with valvate or induplicate-

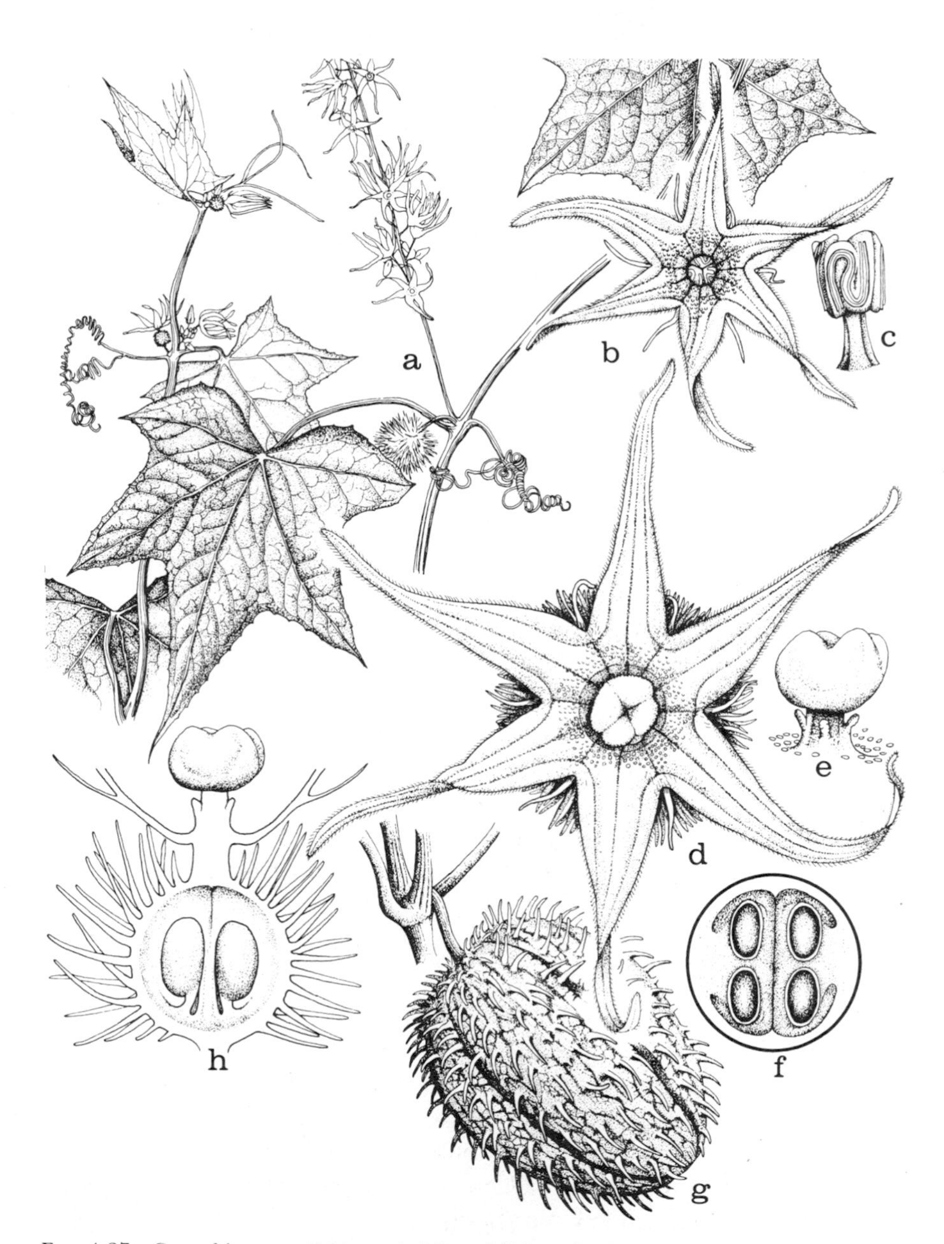

FIG. 4-27 Cucurbitaceae. *Echinocystis lobata* (Michx.) T. & G. a, habit, $\times\frac{1}{2}$; b, top view of staminate flower, ×4; c, androecium, ×8; d, top view of pistillate flower, ×4; e, style and stigma, ×8; f, schematic cross-section of ovary, ×8; g, fruit, ×1; h, portion of pistillate flower, showing ovary in long-section, ×8.

valvate lobes, the corolla sometimes unlike in staminate and pistillate flowers; stamens basically 5, attached to the hypanthium or seldom around the summit of the ovary, alternate with the corolla-lobes, but in nearly all species showing one or another stage in displacement, reduction, and fusion, often apparently 3, two double with dithecal (tetrasporangiate) anthers, and one single, with a monothecal (bisporangiate) anther; individual (unmodified) stamens with monothecal (bisporangiate) anthers except in *Telfairia*; filaments distinct or connate, as also the anthers; anthers straight or often variously bent or folded, extrorse, opening by longitudinal slits; pollen-grains binucleate, variously tricolporate or tricolpate to multicolpate or pantoporate; gynoecium mostly of (2) 3 (–5) carpels united to form a compound, inferior (rarely only half-inferior) ovary with intruded, often much enlarged parietal placentas, or these seldom joined in the center and the ovary thus plurilocular; in *Cyclanthera* (15 spp.) and related genera (tribe Cyclanthereae) the ovary monomerous, with only a single carpel-primordium and a single marginal placenta; nectary in the staminate flowers central on the receptacle, in the pistillate ones commonly on the summit of the ovary, or sometimes the nectaries wanting; style solitary, with 1–3 (–5) usually bilobed stigmas or stigma-lobes, or the styles sometimes 2 or 3, each with a bilobed stigma; ovules (1–) numerous, anatropous, bitegmic, crassinucellar; endosperm-development nuclear; chalazal end of the embryo-sac commonly developing into an elongate, tubular haustorium. FRUIT usually a berry or a pepo, less often a dry or fleshy capsule opening in various fashions, or rarely samaroid; seeds (1–) many, large, commonly compressed and sometimes even winged; embryo oily, straight, with 2 large, flat cotyledons; endosperm wanting or nearly so. X = 7–14.

The family Cucurbitaceae consists of about 90 genera and 700 species. The group is widespread in tropical and subtropical regions, with relatively few species occurring in temperate and cool climates; the aerial parts of all species are sensitive to frost. None of the genera has as many as a hundred species. Various species of *Cucurbita* (pumpkin, squash), *Citrullus* (watermelon), *Cucumis* (cucumber, muskmelon), *Lagenaria* (gourd), *Luffa* (vegetable sponge) and other genera are familiar in cultivation.

In the Engler system the Cucurbitaceae were considered to form a distinct order of Sympetalae, but in that group they have no allies and are aberrant in their bitegmic, crassinucellar ovules. Many other authors, both pre- and post-Engler, have thought that the Cucurbitaceae are allied to some of the families here included in the Violales, most notably the Passifloraceae. A position near the Passifloraceae is now widely accepted, even in the current (1964) Engler Syllabus. Opinion remains divided as to whether the Cucurbitaceae should be included in the same order as the Passifloraceae, or whether the Cucurbitales should be retained as a unifamilial order. I find it useful to retain the Cucurbitaceae in the Violales until the questions about the affinities of the Begoniaceae and Loasaceae are resolved. If these two families are eventually removed from the Violales, then the purposes of classification might be better served by removal of the Cucurbitaceae as well. The Cucurbitaceae are in any case absolutely sharply defined. There are no doubtful genera or species that connect them to other families.

Jeffrey (personal communication) thinks that the Cucurbitaceae may be misplaced in the Violales, and that they may eventually prove to be related to the Grossulariaceae (as here defined) or the Sapindaceae.

Fossil leaves considered to represent the Cucurbitaceae occur in Paleocene and more recent deposits, but identifiable pollen dates only from the Oligocene.

22. Family DATISCACEAE Lindley 1830 nom. conserv., the Datisca Family

Perennial herbs or large trees, not cyanogenic and not tanniferous, lacking both proanthocyanins and ellagic acid, and without chemical-bearing idioblasts; calcium oxalate present only as small crystals; wood (in the 2 woody genera) light in weight and not durable; vessel-segments large, with simple perforations; imperforate tracheary elements with a large lumen, thin walls, and small, simple pits; wood-rays heterocellular, up to 4–7 cells wide, with few uniseriates; wood-parenchyma paratracheal, scanty, mostly restricted to 1 or 2 layers. LEAVES alternate, simple or pinnately compound; stomates anomocytic; stipules wanting. FLOWERS in axillary spikes or racemes or panicles, or in slender, terminal inflorescences that are leafy-bracteate below, more or less regular, mostly unisexual and the plants dioecious, but in *Datisca* sometimes polygamous or androdioecious. Staminate flowers: sepals 4–8, distinct and unequal, or connate into a lobed tube; petals in *Octomeles* 6–8, small, and distinct, in the other genera wanting; stamens of the same number as the sepals and opposite them, or more numerous, up to about 25; anthers tetrasporangiate and dithecal, short or (*Datisca*) elongate, opening by longitudinal slits; a vestigial gynoecium sometimes present.

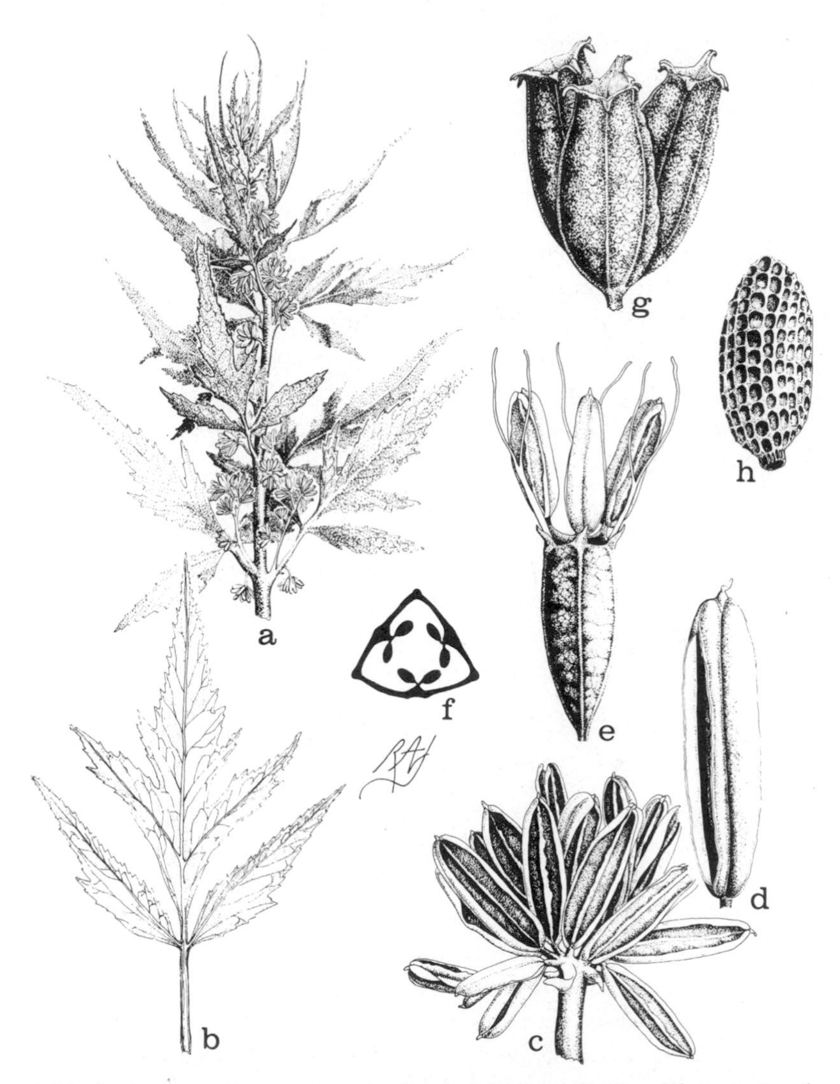

FIG. 4-28 Datiscaceae. *Datisca glomerata* (Presl) Baillon. a, habit, × ½; b, leaf, × ½; c, staminate flower, × 5; d, stamens, × 10; e, perfect flower, × 5; f, schematic cross-section of ovary, × 10; g, cluster of fruits, × 3; h, seed, × 25.

Pistillate (and perfect) flowers: sepals 3–8, short, seated around the periphery of the summit of the ovary; petals none; more or less functional stamens sometimes present around the summit of the ovary; staminodes generally wanting; gynoecium of 3–8 carpels united to form a compound, unilocular, inferior ovary with parietal placentas; styles distinct, sometimes (*Datisca*) bifid; ovules anatropous, bitegmic, crassinucellar; endosperm-development nuclear. FRUIT capsular, opening apically between the persistent styles; seeds numerous, very small, with straight, cylindric, oily, dicotyledonous embryo and very scanty or no endosperm. N = 11 (*Datisca*). (Tetramelaceae)

The family Datiscaceae consists of 3 genera and 4 species. *Octomeles* and *Tetrameles*, each with a single species, are simple-leaved trees of the Malesian region. *Datisca*, a coarse herb with pinnately compound leaves, has one species on the Asiatic mainland and another in western North America.

The Datiscaceae have generally been considered to be related to the Begoniaceae, and Brown (1938, in initial citations) even suggested that these two families might properly constitute a distinct order. On the presently available information, the Datiscaceae are perfectly at home in the Violales, in contrast to the supposedly closely allied Begoniaceae, which are aberrant in their plurilocular ovary and centripetal stamens. The sequence of initiation in the Datiscaceae does not appear to have been investigated. Davidson (1976) considers that the Datiscaceae are related to but more advanced than the Flacourtiaceae, comparable anatomically to the Cistaceae, Violaceae, and Scyphostegiaceae.

Fossil wood thought to represent the Datiscaceae occurs in Eocene deposits in India.

23. Family BEGONIACEAE C. A. Agardh 1825 nom. conserv., the Begonia Family

Rather succulent, glabrous or often coarsely shaggy-hairy herbs or soft shrubs, sometimes climbing, often with crassulacean acid metabolism, accumulating free organic acids (including oxalic and malic acids) in the cell sap, tanniferous, producing proanthocyanins but neither ellagic acid nor iridoid substances, not cyanogenic, often with secretory idioblasts in the parenchymatous tissues, and with solitary or clustered crystals of calcium oxalate in some of the cells; nodes trilacunar; vessel-segments with simple and/or scalariform perforations; ground-tissue of the secondary xylem generally composed chiefly of delicately septate prosenchymatous cells with simple pits. LEAVES alternate, simple or seldom compound, mostly palmately veined and often palmately lobed, commonly asymmetrical; stomates surrounded by 3–6 subsidiary cells, these often arranged in 2 cycles; stipules present, commonly large and persistent, free, sometimes remaining even after the leaf has fallen; cystoliths often present in the mesophyll or in the hair-bases. FLOWERS commonly in axillary, cymose, sometimes long-pedunculate inflorescences, unisexual (the plants usually monoecious) and often irregular; tepals all petaloid, sometimes in 2 sets of 5, the outer larger, but more often less than 10 in all, often 2 unlike, valvate sets of 2 in the staminate flowers and a single imbricate set of 5 in the pistillate flowers, usually all distinct, seldom connate at the base; stamens 4 (*Begoniella*) to usually more or less numerous, originating in centripetal (!) sequence, regularly arranged or all on one side of the flower, the filaments distinct or sometimes connate below; anthers tetrasporangiate and dithecal, opening by longitudinal slits or seldom by terminal pores, the connective sometimes expanded so that the pollen-sacs are well separated; pollen-grains binucleate, tricolporate; gynoecium of (2) 3 (–6) carpels united to form a compound, inferior (seldom not fully inferior), plurilocular ovary with axile placentas, or seldom the intruded placental partitions not meeting and the ovary thus unilocular; ovary very often with (1–) 3 (–6) prominent, equal or unequal wings; styles distinct or sometimes connate at the base, bifid; ovules numerous, anatropous, bitegmic, weakly crassinucellar; endosperm-development nuclear. FRUIT mostly a loculicidal capsule, or seldom a berry; seeds very numerous and small, with tiny, straight, dicotyledonous embryo and virtually no endosperm. X = 10–21+.

The family Begoniaceae consists of the very large genus *Begonia*, with perhaps as many as a thousand species, and 2–4 small satellite genera (according to taxonomic interpretation) with fewer than 20 species in all. The family is widespread in tropical regions except for most of Polynesia and Australia, but best-developed in northern South America. Many species and hybrid cultigens are cultivated as house-plants and tender garden-ornamentals.

The relationships of the Begoniaceae are not entirely clear. Recent authors mostly associate them with the Datiscaceae, in or near the Violales (under whatever ordinal name). This assessment may well be correct, but the usually plurilocular ovary of the Begoniaceae is uncommon in the Violales, and the centripetal development of the stamens is aberrant (though not unprecendented) for the whole subclass Dilleniidae. Given these facts, it may be significant that the several tested members of the Begoniaceae differ from all other tested Violales in some chemical tests reported by Gibbs (1974, in initial citations), but it is not clear what substances are responsible for the observed reactions. On the other hand, Leo Hickey tells me that the structure of the marginal teeth of the leaves is wholly consonant with the Dilleniidae, but not with the Rosidae. Further study is obviously in order.

24. Family LOASACEAE Dumortier 1822 nom. conserv., the Loasa Family

Annual or more often perennial, sometimes climbing herbs, less often shrubs or even small trees, provided with coarse, silicified and often calcified, sometimes stinging, often gland-tipped hairs that often have basal cystoliths, commonly producing iridoid substances, but not cyanogenic and not tanniferous, lacking both ellagic acid and proanthocyanins; clustered crystals of calcium oxalate sometimes present in some of the cells of the parenchymatous tissues; vessel-segments broad, with simple perforations; imperforate tracheary elements with simple or bordered pits. LEAVES alternate or opposite, simple but often lobed; stomates anomocytic; stipules wanting. FLOWERS solitary or more often in cymose inflorescences, perfect, regular, epigynous; sepals (4) 5 (–7), convolute or imbricate, persistent; petals (4) 5 (–7) or 10 by development of petaloid staminodes, induplicate-valvate in bud, distinct or sometimes basally adnate to the androecial tube or forming a distinctly sympetalous, lobed corolla; stamens 10 to more often numerous, centripetal (Mentzelioideae) or centrifugal (Loasoideae), or only 5 (Gronovioideae), distinct or basally connate into a short tube or into antepetalous bundles, sometimes with nearly sessile anthers arising from the corolla-tube, some of the stamens often modified into petaloid or scale-like or nectariferous staminodia; anthers tetrasporangiate and dithecal, opening by longitudinal slits; pollen-grains binucleate or trinucleate, tricolporate; gynoecium of 3–5 (–7) carpels united to form a compound, inferior, unilocular ovary with parietal, often more or less deeply intruded placentas, or rarely fully partitioned and plurilocular, with axile placentas, or (Gronovioideae) the ovary pseudomonomerous with a single pendulous, apical ovule; style solitary; ovules 1-many on each placenta, anatropous or hemitropous, unitegmic, tenuinucellar, with a more or less well developed integumentary tapetum; endosperm-development cellular, with micropylar and chalazal haustoria. FRUIT a loculicidal or septicidal capsule, or seldom dry and indehiscent; seeds with straight or curved, spatulate, dicotyledonous embryo embedded in the abundant, oily, softly fleshy endosperm, or (Gronovioideae) the endosperm scanty or wanting. X = 7–15+. (Gronoviaceae)

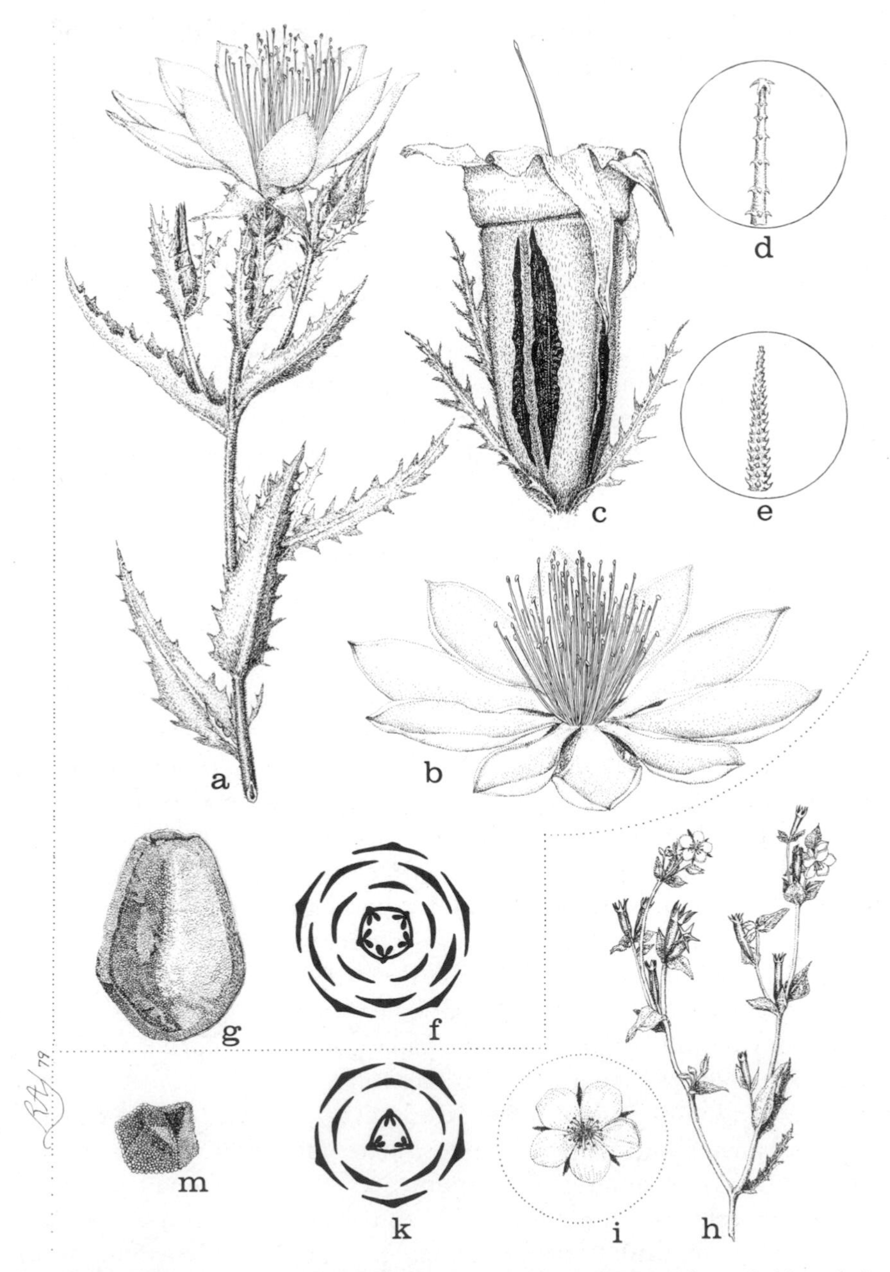

FIG. 4-29 Loasaceae. a–g, *Mentzelia decapetala* (Pursh) Urban & Gilg. a, habit, ×¼; b, flower, ×½; c, opening fruit, ×1; d, e, trichomes, ×40; f, diagram of perianth and pistil, the stamens omitted; g, seed, ×7; h–m, *Mentzelia albicaulis* Dougl. h, habit, ×½; i, flower, ×1½; k, diagram of perianth and pistil, the stamens omitted; m, seed, ×5.

The family Loasaceae as traditionally (and here) defined consists of some 14 genera and a little more than 200 species, widespread in temperate and tropical North and South America. Only one small and unusual genus, *Fissenia* (*Kissenia*), with one or two species in Africa and Arabia, occurs outside the New World. There are 3 well marked subfamilies, and it is conceivable that the Mentzelioideae, with centripetal stamens, do not properly go with the Loasoideae, which have centrifugal stamens. The largest genera are *Loasa* (75), *Mentzelia* (50) and *Caiophora* (50). The small subfamily Gronovioideae has only 3 genera and 8 species.

The affinities of the Loasaceae are uncertain. The unitegmic, tenuinucellar ovules with chalazal and micropylar haustoria are aberrant in the Dilleniidae and suggest the Asteridae, as does the frequent present of iridoid compounds. On the other hand, the numerous stamens and mostly polypetalous corolla are at least as aberrant in the Asteridae, which as here defined are a morphologically coherent and relatively well defined group. Within the Asteridae, Takhtajan uses the multicellular, glandular hairs with basal cystoliths to support a relationship to the Boraginaceae and Hydrophyllaceae, but similar hairs also occur in the Cucurbitaceae. The Loasaceae have some embryological features in common with the Ericaceae (unitegmic, tenuinucellar ovules; cellular endosperm with terminal haustoria), and they also resemble the Ericaceae in having iridoid compounds. On the other hand they do not have the remainder of the ericoid embryological syndrome, and they do not at all resemble the Ericaceae in aspect. The similarities between these two families are here ascribed to a degree of evolutionary parallelism within the Dilleniidae. On balance, I prefer to leave the Loasaceae in the Violales, where they have most often been put in the past, until the evidence on their affinities is more conclusive. The sequence of amino acids in cytochrome *c* might eventually help to resolve the problem.

7. Order SALICALES Lindley 1833

The order consists of the single family Salicaceae.

SELECTED REFERENCES

Binns, W. W., G. Blunden, & D. L. Woods. 1968. Distribution of leucoanthocyanidins, phenolic glycosides and imino-acids in leaves of *Salix* species. Phytochemistry 7: 1577–1581.
Fisher, M. J. 1928. The morphology and anatomy of the flowers of the Salicaceae. Amer. J. Bot. 15: 307–326; 372–394.
Holm, L. 1969. An uredinological approach to some problems in angiosperm taxonomy. Nytt Mag. Bot. 16: 147–150.
Kimura, C. 1963. On the embryo sac formation of some members of the Salicaceae. Sci. Rep. Tôhoku Imp. Univ. Ser. 4, Biol. 29: 393–398.
Meeuse, A. D. J. 1975. Taxonomic relationships of Salicaceae and Flacourtiaceae. Acta Bot. Neerl. 24: 437–457.
Nagaraj, M. 1952. Floral morphology of *Populus deltoides* and *P. tremuloides*. Bot. Gaz. 114: 222–243.
Risch, C. 1960. Die Pollenkörner der Salicaceen. Willdenowia 2: 402–409.
Rowley, J. R., & G. Erdtman. 1967. Sporoderm in *Populus* and *Salix*. Grana Palynol. 7: 517–567.
Suda, Y. 1963. The chromosome numbers of salicaceous plants in relation to their taxonomy. Sci. Rep. Tôhoku Imp. Univ., Ser. 4, Biol. 29: 413–430.

Гзырян, М. С. 1952. Семейство Salicaceae и его положение в системе покрытосеменных по данным анатомии древесины. Автореф. Дисс., Ереван.
Гзырян, М. 1955. Внутрисемейственные взаимоотношения у ивовых. Докл. АН СССР, 105: 832-834.
Малютина, Е. Т. 1972. О морфологической природе частей цветка некоторых видов рода *Salix* L. и возможные пути их эволюции. Бот. Ж. 57: 524-530.
Скворцов, А. К., & М. Д. Голышева. 1966. Исследование анатомии листа ив (*Salix*) в связи с систематикой рода. Acta Bot. Acad. Sci. Hung. 12: 125-174.
Ярмоленко, А. В. 1949. Новый род из семейства ивовых—Salicaceae. Бот. Матер. Герб. Бот. Инст. АН СССР 11: 67-73.

1. Family SALICACEAE Mirbel 1815 nom. conserv., the Willow Family

Trees, shrubs, or sometimes depressed shrublets, characteristically producing special phenolic heterosides such as salicin and populin, commonly with tanniferous idioblasts containing proanthocyanins but not ellagic acid, only seldom cyanogenic and seldom with alkaloids, without iridoid substances; nodes trilacunar; solitary or clustered crystals of calcium oxalate commonly present in some of the cells of the parenchymatous tissues; vessel-segments with simple perforations; im-

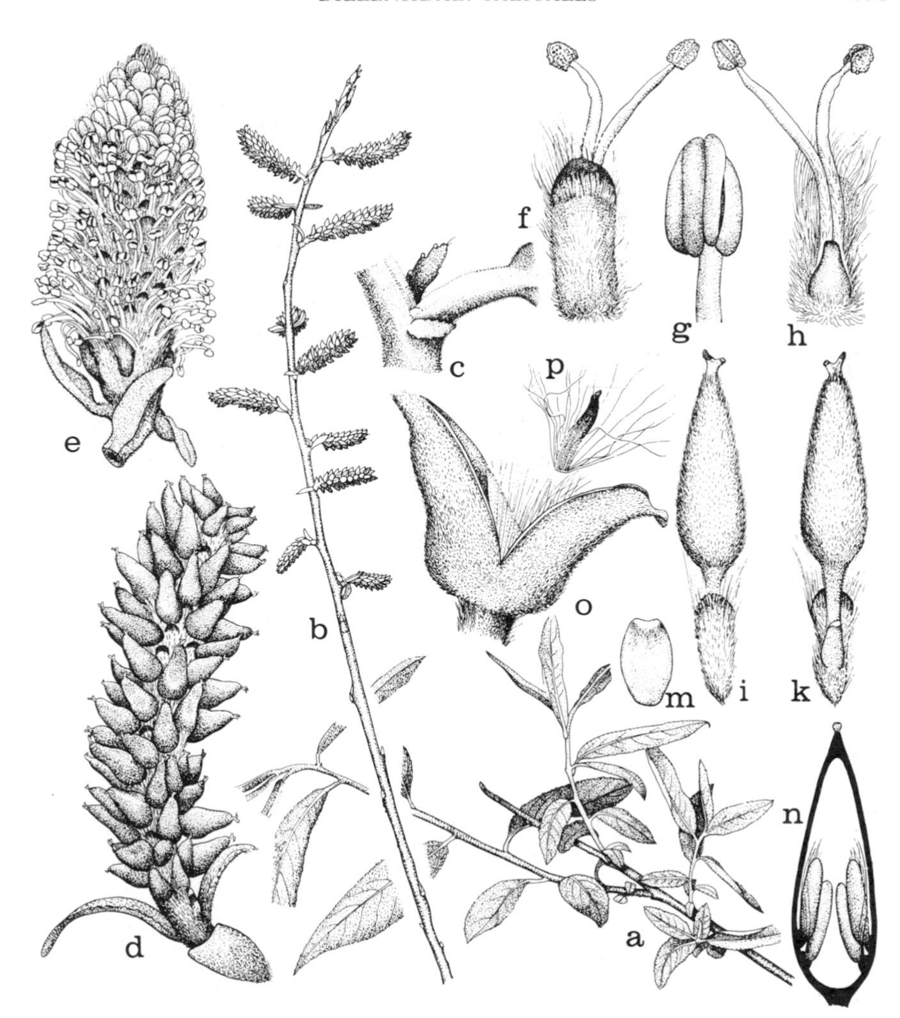

FIG. 4-30 Salicaceae. *Salix sericea* Marshall. a, leafy twig, ×½; b, twig with pistillate catkins, ×½; c, node, base of leaf, and stipules, ×3; d, pistillate catkin, ×3; e, staminate catkin, ×3; f, h, dorsal and ventral view of bract and staminate flower, ×9, the nectary showing in h; g, anther, ×18; i, k, dorsal and ventral view of bract and pistillate flower, ×9, the nectary showing in k; m, nectary of pistillate flower, ×18; n, long-section of ovary, ×12; o, opened fruit, ×18, p, seed, ×18.

perforate tracheary elements with small, mostly simple pits; wood-rays heterocellular (most spp. of *Salix*) or homocellular (most spp. of *Populus*), mostly uniseriate; wood-parenchyma scanty, variously terminal, vasicentric, or diffuse; roots commonly forming ectotrophic mycorhizae. LEAVES alternate, simple, deciduous; stomates paracytic; distal end of the petiole with one or more closed rings of xylem and phloem; stipules present and often conspicuous, but frequently caducous. FLOWERS anemophilous or (most spp. of *Salix*) secondarily entomophilous, borne in unisexual, often precocious catkins (the plants dioecious), each flower axillary to a bract, but without bracteoles, apetalous, but with a vestigial calyx, which in *Populus* is more or less disk-shaped in the staminate flowers and saucer-shaped to cupulate in the pistillate flowers, but in *Salix* is reduced to one or two (3–5) small, commonly fringed, often unequal nectaries that may be united into a lobed ring; stamens (1) 2-many (up to ca 60), usually several or many in *Populus* and 2 in *Salix*, distinct or the filaments sometimes connate below (rarely throughout); anthers tetrasporangiate and dithecal, opening by longitudinal slits; pollen-grains binucleate or sometimes (spp. of *Populus*) trinucleate, inaperturate in *Populus*, tricolpate or tricolporate in *Salix* (sometimes 2-, 4-, or 6-colpate in some grains of some spp.); gynoecium of 2 (less often 3 or 4) carpels united to form a compound ovary with parietal placentas, the stigmas distinct, sessile or often on a short common style, often bifid or irregularly lobed; ovules (2–) more or less numerous on each placenta, anatropous, crassinucellar, unitegmic (the inner integument suppressed) or in a few species of *Populus* weakly bitegmic; endosperm-development nuclear. FRUIT a 2- to 4-valved capsule; seeds tiny, distributed by wind, enveloped by a tuft of long hairs originating from the funiculus or from the placenta; embryo small, straight, with 2 cotyledons; endosperm none, or very scanty and oily. X = 11, 12, 19.

The family Salicaceae consists of 2 genera, *Populus*, with about 40 species, and *Salix*, with perhaps as many as 300. Both genera are widely distributed, most commonly in North Temperate (in *Salix* also Arctic) regions, but wanting from Australia and the Malay Archipelago. Both *Salix* (willow) and *Populus* (poplar) are partial to moist habitats, often growing along streams. A number of species are well known in cultivation.

In the Englerian system the Salicaceae were considered to be very primitive dicotyledons and were placed in the Amentiferae. By 1926 Engler himself admitted that the flowers of Salicaceae are reduced rather than primitively simple, but he put forward the curious argument that the great floral reduction of the family indicates its great age and therefore supports its assignment to a position at the beginning of the taxonomic sequence. Such a view makes hash of efforts to integrate phylogenetic thinking into taxonomy. All modern groups are of the same age, if their remote ancestors are taken into account. Differing ages can be assigned only by considering the time when the essential features of the modern group were acquired. A phylogenetic arrangement properly takes as basal those taxa that are least modified from the common ancestral type, instead of starting with those that are most modified.

In keeping with the advanced floral structure, the wood of Salicaceae is moderately specialized, rather than primitive. The vessels have simple perforations, the imperforate tracheary elements have small, mostly simple pits, and the wood-rays are uniseriate.

If one extrapolates backward from the Salicaceae to the hypothetical ancestral type with small but complete flowers, numerous stamens, and a several-carpellate ovary with parietal placentation and separate styles, one arrives at a group that might reasonably be included in the Violales, near the Flacourtiaceae. The Flacourtiaceae have indeed often been cited as possible or probable allies of the Salicaceae, and Takhtajan emphasizes that *Idesia* and related genera of the Flacourtiaceae are particularly suggestive of the Salicaceae. It is interesting and perhaps significant that salicin, which is widespread in the Salicaceae, also occurs in *Idesia* and a few other Flacourtiaceae but is known almost nowhere else. Furthermore, *Idesia* and the Salicaceae play host to similar rust fungi, belonging to the genus *Melampsora.* The small family Stachyuraceae may also be mentioned as possibly standing near to the ancestry of the Salicaceae.

Fossil leaves credibly considered to represent the genus *Salix* occur in Eocene deposits in North Dakota, but identifiable salicaceous pollen dates only to the Oligocene. There are leaves, twigs, and fruiting racemes of *Populus,* as also leaves of *Salix,* reported from the Middle Eocene in Utah.

8. Order CAPPARALES Hutchinson 1926

Trees, shrubs, lianas, or most often herbs, consistently producing glucosinolates (mustard-oil glucosides) and with specialized myrosin-cells, sometimes with alkaloidal amines and sometimes cyanogenic, occasionally producing mucilage, but without iridoid compounds and mostly not tanniferous, lacking ellagic acid and usually also lacking proanthocyanins; vessel-segments with simple perforations; imperforate tracheary elements with simple or bordered pits; roots only seldom forming mycorhizae. LEAVES alternate or rarely opposite, simple or often variously compound or dissected, with or without stipules; stomates anisocytic (Brassicaceae) or anomocytic. FLOWERS mostly in terminal racemes, seldom in panicles or mixed panicles, perfect or seldom unisexual, hypogynous or much less often perigynous, regular or somewhat irregular; nectaries more or less obviously of receptacular origin; sepals distinct, 2–8, most commonly 4 in 2 opposite, decussate pairs; petals distinct, 2–8, most commonly 4, imbricate or convolute or valvate, very often with a basal claw, or rarely wanting; stamens (2–) 4-many, generally arising from a limited number (most often 4) of primordia, centrifugal when in more than one series; pollen-grains binucleate or trinucleate, mostly tricolporate; gynoecium of 2 or less often 3–6 (–12) carpels, nearly always united to form a compound ovary with typically parietal (seldom axile) placentas, but the ovary and fruit in the large family Brassicaceae partitioned by the replum; stigma 1, simple or lobed, usually elevated on a short to elongate style, or in the Resedaceae the stigmas separate and sessile; ovules (1–) usually more or less numerous, anatropous or more often campylotropous, bitegmic, crassinucellar to tenuinucellar, often with a zigzag micropyle; endosperm-development nuclear. FRUIT most often a specialized type of capsule with a distinct replum, but sometimes a more ordinary capsule or a berry, or even a drupe or a nut; seeds with large, dicotyledonous, usually curved or folded embryo (straight in Moringaceae) and usually scanty or no endosperm (endosperm fairly well developed in Tovariaceae).

The order Capparales as here defined consists of 5 families and nearly 4000 species. By far the largest family is the Brassicaceae, with about 3000 species. Next come the Capparaceae, with about 800. The other 3 families have fewer than a hundred species in all. The Papaveraceae and Fumariaceae, often in the past associated with the Capparales in a collective order Rhoeadales, are here treated as a distinct order Papaverales in the subclass Magnoliidae. The numerous features that require this separation are discussed under the Papaverales.

The most characteristic feature of the Capparales is the production of one or another sort of mustard-oil. Mustard-oils are isothiocyanates, mostly derived by hydrolysis of glucosinolates through the intervention of the enzyme myrosin. Typically the myrosin is stored in scattered idioblasts called myrosin-cells.

In addition to the Capparales, mustard-oils are known in a thin scattering of other groups, a little more than 300 species in all. These are the Bataceae and Gyrostemonaceae (collectively making up the order Batales), Limnanthaceae and Tropaeolaceae (both in the Geraniales), *Bretschneidera* and some Zygophyllaceae (Sapindales), some Salvadoraceae (Celastrales), *Drypetes* (Euphorbiaceae), and some of the Caricaceae (Violales). Thus 7 orders, in 2 subclasses, contain at least some members that produce mustard-oils. There may be a chemical relationship between the unusual method of cyanogenesis in the Violales and the production of mustard-oils by the Capparales.

One may reasonably speculate that there is a functional correlation between the presence of mustard-oils in the Capparales and the absence or restricted occurrence of other repellants such as tannins, alkaloids, and iridoid compounds. Such specialization in repellants is of course an invitation to the evolution of specialized groups of predators that can tolerate (and may even require) the critical substances. Possibly the cabbage-butterflies represent such a group. The white rusts of crucifers may also be especially adapted to mustard-oil. On the other hand, mustard-oils evidently do not discourage herbarium beetles, which feed happily on crucifers as well as on a wide range of other families in other orders.

Aside from the mustard-oils and some obvious features of habit, most of the characters that mark the families of Capparales and the order as a whole are of doubtful selective significance. Standard evolutionary theory would seem to require that the replum has some adaptive value, and that the difference in the replum between the Capparaceae and Brassicaceae must also be adaptive, but I cannot come up with a plausible explanation. What difference does it make to members of these two families whether the replum has a partition or not? How does the gynophore of the Capparaceae confer any advantage or help fit the plants to any particular habitat? What is the value to the Resedaceae in having the ovary open at the top so that the ovules are exposed?

Obviously all these variations in structure function well enough to permit the plants to survive and reproduce and carry on their kind through long periods of time. But why change? It does not appear that the changes have opened up any new or better way of exploiting the habitat, or enabled the plants to occupy any previously unavailable territory, or given them any advantage in the competition for scarce resources. My imagination is inadequate to the task of providing a Darwinian interpretation.

SELECTED REFERENCES

Abdallah, M. S., & H. C. D. de Wit. 1967, 1978. The Resedaceae. A taxonomic revision of the family. Belmontia. New Series. 8(26 A & B): 1–416, + 91 figs. (Pp. 1–98, figs. 1–17, 1967. Pp. 99–416, figs. 18–91, 1978.)

Alexander, I. 1952. Entwicklungsstudien an Blüten von Cruciferen und Papaveraceen. Planta 41: 125–144.

Aleykutty, K. M., & J. A. Inamdar. 1978. Studies in the vessels of some Capparaceae. Flora B 167: 103–109.

Aleykutty, K. M. & J. A. Inamdar. 1978. Structure, ontogeny and taxonomic significance of trichomes and stomata in some Capparidaceae. Feddes Repert. 89: 19–30.

Arber, A. 1931. Studies in floral morphology. I. On some structural features of the cruciferous flower. New Phytol. 30: 11–41.

Arber, A. 1942. Studies in flower structure. VII. On the gynaeceum of *Reseda*, with a consideration of paracarpy. Ann. Bot. (London) II. 6: 43–48.

Behnke, H.-D. 1977. Dilatierte ER-Zisternen, ein mikromorphologisches Merkmal der Capparales? Ber. Deutsch. Bot. Ges. 90: 241–251.

Crété, P. 1951. Embryogénie des Capparidacées. Développement de l'embryon chez le *Cleome graveolens* Rafin. Compt. Rend. Hebd. Séances Acad. Sci. 233: 562–564.

Delaveau, P., B. Koudogbo, & J.-L. Pousset. 1973. Alcaloïdes chez les Capparidaceae. Phytochemistry 12: 2983–2985.

Dutt, B. S. M., L. L. Narayana, & A. Parvathi. 1978. Floral anatomy of *Moringa concanensis* Nimmo. Indian J. Bot. 1: 35–39.

Dvořák, F. 1968. A contribution to the study of the variability of the nectaries. Preslia 40: 13–17.

Dvořák, F. 1971. On the evolutionary relationship in the family Brassicaceae. Feddes Repert. 82: 357–372.

Dvořák, F. 1973. The importance of the indumentum for the investigation of evolutional relationship in the family Brassicaceae. Oesterr. Bot. Z. 121: 155–164.

Eames, A. J., & C. L. Wilson. 1930. Crucifer carpels. Amer. J. Bot. 17: 638–656.

Eggers, O. 1935. Über die morphologische Bedeutung des Leitbündelverlaufes in den Blüten der Rhoeadalen und über das Diagramm der Cruciferen und Capparidaceen. Planta 24: 14–58.

Ernst, W. R. 1963. The genera of Capparaceae and Moringaceae in the southeastern United States. J. Arnold Arbor. 44: 81–95.

Ezelarab, G. E., & K. J. Dormer. 1966. The organization of the primary vascular system in the Rhoeadales. Ann. Bot. (London) II. 30: 123–132.

Feeny, P. 1977. Defensive ecology of the Cruciferae. Ann. Missouri Bot. Gard. 64: 221–234.
Frohne, D. 1962. Das Verhältnis von vergleichender Serobotanik zu vergleichender Phytochemie, dargestellt an serologischen Untersuchungen im Bereich der "Rhoeadales." Pl. Med. 10: 283–297.
Gibson, A. C. 1979. Anatomy of *Koeberlinia* and *Canotia* revisited. Madroño 26: 1–12.
Hennig, L. 1929. Beiträge zur Kentniss der Resedaceen-Blüte und -Frucht. Planta 9: 507–563.
Iltis, H. H. 1957. Studies in the Capparidaceae. III. Evolution and phylogeny of the western North American Cleomoideae. Ann. Missouri Bot. Gard. 44: 77–119.
Iversen, T.-H. 1970. The morphology, occurrence, and distribution of dilated cisternae of the endoplasmic reticulum in tissues and plants of the Cruciferae. Protoplasma 71: 467–477.
Janchen, E. 1942. Das Systeme der Cruciferen. Oesterr. Bot. Z. 91: 1–28.
Jørgensen, L. B., H.-D. Behnke, & T. J. Mabry. 1977. Protein-accumulating cells and dilated cisternae of the endoplasmic reticulum in three glucosinolate-containing genera: *Armoracia, Capparis, Drypetes.* Planta 137: 215–224.
Kjaer, A. 1968. Glucosinolates in Tovariaceae. Phytochemistry 7: 131–133.
Leins, P., & G. Metzenauer. 1979. Entwicklungsgeschichtliche Untersuchungen an Capparis-Blüten. Bot. Jahrb. Syst. 100: 542–554.
Leins, P., & U. Sobick. 1977. Die Blütenentwicklung von *Reseda lutea.* Bot. Jahrb. Syst. 98: 133–149.
Mauritzon, J. 1935. Die Embryologie einiger Capparidaceen sowie von *Tovaria pendula.* Ark. Bot. 26A(No. 15): 1–14.
Murty, Y. S. 1953. A contribution to the anatomy and morphology of normal and some abnormal flowers of *Gynandropsis gynandra* (L.) Briq. J. Indian Bot. Soc. 32: 108–122.
Narayana, H. S. 1962, 1965. Studies in the Capparidaceae. I. The embryology of *Capparis decidua* (Forsk.) Pax. II. Floral morphology and embryology of *Cadaba indica* Lamk. and *Crataeva nurvala* Buch.-Ham. Phytomorphology 12: 167–177, 1962. 15: 158–175, 1965.
Norris, T. 1941. Torus anatomy and nectary characteristics as phylogenetic criteria in the Rhoeadales. Amer. J. Bot. 28: 101–113.
Puri, V. 1941. Life-history of *Moringa oleifera* Lamk. J. Indian Bot. Soc. 20: 263–284.
Puri, V. 1941, 1942, 1950. Studies in floral anatomy. I. Gynaeceum constitution in the Cruciferae. Proc. Indian Acad. Sci. B 14: 166–187, 1941. II. Floral anatomy of the Moringaceae with special reference to gynaeceum constitution. Proc. Natl. Inst. Sci. India 8: 71–88, 1942. VI. Vascular anatomy of the flower of *Crataeva religiosa* Forst., with special reference to the nature of the carpels in the Capparidaceae. Amer. J. Bot. 37: 363–370, 1950.
Puri, V. 1952. Floral anatomy in relation to taxonomy. Agra Univ. J. Res., Sci. 1: 15–35.
Raghavan, T. S. 1937, 1939. Studies in the Capparidaceae. I. The life-history of *Cleome chelidonii* Linn. fil. II. Floral anatomy and some structural features of the capparidaceous flower. J. Linn. Soc., Bot. 51: 43–72, 1937. 52: 239–257.
Raghavan, T. S., & K. R. Venkatasubban. 1941. Studies in the Capparidaceae. V. The floral morphology of *Crataeva religiosa* Forst. VII. Floral ontogeny

and anatomy of *Crataeva religiosa* with special reference to the morphology of the carpel. Beih. Bot. Centralbl. 60(A): 388:396; 397–416.
Rollins, R. C. 1956. Some new primitive Mexican Cruciferae. Rhodora 58: 148–157.
Spratt, E. R. 1932. The gynoecium of the family Cruciferae. J. Bot. 70: 308–314.
Stoudt, H. N. 1941. The floral morphology of some of the Capparidaceae. Amer. J. Bot. 28: 664–675.
Vaughan, J. G., A. J. McLeod, & B. M. G. Jones, eds. 1976. The biology and chemistry of the Cruciferae. Academic Press. London, New York, San Francisco.
Vaughan, J. G., & J. M. Whitehouse. 1971. Seed structure and the taxonomy of the Cruciferae. J. Linn. Soc., Bot. 64: 383–409.
Yen, C. 1959. On a new view of carpel morphology in *Brassica*. Acta Bot. Sin. 8: 271–280. (In Chinese, with English summary.)

Аветисян, В. Е. 1976. Некоторые модификации системы семейства Brassicaceae. Бот. Ж. 61: 1198-1203.
Козо-Полянский, Б. М. 1945. Происхождение цветка Cruciferae в тератологическом освещении. Тератология цветка и новые вопросы его теории. III. Бот. Ж. 30: 14-30.
Чабан, И. А., & М. С. Яковлев. 1974. Эмбриология *Reseda lutea* L. I. Мегаспорогенез и развитие зародышевого мешка. Бот. Ж. 59: 24-37.
Чигуряева, А. А. 1973. Морфология пыльцы семейства Cruciferae. В сб.: Морфология пыльцы и спор современных растений. Труды III Международной Палинологической Конференции, СССР, Новосибирск, 1971.: 93-98. Изд. Наука, Ленинградское Отд. Ленинград.

SYNOPTICAL ARRANGEMENT OF THE FAMILIES OF CAPPARALES

1 Seeds with fairly well developed endosperm; ovary plurilocular, with typically 6 carpels and proliferating axile placentas; sepals, petals, and stamens mostly 8 each, coarse herbs or soft shrubs with trifoliolate leaves .... 1. TOVARIACEAE.

1 Seeds with very scanty or no endosperm; placentas parietal (sometimes intruded), only very rarely axile (or in one small genus of Resedaceae the carpels distinct).

  2 Stigma 1, capitate or slightly lobed, usually elevated above the ovary on a style; ovary closed.

    3 Flowers hypogynous or seldom slightly perigynous, regular or somewhat irregular; sepals (2) 4 (–6); petals (0–) 4 (–6); carpels 2 (–12); fruit very often with a definite replum.

      4 Ovary and fruit typically unilocular with parietal (sometimes intruded) placentas, or rarely plurilocular with axile placentas, only very rarely cross-partitioned by a septum; stamens 4-many, but never tetradynamous; flowers generally with an evident gynophore or androgynophore; pollen-grains mostly binucleate; leaves simple or trifoliolate or palmately compound, but not much dissected; shrubs, or

less often herbs or trees, most species tropical or subtropical .... 2. CAPPARACEAE.

4 Ovary and fruit nearly always cross-partitioned by a septum; placentas parietal; stamens 6 and usually tetradynamous (the 2 outer shorter than the 4 inner), seldom fewer or more numerous; flowers only seldom with a gynophore, and never with an androgynophore; pollen-grains mostly trinucleate; leaves simple to often pinnately more or less dissected, only seldom with definite leaflets; herbs, very seldom shrubs, most species of temperate or warm-temperate or boreal regions .... 3. BRASSICACEAE.

3 Flowers perigynous, irregular; sepals, petals, and functional stamens 5 each; carpels (2) 3 (4); no replum; trees with pinnately decompound leaves and definite leaflets .... 4. MORINGACEAE.

2 Stigmas as many as the carpels, sessile and well separated; ovary usually open at the top; flowers irregular; sepals 4–8; carpels (2) 3–6 (7); no replum; herbs or seldom shrubs, with entire to deeply pinnatifid leaves .... 5. RESEDACEAE.

1. Family TOVARIACEAE Pax in Engler & Prantl 1891 nom. conserv., the Tovaria Family

Strongly odorous coarse herbs or soft shrubs or half-shrubs, sometimes lax and somewhat scrambling, producing myrosin and glucosinolates (mustard-oil glucosides); ER-dependent vacuoles, but not dilated cisternae, present in companion cells and parenchyma cells. LEAVES alternate, trifoliolate; stomates paracytic; stipules none. FLOWERS in elongate terminal racemes, perfect, regular, hypogynous; sepals, petals, and stamens commonly 8, seldom 6, 7, or 9; sepals imbricate, deciduous; petals sessile; stamens borne internally to a lobed nectary-disk; filaments thickened toward the base, short-hairy; anthers sagittate, tetrasporangiate and dithecal, opening by longitudinal slits; pollen-grains (2) 3-colporate, binucleate; gynoecium of (5) 6 (–8) carpels, united to form a compound, plurilocular ovary that is borne on a very short gynophore or virtually sessile; style short, stout, beak-like, with spreading stigmas; ovules numerous, clothing the expanded, proliferating dissepiments, more or less distinctly campylotropous, crassinucellar, bitegmic, with a zigzag micropyle; endosperm-development nuclear. FRUIT a berry, the soft and juicy central placental mass easily separating from the pericarp, to which it is joined by the thin and fragile, sterile peripheral portions of the partitions; seeds very numerous and small; embryo dicotyledonous, embedded in a rather thin sheath of oily endosperm, curved around the periphery of the seed. N = 14.

The family Tovariaceae consists of the single genus *Tovaria*, with 2 tropical American species. *Tovaria* has in the past often been included in the Capparaceae, but modern authors are agreed that it should form a separate family. The embryology and pollen-morphology, as well as the more obvious morphological features, are generally considered to support a relationship to the Capparaceae, and the two families are kept in the same order. In addition to the high number of petals, sepals, and carpels, *Tovaria* differs from most Capparaceae in its fairly well developed endosperm and especially in the distinctly plurilocular ovary with proliferating axile placentation. The endospermous seeds and axile placentation appear to be primitive features, but the number of sepals, petals, and carpels may well reflect a secondary increase. Although the Tovariaceae are here placed at the beginning of the linear sequence of families of the Capparales, they are not to be considered ancestral to the other families. Instead they represent a short, early side-branch. The ecological significance of the characters that distinguish the Tovariaceae from other Capparales is obscure.

2. Family CAPPARACEAE A. L. de Jussieu 1789 nom. conserv., the Caper Family

Shrubs, or less often annual or perennial herbs or seldom trees, sometimes scandent, producing glucosinolates (mustard-oil glucosides) and (in the Capparoideae) pyrrolidine alkaloids, sometimes cyanogenic, but without iridoid substances, and not tanniferous, lacking both proanthocyanins and ellagic acid; endoplasmic reticulum with dilated cisternae; indument of various types of unicellular, uniseriate, stellate, or peltate hairs; scattered myrosin-cells present in all organs; calcium oxalate crystals of various sorts commonly present in some of the cells of the parenchymatous tissues; stems sometimes with anomalous secondary growth of concentric type; vessel-segments with simple perforations; imperforate tracheary elements with small, simple pits (bordered in *Koeberlinia*); wood-rays homocellular or less often heterocellular, usually only a few of them uniseriate, the others 2–5 (6) cells wide; wood-parenchyma paratracheal, commonly scanty and vasicentric (apotracheal in *Koeberlinia*); sieve-tubes with S-type or P-type plastids; roots sometimes forming endotrophic mycorhizae, sometimes nonmycorhizal. LEAVES alternate or rarely opposite, simple or trifoliolate or often palmately compound (reduced and scale-like in *Koeberlinia*); stomates anomocytic; stipules when present small, sometimes modified into glands or spinules. FLOWERS mostly in bracteate (seldom ebracteate) racemes, or occasionally solitary and axillary, perfect or seldom unisexual, regular or often somewhat irregular, hypogynous or seldom slightly perigynous; receptacle usually prolonged into a gynophore (the ovary appearing to be stipitate) or androgynophore; sepals (2–) 4 (–6), often in 2 opposite, decussate pairs, distinct or connate below; petals (2–) 4 (–6), mostly distinct and alternate with the sepals, often with a basal claw, or seldom wanting; androecium (at least typically) derived from 4 staminal primordia and sometimes consisting of 4 stamens alternate with the petals, but often 2 or all 4 of the primordia proliferating to produce a total of 6-many centrifugal stamens, some of which may be staminodial, the stamens however not tetradynamous as in the Brassicaceae; anthers tetrasporangiate and dithecal, opening by longitudinal slits; pollen-grains binucleate or seldom trinucleate, (2) 3 (4)-colporate; nectary varying from an extrastaminal ring to a mere adaxial protrusion of the receptacle; gynoecium consisting of 2 (–12) carpels united to form a compound pistil with a single terminal style and a capitate or bilobed, dry stigma, or the stigma sometimes sessile; ovary unilocular, with parietal, sometimes more or less deeply intruded placentas, or the

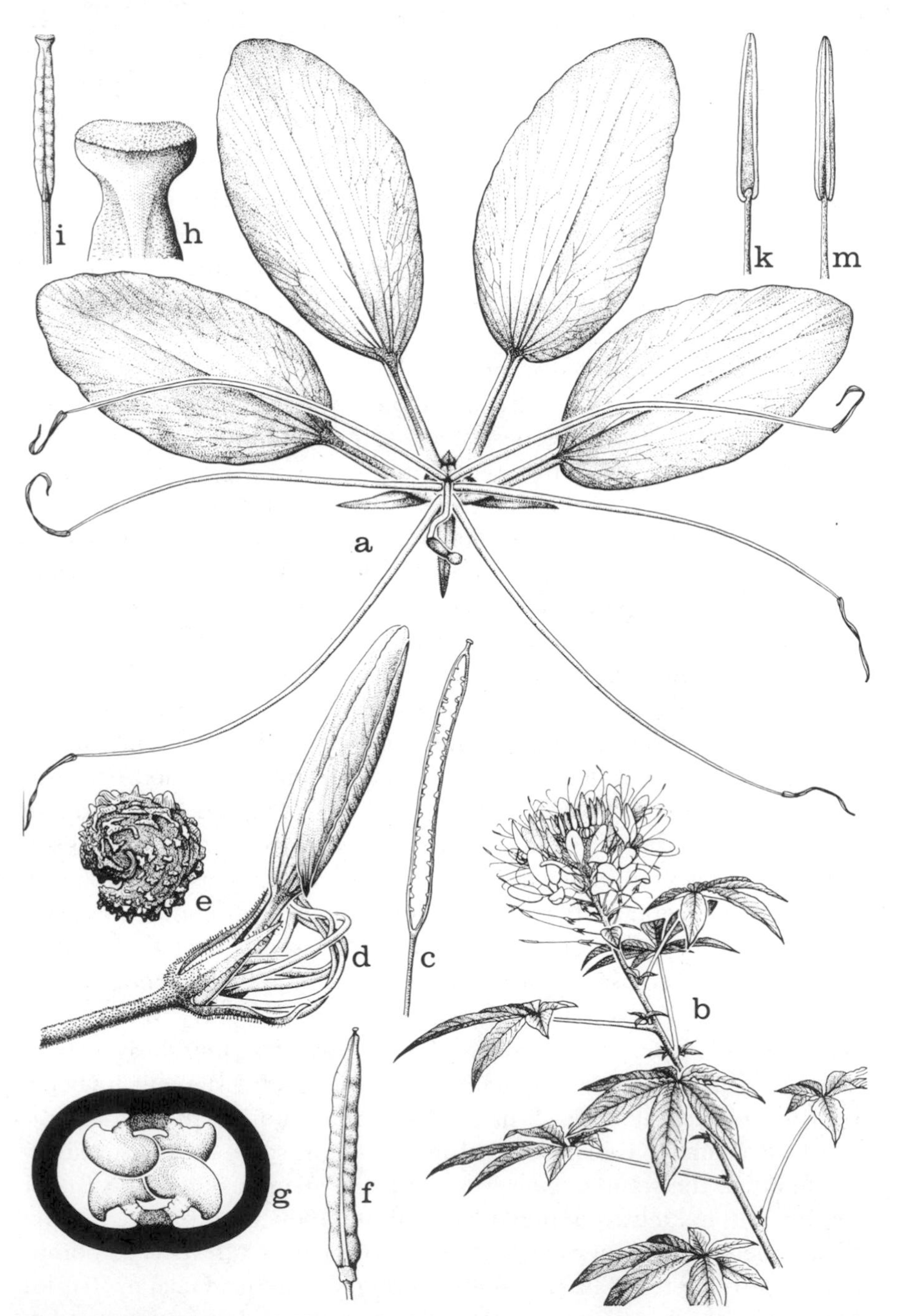

Fig. 4-31 Capparaceae. *Cleome spinosa* L. a, flower, ×2, b, habit, ×¼; c, replum of fruit, ×1; d, flower, just beginning to open, ×2; e, seed, ×8, f, fruit, ×1; g, schematic cross-section of ovary, ×32; h, top of ovary, with stigma, ×12; i, stipitate ovary, ×2; k, m, anthers ×2.

placental partitions rarely meeting in the center and joined so that the ovary is plurilocular, or sometimes a partition present but ephemeral; ovules (1–) numerous on each placenta, campylotropous or seldom anatropous (or anatropous before fertilization but becoming campylotropous thereafter), crassinucellar, bitegmic, with a zigzag micropyle; endosperm-development nuclear. FRUIT usually stipitate, often a berry, or often dry and siliquiform, then either indehiscent or with the valves falling away at maturity, leaving the persistent, frame-like replum (which with rare exceptions differs from that of the Brassicaceae in not bearing a partition), seldom an ordinary capsule or a nut or a drupe; seeds commonly reniform, with a more or less curved or folded, dicotyledonous, oily embryo, sometimes arillate; endosperm scanty or none; a persistent perisperm sometimes present. X = 8–17. (Cleomaceae, Koeberliniaceae, Oceanopapaveraceae, Pentadiplandraceae)

The family Capparaceae as here defined consists of about 45 genera and some 800 species, widespread in tropical and subtropical regions, with a few temperate-zone species, these mostly in arid climates. The traditional spelling Capparidaceae was accepted by scholarly botanists of past generations, but the spelling Capparaceae, without the *-id*, has been enshrined in the list of conserved family names in the International Code of Botanical Nomenclature. Fully two-thirds of the species belong to only two genera, *Capparis* (350) and *Cleome* (200). A few species are cultivated for ornament. Notable among these is *Cleome spinosa* L., the giant spider-flower. The pickled flower-buds of *Capparis spinosa* L., called capers, are used for garnish or seasoning.

Nearly all of the genera can be accommodated in two subfamilies: the Capparoideae, usually woody, the fruits usually indehiscent and in any case lacking a replum; and the Cleomoideae, more often herbaceous, the fruits dehiscent, with a replum. The Capparoideae are reminiscent of the Flacourtiaceae, whereas the Cleomoideae approach the Brassicaceae.

It is now generally believed that the Capparaceae stand near to the base of the order Capparales, and that the Capparales are allied to the Violales and Theales. Some authors, such as Takhtajan, see a close relationship of the Capparaceae to the Flacourtiaceae. This may well be correct, but I believe that the phyletic connection between these two families (and their respective orders) lies in their common origin in the Theales.

As in other groups with centrifugal stamens, it is still open to question whether the fascicled stamens of the Capparaceae represent a stage in androecial reduction or a secondary increase in the number of stamens.

I incline toward the former interpretation. It is likewise debatable whether the few Capparaceae with a plurilocular ovary are primitive or secondary in this regard. I incline toward the view that they are primitive. On the other hand, numbers of 6 or more sepals, petals, and carpels may well reflect a secondary increase, as several authors have supposed. Clarification of these matters awaits more detailed studies of the Capparaceae and a number of other families.

Canotia, sometimes associated with *Koeberlinia*, is here referred to the Celastraceae. *Emblingia*, another vexatious genus sometimes included in the Capparaceae, is here treated under the Polygalaceae.

Fossils thought to represent the Capparaceae occur in Eocene and more recent deposits.

3. Family BRASSICACEAE Burnett 1835 nom. conserv., the Mustard or Crucifer Family

Annual, biennial, or perennial herbs, or very seldom shrubs, producing glucosinolates (mustard-oil glucosides), and often cyanogenic, but without iridoid substances, mostly not tanniferous, lacking both ellagic acid and proanthocyanins (or the latter sometimes present in the seed-coat), and only seldom with alkaloids, cardenolides, or cucurbitacins; endoplasmic reticulum commonly with dilated cisternae; indument diverse, often of malpighian or branched or stellate, sometimes calcified hairs; scattered myrosin-cells present in all organs; calcium oxalate crystals usually wanting; nodes variously unilacunar, trilacunar, or multilacunar; stem occasionally with anomalous structure; vessel-segments with simple perforations; imperforate tracheary elements with small, bordered pits; sieve-tubes with S-type or P-type plastids; roots only seldom forming mycorhizae. LEAVES alternate or rarely opposite, simple to often pinnately more or less dissected, but only seldom with distinct, articulated leaflets; stomates anisocytic; stipules wanting. FLOWERS in racemes (these occasionally with a few long branches); usually bractless, very rarely solitary and terminal on a scape, perfect, regular or very seldom somewhat irregular, hypogynous; receptacle only seldom prolonged into a gynophore, but often giving rise to diversely arranged nectaries that are sometimes confluent into a ring around the base of the ovary or just outside the stamens; sepals 4, in 2 opposite, decussate pairs, the outer pair often gibbous at the base; petals 4, diagonal to the sepals, imbricate or convolute, commonly with an elongate claw and abruptly spreading blade above the top of the cylindrical calyx, forming a cross

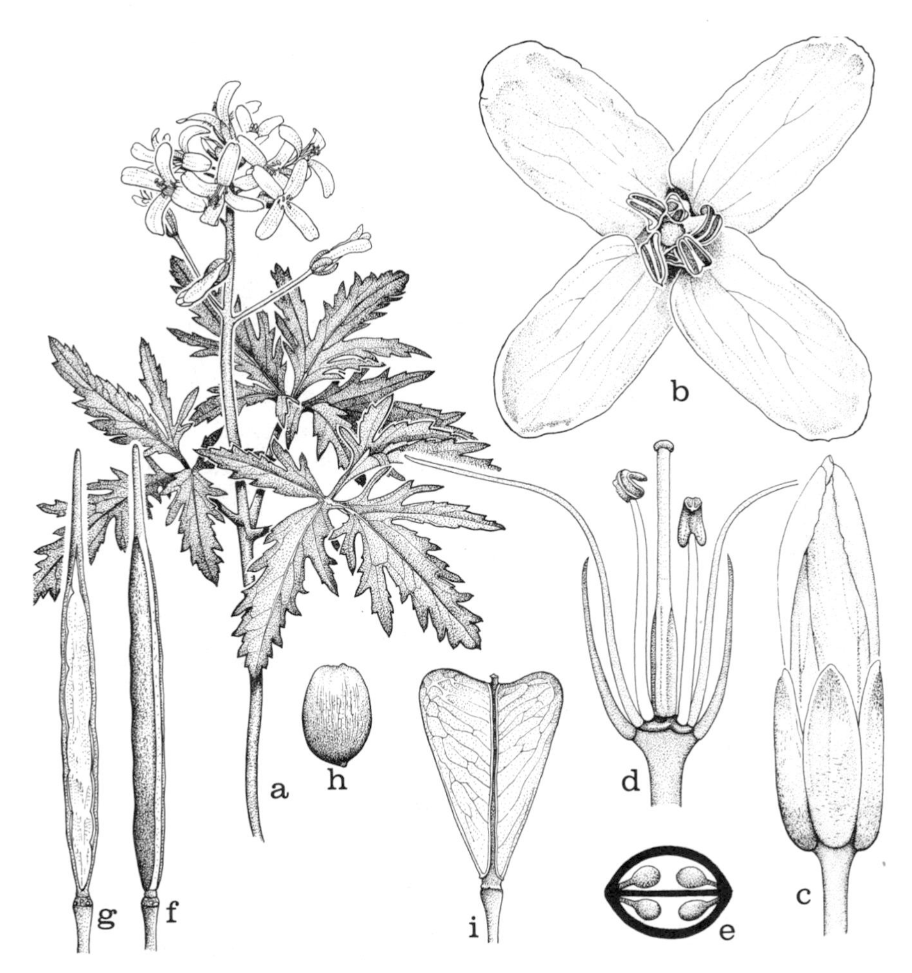

FIG. 4-32 Brassicaceae. a–h, *Dentaria laciniata* Muhl. a, habit, × ½; b, flower, from above, ×4; c, flower bud, ×4; d, schematic long-section of flower, ×4; e, schematic cross-section of ovary, ×16; f, fruit, ×2; g, fruit, after the valves have fallen, ×2; h, seed, ×4. i, *Capsella bursa-pastoris* L., fruit, ×4.

(hence the name Cruciferae), or seldom wanting; stamens commonly 6, the 2 outer shorter than the 4 inner, the androecium said to be tetradynamous; the 4 inner stamens derived from only 2 initial primordia, and sometimes connate below in pairs, rarely the stamens only 2–4, as in spp. of *Lepidium*, or up to 16, as in *Megacarpaea*; anthers tetrasporangiate and dithecal, opening by longitudinal slits; pollen-grains trinucleate or seldom binucleate, 3 (–10)-colporate; gynoecium consisting of 2 carpels united to form a compound pistil, with or without a style, the stigma dry, papillate, capitate, entire or bilobed; ovary divided into 2 chambers by a thin, unvasculated partition connecting the 2 parietal placentas; ovules (1–) several or many on each placenta, in 2 rows separated by the partition, anatropous to more often campylotropous, bitegmic, crassinucellar or weakly crassinucellar to tenuinucellar; endosperm-development nuclear. FRUIT dry and usually dehiscent, generally a silique (elongate) or a silicle (short), the valves falling away at maturity to expose the persistent replum, which nearly always bears a thin partition, or seldom the fruit indehiscent and sometimes transversely jointed; seeds attached to the frame of the replum, the funiculus often persisting after the seeds have been shed; embryo large, oily, commonly containing large amounts of erucic and/or lesquerolic acid in its fats, often also containing raffinose and/or stachyose, folded, the cotyledons commonly lying against the radicle (either edgewise or flatwise); endosperm very scanty, commonly hardly more than a single layer of cells, or none. X = 5–12+. (Cruciferae)

The family Brassicaceae consists of about 350 genera and 3000 species, found mainly in temperate (or cold) and warm-temperate parts of both the Northern and Southern Hemispheres. The greatest concentration of genera and species is in the area from the periphery of the Mediterranean to central Asia. There is a lesser but substantial center in western North America, and there are subsidiary centers in temperate South America, southern Africa, and Australia. Crucifers are partial to dry climates, but some occur in moist regions or habitats, and there are even some submerged aquatics (e.g., *Subularia*). The largest genera are *Draba* (nearly 300), *Cardamine* (150), *Lepidium* (130), *Alyssum* (100), *Arabis* (100), *Heliophila* (100), and *Lesquerella* (70). Many species are cultivated for food or ornament. Among the most familiar of these are *Armoracia rusticana* P. Gaertn., B. Mey., & Scherb. (horse-radish), *Brassica nigra* L., (black mustard), *B. oleracea* L. (broccoli, cabbage, cauliflower, Brussels sprouts), *B. rapa* L. (turnip), *Cheiranthus cheiri* L. (wallflower), *Hesperis matronalis* L. (rocket, dame's violet), *Iberis amara* L. (candytuft), *Lobularia maritima* (L.) Desv. (sweet alyssum), *Lunaria annua* L. (honesty), *Nasturtium officinale* R. Br. (water-cress), and *Raphanus sativus* L. (radish).

The Brassicaceae are one of the more sharply defined and readily recognizable large families of plants. In counterpoint, the genera are ill-defined and frequently confluent.

The gynoecium of the Brassicaceae is unique, although species in which the partition fails to develop compare closely with some of the

Capparaceae. The morphology of the cruciferous gynoecium has occasioned much controversy and is still not settled to the satisfaction of all concerned. I prefer the 2-carpellary interpretation of Puri (1952) to the 4-carpellary interpretation of Eames and Wilson. The cruciferous gynoecium is surely homologous with the gynoecium of those Capparaceae that have a replum but lack a partition, and there is no reason to suppose that these latter genera are anything but bicarpellate as they appear to be.

Botanists are agreed that the Brassicaceae are related to the Capparaceae but more advanced in at least some respects. Some authors (e.g., Janchen, Takhtajan) consider the western American genus *Stanleya*, with the ovary on an elongate gynophore, to be very archaic within the family, and they compare *Stanleya* to members of the subfamily Cleomoideae of the Capparaceae. Others (e.g., Rollins, personal communication) think that the similarities between *Stanleya* (a selenophile) and genera such as *Cleome* reflect parallelism or convergence rather than close relationship. I incline toward the latter view. The connection between the two families might perhaps be better sought among the Old-World genera.

Fossils thought to represent the Brassicaceae occur in Oligocene and more recent deposits.

4. Family MORINGACEAE Dumortier 1829 nom. conserv., the Horse-radish tree Family

Deciduous trees of *Acacia*-like habit, commonly with a stout trunk and gummy bark, with a large, central mucilage-canal in the pith, bearing scattered myrosin-cells and producing glucosinolates (mustard-oil glucosides) and sometimes alkaloidal amines, sometimes cyanogenic but without iridoid substances and not tanniferous, lacking ellagic acid and proanthocyanins; endoplasmic reticulum probably with dilated cisternae; hairs unicellular; clustered crystals of calcium oxalate commonly present in some of the cells of the parenchymatous tissues; vessel-segments with simple perforations; imperforate tracheary elements storied, with simple pits; wood-rays homocellular, Kribs types I and II, up to 3 cells wide, the uniseriate ones sometimes very few; wood-parenchyma paratracheal, vasicentric or somewhat wing-like. LEAVES alternate, 2–3 times pinnately compound, with opposite leaflets; epidermis often mucilaginous; stomates anomocytic; stipules wanting, or represented by glands at the base of the petiole, similar glands also present at the base of the pinnae. FLOWERS in axillary panicles or mixed panicles, red or white, perfect, irregular (seldom essentially regular), perigynous, with a saucer-shaped to shortly tubular, often oblique hypanthium, this lined below with a nectary-disk that has a short free margin; sepals 5, spreading or reflexed, imbricate, unequal, much like the petals; petals 5, imbricate, usually the outermost one the largest and the 2 inner ones the smallest; functional stamens 5, declined, antepetalous alternating with as many (or fewer) staminodia, these setiform or resembling antherless filaments; filaments distinct, inserted on the hypanthium around the margin of the disk; anthers dorsifixed, mostly bisporangiate and monothecal, seldom tetrasporangiate and dithecal, opening by longitudinal slits; pollen-grains binucleate or trinucleate, (2) 3 (4)-colporate; gynoecium of (2) 3 (4) carpels united to form a compound, unilocular ovary that is seated on a short gynophore and bears a slender, terminal, hollow, apically truncate style; placentas parietal, each with 2 rows of ovules, these anatropous, crassinucellar, and bitegmic, with a zigzag micropyle; endosperm-development nuclear. FRUIT a large, elongate, woody, torulose, unilocular, explosively dehiscent capsule, without a replum; seeds attached medially along the valves, numerous, 3-winged or less often wingless, with a straight, oily, dicotyledonous embryo, nearly or quite without endosperm. X = 14.

The family Moringaceae consists of the single genus *Moringa*, with about 10 species of xerophytic trees occurring in Africa and Madagascar and across the Middle East to India. Ben oil, used in perfumery and lubrication, is extracted from the seeds of *Moringa pterygosperma* C. F. Gaertn., the horse-radish tree, a native of India that is widely cultivated in tropical countries. The leaves, young fruits, and roots of this species are eaten.

5. Family RESEDACEAE S. F. Gray 1821 nom. conserv., the Mignonette Family

Annual or perennial herbs or seldom shrubs or half-shrubs, with unicellular hairs, producing glucosinolates (mustard-oil glucosides), and sometimes cyanogenic, mostly without crystals of calcium oxalate (except in the seed coat), without iridoid compounds, without anthocyanins, and not tanniferous, lacking both proanthocyanins and ellagic acid; scattered myrosin-cells present in all organs; endoplasmic reticulum with ER-dependent vacuoles; nodes unilacunar; vessel-segments with simple perforations; imperforate tracheary elements with simple pits; roots only seldom forming mycorhizae. LEAVES alternate, entire to deeply pinnatifid; epidermis often mucilaginous; stomates anomocytic; stipules small, modified into glands. FLOWERS in racemes or spikes (sometimes branched), perfect or seldom unisexual, more or less strongly irregular, hypogynous to seldom somewhat perigynous, with a short androgynophore or gynophore; sepals (4–) 6 (–8), valvate or slightly imbricate in bud; petals (0 or 2) 4–8, most often 6, white to yellow, valvate, distinct, unequal, the uppermost (innermost) one the largest and usually conspicuously fringed-appendiculate from the upper part of the back, the outer ones progressively smaller and with less well developed appendages; androgynophore commonly with an adaxially dilated nectary-disk below the stamens; stamens 3-50+, in *Reseda lutea* L. 17–22 and originating in a single ring beginning at the adaxial side of the flower and progressing in both directions to the abaxial side, the circular bulge from which these stamens arise considered by Leins and Sobick to be homologous with a similar bulge in *Capparis*, on which numerous stamens originate centrifugally; anthers tetrasporangiate and dithecal, opening by longitudinal slits; pollen-grains binucleate, tricolpate or tricolporate; gynoecium of (2) 3–6 (7) carpels, these generally united to form a superior, compound, unilocular ovary that is evidently open at the top and bears small, sessile, well separated, dry stigmas around the rim; in the monotypic genus *Sesamoides* the carpels wholly distinct (considered to be only secondarily so) and bearing only a single ovule (rarely 2 ovules), and in *Caylusea* (3 spp.) the carpels united only near the base, with the several ovules crowded on short, axile placentas; ovules, except as noted, on parietal placentas, campylotropous, bitegmic, crassinucellar or tenuinucellar, endosperm-development nuclear. FRUIT a usually gaping capsule, or sometimes a berry or in *Sesamoides* of distinct, radiating carpels; seeds reniform, with large, curved or folded, oily embryo and little or no endosperm. X = 6–15.

The family Resedaceae consists of 6 genera and about 70 species. It is restricted to the Northern Hemisphere, mostly in the Old World, especially in the Mediterranean region and eastward. A few species occur in southwestern United States and northern Mexico. *Reseda*, with 55 species, is by far the largest genus. *Reseda odorata* L., mignonette, is a common garden-ornamental, grown especially for its fragrance. The distally open ovary of *Reseda* and its allies is unique.

The commonly accepted assignment of the Resedaceae to the Capparales is not currently in dispute.

9. Order BATALES Skottsberg 1940

Trees, shrubs, or half-shrubs, producing glucosinolates (mustard-oil glucosides), and with specialized myrosin-cells, but lacking iridoid compounds, ellagic acid, proanthocyanins, anthocyanins, and betalains; vessel-segments with simple perforations; imperforate tracheary elements with simple or bordered pits; sieve-tubes with S-type plastids. LEAVES alternate or opposite, simple and entire, more or less succulent; stipules minute or wanting. FLOWERS unisexual, small, solitary in the axils of the upper leaves or grouped into terminal or axillary (sometimes ament-like or strobiloid) racemes or spikes; calyx small, of distinct or connate members; petals none (unless the tepals of the staminate flowers of *Batis* are so interpreted); stamens 4-many; anthers dithecal, opening by longitudinal slits; pollen-grains mostly tricolpate or tricolporate, with solid exine; gynoecium of (1) 2-numerous carpels joined to form a compound ovary with as many (or twice as many) locules as carpels; stigmas sessile or nearly so; ovules solitary in each locule, bitegmic, crassinucellar. FRUIT drupaceous, or dry and dehiscent or nutlike; embryo dicotyledonous, endosperm present or absent; perisperm absent.

The order Batales as here defined consists of 2 small families, the Bataceae and Gyrostemonaceae, only 6 genera and about 19 species in all. The Gyrostemonaceae have often been included in the Phytolaccaceae. The Bataceae have been variously treated, often being included in or associated with the Caryophyllales.

The Batales differ from the Caryophyllales in their unusual pollen with solid exine, in their S-type sieve-tube plastids, in having mustard oil, and in lacking both betalains and anthocyanins. The food-reserve in the seeds of the Gyrostemonaceae is thought to be endosperm rather than perisperm, but this interpretation needs to be confirmed. The Bataceae lack both endosperm and perisperm. These several features are now generally considered to require the removal of the Bataceae and Gyrostemonaceae from the Caryophyllales, and it is rapidly becoming customary to associate the two families in some way in the general system of classification. As long ago as 1965, before the chemical features and the sieve-tube plastids had been studied, Kuprianova emphasized external pollen-morphology in assigning the Gyrostemonaceae to the Batales along with *Batis*.

The affinities of the Batales are now uncertain and disputed. Perhaps the best lead is provided by the fact that they have mustard-oils.

Mustard-oils are characteristically produced by members of the fairly large and successful order Capparales, and by some members of the small family Caricaceae (Violales). Outside the Dilleniidae mustard-oils are rare, although they do occur in scattered members of the Geraniales, Sapindales, Celastrales, and Euphorbiales. *Batis* is also reminiscent of the Brassicaceae in having parietal placentation in a partitioned ovary, although the details of structure are different. The recent discovery of special protein-containing vacuoles in the companion-cells of *Gyrostemon* may also be significant. These have been compared to (and may be homologous with) the dilated ER-cisternae of the Capparaceae and Brassicaceae. At the present state of knowledge, I think it most useful to associate the Batales with the Capparales. Inclusion of the Batales in the Dilleniidae brings no important new character to the subclass, except for the solid exine of the pollen-grains, which would be equally anomalous in all the other subclasses except the Magnoliidae.

SELECTED REFERENCES

Behnke, H.-D. 1977. Phloem ultrastructure and systematic position of Gyrostemonaceae. Bot. Not. 130: 255–260.

Behnke, H.-D. 1977. Zur Skulptur der Pollen-Exine bei drei Centrospermen (*Gisekia, Limeum, Hectorella*), bei Gyrostemonaceen und Rhabdodendraceen. Pl. Syst. Evol. 128: 227–235.

Behnke, H.-D. & B. L. Turner. 1971. On specific sieve-tube plastids in Caryophyllales. Further investigations with special reference to the Bataceae. Taxon. 20: 731–737.

Carlquist, S. 1978. Wood anatomy and relationships of Bataceae, Gyrostemonaceae, and Stylobasiaceae. Allertonia 1: 297–330.

Eckardt, Th. 1959 (1960). Das Blütendiagramm von *Batis* P. Br. Ber. Deutsch. Bot. Ges. 72: 411–418.

Eckardt, Th. 1971. Anlegung und Entwicklung der Blüten von *Gyrostemon ramulosus* Desf. Bot. Jahrb. Syst. 90: 434–446.

Goldblatt, P. 1976. Chromosome number and its significance in *Batis maritima* (Bataceae). J. Arnold Arbor. 57: 526–530.

Goldblatt, P., J. W. Nowicke, T. J. Mabry, & H.-D. Behnke. 1976. Gyrostemonaceae: status and affinity. Bot. Not. 129: 201–206.

Johnson, D. S. 1935. The development of the shoot, male flower and seedling of *Batis maritima* L. Bull. Torrey Bot. Club 62: 19–31.

Keighery, G. J. 1975. Chromosome numbers in the Gyrostemonaceae Endl. and the Phytolaccaceae Lindl.: a comparison. Austral. J. Bot. 23: 335–338.

Mabry, T. J., & B. L. Turner. 1964. Chemical investigations of the Batidaceae. Taxon 13: 197–200.

McLaughlin, L. 1959. The woods and flora of the Florida Keys. Wood anatomy and phylogeny of Batidaceae. Trop. Woods 110: 1–15.

Prijanto, B. 1970. Gyrostemonaceae. World Pollen Flora 2.

Prijanto, B. 1970. Batidaceae. World Pollen Flora 3: 1–11.

Royen, P. van. 1956. A new Batidacea, *Batis argillicola.* Nova Guinea 7: 187–195.
Uphof, J. C. 1930. Biologische Beobachtungen an *Batis maritima* L. Oesterr. Bot. Z. 79: 355–367.
van Heel, W. A. 1958. Additional investigations on *Batis argillicola* van Royen. Nova Guinea 9: 1–7.

Куприянова, Л. А. 1965. Палинология сережкоцветных. Изд. Наука. Москва–Ленинград.

SYNOPTICAL ARRANGEMENT OF THE FAMILIES OF BATALES

1 Leaves alternate; stamens 6-many; placentation axile; ovary with (1–) usually more or less numerous carpels and as many locules; fruit dry, dehiscent or less often indehiscent; seeds with endosperm .... 1. GYROSTEMONACEAE.
1 Leaves opposite; stamens 4; placentation parietal-basal; ovary with 2 carpels and 4 locules; fruit drupaceous, with 4 pyrenes; seeds without endosperm .... 2. BATACEAE.

1. Family GYROSTEMONACEAE Endlicher 1841 nom. conserv., the Gyrostemon Family

Trees, shrubs, or half-shrubs with normal secondary growth, producing glucosinolates (mustard-oil glucosides), and with myrosin-cells, but lacking iridoid compounds, ellagic acid, proanthocyanins, anthocyanins, betalains, and calcium oxalate crystals; vessel-segments with simple perforations; imperforate tracheary elements with bordered pits, considered to be tracheids; wood-rays nearly homocellular, Kribs type I, mostly pluriseriate, with only a few uniseriates; sieve tubes with S-type plastids; wood parenchyma chiefly paratracheal, but sometimes also diffuse; companion cells (at least in *Gyrostemon*) with protein-containing, ER-dependent vacuoles that may be homologous with the dilated ER-cisternae of the Capparales. LEAVES alternate, simple, entire, containing some mucilage-cells, often more or less succulent; stomates anomocytic; stipules minute or wanting. FLOWERS solitary in the axils of the upper leaves or in terminal or axillary racemes or spikes, bracteolate, small, unisexual (the plants mostly dioecious), with a notably expanded, disk-like or convex receptacle; calyx more or less synsepalous, with imbricate lobes, persistent in fruit; petals none; stamens 6-many, in one or more cycles around the edge of the receptacle, the cycles developing in centripetal sequence; filaments very short or virtually none; anthers dithecal, opening by longitudinal slits; pollen-grains tricolpate, smooth or a little rough, with solid exine; gynoecium of (2–4) more or less numerous carpels, these adnate to a central column but otherwise distinct or only lightly connate, forming an unusual sort of compound ovary with as many locules as carpels (carpel solitary in the monotypic genus *Cypselocarpus*); central column often expanded at the top into a flat disk, bearing around its margin the short, distinct styles with decurrent stigma; placentation axile; ovules solitary in each locule, campylotropous, with a thickened funicle. FRUIT dry, each carpel usually opening dorsally or ventrally or both and separating from the central column, or indehiscent (*Tersonia*, *Cypselocarpus*); seeds with a basal aril; embryo dicotyledonous, peripheral, curved around the copious, oily and fleshy endosperm. X = 14, 15.

The family Gyrostemonaceae consists of 5 genera (*Codonocarpus*, *Cypselocarpus*, *Didymotheca*, *Gyrostemon*, and *Tersonia*) and about 17 species, native to Australia. None of the species is economically important or familiar to botanists in general.

2. Family BATACEAE Martius ex Meissner 1842 nom. conserv., the Saltwort Family

Rather small, maritime shrubs up to ca 1.5 m tall, with or without crystals of calcium oxalate in the parenchymatous tissues, producing mustard-oil glucosides and with myrosin-bearing guard-cells, but lacking iridoid compounds, ellagic acid, proanthocyanins, anthocyanins, and betalains; twigs commonly (3) 4 (5)-angled, with a group of vascular bundles under each angle, the groups separated by narrow to broad medullary rays; vessel-segments very small, with simple, horizontal to somewhat oblique perforations; imperforate tracheary elements small, thick-walled, with minute, distinctly bordered pits, considered to be fiber-tracheids; wood-rays somewhat heterocellular, Kribs type IIA, mixed uniseriate and pluriseriate, the latter up to 6 (–8) cells wide; wood-parenchyma mainly paratracheal, but also apotracheal in bands; sieve-tubes with S-type plastids. LEAVES opposite, sessile, simple and entire, slender, succulent, with more or less distinctly paracytic stomates; stipules minute, caducous. Plants monoecious (*Batis argillicola*) or dioecious (*B. maritima*), the FLOWERS small and much-reduced, unisexual, borne in small, strobiloid, axillary or in part terminal spikes, the staminate and pistillate flowers intermingled in the same spike in *B. argillicola*; staminate flowers initially enclosed in a membranous, saccate organ, possibly representing the calyx or a pair of bractlets, which eventually splits near the top into 2 or 4 lobes (*B. maritima*) or along one side only (*B. argillicola*); tepals (sometimes considered to be staminodes) 4, distinct; stamens 4, alternating with the tepals; filaments distinct; anthers tetrasporangiate and dithecal, opening by longitudinal slits; pollen-grains binucleate, 3 (4)-colporate, with solid exine; a vestigial gynoecium present in the staminate flowers; pistillate flowers without perianth, composed essentially of a naked ovary, this bicarpellate but 4-locular, the 2 primary locules each divided by a partition extending from the carpellary midrib to the central axis; stigmas 2, sessile; each of the 4 locules bearing a solitary parietal-basal, anatropous, bitegmic, crassinucellar, epitropous ovule. FRUIT adapted to dispersal by flotation in salt water, drupaceous, with 4 pyrenes; seeds with straight or slightly curved, dicotyledonous embryo, lacking both endosperm and perisperm. $2n = 18$.

The family Bataceae consists of the single genus *Batis*, with 2 species, one along the coast of tropical and subtropical America and also in the

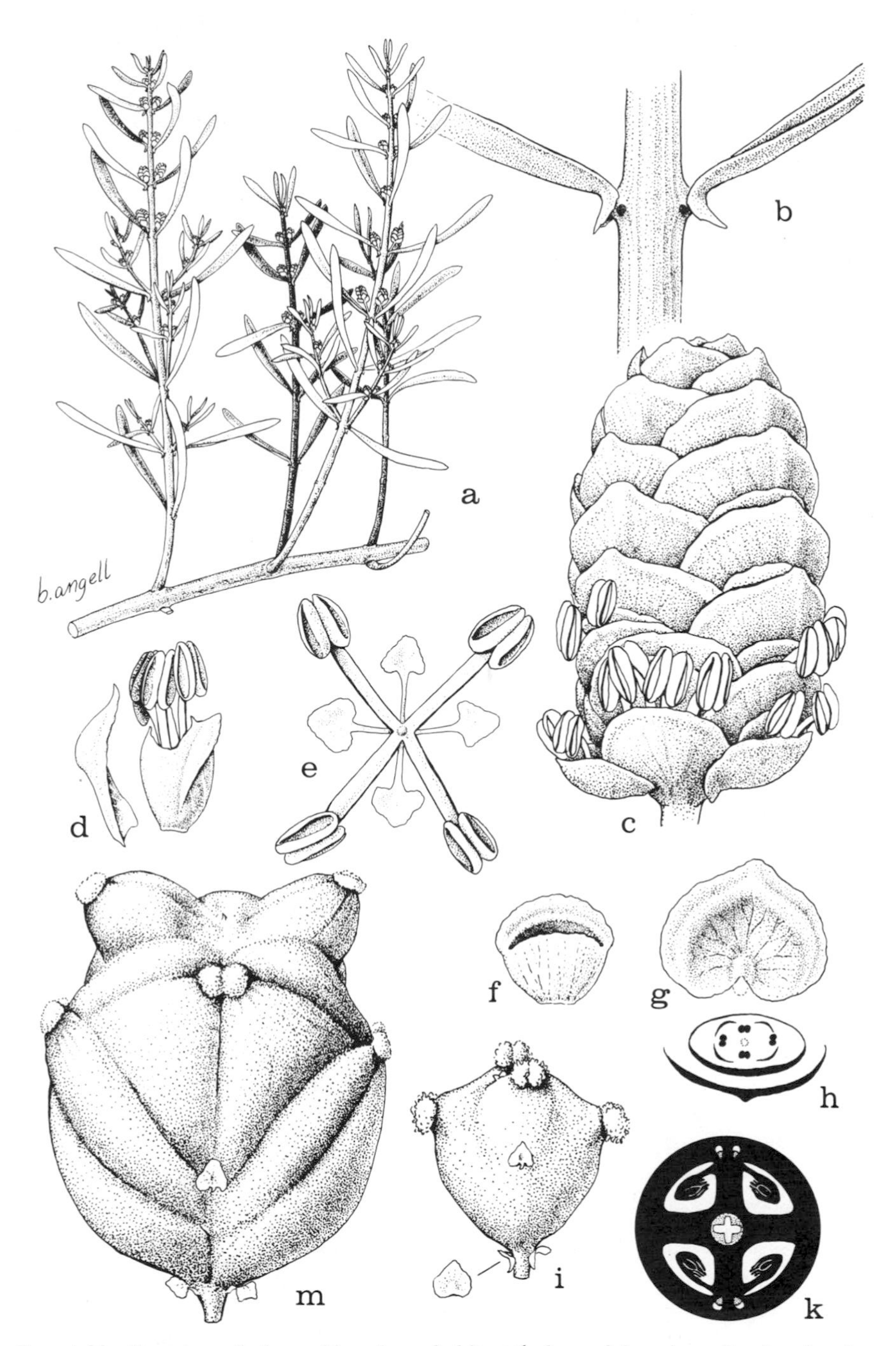

FIG. 4-33 Bataceae. *Batis maritima* L. a, habit, $\times\frac{1}{2}$; b, nodal region of twig, showing stipules and leaf-bases, $\times 5$; c, staminate inflorescence, $\times 8$; d, side view of staminate flower, with its enclosing saccate organ and subtending bract, $\times 8$; e, staminate flower, from above, $\times 8$; f, dorsal view of saccate organ that encloses the staminate flower, $\times 8$; g, dorsal view of bract of staminate inflorescence, $\times 8$; h, diagrammatic cross-section of staminate flower and bract; i, pistillate inflorescence, at anthesis, $\times 8$; k, diagrammatic cross-section of ovary (after Eckardt); m, multiple fruit, $\times 8$.

Galapagos and introduced into the Hawaiian Islands, the other along the coast of New Guinea and northeastern Australia.

Pollen considered to represent *Batis* is known from Maestrichtian deposits in California. Thus the Batales, on the record, appear to be as old as any identifiable order in either the Dilleniidae or the Caryophyllidae.

10. Order ERICALES Lindley 1833

Shrubs or rather small trees, or less often perennial herbs, often more or less strongly mycotrophic, and sometimes even without chlorophyll, without raphides and largely or wholly without alkaloids, commonly tanniferous, producing proanthocyanins but usually not ellagic acid, often producing iridoid compounds, methyl salicylate, 5-methoxy-flavonols, triterpernoid compounds such as ursolic acid, and phenolic heterosides such as arbutin, but only seldom cyanogenic; nodes unilacunar except in Epacridaceae; vessel-segments commonly with scalariform, less often simple perforations; imperforate tracheary elements with bordered pits; wood-parenchyma mostly scanty or wanting, seldom abundant and diffuse. LEAVES alternate, or less commonly opposite or whorled, simple, very often small, firm, and perennial ("ericoid"), but also often of ordinary form and appearance; stomates most commonly anomocytic, less often paracytic or of other types; stipules wanting. FLOWERS most commonly in racemes, perfect or sometimes (notably the Empetraceae) unisexual, hypogynous or sometimes epigynous, regular or slightly irregular, often containing anthocyanin 3-galactosides; sepals 3–7, most often 5, less often 4, distinct or connate below, imbricate or valvate; petals 3–7, most often 5, less often 4, distinct or usually connate at least below to form a sympetalous corolla with imbricate or convolute (less often valvate) lobes, or rarely the petals wanting; stamens usually as many or twice as many as the petals or corolla-lobes, rarely only 2 or as many as 20, attached directly to the receptacle or less often to the corolla-tube, the anthers commonly (except in Cyrillaceae) becoming inverted during ontogeny so that the morphological base is apical, typically without a fibrous layer in the wall, opening lengthwise or much more often by apical (morphologically basal) pores or slits, often provided with 2 (or more) slender appendages (whence the old ordinal name Bicornes); rarely these appendages borne on the upper part of the filament instead of on the anther; pollen-grains in monads or very often in tetrads, binucleate or seldom trinucleate, most commonly tricolporate; intrastaminal nectary-disk commonly present, sometimes forming a cap over an inferior ovary, or represented by scale-like nectaries alternating with the stamens, or the nectaries rarely wanting or (Clethraceae, some Monotropaceae) the basal part of the ovary nectariferous; gynoecium of 2–10 carpels united to form a compound, plurilocular ovary with axile placentas, or not infrequently the partitions failing to meet in the center, so that the ovary is unilocular with intruded parietal placentas, rarely (Grubbiaceae and a few Ericaceae) the partitions evanescent so that the placentation is eventually free-central; rarely (a few Ericaceae) only one carpel fully developed; style often emerging from an apical depression in the ovary, characteristically hollow, the cavity fluted in alignment with the locules; stigma subtruncate to capitate or lobed, or the style seldom cleft, with separate stigmas; ovules (1–) more or less numerous, anatropous to sometimes almost campylotropous, unitegmic, tenuinucellar, commonly with an integumentary tapetum; endosperm-development cellular, typically with haustoria at both ends. FRUIT of various types, often capsular; seeds (1–) more or less numerous, mostly small, with a thin testa consisting of a single layer of cells, or (Cyrillaceae) without a testa; embryo small, embedded in the copious to scanty endosperm, sometimes not differentiated into parts.

The order Ericales as here defined consists of 8 families and about 4000 species. About seven-eighths of the species belong to the Ericaceae (3500), and most of the remainder to the Epacridaceae (400). The other 6 families have only a little more than a hundred species in all.

Most members of the Ericales have a long series of embryological features in common. Among these are: 1) ovules unitegmic, tenuinucellar, the nucellus evanescent; 2) endosperm-development cellular; 3) endosperm-haustoria formed at both ends of the embryo-sac, micropylar as well as chalazal; 4) testa a single layer of cells formed from the outermost layer of the integument; 5) style hollow, the cavity fluted in alignment with the locules of the ovary; 6) anthers introrse at maturity, the position being achieved at some time during ontogeny by inversion of a basically extrorse structure, so that the morphological base of the anther appears to be apical; 7) anthers without a fibrous layer in the wall; 8) anthers dehiscing by terminal pores; 9) anther-tapetum of the glandular type, with multinucleate cells; 10) pollen-grains borne in tetrads. Most of these features are subject to some exception, and most or all of them occur in certain other orders. It is the combination of features, rather than any one, that distinguishes the bulk of the Ericales from other orders.

The mutual affinity among the Ericaceae, Pyrolaceae, Monotropaceae, Epacridaceae, and Clethraceae has for many years been evident to all, and a consensus that the Empetraceae also belong to the group now seems firmly established. A similar consensus about the Cyrillaceae seems to be developing, although they are admittedly somewhat removed from the other families. The position of the Grubbiaceae is more debatable, as noted in the discussion under that family.

The Ericaceae, Pyrolaceae, and Monotropaceae represent progressive stages in dependence on a mycorhizal fungus. The Ericaceae require the fungus for successful growth, but they have green leaves, and the morphology of their aerial parts in general does not obviously reflect their mycotrophic habit. The Pyrolaceae and Monotropaceae have the reduced embryo that so often accompanies parasitism or extreme mycotrophy in other groups. The Pyrolaceae usually have green leaves, although some species of *Pyrola* have leafless as well as leafy forms. The Monotropaceae have reduced, scale-like leaves and are wholly without chlorophyll. The degree of mycotrophy in the other families of the order awaits detailed study.

The Pyrolaceae and Monotropaceae have often been included in the Ericaceae. This is purely a matter of taste. Neither the close affinity of the 3 groups nor the sharpness of the distinctions among them are in dispute. I find it easier to make a synoptical arrangement of the families of the order if the Pyrolaceae and Monotropaceae are held as distinct families.

Vaccinium and its allies also stand off to some degree from the remainder of the Ericaceae and have sometimes been treated as a distinct family Vacciniaceae. This too is mainly a matter of taste. In spite of their inferior ovary, the flowers of *Vaccinium* look so much like those of many other Ericaceae (e.g., *Erica, Chamaedaphne, Pieris*) that I find it useful to recognize the distinction at the subfamilial rather than the familial level. This causes no difficulty in the synoptical arrangement. If the Vacciniaceae were recognized as a separate family, they would stand alongside the Ericaceae in the synopsis here presented. Some modern students of the Ericaceae treat the subfamily Vaccinioideae in an expanded sense to include *Arbutus, Arctostaphylos, Gaultheria, Oxydendrum, Pieris* and a number of other genera with a superior ovary, which have not usually been associated with the Vaccinioideae (or Vacciniaceae) in the past.

In addition to the 8 families here assigned to the Ericales, Takhtajan refers the Actinidiaceae (together with the segregate family Sauraujaceae) to the Ericales. This is defensible on grounds of evolutionary relationships, but it complicates the morphological definition of the order and is in my opinion unnecessary. The Actinidiaceae would be aberrant in the Ericaceae in their numerous stamens, in their cymose inflorescence, and in having raphides. All of these features can easily be accommodated in the Theales. As Takhtajan points out, the Actinidiaceae are closely related to the Theaceae as well as to the Clethraceae and other Ericales. Some members of the Theaceae even have the anthers dehiscent by terminal pores, as in the Actinidiaceae and the bulk of the Ericales. The Actinidiaceae have not been subjected to the same sort of detailed embryological and chemical study as many of the Ericales, so it is uncertain how much of the ericoid syndrome they share beyond the unitegmic, tenuinucellar ovules, poricidal anthers, cellular endosperm, often tetradinous pollen, and the presence in at least some members of iridoid compounds. On balance, the Actinidiaceae are somewhat more archaic than the Ericales, and are in my opinion more usefully included in the Theales.

SELECTED REFERENCES

Bell, H. P., & E. C. Giffin. 1957. The lowbush blueberry: the vascular anatomy of the ovary. Canad. J. Bot. 35: 667–673.

Carlquist, S. 1977. A revision of Grubbiaceae. J. S. African Bot. 43: 115–128.

Carlquist, S. 1977. Wood anatomy of Grubbiaceae. J. S. African Bot. 43: 129–144.

Carlquist, S. 1978. Vegetative anatomy and systematics of Grubbiaceae. Bot. Not. 131: 117–126.

Collinson, M. E., & P. R. Crane. 1978. *Rhododendron* seeds from the Palaeocene of southern England. J. Linn. Soc., Bot. 76: 195–205.

Copeland, H. F. 1933. The development of seeds in certain Ericales. Amer. J. Bot. 20: 513–517.

Copeland, H. F. 1939. The structure of *Monotropsis* and the classification of the Monotropoideae. Madroño 5: 105–119.

Copeland, H. F. 1941. Further studies on Monotropoideae. Madroño 6: 97–119.

Copeland, H. F. 1943. A study, anatomical and taxonomic, of the genera of Rhododendroideae. Amer. Midl. Naturalist 30: 533–625.

Copeland, H. F. 1947. Observations on the structure and classification of the Pyroleae. Madroño 9: 65–102.

Copeland, H. F. 1953. Observations on the Cyrillaceae particularly on the reproductive structures of the North American species. Phytomorphology 3: 405–411.

Copeland, H. F. 1954. Observations on certain Epacridaceae. Amer. J. Bot. 41: 215–222.

Cox, H. T. 1948. Studies in the comparative anatomy of the Ericales. I. Ericaceae—subfamily Rhododendroideae. II. Ericaceae—subfamily Arbutoideae. Amer. Midl. Naturalist 39: 220–245. 40: 493–516.

Dormer, K. J. 1944. Morphology of the vegetative shoot in Epacridaceae. New Phytol. 44: 149–151.

Fagerlind, F. 1947. Die systematische Stellung der Familie Grubbiaceae. Svensk Bot. Tidskr. 41: 315–320.

Franks, J. W., & L. Watson. 1963. The pollen morphology of some critical Ericales. Pollen & Spores 5: 51–68.

Ganapathy, P. S., & B. F. Palser. 1964. Studies of floral morphology in the Ericales. VII. Embryology in the Phyllodoceae. Bot. Gaz. 125: 280–297.

Gibbs, R. D. 1958. Biochemistry as an aid in establishing the relationships of some families of dicotyledons. Proc. Linn. Soc. London 169: 216–230.
Giebel, K. P., & W. C. Dickison. 1976. Wood anatomy of Clethraceae. J. Elisha Mitchell Sci. Soc. 92: 17–26.
Hagerup, O. 1946. Studies on the Empetraceae. Biol. Meddel. Kongel. Danske Vidensk. Selsk. 20(5): 1–49.
Hagerup, O. 1953. The morphology and systematics of the leaves in Ericales. Phytomorphology 3: 459–464.
Hara, N. 1958. Structure of the vegetative shoot apex and development of the leaf in the Ericaceae and their allies. J. Fac. Sci. Univ. Tokyo, Sec. 3, Bot. 7: 367–450.
Harborne, J. B., & C. A. Williams. 1973. A chemotaxonomic survey of flavonoids and simple phenols in leaves of the Ericaceae. J. Linn. Soc., Bot. 66: 37–54.
Jarman, S. J., & R. K. Crowden. 1974. Anthocyanins in the Epacridaceae. Phytochemistry 13: 743–750.
Jarman, S. J., & R. K. Crowden. 1977. The occurrence of flavonol arabinosides in the Epacridaceae. Phytochemistry 16: 929–930.
Kavaljian, L. G. 1952. The floral morphology of *Clethra alnifolia* with some notes on *C. acuminata* and *C. arborea*. Bot. Gaz. 113: 392–413.
Leins, P. 1964. Entwicklungsgeschichtliche Studien an Ericales-Blüten. Bot. Jahrb. Syst. 83: 57–88.
Matthews, J. R., & E. M. Knox. 1926. The comparative morphology of the stamen in the Ericaceae. Trans. & Proc. Bot. Soc. Edinburgh 29: 243–281.
Moore, D. M., J. B. Harborne, & C. A. Williams. 1970. Chemotaxonomy, variation and geographical distribution of the Empetraceae. J. Linn. Soc., Bot. 63: 277–293.
Nowicke, J. W. 1966. Pollen morphology and classification of the Pyrolaceae and Monotropaceae. Ann. Missouri Bot. Gard. 53: 213–219.
Olson, A. R. 1980. Seed morphology of *Monotropa uniflora* L. (Ericaceae). Amer. J. Bot. 67: 968–974.
Palser, B. F. 1951–1961. Studies of floral morphology in the Ericales. I. Organography and vascular anatomy in the Andromedeae. Bot. Gaz. 112: 447–485, 1951. II. Megasporogenesis and megagametophyte development in the Andromedeae. Bot. Gaz. 114: 33–52, 1952. III. Organography and vascular anatomy in several species of the Arbuteae. Phytomorphology 4: 335–354, 1954. IV. Observations on three members of the Gaultherieae. Trans. Illinois Acad. Sci. 51: 24–34, 1958. V. Organography and vascular anatomy in several United States species of the Vacciniaceae. Bot. Gaz. 123: 79–111, 1961.
Palser, B. F., & Y. S. Murty, 1967. Studies of floral morphology in the Ericales. VIII. Organography and vascular anatomy in *Erica*. Bull. Torrey Bot. Club 94: 243–320.
Paterson, B. R. 1961. Studies of floral morphology in the Epacridaceae. Bot. Gaz. 122: 259–279.
Philipson, W. R., & M. N. 1968. Diverse nodal types in *Rhododendron*. J. Arnold Arbor. 49: 193–224.
Samuelsson, G. 1913. Studien über Entwicklungsgeschichte der Blüten einiger Bicornes-Typen. Ein Beitrag zur Kenntnis der systematischen Stellung der Diapensiaceen und Empetraceen. Svensk Bot. Tidskr. 7: 97–188.
Sleumer, H. 1941. Vaccinioideen-Studien. Bot. Jahrb. Syst. 71: 375–510.
Sleumer, H. 1967. Monographia Clethracearum. Bot. Jahrb. Syst. 87: 36–175.

Smith-White, S. 1955. Chromosome numbers and pollen types in the Epacridaceae. Austral. J. Bot. 3: 48–67.
Stevens, P. F. 1971. A classification of the Ericaceae: subfamilies and tribes. J. Linn. Soc., Bot. 64: 1–53.
Stushnoff, C., & B. F. Palser. 1969 (1970). Embryology of five *Vaccinium* taxa including diploid, tetraploid and hexaploid species or cultivars. Phytomorphology 19: 312–331.
Thomas, J. L. 1960. A monographic study of the Cyrillaceae. Contr. Gray Herb. 186: 1–114.
Thomas, J. L. 1961. The genera of the Cyrillaceae and Clethraceae of the southeastern United States. J. Arnold Arbor. 42: 96–106.
Thomas, J. L. 1961. *Schizocardia belizensis*: a species of *Purdiaea* (Cyrillaceae) from Central America. J. Arnold Arbor. 42: 110, 111.
Veillet-Bartoszevska, M. 1960. Embryogénie des Cléthracées. Développement de l'embryon chez le *Clethra alnifolia* L. Compt. Rend. Hebd. Séances Acad. Sci. 251: 2572–2574.
Veillet-Bartoszewska, M. 1961. Embryogénie des Épacridacées. Développement de l'embryon chez le *Dracophyllum secundum* R. Br. Compt. Rend. Hebd. Séances Acad. Sci. 253: 1000–1002.
Veillet-Bartoszewska, M. 1963. Recherches embryogéniques sur les Ericales. Comparaison avec les Primulales. Rev. Gén. Bot. 70: 141–230.
Venkata Rao, C. 1961. Pollen types in the Epacridaceae. J. Indian Bot. Soc. 40: 409–423.
Vijayaraghavan, M. R. 1969. Studies in the family Cyrillaceae. I. Development of male and female gametophytes in *Cliftonia monophylla* (Lam.) Britton ex Sarg. Bull. Torrey Bot. Club 96: 484–489.
Vijayaraghavan, M. R., & U. Dhar. 1978. Embryology of *Cyrilla* and *Cliftonia* (Cyrillaceae). Bot. Not. 131: 127–138.
Wallace, G. D. 1975. Studies of the Monotropoideae (Ericaceae): Taxonomy and distribution. Wasmann J. Biol. 33: 1–88.
Wallace, G. D. 1975. Interrelationships of the subfamilies of the Ericaceae and derivation of the Monotropoideae. Bot. Not. 128: 286–298.
Wallace, G. D. 1977. Studies of the Monotropoideae (Ericaceae). Floral nectaries: Anatomy and function in pollination ecology. Amer. J. Bot. 64: 199–206.
Watson, L. 1964. The taxonomic significance of certain anatomical observations on Ericaceae. The Ericoideae, *Calluna* and *Cassiope*. New Phytol. 63: 274–280.
Watson, L. 1965. The taxonomic significance of certain anatomical variations among Ericaceae. J. Linn. Soc., Bot. 59: 111–125.
Watson, L. 1967. Taxonomic implications of a comparative anatomical study of Epacridaceae. New Phytol. 66: 495–504.
Watson, L. 1976. Ericales revisited. Taxon 25: 269–271.
Watson, L., W. T. Williams, & G. N. Lance. 1967. A mixed-data approach to angiosperm taxonomy: the classification of Ericales. Proc. Linn. Soc. London 178: 25–35.
Wood, C. E. 1961. The genera of Ericaceae in the southeastern United States. J. Arnold Arbor. 42: 10–80.
Yang, B.-Y. 1952. Pollen grain morphology in the Ericaceae. Quart. J. Taiwan Mus. 5: 1–24.

Сладков, А. Н. 1953. О морфологических признаках пыльцевых зерен верескоцветных. (Ericaceae). Докл. АН СССР 92: 1065-1068.
Сладков, А. Н. 1954. Морфологическое описание пыльцы грушанковых,

вертляницевых, вересковых, брусничных и вороников европейской части Союза ССР. Труды Инст. Геогр. АН СССР, 61, Матер. по геоморф. и палеогеогр. СССР, 11:119–156.

SYNOPTICAL ARRANGEMENT OF THE FAMILIES OF ERICALES

1 Embryo normally developed, with 2 cotyledons; plants more or less woody, always chlorophyllous.

2 Pollen-grains borne in monads; petals, when present, distinct or only very shortly connate at the base.

3 Ovary superior; sepals 5 (–7); petals as many as the sepals; carpels 2–5; leaves alternate.

4 Ovules 1–3 per locule; intrastaminal nectary-disk present, fruit indehiscent, though often capsule-like in appearance; seed-coat none ....1. Cyrillaceae.

4 Ovules numerous; nectary-disk wanting, although the basal part of the ovary is sometimes nectariferous; fruit a loculicidal capsule; seed-coat present, consisting of a single layer of cells ....2. Clethraceae.

3 Ovary inferior; sepals 4; petals none; carpels 2; leaves opposite ....3. Grubbiaceae.

2 Pollen-grains mostly borne in tetrads, but if in monads or pseudomonads then the corolla evidently sympetalous; petals, when present, very often united to form a sympetalous corolla.

5 Perianth weakly or scarcely differentiated into calyx and corolla, consisting of 3–6 separate, distinct members arranged in 1 or 2 cycles; ovules solitary in each locule; many or all of the flowers generally unisexual .... 4. Empetraceae.

5 Perianth clearly differentiated into calyx and corolla, usually with more than 3 (typically 5) members in each of the 2 cycles, the corolla sympetalous or seldom polypetalous.

6 Stamens mostly of the same number as the corolla-lobes, often attached to the corolla-tube; anthers mostly bisporangiate, monothecal, and opening by a single longitudinal slit, not appendiculate, or seldom with a single appendage; leaves mostly with palmate or nearly parallel venation ....5. Epacridaceae.

6 Stamens mostly twice as many as the corolla-lobes (or petals), seldom of the same number or more than twice as many, usually free from the corolla-tube or attached only at its very base; anthers tetrasporangiate and dithecal, opening by a pair of pores at the apparent apex, or by more or less elongate slits extending downward from it, only rarely opening longitudinally for their whole length, often pro-

vided with 2 (or more) slender appendages; leaves pinnately veined, or seldom pinnipalmately plinerved 6. ERICACEAE.

1 Embryo very small and undifferentiated, without cotyledons; herbs or (*Chimaphila*, in the Pyrolaceae) half-shrubs, often without chlorophyll.

7 Plants usually with green leaves; anthers opening by seemingly apical pores; pollen-grains usually in tetrads; petals separate ....7. PYROLACEAE.

7 Plants without chlorophyll, the leaves reduced to mere scales; anthers generally opening by elongate, usually longitudinal slits, or rarely by seemingly apical pores; pollen-grains borne in monads; petals separate or united (or wanting) ....8. MONOTROPACEAE.

1. Family CYRILLACEAE Endlicher 1841 nom. conserv., the Cyrilla Family

Glabrous, evergreen or deciduous shrubs or small trees, tanniferous, producing proanthocyanins and ellagic acid or other tannins, but without iridoid substances, and not known to be cyanogenic; nodes unilacunar; solitary or clustered crystals of calcium oxalate commonly present in some of the cells of the parenchymatous tissues; vessel-

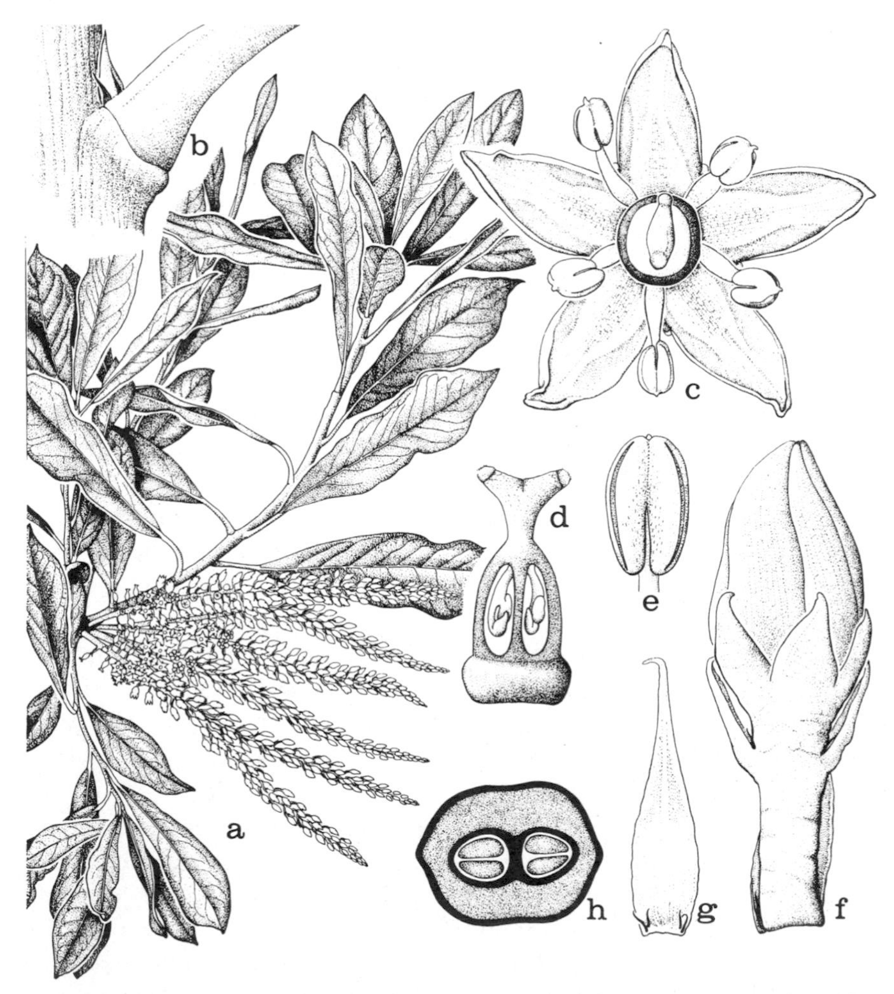

FIG. 4-34 Cyrillaceae. *Cyrilla racemiflora* L. a, habit, $\times\frac{1}{2}$; b, node, with stipules. $\times 6$; c, flower, from above, $\times 12$; d, pistil, in partial long-section, $\times 12$; e, anther, $\times 32$; f, flower bud, $\times 12$; g, bract $\times 12$; h, schematic cross-section of ovary, $\times 20$.

segments with scalariform perforations that commonly have 30–40 (–70) cross-bars; imperforate tracheary elements with small, evidently bordered pits; wood-rays heterocellular with short ends, the uniseriates few, the others mostly 2–4 (–8) cells wide; wood-parenchyma diffuse, or partly but scantily paratracheal. LEAVES alternate, simple, entire; epidermis more or less mucilaginous; stomates anomocytic; petiole with complex vascular anatomy, or the leaves sometimes sessile; stipules wanting. FLOWERS in terminal or axillary racemes, individually axillary to a caducous or persistent bract and often with 2 bractlets on the pedicel, perfect, regular, hypogynous; sepals 5 (–7), connate at the base, imbricate, persistent and often accrescent in fruit; petals as many as the sepals and alternate with them, white or anthocyanic, distinct or shortly connate at the base, imbricate or convolute; stamens twice as many as the petals, or (*Cyrilla*) as many as and alternate with the petals; filaments sometimes petaloid-flattened below; anthers versatile, or so deeply and narrowly sagittate as to appear so, tetrasporangiate and dithecal, not obviously inverted, opening by longitudinal slits or apical pores; pollen-grains in monads, binucleate, 3 (4)-colporate; intrastaminal nectary-disk present around the base of the ovary; gynoecium of 2–5 carpels united to form a compound, plurilocular ovary with axile placentas; style short or nearly suppressed; stigma lobed or lobeless; ovules 1–3 in each locule, pendulous from near the tip (sometimes from a pendulous, stalk-like placenta), anatropous, unitegmic, tenuinucellar; embryo-sac developing endosperm-haustoria at both ends. FRUIT indehiscent, dry and capsule-like, or samaroid or drupaceous; seed-coat wanting; embryo slender, straight, with small, slightly expanded cotyledons, embedded in the copious, fleshy endosperm. X = 10.

The family Cyrillaceae consists of 3 genera and 14 species, occurring in northern South America, Central America, and the West Indies, and on the coastal plain of southeastern United States. *Cyrilla* and *Cliftonia* reach the United States. They are monotypic; the remaining species belong to the genus *Purdiaea*.

According to Thomas (1961), the "Cyrillaceae have a well documented fossil record, with cyrillaceous pollen grains occurring as far back as the Upper Cretaceous. . . . Although these pollen grains cannot be definitely assigned to any of the living members of the Cyrillaceae, they are quite similar to the pollen of *Cyrilla* and *Cliftonia*. The best documented fossil record is in the Brandon Lignite of Vermont. The most abundant wood and second most abundant pollen in this fossil flora is that of *Cyrilla*. . . . The wood, and particularly the pollen, from this deposit is very similar to that of living material—remarkably so in view of the age of the deposit, estimated as late Upper Oligocene by Barghoorn." Fossil wood assigned to the Cyrillaceae is also known from Eocene deposits in Yellowstone National Park.

There has been a long-standing difference of opinion as to whether the Cyrillaceae should be associated with the Celastraceae or with the Ericaceae. Opinion in recent years has shifted strongly to the latter view, based partly on the accumulation of palynological and embryological evidence. *Clethra* and *Cyrilla* are also very similar in aspect.

2. Family CLETHRACEAE Klotzsch 1851 nom. conserv., the Clethra Family

Deciduous or evergreen shrubs or small trees, usually stellate-hairy, storing carbohydrate as fructose oligosaccharides, sometimes accumulating cobalt, tanniferous, producing proanthocyanins but not ellagic acid, and without iridoid compounds; parenchymatous tissues with abundant, often clustered crystals of calcium oxalate in some of the cells, and also with scattered secretory cells; nodes unilacunar; vessel-segments elongate, with scalariform perforations that commonly have 20–50 cross-bars; imperforate tracheary elements with distinctly bordered pits; wood-rays strongly heterocellular, mixed uniseriate and pluriseriate, the latter mostly up to 5 or 6 cells wide, with elongate ends; wood-parenchyma apotracheal, mainly diffuse or diffuse-in-aggregates. LEAVES alternate, simple, toothed or sometimes entire; stomates mostly paracytic, but sometimes anisocytic or anomocytic; stipules wanting. FLOWERS in terminal racemes or panicles, perfect, regular, hypogynous, pentamerous or rarely hexamerous; sepals 5 (6), connate below into a tube, imbricate; persistent; petals 5 (6), distinct or nearly so, imbricate; stamens 10 (12), bicyclic, free and distinct, or barely adnate to the base of the petals; anthers tetrasporangiate and dithecal, deeply sagittate, pointed or shortly tailed at the morphological tip, becoming deeply inverted during ontogeny so that the morphological base is apical, provided with a fibrous layer in the wall, opening by slit-like pores at the apparent tip; pollen-grains borne in monads, binucleate, tricolporate; disk wanting, but the basal part of the ovary often nectariferous; gynoecium of 3 carpels united to form a compound, trilocular ovary with a fluted, hollow, apically trilobed style; stigmas dry; ovules numerous on axile placentas, anatropous, unitegmic, tenuinucellar, with an integumentary tapetum on the upper two-thirds;

FIG. 4-35 Clethraceae. *Clethra alnifolia* L. a, habit, ×½; b, portion of inflorescence, ×1½; c, d, flower, ×4; e, stamens, ×10; f, opened fruit, ×4; g, schematic cross-section of ovary, ×10; h, pistil, ×10.

endosperm-development cellular; embryo-sac developing weak terminal haustoria at both ends. FRUIT a loculicidal capsule; seeds numerous, often winged, with a short, cylindrical, dicotyledonous embryo and well developed, fleshy, oily endosperm; seed-coat very thin, consisting of a single layer of cells. X = 8.

The family Clethraceae consists of the single genus *Clethra*, with about 65 species, occurring from tropical South America north to Mexico and southeastern United States, and also in tropical and subtropical southeastern Asia and the East Indies, with a single endemic species on the island of Madeira. *Schizocardia* A. C. Smith & Standley, described in 1932 as a new genus of Clethraceae from British Honduras, turns out to be a species of *Purdiaea*, in the related family Cyrillaceae.

Fossils presently attributed to the Clethraceae need careful reconsideration.

Giebel & Dickison consider on the basis of wood-anatomy that *Clethra* belongs in the Ericales, as a genus transitional toward the Theales. "The primitive position generally accorded the Clethraceae in the Ericales is strongly supported by wood anatomical characters. Similarly, the transitional nature of the Clethraceae, Cyrillaceae, and *Saurauia*, as links between the Theales and Ericales is apparent."

The Clethraceae and Cyrillaceae are presumably more primitive than other Ericales in having monadinous pollen. The polypetalous corolla of *Clethra* has also generally been regarded as primitive in comparison to the sympetalous corolla of the Ericaceae, but some recent authors have suggested that polypetaly in *Clethra* may be secondary. Regardless of whether these and other features of *Clethra* are considered to be more primitive or more advanced than comparable features in the Ericaceae, botanists are agreed that *Clethra* is allied to the Ericaceae and should be included in the Ericales.

3. Family GRUBBIACEAE Endlicher 1839 nom. conserv., the Grubbia Family

Ericoid shrubs with trilacunar nodes; plants tanniferous, and commonly with single or clustered crystals of calcium oxalate in some of the cells of the parenchymatous tissues; wood containing gummy deposits, vessel-segments elongate, with scalariform perforation-plates that have numerous cross-bars; imperforate tracheary elements elongate, with bordered pits, considered to be tracheids; wood-rays homocellular, mixed uniseriate and pluriseriate, the latter only 2–3 cells wide; wood-parenchyma very scanty, diffuse. LEAVES opposite, simple, narrow, leathery, strongly revolute-margined; stomates anomocytic; stipules wanting, but the leaf-bases connected by a transverse ridge. FLOWERS tiny, arranged in small, axillary, bracteate, compact, often woolly-villous cymes, perfect,

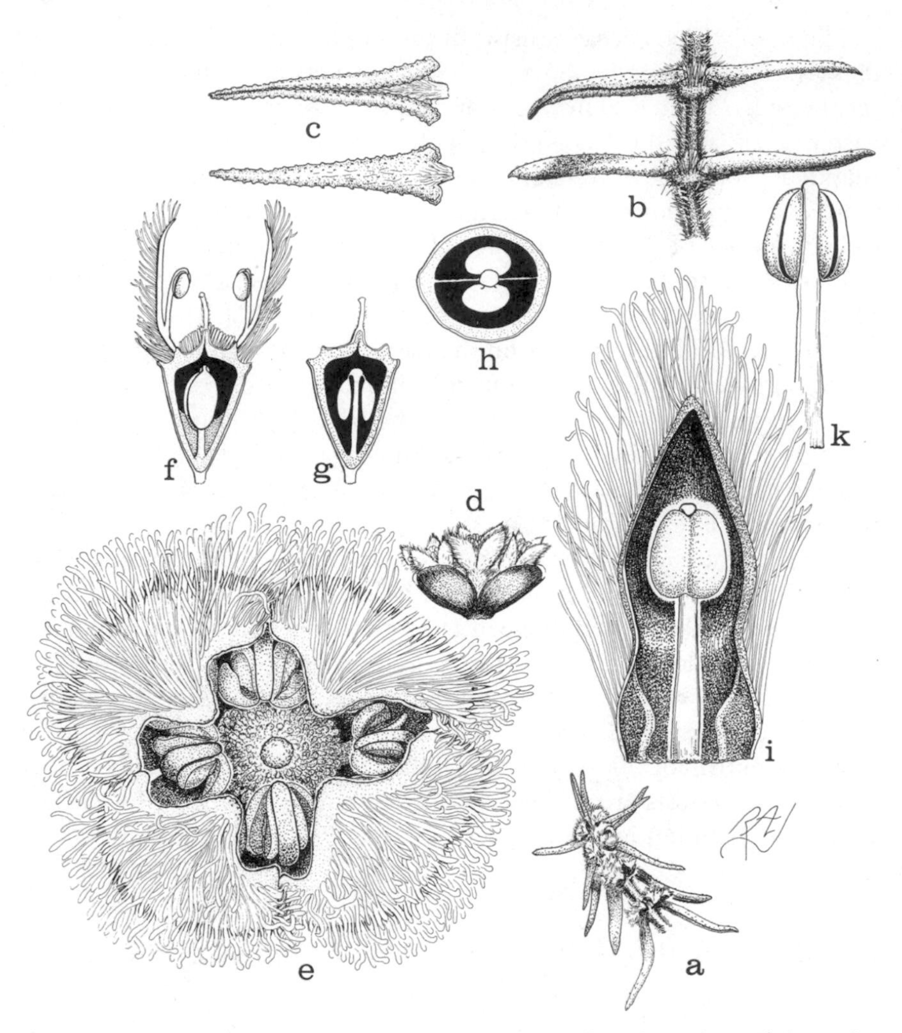

FIG. 4-36 Grubbiaceae. *Grubbia rosmarinifolia* Bergius. a, flowering twig, × 1½; b, portion of twig, with leaves, × 4; c, leaf, from above (below), and from beneath, × 5; d, flower cluster, with basal bracts, × 5; e, flower, from above, × 40, f, schematic long-section of flower, the partition in the lower part of the ovary stippled, × 18; g, schematic long-section of ovary, seen at right angles to the view in f, × 18; h, schematic cross-section of ovary, × 30, i, sepal, with attached stamen, × 40; k, abaxial view of stamen, × 40.

regular, epigynous, apetalous; sepals 4, small, valvate; stamens 8, the antesepalous ones a little longer and slightly adnate to the base of the sepals; anthers minute, dithecal, inverted and adnate their whole length to the distal part of the filament, thus appearing to be extrorse, each pollen-sac opening by a longitudinal slit, said (Carlquist) to have one sterile and one functional microsporangium; pollen-grains in monads, tricolporate; nectary-disk atop the ovary pubescent; gynoecium of 2 carpels united to form a compound, inferior overy capped by a simple style with a simple or 2-lobed stigma; ovary with an ephemeral partition, at first more or less bilocular, but becoming unilocular, with a free-standing columnar central placenta that bears a pair of pendulous ovules from its summit, these anatropous, unitegmic, tenuinucellar, with an integumentary tapetum; endosperm-development probably cellular, with haustoria at both ends. FRUIT a 1-seeded drupe, angular by compression, those of an inflorescence forming a compact cluster suggesting a small cupressaceous cone; seeds with thin testa and long, straight, cylindrical, dicotyledonous embryo embedded in the oily and proteinaceous endosperm.

The family Grubbiaceae consists of the single genus *Grubbia*, with 3 species limited to the Cape Province of South Africa. None of the species is familiar in cultivation or commerce.

The affinities of *Grubbia* are uncertain. It has most commonly been referred to the Santalales, probably because of its very santalalean gynoecium. In other respects, however, it is anomalous in that order. It is wholly autotrophic, and there is nothing to suggest a close alliance with any one family of the Santalales. Metcalfe & Chalk have noted a series of anatomical differences between *Grubbia* and the Santalaceae, which some authors have thought to be related.

The distinctly ericoid habit of *Grubbia* suggests a possible position in the Ericales, although here again there is no one family to which it seems closely allied. The ericoid option is supported to some extent by the sculpture of the pollen grains (fide Erdtman), and more especially by a series of embryological features that have been elucidated by Fagerlind (1947). Furthermore, the inverted position of the anthers is very suggestive of the Ericales, even though the anthers are not exactly like those of any other family. This feature has been generally overlooked, apparently even by Carlquist (1977). It may be noted that Takhtajan (1966) has accepted Fagerlind's suggestion that *Grubbia* belongs to the Ericales.

Carlquist (1977) considers the wood of *Grubbia* to be very primitive, similar to that of the Bruniaceae and Geissolomataceae, and suggests a generalized rosalean affinity for all of these groups. I believe that in so doing he dismisses the embryological features of *Grubbia* too lightly. It is true, as he suggests, that the embryology does not require that *Grubbia* be assigned to the Ericales—indeed the features reported by Fagerlind would be equally compatible with the Asteridae—but it does militate against an assignment to the Rosales.

A study of the secondary metabolites and cytochrome *c* of *Grubbia* might be instructive.

4. Family EMPETRACEAE S. F. Gray 1821 nom., conserv., the Crowberry Family

Small, evergreen shrubs or shrublets, strongly mycotrophic, provided with simple, unicellular hairs and stalked, unicellular or multicellular glands, producing ursolic acid (a triterpenoid) and proanthocyanins, but little or no ellagic acid, not cyanogenic, often with clustered crystals of calcium oxalate in some of the cells of the parenchymatous tissues; nodes unilacunar; vessel-segments with scalariform perforations that have up to 10 cross-bars (*Empetrum*) or with mostly simple perforations; imperforate tracheary elements with numerous small, bordered pits; wood-rays homocellular, 1–2 cells wide; wood-parenchyma very scanty. LEAVES alternate, crowded (sometimes in pseudowhorls), with a pulvinus at the base, small, firm and ericoid, deeply furrowed on the lower side, the groove more or less closed outside (or occluded by the pubescence) seeming to form a central longitudinal cavity, the leaf seeming to be rolled; inner walls of the epidermal cells tending to be mucilaginous; stomates anomocytic; stipules wanting. FLOWERS 1–3 in the axils of the leaves, or in few-flowered terminal heads, usually subtended by one or more bracts, small and inconspicuous, regular, hypogynous, ordinarily unisexual and the plants dioecious, seldom both sexes on the same plant, or some or all of the flowers perfect; perianth of 3–6 distinct tepals, these more or less evidently biseriate and differentiated into 2 or 3 imbricate sepals and as many imbricate petals, or 3–4 and all about alike in a single imbricate series (*Corema*); stamens 2 (*Ceratiola*) or 3 (4), alternate with the petals when the petals are differentiated from the sepals; filaments distinct; anthers small, tetrasporangiate and dithecal, without a fibrous layer in the wall, becoming inverted during ontogeny so that the morphological base is apical, opening by longitudinal slits, without appendages; pollen-grains borne in tetrads, binucleate, tricolporate; nectary-disk wanting; gynoecium of 2–9 carpels united to form a compound, plurilocular ovary with axile-basal placentas; style short, hollow (the cavity fluted in alignment with the locules), more or less deeply cleft; ovules solitary in each locule, erect, anatropous to almost campylotropous, unitegmic, tenuinucellar, with an integumentary tap-

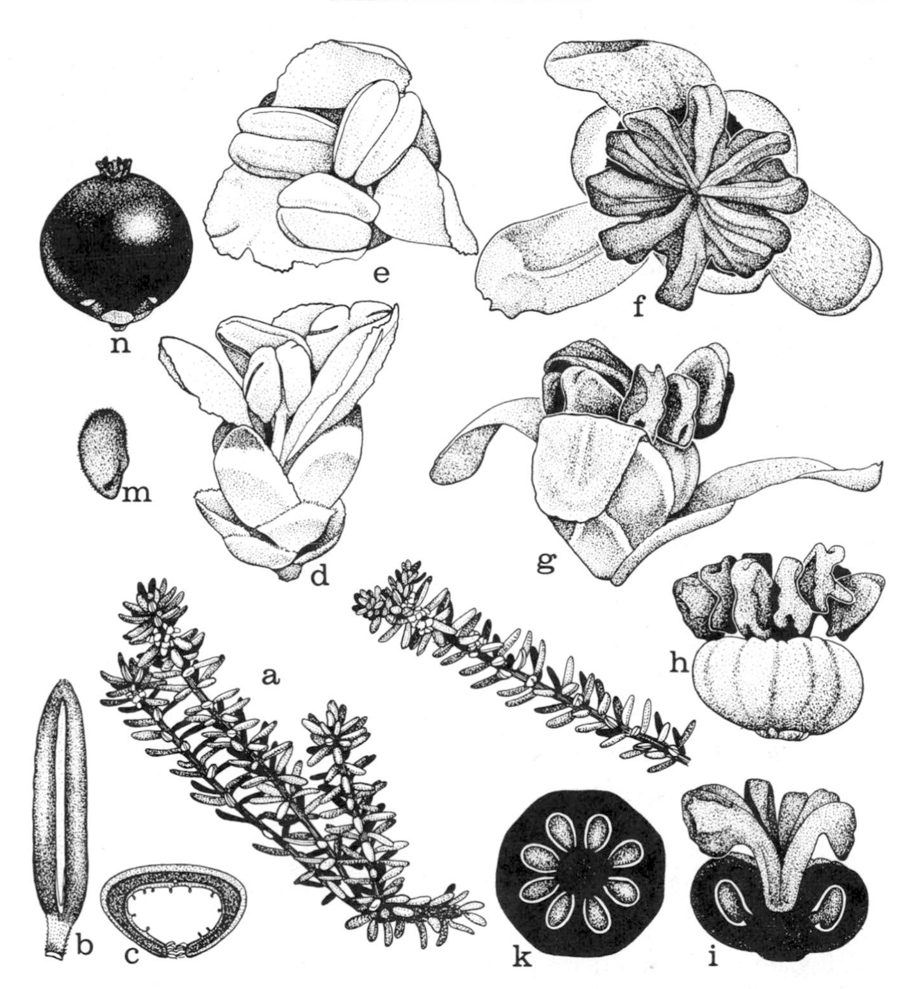

FIG. 4-37 Empetraceae. *Empetrum nigrum* L. a, habit, ×1; b, leaf, from beneath, ×6; c, schematic cross-section of leaf, ×12; d, e, staminate flower, ×12; f, g, pistillate flower, ×12; h, pistil, ×12; i, pistil, in partial long-section, ×12; k, schematic cross-section of ovary, ×12; m, seed, ×6; n, fruit, ×3.

etum; endosperm-development cellular, with haustoria at both ends. FRUIT a juicy or dry drupe with 2–9 stones; seeds with thin testa and straight, slender, elongate, axile, dicotyledonous embryo surrounded by the abundant, fleshy, oily endosperm. X = 13.

The family Empetraceae consists of 3 genera and probably only 5 species. *Empetrum* (2 or perhaps several species) is widespread in the colder parts of the Northern Hemisphere, and occurs also in southern South America and some nearby islands. *E. rubrum* Vahl is notable for its bicentric distribution, with one variety in southern South America, and two others in northeastern United States and southeastern Canada. *Ceratiola* (one sp.) is restricted to the coastal plain of southeastern United States. *Corema* (2) occurs near the coast from Newfoundland to New Jersey and on the Iberian peninsula and in the Azores. Crowberry (*Empetrum nigrum* L.) is gathered and eaten in Europe.

The Empetraceae have variously been referred to the Ericales, Sapindales, and Celastrales in past systems of classification. Aside from the habital resemblance, the Empetraceae resemble the Ericaceae in a series of embryological features and in the morphology of the pollen. Botanists who have paid particular attention to the Empetraceae in the present century, from Samuelsson (1913) and Hagerup (1946) to Maheshwari (1950, initial citations) are agreed that the affinity of the Empetraceae is with the Ericaceae. Erdtman (1952, initial citations) pointed out that the pollen is similar to that of the Ericaceae, especially to that of *Ledum* and other members of the subfamily Rhododendroideae. Even the shared susceptibility to particular rusts links the Empetraceae and Ericaceae (Saville 1979, in general citations). In the 12th (1964) edition of the Engler Syllabus the Empetraceae were moved to a position in the Ericales. Among modern authors the principal holdout has been Hutchinson, who in 1973 still maintained the Empetraceae in the Celastrales.

5. Family EPACRIDACEAE R. Brown 1810 nom. conserv., the Epacris Family

Shrubs (sometimes scandent) or small trees or seldom trailing shrublets, often of ericoid aspect, strongly mycotrophic, glabrous or with scattered, mostly unicellular hairs, producing ursolic acid and sometimes methyl salicylate, tanniferous, with proanthocyanins but not ellagic acid, only seldom cyanogenic; flavonols include the otherwise rare foeniculin, a quercetin 3-arabinoside; parenchymatous tissues often with solitary or clustered crystals in some of the cells, and also commonly with scattered secretory (tanniferous) idioblasts; nodes multilacunar; vessel-segments with scalariform and/or simple perforations; imperforate tracheary elements with evidently bordered pits; wood-rays heterocellular, up to about 10 cells wide, with short ends; wood-parenchyma diffuse or in slender bands. LEAVES xeromorphic, alternate or rarely opposite, often crowded, simple, commonly small and ericoid, sessile or seldom petiolate, sometimes sheathing at the base, entire or seldom toothed, with

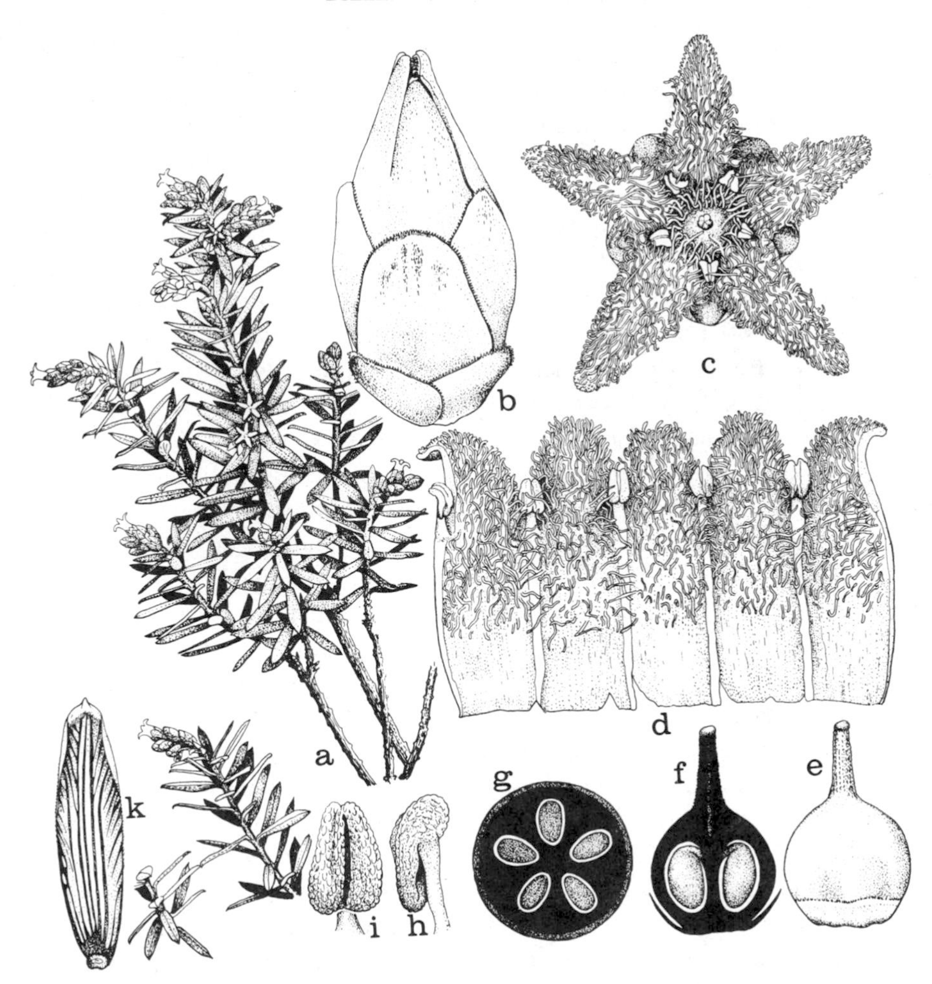

FIG. 4-38 Epacridaceae. *Styphelia suaveolens* (Hook. f.) Warb. a, habit, ×1; b, flower bud, ×12; c, flower, from above, ×12; d, corolla, opened out, with attached stamens, ×12; e, pistil, ×12; f, schematic long-section of pistil, ×12; g, schematic cross-section of ovary, ×16; h, i, anther, ×24; k, lower side of leaf, ×4.

palmate or subparallel venation (pinnate in *Prionotes* and *Wittsteinia*), in *Dracophyllum* and *Richea* parallel-veined, sessile by a broad base and very monocotyledonous in aspect; conspicuous groups of fibers typically associated with the veins; epidermis not mucilaginous; stomates anomocytic, or sometimes paracytic or tetracytic or encyclocytic; stipules wanting. FLOWERS mostly in bracteate racemes or solitary in the axils, bibracteolate, and sometimes with additional sepaloid bractlets on the pedicel, commonly perfect, regular, hypogynous or rarely (*Wittsteinia*) epigynous, most commonly pentamerous throughout; sepals (4) 5, distinct, persistent; corolla commonly with anthocyanidin 3-galactosides and 3-arabinosides, sympetalous, tubular, with (4) 5 imbricate or valvate lobes; stamens unicyclic, as many as the corolla-lobes and alternate with them (rarely only 2); filaments free from the corolla or more often attached to the corolla-tube, sometimes well up in the tube; anthers bisporangiate and usually monothecal, or seldom (*Prionotes* and *Wittsteinia*) tetrasporangiate and dithecal, becoming inverted during ontogeny so that the morphological base is apical, without a fibrous layer in the wall but often with a thick-walled epidermis (exothecium), opening by longitudinal slits, without appendages, or seldom with a single slender appendage; pollen-grains diverse in developmental pattern, borne in tetrads or sometimes in pseudomonads, binucleate or trinucleate, 3 (4)-colporate; small nectary-scales (presumably staminodial) generally alternating with the stamens, or the nectaries sometimes connate to form an intrastaminal ring, rarely (*Sprengelia*) the nectaries wanting; gynoecium of 4–5 (10) carpels united to form a compound, plurilocular ovary with axile placentas, or a technically unilocular ovary with deeply intruded parietal placentas, sometimes some of the locules empty and only 1–4 carpels fertile; style slender, generally hollow, the cavity fluted in alignment with the locules; stigma wet, capitate; ovules (1) several or numerous on each placenta anatropous, unitegmic, tenuinucellar, with an integumentary tapetum on the lower two-thirds; endosperm-development cellular, with haustoria at one or the other or usually both ends. FRUIT a loculicidal capsule with 1-many seeds, or a fleshy or sometimes dry drupe with 1–5 stones; seeds with a very thin testa, consisting of a single layer of cells, and with a straight, cylindrical, dicotyledonous embryo embedded in the copious, oily endosperm. X = 4–14. (Prionotaceae, Stypheliaceae)

The family Epacridaceae as here defined consists of about 30 genera and 400 species, found mostly in Australia, New Zealand, and the East Indies, but reaching also the Philippine Islands and the mainland of southeastern Asia, and extending even to the Hawaiian Islands and Patagonia. The Epacridaceae may be considered the Australian correlatives of the Ericaceae. The largest genera are *Leucopogon* (140), *Styphelia* (40), *Epacris* (40), and *Dracophyllum* (35).

The Epacridaceae and Ericaceae have followed somewhat different paths of advancement from their common ancestry. The Epacridaceae are more advanced than the Ericaceae in their unicyclic, monothecal stamens attached to the corolla-tube, but the Ericaceae are more advanced in their poricidal, often variously ornamented anthers.

Prionotes (including *Lebetanthus*) and *Wittsteinia*, here included in the Epacridaceae, are transitional toward the Ericaceae and have sometimes been included in that family or taken as a separate family Prionotaceae. They have a single cycle of stamens, and the anthers open longitudinally, as in other Epacridaceae. On the other hand, the filaments are attached directly to the receptacle, and the anthers are tetrasporangiate and dithecal, as in the Ericaceae. *Sprengelia* and several other genera of Epacridaceae also have the stamens attached directly to the receptacle, however. The anthers of *Sprengelia*, although bisporangiate, are dithecal and open by two longitudinal slits, in contrast to other Epacridaceae, which are bisporangiate and monothecal and open by a single longitudinal slit. The pinnate venation of the leaves of *Prionotes* and *Wittsteinia* is not matched by other Epacridaceae; in this respect these genera resemble the Ericaceae.

6. Family ERICACEAE A. L. de Jussieu 1789 nom. conserv., the Heath Family

Evergreen or sometimes deciduous shrubs or less often half-shrubs, small trees, or lianas, rarely almost herbaceous, more or less strongly mycotrophic and often growing in acid soils, or not infrequently epiphytic, with various sorts of indument, producing iridoid compounds, phenol heterosides (notably arbutin), triterpenoid compounds (including ursolic acid), proanthocyanins, 5-methoxy-flavonols and often 3-galactoside and 3-arabinoside, often toxic diterpenes (such as andromedotoxin), and occasionally ellagic acid, only seldom cyanogenic; parenchymatous tissues with clustered or less often solitary crystals of calcium oxalate in some of the cells, and with scattered secretory idioblasts; nodes unilacunar or sometimes trilacunar; vessel-segments with scalariform perforations, or some (seldom all) of them with simple perforations, imperforate tracheary elements with large, bordered pits, sometimes with a persistent protoplast; wood-rays heterocellular, sometimes all uniseriate, but more often mixed uniseriate and pluriseriate, the latter with short or less commonly elongate ends, up to 10 cells wide; wood parenchyma mostly scanty or wanting, in a few genera fairly abundant and diffuse. LEAVES alternate or sometimes opposite or even whorled, simple, often small and firm, ("ericoid"); epidermis, or the

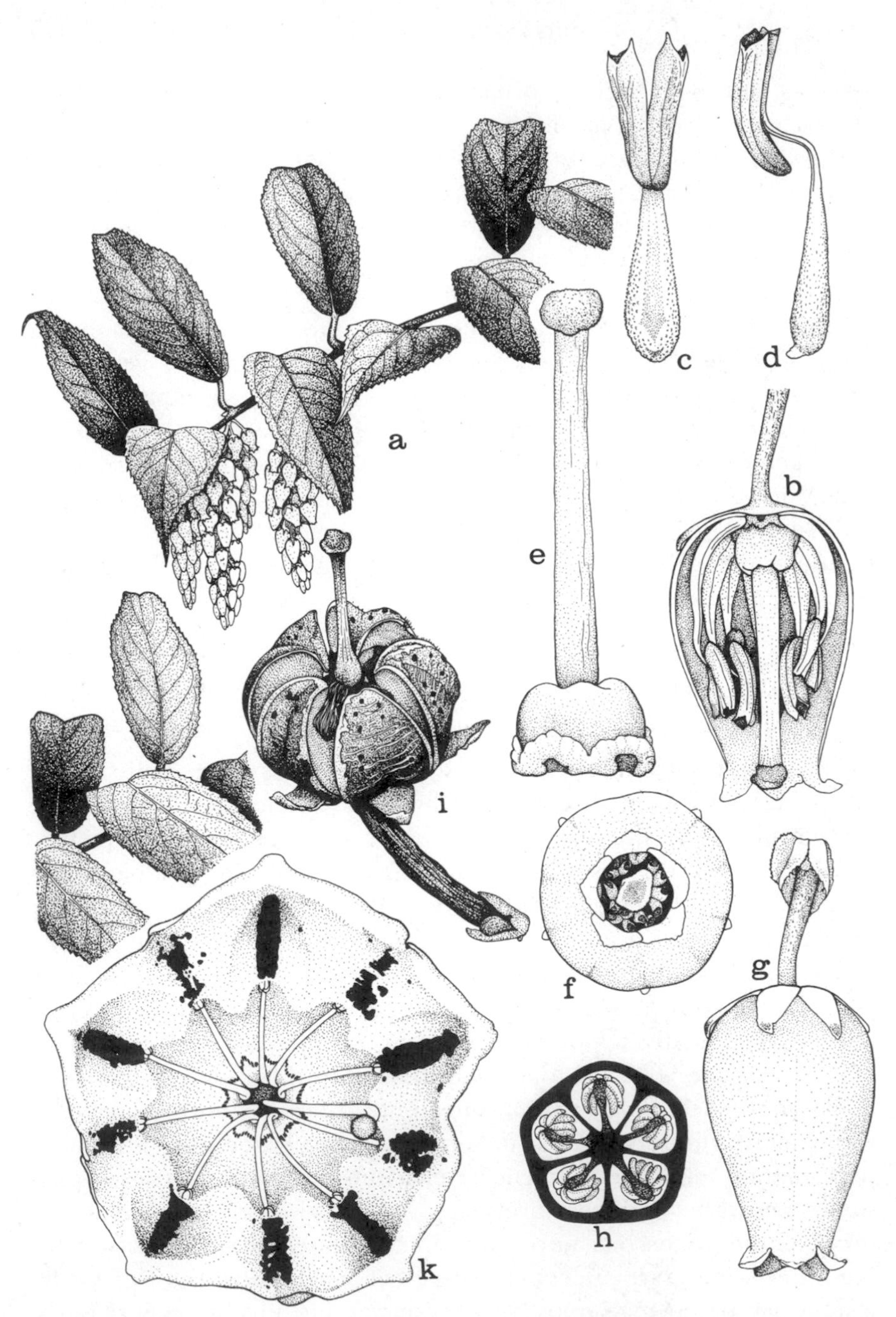

FIG. 4-39 Ericaceae. a–i, *Leucothoe axillaris* (Lam.) D. Don a, habit, ×½; b, flower, in partial long-section, ×6; c, d, stamen, ×12; e, pistil, ×12; f, g, flower, from above and from the side, ×6; h, schematic cross-section of ovary, ×12; i, fruit, ×6. k, *Kalmia latifolia* L., flower, from above, ×3.

inner wall of the epidermal cells, often mucilaginous; stomates mostly anomocytic, except for the Vaccinioideae, in which they are mostly paracytic; stipules wanting. FLOWERS mostly in bracteate racemes, sometimes solitary and terminal or axillary, usually subtended by 2 bracteoles, perfect or seldom functionally unisexual, regular or slightly irregular, hypogynous to sometimes semi-epigynous or (chiefly in Vaccinioideae) epigynous; sepals 3–7, most often 5, less often 4, valvate or imbricate, persistent; petals 3–7, most often 5, less often 4, convolute or imbricate, generally connate at least below to form a short to conspicuous corolla-tube, the corolla often urceolate with short lobes, but also often of other shapes, in a few genera the petals distinct *ab initio* or (*Ledum*) arising from a continuous circular primordium but becoming distinct during ontogeny; stamens mostly bicyclic, twice as many as the corolla-lobes, seldom more numerous (up to 20) or some or all of the members of the antepetalous cycle missing; filaments free from the corolla, or attached at its very base, rarely adnate for much of their length, distinct or rarely connate; anthers dorsifixed or seldom basifixed, becoming inverted during ontogeny so that the morphological base is apical, tetrasporangiate and dithecal, usually without a fibrous layer in the wall, opening by pores at the apparent apex, or by more or less elongate slits extending downward from the apex, rarely opening longitudinally for their whole length, the tip sometimes drawn out into a pair of slender tubules (these sometimes connate), and the body often with two (or more) evident, slender appendages (horns); pollen-grains borne in tetrahedral tetrads or seldom (e.g., *Enkianthus*) in monads, binucleate or often trinucleate, commonly tricolporate, sometimes producing viscin threads on the distal polar surface; intrastaminal nectary-disk present around the base of (and often attached to) the ovary, or covering the top of an inferior ovary; gynoecium of 2–10, most often 5, less often 4 carpels united to form a compound, plurilocular ovary, seldom with twice as many locules as carpels; rarely, as in *Scyphogyne* and related South African genera, the ovary unilocular, with a single pendulous ovule; style slender, hollow, the cavity fluted in alignment with the locules; stigma wet, obtuse or capitate or shallowly lobed; placentation axile, or the partitions sometimes imperfect toward the top of the ovary, so that the placentation is axile below and parietal (on intruded placentas) above; ovules (1–) several or more often numerous on each placenta, anatropous or hemitropous to more or less campylotropous, unitegmic, tenuinucellar, with an integuentary tapetum, at least at the chalazal end; endosperm-development cellular, with haustoria at both ends. FRUIT a septicidal or loculicidal capsule, or (especially the Vaccinioideae) a berry or (Arbutoideae) a drupe, seldom a small nut; seeds commonly more or less numerous, mostly small, sometimes winged, with a very thin testa consisting of a single layer of cells; embryo straight, dicotyledonous, normally developed but often very short, cylindric or sometimes spatulate, embedded in the well developed, oily and proteinaceous, fleshy-firm endosperm. X = 8–23, most often 12 or 13. (Vacciniaceae)

The family Ericaceae as here defined consists of about 125 genera and 3500 species, widespread in temperate, cool, and subtropical regions and in tropical mountains. The ancestry of the group evidently lies in tropical latitudes, but perhaps at high elevations rather than in tropical forests. The largest genera are *Rhododendron* (850) and *Erica* (600), followed by *Vaccinium* (450), *Gaultheria* (150), and *Cavendishia* (150). *Erica* is especially well developed in South Africa, and *Rhododendron* in the Himalayan and Malesian regions. Species of *Arctostaphylos* (manzanita) are well known constituents of the chaparral vegetation in the southwestern United States. The family contains many familiar ornamentals, including species of *Arbutus* (madrone), *Calluna* (heather), *Epigaea* (trailing arbutus), *Erica* (heath), *Kalmia* (mountain laurel), *Oxydendrum* (sourwood), *Pieris*, and *Rhododendron*. Species of *Vaccinium* are cultivated for their edible fruits. *Gaultheria procumbens* L. is the original source of oil of wintergreen.

Organization of the Ericaceae into subfamilies is presently in dispute. It has been customary to recognize 4 subfamilies, the Ericoideae, Rhododendroideae, Arbutoideae, and Vaccinioideae, but some more recent authors would recognize an additional small subfamily Epigaeoideae, and submerge the Arbutoideae in the Vaccinioideae. I take no position on this matter.

The Vaccinioideae have sometimes been taken as a separate family, characterized by their consistently inferior ovary and usually fleshy fruit, in contrast to the nearly always superior ovary and usually capsular fruit of the other groups. Whether to split or lump in this case is purely a matter of taste. I have followed the usual historical practice in accepting only subfamilial status for the Vaccinioideae.

Although individual families of angiosperms are commonly more or less heterobathmic (to use Takhtajan's term), the Ericaceae present a particularly noteworthy example of this mixture of primitive and advanced features. The sympetalous corolla and the unitegmic, tenuinucellar ovules are clearly advanced, along the lines of general evolutionary tendencies in the angiosperms as a whole. The inverted, poricidal, often appendiculate stamens are also clearly advanced, but along a track separate from that of most other angiosperms. On the other hand, the number of stamens (usually twice as many as the corolla-lobes) is only moderately advanced, and their attachment directly to the receptacle instead of to the corolla-tube is a primitive feature. The wood is also relatively primitive; the vessels commonly have scalariform perforation-plates, and the rays are heterocellular, often with elongate ends.

Fossil seeds attributed to *Rhododendron* and *Vaccinium* have been found in Paleocene deposits in England. Pollen representing either the Ericaceae, Epacridaceae, or Clethraceae dates from the Maestrichtian in the Northern Hemisphere, and from the Eocene in Australia.

7. Family PYROLACEAE Dumortier 1829 nom. conserv., the Shinleaf Family

Strongly mycotrophic perennial herbs or half-shrubs from creeping rhizomes (these very slender in *Moneses*), mostly evergreen (but not in *Moneses* and the leafless forms of *Pyrola*), commonly growing in acid soils, storing carbohydrate as fructose oligosaccharides, producing chimaphilin (2, 7-dimethyl–1,4-naphthaquinone), phenol heterosides (notably arbutin), triterpenoid compounds (including ursolic acid), iridoid compounds, and proanthocyanins, but not ellagic acid; nodes unilacunar; secondary tissues progressively reduced, from *Chimaphila* through *Pyrola* to *Moneses*. LEAVES alternate to sometimes opposite or subverticillate, all basal or nearly so except in *Chimaphila*, simple, entire or toothed, in some individuals of certain species of *Pyrola* much-reduced or wholly suppressed; stomates anomocytic; petiole with a single trough-shaped vascular strand; stipules wanting. FLOWERS in bracteate racemes (*Pyrola* and *Orthilia*), umbels or corymbs (*Chimaphila*), or solitary and terminal (*Moneses*), without bracteoles, perfect, regular, hypogynous; sepals (4) 5, distinct or shortly connate at the base, persistent; petals (4) 5, distinct but originating from an annular primordium, imbricate; stamens twice as many as the petals, distinct, attached directly to the receptacle; anthers tetrasporangiate and dithecal, becoming inverted during ontogeny so that the morphological base is apical, without a fibrous layer in the wall, opening by pores at the apparent apex, the thecae sometimes produced into short tubes; pollen-grains tetradinous (except in *Orthilia*), binucleate, commonly tricolporate; intrastaminal nectary-disk present or absent; gynoecium of 5 carpels united to form

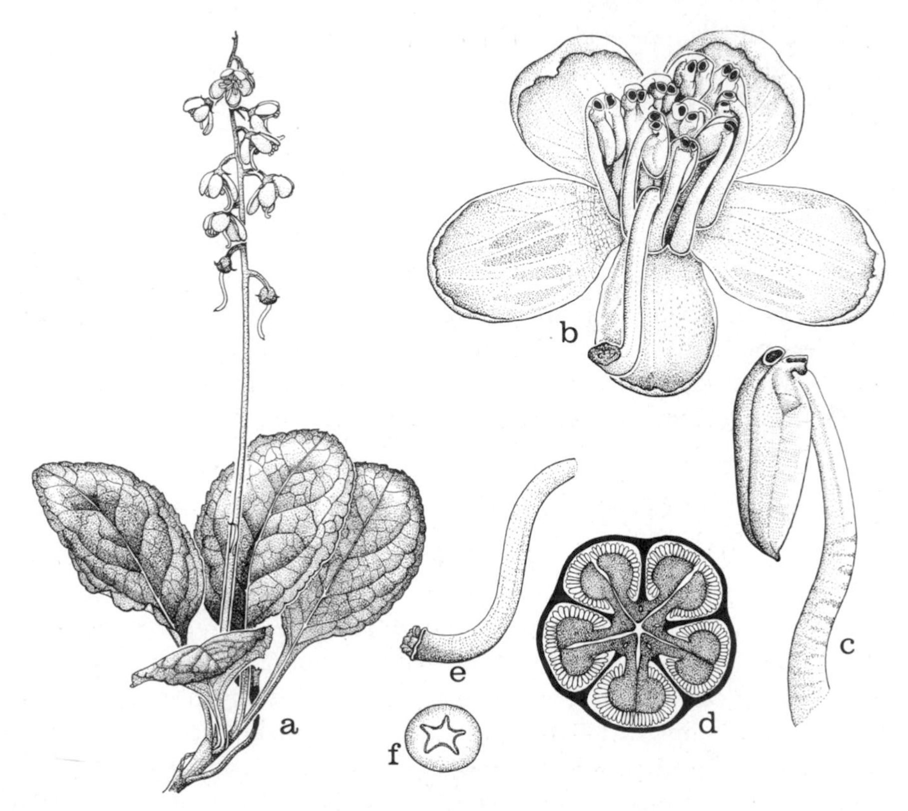

FIG. 4-40 Pyrolaceae. *Pyrola elliptica* Nutt. a, habit, $\times\frac{1}{2}$; b, flower, $\times 4$; c, stamen, $\times 8$; d, schematic cross-section of ovary, $\times 8$; e, style, $\times 4$; f, cross-section of style, $\times 8$.

a compound, often externally furrowed or lobed, imperfectly 5-locular ovary, the intruded parietal placentas not joined in the center; style short or elongate, often declined, hollow, the cavity fluted in alignment with the locules; stigma wet, capitate or peltate, more or less distinctly 5-lobed; ovules numerous, anatropous, unitegmic, tenuinucellar; endosperm-development cellular, with short haustoria at both ends. FRUIT a loculicidal capsule; seeds numerous and small, the testa very thin, only one cell thick; embryo tiny, undifferentiated, embedded in the copious endosperm. X = 8, 11, 13, 16, 19, 23.

The family Pyrolaceae as here defined consists of 4 genera, *Chimaphila*, *Moneses*, *Orthilia*, and *Pyrola*, about 45 species in all. *Orthilia* is often treated as a section of *Pyrola*, and sometimes *Moneses* is included in *Pyrola* as well. Both of these segregate genera are monotypic. *Chimaphila* has about 4 or 5 species. The remainder of the species belong to *Pyrola*. The Pyrolaceae are confined to the Northern Hemisphere, and are most abundant and diversified in temperate and boreal regions.

8. Family MONOTROPACEAE Nuttall 1818 nom. conserv., the Indian Pipe Family

Strongly mycotrophic perennial herbs, often fleshy, variously white to pink, red, purple, yellow, or brown in color, but not green, lacking chlorophyll, storing carbohydrate as fructose oligosaccharides, tanniferous, but without ellagic acid, producing iridoid substances, phenol heterosides (notably arbutin and monotropitoside), and sometimes methyl salicylate, but not cyanogenic; nodes unilacunar; xylem highly reduced, the tracheary elements prosenchymatous or none. LEAVES alternate, much-reduced, scale-like; stomates few or none. FLOWERS solitary or in a short or elongate, bracteate raceme (or a raceme of cymules), usually without bracteoles, perfect, regular, hypogynous; sepals (2–) 4–5 (6), distinct or sometimes shortly connate, imbricate, or rarely obsolete; petals (3) 4–5 (6), mostly of about the same color as the stem, distinct or connate into a lobed tube, imbricate or more or less distinctly convolute, or rarely (*Allotropa*) wanting; stamens 6–12, biseriate, commonly twice as many as the sepals or petals, distinct or shortly connate at the base, attached directly to the receptacle; anthers tetrasporangiate and dithecal, often becoming inverted during ontogeny so that the morphological base is apical, without a fibrous layer in the wall, opening lengthwise or sometimes by terminal pores or by slits across the broad thecal summits, sometimes spurred at the morphological base; pollen-grains borne in monads but often forming a coherent mass, binucleate, (2) 3–4 (5)-colporate or -porate; nectaries alternating with the stamens, sometimes united to form a lobed intrastaminal disk, sometimes reduced to low ridges on the lower part of the ovary-wall; gynoecium of (4) 5 (6) carpels united to form a compound ovary that is plurilocular, with axile placentas, or unilocular with strongly intruded, parietal placentas; style mostly short, hollow, the cavity fluted in alignment with the locules; stigma capitate or peltate, often lobed; ovules numerous, anatropous, unitegmic, tenuinucellar; endosperm-development cellular, at least sometimes (*Sarcodes*) with haustoria at both ends. FRUIT a loculicidal or septicidal capsule, or baccate and indehiscent; seeds numerous and very small, with a very thin testa

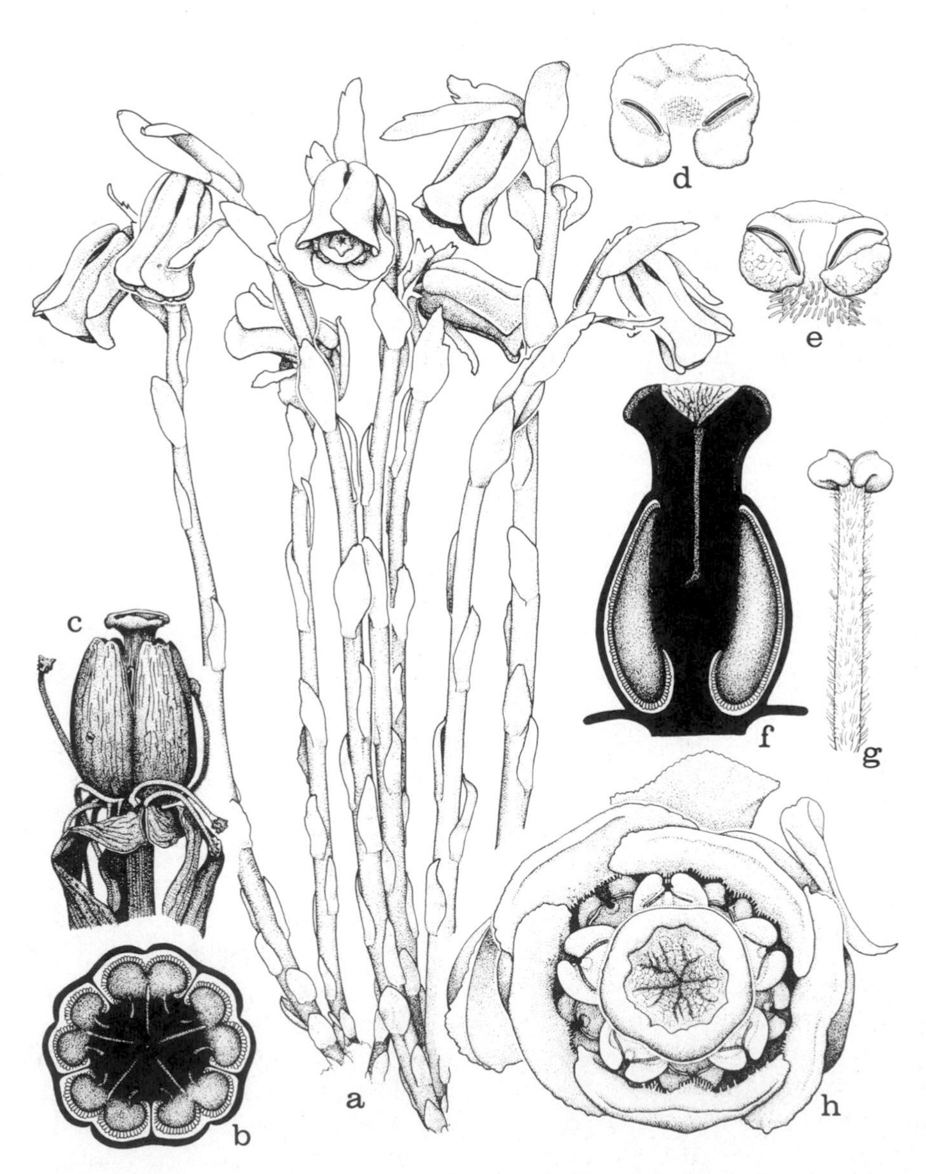

FIG. 4-41 Monotropaceae. *Monotropa uniflora* L. a, habit, ×1; b, schematic cross-section of ovary, ×4; c, opened fruit, ×2; d, e, anther, ×8; f, schematic long-section of pistil, ×4; g, stamen, ×4; h, flower, ×4.

composed of a single layer of cells; embryo tiny, undifferentiated, sometimes consisting of only 2–4 cells, embedded in the well developed to very scanty endosperm. X = 8. (Hypopithydaceae, Semicirculaceae)

The family Monotropaceae consists of 10 genera and 12 species, chiefly of temperate and cool regions in the Northern Hemisphere, but extending south to Colombia in the New World and the Malay Peninsula in the Old. The close relationship of the Monotropaceae to the Pyrolaceae and Ericaceae is not in question. Copeland (1947) considered that the Monotropaceae and Pyrolaceae (which he included as parts of the Ericaceae "represent parallel lines of descent from a common origin, presumably the tribe Andromedeae."

The Monotropaceae represent the extreme of mycotrophic specialization within the Ericales, having become dependent on their associated fungus for food as well as for water and minerals. The fungus commonly forms a mycorhizal association not only with the monotropid but also with the roots of forest trees. Thus the fungus permits the monotropid to parasitize the trees indirectly. The value to the fungus of such an association with the monotropid is obscure.

11. Order DIAPENSIALES Engler & Gilg 1924

The order consists of the single family Diapensiaceae.

SELECTED REFERENCES

Baldwin, J. T. 1939. Chromosomes of the Diapensiaceae. A cytological approach to a phylogenetic problem. J. Heredity 30: 169–171.

Palser, B. F. 1963. Studies of floral morphology in the Ericales. VI. The Diapensiaceae. Bot. Gaz. 124: 200–219.

Wood, C. E., & R. B. Channell. 1959. The Empetraceae and Diapensiaceae of the southeastern United States. J. Arnold Arbor. 40: 161–171.

Yamazaki, T. 1966. The embryology of *Shortia uniflora* with a brief review of the systematic position of the Diapensiaceae. J. Jap. Bot. 41: 245–251.

1. Family DIAPENSIACEAE Lindley 1836 nom. conserv., the Diapensia Family

Evergreen perennial herbs or half-shrubs, strongly mycotrophic, accumulating aluminum, tanniferous, at least sometimes producing both proanthocyanins and ellagic acid, but not cyanogenic, sometimes with crystals (especially clustered crystals) of calcium oxalate in some of the cells of the parenchymatous tissues; nodes unilacunar; vessel-segments commonly with simple perforations; imperforate tracheary elements with bordered pits. LEAVES alternate, simple; stomates mostly anomocytic, but sometimes some of them anisocytic; stipules wanting. FLOWERS solitary or in compact racemes, perfect, regular, hypogynous, white or anthocyanic; sepals 5, each with (1–) 3 (–7) vascular traces, imbricate, distinct or connate to form a lobed tube; corolla sympetalous, with 5 imbricate or convolute lobes, or (*Galax*) the 5 petals virtually distinct; petals or corolla-lobes each with a single vascular trace; androecium most commonly of 5 functional stamens attached to the corolla-tube alternate with the lobes, and 5 staminodes opposite the lobes, the stamens and staminodes sometimes connivent or connate to form a tube, in *Galax* the stamen-staminode tube adnate to the petals at the base and falling with the petals, in *Diapensia* and *Pyxidanthera* the staminodes wanting; anthers tetrasporangiate and dithecal or (*Galax*) bisporangiate and monothecal, provided with a fibrous layer in the wall, sometimes awned at the base, opening by longitudinal or (*Pyxidanthera*) transverse slits; pollen-grains in monads, trinucleate, tricolporate or tricolpate; nectary a poorly developed ring on the base of the ovary, or

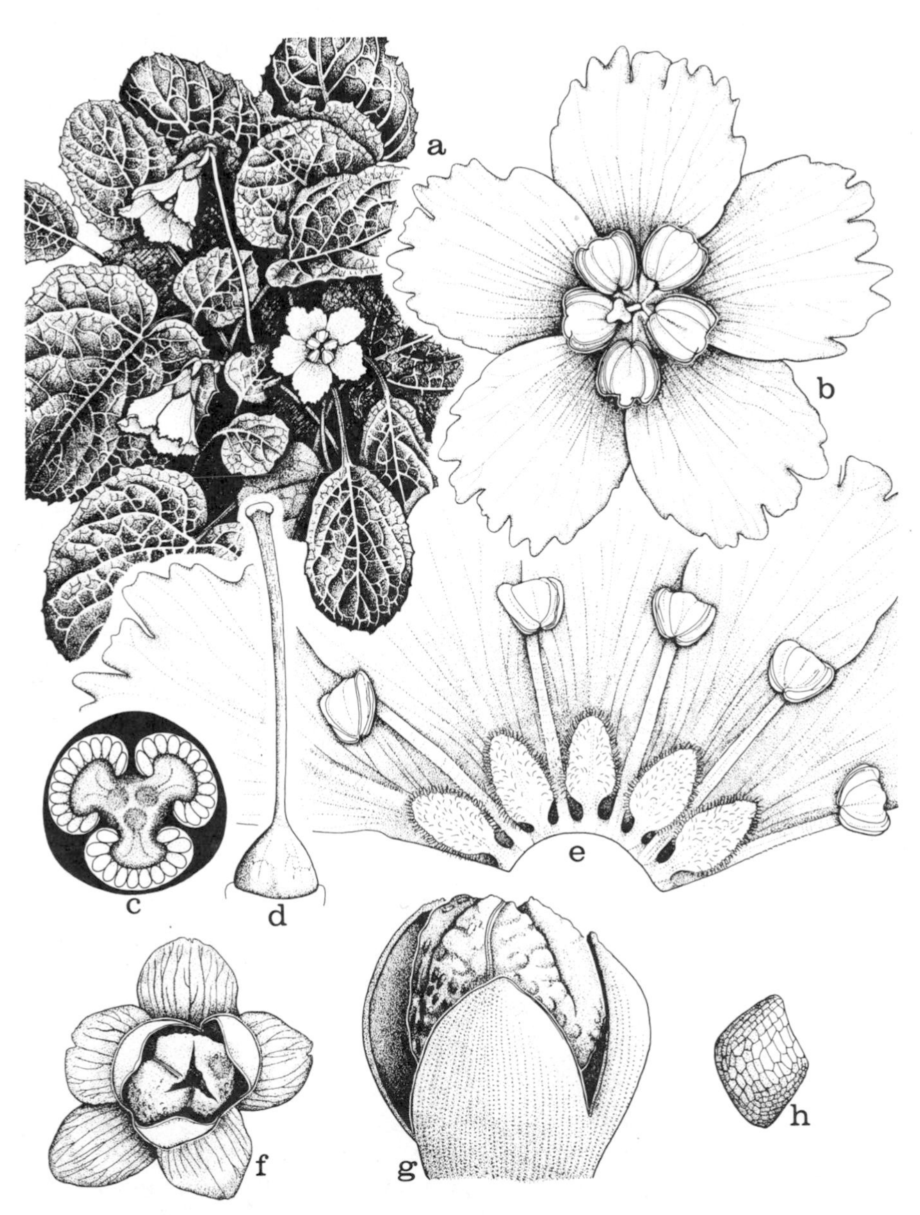

FIG. 4-42 Diapensiaceae. *Shortia galacifolia* T. & G. a, habit, $\times\frac{1}{2}$; b, flower, $\times 3$; c, schematic cross-section of ovary, $\times 9$; d, pistil, $\times 3$; e, corolla, opened out, showing stamens and staminodes; f, mature fruit, after dehiscence, with persistent calyx, $\times 3$; g, opened fruit, with persistent placenta, $\times 3$; h, seed, $\times 20$.

wanting; gynoecium of 3 carpels united to form a compound, trilocular ovary capped by a hollow, internally fluted style ending in a trilobed stigma; ovules several or numerous on axile placentas, anatropous to hemitropous or campylotropous, unitegmic, tenuinucellar, often with an integumentary tapetum; endosperm-development cellular, without haustoria. FRUIT a 3-valved loculicidal capsule; seeds small, with copious, fleshy endosperm around a straight or slightly curved, dicotyledonous embryo. X = 6.

The family Diapensiaceae consists of 6 genera and 18 species, occurring in Arctic and North Temperate regions, and south as far as the Himalaya Mts. *Diapensia lapponica* L. is a well known circumpolar species of the tundra and higher mountains. *Shortia galacifolia* Torrey & Gray, Oconee Bells, and some other species of *Shortia* are sometimes cultivated for ornament.

The Diapensiaceae have often been included in the Ericales, but they differ from typical members of that order in a rather long list of embryological and androecial characters. There is a single set of functional stamens, attached well up in the corolla-tube, with or without an alternating set of staminodes. The anthers have a fibrous layer in the wall, they mostly lack appendages, and they mostly open by longitudinal slits, never by terminal pores. The pollen-grains are borne singly. The characteristic nectary-disk about the base of the ovary, seen in most Ericales, is almost or quite wanting in the Diapensiaceae. The Diapensiaceae lack endosperm-haustoria, whereas the Ericales have typically endosperm-haustoria at both ends of the embryo-sac. These and other differences have led modern students of the group, such as Palser, to exclude the Diapensiaceae from the Ericales.

In spite of the differences, the Diapensiales may well be related to the Ericales. The unitegmic, tenuinucellar ovules and integumentary tapetum suggest that the Diapensiales belong either with the Asteridae or with the more advanced orders of Dilleniidae. The frequent presence of a set of staminodes in addition to the functional stamens suggests that the group would be more at home among the advanced Dilleniidae than among the Asteridae. This leads us back to the Ericales as the most likely allies, and it may not be without significance that two genera of the Diapensiaceae have appendages on the anthers that resemble those of some of the Ericales. The wood-anatomy is perfectly compatible with a relationship between the two orders, and there are some similarities in floral anatomy as well. Any relationship must of course involve the more archaic members of both groups, rather than the more advanced ones. It seems probable that the two orders have a common ancestry in the Theales.

The adaptive significance of the characters that mark the Diapensiales is dubious.

12. Order EBENALES Engler 1892

Trees or shrubs, without raphides, tanniferous, commonly producing proanthocyanins and sometimes also ellagic acid, sometimes producing one or another sort of alkaloid and sometimes cyanogenic, but without iridoid compounds; vessel-segments with scalariform or simple perforations; imperforate tracheary elements with simple or bordered pits; wood-rays mostly heterocellular, mixed uniseriate and pluriseriate, or seldom all uniseriate; wood-parenchyma mostly apotracheal. LEAVES alternate or rarely opposite, simple, often firm and leathery, entire or seldom toothed, usually without stipules; stomates mostly anomocytic, but paracytic in the Symplocaceae and encyclocytic in the Lissocarpaceae. FLOWERS in various sorts of inflorescences that are often axillary or borne at the nodes on old wood, regular, perfect or unisexual, hypogynous to less often epigynous; calyx of 4–6 (–12) more or less distinct sepals, sometimes in 2 series, or tubular or cupulate with (2–) 4–5 (–7) lobes, or the lobes seldom obsolete; corolla sympetalous and (3) 4–6 (–8) lobed, or rarely of distinct petals; stamens mostly 2–3 (4) times as many as the corolla lobes, in 1–3 (4) cycles, sometimes one cycle staminodial, or with the filaments sometimes basally connate to form antesepalous clusters, or less often isomerous with the corolla-lobes, then variously opposite or alternate with the corolla-lobes, most commonly attached to the corolla-tube, but sometimes attached directly to the receptacle; anthers tetrasporangiate and dithecal, opening by longitudinal slits or rarely by apical pores; pollen-grains binucleate or trinucleate, most commonly tricolporate; gynoecium of (2–) 4–5 (–30) carpels united to form a compound, nearly always plurilocular ovary with axile placentas (but the partitions frequently not reaching the summit of the ovarian cavity); style with capitate or merely lobed stigma, or (Ebenaceae) more or less deeply cleft or the styles nearly distinct; ovules 1–4 (–6, rarely more) in each locule, anatropous to sometimes hemitropous, bitegmic or unitegmic, tenuinucellar, sometimes with an integumentary tapetum; endosperm nuclear or cellular; embryo sac not known to produce endosperm-haustoria. FRUIT variously dry or fleshy, dehiscent or indehiscent; seeds few and large, with a well developed embryo that often has large, flat or thickened cotyledons, with or without endosperm at maturity.

The order Ebenales as here defined consists of 5 families and about 1750 species, chiefly of tropical and subtropical regions, with relatively few species in temperate climates. The largest family is the Sapotaceae, with about 800 species, followed by the Ebenaceae, with about 450.

Most authors are agreed that the Ebenales as here constituted form a natural group, and there is a developing consensus that the group takes its origin in the Theales. The Ebenales also resemble the Ericales in some respects, but the relationship is collateral rather than ancestral. The Ebenales are more advanced than the bulk of the Ericales in the reduced number of ovules and in having the stamens attached to the corolla-tube, and also in having only a very few polypetalous members, but so far as known they lack the set of specialized embryological features that mark the Ericales. The bitegmic ovules of the Ebenaceae and Styracaceae are not likely to have been derived from the unitegmic ovules of the Ericales. The two orders may be regarded as having undergone certain parallel and other divergent changes from a common ancestry in the Theales.

Aside from the features associated with secondary metabolites, there is nothing ecologically outstanding about the Ebenales or any of the included families. They are all woody and chiefly tropical, but these are typical angiospermous features. The differences in fruit-type that characterize some of the families are obviously related to seed-dispersal, but none of the families has any special adaptation for seed-dispersal that is not well known in other groups as well. The exserted water-cells of the leaves of Symplocaceae are curious, but their functional significance remains to be elucidated. The importance of the differences in pubescence, nodal anatomy, and ovular structure is equally obscure, and if the genes governing these characters are important for other effects, the nature of these other effects is still wholly unknown.

In contrast to the classical morphological characters, the latex of the Sapotaceae and the naphthaquinones of the Ebenaceae are good candidates for interpretation in terms of survival value. Both probably discourage attacks by insects or other predators. The accumulation of aluminum in Symplocaceae may also have a protective function. The tannins found throughout the order are presumably also protective, but so far as I know there is nothing distinctive about them as compared to the tannins of many other orders.

No one family of the Ebenales is likely to be directly ancestral to any of the others. Each has a combination of relatively primitive and relatively advanced features that bespeaks collateral relationships with the other families. The nearest relatives of the Sapotaceae are the

Ebenaceae. The Ebenaceae are allied to the Styracaceae as well as to the Sapotaceae. The Lissocarpaceae and Symplocaceae are allied inter se, and these two families collectively are allied to the Styracaceae. The Lissocarpaceae are closer to the Styracaceae than are the Symplocaceae. The linear sequence of families here adopted is intended to express these relationships, but it should not be interpreted to mean that the Sapotaceae are more archaic than other families in the order. Indeed, aside from the nodal anatomy, the features that set the Sapotaceae apart appear to be advanced rather than primitive within the order.

SELECTED REFERENCES

Aubréville, A. 1963. Notes sur les Sapotacées. Adansonia sér 2. 3: 19–42.

Baehni, C. 1938, 1965. Mémoires sur les Sapotacées. I. Système de classification. Candollea 7: 394–508, 1938. III. Inventaire des genres. Boissiera 11: 1–262, 1965.

Copeland, H. F. 1938. The *Styrax* of northern California and the relationships of the Styracaceae. Amer. J. Bot. 25: 771–780.

Cronquist, A. 1946. Studies in the Sapotaceae. II. Survey of the North American genera. Lloydia 9: 241–292. VI. Miscellaneous notes. Bull. Torrey Bot. Club 73: 465–471.

Gunasekera, S. P., V. Kumar, M. U. S. Sultanbawa, & S. Balasubramaniam. 1977. Triterpenoids and steroids of some Sapotaceae and their chemotaxonomic significance. Phytochemistry 16: 923–926.

Kukachka, B. F. 1978, 1979, 1980. Wood anatomy of the neotropical Sapotaceae. I–XIX. U. S. Forest Service Res. Pap. FPL 325–331, 1978. 349–354, 1979. 358–363, 1980.

Lam, H. J. 1925. The Sapotaceae, Sarcospermaceae and Boerlagellaceae of the Dutch East Indies and surrounding countries (Malay Peninsula and Philippine Islands). Bull. Jard. Bot. Buitenzorg III. 7: 1–289.

Lam, H. J. 1939. On the system of the Sapotaceae, with some remarks on taxonomical methods. Recueil Trav. Bot. Néerl. 36: 509–525.

Lam, H. J., & W. W. Varossieau. 1938. Revision of the Sarcospermataceae. Blumea 3: 183–200.

Nooteboom, H. P. 1975. Revision of the Symplocaceae of the Old World, New Caledonia excepted. Leiden Bot. Ser. 1.

Record, S. J. 1939. American woods of the family Sapotaceae. Trop. Woods 59: 21–51.

Schadel, W. E., & W. C. Dickison. 1979. Leaf anatomy and venation patterns of Styracaceae. J. Arnold Arbor. 60: 8–27.

Veillet-Bartoszewska, M. 1960. Embryogénie des Styracacées. Développement de l'embryon chez le *Styrax officinalis* L. Compt. Rend. Hebd. Séances Acad. Sci. 250: 905–907.

White, E. B. 1956–1963. Notes on Ebenaceae. I, II. Bull. Jard. Bot. État 26: 237–246; 277–307, 1956. III. Bull. Jard. Bot. État 27: 515–531, 1957. IV. Bol. Soc. Brot. II, 36: 97–100, 1962. V. Bull. Jard. Bot. État 33: 345–367, 1963.

Wood, C. E., & R. B. Channell. 1960. The genera of the Ebenales in the southeastern United States. J. Arnold Arbor. 41: 1–35.

Yamazaki, T. 1970–1972. Embryological studies in Ebenales. 1. Styracaceae. 2. Symplocaceae. 3. Sapotaceae. 4. Ebenaceae. J. Jap. Bot. 45: 267–273; 353–358, 1970. 46: 161–165, 1971. 47: 20–28, 1972.

SYNOPTICAL ARRANGEMENT OF THE FAMILIES OF EBENALES

1 Plants with a well developed latex-system; pubescence of 2-armed (malpighian) hairs (one arm sometimes reduced or obsolete); nodes mostly trilacunar; vessels with simple perforations; ovules unitegmic .... 1. SAPOTACEAE.

1 Plants without a latex-system; pubescence not of 2-armed hairs except in some Ebenaceae; nodes unilacunar (nodal anatomy of Lissocarpaceae unknown); ovules and vessels various.

  2 Flowers mostly unisexual, rarely perfect; style more or less deeply cleft, or the styles almost distinct; vessels with simple perforations, ovules bitegmic .... 2. EBENACEAE.

  2 Flowers mostly perfect, rarely unisexual; style simple with a capitate or merely lobed stigma; vessels with scalariform perforations, or (Lissocarpaceae) some of the perforations scalariform and others simple.

    3 Pubescence characteristically of stellate hairs or peltate scales; fruit mostly dry, seldom fleshy; stamens all in a single series; ovary superior to inferior; anthers more or less linear; ovules bitegmic or unitegmic .... 3. STYRACACEAE.

    3 Pubescence neither of stellate hairs nor of peltate scales, sometimes wanting; fruit more or less fleshy; ovary inferior, or seldom only half-inferior.

      4 Anthers linear; flowers with a corona; stamens 8, twice as many as the corolla-lobes, all in a single series .... 4. LISSOCARPACEAE.

      4 Anthers broadly ovate or rotund; flowers without a corona; stamens (4–) 12 to rather numerous, in more than one series, or grouped into fascicles; ovules unitegmic .... 5. SYMPLOCACEAE.

1. Family SAPOTACEAE A L. de Jussieu 1789 nom. conserv., the Sapodilla Family

Trees or shrubs with trilacunar or seldom unilacunar nodes, with well developed, elongate, commonly parallel latex-sacs in the leaves, bark, and pith, and also with scattered secretory cells, tanniferous, producing proanthocyanins but not ellagic acid, often accumulating steroids (commonly including α-spinasterol) and triterpenoids (of the oleanane series), and often cyanogenic, but without iridoid compounds; vesture of 2-armed (malpighian) hairs, one arm often more or less reduced or sometimes even obsolete; calcium oxalate crystals of various forms often present in some of the cells of the parenchymatous tissues, and some of the cells often containing silica-bodies; vessel-segments with simple perforations; imperforate tracheary elements with simple or seldom more or less distinctly bordered pits; wood-rays mostly heterocellular, Kribs types IIA, IIB, and III, mixed uniseriate and pluriseriate, the latter 2–4 (–6) cells wide, or rarely all uniseriate; wood-parenchyma generally abundant and apotracheal, in more or less well defined ribbons or often reticulate, seldom diffuse or vasicentric. LEAVES alternate or seldom opposite, simple, mostly entire; stomates commonly anomocytic; stipules usually wanting but sometimes well developed, as in *Ecclinusa*. FLOWERS mostly rather small, borne (singly or) in small cymose clusters in the axils or at the nodes on old wood, or seldom (*Sarcosperma*) in large, paniculately branched, many-flowered axillary inflorescences, nearly always perfect, regular, hypogynous, often bracteolate; sepals distinct or nearly so, (4) 5 (–12) and imbricate, or sometimes in 2 cycles of 2, 3, or 4; corolla sympetalous, with 4–8 imbricate lobes, the lobes usually of the same number as the sepals (fewer in some of the species with relatively numerous sepals), and sometimes each bearing a pair of lateral or dorsolateral appendages; stamens epipetalous, in 1–3 whorls, often some of them reduced to staminodes (these commonly alternate with the corolla-lobes, often petaloid), at least one antheriferous whorl opposite the corolla-lobes; anthers tetrasporangiate and dithecal, opening by longitudinal slits; pollen-grains binucleate or trinucleate, 3–5 (–6)-colporate; gynoecium of 2–14 (–30) carpels united to form a compound, usually hairy, plurilocular ovary with axile or axile-basal placentation, or rarely (*Diploon*) the ovary unilocular by failure of the partition and the ovules basal; style single, with a capitate or slightly lobed stigma; ovules 1 per carpel, anatropous to hemitropous, apotropous, unitegmic, tenuinucellar; endosperm-development nuclear. FRUIT fleshy, indehiscent; seeds

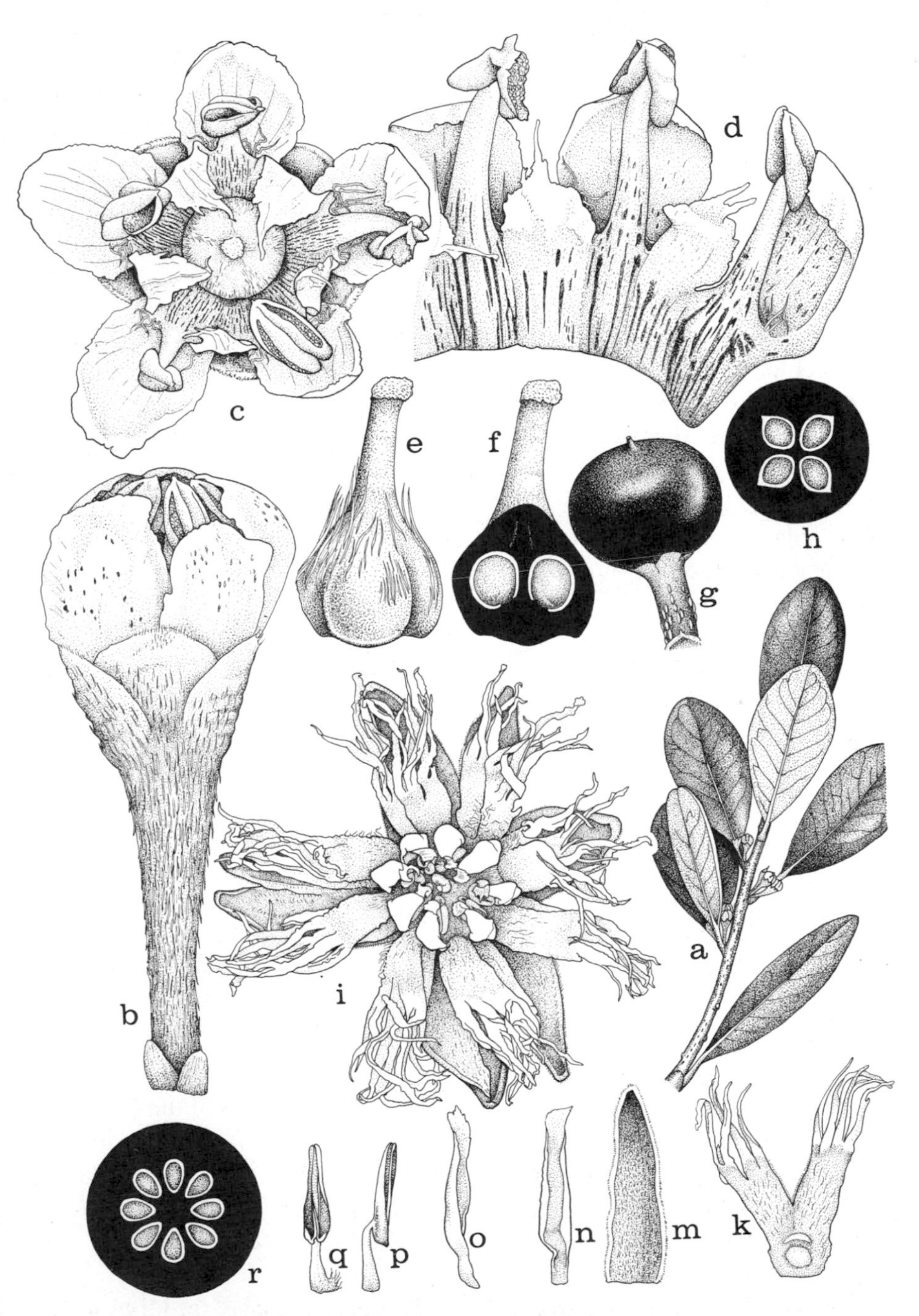

FIG. 4-43 Sapotaceae. a–h, *Sideroxylon inerme* L. a, habit, ×½; b, flower, just beginning to open, ×12; c, flower, from above, ×12; d, corolla, opened out to show stamens and staminodes, ×12; e, pistil, ×12; f, pistil, in partial long-section, ×12; g, fruit, ×2; h, schematic cross-section of ovary, ×12. i–r, *Mimusops commersonii* (G. Don) Engl. i, flower, from above, ×3; k, portion of tubular corolla, torn loose in the lower portion, showing a pair of fimbriate dorsolateral appendages, the blade of the corolla-lobe removed and shown as a crescentic scar, ×3; m, sepal, ×3; n, o, corolla-lobe, removed from the tube and appendages, ×3; p, q, stamen, ×3; r, schematic cross-section of ovary, ×16.

large, commonly with a relatively large, excavated scar of attachment, this lateral and elongate to basilateral or nearly basal and short but often broad, the seed-coat otherwise shiny and usually thick and hard; embryo large, with thin, flat cotyledons, enclosed in a more or less well developed, oily, fleshy or hard endosperm, or with thickened cotyledons and without endosperm at maturity. X = 7, 9–13. (Achraceae, Boerlagellaceae, Bumeliaceae, Sarcospermataceae)

The family Sapotaceae as here defined consists of some 70 genera and about 800 species, widespread in tropical parts of both the Old and the New World, with a relatively few species extending into temperate regions. About three-fourths of the species belong to only 6 genera, *Pouteria* (150), *Palaquium* (115), *Planchonella* (100), *Chrysophyllum* (90), *Manilkara* (75), and *Madhuca* (75). *Manilkara zapota* (L.) Van Royen, the Sapodilla, a Central American and West Indian species, is well known as the classical source of chicle, the essential ingredient of chewing gum. Other gums are produced from other members of the family. A number of species have edible fruits, such as the star-apple, *Chrysophyllum cainito* L., and the Sapote, *Pouteria sapota* (Jacq.) H. E. Moore & Stearn, both of tropical America.

The delimitation of genera in the Sapotaceae is extraordinarily controversial. The shape of the seed-scar and the presence or absence of endosperm are considered by most students of the family to provide highly useful generic characters, but the value of these features is denied by others. In contrast to Baehni, I consider the seed-characters to be useful at the generic level. On the other hand I cannot accept the very narrowly limited genera of Pierre and of Aubreville.

The tropical East Asian genus *Sarcosperma* (6 spp.) is sometimes taken as a separate family, mainly because of its large, branching, many-flowered, panicle-like inflorescence. Its close relationship to the Sapotaceae is not in dispute.

The Sapotaceae stand off sharply from other members of the Ebenales in their well developed latex-system, malpighian hairs, and mostly trilacunar nodes. In spite of these differences, botanists are agreed that the family belongs here. The wood-anatomy of the Ebenaceae is much like that of the Sapotaceae, and some members of the Ebenaceae also have malpighian hairs.

It seems likely that some of the Sapotaceae have undergone an increase in the number of floral parts of a kind, in contrast to the general trend toward reduction among the angiosperms as a whole. Some species of *Pouteria* have as many as 12 sepals in otherwise pentamerous flowers. Some species of *Dipholis* likewise have more sepals than other floral parts. *Manilkara* has 6 sepals in 2 cycles, 6 corolla-lobes, 6 stamens, 6 staminodia, and up to as many as 14 carpels. *Mimusops* has 8 sepals in 2 cycles, 8 corolla-lobes, 8 stamens, 8 staminodes, and a variable number of carpels. The bicyclic calyx appears to be secondary within the family, as do all numbers of more than 5 parts in a cycle.

Fossil pollen attributed to the Sapotaceae occurs in Maestrichtian (late Upper Cretaceous) and more recent deposits.

2. Family EBENACEAE Gürke in Engler & Prantl 1891 nom. conserv., the Ebony Family

Trees or shrubs with unilacunar nodes, commonly with hard, dark or black wood, without latex, usually with scattered tanniferous cells, often or usually producing proanthocyanins, but without ellagic acid and lacking iridoid compounds, sometimes cyanogenic and sometimes producing methyl salicylate, commonly producing black or dark-colored naphthaquinones or their derivatives in the leaves, young stems, and mature wood; vesture commonly of simple hairs, or partly or wholly of 2-armed hairs, or branched hairs, or tufts of hairs, or gland-tipped hairs; large, solitary crystals of calcium oxalate commonly present in some of the cells of the parenchymatous tissues, clustered crystals sometimes present as well; vessel-segments with simple perforations; imperforate tracheary elements with small, bordered pits; wood-rays heterocellular, Kribs types IIB and III, sometimes all uniseriate, sometimes some of them 2 (–4) cells wide; wood-parenchyma typically in numerous uniseriate, apotracheal ribbons, sometimes also diffuse or vasicentric. Leaves alternate or rarely opposite, simple, entire; stomates anomocytic; stipules wanting. Flowers mostly rather small, borne singly (especially the pistillate ones) or in small cymose clusters in the axils, regular, hypogynous, mostly unisexual (the plants commonly dioecious), the staminate flowers with a vestigial ovary, and the pistillate ones often with staminodes, but the flowers perfect in the African genus *Royena*; calyx 3- to 7-lobed, persistent, often accrescent in fruit; corolla sympetalous, 3- to 7-lobed, regular, the lobes imbricate, convolute, or valvate; stamens attached to the base of the corolla-tube or sometimes directly to the receptacle, usually twice as many as the corolla-lobes and in 2 cycles, but sometimes up to 4 or even 5 times as many as the corolla-lobes, often paired and sometimes with the filaments connate in pairs,

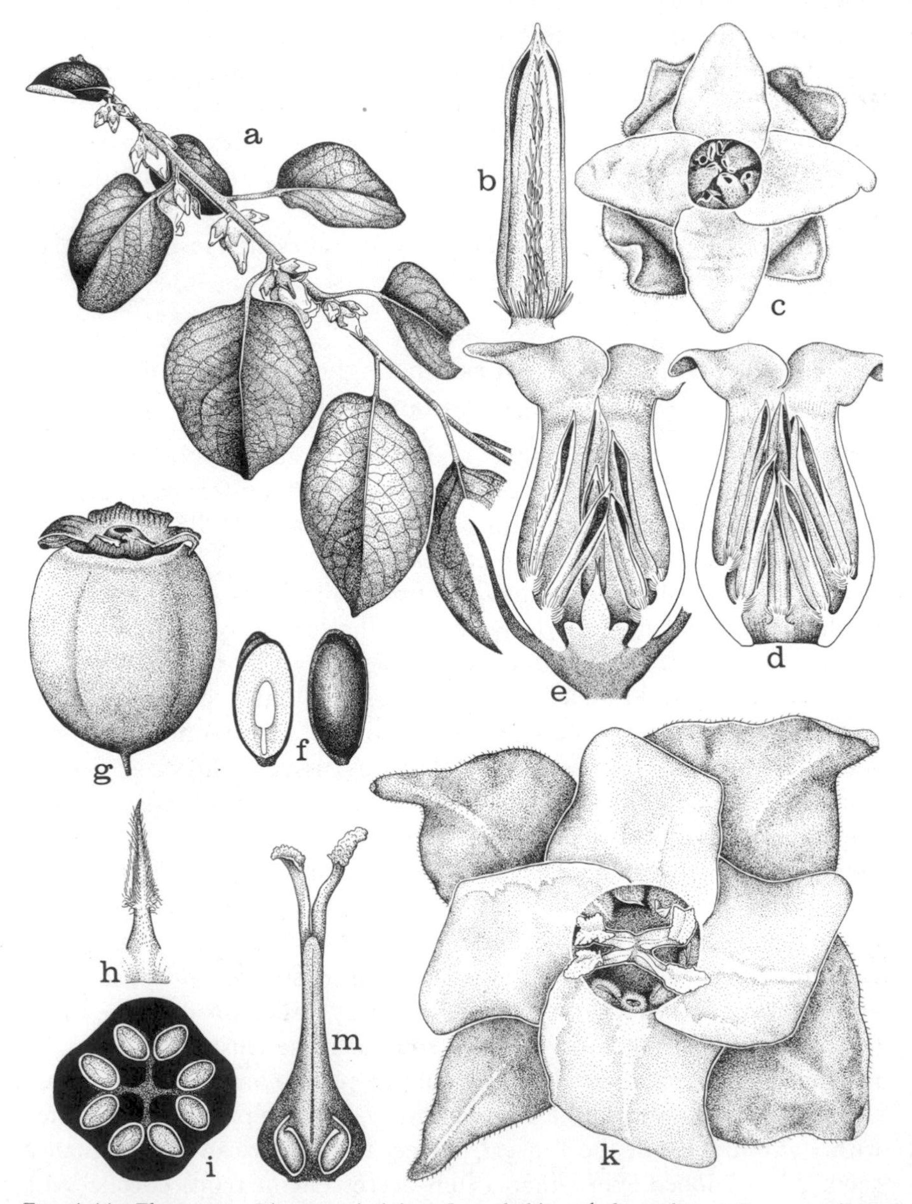

FIG. 4-44 Ebenaceae. *Diospyros virginiana* L. a, habit, ×½; b, anther, ×8; c, staminate flower, from above, ×4; d, corolla of staminate flower in partial long-section, showing stamens in 2 cycles, ×4; e, staminate flower in partial long-section, ×4; f, seed, external and in long-section, ×1; g, fruit, ×1; h, staminode from pistillate flower, ×4; i, schematic long-section of ovary, ×6; k, pistillate flower, from above, ×4; m, pistil in partial long-section.

or seldom of the same number as the corolla-lobes and alternate with them, only 2–3 in *Rhaphidanthe*; anthers tetrasporangiate and dithecal, opening by longitudinal slits or rarely (some spp. of *Diospyros*) by apical pores; pollen-grains binucleate, tricolporate; gynoecium of (2) 3–8 (–10), often 4 carpels, these united to form a compound, plurilocular ovary with apical-axile placentation, each carpel often divided by a median partition so that there are twice as many locules (or locelli) as carpels; style more or less deeply cleft, or the styles (most often 4) almost distinct; ovules solitary (2) in each locule, pendulous, apotropous, anatropous, bitegmic, tenuinucellar, with an integumentary tapetum, without an obturator; endosperm-development variously reported to be nuclear or cellular. FRUIT usually a juicy to somewhat leathery berry, rarely capsular; seeds with thin testa and abundant, hard, often ruminate endosperm that has reserves of hemicellulose and oil; embryo straight or slightly curved, with flat cotyledons. X = 15.

The family Ebenaceae as here defined consists of 5 genera and about 450 species, widespread in tropical and subtropical regions of both the Old and the New World, with only a few species extending into temperate climates. The great bulk of the family belongs to the single genus *Diospyros*, with perhaps 400 species. The other genera are *Euclea*, *Rhaphidanthe*, *Royena*, and *Tetraclis*, the first 3 mainly African, the fourth from Madagascar. The family is the source of several economically important woods, including most notably ebony, the hard, very heavy, black heartwood of *Diospyros ebeneum* König and related species. The fruits of some species are edible. Those of *D. virginianum* L., persimmon, are much esteemed after frost in rural southeastern United States. The rather similar *D. kaki* L.f., Japanese persimmon, is favored in the Orient.

Leaves thought to represent *Diospyros* are found in Upper Cretaceous and more recent deposits, but the identification needs to be re-examined on modern criteria. Wood considered to belong to the Ebenaceae is known from Oligocene and later deposits, and pollen attributed to the group dates from the Lower Eocene.

3. Family STYRACACEAE Dumortier 1829 nom. conserv., the Storax Family

Trees or shrubs with unilacunar nodes, often with resiniferous intercellular canals in the bark and wood but without latex, tanniferous, often or regularly producing proanthocyanins but not ellagic acid, not cyanogenic, and without iridoid compounds; vesture typically of stellate

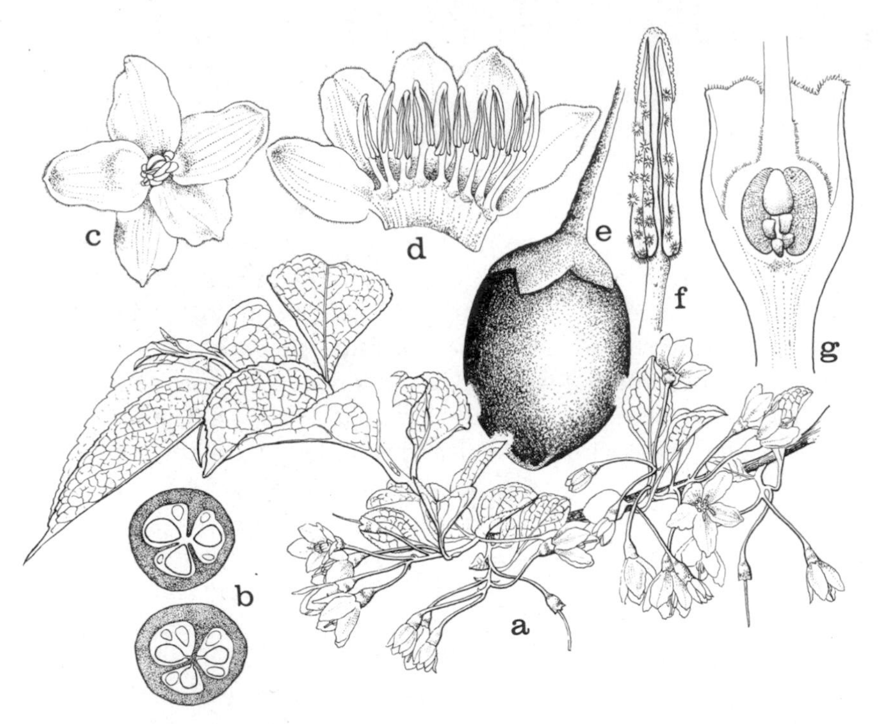

FIG. 4-45 Styracaceae. *Styrax* hybrid. a, habit, ×½; b, schematic cross-sections of ovary, ×5, near the base (below) and above the middle (above); c, flower, from above, ×2; d, corolla, opened out to show stamens, ×2; e, fruit, ×2; f, anther, ×8; g, schematic partial long-section of calyx and pistil, one fertile ovule (above) and several sterile ovules (below), in one locule, ×5.

hairs or peltate scales, usually brown or rufous, often present on the outside of the calyx and corolla as well as on the leaves and twigs; solitary or clustered crystals of calcium oxalate often present in some of the cells of the parenchymatous tissues, and grains of silica sometimes present in the wood; vessel-segments with scalariform perforations that have fewer than 20 cross-bars, or seldom some of the perforations simple; imperforate tracheary elements with bordered pits; wood-rays heterocellular, Kribs type IIA and B, mixed uniseriate and pluriseriate, the latter 2–4 (–6) cells wide, with short ends; wood-parenchyma apotracheal, diffuse and in regular, uniseriate lines, or seldom diffuse and scanty-paratracheal. LEAVES alternate, simple, entire or toothed; stomates anomocytic; stipules wanting. FLOWERS variously in panicles, racemes, or cymes, or seldom solitary, perfect or seldom (*Bruinsmia*) some of them unisexual (the plants then polygamo-dioecious), regular, hypogynous to epigynous, without bracteoles; calyx tubular, with (2–)

4–5 (–7) short, valvate or open lobes or teeth, or the teeth obsolete; corolla of 2–5 (–7) petals united below to form a tube generally much shorter than the imbricate or valvate lobes, rarely the petals distinct (*Bruinsmia*) or the lobes distinctly shorter than the tube (*Halesia*); stamens commonly twice as many (or up to 4 times as many) as the corolla-lobes but all in one row, or seldom (*Pamphilia*) only 5 and alternate with the corolla-lobes; filaments adnate to the corolla-tube, or less often attached directly to the receptacle, usually connate below into a short or more or less elongate tube; anthers linear, tetrasporangiate, dithecal, opening by longitudinal slits, the connective sometimes shortly prolonged; pollen-grains binucleate, tricolporate; gynoecium of (2) 3–5 carpels united to form a compound, superior to often inferior ovary that is plurilocular below but commonly unilocular above; style slender, with a capitate or merely lobed stigma; placentation axile, with 1-many (most often 4–6) erect or pendulous ovules on each placenta, but not more than 2 ovules per locule maturing into seeds; ovules apotropous, anatropous to hemitropous, bitegmic (*Styrax*) or unitegmic (*Halesia*), tenuinucellar, sometimes with an obturator; endosperm-development cellular. FRUIT mostly dry and usually capsular, varying to indehiscent and sometimes samaroid, or seldom (*Parastyrax*) drupaceous, subtended by the persistent calyx; seeds with thin to indurated testa, abundant, oily endosperm, and large, straight or sometimes slightly curved embryo with broad cotyledons. X = 8, 12.

The family Styracaceae as here defined consists of about 10 genera and 150 species; about 120 species belong to the genus *Styrax*. The family occurs in 3 widely disjunct areas: the New World from central U.S.A. to northern Argentina; the Mediterranean region; and southeastern Asia and adjacent islands. *Styrax benzoin* Dryand. is the source of gum benzoin, and *S. officinalis* L. is the source of storax, a gummy resin used in incense. Species of *Styrax* and *Halesia* (silver-bell tree) are sometimes cultivated for ornament.

The tropical African genus *Afrostyrax*, sometimes included in the Styracaceae, is here referred to the family Huaceae in the Violales.

Fossils of the Styracaceae are not known from deposits older than the Tertiary.

4. Family LISSOCARPACEAE Gilg in Engler & Gilg 1924 nom. conserv., the Lissocarpa Family

Rather small trees, without stellate or peltate hairs; vessel-segments with partly simple and partly scalariform perforations; imperforate tracheary elements with very small bordered pits; wood-rays heterocellular; wood-parenchyma in ribbons. LEAVES alternate, simple, pinnately veined, entire; stomates encyclocytic or anomocytic; stipules wanting. FLOWERS in small axillary cymes, bibracteolate, perfect, regular, epigynous; calyx 4-lobed, the lobes imbricate; corolla 4-lobed, the lobes convolute in bud; corolla-throat bearing a short, tubular, 8-toothed corona; stamens 8, in a single series, the filaments shortly connate, attached to the corolla-tube below its middle; anthers linear, tetrasporangiate and dithecal, with apiculate-prolonged connective, opening by longitudinal slits; pollen-grains triporate; gynoecium of 4 carpels united to form a compound, plurilocular, inferior ovary with axile placentas and a terminal, clavate style with a capitate or shallowly 4-lobed stigma; ovules 2 in each locule, pendulous. FRUIT indehiscent, somewhat fleshy, 1- or 2-seeded; seeds with abundant endosperm and rather large, straight embryo.

The family Lissocarpaceae consists of the single genus *Lissocarpa*, with 2 species native to tropical South America. *Lissocarpa* has been insufficiently studied, but most recent authors are agreed that it forms a family in the Ebenales.

5. Family SYMPLOCACEAE Desfontaines 1820 nom. conserv., the Sweetleaf Family

Evergreen or rarely deciduous trees or shrubs with unilacunar nodes, glabrous or with multicellular hairs or stalked glands, but without malpighian or stellate hairs, commonly accumulating aluminum (often in distinctive intercellular bodies), tanniferous, producing proanthocyanins and also ellagic and gallic acids, sometimes producing methyl salicylate or indole alkaloids of the carboline group, seldom benzyl-isoquinoline alkaloids, but not cyanogenic and without iridoid compounds; solitary (seldom clustered) crystals of calcium oxalate commonly present in some of the cells of the parenchymatous tissues; vessels small, solitary, angular, with scalariform perforations that commonly have more than 20 cross-bars, the walls often spirally thickened; imperforate tracheary elements with evidently bordered pits; wood-rays heterocellular, Kribs types I and IIA, varying to almost homocellular, mixed uniseriate and pluriseriate, the latter up to 3–5 (–12) cells wide, with long or short ends; wood-parenchyma apotracheal, diffuse or in short, uniseriate lines, sometimes also scanty-paratracheal. LEAVES alternate, simple, firm, often sweet-tasting, often with bladder-like water-cells

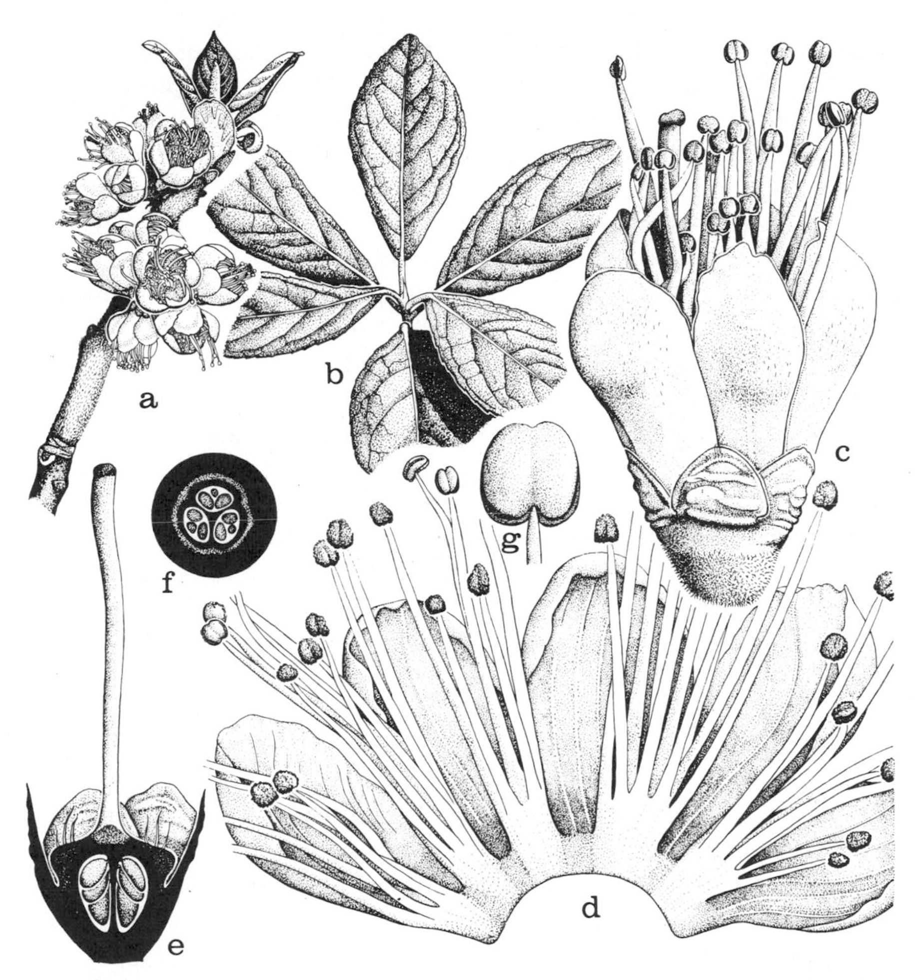

FIG. 4-46 Symplocaceae. *Symplocos tinctoria* (L.) L'Herit. a, flowering branch, ×1; b, leafy twig, ×½; c, flower, ×6; d, corolla, opened out to show stamens, ×6; e, pistil in partial long-section, with calyx attached, ×6; f, schematic cross-section of ovary, ×6; g, anther, ×18.

exserted from the epidermis on one or both sides; stomates mostly paracytic; stipules wanting. FLOWERS in racemes, or less often in spikes, panicles, or fascicles, or solitary, each axillary to a bract and provided with 2 bracteoles, regular, perfect or sometimes some of them unisexual (the plants then polygamous), epigynous or less often half-epigynous; calyx of (3–) 5 basally connate sepals, the lobes valvate or imbricate; corolla of (3–) 5 (–11) petals connate below to form a usually short tube, the lobes imbricate, sometimes more or less distinctly in 2 rows;

stamens (4–) 12 to rather numerous, attached to the base of the corolla-tube, usually in more than one series, or in fascicles alternate with the corolla-lobes; anthers short and broad, mostly rotund-ovate, opening by longitudinal slits; pollen-grains binucleate, mostly oblate, with 3 very short colpi, the ornamentation of the exine rather diverse; gynoecium of 2–5 carpels united to form a compound, usually fully plurilocular, inferior or seldom only half-inferior ovary, the slender style often surrounded by a nectary-disk; stigma capitate or merely lobed; ovules 2–4 in each locule, pendulous on axile (or deeply intruded) placentas, epitropous, anatropous, unitegmic, tenuinucellar, with an integumentary tapetum; endosperm-development cellular. FRUIT crowned by the persistent calyx-lobes, more or less fleshy, most commonly a single-stoned drupe, less often baccate; endocarp with an apical germinal pore, or the fruit (when pluricarpellate) with one such pore per fertile carpel; seeds with abundant, proteinaceous and oily (sometimes also starchy) endosperm and a rather large, straight or curved embryo with very short cotyledons. X = 11–14.

The family Symplocaceae consists of the single genus *Symplocos*, with 300–400 species, widespread in moist tropical and subtropical regions of America, southern and eastern Asia, Australia, and the East Indies, but absent from Africa, western Asia, and Europe. Only a few species grow in temperate regions. A yellow dye is extracted from the leaves and bark of several species, notably *S. tinctoria* (L.) L'Her., of the southeastern United States.

Pollen attributed to the Symplocaceae is known from Maestrichtian (uppermost Cretaceous) and more recent deposits, and fruits date from the Lower Eocene.

Recent authors (with the exception of Nooteboom) are agreed that the Symplocaceae are related to the Styracaceae. The fascicled stamens of some species are also reminiscent of the Theales, to which the Ebenales are now usually considered to be related.

13. Order PRIMULALES Lindley 1833

Herbs or woody plants with a well developed (Myrsinaceae, some Primulaceae) to rudimentary or vestigial (Theophrastaceae, many Primulaceae) schizogenous secretory system containing red to yellow or brownish material, generally tanniferous (except in many Theophrastaceae), often containing proanthocyanins but only seldom ellagic acid, commonly producing triterpenoid saponins, but not cyanogenic and without iridoid compounds; trichomes of various sorts, often some of them capitate-glandular or peltate-glandular, sometimes sunken in the epidermis; nodes unilacunar; vessel-segments mostly or all with simple perforations; imperforate tracheary elements in woody plants with simple pits. LEAVES alternate to less often opposite or whorled, mostly simple and entire or merely toothed, seldom pectinately pinnatifid; stomates anomocytic or anisocytic; vascular anatomy of the petiole often complex; stipules wanting. FLOWERS in various sorts of indeterminate inflorescences, perfect or seldom unisexual, regular or rarely somewhat irregular, hypogynous or rarely half-epigynous, most often pentamerous; calyx synsepalous and toothed or lobed, or less often of distinct sepals; corolla sympetalous, rarely polypetalous or wanting; functional stamens as many as and opposite the lobes of the corolla, the filaments attached to the corolla-tube or rarely almost free; petaloid or vestigial staminodes sometimes alternating with the stamens; anthers tetrasporangiate and dithecal, opening by longitudinal slits or seldom by apical pores; pollen-grains binucleate, tricolporate, varying to sometimes pantoporate; gynoecium of (3–) 5 (6) carpels united to form a compound, unilocular ovary with a free-central or sometimes basal placenta; style solitary, with a simple or merely lobed, often capitate stigma; ovules few to usually more or less numerous, anatropous to hemitropous or nearly campylotropous, bitegmic (rarely unitegmic), tenuinucellar, often with an integumentary tapetum; endosperm-development nuclear, without haustoria so far as known. FRUIT usually a capsule, berry, or drupe, with 1-many small to large seeds; embryo small to large, straight or slightly curved, dicotyledonous (monocotyledonous in *Cyclamen*, of the Primulaceae), embedded in the copious endosperm, or the endosperm rarely wanting.

The order Primulales as here defined consists of 3 families and about 1900 species. The Myrsinaceae have about 1000 species, the Primulaceae about 800, and the Theophrastraceae about 100.

The Primulales are sympetalous dicotyledons with the functional stamens opposite the corolla-lobes (with or without an alternating set of staminodes), and with a compound ovary that has a single style and several to numerous tenuinucellar, mostly bitegmic ovules on a free-central or basal placenta. No other group has this combination of characters. The Plumbaginaceae, which have often been referred to the Primulales, differ among other respects in their solitary ovules and partly or wholly distinct styles, and are here treated as an order of the subclass Caryophyllidae.

Most taxonomists agree that the three families of Primulales are closely allied. The Theophrastaceae have indeed often been included in the Myrsinaceae, but they stand somewhat apart on anatomical as well as floral characters. *Lysimachia* is often cited as an archaic member of the Primulaceae that suggests a connection to the Myrsinaceae. Hutchinson disagrees, of necessity, assigning the Primulaceae to his Herbaceae and the other two families to his Lignosae.

No one family of the Primulales is likely to be directly ancestral to either of the others. The Theophrastaceae are more primitive than the other families in consistently having a set of staminodes in addition to the functional stamens, but they stand between the Myrsinaceae and the Primulaceae in the vascular anatomy of the stem, and their leaf-architecture is specialized in a way unlike either of the other two families. It might be possible to consider that the well developed secretory system of the Myrsinaceae has been reduced in the Primulaceae. *Lysimachia* can then be considered to be a connecting form in this respect, but some species of *Lysimachia* are more primitive than any of the Myrsinaceae in having evident staminodes.

It is uncertain whether the characteristic schizogenous secretory system of the Myrsinaceae is a primitive or an advanced feature in the order. If it is primitive, then it has been almost completely lost in the Theophrastaceae, being represented only by the glandular dots or lines in the calyx and corolla. If it is advanced, then it has been independently developed to varying degree in parallel fashion in the three families—only slightly in the Theophrastaceae, somewhat more, but inconsistently, in the Primulaceae, and to the full extent only in the Myrsinaceae. Take your pick.

The Primulales are probably related to the Ebenales. Takhtajan has suggested a possible homology of the secretory system of the Myrsinaceae with that of the Sapotaceae. I express no opinion on this concept here. One of the unusual features of both the Primulales and the Ebenales is the combination of sympetalous flowers and bitegmic ovules.

Furthermore, these bitegmic ovules are tenuinucellar. As Philipson (1974, in initial citations) has pointed out, bitegmic, tenuinucellar ovules are very common among the Theales and some other Dilleniidae, but uncommon elsewhere.

Neither the Primulales nor the Ebenales can be considered ancestral to the other, but they might have a common ancestry just short of the Theales. Such an ancestor would be a tropical tree with hypogynous, sympetalous flowers that have two or three sets of epipetalous stamens, a compound ovary, separate styles, and numerous bitegmic, tenuinucellar ovules on axile placentas. Only the sympetalous condition would be at odds with the characters of the Theales. Alternatively, one could suppose that the Ebenales and Primulales achieved the sympetalous condition independently, in which case the nearest common ancestor would be referable to the Theales. The latter assumption would be more nearly in accord with the results of recent serological studies by Frohne and John (1978), which suggest a mutual affinity among the Primulales, Theales, and Ericales, and a less close affinity of the Ebenales to these three.

Botanists once entertained the notion that the Primulales might be derived from the Caryophyllales, but further study has made such a relationship seem highly unlikely. The Primulales have none of the special features of the Caryophyllales other than free-central placentation, and furthermore the Caryophyllales appear to be primitively herbaceous rather than primitively woody. It is hard to see how any possible ancestor of the Primulales could be included in the Caryophyllidae. In seeking an ancestor for the Primulales one must look for something that could give rise to one or the other of the woody families, not something that suggests the obviously advanced, herbaceous family Primulaceae.

The Primulales as a whole do not appear to be adapted to any particular ecological niche, nor are the characters that mark the order of any obvious selective significance. The features of growth-habit and fruit-type that distinguish the Primulaceae from the other two families are evidently adaptive, but there is nothing distinctive about them as compared to other angiosperms. The differences in the fruits are doubtless correlated with different strategies for seed-dispersal, but there is nothing unusual about these strategies as compared with those of other angiosperms. The biological significance of the staminodes of the Theophrastaceae is obscure. The chemical defenses are doubtless adaptive. The special secretory system of the Myrsinaceae may be unlike anything in any other order, but the chemistry of the compounds remains to be elucidated. All 3 families have triterpenoid saponins, but compounds of these groups occur in a number of other orders as well.

SELECTED REFERENCES

Channell, R. B., & C. E. Wood. 1959. The genera of the Primulales of the southeastern United States. J. Arnold Arbor. 40: 268–288.

Dickson, J. 1936. Studies in floral anatomy. III. An interpretation of the gynaeceum in the Primulaceae. Amer. J. Bot. 23: 385–393.

Douglas, G. E. 1936. Studies in the vascular anatomy of the Primulaceae. Amer. J. Bot. 23: 199–212.

Frohne, D., & J. John. 1978. The Primulales: serological contributions to the problem of their systematic position. Biochem. Syst. Ecol. 6: 315–322.

Lys, J. 1955. Sur la nature et l'évolution biochimique des glucides des Lysimaques. Compt. Rend. Hebd. Séances Acad. Sci. 241: 1842–1844.

Lys, J. 1956. Les glucides de quelques Primulacées. Rev. Gén. Bot. 63: 95–100.

Roth, I. 1959. Histogenese und morphologische Deutung der Kronblätter von *Primula*. Bot. Jahrb. Syst. 79: 1–16.

Roth, I. 1959. Histogenese und morphologische Deutung der Plazenta von *Primula*. Flora 148: 129–152.

Sattler, R. 1962. Zur frühen Infloreszenz- und Blütenentwicklund der Primulales sensu lato mit besonderer Berücksichtigung der Stamen-Petalum-Entwicklung. Bot. Jahrb. Syst. 81: 358–396.

Schaeppi, H. 1937. Vergleichend-morphologische Untersuchungen am Gynoeceum der Primulaceen. Z. Gesamte Naturwiss. (Brunswick) 3: 239–250.

Souèges, R. 1937. Embryogénie des Primulacées. Développement de l'embryon chez le *Samolus Valerandi* L. Compt. Rend. Hebd. Séances Acad. Sci. 204: 145–147.

Spanowsky, W. 1962. Die Bedeutung der Pollenmorphologie für die Taxonomie der Primulaceae-Primuloideae. Feddes Repert. Spec. Nov. Regni Veg. 65: 149–214.

Walker, E. H. 1940. A revision of the eastern Asiatic Myrsinaceae. Philipp. J. Sci. 73: 1–258.

Свешникова, И. Н. 1951. К морфологии соцветия рода *Primula* L. Бот. Ж. 36: 160–174.

SYNOPTICAL ARRANGEMENT OF THE FAMILIES OF PRIMULALES

1 Plants nearly always woody, often arborescent, largely tropical and subtropical; fruits mostly fleshy, often 1-seeded (even though there are several or many ovules in the ovary).

2 Flowers with staminodes alternate with the corolla-lobes; plants without an evident secretory system in the stems and leaves, the leaves not gland-dotted; leaves commonly with elongate subepidermal fibers and often with a submarginal fibrous strand .... 1. THEOPHRASTACEAE.

2 Flowers without staminodes; plants with a well developed schizogenous secretory system in the stem and leaves, the leaves gland-dotted; leaves usually without subepidermal fibers and usually without a submarginal fibrous strand .... 2. MYRSINACEAE.

1 Plants herbaceous or rarely half-shrubby, chiefly of temperate or cold regions or altitudes; fruit dry, mostly capsular, with (1-) several or many seeds ....3. PRIMULACEAE.

1. Family THEOPHRASTACEAE Link 1829 nom. conserv., the Theophrasta Family

Trees or low to tall shrubs, often of palm-like habit, commonly producing triterpenoid saponins and sometimes containing emetic substances in the roots, sometimes more or less tanniferous, but usually or always lacking both proanthocyanins and ellagic acid, also lacking iridoid compounds and not cyanogenic; resin-ducts wanting; solitary or clustered crystals of calcium oxalate commonly present in some of the cells of the parenchymatous tissues; vascular system of the stem more dissected than in the Myrsinaceae, with broad medullary rays; nodes unilacunar; vessel-segments with simple perforations; imperforate tracheary elements with simple pits, not septate; wood-rays more or less distinctly heterocellular; pluriseriate; wood-parenchyma scanty or none. LEAVES alternate, but often closely clustered at the end of the main stem or its branches and appearing almost whorled, simple, often pungent-tipped, entire or often spiny-toothed, often beset with impressed, capitate-glandular hairs, pinnately veined, very often with an immediately inframarginal sclerenchymatous strand (or fibrovascular bundle), and commonly with variously arranged elongate fibers just internal to both the upper and the lower epidermis; petiole often with complex vascular anatomy; stipules wanting. FLOWERS mostly rather large, white to yellow or pink, in terminal racemes, corymbs, or panicles, or seldom in lateral inflorescences or solitary, regular, hypogynous, perfect or sometimes (*Clavija*) unisexual, the plants then polygamodioecious; calyx of (4) 5 distinct sepals, or the sepals sometimes (*Clavija*) connate at the base; corolla fleshy, sympetalous, with a short tube and (4) 5 imbricate lobes; both calyx and corolla marked with glandular dots or pits or lines; functional stamens as many as and opposite the corolla-lobes, attached to the corolla-tube near its base, the filaments distinct or sometimes connate below to form a tube; anthers tetrasporangiate and dithecal, introrse or (*Jacquinia*) extrorse, opening by longitudinal slits, the connective often shortly apiculate-prolonged; pollen-grains binucleate, tricolporate; a cycle of petaloid or glandular staminodes alternating with the functional stamens and inserted a little higher in the corolla-tube or at the sinuses; gynoecium of 5 carpels united to form a compound, unilocular, superior ovary with a terminal style; stigma punctate or discoid or craterform, sometimes shallowly lobed; ovules more or less numerous, anatropous to sometimes hemitropous or almost campylotropous, bitegmic, tenuinucellar, attached

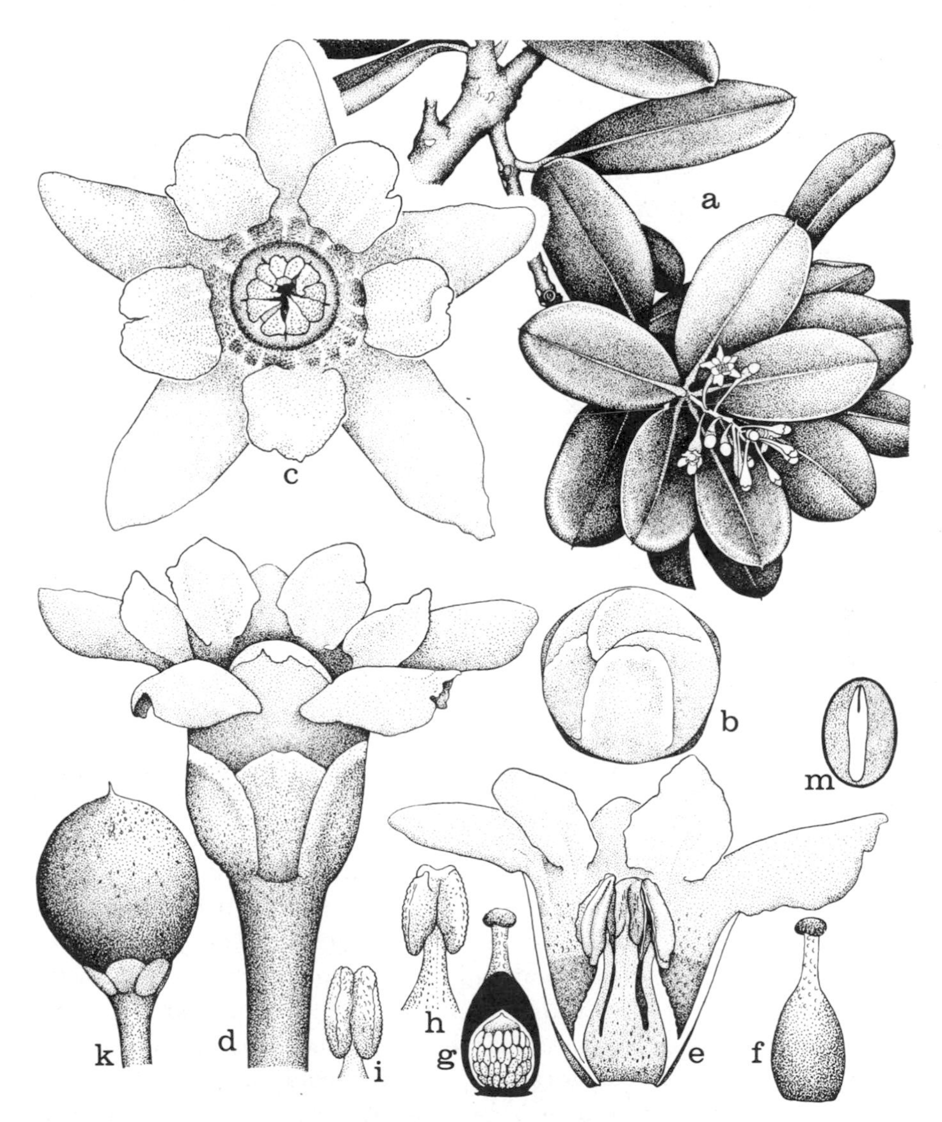

FIG. 4-47 Theophrastaceae. *Jacquinea arborea* Vahl. a, habit, ×½; b, flower bud, ×6; c, d, flower, ×6; e, corolla in partial long-section, with attached stamens and staminodes, ×6; f, pistil, ×6; g, pistil, in partial long-section, ×6; h, i, extrorse anther, ×6; k, fruit, ×2; m, schematic long-section of seed, ×2.

around the periphery of a columnar, apically sterile, free-central placenta, or rarely to a basal placenta (i.e., the placental column sometimes obsolete), embedded in a mucilaginous matrix that fills the ovarian cavity; endosperm-development nuclear. FRUIT a large, several-seeded, commonly yellow to orange-red, often rather dry berry, or seldom a 1-seeded drupe; seeds rather large, yellow to orange-red, with copious, oily endosperm and a rather large, straight embryo with well developed cotyledons. N = 18.

The family Theophrastaceae consists of 4 genera and about a hundred species of essentially New World tropical distribution, occurring from Mexico and southern Florida to northern Paraguay. A few of the species are cultivated for ornament in tropical countries, but none is of any great economic importance. The West Indian species *Jacquinia barbasco* (Loefl.) Mez is said to be used as a fish-poison. The two largest genera are *Clavija* and *Jacquinia*, with about 50 species each. *Theophrasta* and *Deherainia* have only 2 or 3 species each.

2. Family MYRSINACEAE R. Brown 1810 nom. conserv., the Myrsine Family

Mostly evergreen trees or shrubs or sometimes woody vines, seldom only half-shrubby, rarely (*Ardisia primulifolia* Gardn. & Champ.) herbaceous and nearly acaulescent, sometimes epiphytic, characteristically with scattered schizogenous secretory ducts or cavities containing a yellowish to brownish resinous material in some of the parenchymatous tissues of the shoot, and commonly with scattered individual secretory cells as well, tanniferous, usually with proanthocyanins but only seldom with ellagic acid, producing benzoquinones and triterpenoid saponins, but without iridoid compounds and not cyanogenic; solitary or scattered crystals of calcium oxalate commonly present in some of the cells of the parenchymatous tissues; young stem with the xylem-cylinder continuous, not broken up into vascular bundles; nodes unilacunar; vessel-segments with simple perforations, or occasionally some of them with scalariform perforation-plates; imperforate tracheary elements with simple pits, usually septate; wood-rays heterocellular to seldom essentially homocellular, typically all multiseriate (4–30 cells wide) and often also very high, but sometimes only 2–4-seriate, or some of them uniseriate; wood-parenchyma paratracheal and generally scanty, seldom diffuse or none. LEAVES alternate, but sometimes so closely crowded as

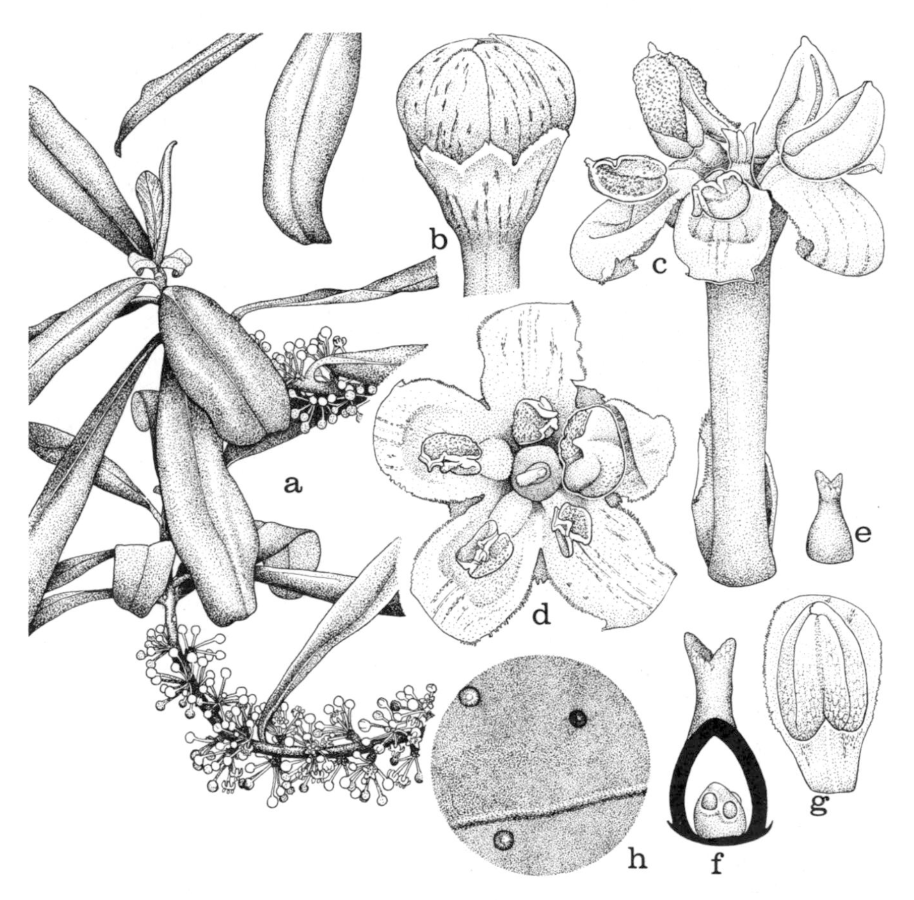

FIG. 4-48 Myrsinaceae. *Myrsine salicina* Heward. a, habit, ×½; b, flower bud, ×9; c, d, flower, ×9; e, pistil, ×9; f, pistil, in partial long-section, ×25; g, stamen and corolla-lobe, ×9; h, lower leaf-surface, with scattered glands, ×25.

to appear whorled, simple, entire or less often toothed, pinnately veined, glandular-punctate or with schizogenous secretory lines, often glandular-hairy, the glandular trichomes often peltate or capitate with a multicellular head, depressed below the surface in *Aegiceras*; epidermis often mucilaginous; stomates anomocytic or anisocytic; sclerenchymatous subepidermal fibers and submarginal strands wanting; petiolar anatomy often complex; stipules wanting. FLOWERS small, in axillary or less often terminal inflorescences of various sorts, in most genera axillary to bracts but without bracteoles (bracts wanting in *Aegiceras*, bracteoles present in *Maesa*), regular or nearly so (at least as to the corolla), hypogynous to half-epigynous, perfect or seldom unisexual (the plants then mostly dioecious); calyx of (3) 4–5 (6) sepals, these distinct or more often connate at the base, variously imbricate, convolute, or valvate, persistent, often glandular-punctate; corolla sympetalous, rotate or tubular, the lobes variously imbricate, convolute, or rarely valvate, or seldom (notably in *Embelia*) the petals distinct; stamens as many as and opposite the corolla-lobes (or petals), adnate to the corolla-tube or rarely almost free, the filaments distinct or sometimes connate below into a short tube; anthers tetrasporangiate and dithecal, introrse, opening by longitudinal slits or sometimes by apical pores, in *Aegiceras* transversely septate; pollen-grains binucleate, 3 (–5)-colporate; staminodes wanting, but often represented by vestigial traces in the receptacle; gynoecium of 3–5 (6) carpels united to form a compound, unilocular, superior or rarely (*Maesa*) half-inferior ovary with a short terminal style and a simple or obscurely lobed, often capitate stigma; ovules few to numerous, anatropous to hemitropous, or nearly campylotropous, bitegmic (unitegmic in *Aegiceras*), tenuinucellar, with an integumentary tapetum, generally more or less embedded in the free-central (or basal) placenta; endosperm-development nuclear. FRUIT usually a berry or drupe (an elongate, 1-seeded capsule in *Aegiceras*), commonly 1-seeded (but with rather numerous seeds in *Maesa*); seeds relatively small, generally dark-colored, with an axile, cylindrical, straight or slightly curved, dicotyledonous embryo and oily, fleshy or hard endosperm, or seldom (*Aegiceras*) the seeds larger and without endosperm at maturity. X = 10–13, 23.

The family Myrsinaceae as here defined consists of more than 30 genera and about a thousand species, widely distributed in tropical and subtropical regions of both the Old and the New World, and also occurring to some extent in temperate regions of the Old World. Well over half of the species belong to only 4 genera, *Ardisia* (250), *Rapanea* (150), *Embelia* (130) and *Maesa* (100). A few species are sometimes cultivated for ornament, but none is of any great economic importance.

Aegiceras, a genus of 2 palaeotropical species, has sometimes been taken as a separate family. It certainly stands apart from the rest of the Myrsinaceae, but the relationship is not in dispute. *Aegiceras* is a mangrove, and some of the principal characters by which it differs from typical Myrsinaceae relate to the adaptation of the seed and fruit to the mangrove habit. The inclusion of *Aegiceras* in the Myrsinaceae creates no problems in the scheme of classification.

3. Family PRIMULACEAE Ventenat 1799 nom. conserv., the Primrose Family

Annual or more often perennial herbs, or rarely half-shrubs, commonly accumulating triterpenoid saponins, notably primulagenin, and mostly with scattered tanniferous cells containing proanthocyanins but not ellagic acid, sometimes also (as in *Lysimachia*) with secretory cells or cavities containing red or reddish materials (even the petals sometimes streaked in lines with secretory canals), but not cyanogenic, and without iridoid compounds; trichomes variously uniseriate, or branched, or multicellular and capitate, the capitate hairs producing a crystalline substance that causes the farinaceous appearance of the leaves of *Primula*; calcium oxalate crystals wanting; stem commonly with a cylinder of separate vascular bundles separated by broad medullary rays, but the bundles sometimes in a more complex, anomalous arrangement, or rather numerous and separated by narrow medullary rays so that the xylem forms a nearly continuous cylinder; nodes unilacunar; vessel-segments with simple perforations. LEAVES alternate or more often opposite or whorled, often all basal, often glandular-punctate or farinaceous, mostly simple and entire or merely toothed or lobed, but in the aquatic genus *Hottonia* pectinately pinnatisect; stomates mostly anomocytic; petiole often with complex vascular anatomy; stipules wanting. FLOWERS variously in panicles, umbels, racemes, or heads or solitary, individually bracteate but without bracteoles, perfect, hypogynous (half-epigynous in *Samolus*), regular (somewhat irregular in *Coris*), (3–) 5 (–9)-merous, often heterostylic; calyx synsepalous, toothed or lobed, commonly persistent; corolla sympetalous, with imbricate lobes shorter to much longer than the tube (polypetalous in the monotypic, trimerous genus *Pelletiera*, wanting in *Glaux*); stamens as many as and opposite the corolla-lobes, the filaments borne in the corolla-tube (alternate with the calyx-lobes in *Glaux*); anthers introrse, tetrasporangiate and dithecal, opening by longitudinal slits or sometimes by terminal pores; pollen-grains binucleate, tricolporate or 5–8-zonocolpate to 3–10-porate; small staminodes sometimes alternating with the functional stamens, with or without a vascular trace in the receptacle; gynoecium mostly of 5 carpels, these united to form a compound, superior or (*Samolus*) half-inferior ovary, the ovary unilocular but often with vestigial partitions at the base; style solitary, terminating in a usually capitate stigma; ovules (5–) more or less numerous on a free-central placenta,

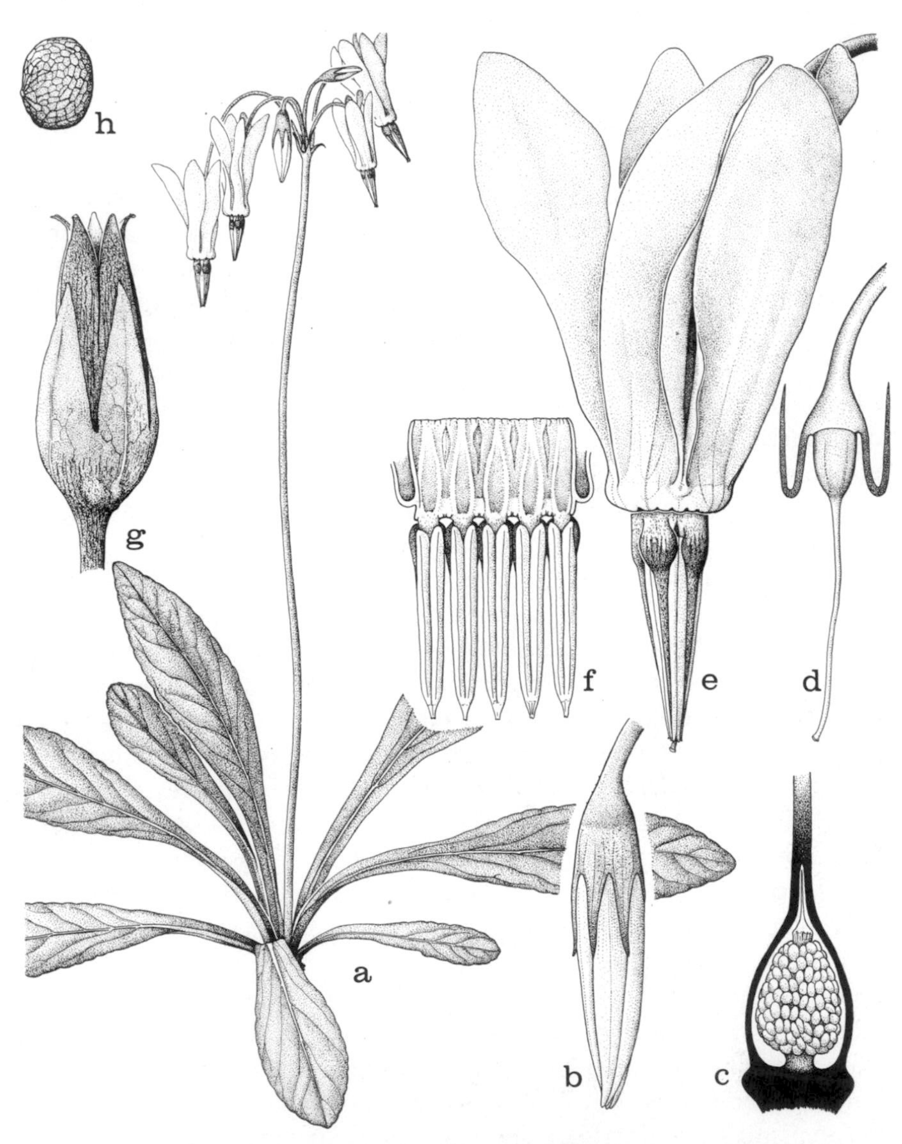

FIG. 4-49 Primulaceae. *Dodecatheon meadia* L. a, habit, ×½; b, flower bud, ×3; c, ovary in long-section, ×6; d, developing pistil, and reflexed sepals, ×3; e, flower, ×3; f, corolla-tube, opened to show attached stamens, ×3; g, opened capsule, with persistent calyx, ×3; h, seed, ×12.

hemitropous or sometimes anatropous, bitegmic, tenuinucellar, with an integumentary tapetum (campylotropous and unitegmic in *Cyclamen);* endosperm-development nuclear, without haustoria. FRUIT capsular, dehiscent by valves or sometimes circumscissile, or rarely indehiscent; seeds (1–) usually more or less numerous; embryo dicotyledonous (monocotyledonous in *Cyclamen*) straight, linear or short, centrally embedded in the copious, mostly rather hard endosperm, which contains reserves of oil, protein, and amylose, but not starch. X = 5, 8–15, 17, 19, 22. (Coridaceae)

The family Primulaceae consists of about 30 genera and a thousand species, occurring mainly in temperate and cold regions of the Northern Hemisphere, and in tropical mountains. About half of the species belong to the single large genus *Primula.* The next largest genera are *Lysimachia* (150) and *Androsace* (100). A number of species, especially of the genera *Primula* and *Cyclamen,* are cultivated for ornament as pot-plants or in rock-gardens. *Primula* is allergenic to some people and is a frequent cause of dermatitis.

V. Subclass ROSIDAE Takhtajan 1966

Woody or herbaceous dicotyledons (seldom vegetatively reduced or modified and thalloid) with diverse sorts of repellents, often including tannins, less often iridoid compounds, and sometimes triterpenoid compounds or various other substances, frequently with one or another sort of alkaloid, but only rarely with benzyl-isoquinoline alkaloids or mustard-oils, and without betalains, vessel-segments with simple or less often scalariform perforations; wood-rays with short or seldom elongate ends; sieve-tubes with S-type or seldom (but notably in the Fabales) P-type plastids. LEAVES simple or often pinnately (less often palmately) compound or dissected, or seldom much-reduced or wanting. FLOWERS hypogynous to perigynous or epigynous, with the petals distinct or occasionally connate at the base, or some petals connate and some distinct, rarely all connate to form a lobed tube, or sometimes the petals much-reduced or wanting; nectaries of various types, often of staminodial origin, frequently forming an intrastaminal or extrastaminal disk; stamens, when numerous, mostly initiated in centripetal sequence (notable exception: Punicaceae); pollen-grains binucleate or less often trinucleate, triaperturate (typically tricolporate) or sometimes with only 2 or more than 3 apertures, or seldom inaperturate; gynoecium apocarpous (or monocarpous) in the Fabales, Proteales, and many Rosales, otherwise usually syncarpous, the ovary variously superior or inferior; placentation various, in compound ovaries most commonly axile (or basal-axile or apical-axile), less often parietal or free-central, or basal or apical in a unilocular ovary; ovary, when plurilocular, very often with only one or 2 ovules per locule, but sometimes with several or many; ovules typically bitegmic and crassinucellar, but sometimes unitegmic, or tenuinucellar, or both, or even ategmic, in scattered families and orders with an integumentary tapetum; endosperm present or absent, but perisperm mostly wanting; embryo tiny to more often fairly large in relation to the size of the seed, usually embedded in the endosperm when endosperm is present, variously straight or curved or even spiral, only rarely peripheral and curved or annular.

The subclass Rosidae as here defined consists of 18 orders, 114 families, and about 58,000 species. It is the largest subclass of angiosperms, in terms of the number of families, but about the same size as the Asteridae in terms of the number of species. Nearly three-fourths of the species in the Rosidae belong to only 5 large orders: Fabales (14,000), Myrtales (9,000), Euphorbiales (7,600), Rosales (6,600), and Sapindales (5,400). The remaining orders have a little more than 15,000 species amongst them.

The 18 orders that make up the Rosidae evidently hang together as a natural group. Only the Euphorbiales and Rafflesiales are obviously debatable, the former because of some similarities to the Malvales in the subclass Dilleniidae, the latter because the morphological reduction in association with parasitism makes their affinities hard to establish.

In general, the Rosidae are more advanced than the Magnoliidae in one or another respect, but less advanced than the Asteridae, and they mostly lack the special features that characterize the Hamamelidae and Caryophyllidae. Those Rosidae that are florally similar to the Hamamelidae are clearly linked to more typical members of their subclass. The Julianiaceae, for example, are obviously related to the Anacardiaceae, in spite of their amentiferous floral organization.

The subclass most likely to be confused with the Rosidae is the Dilleniidae. These likewise are more advanced than the Magnoliidae and less advanced than the Asteridae. Compound leaves with distinct, articulated leaflets are much less common in the Dilleniidae than in the Rosidae. Most of the Dilleniidae are polypetalous (as are the vast majority of the Rosidae), but a considerable number are sympetalous, and some are apetalous or even amentiferous. In contrast to the Rosidae, the Dilleniidae usually have centrifugal stamens (when the stamens are numerous), and only a few families of Dilleniidae (notably the Brassicaceae) have trinucleate pollen. Parietal placentation is common in the Dilleniidae (in contrast to its relative rarity in the Rosidae), but other types are also well represented. Uniovulate or biovulate locules are much less common in the Dilleniidae than in the Rosidae, but are well represented in the Malvales, whose position in the Dilleniidae is well established. Not many of the Dilleniidae have a typical nectary-disk of the sort so common in the Rosidae, but other types of nectaries are common. Studies now under way by Leo Hickey suggest that there may be a fairly consistent difference between the Rosidae and Dilleniidae in the architecture and supporting venation of the leaf-teeth. I have not yet fully grasped the distinction, and have made no effort to incorporate it into the descriptions.

In the last analysis, the Rosidae and Dilleniidae are kept apart as subclasses because each seems to constitute a natural group separately derived from the ancestral Magnoliidae, rather than because of any definitive distinguishing characters. The same sorts of evolutionary advances have occurred in both groups, but with different frequencies. In spite of the lack of solid distinguishing criteria, I believe that it is conceptually more useful to hold the two as separate subclasses than to combine them into one or to abandon any attempt at organization of the Magnoliopsida into subclasses.

The Rosidae show no ecologic unity whatsoever. They exploit a wide range of habitats in a wide variety of ways, and they do not occupy any major habitat or adaptive zone to the exclusion of members of other subclasses.

The Rosidae are evidently derived from the Magnoliidae, as indicated by the fossil record as well as by comparative morphology. None of the especially primitive characters that we have noted in some of the Magnoliidae is known in the Rosidae. Every significant respect in which any member of the Rosidae differs from the more archaic members of the Magnoliidae appears to represent a phyletic advance, with the possible exception of the nature of the chemical defenses, on which the phyletic progression among the subclasses is still debatable.

The fossil record suggests that the Rosidae began to diverge from the ancestral Magnoliidae near the middle of the Albian stage of the Lower Cretaceous, as a group with tricolpate pollen and pinnatifid (then compound) leaves. *Sapindopsis* is a good candidate for the earliest recognizable antecedent of the subclass. The implication here is that the pinnate-leaved habit, so common in the Rosidae, is basic to the group, and that modern families of Rosidae with simple leaves have a pinnate ancestry (cf. Hickey & Doyle, 1977, in general citations). On the basis of comparative morphology of modern species, there is no doubt that simple leaves have repeatedly been derived from compound leaves in several families of the Rosidae, with unifoliolate compound leaves as an intermediate stage. Extension of this interpretation to all the simple-leaved members of the subclass may seem a bit drastic, but it is entirely compatible with the phyletic arrangements I have previously proposed for the group. On the other hand, the occurrence of fossil pollen of the Aquifoliaceae in early Upper Cretaceous (Turonian) deposits suggests that simple leaves in the Rosidae have a long evolutionary history.

Two caveats should be entered at this point. One is that in the absence of fossil flowers it is not possible to identify *Sapindopsis* with any modern family or order of Rosidae. Indeed the tricolpate pollen associated with *Sapindopsis* militates against such an identification. In the modern members of the subclass truly tricolpate pollen is relatively rare, and triaperturate types are generally tricolporoidate or tricolporate or triporate. Hickey & Doyle (1977, in general citations) consider that the most primitive pollen-type among modern Rosidae is tricolporoidate, and that the few tricolpates in the group are secondary. Tricolporoidate pollen makes its first appearance in the Late Albian, some time after the first appearance of *Sapindopsis* leaves. The second caveat is that these ages are based on fossils of the Potomac Group, which may reflect immigration after a somewhat earlier divergence in Gondwanaland.

The Rosales are the most archaic order of their subclass. All other orders of the Rosidae appear to be derived directly or indirectly from the Rosales. If the other orders of the subclass were wiped out of existence, the Rosales could be accommodated without great difficulty as a somewhat isolated order of the Magnoliidae. A large proportion of the Rosales are apocarpous, whereas apocarpy is rare in the rest of the group (except for the basically monocarpous orders Fabales and Proteales). On the other hand, the relationship between the Rosales and some of the more advanced orders of the subclass is so obvious, and the

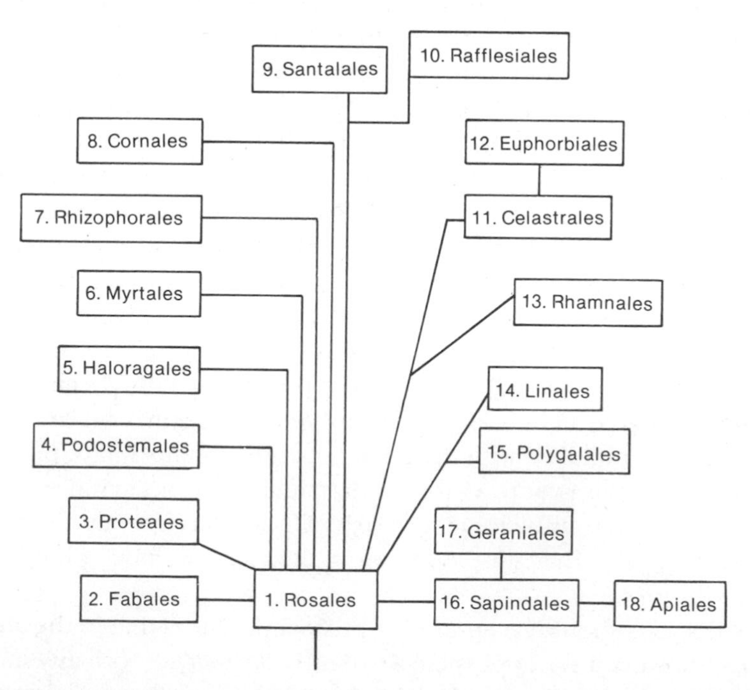

FIG. 5.1 Putative evolutionary relationships among the orders of Rosidae. The length of the lines reflects the exigencies of two-dimensional representation.

difficulty of delimiting these from the Rosales is so great, that it seems necessary to associate the Rosales with their descendents rather than with their collateral ancestors.

SELECTED REFERENCES

Heimsch, C. H. 1942. Comparative anatomy of the secondary xylem in the "Gruinales" and "Terebinthales" of Wettstein with reference to taxonomic grouping. Lilloa 8: 83–198.

Mauritzon, J. 1939. Contributions to the embryology of the orders Rosales and Myrtales. Acta Univ. Lund. N.S. Avd. 2, 35(2): 1–121.

Rullmann, H.-D. 1978. Phytoserologische Untersuchungen zur Gliederung der Geraniales. Dissertation zur Erlangung des Doktorgrades des Fachbereichs Math.-Naturwiss. der Christian-Albrechts-Univ. Kiel.

SYNOPTICAL ARRANGEMENT OF THE ORDERS OF ROSIDAE

1 Flowers of relatively primitive structure, commonly either apocarpous (with 1-many carpels), or with more or less numerous stamens, often with more than one of these features; plants never with internal phloem, never parasitic (though sometimes insectivorous), and never highly modified aquatics.

2 Gynoecium mostly of 2-many carpels, only seldom (but notably in some Connaraceae and in the subfamily Prunoideae of the Rosaceae) of a single carpel, and then the plants not with the special features of the Fabales or Proteales; endosperm variously present or absent .... 1. ROSALES.

2 Gynoecium nearly always of a single carpel; endosperm usually none or scanty, seldom well developed.

3 Petals usually present; perianth most commonly 5-merous, only seldom 4-merous or of other numbers; stamens most commonly 10, less often 9, seldom fewer or more numerous; leaves mostly compound, seldom simple, mostly stipulate and with a basal pulvinus; fruit of various types, but most commonly dry and dehiscent along both sutures, usually with 2 or more seeds; ovules (1) 2-many .... 2. FABALES.

3 Petals wanting, or represented only by scales or glands; flowers mostly 4-merous, with not more than 8 stamens; leaves variously simple or compound, but exstipulate and without a pulvinus; fruit of diverse types, very often with only a single seed, dehiscent (if at all) along only one suture; ovules 1-several or seldom many .... 3. PROTEALES.

1 Flowers more advanced in one or another respect, mostly syncarpous (seldom pseudomonomerous or with distinct carpels), typically with not more than twice as many stamens as sepals or petals

(stamens numerous in some Myrtales—notably the Myrtaceae, which have internal phloem, only rarely numerous in other groups), very often with only 1 or 2 ovules per carpel; plants variously autotrophic or sometimes parasitic, sometimes highly modified aquatics.

4 Stem commonly with internal phloem; flowers usually strongly perigynous or epigynous, often tetramerous, often with numerous stamens and numerous ovules; embryo often of unusual structure, but also often of perfectly ordinary type ....6. MYRTALES.

4 Stem nearly always without internal phloem; flowers and embryo various.

5 Plants mostly parasitic or hemiparasitic (wholly autotrophic in some Santalales); gynoecium very often unusual in either the number or the arrangement of the ovules, which are often much-reduced in structure.

6 Ovules few, commonly 1–8 or up to about 12, often on a free-central or basal placenta, or pendulous from near the summit of a columnar placenta that extends upward beyond the basal partial partitions; seeds solitary, or rarely 2–3; plants with or without chlorophyll ....9. SANTALALES.

6 Ovules and seeds very numerous, commonly in the thousands, on parietal, often deeply intruded placentas or in an irregularly plurilocular ovary; plants without chlorophyll ....10. RAFFLESIALES.

5 Plants (with few exceptions, such as the Krameriaceae) autotrophic.

7 Flowers generally more or less reduced (not very much so in some Euphorbiaceae) and often unisexual, the perianth mostly poorly developed or wanting; styles generally distinct or only basally connate.

8 Endosperm wanting; flowers perfect, hypogynous, generally with numerous ovules; highly modified aquatics ....4. PODOSTEMALES.

8 Endosperm more or less well developed; flowers with only 1 or 2 ovules per locule, otherwise with diverse structure; habitat various.

9 Mostly herbs, often aquatic, with an inferior ovary and without milky juice; flowers not forming pseudanthia ....5. HALORAGALES.

9 Herbs, shrubs, or trees, with superior or naked ovary and often with milky juice; flowers often grouped into pseudanthia .... 12. EUPHORBIALES.

7 Flowers mostly of ordinary type, generally not much reduced except in a few small families such as the Garryaceae and Julianiaceae, very often with a nectary-disk and very often with only one or two ovules per locule; styles distinct, or very often connate for much or all of their length into a single compound style.

10 Leaves mostly simple and entire or merely toothed, only seldom pinnatifid (as in *Aralidium*, of the Cornaceae), or compound (as in many Vitaceae, order Rhamnales).

11 Flowers usually epigynous or strongly perigynous.

12 Ovules bitegmic, 2 (or more) per carpel; leaves stipulate; fruit mostly a berry, seldom a capsule; plants not known to have iridoid compounds, sometimes of mangrove habit .... 7. RHIZOPHORALES.

12 Ovules unitegmic, one per carpel (2 ovules present in the single locule of the bicarpellate ovary of Garryaceae); leaves nearly always without stipules; fruit usually a drupe, seldom a berry; plants often producing iridoid compounds, never of mangrove habit .... 8. CORNALES.

11 Flowers hypogynous, or sometimes half-epigynous or somewhat perigynous.

13 Flowers regular or nearly so; anthers generally opening by longitudinal slits, only seldom by transverse slits or short distal slits or terminal pores.

14 Stamens distinct, seldom more than 5; flowers usually with a solitary (sometimes cleft) style, seldom with distinct styles.

15 Stamens alternate with the petals, or rarely both alternate and opposite the petals ....11. CELASTRALES.

15 Stamens opposite the petals ....13. RHAMNALES.

14 Stamens usually connate by their filaments, at least toward the base, often more than 5; styles distinct, or style sometimes solitary and then variously entire or cleft .... 14. LINALES.

13 Flowers either more or less strongly irregular, or with poricidal anthers, or often both .... 15. POLYGALALES.

10 Leaves mostly compound or conspicuously lobed or cleft, only seldom (as in the Balsaminaceae) simple and unlobed.

16 Ovary superior; pollen-grains binucleate or less often trinucleate.
17 Plants mostly woody, seldom herbaceous .... 16. SAPINDALES.
17 Plants mostly herbaceous or nearly so, seldom shrubby or even arborescent .... 17. GERANIALES.
16 Ovary (with the rarest of exceptions) inferior; pollen-grains mostly trinucleate .... 18. APIALES.

1. Order ROSALES Lindley 1833

Autotrophic (sometimes insectivorous), terrestrial or occasionally semi-aquatic plants, woody or herbaceous, chemically and anatomically diverse, but without milky juice, and without internal phloem; nodes trilacunar, or less often unilacunar or multilacunar. LEAVES variously alternate or opposite or whorled, simple or compound, with or without stipules. FLOWERS perfect or seldom unisexual, hypogynous to more often perigynous or sometimes epigynous, regular to somewhat irregular, generally with a well developed (most commonly pentamerous) perianth that is differentiated into sepals and petals; petals distinct or seldom more or less connate towards the base, or sometimes wanting; stamens numerous and developing in centripetal sequence, or in 1-several more or less definite cycles, or rarely the stamen solitary; pollen-grains triaperturate or of triaperturate-derived type; gynoecium of 1-many distinct carpels, or the carpels united to form a compound pistil with separate styles or a single style (or the stigmas sessile); placentation variously marginal, axile, parietal, or basal; ovules (1–) 2-several or many, bitegmic or less often unitegmic, crassinucellar or less often tenuinucellar; endosperm-development nuclear or cellular. FRUIT of various sorts; seeds with or without endosperm; embryo with 2 cotyledons. (Connarales, Cunoniales, Grossulariales, Pittosporales, Saxifragales)

The order Rosales as here defined consists of 24 families and about 6600 species, widely distributed throughout the world. The Rosaceae, with about 3000 species, are by far the largest family, followed by the Crassulaceae (900) and Saxifragaceae (700). These 3 well known families together make up about seven-tenths of the order. Another 6 families of moderate size are familiar to botanists generally. These are the Chrysobalanaceae (400), Grossulariaceae (350), Cunoniaceae (350), Connaraceae (350), Pittosporaceae (200), and Hydrangeaceae (170). The remaining 15 small, much less familiar families have only about 225 species in all and collectively constitute less than 4% of the order. These are the Bruniaceae (75), Brunelliaceae (50), Anisophylleaceae (40), Alseuosmiaceae (12), Neuradaceae (10), Crossosomataceae (8–9), Surianaceae (6), Eucryphiaceae (5), Byblidaceae (4), Columelliaceae (4), Rhabdodendraceae (3), Greyiaceae (3), Cephalotaceae (1), Davidsoniaceae (1), and Dialypetalanthaceae (1).

The Rosales form an exceedingly diverse order, standing at the evolutionary base of their subclass. In effect, they are what is left over after all the more advanced, specialized orders have been delimited.

Aside from internal phloem and a parasitic or highly modified aquatic habitat, most of the features that mark the more advanced orders of the subclass Rosidae (and indeed even some of the features of the Asteridae) can be found individually within the order Rosales, but in the Rosales these features do not occur in the combinations that mark the more advanced groups. Two characters that are very common in the Rosales are much less common among the other orders of the subclass. These are a polystemonous androecium and an apocarpous gynoecium (although the gynoecium is monocarpous in the large order Fabales). Furthermore, a great many of the Rosales have more or less numerous ovules per carpel, and relatively few have only one or two. In contrast, many of the other orders of the Rosidae have only one or two ovules per carpel, and most of those that have numerous ovules are well differentiated from the Rosales in other respects.

In spite of the diversity of the order, and in spite of the lability of the critical characters even within some of the larger families, the 24 families all appear to be interrelated. None of the families is notably archaic within the order. Each family has its own combination of relatively primitive and more advanced features. The subordinal groupings are, I think, for the most part useful and reasonably in accord with probable phylogenetic relationships, but the Connaraceae may be as closely allied to the Rosineae and even to the Fabales as they are to the Cunoniineae. It would be possible to elevate the 5 suborders here recognized to full ordinal status, but I would be reluctant to see the Rosaceae and Saxifragaceae referred to separate orders, and if these two are kept together it is difficult to extricate any of the other suborders. Conceptually, I find it more useful to delimit the order broadly and to recognize suborders within it, than to fragment the order and lose sight of the interrelationships among its parts.

SELECTED REFERENCES

Baehni, C., & P. Dansereau. 1939. *Polygonanthus*, genre de Saxifragacées. Bull. Soc. Bot. France 86: 183–186.

Bange, G. G. J. 1952. A new family of dicotyledons: Davidsoniaceae. Blumea 7: 293–296.

Banwar, S. C. 1970. Fossil leaves of *Lyonothamnus*. Madroño 20: 359—364.

Bates, J. C. 1933. Comparative anatomical research within the genus *Ribes*. Univ. Kansas Sci. Bull. 21: 369–398.

Bate-Smith, E. C. 1961. Chromatography and taxonomy in the Rosaceae, with special reference to *Potentilla* and *Prunus*. J. Linn. Soc., Bot. 58: 39–54.

Bate-Smith, E. C. 1977. Chemistry and taxonomy of the Cunoniaceae. Biochem. Syst. Ecol. 5: 95–105.

Bausch, J. 1938. A revision of the Eucryphiaceae. Kew Bull. Misc. Inform. 1938: 317–349.

Becker, H. F. 1963. The fossil record of the genus *Rosa*. Bull. Torrey Bot. Club 90: 99–110.

Bensel, C. R., & B. F. Palser. 1975. Floral anatomy in the Saxifragaceae sensu lato. I. Introduction, Parnassioideae and Brexioideae. II. Saxifragoideae and Iteoideae. III. Kirengeshomoideae, Hydrangeoideae and Escallonioideae. IV. Baueroideae and conclusions. Amer. J. Bot. 62: 176–185; 661–675; 676–687; 688–694.

Brizicky, G. K. 1961. A synopsis of the genus *Columellia* (Columelliaceae). J. Arnold Arbor. 42: 363–372.

Carlquist, S. 1976. Wood anatomy of Byblidaceae. Bot. Gaz. 137: 35–38.

Carlquist, S. 1976. Wood anatomy of Roridulaceae: ecological and phylogenetic implications. Amer. J. Bot. 63: 1003–1008.

Carlquist, S. 1978. Wood anatomy of Bruniaceae: correlations with ecology, phylogeny, and organography. Aliso 9: 323–364.

Challice, J. S. 1974 (1975). Rosaceae. Chemotaxonomy and the origins of the Pomoideae. J. Linn. Soc., Bot. 69: 239–259.

Croizat, L. 1939. Une nouvelle sous-famille des Olacacées au Brésil. Bull. Soc. Bot. France 86: 5–7.

Cuatrecasas, J. 1970. Brunelliaceae. Flora Neotropica Monograph No. 2. Hafner. Darien, Conn.

Dahlgren, K. V. O. 1930. Zur Embryologie der Saxifragoideen. Svensk Bot. Tidskr. 24: 429–448.

Dahlgren, R., S. R. Jensen, & B. J. Nielsen. 1977. Seedling morphology and iridoid occurrence in *Montinia caryophyllacea* (Montiniaceae). Bot. Not. 130: 329–332.

Dandy, J. E. 1927. The genera of Saxifragaceae. Pp. 107–118 *in*: J. Hutchinson, Contributions towards a phylogenetic classification of flowering plants, VI. Kew Bull. Misc. Inform. 1927: 100–118.

DeBuhr, L. E. 1978. Wood anatomy of *Forsellesia* (*Glossopetalon*) and *Crossosoma* (Crossosomataceae, Rosales). Aliso 9: 179–184.

Dickinson, T. A., & R. Sattler. 1974. Development of the epiphyllous inflorescence of *Phyllonoma integerrima* (Turcz.) Loes.: implications for comparative morphology. J. Linn. Soc., Bot. 69: 1–13.

Dickison, W. C. 1971–1974. Anatomical studies in the Connaraceae. I. Carpels. II. Wood Anatomy. III. Leaf anatomy. IV. The bark and young stem. J. Elisha Mitchell Sci. Soc. 87: 77–86, 1971. 88: 120–136. 89: 121–138, 1973; 166–171, 1973 (1974).

Dickison, W. C. 1975. Floral morphology and anatomy of *Bauera*. Phytomorphology 25: 69–76.

Dickison, W. C. 1975. Studies on the floral anatomy of the Cunoniaceae. Amer. J. Bot. 62: 433–447.

Dickison, W. C. 1975. Leaf anatomy of Cunoniaceae. J. Linn. Soc., Bot. 71: 275–294.

Dickison, W. C. 1977. Wood anatomy of *Weinmannia* (Cunoniaceae). Bull. Torrey Bot. Club 104: 12.

Dickison, W. C. 1978. Comparative anatomy of Eucryphiaceae. Amer. J. Bot. 65: 722–735.

Dickison, W. C. 1979. A survey of pollenmorphology of the Connaraceae. Pollen & Spores 21: 31–79.
Dickison, W. C. 1980. Diverse nodal anatomy of the Cunoniaceae. Amer. J. Bot. 67: 975–981.
Ensign, M. 1942. A revision of the celastraceous genus *Forsellesia* (*Glossopetalon*). Amer. Midl. Naturalist 27: 501–511.
Erdtman, G. 1946. Pollen morphology and plant taxonomy. VII. Notes on various families. Svensk Bot. Tidskr. 40: 77–84.
Ferguson, I. K., & D. A. Webb. 1970. Pollen morphology in the genus *Saxifraga* and its taxonomic significance. J. Linn. Soc., Bot. 63: 295–311.
Fryns-Claessens, E., & W. Van Cotthem. 1966. L'appareil stomatique des Pandacées et Davidsoniacées. Rev. Gén. Bot. 63: 783–789.
Gardner, R. O. 1978. Systematic notes on the Alseuosmiaceae. Blumea 24: 138–142.
Gelius, L. 1967. Studien zur Entwicklungsgeschichte an Blüten der Saxifragales sensu lato mit besonderer Berücksichtigung des Androeceums. Bot. Jahrb. Syst. 87: 253–303.
Goldblatt, P. 1976. Cytotaxonomic studies in the tribe Quillajeae (Rosaceae). Ann. Missouri Bot. Gard. 63: 200–206.
Gorenflot, R. 1971. Intérêt taxonomique et phylogénique des caractères stomatiques (application à la tribu des Saxifragées). Boissiera 19: 181–192.
Gorenflot, R., & F. Moreau. 1970. Types stomatiques et phylogénie des Saxifraginées (Saxifragacées). Compt. Rend. Hebd. Séances Acad. Sci. 270 Ser. D.: 2802–2805.
Gutzwiller, M.-A. 1961. Die phylogenetische Stellung von *Suriana maritima* L. Bot. Jahrb. Syst. 81: 1–49.
Hamel, J. 1953. Contribution a l'étude cyto-taxinomique des Saxifragacées. Rev. Cytol. Biol. Vég. 14: 113–311.
Herr, J. M. 1954. The development of the ovule and female gametophyte in *Tiarella cordifolia*. Amer. J. Bot. 41: 333–338.
Hideux, M. J., & I. K. Ferguson. 1976. The stereostructure of the exine and its evolutionary significance in Saxifragaceae sensu lato. *In*: I. K. Ferguson & J. Muller, eds., The evolutionary significance of the exine, pp. 327–377. Linn. Soc. Symposium Ser. No. 1. Academic Press. London and New York.
Hilger, H. H. 1978. Der multilakunäre Knoten einiger *Melianthus*- und *Greyia*-Arten im Vergleich mit anderen Knotentypen. Flora B 167: 165–176.
Hoogland, H. G. 1960. Studies in the Cunoniaceae. 1. The genera *Ceratopetalum, Gillbeea, Aistopetalum,* and *Calycomis*. Austral. J. Bot. 8: 318–341.
Huber, H. 1963. Die Verwandtschaftsverhältnisse der Rosifloren. Mitt. Bot. Staatssamml. München 5: 1–48.
Hutchinson, J. 1927. Contributions towards a phylogenetic classification of flowering plants. VI. Hydrangeaceae and Saxifragaceae. A. The genera of Hydrangeaceae. Kew Bull. Misc. Inform. 1927: 100–107.
Ingle, H. D., & H. E. Dadswell. 1956. The anatomy of the timbers of the Southwest Pacific area. IV. Cunoniaceae, Davidsoniaceae, and Eucryphiaceae. Austral. J. Bot. 4: 125–151.
Jackson, G. 1934. The morphology of the flowers of *Rosa* and certain closely related genera. Amer. J. Bot. 21: 453–466.
Jay, M. 1967. Recherches chimiotaxinomiques sur les plantes vasculaires. Distribution des flavonoides chez les Saxifragacées. Compt. Rend. Hebd. Séances Acad. Sci. 264 Ser. D.: 1754–1756.

Jay, M. 1968. Distribution des flavonoides chez les Bruniacées. Taxon 17: 484–488.
Jay, M. 1968. Distribution des flavonoides chez les Cunoniacées. Taxon 17: 489–495.
Jay, M. 1969. Chemotaxonomic researches on vascular plants XIX. Flavonoid distribution in the Pittosporaceae. J. Linn. Soc., Bot. 62: 423–429.
Jay, M. 1970 (1971). Quelques problèmes taxinomiques et phylogénétiques des Saxifragacées vus à la lumière de la biochemie flavonique. Bull. Mus. Hist. Nat. (Paris) II. 42: 754–775.
Jay, M., & P. Lebreton. 1965. Recherches chimiotaxinomiques sur les plantes vasculaires. III. Distribution des flavonoides dans le genre *Saxifraga*. Bull. Soc. Bot. France, Mém. 112: 125–140.
Jensen, L. C. W. 1968. Primary stem vascular patterns in three subfamilies of the Crassulaceae. Amer. J. Bot. 55: 553–563.
Kania, W. 1973. Entwicklungsgeschichtliche Untersuchungen an Rosaceenblüten. Bot. Jahrb. Syst. 93: 175–246.
Kapil, R. N., & R. S. Vani. 1963. Embryology and systematic position of *Crossosoma californicum*. Curr. Sci. 32: 493–495.
Klopfer, K. 1968, 1970, 1972. Beiträge zur floralen Morphogenese und Histogenese der Saxifragaceae. 1. Die Infloreszenz-Entwicklung von *Tellima grandiflora*. 2. Die Blütenentwicklung von *Tellima grandiflora*. 4. Die Blütenentwicklung einiger *Saxifragen*-Arten. 7. *Parnassia palustris* und *Francoa sonchifolia*. Flora B 157: 461–476, 1968. 158: 1–21, 1968. 159: 347–365, 1970. 161: 320–332, 1972.
Klopfer, K. 1973. Florale Morphogenese und Taxonomie der Saxifragaceae sensu lato. Feddes Repert. 84: 475–516.
Krach, J. E. 1976. Die Samen der Saxifragaceae. Bot. Jahrb. Syst. 97: 1–60.
Krach, J. E. 1977. Seed characters in and affinities among the Saxifragineae. Pl. Syst. Evol. Suppl. 1: 141–153.
Kumazawa, M. 1970. On the pleiocotyly in the genus *Pittosporum*. A preliminary note. Bot. Mag. (Tokyo) 83: 119–124. (In Japanese, with English abstract.)
Leinfellner, W. 1964. Uber die falsche Sympetalie bei *Lonchostoma*. Oesterr. Bot. Z. 111: 345–353.
Leinfellner, W. 1970. Über die Karpelle der Connaraceen. Oesterr. Bot. Z. 118: 542–559.
Lemesle, R. 1948. Position phylogénétique de l'*Hydrastis canadensis* L. et du *Crossosoma californicum* Nutt., d'après les particularités histologiques du xylème. Compt. Rend. Hebd. Séances Acad. Sci. 227: 221–223.
Lopez Naranjo, H., & H. Huber. 1971. Anatomia comparativa de las semillas de *Brunellia* y *Weinmannia* con respecto a su posicion sistematica. Pittieria 3: 19–28.
Mason, C. T. 1975. *Apacheria chiracahuensis*: a new genus and species from Arizona. Madroño 23: 105–108.
Mitchell, R. E., & T. A. Geissman. 1971. Constituents of *Suriana maritima*: A triterpene diol of novel structure and a new flavonol glycoside. Phytochemistry 10: 1559–1567.
Moffett, A. A. 1931. The chromosome constitution of the Pomoideae. Proc. Roy. Soc. London, Ser. B, Biol. Sci. 108: 423–446.
Moreau, F. 1971. Apport des caractères stomatiques à la taxinomie et à la phylogenie des Saxifragées. Bull. Soc. Bot. France 118: 381–427.

Morf, E. 1950. Vergleichend-morphologische Untersuchungen am Gynoeceum der Saxifragaceen. Ber. Schweiz. Bot. Ges. 60: 516–590.

Murbeck, S. 1916. Über die Organisation, Biologie und verwandtschaftlicher Beziehungen der Neuradoideen. Acta Univ. Lund. n.s. 12(6): 1–28.

Narayana, L. L., & M. Radhakrishnaiah. 1976. Floral anatomy of the Pittosporaceae 1. J. Jap. Bot. 51: 278–282.

Narayana, L. L., & K. T. Sundari. 1977. Embryology of Pittosporaceae (1). J. Jap. Bot. 52: 204–209.

Paliwal, G. S., & L. M. Srivastava. 1969 (1970). The cambium of *Alseuosmia*. Phytomorphology 19: 5–8.

Pastre, A., & A. Pons. 1973. Quelques aspects de la systématique des Saxifragacées a la lumière des données de la palynologie. Pollen & Spores 15: 117–133.

Pillans, N. S. 1947. A revision of Bruniaceae. J. S. African Bot. 13: 121–206.

Pires, J. M., & W. A. Rodriguez. 1971. Notas sôbre os gêneros *Polygonanthus* e *Anisophyllea*. Acta Amazonica 1(2): 7–15.

Plouvier, V. 1965. Études chimiotaxinomiques sur les Saxifragacées. Bull. Soc. Bot. France, Mém. 1965: 150–161.

Prance, G. T. 1965. The systematic position of *Stylobasium* Desf. Bull. Jard. Bot. État 35: 435–448.

Prance, G. T. 1968. The systematic position of *Rhabdodendron* Gilg & Pilg. Bull. Jard. Bot. Nat'l. Belg. 38(2): 127–146.

Prance, G. T. 1970. The genera of Chrysobalanaceae in the southeastern United States. J. Arnold Arbor. 51: 521–528.

Prance, G. T. 1972. Chrysobalanaceae. Flora Neotropica Monogr. 9.

Prance, G. T. 1972. Rhabdodendraceae. Flora Neotropica Monogr. 11.

Prance, G. T., D. J. Rogers, & F. White. 1969. A taximetric study of an angiosperm family: generic delimitation in the Chrysobalanaceae. New Phytol. 68: 1203–1234.

Puff, C., & A. Weber. 1976. Contributions to the morphology, anatomy, and karyology of *Rhabdodendron*, and a reconsideration of the systematic position of the Rhabdodendraceae. Pl. Syst. Evol. 125: 195–222.

Raghavan, T. S., & V. K. Srinivasan. 1942. A contribution to the life history of *Vahlia viscosa*, Roxb. and *V. oldenlandioides*, Roxb. Proc. Indian Acad. Sci. B. 15: 83–105.

Richardson, P. E. 1970. The morphology of the Crossosomataceae: I. Leaf, stem and node. Bull. Torrey Bot. Club 97: 34–39.

Rizzini, C. T., & P. Occhioni. 1949. Dialypetalanthaceae. Lilloa 17: 243–286.

Robertson, K. R. 1974. The genera of Rosaceae in the southeastern United States. J. Arnold Arbor. 55: 303–332; 344–401; 611–662.

Savile, D. B. O. 1975. Evolution and biogeography of Saxifragaceae with guidance from their rust parasites. Ann. Missouri Bot. Gard. 62: 354–361.

Sax, K. 1931. The origin and relationships of the Pomoideae. J. Arnold Arbor. 12: 3–22.

Sax, K. 1932. Chromosome relationships in the Pomoideae. J. Arnold Arbor. 13: 363–367.

Saxena, N. P. 1964. Studies in the family Saxifragaceae. I. A contribution to the morphology and embryology of *Saxifraga diversifolia* Wall. II. Development of ovule and megagametophyte in *Parnassia nubicola* Wall. Proc. Indian Acad. Sci. B 60: 38–51; 196–202.

Schaeppi, H. 1970. Untersuchungen über den Habitus von *Aruncus, Astilbe* und einiger ähnlicher Pflanzen. Beitr. Biol Pflanzen 46: 371–387.

Schaeppi, H. 1971. Zur Gestaltung des Gynoeceums von *Pittosporum tobira*. Ber. Schweiz. Bot. Ges. 81: 40–51.
Schaeppi, H., & F. Steindl. 1950. Vergleichend-morphologische Untersuchungen am Gynoecium der Rosoideen. Ber. Schweiz. Bot. Ges. 60: 15–50.
Sharma, V. K. 1968. Morphology, floral anatomy and embryology of *Parnassia nubicola* Wall. Phytomorphology 18: 193–204.
Sherwin, P. A., & R. L. Wilbur. 1971. The contributions of floral anatomy to the generic placement of *Diamorpha smallii* and *Sedum pusillum*. J. Elisha Mitchell Sci. Soc. 87: 103–114.
Sleumer, H. 1968. Die Gattung *Escallonia* (Saxifragaceae). Verh. Kon. Ned. Akad. Wetensch., Afd. Natuurk., Tweede Sect. 58(2): 1–146.
Spongberg, S. A. 1972. The genera of Saxifragaceae in the southeastern United States. J. Arnold Arbor. 53: 409–498.
Spongberg, S. A. 1978. The genera of Crassulaceae in the southeastern United States. J. Arnold Arbor. 59: 197–248.
Sterling, C. 1969. Comparative morphology of the carpel in the Rosaceae. X. Evaluation and summary. Oesterr. Bot. Z. 116: 46–54.
Stern, W. L. 1974, 1978. Comparative anatomy and systematics of woody Saxifragaceae. *Escallonia*. J. Linn. Soc., Bot. 68: 1–20, 1974. *Hydrangea*. J. Linn. Soc., Bot. 76: 83–113, 1978.
Stern, W. L., G. K. Brizicky, & R. H. Eyde. 1969. Comparative anatomy and relationships of Columelliaceae. J. Arnold Arbor. 50: 36–75.
Stern, W. L., E. M. Sweitzer, & R. E. Phipps. 1970. Comparative anatomy and systematics of woody Saxifragaceae. *Ribes*. *In*: N. K. B. Robson, D. F. Cutler, & M. Gregory, eds., New research in plant anatomy, pp. 215–237. J. Linn. Soc., Bot. 63. Suppl. 1. Academic Press. London & New York.
Steyn, E. 1974. Leaf anatomy of *Greyia* Hooker & Harvey (Greyiaceae). J. Linn. Soc., Bot. 69: 45–51.
Swamy, B. G. L. 1954. Morpho-taxonomical notes on the Escallonioideae, Part A. Nodal and petiolar vasculature. J. Madras Univ. 24B: 299–306.
Tatsuno, A., & R. Scogin. 1978. Biochemical profile of Crossosomataceae. Aliso 9: 185–188.
Thorne, R. F., & R. Scogin. 1978. *Forsellesia* Greene (*Glossopetalon* Gray), a third genus in the Crossosomataceae, Rosineae, Rosales. Aliso 9: 171–178.
Tillson, A. H. 1940. The floral anatomy of the Kalanchoideae. Amer. J. Bot. 27: 595–600.
Uhl, C. H. 1963. Chromosomes and phylogeny of the Crassulaceae. Cact. Succ. J. (Los Angeles) 35: 80–84.
Vani-Hardev [no initial] 1972. Systematic embryology of *Roridula gorgonias* Planch. Beitr. Biol. Pflanzen 48: 339–351.
Wakabayashi, M. 1970. On the affinity in Saxifragaceae s. lato with special reference to the pollen morphology. (In Japanese; English summary). Acta Phytotax. Geobot. 24: 128–145.
Watari, S. 1939. Anatomical studies on the leaves of some saxifragaceous plants, with special reference to the vascular system. J. Fac. Sci. Univ. Tokyo, Sect. 3, Bot. 5: 195–316.
Weberling, F. 1976. Weitere Untersuchungen zur Morphologie des Unterblattes bei den Dikotylen. IX. Saxifragaceae s.l., Brunelliaceae und Bruniaceae. Beitr. Biol. Pflanzen 52: 163–181.
Yen, T. K. 1936. Floral development and vascular anatomy of the fruit of *Ribes aureum*. Bot. Gaz. 98: 105–120.

Агабабян, В. Ш. 1960. К палиносистематике семейства Iteaceae. Изв. АН Армянской ССР, Биол. Науки, 13: 99-102.

Агабабян, В. Ш. 1961. Материалы к палиносистематическому изучению семейства Saxifragaceae s. l. Изв. АН Армянской ССР, Биол. Науки, 14 (2): 45-61.

Агабабян, В. Ш. 1961. К палиноморфологии семейства Hydrangeaceae Dum. Изв. АН Армянской ССР, Биол. Науки, 14 (11): 17-26.

Агабабян, В. Ш. 1963. К палиноморфологии рода *Ribes* L. Изв. АН Армянской ССР 16 (4): 93-98.

Агабабян, В. Ш. 1964. Эволюция пыльцы в порядках Cunoniales и Saxifragales в связи с некоторыми вопросами их систематики и филогении. Изв. АН Армянской ССР, Биол. Науки, 17: 59-72.

Борисовская, Г. М. 1960. Анатомо-систематическое исследование некоторых представителей семейства Crassulaceae DC. Вестн. Ленинградск. Унив., сер. биол., 21(4): 159-162.

Гладкова, В. Н. 1972. О присхождении подсемейства Maloideae. Бот. Ж. 57: 42-49.

Демченко, Н. И. 1966. Морфология пыльцы сем. Neuradaceae. Бот. Ж. 51: 559-562.

Демченко, Н. И. 1973. О морфологии пыльцы семейства Chrysobalanaceae. В сб.: Морфология и происхождение современных растений. Труды III Международной Палинологической Конференции. СССР, Новосибирск, 1971: 69-73. Изд. Наука Ленинградск. Отд. Ленинград.

Комар, Г. А. 1967. О природе нижней завязи крыжовниковых (Grossulariaceae). Бот. Ж. 52: 1611-1629.

Комар, Г. А. 1970. Развитие цветка и соцветия некоторых представителей сем. крыжовниковых (Grossulariaceae). Бот. Ж. 55: 954-971.

Мандрик, В. Ю ., & Л. В. Голышкин. 1973. Змбриологическое исследование некоторых видов семейства Crassulaceae. Бот. Ж. 58: 263-272.

Шапошников, Г. Х. 1951. Эволюция некоторых групп тлей в связи с эволюцией розоцветных. В сб.: Чтения памяти Н. А. Холодковского 1950: 28-60. Изд. Наука, Москва-Ленинград.

SYNOPTICAL ARRANGEMENT OF THE FAMILIES OF ROSALES

1 Seeds mostly with more or less well developed (seldom rather scanty) endosperm, or the endosperm wanting in a few Connaraceae and a few Saxifragaceae.

2 Plants more or less strongly woody, and not succulent (herbaceous only in some Byblidaceae and some Hydrangeaceae).

3 Leaves mostly pinnately compound or trifoliolate, seldom unifoliolate or simple, mostly stipulate except in Connaraceae (simple but with prominent stipules in Dialypetalanthaceae); flowers mostly hypogynous or only slightly perigynous (epigynous in Dialypetalanthaceae, half-epigynous in a few Cunoniaceae) (Cunoniineae).

4 Carpels distinct, or united only at the very base, or the carpel solitary; ovules 2 in each carpel, collateral.

5 Leaves opposite or sometimes ternate, with stipules and stipellules; stigma elongate, decurrent on the style .... 1. BRUNELLIACEAE.

5 Leaves alternate, without stipules or stipellules; stigma terminal, capitate .... 2. CONNARACEAE.

4 Carpels with few exceptions united into a compound ovary with axile placentas; ovules (1) 2-many in each locule.

6 Flowers hypogynous or rarely (some Cunoniaceae) half-epigynous; styles as many as the carpels, mostly distinct.

7 Carpels 4–14 (–18), pluriovulate; stamens numerous; ovules epitropous; flowers solitary, axillary, large, with showy petals .... 3. EUCRYPHIACEAE.

7 Carpels 2 (3–5), (1) 2- to several-ovulate; stamens 8–10, or seldom numerous; ovules mostly apotropous; flowers in various sorts of inflorescences (only seldom solitary and axillary), mostly small, the petals mostly shorter than the sepals, or the petals sometimes wanting .... 4. CUNONIACEAE.

6 Flowers epigynous, bicarpellate, with a single style and numerous ovules; stamens 16–25 .... 6. DIALYPETALANTHACEAE.

3 Leaves simple, entire to deeply cleft, exstipulate or nearly so; flowers variously hypogynous to strongly perigynous or epigynous, with united carpels and axile or parietal placentas.

8 Flowers hypogynous; plants (at least the Pittosporaceae) with well developed schizogenous secretory canals; ovules tenuinucellar, unitegmic at least in Pittosporaceae; leaves alternate, but sometimes closely clustered at the tips of the branches. (Pittosporineae).

9 Leaves without insect-catching hairs; mostly shrubs or trees or woody-vines, seldom half-shrubs; endosperm-development nuclear; flowers often with a more or less definite corolla-tube .... 7. PITTOSPORACEAE.

9 Leaves with insect-catching hairs; herbs or half-shrubs; endosperm-development cellular; petals very nearly distinct .... 8. BYBLIDACEAE.

8 Flowers usually half to fully epigynous, or seldom merely perigynous or (Greyiaceae, some Hydrangeaceae) virtually hypogynous; plants without secretory canals. (Grossulariineae)

10 Leaves opposite or seldom whorled, only rarely alternate; ovules unitegmic, tenuinucellar.

11 Petals distinct; stamens (4) 8-many; styles various, sometimes as in the next group, but more often as many

as the carpels and distinct, or connate below into a common style with distinct stigmatic branches .... 9. HYDRANGEACEAE.

11 Petals shortly connate below; stamens 2; style solitary, short and thick, with a merely lobed stigma .... 10. COLUMELLIACEAE.

10 Leaves mostly alternate, only seldom opposite; ovules various.

12 Petals distinct (rarely shortly connate at the base), or sometimes wanting.

13 Ovules mostly numerous; leaves normally developed, the plants not ericoid in aspect.

14 Flowers perigynous to more often epigynous; gynoecium of 2–3 (–7) carpels; usually some or all of the vessel-segments with scalariform perforations; imperforate tracheary elements with bordered pits, often septate .... 11. GROSSULARIACEAE.

14 Flowers hypogynous; gynoecium of 5 (6) carpels; vessel-segments with simple perforations; imperforate tracheary elements with simple pits, not septate .... 12. GREYIACEAE.

13 Ovules 1–2 (–8) in each locule of the 2–3-locular (seldom pseudomonomerous) ovary; leaves small, closely set, commonly imbricate, the plants typically of ericoid aspect .... 13. BRUNIACEAE.

12 Petals united below to form a corolla-tube .... 15. ALSEUOSMIACEAE.

2 Plants either herbaceous, or succulent, or both. (Saxifragineae)

15 Plants distinctly succulent; carpels as many as the petals, distinct or united only at the base; flowers hypogynous or seldom slightly perigynous .... 16. CRASSULACEAE.

15 Plants only slightly or not at all succulent; carpels seldom of the same number as the petals, variously distinct or united; flowers almost hypogynous in *Parnassia* (Saxifragaceae), otherwise perigynous to often partly or wholly epigynous.

16 Carpels 6, distinct, mostly with a single basal ovule; sepals 6; petals none; stamens 12; some of the leaves modified into small pitchers .... 17. CEPHALOTACEAE.

16 Carpels 2–4 (–7), usually more or less united at least below (seldom wholly distinct), and usually with numerous ovules on parietal or axile placentas; flowers mostly 4–5-merous, with 1 or 2 cycles of stamens and usually (not always) with petals; leaves not modified into pitchers .... 18. SAXIFRAGACEAE.

1 Seeds mostly with very scanty or no endosperm (endosperm copious in Crossosomataceae and a few Rosaceae) (Rosineae, except that the Davidsoniaceae belong with the Cunoniineae, and the Anisophylleaceae belong with the Grossulariineae).

17 Style or styles not gynobasic, sometimes (some Rosaceae) basiventral, but then the separate carpels usually more than 5.

18 Flowers hypogynous; gynoecium of 2 carpels united to form a compound, bilocular ovary with distinct styles; leaves very large, pinnately compound; trees .... 5. DAVIDSONIACEAE.

18 Flowers perigynous to epigynous; other characters various, but not combined as above.

19 Stamens mostly 8; ovary inferior, with mostly 4 carpels; woody plants with simple leaves .... 14. ANISOPHYLLEACEAE.

19 Stamens mostly 10 or more, but if less than 10 then the other features not as in the foregoing group.

20 Seeds arillate, with well developed endosperm; desert shrubs with small leaves .... 21. CROSSOSOMATACEAE.

20 Seeds not arillate, usually with little or no endosperm; plants of various habit and habitat.

21 Gynoecium apocarpous, or syncarpous with 2–5 styles; stamens usually more than 10, but sometimes as few as 5 or even only 1; flowers perigynous or less often epigynous .... 19. ROSACEAE.

21 Gynoecium syncarpous, with 10 styles; stamens 10; flowers epigynous ....20. NEURADACEAE.

17 Style or styles gynobasic or basiventral; carpels 1–5, distinct or united only by a common style.

22 Leaves not punctate; ovules 2 (–5); stigma short.

23 Flowers with a nectary-disk lining the hypanthium below the stamens; gynoecium ancestrally (and in a few modern species) tricarpellate, with the carpels united only by their common gynobasic style, but 2 of the carpels generally more or less reduced, so that the ovary may appear to be monomerous with a basal style; plants commonly with silica-bodies in some of the cells of the parenchyma and epidermis, and often with some of the cell-walls silicified .... 22. CHRYSOBALANACEAE.

23 Flowers without a nectary-disk; gynoecium of 1–5 distinct carpels, each with a basiventral style; plants without silica-bodies, and the cell-walls not silicified .... 23. SURIANACEAE.

22 Leaves pellucid-punctate; ovule solitary; stigma long-decurrent on the style; nectary-disk wanting .... 24. RHABDODENDRACEAE.

1. Family BRUNELLIACEAE Engler in Engler & Prantl 1897 nom. conserv., the Brunellia Family

Evergreen trees, often densely tomentose with unicellular, thick-walled hairs, lacking raphides but often with crystals of calcium oxalate in some of the cells of the parenchymatous tissues; twigs angular, with a rather large pith; nodes trilacunar or pentalacunar; vessel-segments elongate, annular, relatively thin-walled, with slanting end-walls, some of these with simple perforations, others with scalariform perforation-plates that have up to ca 35 cross-bars; imperforate tracheary elements with small, simple or obscurely bordered pits, often septate; wood-rays heterocellular, Kribs types I and III, exclusively uniseriate, or some of them 2–6 cells wide and with elongate ends; wood-parenchyma wanting. LEAVES opposite or less often ternate, pinnately compound or trifoliolate to seldom unifoliolate or simple, the leaflets opposite, stipellate, pinnately veined, entire or often toothed, sometimes doubly toothed; stomates anomocytic; stipules small, caducous, sometimes more than 2. FLOWERS in axillary or terminal, branching cymes, small, hypogynous or nearly so, regular, apetalous, perfect to more often unisexual, the plants commonly dioecious or gyno-dioecious; staminate flowers with a vestigial gynoecium, and pistillate flowers with a vestigial androecium; sepals (4) 5 (–8), connate at the thickened base, valvate, persistent under the fruits; stamens in 2 series and usually twice as many as the sepals, or seldom some or all of the members of the antesepalous (inner) set paired, so that the inner set may have up to twice as many members as the outer; filaments slender, inserted in the notches of an annular, hairy, intrastaminal nectary-disk; anthers introrse, tetrasporangiate and dithecal, opening by longitudinal slits; pollen-grains tricolporate; gynoecium of distinct carpels, these commonly as many as the sepals, but sometimes only 2 or 3, more or less adnate to the disk on the outer side toward the base, 5-veined by early division of the primary lateral veins, each carpel tapering to a slender, elongate, recurved or almost circinate style with a linear stigma decurrent along the sulcus; ovules 2 in each carpel, collateral, attached near the middle of the ventral suture, pendulous with the micropyle directed upwards, epitropous, anatropous, bitegmic and presumably crassinucellar. FRUITS follicular, becoming abaxially deformed in growth so that the style points outward or even downward; exocarp subcoriaceous, generally more or less tomentose with a dense, short, yellowish to reddish indument, and also beset with longer, straight, lignified, thick-walled, pointed trichomes; endocarp more or less lignified, separating from the exocarp at maturity; seeds 1–2 per carpel, with thick, hard, smooth, shiny testa and corky-thickened, subarillate raphe, exserted from the opened ripe fruit and more or less persistent on a funiculus that is continuous with a divergent placental strip; embryo large, straight, with 2 flattened cotyledons, embedded in the abundant, mealy endosperm.

The family Brunelliaceae consists of the single genus *Brunellia,* with about 50 species native to tropical America, from southern Mexico and the West Indies to Venezuela and the Andes of Bolivia and Peru. None of the species is of any economic importance.

It is now widely agreed that the Brunelliaceae are allied to the Cunoniaceae and also to the Rosaceae. In addition to their distinctive features of pubescence and fruit, the Brunelliaceae differ from the Cunoniaceae notably in their elongate, decurrent stigma and in the absence of wood-parenchyma. They also differ from the bulk of the Cunoniaceae in their distinct carpels and epitropous ovules. In spite of the evident relationship, familial status for the Brunelliaceae seems well warranted.

2. Family CONNARACEAE R. Brown, in Tuckey 1818 nom. conserv., the Connarus Family

Shrubs, woody vines, or seldom rather small trees with trilacunar (seldom 5–7-lacunar) nodes, commonly with solitary crystals of calcium oxalate in some of the cells of the parenchymatous tissues, sometimes with intracellular grains of silica in the secondary xylem, and very often with mucilage-canals or tanniferous secretory cavities or both, probably producing proanthocyanins, but lacking ellagic acid and without saponins; bark, fruit, and seeds often highly poisonous, the poisonous principle yet to be characterized; vessel-segments with simple perforations; imperforate tracheary elements with simple pits, mostly septate; wood-rays homocellular or heterocellular, all uniseriate (or sometimes some of them biseriate for part of their length), or seldom some of them up to 5 cells wide; wood-parenchyma diffuse or scanty-paratracheal or wanting; sieve-tube elements with oblique or highly oblique compound sieve-plates. LEAVES alternate, pinnately compound or sometimes unifoliolate, with entire leaflets containing sclereids; epidermis often with many of the cells mucilaginous; stomates of diverse types; stipules wanting. FLOWERS in terminal, pseudoterminal, or axillary panicles or racemes, small, perfect or rarely unisexual (the plants then dioecious), regular, hypogynous or slightly perigynous, pentamerous or seldom tetramerous; sepals distinct or sometimes basally connate, imbricate or valvate, very often becoming indurated and persistent around the base of the fruit; petals distinct or sometimes basally connate, imbricate or rarely valvate; stamens bicyclic, the inner (antepetalous) set sometimes staminodial; anthers dithecal, opening by longitudinal slits; pollen-grains predominantly tricolporate, sometimes tricolpate (tetracolpate in *Jollydora*); nectary-disk poorly developed and usually extrastaminal, or wanting, but the receptacle sometimes nectariferous at the center, internal to the carpels, or also between the carpels; gynoecium of 1 or 5 (rarely 3, 7, or 8) carpels, often 5 but 4 of them abortive, distinct or connate only at the base where they join the receptacle, often not fully sealed, each with (3) 5 primary vascular traces; each carpel with a terminal style and capitate stigma, the style often hollow and with an evident ventral suture; ovules 2 in each carpel, marginal, collateral, ascending from near the base, anatropous to hemitropous or often orthotropous, bitegmic, crassinucellar, usually only 1 maturing; endosperm-development nuclear. FRUITS dry, opening along the ventral suture or seldom along both sutures, or rarely indehiscent and nut-like; seeds often with an aril; endosperm copious and oily to scanty or none. X = 13, 14.

The family Connaraceae consists of some 16–24 genera (depending on generic concepts) and 300–400 species, widespread in tropical regions especially in the Old World. The largest genera, *Connarus* and *Rourea* (including *Santaloides*), are pantropical and have about a hundred species each. The paleotropical genera *Agelaea* and *Cnestis* have about 40–45 species each. The Malesian species *Cnestis platantha* Griff. has edible seeds.

It now seems to be fairly well established that the Connaraceae are related to the Rosales, Fabales, and Sapindales, lying in the nebulous area where these 3 orders join at the base. They would be anomalous in the Sapindales because of their essentially distinct carpels, and would stand out like a sore thumb from the otherwise closely knit order Fabales because of their exstipulate leaves, frequently pleiomerous gynoecium, and usually follicular fruits. On the other hand, they can be accommodated in the Rosales without undue strain. After careful morphological and anatomical study, Dickison (1971, 1972, 1973) thinks the relationship to the Rosales is the closest one.

The Connaraceae do not fit comfortably into any of the 5 suborders of Rosales here recognized. Those botanists who would make several orders out of the Rosales as here defined thus have some basis for also treating the Connaraceae as a distinct order, as Takhtajan has done. I find the broader definition of the Rosales more appealing and useful, and I think it will do no harm to associate the Connaraceae with the families of the suborder Cunoniinae.

3. Family EUCRYPHIACEAE Endlicher 1841 nom. conserv., the Eucryphia Family

Evergreen trees or shrubs with simple, unicellular hairs, producing gum or mucilage and also tannin, often with tanniferous cells, producing proanthocyanins but not ellagic acid, without iridoid compounds and not cyanogenic; solitary or clustered crystals of calcium oxalate often present in some of the cells of the parenchymatous tissues; cortex of young stems spongy; pith quadrangular and heterocellular; nodes trilacunar; vessel-segments with oblique, scalariform perforation-plates that mostly have 5–20 (–40) cross bars, or sometimes with simple perforations; imperforate tracheary elements with bordered pits; wood-rays weakly heterocellular, uniseriate or some of them 2–3-seriate; wood-parenchyma apotracheal, mostly diffuse, but with some tendency toward banding in later wood; sieve-tubes with P-type plastids containing a single polygonal protein-crystalloid. LEAVES opposite, simple or pinnately compound; stomates paracytic; petiolar anatomy complex; stipules interpetiolar, caducous, with large colleters, the terminal bud sticky-resinous. FLOWERS large, solitary in the axils of the leaves, perfect, regular, hypogynous; sepals 4 (5), leathery, imbricate, coherent at the top and deciduous as a calyptra; petals 4 (5), imbricate, white; androecium served by a limited number of trunk-bundles; stamens very numerous, multiseriate and originating in centripetal sequence on an enlarged, often raised and dome-like receptacle; filaments slender; anthers small, rounded, versatile, tetrasporangiate and dithecal, latrorse, opening by longitudinal slits; pollen-grains very small, binucleate, dicolpate; gynoecium of 4–14 (–18) carpels united to form a compound, plurisulcate, plurilocular ovary with slender, distinct or basally united, persistent styles that have a terminal, dry, non-papillate stigma; ovules on an axile placenta, in each locule biseriate and several or numerous, anatropous, epitropous, bitegmic. FRUIT a leathery or woody, septicidal capsule, the carpellary segments keeled on the back, separating from the axis of the fruit and opening ventrally; seeds few, pendulous, elongate, flattened, winged, with more or less copious endosperm;

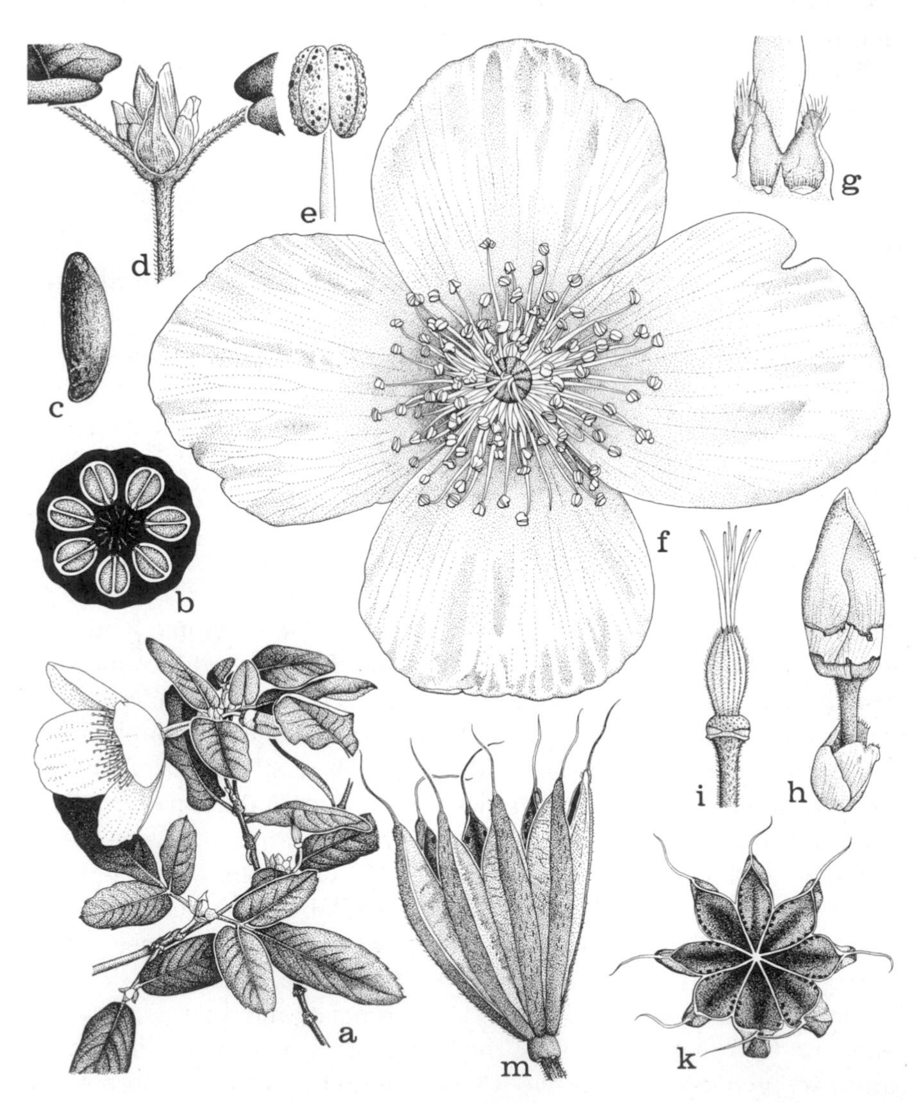

FIG. 5.2 Eucryphiaceae. *Eucryphia* × *nymansensis* Bausch. a, habit, x½; b, schematic cross-section of ovary, ×8; c, seed, ×4; d, node, with interpetiolar stipule, ×2; e, anther, ×12; f, flower, ×2; g, base of stamen, with nectary, ×12; h, flower bud, with the calyx coming off as a calyptra, ×2; i, pistil and receptacle, ×2; k, opened fruit, from above, ×2; m, opened fruit, from the side, ×2.

embryo with 2 large, foliaceous cotyledons and a very short hypocotyl. N = 15, 16.

The family Eucryphiaceae consists of the single genus *Eucryphia*, with about 6 species native to eastern Australia, Tasmania, and Chile. *Eucryphia cordifolia* Cav., of Chile, is a stately tree sometimes cultivated in mild climates.

The Eucryphiaceae were included in a very broadly defined order Parietales in the original Englerian system, and the current edition of the Engler Syllabus notes a certain habital similarity to the Theaceae. In some other systems, including those of Hallier and Takhtajan, *Eucryphia* has been associated with the Cunoniaceae. The vegetative and floral anatomy and pollen-morphology are considered by recent authors to support the cunoniaceous affinity, and the centripetal stamens also suggest a position in the Rosidae rather than in the Dilleniidae.

4. Family CUNONIACEAE R. Br. in Flinders 1814 nom. conserv., the Cunonia Family

Shrubs or trees, or sometimes (as in *Aphanopetalum*) woody climbers, often accumulating aluminum, strongly tanniferous, commonly with scattered tanniferous and/or mucilaginous secretory cells, producing proanthocyanins and very often also ellagic acid, but lacking alkaloids and iridoid compounds, and only seldom cyanogenic; solitary or clustered crystals of calcium oxalate commonly present in some of the cells of the parenchymatous tissues; cortex of young stems commonly spongy; pith typically quadrangular and heterocellular; nodes of diverse types, variously trilacunar or sometimes multilacunar or (as in *Bauera*) unilacunar; wood-anatomy very diverse; vessel-segments with scalariform perforation-plates that have more or less numerous cross-bars, or sometimes some or all of the perforations simple; imperforate tracheary elements with simple or bordered pits; wood-rays slightly to strongly heterocellular, mixed uniseriate and pluriseriate, with short or elongate ends; wood-parenchyma apotracheal, diffuse and in bands 1–4 cells wide. LEAVES firm, opposite or sometimes whorled, pinnately veined, mostly pinnately compound or trifoliolate, seldom simple, the leaflets often glandular-serrate; cells of the epidermis and hypodermis commonly with mucilaginous inner walls; stomates small, with nearly orbicular guard-cells, variously paracytic or anomocytic or anisocytic or encyclocytic; petiole typically with a nearly complete, usually adaxially

FIG. 5.3 Cunoniaceae. *Callicoma serratifolia* Andr. a, habit, $\times\frac{1}{2}$; b, node, with interpetiolar stipule, $\times 2$; c, flower, from above, $\times 4$; d, flower, from the side, $\times 4$; e, anther, $\times 16$; f, schematic cross-section of ovary, $\times 16$; g, pistil, $\times 8$; h, fruit, from the side, $\times 8$; i, fruit, from above, $\times 8$.

flattened, medullated vascular cylinder with the flat adaxial segment separated from the abaxial arc; stipules present (except notably in *Bauera*), sometimes large and conspicuous, often interpetiolar and connate in pairs, commonly with small colleters. FLOWERS mostly small, regular, variously in panicles or mixed panicles or racemes or heads, rarely solitary in the axils, perfect or seldom (as in *Pancheria*) unisexual, the plants then dioecious or polygamo-dioecious, regular, hypogynous or rarely half-epigynous; sepals (3) 4–5 (–10), imbricate or valvate, distinct or basally connate; petals alternate with and mostly smaller than the sepals, or sometimes wanting; stamens mostly biseriate and 8–10, or less often uniseriate and opposite the sepals, seldom more or less numerous (more than 20); filaments slender, surpassing the petals; anthers at least sometimes versatile and inverted, tetrasporangiate and dithecal, opening by longitudinal slits; pollen-grains very small, binucleate, mostly dicolporate or dicolpate, sometimes tricolporate, smooth or nearly so; a saucer-shaped or annular nectary disk commonly but not always developed around the gynoecium; gynoecium of 2 or less often 3–5 carpels, these united to form a plurilocular, superior or seldom partly (up to half) inferior ovary with separate styles that have a terminal stigma, or seldom the carpels more or less distinct; carpels often with 5 vascular traces; carpellary margins inrolled, except in *Bauera*; ovules (1) 2-many in each locule or carpel on axile or apical-axile placentas, mostly apotropous (but sometimes epitropous, as in *Acsmithia* and *Spiraeanthemum*), anatropous to sometimes hemitropous or campylotropous, bitegmic, crassinucellar, with zigzag micropyle; endosperm-development nuclear. FRUIT mostly capsular, the carpels at least sometimes separating above and then opening ventrally, or seldom the fruit nutlike or drupaceous or follicular; seeds small, winged or hairy, with thin testa, commonly with a small, straight, dicotyledonous embryo embedded in the well developed, oily endosperm. N = 12, 15, 16. (Baueraceae)

The family Cunoniaceae as here defined consists of about 25 genera and 350 species, mostly native to the Southern Hemisphere, especially in Australia, New Guinea, and New Caledonia, but in the New World extending north to Mexico and the West Indies. The largest genus is *Weinmannia*, which is pantropical and has about 170 species, some of them used for lumber or for tanning. The type-genus, *Cunonia*, has a notably bicentric distribution, with a dozen or more species in New Caledonia and just one in South Africa. *Cunonia* and *Weinmannia* might perhaps better be combined.

The Australian and Tasmanian genus *Bauera* (3 spp.), here included in the Cunoniaceae, has variously been referred to the Saxifragaceae (sens. lat.), Grossulariaceae, Hydrangeaceae, or even the Tremandraceae, or taken as a separate family. Mauritzon (1933, 1939) and more recently Bensel & Palser (1975) have suggested that on embryological grounds *Bauera* belongs with the Cunoniaceae. Erdtman (1946) came to the same conclusion on the basis of pollen-morphology. Although it is not one of the most typical members of the family, I see no reason why *Bauera* cannot be accommodated in the Cunoniaceae.

Leaf-fossils that have credibly been referred to the Cunoniaceae are known from Eocene deposits in North Dakota. *Cunonioxylon*, from the Upper Oligocene of Europe, may belong to the Cunoniaceae. Some Paleocene wood from Patagonia may also belong here. Pollen referred to *Weinmannia* is known from Oligocene and more recent deposits in New Zealand.

5. Family DAVIDSONIACEAE Bange 1952, the Davidsonia plum Family

Small trees with scattered tanniferous cells in the parenchymatous tissues, probably producing proanthocyanins, but lacking ellagic acid and iridoid compounds; vessel-segments with simple and scalariform perforation-plates; imperforate tracheary elements very thick-walled, with narrowly bordered pits; wood-rays weakly heterocellular, 1–3 (4) cells wide; wood-parenchyma apotracheal, from diffuse to diffuse-in-aggregates. LEAVES alternate, very large, up to about 1 m long, pinnately compound with toothed leaflets and distally tooth-winged rachis, covered with bright red, pungent, somewhat irritating hairs when young, the hairs later fading and eventually deciduous; stomates paracytic; stipules large and conspicuous, palmately veined, up to ca 2 cm wide. FLOWERS in large, axillary or supra-axillary open panicles with spike-like branches, perfect, regular, hypogynous, apetalous; sepals 4 or 5, connate half their length, valvate; stamens 8 or 10, alternating with as many nectary-scales; filaments distinct, becoming shortly exserted, abruptly inflexed near the tip, so that the fundamentally extrorse, dorsifixed, versatile anthers appear to be introrse in bud; anthers tetrasporangiate and dithecal, appearing somewhat sagittate at the morphological apex, opening by apical pores that later elongate into longitudinal slits; pollen-grains tricolporate; gynoecium of 2 carpels united to form a compound, bilocular ovary with distinct styles, the locules confluent at the top; ovules apical-axile, several (ca 5–7) in each locule, pendulous, epitropous, anatropous, bitegmic. FRUIT red-velvety when young, pruinose-glaucous and plum-like at maturity, indehiscent, (1) 2-seeded, with fleshy mesocarp, the leathery endocarp forming (1) 2 flattened, fimbriate-laciniate pyrenes; embryo with 2 large cotyledons and a very short radicle; endosperm wanting.

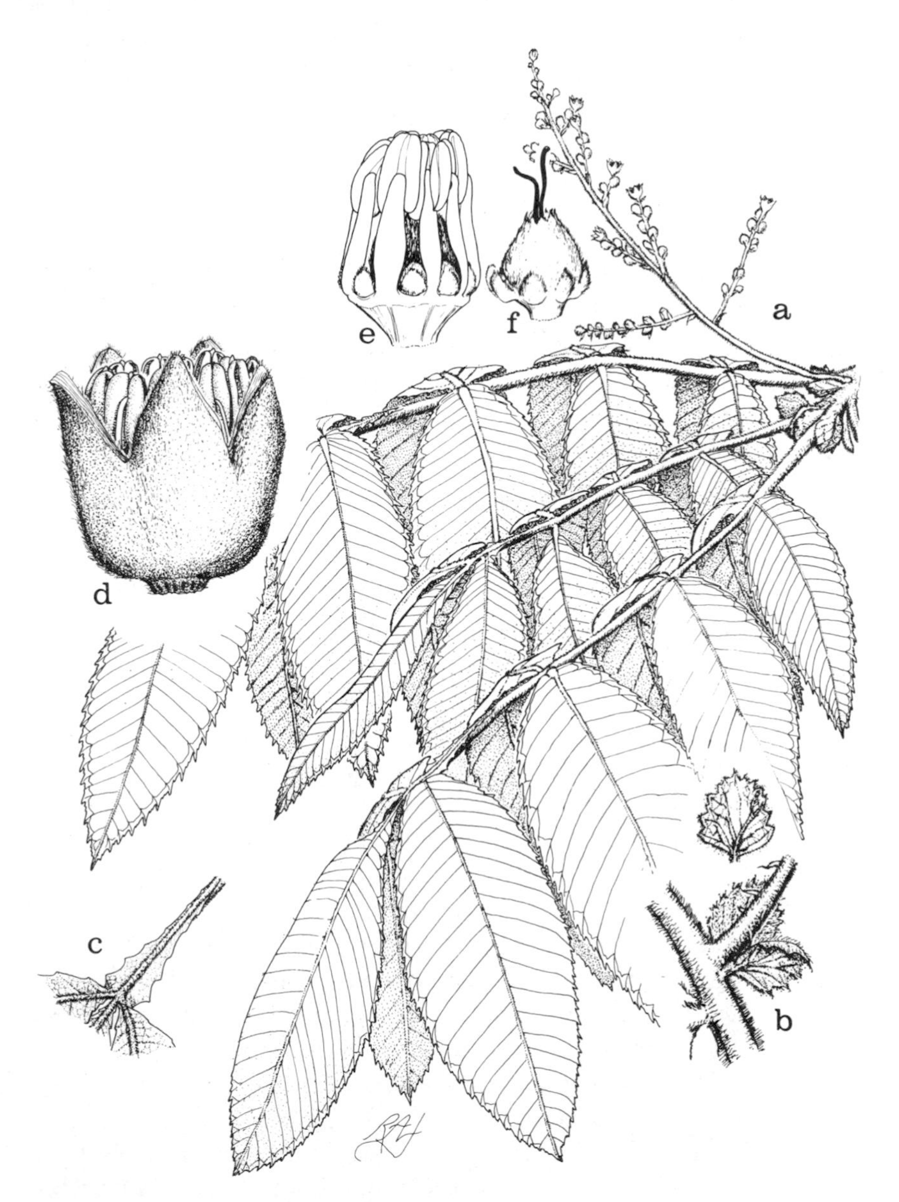

FIG. 5.4 Davidsoniaceae. *Davidsonia pruriens* F. Muell. a, flowering twig, ×¼; b, nodal region of twig, with stipules, ×1; c, distal part of leaf-rachis, ×¼; d, flower, at early anthesis before the anthers have been exserted, ×6; e, flower, with sepals removed, ×6; f, pistil, with subtending nectary-scales, ×6.

The family Davidsoniaceae consists of the single species *Davidsonia pruriens* F. Muell., native to northeastern Australia. The fruits are tart and are used locally to make jam. *Davidsonia* is generally considered to be allied to the Cunoniaceae, and has often been included in that family. It seems adequately distinct, however, in its alternate leaves, epitropous ovules, and nonendospermous seeds.

6. Family DIALYPETALANTHACEAE Rizzini & Occhioni 1949 nom. conserv., the Dialypetalanthus Family

Trees with white-hairy twigs; vessel-segments with simple perforations and vestured pits; imperforate tracheary elements septate; wood rays heterocellular, mixed uniseriate and pluriseriate, the latter 2–7 cells wide; wood-parenchyma scanty-paratracheal, or virtually none. LEAVES opposite, simple, entire, pinnately veined, provided with large intrapetiolar stipules that are laterally connate toward the base in pairs. FLOWERS in a terminal cymose panicle or thyrse, showy, perfect, regular, epigynous, each subtended by a pair of bracteoles; calyx of 4 distinct sepals decussate in 2 sets of 2; petals 4, distinct, bicyclic, white; stamens 16–25, most commonly 18, bicyclic, free from the corolla, the short filaments basally connate into a tube; anthers elongate, tetrasporangiate and dithecal, with separate, linear thecae on the triquetrous connective, introrse, opening by terminal pores; pollen tricolporate; nectary disk forming a fimbriate ring atop the ovary; gynoecium of 2 carpels united to form a compound, inferior, bilocular ovary with a single elongate style and very shortly 2-lobed stigma, the summit of the ovary shortly protruding beyond the base of the calyx, especially in fruit; ovules numerous, anatropous, bitegmic, borne on axile placentas. FRUIT a septifragal capsule crowned by the persistent calyx; seeds numerous, elongate, slender, fusiform, somewhat sigmoid, with straight, terete embryo, 2 short cotyledons, and thin, oily endosperm.

The family Dialypetalanthaceae consists of a single species, *Dialypetalanthus fuscescens* Kuhlmann, of Brazil. *Dialypetalanthus* has sometimes been referred to or associated with the Rubiaceae, which it resembles in its opposite leaves with more or less connate stipules, and with which it is at least compatible in wood-anatomy. The separate petals and relatively numerous stamens, on the other hand, are quite out of harmony with the Rubiaceae, as is the insertion of the stamens atop the ovary, free from the corolla. It has also been referred to the Myrtales, where it is anomalous in its well developed stipules and lack of internal phloem, although the floral features and vestured pits would be perfectly appropriate. Admitting that *Dialypetalanthus* has no obviously close relatives anywhere, I believe it can best be accommodated in the rather amorphous order Rosales. More detailed studies are necessary to permit a more confident assignment of the family.

7. Family PITTOSPORACEAE R. Brown in Flinders 1814 nom. conserv., the Pittosporum Family

Rather small trees or more often shrubs or sometimes half-shrubs, sometimes spiny, often twining or with flexuous or twining branches, glabrous or with simple or T-shaped or clavate-glandular hairs, but not catching insects, with well developed schizogenous secretory canals in some of the parenchymatous tissues of both the shoot and the root, chemically distinctive within the order, not cyanogenic, without iridoid compounds, and without ellagic acid, but often with proanthocyanins, and commonly with essential oils, chlorogenic and quinic acids, polyacetylenes, and often triterpenoid saponins; solitary or clustered crystals of calcium oxalate commonly present in some of the cells of the parenchymatous tissues; nodes trilacunar; vessel-segments with simple but oblique perforations; imperforate tracheary elements septate, with small, simple or very narrowly bordered pits; wood-rays nearly homocellular, a few of them uniseriate, the rest mostly 3–7 cells wide; wood-parenchyma scanty, paratracheal. LEAVES alternate, sometimes so closely crowded at the branch-tips as to appear almost whorled, simple and usually entire (sometimes wavy-margined), leathery; stomates paracytic; stipules none. FLOWERS solitary or in short corymbs or cymose panicles, perfect or seldom some of them unisexual and the plants polygamous, regular or seldom (*Cheiranthera*) slightly irregular, hypogynous, with 2 bracteoles; sepals 5, imbricate, distinct or sometimes connate below; petals 5, imbricate, distinct or more often weakly connate toward the base to form a more or less definite corolla-tube; stamens 5, alternate with the petals, distinct or slightly connate at the base; anthers tetrasporangiate and dithecal, opening by longitudinal slits or terminal pores; pollen-grains binucleate or trinucleate, tricolporate; gynoecium of 2 (3–5) carpels united to form a compound, unilocular or seldom plurilocular ovary with a simple style and a capitate or slightly lobed stigma; carpels at least sometimes with 5 traces, as in Rosaceae; ovules several to usually more or less numerous on each of the intruded parietal or basal-parietal (seldom axile) placentas, anatropous or almost campylo-

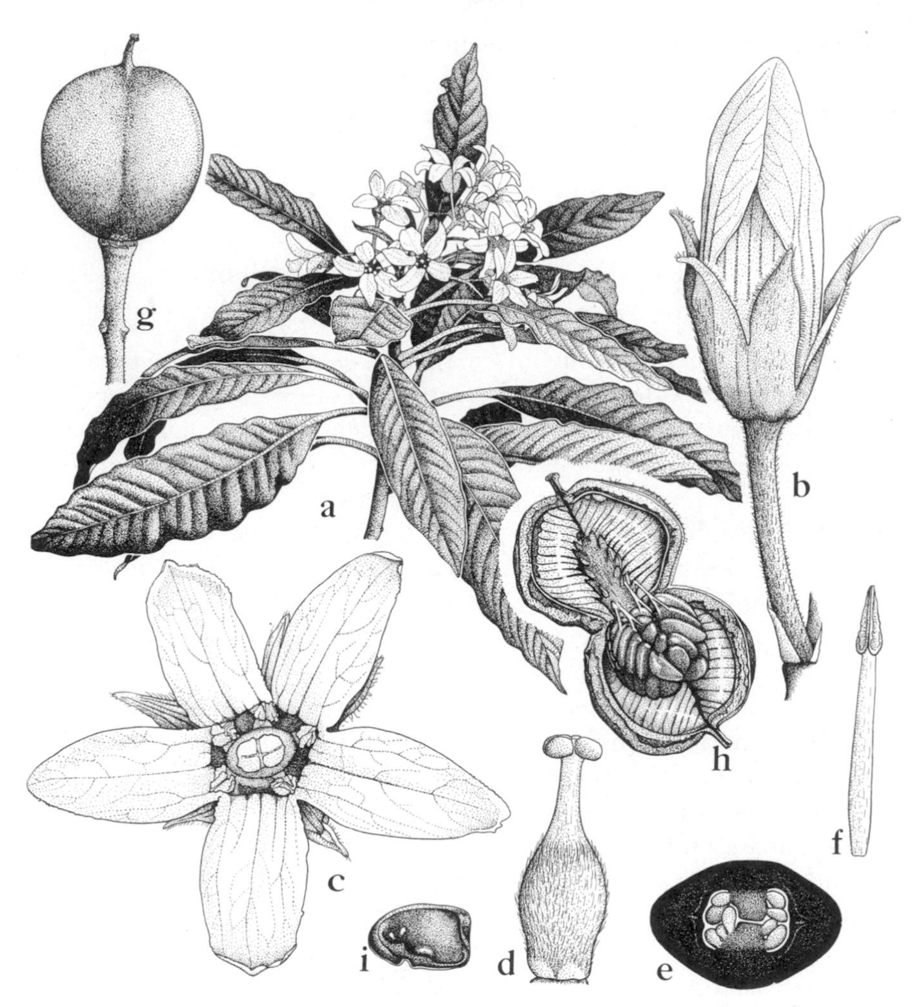

FIG. 5.5 Pittosporaceae. *Pittosporum undulatum* Vent. a, habit, ×½; b, flower bud, ×4; c, flower, from above, ×4; d, pistil, ×4; e, schematic cross-section of ovary, ×8; f, stamen, ×4; g, fruit, ×2; h, fruit, after dehiscence, ×2; i, seed, ×4.

tropous, unitegmic, tenuinucellar; endosperm-development nuclear. FRUIT a loculicidal (or loculicidal and septicidal) capsule or a berry; seeds often immersed in a viscid pulp; embryo tiny, with 2 or in some species 3–5 cotyledons, situated at the base of the well developed, fleshy-firm, oily and proteinaceous endosperm. X = 12.

The family Pittosporaceae consists of 9 genera and about 200 species, widely distributed in tropical and warm-temperate parts of the Old

World, but best developed in Australia. About 150 of the species belong to the genus *Pittosporum*. Few members of the family are of any economic interest. *Pittosporum tobira* (Thunb.) Ait., sometimes called Australian laurel, is cultivated as an ornamental evergreen shrub in warm-temperate regions, and is said to do well especially near the seacoast. Some other species of the genus are used for lumber in Australia.

In most systems of classification the Pittosporaceae are referred to a broadly defined order Rosales or to some smaller order associated with a more narrowly defined order Rosales. Nevertheless, they occupy a relatively isolated position because of their unitegmic, tenuinucellar ovules, schizogenous secretory canals, usually more or less sympetalous corolla, and distinctive chemical features. The only closely allied family appears to be the Byblidaceae.

The secretory canals and some of the chemical features, notably the presence of polyacetylenes, have led some authors to propose a close relationship between the Pittosporaceae and the Araliales. The ovular structure would also be consistent with such a relationship. On the other hand, the hypogynous, multiovulate flowers of the Pittosporaceae, with a well developed calyx, are obviously much more primitive than the epigynous, pauciovulate flowers of the Araliales, with reduced calyx. The Araliales terminate a long line of compound-leaved members of the Rosidae, and it is not likely that the simple-leaved Pittosporaceae stand very close to their immediate ancestry. Rather we must suppose that the anatomical and chemical similarities between the Pittosporaceae and Araliales illustrate the pervasive parallelism that besets efforts to establish phylogenetic relationships among the angiosperms. Both groups belong to the Rosidae, but the Pittosporaceae must be considered to be an early, simple-leaved offshoot that independently developed the secretory system and some of the special chemicals of some of the more advanced Rosidae.

The unitegmic, tenuinucellar ovules and commonly more or less sympetalous corolla of the Pittosporaceae also suggest that the family may stand near to the ancestry of the Asteridae. Here again the correspondence is less than complete, since the Pittosporaceae lack the stipules and the iridoid compounds found in the more archaic members of the Asteridae.

The Pittosporaceae are not well represented in the fossil record, but flowers of *Billarderites* from Miocene amber in Europe appear to belong here.

8. Family BYBLIDACEAE Domin 1922 nom. conserv., the Byblis Family

Perennial herbs or half-shrubs, insectivorous (at least in *Byblis*), beset with long-stalked, oily or mucilaginous glands, and in *Byblis* with sessile glands as well, the former trapping insects, the latter said to digest them; plants without naphthaquinones (differing from the Droseraceae in this regard), also lacking proanthocyanins, and not cyanogenic; vessel-segments with scalariform perforation-plates (*Roridula*) or with mainly simple perforations (*Byblis*); imperforate tracheary elements at least in *Byblis* consisting of true tracheids, in both genera with bordered pits; wood-rays almost exclusively uniseriate in *Roridula*, narrowly pluriseriate in *Byblis*; wood-parenchyma diffuse. LEAVES alternate, simple, linear or nearly so; stomates anomocytic (*Roridula*) or paracytic (*Byblis*); stipules wanting. FLOWERS solitary in the axils of the leaves (*Byblis*) or borne in terminal racemes (*Roridula*), perfect, hypogynous, regular; sepals 5, imbricate, distinct or very shortly connate at the base; petals 5, very shortly connate at the base, imbricate or convolute; stamens 5, alternating with the petals and adnate to the base of the corolla; anthers tetrasporangiate and dithecal, opening by terminal pores or short, pore-like slits; pollen-grains tricolporate or 3–4-colpate; gynoecium of 2 (*Byblis*) or 3 (*Roridula*) carpels united to form a compound, 2–3-locular ovary with a simple style and a terminal, capitate, sometimes lobed stigma; ovules more or less numerous on axile placentas (*Byblis*) or 1-several in each locule and apical-axile (*Roridula*), anatropous, tenuinucellar, with an integumentary tapetum, variously reported as bitegmic or unitegmic; endosperm-development at least in *Byblis* cellular and with both micropylar and chalazal haustoria. FRUIT a loculicidal capsule; seeds with small, straight, dicotyledonous embryo and abundant endosperm. (Roridulaceae)

The family Byblidaceae as here defined consists of only 2 genera and 4 species. *Byblis* (2) is native to Australia, and *Roridula* (2) to South Africa. The two genera differ sharply in several features, but each appears to be the closest ally of the other. Both genera were at one time included in the Droseraceae, but they stand apart collectively from that family in several features. Current opinion emphasizes the similarity of the Byblidaceae to the Pittosporaceae in features other than the specialized, insectivorous habit. *Byblis gigantea* Lindl., in particular, is compared to *Cheiranthera* in the Pittosporaceae.

The existence of true insectivory in *Roridula* has been questioned in recent years. There is no doubt that insects become entangled amongst the long, gland-tipped hairs, but it is not clear that the plant extracts any nutrients from them. Insectivory in *Byblis*, on the other hand, seems to be clearly established.

There is some confusion in the literature as to whether the ovules in *Byblis* and *Roridula* are bitegmic or unitegmic, and I do not know of any recent study directed toward the clarification of this point. If the ovules are in fact bitegmic and tenuinucellar, then the possible relationship of the Byblidaceae to the Nepenthales might warrant reconsideration. Bitegmic, tenuinucellar ovules are common in the Dilleniidae, and occur in some of the Nepenthales, but they are rare in the Rosidae.

9. Family HYDRANGEACEAE Dumortier 1829 nom. conserv., the Hydrangea Family

Shrubs, small trees, lianas, or seldom half-shrubs or even (*Cardiandra, Deinanthe, Kirengeshoma*) rhizomatous herbs, very often with pointed, unicellular hairs, often accumulating aluminum, commonly but not always with raphide-sacs that also contain mucilage, and often with other sorts of secretory cells as well, commonly tanniferous, containing proanthocyanins but not ellagic acid, often producing saponins, sometimes cyanogenic, and at least sometimes (as in *Hydrangea*) with iridoid compounds; nodes trilacunar or sometimes multilacunar; vessel-segments commonly with scalariform perforation-plates that have numerous cross-bars, or sometimes some of them with simple or nearly simple perforations; imperforate tracheary elements commonly with bordered pits, often septate, some of them often true tracheids (these not septate); wood-rays heterocellular or homocellular, mixed uniseriate and pluriseriate, the latter 2–11 cells wide, with elongate or less often short ends; wood-parenchyma often wanting. LEAVES opposite, seldom whorled or (*Cardiandra*) alternate, simple but sometimes lobed; stomates variously anomocytic or paracytic; petiole often with complex anatomy; stipules none. FLOWERS in complex, corymbiform to paniculiform, basically cymose inflorescences, mostly half to fully epigynous with an only shortly or scarcely prolonged hypanthium, but hypogynous or nearly so in several small genera (*Carpenteria, Fendlera, Fendlerella, Jamesia, Whipplea*), perfect, regular, or the marginal ones often neutral and irregular, with enlarged, petaloid sepals, or seldom (*Broussaisia*) many of the flowers unisexual and the plants polygamo-dioecious; calyx-lobes 4–5 (–12), valvate or imbricate; petals 4–5 (–12), distinct, valvate,

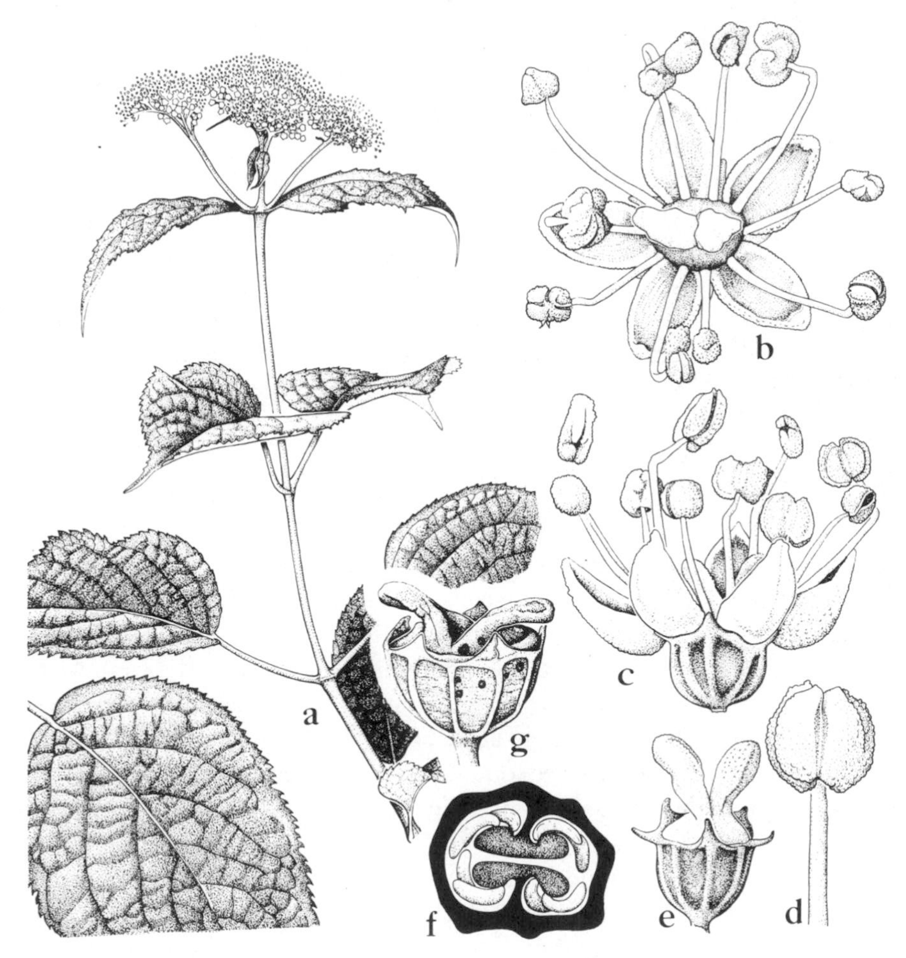

FIG. 5.6 Hydrangeaceae. *Hydrangea arborescens* L. a, habit, ×½; b, c, flower, from above and from the side, ×8; d, stamen, ×16; e, ovary and sepals, ×8; f, schematic cross-section of ovary, ×16; g, fruit, ×8.

imbricate, or convolute; stamens (1) 2-several times as many as the petals, sometimes as many as 50 or even (*Carpenteria*) 200, when biseriate the outer series arising first, when numerous arising centripetally from a limited number of branching primordia; filaments distinct or slightly connate at the base; pollen-grains tricolpate or tricolporate; gynoecium of (2) 3–5 (–12) carpels united to form a compound, half to fully inferior (seldom superior), plurilocular ovary, or sometimes the ovary unilocular and with intruded parietal placentas; styles distinct, or shortly connate below, the branches often stigmatic for much of their length, or seldom the style solitary, short and stout, with distinct stigmas or a capitate, merely lobed stigma; stigmas dry, papillate; an intrastaminal nectary-disk usually present atop the ovary; ovules (1–) several or numerous on each of the axile or intruded parietal placentas, anatropous, unitegmic, tenuinucellar, with an integumentary tapetum; endosperm-development cellular or seldom nuclear. FRUIT a septicidal or loculicidal capsule, or seldom a berry; seeds with a straight, linear, dicotyledonous embryo embedded in the fleshy endosperm. X = 13–18+. (Kirengeshomaceae, Philadelphaceae)

The family Hydrangeaceae as here defined consists of some 17 genera and perhaps 170 species, widespread in temperate and subtropical parts of the Northern Hemisphere, with a few species extending into southeast Asia and Malesia. Most of the species belong to only 3 genera, *Deutzia* (50), *Philadelphus* (50), and *Hydrangea* (24). Various species of all 3 genera are commonly cultivated as ornamentals.

The affinity of the Hydrangeaceae with the other woody families that are generally associated with the Saxifragaceae is widely accepted (though not by Dahlgren, 1977, initial citations). Many botanists have also remarked on a certain similarity of aspect between *Hydrangea* and *Viburnum*, of the Caprifoliaceae, but *Viburnum* is more advanced in so many features (sympetalous corolla, isomerous, epipetalous stamens, drupaceous fruit, tricarpellate pistil with 2 sterile locules and one uniovulate fertile locule) that it is difficult to see any direct relationship between these two genera. Furthermore, *Viburnum* and other Caprifoliaceae lack raphides, which are very common in *Hydrangea* and some other genera of Hydrangeaceae.

10. Family COLUMELLIACEAE D. Don 1828 nom. conserv., the Columellia Family

Bitter shrubs or trees, tanniferous (though scarcely so in the floral parts), probably producing proanthocyanins, but not cyanogenic and without iridoid compounds; clustered crystals of calcium oxalate present in a few of the cells of the parenchymatous tissues, but raphides wanting; nodes unilacunar, with a single trace; vessel-segments elongate, narrow, with slanting, scalariform perforation-plates; imperforate tracheary elements with bordered pits; wood-rays homocellular, uniseriate; wood-parenchyma mostly scanty and vasicentric. LEAVES opposite, rather small, simple, entire or toothed; stomates anomocytic; stipules wanting. FLOWERS in few-flowered terminal cymes, or solitary and terminal, epigynous, perfect, slightly irregular, mostly (4) 5 (–8)-merous as to the calyx and corolla; sepals valvate or slightly imbricate; corolla nearly rotate, with very short tube and imbricate lobes; stamens 2, attached to the corolla near its base; filaments short and stout; anthers with broad connective and 2 plicate, twisted pollen-sacs that open by longitudinal slits; pollen-grains tricolporate; nectary-disk wanting; gynoecium of 2 carpels united to form a compound, inferior ovary with a short, thick style and broadly 2- or 4-lobed stigma; ovary incompletely bilocular, the intruded parietal placentas (partial partitions) nearly meeting in the center; ovules numerous, anatropous, unitegmic, tenuinucellar. FRUIT a septicidal capsule crowned by the persistent calyx; seeds numerous, with small, straight, dicotyledonous embryo surrounded by the abundant, fleshy endosperm.

The family Columelliaceae consists of the single genus *Columellia*, with 4 species native to the Andean region from Colombia to Bolivia. Because of its distinctly (though shortly) sympetalous corolla, *Columellia* has usually been associated with the families of the Asteridae, and the unitegmic, tenuinucellar ovules are compatible with such a position. On the other hand, the relatively primitive wood (vessel-segments with scalariform perforations, imperforate tracheary elements with bordered pits) militates against a position in the Asteridae. Although *Columellia* is so isolated taxonomically that it must stand as a separate family, its individual characters are readily compatible with those of the Rosales, especially the woody families that have been associated with (or included in) the Saxifragaceae. I follow Stern et al. (1969) in this assignment.

11. Family GROSSULARIACEAE A. P. de Candolle in Lamarck & de Candolle 1805 nom. conserv., the Currant Family

Shrubs (or seldom half-shrubs) or sometimes trees, sometimes spiny, very often with unicellular, pointed hairs, sometimes accumulating aluminum, commonly with tanniferous secretory cells in the parenchymatous tissues and producing proanthocyanins and sometimes also ellagic acid, sometimes with iridoid compounds (as in *Escallonia* and *Montinia*) and sometimes cyanogenic; solitary or often clustered crystals of calcium oxalate generally present in some of the cells of the parenchymatous tissues; but raphides wanting; nodes unilacunar or trilacunar, seldom pentalacunar; vessel-segments with scalariform perforation-plates, or less often some or even all of them with simple

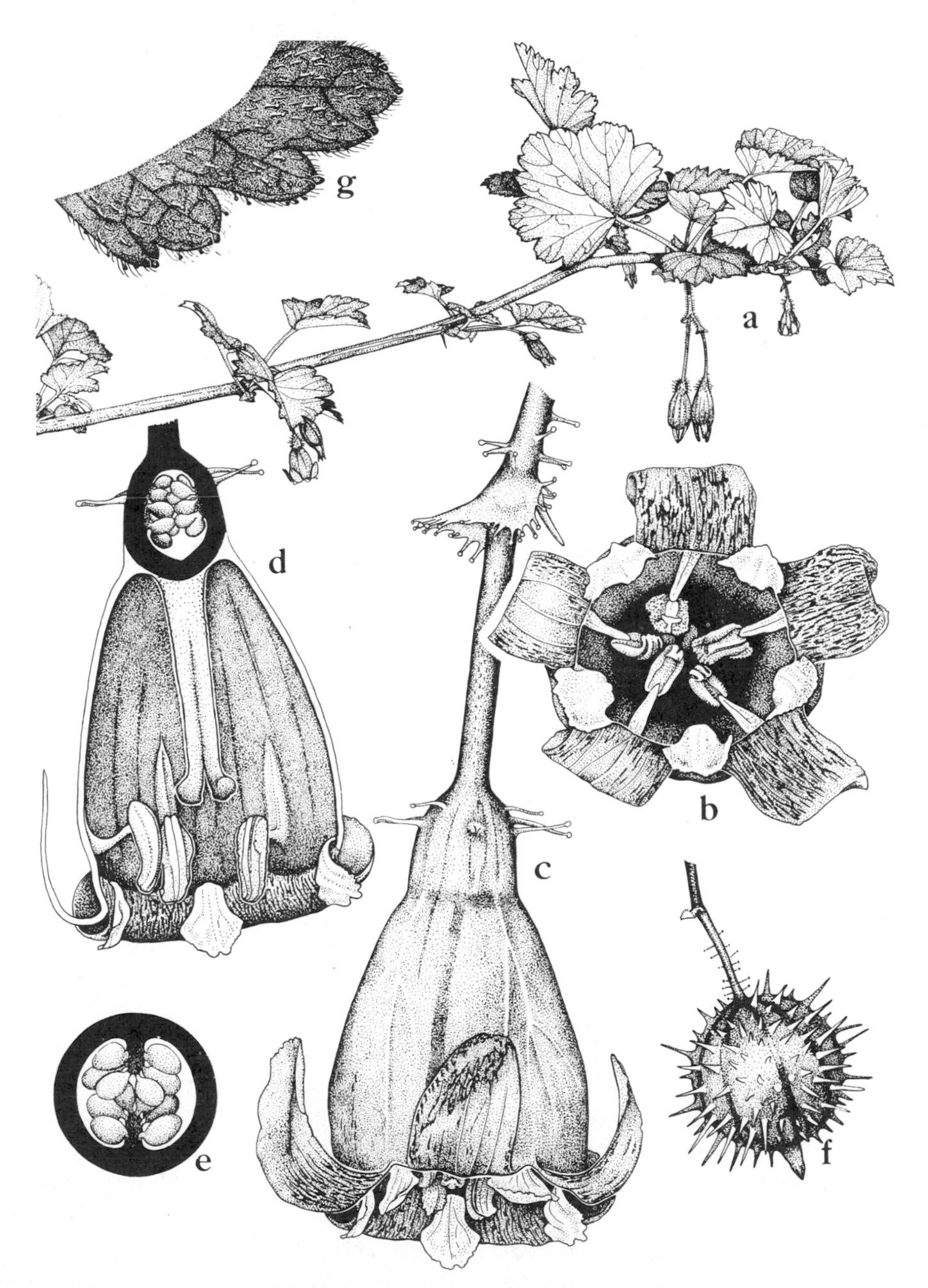

FIG. 5.7 Grossulariaceae. *Ribes cynosbati* L. a, habit, ×1; b, c, two views of flower, ×12; d, flower in long-section, ×12; e, schematic cross-section of ovary, ×24; f, fruit, ×2; g, portion of leaf, ×6.

perforations; imperforate tracheary elements with bordered pits, often septate; vasicentric tracheids often present; wood-rays heterocellular or less often homocellular, mixed uniseriate and pluriseriate, the latter sometimes as much as 15 cells wide; wood-parenchyma apotracheal and diffuse, sometimes also paratracheal, or wanting. LEAVES alternate (seldom opposite, as in *Polyosma*), simple, but sometimes so deeply cleft as to be almost compound, pinnately or palmately veined, often glandular-toothed, very often with hydathodes; stomates paracytic or anomocytic or sometimes almost encyclocytic; petiolar anatomy often complex; stipules most often wanting, or sometimes present (as in *Brexia, Itea, Phyllonoma, Pterostemon*) but then generally small and soon deciduous, seldom well developed and basally adnate to the petiole, as in some spp. of *Ribes*. FLOWERS commonly in terminal or axillary, bracteate or bractless racemes, less often in panicles or small umbels, or solitary in the upper axils, seldom (*Phyllonoma*) the flower-cluster arising from the upper surface of the leaf, regular or nearly so, perfect or seldom unisexual (the plants then sometimes dioecious), perigynous to more often epigynous with a prolonged, saucer-shaped to tubular hypanthium, the (3–) 5 (–9) persistent, imbricate or valvate sepals appearing as lobes on the hypanthium or sometimes forming a calyx-tube that extends beyond the hypanthium, sometimes more or less petaloid and more showy than the proper petals; petals as many as and alternate with the sepals, imbricate, valvate, or convolute, or sometimes wanting; stamens mostly isomerous with and opposite the sepals, but a second, functional or staminodial set sometimes present and alternating with the sepals; anthers tetrasporangiate and dithecal, opening by longitudinal slits; pollen-grains binucleate, (2) 3 (–5)-colporate or -porate, or up to 11-porate in *Ribes*; a lobed nectary disk often present internal to the stamens; gynoecium of 2–3 (–7) carpels united to form a compound, superior to usually partly or wholly inferior ovary, this plurilocular and with axile placentas, or unilocular and with more or less intruded parietal placentas, rarely (*Tetracarpaea*) the carpels distinct; styles distinct, or connate at the base or more or less throughout to form a single style with a wet or dry, lobed to capitate or peltate stigma; ovules numerous, anatropous, bitegmic (*Ribes*) or unitegmic, crassinucellar or tenuinucellar, at least sometimes with an integumentary tapetum; endosperm-development variously nuclear, cellular, or even helobial. FRUIT a capsule or a berry (of follicles in *Tetracarpaea*); seeds numerous, often arillate; embryo dicotyledonous, small to seldom large, the endosperm accordingly copious or scanty; endosperm with reserves of oil and protein, sometimes also hemicellulose. X = 8, 9, 11, 12, 17, 30. (Brexiaceae, Dulongiaceae, Escalloniaceae, Iteaceae, Montiniaceae, Phyllonomaceae, Pterostemonaceae, Tetracarpaeaceae, Tribelaceae)

The family Grossulariaceae as here broadly defined consists of about 25 genera and less than 350 species, of cosmopolitan distribution. About three-fourths of the species belong to only 3 genera, *Ribes* (150), *Polyosma* (60), and *Escallonia* (50). The next largest genus is *Itea* (20), followed by *Argophyllum* (10) and *Brexia* (10). *Ribes* is familiar in cultivation as the source of currants and gooseberries. It is also the alternate host for white pine blister rust.

The two dozen genera here included in the Grossulariaceae are by some authors (e.g., Takhtajan) segregated into as many as 8 families, 5 of them with only a single genus each. At the other extreme, all are included in an amorphous family Saxifragaceae in the Englerian system. The family Grossulariaceae as here defined certainly encompasses a great deal of diversity, and even some of the larger genera are highly variable, but the mutual relationships among the genera are also widely admitted. By drawing the boundaries as I have done, we come up with a fairly well defined group of reasonable size that can usefully be compared with the other woody families related to the Saxifragaceae.

Fossil leaves considered to represent *Ribes* are found in Tertiary deposits of various ages. *Riboidoxylon*, from late Upper Cretaceous deposits in California, some 70 to 80 million years before present, is plausibly compared with the wood of modern *Ribes*. Pollen considered to represent the "Escalloniaceae" dates from the upper Eocene, as does the distinctive pollen of *Itea*.

12. Family GREYIACEAE Hutchinson 1926 nom. conserv., the Greyia Family

Coarse, soft-wooded shrubs or small, scrubby trees, tanniferous and producing ellagic acid, but apparently without proanthocyanins, not cyanogenic, and not saponiferous; raphides and/or clustered crystals of calcium oxalate present in some of the cells of the parenchymatous tissues; nodes multilacunar; vessel-segments with simple perforations; imperforate tracheary elements very short, with simple pits; wood-rays heterocellular, mixed uniseriate and pluriseriate, the latter up to 7 or 8 cells wide, with short ends; wood-parenchyma scanty-paratracheal and also with a few diffusely scattered cells. LEAVES deciduous, alternate, simple, lobulate and toothed, currant-like; stomates anomocytic; petiole

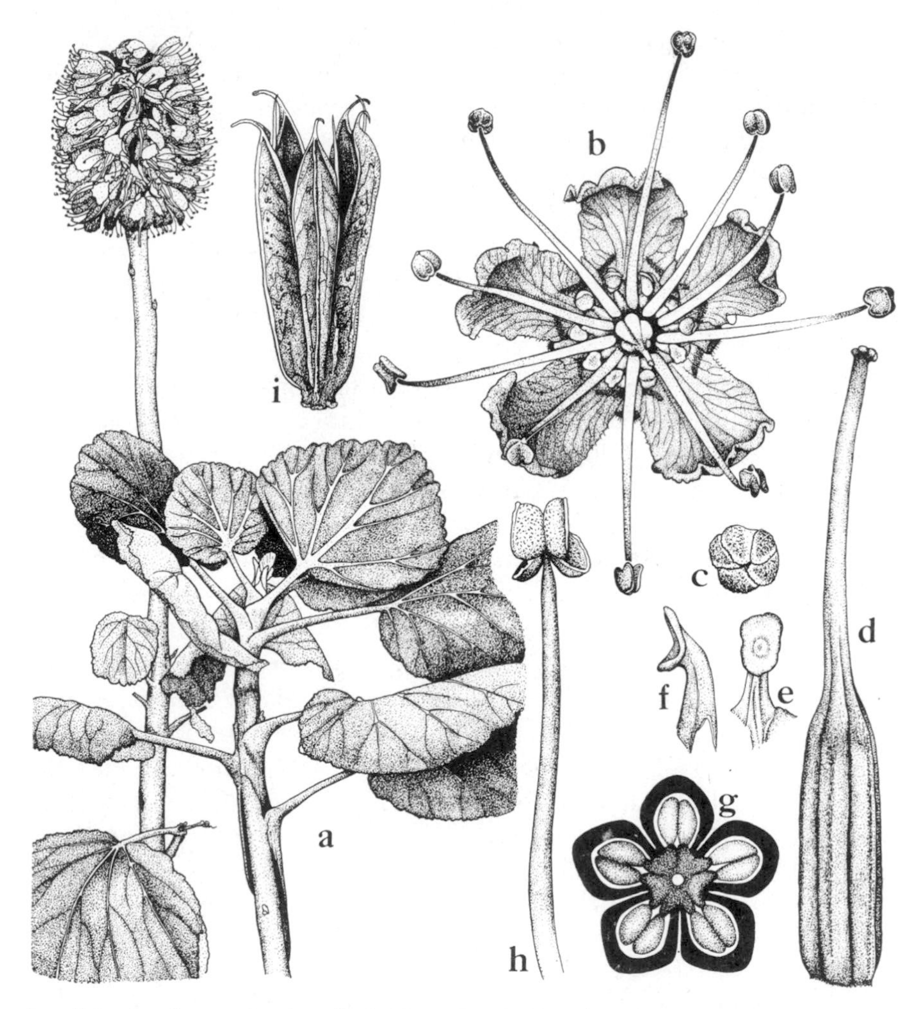

FIG. 5.8 Greyiaceae. *Greyia radlkoferi* Szyszyl. a, habit, ×¼; b, flower, from above, ×2; c, stigma, from above, ×8; d, pistil, ×4; e, f, staminodes, ×4; g, schematic cross-section of ovary, ×8; h, stamen, ×8; i, opened fruit, ×2.

somewhat sheathing at the base; stipules wanting. FLOWERS in racemes, perfect, regular, hypogynous, pentamerous (or the ovary casually with 4 or 6 carpels); sepals distinct, imbricate, persistent; petals red, distinct, imbricate; nectary-disk well developed, extrastaminal, cupulate, crowned by 10 small staminodes alternate with the stamens; stamens 5 + 5, distinct, seemingly in a single series both opposite and alternate with the petals; anthers long-exserted, latrorse, tetrasporangiate and dithecal, opening by longitudinal slits; pollen-grains tricolporate, said to be rather similar to those of *Bersama*, in the Melianthaceae, and very different from those of *Ribes* (Hideux, personal communication), but also said to be similar to those of *Grevea madagascariensis* Baill. (Pastre & Pons, 1973), in the Grossulariaceae as here defined; gynoecium of (4) 5 (6) carpels united to form a deeply 5-lobed ovary with a columnar or cylindric, solid or hollow, axile placenta that is prolonged as a terminal style with a minutely lobed, wet, non-papillate stigma, the carpellary lobes of the ovary distinct except for their common placental column or united by their ventral margins around a central hollow; ovules numerous in 2 rows in each locule, anatropous, bitegmic, crassinucellar. Mature carpels separating from the central axis (or hollow) and opening ventrally; seeds numerous, small, with copious endosperm and small, straight, dicotyledonous embryo. X = 16 or 17.

The family Greyiaceae consists of the single genus *Greyia*, with 3 species confined to South Africa. *Greyia* has often been included in the Melianthaceae, with which it has little in common besides geography. The simple leaves and especially the numerous ovules are out of harmony not only with the Melianthaceae but with the Sapindales as a whole. In spite of its relatively specialized wood, I must agree with Hutchinson that *Greyia* properly forms a family to be associated with the families that he includes in the order Cunoniales.

13. Family BRUNIACEAE A. P. de Candolle 1825, the Brunia Family

Shrubs or under-shrubs, or seldom small trees, commonly of ericoid habit, mostly with long, slender, unicellular hairs, commonly producing proanthocyanins but not ellagic acid, and without iridoid compounds; solitary and clustered crystals of calcium oxalate commonly present in some of the cells of the parenchymatous tissues; wood with deposits of amorphous, dark-staining substances; vessel-segments slender, elongate, with scalariform perforation-plates that have numerous cross-bars; imperforate tracheary elements elongate, thick-walled, with bordered pits, considered to be tracheids; wood-rays heterocellular, Kribs heterogeneous type I, mixed uniseriate and pluriseriate, the latter mostly 2–7 cells wide, with tapering ends; wood-parenchyma mostly diffuse. LEAVES alternate, small, simple, entire, closely set and commonly imbricate, very often with centric structure; stomates of various types, often surrounded by 4–7 relatively small epidermal cells; stipules wanting or vestigial. FLOWERS sessile in spikes or more often heads (the latter sometimes involucrate in a manner reminiscent of the Asteraceae), the heads sometimes arranged in racemes or panicles that may (as in

Berzelia), have characteristic swellings at the nodes, or seldom the flowers solitary and terminal or axillary, mostly small, perfect, regular, almost always partly or wholly epigynous, very seldom merely perigynous with an essentially superior ovary; sepals (or calyx-lobes) (4) 5, imbricate; petals as many as and alternate with the sepals, imbricate, distinct and often clawed, or seldom (*Lonchostoma*) united at the base, forming a short corolla-tube; stamens as many as and alternate with the petals, the filaments free or sometimes adnate to the claws of the petals (in *Lonchostoma* the anthers subsessile on the short corolla-tube); anthers dithecal, opening by longitudinal slits, sometimes with a prolonged connective; pollen-grains tricolporate or 6–10-colporate; an intrastaminal nectary-disk sometimes present; gynoecium usually of 2, rarely (*Audouinia*) 3 carpels united to form a compound ovary with as many locules as carpels but in *Mniothamnea* and *Berzelia* the gynoecium apparently of a single carpel, with 1 locule, 1 ovule, and 1 style; styles terminal, tending to be somewhat connate at least below, each with a small, terminal stigma; ovules 1–2 (–12) per locule, pendulous from near the summit of the partition, apotropous, anatropous, unitegmic, crassinucellar. FRUIT dry, commonly crowned by the persistent calyx, often indehiscent and achene-like or nut-like, with only a single seed, or the 1–2-seeded carpels separating and opening along the ventral suture; seeds small, occasionally arillate, with very small, straight, dicotyledonous embryo at the base of the abundant, fleshy endosperm. 2n = 16 (*Staavia*). (Berzeliaceae)

The well defined family Bruniaceae consists of 12 genera and about 75 species, native to South Africa, almost entirely in the Cape Province, with a single species in Natal. The Bruniaceae have often in the past been associated with the Hamamelidaceae, but most modern authors (with the notable exception of Hutchinson) agree that the family is better placed among the woody relatives of the Saxifragaceae. Carlquist (1978) considers the wood to be primitive and compatible with a rosalean affinity.

14. Family ANISOPHYLLEACEAE Ridley 1922, the Anisophyllea Family

Trees or shrubs, commonly accumulating aluminum; vessel-segments with simple perforations; imperforate tracheary elements with bordered pits; wood-rays Kribs type II, heterocellular, mixed uniseriate and pluriseriate, the latter up to 20 cells wide; wood-parenchyma apotracheal and often banded, varying to paratracheal and irregular; lysigenous secretory cavities sometimes present in the parenchymatous tissues. Leaves alternate, simple, often 3–5-plinerved, often 2-ranked, sometimes pellucid-punctate; stipules wanting. Flowers in axillary spikes or racemes, or in panicles of catkin-like spikes on leafless shoots, perfect or unisexual, regular, mostly 4-merous, epigynous and with a slightly prolonged hypanthium that has a lobed disk at the base, surrounding the top of the ovary; petals 4 and distinct, alternating with the calyx-lobes, variously entire to bifid or dissected, or sometimes wanting; stamens bicyclic, twice as many as the petals, tetrasporangiate and dithecal, opening by longitudinal slits; pollen-grains tricolporate, at least sometimes borne in tetrads; gynoecium of (3) 4 carpels united to form a compound, inferior ovary with separate, subulate styles and as many locules as carpels; ovules 1–2 in each locule, pendulous from the axile placenta. Fruit indehiscent, woody or coriaceous to drupaceous, sometimes winged, 1–4-seeded; cotyledons small, endosperm none. (Polygonanthaceae)

The family Anisophylleaceae consists of 4 genera (*Anisophyllea, Combretocarpus, Poga* and *Polygonanthus*) and about 40 species of tropical or subtropical forests, mostly of Africa and Indomalaysia, but with a few species in South America. *Poga* and *Combretocarpus* are monotypic, and *Polygonanthus* has only 2 described species. The remaining species belong to *Anisophyllea*, which occurs throughout most of the range of the family.

The affinities of the Anisophylleaceae are in dispute. *Polygonanthus* has sometimes been referred to the Escalloniaceae, Euphorbiaceae, Hydrangeaceae, Olacaceae, and Rhizophoraceae. The other genera have often been associated with the Rhizophoraceae, with which they are compatible in wood anatomy, but from which they differ in their alternate, exstipulate leaves, distinct styles, and features of pollen-morphology. In my opinion the group is best treated as a distinct family and referred to the Rosales. Within the Rosales the Anisophylleaceae appear to belong to the suborder Grossulariinae, in spite of their nonendospermous seeds. The pollen of *Polygonanthus*, at least, is highly compatible with this assignment.

15. Family ALSEUOSMIACEAE Airy Shaw 1965, the Alseuosmia Family

Shrubs, often suggesting the Pittosporaceae in aspect; vessel-segments mostly or all with scalariform perforation-plates; imperforate tracheary elements septate, with simple pits; wood-rays absent; wood-parenchyma cells often occurring in radial files, and terminally at the growth-rings. LEAVES alternate or subopposite, simple, entire or sinuately toothed; stomates anomocytic; stipules wanting. FLOWERS borne in large or small axillary cymes, or solitary in the axils, or on old wood, regular, perfect or sometimes unisexual, (4–) 5 (–7)-merous as the calyx, corolla, and androecium; sepals distinct, valvate; corolla sympetalous, with valvate lobes; stamens alternate with the corolla-lobes, attached to the corolla-tube, or free from it; anthers introrse, dorsifixed, tetrasporangiate and dithecal, opening by longitudinal slits; pollen-grains tricolporate; gynoecium of 2 carpels united to form a bilocular, half to fully inferior ovary with axile placentas, the hypanthium sometimes prolonged beyond the ovary; ovary crowned by a more or less well developed nectary-disk; style slender to stout, with an enlarged, more or less 2-lobed stigma; ovules 2-many in each locule. FRUIT a bilocular berry with 1-many seeds; seeds with a dicotyledonous embryo embedded in the copious endosperm.

The family Alseuosmiaceae consists of about a dozen species and 3 genera, *Alseuosmia, Memecylanthus,* and *Periomphale* (*Pachydiscus*), native to New Zealand and New Caledonia. They have usually been referred to the Caprifoliaceae, but they differ notably from that family in their alternate leaves, valvate corolla lobes, and stamens that are often free from the corolla. Furthermore the pollen is said to be unlike that of the Caprifoliaceae. Several authors have noted that the plants are often habitally much like the Pittosporaceae, but they differ from the Pittosporaceae in having the ovary inferior or half-inferior. They are also much like those members of the Grossulariaceae that have sometimes been segregated as a family Escalloniaceae, differing most obviously in their sympetalous corolla. On the basis of present information, the Alseuosmiaceae are best accommodated in the Rosales, near the Grossulariaceae and Pittosporaceae. See also Blumea 29: 387–394. 1984.

16. Family CRASSULACEAE A. P. de Candolle in Lamarck & de Candolle 1805 nom. conserv., the Stonecrop Family

Succulent herbs, half-shrubs, and small shrubs, rarely even arborescent, commonly producing considerable amounts of sedoheptulose and isocitric acid in association with crassulacean acid metabolism, tanniferous and usually producing proanthocyanins, and very often with localized or pervasive red anthocyanin, the root-tips in particular commonly red; plants often with pyridine alkaloids, especially sedamine, and sometimes

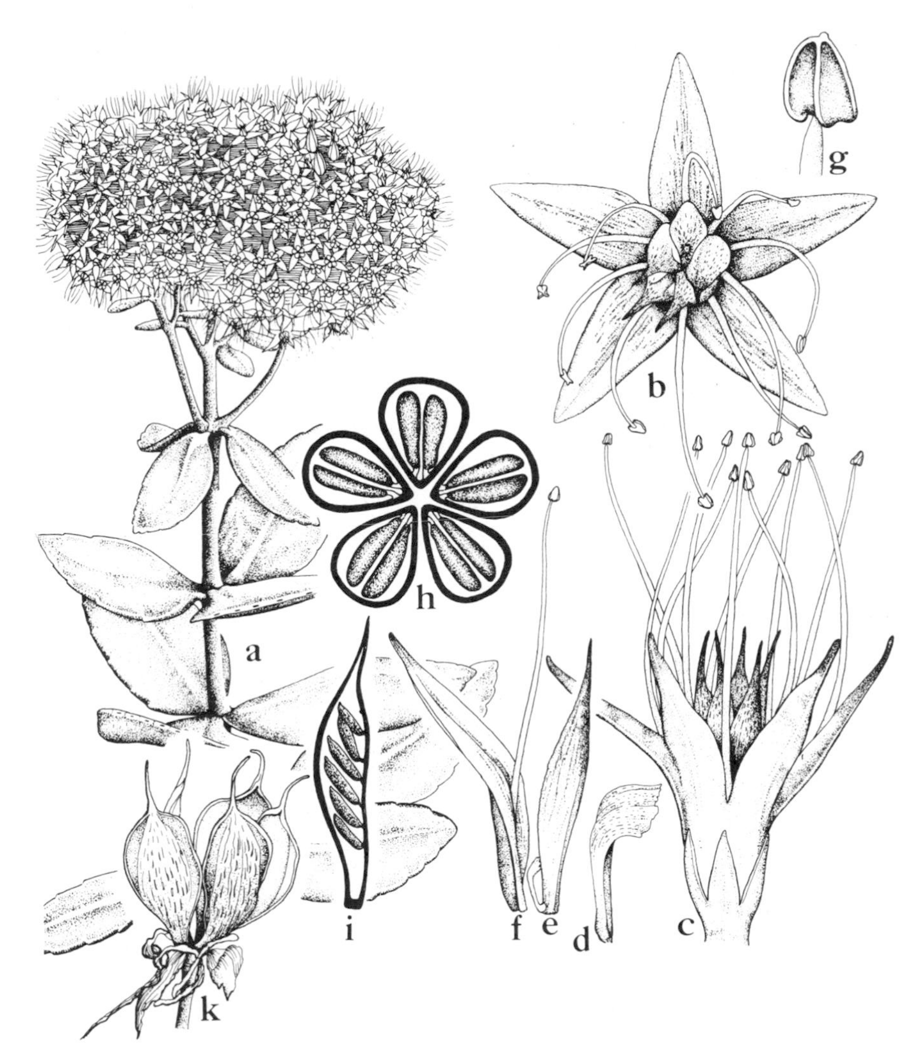

FIG. 5.9 Crassulaceae. *Sedum spectabile* Boreau. a, habit, ×½; b, c, two views of flower, ×5; d, staminode, ×20; e, carpel, with nectary, ×5; f, petal, with stamen, ×10; g, anther, ×20; h, schematic cross-section of gynoecium, ×10; i, long-section of carpel, ×5; k, postmature flower, after dehiscence of fruits, ×5.

cyanogenic, but without ellagic acid and iridoid compounds; various sorts of crystals of calcium oxalate commonly present in some of the cells of the parenchymatous tissues, but raphides wanting; nodes unilacunar or trilacunar; stem often with one or another sort of anomalous structure, including cortical and/or medullary vascular bundles; xylem usually forming a more or less continuous cylinder in the stem, only seldom dissected by wide rays into distinct vascular bundles; vessel-segments with simple perforations; imperforate tracheary elements with simple pits. LEAVES alternate, opposite, or sometimes whorled, simple and commonly entire, often with Kranz anatomy, and often with hydathodes; stomates almost always anisocytic; petiole (when present) usually with a single vascular bundle, but often with complex vascular anatomy; stipules wanting. FLOWERS mostly in one or another sort of cymose (often sympodial) inflorescence, or sometimes solitary, perfect or rarely unisexual, hypogynous or seldom slightly perigynous, regular, most commonly pentamerous, seldom 3-, 4-, or 6-merous or even polymerous throughout, rarely with numerous and unstable numbers of parts (up to more than 50) of a kind; sepals commonly distinct or nearly so, less often connate below, seldom nearly to the tip; petals distinct, or sometimes connate near the base, less often forming a definite corolla-tube of some length; stamens mostly twice as many as the petals, in 2 cycles, less often isomerous and alternate with the petals; filaments distinct or rarely connate below, in sympetalous flowers borne on the corolla-tube, the antepetalous ones a little higher than the antesepalous ones; anthers tetrasporangiate and dithecal, opening by longitudinal slits; pollen-grains binucleate, commonly tricolporate; carpels as many as the sepals or petals, distinct or connate only at the base, rarely (as in *Diamorpha* and *Pagella*) connate nearly to the middle, each tapering into a short or elongate style with a wet stigma; a small, nectariferous appendage borne externally on each carpel near the base (the appendate larger and petaloid in *Monanthes* and some spp. of *Sedum*); ovules (1–) few to more often numerous in each carpel, on a submarginal (or proximally axile) placenta, anatropous, bitegmic, crassinucellar; endosperm-development cellular, sometimes with chalazal haustoria. FRUIT mostly of separate follicles, in *Diamorpha* capsular with the carpels opening dorsally; seeds small; embryo dicotyledonous; endosperm oily and proteinaceous, copious to scanty. X = 4–22+.

The family Crassulaceae consists of about 25 genera and 900 species, of nearly cosmopolitan distribution (except Australia and Polynesia), but most common in arid, temperate or warm-temperate regions. They are especially abundant and diversified in South Africa, but also well developed in the mountains of Mexico and Asia. The pervasive succulent habit and associated crassulacean acid metabolism of the family reflect adaptation to dry habitats, but many species occur in mesic or even wet places, often side by side with typical mesophytes. More than two-thirds of the species belong to only 3 genera, *Sedum* (300), *Crassula* (250), and *Kalanchoe* (120). Many species of these and other genera are familiar in cultivation, either as potted plants or in rock-gardens.

The relationship of the Crassulaceae to the Saxifragaceae has long been plain for all to see. There is even a transitional genus, *Penthorum* (1–3 spp.), that has variously been included in one or the other family or treated as a separate family Penthoraceae. *Penthorum* resembles the Crassulaceae in having as many carpels as sepals, and Agababyan (1961) considers that in pollen-structure it is intermediate between the two families. In other respects *Penthorum* is more like the Saxifragaceae. It is not succulent; the flowers are somewhat perigynous; the carpels are connate up to about the middle, and they lack the nectariferous appendage of the Crassulaceae. Mauritzon (1939) considered that in embryological features *Penthorum* is more like the Saxifragaceae than the Crassulaceae. In my opinion it is most useful to include *Penthorum* in the Saxifragaceae, an already highly varied family. It is not sufficiently different from the Saxifragaceae and Crassulaceae to warrant being treated as a distinct family, and it would be a discordant element in the Crassulaceae.

Fossil pollen referable to the Crassulaceae is known from Miocene and more recent deposits.

17. Family CEPHALOTACEAE Dumortier 1829 nom. conserv., the Australian Pitcher-plant Family

Small, insectivorous bog-herbs from a short rhizome, tanniferous, with ellagic and gallic acids and probably proanthocyanins, not cyanogenic, without iridoid compounds, and without raphides; vascular bundles borne in a ring; vessel-segments with scalariform perforation-plates. LEAVES all basal, small, alternate, closely set in a sort of rosette, of two sorts, the inner with flat, elliptic, entire blade, the outer modified into small pitchers at the ground-level, the pitcher perhaps representing an ascidiate petiole, at first closed by a lid, but later opening; distal inner surface of the pitcher and lower-surface of the lid slippery, coated with overlapping, downwardly directed projections from the epidermal cells; multicellular, embedded, solid glands present on both the inner and outer surface of the pitcher and also on the petiole and the lower surface of the vegetative leaves; flask-shaped embedded glands also present within the pitcher, especially in brightly colored, cushion-like projections from the surface. Scape arising from the center of the rosette, distally bearing racemosely arranged dichasia of small flowers; FLOWERS perfect, regular, perigynous, the hypanthium appearing as a calyx-tube bearing the 6 valvate, hooded, colored sepals as lobes; petals none; stamens in 2 unequal sets of 6 each, borne at the summit of the hypanthium above a glandular, setose disk; anthers tetrasporangiate and dithecal; connective with a swollen, glandular tip; pollen-grains tricolporate, smooth; gynoecium of 6 distinct carpels, each topped by a more or less circinately recurved style; ovules solitary (rarely 2) in each carpel, basal, erect, anatropous, bitegmic, crassinucellar, with an integumentary tapetum. FRUITS follicular, hairy, collectively surrounded at the base by the accrescent hypanthium; seeds with very small, straight embryo surrounded by the copious, fleshy endosperm.

The family Cephalotaceae consists of the single species *Cephalotus follicularis* Labill., of southwestern Australia. Authors are agreed that the family is related to the Crassulaceae and Saxifragaceae.

18. Family SAXIFRAGACEAE A. L. de Jussieu 1789 nom. conserv., the Saxifrage Family

Perennial (seldom annual) herbs, or seldom subshrubs, often with multicellular hairs, sometimes slightly succulent and producing sedoheptulose, often with red (anthocyanic) root-tips, often with scattered tanniferous secretory cells containing proanthocyanins and/or ellagic acid, and sometimes cyanogenic, but without iridoid compounds; crystals uncommon, clustered when present; stem with the vascular bundles in a ring, or with a more or less continuous cylinder of xylem, sometimes also with cortical and/or medullary bundles; nodes most commonly trilacunar, but sometimes unilacunar (as in *Parnassia* and *Penthorum*) or seldom multilacunar (as in *Astilbe*); vessel-segments with simple or seldom (as in *Penthorum*) scalariform perforations; imperforate tracheary elements, when present, small, with bordered pits. LEAVES alternate (sometimes all basal) or less often opposite, simple and pinnately or often palmately veined, or pinnately or palmately compound or decompound, often with hydathodes, these sometimes functioning as chalk-glands; stomates most commonly anomocytic, but sometimes anisocytic or diacytic; petiole often (especially in *Saxifraga*) with 1–3 concentric bundles; stipules wanting, or represented only by the expanded margins of a sheathing petiolar leaf-base, seldom more clearly developed, as in *Francoa*. FLOWERS in various sorts of cymose to racemose inflorescences, or solitary, regular or somewhat irregular, perfect or seldom some or

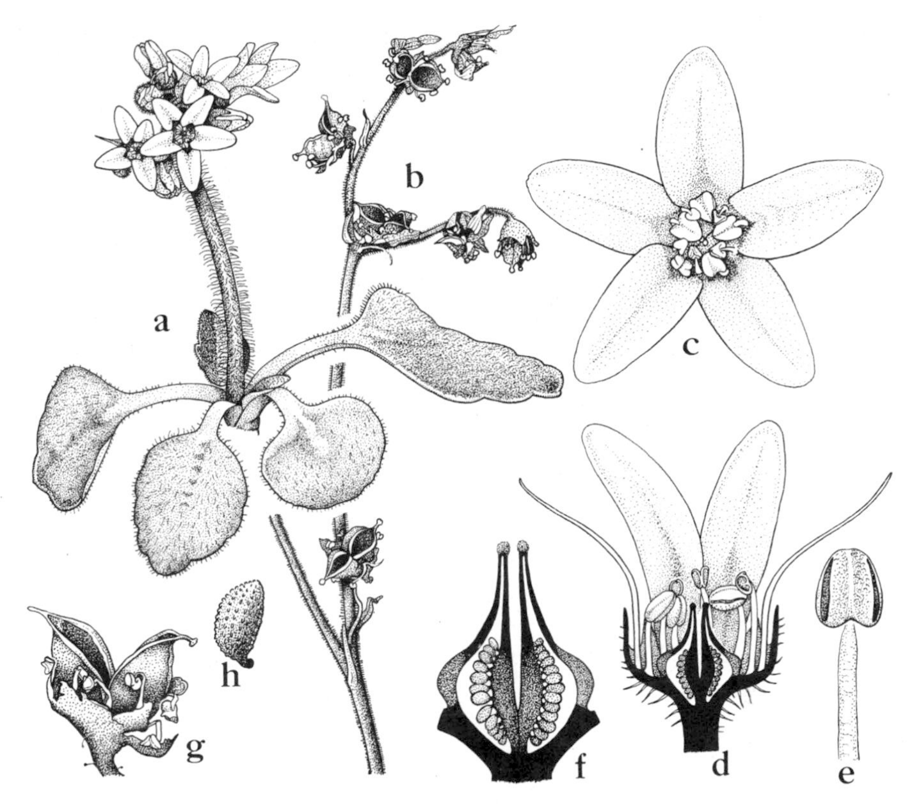

FIG. 5.10 Saxifragaceae. *Saxifraga virginiensis* Michx. a, habit, ×2; b, portion of fruiting inflorescence, ×2; c, flower, from above, ×6; d, flower, in partial section, ×6; e, stamen, ×12; f, schematic long-section of pistil, ×12; g, opened fruit, ×6; h, seed, ×36.

all of them unisexual, perigynous (almost hypogynous in *Parnassia*) to often partly or wholly epigynous, the (3–) 5 (–10) imbricate or valvate sepals commonly appearing as lobes of the hypanthium; petals typically as many as and alternate with the sepals, imbricate or convolute, often clawed, sometimes cleft or dissected, well developed or often relatively small and inconspicuous (and then sometimes irregularly developed and fewer than the sepals), sometimes soon deciduous, or even completely wanting; stamens bicyclic or often unicyclic (then either opposite or alternate with the sepals), one set sometimes staminodial; anthers tetrasporangiate and dithecal, opening by longitudinal slits; pollen-grains 2–3-nucleate, tricolporate or tricolpate, varying to 6–9-porate; an intrastaminal nectary-disk or annulus often present; gynoecium of 2–4 (–7) carpels, these sometimes essentially distinct, but more often more or less connate at least toward the base to form a compound, distally often more or less deeply lobed ovary, each lobe commonly prolonged into a hollow or solid stylar beak that is terminated by a usually capitate (seldom somewhat decurrent) stigma (in *Francoa* and *Parnassia* the stigmas commisural and sessile or nearly so at the top of the ovary); placentation variously marginal, axile, or parietal, or axile below and parietal above, in *Vahlia* 2 elongate, multiovulate placentas pendulous from the top of a compound, unilocular ovary, and in *Penthorum* a single pendulous, marginal placenta produced from the distal (free) part of each carpel; ovules several to usually numerous on each placenta (solitary in each locule in *Eremosyne*), anatropous, bitegmic or sometimes unitegmic, sometimes with a zigzag micropyle, crassinucellar to less often (as in *Parnassia* and *Vahlia*) tenuinucellar, sometimes (as in *Vahlia*) with an integumentary tapetum; endosperm-development variously cellular, helobial, or (*Francoa*) nuclear. FRUIT dry, dehiscent, most often septicidal or dehiscent along the ventral sutures of the carpels above their level of union; seeds generally more or less numerous, small, with small to fairly large, straight, dicotyledonous embryo embedded in the copious, oily endosperm, or the endosperm rarely scanty or even wanting. X = 6–15, 17, perhaps typically 7. (Eremosynaceae, Francoaceae, Lepuropetalaceae, Parnassiaceae, Penthoraceae, Vahliaceae)

The family Saxifragaceae as here defined consists of about 40 genera and 700 species, almost cosmopolitan in distribution, but best developed in temperate and cold, often mountainous parts of the Northern Hemisphere. About half of the species belong to the single large, highly variable genus *Saxifraga. Astilbe* is frequently cultivated as a garden-ornamental, and species of *Saxifraga* and several other genera are often grown in rock-gardens.

The Saxifragaceae have been variously defined in different systems of classification. Engler and his followers took a very broad view, and included not only herbaceous genera but also the woody genera here segregated into the families Hydrangeaceae and Grossulariaceae. *Bauera,* here referred to the Cunoniaceae, is also included in the Saxifragaceae sensu Engler. Many recent authors have found the Englerian concept too broad to be useful, and have therefore defined the family more narrowly. In 1966 Takhtajan distributed the Englerian Saxifragaceae among no less than 15 families, in two orders. (In 1980 he reduced the number to 10, all in the order Saxifragales.) Most authors (Hutchinson excepted) agree that all of these groups are more or less closely related.

The question is how best to organize them into a taxonomic system. I have found it useful to restrict the Saxifragaceae to the herbaceous genera of the Englerian group. The 5 (now 4) additional herbaceous families that Takhtajan would recognize have only about 65 species in all, 50 of them in the single genus *Parnassia*.

Even if one takes the narrow view of Takhtajan, the Saxifragaceae remain a morphologically diversified group. The single genus *Mitella* includes species with 10 stamens, species with 5 stamens alternate with the sepals, and species with 5 stamens opposite the sepals. *Saxifraga* encompasses species with the carpels essentially distinct and species with the carpels united almost up to the stigmas. Furthermore the ovary in syncarpous species of *Saxifraga* ranges from superior to inferior, and the placentation within the genus varies from essentially marginal, through axile below and marginal above, to wholly parietal.

Given this wide range of floral diversity that must perforce be included within the family, I have no difficulty in extending the boundaries a little farther to include the small but closely related families Eremosynaceae, Francoaceae, Lepuropetalaceae, Parnassiaceae, Penthoraceae, and Vahliaceae. There is on the other hand a clear and readily recognizable gap between the herbaceous family Saxifragaceae as here defined, and the essentially woody families Hydrangeaceae and Grossulariaceae. Conceptually these woody families are more usefully associated with the other woody families of the order, such as the Cunoniaceae and Pittosporaceae.

Savile (1975) thinks that the Saxifragaceae originated in eastern or southeastern Asia from an "ancestral plant that may have looked much like *Astilbe*." On the basis of the heteroecious rusts, he considers that "it is difficult to believe that the Saxifragaceae originated before the Oligocene period." It may be desirable to reconsider the status of some Eocene fossils from Europe that have been referred to the Saxifragaceae. If *Astilbe* is indeed similar to the ancestral prototype of the Saxifragaceae, then the strong habital similarity between *Astilbe* and *Aruncus* (one of the more archaic genera of Rosaceae) may be less difficult to explain than has been supposed.

19. Family ROSACEAE A. L. de Jussieu 1789 nom. conserv., the Rose Family

Trees, shrubs, or herbs, generally tanniferous, producing proanthocyanins and often (Rosoideae) also ellagic and gallic acids, commonly

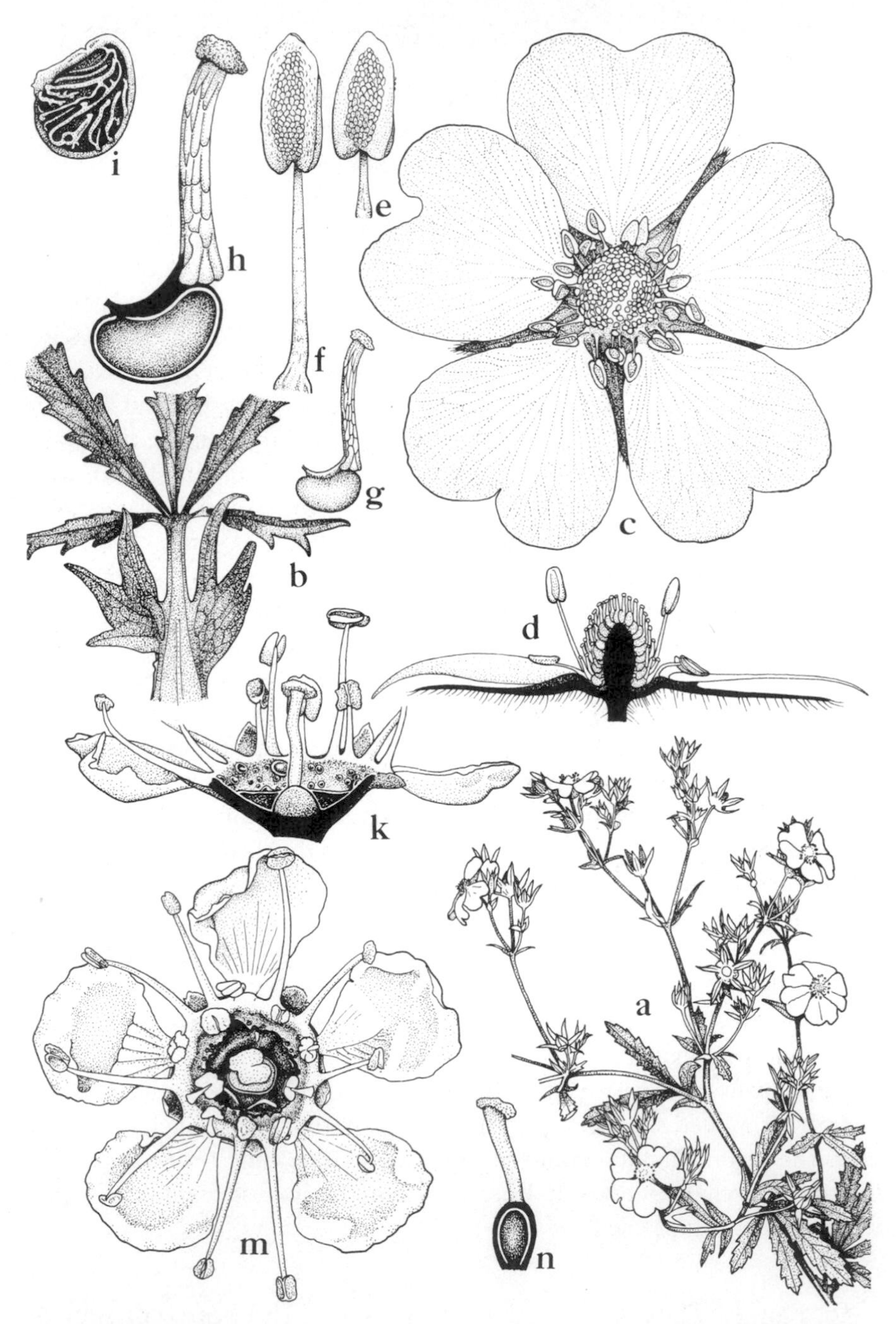

FIG. 5.11 Rosaceae. a–i, *Potentilla recta* L. a, habit, $\times\frac{1}{2}$; b, leaf, with petiolar stipules, $\times 1$; c, flower, from above, $\times 3$; d, flower, in long-section, $\times 3$; e, f, stamens, $\times 12$; g, carpel, $\times 12$; h, carpel, in partial section, $\times 24$; i, achene, $\times 12$. k–n, *Prunus serotina* Ehrh. k, flower, in long-section, $\times 6$; m, flower, from above, $\times 6$; n, pistil, in partial long-section, $\times 6$.

accumulating triterpenoid saponins and glucitol (sorbitol), and often cyanogenic through a pathway based on phenylalanine or leucine, but without iridoid compounds and only seldom with alkaloids; solitary or clustered crystals of calcium oxalate often present in some of the cells of the parenchymatous tissues, but raphides wanting; nodes trilacunar, or sometimes multilacunar or unilacunar; vessel-segments commonly with simple perforations, less often some of them with scalariform perforations; imperforate tracheary elements with bordered pits; wood-rays heterocellular to less often homocellular, mostly 2–5 (–10) cells wide, with short ends, or seldom uniseriate; wood-parenchyma apotracheal and diffuse or partly in short bands, less often some of it paratracheal. LEAVES alternate or rarely opposite, simple or variously compound or even dissected; stomates anomocytic; petiolar anatomy often complex; stipules usually present, sometimes adnate to the petiole, seldom (as in *Spiraea*) wanting. FLOWERS solitary or more often in various sorts of usually more or less cymose inflorescences, often cantharophilous, mostly regular and perfect, perigynous, with a definite hypanthium, or sometimes (mainly in the Maloideae) epigynous; sepals (3–) 5 (–10), imbricate, often appearing as lobes of the hypanthium (or the hypanthium appearing as a calyx-tube); petals (3–) 5 (–10), distinct, imbricate, often large and showy, equal or rarely unequal, or seldom wanting; stamens mostly more or less numerous but tending to be in more or less definite sets of 5 or 10, often 20 in all, seldom as few as 5 or even (*Aphanes*) only 1, the sets at least sometimes originating in centripetal sequence; filaments slender, distinct or seldom connate, attached to the hypanthium; anthers tetrasporangiate and dithecal, opening by longitudinal slits or rarely by terminal pores; pollen tricolporate or rarely triporate or (*Sanguisorba*) pantoporate; inner surface of the hypanthium commonly nectariferous; gynoecium of 1 (mainly in Prunoideae) to usually several or many distinct carpels, or sometimes (mainly in Maloideae) the carpels united to form a compound, mostly inferior ovary with 2–5 mostly separate styles and axile placentas; stigmas wet or dry; carpels very often with 5 vascular traces (considered to be the primitive condition in the family); ovules mostly several or numerous on a marginal placenta in the Spiraeoideae, only 1 or 2 per carpel in the other subfamilies, anatropous to less often hemitropous or campylotropous, crassinucellar, bitegmic or less often unitegmic; endosperm-development mostly nuclear; an obturator commonly present. FRUIT various, most commonly of separate follicles or achenes (these sometimes seated on an enlarged, fleshy receptacle, as in *Fragaria*, or enclosed in a swollen, narrow-mouthed hypanthium, as in *Rosa*), or of laterally coherent drupelets (as in *Rubus*) drupe in Prunoideae, or sometimes (Maloideae) a pome, or (Prunoideae) a drupe, seldom (*Lindleya*) a capsule; seeds without endosperm, or with very scanty endosperm, very seldom with copious, fleshy endosperm (as in *Physocarpus*); embryo straight or bent, with 2 expanded, flat cotyledons. X = 7–9, 17+. (Amygdalaceae, Drupaceae, Malaceae, Pomaceae)

The family Rosaceae as here defined consists of about a hundred genera and 3000 species, nearly cosmopolitan in distribution, but most common in temperate and subtropical parts of the Northern Hemisphere. Many genera are commonly cultivated, either for ornament (e.g., *Crataegus, Kerria, Pyracantha, Rosa, Sorbus, Spiraea*) or for their edible fruits (e.g., *Fragaria, Prunus, Pyrus, Rubus*). The taxonomy of some of the larger genera, such as *Rubus* and *Crataegus*, is complicated by polyploidy and apomixis; estimates of the number of species in such groups are notoriously subjective. Some other large genera are *Potentilla* (300), *Prunus* (200), *Rosa* (100+), and *Spiraea* (100).

The Rosaceae consist of 4 fairly well marked subfamilies. The following synoptical arrangement of the subfamilies is subject to a considerable number of exceptions, but will hold for most species of most genera.

Pistil or pistils simple, each composed of a single carpel (but these sometimes laterally coherent); ovary or ovaries superior.
- Pistils 2 or more.
  - Pistils each with several or many ovules, ripening into follicles .... SPIRAEOIDEAE.
  - Pistils each with a single ovule, ripening into achenes or coherent drupelets .... ROSOIDEAE.
- Pistil 1, with a single ovule, ripening into a drupe .... PRUNOIDEAE.

Pistil compound, composed of 2–5 carpels; ovary inferior, ripening into a pome .... MALOIDEAE.

According to Sax (1930), some of the Maloideae (x = 17) probably arose through polyploidy following hybridization between a member of the Spiraeoideae (x = 8, 9) and a member of the Rosoideae (x = 7, 8, 9). Subsequent authors have extended this interpretation to cover the whole subfamily Maloideae. This view has received considerable acceptance, but as Gladkova (1972) has pointed out, there is no need to invoke intersubfamilial hybridization here. *Quillaja* and the closely allied *Kageneckia*, which are morphologically at home in the Spiraeoideae, have x = 14 and 17, respectively, and *Lindleya*, which is morphologically somewhat transitional between the Spiraeoideae and Maloideae, also has x = 17. Gladkova's view, that the Maloideae probably arose from members of the Spiraeoideae more or less similar to *Quillaja*, is here considered the more likely alternative. The number x = 17 doubtless reflects allopolyploidy, but there is no obvious need to look outside the Spiraeoideae for the likely antecedents. It is interesting and perhaps significant that the rust fungi correlate well with the 4 traditional subfamilies and provide some support for the view that the Maloideae originated from the Spiraeaoideae (Savile 1979, in general citations).

The Rosaceae may well be a fairly old family, but the fossil record as presently known is ambiguous. The readily recognizable advanced genus *Rosa* appears in the Paleocene or Eocene of North America, and is widespread by the Oligocene. Pollen assigned to the Rosaceae occurs in Oligocene and more recent deposits, and wood considered to represent *Prunus* occurs in Eocene deposits in Yellowstone National Park.

20. Family NEURADACEAE J. G. Agardh 1858, the Neurada Family

Prostrate, densely pubescent annual herbs with sympodial stems that have lysigenous mucilage-ducts in the pith; raphides wanting; vessel-segments with simple perforations. LEAVES alternate, toothed or pinnately lobed or pinnatifid; stomates anomocytic; stipules wanting from most leaves, present on some others. FLOWERS solitary, upturned on horizontally spreading, seemingly axillary peduncles, perfect, regular, more or less epigynous; sepals 5, distinct, valvate; petals 5, distinct, imbricate or convolute; stamens 10, borne on the slightly prolonged hypanthium; anthers tetrasporangiate and dithecal; pollen trinucleate, of a unique, bipolar type, with 3 (4) pores at each end; gynoecium of 10 carpels (thought to have evolved by splitting or doubling of 5), united to form a compound, more or less inferior, plurilocular ovary, but 2–4 of the locules on the side toward the peduncle more or less reduced, or their ovules not maturing; styles distinct, eventually becoming indurated; stigmas capitate; ovules solitary in each of the fully developed locules, apical-axile, pendulous, apotropous, anatropous, bitegmic, crassinucellar. FRUIT dry, indehiscent; seeds without endosperm; embryo with 2 cotyledons. N = 6.

The family Neuradaceae consists of 3 small genera, *Grielum* (6), *Neurada* (1), and *Neuradopsis* (3), native to deserts in Africa and across the Middle East to India. None of the species is of any economic importance. Most authors agree that the Neuradaceae are closely allied to the Rosaceae, in which they have often been included. Aside from their distinctive gynoecium, the Neuradaceae have very unusual pollen. All of the features in which the Neuradaceae differ from typical Rosaceae represent phyletic advances.

21. Family CROSSOSOMATACEAE Engler in Engler & Prantl 1897 nom. conserv., the Crossosoma Family

Glabrous, xerophytic shrubs, without secretory idioblasts, producing tannins (including ellagic and gallic acids) and syringin, but not proanthocyanins, not cyanogenic, and without saponins; nodes trilacunar to unilacunar; vessel-segments of the secondary wood small and short, with simple perforations; imperforate tracheary elements short, with bordered pits, considered to be tracheids or fiber-tracheids; wood-rays heterocellular, mixed uniseriate and pluriseriate, the latter up to 6 cells wide, with short ends; wood-parenchyma scanty, mostly apotracheal and diffuse; calcium oxalate crystals wanting; minute, yellow, acicular crystals closely packed in some of the cells of the parenchymatous tissues of the shoot, or even in the epidermis. LEAVES deciduous, alternate or (*Apacheria*) opposite, small, simple, entire or some of them apically tridentate, with Kranz anatomy, commonly containing masses of acicular yellow crystals; stomates anomocytic; stipules minute or absent. FLOWERS solitary and axillary or terminal, lacking flavones and flavonols, perfect or some of them unisexual, shortly perigynous, the hypanthium forming a thickened nectary-disk to which the stamens are attached, or produced at the base into an annular nectary-disk around which the stamens are seated; sepals (3) 4–5 (6); petals (3) 4–5 (6), white, imbricate; stamens in *Crossosoma* numerous (up to ca. 50) in 3 or 4 cycles and associated with about 10 trunk bundles, originating in centrifugal sequence (fide Eames, 1953) or in centripetal sequence (fide Thorne, 1978 and personal communication), in *Apacheria* and some species of *Glossopetalon* bicyclic, in other species of *Glossopetalon* more or less definitely unicyclic through reduction or suppression of the antepetalous cycle; anthers tetrasporangiate and dithecal, opening by longitudinal slits; pollen-grains binucleate, (2) 3-colporate; gynoecium of

FIG. 5.12 Crossosomataceae. *Crossosoma californicum* Nutt. a, habit, $\times\frac{3}{4}$; b, nodal region of twig, $\times 2$; c, flower, from above, $\times 4$; d, side view of flower, some stamens removed, $\times 2$; e, fruits, $\times 1$; f, seed, $\times 5$.

1–5 (–9) distinct carpels, each with a short, stout style and terminal, expanded stigma, or (*Apacheria*) the stigma ventrally decurrent on the short, stout style; ovules (1) 2-many on a marginal placenta, amphitropous or campylotropous, bitegmic and crassinucellar; endosperm-development nuclear. FRUIT follicular; seeds arillate; embryo with 2 cotyledons; endosperm thin to copious, oily. N = 12. (*Crossosoma*)

The family Crossosomataceae as here defined consists of 3 genera and less than 10 species, all native to arid parts of western United States and adjacent Mexico. *Crossosoma* has 2 species, *Glossopetalon* (*Forsellesia*) has 5 or 6 species, and *Apacheria* only one. *Glossopetalon* has usually been included in the Celastraceae, but Thorne & Scogin (1978) present a convincing case for transferring it to the Crossosomataceae. None of the species of Crossosomataceae is familiar in cultivation or has any great economic importance.

The affinities of the Crossosomataceae are uncertain. The two most obvious possible positions for the family are in the Dilleniales and in the Rosales. The well developed aril suggests the Dilleniales, but the floral perigyny suggests the Rosales. Conflicting reports on the sequence of initiation of the stamens must be reconciled before this character can be brought to bear on the question. In either the Dilleniales or the Rosales the family must occupy an isolated position.

The three genera of Crossosomataceae have a certain ecogeographic unity in that they are all microphyllous xerophytic shrubs of western United States and adjacent Mexico. There is nothing unique or even very unusual about this combination of features, which the Crossosomataceae share with various genera in other families of dicotyledons.

22. Family CHRYSOBALANACEAE R. Brown in Tuckey 1818 nom. conserv., the Cocoa-plum Family

Trees or shrubs, seldom half-shrubs, mostly tanniferous, commonly producing proanthocyanins and ellagic and gallic acids, but lacking saponins and iridoid compounds, not known to produce alkaloids, and not cyanogenic in the few members tested; silica-bodies commonly present in some of the parenchymatous and epidermal cells, and often some of the cell-walls silicified; crystals of calcium oxalate sometimes present in some of the cells of the parenchymatous tissues, but raphides wanting; vessels typically arranged in oblique lines, the segments with exclusively simple perforations; imperforate tracheary elements with evidently bordered pits; wood-rays mainly or wholly uniseriate, hetero-

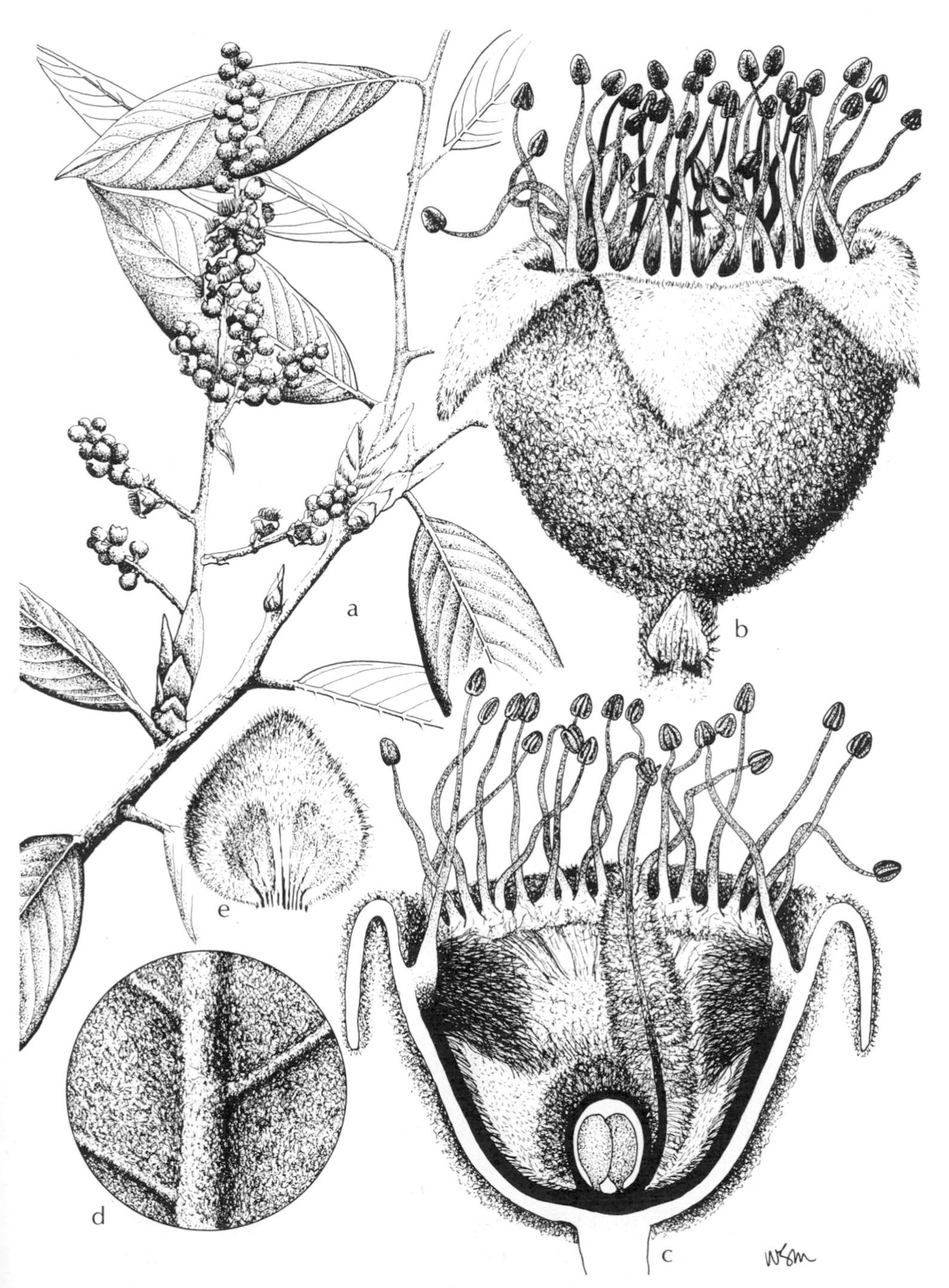

FIG. 5.13 Chrysobalanaceae. *Licania cabrerae* Prance. a, habit, ×½; b, flower, ×9; c, flower in long-section, ×9; d, detail of lower leaf-surface, ×5; e, petal, from flower bud, ×9. From Brittonia 28: 211. 1976; courtesy of G. T. Prance and the New York Botanical Garden.

cellular to rarely almost homocellular; wood-parenchyma apotracheal, in numerous narrow (1–3 cells wide) but elongate (up to 16 cells long) bands. LEAVES alternate, simple, entire, pinnately veined; stomates paracytic; petiolar anatomy often complex; stipules present. FLOWERS in various sorts of cymose to racemose, spicate, or paniculate, terminal or axillary inflorescences, or seldom solitary, mostly rather small and inconspicuous, regular to usually more or less strongly irregular, perfect or seldom some of them imperfect and the plants then polygamous, distinctly perigynous, with a well developed hypanthium that is lined with an annular nectary below the insertion of the stamens; sepals 5, imbricate; petals 5, distinct, imbricate, or sometimes wanting; stamens (2–) 8–20 or numerous (–300), all antheriferous or some of them reduced to long or short staminodes, the filaments elongate, distinct or all connate or connate into groups for up to half their length, regularly arranged around the hypanthium, or in the more zygomorphic members of the family all on one side of the hypanthium and sometimes ligulately connate; anthers tetrasporangiate and dithecal, opening by longitudinal slits; pollen 3 (4)-colpate or -colporate; gynoecium ancestrally (and in a few modern species) tricarpellate, with the carpels united only by the common gynobasic style, but 2 of the carpels commonly more or less reduced, so that the ovary may appear to be monomerous with a basal style; ovary often excentrically placed in the hypanthial cup or tube; style with a simple or 3-lobed stigma; ovules 2 in the single fertile locule, or seldom 2 in each of the 3 fertile locules, or sometimes the single carpel divided by a median partition, so that there appear to be 2 uniovulate locules, the ovules erect, anatropous, epitropous. FRUIT a fleshy or sometimes dry, 1-seeded drupe; seeds with large embryo and 2 commonly thickened cotyledons; seed-fats often with large amounts of licanic or parinaric acid; endosperm none. N = 10, 11.

The family Chrysobalanaceae as here defined consists of 17 genera and about 450 species, basically pantropical in distribution, but best developed in the New World. A third of the species belong to the single genus *Licania* (160), which like the family as a whole is primarily Neotropical but also occurs in the Paleotropics (1 African and 2 Asian spp.). The next largest genera are *Hirtella* (85), *Couepia* (55), and *Parinari* (50). The fruit of *Chrysobalanus icaco* L., the cocoa-plum, is bottled in syrup and sold in northern South America. No other members of the family are familiar in commerce or cultivation.

The Chrysobalanaceae have traditionally been considered to form a tribe or subfamily of the Rosaceae, but there is a growing consensus that the group is better treated as a separate family. The Chrysobalanaceae are anatomically so unlike the Rosaceae that Metcalfe & Chalk (1950, initial citations) took the unusual step of describing the Chrysobalanaceae separately even while formally including them in the Rosaceae. On the basis of chemical evidence Gibbs (1974, initial citations) also feels "quite happy to separate this group from the true Rosaceae." The Chrysobalanaceae are morphologically well set off from the Rosaceae, and Demchenko (1973) considers that the pollen is also distinctive.

23. Family SURIANACEAE Arnott in Wight & Arnott 1834 nom. conserv., the Suriana Family

Shrubs or trees, tanniferous, at least sometimes with proanthocyanins, but not known to contain ellagic acid, not saponiferous, not cyanogenic; crystals of calcium oxalate present in some of the cells of the parenchy-

FIG. 5.14 Surianaceae. *Suriana maritima* L. a, habit, ×½; b, leaf, ×3; c, portion of twig, with leaf-bases, ×4; d, inflorescence, ×2; e, flower, in partial long-section, ×5; f, fruit, ×2.

matous tissues; wood diffuse-porous, the vessels solitary or in groups of 2–9, with simple perforations; imperforate tracheary elements with simple pits, sometimes septate; rays heterocellular (homocellular in *Cadellia*), all uniseriate or a few of them biseriate; wood-parenchyma scanty, mostly paratracheal, sometimes also diffuse. LEAVES alternate, simple; stomates more or less distinctly anomocytic, varying to anisocytic; stipules small and deciduous, or wanting. FLOWERS borne in axillary and terminal, small to large mixed panicles, or solitary in the axils, perfect or sometimes unisexual, hypogynous, pentamerous, anemophilous in *Stylobasium*; sepals distinct or nearly so, imbricate; petals distinct, imbricate, or sometimes (*Stylobasium*) wanting; androecium bicyclic, but in *Suriana* some or all of the members of the antepetalous cycle often reduced and staminodial or obsolete; anthers tetrasporangiate and dithecal, opening by longitudinal slits; pollen-grains tricolporate, in *Suriana* and *Stylobasium* triangular with apertures at the angles, in *Cadellia* and *Guilfoylia* ellipsoid; nectary-disk wanting; gynoecium of a solitary carpel in *Stylobasium*, one or seldom 2 distinct carpels in *Guilfoylia*, 5 distinct carpels in *Suriana* and *Cadellia*, each carpel with a ventral-basal style; ovules anatropous, bitegmic, crassinucellar, 2 in each carpel, collateral on a basal-marginal placenta, or up to 5 and marginal in *Cadellia*; endosperm-development presumably nuclear. FRUIT indehiscent, baccate or drupaceous or nutlike; seeds with curved or folded embryo and very little or no endosperm; cotyledons thickened (except in *Cadellia*), variously starchy or oily. (Stylobasiaceae)

The family Surianaceae as here defined consists of 4 small genera, *Suriana* (1, tropical maritime), *Cadellia* (2, subtropical Australia), *Guilfoylia* (1, northeastern and east-central Australia), and *Stylobasium* (2, southwestern Australia). None of the species is familiar in cultivation or commerce.

Stylobasium has commonly been included in the Chrysobalanaceae, but Prance (1965) showed that it is out of place there in a series of anatomical and palynological features as well as some features of floral morphology. In my opinion the characters that separate *Stylobasium* from the Chrysobalanaceae bring it close to *Suriana*, which it also resembles in aspect. *Suriana, Cadellia,* and *Guilfoylia* have often been included in the Simaroubaceae, but they are anomalous there in their simple leaves and distinct carpels. Furthermore *Suriana*, at least, lacks characteristic terpenoid lactones that make the bark in true Simaroubaceae bitter.

Although I am reasonably satisfied that these 4 genera properly constitute a distinct family, the ordinal position of the family might be debated. By tradition, it ought to be placed in the Sapindales near the Simaroubaceae, but the simple leaves, apocarpous flowers, and absence of both a nectary-disk and a gynophore are out of harmony with the order as well as the family. *Recchia* (*Rigiostachys*) and a few other Simaroubaceae have distinct carpels, but they have compound leaves and produce the characteristic terpenoid lactones of the family. On the other hand, there is nothing in the features of the Surianaceae that is out of harmony with the rather amorphous order Rosales, and the characterization of the two orders is facilitated by associating the Surianaceae with the latter group.

24. Family RHABDODENDRACEAE Prance 1968, the Rhabdodendron Family

Shrubs, two species with anomalous secondary growth; nodes multilacunar, parenchymatous tissues containing scattered secretory cavities with resinous contents; vessel-segments with simple perforations; imperforate tracheary elements with many small, simple or bordered pits; wood-rays heterocellular to sometimes homocellular, variously 1–10 cells wide, at least sometimes containing intracellular grains of silica; wood-parenchyma paratracheal and very scanty-diffuse, or largely in apotracheal bands; sieve-tube plastids with a single polygonal proteinaceous crystalloid as well as some starch grains, but without the peripheral proteinaceous fibrils that mark the Caryophyllales. LEAVES alternate, simple, entire, leathery, provided with short-stalked, peltate hairs, pellucid-puncate with lysigenous secretory cavities, the mesophyll also with scattered fatty bodies and traversed by fiber-like, simple or branched sclereids forming prolongations of the vein-ends, many of the cells of the mesophyll with silicified walls; lower epidermis with peltate, multicellular trichomes that have siliceous inclusions; stomates anomocytic; stipules minute and caducous, or wanting. FLOWERS borne in axillary or supra-axillary racemes or raceme-like cymes that have a terminal flower, regular, slightly perigynous, mostly perfect; calyx very short, 5-lobed (the lobes imbricate) or nearly entire; petals 5, sepal-like, glandular-punctate, imbricate or cochlear, caducous; stamens numerous (ca 25–50) tending to be in 3 whorls, originating simultaneously or weakly centripetally; filaments very short, flattened, persistent; anthers tetrasporangiate and dithecal, elongate, linear, erect, basifixed, opening by longitudinal slits, caducous; pollen-grains trinucleate, tricolporate, with finely reticulate exine; nectary-disk wanting; gynoecium of a single carpel with an elongate virtually basal style that is stigmatic for half to all of its length on the outer side; ovule solitary (a second abortive one sometimes present), basal, unitegmic, hemitropous and epitropous, the micropyle directed upwards. FRUIT a small drupe with thin, eventually crustaceous exocarp and slightly woody endocarp, shortly stipitate within the cupular hypanthium; seed with 2 thick, fleshy cotyledons, the radicle small and bent inwards toward the hilum; endosperm none; germination hypogaeous. X = 10.

The family Rhabdodendraceae consists of the single tropical genus *Rhabdodendron*, with 3 species native to tropical South America. *Rhabdodendron* has often been included in the Rutaceae, or sometimes in the Chrysobalanaceae or Phytolaccaceae. Aside from the secretory cavities it has little in common with the Rutaceae. Because of its simple leaves, perigynous flowers, numerous stamens, single carpel with a basal style and long-decurrent stigma, unitegmic ovules, and absence of a nectary disk it would be discordant not only in the Rutaceae but also in the Sapindales as a whole. The anomalous secondary growth of *Rhabdodendron amazonicum* played some role in Prance's assignment of the genus to the Phytolaccaceae. Some of the embryological and chemical features that would be critical to such an affinity have not been investigated in *Rhabdodendron*, but it has neither the seeds nor the sieve-tube plastids of the Caryophyllales. I can agree with Prance's exclusion of *Rhabdodendron* from the Chrysobalanaceae, but I think it most comfortably stands somewhere near that family in the order Rosales.

2. Order FABALES Bromhead 1838

Trees, shrubs, herbs, or vines, very often bearing root-nodules that harbor nitrogen-fixing bacteria, very often with non-protein amino-acids in the seeds and/or vegetative parts, commonly with scattered tanniferous cells and sometimes also other sorts of secretory cells or cavities, generally producing proanthocyanins, and sometimes cyanogenic, but without ellagic acid and iridoid compounds, often producing one or another sort of alkaloid, especially of the pyridine, quinolizidine, and indole groups; nodes trilacunar or less often pentalacunar; vessel-segments with simple perforations; imperforate tracheary elements mostly or all with small, simple pits, sometimes septate; sieve-tube plastids containing irregular protein crystalloids and generally also starch grains, or seldom with starch grains only. LEAVES alternate or rarely opposite, mostly pinnately once or twice compound, less often palmately compound or trifoliolate, seldom unifoliolate or simple, the petiole and the individual leaflets commonly each with a basal pulvinus that governs its orientation, or the pulvinus present but not functional, or wanting; petiole often with complex vascular anatomy; stipules mostly present, sometimes developed into prickles or spines. FLOWERS mostly in racemes or corymbs or spikes or heads, hypogynous to somewhat perigynous, perfect or rarely unisexual, regular to often strongly irregular; calyx of (3–) 5 (6) sepals, these distinct or connate into a lobed tube that is often somewhat bilabiate, or seldom the calyx much-reduced or virtually obsolete; corolla of (0–) 5 (6) petals, these usually isomerous with the sepals, seldom fewer or none, distinct (and very often highly differentiated inter se), or connate to form a lobed tube, or the two lower (abaxial) ones often connate and the others distinct; stamens most commonly 10, less often 9, sometimes fewer or more numerous, very often connate by their filaments to form a closed or open sheath around the ovary; pollen-grains 2–3-nucleate, commonly triaperturate; gynoecium nearly always of a solitary carpel (rarely 2 or more distinct carpels) with a terminal style and stigma; ovules 2-many on a marginal placenta (ovule solitary in *Andira*, which has a drupaceous fruit), variously anatropous or hemitropous or very often campylotropous, consistently bitegmic and crassinucellar, often with a zigzag micropyle; endosperm-development nuclear. FRUIT commonly dry and dehiscent along both sutures (i.e., a typical legume), but sometimes indehiscent (and then sometimes winged) or breaking transversely into 1-seeded joints; seeds typically with a hard, often impervious testa, often very long-lived; embryo large, with 2 cotyledons; endosperm mostly wanting or very scanty, seldom more or less copious.

The order Fabales consists of 3 families and about 17,000 species, widely distributed throughout the world. A little more than two-thirds of the species belong to the Fabaceae. The remainder are nearly equally divided between the Mimosaceae and Caesalpiniaceae. The three families have often been considered to be subfamilies of a single large family Leguminosae. Under such a definition the name Fabaceae is conserved over the older names Caesalpiniaceae and Mimosaceae, for use by those who prefer to base all names of families on genera.

The existence of the three major groups (here called families), which collectively constitute a larger group (here called an order), is widely admitted. It is only the taxonomic rank of the groups on which opinion remains sharply divided. I prefer the treatment here presented as being more in harmony with customary definitions of families of angiosperms. Consider, for example, Brassicaceae-Capparaceae, Apiaceae-Araliaceae, and Apocynaceae-Asclepiadaceae.

Many authors have treated the Leguminosae as a family of a broadly defined order Rosales. This is not really wrong, but I believe that the organization here presented is conceptually more useful. The legumes, whether treated as one family or three, form such a coherent group with such an abundance of genera and species that the assignment of this group to any other order raises questions of what is tail and what is dog.

The Fabales are by all accounts closely related to the Rosales as here defined. The Connaraceae, here referred to the Rosales, are often cited as a family having much in common with the Fabales. The rusts and smuts of the Fabales emphasize their distinction from the Rosaceae, with which they have often been associated (Savile, 1979).

The Fabaceae (sens. strict.) are obviously the most advanced family of the Fabales. The Mimosaceae and Caesalpiniaceae apparently diverge from a common base, with neither family being ancestral to the other. In a linear sequence it is customary and helpful to start with the Mimosaceae and finish with the Fabaceae. The Caesalpiniaceae connect to both of the other families, and the Fabaceae may reasonably be regarded as derived from the Caesalpiniaceae.

SELECTED REFERENCES

Bandel, G. 1974. Chromosome numbers and evolution in the Leguminosae. Caryologia 27: 17–32.

Barneby, R. C. 1964. Atlas of North American *Astragalus*. Mem. New York Bot. Gard. 13: 1–1188.

Bell, E. A., J. A. Lackey, & R. M. Polhill. 1978. Systematic significance of canavanine in the Papilionoideae (Faboideae). Biochem. Syst. Ecol. 6: 201–212.

Birdsong, G. A., R. Alston, & B. L. Turner. 1960. Distribution of canavanine in the family Leguminosae as related to phyletic groupings. Canad. J. Bot. 38: 499–505.

Buss, P. A., & N. R. Lersten. 1975. Survey of tapetal nuclear number as a taxonomic character in Leguminosae. Bot. Gaz. 136: 388–395.

Crepet, W. L., & D. L. Dilcher. 1977. Investigations of angiosperms from the Eocene of North America: A mimosoid inflorescence. Amer. J. Bot. 64: 714–725.

Daghlian, C. P., W. L. Crepet, & T. Delevoryas. 1980. Investigations of Tertiary angiosperms: A new flora including *Eomimosoidea plumosa* from the Oligocene of eastern Texas. Amer. J. Bot. 67: 309–320.

Dnyansagar, V. R. 1955. Embryological studies in the Leguminosae. XI. Embryological feature and formula and taxonomy of the Mimosaceae. J. Indian Bot. Soc. 34: 362–374.

Dormer, K. J. 1945. An investigation of the taxonomic value of shoot structure in angiosperms with especial reference to Leguminosae. Ann. Bot. (London) II. 9: 141–153.

Dormer, K. J. 1946. Vegetative morphology as a guide to the classification of the Papilionatae. New Phytol. 45: 145–161.

El-Gazzar, A., & M. A. El-Fiki. 1977. The main subdivisions of Leguminosae. Bot. Not. 129: 371–375.

Elias, T. S. 1974. The genera of Mimosoideae (Leguminosae) in the southeastern United States. J. Arnold Arbor. 55: 67–118.

Guinet, P. 1969. Les Mimosacées. Étude de palynologie fondamentale, corrélations, évolution. Trav. Sect. Sci. Techn. Inst. Franç. Pondichery [India] 9: 1–293.

Harborne, J. B., D. Boulter, & B. L. Turner, eds. 1971. Chemotaxonomy of the Leguminosae. Academic Press. London, New York, San Francisco.

Kopooshian, H., & D. Isely. 1966. Seed character relationships in the Leguminosae. Proc. Iowa Acad. Sci. 73: 59–67.

Lackey, J. A. 1977. A revised classification of the tribe Phaseoleae (Leguminosae: Papilionoideae), and its relation to canavanine distribution. J. Linn. Soc., Bot. 74: 163–178.

Leinfellner, W. 1970. Zur kenntnis der Karpelle der Leguminosen. 2. Caesalpiniaceae and Mimosaceae. Oesterr. Bot. Z. 118: 108–120.

Pettigrew, C. J., & L. Watson. 1977. On the classification of Caesalpinioideae. Taxon 26: 57–64.

Picklum, W. E. 1954. Developmental morphology of the inflorescence and flower of *Trifolium pratense* L. Iowa State Coll. J. Sci. 28: 477–495.

Rao, V. S., K. Sirdeshmukh, & M. G. Sardar. 1958. The floral anatomy of the Leguminosae. J. Univ. Bombay 26 (New Series) Part 5B: 65–138.

Rau, M. A. 1953. Some observations on the endosperm in Papilionaceae. Phytomorphology 3: 209–222.

Rau, M. A. 1954. The development of the embryo of *Cyamopsis, Desmodium* and *Lespedeza*, with a discussion on the position of the Papilionaceae in the system of embryogenic classification. Phytomorphology 4: 418–430.

Rembert, D. H. 1971 (1972). Phylogenetic significance of megaspore tetrad patterns in Leguminales. Phytomorphology 21: 1–9.

Robertson, K. R., & Y.-T. Lee. 1976. The genera of Caesalpinioideae (Leguminosae) in the southeastern United States. J. Arnold Arbor. 57: 1–53.

Van Campo, M., & P. Guinet. 1961. Les pollens composés. L'exemple des Mimosacées. Pollen & Spores 3: 201–218.
Vishnu-Mittre & B. D. Sharma. 1962. Studies of Indian pollen grains. I. Leguminosae. Pollen & Spores 4: 5–45.

Анели, Н. А. 1953. Материалы к вопросу об анатомическом родстве древесных бобовых растений. (In Georgian. Russian summary.) Вестн. Тбилисск. Бот. Сада, 61: 95-116.
Федоров, Ан. А. 1937. К морфологии цветков некоторых видов рода *Acacia*. Труды прикл. Бот., Генет. и Селекц., сер. I, 2: 233-240.
Яковлев, Г. П. 1972. Дополнения к системе порядка Fabales Nakai (Leguminales Jones). Бот. Ж. 57: 585-595.

SYNOPTICAL ARRANGEMENT OF THE ORDERS OF FABALES

1 Flowers hypogynous or slightly perigynous, regular (at least as to the corolla); petals mostly valvate, often connate below to form a tube; stamens often more than 10, the filaments often colored and long-exserted and forming the conspicuous part of the inflorescence; ovules anatropous or sometimes hemitropous; leaves mostly bipinnately compound (sometimes reduced to phyllodes); plants mainly tropical and woody .... 1. MIMOSACEAE.

1 Flowers slightly to evidently perigynous, only rarely essentially hypogynous, the corolla usually more or less strongly irregular; petals imbricate, distinct or only the 2 lower ones connate; stamens most commonly 10, less often 9, or sometimes fewer, but only seldom more numerous, not forming the most conspicuous part of the inflorescence; ovules anatropous or hemitropous to often amphitropous or campylotropous; leaves and habit various.

2 Corolla not papilionaceous; adaxial (upper) petal usually borne internally to the lateral petals and smaller than them; sepals mostly distinct; filaments distinct or variously connate, but not usually forming a definite sheath around the pistil; plants mainly tropical and woody, with pinnately or bipinnately compound (seldom unifoliolate or simple) leaves .... 2. CAESALPINIACEAE.

2 Corolla mostly papilionaceous, the adaxial petal (called the banner or standard) borne externally to the others and generally the largest, folded along the midline so as to embrace the other petals in bud; 2 lateral petals (called wings) similar inter se and mostly distinct; 2 lower petals innermost, similar inter se, mostly connate distally to form a keel enfolding the androecium and gynoecium; sepals mostly connate below to form a tube (beyond the hypanthial base); stamens mostly 10 (seldom fewer, but not

more), usually connate by their filaments to form an open or closed sheath around the pistil, the uppermost one often more or less separate from the others so that the androecium is diadelphous (9 + 1), or the uppermost stamen sometimes obsolete, or sometimes the filaments all distinct; plants of varying habit and habitat, often herbaceous and extratropical, the leaves pinnately or less often palmately once compound or trifoliolate, seldom unifoliolate or simple ....3. FABACEAE.

1. Family MIMOSACEAE R. Brown in Flinders 1814 nom. conserv., the Mimosa Family

Trees or shrubs (sometimes lianoid), rarely herbs, sometimes spiny, provided with glandular and often also eglandular hairs, but without malpighian hairs, commonly bearing root-nodules that harbor nitrogen-fixing bacteria, very often with non-protein amino-acids in the seeds and/or vegetative parts, frequently with extrafloral nectaries, often with solitary crystals of calcium oxalate in some of the parenchyma-cells, but without raphides, strongly tanniferous, with scattered tanniferous cells and commonly also with other sorts of secretory cells or sacs (these often containing gum or mucilage) in the parenchymatous tissues; commonly but not always producing proanthocyanins, and sometimes also cyanogenic, but lacking ellagic acid and iridoid compounds, often producing one or another sort of alkaloid, especially of the pyridine and indole groups; nodes trilacunar or less often pentalacunar; vessel-segments with simple perforations; imperforate tracheary elements with small, simple pits, sometimes septate; wood-rays homocellular, most commonly 2–5 cells wide, but sometimes up to 9 cells wide, with short ends, or some or all of them uniseriate; wood-parenchyma usually abundant and predominantly paratracheal; sieve-tube plastids containing irregular protein crystalloids and generally also starch grains. Leaves alternate or rarely opposite (as in *Parkia oppositifolia* Spruce ex Bentham), usually bipinnately compound, often with very numerous and small leaflets, seldom only once pinnate or (especially in spp. of *Acacia*) modified into narrow phyllodia, the petiole, individual pinnae, and ultimate leaflets each commonly with a swollen, basal pulvinus that governs its orientation, or the pulvinus present but not functional; stomates usually paracytic; petiole often glandular, and commonly with complex vascular anatomy, the xylem often forming a hollow cylinder; stipules mostly present, sometimes developed into prickles or spines, which in spp. of *Acacia* are much-enlarged and harbor ants. Flowers in racemes, spikes, or heads, individually generally rather small (the inflorescence rather than the individual flowers being showy), hypogynous or slightly perigynous, perfect or rarely unisexual or some of them neutral (as in *Parkia* and *Neptunia*), regular, or seldom with a somewhat irregular calyx; calyx of (3–) 5 (6) sepals united to form a tube with valvate (rarely imbricate) lobes, sometimes much-reduced or virtually obsolete; corolla of (3–) 5 (6) petals (commonly isomerous with the sepals), these distinct or often connate below to form a tube, valvate

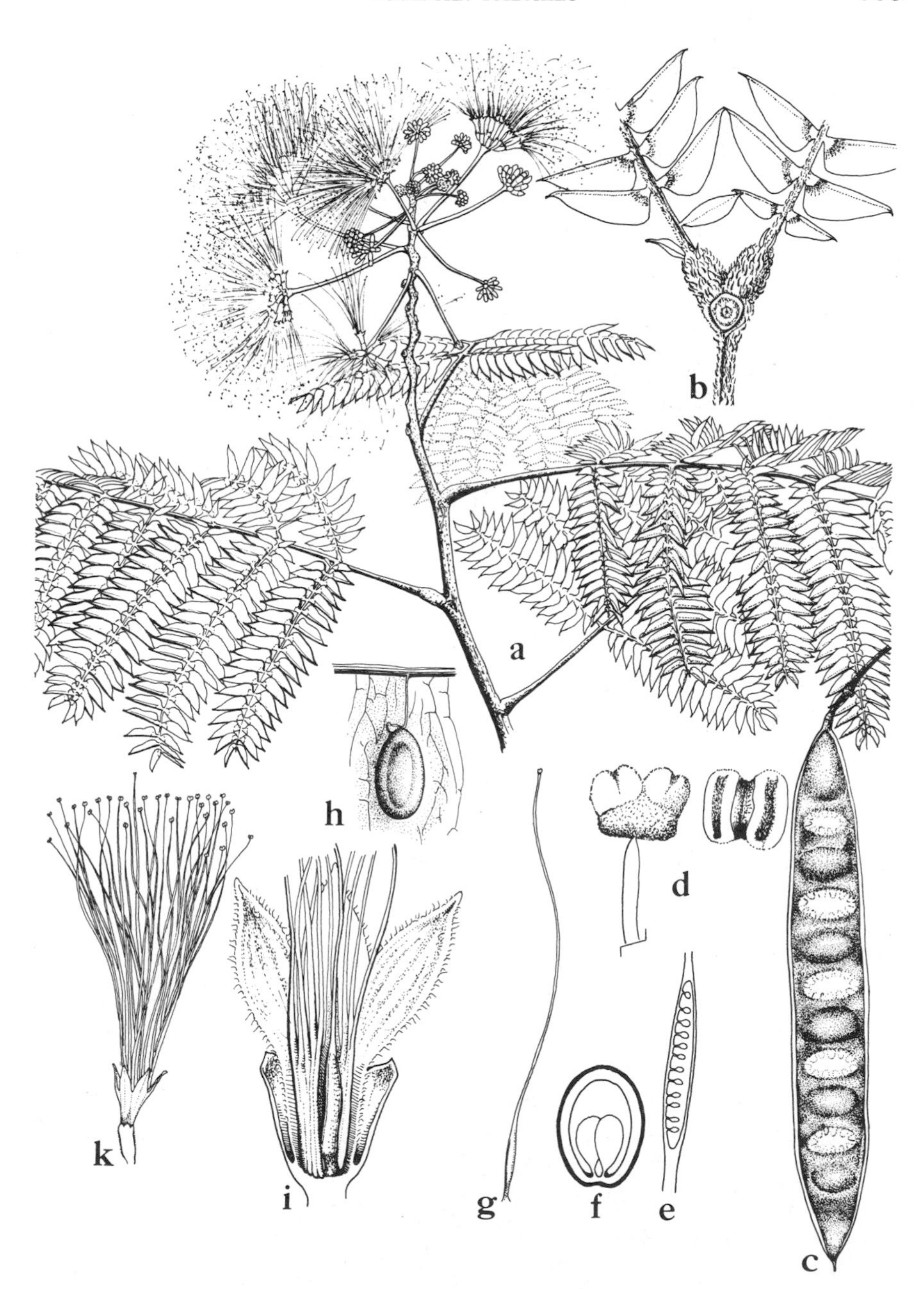

FIG. 5.15 Mimosaceae. *Albizia julibrissin* Durazz. a, habit, ×½; b, distal pinnae, ×1½; c, fruit, ×½; d, two views of anther, ×20; e, schematic long-section of ovary, ×5; f, schematic cross-section of ovary, ×12; g, pistil, ×1½; h, seed and funiculus, ×1; i, flower, in long-section, ×5; k, flower, ×1½.

or very rarely imbricate; stamens usually twice as many as the sepals or petals, or numerous, seldom only isomerous; filaments distinct or connate below, commonly colored and long-exserted, and collectively forming the most conspicuous part of the inflorescence; anthers small, tetrasporangiate and dithecal, opening by longitudinal slits, often with a deciduous gland at the tip; tapetal cells of the anther uninucleate; pollen-grains binucleate, commonly tricolporate or triporate, borne in monads or very often in tetrads or polyads; gynoecium of a single carpel, or in a few small genera of 2-several (–16) distinct carpels, each with a terminal style and stigma; ovules 2-many on a marginal placenta, anatropous or sometimes hemitropous, crassinucellar, bitegmic, sometimes with a zigzag micropyle; endosperm-development nuclear. FRUIT commonly dry and dehiscent along both sutures (i.e., a typical legume), but sometimes indehiscent or breaking transversely into 1-seeded joints; seeds mostly flattened, often with an elongate funiculus or a short funicular aril, or sometimes with a sarcotesta, more often the seed-coat very hard and impervious to water and air; seed coat commonly with a horseshoe-shaped groove that follows the curve of the margin on the flat surface and is usually open at the hilar end; embryo large, generally straight, the short, thick radicle never folded, the cotyledons basally cordate and commonly thickened; endosperm mostly wanting or very scanty, well developed in a few genera such as *Prosopis*. X = 8, 11–14, most often 13 or 14.

The family Mimosaceae consists of about 50–60 genera and 3000 or more species, widespread in tropical and subtropical regions, especially in arid or semi-arid climates; only a few species extend into distinctly temperate climates. The largest genera are *Acacia* (700–800), and *Mimosa* (450–500). Other genera with more than a hundred species are *Albizia* (100–150), *Calliandra* (150), *Inga* (250), and *Pithecellobium* (150–200). *Acacia* is familiar on tropical and subtropical savannas, especially in Africa and Australia. Gum arabic is obtained from *Acacia senegal* Willd. and other African species. *Prosopis* (mesquite) is well known in warm desert regions of both North and South America. *Albizia julibrissin* Durazz. is familiar in cultivation in the southeastern United States under the name Mimosa. *Mimosa pudica* L. is the sensitive plant; other species of *Mimosa* and some other genera respond in similar but less dramatic fashion to being jostled. Species of several genera produce valuable timber.

Fossil inflorescences and tetradinous pollen considered to represent fairly well advanced members of the Mimosaceae are known from middle Eocene deposits in Tennessee and from Oligocene deposits in Texas. Several kinds of mimosaceous pollen occur in upper Eocene and more recent deposits in Africa.

2. Family CAESALPINIACEAE R. Brown in Flinders 1814 nom. conserv., the Caesalpinia Family

Trees or shrubs (sometimes lianoid), or less often herbs, sometimes spiny, very often with non-protein amino-acids in the seeds and/or vegetative parts, and often with various sorts of glandular or eglandular hairs (the latter sometimes malpighian or stellate), but only rarely with hairs that have short basal cells and an elongate distal cell; plants sometimes bearing root-nodules that harbor nitrogen-fixing bacteria, but more often not, seldom with extrafloral nectaries, commonly with both solitary and clustered crystals of calcium oxalate in some of the parenchyma-cells, but without raphides, usually with scattered tanniferous or other secretory cells, and sometimes with secretory canals or lined cavities with diverse sorts of (sometimes resinous or tanniferous) contents, commonly producing proanthocyanins, and sometimes also cyanogenic, but lacking ellagic acid and iridoid compounds, often producing one or another sort of alkaloid, especially of the pyridine group; nodes trilacunar or less often pentalacunar; vessel-segments with simple perforations; imperforate tracheary elements with small, simple pits, sometimes septate; wood-rays homocellular or heterocellular, mostly 2–3 cells wide, but sometimes up to 7 cells wide, or some or all of them uniseriate; wood-parenchyma usually abundant and predominantly paratracheal, but sometimes scanty; sieve-tube plastids containing irregular protein crystalloids and generally also starch grains. LEAVES alternate, usually pinnately or less often bipinnately compound, seldom unifoliolate or simple (simple and bifid in *Bauhinia,* these sometimes modified into coiled tendrils), the petiole, individual pinnae, and secondary pinnae (when present) commonly with a swollen, basal pulvinus that governs their orientation, or the pulvinus present but not functional; stomates of diverse sorts, but most commonly paracytic or anomocytic; stipules present, but stipellules mostly absent. FLOWERS in racemes or spikes or sometimes cymes, small and individually inconspicuous to more often larger and more or less showy, slightly to evidently perigynous, only seldom hypogynous, usually perfect, and more or less irregular, but not papilionaceous (nearly so in *Cercis,* but the banner internal to the wings); sepals mostly 5, distinct or nearly

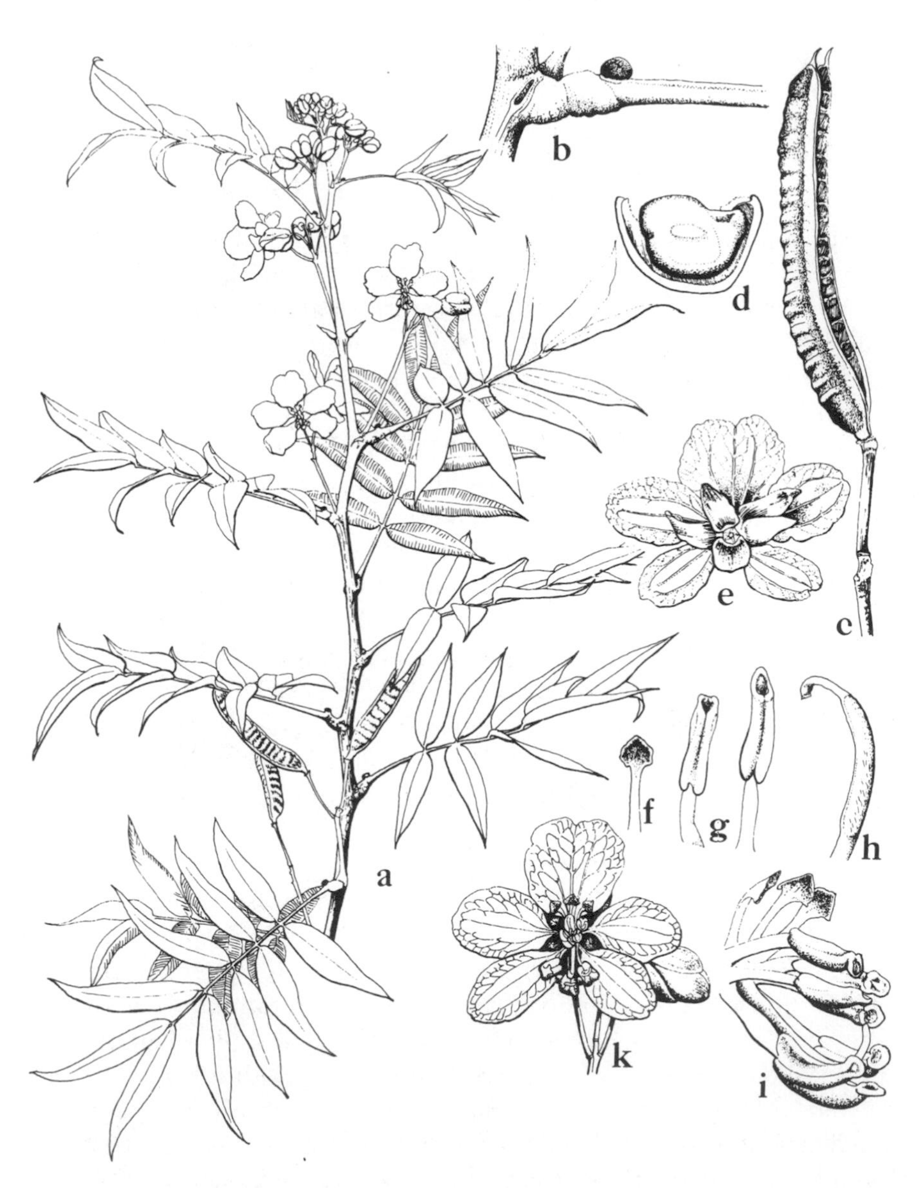

FIG. 5.16 Caesalpiniaceae. *Senna sophera* (L.) Roxb. a, habit, $\times\frac{1}{3}$; b, node and leaf-base, showing pulvinus and petiolar gland, $\times 1\frac{1}{4}$; c, opening fruit, $\times 1$; d, portion of opened fruit, with seed, $\times 2\frac{1}{2}$; e, flower, from beneath, $\times 1$; f, staminode, $\times 2\frac{1}{2}$; g, two forms of stamens, $\times 2\frac{1}{2}$; h, pistil, $\times 2\frac{1}{2}$; i, flower, with the perianth removed, $\times 2\frac{1}{2}$; k, flower, from above, $\times 1$.

so, or the 2 upper ones more or less connate, imbricate or rarely valvate, or all united into a 5-toothed cup (as in *Dimorphandra*), or sometimes the calyx spathaceous or splitting irregularly, or more or less reduced or even obsolete; corolla of (0–) 5 distinct, imbricate petals, irregular (regular in *Gleditsia*), the uppermost petal generally internal to the 2 adjacent ones and often smaller than them; stamens (1–) 10 (-numerous), distinct or the filaments sometimes variously connate, but not long-exserted and not forming the showy part of the flower, all alike or variously heteromorphic, sometimes some of them staminodial; anthers tetrasporangiate and dithecal, opening lengthwise or sometimes by terminal or basal pores; in *Dicorynia* the stamens only 2, one or both of them with the microsporangia deeply cleft so that the anther shows 8 or 10 microsporangia at and above midlength; tapetal cells of the anther nearly always with 2 or more nuclei; pollen-grains binucleate, commonly tricolporate, but variously porate or colpate or inaperturate in some genera, generally borne in monads; nectary commonly a ring on the receptacle, around the ovary; gynoecium of a single carpel with a terminal (or excentric) style and stigma; ovules 2-many on a marginal placenta, anatropous to sometimes hemitropous or somewhat campylotropous, crassinucellar, bitegmic, often with a zigzag micropyle; endosperm-development nuclear. FRUIT commonly dry and dehiscent down both sutures (i.e., a typical legume), but sometimes indehiscent (and then drupaceous or samaroid), or breaking transversely into 1-seeded joints; seeds often with an elongate funiculus, sometimes arillate, rarely winged (*Batesia*), and commonly with a hard, impervious coat; embryo straight, with 2 thickened, commonly cordate cotyledons, the radicle usually short and thick, never folded, the axis of the embryo commonly short and straight; endosperm mostly wanting or very scanty, seldom copious. X = 6–14.

The family Caesalpiniaceae consists of about 150 genera and 2200 species, widespread in tropical and subtropical regions; only a few species grow in distinctly temperate climates. The largest genera are *Bauhinia, Chamaecrista,* and *Senna,* with about 250 species each, followed by *Caesalpinia* and *Swartzia,* with about 125 each. *Swartzia,* with usually only a single petal, is sometimes referred to the Fabaceae. *Cassia,* formerly interpreted to include *Senna* and *Chamaecrista,* is now restricted to a group of about 30 species. Senna is obtained from some Old-World species of *Senna,* such as *S. angustifolia* Vahl. *Delonix regia* (Bojer) Raf., the royal poinciana, and *Amherstia nobilis* Wall. are familiar as ornamental trees in the tropics. *Copaifera, Haematoxylon,* and *Tamarindus* are some other familiar tropical genera. *Cercis, Gymnocladus,* and *Gleditsia* are among the more frost-hardy genera, and *Gleditsia triacanthos* L., the honey locust, is a common street-tree in temperate regions. The resin of *Copaifera* has been used medicinally, and some species are exciting interest as a possible source of commercial sesquiterpene resins.

Pollen very similar to that of *Sindora* occurs in Maestrichtian deposits in such diverse places as Siberia, Canada, and Colombia. Pollen referred to *Crudia* dates from the Paleocene. Thus on the basis of the fossil record the Caesalpiniaceae appear to be the oldest family of legumes.

3. Family FABACEAE Lindley 1836 nom. conserv., the Pea or Bean Family

Herbs (sometimes twining, or climbing by tendrils), or less often shrubs, trees, or woody vines, seldom spiny (but notably so in some Old World species of Genisteae and *Astragalus*), sometimes with anomalous stem-structure, often provided with various sorts of glandular or eglandular hairs, the eglandular ones often malpighian or uniseriate with short basal cells and an elongate terminal cell, or sometimes stellate; plants commonly bearing root-nodules that harbor nitrogen-fixing bacteria, very often with non-protein amino-acids in the seeds and/or vegetative parts, and commonly with solitary crystals of calcium oxalate in some of the epidermal and parenchymatous cells, but without raphides, tanniferous, commonly with scattered secretory cells (sometimes also sacs or canals) containing tannin and often also gum or other substances, sometimes producing proanthocyanins, and sometimes cyanogenic, but without ellagic acid and without iridoid compounds, often producing one or another sort of alkaloid, especially of the pyridine, quinolizidine, and indole groups; nodes trilacunar or less often pentalacunar; vessel-segments with simple perforations; imperforate tracheary elements mostly or all with small, simple pits, only very rarely septate, but some vasicentric tracheids sometimes also present; wood-rays homocellular or heterocellular, in woody species uniseriate or more often 2–3 cells wide, or sometimes up to 12 or even 20 cells wide; wood-parenchyma usually abundant, variously disposed; sieve-tube plastids containing irregular protein crystalloids and generally also starch grains, or sometimes with starch grains only. LEAVES alternate, or rarely opposite, as in *Platymiscium,* pinnately or less often palmately compound or trifoliolate, or sometimes unifoliolate or even simple, the petiole and individual leaflets commonly with a swollen, basal pulvinus that governs their

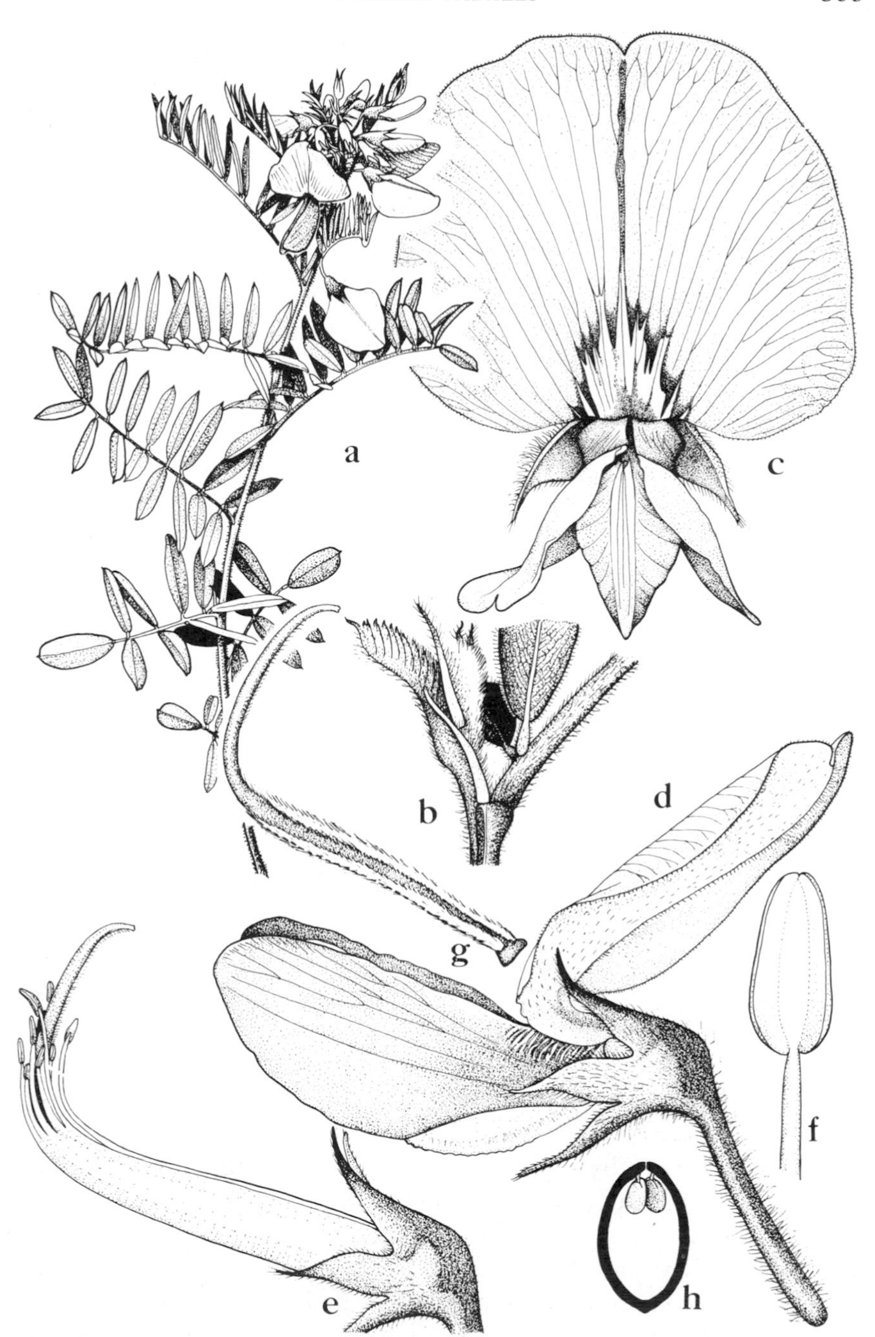

FIG. 5.17 Fabaceae. *Tephrosia virginiana* (L.) Pers. a, habit, ×½; b, shoot-tip, with stipules and developing leaves, ×4; c, d, two views of flower, ×4; e, flower, with the corolla removed, ×4; f, anther, ×16; g, pistil, ×4; h, schematic cross-section of ovary, ×8.

orientation, or the pulvinus present but not functional, or wanting; leaflets often provided with stipellules; stomates of diverse types; petiole often with complex vascular anatomy; stipules present, seldom modified into spines or prickles. FLOWERS mostly in racemes, spikes, or heads, commonly more or less showy, slightly to seldom strongly perigynous, perfect, usually strongly irregular; sepals mostly 5, more or less connate into a lobed tube that is often somewhat bilabiate; corolla typically papilionaceous and consisting of 5 petals, the uppermost (adaxial) one, called the *banner* or *standard,* borne externally to the others and usually the largest, folded along the midline and more or less enfolding the other petals in bud; rarely (as in spp. of *Canavalia, Centrosema,* etc.) the flower resupinate, with the banner lowermost; 2 lateral petals, called *wings,* usually similar inter se and mostly distinct, but sometimes connivent distally or lightly adherent in a small area to the keel-petals; 2 lower petals innermost, similar inter se, mostly connate (or imbricate) distally to form a *keel* enfolding the androecium and gynoecium; petals seldom fewer than 5, seldom all but the banner suppressed, or rarely all the petals completely wanting or (*Etaballia*) all alike and distinct; stamens mostly 10, seldom only 5–9, 9 of the filaments commonly connate into an open sheath around the pistil, the 10th (adaxial) one usually partly or wholly separate from the other 9 (the stamens then said to be diadelphous), or less often all 10 stamens connate by their filaments to form a closed (or rarely adaxially open) sheath (the stamens then said to be monadelphous), or the 10th stamen suppressed and the other 9 monadelphous in an adaxially open sheath, or the filaments sometimes all distinct; exceptionally the stamens polyadelphous (1, 4, 1, 4), as in *Discolobium*; anthers tetrasporangiate and dithecal, mostly opening by longitudinal slits, all alike or sometimes dimorphic and alternating; tapetal cells of the anther uninucleate; pollen-grains 2–3-nucleate, commonly tricolporate or sometimes tricolpate or triporate, generally borne in monads; filaments sometimes nectariferous toward the base, but the nectary more often a ring (often lopsided) on the receptacle around the overy; gynoecium of a single carpel with a terminal style and stigma; stigma wet, papillose; ovules (1) 2-many on a marginal placenta, variously anatropous, hemitropous, or much more often campylotropous, crassinucellar, bitegmic, often with a zigzag micropyle; endosperm-development nuclear. FRUIT commonly dry and dehiscent down both sutures (i.e., a typical legume), but sometimes follicular, or indehiscent and then sometimes winged or breaking transversely into 1-seeded joints (i.e., a loment), rarely bladdery-inflated or more or less drupaceous (*Andira*) or otherwise fleshy, or nut-like or achene-like; seldom (as in spp. of *Astragalus*) the dorsal suture of the fruit giving rise to a partial or even complete partition; seeds with a short funiculus and usually a hard, impervious seed-coat, often very long-lived, generally with a complex, specialized hilum and usually a small strophiolar swelling between the hilum and the chalaza; embryo commonly with thickened, noncordate cotyledons, the radicle very often folded, sometimes elongate (i.e., the embryo typically curved, only seldom straight); endosperm mostly wanting or very scanty, seldom copious; seeds very often containing considerable amounts of canavanine, a non-protein amino-acid apparently restricted to this family. X = 5–13. (Papilionaceae)

The family Fabaceae as here narrowly defined consists of about 440 genera and 12,000 species, widespread in temperate and cold as well as tropical regions. The largest genus is *Astragalus*, with perhaps as many as 2000 species; some species are loco-weeds, some are selenium-accumulators, and some are both, but most are neither. Some other large genera are *Indigofera* (500), *Crotalaria* (500), *Trifolium* (300, including *T. pratense* L., red clover, and *T. repens* L., white clover), *Dalea* (160+), *Phaseolus* (200, including *P. vulgaris* L., kidney bean, and *P. limensis* Macfady, lima bean), *Lupinus* (200), *Dalbergia* (200), *Vicia* (150, including *V. faba* L., broad bean, and *V. sativa* L., common vetch), *Lathyrus* (150, including *L. odoratus* L., sweet pea), *Onobrychis* (170), *Hedysarum* (160), *Psoralea* (150), *Lotus* (150), *Dolichos* (120, including *D. lablab* L., hyacinth bean), *Medicago* (110, including *M. sativa* L., alfalfa), and *Lespedeza* (100, bush clover). Some other well known members of the family are *Abrus precatorius* L., rosary pea; *Arachis hypogaea* L., peanut; *Cytisus* and *Genista,* broom; *Derris,* tuba root; *Erythrina; Glycine max* (L.) Merrill, soy bean; *Glycyrrhiza glabra* L., licorice; *Lens culinaris* Medic., lentil; *Lonchocarpus utilis* HBK., cubé root; *Melilotus,* sweet clover; *Pisum sativum* L., common pea; *Pueraria thunbergiana* Benth., kudzu vine; *Robinia pseudoacacia* L., black locust; *Ulex europaeus* L., gorse; and *Wisteria*. The Fabaceae rank second only to the Poaceae in agricultural importance.

Bell, Lackey, & Polhill (1978) suggest that canavanine, which acts as an antimetabolite of arginine, plays an important role in chemical defense, as well as in the storage of bound nitrogen.

Pollen identifiable with the Fabaceae enters the fossil record only in the middle part of the Upper Miocene. Thus the Fabaceae appear to be the youngest of the 3 families of the order. That assessment is in harmony with relationships postulated on the basis of the modern members of the Fabales.

3. Order PROTEALES Lindley 1833

Woody plants, or seldom almost herbaceous, tanniferous, producing proanthocyanins and/or ellagic acid, but without iridoid compounds and only seldom with alkaloids; vessel-segments nearly always with simple perforations; imperforate tracheary elements commonly with bordered pits; wood-rays often very broad; phloem sometimes stratified into hard and soft layers, or with irregular strands of fibers; no internal phloem. LEAVES alternate to seldom opposite or whorled, simple to pinnately compound, exstipulate. FLOWERS perfect or seldom unisexual, hypogynous to strongly perigynous, regular or irregular, most commonly 4-merous; sepals valvate, usually connate at least below to form a tube, or appearing as lobes on a hypanthium, seldom distinct and hypogynous; petals wanting or represented by distinct or variously connate scales or glands; stamens unicyclic or seldom bicyclic, sometimes some of them reduced or staminodial, generally borne on the calyx-tube or the hypanthium or adnate to the base of the sepals, seldom strictly hypogynous; pollen-grains binucleate, mostly tricolporate or triporate; gynoecium of a single carpel with an elongate, terminal style; ovules 1-several or seldom numerous, marginal or basal, bitegmic or seldom unitegmic, crassinucellar; endosperm-development nuclear. FRUIT of diverse types, very often containing only a single seed; embryo usually dicotyledonous; endosperm scanty or more often none.

The order Proteales as here defined consists of only 2 families, the Proteaceae and Elaeagnaceae. Each of these families has sometimes been considered to form a distinct, unifamilial order, and each has by some authors been associated with one or another order of the Rosidae, or less often of the Dilleniidae.

In spite of the impressive number of technical features shared by the Proteaceae and Elaeagnaceae, it is not at all certain that the two groups are closely related. The Elaeagnaceae do not really look like the Proteaceae, but they do look remarkably like some of the Thymelaeaceae.

The pistil in the Proteaceae clearly consists of a single carpel, which is often not fully sealed. Fossil pollen regarded as proteaceous is known from Santonian (i.e., mid-Upper Cretaceous) deposits in Australia, laid down about 82 million years ago, and from somewhat younger (Maestrichtian, i.e., latest Cretaceous) deposits in South America. No order of Rosidae more advanced than the Rosales is a likely ancestor for the Proteaceae.

The pistil in the Elaeagnaceae also appears to consist of a single carpel, but it has not been sufficiently studied to exclude the possibility that it is only pseudomonomerous. The Thymelaeaceae include both pluricarpellate and pseudomonomerous types, and a little further reduction might conceivably lead to something like the Elaeagnaceae. The pollen of the Elaeagnaceae dates only from the Upper Miocene, so there is no problem of timing. Yamazaki (1975) compares the embryogeny of *Elaeagnus umbellata* to that of *Myrtus,* and says it is unlike that of the Proteaceae. As a putative member of the Myrtales, the Elaeagnaceae would stand out like a sore thumb on anatomical as well as floral morphological grounds, but an evolutionary relationship via the Thymelaeaceae cannot be ruled out on the basis of present evidence.

The position of the Elaeagnaceae cannot be resolved by supposing a close relationship between the Proteaceae and Thymelaeaceae. The Thymelaeaceae clearly have a basically compound pistil, in contrast to the simple pistil of the Proteaceae. According to Axelrod & Raven (1974, in initial citations) Doyle considers that on the basis of the pollen the most recent common ancestor of the Proteaceae and Thymelaeaceae could scarcely be less than about 110 million years old, that is, somewhere near the Albian-Cenomanian border. At that time the Rosidae were probably only in process of differentiating from the ancestral Magnoliidae.

For the present it will do no harm to retain the Proteaceae and Elaeagnaceae in the same order. When more evidence is available it may become necessary to restore the order Elaeagnales and insert it in a position following the Myrtales.

SELECTED REFERENCES

Bond, G., J. T. MacConnell, & A. H. McCallum. 1956. The nitrogen-nutrition of *Hippophaë rhamnoides* L. Ann. Bot. (London) II. 20: 501–512.

Cooper, D. C. 1932. The development of the peltate hairs of *Shepherdia canadensis.* Amer. J. Bot. 19: 423–428.

Elsworth, J. F., & K. R. Martin, 1971. Flavonoids of the Proteaceae, Part 1. A chemical contribution to studies on the evolutionary relationships in the S. African Proteoideae. J. S. African Bot. 37: 199–212.

Gardner, I. C. 1958. Nitrogen fixation in *Elaeagnus* root nodules. Nature 181: 717–718.

Gardner, I. C., & G. Bond. 1957. Observations on the root nodules of *Shepherdia.* Canad. J. Bot. 35: 305–314.

Graham, S. A. 1964. The Elaeagnaceae in the southeastern United States. J. Arnold Arbor. 45: 274–278.

Haber, J. M. 1959–1966. The comparative anatomy and morphology of the flowers and inflorescences of the Proteaceae. I. Some Australian taxa. II. Some American taxa. III. Some African taxa. Phytomorphology 9: 325–358, 1959 (1960). 11: 1–16, 1961. 16: 490–527, 1966.

Hawker, L. E., & J. Fraymouth. 1951. A re-investigation of the root-nodules of species of *Elaeagnus, Hippophae, Alnus,* and *Myrica,* with special reference to the morphology and life histories of the causative organisms. J. Gen. Microbiol. 5: 369–386.

Johnson, L. A. S., & B. G. Briggs. 1963. Evolution in the Proteaceae. Austral. J. Bot. 11: 21–61.

Johnson, L. A. S., & B. G. Briggs. 1975. On the Proteaceae—the evolution and classification of a southern family. J. Linn. Soc., Bot. 70: 83–182.

Kausik, S. B. 1940. Vascular anatomy of the flower of *Macadamia ternifolia* F. Muell (Proteaceae). Curr. Sci. 9:22–25.

Kausik, S. B. 1941. Studies in the Proteaceae. V. Vascular anatomy of the flower of *Grevillea robusta* Cunn. Proc. Natl. Inst. Sci. India 7: 257–266.

Leins, P. 1967. Morphologische Untersuchungen an Elaeagnaceen–Pollenkörnern. Grana Palyn. 7: 390–399.

Purnell, H. M. 1960. Studies of the family Proteaceae. I. Anatomy and morphology of the roots of some Victorian species. Austral. J. Bot. 8: 38–50.

Rao, V. S. 1974. The nature of the perianth in *Elaeagnus* on the basis of floral anatomy, with some comments on the systematic position of Elaeagnaceae. J. Indian Bot. Soc. 53: 156–161.

Rourke, J., & D. Wiens. 1977. Convergent floral evolution in South African and Australian Proteaceae and its possible bearing on pollination by nonflying mammals. Ann. Missouri Bot. Gard. 64: 1–17.

Venkata Rao, C. 1957. Cytotaxonomy of the Proteaceae. Proc. Linn. Soc. New South Wales 82: 257–271.

Venkata Rao, C. 1960–1971. Studies in the Proteaceae. I–IV, XIII. Proc. Natl. Inst. Sci. India 26B: 300–337, 1960. 27B: 126–151, 1961. 29B: 489–510, 1963. 30B: 197–244, 1964. 35B: 471–486, 1969. V, VI. J. Indian Bot. Soc. 44: 244–270; 479–494, 1965. XIV. Proc. Indian Natl. Sci. Acad. 36B: 345–363, 1971.

Venkata Rao, C. 1965. Pollen grains of Proteaceae. J. Palynol. (Lucknow) 1: 1–9.

Venkata Rao, C. 1967. Morphology of the nectary in Proteaceae. New Phytol. 66: 99–107.

Venkata Rao, C. 1967. Origin and spread of the Proteaceae. Proc. Natl. Inst. Sci. India 33B: 219–251.

Venkata Rao, C. 1971. Proteaceae. Bot. Monogr. 6. Council of Sci. & Indus. Res. New Delhi.

Vickery, J. R. 1971. The fatty acid composition of the seed oils of Proteaceae: a chemotaxonomic study. Phytochemistry 10: 123–130.

Yamazaki, T. 1975. Embryogeny of *Elaeagnus umbellata* Thunb. J. Jap. Bot. 50: 281–284.

Давтян, А. Г. 1950. Сравнительно-анатомическое исследование древесины дикорастущих и культивируемых на Кавказе видов рода *Elaeagnus.* Труды Бот. Инст. АН Армянской ССР, 7: 133-144.

Массагетов, П. С. 1946. Алкалоиды в растениях семейства Elaeagnaceae. Ж. Общ. Химии 16: 139-140.

SYNOPTICAL ARRANGEMENT OF THE FAMILIES OF PROTEALES

1 Plants provided with a vesture of peltate scales or stellate hairs; fruit typically a pseudodrupe, the dry achene surrounded by the

persistent, thickened, externally fleshy or mealy base of the hypanthium; stamens as many as and alternate with the sepals, or twice as many as the sepals; nodes unilacunar 1. ELAEAGNACEAE.

1 Plants without peltate scales or stellate hairs; fruits diverse, but not as in the Elaeagnaceae; stamens as many as and opposite the sepals; nodes trilacunar ....2. PROTEACEAE.

1. Family ELAEAGNACEAE A. L. de Jussieu 1789 nom. conserv., the Oleaster Family

Shrubs or sometimes rather small trees, often thorny, commonly harboring symbiotic nitrogen-fixing bacteria in nodules on the roots, copiously provided with lepidote or stellate trichomes, strongly tanniferous, generally with scattered tanniferous cells in the parenchymatous tissues, accumulating ellagic acid and often also proanthocyanins, commonly producing quebrachitol and sometimes also saponins and/or indole alkaloids, but not cyanogenic and without iridoid compounds; nodes unilacunar; calcium-oxalate crystals of various sorts commonly present in some of the cells of the parenchymatous tissues; vessel-segments with simple perforations; imperforate tracheary elements consisting of fiber-tracheids and true tracheids, with bordered pits; wood-rays homocellular or heterocellular, mixed uniseriate and pluriseriate, the uniseriate ones rather few, the others 2–20 cells wide, with short ends; wood-parenchyma diffuse, sometimes very scanty; phloem commonly stratified tangentially into hard and soft layers; no internal phloem. LEAVES alternate or (*Shepherdia*) opposite, simple and entire, pinnately veined; stomates anomocytic; stipules wanting. FLOWERS in racemes or small umbels or solitary in the axils of the leaves, perfect or sometimes unisexual (the plants then dioecious or polygamo-dioecious), regular, mostly 4-merous, strongly perigynous, apetalous; hypanthium in perfect and pistillate flowers tubular, commonly constricted just above (but free from) the ovary, in staminate flowers generally cupulate or almost flat; sepals appearing as lobes on the hypanthium, commonly 4, seldom 2 or 6, valvate, often somewhat petaloid; stamens borne in the throat of the hypanthium, as many as and alternate with the sepals (*Elaeagnus*) or twice as many as the sepals and both alternate with and opposite to them (*Hippophae* and *Shepherdia*); filaments very short; anthers tetrasporangiate and dithecal, opening by longitudinal slits; pollen-grains binucleate, (2) 3 (4)-colporate; a more or less well developed, often lobulate nectary-disk commonly borne on the inner surface of the hypanthium, in *Elaeagnus* at or just above the constriction, in *Shepherdia* near the summit, the lobes alternating with the stamens in *Shepherdia*; gynoecium of a single carpel, with an elongate, slender style terminating in a linear to capitate stigma; ovule solitary, basal, anatropous, bitegmic, crassinucellar, with a funicular obturator; endosperm-development nuclear. FRUIT drupe-like or berry-like, the dry achene enveloped by (but free from) the persistent base of the hypanthium, which becomes

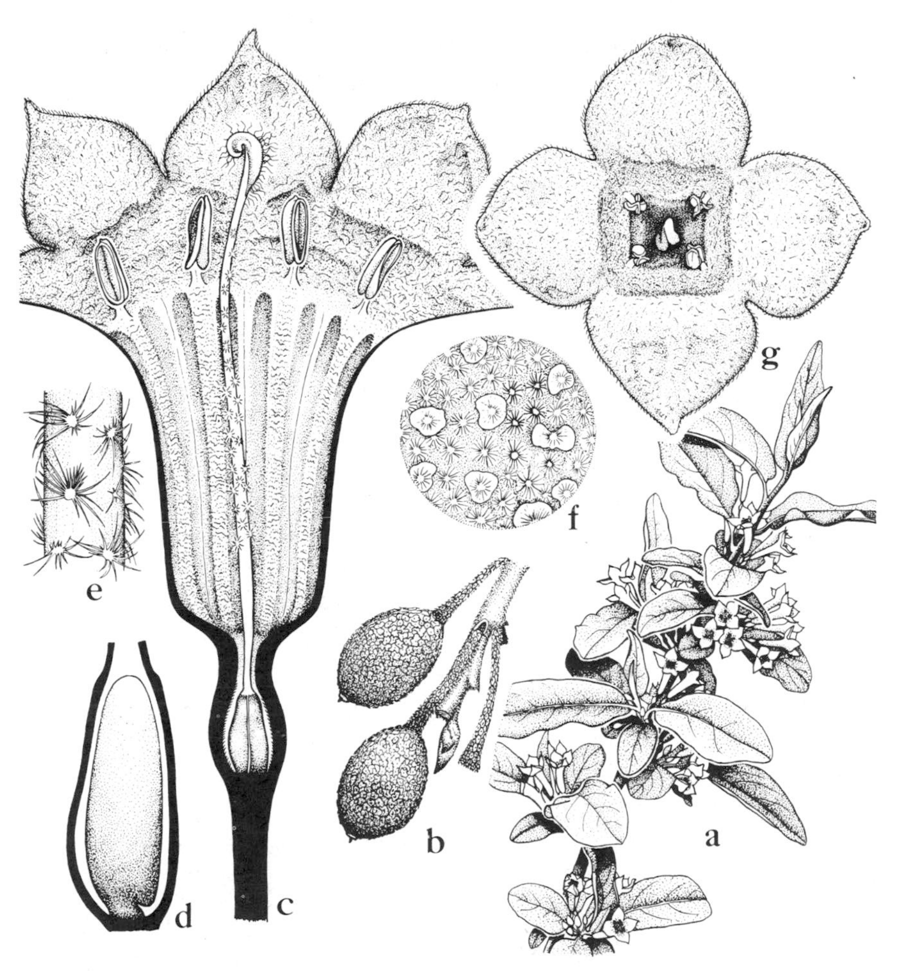

FIG. 5.18 Elaeagnaceae. *Elaeagnus umbellata* Thunb. a, habit, $\times\frac{1}{2}$; b, fruits, $\times 1\frac{1}{2}$; c, flower, slit lengthwise and laid out, $\times 6$; d, ovary in partial long-section, $\times 18$; e, pubescence of style, $\times 18$; f, portion of lower surface of leaf, with peltate scales and stellate hairs, $\times 12$; g, flower, from above, $\times 6$.

mealy or fleshy, very often with a bony inner layer; embryo straight, with 2 expanded, fleshy-thickened, oily and proteinaceous (sometimes also starchy) cotyledons; endosperm scanty or none. X = 6, 10, 11, 13, 14.

The family Elaeagnaceae consists of 3 genera and about 50 species, occurring mostly in temperate and subtropical regions of the Northern Hemisphere, but extending also to tropical Asia and even to northern Australia. *Elaeagnus*, with about 45 species, is by far the largest genus.

Elaeagnus angustifolia L., the Russian olive, is sometimes planted for ornament, especially in temperate and warm-temperate dry regions.

Fossil pollen considered to represent the Elaeagnaceae dates only from the Paleocene. Thus the family appears to be considerably younger than the Proteaceae.

2. Family PROTEACEAE A. L. de Jussieu 1789 nom. conserv., the Protea Family

Evergreen shrubs or less often trees, seldom almost herbaceous, glabrous or more often provided with characteristic, 3-celled, often thick-walled trichomes (the basal cell embedded in the epidermis, the stalk-cell short, the terminal cell elongate and sometimes equally or unequally bifid), occasionally with glandular hairs as well, commonly with scattered tanniferous cells in the parenchymatous tissues, producing proanthocyanins but apparently not ellagic acid, often accumulating aluminum, sometimes cyanogenic, but without saponins or iridoid compounds; roots not forming mycorhizae, commonly producing clusters of specialized short lateral roots ("proteoid roots"); nodes trilacunar; crystals seldom present in any of the cells, except often in the endocarp; vessel-segments largely or usually wholly with simple perforations; imperforate tracheary elements commonly with small bordered pits; vasicentric tracheids sometimes present; wood-rays nearly or quite homocellular, all uniseriate, or often mixed uniseriate and pluriseriate, the latter often 10–30 cells wide and high, or even larger; wood-parenchyma variously paratracheal or apotracheal, sometimes banded; phloem sometimes more or less stratified into hard and soft layers, or with irregular strands of fibers; no internal phloem. LEAVES alternate to seldom opposite or whorled, simple and entire or toothed to often pinnatifid or pinnately or bipinnately compound, diverse in form and structure, very often xeromorphic; stomates commonly paracytic; petiole with complex vascular anatomy; stipules wanting. FLOWERS pollinated by insects, birds, mice, or small marsupials, solitary or paired (seldom several) in the axils of bracts, arranged in racemes, umbels, or cone-like inflorescences, or often in involucrate heads that suggest those of the Asteraceae, the primary inflorescences often reduced to a pair of bracteate, pedicellate flowers, which may be arranged into secondary racemes, usually perfect, seldom unisexual (the plants then monoecious or dioecious) protandrous, regular or more or less irregular, 4-merous, hypogynous to more or less definitely perigynous; sepals valvate, commonly petaloid, distinct

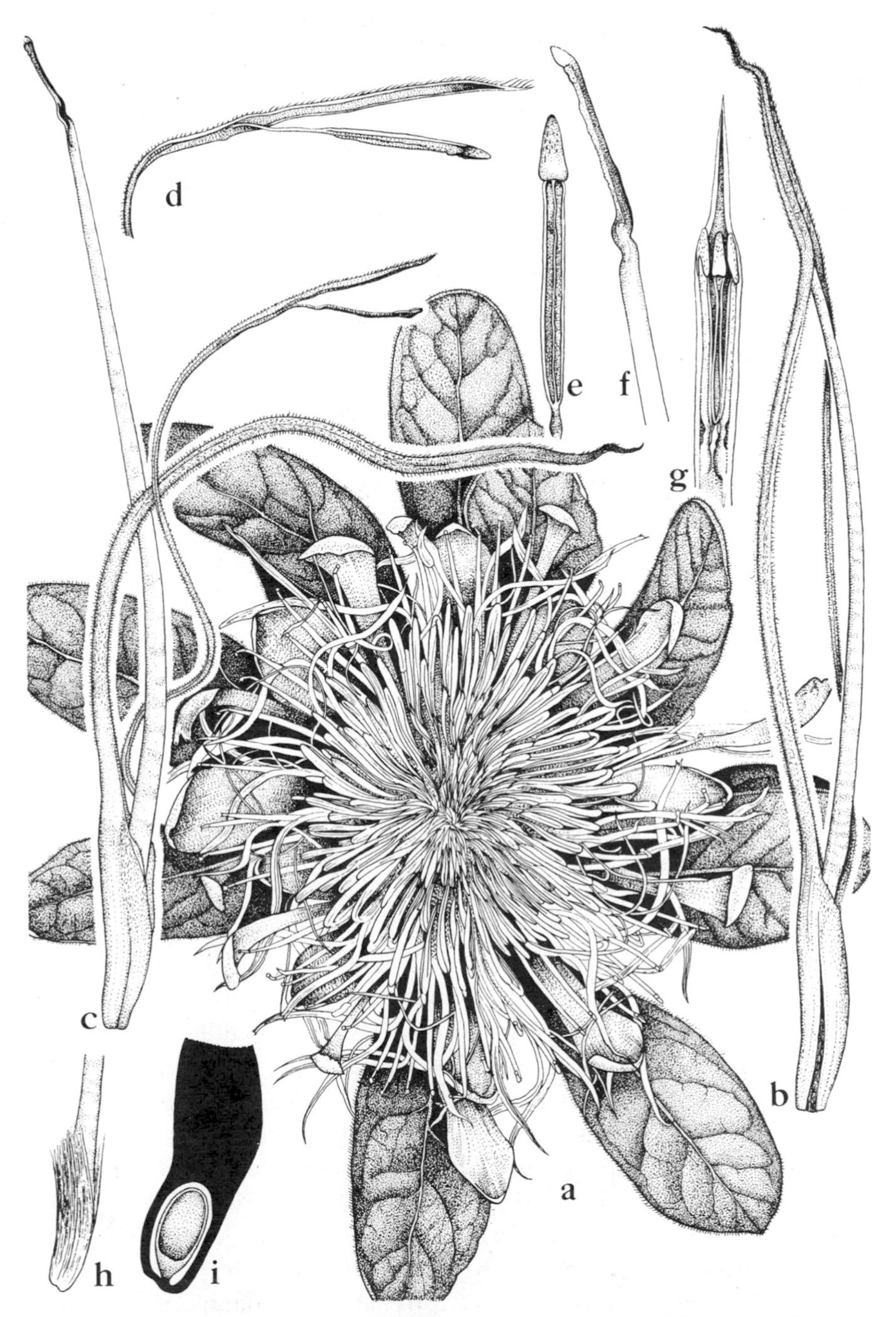

FIG. 5.19 Proteaceae. *Protea susannae* Phillips. a, inflorescence and leaves, ×1; b, young flower, ×2; c, fully opened flower, ×2; d, sterile stamen and its associated calyx-lobe, ×4; e, fertile stamen, ×6; f, stigma, ×4; g, three fertile stamens and associated calyx-lobes, ×4; h, base of pistil, ×2; i, schematic long-section of ovary, ×8.

or more often united below to form a calyx-tube that may be more deeply cleft on one side, or 3 sepals connate and the fourth distinct; corolla apparently represented only by an annular or horseshoe-shaped, often 4-lobed, hypogynous nectary-disk, or by 4 (less often 3 or 2) distinct or variously connate, hypogynous nonvasculated scales or glands alternating with the sepals, or completely wanting (these scales or glands probably best considered to be mere enations or otherwise noncorolline); stamens 4, antesepalous, the filaments broad, adnate to the base of the sepals or to the calyx-tube to varying heights (in *Franklandia* borne on a well developed, tubular hypanthium), or rarely (*Bellendena*) free from the calyx and hypogynous; anthers opening lengthwise, tetrasporangiate and dithecal (rarely bisporangiate, as in *Conospermum* and *Synaphea*), or sometimes one or more of them more or less reduced and monothecal or staminodial; connective often prolonged as an evident, terminal appendage; pollen-grains binucleate, (2) 3 (–8)-porate or sometimes -colporate; gynoecium of a single conduplicate, often stipitate carpel, sealed to varying degrees, with an elongate style and a terminal or lateral stigma, the style sometimes somewhat modified to serve in pollen-presentation; ovules 1 or 2 or less often several or more or less numerous, marginal, anatropous to more often hemitropous or amphitropous, or sometimes orthotropous, bitegmic, crassinucellar; endosperm-development nuclear. FRUIT a follicle, nut, achene, or drupe, when indehiscent often 1-seeded even though the ovary may have 2 or more ovules; seeds often winged; embryo straight, oily, lacking a suspensor; cotyledons mostly 2, in *Persoonia* 3–8; endosperm usually wanting, but present in *Bellendena*; germination usually epigaeous. X = 5, 7, 10–13, perhaps primitively 7; chromosomes sometimes very large.

The well defined family Proteaceae consists of about 75 genera and more than 1000 species, widespread in tropical and subtropical regions and especially in the warmer parts of the Southern Hemisphere. The greatest centers of distribution are in South Africa and Australia. A large proportion of the species grow in regions with alternating wet and dry seasons, often in nutrient-poor soil, but many genera, such as *Helicia,* are trees of the rain-forest. The largest genera are *Grevillea* (250), and *Protea* and *Hakea,* each with more than a hundred species. *Protea* is African, and *Hakea* is Australian. Macadamia nuts, from *Macadamia integrifolia* Maiden & Betche and *M. tetraphylla* L. A. S. Johnson, have become an important article of commerce in recent years.

Johnson and Briggs (personal communication) consider that the primitive Proteaceae were trees of mesothermic closed forest, with entomophilous flowers of relatively unspecialized form.

Leaves considered to represent the Proteaceae are known from Paleocene and more recent deposits in the Southern Hemisphere. Raven and Axelrod (1974, in initial citations) think to see unequivocal proteaceous pollen from the Santonian of Australia, ca. 82 million years before present, and they consider that it reaches South America in mid-Maestrichtian, about 68 million years before present. On the other hand, Muller (Bot. Rev. 1981 in press) considers that the oldest clearly proteaceous pollen is Maestrichtian.

4. Order PODOSTEMALES Lindley 1833

The order consists of the single family Podostemaceae.

SELECTED REFERENCES

Battaglia, E. 1971. The embryo sac of Podostemaceae—an interpretation. Caryologia 24: 403–420.

Bezuidenhout, A. 1964. Pollen of the African Podostemaceae. Pollen & Spores 6: 463–478.

Govindappa, D. A., & C. R. Nagendran. 1975. Is there a Podestemum type of embryo sac in the genus *Farmeria*? Caryologia 28: 229–235.

Graham, S. A., & C. E. Wood. 1975. The Podostemaceae in the southeastern United States. J. Arnold Arbor. 56: 456–465.

Grubert, M. 1974 (1975). Podostemaceen-Studien. Teil I. Zur Ökologie einiger venezolanischer Podostemaceen. Beitr. Biol. Pflanzen 50: 321–391.

Jäger-Zürn, I. 1967. Embryologische Untersuchungen an vier Podostemaceen. Oesterr. Bot. Z. 114: 20–45.

Jäger-Zürn, I., 1970. Morphologie der Podostemaceae. I. *Tristicha trifaria* (Bory ex Willd.) Spreng. Beitr. Biol. Pflanzen 47: 11–52.

Razi, B. A. 1949. Embryological studies of the two members of Podostemonaceae. Bot Gaz. 111: 211–218.

Royen, P. van. 1951–1955. The Podostemaceae of the New World. Meded. Bot. Mus. Herb. Rijks Univ. Utrecht 107: 1–151, 1951. 115: 1–21, 1953. 119: 215–263, 1955.

Schnell, R. 1967. Études sur l'anatomie et la morphologie des Podostémacées. Candollea 22: 157–225.

Schnell, R. 1969. Contribution à l'étude des Podostémacées de Guyane. Adansonia sér. 2. 9: 249–271.

Schnell, R., & G. Cusset. 1963. Remarques sur la structure des plantules des Podostémonacées. Adansonia sér. 2. 3: 358–369.

1. Family PODOSTEMACEAE L. C. Richard ex C. A. Agardh 1822 nom. conserv., the River-weed Family

Aquatic herbs of fast rivers with a stony bed, often annual, mostly submerged or with some of the parts floating, but producing aerial flowers and fruits at times of low water; plants vegetatively highly modified, diverse in form and organization but anatomically relatively simple, commonly more or less thalloid and often lichen-like or fucoid in appearance, or the parts sometimes recognizable with some difficulty as modified roots, stems, and leaves; silica-bodies very often present in such quantities in the superficial tissues as to maintain the size and form of the thallus under conditions of desiccation; scattered resin-cells or latex-cells or latex-channels present in many species; xylem very much reduced, often central, lacking vessels, commonly represented only by a few tracheids with spiral or annular thickenings, or completely wanting; epidermis chlorophyllous, seldom well differentiated from the adjacent internal, also chlorophyllous tissues; primary root suppressed, but plagiotrophic, rootlike (but chlorophyllous) branches often produced at the base of the thallus, commonly attached to the substrate by numerous epidermal hairs or by specialized branches called hapters. LEAVES, when recognizable, alternate, entire to more or less dissected, without axillary buds, the branches of the thallus (or shoot) not arising from their axils. FLOWERS solitary or in cymose (often spiciform) inflorescences, very small, perfect, regular or irregular, hypogynous, apetalous, variously entomophilous, anemophilous, or cleistogamous; perianth of 2–3 (–5) distinct or more or less connate, petaloid sepals (subfamily Tristichoideae) or (subfamily Podostemoideae, by far the larger group) of 2-many distinct sepals or a small annular scale or completely wanting; young flowers or flower-clusters subtended or enclosed by a pair of bracteoles, which in the Podostemoideae are modified to form a small spathe, called a spathella, that may enclose up to 20 flowers; stamens in 1-several whorls (rarely the androecium reduced to a solitary stamen), the filaments distinct or more often connate at the base; anthers tetrasporangiate and dithecal; pollen-grains borne in diads or monads, binucleate, tricolpate or tricolporate, or rarely pantoporate or inaperturate; gynoecium of (1) 2 (3) carpels united to form a compound ovary with as many locules as carpels, the styles distinct or basally connate; ovules (2-) more or less numerous on a thickened, axile placenta (when there is more than one carpel), anatropous, bitegmic, tenuinucellar or nearly so; embryo-sac monosporic or bisporic, 4–6-nucleate, without "polar" nuclei, and not undergoing double fertilization; endosperm not produced; developing embryo with a well developed suspensor-haustorium. FRUIT a septicidal capsule with usually numerous and very small seeds, these often with a mucilaginous testa; embryo straight, dicotyledonous; germinating seeds not producing a primary root. X apparently = 10. (Tristichaceae)

The family Podostemaceae consists of some 40 genera and a little more than 200 species, widespread in tropical regions, especially in Asia and America, with only a few species in temperate climates. The largest genera are *Apinagia* (50), *Marathrum* (25), *Rhyncholacis* (25), *Inversodicraea* (20), *Podostemum* (17), and *Oenone* (17).

The Podostemaceae are taxonomically isolated, but most authors agree that they are related to the Saxifragaceae and Crassulaceae. The

Crassulaceae might seem an odd starting-point for a group of aquatic plants, but *Tillaea aquatica* L. (Crassulaceae) is semi-aquatic. *Tillaea* is not to be regarded as on the direct line of evolution to the Podostemaceae; it merely indicates the potentiality within the Crassulaceae to adapt to aquatic habitats.

The family Hydrostachyaceae, which has in the past often been associated with the Podostemaceae, is now generally considered to belong to the subclass Asteridae. It is here referred to the order Callitrichales.

5. Order HALORAGALES Novák 1954[1]

Mostly herbs, often aquatic, sometimes with crystals of calcium oxalate in some of the tissues, but without raphides, sometimes producing mucilage, but without milky juice; vessel-segments with simple perforations; imperforate tracheary elements with simple pits. LEAVES variously alternate, opposite, or whorled, entire to dissected, exstipulate, or with an axillary scale that has been interpreted as stipular. FLOWERS mostly small and individually inconspicuous, commonly anemophilous, variously solitary and axillary, or in axillary dichasia, or in terminal spikes, racemes, or panicles, epigynous, regular, perfect or often unisexual; perianth usually minute, the sepals 2–4, valvate, or nearly obsolete; petals 2–4, or wanting; stamens 1–4, or in 2 cycles of 4; filaments short; anthers tetrasporangiate and dithecal; pollen-grains binucleate or trinucleate, with 3 or more colpi or pores; carpels 2–4, united to form a compound, inferior ovary with 1–4 locules and 2–4 distinct styles; each locule with a solitary, anatropous or hemitropous, bitegmic, crassinucellar ovule pendulous from the apex; endosperm-development cellular or seldom nuclear. FRUIT small, nutlike or drupaceous, or separating into 1-seeded mericarps; seeds with a tiny or straight and cylindrical, dicotyledonous embryo embedded in the usually copious, oily endosperm.

The order Haloragales as here defined consists of 2 families, the Haloragaceae and Gunneraceae, probably less than 200 species in all. Two other families that have often been associated with these 2 are now considered to belong elsewhere. The Theligonaceae are allied to the Rubiaceae, and the Hippuridaceae are associated with the Callitrichaceae and Hydrostachyaceae to form an order (Callitrichales) of the Asteridae.

The Haloragales have often been considered to be allied to the Myrtales, or even included in that order. In addition to their reduced flowers, however, the Haloragales differ from the Myrtales in having distinct styles (a primitive feature) rather than a common style, and

[1] A Latin diagnosis, not given by Novák, is provided here:
Pleraeque herbae, saepe aquaticae; flores parvi inconspicui regulares epigyni plerumque anemophili; perianthium plerumque minimum, sepalis 2–4 vel fere obsoletis, petalis 2–4 vel 0; stamina 1–4, vel 4 + 4; carpella 2–4, connata, ovarium compositum 1–4-locularem stylis distinctis componientia, in quoque loculo ovulo solitario pendulo; semina plerumque albumine copioso instructa, embryone cotyledonibus 2. Type: Haloragaceae R. Brown in Flinders, Voy. Terra Austr. 2: 549. 1814.

they further differ from characteristic members of the Myrtales in having a well developed endosperm and in lacking internal phloem. The embryo of the Gunneraceae is very small in relation to the endosperm, as in archaic angiosperms in general. Thus it does not seem likely that the Haloragales are florally reduced, aquatic derivatives of the Myrtales. Both orders must instead by derived from a generalized Rosalean ancestry.

SELECTED REFERENCES

Bader, F., & J. Walter. 1961. Das Areal der Gattung *Gunnera* L. Bot. Jahrb. Syst. 80:281–293.

Bawa, S. B. 1969. Embryological studies on the Haloragidaceae. II. *Laurembergia brevipes* Schindl. and a discussion of systematic considerations. Proc. Natl. Inst. Sci. India 35B: 273–290.

Gruas-Cavagnetto, C., & J. Praglowski. 1977. Pollen d'Haloragacées dans le Thanétien et le Cuisien du bassin de Paris. Pollen & Spores 19: 299–308.

Jarzen, D. M., 1980. The occurrence of *Gunnera* pollen in the fossil record. Biotropica 12: 117–123.

Kapil, R. N., & S. B. Bawa. 1968. Embryological studies on the Haloragidaceae. I. *Haloragis colensoi* Skottsb. Bot. Not. 121: 11–28.

Nagaraj, M., & B. H. M. Nijalingappa. 1974. Embryological studies in *Laurembergia hirsuta*. Bot. Gaz. 135: 19–28.

Orchard, A. E. 1975. Taxonomic revisions in the family Haloragaceae. I. The genera *Haloragis, Haloragodendron, Glischrocaryon, Meziella* and *Gonocarpus*. Bull. Auckland Inst. Mus. 10.

Praglowski, J. 1969. Pollen types in species of *Haloragis*. Svensk Bot. Tidskr. 63: 486–490.

Praglowski, J. 1970. The pollen morphology of the Haloragaceae with reference to taxonomy. Grana 10: 159–239.

SYNOPTICAL ARRANGEMENT OF THE FAMILIES OF HALORAGALES

1 Ovary 2- to 4-locular, with as many styles as locules and with a single ovule in each locule; stamens (3) 4 or 8; embryo-sac monosporic, 8-nucleate; embryo straight, cylindric, flowers individually subtended by a pair of bracteoles; stem monostelic; plants commonly cyanogenic and producing proanthocyanins, aquatic or amphibious or of marshes, with small to medium-sized, alternate or often opposite or whorled leaves lacking stipules and axillary scales; leaf-venation tending to be pinnate .... 1. HALORAGACEAE.

1 Ovary unilocular, with a single pendulous ovule and 2 styles; stamens 1–2; embryo-sac tetrasporic, 16-nucleate; embryo tiny, obcordate; flowers not bracteolate; stem polystelic; plants neither cyanogenic nor with proanthocyanins, terrestrial, with alternate, palmately veined, often very large leaves which have an axillary scale that is sometimes considered to be stipular ....2. GUNNERACEAE.

1. Family HALORAGACEAE R. Brown in Flinders 1814 nom. conserv., the Water Milfoil Family

Submerged to emergent, aquatic or amphibious herbs or plants of marshes, less often distinctly terrestrial, seldom (*Haloragodendron*) shrubs or even small trees, glabrous or with uniseriate hairs, commonly tanniferous, producing both proanthocyanins and ellagic acid and also cyanogenic, generally with crystals of calcium oxalate in some of the parenchymatous tissues, often in hairlike cortical cells; cortex commonly with numerous air-cavities; vascular system generally more or less reduced, often consisting of a central fibrovascular strand, without a pith, but in some of the more or less terrestrial species the stem with a well developed cambium and a continuous cylinder of secondary xylem with narrow rays; vessel-segments slender, with simple perforations; imperforate tracheary elements thick-walled, with a narrow lumen and simple pits. LEAVES alternate, opposite, or whorled, very diverse in form and size, tending to be pinnately veined; stomates, when present, mostly anomocytic; stipules wanting. FLOWERS mostly small and anemophilous (relatively large in *Loudonia*), solitary and axillary, or in terminal spikes, racemes, or panicles, perfect or more often unisexual, epigynous, regular, commonly 4-merous, less often 3-merous, individually subtended by a pair of bracteoles; sepals valvate, persistent in fruit; petals often somewhat surpassing the sepals, commonly deciduous, or sometimes wanting; stamens most commonly 8 in 2 cycles, the outer set opposite the sepals, or less often 4, seldom only 3; filaments mostly short or very short; anthers relatively large (sometimes some of them smaller and more or less staminodial), tetrasporangiate and dithecal, opening by longitudinal slits; pollen-grains binucleate or trinucleate, with 3 or more apertures, often with 4–6 (or more) short colpi or pores, sometimes dimorphic; gynoecium of (2) 3 or 4 carpels united to form a compound, inferior ovary with as many locules as carpels (or the partitions sometimes feebly developed) and with distinct, feathery styles; each locule with a solitary, anatropous or hemitropous, bitegmic, crassinucellar ovule pendulous from the apex, with a weakly developed funicular obturator; embryo-sac monosporic, 8-nucleate; endosperm-development cellular or seldom nuclear; developing embryo with a haustorial suspensor. FRUIT small, nutlike or drupaceous, sometimes (*Myriophyllum* and *Vinkia*) separating into 1-seeded mericarps; seeds with a well developed, straight, cylindrical, dicotyledonous embryo embedded in the usually more or less copious, softly fleshy, oily endosperm. X most commonly = 7. (Myriophyllaceae)

The family Haloragaceae as here defined consists of 8 genera and about 100 species, widely distributed throughout the world, but best developed in the Southern Hemisphere, especially in Australia. The largest genera are *Gonocarpus* (36), *Haloragis* (26), and *Myriophyllum* (20+). *Myriophyllum aquaticum* (Vell.) Verdc., called parrot's feather, is a well known aquarium-plant.

Pollen that fairly clearly represents the Haloragaceae is known from Paleocene and more recent deposits. Pollen more doubtfully of this affinity extends well back into the Upper Cretaceous. Fruits of *Proserpinaca* date from the Miocene epoch.

2. Family GUNNERACEAE Meissner 1841 nom. conserv., the Gunnera Family

Terrestrial, often megaphytic, perennial herbs, often with a coarse, procumbent, only apically assurgent stem, harboring symbiotic colonies of *Nostoc* in the superficial tissues of the stem and/or adventitious roots in association with lysigenous, glandular cavities filled with tanniferous mucilage; hairs unicellular; plants not cyanogenic, and lacking proanthocyanins; vascular system of the shoot polystelic, with a small to large number of separate, variously oriented, anastomosing steles, each with its own endodermis; vessel-segments with simple perforations; imperforate tracheary elements with simple pits; sieve-tubes with P-type plastids. LEAVES alternate, all radical, often very large, long-petiolate, and provided with a large, median axillary scale that has sometimes been considered to be stipular, the stipules otherwise wanting; blade palmately veined, commonly orbicular or ovate or sometimes peltate, often large and coarse. FLOWERS epigynous, individually small, borne in a commonly very large, erect, panicle (this terminal or from an upper axil), not individually bracteolate, typically the lower flowers pistillate and the upper ones staminate, the middle ones sometimes perfect, or seldom all the flowers perfect or the flowers all unisexual and the plants dioecious; sepals 2 (3) valvate, very small, often nearly obsolete; petals 2, mitre-shaped, somewhat surpassing the sepals, or more often wanting; stamens 1 or 2, with short filaments; anthers tetrasporangiate and dithecal, opening by longitudinal slits; pollen-grains binucleate, with 3 (–5) long, deep colpi, said to be of a very distinctive type; gynoecium of 2 carpels united to form a compound, inferior, unilocular ovary with 2 distinct, terminal styles; stigmas dry, papillate; ovule solitary, pendulous from the apex of the locule, anatropous, bitegmic, crassinucellar, without an obturator; embryo-sac tetrasporic, 16-nucleate; endosperm-development cellular; suspensor not producing a haustorium. FRUIT drupaceous; seeds with a very small, dicotyledonous, obcordate embryo embedded in the copious, oily endosperm. X = 12 (or 11?), 17, 18.

The family Gunneraceae consists of the single genus *Gunnera,* with perhaps as many as 50 species, occurring mainly in scattered parts of the Southern Hemisphere; a few species extend north to southern Mexico. *Gunnera chilensis* Lam. and some related species are occasionally cultivated as specimen-plants in moist, mild climates because of their spectacularly large leaves. A few of the smaller species are sometimes grown in rock-gardens.

Gunnera has often been included in the family Haloragaceae, to which it appears to be related. There is, however, a long list of differences, including features of the habit, inflorescence, androecium, pollen, gynoecium, embryology, chemistry, and vascular anatomy. *Gunnera* is such a discordant element in the Haloragaceae that it seems better treated as a distinct family.

Pollen considered to represent *Gunnera* dates from the Turonian. In the absence of comparable macrofossils, however, the identity of these Upper Cretaceous pollen-grains must be regarded with some suspicion.

6. Order MYRTALES Lindley 1833

Plants woody or herbaceous, terrestrial or sometimes aquatic, tanniferous, commonly accumulating ellagic acid and often also proanthocyanins, but lacking iridoid compounds and only seldom with alkaloids; nodes usually or always unilacunar; clustered or solitary crystals of calcium commonly present in some of the cells of the parenchymatous tissues, often in distinctive idioblasts, but raphides nearly always wanting except in the Onagraceae; vessel-segments with simple or rarely scalariform perforations and vestured pits; imperforate tracheary elements with simple or bordered pits, often septate; wood-rays heterocellular to homocellular, in most families mixed uniseriate and pluriseriate, but the latter usually only 2 or 3 cells wide, many of the ray-cells generally containing amorphous gummy deposits; some of the axial parenchyma generally consisting of vertical crystalliferous strands; phloem of young twigs often tangentially stratified into hard and soft layers; internal phloem characteristically present, next to the pith; interxylary phloem sometimes also present. LEAVES simple and most commonly entire, or seldom (some Onagraceae) lyrate-pinnatifid, opposite or less often alternate or whorled; stipules vestigial or none, or rarely well developed. FLOWERS in variously racemose or cymose inflorescences, perfect or seldom unisexual, regular or sometimes somewhat irregular, often tetramerous, generally strongly perigynous or epigynous, the hypanthium often prolonged well beyond the ovary, only rarely (some Thymelaeaceae) nearly hypogynous; sepals often appearing as lobes on the hypanthium, sometimes much-reduced or obsolete; petals distinct, alternating with the sepals, or sometimes wanting; stamens numerous and developing in centripetal or less often centrifugal sequence, sometimes clustered on a limited number of primordia associated with trunk-bundles, or often in only 1 or 2 cycles, or even reduced to 1; pollen-grains triaperturate or of triaperturate-derived type; gynoecium of 2-many carpels united to form a compound ovary with as many locules as carpels and with axile placentation, or less often the ovary unilocular with basal, apical, or even parietal placentation, sometimes (Thymelaeaceae) pseudomonomerous; style apical, with a punctate to capitate or lobed stigma; ovules (1) 2-many per carpel, anatropous or seldom hemitropous or campylotropous, bitegmic, crassinucellar or rarely tenuinucellar; embryo-sac sometimes with unusual ontogeny; endosperm nuclear (endosperm-nucleus soon degenerating in *Trapa*). FRUIT of various sorts; seeds mostly with very little or no endosperm; embryo often of unusual structure, with twisted or folded or very unequal cotyledons, but also often of perfectly ordinary type.

The order Myrtales are here defined consists of 12 families and more than 9000 species. About three-fourths of the species belong to only two large families, the Melastomataceae (4000) and the Myrtaceae (3000). Another 4 families, the Onagraceae, Combretaceae, Lythraceae, and Thymelaeaceae, have 400 to 650 species each. The remaining 6 families have only about 60 species in all.

As so defined, the order is fairly homogeneous, without any strongly discordant elements. The most distinctive family, in contrast to all the others, is the Thymelaeaceae, marked by its usually pseudomonomerous ovary and often crotonoid pollen. Furthermore, some few members of the family are unusual in the Myrtales in having essentially hypogynous flowers. On the other hand, more ordinary kinds of pollen and gynoecium, with transitional types, also occur within the Thymelaeaceae. Thus it is unnecessary to seek the placement of the Thymelaeaceae in any other order. Indeed the internal phloem and strongly perigynous, polypetalous to apetalous flowers of characteristic members of the Thymelaeaceae would be out of harmony with any other order that might be suggested as a haven for the family. Furthermore, the characteristic obturator of the Thymelaeaceae, though not identical in detail, might be compared with the obturator of the Combretaceae; and the glandular-punctate leaves of some Thymelaeaceae recall those of the Myrtaceae.

Several families that have been referred to the Myrtales in some systems of classification are placed elsewhere in the present system. Among these are the Lecythidaceae, Rhizophoraceae, Haloragaceae, and Theligonaceae. All of these lack internal phloem and would also be anomalous in other respects within the Myrtales.

An ecological interpretation of the Myrtales is difficult and unsatisfying. Tannins, including ellagic acid and often proanthocyanins, play a major role, for most species, in discouraging predators. A number of other orders rely on these same substances. The habitat and growth-habit of the Myrtales embrace most of the possibilities open to angiosperms. Only some of the smaller, highly specialized families, such as the Trapaceae, are ecologically distinctive. Members of the Myrtales seem to be peculiarly susceptible to fixation of disturbances in the development of the embryo-sac, embryo, and endosperm, but the adaptive significance of these unusual types of development remains to be elucidated. The adaptive significance of perigyny and epigyny in protecting the ovules has been inconclusively debated. The functional importance of internal phloem is wholly obscure, especially inasmuch as it occurs in plants of diverse habit in diverse habitats, which grow intermingled with plants of similar aspect belonging to orders that do not have internal phloem.

SELECTED REFERENCES

Baas, P., & R. C. V. J. Zweypfenning. 1979. Wood anatomy of the Lythraceae. Acta Bot. Neerl. 28: 117–155.

Baehni, C., & C. E. B. Bonner. 1949. La vascularisation du tube floral chez les Onagracées. Candollea 12: 345–359.

Beusekom-Osinga, R. van, & C. F. van Beusekom. 1975. Delimitation and subdivision of the Crypteroniaceae (Myrtales). Blumea 22: 255–266.

Bridgwater, S. D., & P. Baas. 1978. Wood anatomy of the Punicaceae. IAWA Bull. 1978 (1): 3–6.

Briggs, B. G., & L. A. S. Johnson. 1979. Evolution in the Myrtaceae—evidence from inflorescence structure. Proc. Linn. Soc. New South Wales 102: 157–256.

Brown, C. A. 1967. Pollen morphology of the Onagraceae. Rev. Palaeobot. Palynol. 3: 163–180.

Bunniger, L. 1972. Untersuchungen über die morphologische Natur des Hypanthiums bei Myrtales– und Thymelaeales–Familien. II. Myrtaceae. III. Vergleich mit den Thymelaeaceae. Beitr. Biol. Pflanzen 48: 79–156.

Bunniger, L., & F. Weberling. 1968. Untersuchungen über die morphologische Natur des Hypanthiums bei Myrtales-Familien I. Onagraceae. Beitr. Biol. Pflanzen 44: 447–477.

Carlquist, S. 1975. Wood anatomy of Onagraceae, with notes on alternative modes of photosynthate movement in dicotyledon woods. Ann. Missouri Bot. Gard. 62: 386–424.

Carlquist, S., & L. DeBuhr. 1977. Wood anatomy of Penaeaceae (Myrtales): comparative, phylogenetic, and ecological implications. J. Linn. Soc., Bot. 75: 211–227.

Carr, S. G. M., & D. J. Carr. 1969, 1970. Oil glands and ducts in *Eucalyptus* L'Hérit. I. The phloem and the pith. II. Development and structure of oil glands in the embryo. Austral. J. Bot. 17: 471–513, 1969. 18: 191–212, 1970.

Carr, S. G. M., D. J. Carr, & L. Milkovits. 1970. Oil glands and ducts in *Eucalyptus* L'Herit. III. The flowers of series Corymbosae (Benth.) Maiden. Austral. J. Bot. 18: 313–333.

Cheung, M., & R. Sattler. 1967. Early floral development of *Lythrum salicaria.* Canad. J. Bot. 45: 1609–1618.

Chrtek, J. 1969. Die Kronblattnervatur in der Familie Lythraceae. Preslia 41: 323–326.

Domke, W., 1934. Untersuchungen über die systematische und geographische Gliederung der Thymelaeaceen nebst einer Neubeschreibung ihrer Gattungen. Biblioth. Bot. 27(Heft 111): 1–151.

Erdtman, G., & C. R. Metcalfe. 1963. Affinities of certain genera *incertae sedis* suggested by pollen morphology and vegetative anatomy. I. The Myrtaceous affinity of *Kania eugenioides* Schltr. II. The Myrtaceous affinity of *Tristania merguensis* Griff. (*Thorelia deglupta* Hance). Kew Bull. 17: 249–250; 251–252.

Exell, A. W. 1931. The genera of Combretaceae. J. Bot. 69: 113–128.

Exell, A. W. 1962. Space problems arising from the conflict between two evolutionary tendencies in the Combretaceae. Bull. Soc. Roy. Bot. Belgique 95: 41–49.

Exell, A. W., & C. A. Stace. 1966. Revision of the Combretaceae. Bol. Soc. Brot. ser. 2, 40: 5–26.

Eyde, R. H. 1977, 1978. Reproductive structures and evolution in *Ludwigia* (Onagraceae). I. Androecium, placentation, merism. II. Fruit and seed. Ann. Missouri Bot. Gard. 64: 644–655, 1977. 65: 656–675, 1978.

Eyde, R. H., & J. T. Morgan. 1973. Floral structure and evolution in Lopezieae (Onagraceae). Amer. J. Bot. 60: 771–787.

Eyde, R. H., & J. A. Teeri. 1967. Floral anatomy of *Rhexia virginica* (Melastomataceae). Rhodora 69: 163–178.

Fagerlind, F. 1940. Zytologie und Gametophytenbildung in der Gattung *Wikstroemia*. Hereditas 26: 23–50.

Fuchs, A. 1938. Beiträge zur Embryologie der Thymelaeaceae. Oesterr. Bot. Z. 87: 1–41.

Fujita, E., K. Bessho, Y. Saeki, M. Ochiai, & K. Fuji. 1971. Lythraceous alkaloids. V. Isolation of ten alkaloids from *Lythrum anceps*. Lloydia 34: 306–309.

Gleason, H. A. 1931. The relationships of certain myrmecophilous melastomes. Bull. Torrey Bot. Club 58: 73–85.

Graham, A., & S. A. Graham. 1971. The geologic history of the Lythraceae. Brittonia 23: 335–346.

Graham, A., S. A. Graham, & D. Geer. 1968. Palynology and systematics of *Cuphea* (Lythraceae). I. Morphology and ultrastructure of the pollen wall. Amer. J. Bot. 55: 1080–1088.

Graham, S. A. 1964. The genera of Lythraceae in the southeastern United States. J. Arnold Arbor. 45: 235–250.

Graham, S. A., & A. Graham. 1971. Palynology and systematics of *Cuphea* (Lythraceae). II. Pollen morphology and infrageneric classification. Amer. J. Bot. 58: 844–857.

Heinig, K. H. 1951. Studies in the floral morphology of the Thymelaeaceae. Amer. J. Bot. 38: 113–132.

Joshi, A. C. 1939. Embryological evidence for the relationships of the Lythraceae and related families. Curr. Sci. 8: 112, 113.

Joshi, A. C., & J. Venkateswarlu. 1935, 1936. Embryological studies in the Lythraceae. I. *Lawsonia inermis* Linn. II. *Lagerstroemia* Linn. III. Proc. Indian Acad. Sci. B. 2: 481–493; 523–534, 1935. 3: 377–400, 1936.

Kausel, E. 1955. Beitrag zur Systematik der Myrtaceen. I, II. Ark. Bot. 3: 491–516; 607–611.

Kurabayashi, M., H. Lewis, & P. H. Raven. 1962. A comparative study of mitosis in the Onagraceae. Amer. J. Bot. 49: 1003–1026.

Leandri, J. 1930. Recherches anatomiques sur les Thyméléacées. Ann. Sci. Nat. Bot. sér 10. 12: 125–237.

Leins, P. 1965. Die Inflorescenz und frühe Blütenentwicklung von *Melaleuca nesophila* F. Muell. (Myrtaceae). Planta 65: 195–204.

Lowry, J. B. 1976. Anthocyanins of the Melastomataceae, Myrtaceae and some allied families. Phytochemistry 15: 513–516.

McVaugh, R. 1956. Tropical American Myrtaceae. Notes on generic concepts and descriptions of previously unrecognized species. Fieldiana: Bot. 29: 145–228.

Mauritzon, J. 1934. Zur Embryologie einiger Lythraceen. Acta Horti Gothob. 9: 1–21.

Mayr, B. 1969. Ontogenetische Studien an MyrtalesBlüten. Bot. Jahrb. Syst. 89: 210–271.
Miki, S. 1959. Evolution of *Trapa* from ancestral *Lythrum* through *Hemitrapa*. Proc. Imp. Acad. Japan 35(6): 289–294.
Mújica, M. B., & D. F. Cutler. 1974. Taxonomic implications of anatomical studies on the Oliniaceae. Kew Bull. 29: 93–123.
Muller, J. 1969. A palynological study of the genus *Sonneratia* (Sonneratiaceae). Pollen & Spores 11: 223–298.
Muller, J. 1975. Note on the pollen morphology of Crypteroniaceae s.l. Blumea 22: 275–294.
Muller, J. 1978. New observations on pollen morphology and fossil distribution of the genus *Sonneratia* (Sonneratiaceae). Rev. Palaeobot. Palyn. 26: 277–300.
Nevling, L. I. 1959. A revision of the genus *Daphnopsis*. Ann. Missouri Bot. Gard. 46: 257–358.
Nevling, L. I. 1962. The Thymelaeaceae in the southeastern United States. J. Arnold Arbor 43: 428–434.
Outer, R. W. den, & J. M. Fundter. 1976. The secondary phloem of some Combretaceae and the systematic position of *Strephonema pseudocola* A. Chev. Acta Bot. Neerl. 25: 481–493.
Pike, K. M. 1956. Pollen morphology of Myrtaceae from the south-west Pacific area. Austral. J. Bot. 4: 13–53.
Plitmann, U., P. H. Raven, & D. E. Breedlove. 1973. The systematics of Lopezieae (Onagraceae). Ann. Missouri Bot. Gard. 60: 478–563.
Prakasa Rao, P. S. 1972. Wood anatomy of some Combretaceae. J. Jap. Bot. 47: 358–377.
Ram, M. 1956. Floral morphology and embryology of *Trapa bispinosa* Roxb. with a discussion on the systematic position of the genus. Phytomorphology 6: 312–323.
Rao, V. S., & R. Dahlgren. 1968. Studies on Penaeaceae. V. The vascular anatomy of the flower of *Glischrocolla formosa*. Bot. Not. 121: 259–268.
Rao, V. S., & R. Dahlgren. 1969. The floral anatomy and relationships of Oliniaceae. Bot. Not. 122: 160–171.
Raven, P. 1963. The Old World species of *Ludwigia* (including *Jussiaea*), with a synopsis of the genus (Onagraceae). Reinwardtia 6: 327–427.
Raven, P. H. 1964. The generic subdivision of Onagraceae, tribe Onagreae. Brittonia 16: 276–288.
Raven, P. H. 1976. Generic and sectional delimitation in Onagraceae, tribe Epilobieae. Ann. Missouri Bot. Gard. 63: 326–340.
Schmid, R. 1972. Floral anatomy of Myrtaceae. I. *Syzygium* s.l. Bot. Jahrb. Syst. 92: 433–489, 1972. II. *Eugenia*. J. Arnold Arbor. 52: 336–363, 1972.
Schmid, R. 1972. A resolution of the *Eugenia-Syzygium* controversy (Myrtaceae). Amer. J. Bot. 59: 423–436.
Seavey, S. R., R. E. Magill, & P. H. Raven. 1977. Evolution of seed size, shape, and surface architecture in the tribe Epilobieae (Onagraceae). Ann. Missouri Bot. Gard. 64: 18–47.
Sinha, S. C., & B. C. Joshi. 1959. Vascular anatomy of the flower of *Punica granatum* L. J. Indian Bot. Soc. 38: 35–45.
Skvarla, J. J., P. H. Raven, W. F. Chissoe, & M. Sharp. 1978. An ultrastructural study of viscin threads in Onagraceae pollen. Pollen & Spores 20: 5–143.
Skvarla, J. J., P. H. Raven, & J. Praglowski. 1976. Ultrastructural survey of Onagraceae pollen. *In:* I. K. Ferguson, & J. Muller, eds., The evolutionary significance of the exine, pp. 447–479. Linn. Soc. Symp. Series No. 1.

Smith, B. B., & J. M. Herr. 1971. Ovule development, megagametogenesis, and early embryogeny in *Ammannia coccinea* Rothb. J. Elisha Mitchell Sci. Soc. 87: 192–199.
Stace, C. A. 1965. The significance of the leaf epidermis in the taxonomy of the Combretaceae. I. A general review of tribal, generic and specific characters. J. Linn. Soc., Bot. 59: 229–252.
Stern, W. L., & G. K. Brizicky. 1958. The comparative anatomy and taxonomy of *Heteropyxis*. Bull. Torrey Bot. Club 85: 111—123.
Subramanyam, K., & L. L. Narayana. 1969. A contribution to the floral anatomy of some members of Melastomataceae. J. Jap. Bot. 44: 6–16.
Thanikaimoni, G., & D. M. A. Jayaweera. 1966. Pollen morphology of Sonneratiaceae. Trav. Sect. Sci. Techn. Inst. Franç. Pondichéry 5(3): 1–12.
Tiagi, Y. D. 1969. Vascular anatomy of the flower of certain species of Combretaceae. Bot. Gaz. 130: 150–157.
Ting, W. S. 1966. Pollen morphology of Onagraceae. Pollen & Spores 8: 9–36.
Venkatesh, C. S. 1955. The structure and dehiscence of the anther in *Memecylon* and *Mouriria*. Phytomorphology 5: 435–440.
Venkateswarlu, F. N. I., & P. S. P. Rao. 1970. The floral anatomy of Combretaceae. Proc. Natl. Acad. Sci. India 36B: 1–20.
Venkateswarlu, J. 1937. A contribution to the embryology of Sonneratiaceae. Proc. Indian Acad. Sci. B 5: 206–223.
Venkateswarlu, J. 1947. Embryological studies in the Thymelaeaceae. II. *Daphne cannabina* Wall. and *Wikstroemia canescens* Meissn. J. Indian Bot. Soc. 26: 13–39.
Venkateswarlu, J. 1952. Contributions to the embryology of Combretaceae. I. *Poivrea coccinea* DC. Phytomorphology 2: 231–240.
Vliet, G. J. C. M. van. 1978. Vestured pits of Combretaceae and allied families. Acta Bot. Neerl. 27(5/6): 273–285.
Vliet, G. J. C. M. van. 1979. Wood anatomy of the Combretaceae. Blumea 25: 141–223.
Vliet, G. J. C. M. van, & P. Baas. 1975. Comparative anatomy of the Crypteroniaceae sensu lato. Blumea 22: 175–195.
Weberling, F. 1956. Untersuchungen über rudimentäre Stipeln bei den Myrtales. Flora 143: 201–218.
Weberling, F. 1960. Weitere Untersuchungen über das Vorkommen rudimentärer Stipeln bei den Myrtales (Combretaceae, Melastomataceae). Flora 149: 189–205.
Weberling, F. 1963. Ein Beitrag zur systematischen Stellung der Geissolomataceae, Penaeaceae und Oliniaceae sowie der Gattung *Heteropyxis* (Myrtaceae). Bot. Jahrb. Syst. 82: 119–128.
Wilson, C. L. 1950. Vasculation of the stamen in the Melastomaceae with some phyletic implications. Amer. J. Bot. 37: 431–444.
Wilson, K. A. 1960. The genera of Myrtaceae in the southeastern United States. J. Arnold Arbor. 41: 270–278.
Wurdack, J. J. 1953. A revision of the genus *Brachyotum* (Tibouchineae-Melastomaceae). Mem. New York Bot. Gard. 8(4): 343–407.
Ziegler, A. 1925. Beiträge zur Kenntnis des Androeceums und der Samenentwicklung einiger Melastomaceen. Bot. Arch. 9: 398–467.

Анели, Н. 1952. Текстура проводящей системы и некоторые вопросы систематики сем миртовых. (In Georgian. Russian summary.) **Вестн. Тбилисск. Бот. Сада АН Грузинск. ССР, 60: 145-160.**

Архангельский, Д. Б. 1966. Пыльцевые зерна семейства Thymelaeaceae п Gonystylaceae. Бот. Ж. 51: 484–494.
Архангельский, Д. Б. 1971. Палинотаксономия Thymelaeaceae s. l. В сб.: Морфология пыльцы Cucurbitaceae, Thymelaeaceae, Cornaceae. Л. А. Куприянова и М. С. Яковлев, редакторы: 104–234. Изд. Наука, Ленинградск. Отд. Ленинград.
Васильев, В. Н. 1960. Водяной орех и перспективы его культуры в СССР. Москва–Ленинград. Бот. Инст. им Комарова, Научно-Популярная Сер.
Зажурило, К. К. 1935. Современные проблемы анатомии в карпологии. Труды Воронежск. Госуд. Унив. 7: 21–42.
Полунина, Н. Н. 1963. Сравнительно-эмбриологическое исследование некоторых представителей семейства миртовых. Бюлл. Главн. Бот. Сада АН СССР, 49: 82–90.
Шилкина, И. А. 1973. К анатомии древесины рода *Punica* L. Бот. Ж. 58: 1628–1630.

SYNOPTICAL ARRANGEMENT OF THE FAMILIES OF MYRTALES

1 Ovary superior both at anthesis and in fruit, only the very base sometimes adnate to the hypanthium.
  2 Ovules 2-many per carpel; ovary generally 2-several-locular; fruit commonly capsular, seldom indehiscent.
    3 Stamens numerous, in more than 2 cycles, seldom only 12, carpels 4–20; trees, often of mangrove habit .... 1. SONNERATIACEAE.
    3 Stamens mostly bicyclic or unicyclic, rarely more than 12, seldom reduced to 1; carpels 2–4 (–6); not mangroves.
      4 Plants with perfect flowers; petals present or absent.
        5 Embryo-sac monosporic, 8-nucleate; filaments more or less elongate; connective not notably expanded; carpels 2–4 (–6), each with (2-) more or less numerous ovules; most commonly (though not always) diplostemonous .... 2. LYTHRACEAE.
        5 Embryo-sac tetrasporic, 16-nucleate; filaments very short; connective strongly expanded, thickly laminar; carpels 4, each with 2–4 ovules; haplostemonous ..3. PENAEACEAE.
      4 Plants polygamo-dioecious; petals none; carpels 2, multiovulate; haplostemonous .... 4. CRYPTERONIACEAE.
  2 Ovule solitary in a pseudomonomerous ovary, or the ovules solitary in each of 2 or more locules; fruit commonly indehiscent, less often capsular ....5. THYMELEACEAE.
1 Ovary half to more often fully inferior, or in a relatively few genera and species of some families superior.
  6 Plants annual aquatics; ovary bilocular, with a single pendulous apical-axile ovule in each locule, half-inferior at anthesis, be-

coming almost wholly inferior in fruit; fruits horned; cotyledons very unequal; haplostemonous .... 6. TRAPACEAE.

6 Plants woody or herbaceous, terrestrial, or less often aquatic; other features not combined as above.

7 Stamens mostly numerous, seldom only as many or twice as many as the sepals or petals.

8 Leaves glandular-punctate; carpels mostly 2–5 (16); fruit various, but only rarely many-seeded and indehiscent, in any case the seeds without a proliferating sarcotesta .... 7. MYRTACEAE.

8 Leaves not glandular-punctate; carpels rather numerous, commonly 7–9 (–15); fruit indehiscent, with a firm rind and many seeds embedded in a pulpy mass representing the proliferated sarcotestas .... 8. PUNICACEAE.

7' Stamens usually not more than twice as many as the sepals or petals.

9 Placentation axile to less often basal or parietal; fruit only seldom 1-seeded and indehiscent; plants woody or herbaceous.

10 Anthers opening by longitudinal slits; connective without appendages; leaves pinnately veined.

11 Embryo-sac 4-nucleate, the endosperm diploid; ovules usually numerous, seldom only 1 or 2 per carpel; fruit usually capsular, less often indehiscent, but not drupaceous; plants with raphides; most species herbaceous .... 9. ONAGRACEAE.

11 Embryo-sac 8-nucleate, the endosperm triploid; ovules 2 (3) per carpel; fruit drupaceous; plants without raphides; shrubs or trees .... 10. OLINIACEAE.

10 Anthers mostly opening by a terminal pore or pores, only seldom by longitudinal slits; connective commonly provided with conspicuous appendages; leaves mostly with 3–9 prominent longitudinal veins, only seldom pinnately veined .... 11. MELASTOMATACEAE.

9 Placentation apical in a unilocular ovary; ovules 2 (–6); fruit indehiscent, 1-seeded; trees, shrubs, or woody vines .... 12. COMBRETACEAE.

1. Family SONNERATIACEAE Engler & Gilg 1924 nom. conserv., the Sonneratia Family

Trees, tanniferous and producing ellagic acid; stem with internal phloem, next to the pith; vessel-segments with simple perforations and vestured pits; imperforate tracheary elements with simple pits, in *Sonneratia* septate; wood-rays 1–2 (3) cells wide, homocellular (*Sonneratia*) or heterocellular (*Duabanga*), many of the cells containing amorphous gummy deposits, others (in vertical strands within the ray) containing each a solitary crystal of calcium oxalate; wood-parenchyma absent (*Sonneratia*) or paratracheal (*Duabanga*). LEAVES opposite (but not decussate) or sometimes whorled, simple, entire, leathery; mesophyll with many large mucilage-cells and with branching sclereids in the central part; stomates encyclocytic, at least in *Sonneratia*; stipules vestigial. FLOWERS 1–3 at the branch tips (*Sonneratia*), or in terminal umbelliform cymes (*Dunabanga*), rather large, pollinated by bats, regular, perfect or seldom unisexual, strongly perigynous; hypanthium campanulate, thick and leathery; sepals 4–8, valvate; petals 4–8, often crumpled in bud, or sometimes wanting; stamens commonly numerous, less often only 12, borne on the hypanthium in several cycles, or in clusters opposite the sepals, often incurved before anthesis, originating from a limited number of primordia associated with trunk-bundles; filaments distinct; anthers reniform, tetrasporangiate and dithecal; pollen-grains binucleate, triporate, with a smooth or warty surface; gynoecium of 4–20 carpels united to form a compound ovary with an elongate, terminal style and a capitate stigma, the style bent in bud; ovary superior, or adnate at the base to the hypanthium, plurilocular, with thin partitions and a thickened, axile placenta; ovules numerous, anatropous, bitegmic, crassinucellar; embryo-sac monosporic, 8-nucleate. FRUIT capsular (*Duabanga*) or an edible acid berry (*Sonneratia*); seeds numerous, without endosperm; embryo straight or curved, with 2 cotyledons. N = 9, 18, 24.

The family Sonneratiaceae consists of only 2 genera, *Sonneratia,* with about 8 species, and *Duabanga,* with only 2, all native to the tropics of the Old World. *Sonneratia* is a mangrove with characteristic pneumatophores arising from the roots, but *Duabanga* is a tree of the monsoon and rain forests. The Sonneratiaceae are generally considered to be closely allied to the Lythraceae. Embryological studies disclose no basis at all for the separation, but the relatively numerous stamens and carpels of the Sonneratiaceae do set them somewhat apart from the Lythraceae.

Pollen considered to represent the Sonneratiaceae dates from the base of the Miocene epoch, and wood very much like that of *Sonneratia* occurs in middle Eocene and more recent deposits. Some flowers and fruits from Eocene deposits in India probably represent the Sonneratiaceae or Lythraceae. Muller (1978) regards the Sonneratiaceae as "a highly specialized offshoot of lythraceous stock." It is possible that the floral form of the Sonneratiaceae evolved in mid-Tertiary in association with bat-pollination.

2. Family LYTHRACEAE Jaume St.-Hilaire 1805 nom. conserv., the Loosestrife Family

Herbs, or seldom shrubs or even trees, tanniferous and producing ellagic acid, but without proanthocyanins and only seldom cyanogenic, often with mucilage-cells in the epidermis, and often producing alkaloids of the piperidine, pyridine, and quinolizidine groups; trichomes sometimes silicified; stem with internal phloem, next to the pith; nodes unilacunar (trilacunar in *Alzatea*); crystals of calcium oxalate commonly present in some of the parenchymatous tissues, including the wood-parenchyma; vessel-segments with simple perforations and vestured pits; imperforate tracheary elements with simple or minutely bordered pits, commonly septate; wood-rays uniseriate or less often some of them 2–3 seriate, homocellular or heterocellular, many of the cells containing amorphous gummy deposits; wood-parenchyma mostly paratracheal, frequently scanty. LEAVES opposite or less often whorled, only seldom alternate, simple and commonly entire; stomates mostly anomocytic; stipules vestigial or none. FLOWERS solitary or fascicled in the axils, or often in terminal racemes or spikes, less often in dichasia or panicles, perfect, often heterostylic, regular or sometimes irregular, strongly perigynous, with a prominent, sometimes spurred hypanthium, commonly 4-, 6-, or 8-merous, seldom only 3-merous or up to 16-merous; sepals appearing like lobes of the hypanthium, valvate, often alternating with external appendages at the sinuses; petals alternate with the sepals, distinct, attached at the summit of the hypanthium or within its tube, commonly pinnately veined, crumpled in bud, or sometimes wanting; stamens most commonly twice as many as the sepals or petals and bicyclic, attached below the summit of the hypanthium, the outer set

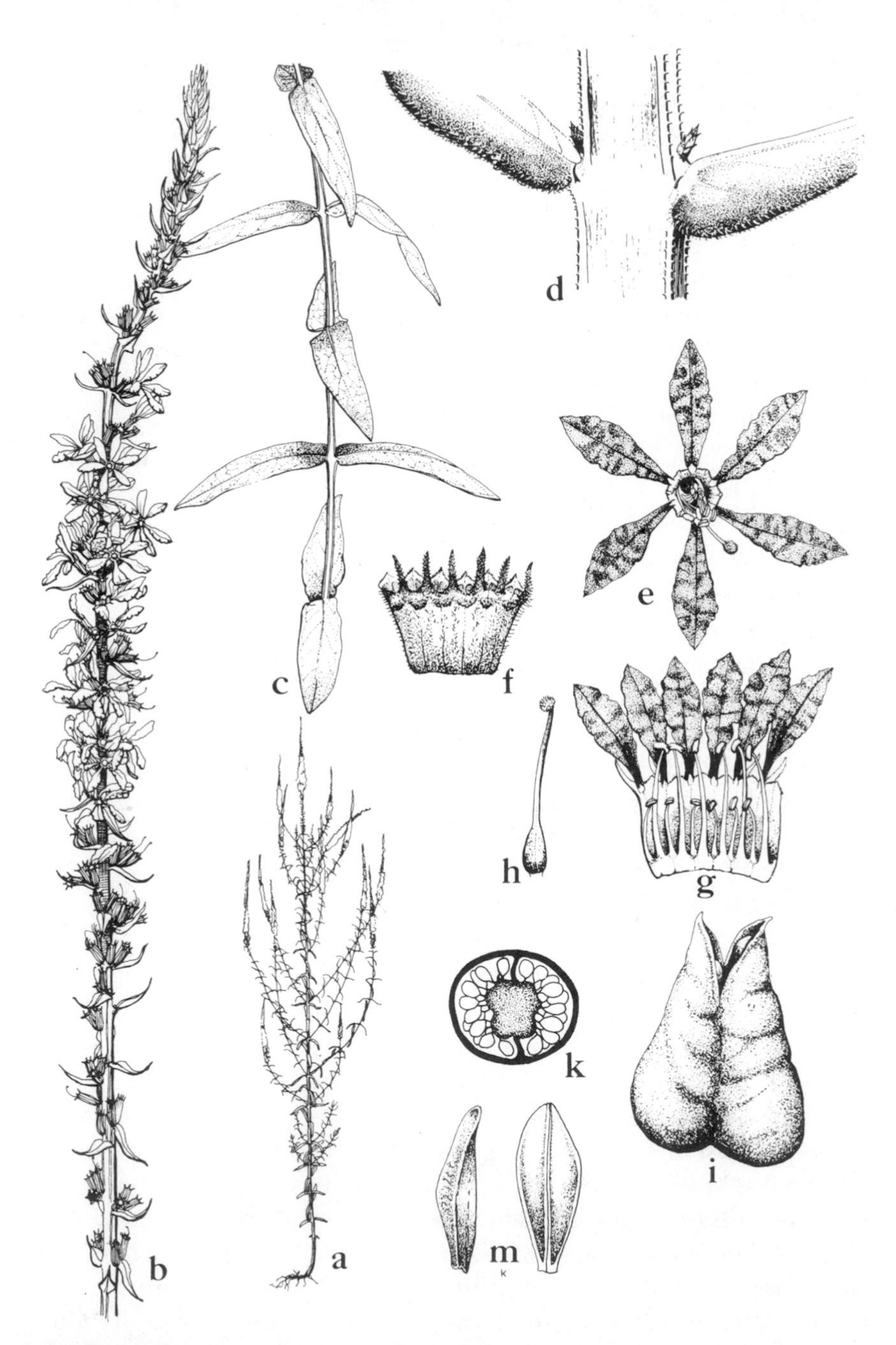

FIG. 5.20 Lythraceae. *Lythrum salicaria* L. a, habit, $\times\frac{1}{16}$; b, inflorescence, $\times\frac{1}{2}$; c, portion of shoot, showing decussate leaves, $\times\frac{1}{2}$; d, node and leaf-base, with minute stipules, $\times 5$; e, flower, from above, $\times 4$; f, external view of hypanthium and calyx, opened out, $\times 2$; g, internal view of flower, opened out, with the pistil removed, $\times 2$; h, pistil, $\times 2$; i, fruit, $\times 10$; k, schematic cross-section of ovary, $\times 10$; m, seeds, $\times 20$.

initiated before the inner, or the inner (lower, antepetalous) cycle sometimes wanting, seldom in more than 2 cycles or more or less numerous, in *Lagerstroemia* numerous and centrifugal, in *Rotala* the stamen solitary; filaments more or less elongate; anthers versatile or seldom basifixed, tetrasporangiate and dithecal, opening by longitudinal slits; pollen-grains binucleate, of diverse architecture, commonly tricolporate to triporate, sometimes with some additional furrows that lack an ora; gynoecium of 2–4 (–6) carpels united to form a compound pistil that is often surrounded at the base by an annular nectary-disk, or the nectary unilateral; style filiform, commonly elongate, sometimes bent in bud, the stigma mostly capitate; ovary superior, plurilocular with as many locules as carpels, but the partitions sometimes not reaching the summit; placentation axile, or rarely free-central by failure of the partitions, or rarely the ovary pseudomonomerous, with a single locule and a single parietal placenta; ovules (2–) more or less numerous in each locule, epitropous, anatropous, crassinucellar or seldom tenuinucellar, bitegmic, with a zigzag micropyle; embryo-sac monosporic 8-nucleate; endosperm-development nuclear. FRUIT dry, commonly capsular, dehiscing by valves or by a transverse slit or irregularly, or seldom indehiscent; seeds (1–) generally more or less numerous in each locule, sometimes winged, nearly or quite without endosperm; embryo oily, straight, with a short radicle and 2 expanded cotyledons. X = 5–11.

The family Lythraceae consists of about 27 genera and nearly 575 species, widespread in tropical countries, with relatively few species in temperate regions. The largest genus by far is *Cuphea,* with about 200 species native to tropical and subtropical America and introduced into the Hawaiian Islands. Some other genera of considerable size are *Rotala* (35), *Lythrum* (30), and *Lagerstroemia* (3). *Lythrum salicaria* L., the purple loosestrife, is native to Eurasia but now well established in eastern United States. *Lagerstroemia indica* L., the crape-myrtle, is native to Asia but widely cultivated for ornament in warm-temperate or subtropical regions elsewhere. Commercial henna is made from the dried leaves of the paleotropical species *Lawsonia inermis* L.

Macrofossils credibly referred to the Lythraceae occur from the Lower Eocene (both Europe and India) to the present, and the modern genus *Lagerstroemia* is recorded from the Middle Eocene. Lythraceous pollen, on the other hand, dates only from the uppermost Eocene. The distinctive pollen of *Cuphea,* one of the more advanced genera, is known from mid-Miocene to the present.

Alzatea, a monotypic Peruvian genus, may or may not be best placed in the Lythraceae. Its flowers are apetalous, and have stamens alternate with the calyx-lobes. The ovary is bicarpellate and unilocular, with 2 parietal placentas. The fruit develops a pair of intruded partial partitions from the carpellary midribs. *Alzatea* is one of the several genera often included in the Lythraceae that van Beusekom-Ozinga and van Beusekom propose to transfer to the Crypteroniaceae. See also Lourteig in Ann. Missouri Bot. Garden 52: 371–378. 1965.

3. Family PENAEACEAE Guillemin 1828 nom. conserv., the Penaea Family

Evergreen shrubs, often of ericoid aspect, typically rather small, varying to sometimes diminutive shrubs or arborescent shrubs, glabrous or seldom with a few small, mostly unicellular hairs; plants tanniferous, and with cluster-crystals of calcium oxalate in some of the cells of the parenchymatous tissues; nodes unilacunar; xylem forming a continuous cylinder traversed by narrow rays; vessel-segments with simple perforations and vestured pits; imperforate tracheary elements generally with typical bordered pits, considered to be tracheids, wood-rays heterocellular, with a predominance of upright cells over procumbent ones, most (or all) of the rays uniseriate, the pluriseriate ones, when present, mostly only 2–3 cells wide; many cells of the rays (and also of the wood-parenchyma) containing amorphous, gummy deposits; wood-parenchyma mostly scanty and diffuse, sometimes consisting in part of vertical strands with a cluster-crystal in each cell; internal phloem characteristically present, next to the pith. LEAVES opposite, decussate, often crowded, simple and entire, commonly sessile; stomates anomocytic; mesophyll containing scattered fibers, usually also some branching sclereids, and in some cells a cluster-crystal of calcium oxalate; stipules wanting, or tiny and vestigial. FLOWERS solitary in the axils of the upper leaves, often closely crowded, perfect, regular, strongly perigynous, apetalous, the hypanthium and sepals commonly colored like a corolla; sepals 4, valvate, appearing as lobes on the hypanthium; stamens 4, attached to the hypanthium and alternate with the sepals; filaments very short; anthers tetrasporangiate and dithecal, opening by longitudinal slits, the connective much-expanded, thickly laminar, often much longer than the frequently well separated pollen-sacs, which are seated on its ventral or ventromarginal surface; pollen-grains 3–5-colporate, the colpi alternating with colpoid grooves, disk wanting; gynoecium of

4 carpels united to form a compound, superior, 4-locular ovary with a terminal style and a capitate or 4-lobed stigma; placentation axile-basal and/or axile-apical, or median-axile; ovules 2–4 in each locule, pendulous or erect, anatropous, bitegmic, crassinucellar; embryo-sac tetrasporic, 16-nucleate; embryo without a suspensor; endosperm-development nuclear. FRUIT a loculicidal capsule, included in the persistent hypanthium; seeds nearly or quite without endosperm, often solitary in the locules; embryo with a large hypocotyl and 2 very small cotyledons. N = 11 or 12.

The family Penaeaceae consists of 7 genera and about 20 species, native to the Cape Province in South Africa. None of the species is well known in cultivation or commerce.

The Penaeaceae have by different authors been associated with the Lythraceae or the Thymelaeaceae. Since both of these families are here included in the same order, neither association need be denied.

4. Family CRYPTERONIACEAE Alphonse de Candolle 1868 nom. conserv., the Crypteronia Family

Polygamo-dioecious trees with quadrangular twigs, commonly accumulating aluminum; nodes unilacunar; vessel-segments with simple perforations and vestured pits; imperforate tracheary elements with bordered pits, considered to be tracheids; wood-rays heterocellular, 1–4 cells wide, often with radial intercellular canals, many of the cells containing amorphous gummy deposits; wood-parenchyma diffuse, with a tendency to form uniseriate bands, some of these bands with the cells containing each a cluster-crystal of calcium oxalate; internal phloem characteristically present, next to the pith. LEAVES opposite, simple, entire, pinnately veined, with a continuous submarginal vein; stomates anomocytic; some of the cells of the mesophyll containing a cluster-crystal of calcium oxalate; stipules vestigial. FLOWERS in axillary racemes or spikes or panicles, numerous and very small, white or green, regular, apetalous, strongly perigynous, with a cupulate or shortly tubular hypanthium; sepals 4–5, appearing as valvate lobes on the hypanthium; stamens as many as and alternate with the sepals; anthers very short, tetrasporangiate and dithecal; pollen-grains very small, syndicolpate, suggesting those of the Eucryphiaceae and Cunoniaceae; gynoecium of 2 carpels united to form a compound, bilocular, superior ovary with a terminal style; ovules numerous, on axile placentas, anatropous. FRUIT a loculicidal capsule, the valves remaining connected by the persistent base of the style; seeds small, elongate, with a cylindrical, dicotyledonous embryo; endosperm wanting.

The family Crypteroniaceae as traditionally defined consists of the single genus *Crypteronia,* with 4 species native to India, the Philippine Islands, and the Malay Archipelago. It has sometimes been associated with the Cunoniaceae and Eucryphiaceae in the suborder Cunoniinae of the Rosales, and except for the internal phloem that disposition is reasonably plausible. On the other hand, *Crypteronia* is perfectly at home in the Myrtales, and some authors have sought to expand the Crypteroniaceae to include as many as 4 other genera (*Alzatea, Axinandra, Dactylocladus,* and *Rhynchocalyx*) that are always included in the Myrtales and often referred to the Lythraceae. We still await the promised argumentation for this broader definition, said in Blumea 22 (2) to be forthcoming in the next issue, but not actually included therein. Although its largely unisexual flowers indicate that it is not on the direct line of descent, *Crypteronia* does strengthen the case for deriving the Myrtales from the suborder Cunoniinae of the Rosales.

5. Family THYMELAEACEAE A. L. de Jussieu 1789 nom. conserv., the Mezereum Family

Highly poisonous shrubs or sometimes trees, seldom herbs or lianas, mostly with unicellular hairs, not strongly tanniferous, seldom producing proanthocyanins, lacking ellagic acid and iridoid compounds, rarely cyanogenic, but sometimes producing glycosides, and characteristically accumulating the simple coumarin daphnin (or allied compounds); epidermis and parenchymatous tissues often with scattered mucilage-cells; solitary or clustered crystals of calcium oxalate, diverse in form, often present in some of the cells of the parenchymatous tissues; nodes unilacunar; vessel-segments with simple perforations and vestured pits; imperforate tracheary elements commonly with bordered pits, at least some of them considered to be tracheids; wood-rays homocellular to heterocellular, either predominantly uniseriate, or up to 4 (–9) cells wide with few uniseriates; wood-parenchyma mostly paratracheal; phloem of young stems with broad, wedge-shaped medullary rays, the truncated triangles of phloem permeated by a network of tough, variously arranged fibers; internal phloem present, next to the pith, except in Gonystyloideae and a few genera of other subfamilies; interxylary phloem sometimes present as well. LEAVES alternate or

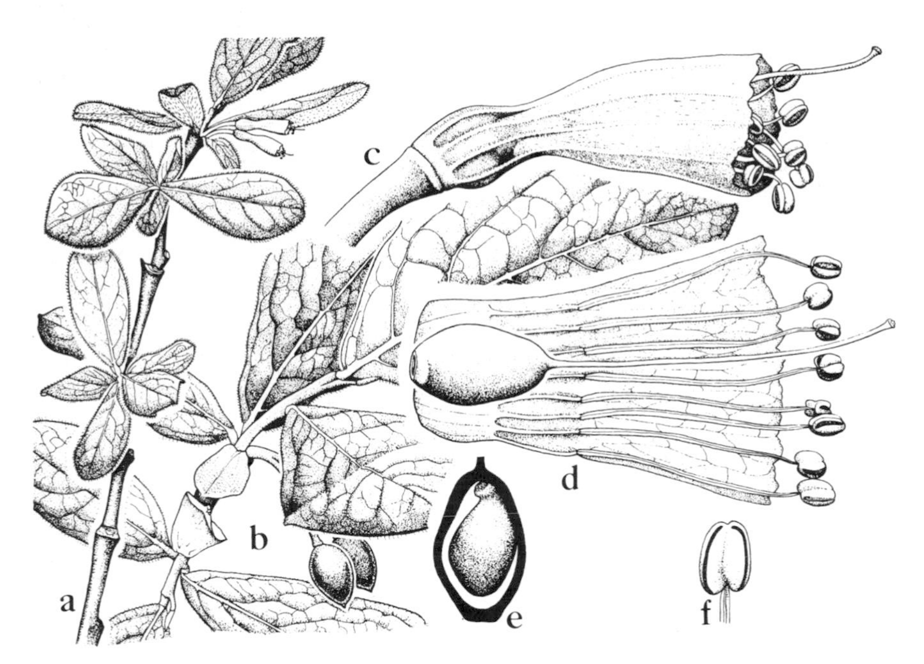

FIG. 5.21 Thymelaeaceae. *Dirca palustris* L. a, habit, in flower, ×1; b, habit, in fruit, ×1; c, flower, ×6; d, flower, opened and laid out, ×6; e, schematic long-section of ovary, ×6; f, anther, ×12.

opposite or in irregular pseudowhorls, simple, entire, pinnately veined, sometimes ericoid, or reduced to a parallel-veined petiolar sheath (as in *Struthiola*), in the Gonystyloideae commonly pellucid-punctate; stomates anomocytic or sometimes more or less actinocytic; mesophyll often with scattered sclereids; stipules wanting or vestigial. Flowers variously in terminal, axillary or extra-axillary, racemose or spicate to capitate or umbelliform inflorescences, or sometimes solitary, perfect or sometimes unisexual, regular or rarely somewhat irregular, (3) 4–5 (6)-merous, generally (at least in the Thymelaeoideae) more or less strongly perigynous, with a well developed, often colored hypanthium on which the more or less petaloid, imbricate or seldom valvate sepals appear as lobes, or the hypanthium sometimes merely erose at the summit and scarcely lobed, or seldom the sepals essentially distinct and the flowers hypogynous; petals small and often scale-like, usually scarcely or not at all corolloid, attached to the throat of the hypanthium, as many as and alternate with the sepals, or twice as many and paired opposite the sepals, or seldom numerous (Gonystyloideae) and not arranged in a clear relationship with the sepals, sometimes connate at the base, or often wholly wanting; stamens as many as and opposite the sepals, or more often in 2 cycles with the upper cycle opposite the sepals, or seldom numerous (*Gonystylus*) or as few as 2 (*Pimelea*); filaments usually short, or even obsolete; anthers tetrasporangiate and dithecal, opening by longitudinal slits; pollen-grains trinucleate, often crotonoid and 6–8-porate, sometimes non-crotonoid and 3–6-porate, or pantoporate with up to ca. 60 pores; a hypogynous, annular to cupular or lobed nectary-disk commonly present, or the disk sometimes represented by separate scales, wanting in the Gonystyloideae and most of the Aquilarioideae; gynoecium of 2–5 (–12) carpels united to form a compound ovary with as many locules as carpels, or more commonly bicarpellate but pseudomonomerous and unilocular, generally with an obturator descending into each locule from near the base of the stylar canal; style commonly elongate and slender, but sometimes nearly obsolete, often excentric, especially in the pseudomonomerous genera; stigma capitate or of various other forms; ovules solitary and pendulous in each locule, anatropous to hemitropous, bitegmic, crassinucellar; endosperm-development nuclear. FRUIT commonly indehiscent, variously dry or fleshy and baccate or drupaceous, or less often (Aquilarioideae and Gonystyloideae) a loculicidal capsule; seeds commonly with a caruncular thickening or a more or less caudate appendage; embryo oily, large, straight, with expanded, flat cotyledons; endosperm generally scanty, seldom copious or none. X most commonly = 9.

The family Thymelaeaceae as here defined consists of about 50 genera and 500 species, of cosmopolitan distribution. The group is best developed in Australia and tropical Africa, with lesser centers in the Mediterranean region and in eastern and southeastern Asia. The largest genus is *Gnidia* (140), mainly of tropical and southern Africa and Madagascar. The most familiar genus in the Northern Hemisphere is *Daphne,* with about 50 species in Eurasia and North Africa. *Daphne mezereum* L. is frequently cultivated for ornament.

The Thymelaeaceae consist of 4 rather distinctive subfamilies of very unequal size, as follows:

a Fruit capsular; disk mostly wanting; stems mostly without internal phloem; carpels often more than 2; hypanthium often scarcely developed.

b Style elongate, with a small stigma; petals mostly numerous, commonly linear, sometimes cleft or dissected; stamens (8-) numerous; leaves commonly pellucid-punctate .... GONYSTYLOIDEAE.

b Style usually very short or none, with a large stigma; petals small and scale-like, or none; stamens 5 or 10; leaves not pellucid-punctate .... AQUILARIOIDEAE.

a Fruit indehiscent; disk mostly present; stems mostly with internal phloem; carpels 2, one of them commonly much reduced so that the ovary is pseudomonomerous.

c Hypanthium scarcely developed .. SYNANDRODAPHNOIDEAE.

c Hypanthium more or less well developed .... THYMELAEOIDEAE.

The Thymelaeoideae, with about 40 genera and more than 400 species, constitute the bulk of the family, to which the other subfamilies are appended. The Synandrodaphnoideae consist of a single genus and species, *Synandrodaphne paradoxa* Gilg of western tropical Africa. The Aquilarioideae and Gonystyloideae have often been considered to form separate small families. This treatment can be defended, but the mutual relationships are not in dispute.

The affinities of the Thymelaeaceae have been much-debated. The Euphorbiaceae, Elaeagnaceae, Flacourtiaceae, and various families of the Myrtales and Malvales are among the groups that have been suggested as possible relatives. The internal phloem and prominent hypanthium of typical members suggest a myrtalean affinity. The fibrous phloem and mucilaginous epidermal cells suggest the Malvalves, but are not really out of harmony with the Myrtales. The possible relationship to the Elaeagnaceae is discussed under that family; if the two families are related, the Elaeagnaceae must be the more advanced group. The pollen and uniovulate locules of the Thymelaeaceae suggest an affinity with the Euphorbiaceae. This suggestion is strengthened by the recent discovery of diterpenoids of the phorbol type in *Pimelea* and *Daphne,* since these substances are otherwise known only in the Euphorbiaceae. The relationship of these two families can scarcely be more than a collateral one, however, since they have different sets of specialized features. Furthermore, the Euphorbiaceae, like the Thymelaeaceae, are in different systems of classification assigned to different subclasses. The assignment of both families to the Rosidae, and of the Thymelaeaceae to the Myrtales in the present system is intuitive and arbitrary. A more definitive placement must await further evidence.

Fossil pollen thought to represent the Gonystyloideae dates from the Oligocene, and that of the Thymelaeoideae from the Lower Eocene. Macrofossils referred to the Thymelaeoideae occur in Oligocene and more recent strata.

6. Family TRAPACEAE Dumortier 1828 nom. conserv., the Water Chestnut Family

Aquatic annual herbs, rooted in the substrate but often breaking off and becoming free-floating, tanniferous but without proanthocyanins; clustered crystals of calcium oxalate commonly present in some of the cells of the parenchymatous tissues, but raphides wanting; vessel-segments with simple perforations; vascular bundles bicollateral, with internal phloem; submersed part of the stem with elongate internodes and opposite (or paired at the same level, but not fully opposite) or ternate, elongate, filiform-dissected green leaves or leaflike organs (these sometimes interpreted as photosynthetic roots or [Vasilev] as stipules), and also producing filiform adventitious roots from some of the nodes; aerial part of the stem short, bearing densely crowded, rosulate LEAVES with small, cleft, deciduous stipules, the blade more or less rhombic, the elongate petiole with an inflated, aerenchymatous float near midlength; stomates anomocytic. FLOWERS solitary in the axils, short-pedicellate, slightly elevated above the water, perfect, regular, tetramerous, slightly perigynous as well as partly epigynous; sepals valvate, joined at the base to form a short hypanthial tube, 2 or all 4 of them persistent and indurate-accrescent in fruit as hornlike or spinelike projections at the distal margin; petals imbricate, distinct, alternate with the sepals; stamens alternate with the petals; filaments short; anthers tetrasporangiate and dithecal, opening by longitudinal slits; pollen-grains binucleate, with 3 prominent meridional crests joined at the poles, alternating with 3 meridionally elongate colpi; gynoecium of 2 carpels united to form a compound, bilocular, partly inferior ovary, the lower part sunken in the receptacle, and surrounded at the base by a cupular, often 8-lobed disk, the ovary becoming almost wholly inferior in fruit; style subulate, with a capitate terminal stigma; each locule of the ovary with a solitary, pendulous, apical-axile, anatropous, bitegmic, crassinucellar ovule, but one locule and its ovule aborting after anthesis, so that the fruit is unilocular and 1-seeded; ovule with a long nucellar beak projecting from the micropyle; embryo-sac 8-nucleate, developing from the chalazal megaspore of a linear tetrad; endosperm-nucleus degenerating soon after its formation; embryo with a prominent haustorial suspensor. FRUIT indehiscent, with a thin, evanscent, fleshy exocarp and a persistent, stony endocarp; cotyledons 2, very unequal, one large, densely packed with starch, and retained within the fruit, the other small and scale-like, growing out (along with the radicle and plumule) through the terminal pore left by the fall of the style; endosperm wanting. N = ca 18, 24. (Hydrocaryaceae)

The family Trapaceae consists of the single genus *Trapa,* with about 15 species native to tropical and subtropical Africa and Eurasia. *Trapa natans* L., the water-chestnut, is used in specialty-dishes, particularly in Chinese cookery.

Trapa has often in the past been included in the Onagraceae, but botanists are now agreed that it should form a separate family. In addition to its specialized habit and fruit, *Trapa* differs from the Onagraceae in its nearly superior ovary surrounded by a cupular disk, in not having raphides, in the structure of the pollen, and in a series of embryological features. Takhtajan considers that *Trapa* is related to the more archaic members of the Onagraceae (tribe Jussieae) and a little less closely to the Lythraceae. On the other hand, Raven (personal communication) sees no link at all between *Trapa* and the Jussieae or other Onagraceae.

Trapa fruits are well represented in the fossil record throughout the Tertiary period and for some distance into the Upper Cretaceous. Miki (1959, 1969) considers that *Trapa* evolved from *Lythrum* through the intermediate fossil genus *Hemitrapa.* In contrast to the fruits, pollen of *Trapa* is known only from Miocene and later deposits.

7. Family MYRTACEAE A. L. de Jussieu 1789 nom. conserv., the Myrtle Family

Trees or shrubs with unicellular or sometimes bicellular or multicellular hairs, bearing ethereal oils (variously monoterpenes, sesquiterpenes, triterpenes, other terpenoids, or polyphenols) in abundant, scattered, small, spherical to sometimes elongate, schizogenous secretory cavities that occur in most or all of the unlignified tissues of the shoot, and also with scattered tanniferous cells; plants nearly always with proanthocyanins and usually also with ellagic and gallic acids, sometimes producing saponins, but only seldom cyanogenic, and lacking iridoid compounds; solitary or clustered crystals of calcium oxalate commonly present in some of the cells of the parenchymatous tissues; nodes typically unilacunar; vessel-segments with simple or seldom scalariform perforations, and usually with some vestured pits; imperforate tracheary elements commonly with evidently bordered (seldom simple) pits, most of them in most species considered to be fiber-tracheids, but those surrounding the vessels considered (in most species) to be true tracheids; wood-rays

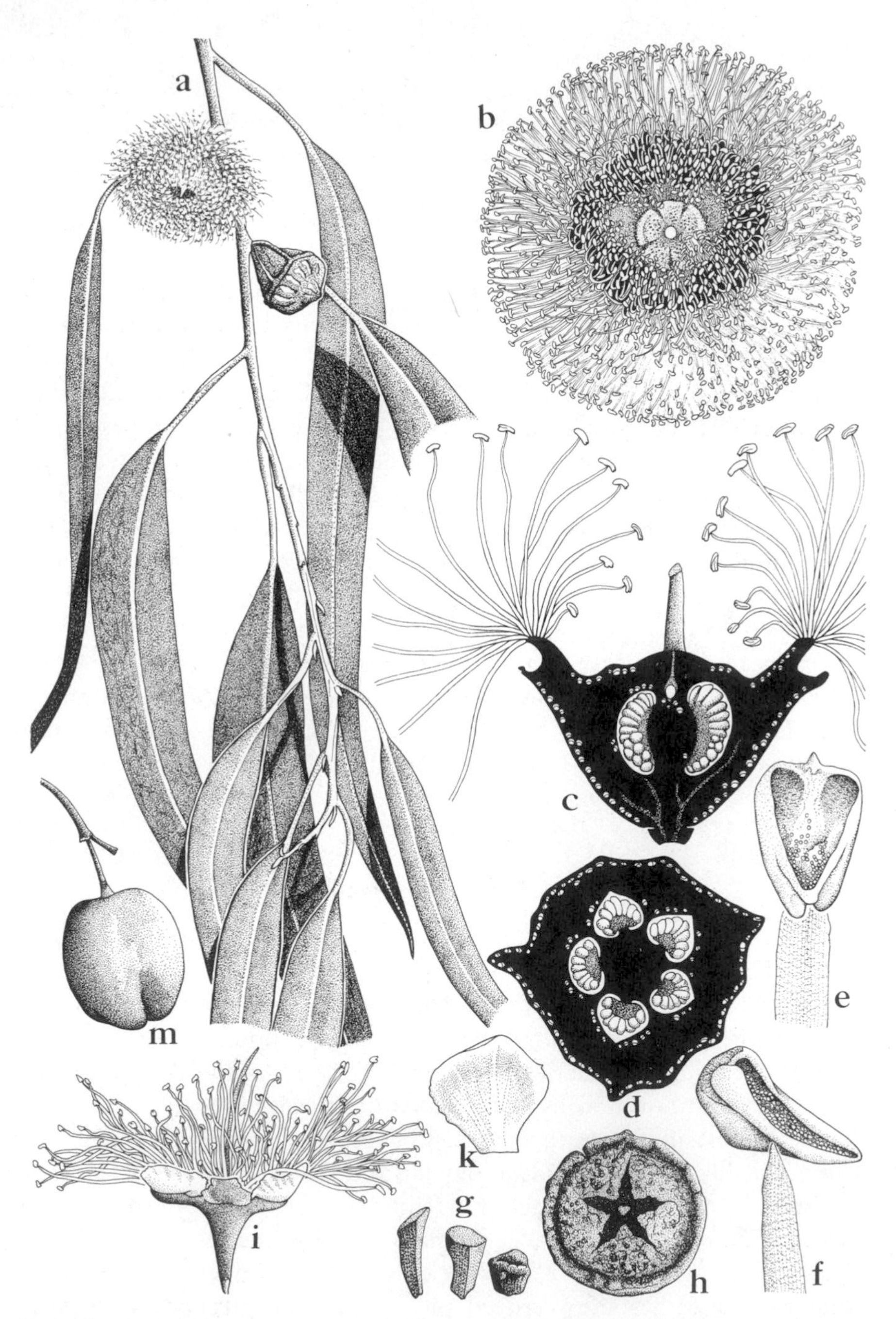

FIG. 5.22 Myrtaceae. a–h, *Eucalyptus globulus* Labill. a, habit, $\times\frac{1}{2}$; b, flower, from above, $\times 1$; c, schematic long-section of flower, $\times 2$; d, schematic cross-section of ovary, $\times 2$; e, f, anthers, $\times 16$; g, seeds, $\times 4$; h, fruit, $\times 1$. i–m, *Eugenia paniculata* Banks. i, flower, $\times 2$; k, petal, $\times 4$; m, fruit, $\times 1$.

heterocellular to sometimes homocellular, all uniseriate, or some of them 2 or 3 (–8) cells wide, many of the cells containing amorphous gummy deposits; wood-parenchyma variously diffuse or in uniseriate ribbons, or partly or largely paratracheal, generally including some vertical crystalliferous strands; wood often with vertical lysigenous canals; internal phloem characteristically present, next to the pith; secondary phloem of young stems commonly stratified tangentially into fibrous and soft layers, seldom wholly soft; most species producing ectotrophic mycorhizae. LEAVES opposite or less often alternate, rarely whorled, simple, commonly coriaceous and always entire, glandular-punctate, often with a continuous intramarginal vein; stomates anomocytic or seldom paracytic; petiole variously with a single more or less evidently bicollateral vascular arc or ring, or with several bicollateral bundles in a line or arc; stipules vestigial or wanting. FLOWERS in various sorts of cymose to racemose, often complex inflorescences (the cymose condition thought to be primitive), or seldom solitary and axillary, commonly bibracteolate at the base, perfect or rarely (as in *Psiloxylon*) unisexual (some flowers staminate in spp. of *Eucalyptus*), regular, epigynous (the hypanthium often prolonged beyond the ovary) to seldom (some Leptospermoideae) only half-epigynous or merely perigynous with the well developed hypanthium free from the ovary, predominantly nectariferous and zoophilous (often pollinated by long-billed, nectar-eating birds); conspicuous floral parts sometimes the corolla, sometimes the calyx, sometimes the stamens (especially in brush-inflorescences) or bracts, or the marginal flowers in a capitate pseudanthium; sepals (3) 4–5 (6), commonly imbricate, or the calyx sometimes undivided in bud and splitting irregularly or deciduous as a calyptra, or the calyx sometimes much-reduced; petals (3) 4–5 (6), imbricate, sometimes (as in spp. of *Eucalyptus*) connivent to form a calyptra, or sometimes wanting; stamens commonly numerous, originating in centripetal sequence, borne at the rim or on the upper surface of the hypanthium, or on a flat disk surrounding the style at the top of the ovary, distinct or united at the base into 4 or 5 bundles developing from as many primordia, each primordium generally supplied by a single vascular trunk bundle, or less often the stamens only twice as many as the sepals or petals and bicyclic, or (as in *Heteropyxis*) reduced to a single antepetalous cycle; anthers tetrasporangiate and dithecal, small, versatile, the connective typically with an evident apical secretory cavity; pollen-sacs opening by longitudinal slits or sometimes by terminal pores; pollen-grains binucleate, commonly tricolporate, seldom dicolpate or triporate, sometimes with a granular instead of columellar infratectal structure; nectary-disk borne on the summit of the ovary or lining the prolonged hypanthium; gynoecium of 2–5 (16) carpels united to form a compound ovary with as many locules as carpels, or the ovary seldom (some Leptospermoideae) pseudomonomerous, with a single locule; style terminal and generally elongate, with a capitate stigma, or seldom (*Psiloxylon*) the stigma sessile and lobed; placentation in plurilocular ovaries axile (basal-axile to apical-axile); ovules 2-many in each locule, anatropous or campylotropous, bitegmic or seldom unitegmic, crassinucellar; endosperm-development nuclear; embryo-sac monosporic, 8-nucleate, developing from the chalazal megaspore of a linear tetrad; embryo commonly formed by nucellar budding after degeneration of the zygote. FRUIT a (1–) few (-many)-seeded berry, or a loculicidal capsule, or sometimes a drupe or a nut; seeds with very little or commonly no endosperm, often polyembryonic in early ontogeny, but usually with only a single embryo at maturity; embryo of diverse forms, the 2 cotyledons small to large, sometimes connate or folded or twisted together, the hypocotyl very short to elongate, often curved or spirally twisted; reserve food in the embryo usually starchy. X = 11, or less often 6–9 or 12. (Heteropyxidaceae, Psiloxylaceae).

The family Myrtaceae consists of about 150 genera and 3600 or more species, found in tropical and subtropical regions throughout the world, and also well developed in temperate Australia. There are 2 well marked but evidently allied subfamilies: the Myrtoideae, widespread but best developed in tropical America, with mostly bilocular ovary and baccate or sometimes drupaceous fruit; and the Leptospermoideae (including Chamaelaucioideae), also widespread, but best developed in Australia, Malesia, and Polynesia, with mostly capsular, 2–5-locular or sometimes multilocular fruits, or a nutlike and 1-seeded fruit. *Heteropyxis*, a small (3 spp.) South African genus here referred to the Leptospermoideae, is sometimes considered to form another subfamily. Some of the larger genera of Myrtaceae are *Eucalyptus* (500) and *Melaleuca* (100) in the Leptospermoideae; and *Eugenia* (600), *Myrcia* (300), *Syzygium* (200+) and *Psidium* (100) in the Myrtoideae.

Myrtus communis L., myrtle, and many species of *Eucalyptus* are cultivated for ornament, and species of *Eucalyptus* are used for commercial timber and as street-trees. *Eucalyptus* forms great forests under warm-temperate climatic conditions in southern Australia; some species attain a height of more than 100 m. The flower-buds of *Syzygium aromaticum* (L.) Merrill & Perry are cloves. *Pimenta dioica* (L.) Merrill is the source of allspice. Guava, cultivated in the tropics for its edible fruit, is *Psidium guavaja* L.

It has been suggested that the number of stamens in the Myrtaceae has increased in association with pollination by birds or bats.

Fossil pollen attributed to the Myrtaceae occurs in deposits of mid-Senonian (Upper Cretaceous) and younger age. Macrofossils attributed to the Myrtaceae are common in Tertiary deposits, and specimens that have been referred to the modern genus *Syzygium* have been found in India.

It is possible that the monotypic Mascarene genus *Psiloxylon* should be excluded from the Myrtaceae and treated as a distinct family Psiloxylaceae Croizat, as recently suggested by Briggs & Johnson. *Psiloxylon* differs from characteristic members of the Myrtaceae in its unisexual flowers, diplostemonous androecium, superior ovary, and large, lobed, sessile stigma.

8. Family PUNICACEAE Horaninow 1834 nom. conserv., the Pomegranate Family

Small trees or shrubs with quadrangular twigs and unilacunar nodes, often thorny, producing pyridine alkaloids and free triterpenes, tanniferous and with ellagic and gallic acids, but without proanthocyanins, and without iridoid compounds; scattered secretory cells present in the cortex and pith; crystals of calcium oxalate (but not raphides) present in some of the cells of the parenchymatous tissues; vessel-segments with simple perforations and vestured pits; imperforate tracheary elements in *Punica granatum* all with simple pits, septate, each chamber bearing a crystal, in *P. protopunica* some vascular tracheids and fiber-tracheids with bordered pits also present; wood-rays all uniseriate and homocellular or some of them biseriate or triseriate and heterocellular, many of the cells containing amorphous gummy deposits; wood parenchyma very scanty paratracheal or virtually wanting; internal phloem present, next to the pith; secondary phloem of young stems with broad, wedge-shaped medullary rays and truncated triangles of phloem. LEAVES mostly opposite or subopposite, sometimes closely crowded at the tips of the twigs, simple and entire, not glandular; stomates commonly anomocytic; stipules vestigial. FLOWERS terminal and axillary, solitary or in clusters terminating axillary branches, perfect, regular, epigynous, the colored hypanthium prolonged well beyond the ovary; sepals 5–8, valvate, appearing as lobes on the hypanthium; petals as many as and

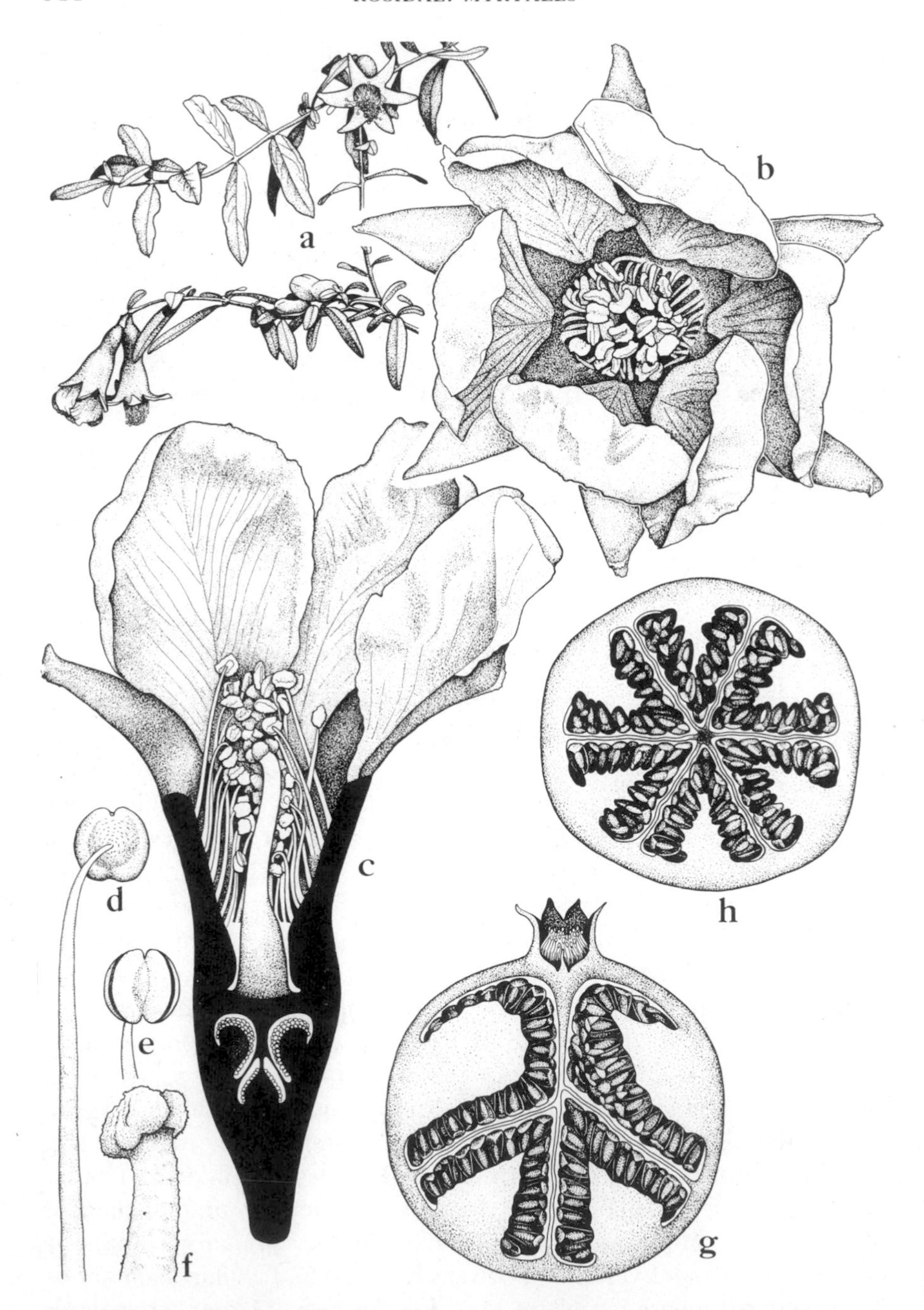

FIG. 5.23 Punicaceae. *Punica granatum* L. a, habit, $\times\frac{1}{2}$; b, flower, from above, $\times 3$; c, flower, in partial long-section, $\times 3$; d, stamen, $\times 6$; e, anther, $\times 6$; f, stigma and top of style, $\times 6$; g, schematic long-section of fruit, $\times\frac{3}{8}$; h, schematic cross-section of fruit, $\times\frac{3}{8}$.

alternate with the sepals, imbricate and crumpled in bud; stamens numerous, produced on the hypanthial tube in centrifugal sequence; filaments slender, distinct; anthers small, dorsifixed, tetrasporangiate and dithecal, opening by longitudinal slits; pollen-grains binucleate, tricolporate; gynoecium of several (commonly 7–9, or up to 15) carpels, those in *P. protopunica* united to form an ordinary compound ovary with axile placentation, those in *P. granatum* undergoing differential growth to become superposed in 2 or 3 layers, the lower layer with axile placentation, the upper appearing parietal because of asymmetric growth; style slender, more or less elongate, with a wet, papillate, capitate stigma; ovules more or less numerous on each placenta, anatropous, bitegmic, crassinucellar; embryo-sac monosporic, 8-nucleate, developing from the chalazal megaspore of a linear tetrad; endosperm-development nuclear. FRUIT with a leathery rind, and crowned by the persistent sepals; seeds embedded in a pulpy mass representing the proliferated sarcotestas; endosperm wanting; embryo oily, straight, with 2 large, spirally rolled cotyledons; seed-fat yielding the otherwise rare punicic acid. X = 8, 9.

The family Punicaceae consists of the single genus *Punica,* with 2 species. *Punica granatum* L., the pomegranate, is a cultigen that also grows wild from the Balkan region to northern India. *Punica protopunica* Balf. f., native to the island of Socotra, may be its ancestor. The discovery of *P. protopunica* makes *P. granatum* seem less distinctive and taxonomically less isolated than before, but *Punica* still does not fit comfortably into any other family. Bridgwater and Baas (1978) consider that the wood anatomy of *Punica* supports a close relationship with the Lythraceae.

9. Family ONAGRACEAE A. L. de Jussieu 1789 nom. conserv., the Evening Primrose Family

Herbs, or sometimes (notably *Fuchsia* and *Hauya*) small to arborescent shrubs, or even trees to 30 m tall, with unicellular or uniseriate hairs, tanniferous, accumulating ellagic and gallic acids, but without proanthocyanins, without saponins, not cyanogenic and not reliably reported to produce alkaloids, often with oil-cells in the epidermis, and generally producing raphides (unique in the order in this respect) in scattered sacs in the parenchymatous tissues, the raphides sometimes embedded in or replaced by mucilage; nodes unilacunar; vessel-segments commonly with transverse to somewhat oblique ends, simple perforations,

FIG. 5.24 Onagraceae. *Oenothera biennis* L. a, habit, ×½; b, seed, ×10; c, flower, in schematic veiw from the top, ×1; d, flower, from above, ×2; e, stamen, ×4; f, side view of flower, ×2; g, pistil, ×2; h, schematic cross-section of ovary, ×6; i, opened fruit, ×2; k, opened fruit, from above, ×2.

and vestured pits; imperforate tracheary elements with simple pits, commonly either septate or persistently nucleated; wood-rays heterocellular, 1-several (–9)-seriate, commonly with amorphous gummy deposits in many of the cells; wood-parenchyma mostly scanty and paratracheal, and including some vertical crystalliferous strands; internal phloem characteristically present; interxylary phloem often present as well. LEAVES variously alternate, opposite, or whorled, considered by Raven (personal communication) to be primitively opposite, commonly simple and entire or toothed, but sometimes lobed or lyrate-pinnatifid; stomates with 3 or more subsidiary cells, often anisocytic as in the Brassicaceae; stipules primitively well developed, but in most genera wanting or small and soon deciduous. FLOWERS solitary in the axils, or in leafy-bracteate or naked spikes, racemes, or panicles, perfect or seldom (spp. of *Fuchsia*) unisexual, epigynous and with the hypanthium generally more or less prolonged and nectariferous toward the base within (not prolonged in *Ludwigia*), regular or less often irregular, most often 4-merous or 2-merous (*Circaea*); about half of the species self-pollinated, the others outcrossed, variously pollinated by birds or by bees, moths, flies, or other insects, but not by bats, beetles, or wind, and not known to be apomictic; sepals valvate, commonly appearing as lobes on the hypanthium; petals generally as many as the sepals, variously imbricate or valvate or convolute, distinct, often clawed or stipitate, or seldom wanting; stamens attached within the hypanthium, or surrounding an epigynous disk, commonly bicyclic, or unicyclic by reduction of the antepetalous series, or rarely reduced to 2 (and one of these sterile in *Lopezia*); anthers tetrasporangiate and dithecal, sometimes cross-partitioned, opening by longitudinal slits; pollen-grains binucleate, borne in tetrads or monads, with viscin threads on the proximal surface and tending to cohere in large groups; exine of complex and unusual structure, (2) 3 (–6)-aperturate (variously colpate, colporate, or often porate); carpels generally as many as the sepals or petals, united to form a compound ovary with as many locules as carpels, or the partitions sometimes imperfect, the placentation accordingly axile or parietal; style with a capitate to 4-lobed, wet or sometimes dry stigma; ovules mostly several or numerous (seldom only 1 or 2) in each locule or on each placenta, anatropous, bitegmic, crassinucellar; embryo-sac monosporic, 4-nucleate (unique in the order), the endosperm diploid and initially nuclear. FRUIT a loculicidal (or both loculicidal and septicidal) capsule, or sometimes a berry (*Fuchsia*) or a small nut (notably *Gaura* and *Circaea*); seeds numerous or seldom few or only one, without endosperm; embryo oily, straight or nearly so, with 2 cotyledons. X = 6–18, perhaps originally 11, now most commonly 7.

The well defined family Onagraceae consists of some 17 genera and about 675 species, mainly of temperate and subtropical regions especially in the New World, most abundant and diversified in western United States. The two least specialized or archaic genera, *Ludwigia* and *Fuchsia*, are heavily concentrated in South America. The family is somewhat isolated within the order, being well set off by its embryological and other features, but its relationships are not in dispute. The largest genera are *Epilobium* (200), *Oenothera* (125), *Fuchsia* (90), and *Ludwigia* (80). Species of these and several other genera (notably *Clarkia*) are often cultivated for ornament.

Onagraceous pollen appears in generalized form in the Maestrichtian, and in gradually increasing diversity from the Eocene onwards. Pollen attributed to the modern genera *Ludwigia* and *Boisduvalia* dates from the Eocene and Oligocene epochs onward respectively, suggesting at least the presence of the tribes to which these genera belong. Fruits attributed to *Circaea,* one of the more specialized modern genera, occur in Oligocene and more recent deposits.

10. Family OLINIACEAE Arnott ex Sonder in Harvey & Sonder 1862 nom. conserv., the Olinia Family

Much-branched, tanniferous shrubs or rather small trees with quadrangular to terete twigs, unilocular nodes, and simple, unicellular hairs; solitary crystals of calcium oxalate present in some of the cells of the parenchymatous tissues; vessel-segments with slanting ends, simple perforations, and vestured pits; imperforate tracheary elements septate, with simple pits; wood-rays heterocellular, most of them biseriate, but some uniseriate, and sometimes some triseriate, many of the cells containing amorphous gummy deposits; wood-parenchyma very sparse and mostly paratracheal, some vertical crystalliferous strands also present; internal phloem present, next to the pith. LEAVES opposite, simple, entire, pinnately veined; stomates both paracytic and anomocytic, verging toward encyclocytic; petiole with a bicollateral, trough-shaped vascular strand that may have incurved edges; stipules vestigial. FLOWERS in terminal or axillary cymose inflorescences, small, perfect, regular, 4–5-merous, epigynous with the hypanthium prolonged beyond the ovary; calyx represented by a narrow rim on the hypanthium, usually with 4 or 5 small, blunt teeth or lobes; petals alternating with the sepals (when these can be distinguished), imbricate, spatulate; a set of incurved, hairy, scale-like, colored or white staminodes alternating with the petals and on the same radii as the fertile stamens, which also alternate with small, antepetalous staminodes (or these virtually obsolete); filaments short, recurved; anthers small, introrse, dithecal, the pollen-sacs small and distinctly separated on a thickened connective, opening by longitudinal slits; pollen-grains tricolporate; gynoecium of (3) 4–5 carpels united to form a compound, inferior ovary with as many locules as carpels and with axile placentation; style terminal, with a capitate stigma; disk none; 2 (3) superposed, hemitropous, apotropous, bitegmic, crassinucellar ovules in each locule; embryo-sac 8-nuclear; endosperm-development nuclear. FRUIT drupaceous, with a single plurilocular stone and 1 seed in each locule; embryo oily, with a short radicle and 2 spirally twisted or irregularly folded cotyledons; endosperm none. X apparently = 10.

The family Oliniaceae consists of the single genus *Olinia,* with 8 species confined to tropical and southern Africa and the Island of St. Helena. *Olinia* is obviously at home in the Myrtales, but it cannot readily be squeezed into any of the other families.

11. Family MELASTOMATACEAE A. L. de Jussieu 1789 nom. conserv. the Melastome Family

Shrubs or herbs, or sometimes lianas or rather small trees, sometimes epiphytic, often with quadrangular stems, and commonly with various sorts of complex hairs, sometimes myrmecophilous, often accumulating aluminum, tanniferous, accumulating ellagic acid and often also proanthocyanins, and producing abundant acylated anthocyanins (these reputedly lacking from the Myrtaceae, the next largest family of the order), but lacking saponins and iridoid substances, only seldom cyanogenic, and only rarely with alkaloids; nodes unilacunar; clustered or solitary crystals of calcium oxalate commonly present in some of the parenchymatous tissues, which often also have scattered secretory cells; vessel-segments with simple perforations and vestured pits; imperforate tracheary elements with clearly bordered pits, or with simple pits and then generally septate; wood-rays uniseriate or some of them 2–5-seriate, heterocellular or less often homocellular, many of the cells containing amorphous gummy deposits; wood-parenchyma commonly paratracheal and rather scanty, but sometimes also in apotracheal ribbons, and including some vertical crystalliferous strands; internal phloem characteristically present, and interxylary phloem sometimes present as well; cortical and/or pith bundles usually present. LEAVES opposite (but often the members of a pair unequal, or one member even suppressed) or seldom whorled, simple and usually entire, commonly with 3–9 prominent, subparallel veins, seldom pinnately veined; stomates mostly anomocytic or anisocytic, but sometimes diacytic or paracytic; mesophyll often containing scattered sclereids; petiolar anatomy complex; stipules mostly wanting or vestigial, seldom well developed. FLOWERS in various sorts of cymose inflorescences, commonly rather large and showy, sometimes subtended by showy bracts, typically without nectar and visited by pollen-gathering insects, but in *Blakea* nectariferous and pollinated by mice, perfect or seldom unisexual,

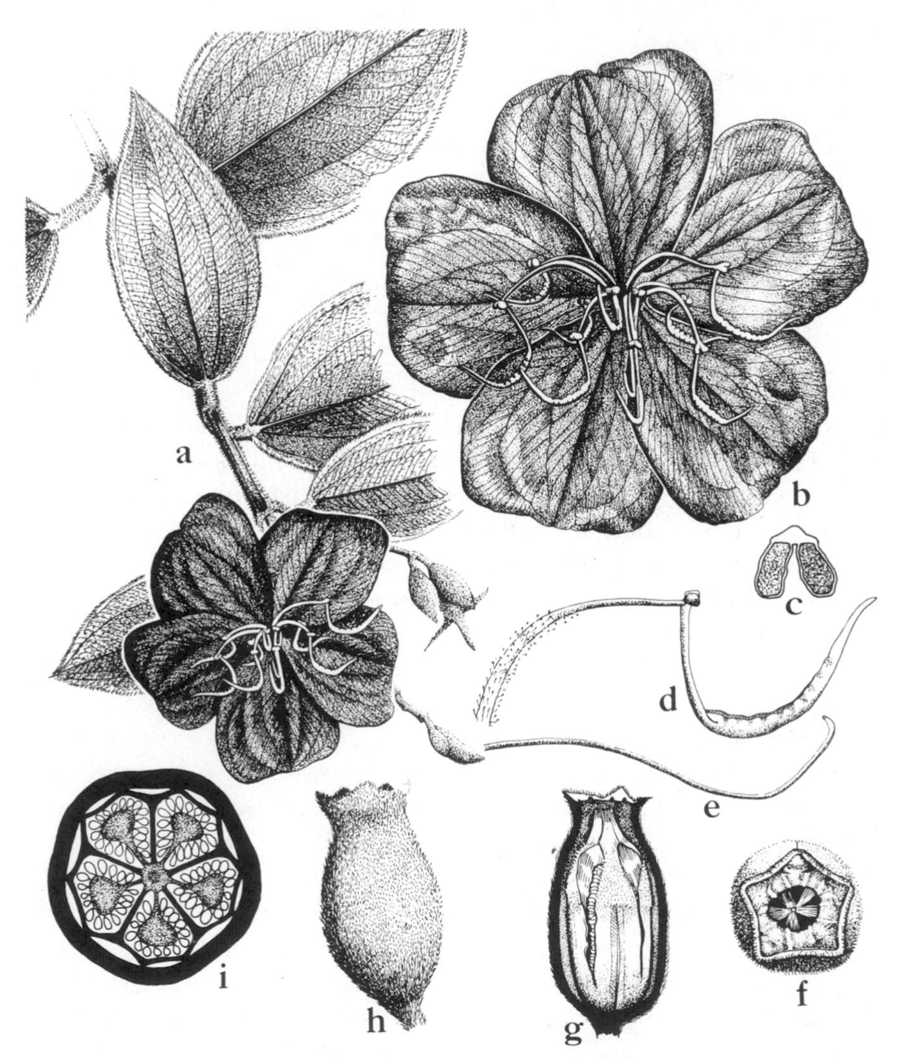

FIG. 5.25 Melastomataceae. *Tibouchina urvilleana* (DC.) Cogn. a, habit, ×½; b, flower, ×1; c, anther in cross-section, ×8; d, stamen, ×2; e, style, ×2; f, g, h, mature fruits, ×2, top view, long-section, and side view; i, schematic cross-section of ovary, ×4.

partly or wholly epigynous, or less often strongly perigynous with the persistent hypanthium enveloping but free from the ovary, or sometimes the hypanthium attached to the ovary only by longitudinal, septum-like ribs, regular except for the stamens, (3) 4–5 (–10)-merous; calyx-lobes valvate, or the calyx sometimes calyptrate or reduced to a mere rim on the hypanthium; petals distinct or rarely connate at the base, convolute in bud; stamens generally in 2 cycles and twice as many as the petals, often dimorphic, or seldom in only one cycle alternate with or opposite to the petals, or seldom more or less numerous; filaments generally inflexed in bud (each anther sometimes occupying a cavity between the hypanthium and the ovary), commonly twisted at anthesis so as to bring all the anthers to one side of the flower; anthers tetrasporangiate and dithecal, or seldom (*Rhexia*) unithecal but still tetrasporangiate, opening by a single terminal pore or seldom by 2 pores or 2 longitudinal slits; connective often thickened at the base, and commonly provided with appendages of diverse form; pollen-grains binucleate, typically tricolporate and with 3 alternating poreless furrows, the surface smooth or nearly so; nectaries mostly wanting; gynoecium of (2) 3–5 (–15) carpels united to form a compound ovary with as many locules, or seldom unilocular through failure of the partitions; style terminal, with a capitate or punctate or seldom lobed stigma; ovules (1–) more or less numerous in each locule, on axile or less often basal or rarely even free-central (as in *Memecylon*) or parietal placentas, anatropous or seldom campylotropous, crassinucellar, bitegmic, with a zigzag micropyle; embryo-sac monosporic, 8-nuclear, developing from the chalazal megaspore of a linear tetrad; endosperm-development nuclear, never becoming cellular. FRUIT a loculicidal capsule or a berry; seeds commonly numerous, mostly small, without endosperm; cotyledons 2, commonly unequal. X = 7–18 or more. (Memecylaceae, Mouririaceae)

The well defined family Melastomataceae consists of about 200 genera and 4000 species, widespread in tropical and subtropical regions, but best developed in South America. Many have very showy flowers, but only a few have been brought into cultivation, and these are not commonly grown. The largest genus by far is *Miconia,* with perhaps as many as 1000 species. *Medinilla* (300), *Tibouchina* (250), *Memecylon* (200), *Leandra* (200), *Clidemia* (175), *Gravesia* (100), and *Microlicia* (100) are some other large genera. Details of the structure of the anthers, doubtless related eventually to mechanisms of pollination, provide many generic and specific characters in the Melastomataceae.

Melastome leaves thought to be related to the modern genus *Hederella* occur in Eocene deposits of North Dakota, but pollen of the family dates only from the Oligocene.

12. Family COMBRETACEAE R. Brown 1810 nom. conserv., the Indian Almond Family

Trees or shrubs, often scandent and sometimes twining, seldom (*Laguncularia, Lumnitzera*) mangroves, with various kinds of hairs (typically

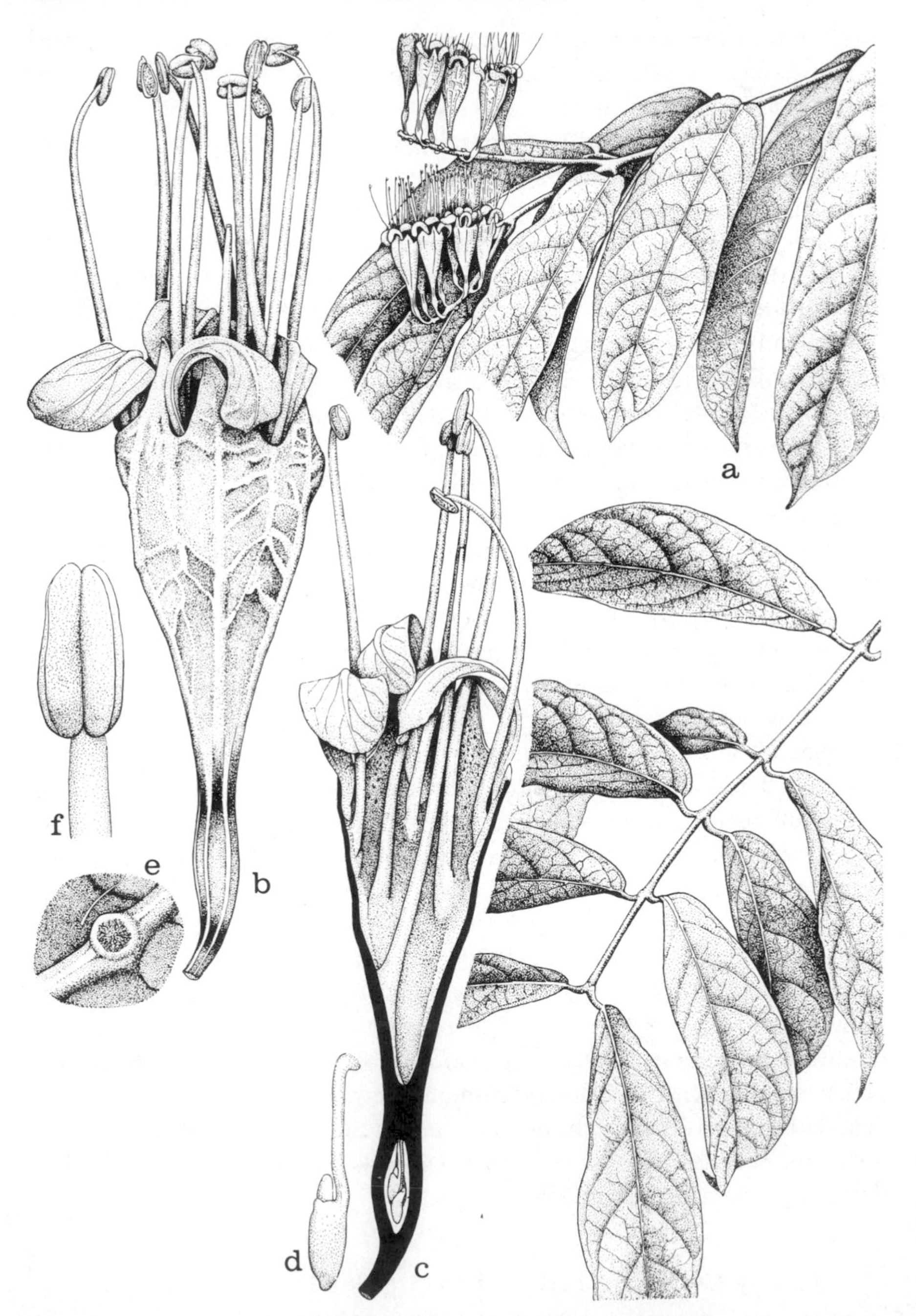

FIG. 5.26 Combretaceae. *Combretum grandiflorum* G. Don. a, habit, ×½; b, flower, ×3; c, flower, in long-section, ×3; d, ovule, ×12; e, lower surface of leaf, showing gland, ×18; f, anther, ×12.

some or all of them long, straight, sharp-pointed, unicellular and very thick-walled, with a conical internal compartment at the base), tanniferous, producing ellagic and gallic acids and often also proanthocyanins, sometimes also cyanogenic, often accumulating triterpenes, especially as saponins, but without iridoid compounds; mucilaginous secretory cells or canals often present in the parenchymatous tissues, sometimes even in the wood; solitary or clustered crystals of calcium oxalate often present in some of the cells of the parenchymatous tissues, those in the leaves often in stellate idioblasts, so that the leaf may appear translucent-punctate; wood mostly diffuse-porous; vessel-segments with simple (oblique to transverse) perforations and vestured pits; very slender vessels commonly intermingled with ordinary ones in *Combretum* and its immediate allies; imperforate tracheary elements with small, mostly simple pits (bordered in *Strephonema*), often some of the elements septate; vascular tracheids and radial vessels sometimes present, notably in *Combretum* and its immediate allies; wood-rays uniseriate, or some of them 2–3 (–6)-seriate, homocellular or heterocellular, generally many of the cells containing amorphous gummy deposits; wood-parenchyma mainly paratracheal, commonly wing-like, and including some vertical crystalliferous strands; some apotracheal parenchyma sometimes also present; internal phloem commonly but not always present, next to the pith, interxylary phloem often present as well. LEAVES alternate, opposite, or less often whorled, simple and entire, pinnately veined; base of the leaf-blade commonly with 2 flask-shaped cavities, each containing a multicellular gland; stomates anomocytic or seldom (*Laguncularia, Lumnitzera*) encyclocytic or (*Strephonema*) paracytic; stipules vestigial or wanting. FLOWERS in terminal and axillary racemes, spikes, or heads, mostly rather small, perfect or less often unisexual, regular or seldom slightly irregular, epigynous or rarely (*Strephonema*) only half-epigynous, the hypanthium only slightly to often conspicuously prolonged beyond the ovary, often nectariferous within; sepals 4–5 (–8), often appearing as lobes on the hypanthium, or sometimes forming a short calyx-tube above the hypanthium, persistent, commonly valvate or sometimes imbricate, or sometimes much-reduced; petals small, alternating with the sepals and imbricate or valvate, or often wanting; stamens commonly twice as many as the sepals and bicyclic, the members of the outer or upper (antepetalous) cycle sometimes reduced or wanting or rarely in pairs or triplets, or rarely the stamens in 3 cycles; filaments inflexed in bud, often long-exserted at anthesis; anthers usually versatile, tetrasporangiate and dithecal, opening by longitudinal slits; pollen-grains tricolporate or triporate, binucleate or sometimes trinucleate; an epigynous disk commonly present; gynoecium of 2–5 carpels united to form a compound, unilocular ovary with an elongate terminal style ending in a punctate or seldom capitate stigma; ovary commonly with as many raised ribs as there are sepals; ovules 2 (–6), pendulous on an elongate, slender funiculus from the tip of the locule, anatropous, bitegmic, crassinucellar, with a zigzag micropyle; embryo-sac monosporic and 8-nuclear, developing from the chalazal megaspore or a linear tetrad, or seldom tetrasporic and 16-nuclear in the manner of *Penaea;* endosperm-development nuclear; an elaborate obturator often produced from the funiculus. FRUIT 1-seeded, generally indehiscent and adapted to dispersal by water, commonly leathery with an aerenchymatous mesocarp, less often drupaceous, generally ribbed, the ribs often produced into conspicuous wings, or rarely the fruit dry and dehiscent; seeds without endosperm; embryo oily, with a small radicle and 2 (3 in some spp. of *Terminalia* from se. Asia) folded or spirally twisted cotyledons (cotyledons massive and hemispheric in *Strephonema,* totally fused in some African spp. of *Combretum*). X = 7, 11, 12, 13. (Strephonemataceae)

The family Combretaceae consists of about 20 genera and nearly 400 species, widespread in tropical and subtropical regions, especially in Africa, very often in savannas. *Combretum* (150+) and *Terminalia* (100+) make up at least two-thirds of the family, *Quisqualis indica* L., Rangoon creeper, a climber with attractive flowers, is cultivated for ornament in the tropics. *Terminalia catappa* L., Indian almond, is widely cultivated in the tropics for its edible seed and tanniferous bark, and other species of *Terminalia* are important timber trees in Africa. *Laguncularia racemosa* (L.) Gaertn. f. is the white mangrove.

The Combretaceae stand somewhat apart from the other Myrtales in their unilocular ovary with pendulous, apical ovules, and some of them lack the characteristic internal phloem of the order. In other respects they fit well enough into the Myrtales, and their inclusion in the order is not in dispute.

Fossil leaves and pollen thought to represent the Combretaceae occur in Eocene deposits in southeastern United States. A fossil flower thought to be allied to *Combretum* was found in Eocene deposits of Tennessee.

7. Order RHIZOPHORALES Van Tieghem & Constantin 1918

The order consists only of the family Rhizophoraceae.

SELECTED REFERENCES

Graham, S. A. 1964. The genera of Rhizophoraceae and Combretaceae in the southeastern United States. J. Arnold Arbor. 45: 285–301.

Lersten, N. R., & J. D. Curtis. 1974. Colleter anatomy in red mangrove, *Rhizophora mangle* (Rhizophoraceae). Canad. J. Bot. 52: 2277–2278.

Marco, H. F. 1935. Systematic anatomy of the woods of the Rhizophoraceae. Trop. Woods 44: 1–20.

Muller, J., & C. Caratini. 1977. Pollen of *Rhizophora* (Rhizophoraceae) as a guide fossil. Pollen & Spores 19: 361–389.

Prance, G. T., et al. 1975. Revisão taxonômica das espécies amazõnicas de Rhizóphoraceae. Acta Amazonica 5: 5–22.

Tomlinson, P. B., & D. W. Wheat. 1979. Bijugate phyllotaxis in Rhizophoreae (Rhizophoraceae). J. Linn. Soc., Bot. 78: 317–321.

van Vliet, G. J. C. M. 1976. Wood anatomy of the Rhizophoraceae. Leiden Bot. Ser. 3: 20–75.

1. Family RHIZOPHORACEAE R. Brown in Flinders 1814 nom. conserv., the Red Mangrove Family

Trees or shrubs, often of mangrove habit, mostly with unicellular hairs, tanniferous (the bark often especially so) and at least sometimes with vertically elongate tanniferous cells in some of the parenchymatous tissues, producing proanthocyanins and sometimes also ellagic acid, and also at least sometimes with alkaloids of the pyrrolidine, pyrrolizidine, and/or tropane groups, but not cyanogenic, and lacking saponins and iridoid compounds; nodes trilacunar; solitary or clustered crystals of calcium oxalate often present in some of the cells of the parenchymatous tissues; vessel-segments with scalariform, or simple, or mixed scalariform and simple perforations; imperforate tracheary elements with simple or bordered pits; wood-rays nearly always heterocellular, usually mixed uniseriate and pluriseriate with short ends, the latter sometimes as much as 10 or 15 cells wide; no internal phloem; "pneumatophores" of dubious function produced by the mangrove genera. LEAVES simple and entire, opposite (but not decussate), with well developed but caducous interpetiolar stipules, these sheathing the terminal bud, and at least sometimes with colleters on the inner surface at the base; stomates of various types, most often anomocytic or encyclocytic. FLOWERS solitary

FIG. 5.27 Rhizophoraceae. *Rhizophora mangle* L. a, habit, ×½; b, terminal bud, × 1; c, stipule, × 1; d, portion of lower inner side of stipule, showing colleters, ×4; e, flower, from above, ×4; f, g, anther, ×4; h, pistil and base of calyx from above, showing insertion of 8 stamens in the perigynous nectary ring, ×4, i, superior portion of ovary, ×4; k, schematic long-section of pistil, ×4; m, germinating fruit, ×1.

and axillary, or in axillary, few-flowered cymes or less often racemes, perfect or seldom some of them unisexual, regular, most often 4–5-merous, perigynous or epigynous (hypogynous in Cassipourea), the hypanthium in epigynous flowers sometimes prolonged beyond the ovary; sepals (3) 4–5 (–16), valvate, thick, commonly fleshy or leathery; petals as many as and alternate with the sepals, distinct, commonly fleshy and often shorter than the sepals, convolute or infolded in bud; stamens twice as many or sometimes 3 or 4 times as many as the sepals and petals, or of indefinite number, often in pairs opposite the petals, generally all borne in a single cycle; filaments distinct or connate at the base, attached to or around the base of a perigynous nectary-disk, or the disk sometimes wanting, or the anthers sessile; anthers tetrasporangiate and bilocular, or (*Rhizophora*) cross-partitioned and dehiscent by a separating longitudinal valve; pollen-grains binucleate, 3 (4)-colporate; gynoecium of 2–5 (–6) carpels united to form a compound ovary with as many locules as carpels, or the ovary sometimes unilocular by failure of the partitions; style terminal, simple, with a shallowly to evidently lobed stigma; ovules 2 (seldom 4 or more) in each locule, apical-axile, pendulous, anatropous to hemitropous, bitegmic, crassinucellar, with a zigzag micropyle; endosperm-development nuclear. FRUIT baccate, 1-seeded or with 1 seed per locule, or seldom capsular; seeds sometimes arillate; embryo dicotyledonous, straight, linear, often green, in the mangrove genera viviparous and with an enlarged hypocotyl; endosperm well developed, fleshy, oily. X = 8, 9.

The family Rhizophoraceae as here defined consists of about 14 genera and a hundred species, widely distributed in tropical regions. The mangrove genera (tribe Rhizophoreae, 4 genera, ca. 17 spp.) are the most familiar members of the family, but the majority of the genera and species are inland plants that are not mangroves. The largest genus is *Cassipourea,* which is pantropical, with about 70 species, best developed in tropical Africa. *Cassipourea* and a few other genera have hypogynous flowers, with a perigynous disk but no hypanthium. Most other members of the family are epigynous. *Anisophyllea* and its immediate allies, which have sometimes been included in the Rhizophoraceae, are here referred to the Rosales as a family Anisophylleaceae.

The proper taxonomic disposition of the Rhizophoraceae presents a difficult problem. Traditionally they have been referred to a broadly defined and poorly characterized order Myrtales. In the present system the Myrtales are a more homogeneous group, and the Rhizophoraceae would be as out of place as a giraffe in a herd of bison. The Myrtales nearly always have internal phloem, which is wanting from the Rhizophoraceae. The Myrtales seldom have well developed stipules, whereas the Rhizophoraceae have large interpetiolar stipules. The Myrtales mostly have seeds with very little or no endosperm, whereas the endosperm is well developed in the Rhizophoraceae. The Myrtales nearly always have vessels with simple perforations, whereas many of the Rhizophoraceae have scalariform vessels. The Myrtales are poor in alkaloids; the Rhizophoraceae have alkaloids of groups unknown in the Myrtales. The fact that both the Combretaceae (*Myrtales*) and the Rhizophoraceae include some mangrove genera must be viewed in the light of the fact that in both families the mangroves are specialized members; neither family has inherited the mangrove habit from the other. The exclusion of the Rhizophoraceae from the Myrtales thus seems well justified.

Another order which might be considered as a refugium for the Rhizophoraceae is the Cornales. I proposed this disposition of the family, with some hesitation, in 1957. The Cornales even without the Rhizophoraceae are a rather diverse group, so that it might be thought that it would do no great harm to expand their limits to include that family. On the other hand, the Rhizophoraceae would bring in several characters that are not otherwise known in the Cornales, to wit: stipulate leaves, perigynous (rather than epigynous) flowers, convolute or infolded petals, two or more ovules per carpel, and bitegmic ovules. Several kinds of alkaloids have been found in members of the Cornales, but these do not belong to the same chemical groups as the alkaloids of the Rhizophoraceae. In contrast, iridoid compounds are widespread in the Cornales, but unreported from the Rhizophoraceae. Furthermore, the fruit of most members of the Rhizophoraceae is baccate, whereas that of most Cornales is drupaceous. Some few members of the Rhizophoraceae have capsular fruits, which are unknown in the Cornales. The Rhizophoraceae cannot be attached to the Cornales as an aberrant small group, because there are nearly as many species of Rhizophoraceae as there are of Cornales. Therefore I withdraw my earlier suggestion that the Rhizophoraceae might usefully be included in the Cornales.

A third conceivable place for the Rhizophoraceae is in the order Rosales. The Rosales are the basal complex of their subclass, and many of the features that characterize the more advanced orders can be found individually in particular members of the Rosales. Thus it should not be surprising that the Rhizophoraceae would bring almost nothing in the way of new characters to the Rosales. On the other hand, the Rhizophoraceae do not appear to be closely allied to any family of Rosales, and they have none of the primitive features which mark that order. They do not have separate carpels, nor do they have separate styles; they do not usually have numerous stamens, and they do not usually have numerous ovules. If one were to try to force the Rhizophoraceae into the Rosales they would have to be put somewhere near the Grossulariaceae and Hydrangeaceae. Their ancestry may indeed lie somewhere near those two families, as does that of the Cornales as well, but I do not believe the interests of taxonomy would be well served by extending the limits of the Rosales to encompass either the Rhizophoraceae or the Cornales.

The fourth alternative, and the one here adopted, is to treat the Rhizophoraceae as a distinct order. When it is reasonably possible I prefer to avoid recognizing monotypic groups, especially when these do not have a long history of taxonomic acceptance, but in this instance the other alternatives are even less attractive.

Pollen considered to represent the tribe Rhizophoreae is known from uppermost Eocene and more recent deposits.

8. Order CORNALES Lindley 1833

Woody plants, or rarely herbs from woody rhizomes, often with iridoid compounds (notably aucubin), tanniferous or not, with or without ellagic acid and proanthocyanins, sometimes with triterpenoid saponins and/or alkaloids (notably of emetine or diterpene type); nodes nearly always trilacunar; vessel-segments with scalariform or less often simple perforations; imperforate tracheary elements with bordered or less often simple pits; wood-rays heterocellular, often with elongate ends; wood-parenchyma generally apotracheal, often diffuse. LEAVES opposite or less commonly alternate, simple, nearly always exstipulate. FLOWERS individually small, regular or nearly so, epigynous, very often 4-merous, perfect or less often unisexual; petals distinct or sometimes wanting; stamens isomerous with the petals or sepals, or less often twice as many or even 3 or 4 times as many, or seldom several and the flowers without perianth; pollen-grains triaperturate, or of triaperturate-derived type; gynoecium of 2–9 carpels united to form a compound, inferior ovary with as many locules as carpels, or the ovary unilocular and bicarpellate or pseudomonomerous; central axis of the ovary usually without vascular bundles; style terminal, simple or lobed, or the individual styles sometimes nearly or quite distinct; ovules solitary in each locule (2 in the single locule of *Garrya*), pendulous from the apex, anatropous, unitegmic, crassinucellar or seldom tenuinucellar, often with an integumentary tapetum; endosperm-development cellular or nuclear. FRUIT usually a drupe with a single unilocular to plurilocular stone (one seed in each locule, or some locules empty), or seldom a 1-seeded or few-seeded berry, or (*Kaliphora*) a drupe with 2 1-seeded pyrenes; seeds with copious, usually or always oily endosperm and small to elongate, dicotyledonous embryo. X = 8–11.

The order Cornales as here defined consists of 4 families and scarcely 150 species. The Cornaceae, with about a hundred species, are the largest family, with which the other families are to be associated individually. Recent serological studies among species of *Cornus* (sens. lat.), *Nyssa, Griselinia, Corokia,* and *Garrya* suggest that *Nyssa* is fairly close to *Cornus,* followed by *Griselinia* and *Corokia,* and then *Garrya.*

The families of Cornales have often been included in a broadly defined order Araliales or Umbellales. The merits of that treatment, as opposed to the one presented here, have been extensively discussed in the taxonomic literature, and the arguments need not be repeated here. In my opinion the Araliales are derived from the Rosales via the

Sapindales, whereas the Cornales are derived directly from the Rosales, in the vicinity of the Hydrangeaceae and Grossulariaceae. Features of gross morphology, wood-anatomy, palynology, and chemistry combine to support this view of the relationships of the Cornales, as does the existence of the non-missing link *Corokia* (q.v., sub Cornaceae), which has with good reason been compared to *Argophyllum,* in the Grossulariaceae (Escalloniaceae).

SELECTED REFERENCES

Adams, J. E. 1949. Studies in the comparative anatomy of the Cornaceae. J. Elisha Mitchell Sci. Soc. 65: 218–244.

Bate-Smith, E. C., I. K. Ferguson, K. Hutson, S. R. Jensen, B. J. Nielsen, & T. Swain. 1975. Phytochemical interrelationships in the Cornaceae. Biochem. Syst. Ecol. 3: 79–89.

Bloembergen, S. 1939. A revision of the genus *Alangium.* Bull. Jard. Bot. Buitenzorg III. 16: 139–235.

Brooks, R. R., J. A. McCleave, & E. K. Schofield. 1977. Cobalt and nickel uptake by the Nyssaceae. Taxon 26: 197–201.

Brunner, F., & D. E. Fairbrothers. 1978. A comparative serological investigation within the Cornales. Serol. Mus. Bull. 53: 2–5.

Chao, C.-Y. 1954. Comparative pollen morphology of the Cornaceae and allies. Taiwania 5: 93–106.

Chopra, R. N., & H. Kaur. 1965. Some aspects of the embryology of *Cornus.* Phytomorphology 15: 353–359.

Dahling, G. V. 1978. Systematics and evolution of *Garrya.* Contr. Gray Herb. 209: 1–104.

Eyde, R. H. 1963. Morphological and paleobotanical studies of the Nyssaceae. I. A survey of the modern species and their fruits. J. Arnold Arbor. 44: 1–59.

Eyde, R. H. 1964. Inferior ovary and generic affinities of *Garrya.* Amer. J. Bot. 51: 1083–1092.

Eyde, R. H. 1966. The Nyssaceae in the southeastern United States. J. Arnold Arbor. 47: 117–125.

Eyde, R. H. 1966. Systematic anatomy of the flower and fruit of *Corokia.* Amer. J. Bot. 53: 833–847.

Eyde, R. H. 1967 (1968). The peculiar gynoecial vasculature of Cornaceae and its systematic significance. Phytomorphology 17: 172–182.

Eyde, R. H. 1968. Flowers, fruits, and phylogeny of Alangiaceae. J. Arnold Arbor. 49: 167–192.

Eyde, R. H. 1972. Pollen of *Alangium:* Toward a more satisfactory synthesis. Taxon 21: 471–477.

Eyde, R. H., & E. S. Barghoorn. 1963. Morphological and paleobotanical studies of the Nyssaceae. II. The fossil record. J. Arnold Arbor. 44: 328–376.

Eyde, R. H., A. Bartlett, & E. S. Barghoorn. 1969. Fossil record of *Alangium.* Bull. Torrey Bot. Club 96: 288–314.

Ferguson, I. K. 1966. The Cornaceae in the southeastern United States. J. Arnold Arbor. 47: 106–116.

Ferguson, I. K. 1977. Angiospermae. Cornaceae Dum. World Pollen and Spore Flora 6.

Goldblatt, P. 1978. A contribution to cytology in Cornales. Ann. Missouri Bot. Gard. 65: 650–655.

Gopinath, D. M. 1945. A contribution to the embryology of *Alangium lamarckii* Thw. with a discussion of the systematic position of the family Alangiaceae. Proc. Indian Acad. Sci. B. 22: 225–231.

Govindarajalu, E. 1961, 1962. The comparative morphology of the Alangiaceae. I. The anatomy of the node and internode. II. Foliar histology and vascularization. III. Pubescence. IV. Crystals. Proc. Natl. Inst. Sci. 27B: 375–388, 1961. 28B: 100–114; 507–517; 518–531, 1962.

Govindarajalu, E., & B. G. L. Swamy. 1956. Petiolar anatomy and subgeneric classification of the genus *Alangium.* J. Madras Univ. 26B: 583–588.

Hallock, F. A. 1930. The relationship of *Garrya.* Ann. Bot. (London) 44: 771–812.

Hegnauer, R. 1969. Chemical evidence for the classification of some plant taxa. *In:* J. B. Harborne & T. Swain, eds., Perspectives in phytochemistry, pp. 121–138. Academic Press. London & New York.

Hohn, M. E., & W. G. Meinschein. 1976. Seed oil fatty acids: evolutionary significance in the Nyssaceae and Cornaceae. Biochem. Syst. Ecol. 4: 193–199.

Kapil, R. N., & P. R. Mohana Rao. 1966. Studies of the Garryaceae. II. Embryology and systematic position of *Garrya* Douglas ex Lindley. Phytomorphology 16: 564–578.

Kubitzki, K. 1963. Zur Kenntnis des unilokularen Cornaceen-Gynözeums (Cornaceen-Studien I). Ber. Deutsch. Bot. Ges. 76: 33–39.

Li, H.-L. 1954 (1955). *Davidia* as the type of a new family Davidiaceae. Lloydia 17: 329–331.

Li, H.-L., & C.-Y. Chao. 1954. Comparative anatomy of the woods of Cornaceae and allies. Quart. J. Taiwan Mus. 7: 119–136.

Maekawa, F. 1965. *Aucuba* and its allies—the phylogenetic consideration on the Cornaceae. J. Jap. Bot. 40: 41–47.

Markgraf, F. 1963. Die phylogenetische Stellung der Gattung *Davidia.* Ber. Deutsch. Bot. Ges. 76: (63)–(69).

Mohana Rao, P. R. 1972 (1973). Embryology of *Nyssa sylvatica* and systematic consideration of the family Nyssaceae. Phytomorphology 22: 8–21.

Moseley, M. F., & R. M. Beeks. 1955. Studies of the Garryaceae. I. The comparative morphology and phylogeny. Phytomorphology 5: 314–346.

Philipson, W. R. 1967. *Griselinia* Forst. fil.—Anomaly or link. New Zealand J. Bot. 5: 134–165.

Reitsma, T. 1970. Pollen morphology of the Alangiaceae. Rev. Palaeobot. Palynol. 10: 249–332.

Sohma, K. 1963, 1967. Pollen morphology of the Nyssaceae. I. *Nyssa* and *Camptotheca.* II. *Nyssa* and *Davidia.* Sci. Rep. Tôhoku Imp. Univ. Ser. IV. 29: 389–392, 1963. 33: 527–532, 1967.

Tandon, S. R., & J. M. Herr. 1971. Embryological features of taxonomic significance in the genus *Nyssa.* Canad. J. Bot. 49: 505–514.

Titman, P. W. 1949. Studies in the woody anatomy of the family Nyssaceae. J. Elisha Mitchell Sci. Soc. 65: 245–261.

Wilkinson, A. M. 1944. Floral anatomy of some species of *Cornus.* Bull. Torrey Bot. Club 71: 276–301.

Ерамян, Е. Н. 1967. Типы оболочки микроспор представителей порядка Cornales и их генетические связи. Бот. Ж. 52: 1287-1294.

Ерамян, Е. Н. 1971. Палинологические данные к систематике и филогении Cornaceae Dumort. и родственных семейств. В сб.: Морфология пыльцы Cucurbitaceae, Thymelaeaceae, Cornaceae. Л. А. Куприянова и М. С. Яковлев, редакторы: 235–273. Изд. Наука. Ленинградск. Отд. Ленинград.

SYNOPTICAL ARRANGEMENT OF THE FAMILIES OF CORNALES

1 Ovules solitary in each of the 1-several locules, or some locules empty; flowers not in aments.
  2 Leaves strictly alternate; stamens often more numerous than the petals; embryo fairly large, about as long as the endosperm.
    3 Petals valvate; flowers perfect or rarely unisexual; plants with articulated laticifers and isoquinoline (emetine) alkaloids, but without iridoid compounds; stone of the drupe not opening by valves; vessel-segments in most species with simple perforations .... 1. ALANGIACEAE.
    3 Petals (when present) imbricate; flowers of 2 sorts, one sort staminate, the other sort pistillate or perfect; plants without laticifers and without isoquinoline alkaloids, but at least sometimes producing iridoid compounds; stone of the drupe opening by subapical, abaxial valves at the time of germination; vessel-segments with scalariform perforations .... 2. NYSSACEAE.
  2 Leaves opposite or less often alternate; stamens isomerous with the petals, these valvate or seldom imbricate; embryo from very small to as large as the preceding group; flowers perfect or sometimes unisexual; plants commonly producing iridoid compounds, but without laticifers and without isoquinoline alkaloids .... 3. CORNACEAE.
1 Ovules two in the single locule of the ovary; flowers in unisexual aments; leaves opposite; plants producing iridoid compounds .... 4. GARRYACEAE.

1. Family ALANGIACEAE A. P. de Candolle 1828 nom. conserv., the Alangium Family

Trees, or less often shrubs or woody vines, sometimes thorny, often with stellate or peltate hairs or with other sorts of hairs, tanniferous, but apparently lacking both proanthocyanins and ellagic acid, not cyanogenic, but producing triterpenoid saponins or similar compounds and commonly also isoquinoline alkaloids of the emetine group; various types of solitary and clustered crystals of calcium oxalate commonly present in some of the cells of the parenchymatous tissues; articulated laticifers present in various tissues of the shoot; nodes trilacunar; vessel-segments mostly with simple, oblique perforations, or sometimes with scalariform perforations and numerous cross-bars; imperforate tracheary elements with simple pits; wood-rays heterocellular, mixed uniseriate and pluriseriate, the latter up to 8 cells wide, with elongate ends; wood-parenchyma apotracheal, diffuse or to some extent in uniseriate lines. LEAVES alternate, simple, entire or lobed, pinnately or palmately veined, often some of the cells of the mesophyll containing cluster-crystals and appearing as translucent dots; stomates usually anomocytic; stipules none. FLOWERS in axillary cymes, perfect or rarely unisexual, epigynous; calyx-lobes or -teeth 4–10, or obsolete; petals 4–10, more or less linear, valvate, hairy within, sometimes connate at the base, reflexed after anthesis; stamens as many as and alternate with the petals, or ca. 2–4 times as many as the petals, in either case arranged in a single cycle around a prominent, pulvinate, epigynous disk; filaments hairy within, distinct or slightly connate at the base, sometimes adnate to the petals at the base; anthers tetrasporangiate, dithecal or tetrathecal, opening by longitudinal slits; pollen-grains binucleate, (2) 3–4 (–8)-colporate or -porate; gynoecium of 2 carpels united to form a compound, bilocular ovary that lacks axile vascularization, or more often pseudomonomerous, with a single locule; style terminal, with a capitate and entire to 2–4-lobed stigma, or the proper style short and with 2 elongate, ventrally stigmatic branches; ovules solitary and pendulous in each locule (or one locule empty), anatropous, unitegmic, crassinucellar, with an integumentary tapetum; endosperm-development cellular or perhaps sometimes nuclear. FRUIT a drupe, crowned by the sepals and disk, containing a single unilocular or less often bilocular stone, each locule with 1 seed, or one locule empty and sometimes more or less reduced; seeds with large, straight embryo, 2 foliaceous cotyledons, and an elongate hypocotyl; endosperm copious, oily and also with reserves of hemicellulose. X = 8, 11.

The family Alangiaceae consists of the single diversified genus *Alangium,* with about 20 species, native to eastern and tropical Asia, eastern Australia, the Pacific islands, Madagascar, and western Africa. The family is distinctive within the Cornales in its alkaloids and in having laticifers, but its flowers and fruits are harmonious with the order, and it has the otherwise unusual combination of unitegmic, crassinucellar ovules found in most other members of the group. No other order can easily accommodate the Alangiaceae, and I see no reason why they should not be retained in the Cornales.

Fossil pollen referred to *Alangium* dates from lower Eocene deposits in Europe. Identifiable fruits are a little more recent, dating from the Oligocene.

Alangium villosum (Blume) Wangerin has domatium-like pits scattered along (but not on) the primary lateral veins of the leaves.

2. Family NYSSACEAE Dumortier 1829 nom. conserv., the Sour Gum Family

Trees or less often shrubs, tanniferous and commonly with ellagic and gallic acids, but without proanthocyanins, at least sometimes with iridoid compounds, often (*Nyssa* and to a lesser extent *Camptotheca,* but not *Davidia*) accumulating cobalt, not cyanogenic; solitary or clustered crystals of calcium oxalate present in some of the cells of the parenchymatous tissues; nodes trilacunar; vessel-segments elongate and slender, with elongate perforation-plates that have numerous (25–100) cross-bars; imperforate tracheary elements with bordered pits, or a few of them with simple pits; wood-rays heterocellular, variously Kribs type I or IIA or B, mixed uniseriate and pluriseriate, the latter mostly 2–4 cells wide, with short or elongate ends; wood-parenchyma sparse, diffuse or partly vasicentric. LEAVES alternate, simple, entire or merely toothed, pinnately or pinnipalmately veined, deciduous; epidermis mucilaginous; stomates paracytic; sclereids often present in the mesophyll; stipules wanting. FLOWERS small, regular or nearly so, some staminate, others perfect or functionally pistillate, borne on the same or different individuals, often in different inflorescences, these consisting of pedunculate heads, compact umbels, or short racemes, or the

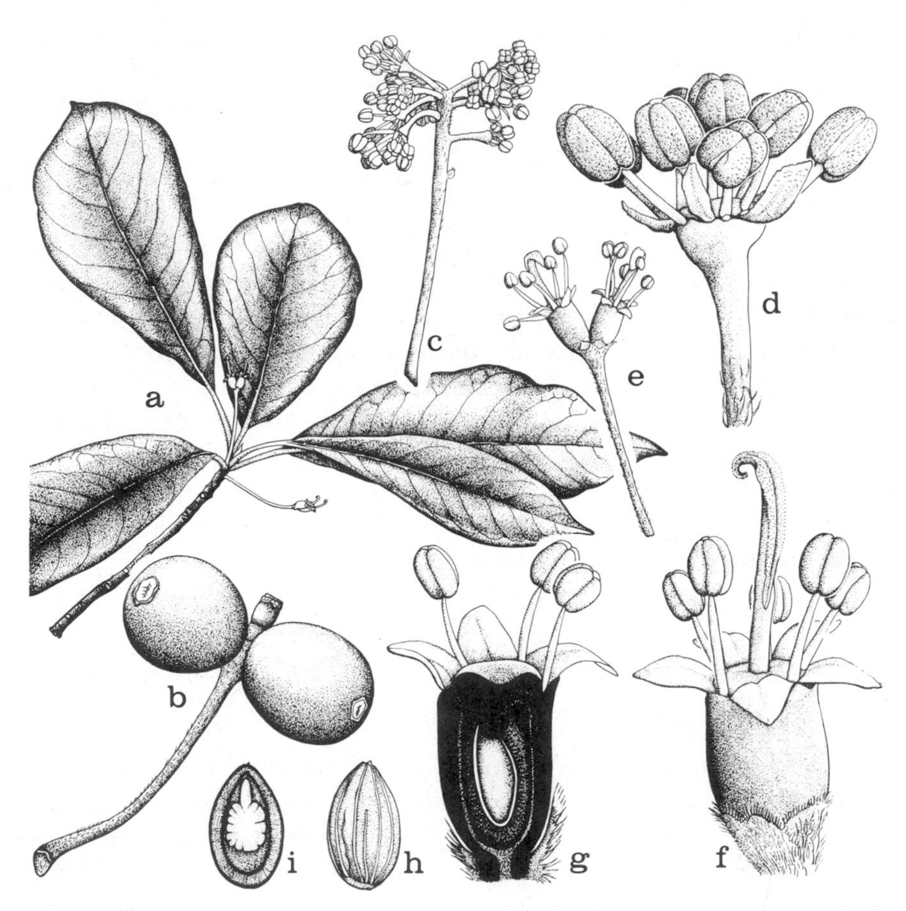

FIG. 5.28 Nyssaceae. *Nyssa sylvatica* Marsh. a, habit, ×½; b, fruits, ×2; c, staminate inflorescence, ×2; d, staminate flower, ×8; e, pair of perfect flowers, ×2; f, perfect flower, ×8; g, perfect flower, the ovary in long-section, ×8; h, i, stone, external and in long-section, ×2.

inflorescence sometimes reduced to a single flower; in *Davidia* the heads pseudanthial and subtended by a pair of large, petaloid bracts; perianth in *Davidia* wanting; in the other genera the calyx represented by a minute, 5-toothed or irregularly toothed rim, or virtually obsolete, and the corolla represented by 5 (–10) small, imbricate petals; a well developed nectary disk present in all flowers in *Nyssa* and *Camptotheca,* but not in *Davidia;* stamens in staminate flowers of *Nyssa* and *Camptotheca* 8–15, most commonly 10 and bicyclic, with the outer cycle antepetalous; in perfect flowers the stamens usually as many as and alternate with the petals, sometimes not fully functional; in *Davidia* the stamens (1–) 5–6

(−12); anthers tetrasporangiate and dithecal, opening by longitudinal slits; pollen grains binucleate, tricolporate; gynoecium in *Nyssa* typically pseudomonomerous and unilocular, with a single style, but sometimes bicarpellate and bilocular, with 2 basally united styles, in *Camptotheca* unilocular but the style with 2 or 3 branches, and in *Davidia* with 6–9 carpels, locules, and style-branches; ovules solitary in each locule, apical-parietal or apical-axile, pendulous, apotropous, anatropous or hemitropous, unitegmic, tenuinucellar or (*Davidia*) more or less crassinucellar, with an integmumentary tapetum; endosperm-development nuclear or (*Davidia*, possibly *Nyssa*) cellular. FRUIT drupaceous, with a single stone that has 1-several locules, each locule opening apically by a triangular abaxial valve at the time of germination; in *Camptotheca* the endocarp thin and hardly stony; seeds solitary in each locule, or some locules empty; embryo rather large, straight, with 2 cotyledons; endosperm moderately copious, oily and also with reserves of hemicellulose. X = 21, 22. (Davidiaceae)

The family Nyssaceae consists of only 3 genera and 7 or 8 species. *Nyssa*, the sour gum, has 5 or 6 species, with a bicentric distribution in eastern North America and eastern Asia (and the adjacent Pacific islands). *Davidia*, the dove-tree, and *Camptotheca* are monotypic Chinese genera. *Davidia* has sometimes been considered to form a distinct family, and this treatment can be defended. On the other hand, although *Davidia* is obviously distinctive, its relationship to *Nyssa* is evident. The same considerations that lead me to define the Cornaceae rather broadly also lead me to include *Davidia* in the Nyssaceae. Recent serological data support a relationship of *Davidia* to the Nyssaceae, but somewhat removed from the other 2 genera. The seed oils of *Davidia* also support a relationship to *Nyssa*.

The Nyssaceae appear to have been widely distributed on the northern (Laurasian) continents from early Eocene until the beginning of the Pleistocene. Fossil fruits of Nyssaceae can be fairly reliably identified because of the characteristic germination valve(s) in the stone. They can be found in numerous deposits dating back to the lower Eocene. Many of the older fossils, in particular, have a 3–4-locular stone. What appear to be nyssaceous leaves have also been found, sometimes in association with the fruits, in deposits from the Eocene to the present. Wood considered to represent *Nyssa* has been found in Eocene deposits in Yellowstone National Park, U.S.A. Nyssaceous pollen has also been collected in deposits as old as the Paleocene. None of the Cretaceous fossils that have been ascribed to the Nyssaceae is securely identified.

3. Family CORNACEAE Dumortier 1829 nom. conserv., the Dogwood Family

Trees, shrubs, or rarely herbs with woody rhizomes, sometimes with calcified trichomes, sometimes (as in *Mastixia*) accumulating aluminum, often producing isokestose and sometimes inulin, commonly with scattered secretory cells in some of the parenchymatous tissues, these at least sometimes tanniferous, containing ellagic acid and/or proanthocyanins and sometimes also gallic acid, but the plants also often lacking all of these substances, commonly (though not in *Helwingia*) producing iridoid substances, notably aucubin, but not cyanogenic and only seldom producing saponins; well developed secretory canals present in the cortex and sometimes also the pith, in *Mastixia* only; crystals of calcium oxalate commonly present in some of the cells of the parenchymatous

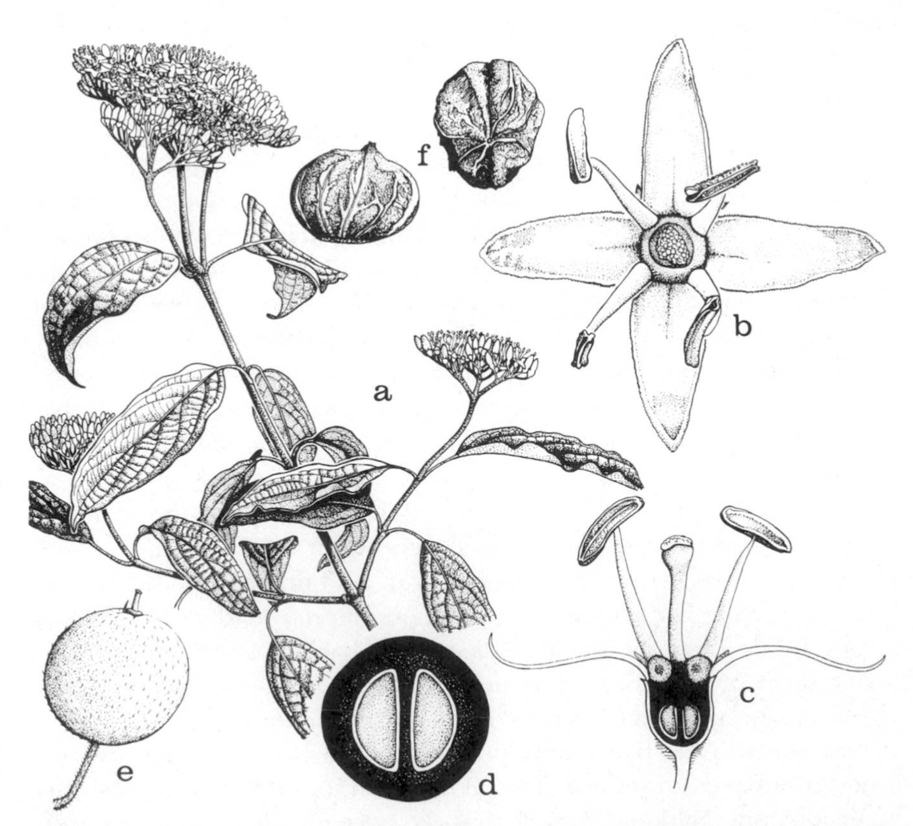

FIG. 5.29 Cornaceae. *Cornus stolonifera* Michx. a, habit, ×½; b, flower, from above, ×6; c, flower, in partial long-section, ×6; d, schematic cross-section of ovary, ×18; e, fruit, ×3; f, stone, ×3.

tissues, either singly, or in clusters, or as crystal sand; nodes mostly trilacunar (multilacunar in *Aralidium* and spp. of *Griselinia*); vessel-segments with scalariform perforations; imperforate tracheary elements with bordered or less often simple pits; wood-rays heterocellular, mixed uniseriate and pluriseriate, the latter 3–8 cells wide, mostly with elongate ends; wood-parenchyma mostly apotracheal and diffuse, often sparse or sometimes even wanting, paratracheal in *Aralidium,* scanty-paratracheal in *Aucuba, Griselinia* and *Mastixia;* cortex generally collenchymatous, at least toward the periphery, except in *Mastixia,* which has cortical vascular bundles. LEAVES opposite or less commonly alternate, simple, entire or sometimes toothed (deeply pinnatifid in *Aralidium*), pinnately or seldom palmately veined; upper epidermis sometimes mucilaginous; stomates mostly anomocytic (anisocytic in *Aralidium*); stipules wanting except in *Helwingia.* FLOWERS perfect or less often unisexual, regular, epigynous, individually small, borne in various types of often cymose or cymose-capitate inflorescences (a raceme in *Melanophylla*), the latter sometimes subtended by a whorl of large, petaloid bracts, the compact inflorescence of *Helwingia* sessile on the upper surface of a leaf; calyx in perfect and pistillate flowers mostly represented by 4, less often 5 (–7) small teeth around the summit of the ovary, or obsolete, but in staminate flowers at least sometimes forming a lobed tube; petals 4 or less often 5, alternate with the sepals, valvate or seldom (*Griselinia* and *Melanophylla*) imbricate, wanting from the pistillate flowers; stamens as many as and alternate with the petals, generally attached to or around the edge of an epigynous, often lobulate disk; anthers tetrasporangiate and dithecal, opening by longitudinal slits; pollen-grains binucleate, (2) 3 (–6)-colporate; gynoecium of 2–4 (5) carpels united to form a compound, inferior ovary with as many locules as carpels (only one locule ovuliferous in *Aralidium* and *Griselinia*), or the ovary seldom unilocular and pseudomonomerous; axis of the ovary without vascular bundles, except in *Curtisia, Helwingia,* and *Corokia*; style terminal, with a capitate or lobed stigma, or the styles sometimes united only toward the base or wholly distinct; stigmas dry, not papillate; ovules solitary in each locule, apical, pendulous, apotropous or seldom (*Curtisia, Mastixia*) epitropous, unitegmic, crassinucellar, with an integumentary tapetum; embryo-sac monosporic or sometimes tetrasporic; endosperm-development cellular. FRUIT most commonly a drupe with a single 1–5 locular, longitudinally grooved stone that has 1 seed in each locule (in *Kaliphora* a drupe with 2 pyrenes), less often (*Aucuba* and *Griselinia*) a berry; seeds with small to often elongate, spatulate, dicotyledonous embryo axially embedded in the copious, oily endosperm; petroselinic acid found only in *Aucuba*. X = 8–13, 19. (Aralidiaceae, Aucubaceae, Curtisiaceae, Griseliniaceae, Helwingiaceae, Mastixiaceae, Melanophyllaceae, Toricelliaceae)

The family Cornaceae as here described consists of 11 genera and about 100 species, widespread in North Temperate regions, and irregularly distributed in tropical and South Temperate regions. Species of *Cornus* and *Aucuba* are often cultivated as ornamentals.

By far the largest genus of the family, with about 50 species, is *Cornus*, here broadly defined to include 8 sections or subgenera. Each of these 8 groups has sometimes been treated as a distinct genus, but their mutual affinity is widely recognized.

Each of the 11 genera of the Cornaceae has sometimes been considered to form a monotypic family, except for *Kaliphora* and *Corokia;* the latter has been booted back and forth between the Cornaceae and the Grossulariaceae sens. lat., and serological data do emphasize its separation from other Cornaceae. Each of the segregate families has in fact some justification, but it seems pointless to fragment the larger group into a series of monotypic families if these are all to be kept in the same order. In my opinion the purposes of classification are better served by defining the Cornaceae broadly.

Leaves attributed to *Cornus* occur in Eocene rocks of North Dakota, and in more recent deposits.

4. Family GARRYACEAE Lindley 1834 nom. conserv., the Silk Tassel Family

Dioecious evergreen shrubs or trees with numerous crown-sprouts, usually provided with various sorts of unicellular trichomes that have ridges and furrows in a counterclockwise orientation (glabrous in *G. glaberrima* Wangerin), often with crystal sand in some of the cells of the parenchymatous tissues, lacking tannins, ellagic acid, and proanthocyanins, and only seldom cyanogenic, but commonly producing iridoid substances (notably aucubin) and highly toxic diterpenoid alkaloids; nodes trilacunar; vessel-segments with oblique, scalariform perforations, the cross-bars relatively few, (1–) 4–7 (–13); imperforate tracheary elements with bordered pits; early wood often with vasicentric tracheids; wood-rays heterocellular, Kribs types II A and B, mixed uniseriate and pluriseriate, the larger ones up to 8 (–14) cells wide, with short ends; wood-parenchyma commonly diffuse or diffuse-in-aggregates. LEAVES

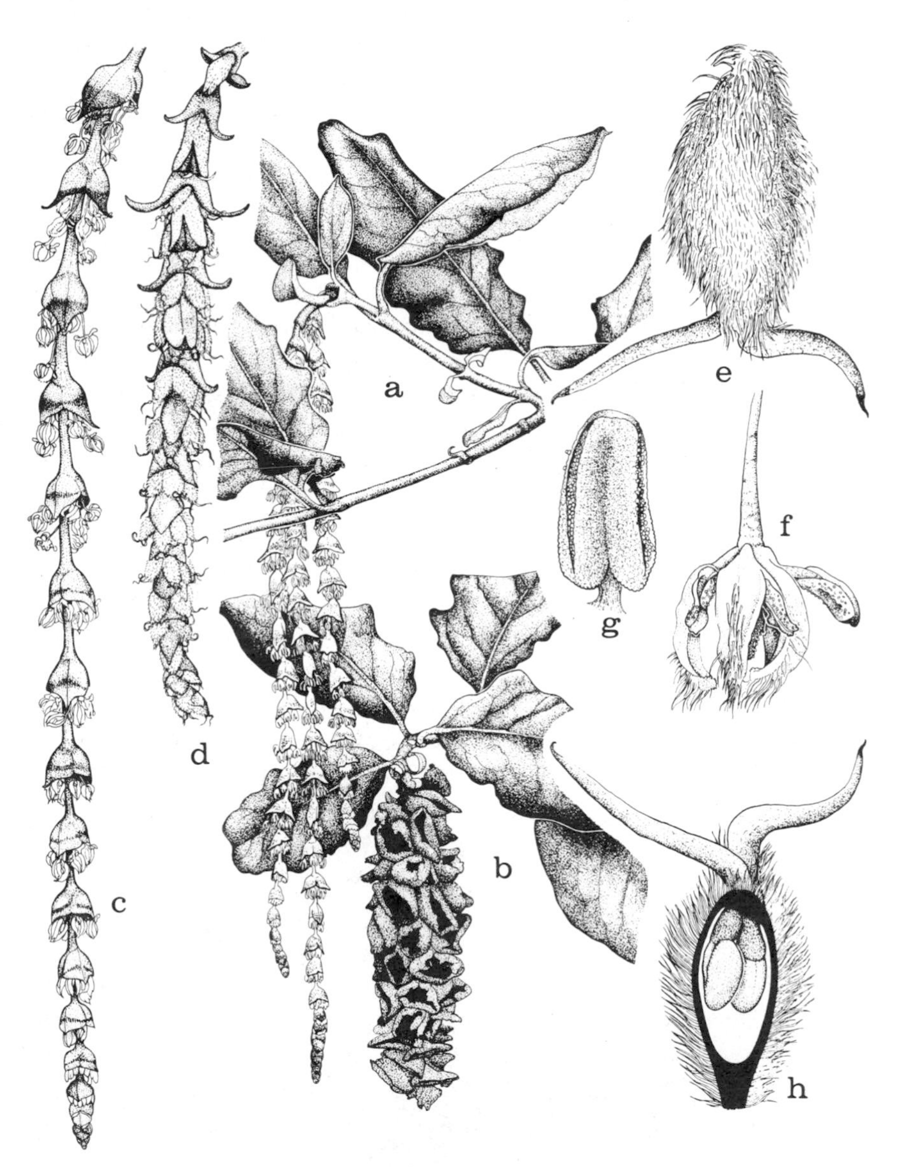

FIG. 5.30 Garryaceae. *Garrya elliptica* Dougl. a, habit, in flower, $\times \frac{1}{2}$; b, habit, in fruit, $\times \frac{1}{2}$; c, staminate catkin, $\times 1$; d, pistillate catkin, $\times 1$; e, pistillate flower, $\times 6$; f, staminate flower, $\times 6$; g, stamen, $\times 12$; h, pistil, the ovary in long-section, $\times 6$.

decussately opposite (but the vascular traces of a pair originating at different levels), simple, entire or subentire, pinnately veined, leathery, evergreen, commonly with sclereids in the mesophyll; stomates paracytic; petioles basally connate; stipules wanting. FLOWERS small, anemophilous, borne singly or 2–3 together in the axils of opposite, decussate bracts in silky catkins (slender, sometimes branched racemes) that are usually pendulous; staminate flowers with a perianth of 4 slender, usually apically connate, bract-like tepals; stamens 4, alternate with the tepals; filaments distinct; anthers basifixed, tetrasporangiate and dithecal, opening by longitudinal slits; pollen-grains binucleate, 3 (–15)-colporate, the exine with a reticulate pattern; pistillate flowers with a vestigial perianth of 2 small appendages near the base of the styles, or nude; gynoecium of 2 (3) carpels united to form a compound, unilocular ovary with distinct, spreading styles; stigmas dry, not papillate; ovules 2 (3), pendulous from the wall of the upper part of the locule, anatropous, with the micropyle toward the placenta, unitegmic, crassinucellar; endosperm-development nuclear; placenta producing an irregular obturator that fills much of the upper part of the locule. FRUIT a long-persistent, 2-seeded berry, becoming dry and thin-walled at maturity, but indehiscent or only tardily dehiscent, crowned by the persistent styles; seed solitary, or sometimes 2; embryo small, linear, dicotyledonous, basal to basal-axile; endosperm abundant, softly fleshy, oily (with a high percentage of petroselinic acid), and with reserves of hemicellulose. X = 11.

The family Garryaceae consists of the single genus *Garrya,* with 13 species native to western North America, from Washington to Panama, plus an outlying species in the Greater Antilles. Several species are occasionally planted as ornamental shurbs.

The affinities of *Garrya* have been disputed in the past. Recent opinion, based on morphological, anatomical, and chemical studies, is that it is allied to the Cornaceae. *Aucuba* and *Griselinia* have been suggested as the cornaceous genera most nearly allied to *Garrya.* Even the pollen of *Garrya* resembles that of *Aucuba,* and these two genera are also alike (and unique in the Cornales) in having petroselinic acid as the major fatty acid in the seed oils.

Fossils of *Garrya,* including leaves, seeds, and an inflorescence, are known from deposits in western United States of Miocene and more recent age. Thus the Garryaceae, which have the most reduced flowers in the order, are also the last to appear in the fossil record.

9. Order SANTALALES Lindley 1833

Plants sometimes autotrophic and arborescent, but usually hemiparasitic (with chlorophyll) or fully parasitic (without chlorophyll) and attaching to the roots or stems of other plants, the parasitic forms variously trees, shrubs, shrublets, or herbs, sometimes fungoid in aspect; plants often tanniferous, sometimes producing proanthocyanins, but not known to produce ellagic acid, sometimes cyanogenic, or saponiferous, or producing polyacetylenes, but without iridoid compounds; laticifers sometimes present; silicified cells (or cells containing siliceous cystoliths) often scattered singly or in groups in the mesophyll and frequently also in the parenchymatous tissues of the stem; vessel-segments with simple or scalariform perforations, or vessels wanting, the vascular tissue often more or less reduced. LEAVES variously alternate, opposite, or sometimes whorled, simple and generally entire, often evergreen, or often reduced to mere scales; small stipules present in Medusandraceae and Dipentodontaceae, the other families exstipulate. FLOWERS from fairly large to exceedingly small, perfect or unisexual, hypogynous to epigynous, regular or seldom somewhat irregular, dichlamydeous (but the calyx often much reduced) to strictly monochlamydeous (the single set of tepals probably representing the petals), or in the most reduced forms the flowers naked; petals (or tepals) commonly valvate (or open), distinct or connate below; stamens commonly isomerous and opposite the petals (alternate in Dipentodontaceae), or sometimes 2–5 times as many but still uniseriate, or reduced in number or variously modified or coalescent; pollen-grains binucleate or sometimes trinucleate, triaperturate or of various triaperturate-derived types; gynoecium of 2–5 carpels united to form a compound, superior to inferior ovary with a single style or sometimes with separate styles, seldom pseudomonomerous, generally unilocular or partitioned only at the base into as many semilocules as carpels, rarely completely partitioned, or sometimes appearing solid, without an evident locule; placentation various, but most often either free-central or basal in a unilocular ovary, or with the ovules pendulous from a columnar placenta that extends upwards beyond the basal partial partitions, the ovules then commonly cupped at the base in the semilocules; ovules mostly few (1–8, or up to 12 in Loranthaceae), often one in each locule or semilocule, from well developed and bitegmic to usually more or less reduced, and then unitegmic or often not divided into nucellus and integument, sometimes even embedded in and hardly differentiated from the placenta and thus seeming to consist essentially of an embryo-sac; embryology often complex and unusual; endosperm-development cellular or sometimes helobial. FRUIT capsular in 2 small peripheral families, otherwise indehiscent and usually fleshy or nutlike, often a berry or drupe; seed solitary, or rarely 2 or 3, frequently surrounded or cupped by viscid tissue, ovary often lacking a differentiated testa; endosperm usually well developed; embryo sometimes of ordinary proportions and with 2 or 3–6 cotyledons, or often small and poorly differentiated. (Balanophorales, Olacales)

The order Santalales as here defined consists of 10 families and about 2000 species. The Medusandraceae and Dipentodontaceae, with a single species each, are somewhat doubtfully included in the order, and the position of the small family Balanophoraceae has been debated. The remaining 7 families, making up the great bulk of the order, are generally considered to form a highly natural group showing progressive adaptation to parasitism. The largest family is the Loranthaceae (900), followed by the Santalaceae (400), Viscaceae (300), and Olacaceae (250). None of the other families has as many as a hundred species.

Onotogenetic and anatomical studies suggest that the inferior ovary in members of the Santalales is at least partly receptacular rather than appendicular; i.e., the ovary is sunken into the receptacle.

The intrepretation of the fossil record of the order is in a state of flux. Macrofossils referred to the Olacaceae and Santalaceae date from the Eocene and Oligocene, respectively. Jarzen (1977) proposes that "the fossil pollen form *Aquilapollenites* evolved from a Santalalean stock during the upper part of the late Cretaceous, along primarily two lines. One line proceeded through the isopolar forms of *Aquilapollenites,* some obviously becoming extinct, others following a trend through lorantha-ceous development. The second line developed the subisopolar and heteropolar forms . . . leading to the *Arjona* form of the Santalaceae and also many extinct forms." Other authors have considered that pollen representing the Olacaceae (indeed the modern genus *Anacolosa*) enters the fossil record in the Maestrichtian, the Santalaceae (*Santalum*) in the late Paleocene, the Loranthaceae in the lower Eocene, and the Balanophoraceae in the Miocene.

Thus the fossil record as now interpreted is at least compatible with, and may give some support to, the view that the Olacaceae are basal to most of the other families of the order, and that the Balanophoraceae are relatively late comers to the scene.

SELECTED REFERENCES

Agarwal, S. 1963. Morphological and embryological studies in the family Olacaceae. I. *Olax* L. II. *Strombosia* Blume. Phytomorphology 13: 185–196; 348–356.

Arekal, G. D., & G. R. Shivamurthy. 1976. "Seed" germination in *Balanophora abbreviata*. Phytomorphology 26: 135–138.

Barlow, B. A. 1964. Classification of the Loranthaceae and Viscaceae. Proc. Linn. Soc. New South Wales II, 89: 268–272.

Barlow, B. A., & D. Wiens, 1971. The cytogeography of the loranthaceous mistletoes. Taxon 20: 291–312.

Barlow, B. A., & D. Wiens. 1973. The classification of the generic segregates of *Phrygilanthus* (= *Notanthera*) of the Loranthaceae. Brittonia 25: 26–39.

Bhatnagar, S. P. 1960. Morphological and embryological studies in the family Santalaceae. IV. *Mida salicifolia* A. Cunn. Phytomorphology 10: 198–207.

Bhatnagar, S. P., & S. Agarwal. 1961. Morphological and embryological studies in the family Santalaceae. VI. *Thesium* L. Phytomorphology 11: 273–282.

Bhatnagar, S. P., & P. C. Joshi. 1965. Morphological and embryological studies in the family Santalaceae. VII. *Exocarpus bidwellii* Hook. f. Proc. Natl. Inst. Sci. India 31B: 34–44.

Brenan, J. P. M. 1952. Plants of the Cambridge Expedition 1947–1948. II. A new order of flowering plants from the British Cameroons. Kew Bull. 1952: 227–236.

Cohen, L. I. 1968. Development of the staminate flower in the dwarf mistletoe, *Arceuthobium*. Amer. J. Bot. 55: 187–193.

Cohen, L. I. 1970. The development of the pistillate flower in the dwarf mistletoes, *Arceuthobium*. Amer. J. Bot. 57: 477–485.

Danser, B. H. 1933. A new system for the genera of Loranthaceae Loranthoideae, with a nomenclator for the Old World species of this subfamily. Verh. Kon. Ned. Akad. Wetensch. Afd. Natuurk., Tweede Sect. 29(6): 1–128.

DeFilipps, R. 1969. Parasitism in *Ximenia* (Olacaceae). Rhodora 71: 439–443.

Dixit, S. N. 1958, 1961. Morphological and embryological studies in the family Loranthaceae. IV. *Amyema* Van-Tiegh. V. *Lepeostegeres gemmiflorus* (Bl.) Bl. VIII. *Tolypanthus* Bl. Phytomorphology 8: 346–364; 365–376, 1958. 11: 335–345, 1961.

Dixit, S. N. 1962. Rank of the subfamilies Loranthoideae and Viscoideae. Bull. Bot. Surv. India 4: 49–55.

Fagerlind, F. 1945. Blüte und Blütenstand der Gattung *Balanophora*. Bot. Not. 1945: 330–350.

Fagerlind, F. 1947. Gynöceummorphologische und embryologische Studien in der Familie Olacaceae. Bot. Not. 1947: 207–230.

Fagerlind, F. 1948. Beiträge zur Kenntnis der Gynäceummorphologie und Phylogenie der Santalales-Familien. Svensk Bot. Tidskr. 42: 195–229.

Fagerlind, F. 1959. Development and structure of the flower and gametophytes in the genus *Exocarpus*. Svensk Bot. Tidskr. 53: 257–282.

Hansen, B. 1972. The genus *Balanophora* J. R. & G. Forster. A taxonomic monograph. Dansk Bot. Ark. 28(1): 1–188.

Hansen, B. 1976. Pollen and stigma conditions in the Balanophoraceae sens. lat. Bot. Not. 129: 341–345.

Hansen, B. 1980. Balanophoraceae. Fl. Neotropica Monograph 23.

Hansen, B. & K. Engell. 1978. Inflorescences in Balanophoroideae, Lophophytoideae and Scybalioideae (Balanophoraceae). Bot. Tidsskr. 72: 177–187.

Hiepko, P. 1971. Die Gattungsabgrenzung bei den Opiliaceae. Ber. Deutsch. Bot. Ges. 84: 661–663.

Jarzen, D. M. 1977. *Aquilapollenites* and some santalalean genera. Grana 16: 29–39.

Johri, B. M., & others. 1957, 1969. Morphological and embryological studies in the family Loranthaceae. I. *Helicanthes elastica* (Desr.) Dans. Phytomorphology 7: 336–354, 1957. XII. *Moquiniella rubra* (Spreng.f.) Balle. Oesterr. Bot. Z. 116: 475–485, 1969.

Johri, B. M., & S. Agarwal. 1965. Morphological and embryological studies in the family Santalaceae. VIII. *Quinchamalium chilense* Lam. Phytomorphology 15: 360–372.

Johri, B. M., & S. P. Bhatnagar. 1960. Embryology and taxonomy of the Santalales. Proc. Natl. Inst. Sci. India 26B, Suppl.: 199–220.

Joshi, P. C. 1960. Morphological and embryological studies in the family Santalaceae. V. *Osyris wightiana* Wall. Phytomorphology 10: 239–248.

Kuijt, J. 1960. Morphological aspects of parasitism in the dwarf mistletoes (*Arceuthobium*). Univ. Calif. Publ. Bot. 30: 337–436.

Kuijt, J. 1968. Mutual affinities of Santalalean families. Brittonia 20: 136–147.

Kuijt, J. 1969. The biology of parasitic flowering plants. Univ. Calif. Press. Berkeley.

Kuijt, J., & F. Weberling. 1972. The flower of *Phthirusa pyrolifolia* (Loranthaceae). Ber. Deutsch. Bot. Ges. 85: 467–480.

Maguire, B., J. J. Wurdack, & Yung-chau Huang. 1974. Pollen grains of some American Olacaceae. Grana 14: 26–38.

Maheshwari, P., B. M. Johri, & S. N. Dixit. 1957. The floral morphology and embryology of the Loranthoideae (Loranthaceae). J. Madras Univ. 27B: 121–136.

Merrill, E. D. 1941. Dipentodontaceae. Pp. 69–73 *in:* Plants collected by Captain F. Kingdon Ward on the Vernay-Cutting Expedition, 1938–39. Brittonia 4: 20–188.

Metcalfe, C. R. 1952. *Medusandra richardsiana* Brenan. Anatomy of the leaf, stem and wood. Kew Bull. 1952: 237–244.

Narayana, R. 1958. Morphological and embryological studies in the family Loranthaceae—II. *Lysiana exocarpi* (Behr.) Van Tieghem. III. *Nuytsia floribunda* (Labill.) R. Br. Phytomorphology 8: 146–169; 306–323.

Paliwal, R. L. 1956. Morphological and embryological studies in some Santalaceae. Agra Univ. J. Res., Sci. 5: 193–284.

Piehl, M. A. 1965. The natural history and taxonomy of *Comandra* (Santalaceae). Mem. Torrey Bot. Club 22: 1–97.

Prakash, S. 1960, 1963. Morphological and embryological studies in the family Loranthaceae—VI. *Peraxilla tetrapetala* (Linn.f.) Van Tiegh. X. *Barathranthus axanthus* (Korth.) Miq. Phytomorphology 10: 224–234, 1960. 13: 97–103, 1963.

Ram, M. 1957, 1959. Morphological and embryological studies in the family Santalaceae. I. *Comandra umbellata* (L.) Nutt. II. *Exocarpus,* with a discussion on its systematic position. III. *Leptomeria* R. Br. Phytomorphology 7: 24–35, 1957. 9: 4–19; 20–33, 1959.

Rao, L. N. 1942. Studies in the Santalaceae. Ann. Bot. (London) II. 6: 151–175.

Reed, C. 1955. The comparative morphology of the Olacaceae, Opiliaceae and Octoknemaceae. Mem. Soc. Brot. 10: 29–79.

Schaeppi, H., & F. Steindl. 1942. Blütenmorphologische und embryologische Untersuchungen an Loranthoideen. Vierteljahrsschr. Naturf. Ges. Zürich 87: 301–372.
Schaeppi, H., & F. Steindl. 1945. Blütenmorphologische und embryologische Untersuchungen an einigen Viscoideen. Vierteljahrsschr. Naturf. Ges. Zürich 90, Beih. 1: 1–46.
Singh, V., & G. Ratnakar. 1974. Contribution to the floral anatomy of the Loranthaceae. I. Subfamily Loranthoideae. J. Indian Bot. Soc. 53: 162–169.
Smith, F. H., & E. C. Smith. 1943. Floral anatomy of the Santalaceae and some related forms. Oregon State Monogr. Stud. Bot. 5.
Stauffer, H. U. 1959, 1961. Santalales-Studien IV. Revisio Anthobolearum, morphologische Studie mit Einschluss der Geographie, Phylogenie und Taxonomie. Mitt. Bot. Mus. Univ. Zürich 213: 1–260, 1959. VIII. Zur Morphologie und Taxonomie der Olacaceae-Tribus Couleae. Vierteljahrsschr. Naturf. Ges. Zürich 106: 412–418.
Swamy, B. G. L. 1949. The comparative morphology of the Santalaceae: node, secondary xylem and pollen. Amer. J. Bot. 36: 661–673.
Werth, C. R., & W. V. Baird. 1979. Root parasitism in *Schoepfia* Schreb. (Olacaceae). Biotropica 11: 140–143.
Wiens, D. 1975. Chromosome numbers in African and Madagascan Loranthaceae and Viscaceae. J. Linn. Soc., Bot. 71: 295–310.
Wiens, D., & B. A. Barlow. 1971. The cytogeography and relationships of the viscaceous and eremolepidaceous mistletoes. Taxon 20: 313–332.
Yoshida, O., & H. Kawaguchi. 1971. Embryology of *Korthalsella japonica* (Thunb.) Engler. J. Coll. Arts Chiba Univ. B-4: 37–47.

Тамамшян, С. Г. 1958. Santalales в системе Spermatophyta. Пробл. Бот., Москва-Ленинград. 3: 67-97.
Терёхин, Э. С., & М. С. Яковлев. 1967. Эмбриология Balanophoraceae (к вопросу о гомологиях "цветка" *Balanophora*). Бот. Ж. 52: 745-758.
Терёхин, Э. С., М. С. Яковлев, & З. И. Никитичева. 1975. Развитие микроспорангия, пыльцевых зерен, семяпочки и зародышевого мешка у *Cynomorium songaricum* Rupr. (Cynomoriaceae). Бот. Ж. 60: 153-162.

SYNOPTICAL ARRANGEMENT OF THE FAMILIES OF SANTALALES

1 Fruit capsular; leaves with small stipules; plants autotrophic; ovules 6–8.
  2 Fertile stamens opposite the petals, alternating with well developed staminodes; petals and sepals well differentiated; flowers in slender, catkin-like racemes .... 1. MEDUSANDRACEAE.
  2 Fertile stamens alternate with the petals, and alternate with a set of nectary-glands; petals and sepals very much alike; flowers in globose umbels .... 2. DIPENTODONTACEAE.
1 Fruit indehiscent, often fleshy; leaves without stipules; plants partly or wholly parasitic, except some Olacaceae; ovules 1–5 in most families, up to 12 in Loranthaceae.
  3 Plants chlorophyllous and photosynthetic; shoot of fairly ordinary construction, with obvious nodes and internodes, the leaves well

developed or reduced to small scales; inflorescence with an ordinary (not an extremely large) number of flowers.

4 Plants terrestrial, (autotrophic or) attached to the roots of their host, except in a few Santalaceae; fruit without viscid tissue, except in some Santalaceae; ovules well differentiated from the placenta (except in *Exocarpos,* of the Santalaceae), variously bitegmic or unitegmic, or not differentiated into nucellus and integument.

5 Perianth mostly dichlamydeous; leaves alternate; ovary superior or less often inferior.

6 Ovary with partitions at the base (rarely throughout), thus partly (rarely wholly) 2–5-locular, with one ovule per locule or semilocule; leaves often with silicified cells borne singly or in groups in the mesophyll, and also often with spicular sclereids, but without a branching system of lignified cells connecting the veins .... 3. OLACACEAE.

6 Ovary strictly unilocular, with one ovule; leaves with a branching system of lignified cells connecting the veins, but without silicified cells or spicular sclereids .... 4. OPILIACEAE.

5 Perianth strictly monochlamydeous; leaves opposite or less commonly alternate; ovary inferior or less often superior .... 5. SANTALACEAE.

4 Plants aerial, attached to the branches of their host (terrestrial in a few Loranthaceae); fruit various, ovules not differentiated into nucellus and integument.

7 Ovules well differentiated from the placental column; fruit dry, without viscid tissues, airborne by 3 strongly accrescent, feathery staminodes; leaves alternate; inflorescence catkin-like .... 6. MISODENDRACEAE.

7 Ovules embedded in and scarcely differentiated from the large, free-central or basal placenta; fruit fleshy, the seed or stone usually surrounded or capped at one end by viscid tissue; leaves and inflorescence various.

8 Flowers perfect, only rarely unisexual, generally dichlamydeous, often large and showy; endosperm compound, usually lacking chlorophyll; ovules several, commonly 4–12; embryo-sac monosporic; leaves opposite or sometimes alternate .... 7. LORANTHACEAE.

8 Flowers strictly unisexual, generally small and inconspicous; perianth monochlamydeous; endosperm simple, chlorophyllous; ovules 2; embryo-sac bisporic.

9 Leaves opposite; inflorescence not catkin-like; placental column present .... 8. VISCACEAE.
9 Leaves alternate; inflorescence catkin-like; placental column absent, the ovules embedded in the base of the ovary .... 9. EREMOLEPIDACEAE.
3 Plants wholly without chlorophyll; shoot often fungoid in aspect; inflorescence massive, with numerous or very numerous, often minute flowers; plants terrestrial ....10. BALANOPHORACEAE.

1. Family MEDUSANDRACEAE Brenan 1952 nom. conserv., the Medusandra Family

Trees with scattered secretory canals (even in the xylem) containing a yellow liquid; vessel-segments elongate, with slanting, scalariform perforation-plates that have (11–) 19–24 (–43) cross-bars; imperforate tracheary elements with bordered pits; wood-rays heterocellular, mostly uniseriate; wood-parenchyma very sparse, paratracheal and diffuse. LEAVES alternate, apparently simple, remotely crenulate, pinnately veined, provided on the lower surface with hairs of a unique, thick-based type; stomates anomocytic; secretory canals associated with the veins; petiole with complex vascular anatomy, and with a swollen pulvinus at the tip, suggesting that the leaves may be unifoliolate rather than truly simple; stipules small, deciduous. FLOWERS perfect, regular, hypogynous, borne in axillary, solitary or paired, pendulous, shortly tomentose, ament-like racemes with tiny, caducous bracts; sepals distinct or connate at the base, persistent, more or less accrescent in fruit, open in bud; petals 5, distinct, small, imbricate; fertile stamens 5, opposite the petals, distinct; anthers tetrasporangiate and dithecal, opening by longitudinally recurving valves; pollen-grains tricolporate; staminodes 5, alternating with the fertile stamens, linear, elongating and much-surpassing the petals, very densely and shortly pubescent above the glabrous base, and with a vestigial anther at the tip; nectary-disk wanting; gynoecium of 3 (4) carpels united to form a compound, superior ovary with distinct, short styles that have a very small, terminal stigma; ovary unilocular, with a slender placental column arising from the base and joined to the top of the ovary; ovules 6 (–8), pendulous from the top of the placental column, anatropous, epitropous. FRUIT capsular, with a single large, pendulous seed; seeds with copious, slightly ruminate endosperm and a small, straight, dicotyledonous embryo.

The family Medusandraceae consists of the single genus and species *Medusandra richardsiana* Brenan, native to rain-forests of tropical west Africa.

Soyauxia, sometimes included in the Medusandraceae, is here referred to the Flacourtiaceae. It differs from *Medusandra* in its very numerous stamens, imbricate sepals, entire leaves without a pulvinus, annular nectary-disk, absence of secretory canals, and other features.

2. Family DIPENTODONTACEAE Merrill 1941 nom. conserv., the Dipentodon Family

Shrubs or small trees without secretory canals; vessel-segments elongate, with scalariform perforation-plates that have numerous (up to 20 or more) cross-bars; imperforate tracheary elements with bordered pits; wood rays homocellular, uniseriate, wood-parenchym wanting. LEAVES alternate, simple, toothed, pinnately veined; stomates anomocytic; stipules small, deciduous. FLOWERS small, perfect, regular, hypogynous, ca 25–30 in pedunculate, axillary, globose umbels that are at first subtended by an involucre of 4–5 small bracts; pedicels jointed at the middle; sepals 5–7, connate at the base, slender, valvate; petals essentially similar to the sepals and alternate with them, valvate; stamens as many as and alternate with the petals; filaments distinct; anthers tetrasporangiate and dithecal, opening by longitudinal slits; pollen tricolporate; nectary-glands (staminodes?) borne opposite the petals; gynoecium of 3 carpels united to form a compound, superior ovary topped by a simple style with a small terminal stigma; ovary essentially unilocular, but with partial partitions at the base; ovules 6, borne at the top of a columnar, free-central placenta. FRUIT a tardily deciduous capsule with a single seed borne on the thickened placenta.

The family Dipentodontaceae consists of the single species *Dipentodon sinicus* Dunn, native to southern China and adjacent Burma. The proper taxonomic placement of the family awaits detailed study. The gynoecium is very santalalean, and for the present it will do no harm to follow Schultze-Motel (in the current Engler Syllabus) and Takhtajan in referring the family to the Santalales. Hutchinson's view of the affinities of the family is similar, since he includes it in his order Olacales, a segregate from the Santalales.

3. Family OLACACEAE Mirbel ex A. P. de Candolle 1824 nom. conserv., the Olax Family

Terrestrial, mostly evergreen trees, shrubs (seldom half-shrubs) or woody vines, green and photosynthetic, but most members hemiparasitic, attaching to the roots of other plants; plants often cyanogenic and sometimes tanniferous, but only seldom saponiferous, and without iridoid compounds, sometimes with schizogenous resin-glands and/or branched, articulated or nonarticulated laticifers, and commonly with solitary or clustered crystals of calcium oxalate in some of the cells of

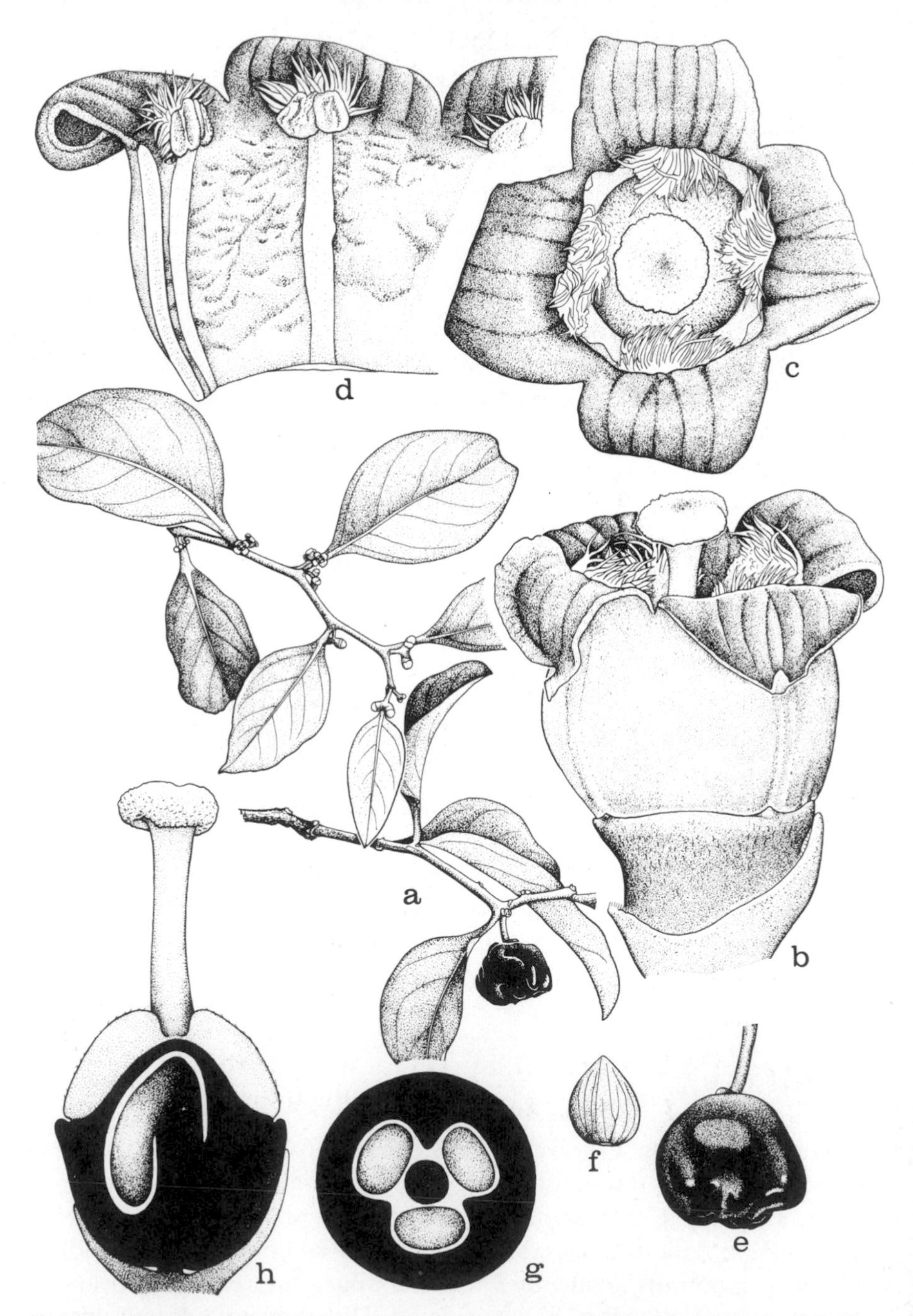

FIG. 5.31 Olacaceae. *Schoepfia schreberi* J. F. Gmelin. a, habit, ×½; b, c, flower, ×20; d, corolla, laid open to show stamens, ×20; e, fruit, ×2; f, seed, ×2; g, schematic cross-section of ovary, ×20; h, pistil, the ovary in long-section, ×20.

the parenchymatous tissues; nodes unilacunar or trilacunar, seldom pentalacunar; vessel-segments with simple or less often scalariform perforations, the latter with mostly 6–10 (seldom 20 or more) cross-bars; imperforate tracheary elements with few and simple pits, or less often with numerous, distinctly bordered pits; wood-rays heterocellular, mixed uniseriate and pluriseriate, or almost exclusively one or the other, the pluriseriate ones mostly 2–4 (–9) cells wide, short or elongate ends; wood-parenchyma commonly apotracheal, often diffuse. LEAVES alternate, simple, entire, pinnately veined, the mesophyll often with silicified cells borne singly or in groups, and also often with spicular sclereids; stomates of diverse types, sometimes paracytic; stipules wanting. FLOWERS mostly rather small, in axillary clusters that may be elaborated into racemes or panicles, regular, perfect or rarely unisexual (the plants then dioecious), hypogynous or semi-epigynous to seldom epigynous or perigynous; calyx small, generally more or less cupular and inconspicuously 3–6-toothed (or the teeth seldom obsolete), often accrescent in fruit; petals 3–6, alternate with the calyx teeth, valvate or rarely imbricate, distinct or connate below, seldom (as in spp. of *Schoepfia*) forming a long tube with short lobes; nectary-disk intrastaminal, surrounding the ovary (which is often more or less sunken into it), or less often surronding the style atop the ovary, or extrastaminal and annular or consisting of glands alternate with the petals; stamens as many as and opposite the petals, or 2–5 times as many, but apparently all in a single cycle, sometimes some of them staminodial; filaments free and distinct, or sometimes adnate to the base of the corolla (or adnate throughout much of its length) or connate into a sheath around the style; anthers tetrasporangiate and dithecal, opening by longitudinal slits or seldom by terminal valves; pollen-grains 2–3-nucleate, of diverse forms, variously tricolpate, tricolporate, or 3–8-porate; gynoecium of (2) 3 (–5) carpels united to form a compound, superior to inferior ovary with a terminal style and 2–5-lobed stigma; ovary 2–5-locular at the base, but mostly unilocular above, the partitions rarely reaching the top of the ovarian cavity; ovules solitary in each locule or semilocule, pendulous from the top of the free-central (or partly or wholly axile) placenta (in *Octoknema* the placenta reaching and joined to the top of the essentially unilocular ovary, as in *Medusandra*), anatropous or seldom orthotropous, bitegmic or unitegmic and tenuinucellar or nearly so, or not divided into nucellus and integuments; embryo-sac monosporic, or seldom bisporic, in either case 8-nucleate; endosperm-development cellular or helobial; embryology of diverse, often unusual types. FRUIT a drupe or a nut, almost always 1-seeded, often included in an accrescent calyx; seeds with thin testa and small or tiny embryo near the tip of the copious, oily and sometimes also starchy endosperm; cotyledons 2–6. X = 19, 20. (Aptandraceae, Cathedraceae, Chaunochitonaceae, Coulaceae, Erythropalaceae, Heisteriaceae, Octoknemaceae, Schoepfiaceae, Scorodocarpaceae, Strombosiaceae, Tetrastylidiaceae)

The family Olacaceae as here rather broadly defined consists of some 25 or 30 genera and 250 species, widespread in tropical and subtropical regions. The largest genera are *Olax*, with about 50 species, and *Schoepfia*, with about 40. The familiar pantropical genus *Ximenia* has about 15 species. *Ximenia americana* L. is known for its valuable wood, edible fruits, and oil-rich seeds. The considerable morphological diversity within the Olacaceae is reflected in the large number of segregate families that have been proposed, but the mutual affinities among the several more narrowly defined groups are generally conceded.

4. Family OPILIACEAE Valeton 1886 nom. conserv., the Opilia Family

Terrestrial, mostly evergreen trees or shrubs, sometimes climbing; plants green and photosynthetic, but usually or always hemiparasitic, attaching to the roots of other plants; parenchymatous tissues of the shoot, even of the secondary wood, containing siliceous cystoliths in scattered specialized cells or groups of cells, but crystals otherwise wanting; vessel-segments with simple perforations; imperforate tracheary elements with bordered or less often simple pits; wood-rays homocellular, up to 5 cells wide, with very few uniseriates; wood-parenchyma apotracheal, commonly diffuse. LEAVES alternate, simple, entire, pinnately veined, with a branching system of lignified cells connecting the veins of the mesophyll, sometimes also with scattered secretory cells, these at least sometimes mucilaginous; stomates paracytic; stipules wanting. Inflorescence of axillary or cauliflorous spikes, racemes, umbels, or panicles, sometimes catkin-like, with the flowers individually subtended by caducous bracts; FLOWERS small, 4–5-merous, mostly perfect, but unisexual in *Agonandra* and *Gjellerupia*, and the plants then dioecious; calyx small and inconspicuous, cupular and entire-margined or with 4–5 small lobes or teeth, not accrescent, sometimes virtually obsolete; petals in perfect and staminate flowers 4 or 5, distinct or connate below, seldom connate for more than half their length, valvate, in pistillate flowers the petals usually wanting; stamens as many as and opposite the petals, free and distinct, or the filaments borne on the petals or on the corolla-tube; anthers tetrasporangiate and dithecal, opening by longitudinal slits; pollen-grains binucleate, tricolporate; intrastaminal nectary-disk consisting of distinct or more or less connate nectaries alternating with the stamens; gynoecium of 2–5 carpels united to form a compound ovary topped by a simple style and a small or capitately expanded stigma, or the stigma sessile; ovary superior, or half-sunken in the disk, strictly unilocular, with a single ovule pendulous from the summit of the columnar free-central placenta, or in *Agonandra* the ovule basal and erect; ovule anatropous, unitegmic and tenuinucellar, or not divided into integument and nucellus; endosperm-development cellular; embryology often complex and unusual. FRUIT a drupe; seeds with rather small embryo and abundant, oily and starchy endosperm; cotyledons (2) 3–4. N = 10 in the only species so-far counted. (Cansjeraceae)

The family Opiliaceae consists of about 9 genera and less than 50 species, widespread in tropical and subtropical regions. The three largest genera are *Opilia, Rhopalopilia*, and *Agonandra*, each with about 10 species. *Opilia* is widespread in the Old World, *Rhopalopilia* is strictly African, and *Agonandra*, the only American genus, is confined to the New World. *Agonandra* is unique in the family in its basal ovule, and is also unusual in having unisexual flowers.

Although the Opiliaceae have often in the past been included in the Olacaceae, most present-day authors consider the two groups to form distinct families. The clear anatomical distinction between the two groups reinforces the difference in the gynoecium.

5. Family SANTALACEAE R. Brown 1810 nom. conserv., the Sandalwood Family

Terrestrial small trees, shrubs, or perennial herbs, green and photosynthetic, but hemiparasitic, attaching to the roots of other plants, or seldom (notably in *Dendrotrophe*) aerial and attached to the branches of the host (and then sometimes with a diffusely branched haustorium as in the Viscaceae), sometimes xeromorphic, sometimes thorny, commonly accumulating polyacetylenes (with acetylenic fatty acids in the seeds and vegetative organs), sometimes tanniferous (with proanthocyanins) but without ellagic acid and without iridoid compounds, not saponiferous and not cyanogenic, often with groups of silicified cells in the parenchymatous tissues of the shoot, and with solitary or clustered crystals of

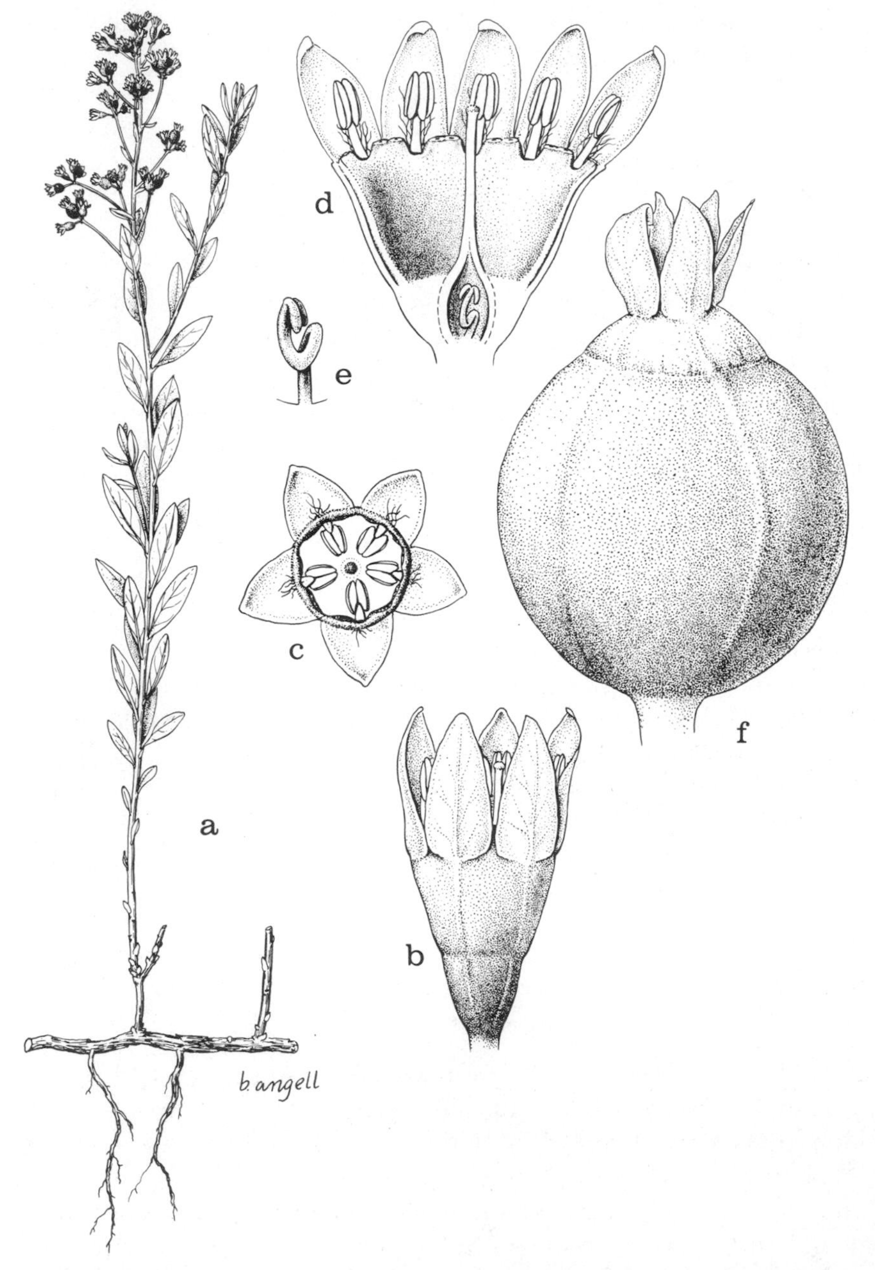

FIG. 5.32 Santalaceae. *Comandra umbellata* (L.) Nutt. a, habit, ×½; b, external view of flower, ×8; c, flower, from above, ×8; d, flower, opened up, the ovary in long-section, ×8; e, placenta, with 2 pendulous ovules, ×16; f, fruit, ×8.

calcium oxalate in some of the parenchyma-cells; nodes unilacunar; vessel-segments with simple perforations; imperforate tracheary elements often including some tracheids or vascular tracheids, the others fibers or fiber-tracheids with simple or bordered pits; wood-rays homocellular to somewhat heterocellular, mixed uniseriate and pluriseriate, the latter 2–6 cells wide, with short or sometimes elongate ends; wood-parenchyma commonly rather scanty, apotracheal and often diffuse to sometimes paratracheal. LEAVES opposite or less commonly alternate, simple, entire, well developed or sometimes reduced to mere scales, commonly with groups of silicified cells in the mesophyll; stomates paracytic or sometimes anomocytic, or of various unusual types; stipules wanting. FLOWERS in diverse sorts of inflorescences, often with a small dichasium axillary to each bract, small, often greenish, perfect or unisexual (the plants then monoecious or dioecious), regular, hypogynous or perigynous or half-epigynous to more often epigynous; perianth strictly monochlamydeous (probably representing the corolla, the calyx obsolete), the tepals distinct or forming a valvately (3) 4–5 (–8)-lobed, often fleshy cup or tube; stamens as many as and opposite the tepals, to the base of which they are often adnate by their filaments; anthers tetrasporangiate and dithecal, opening by longitudinal slits or sometimes by a single apical pore; pollen-grains 2–3-nucleate, triaperturate, with various types of apertures; intrastaminal, lobed nectary-disk commonly surrounding or seated on the ovary, or lining the lower part of the perianth-tube; gynoecium of (2) 3 (–5) carpels united to form a compound, superior or half-inferior to more often wholly inferior ovary with simple, terminal style and capitate or lobed stigma; ovary completely unilocular, or partitioned at the base only, the erect, free-central placenta bearing 1–4 pendulous ovules (ovule solitary, embedded, and scarcely differentiated in *Exocarpos*, so that the placental column might be taken for a single massive ovule); ovules anatropous or less often hemitropous, unitegmic and tenuinucellar, or not differentiated into nucellus and integument; embryo-sac monosporic or seldom bisporic, 8-nucleate; endosperm-development cellular or less often helobial; embryology of diverse, often unusual types. FRUIT a nut or a drupe, usually without any viscid tissue; seed solitary, without a differentiated testa; embryo axile, straight, with 2 cotyledons, surrounded by the copious, fleshy, oily or (*Thesium*) starchy endosperm. X = 5, 6, 7, 12, 13, and higher numbers. (Anthobolaceae, Canopodaceae, Exocarpaceae, Osyridaceae, Podospermaceae)

The family Santalaceae as here defined consists of about 35 genera and 400 species, nearly cosmopolitan in distribution, but most common in tropical and subtropical, often arid climates. About half of the species belong to the single genus *Thesium*, native chiefly to Africa and the Mediterranean region. Sandalwood (*Santalum album* L. and related species) has long been prized for ceremonial burning rites in some Eastern societies. Sandal oil, a yellow aromatic oil used ceremonially and in soaps and cosmetics, is distilled from sandalwood.

6. Family MISODENDRACEAE J. G. Agardh 1858 nom. conserv., the Feathery Mistletoe Family

Dioecious shrublets, hemiparasitic on the branches of *Nothofagus*, more or less green and chlorophyllous; plants with thickened haustoria that promote overgrowth of the host at the contact-zone; twigs stout, the apex aborting at the end of each growing season, the next season's growth coming from one or more lateral branches; crystals of calcium oxalate present in some of the cells of the parenchymatous tissues; vessel-segments small, with simple perforations; imperforate tracheary elements very short and broad, with bordered pits; wood-rays homocellular, uniseriate; wood-parenchyma vasicentric. LEAVES alternate, small and simple, sometimes reduced and scale-like; mesophyll often with scattered groups of silicified cells; stipules wanting. FLOWERS small, unisexual, in catkin-like compound racemes or spikes, staminate flowers without perianth, consisting of 2–3 stamens seated around a small, lobed nectary-disk; anthers bisporangiate and monothecal, opening by a terminal slit; pollen-grains 4–12-colporate; pistillate flowers provided with a monochlamydeous perianth of 3 members that are adnate (except along the margins) to the ovary, but separate from each other except at the very base, and projecting beyond the ovary as very short free lobes; a linear, accrescent staminode borne at each sinus of the perianth, or adnate to the side of the ovary between the perianth-members; gynoecium of 3 carpels united to form a compound, unilocular ovary capped by a very short, stout style with 3 stigmas; ovules 3, pendulous from the top of the free-central placental column, not differentiated into nucellus and integument, each hanging down into a basal pocket of the ovarian cavity; endosperm-development cellular; embryology complex and unusual, resembling that of some members of the Olacaceae. Fruit an achene or a small nut, crowned with the strongly accrescent, feathery staminodes, by which it is airborne; seed solitary, without a testa, the oily green endosperm surrounding a straight embryo.

The family Misodendraceae consists of the single genus *Misodendrum*, with about 10 species native to temperate, forested regions of South America, from the Straits of Magellan north to about 33 degrees South Latitude. The spelling Misodendraceae has been conserved over Myzodendraceae.

7. Family LORANTHACEAE A. L. de Jussieu 1808 nom. conserv., the Showy Mistletoe Family

Chlorophyllous, photosynthetic, evergreen, brittle shrublets hemiparasitic on the branches of trees (or on epiphytes growing on the branches of trees), or seldom terrestrial shrubs or vines or even small trees to 12 m tall (*Nuytsia*) hemiparasitic on the roots of other plants, with a single haustorium, or often producing epicortical roots that grow along a branch of the host and form haustoria at intervals, occasionally almost dodder-like in habit; haustoria typically fairly massive and more or less woody, promoting overgrowth of the host in the contact-region, or seldom diffuse and internal as in the Viscaceae; plants often with scattered tanniferous cells in some of the parenchymatous tissues, but without proanthocyanins, and without iridoid compounds; solitary or clustered crystals of calcium oxalate commonly present in some of the cells of the parenchymatous tissues; stem often dichasially branched, but without nodal constrictions; nodes unilacunar; vessel-segments with simple perforations; imperforate tracheary elements very short, with simple or evidently bordered pits; wood-rays slightly to strongly heterocellular, variously 1–4 cells wide, or much broader (to 12 cells wide) and then with few or no uniseriates; wood-parenchyma apotracheal, diffuse or in short, uniseriate lines. Leaves well developed or sometimes reduced to small scales, opposite or sometimes ternate, simple, entire; mesophyll commonly with scattered groups of silicified cells, and often also with scattered sclereids; stomates mostly paracytic; stipules wanting. Flowers entomophilous or often ornithophilous, often red or yellow, frequently rather large and showy, seldom less than 1 cm long, perfect or rarely unisexual, epigynous, regular or somewhat irregular, borne in various types of inflorescences that may appear to be racemes, umbels, spikes, or heads, but the basic unit of the inflorescence generally a dichasium; perianth more or less distinctly dichlamydeous, the calyx represented by a usually toothed or shortly lobed or entire rim or shallow cup around the summit of the ovary, the teeth or lobes sometimes vascularized, sometimes not; petals (3–) 5–6 (–9), sometimes nectariferous at the base within, valvate, distinct or often connate below to form an evident corolla-tube that is equally or unequally cleft; stamens as many as and opposite the petals, to which they are often adnate by their filaments, sometimes subsessile; anthers tetrasporangiate and dithecal (the microsporangia sometimes cross-partitioned), or seldom with only 2 or 3 sporangia and a single theca, opening by longitudinal slits; pollen-grains binucleate, with 3 or seldom 4 apertures (inaperturate in *Atkinsonia*), usually trilobate or triangular, seldom spherical; nectary-disk present or not; gynoecium of 3–4 carpels united to form a compound, inferior, usually unilocular ovary (or the ovary sometimes solid, without a locule) with a very short to much-elongate style and small stigma, or the stigma sessile; ovules several, commonly 4–12, embedded in the erect, free-central placental column (or this scarcely developed and the ovules embedded in the basal tissue of the ovary), or rarely (*Lysiana*) the ovary 4-locular with an axile placenta; each ovule consisting essentially of a monosporic, 8-nucleate embryo-sac, without clearly defined nucellus or integument; embryo with a highly elongate suspensor; embryology complex and unusual. FRUIT usually a laticiferous berry or drupe with 1 seed, seldom with 2–3 seeds, or rarely the fruit dry and indehiscent; seed without a testa, surrounded or capped at one end by viscid tissue, often containing more than one embryo, embryo rather large, axile, at least sometimes without an evident radicle; cotyledons initially 2, but often becoming fused during ontogeny; endosperm copious, starchy, compound, derived from several primary endosperm-nuclei, usually without chlorophyll, or seldom wanting at maturity. X = 8–12, perhaps primitively 12.

The family Loranthaceae as here delimited consists of about 60–70 genera and perhaps 700 species, largely tropical and subtropical in distribution, especially in the Southern Hemisphere. The largest American genera are *Psittacanthus* and *Struthanthus* (sens. lat.) each estimated at about 75 species. *Loranthus*, broadly defined in the past, is now restricted by specialists to *L. europaeus* Jacq., and the other several hundred species are divided among many genera.

8. Family VISCACEAE Miers 1851, the Christmas Mistletoe Family

Brittle shrublets hemiparasitic on the branches of trees, more or less green and photosynthetic, producing haustoria that penetrate the host and ramify into slender branches (the haustoria sometimes almost like a mycelium), in *Arceuthobium* stimulating the host to produce witches' brooms; shoot often dichasially branched, usually with nodal constrictions, often with scattered tanniferous cells in some of the parenchymatous tissues, with or without proanthocyanins, but lacking iridoid compounds and not cyanogenic; solitary or clustered crystals of calcium

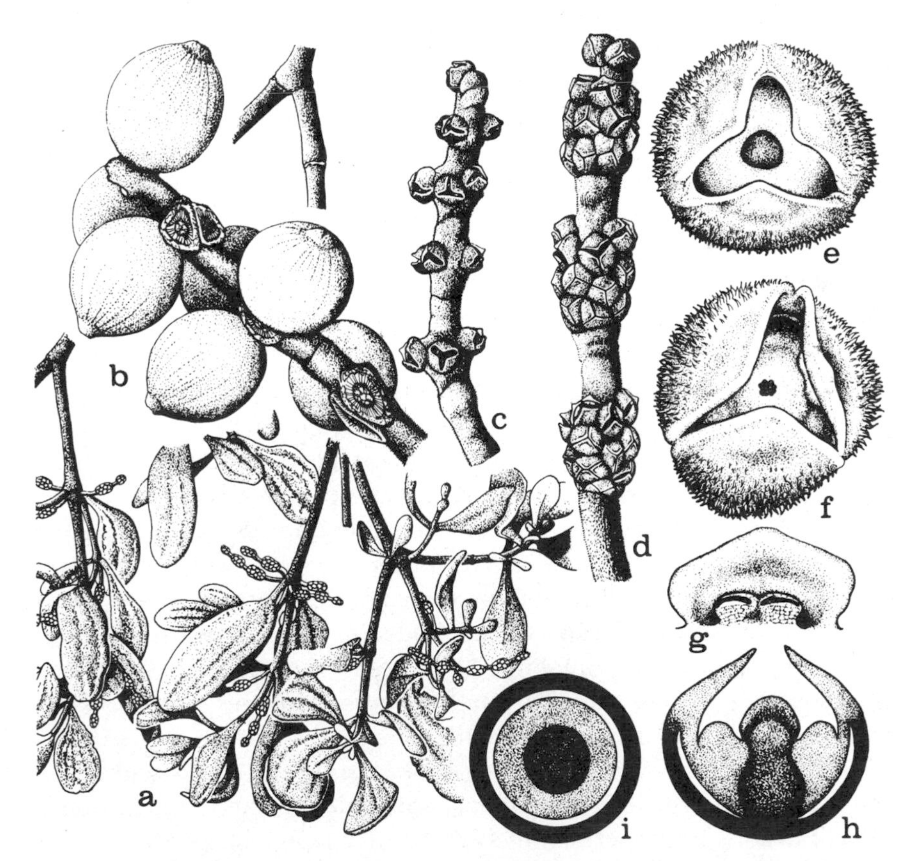

FIG. 5.33 Viscaceae. *Phoradendron serotinum* (Raf.) M. C. Johnston. a, habit, ×½; b, mature fruits, ×3; c, pistillate inflorescence, ×3; d, staminate inflorescence, ×3; e, pistillate flower, from above, ×15; f, staminate flower, from above, ×15; g, single tepal of staminate flower, with attached stamen, ×15; h, schematic long-section of pistillate flower, ×15; i, schematic cross-section of pistillate flower, ×15.

oxalate commonly present in some of the cells of the parenchymatous tissues; vessel-segments with simple perforations; imperforate tracheary elements very short, with simple or evidently bordered pits; wood-rays slightly to strongly heterocellular, variously 1–4 cells wide, or much broader (up to about 21 cells wide) and then with few or no uniseriates; wood-parenchyma apotracheal, diffuse or in short, uniseriate lines. LEAVES well developed or often reduced to small scales, opposite, simple, entire; mesophyll commonly with scattered groups of silicified cells, and often also with scattered sclereids; stomates mostly paracytic; stipules wanting. FLOWERS small, mostly greenish or yellow, entomophilous or anemophilous, strictly unisexual (the plants monoecious or dioecious), epigynous, regular, individually usually sessile or nearly so, mostly borne in spike-like, often branching inflorescences with a 3-flowered (or by reduction 2-flowered or even 1-flowered) dichasium in the axil of each small bract; perianth strictly monochlamydeous, seldom with even a vague suggestion of an outer calycine rim; tepals valvate, small, seldom more than 4 in staminate flowers or 3 in pistillate flowers, often reduced to mere teeth or points around the rim of the ovary; stamens in staminate flowers as many as and opposite the tepals and commonly adnate to them at the base, the filaments often very short; anthers sometimes tetrasporangiate, dithecal, and opening by longitudinal slits, but often more or less reduced and modified, sometimes with only one or two sporangia and opening irregularly by terminal pores or slits, or transversely partitioned and opening by a number of transverse slits, or the anthers even confluent into a synandrium; pollen-grains binucleate or trinucleate, triaperturate, mostly spherical; gynoecium of 3–4 carpels united to form a compound, inferior, unilocular ovary (or the ovary sometimes solid, without a locule) with a very short terminal style and small stigma, or the stigma essentially sessile; ovary containing a massive, placenta-like body (mamelon) nearly or quite filling the locule and containing 2 bisporic, 8-nucleate embryo-sacs; embryology complex and unusual; embryo with very short or no suspensor. FRUIT a shining, sometimes explosive berry; seed solitary (2), without a testa, surrounded or capped at one end with viscid tissue; embryo rather large, dicotyledonous, embedded in the simple, starchy, chlorophyllous endosperm, which is derived from the primary endosperm-nucleus of a single embryo-sac. X = 10–15, perhaps primitively 14.

The family Viscaceae as here defined consists of 7 or 8 genera and perhaps 350 species, of cosmopolitan distribution, but most abundant in tropical regions. By far the largest genus is *Phoradendron*, with about 200 species. *Viscum album* L. is the traditional Christmas mistletoe in Europe; in North America its place is commonly taken by *Phoradendron serotinum* (Raf.) M. C. Johnston.

The Viscaceae have traditionally been included in the Loranthaceae, often as a separate subfamily. Embryological studies of the past two or three decades have emphasized the distinction between the two groups, which now appear to be better treated as distinct families. The monotypic western South American genus *Lepidoceras* is not certainly placed. It suggests the Viscaceae in some features and the Loranthaceae in others.

9. Family EREMOLEPIDACEAE Van Tieghem 1910, the Catkin-mistletoe Family

Shrublets hemiparasitic on the branches of trees, more or less green and photosynthetic, with thickened haustoria that promote overgrowth of the host in the contact-zone, and often producing epicortical roots. LEAVES alternate, simple, entire, exstipulate. FLOWERS unisexual (the plants dioecious or sometimes monoecious), small, in catkin-like inflorescences, subtended by imbricate, often deciduous bracts, mostly sessile, epigynous or semi-epigynous, strictly monochlamydeous; tepals small, 3 or 4 (or none) in staminate flowers, 2 or 3 in pistillate flowers; stamens distinct, opposite the tepals when these are present, tetrasporangiate, dithecal, opening by terminal slits; pollen-grains tricolporate; gynoecium of 3 or 5 carpels united to form a compound, inferior (half-inferior in *Antidaphne*), unilocular ovary; ovules 2, embedded in the base of the ovary, each consisting of a bisporic, 8-nucleate embryo-sac, without clearly defined nucellus and integument; embryology complex and unusual; embryo with short or no suspensor. FRUIT a berry, the single seed lacking a testa, surrounded or capped at one end by viscid tissue; embryo rather large, dicotyledonous, embedded in the simple, chlorophyllous endosperm. N = 10, 13.

The family Eremolepidaceae consists of 3 small tropical American genera, *Eremolepis, Antidaphne*, and *Eubrachion*, about a dozen species in all. These genera have usually been included in the Viscaceae (or in the subfamily Viscoideae of the Loranthaceae), but Kuijt (1968) considers them to be more closely allied to the group here treated as the family Opiliaceae. More recently Wiens & Barlow (1971) have suggested that the Eremolepidaceae represent diverse aerial-parasitic members of the Santalaceae. More study is in order.

10. Family BALANOPHORACEAE L. C. & A. Richard 1822 nom. conserv., the Balanophora Family

Fleshy root-parasites without chlorophyll, attaching to the root of the host by an amorphous, highly modified root, commonly called a tuber, which very often incorporates tissues of the host as well as the parasite; tubers in some genera accumulating balanophorin, a waxy food-reserve, in others starch; slender rhizomes (roots?) produced from the tuber grow horizontally through the ground and attack new hosts; plants commonly tanniferous, and producing solitary or clustered crystals of calcium oxalate in some of the cells of the parenchymatous tissues; vascular system much-reduced, but at least sometimes including some vessels, these with simple perforations. LEAVES alternate (sometimes very closely spiralled) or whorled, scale-like, without stomates (except in *Cynomorium*), or wanting. Inflorescence terminal, erect, generally massive, often fungoid in appearance, often arising endogenously from the tuber; FLOWERS numerous or very numerous and often very small (those of some genera among the smallest of all angiosperms), at least sometimes entomophilous, monochlamydeous or naked, unisexual, the plants monoecious or dioecious; staminate flowers very diverse, in the least-reduced forms with 3–4 (–8) distinct or basally connate, valvate tepals with a stamen opposite each tepal, the stamen with a tetrasporangiate, dithecal anther and opening by longitudinal slits, but the stamens often reduced in various ways, sometimes only 1 or 2 (one on each edge of the 2-lobed upper lip in *Mystropetalon*), sometimes with a monothecal anther opening by a terminal pore, sometimes coalescent to form a synandrium opening by numerous transverse slits, and the perianth sometimes reduced to only 2 tiny, filamentous tepals; pollen-grains 3–5-colpate or 3-many-porate, or inaperturate, variously binucleate or trinucleate, the binucleate types associated with a wet stigma, the trinucleate ones with a dry one; pistillate flowers without perianth, or in 2 genera with minute tepals, these hypogynous and united into a cup in *Mystropetalon*, epigynous in *Cynomorium*; gynoecium of 2 or 3 carpels united to form a compound ovary with distinct styles or a single trifid style, or sometimes the stigma sessile and discoid, or (*Balanophora*) the gynoecium pseudomonomerous and with a single undivided style; ovary typically solid, without a locule, and containing 1 or 2 embryo-sacs (very much reduced ovules) without recognizable nucellus or integuments; early ontogenetic stages of the ovary sometimes showing a massive central placental column that later fuses with the ovary wall;

in *Cynomorium* the ovary unilocular and with a single unitegmic, crassinucellar ovule pendulous from the summit; embryo-sac monosporic or bisporic; endosperm-development cellular. FRUIT indehiscent, often very tiny, in *Mystropetalon* surrounded by the swollen perianth-tube; numerous individual fruits sometimes aggregated to form a fleshy multiple fruit; seed solitary, with a very small, undifferentiated embryo embedded in the endosperm. X = 8, 9, 12, and more. (Cynomoriaceae)

The family Balanophoraceae as here defined consists of some 19 genera and about 45 species, widespread in tropical and subtropical regions. The largest genus is *Balanophora*, with about 15 (or according to some authors as many as 70 or 80) species. The small Mediterranean and Asian genus *Cynomorium* (2 spp) is often taken as a distinct family. This treatment can be defended, but in a group that already shows such diversity in floral structure it hardly seems necessary.

The relationship of the Balanophoraceae to the rest of the Santalales has sometimes been questioned, but the family has no other evident allies. Fagerlind (1945, 1948) found the Balanophoraceae to be embryologically highly compatible with the Santalaceae. It is of course possible that this specialized, achlorophyllous family has an origin entirely apart from the Santalales, and that the similarities reflect convergent adaptations to parasitism. On the other hand, it seems more logical to seek the origin of the Balanophoraceae in some less reduced parasitic group. The Santalales are an obvious candidate for such an ancestral group. I see no reason why the Balanophoraceae might not have originated from the Olacaceae or something much like that family.

10. Order RAFFLESIALES Kerner 1891

Parasitic, fleshy herbs without chlorophyll, vegetatively much reduced and modified, often with only the flowers or the inflorescence emergent from the host. LEAVES much-reduced and scale-like, or none. FLOWERS or compact flowering shoots arising endogenously from the host or endogenously from a rhizome-like root, the flowers small to very large, generally fleshy, commonly malodorous, entomophilous (often cantharophilous), regular, monochlamydeous, perfect or unisexual, hypogynous to epigynous; tepals 3–5, or sometimes more or less numerous (apparently by splitting), often connate toward the base; androecium variously modified, never of simple, ordinary type; pollen-grains smooth, variously monosulcate, disulcate, triaperturate, or of triaperturate-derived type; gynoecium of 3 or more carpels united to form a compound pistil, this typically unilocular but often with deeply intruded placentas, sometimes irregularly multilocular; ovules very numerous, anatropous to more often orthotropous, bitegmic or unitegmic, tenuinucellar, or scarcely differentiated from the placenta except as embryo-sacs. FRUITS often fleshy; seeds very numerous and tiny, with a hard testa and a minute, undifferentiated embryo surrounded by endosperm, sometimes also with perisperm.

The order Rafflesiales as here defined consists of 3 families and perhaps 60 species. The Mitrastemonaceae are obviously allied to the Rafflesiaceae and have often been included in that family. The Hydnoraceae are clearly distinctive, but their relationship to the other families is generally admitted. Aside from the features common to the order, *Prosopanche* (Hydnoraceae) resembles *Mitrastemon* in its dome-like synandrium covering the gynoecium, although the details of structure are different.

The relationships of the Rafflesiales are disputed and doubtful. Traditionally they have been associated with the Aristolochiaceae, presumably because of some vague similarity in the perianth. In other respects, however, the two groups are very different. The Rafflesiales are so highly specialized that they must have a long history of parasitism, of which there is not a whisper in the Aristolochiaceae. The ethereal oil cells of the Aristolochiaceae and other Magnoliales are wanting from the Rafflesiales. Syncarpous members of the Aristolochiaceae have axile placentation, whereas in the Rafflesiales the placentation is parietal or sometimes apical, or highly modified. In my opinion the Rafflesiales are singularly misplaced in the Aristolochiales.

Ongoing (1980) studies carried out under the direction of Armen Takhtajan disclose that the pollen of *Hydnora* is monosulcate (personal communication). The pollen of *Prosopanche* and *Mitrastemon*, on the other hand, is bisulcate, and that of the Rafflesiaceae varies from multiaperturate to 2–3-porate or inaperturate. Aside from *Hydnora*, all pollen of the Rafflesiales can readily enough be interpreted as belonging to the triaperturate or triaperturate-derived series that characterizes most dicotyledons.

Monosulcate pollen is now uniformly regarded as primitive among the angiosperms, with triaperturate and other types being derived. Aside from *Hydnora*, all dicotyledons with monosulcate pollen are included in the Magnoliidae. These groups with monosulcate pollen have other putatively primitive characters and are generally regarded as archaic.

I am reluctant to apply this concept to the Rafflesiales. There is no group of Magnoliidae to which they seem to be at all allied. Instead I suggest that the monosulcate pollen of *Hydnora* may represent a reversion to a primitive type, possibly fostered by the overall simplification of structure associated with the parasitic habit. Virtually all other evolutionary advances within the angiosperms are subject to reversion, and I see no a priori reason why the structure of the pollen should be a unique exception.

If one can bypass the problem of a monosulcate pollen in *Hydnora*, then in my opinion the most likely relatives of the Rafflesiales are the Santalales. If one mentally extrapolates backward from existing Rafflesiales to suppose what their less specialized ancestors may have been like, one visualizes green, tanniferous root-parasites with perfect, hypogynous, regular, monochlamydeous flowers that have more or less numerous bitegmic, tenuinucellar ovules in a compound ovary on deeply intruded parietal placentas (or possibly on axile placentas), and with fleshy fruits that have more or less numerous seeds with a well developed endosperm. The only features in the list that are out of harmony with the Santalales are the nature of the placentas and the number of ovules and seeds. The very large number of ovules and seeds in the Rafflesiales is clearly a specialization derived from some smaller number, and the complex placental organization of the group is obviously associated with the increase in number of ovules. Thus these features by themselves do not seem to present an insurmountable barrier to a relationship with the Santalales. The tendency toward simplification of the ovules and the embryo in both orders may simply reflect parallel adaptations to a parasitic habit, but not all groups of parasites show these changes. The endotrophic, mycelium-like haustorial body of the Rafflesiaceae and Mitrastemonaceae is very reminiscent of that in many of the Viscaceae. Here we are clearly dealing with parallel adaptations to parasitism, but no other parasitic group has developed a similar structure. As I pointed out in 1968 and in other publications, parallelism in a number of features is in itself some indication of relationship, to be considered along with other evidence.

At the same time, it is clear that Rafflesiales cannot be derived from any of the more advanced members of the Santalales. Some of the Rafflesiales have bitegmic ovules, and nothing in the Santalales any more advanced than the Olacaceae has more than one integument—indeed most of them have none at all. Furthermore, all of the branch-parasites in the Santalales are immediately excluded as possible ancestors; only root-parasites will do.

If any other families or orders merit serious consideration as possible ancestors of the Rafflesiales, the fact is not immediately evident. All of the other parasitic groups are clearly of very different affinity, and none of the autotrophic families with numerous ovules on parietal placentas presents itself to mind as a likely ancestor.

The familiar name Rafflesiales is here preferred over the older name Cytinales G. T. Burnett 1835.

SELECTED REFERENCES

Cocucci, A. E. 1965–1976. Estudios en el genero *Prosopanche* (Hydnoraceae). I, II, III. Kurtziana 2: 53–74, 1965. 8: 7–15, 1975. 9: 19–39, 1976.

Matuda, E. 1947. On the genus *Mitrastemon*. Bull. Torrey Bot. Club 74: 133–141.

Meijer, W. 1958. A contribution to the taxonomy and biology of *Rafflesia arnoldi* in west Sumatra. Ann. Bogor. 3: 33–44.

Olah, L. von. 1960. Cytological and morphological investigations in *Rafflesia arnoldi* R. Br. Bull. Torrey Bot. Club 87: 406–416.

Royen, P. van. 1963. Sertulum Papuanum 8. Rafflesiaceae. Nova Guinea Bot. 14: 243–245.

SYNOPTICAL ARRANGEMENT OF THE FAMILIES OF RAFFLESIALES

1 Plants essentially ectoparasitic, the vegetative body consisting of a coarse, rhizome-like pilot-root from which the numerous slender, unbranched haustorial roots emerge; leaves strictly wanting; style none; flowers perfect, ovary inferior .... 1. HYDNORACEAE.

1 Plants essentially endoparasitic, the vegetative body much branched and mycelium-like, permeating the tissues of the host, no external

roots or rhizome-like structures; scale leaves present on the emergent flowering shoot; stylar column more or less well developed, often large and conspicuous.

2 Ovary superior; flowers perfect; stamen-tube free from the pistil and enclosing it, except for a small apical opening, but deciduous after the pollen has been shed ....2. MITRASTEMONACEAE.

2 Ovary inferior or half-inferior; flowers mostly unisexual; stamens adnate to the stylar column, or sometimes merely forming a sheath around it, the androecium not deciduous as a unit ....3. RAFFLESIACEAE.

1. Family HYDNORACEAE C. A. Agardh 1821 nom. conserv., the Hydnora Family

Strictly leafless, highly modified, terrestrial parasitic herbs without chlorophyll; vegetative body consisting of a coarse, rhizome-like pilot-root from which numerous slender, unbranched haustorial roots emerge and parasitize the roots of various other plants; pilot-root provided with a root-cap, not articulated, angular or less often cylindric, covered externally with tubercles or vermiform projections, or these concentrated along the angles, provided internally with variously arranged vascular bundles that at least sometimes have vessels with simple

FIG. 5.34 Hydnoraceae. Above, monosulcate pollen of *Hydnora africana* Thunb., in surface view (left) and in section (right). Electron micrographs courtesy of Armen Takhtajan. Below, mainly disulcate pollen of *Prosopanche americana* (R. Br.) Kuntze, left, courtesy of Armen Takhtajan, right, courtesy of Beryl Simpson.

perforations; parenchymatous tissues of the pilot-root with scattered tanniferous mucilage-cells (or sometimes lysigenous mucilage-canals or cavities) containing catechin and probably proanthocyanins. FLOWERS arising individually and endogenously from the pilot-roots, short-stalked, barely reaching the surface of the ground (the lower part commonly remaining buried), rather large, fleshy, malodorous, cantharophilous, perfect, epigynous with a shortly prolonged hypanthium, regular, monochlamydeous; tepals 3 or 4 (5), thick and fleshy, valvate, connate below, the outer surface coarse, cracked, brown, the inner surface showy, white to partly or wholly pink or red, often bearing retrorse bristles; androecium highly modified and synandrial, nearly or quite without filaments, consisting basically of as many stamens as tepals (opposite the tepals), but with numerous elongate, extrorse, bisporangiate pollen-sacs that open by longitudinal slits; in *Hydnora* the androecium forming a lobed ring on the hypanthium, the lobes opposite the tepals and sometimes with an upright free tip, in *Prosopanche* the very short filaments arising from the hypanthium, and the extrorse anthers connate to form a dome or cap with a small central opening; pollen-grains spherical to oblong, binucleate, monosulcate (*Hydnora*) or bisulcate (*Prosopanche*), not sculptured; small, fleshy staminodes present below and alternate with the stamens in *Prosopanche*; gynoecium of 3 (4) carpels united to form a compound, inferior, unilocular ovary that eventually becomes filled by the growth of the accrescent placentas; stigma sessile, in *Hydnora* mostly 3-lobed, in *Prosopanche* consisting essentially of the slightly protruding tips of the placental lamellae; placentas lamellar, covered with ovules, in *Prosopanche* numerous in 3 groups, parietal, deeply intruded but not joined in the center, in *Hydnora* numerous, suspended from the top of the ovary, and branched; ovules very numerous, orthotropous, with a massive single integument, tenuinucellar, in *Prosopanche* embedded and scarcely differentiated from the placenta except as embryo-sacs, the single integument recognizable only in the micropylar area; embryo-sac bisporic in *Prosopanche*, tetrasporic in *Hydnora*; endosperm-development cellular. FRUIT with a more or less woody pericarp and fleshy, edible interior, in *Prosopanche* bursting in circumscissile fashion when ripe; seeds very numerous (in the thousands), tiny, with a hard testa, a thin layer of perisperm, and a well developed endosperm with polysaccharide (arabinose) food-reserves; embryo minute, undifferentiated, enclosed in the endosperm.

The family Hydnoraceae consists of 2 genera and perhaps 10 species. *Hydnora* occurs in the drier parts of Africa from Ethiopia south to the

Cape, and on Madagascar. *Prosopanche* (2 spp.) occurs in grasslands and other dry regions from Paraguay and northern Argentina to Patagonia.

2. Family MITRASTEMONACEAE Makino 1911 nom. conserv., the Mitrastemon Family

Plants without chlorophyll, endoparasitic in the roots of other plants, often causing broomlike overgrowths of the host-root; vegetative body much-dissected and largely filamentous, resembling a fungal mycelium, permeating the tissues of the root of the host, but not extending into the apical meristem. FLOWERS arising endogenously from the roots of the host, often forming a sort of fairy-ring on the ground, each flower terminal to a short, fleshy stem that bears reduced, opposite, scale-like but somewhat fleshy leaves without stomates, the uppermost leaves tending to be somewhat cupped and accumulating nectar. FLOWERS perfect, hypogynous, regular, monochlamydeous, protandrous; tepals several, not strongly petaloid, connate below to form a cup, the rim of the cup 4-lobed in the Pacific species, irregularly undulate in the American species; stamens united into a tube surrounding but free from the pistil, the tube domelike, nearly closed, with only a small apical aperture, bearing numerous sessile anthers in a series of rings shortly below the summit, the anthers extrorse, dithecal, and longitudinally dehiscent; pollen-grains dicolpate; stamen-tube deciduous after the pollen has been shed, exposing the gynoecium; gynoecium apparently of 9–15 carpels, these united to form a compound, superior ovary with a short stylar column and a somewhat expanded, fleshy stigma; placentas parietal, deeply intruded, but not joined in the center, covered with the numerous ovules, these anatropous, unitegmic, tenuinucellar. FRUITS baccate, or capsular and tardily opening by a horizontal slit; seeds very numerous and tiny, with a hard testa and a minute, undifferentiated embryo surrounded by endosperm. X = 10.

The family Mitrastemonaceae consists of the single genus *Mitrastemon*, with 2 widely disjunct species, one occurring from Borneo and Sumatra to Indochina and Japan, the other in Mexico and Central America. *Mitrastemon* has often been included in the Rafflesiaceae, to which it is evidently related. It is more primitive than the Rafflesiaceae in its hypogynous, perfect flowers, and its dome-like synandrium is more nearly comparable to that of *Prosopanche*, in the Hydnoraceae, than to anything in the Rafflesiaceae. Whether *Mitrastemon* should be included in the Rafflesiaceae or kept as a separate family is largely a matter of taste.

3. Family RAFFLESIACEAE Dumortier 1829 nom. conserv., the Rafflesia Family

Plants without chlorophyll, endoparasitic in the roots or less often the shoots of other plants, only the flowers or the short flowering stem exserted; vegetative body much-dissected and largely filamentous, resembling a fungal mycelium, permeating the tissues of the host and sometimes extending even into the apical meristem; flowering shoots short, fleshy, tanniferous, originating endogenously from undifferentiated tissue of the parasite within the host, and emerging endogenously from the host, terminating in a single flower or in a short, fleshy spike; shoot in the less reduced forms with a vestigial vascular system; plastids

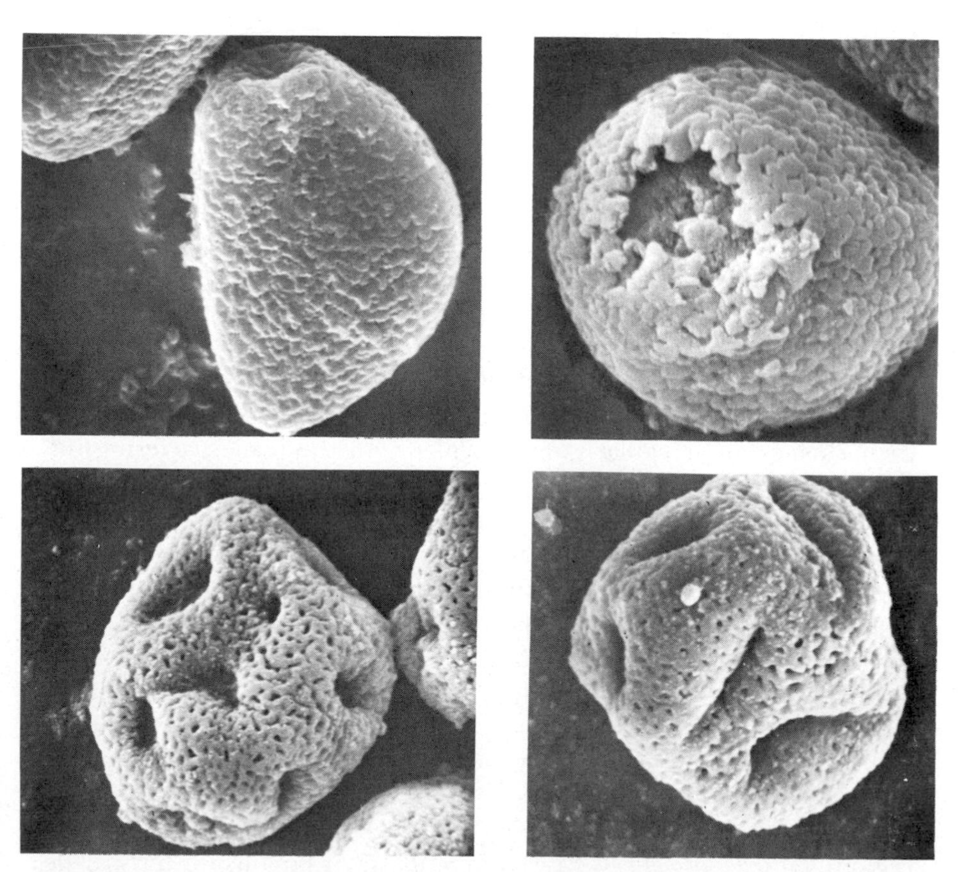

FIG. 5.35 Rafflesiaceae. Above, diporate pollen of *Cytinus ruber* Fritsch, × about 3300. Below, multiaperturate pollen of *Cytinus dioicus* Juss., × about 1300. All courtesy of Armen Takhtajan.

of the sieve-tubes lacking both protein and starch. Reduced, scale-like LEAVES present on the flowering stem or around the base of the solitary flower, usually whorled, sometimes opposite or even alternate, without normal stomates. FLOWERS sometimes small (as in *Cytinus*) and partly embedded in the fleshy axis of the inflorescence, but more often large or very large and arising individually, commonly fleshy, often malodorous, often brightly colored, unisexual or rarely perfect, regular, monochlamydeous; tepals usually 4 or 5, sometimes up to 10 or even more, the larger numbers perhaps reflecting segmentation of the ancestral 4 or 5, distinct or often connate below to form a tube, imbricate or rarely valvate, petaloid or not; stamens in staminate flowers 5 to numerous, connate by their filaments into a tube surrounding the stylar column, or more often adnate to the column, from which the anthers originate in 1-several cycles; stylar column in the larger-flowered species expanded above the anthers to form an often large and complex disk; anthers in the less modified forms tetrasporangiate, dithecal, and opening by longitudinal slits, but in others monothecal or irregularly polythecal and opening by transverse slits or an apical pore; pollen-grains binucleate, multiaperturate to 2–3-porate or inaperturate, not sculptured; pistillate flowers epigynous or semi-epigynous, with 4–8 carpels united to form a compound, inferior or half-inferior ovary crowned by a stout, columnar style that is distally expanded into an often large and complex disk; ovary variously unilocular and with a limited number of parietal placentas, or with the ovules covering the whole inner wall, or with as many as 8–14 more or less deeply intruded, sometimes branched parietal placentas that bear ovules over most of their surface, or irregularly multilocular with the ovules covering the surfaces of the partitions; ovules very numerous, anatropous to more often orthotropous, bitegmic or unitegmic, tenuinucellar; endosperm-development nuclear or cellular. FRUITS indehiscent or irregularly dehiscent, often fleshy, sometimes multiple; seeds very numerous (commonly in the thousands) and tiny, with an undifferentiated, often few-celled embryo surrounded by endosperm. N = 12. (Cytinaceae)

The family Rafflesiaceae as here defined consists of about 7 genera and perhaps as many as 50 species, widespread in tropical and subtropical regions, but apparently never abundant. Only a few species occur in temperate climates, as for example the well known *Cytinus hypocistis* L., parasitic on the roots of Cistaceae in the Mediterranean region. *Rafflesia arnoldii* R. Br., from Sumatra, is famous for having the largest flower in the world, as much as 1 m across.

11. Order CELASTRALES Wettstein 1907

Autotrophic, terrestrial plants, woody or seldom herbaceous, often producing one or another sort of alkaloid, with or more often without proanthocyanins, and only seldom with ellagic acid or iridoid compounds; internal phloem wanting, but interxylary phloem sometimes present. LEAVES opposite or alternate, simple, entire or toothed; stipules well developed in the Dichapetalaceae, otherwise wanting or very small. FLOWERS mostly rather small, perfect or less often unisexual, hypogynous to seldom epigynous or evidently perigynous, dichlamydeous or seldom monochlamydeous, commonly 4–5-merous, haplostemonous with antesepalous stamens (some members of the cycle sometimes wanting, and the remaining members then sometimes not aligned with the sepals), or seldom diplostemonous; sepals or calyx-lobes imbricate or rarely valvate; petals distinct or sometimes connate at the base, seldom connate for much of their length, imbricate or less often valvate, rarely convolute; stamens distinct, or the filaments rarely connate below; nectary-disk present and often conspicuous, or often wanting; carpels 2-several, united to form a compound ovary with as many locules as carpels, or functionally unilocular in a few families, capped by a single, often short style (or the stigmas sometimes sessile), or seldom by 2–4 distinct styles; placentation axile, or often axile-basal or axile-apical, or simply apical or subapical when there is only one functional locule; ovules mostly anatropous, commonly apotropous, usually 1–2 in each locule at the base or the summit, or less often several and superposed in 2 rows in each locule, bitegmic or less often unitegmic, crassinucellar or less often tenuinucellar; endosperm-development nuclear or (Aquifoliaceae) cellular. FRUIT variously capsular, drupaceous, baccate, or samaroid; seeds with or without endosperm; embryo with 2 cotyledons.

The order Celastrales as here defined consists of 11 families and a little more than 2000 species. The Celastraceae (800), Hippocrateaceae (300), Aquifoliaceae (300+) and Icacinaceae (400) are fairly closely related and make up the bulk of the order. The Dichapetalaceae (200) stand somewhat apart from the other families and are by some authors excluded from the Celastrales. The remaining 6 families have fewer than 50 species altogether. The position of the Salvadoraceae (12) and Stackhousiaceae (20–25) as members of the Celastrales is generally accepted, but the affinities of the Corynocarpaceae (5), Cardiopteridaceae (3), Aextoxicaceae (1) and Geissolomataceae (1) are more debatable.

There is nothing ecologically distinctive about the Celastrales, nor (aside from the ensemble of secondary metabolites) any of the major families of the order. If the order and its constituent families exploit the habitat in any way notably different from the ways of numerous other angiosperms, the fact is not immediately evident.

The new family Tepuianthaceae Maguire & Steyermark, which is awaiting publication as this manuscript is being completed, might perhaps best be referred to the Celastrales. It may be characterized as follows:

Tepuianthaceae Maguire & Steyermark ined.

Trees or shrubs with somewhat bitter bark; indument of simple, more or less appressed hairs; vessel-segments with simple perforations; imperforate tracheary elements with distinctly bordered pits; wood-rays homocellular, most of them uniseriate, a few biseriate for part of their length; wood-parenchyma scanty-vasicentric. Leaves simple, entire, pinnately veined, exstipulate, alternate or opposite. Inflorescences cymose, terminal or in the uppermost axils, the plants androdioecious; flowers hypogynous, regular, calyx of 5 distinct, imbricate sepals; corolla of 5 distinct, imbricate petals that tend to be clawed at the base; disk extrastaminal, consisting of 5–10 discrete but contiguous fleshy glands; stamens borne in 1 or more often 2 or 3 cycles, when unicyclic alternate with the petals; anthers sagittate, tetrasporangiate and dithecal; pollen-grains spheroidal, tricolporate, prominently sculptured; gynoecium of 3 carpets united to form a compound, trilocular, superior ovary with distinct, bifid styles; ovules solitary in each locule, anatropous, pendulous with ventral raphe, the micropyle directed upward and outward. Fruit a densely sericeous, trilocular, bony, loculicidal capsule with a single seed in each locule; endosperm copious; embryo small, with 2 poorly differentiated cotyledons.

The family consists of the single genus *Tepuianthus*, with 5 species native to the Guayana Highlands and some of the nearby lowlands in northern South America. The group is anomalous in the Celastrales in that 4 of the 5 species have a bicyclic or tricyclic androecium. Comparison with any member of the order discloses additional significant differences. Possibly the family should be regarded as a simple-leaved member of the Sapindales, but here again no existing family can readily accommodate it.

SELECTED REFERENCES

Baas, P. 1973. The wood anatomical range in *Ilex* (Aquifoliaceae) and its ecological and phylogenetic significance. Blumea 21: 193–258.

Baas, P. 1975. Vegetative anatomy and the affinities of Aquifoliaceae, *Sphenostemon, Phelline,* and *Oncotheca.* Blumea 22: 311–407.

Bailey, I. W., & R. A. Howard. 1941. The comparative morphology of the Icacinaceae. I–IV. J. Arnold Arbor. 22: 125–132; 171–187; 432–442; 556–568.

Bailey, I. W., & B. G. L. Swamy. 1953. The morphology and relationships of *Idenburgia* and *Nouhuysia.* J. Arnold Arbor. 34: 77–85.

Barker, W. R. 1977. Taxonomic studies in *Stackhousia* Sm. (Stackhousiaceae) in South Australia. J. Adelaide Bot. Gard. 1: 69–82.

Berkeley, E. 1953. Morphological studies in the Celastraceae. J. Elisha Mitchell Sci. Soc. 69: 185–206.

Bernardi, L. 1964. La position systématique du genre *Sphenostemon* Baillon sensu van Steenis. Candollea 19: 199–205.

Brizicky, G. K. 1964. The genera of Celastrales in the southeastern United States. J. Arnold Arbor. 45: 206–234.

Carlquist, S. 1975. Wood anatomy and relationships of Geissolomataceae. Bull. Torrey Bot. Club 102: 128–134.

Copeland, H. F. 1963. Structural notes on hollies (*Ilex aquifolium* and *I. cornuta,* family Aquifoliaceae). Phytomorphology 13: 455–464.

Copeland, H. F. 1966. Morphology and embryology of *Euonymus japonica.* Phytomorphology 16: 326–334.

Dahl, A. O. 1952. The comparative morphology of the Icacinaceae. VI. The pollen. J. Arnold Arbor. 33: 252–295.

Dahlgren, R., & V. S. Rao. 1969. A study of the family Geissolomataceae. Bot. Not. 122: 207–227.

David, E. 1938. Embryologische Untersuchungen an Myoporaceen, Salvadoraceen, Sapindaceen und Hippocrateaceen. Planta 28: 680–703.

Fagerlind, F. 1945. Bau des Gynöceums, der Samenanlage und des Embryosackes bei einigen Repräsentanten der Familie Icacinaceae. Svensk Bot. Tidskr. 39: 346–364.

Hartog, R. M. den, & P. Baas, 1978. Epidermal characters of the Celastraceae sensu lato. Acta Bot Neerl. 27(5/6): 355–388.

Heintzelmann, C. E., & R. A. Howard, 1948. The comparative morphology of the Icacinaceae V. The pubescence and the crystals. Amer. J. Bot. 35: 42–52.

Herr, J. M. 1959. The development of the ovule and megagametophyte in the genus *Ilex* L. J. Elisha Mitchell Sci. Soc. 74: 107–128.

Herr, J. M. 1961. Endosperm development and associated ovule modifications in the genus *Ilex.* J. Elisha Mitchell Sci. Soc. 77: 26–32.

Hou, D. 1969. Pollen of *Sarawakodendron* (Celastraceae) and some related genera, with notes on techniques. Blumea 17: 97–120.

Hu, S-y. 1967. The evolution and distribution of the species of Aquifoliaceae in the Pacific area. J. Jap. Bot. 42: 13–32; 49–59.

Inamdar, J. A. 1969. The stomatal structure and ontogeny in *Azima* and *Salvadora.* Flora B 158: 519–525.

Ingle, H. D. 1956. A note on the wood anatomy of the genus *Corynocarpus.* Trop. Woods 105: 8–12.

Johnston, M. C. 1975. Synopsis of *Canotia* (Celastraceae) including a new species from the Chihuahuan desert. Brittonia 27: 119–122.

Kshetrapal, S. 1970. A contribution to the vascular anatomy of the flower of certain species of the Salvadoraceae. J. Indian Bot. Soc. 49: 92–99.

Lobreau, D. 1969. Les limites de l'"ordre" des Célastrales d'après le pollen. Pollen & Spores 11: 499–555.

Lobreau-Callen, D. 1972, 1973. Pollen des Icacinaceae. I. Atlas (1). II. Observations en microscopie electronique, correlations, conclusions (1). Pollen & Spores 14: 345–388, 1972. 15: 47–89, 1973.
Lobreau-Callen, D. 1975. Les pollens colpés dans les Célastrales: interprétation nouvelle de l'aperture simple. Compt. Rend. Hebd. Séances Acad. Sci. 280: 2547–2550.
Lobreau-Callen, D. 1977. Nouvelle interpretation de l'"ordre" des Celastrales à l'aide de la palynologie. Compt. Rend. Hebd. Séances Acad. Sci. 284D: 915–918.
Lobreau-Callen, D., & B. Lugardon. 1972–1973. L'Aperture a repli du pollen des Celastraceae. Naturalia Monspel., Sér. Bot., Fasc. 23–24: 205–210.
Mauritzon, J. 1936. Embryologische Angaben über Stackhousiaceae, Hippocrateaceae und Icacinaceae. Svensk Bot. Tidskr. 30: 541–550.
Merrill, E. D., & F. L. Freeman. 1940. The old world species of the celastraceous genus *Microtropis* Wallich. Proc. Amer. Acad. Arts 73: 271–310.
Metcalfe, C. R. 1956. The taxonomic affinities of *Sphenostemon* in the light of the anatomy of its stem and leaf. Kew Bull. 1956: 249–253.
Narang, N. 1953. The life-history of *Stackhousia linariaefolia* A. Cunn. with a discussion on its systematic position. Phytomorphology 3: 485–493.
Prance, G. T. 1972. Dichapetalaceae. Flora Neotropica Monograph 10.
Record, S. J. 1938. The American woods of the orders Celastrales, Olacales, and Santalales. Trop. Woods 53: 11–38.
Robson, N. 1965. Taxonomic and nomenclatural notes on Celastraceae. Bol. Soc. Brot., ser. 2, 39: 5–55.
Sleumer, H. 1969. Materials towards the knowledge of the Icacinaceae of Asia, Malesia, and adjacent areas. Blumea 17: 181–264.
Smith, A. C. 1940. The American species of Hippocrateaceae. Brittonia 3: 341–555.
Smith, A. C., & I. W. Bailey. 1941. *Brassiantha*, a new genus of Hippocrateaceae from New Guinea. J. Arnold Arbor. 22: 389–394.
Stant, M. Y. 1952. Notes on the systematic anatomy of *Stackhousia.* Kew Bull. 1951: 309–318.
Van Campo, M., & N. Hallé. 1959. Les pollens des Hippocratéacées d'Afrique de l'Ouest. Pollen & Spores 1: 191–272.
Van Staveren, M. G. C., & P. Baas. 1973. Epidermal leaf characters of the Malesian Icacinaceae. Acta Bot. Neerl. 22: 329–359.
Waanders, G. L., J. J. Skvarla, & C. C. Pyle. 1968. Fine structure of Hippocrateaceae pollen walls. Pollen & Spores 10: 189–196.

SYNOPTICAL ARRANGEMENT OF THE FAMILIES OF CELASTRALES

1 Flowers diplostemonous, with 4 + 4 functional stamens; petals none; disk none; habit ericoid .... 1. GEISSOLOMATACEAE.

1 Flowers haplostemonous (diplostemonous in a few Celastraceae), sometimes (Corynocarpaceae) with a set of staminodes in addition to the functional stamens; petals present or rarely absent; disk often present; habit not ericoid.

2 Ovules mostly erect and basal-axile, or several and superposed in 2 rows on the axile placenta in each locule (seldom apical-axile

in Celastraceae), bitegmic; disk well developed except in Salvadoraceae; leaves opposite or less often alternate.

3 Flowers with a well developed nectary-disk, pentamerous or less often tetramerous; carpels mostly 3–5, seldom only 2.

4 Woody plants; flowers hypogynous or only slightly perigynous (but the disk sometimes enveloping the ovary); leaves opposite or alternate.

5 Disk intrastaminal, or the stamens seated on the disk, only rarely the disk extrastaminal; stamens 4–5 (–10); seeds mostly with endosperm; plants with or often without a latex-system; seeds mostly arillate .... 2. CELASTRACEAE.

5 Disk extrastaminal; stamens (2) 3 (–5); seeds without endosperm; plants generally with a well developed latex-system .... 3. HIPPOCRATEACEAE.

4 Herbs, flowers distinctly perigynous, the disk lining the hypanthium; leaves alternate; seeds with endosperm .... 4. STACKHOUSIACEAE.

3 Flowers without a fully developed nectary-disk (but the stamens sometimes alternating with nectariferous glands), mostly tetramerous; carpels 2; seeds without endosperm .... 5. SALVADORACEAE.

2 Ovules pendulous, apical or apical-axile, 1–2 per locule, bitegmic or more often unitegmic; disk mostly wanting, or represented only by nectary-glands alternating with the stamens, fully developed only in some Dichapetalaceae; leaves alternate or rarely opposite.

6 Stipules wanting or vestigial; seeds with endosperm except in some Icacinaceae.

7 Woody plants, without milky juice; fruit drupaceous or seldom samaroid; corolla polypetalous or sometimes more or less sympetalous.

8 Flowers not enclosed by a calyptrate bracteole; ovules unitegmic; no disk or nectary-glands except in some Icacinaceae; endosperm not ruminate.

9 Locules (2–) 4–6 or more, each with a solitary (rarely 2) ovule; pedicel not articulated at the summit ....6. AQUIFOLIACEAE.

9 Fertile locule solitary (rarely 3), with (1) 2 ovules; pedicel articulated at the summit ....7. ICACINACEAE.

8 Flowers enclosed in bud by a calyptrate bracteole; ovules bitegmic; nectary-glands well developed, alternating with the stamens; endosperm ruminate; ovary bilocular, one locule with 2 ovules, the other empty 8. AEXTOXICACEAE.

7 Climbing herbs with milky juice; fruit samaroid; corolla distinctly sympetalous .... 9. CARDIOPTERIDACEAE.

6 Stipules present; endosperm wanting; disk present and intrastaminal, or more often represented by nectary-glands alternating with the stamens.

10 Flowers with petaloid staminodes alternating with the functional stamens, which are opposite the petals; ovary bicarpellate, but one carpel usually more or less reduced; ovule solitary, or solitary in each of the 2 locules; ovules crassinucellar; pollen grains dicolporate 10. CORYNOCARPACEAE.

10 Flowers without staminodes; stamens alternate with the petals; ovary with 2–3 (4) equal carpels and locules, each locule with 2 tenuinucellar ovules; pollen grains tricolporate .... 11. DICHAPETALACEAE.

1. Family GEISSOLOMATACEAE Endlicher 1841 nom. conserv., the Geissoloma Family

Xerophytic evergreen shrubs with somewhat quadrangular twigs that are at first covered with unicellular, malpighian hairs; plants accumulating aluminum; solitary or clustered crystals of calcium oxalate present in some of the cells of the parenchymatous tissues, including the mesophyll; vessel-segments with slanting, scalariform perforation-plates that have rather numerous (ca 12–23) cross-bars; imperforate tracheary elements with bordered pits; wood-rays heterocellular, mixed uniseriate and pluriseriate, the latter up to 6 or 8 cells wide, with elongate ends; wood-parenchyma scanty, diffuse; cortex containing scattered stone-cells. LEAVES opposite, subsessile, leathery, simple, entire, pinnately veined; epidermis containing some mucilaginous cells; stomates anomocytic; mesophyll without sclereids; stipules petiolar, minute and vetigial. FLOWERS terminal on short, axillary branches that have 3 pairs of bracts, sometimes also axillary to these bracts, perfect, regular, tetramerous throughout, monochlamydeous; tepals and stamens adnate to the base of the ovary; tepals 4, petaloid, rose-colored; nectary-disk wanting; stamens in 2 cycles of 4 each, the outer cycle alternating with the tepals; filaments slender; anthers tetrasporangiate and dithecal, opening by longitudinal slits; pollen-grains tricolporate; gynoecium of 4 carpels united to form a compound, 4-locular, largely superior ovary that is 4-ridged and -grooved and 4-lobed at the tip, each lobe prolonged into a slender style with a terminal, punctate stigma, the styles and stigmas distally connivent; ovules 2 in each locule, pendulous from near the top of the central axis of the ovary, anatropous, bitegmic, sometimes with an integumentary tapetum. FRUIT a loculicidal capsule enveloped by the persistent perianth; seeds solitary in each locule, with an elongate, dicotyledonous, central embryo and rather thin endosperm.

The family Geissolomataceae consists of the single species *Geissoloma marginatum* (L.) A. Juss., native to the Langeberg Mts., Cape province, South Africa. The family has often in the past been associated with the Penaeaceae (Myrtales), which occur in the same region, and to which it has some habital resemblance. It differs from the Penaeaceae and other Myrtales in lacking internal phloem, and recent studies have shown a series of other differences. *Geissoloma* would be a highly discordant element in the otherwise relatively homogeneous order Myrtales.

The least uncomfortable position for the Geissolomataceae is at the beginning of the order Celastrales. The pollen of *Geissoloma* is much like that of some of the Celastraceae and Hippocrateaceae. *Geissoloma* differs from the bulk of the Celastraceae in having monochlamydeous flowers with 2 cycles of stamens, and in lacking a disk, but each of these features can be found individually in undoubted Celastrales. The apically lobed ovary and distinct styles of *Geissoloma* are more primitive than anything in the Celastrales, however, and suggest a connection to the ancestral Rosales. Inasmuch as both the Celastrales and the Myrtales are here considered to be derived directly from the Rosales, the transfer of the Geissolomataceae from a position in the Myrtales to a position at the beginning of the Celastrales is perhaps not so significant as the change in the linear sequence of families might suggest.

2. Family CELASTRACEAE R. Brown in Flinders 1814 nom. conserv., the Bittersweet Family

Trees or shrubs, sometimes scandent, usually glabrous, commonly with scattered tanniferous cells (these containing proanthocyanins but only seldom ellagic acid) and sometimes saponiferous, often with alkaloidal amines such as cathine, and rarely with benzyl-isoquinoline alkaloids, only rarely cyanogenic, and without iridoid compounds; latex-sacs or -canals often present in the phloem of the stem and leaves; some of the cells of the parenchymatous tissues commonly with solitary or clustered crystals of calcium oxalate or containing crystal sand; nodes unilacunar; vessel-segments with simple or seldom scalariform perforations; imperforate tracheary elements with simple or bordered pits, sometimes scattered or in bands, often some of them septate; wood-rays heterocellular or homocellular, of various types, often some of them uniseriate and others 2–8 cells wide; wood-parenchyma very diverse, commonly scanty or wanting. LEAVES opposite or alternate, simple, seldom (as in *Canotia*) much reduced or virtually obsolete; stomates variously laterocytic or paracytic, less commonly anisocytic or anomocytic; petiolar anatomy very diverse, often complex; stipules mostly small and caducous, or none. FLOWERS in terminal or axillary, cymose or less often racemose inflorescences (some of the inflorescences modified into coiled tendrils in *Lophopyxis*), or seldom solitary in the axils, mostly rather small, commonly greenish or white, perfect or less often unisexual, regular, 4–5-merous, hypogynous or shortly perigynous or sometimes half-epigynous; sepals small, imbricate or seldom valvate, distinct or

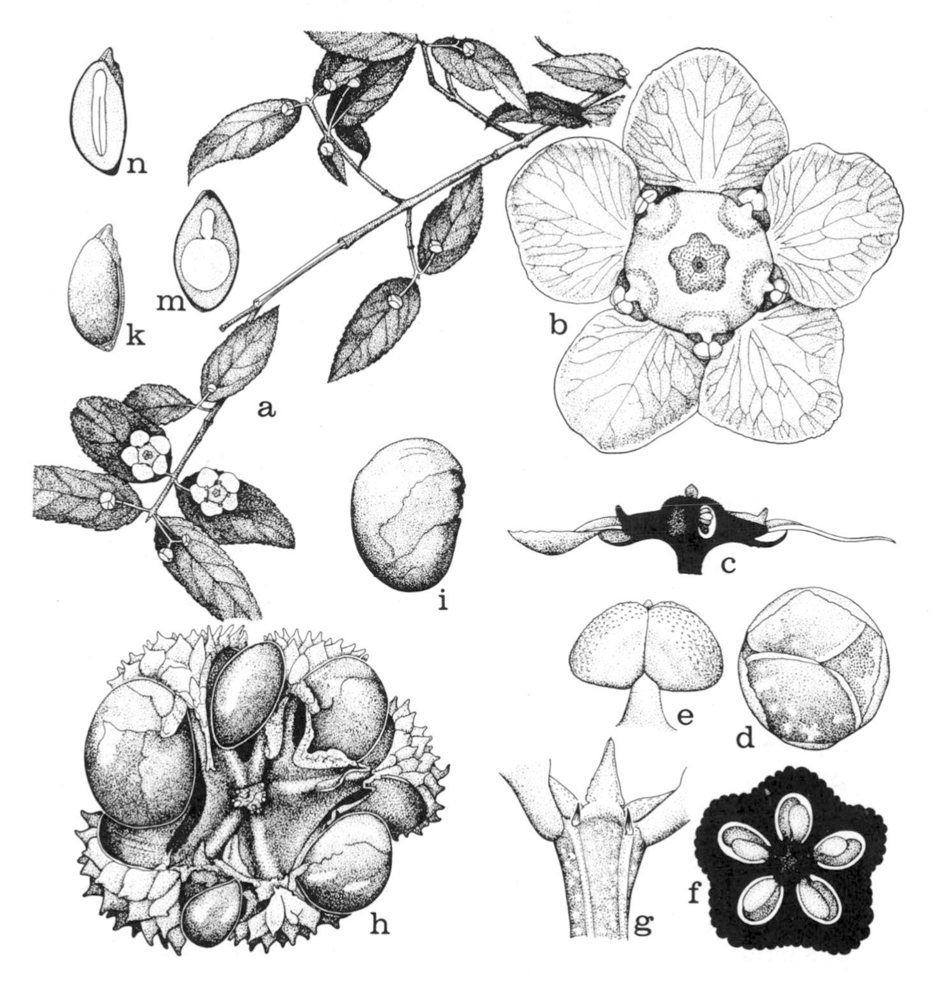

FIG. 5.36 Celastraceae. *Euonymus americanus* L. a, habit, ×½; b, flower, ×4; c, schematic long-section of flower, ×4; d, flower bud, ×4; e, stamen, ×12; f, schematic cross-section of ovary, ×12; g, node, with stipules, ×8; h, opening fruit, from above, ×3; i, seed, with aril, ×3; k, seed, stripped of aril, ×3; m, n, long sections of seed, ×3.

connate at the base or even to above the middle; petals imbricate, rarely convolute or valvate, distinct, or rarely wanting; stamens seated on or outside (seldom inside) the nectary-disk, typically unicyclic and alternate with the petals, seldom bicyclic, in *Lophopyxis* unicyclic and with a set of alternisepalous staminodes; anthers tetrasporangiate and dithecal, or seldom bisporangiate, commonly introrse (extrorse in spp. of *Euonymus*; pollen-grains binucleate or trinucleate, tricolporate or sometimes (as in *Siphonodon*) triporate, often borne in tetrads or polyads; nectary-disk generally well developed, often adnate to the ovary, which may appear to be more or less sunken into it (much-enlarged and almost enclosing the ovary in *Siphonodon*, poorly developed in *Schaefferia*; gynoecium of 2–5 carpels united to form a compound, superior or rarely half-inferior ovary with as many locules as carpels (rarely all the locules but one abortive); style terminal, generally short, with capitate or 2–5 lobed stigma, or seldom (*Goupia*) the styles distinct; stigmas dry; ovules (1) 2 or sometimes up to 6 (numerous in *Goupia*) in each locule, erect or less often pendulous from the axile placenta, apotropous or very rarely epitropous, anatropous, bitegmic, crassinucellar or tenuinucellar, with an integumentary tapetum; endosperm-development nuclear. FRUIT variously a capsule, samara, berry, or drupe; seeds commonly more or less completely enveloped by a well developed aril originating from the integument near the funiculus; endosperm copious and more or less oily, or rarely wanting; embryo with a short radicle and 2 large, flat, foliaceous cotyledons. X = 8, 12, or more. (Canotiaceae, Chingithamnaceae, Goupiaceae, Lophopyxidaceae, Siphonodontaceae)

The family Celastraceae as here defined consists of about 50 genera and 800 species, pantropical and with a lesser number of genera and species in temperate regions. The largest genus by far is *Euonymus*, with more than 200 species, more or less cosmopolitan, but best developed in southeast Asia. A number of species of *Euonymus* are cultivated in temperate regions for their ornamental foliage. The family is rather diversified and loosely knit. Five of the more aberrant genera, *Canotia, Chingithamnus, Goupia, Lophopyxis,* and *Siphonodon*, have been taken as separate families.

Pollen referred to the Celastraceae is known from Oligocene and more recent deposits. Cretaceous macrofossils referred to the family need to be re-examined.

3. Family HIPPOCRATEACEAE A. L. de Jussieu 1811 nom. conserv., the Hippocratea Family

Shrubs, rather small and slender trees, or woody vines with slender, succulent branches, usually glabrous, tanniferous but apparently without proanthocyanins, only seldom saponiferous, sometimes with one or another sort of included phloem, generally (not always) with well developed latex-canals in the parenchymatous tissues of the stem, sometimes also in the leaves, and with solitary or clustered crystals of calcium oxalate in some of the cells; vessel-segments mostly with simple perforations; imperforate tracheary elements with simple or bordered

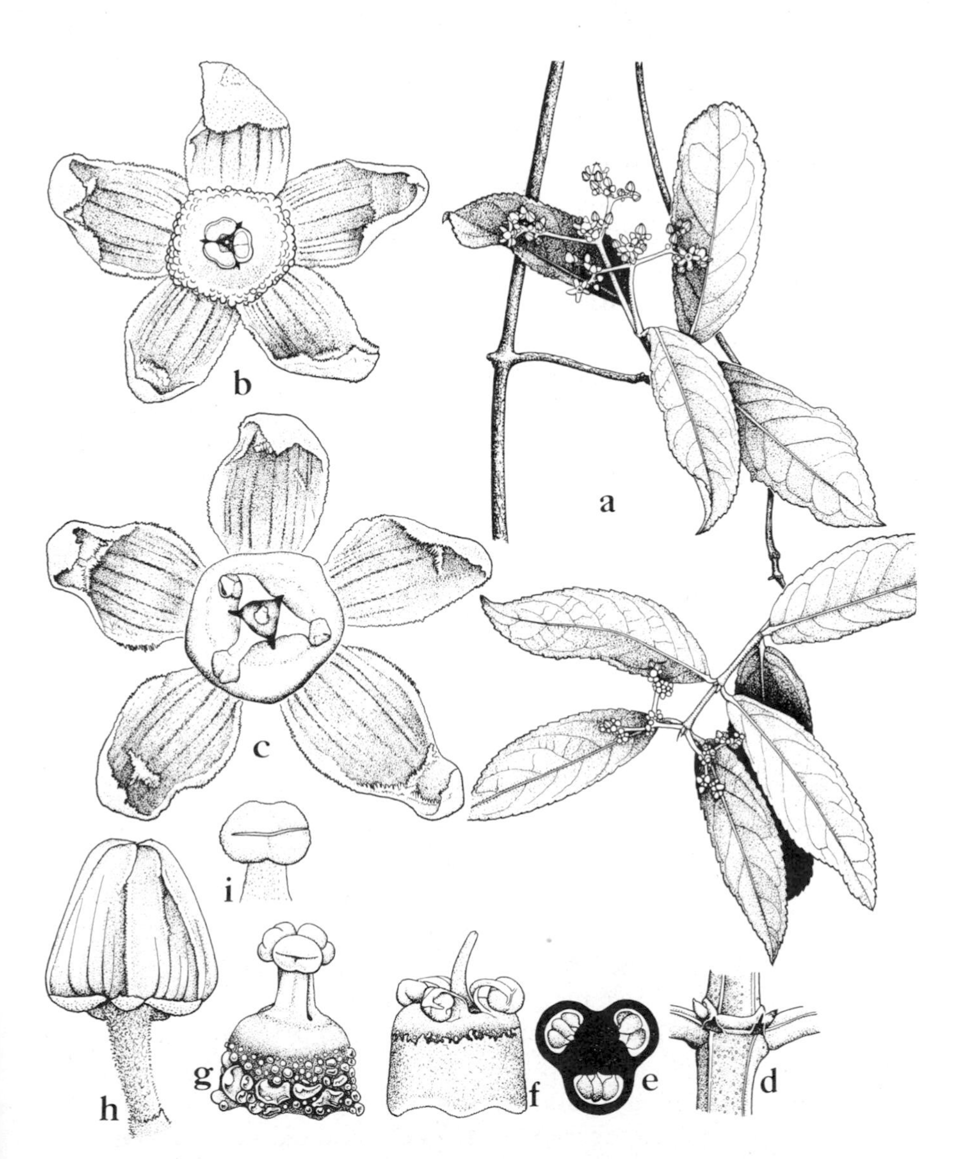

FIG. 5.37 Hippocrateaceae. *Hippocratea volubilis* L. a, habit, $\times\frac{1}{2}$; b, flower at early anthesis, the stamens surrounding the style, ×9; c, flower at late anthesis, with the stamens recurved, revealing the ovary and style, ×9; d, node, with minute stipules, ×3; e, schematic cross-section of ovary, ×15; f, stamens, style, and nectary disk at late anthesis, ×9; g, stamens and nectary disk at early anthesis, ×9; h, flower bud, ×9; i, anther, ×15.

pits; wood-rays heterocellular, uniseriate or of mixed sizes, the larger ones sometimes up to 30 cells wide; wood-parenchyma rather scanty and paratracheal, but thin-walled septate fibers scattered in parenchyma-like fashion amongst ordinary fibers. LEAVES opposite or rarely alternate, simple; stomates anomocytic or encyclocytic (laterocytic); petiolar anatomy very diverse and often complex; stipules small or wanting. FLOWERS mostly rather small, in various sorts of mostly cymose inflorescences, bracteolate, generally perfect, regular, hypogynous; sepals 5, rarely only 2 or 3, imbricate, generally connate at the base; petals 5, or rarely only 2, distinct, imbricate or valvate; nectary-disk well developed and extrastaminal, often cupular, or sometimes forming a short, stout androgynophore, or rarely wanting (as in *Campylostemon*); stamens mostly 3, rarely 2, 4, or 5, usually or always aligned with the sides of the ovary when 3, alternate with the petals when 5; filaments often expanded and shortly connate at the base; anthers tetrasporangiate and dithecal, extrorse or seldom introrse, commonly opening by transverse slits; pollen-grains binucleate, tricolporate, often borne in tetrads or polyads; gynoecium of 3 (rarely 2 or 5) carpels united to form a compound, superior, commonly triangular ovary with as many locules as carpels and with axile placentation; style terminal, with as many stigmas as carpels; ovules 2–10 (–15) in each locule, anatropous, bitegmic, crassinucellar (often weakly so) or seldom tenuinucellar, with an integumentary tapetum at least in *Hippocratea*; endosperm-development nuclear. FRUIT a 1–3-locular drupe or a berry, or often a strongly 3-lobed capsule, the free lobes extending far above the short, axile placenta; seeds mostly compressed or angular, often winged, without endosperm; embryo large, oily, with a minute radicle and 2 thickened, often connate cotyledons. X = 14.

The family Hippocrateaceae consists of more than 300 species, widespread in tropical regions. The vast majority of the species have traditionally been referred to only 2 genera, *Hippocratea* (100) and *Salacia* (200). In another view, these genera are subdivided (as many as 12 for *Hippocratea* alone), and *Hippocratea* sens. strict. becomes a monotype. None of the species of Hippocrateaceae is of any great economic importance.

Authors are agreed that the Hippocrateaceae are closely allied to the Celastraceae, and some would combine the two families. Robson (1965) considers that the Hippocrateaceae are not a natural group, being derived in two separate lines from the Celastraceae.

Pollen considered to represent the Hippocrateaceae is known from Oligocene and more recent deposits.

4. Family STACKHOUSIACEAE R. Brown in Flinders 1814 nom. conserv., the Stackhousia Family

More or less xeromorphic, rhizomatous perennial herbs, seldom half-shrubby or annual, with scattered tanniferous cells in the parenchymatous tissues, producing proanthocyanins, apparently lacking crystals; xylem in the woody spp forming a continuous cylinder with uniseriate rays; vessel-segments with simple perforations; imperforate tracheary elements with bordered pits. LEAVES alternate, simple, entire, sessile, fleshy or leathery, or sometimes much-reduced; stomates anomocytic; stipules tiny, often deciduous, or wanting. FLOWERS variously in racemose or cymose inflorescences, rather small, perfect, regular (except for the usually unequal stamens), perigynous, with a well developed cupular hypanthium, pentamerous, haplostemonous, with 2 bracteoles except in *Macgregoria*; sepals 5, appearing as lobes of the hypanthium, imbricate; petals 5, imbricate, clawed, the claws distinct, but the blades (except in *Macgregoria*) connate into a tube with imbricate lobes; nectary-disk thin, lining the hypanthium; stamens alternate with the petals, 3 longer than the other 2 (except in *Macgregoria*); anthers tetrasporangiate and dithecal, opening longitudinally; pollen-grains 2–3-nucleate, tricolporate; gynoecium of (2) 3 or less often 5 carpels united to form a compound, plurilocular ovary that is laterally and usually also apically lobed; style terminal, but commonly sunken between the lobes; stigmas or style-branches as many as the carpels; locules as many as the carpels; each with a solitary, erect, axile-basal ovule, these apotropous to sometimes epitropous, anatropous, bitegmic, tenuinucellar; endosperm-development nuclear; suspensor very short. FRUIT dry, ripening into indehiscent mericarps; seeds with large, straight, dicotyledonous embryo and oily endosperm. X = 9, 10, 15.

The family Stackhousiaceae consists of 3 genera and some 20–25 species, native principally to Australia and New Zealand, but with one species extending to many of the tropical islands of the Southwest Pacific. *Macgregoria* and *Tripterococcus* are monotypic; the remaining species belong to *Stackhousia*.

5. Family SALVADORACEAE Lindley 1836 nom. conserv., the Mustard-tree Family

Shrubs (sometimes scandent) or rather small trees, sometimes producing mustard-oils, and sometimes piperidine alkaloids, but not tanniferous, lacking both proanthocyanins and ellagic acid; clustered crystals of calcium salts (not oxalate) present in some of the cells of the parenchymatous tissues; vessel-segments with simple perforations; imperforate tracheary elements very short, with simple pits; wood-rays homocellular or slightly heterocellular, mostly 2–5 cells wide, with only a few uniseriates; wood-parenchyma scanty-paratracheal to vasicentric; interxylary phloem often present. LEAVES opposite, simple, commonly leathery, sometimes with axillary thorns; stomates paracytic or sometimes anisocytic or anomocytic; stipules very small or wanting. FLOWERS in various sorts of axillary or terminal, mostly indeterminate inflorescences, small, perfect or unisexual (the plants then variously polygamous or dioecious), regular, hypogynous, haplostemonous; calyx of 2–4 (5) sepals united into a lobed tube, the lobes commonly imbricate; petals 4 (5), imbricate, distinct or (*Salvadora*) shortly connate at the base; stamens alternate with the petals, the filaments variously free and distinct (*Azima*), or connate below into a tube (*Dobera*), or adnate at the base to the corolla-tube (*Salvadora*); anthers tetrasporangiate and dithecal, opening by longitudinal slits; pollen-grains commonly tricolporate; nectary-disk wanting, or sometimes represented by small, distinct glands alternating with the stamens; gynoecium of 2 carpels united to form a compound ovary with a terminal, short or very short style and entire or 2-lobed stigma; ovary bilocular in *Azima*, unilocular in *Dobera* and *Salvadora*; ovules 1 or 2 in each locule, basal or axile-basal, erect, apotropous, anatropous, bitegmic, crassinucellar; endosperm-development nuclear. FRUIT a mostly 1-seeded berry or drupe; seeds without endosperm; embryo with 2 thickened, oily, cordate cotyledons. X = 12.

The family Salvadoraceae consists of 3 small genera, *Azima* (4), *Dobera* (3), and *Salvadora* (5), native to Africa and Madagascar, and across the Middle East to India, Ceylon, and Southeast Asia, usually in fairly dry regions. *Salvadora persica* L., the mustard-tree, occurs in shrub-savannas from northwestern India to Africa. It has tough wood and pungent, edible fruits.

6. Family AQUIFOLIACEAE Bartling 1830 nom. conserv., the Holly Family

Shrubs or trees (the trees mostly rather small), evergreen or less often deciduous, with unilacunar or sometimes trilacunar nodes, sometimes saponiferous and sometimes producing purine or pyridine base alkaloids, tanniferous or not, sometimes accumulating proanthocyanins but not ellagic acid, only seldom cyanogenic, and without iridoid com-

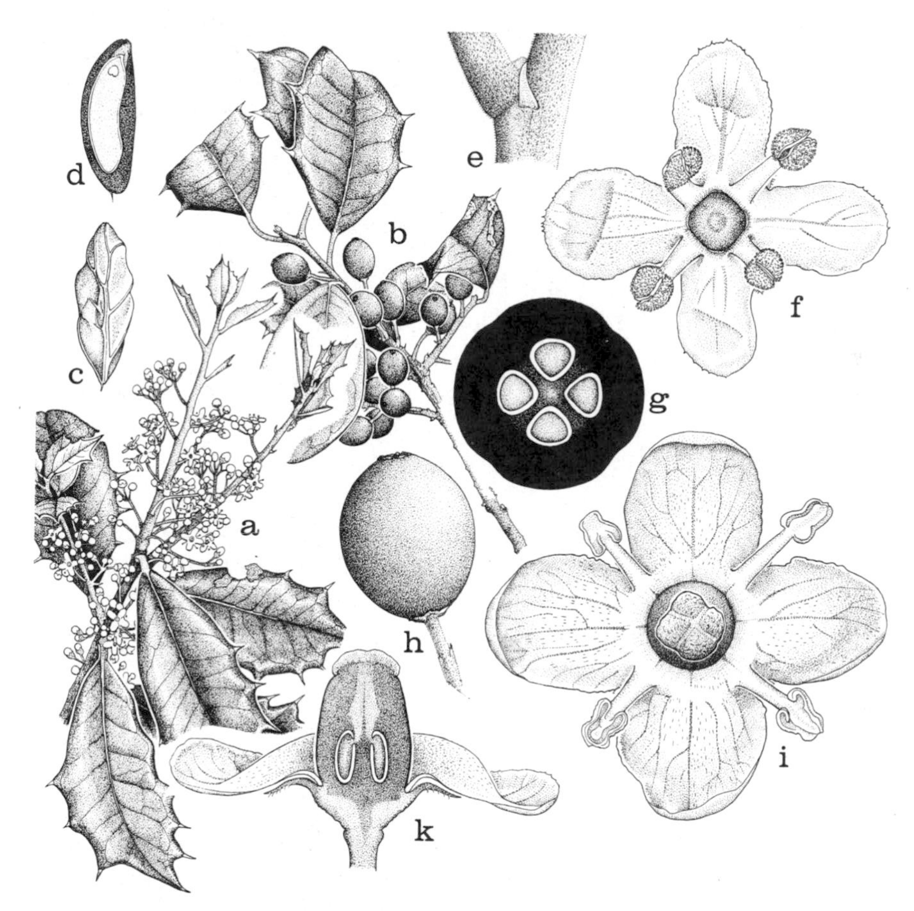

FIG. 5.38 Aquifoliaceae. *Ilex opaca* Aiton. a, b, habit, ×½; c, pyrene, ×3; d, pyrene and seed in schematic long-section, ×3; e, node, with stipule, ×6; f, staminate flower, ×6; g, schematic cross-section of ovary, ×12; h, fruit, ×2; i, perfect flower, from above, ×6; k, perfect flower, in schematic long-section, ×6.

pounds; clustered and sometimes also solitary crystals of calcium oxalate present in some of the cells of the parenchymatous tissues; vessel-segments with more or less oblique, scalariform perforations, commonly with numerous cross-bars; imperforate tracheary elements with evidently bordered pits, often septate; wood-rays heterocellular, mixed uniseriate and pluriseriate, the latter (2–) 5–15 (–25) cells wide, with elongate or (*Phelline*) short ends; wood-parenchyma diffuse or diffuse-in-aggregates, or (*Phelline*) scanty-paratracheal. LEAVES simple, alternate or rarely opposite, sometimes (as in *Phelline*) in well spaced pseudo-whorls; epidermis commonly mucilaginous; stomates of diverse types;

mesophyll and parenchymatous tissues of the stem commonly containing resiniferous and laticiferous idioblasts, the contents of which are the "fat bodies" of the older literature; stipules very small and often caducous, or wanting. FLOWERS in small, axillary, cymose inflorescences or (*Phelline*) in axillary racemes or panicles or mixed panicles, or (*Sphenostemon*) in terminal or subterminal racemes, or sometimes solitary and axillary, in any case small, regular, and hypogynous, generally unisexual (the plants dioecious) but in *Sphenostemon* and spp. of *Ilex* sometimes perfect, 4-merous or less often 5–6-merous, seldom 7–8-merous; sepals small, imbricate, typically more or less connate at the base but distinct and caducous in *Sphenostemon,* in *Nemopanthus* wanting from the pistillate flowers and caducous in the staminate flowers; petals shortly connate at the base in *Ilex,* distinct in the other genera, imbricate or (*Phelline*) valvate, or rarely wanting; stamens usually as many as and alternate with the petals, commonly adnate to the base of the corolla in *Ilex,* sometimes more numerous than the petals in polycarpous species of *Ilex,* and up to 12 (but still apparently uniseriate) in *Sphenostemon*; anthers tetrasporangiate and generally dithecal, opening by longitudinal slits, but in *Sphenostemon* the stamens more or less laminar, each with 4 embedded microsporangia; pollen-grains binucleate, 3–4 colporate or -porate, or inaperturate; nectary-disk wanting; gynoecium of (2–) 4–6 (–8 or even 24) carpels united to form a compound, superior ovary with as many locules as carpels; style terminal and short, or wanting, the stigma lobed or capitate; ovules solitary (2) in each locule, apical-axile, pendulous, apotropous, anatropous to sometimes (*Phelline*) hemitropous or somewhat campylotropous, unitegmic, crassinucellar or sometimes tenuinucellar, with an integumentary tapetum, the funiculus often with a ventral protuberance that may represent a suppressed second ovule; endosperm-development cellular. FRUIT a drupe with usually as many stones as carpels; seeds with a very small, dicotyledonous embryo near the micropyle; endosperm abundant, oily and proteinaceous, without starch. X = 9, 10. (Phellineaceae, Sphenostemonaceae)

The cosmopolitan family Aquifoliaceae as here defined consists of 4 genera, *Ilex* (including *Byronia*), *Nemopanthus, Phelline,* and *Sphenostemon* (including *Nouhuysia*). *Ilex* is by far the largest genus, with 300–400 species, of irregularly cosmopolitan distribution. *Phelline* has about 10 species, native to New Caledonia, and *Sphenostemon* has about 7, native to the southwestern Pacific. *Nemopanthus* has a single species, native to eastern North America. *Phelline* and *Sphenostemon* have sometimes been taken as separate families, but the relationships of the former, at least, are not in dispute. A number of species of *Ilex* (holly) are cultivated for their ornamental evergreen foliage. Leaves of *Ilex paraguariensis* A. St.-Hil. are used in South America to brew maté, a popular tea.

Fossil pollen attributed to the genus *Ilex* first appears in Turonian (early Upper Cretaceous) deposits in Australia. By Coniacian time it is also in Africa. By the beginning of the Tertiary the type is cosmopolitan. Hu (1967) considers that "The evolutionary lines that involve the widespread groups of the family were well established in the Oligocene."

7. Family ICACINACEAE Miers 1851 nom. conserv., the Icacina Family

Trees, shrubs, or scrambling or twining woody vines with various types of hairs, chemically diverse, sometimes accumulating aluminum, sometimes producing purine-base or emetine alkaloids, sometimes with iridoid compounds (not known in other families of the order), sometimes cyanogenic, sometimes saponiferous, and at least sometimes accumulating proanthocyanins, but without ellagic acid; mucilage-canals sometimes present in the parenchymatous (or other living) tissues; clustered (and often also solitary) crystals of calcium oxalate usually present in some of the cells of the parenchymatous tissues, but crystal sand only seldom present; stem often with anomalous structure, sometimes with interxylary phloem; nodes trilacunar or unilacunar; vessels variously with scalariform or simple perforations, the two types sometimes occurring together; imperforate tracheary elements with bordered or simple pits, variously consisting of tracheids, fiber-tracheids, or libriform fibers; wood-rays heterocellular, mixed uniseriate and pluriseriate, the latter 3–10 cells wide (or even wider), with short or elongate ends; wood-parenchyma varying from apotracheal and diffuse to vasicentric. LEAVES alternate or rarely (*Iodes*) opposite, simple, entire or occasionally toothed; stomates of diverse types, most often encyclocytic; stipules wanting. FLOWERS in various sorts of usually axillary inflorescences, hypogynous, regular, perfect or less often unisexual (the plants then from polygamous to dioecious), (3) 4–5 (6)-merous; pedicel articulated at the summit; sepals small, connate into a tube with imbricate or rarely valvate lobes, not accrescent; petals valvate or rarely subimbricate, distinct or connate at the base or even for much of their length, or rarely wanting; stamens antesepalous, isomerous with the petals (or sepals), filaments free or borne on the corolla alternate with its lobes, often hairy near the tip; anthers tetrasporangiate and dithecal, opening by longitudinal slits or less often by apical pores; pollen-grains binucleate, of diverse, mostly triaperturate types, or sometimes inaperturate; nectary-disk usually wanting, but sometimes present or represented by distinct glands; gynoecium of (2) 3 (–5) carpels united to form a compound ovary with a short, terminal style and simple to evidently lobed stigma, pseudomonomerous in *Gomphandra, Gonocaryum,* and *Phytocrene*; ovary usually functionally unilocular, only one locule fully developed and ovuliferous, the others empty and reduced or vestigial or obsolete, only rarely (*Emmotum*) all 3 locules ovuliferous; ovules (1) 2 in the fertile locule(s), hanging back to back from the top of the ovary, with the micropyle turned outwards, anatropous, unitegmic, crassinucellar to tenuinucellar, with a funicular thickening near the micropyle; endosperm-development nuclear. FRUIT commonly a 1-seeded drupe, seldom dry and samaroid; seeds with a straight or curved, dicotyledonous embryo and well developed, oily endosperm, or sometimes without endosperm. X = 10, 11. *Metteniusa* seems unicarpellate.

The family Icacinaceae consists of about 50 genera and some 400 species, of pantropical distribution, with relatively few species in temperate regions. The largest genus is *Gomphandra,* with about 50 species native to Indomalesia and some of the islands of the southwestern Pacific. A few species are of minor economic importance. The seeds and tubers of *Icacina senegalensis* Juss., a tropical African species, provide a starchy flour. Species of the tropical South American genus *Poraqueiba* have a locally esteemed edible fruit.

Fossil pollen of the Icacinaceae occurs in rocks as old as the Paleocene, and is perhaps more doubtfully reported from the Upper Cretaceous of New York. Fossil fruits are known from the Paleocene of Egypt and the Eocene of Europe and Oregon.

8. Family AEXTOXICACEAE Engler & Gilg 1919 nom. conserv., the Aextoxicon Family

Trees with the twigs, lower side of the leaves, and inflorescences covered with peltate scales, tanniferous but not saponiferous; vessels with scalariform perforations that have rather numerous cross-bars; imperforate tracheary elements with bordered pits; wood-rays heterocellular, with elongate ends; wood-parenchyma diffuse. LEAVES alternate or subopposite, simple, entire, pinntely veined; stomates encyclocytic, deeply impressed; stipules wanting. FLOWERS in axillary racemes, small, unisexual (the plants dioecious), (4) 5 (6)-merous, hypogynous, regular, completely enveloped in bud by a calyptrate bracteole; sepals distinct, strongly imbricate, thin, deciduous; petals distinct, broadly clawed, imbricate; stamens distinct, antesepalous, alternating with well developed reniform nectary-glands; filaments thick and fleshy; anthers dithecal, opening by short slits toward the summit, pollen-grains tricolporate; staminate flowers with a vestigial gynoecium, and pistillate flowers with short but evident, linear, fleshy, antesepalous staminodia alternating with the nectary-glands; gynoecium of 2 carpels united to form a compound, bilocular ovary, the short style recurved and appressed to the ovary, bilobed at the tip; one locule of the ovary empty, the other with 2 anatropous, bitegmic, crassinucellar, apotropous ovules pendulous from the summit; nucellus with a massive beak protruding beyond the integuments. FRUIT a dry drupe with a single stone and seed; seeds with ruminate endosperm and a well developed embryo with a well developed radicle and 2 flattened, cordate-orbicular cotyledons. N = 16.

The family Aextoxicaceae consists of the single species *Aextoxicon punctatum* Ruiz & Pavon, native to Chile. *Aextoxicon* has variously been included in the Euphorbiaceae or treated as a family in the Celastrales, Euphorbiales, Laurales, or Sapindales. It is morphologically at home in the Celastrales, but the massive nucellar beak is reminiscent of that found in several genera of Euphorbiaceae.

9. Family CARDIOPTERIDACEAE Blume 1843 nom. conserv., the Peripterygium Family

Glabrous, twining herbs with articulated laticifers, bearing a milky juice in the leaves and stems; vessel-segments short and very broad, with simple perforations; imperforate tracheary elements with small bordered pits. LEAVES alternate, petiolate, with cordate, entire or lobed, palmately veined blade; stomates anomocytic; stipules wanting. FLOWERS in bractless, axillary cymes with elongate, somewhat secund branches, very small, perfect or unisexual (the plants then polygamous), 5-merous, hypogynous, regular, calyx 5-lobed, the lobes imbricate; corolla shortly sympetalous, 5-lobed, the lobes imbricate; stamens 5, the short filaments attached to the corolla-tube alternate with the lobes; anthers dithecal, opening by longitudinal slits; pollen-grains tricolporate; nectary-disk wanting; gynoecium of 2 carpels united to form a compound, superior ovary with 2 distinct, dissimilar styles, one elongate and persistent in fruit, the other very short, with an evident, capitate stigma; ovary unilocular, with 2 ovules pendulous from the summit, or one of the ovules abortive. FRUIT a broadly 2-winged samara; seed with a very small embryo at the tip of the fleshy endosperm. (Peripterygiaceae)

The family Cardiopteridaceae consists of the single genus *Peripterygium (Cardiopteris)*, with 3 species native from southeastern Asia to New Guinea and Queensland, Australia. The genus is insufficiently studied, and may prove to be misplaced in the Celastrales. The habit, leaves, and milky juice are reminiscent of the Convolvulaceae, and the sympetalous corolla with isomerous stamens attached to the corolla-tube alternate with the lobes, would at least be compatible with the placement of the Cardiopteridaceae in the Asteridae. On the other hand, there are numerous differences between *Peripterygium* and the Convolvulaceae or any other family of Asteridae, and many similarities with the Icacinaceae, which include some genera with a distinctly sympetalous corolla. The latex-system of *Peripterygium* has been compared with that of some Olacaceae, but the two groups do not seem very close in other respects, and it has never been suggested that *Peripterygium* might be parasitic. The Hippocrateaceae, one of the central families in the Celastrales, also have a well developed latex-system. It seems reasonable to retain the Cardiopteridaceae in the Celastrales, at least until the plants have been more fully studied.

10. Family CORYNOCARPACEAE Engler in Engler & Prantl 1897 nom. conserv., the Karaka Family

Trees, containing very toxic, bitter glucosides, at least in the bark and seeds, and also tanniferous, accumulating ellagic acid (at least in the bark) but not proanthocyanins; solitary or clustered crystals of calcium oxalate present in some of the cells of the parenchymatous tissues; vessel-segments short, with simple perforations; imperforate tracheary elements with simple or faintly bordered pits; wood-rays heterocellular, all pluriseriate and notably broad, mostly 6–10 (–16) cells wide; wood-parenchyma vasicentric and in broad paratracheal strips. LEAVES alternate, simple, leathery, entire; stomates paracytic; petiole with 3 traces entering at the base; stipules intrapetiolar, crescent-shaped, subtending the axillary bud, deciduous. FLOWERS in terminal panicles (or mixed panicles), essentially hypogynous, regular, perfect; sepals 5, distinct, strongly imbricate; petals 5, imbricate; functional stamens 5, opposite the petals and adnate to them at the base, alternating with 5 petaloid staminodes; anthers tetrasporangiate and dithecal, the pollen-sacs open-

FIG. 5.39 Corynocarpaceae. *Corynocarpus laevigata* J. R. & G. Forster. a, habit, ×½; b, c, flower, ×8; d, pistil and nectaries, ×8; e, pistil, ×8; f, pistil, in partial long-section, ×12; g, petal and stamen, ×8; h, anther, ×12; i, staminode and nectary, ×8.

ing by longitudinal slits; pollen grains nearly smooth, oblate-ellipsoid, dicolporate, with a short colpus at each end; 5 well developed nectaries borne opposite (and internal to) the staminodes; gynoecium primitively of 2 carpels, with separate styles, capitate stigmas, and a bilocular, compound ovary, but one carpel commonly more or less reduced, or vestigial so that the ovary is pseudomonomerous; each functional locule with a single pendulous, anatropous, bitegmic, crassinucellar ovule, the micropyle directed upwards; endosperm-development nuclear. FRUIT a drupe; seeds very poisonous, with a straight, oily and starchy embryo, 2 large, thickened cotyledons, a minute radicle and plumule, and no endosperm. 2n = 44.

The family Corynocarpaceae consists of the single genus *Corynocarpus*, with about 5 species native to New Zealand, northeastern Australia, New Guinea, and some other islands on the Gondwanaland plate.

The Corynocarpaceae do not fit well into any of the orders of angiosperms, and their taxonomic position has long been uncertain. Most commonly they have been included in the Sapindales or Celastrales, often with an expression of doubt. At one time I referred them to the Ranunculales, but that position has become less attractive in the light of newer information on the chemistry and the pollen. Only the wood-anatomy remains to suggest a position in the Ranunculales, near the Berberidaceae (Heimsch, 1942, under Sapindales; Metcalfe & Chalk). The presence of ellagic acid in *Corynocarpus* is out of harmony with the Magnoliidae as a whole, but not entirely unprecedented, since it also occurs in *Nuphar* (Nymphaeaceae) and the more doubtfully placed family Coriariaceae. The bitter principle of *Corynocarpus*, called karakin, is reported to be identical with hiptagin, of the otherwise very different genus *Hiptage*, in the Malpighiaceae. Walker & Doyle (1975, in general citations) consider that the pollen-morphology of *Corynocarpus* militates against a position in the Ranunculales and argues instead for a place in the Rosidae. Lobreau-Callen (1977) more specifically favors a position in the Myrtales on the basis of pollen-morphology, but *Corynocarpus* has hypogynous flowers and does not have the characteristic internal phloem of the Myrtales. The well developed, intrastaminal nectaries, apparently equivalent to disk-glands, suggest a position in the Rosidae, but not in the Myrtales.

Within the Rosidae, the simple leaves, antepetalous stamens, and other classical morphological features of *Corynocarpus* might at first suggest a position in the Rhamnales. Unfortunately, the well developed staminodes and subapical ovules are out of harmony with that tightly knit order, as is the wood anatomy. It therefore seems most useful to insert the Corynocarpaceae in the more heterogeneous order Celastrales. For whatever reasons, Hutchinson, Takhtajan, and Scholz (in the 12th Engler Syllabus) also find the Celastrales to be the least obnoxious place for the Corynocarpaceae.

11. Family DICHAPETALACEAE Baillon in Martius 1886, nom. conserv. the Dichapetalum Family

Shrubs, rather small trees, or lianas, often highly poisonous, containing fluoracetic acid and pyridine alkaloids, commonly with a characteristic indument consisting of unicellular hairs that have conical or wart-like

Fig. 5.40 Dichapetalaceae. *Dichapetalum gentryi* Prance. a, habit, ×½; b, fruit, ×1; c, flower, ×10; d, flower, in long-section, ×10; e, petal and nectary-gland, ×10; f, detail of lower surface of leaf, ×3. From Brittonia 29: 157. 1977. Courtesy of G. T. Prance and the New York Botanical Garden.

papillae; solitary and clustered crystals of calcium oxalate present in some of the cells of the parenchymatous tissues; mucilage cells sometimes present in the parenchymatous tissues; vessel-segments small, with simple or mixed simple and scalariform perforations; imperforate tracheary elements with very small bordered pits; wood-rays heterocellular, mixed uniseriate and pluriseriate, the latter up to 5 (–10) cells wide; wood-parenchyma both paratracheal and diffuse, but mainly the former. LEAVES alternate, simple, entire, pinnately veined, often with mucilage cells in the epidermis and hypodermis, often bearing a few glands, especially on the lower side towards the base; stomates paracytic; stipules present, usually caducous. FLOWERS small, in axillary to petiolar or epiphyllous cymes, perfect or rarely unisexual, regular or seldom (*Tapura*) somewhat irregular, hypogynous to epigynous (4) 5-merous; pedicels often articulated; sepals distinct or connate below, imbricate; petals mostly bilobed or bifid, distinct and imbricate or seldom connate to form a lobed tube; stamens alternate with the petals or corolla-lobes, free or borne on the corolla-tube, seldom only 3 fertile and 2 staminodial; anthers dithecal, opening longitudinally; connective often thickened on the back; pollen-grains tricolporate; a basal nectary gland borne opposite each petal, or the glands confluent to form a ring when the corolla is sympetalous; gynoecium of 2–3 (4) carpels united to form a compound, superior to inferior, plurilocular ovary; style terminal, simple, apically lobed or with a lobed stigma, or less often the styles distinct; ovules 2 in each locule, apical-axile, pendulous, anatropous, tenuinucellar. FRUIT a dry or seldom fleshy drupe with a unilocular or sometimes 2–3-locular stone that generally has only one seed in each locule, the exocarp sometimes splitting; seeds with a large, straight, oily dicotyledonous embryo and no endosperm, often carunculate. $2n = 20, 24$. (Chailletiaceae)

The family Dichapetalaceae consists of the large, pantropical but mainly African genus *Dichapetalum*, with perhaps 200 species, and 2 additional small satellite genera (*Tapura* and *Stephanopodium*) with only about 35 species in all. The affinities of the Dichapetalaceae are debatable. Morphologically they are most at home in the Celastrales, and Prance (1972) considers them to be closely allied to the Icacinaceae and Celastraceae. Takhtajan (1966) refers them to the Euphorbiales, partly because of some palynological similarities with some of the Euphorbiaceae, but their flowers are not so reduced and modified as in that order. Inasmuch as the Euphorbiales are here regarded as derived from the Celastrales, it seems reasonable to treat the Dichapetalaceae as the final family of the Celastrales.

12. Order EUPHORBIALES Lindley 1833

Autotrophic, terrestrial plants, woody or less often herbaceous, chemically diverse, often producing alkaloids of various sorts, and often with milky or colored juice, but only rarely with ellagic acid or iridoid compounds, and rarely (*Drypetes*) with mustard oils. LEAVES simple or sometimes compound, with or without stipules. FLOWERS hypogynous, unisexual, mostly small, with reduced, often monochlamydeous perianth, or without perianth, only seldom with an evident, even showy perianth of ordinary size; stamens as many or twice as many as the sepals or tepals, or occasionally more numerous, or sometimes reduced (even to 1); pollen of diverse structure, most commonly tricolporate; disk present or absent; gynoecium of 2–5 (-numerous) carpels united to form a compound, plurilocular ovary with as many styles or primary style-branches or sessile stigmas as carpels, very rarely the ovary pseudomonomerous; ovules 1–2 in each locule, apical-axile, pendulous, bitegmic, crassinucellar. FRUIT most commonly an elastically dehiscent capsular schizocarp, less often drupaceous, rarely samaroid or baccate; seeds with copious, oily endosperm, or the endosperm rarely wanting.

The order Euphorbiales as here defined consists of 4 families, dominated by the very large family Euphorbiaceae, to which the other families are attached as small satellites. Of these satellite families, only the Pandaceae are without question closely allied to the Euphorbiaceae. The Buxaceae have usually been associated with the Euphorbiaceae, and both the pollen-morphology and the floral anatomy have been interpreted to favor such an affinity, but an alternative position in the Hamamelidae cannot yet be ruled out. The affinities of *Simmondsia* are uncertain, but its inclusion in the Euphorbiales brings no significant new characters to the order. *Daphniphyllum*, often included in the Euphorbiales or even in the Euphorbiaceae, is here considered to form a distinct family and order in the Hamamelidae.

Leaving aside the satellite families, the position of the Euphorbiaceae in the general scheme of classification is also debatable. In the twelfth edition of the Engler Syllabus they are included in the Geraniales. Takhtajan allies them with the Malvales, in the subclass Dilleniidae. Each of these assignments has a substantial body of tradition and respectable opinion behind it. Webster (1967) cautiously favors a rosid rather than a dilleniid affinity for the family.

The Euphorbiaceae are so diverse in vegetative and chemical features and in pollen-morphology that one could compare some member of the group to any of a wide range of families and orders of dicotyledons.

The euphorbiad gynoecium, on the other hand, is much more stable, as is also the nectary-disk in flowers that are not highly reduced and aggregated into pseudantha. These features should therefore provide a better guide to the affinities of the group, and in these respects the Euphorbiaceae compare well with the Celastrales and Sapindales. Similar gynoecia are much less common in the Dilleniidae, and occur mainly in groups that have other specialized features that militate against a close relationship with the Euphorbiaceae. The malvalean gynoecium is more or less compatible with that of the Euphorhiaceae, but the nectaries in the two groups are wholly different. Therefore I prefer to assign the Euphorbiaceae to the Rosidae rather than to the Dilleniidae. Although some euphorbiads have compound leaves, simple leaves appear to be basic for the group. Thus it seems better to associate the Euphorbiaceae more closely with the Celastrales than with the Sapindales or Geraniales. It may also be noted that the tendency toward dioecism is well established in the Celastrales, and that two celastralean families, the Aextoxicaceae and Dichapetalaceae, have often been thought to be related to the Euphorbiaceae.

SELECTED REFERENCES

Bailey, D. C. 1980. Anomalous growth and vegetative anatomy of *Simmondsia chinensis*. Amer. J. Bot. 67: 147–161.

Croizat, L. 1940. On the phylogeny of the Euphorbiaceae and some of their presumed allies. Revista Univ. (Santiago) 25: 205–220.

Dehay, C. 1935. L'Appareil libéro-ligneux foliaire des Euphorbiacées. Ann. Sci. Nat. Bot. sér. 10, 17: 147–295.

Forman, L. L. 1966. The reinstatement of *Galearia* Zoll. & Mor. and *Microdesmis* Hook.f. in the Pandaceae. With appendices by C. R. Metcalfe and N. Parameswaran. Kew Bull. 20: 309–321.

Forman, L. L. 1968. The systematic position of *Panda* Pierre. Proc. Linn. Soc. London 179: 269–270.

Gentry, H. S. 1958. The natural history of jojoba (*Simmondsia chinensis*) and its cultural aspects. Econ. Bot. 12: 261–295.

Hayden, W. J. 1977. Comparative anatomy and systematics of *Picrodendron*, genus incertae sedis. J. Arnold Arbor. 58: 257–278.

Köhler, E. 1965, Die Pollenmorphologie der biovulaten Euphorbiaceae und ihre Bedeutung für die Taxonomie. Grana Palynol. 6: 26–120.

Mathou, T. 1939. Recherches sur la famille des Buxacées; étude anatomique, microchimique et systématique. Thése Fac. Sci. Toulouse. 1–448.

Nair, N., & V. Abraham. 1962. Floral morphology of a few species of Euphorbiaceae. Proc. Indian Acad. Sci. B, 56: 1–12.

Punt, W. 1962. Pollen morphology of the Euphorbiaceae with special reference to taxonomy. Wentia 7: 1–116.

Raju, V. S., & P. N. Rao. 1977. Variation in the structure and development of foliar stomata in the Euphorbiaceae. J. Linn. Soc., Bot. 75: 69–97.

Record, S. J. 1938. The American woods of the family Euphorbiaceae. Trop. Woods 54: 7–40.
Singh, R. P. 1954. Structure and development of seeds in Euphorbiaceae: *Ricinus communis* L. Phytomorphology 4: 118–123.
Vaughan, J. G., & J. A. Rest. 1969. Note on the testa structure of *Panda* Pierre, *Galearia* Zoll. & Mor. and *Microdesmis* Hook. f. (Pandaceae). Kew Bull. 23: 215–218.
Webster, G. L. 1967. The genera of Euphorbiaceae in the southeastern United States. J. Arnold Arbor. 48: 303–430.
Webster, G. L., & E. A. Rupert. 1973. Phylogenetic significance of pollen nuclear number in the Euphorbiaceae. Evolution 27: 524–531.
Wunderlich, R. 1967 (1968). Some remarks on the taxonomic significance of the seed coat. Phytomorphology 17: 301–311.

Баранова, М. А. 1980. Сравнительно-стоматографическое исследованне семейств Buxaceae и Simmondsiaceae. В кн.: Систематика и эволюция высших растений: 68-75. Наука, Ленинград.
Меликян, А. П. 1968. Положение семейств Buxaceae и Simmondsiaceae в системе. Бот. Ж. 53: 1043-1047.

SYNOPTICAL ARRANGEMENT OF THE FAMILIES OF EUPHORBIALES

1 Ovules apotropous, the raphe dorsal.
  2 Ovules 2 per carpel; endosperm copious; vessels nearly always with scalariform perforations; styles borne on or forming continuations of the apical margin of the ovary; secondary growth of normal type .... 1. BUXACEAE.
  2 Ovules one per carpel; endosperm very scanty or none; vessels with simple perforations; stigmas sessile, deciduous, clustered at the summit of the ovary; secondary growth anomalous, concentric .... 2. SIMMONDSIACEAE.
1 Ovules epitropous, the raphe ventral, or rarely (*Panda*) the ovule orthotropous; ovary with a single common style, or with the styles or stigmas clustered at the summit of the ovary.
    3 Obturator wanting; fruit drupaceous; flowers without a disk; seeds without a caruncle; plants without milky juice; leaves arranged in flat sprays that simulate pinnately compound leaves .... 3. PANDACEAE.
    3 Obturator well developed; fruit usually a capsular schizocarp, only seldom drupaceous; flowers usually with a disk; seeds often with a caruncle; plants often with milky juice; leaves simple or less often compound, or much reduced, only seldom arranged in flat sprays that simulate pinnately compound leaves .... 4. EUPHORBIACEAE.

1. Family BUXACEAE Dumortier 1822 nom. conserv., the Boxwood Family

Evergreen or seldom deciduous woody plants or (*Pachysandra*) suffrutescent herbs, commonly producing steroid alkaloids, but not cyanogenic, not saponiferous, without iridoid compounds, and generally not tanniferous, lacking ellagic acid and only seldom with proanthocyanins; secretory cells scattered in the parenchymatous tissues or arranged in long, branching rows; solitary or clustered crystals of calcium oxalate commonly present in some of the cells of the parenchymatous tissues; nodes unilacunar; vessel-segments very small, nearly always with scalariform perforations; imperforate tracheary elements with obscurely to sometimes evidently bordered pits; wood-rays heterocellular, some uniseriate, others 2 (–4)-seriate, with short or less often elongate ends; wood-parenchyma apotracheal, commonly diffuse. LEAVES opposite or

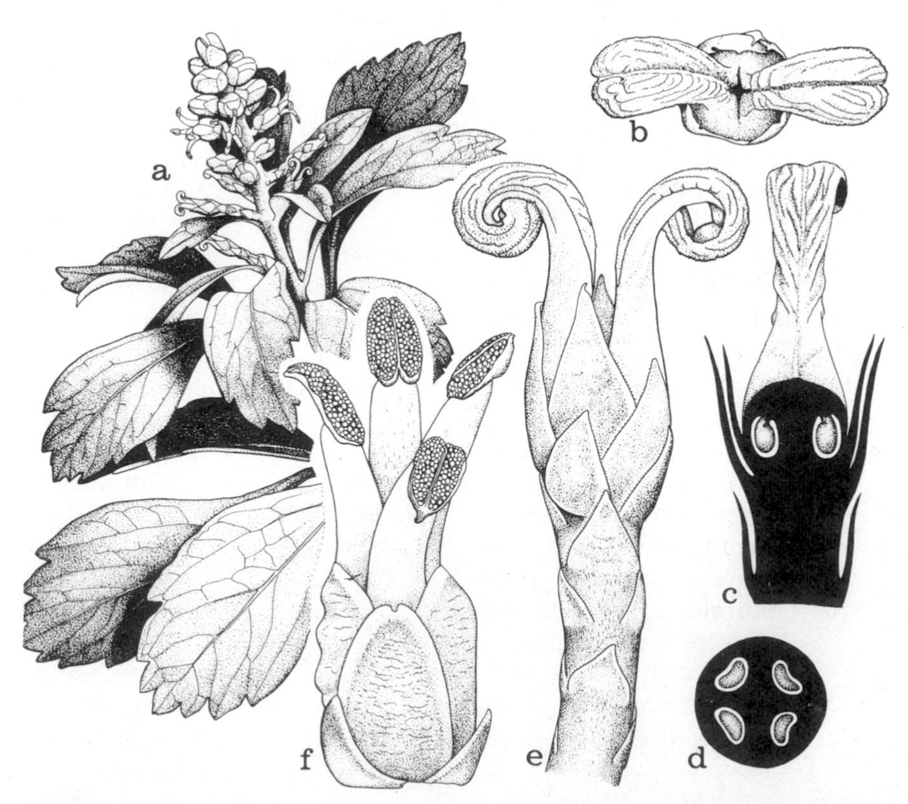

FIG. 5.41 Buxaceae. *Pachysandra terminalis* Sieb. & Zucc. a, habit, ×1; b, pistillate flower, from above, ×8; c, pistillate flower, in partial long-section, ×8; d, schematic cross-section of ovary, ×8; e, pistillate flower, from the side, ×8; f, staminate flower, ×8.

less often alternate, leathery, simple, entire or toothed, mostly penniveined; stomates laterocytic, encyclocytic, or sometimes anomocytic or of an intermediate type; stipules wanting. FLOWERS commonly in spikes or dense racemes or heads, small, regular, hypogynous, unisexual (the plants monoecious or seldom dioecious) or rarely a few of them perfect; tepals small, not petaloid, commonly 2 + 2 (sometimes 5 or 3 + 3, especially in pistillate flowers, wanting in staminate flowers of *Styloceras*); stamens 4 and opposite the tepals, or 2 opposite the outer tepals and 4 in 2 pairs opposite the inner tepals (in *Notobuxus*), or 6–30 (in *Styloceras*); filaments distinct, often broad; anthers large, often dorsifixed, introrse, with a usually broad back, tetrasporangiate and dithecal, opening by longitudinal slits; pollen-grains binucleate, tricolporate to more often multiporate; disk wanting; gynoecium of 3, seldom 2 or casually 4 carpels united to form a compound, plurilocular ovary with the persistent styles borne on or forming continuations of the apical margin; primary locules divided into uniovulate locelli in *Styloceras* and *Pachysandra;* stigma dry, decurrent on the inner side of the style, commonly with a median furrow; ovules 2 in each primary locule, pendulous, apical-axile, apotropous (with dorsal raphe), anatropous, bitegmic, crassinucellar, provided with an obturator at least in *Buxus* and *Sarcococca;* endosperm-development cellular or less often nuclear. FRUIT a loculicidal, elastically dehiscent capsule, or less often a drupe; seeds black and shining, commonly with a straight, axile, dicotyledonous embryo and abundant, firm, oily endosperm, usually carunculate. X = 10, 14. (Pachysandraceae, Stylocerataceae)

The family Buxaceae as here defined consists of 5 genera and about 60 species, nearly cosmopolitan in distribution. Well over half of the species belong to the genus *Buxus*. *Buxus sempervirens* L., boxwood, is often used in hedges because of its dense growth and ornamental evergreen foliage. *Pachysandra terminalis* Sieb. & Zucc. is a familiar ground-cover for shady sites.

Pollen thought to be related to *Pachysandra* is known from Campanian and more recent deposits. Leaves attributed to *Buxus* occur in Upper Miocene deposits of Japan.

2. Family SIMMONDSIACEAE Van Tieghem 1898, the Jojoba Family

Freely branched, evergreen, xerophytic, dioecious shrubs with anomalous secondary growth of concentric type, producing proanthocyanins

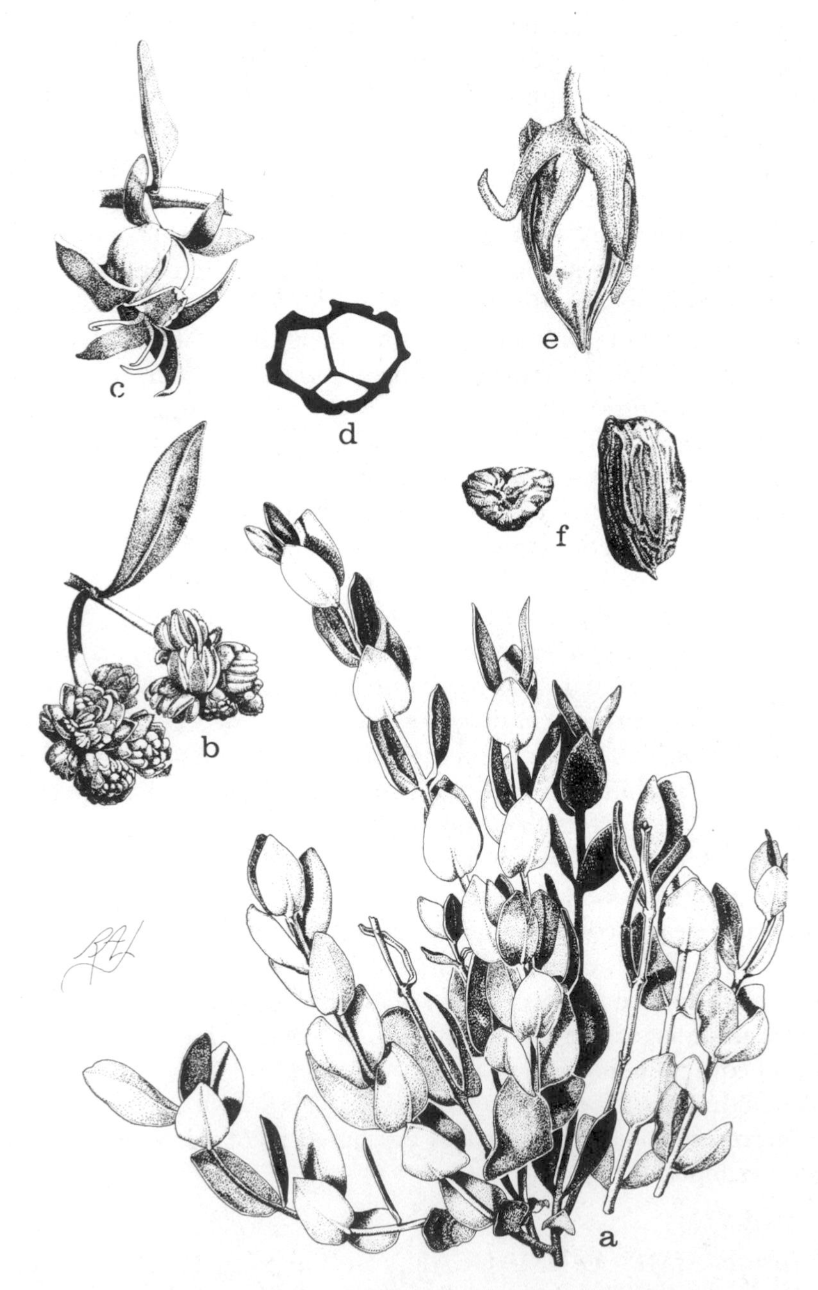

FIG. 5.42 Simmondsiaceae. *Simmondsia chinensis* (Link) C. K. Schneider. a, habit, ×½; b, staminate inflorescences, ×2; c, pistillate flower, ×3; d, schematic cross-section of developing ovary, ×5; e, fruit, ×2; f, two views of seed, ×2.

but not alkaloids; vessel-segments with simple perforations; imperforate tracheary elements with bordered pits; sieve-tube plastids of S-type; wood-parenchyma very scanty. LEAVES small, opposite, sessile, simple but jointed at the base, entire, leathery, pinnately veined; mesophyll consisting wholly of palisade tissue, some of the cells tanniferous, some others containing solitary or clustered crystals of calcium oxalate; stomates laterocytic or some of them anomocytic; stipules wanting. FLOWERS small, regular, hypogynous, apetalous, unisexual, the staminate ones displostemonous, borne in pedunculate or subsessile, head-like, axillary clusters, the pistillate ones mostly solitary and pedicellate in the axils; sepals (4) 5 (6), imbricate, those of the pistillate flowers foliaceous-accrescent; disk wanting; stamens (8) 10 (12), distinct; anthers elongate, basifixed, articulated to the short, stout filament, tetrasporangiate and dithecal, latrorse, opening by longitudinal slits; pollen-grains with 3 poorly defined pores; gynoecium of 3 carpels united to form a compound, trilocular ovary with 3 elongate, slender, papillate-hairy, sessile, deciduous stigmas clustered at the top; ovules solitary in each locule, pendulous, apical-axile, apotropous (with dorsal raphe), anatropous, obturator wanting; endosperm-development nuclear. FRUIT a loculicidal capsule with a single large seed, 2 of the locules narrow and empty; seed glandular and short-hairy, not carunculate; embryo straight, with 2 fleshy-thickened cotyledons, containing a cyanogenic glucoside (simmondsin) and a high proportion of an unusual liquid wax; endosperm very scanty or none.

The family Simmondsiaceae consists of the single species *Simmondsia chinensis* (Link) C. K. Schneider, which in spite of its name is native to southern California, western Arizona, and northern Baja California, not China. *Simmondsia* has usually been included in the Buxaceae, but it differs in so many ways that it seems prudent to follow Takhtajan and Melikyan in treating it as a distinct family until more information is available. Recent studies by Margarita Baranova (1980) show a strong stomatographic similarity between *Simmondsia* and some members of the Buxaceae, notably *Styloceras*.

3. Family PANDACEAE Engler & Gilg 1912 nom. conserv., the Panda Family

Dioecious trees or shrubs, sometimes saponiferous and at least sometimes producing alkaloidal peptides; solitary or clustered crystals of calcium oxalate present in some of the cells of the parenchymatous tissues of the shoot; vessel-segments with scalariform or both scalariform and simple perforations; imperforate tracheary elements with bordered pits, sometimes some of them true tracheids; wood-rays heterocellular, some uniseriate, others 2–4 (–6) cells wide; wood-parenchyma mainly diffuse or in uniseriate apotracheal bands. LEAVES pinnately veined, sometimes pellucid-punctate, alternate, simple, and entire, but seemingly pinnately compound, arranged in flat sprays along lateral branches of limited growth, a bud present in the axil of the leafy branch, but often not in the axil of the leaf itself; stomates variously paracytic (*Galearia*), anomocytic (*Microdesmis*) or more or less encyclocytic (*Panda*); stipules small, unequally inserted. FLOWERS in terminal or axillary or cauline racemes or raceme-like inflorescences, or solitary or clustered in the leaf-axils, small, unisexual, hypogynous, normally 5-merous; sepals distinct, or connate into a toothed or lobed cup; petals valvate or imbricate, distinct; stamens in one or two sets of 5 each, or 10 or 15 in a single series; anthers opening by longitudinal slits; pollen-grains tricolporate; disk wanting; gynoecium of 2–5 carpels united to form a compound, plurilocular ovary terminating in distinct or basally connate styles, or the stigmas virtually sessile; ovary with a single pendulous, bitegmic ovule in each locule, lacking obturators; ovules orthotropous in *Panda,* otherwise anatropous and epitropous, with ventral raphe. FRUIT drupaceous, the thin-walled to thick-walled stone with as many locules and seeds as carpels; seeds not caruncPulate; embryo with 2 thin, flat cotyledons; endosperm abundant, oily. 2n = 30. (*Microdesmis*)

The family Pandaceae as here defined (following Forman) consists of 3 genera, native to tropical Africa, Asia, and New Guinea. *Galearia* is the largest genus, with about 15 species; *Microdesmis* has about 10 species, and *Panda* only one.

4. Family EUPHORBIACEAE A. L. de Jussieu 1789 nom. conserv., the Spurge Family

Trees, shrubs, or less often herbs, sometimes vines, sometimes succulent, very diverse chemically and in vegetative morphology, but more consistent in some gynoecial features, provided with various sorts of hairs (these sometimes stinging) and very often producing one or another sort of alkaloid (benzylisoquinoline alkaloids in *Croton*), often tanniferous, sometimes with proanthocyanins and/or ellagic acid, sometimes accumulating aluminum, but without iridoid compounds, only seldom saponiferous, and seldom (*Drypetes*) with mustard oils; solitary or clustered crystals of calcium oxalate commonly present in some of the

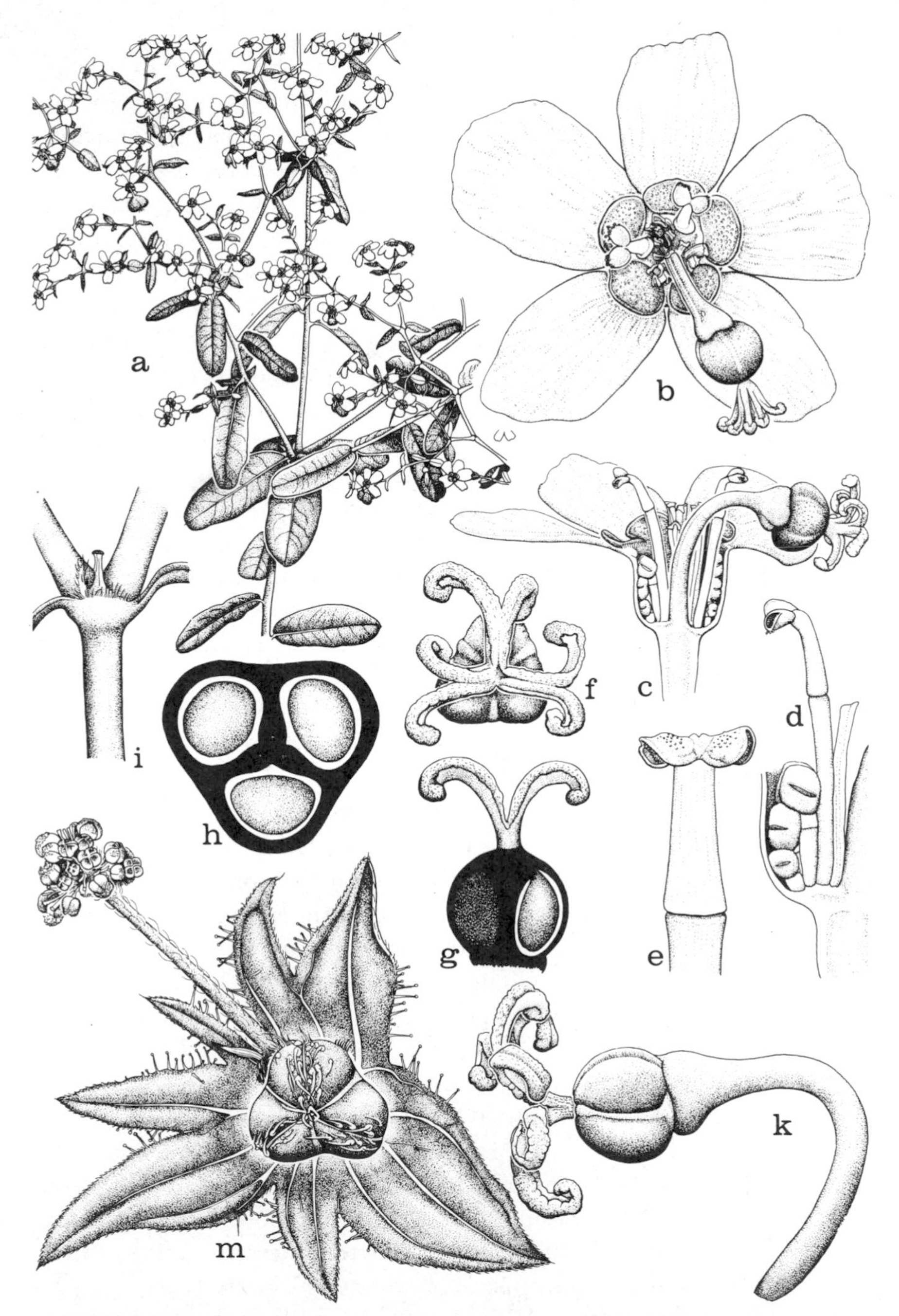

FIG. 5.43 Euphorbiaceae. a–k, *Euphorbia corollata* L. a, habit, ×½; b, cyathium, ×6; c, cyathium, in long section, ×6; d, successive staminate flowers in a cyathium, ×12; e, staminate flower, ×24; f, pistil, from above, ×12; g, pistil, in partial long-section, ×12; h, schematic cross-section of ovary, ×18; i, nodal region of stem, with axillary branches, reduced stipules, and foreshortened, abortive axis above the node, ×6; k, pistillate flower, ×12. m, *Acalypha rhomboidea* Raf., inflorescence, the staminate portion segregated from the pistillate, ×6.

cells of the parenchymatous tissues; stem and leaves very often with specialized cells or tubes (articulated or nonarticulated, sometimes branched) containing diverse sorts of milky or colored latex, or less often containing tanniferous or mucilaginous substances; vessel-segments with simple or less often scalariform perforations; imperforate tracheary elements commonly with simple pits, sometimes septate; wood-rays heterocellular, mixed uniseriate and pluriseriate, the latter 2–17 cells wide, sometimes some of the cells containing silica-bodies; wood-parenchyma commonly abundant and apotracheal, but sometimes wanting; internal phloem sometimes present. LEAVES alternate or less commonly opposite or seldom whorled, simple or less often compound, with pinnate or palmate venation, sometimes much-reduced; stomates paracytic, or less often anisocytic, anomocytic, or diacytic; stipules usually present, sometimes large and protecting the terminal bud, or sometimes reduced to mere glands, or even wanting. FLOWERS in very diverse types of inflorescences of eventually cymose origin, commonly regular, always unisexual (the plants monoecious or dioecious), sometimes individually very much reduced and grouped into bisexual pseudanthia; perianth mostly inconspicuous; mono- or dichlamydeous, or wanting, the tepals distinct or connate toward the base; stamens 5-many, or sometimes fewer or even solitary, distinct or variously connate, anthers most commonly tetrasporangiate and dithecal, opening by longitudinal slits or seldom by apical pores; pollen-grains 2–3-nucleate, of very diverse types, but most often tricolporate, sometimes of a special inaperturate type called crotonoid (from the genus *Corton*); an extrastaminal or intrastaminal nectary-disk (cupulate or of separate segments) commonly present in flowers that are not highly reduced; gynoecium of (1) 3 (4-many) carpels united to form a compound, plurilocular ovary with distinct styles, or these united below into a common style whose branches may be bifid or further divided, rarely the ovary pseudomonomerous; ovules 1 or 2 in each locule, pendulous, apical-axile, epitropous, with ventral raphe, anatropous or hemitropous, bitegmic and crassinucellar, the nucellus often expanded, protruding from the micropyle, and coming into contact with the obturator, a proliferation of the placenta tht forms a roof over the micropyle and provides a passageway for the pollen-tube; embryo-sac of diverse types, but predominantly of the ordinary *Polygonum*-type; endosperm-development nuclear. FRUIT typically a capsular schizocarp, the mericarps separating elastically from the persistent columella and opening ventrally to release the seeds, or the fruit seldom drupaceous or even samaroid or baccate; seeds commonly with a micropylar caruncle; embryo straight or curved, with 2 cotyledons that are usually wider than the radicle; endosperm usually copious, oily and often containing poisonous proteins, or rarely wanting. X = 6 – 14 +. (Androstachydaceae, Hymenocardiaceae, Picrodendraceae, Putranjivaceae, Scepaceae, Stilaginaceae, Uapacaceae)

The family Euphorbiaceae consists of about 300 genera and 7500 species, of cosmopolitan distribution, but best developed in tropical and subtropical regions. By far the largest genus is *Euphorbia* sens. lat., with about 1500 species, found in all parts of the world. Some other large genera are *Croton* (700), *Phyllanthus* (400), *Acalypha* (400), *Macaranga* (250), *Antidesma* (150), *Drypetes* (150), *Tragia* (150), *Jatropha* (150), and *Manihot* (150).

In *Euphorbia* the flowers are individually much-reduced and aggregated into bisexual pseudanthia commonly called cyathia. The cyanthium has a solitary, terminal pistillate flower composed essentially of a naked pistil, surrounded by (4) 5 staminate cymes. The bracts subtending these staminate cymes are connate into a campanulate or hemispheric involucre (simulating a hypanthium), and the tips of the bracts alternate with nectary-glands that may have each a small petaloid appendage. Each male cyme produces 1–10+ closely aggregated flowers; each such flower consists of a single stamen that may or may not be seated in its own small cup (vestigial perianth). The pistillate flower at the center of the hypanthium may at first be sessile or nearly so, but its pedicel commonly elongates with increasing maturity, so that the trilocular ovary hangs out of the cyathium on a stalk.

Many members of the Euphorbiaceae are economically important. *Hevea brasiliensis* (Willd.) Muell.-Arg. is the source of Para rubber. Castor-oil comes from the seeds of the castor-bean, *Ricinus communis* L. The tuberous roots of *Manihot esculenta* Crantz are a staple food in tropical regions. *Euphorbia pulcherrima* Willd. (poinsettia) and a number of other species of *Euphorbia* are cultivated for ornament.

The Euphorbiaceae are highly diversified ecologically as well as morphologically and chemically. Some of the African species of *Euphorbia* are spiny, essentially leafless, arborescent succulents that are ecologically comparable to some of the cacti. The milky juice and the alkaloids found in so many members of the family can readily be believed to help protect the plants from predators. Otherwise it is difficult or impossible to find any unifying ecological features in the group. We still await a convincing Darwinian explanation of the progressive reduction of the euphorbiaceous flower, culminating unexpectedly and very successfully in pseudanthia that are functionally comparable to ordinary flowers in other families.

Macrofossils considered to represent the Euphorbiaceae are known from Eocene and more recent deposits. Euphorbiaceous pollen can be traced back one epoch farther, to the Paleocene.

13. Order RHAMNALES Lindley 1833

Trees, shrubs, or woody vines, seldom herbs, producing mucilage, tanniferous, often with proanthocyanins but without ellagic acid, seldom cyanogenic, lacking iridoid compounds, commonly with crystals of calcium oxalate in some of the cells of the parenchymatous tissues, sometimes also with raphides; vessel-segments with simple perforations; imperforate tracheary elements with simple or seldom bordered pits; wood-rays heterocellular. LEAVES alternate or sometimes opposite, simple or sometimes compound; stipules mostly deciduous or wanting. FLOWERS perfect or sometimes unisexual, mostly rather small, regular, hypogynous or perigynous to sometimes epigynous by adnation of the disk to both ovary and hypanthium, (3) 4–5 (–7)-merous; petals distinct or sometimes connate at the base, sometimes distally coherent, or rarely wanting; stamens as many as and opposite the petals (or as many as and alternate with the sepals, when the petals are wanting); pollen-grains mostly tricolporate; nectary-disk intrastaminal, generally well developed (except often in Leeaceae); gynoecium of 2-several carpels united to form a compound, (often imperfectly) plurilocular ovary with a single, often lobed or cleft style (or the stigma seldom sessile); ovules 1–2 in each locule, basal, erect, apotropous or epitropous or pleurotropous, anatropous, bitegmic, crassinucellar; endosperm-development nuclear. FRUIT drupaceous or baccate, or sometimes schizocarpic.

The order Rhamnales as here defined consists of 3 families and about 1700 species. Most members of the order are readily recognized by the combination of dichlamydeous, polypetalous, haplostemonous flowers with the stamens opposite the petals, and with 1 or 2 ovules in each chamber of a compound ovary. There is nothing ecologically distinctive about the group, and there is considerable ecological diversity among its members.

The close relationship of the Leeaceae to the Vitaceae is not in dispute. The Rhamnaceae are generally thought to be a little more distantly related, and some modern students would remove them from any association with the Vitaceae and Leeaceae. Further study is in order.

The origin of the Rhamnales is to be sought in the Rosales, in a common complex with the ancestors of the Celastrales and Sapindales. There may well have been a diplostemonous common ancestor of the Rhamnales and Celastrales (a very few of the Celastrales are in fact diplostemonous). Differentiation of the two orders from this common ancestor involved loss of the antesepalous stamens by the Rhamnales, and loss of the antepetalous stamens by the Celastrales. Unpublished studies by Piechura show strong serological affinities among tested members of the Rhamnaceae, Celastraceae, Hydrangeaceae, and Grossulariaceae.

SELECTED REFERENCES

Ali, M. A. 1955. Studies on the nodal anatomy of six species of the genus *Vitis* (Tourn.) L. Pakistan J. Sci. Res. 7: 140–152.

Behnke, H.-D. 1974. P- und S-Typ Siebelement-Plastiden bei Rhamnales. Beitr. Biol. Pflanzen 50: 457–464.

Bennek, C. 1958. Die morphologische Beurteilung der Staub- und Blumenblätter der Rhamnaceen. Bot. Jahrb. Syst. 77: 423–457.

Brizicky, G. K. 1964. The genera of Rhamnaceae in the southeastern United States. J. Arnold Arbor. 45: 439–463.

Brizicky, G. K. 1965. The genera of Vitaceae in the southeastern United States. J. Arnold Arbor. 46: 48–67.

Jones, J. H., & D. L. Dilcher, 1980. Investigations of angiosperms from the Eocene of North America: *Rhamnus marginatus* (Rhamnaceae) reexamined. Amer. J. Bot. 67: 959–967.

Kashyap, G. 1955–1958. Studies in the family Vitaceae. I. Floral morphology of *Vitis trifolia* L. Agra Univ. J. Res., Sci. 4 (Suppl.): 777–783, 1955. II. Floral anatomy of *Vitis trifolia* Linn., *Vitis latifolia* Roxb., and *Vitis himalayana* Brandis. III. Floral morphology of *Vitis latifolia* Roxb., *Vitis himalayana* Brandis and *Vitis trifolia* Linn. J. Indian Bot. Soc. 36: 317–323, 1957. 37: 240–248, 1958.

Nair, N. C. 1968. Contribution to the floral morphology and embryology of two species of *Leea* with a discussion on the taxonomic position of the genus. J. Indian Bot. Soc. 47: 193–205.

Nair, N. C., & K. V. Mani. 1960. Organography and floral anatomy of some species of Vitaceae. Phytomorphology 10: 138–144.

Nair, N. C., & P. N. N. Nambisan. 1957. Contribution to the floral morphology and embryology of *Leea sambucina* Wild. Bot. Not. 110: 160–172.

Nair, N. C., & V. S. Sarma. 1961. Organography and floral anatomy of some members of the Rhamnaceae. J. Indian Bot. Soc. 40: 47–55.

Prichard, E. C. 1955. Morphological studies in Rhamnaceae. J. Elisha Mitchell Sci. Soc. 71: 82–106.

Reille, M. 1967. Contribution à l'étude palynologiques de la famille des Vitacées. Pollen & Spores 9: 279–303.

Shah, J. J., & Y. S. Dave. 1970. Morpho-histogenetic studies on the tendrils of Vitaceae. Amer. J. Bot. 57: 363–370.

Souèges, R. 1957. Embryogénie des Ampélidacées (Vitacées). Developpement de l'embryon chez l'*Ampelopsis hederaceae* DC. Compt. Rend. Hebd. Séances Acad. Sci. 244: 2446–2450.

Tiffney, B. H., & E. S. Barghoorn. 1976. Fruits and seeds of the Brandon Lignite I. Vitaceae. Rev. Paleobot. Palynol. 22: 169–191.

Баранов, П. А. 1949. Приспособительная эволюция виноградной лозы. Труды Главного Бот. Сада, 1: 5-26.

Вихирева, В. В. 1951. Развитие семени у семейства крушинных (Rhamnaceae). Труды Бот. Инст. АН СССР, сер. 7, 2: 221-227.

Вихирева, В. В. 1952. Морфолого-анатомическое исследование плодов крушиновых. Труды Бот. Инст. АН СССР, сер, VII, 3: 240–292.

Зубкова, И. Г. 1965. Анатомическое строение черешка в сем. Vitaceae Juss., его таксономическое и эволюционное значение. Бот. Ж. 50: 1556–1567.

Зубкова, И. Г, 1966. Эпидерма листа Vitaceae и её систематическое значение. Бот. Ж. 51: 278–283.

Кравченко, Л. К. 1961. Некоторые сведения о морфологии, типе цветка и биологии цветения диких видов рода *Vitis* L. Докл. АН Уз. ССР, 7: 56–58.

Никитин, А. А. 1938. Сравнительно-анатомическое исследование вторичной древесины сем. Rhamnaceae флоры СССР. Труды Бот. Инст. АН СССР, сер 5, 1: 215–288. (Journal title also in Latin: Acta Institute Botanici Academiae Scientiarum URSS).

SYNOPTICAL ARRANGEMENT OF THE FAMILIES OF RHAMNALES

1 Flowers perigynous to sometimes essentially epigynous; fruits mostly drupaceous or separating into mericarps, not baccate; plants without raphides; embryo rather large; sieve-tubes with S-type plastids; ovules mostly solitary in each chamber of the ovary; mostly shrubs or trees, often thorny, but sometimes scrambling or tendriliferous vines .... 1. RHAMNACEAE.

1 Flowers hypogynous; fruits baccate (though sometimes rather dry); plants with raphide-sacs in the parenchymatous tissues; embryo small; sieve-tubes with P-type plastids.

  2 Ovules solitary in each chamber of the ovary; filaments connate into a tube; plants erect, without tendrils ....2. LEEACEAE.

  2 Ovules paired in each chamber of the ovary; filaments distinct; plants mostly (not always) tendriliferous vines .....3. VITACEAE.

1. Family RHAMNACEAE A. L. de Jussieu 1789 nom. conserv., the Buckthorn Family

Trees or shrubs, often thorny, sometimes scrambling or twining or with coiled tendrils, or rarely (*Crumenaria*) half-shrubs or even herbs, in spp. of *Ceanothus* harboring nitrogen-fixing actinomycetes in the roots, commonly with scattered mucilage-cells (and sometimes also mucilage-cavities) in the parenchymatous tissues, often with tanniferous secretory cells as well, often producing proanthocyanins but not ellagic acid,

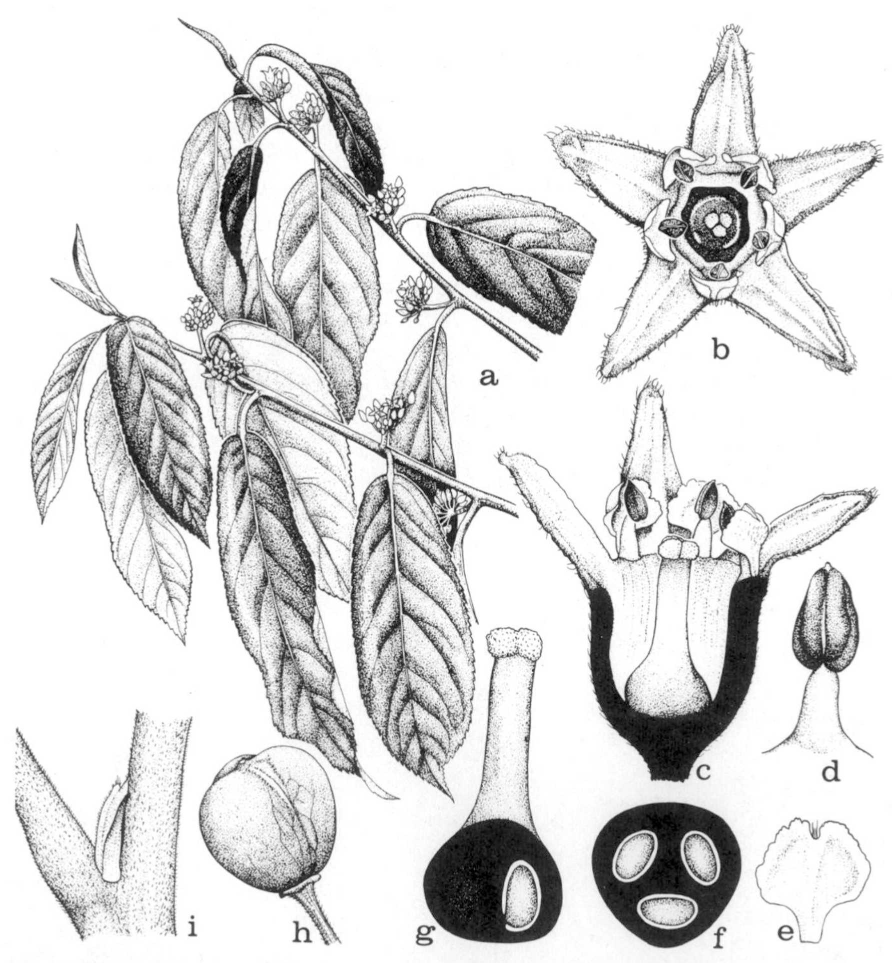

FIG. 5.44 Rhamnaceae. *Rhamnus carolinianus* Walter. a, habit, ×½; b, flower, from above, ×8; c, flower, in long-section, ×8; d, stamen, ×16; e, petal, ×16; f, schematic cross-section of ovary, ×16; g, pistil, in partial long-section, ×16; h, fruit, ×3; i, node, with stipules, ×4.

commonly accumulating anthraquinone glycosides, often saponiferous and often with one or another sort of alkaloid (seldom benzyl-isoquinoline alkaloids), but only seldom cyanogenic, and without iridoid compounds; solitary or clustered crystals of calcium oxalate commonly present in some of the cells of the parenchymatous tissues, but raphides wanting; nodes trilacunar; vessel-segments with simple perforations; imperforate tracheary elements mostly with simple pits, but sometimes some of them with bordered pits and considered to be true tracheids; wood-rays heterocellular to seldom homocellular, usually some uniseriate and others 2–5 (-many)-seriate, seldom all uniseriate; wood-parenchyma varying from mainly paratracheal to mainly apotracheal, sometimes very scanty; sieve-tubes with S-type plastids. LEAVES alternate or less often opposite, simple, pinnately veined or with several main veins from the base, sometimes much-reduced, the stem then strongly photosynthetic; epidermis commonly mucilaginous or with some scattered mucilaginous cells; stomates most commonly anomocytic, sometimes paracytic, seldom anisocytic; stipules generally present but very small, sometimes modified into spines, wanting in most spp. of *Phylica*. FLOWERS in terminal or axillary cymes or mixed panicles or apparent umbels or racemes, or in axillary fascicles that may be reduced to a single flower, mostly rather small, perfect or rarely unisexual, 5-merous or sometimes 4-merous, haplostemonous, rather shortly perigynous (the hypanthium resembling a calyx-tube) to essentially epigynous by adnation of the nectary-disk to both the ovary and the hypanthium; hypanthium often circumscissile and deciduous above the middle, or the calyx-lobes individually deciduous, seldom persistent; calyx-lobes triangular, valvate in bud, sometimes covered on the inner side by a fleshy, glossy layer, histologically similar to the disk, which ends as a tubercle near the sepal-tip, or less modified and merely raised and fleshy along the mid-line on the inner surface; petals commonly more or less concave or hooded, with the anther fitting into the concavity, often clawed at the base, or rarely the petals wanting; stamens distinct, alternate with the sepals, the filament adnate to the base of the petal; anthers generally dithecal, opening by longitudinal slits; pollen-grains binucleate, commonly tricolporate, often more or less triangular; intrastaminal nectary-disk generally well developed, usually adnate to the hypanthium, free from the ovary or adnate around the base or even for the whole length of the ovary; gynoecium of 2–3 (–5) carpels united to form a compound, fully or imperfectly plurilocular ovary (rarely pseudomonomerous) with a terminal, merely lobed to often deeply cleft style; ovules solitary (2 in *Karwinskia*) in each locule, anatropous, bitegmic, and crassinucellar, basal (derived from axile) and erect, usually pleurotropous, with lateral raphe, less often epitropous; endosperm-development nuclear. FRUIT basically drupaceous, with separate stones or a plurilocular stone, but varying to dry and eventually dehiscent, or separating into mericarps; seeds sometimes with a single groove on the back; embryo large, oily, straight or seldom bent, with 2 cotyledons; endosperm usually rather scanty (ruminate in *Reynosia*), or sometimes wanting. X = 9–13, 23, most often 12. (Camarandraceae, Frangulaceae, Phylicaceae)

The family Rhamnaceae consists of about 55 genera and 900 species, cosmopolitan in distribution, but most common in tropical and subtropical regions. More than half of the species belong to only 5 large genera: *Rhamnus* (150, widespread in North Temperate regions and in North Africa); *Phylica* (150, Africa and Madagascar); *Zizyphus* (100, tropics, especially Indomalaysia); *Gouania* (60, tropics); and *Ceanothus* (40, North America, especially California). The fruits of *Rhamnus cathartica* L. and the bark of *R. purshiana* DC. have traditionally been used as potent laxatives.

Leaves assigned to the Rhamnaceae occur in Middle Eocene deposits of Kentucky and Tennessee. Jones & Dilcher (1980) maintain that these leaves "support the hypothesis that modern tribes and possibly genera of Rhamnaceae had evolved by the Middle Eocene." On the other hand, pollen identified as rhamnaceous dates only from the Oligocene epoch.

2. Family LEEACEAE Dumortier 1829 nom. conserv., the Leea Family

Erect herbs or shrubs with scattered tanniferous cells, mucilage-cells, and raphide-sacs (which commonly contain mucilage as well) in the parenchymatous tissues, commonly accumulating proanthocyanins and at least sometimes cyanogenic, but lacking iridoid compounds and not known to be saponiferous; solitary and clustered crystals of calcium oxalate commonly present in some of the cells of the parenchymatous tissues; vessel-segments small, with simple perforations; imperforate tracheary elements septate, with simple or sometimes bordered pits; wood-rays heterocellular, mixed uniseriate and pluriseriate, the latter often very high and wide, with elongate ends; wood-parenchyma mainly paratracheal; sieve-tubes with P-type plastids that contain several polygonal protein crystalloids. LEAVES alternate or rarely opposite, simple or more often pinnately to tripinnately compound, often bearing specialized, multicellular, stalked, deciduous "pearl glands," stomates actinocytic, encyclocytic, or rarely anomocytic; stipules wanting, but the petiole sometimes with a pair of auricles or narrow flanges at the base. FLOWERS in many-flowered, often very large, cymose (frequently umbel-like) inflorescences that are usually borne opposite the leaves (the stem sympodial), individually rather small, perfect, regular, hypogynous, haplostemonous, pentamerous or rarely tetramerous; calyx a shortly toothed cup or saucer, the teeth valvate; petals and stamens basally adnate to form a shortly tubular common structure free from the calyx; petals valvate, often reflexed at anthesis; stamens opposite the petals, the filaments connate (above the common structure with the corolla) for some distance into a tube that may bear lobes alternate with the anthers and that sometimes proliferates near the middle to form a pendulous, tubular-obconic membrane; anthers introrse, tetrasporangiate and dithecal; pollen-grains tricolporate or triporate, smooth, trinucleate or seldom binucleate; ovary superior but sometimes partly sunken in the disk (or the disk wanting), incompletely divided into 4 or 6 (8) apparent locules (the partitions not meeting at midlength), but sometimes interpreted as consisting of only 2 or 3 (4) carpels, with each primary locule being divided lengthwise by a "false" partition from the carpellary midrib; style solitary, terminal, with a dry, papillate, capitate stigma; ovules solitary in each chamber of the ovary, basal, erect, anatropous, bitegmic, crassinucellar, said to be apotropous; endosperm-development nuclear. FRUIT a plurilocular berry with rather thin (and sometimes dry) flesh; embryo small, with 2 cotyledons; endosperm ruminate, oily and proteinaceous.

The family Leeaceae consists of the single paleotropical genus *Leea*, with about 70 species, mostly in southern and eastern Asia. *Leea* has often been included in the Vitaceae, to which it is obviously related, but it is becoming customary to recognize it as a distinct family. *Leea* differs from most of the Vitaceae in its erect habit and absence of tendrils, but the most important difference lies in the structure of the ovary. The ovary is usually interpreted to have as many carpels as locules, with a single ovule in each locule (in contrast to the biovulate locules of the Vitaceae), but on the basis of the vascular anatomy Nair considers that each of the primary, biovulate locules is divided into two uniovulate locelli by a "false" partition originating from the carpellary midrib. Nair's interpretation may well be correct, but it does not provide for the 3-locular ovaries implied in standard descriptions in which the ovary is said to have 3 to 8 carpels or locules. I have not myself seen any specimens of *Leea* with a three-chambered ovary.

3. Family VITACEAE A. L. de Jussieu 1789 nom. conserv., the Grape Family

Climbing, woody vines, commonly with leaf-opposed tendrils, or seldom erect, subsucculent small trees or erect herbs, usually unarmed, with scattered tanniferous cells, mucilage-cells, and raphide-sacs (which commonly contain mucilage as well) in the parenchymatous tissues, often with other sorts of crystals as well, usually accumulating proanthocyanins, but generally not ellagic acid, only seldom cyanogenic, rarely if ever saponiferous, and lacking anthraquinones, alkaloids, and iridoid compounds; nodes 3–7-lacunar, often swollen; vessel-segments commonly rather large, with simple perforations; imperforate tracheary elements septate, with simple pits; wood-rays heterocellular, commonly broad (even up to 20 cells wide) and high; uniseriate rays mostly wanting except in the erect species; wood-parenchyma paratracheal, commonly rather scanty; sieve-tubes with P-type plastids that contain several polygonal protein crystalloids. LEAVES alternate, simple or less often palmately or even pinnately compound, very often palmately lobed or palmately veined, often bearing specialized, multicellular, stalked, deciduous "pearl glands" and sometimes also peltate scales; stomates encyclocytic or actinocytic, or sometimes (spp. of *Cissus*) anomocytic; petiole variously with a complete cylinder of vascular tissue, or with a ring of vascular bundles connected by parenchyma or by more or less lignified fibers; stipules commonly deciduous. FLOWERS in leaf-opposed or terminal, seldom axillary, more or less cymose or sometimes paniculate inflorescences, individually rather small, perfect or not infrequently unisexual, regular, hypogynous, haplostemonous, 4–5-merous, seldom 3-merous or 6–7-merous; calyx small, often only indistinctly toothed or lobed, sometimes reduced to a mere collar; petals valvate, distinct or very rarely connate at the base, but sometimes (notably in *Vitis*) distally coherent and calyptrately deciduous by the growth of the androecium; stamens distinct, opposite the petals; filaments slender; anthers introrse, tetrasporangiate or seldom bisporangiate, dithecal; pollen 2–3-nucleate, tricolporate; nectary-disk intrastaminal, commonly annular or cupulate, or sometimes of 5 distinct glands; gynoecium of 2 carpels united to form a compound (often incompletely)

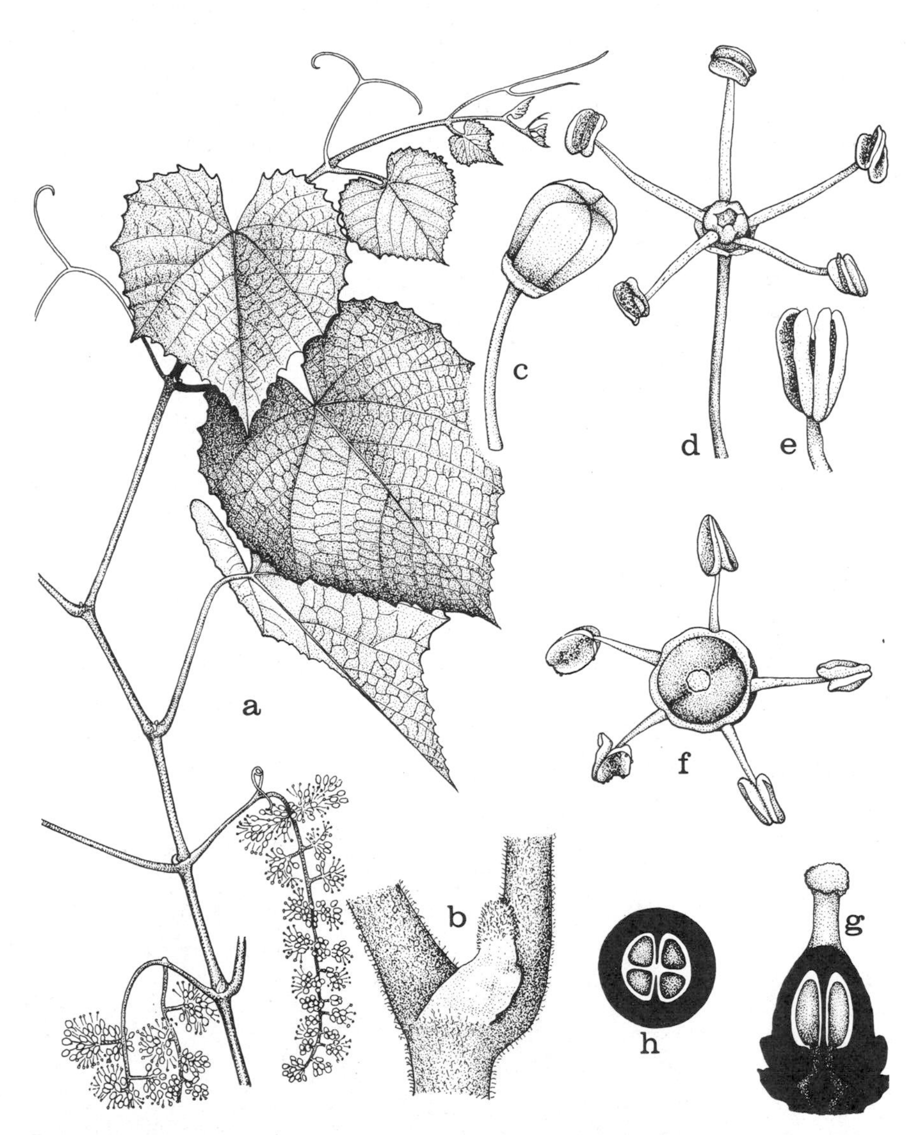

FIG. 5.45 Vitaceae. *Vitis aestivalis* Michx. a, habit, ×½; b, node, with stipules, ×3; c, staminate flower bud, ×8; d, staminate flower, after abscission of the petals, ×8; e, anther, ×16; f, perfect flower, after abscission of the petals, ×12; g, pistil, in partial long section, ×12; h, schematic cross-section of ovary, ×12.

bilocular ovary with a simple style and discoid or capitate, seldom (*Tetrastigma*) quadrifid stigma, or the stigma seldom sessile; ovary superior, but commonly somewhat sunken in the disk; ovules 2 in each locule, ascending from the carpellary margins near the base of the locule, apotropous, anatropous, bitegmic, crassinucellar, sometimes with a placental obturator; endosperm-development nuclear. FRUIT a 1–2-locular (sometimes rather dry) berry, with typically 2 seeds in each locule; seeds with a conspicuous abaxial chalazal knot and 2 deep adaxial grooves, one on each side of the raphe; embryo small, straight, spatulate, with 2 cotyledons, surrounded by the copious, commonly 3-lobed, often ruminate, oily and proteinaceous endosperm. X = 11–20.

The family Vitaceae as here defined consists of about 11 genera and 700 species, mainly of tropical and subtropical regions, with relatively few members in temperate climates. By far the largest genus is *Cissus,* with about 300 species, *Vitis vinifera* L. and other species of grape are well known in cultivation. *Parthenocissus quinquefolia* (L.) Planchon is the Virginia creeper, and *P. tricuspidata* (Sieb. & Zucc.) Planchon is the Boston ivy.

The position of the tendrils opposite the leaves in the Vitaceae suggests that the stem is sympodial. *Parthenocissus* appears to have a very ordinary sympodial structure; leaves that are not opposed to tendrils have an obvious axillary bud, whereas tendril-opposed leaves have none. *Cissus* and *Vitis* have a more complex structure; even the tendril-opposed leaves have an evident axillary bud, which may be somewhat displaced to one side.

Fossil seeds of Eocene and more recent age are credibly attributed to the Vitaceae, and vitaceous pollen dates from the Oligocene. Some upper Cretaceous fossils have also been thought to belong to this family.

14. Order LINALES Cronquist 1957

Plants woody or herbaceous, the herbaceous forms commonly with the vascular bundles becoming confluent through the activity of an interfascicular cambium, lacking iridoid compounds, but sometimes tanniferous, with ellagic acid and proanthocyanins, and sometimes producing alkaloids, especially of the tropane group; vessel-segments with scalariform or simple perforations; imperforate tracheary elements with bordered or seldom simple pits; wood-rays heterocellular or homocellular, mixed uniseriate and pluriseriate, the latter 2–5 cells wide, or all uniseriate; wood-parenchyma of diverse types. LEAVES alternate or less often opposite, simple and often entire, frequently sessile; stomates most commonly paracytic; stipules mostly small or wanting, sometimes modified into glands. FLOWERS in axillary or terminal, mostly more or less cymose inflorescences, hypogynous, perfect or seldom unisexual, regular or nearly so (or the sepals sometimes dissimilar), (4) 5-merous; petals distinct, often clawed; nectary-disk intra- or extrastaminal, sometimes cupulate, sometimes consisting of only 2–5 extrastaminal glands, wanting in the Erythroxylaceae, but the petals then with a ligular appendage near the base; stamens mostly in one or two cycles (the second cycle often staminodial), or sometimes in 3 or 4 cycles, or even numerous, the filaments generally connate below to form a tube, except in the Ixonanthaceae; pollen-grains triaperturate or multiaperturate, trinucleate or seldom binucleate; gynoecium of 2–5 carpels united to form a compound, plurilocular ovary with axile or apical-axile placentas, sometimes with a more or less well developed "false" partition between the two ovules of a primary locule; styles distinct, or connate below, or the single style more or less deeply cleft or undivided, the stigma(s) in any case terminal; ovules 1–2 per locule (only one locule ovuliferous in Erythroxylaceae), pendulous, epitropous, with the micropyle directed upwards and outwards, anatropous to hemitropous, bitegmic, crassinucellar or sometimes tenuinucellar, often with an integumentary tapetum; endosperm-development nuclear or sometimes helobial. FRUIT usually drupaceous or capsular, sometimes a nut or schizocarp; seeds with straight or slightly curved, dicotyledonous embryo, sometimes arillate or winged; endosperm copious to scanty or none.

The order Linales as here defined consists of 5 families and about 550 species. The two largest families are the Linaceae (220) and Erythroxylaceae (200). The other 3 families are distinctive small groups that have often been included in the Linaceae. Among these three, the

Humiriaceae in particular form a well marked, coherent group that seem to be most appropriately treated as a distinct family. Once the Humiriaceae are removed from the Linaceae, then the Ixonanthaceae and Hugoniaceae also appear as aberrant groups that might more reasonably be treated as separate families. Each of the five families of the order then appears as a well defined group. The mutual interrelationships among the five families are not in serious dispute.

The families of the Linales have often been included in a broadly defined order Geraniales, distinguished from the Sapindales by the epitropous rather than apotropous ovules. Under the concepts that I have developed in this and previous works, the orientation of the ovules is considered to be a feature of less than ordinal significance, and the more narrowly defined order Geraniales is considered to be a largely herbaceous offshoot of the characteristically woody order Sapindales. Both the Sapindales and the Geraniales characteristically have compound or evidently lobed leaves, and the forms with simple, entire or merely toothed leaves are considered to be secondary within the group. The Linales, on the other hand, all have simple, entire or merely toothed leaves, and the habit ranges from distinctly woody (and sometimes with scalariform vessels) in the more archaic forms to herbaceous in the more advanced ones. Thus the Linales cannot be included in or derived from the Geraniales as here conceived, but they might possibly be a simple-leaved offshoot from archaic, extinct members of the Sapindales with scalariform vessels, well developed stipules and distinct styles. A perhaps more likely alternative is that the Linales are derived more directly from simple-leaved members of the Rosales, parallel to the compound-leaved Sapindales. In any case it seems clear enough that the Sapindales, Geraniales, and Linales are all derived eventually from the Rosales.

Pollen attributed to the Linales (Linaceae, sens. lat.) is known from Maestrichtian and more recent deposits.

SELECTED REFERENCES

Cuatrecasas, J. 1961. A taxonomic revision of the Humiriaceae. Contr. U.S. Natl. Herb. 35: 25–214.

Forman, L. L. 1965. A new genus of Ixonanthaceae with notes on the family. Kew Bull. 19: 517–526.

Heimsch, C., & E. E. Tschabold. 1972. Xylem studies in the Linaceae. Bot. Gaz. 133: 242–253.

Leinfellner, W. 1955. Beiträge zur Kronblattmorphologie. VII. Die Kronblätter einiger Linaceen. Oesterr. Bot. Z. 102: 322–338.

Narayana, L. L. 1964. A contribution to the floral anatomy and embryology of Linaceae. J. Indian Bot. Soc. 43: 343–357.
Naryana, L. L., & D. Rao. 1966. Floral morphology of Linaceae. J. Jap. Bot. 41: 1–10.
Narayana, L. L., & D. Rao. 1969–1977. Contributions to the floral anatomy of Humiriaceae 1–6. J. Jap. Bot. 44: 328–335, 1969. 48: 143–146; 242–276, 1973. 51: 12–15; 42–44, 1976. 52: 145–153, 1977.
Narayana, L. L., & D. Rao. 1969–1978. Contributions to the floral anatomy of Linaceae. [13 parts.] J. Jap. Bot. 44: 289–294, 1969. 48: 205–208, 1973. 51: 92–96; 349–352, 1976. 52: 56–59; 231–234; 315–317, 1977. 53: 12–14; 161–163; 213–218, 1978. Phytomorphology 21: 64–67, 1971 (1972). Curr. Sci. 43: 226–227: 391–393, 1974.
Nooteboom, H. P. 1967. The taxonomic position of Irvingioideae, *Allantospermum* Forman and *Cyrillopsis* Kuhlm. Adansonia sér. 2. 7: 161–168.
Oltmann, O. 1968 (1969). Die Pollenmorphologie der Erythroxylaceae und ihre systematische Bedeutung. Ber. Deutsch. Bot. Ges. 81: 505–511.
Rao, D., & L. L. Narayana. 1965. Embryology of Linaceae. Curr. Sci. 34: 92, 93.
Robertson, K. R. 1971. The Linaceae in the southeastern United States. J. Arnold Arbor. 52: 649–665.
Robson, N. K. B., & H. K. Airy Shaw. 1962. A note on the taxonomic position of the genus *Cyrillopsis* Kuhlmann. Kew Bull. 15: 387–388.
Rogers, C. M., & K. S. Xavier. 1972. Parallel evolution in pollen structure in *Linum*. Grana 12: 41–46.
Saad, S. I. 1961. Pollen morphology and sporoderm stratification in *Linum*. Grana Palynol. 3(1): 109–129.
Saad, S. I. 1962. Pollen morphology of *Ctenolophon*. Bot. Not. 115: 49–57.
Saad, S. I. 1962. Palynological studies in the Linaceae. Pollen & Spores 4: 65–82.
Suryakanta. [no initial] 1974. Pollen morphological studies in the Humiriaceae. J. Jap. Bot. 49: 112–122.

SYNOPTICAL ARRANGEMENT OF THE FAMILIES OF LINALES

1 Petals usually with a ligular appendage toward the base on the inner side; usually only one locule of the ovary ovuliferous; disk and nectary-glands wanting; trees or shrubs; stamens 10; fruit a drupe; vessel-segments with simple perforations .. 1. ERYTHROXYLACEAE.

1 Petals not appendaged; all locules ovuliferous; disk or nectary-glands present.

2 Style solitary, simple, only the stigmas sometimes separate; disk intrastaminal, well developed or sometimes represented only by 2–5 glands; trees or shrubs.

3 Fruit a drupe; seeds neither arillate nor winged; filaments connate for much of their length into a tube; anthers often of unusual structure; sepals connate into a lobed or toothed cup or tube; vessel-segments with scalariform perforations; stamens 10-many ....2. HUMIRIACEAE.

3 Fruit a capsule with arillate or winged seeds; filaments distinct; anthers of normal structure; sepals distinct or only basally connate; vessel-segments with simple perforations; stamens 5–20 .... 3. IXONANTHACEAE.

2 Style more or less deeply cleft, or often the styles distinct; disk extrastaminal, often represented only by 2–5 glands; sepals distinct or only basally connate.

4 Trees, shrubs, or often woody vines; fruit a drupe or a nut; vessel-segments with simple or scalariform perforations; stamens (5) 10 (15) .... 4. HUGONIACEAE.

4 Herbs, or seldom shrubs; fruit usually a capsule (of indehiscent mericarps in *Anisadenia*); vessel-segments with simple perforations; functional stamens (4) 5 .... 5. LINACEAE.

1. Family ERYTHROXYLACEAE Kunth in H. B. K. 1822 nom. conserv., the Coca Family

Rather small, hard-wooded, glabrous trees and shrubs, often producing alkaloids, especially of the tropane group (including cocaine) seldom cyanogenic, but sometimes saponiferous and sometimes tanniferous, sometimes producing proanthocyanins, but not ellagic acid, and sometimes accumulating ethereal oils in the wood, commonly with solitary crystals of calcium oxalate in the parenchymatous tissues and epidermis; vessel-segments with simple perforations; imperforate tracheary ele-

FIG. 5.46 Erythroxylaceae. *Erythroxylum areolatum* L. a, b, habit, ×½; c, portion of young shoot, showing stipules, ×3; d, flower bud, ×9; e, flower, from above, ×9; f, petal, with internal appendage, ×9; g, androecium, ×9; h, pistil, ×9; i, submature fruit, ×3; k, schematic cross-section of ovary, ×18; m, schematic long-section of ovary, ×18.

ments with strongly bordered to sometimes simple pits; wood-rays heterocellular, mixed uniseriate and pluriseriate, the latter mostly 2–5 cells wide; grains of silica sometimes present in some of the cells of the wood-rays; wood-parenchyma partly apotracheal (mostly diffuse) and partly paratracheal. LEAVES alternate or rarely (*Aneulophus*) opposite, simple and entire; epidermis often mucilaginous; stomates paracytic; mesophyll with scattered sclereids; stipules intrapetiolar, often caducous. FLOWERS in axillary fascicles or solitary in the axils, small, hypogynous, regular, perfect or seldom unisexual, pentamerous, commonly heterostylic; sepals connate below, forming a tube with imbricate or valvate lobes; petals distinct, imbricate, caducous, usually with a ligular appendage toward the base on the inner side; nectary-disk wanting; stamens 10, the filaments united at least at the base, usually forming an evident tube; anthers tetrasporangiate and dithecal, opening by longitudinal slits; pollen-grains trinucleate, tricolporate; gynoecium of 3 or less often only 2 carpels united to form a compound ovary with as many locules as carpels; styles distinct, or more or less connate into a common style; usually only one locule of the ovary ovuliferous, the others empty; ovules solitary (less often 2) in the fertile locule, axile, pendulous, epitropous, with the micropyle directed upward and outward, anatropous to hemitropous, bitegmic, crassinucellar, with an integumentary tapetum; obturator wanting; endosperm-development nuclear. FRUIT a usually single-seeded drupe; seed with a straight, dicotyledonous embryo and copious, starchy endosperm, or seldom without endosperm. X = 12. (Nectaropetalaceae)

The family Erythroxylaceae consists of 4 genera and about 200 species, pantropical but most abundant in the New World. Nearly all of the species belong to the single large genus *Erythroxylum*. The other three genera (*Aneulophus, Nectaropetalum* and *Pinacopodium*) have only about 10 species altogether. The Andean species *Erythroxylum coca* Lam. is well known as the source of cocaine. Fossils attributed to the Erythroxylaceae are known from Eocene deposits in Argentina.

2. Family HUMIRIACEAE A. H. L. de Jussieu in St.-Hil. 1829 nom. conserv., the Humiria Family

Hard-wooded evergreen trees or shrubs, at least sometimes tanniferous, with ellagic acid and proanthocyanins, often with an aromatic, "balsamic" juice, but without prominent secretory structures except often in the fruit; solitary or clustered crystals of calcium oxalate often present in some of the cells of the parenchymatous tissues; vessel-segments elongate, with slanting, scalariform perforation plates that have mostly 15–25 cross-bars; imperforate tracheary elements elongate, with evidently bordered pits; wood-rays heterocellular, mixed uniseriate and pluriseriate, the latter 2–3 (4) cells wide, with short ends; grains of silica sometimes present in some of the cells of the wood-rays; wood-parenchyma both paratracheal and apotracheal, most of it usually diffuse. LEAVES alternate, often distichous, coriaceous, simple, entire or merely toothed, pinnately veined; stomates anomocytic or paracytic; stipules tiny and caducous, or wanting. FLOWERS in axillary or terminal cymes, perfect, regular or nearly so, hypogynous, pentamerous; sepals 5, persistent, usually more or less connate into a thickened tube or cup with imbricate lobes, the 2 outer ones sometimes smaller than the others, or rarely the lobes suppressed; petals 5, deciduous or sometimes persistent, distinct, usually 3–5-nerved, thick, convolute or imbricate, mostly white to ochroleucous or chloroleucous, rarely red; stamens in *Vantanea* numerous and fascicled, associated with trunk bundles; in the other genera the stamens 10–30, sometimes 5 antesepalous clusters of 3, plus 5 antepetalous singles; filaments connate for much of their length into a tube; some of the stamens sometimes staminodial, without anthers; anthers dorsifixed or attached near the base, with an expanded, prolonged connective and relatively small, often well separated pollen-sacs, in *Vantanea* the pollen-sacs 2 and distinct but adjacent at the base of the connective, each with 2 confluent microsporangia, opening longitudinally; in the other genera the pollen-sacs all monosporangiate, either 2 or 4, one near each margin of the connective toward the base, sometimes also one near the middle toward each side; pollen-grains 3 (4)-colporate, or seldom 3 (4)-porate; nectary-disk intrastaminal, free from the ovary and stamens, or adnate to the base of the ovary or to the base of the filament-tube, usually cupulate or tubular and toothed or lobed, or sometimes of 10–20 distinct scales; gynoecium of (4) 5 (–7) carpels united to form a compound, plurilocular ovary with axile placentas and a single style with a lobed or cleft stigma; ovary sometimes unilocular above, the partitions not reaching the summit; ovules solitary in each locule, or 2 and superposed, pendulous, epitropous, with the micropyle directed upward and outward, anatropous, bitegmic, crassinucellar; endosperm-development probably nuclear. FRUIT drupaceous, with a single, usually plurilocular stone, the endocarp sometimes of uniformly dense texture, sometimes with numerous resinous secretory cavities, the stone then buoyant and adapted to distribution by water;

stone with as many longitudinal valves or opercula as carpels, one or more of these pushed out at the time of germination; seeds usually only 1 or 2, with straight or slightly curved embryo and copious, oily endosperm. X = 12.

The family Humiriaceae consists of 8 genera and about 50 species, mainly of tropical South America (and north to Costa Rica), but with one species in tropical West Africa. Most of the species grow in rainforests, some of them characteristically along rivers. The largest genera are *Vantanea,* with 14 species, and *Humiriastrum,* with 13. *Vantanea* is notable (and probably primitive) in the family in having numerous (ca. 50–180) stamens; the other genera have 30 or fewer. Fossils of the Humiriaceae are found in South American deposits of Eocene and later age.

None of the species of Humiriaceae is of much economic importance. The wood is often used locally for construction. The bark and wood of some species produce a locally used balsam of undetermined chemical composition, and the fatty oil from the seeds and exocarp of some species is used domestically in the Amazon region.

3. Family IXONANTHACEAE Exell & Mendonça 1951 nom. conserv., the Ixonanthes Family

Trees or shrubs, at least sometimes tanniferous, with ellagic acid and proanthocyanins; crystals of calcium oxalate often present in some of the cells of the parenchymatous tissues; vessel-segments with simple perforations; imperforate tracheary elements with bordered pits; vasicentric tracheids present; wood-rays heterocellular, mixed uniseriate and pluriseriate, the latter 2–4 cells wide; wood-parenchyma apotracheal, commonly in bands 2–3 cells wide. LEAVES alternate, simple, entire or merely toothed; stomates paracytic; stipules small or wanting. FLOWERS in axillary or terminal cymes, thyrses, or racemes, small, mostly perfect, hypogynous or slightly perigynous, regular or nearly so, commonly pentamerous; sepals imbricate, distinct or connate only at the base; petals imbricate or convolute, distinct; stamens 5–20, the filaments expanded at the base but not connate, free or basally adnate to the well developed, annular to cupular intrastaminal nectary-disk; anthers short, tetrasporangiate and dithecal, opening by longitudinal slits; pollen-grains binucleate, tricolporate; gynoecium of (2–) 5 carpels united to form a compound, plurilocular ovary with axile-apical placentas and a terminal style with a more or less capitate stigma; ovary sometimes divided into locelli in the manner of the Linaceae, and sometimes unilocular at the very top, the partitions not reaching the summit; style and filaments folded in bud; ovules 1 or 2 in each locule, pendulous, epitropous, with ventral raphe, the micropyle directed upwards and outwards, anatropous, bitegmic and presumably crassinucellar, provided with a placental obturator; endosperm-development presumably nuclear. FRUIT a septicidal (sometimes also loculicidal) capsule, with or without a persistent central column; seeds arillate or winged, and with little or no endosperm.

The family Ixonanthaceae as here defined consists of 5 genera (*Allantospermum, Cyrillopsis, Ixonanthes, Ochthocosmus,* and *Phyllocosmus*) and about 30 species, pantropical in distribution. None of the species is of much economic importance or well known to botanists generally.

4. Family HUGONIACEAE Arnott 1834, the Hugonia Family

Trees, shrubs, or more commonly woody vines, at least sometimes saponiferous; vessel-segments with scalariform perforations, or more often the perforations mainly or exclusively simple; imperforate tracheary elements with evidently bordered pits; wood-rays heterocellular, pluriseriate, in *Indorouchera* with elongate ends; wood-parenchyma of diverse types. LEAVES alternate (opposite in *Ctenolophon*), simple, entire; epidermis often somewhat mucilaginous; stomates paracytic; stipules deciduous. FLOWERS in terminal or axillary racemes, spikes, or panicles, large to small, perfect, hypogynous, regular (except as to the sepals), commonly pentamerous; sepals imbricate, distinct, commonly more or less dissimilar; petals distinct, convolute; nectary-disk extrastaminal, well developed in *Ctenolophon*, otherwise mostly represented by 2–5 glands; stamens (5) 10 (15), mostly unequal; filaments connate for much of their length to form a tube; anthers dithecal, opening by longitudinal slits; pollen-grains tricolporate; gynoecium of 2–5 carpels united to form a compound, plurilocular ovary with apical-axile placentas and without false partitions; ovary sometimes unilocular at the very top, the partitions not reaching the summit; styles distinct, or the single style evidently cleft; ovules 2 in each locule, pendulous, epitropous, with ventral raphe and with the micropyle directed upwards and outwards, anatropous, bitegmic, crassinucellar, provided with a placental obturator. FRUIT a drupe or (*Ctenolophon*) a nut; seeds with straight or slightly curved, dicotyledonous embryo and thick (*Indorouchera*) or more often scanty or no endosperm. N = 12, 13. (Ctenolophonaceae)

The family Hugoniaceae as here defined consists of about 7 genera and 60 species, widespread in tropical regions. The largest genus is *Hugonia*, with about 32 species confined to the Old World. The roots and bark of the Indian species *Hugonia mystax* L. are used locally in medicine. Pollen considered to represent *Ctenolophon* is found in Maestrichtian and more recent deposits, but as Saad (1962) has pointed out, the pollen of *Ctenolophon* is remarkably similar to that of some Malpighiaceae.

5. Family LINACEAE S. F. Gray 1821 nom. conserv., the Flax Family

Herbs, or seldom half-shrubs, shrublets, or (*Tirpitzia*) shrubs, often cyanogenic, but generally not saponiferous and only seldom tanniferous, lacking both proanthocyanins and ellagic acid; nodes trilacunar; vascular bundles of the stem commonly becoming confluent through the activity of an interfascicular cambium, forming a continuous ring of xylem and phloem with narrow medullary rays; pericycle commonly with vertical fibrous strands that may be confluent to form a nearly continuous sheath; vessel-segments with simple perforations; imperforate tracheary elements with bordered pits; wood-rays mostly homocellular and uniseriate, sometimes some of them 2–3-seriate; axial parenchyma in *Tirpitzia* vasicentric, otherwise mostly scanty-paratracheal or very often wanting. LEAVES alternate or opposite, simple, entire, very often narrow and sessile; epidermis often somewhat mucilaginous; stomates commonly paracytic; stipules small and inconspicuous, sometimes modified into glands, or wanting. FLOWERS in small to diffusely branched, basically cymose, sometimes raceme-like or spike-like inflorescences, hypogynous, perfect, regular, 5-merous or rarely (*Radiola*) 4-merous; sepals imbricate, distinct or basally connate; petals convolute, distinct, commonly clawed, usually caducous; small nectary-glands commonly present external to the filaments (or to some of them), or at the inner base of the petals; stamens isomerous and alternate with the petals (opposite the petals in *Anisadenia*), sometimes alternating with filamentous or toothed staminodia; filaments expanded below and connate to form an evident tube; anthers tetrasporangiate and dithecal, opening by longitudinal slits; pollen-grains trinucleate, variously tricolpate, polycolpate, or multiporate; gynoecium of (2) 3–5 carpels united to form a compound, plurilocular ovary with axile or apical-axile placentas; ovary sometimes unilocular at the very top, the partitions not reaching the

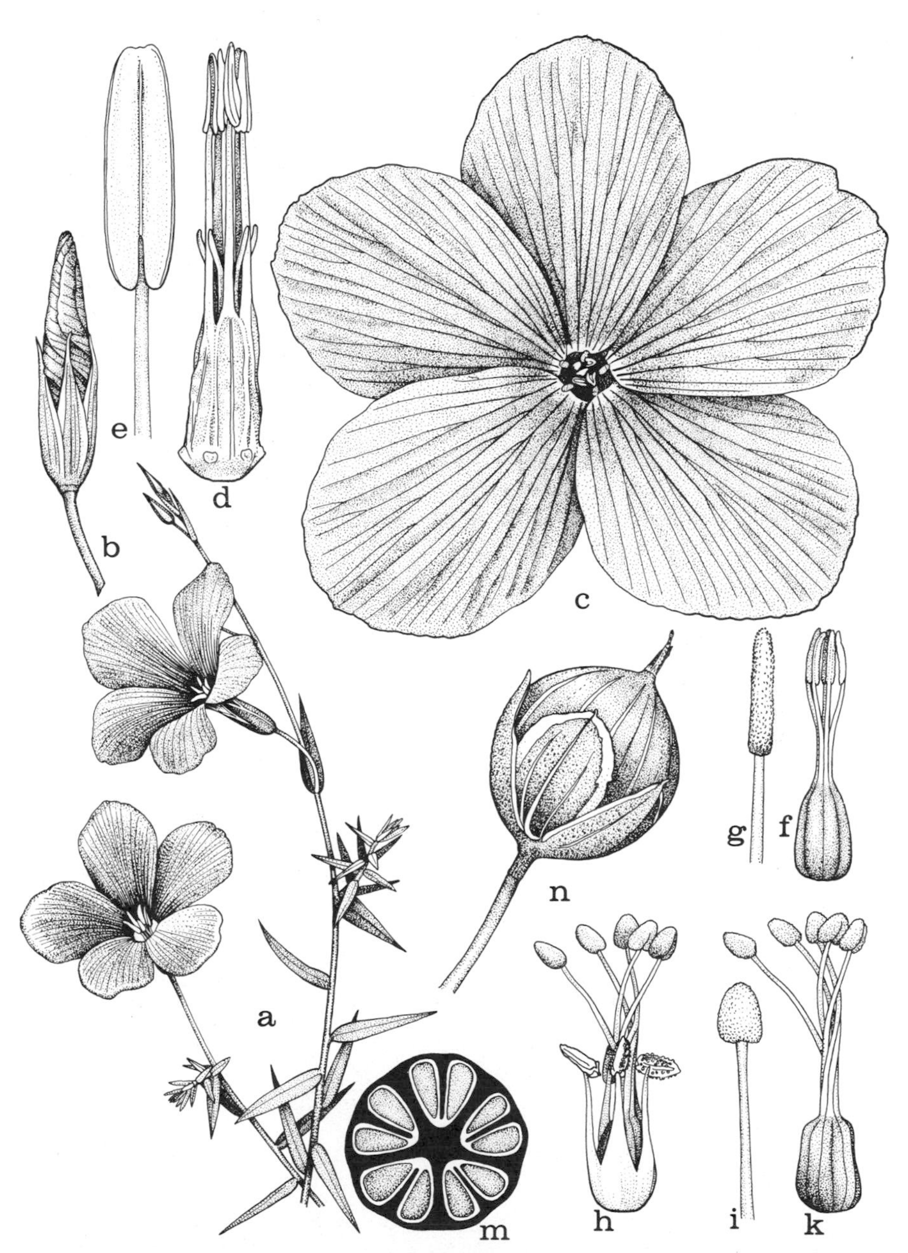

FIG. 5.47 Linaceae. *Linum narbonense* L. a, habit, $\times\frac{1}{2}$; b, flower bud, $\times 2$; c, flower, from above, $\times 2$; d, androecium, with styles protruding, $\times 4$; e, anther, $\times 8$; f, pistil at early anthesis, $\times 4$; g, style and stigma, at early anthesis, $\times 8$; h, androecium and gynoecium, at late anthesis, $\times 4$; i, style and stigma at late anthesis, $\times 8$; k, pistil, at late anthesis, $\times 4$; m, schematic cross-section of ovary, $\times 12$; n, fruit, $\times 2$.

summit; styles distinct, or evidently connate below; stigmas terminal; ovules 2 in each locule, collateral, commonly (but not in *Anisadenia*) separated by an incomplete "false" septum extending inward from the ovary wall, pendulous, epitropous, with the micropyle directed upwards and outwards, anatropous, bitegmic, crassinucellar or tenuinucellar, with an integumentary tapetum and commonly provided with a placental obturator; endosperm-development nuclear or less often helobial. FRUIT a septicidal capsule, or in *Anisadenia* of 2 indehiscent, 1-seeded mericarps; seeds with a straight, spatulate, dicotyledonous, oily embryo and scanty endosperm. X = 6–11+.

The family Linaceae as here narrowly defined consists of 6 genera and about 220 species, widely distributed throughout the world, but most common in temperate and subtropical regions. By far the largest genus is *Linum,* with perhaps 200 species. Common flax, a traditional source of fiber and oil, is *Linum usitatissimum* L.

15. Order POLYGALALES Bentham & Hooker 1862

Plants woody or herbaceous, but even the herbaceous forms with a continuous cylinder of xylem, or with the vascular bundles soon becoming confluent through the activity of an interfascicular cambium, tanniferous or not, with or without proanthocyanins and ellagic acid, frequently accumulating aluminum, but lacking iridoid compounds and lacking alkaloids except in some Malpighiaceae; vessel-segments, or most of them, with simple perforations; imperforate tracheary elements with simple or bordered pits, often septate; wood-rays narrow, many or all of them uniseriate, the others, when present, seldom more than 4 cells wide. LEAVES alternate, opposite, or whorled, simple, and often entire (trifoliolate in one species of *Krameria*); stipules present or absent, sometimes modified into glands or spines. FLOWERS in various sorts of inflorescences, mostly hypogynous or nearly so (perigynous in some Polygalaceae), seldom epigynous, perfect, regular (Tremandraceae) or more often more or less strongly irregular, sometimes (Malpighiaceae) only obliquely so, sometimes papilionaceous in aspect, basically 5-merous or less often 4-merous, but often with some members of some kinds suppressed; petals distinct and often clawed, or connate at the base, or only some of them connate; nectary-disk extrastaminal in *Tremandra,* otherwise intrastaminal or opposed to the stamens or wanting; stamens 1–10 (–15), often some of them staminodial; filaments distinct or often connate into a cleft tube or into 2 groups; anthers opening by longitudinal slits or by terminal or subterminal pores or short slits; pollen-grains very diverse, triaperturate to multiaperturate, binucleate or trinucleate; gynoecium of 2–5 (–8) carpels, generally united to form a compound, usually plurilocular ovary with axile placentas, seldom unilocular (even pseudomonomerous) and with 2 parietal placentas or with one or 2 pendulous subapical ovules; style terminal and solitary, or (Malpighiaceae) the styles distinct or connate only at the base; ovules mostly 1–2 per locule, seldom more or less numerous, epitropous, anatropous to hemitropous, bitegmic, crassinucellar; endosperm-development nuclear, so far as known. FRUIT of various types; seeds sometimes winged or arillate or pubescent; embryo with 2 cotyledons, straight to sometimes curved or even circinate; endosperm copious to wanting.

The order Polygalales as here defined consists of 7 families and about 2300 species. Most of the species belong to only 3 families, the Malpighiaceae (1200+), Polygalaceae (750), and Vochysiaceae (200).

The other 4 families have only a little more than a hundred species in all.

The mutual affinity of 5 of these families has been widely accepted. These are the Polygalaceae, Xanthophyllaceae, Vochysiaceae, Trigoniaceae, and Tremandraceae. The Tremandraceae are unusual in the order in having regular flowers. In other respects, however, including the wood-anatomy and the poricidal anthers, they fit well into the Polygalales, and they do not appear to be closely allied to any other order.

The other two families, the Malpighiaceae and Krameriaceae, are more controversial. The affinities of the Krameriaceae are discussed under that family. The Malpighiaceae have sometimes been associated with the families here grouped as the Linales. Anatomically the Linales and Polygalales are somewhat similar, and they share certain common differences from the Sapindales (see Heimsch, 1942), but each order also hangs together as a group, and the Malpighiaceae go with the Polygalales. The pollen-morphology suggests some affinity of the Malpighiaceae to the Trigoniaceae and Tremandraceae, and also to the Humiriaceae and Zygophyllaceae in allied orders. The Malpighiaceae are in some respects the most archaic family of the Polygalales, and they form a sort of link between this order and the Linales. They (the Malpighiaceae) cannot be considered as directly ancestral to the rest of the Polygalales, however, because their seeds lack endosperm, whereas the Polygalaceae and some of the other families have endospermous seeds. See further discussion under Malpighiaceae.

The Polygalales and Linales are here regarded as a pair of closely related, basically simple-leaved offshoots from the Rosales, parallel in some respects to the compound-leaved Sapindales on the one hand, and to the simple-leaved Celastrales and Rhamnales on the other. The Polygalales differ from the other simple-leaved orders in having irregular flowers, or poricidal anthers, or both. These are of course advanced features. Primitively woody, the order has given rise to herbaceous forms in the Polygalaceae, Vocyhsiaceae, and Krameriaceae. These three families include both woody and herbaceous species, but only the Polygalaceae have a large proportion of herbs.

Ecologically the families of Polygalales have little in common beyond the possible similarities in pollinating mechanisms as suggested by the structure of the flowers. The family Polygalaceae is indeed highly diverse ecologically, as it includes small trees, shrubs, ordinary herbs, climbers, and even a few non-green mycoparasites. The several families (and sometimes genera within the families) differ in the number of carpels, number of ovules per locule, and presence or absence of stipules and endosperm, but these characters are difficult to assess in terms of ecological adaptation or survival value. The differences in habit and type of fruit that mark some of the genera are apparently adaptive but are duplicated in many other groups of dicotyledons.

SELECTED REFERENCES

Anderson, W. R. 1977 (1978). Byrsonimoideae, a new subfamily of the Malpighiaceae. Leandra 7: 5–18.

Busse-Jung, F. 1979. Phytoserologische Untersuchungen zur Frage der systematischen Stellung von *Krameria triandra* Ruiz et Pav. Dissertation Christian-Albrechts-Universität zu Kiel.

Carlquist, S. 1977. Wood anatomy of the Tremandraceae: Phylogenetic and ecological implications. Amer. J. Bot. 64: 704–713.

Dickison, W. C. 1973. Nodal and leaf anatomy of *Xanthophyllum* (Polygalaceae). J. Linn. Soc., Bot. 67: 103–115.

Dube, V. P. 1962. Morphological and anatomical studies in Polygalaceae and its allied families. Agra Univ. J. Res., Sci. 11: 109–112.

Erdtman, G. 1944. The systematic position of the genus *Diclidanthera* Mart. Bot. Not. 1944: 80–84.

Erdtman, G., P. Leins, R. Melville, & C. R. Metcalfe. 1969. On the relationships of *Emblingia.* J. Linn. Soc., Bot. 62: 169–186.

Fouët, M. 1966. Contribution à l'étude cyto-taxinomique des Malpighiacées. Adansonia sér 2. 6: 457–505.

Leinfellner, W. 1971. Das Gynözeum von *Krameria* und sein Vergleich mit jenem der Leguminosae und der Polygalaceae. Oesterr. Bot. Z. 119: 102–117.

Leinfellner, W. 1972. Zur Morphologie des Gynözeums der Polygalaceen. Oesterr. Bot. Z. 120: 51–76.

Lleras, E. 1975. A preliminary monograph of the family Trigoniaceae. Doctoral thesis, City Univ. of New York, 1975.

Milby, T. H. 1971. Floral anatomy of *Krameria lanceolata.* Amer. J. Bot. 58: 569–576.

Miller, N. G. 1971. The Polygalaceae in the southeastern United States. J. Arnold Arbor. 52: 267–284.

Morton, C. V. 1968. A typification of some subfamily, sectional, and subsectional names in the family Malpighiaceae. Taxon 17: 314–324.

O'Donnell, C. A. 1941. La posicion sistematica de "*Diclidanthera*" Mart. Lilloa 6: 207–212.

Rao, A. N. 1964. An embryological study of *Salomonia cantoniensis* Lour. New Phytol. 63: 281–288.

Robertson, K. R. 1972. The Malpighiaceae in the southeastern United States. J. Arnold Arbor. 53: 101–112.

Robertson, K. R. 1973. The Krameriaceae in the southeastern United States. J. Arnold Arbor. 54: 322–327.

Simpson, B. B., & J. L. Neff. 1978. Dynamics and derivation of the pollination syndrome of *Krameria* (Krameriaceae). Bot. Soc. Amer. Misc. Ser. Publ. 156: 14.

Singh, B. 1959, 1961. Studies in the family Malpighiaceae. I. Morphology of *Thryallis glauca* Kuntze. II. Morphology of *Malpighia glabra* Linn. III. Development and structure of seed and fruit of *Malpighia glabra* Linn. Hort. Advance 3: 1–19, 1959. 5: 83–96; 145–155, 1961.

Stafleu, F. A. 1948–1954. A monograph of the Vochysiaceae. 1–4. Recueil Trav. Bot. Néerl. 41: 397–540, 1948. Acta Bot. Neerl. 1: 222–242, 1952. 2: 144–217, 1953. 3: 459–480, 1954.

Stenar, H. 1937. Zur Embryosackentwicklung einiger Malpighiazeen. Bot. Not. 1937: 110–118.

Styer, C. H. 1977. Comparative anatomy and systematics of Moutabeae (Polygalaceae). J. Arnold Arbor. 58: 109–145.

Subra Rao, A. M. 1940, 1941. Studies in the Malpighiaceae. I. Embryo-sac development and embryogeny in the genera *Hiptage, Banisteria* and *Stigmatophyllum*. J. Indian Bot. Soc. 18: 145–156. 2. Structure and development of the ovules and embryo-sacs of *Malpighia coccifera* Linn. and *Tristellateia australis* Linn. Proc. Natl. Inst. Sci. India, Pt. B, Biol. Sci. 7: 393–404.

Turner, B. L. 1958. Chromosome numbers in genus *Krameria*: evidence for familial status. Rhodora 60: 101–106.

Weberling, F. 1974. Weitere Untersuchungen zur Morphologie des Unterblattes bei den Dikotylen. VII. Polygalales. VIII. *Koeberlinia* Zucc. Beitr. Biol. Pflanzen 50: 277–289.

Тамамшян, С. Г. 1940. К вопросу о происхождении крыльев (alae) у *Polygala* Tourn. Уч. зап. Моск. Гос. унив., Бот., 36: 209-222.

SYNOPTICAL ARRANGEMENT OF THE FAMILIES OF POLYGALALES

1 Anthers mostly opening by longitudinal slits, seldom by terminal pores; gynoecium of (2) 3 (–5) carpels, rarely pseudomonomerous; stipules usually present and more or less well developed, but small or wanting in some Vochysiaceae and Malpighiaceae.

2 Styles generally distinct or connate only at the base, only rarely (as in spp. of *Bunchosia*) fully connate; plants generally with malpighian hairs; sepals usually with a pair of prominent abaxial glands near the base; embryo-sac often tetrasporic and 16-nucleate; flowers mostly irregular, the 5 petals commonly clawed, usually with ciliate, toothed, or fringed margins; ovules solitary in each locule .... 1. MALPIGHIACEAE.

2 Style solitary, with a capitate or lateral stigma; plants seldom with malpighian hairs; sepals without abaxial glands; embryo-sac not known to be tetrasporic or 16-nucleate; ovules often more than 1 per locule.

3 Flowers obliquely irregular, but not papilionaceous, stamens distinct or connate into 2 groups, very often only one stamen antheriferous; plants commonly accumulating aluminum .... 2. VOCHYSIACEAE.

3 Flowers papilionaceous, with a 2-petaled keel, 2 wings, and a basally saccate or spurred standard; stamens collectively uni-

lateral on the anterior side of the flower, monadelphous by their filaments, 5–8 of them with anthers; plants not accumulating aluminum .... 3. TRIGONIACEAE.

1. Anthers mostly opening by terminal or subterminal pores or short slits, less frequently (Xanthophyllaceae, some Polygalaceae) by longitudinal slits; gynoecium of 2 (less often 3–5 or even 8) carpels, or sometimes pseudomonomerous; stipules mostly wanting or poorly developed.
   4 Flowers regular, with distinct, induplicate-valvate petals; pollen-grains tricolporate; plants mostly of ericoid aspect .... 4. TREMANDRACEAE.
   4 Flowers usually evidently irregular, commonly papilionaceous in aspect; plants not ericoid.
      5 Ovules solitary in each of the (1) 2–5 (–8) locules of the ovary; filaments generally forming a cleft tube; seeds with or less often without endosperm; pollen-grains polycolporate ....5. POLYGALACEAE.
      5 Ovules 2 or more in the single locule of the ovary; filaments not forming a tube; seeds without endosperm.
         6 Ovary with 2 parietal placentas and 2–16 ovules in all; anthers opening by elongate longitudinal slits; pollen-grains polycolporate; fruit unarmed; small trees, commonly accumulating aluminum .... 6. XANTHOPHYLLACEAE.
         6 Ovary with a pair of pendulous subapical ovules; anthers opening by terminal pores or short slits; pollen-grains with 3 (4) apertures; fruit generally covered with bristles or spines that are often retrorsely barbed; hemiparasitic shrubs and perennial herbs ....7. KRAMERIACEAE.

1. Family MALPIGHIACEAE A. L. de Jussieu 1789 nom. conserv., the Barbados cherry Family

Woody vines, erect shrubs or small trees, often with one or another sort of anomalous secondary growth, sometimes storing carbohydrates as inulin, commonly provided with characteristic malpighian hairs, the hairs always unicellular, nearly always 2-armed, never glandular; some of the cells of the parenchymatous tissues containing solitary or more often clustered crystals of calcium oxalate, some others secretory, the plants sometimes tanniferous and sometimes producing proanthocyanins, but without ellagic acid, only seldom saponiferous and seldom cyanogenic, sometimes producing alkaloids of the indole group (especially of the β-carboline and tryptophane subgroups); nodes trilacunar or sometimes unilacunar; vessel-segments with simple perforations; imperforate tracheary elements with simple pits, often septate; wood-rays heterocellular, mixed uniseriate and pluriseriate, the latter mostly 2–3 cells wide; wood parenchyma of diverse sorts, most often scanty and paratracheal, sometimes mainly in apotracheal bands. LEAVES mostly opposite, seldom ternate or subopposite, simple and commonly entire (lobed in *Stigmaphyllon*), mostly pinnately veined, very often with a pair of large, complex, fleshy glands on the petiole or on the abaxial surface of the blade; epidermis often mucilaginous; stomates mostly paracytic; petiole with a central, trough-shaped vascular strand, sometimes accompanied by smaller lateral ones; stipules usually present, sometimes large and connate. FLOWERS in terminal or axillary racemes, panicles, or cymes, on jointed, bibracteolate pedicels, often large, hypogynous, perfect or seldom unisexual, somewhat irregular (tending to be bilaterally symmetrical), varying to sometimes virtually regular, pentamerous, seldom dimorphic and some of them cleistogamous; receptacle convex, or sometimes produced into a pyramidal structure; sepals distinct or slightly connate at the base, imbricate, very often with a pair of prominent abaxial glands near the base, sometimes one of the sepals with only one large gland or without a gland, but the others biglandular, or seldom all the sepals glandless; petals distinct, imbricate, sometimes irregularly so, commonly clawed, with ciliate, toothed, or fringed margins; stamens mostly bicyclic, often some of them or one set without anthers or with abortive anthers, seldom unicyclic or tricyclic, filaments more or less connate into a tube at the base, or seldom distinct; anthers tetrasporangiate and dithecal or some of them monothecal, dehiscent by longitudinal slits or seldom by terminal pores, sometimes

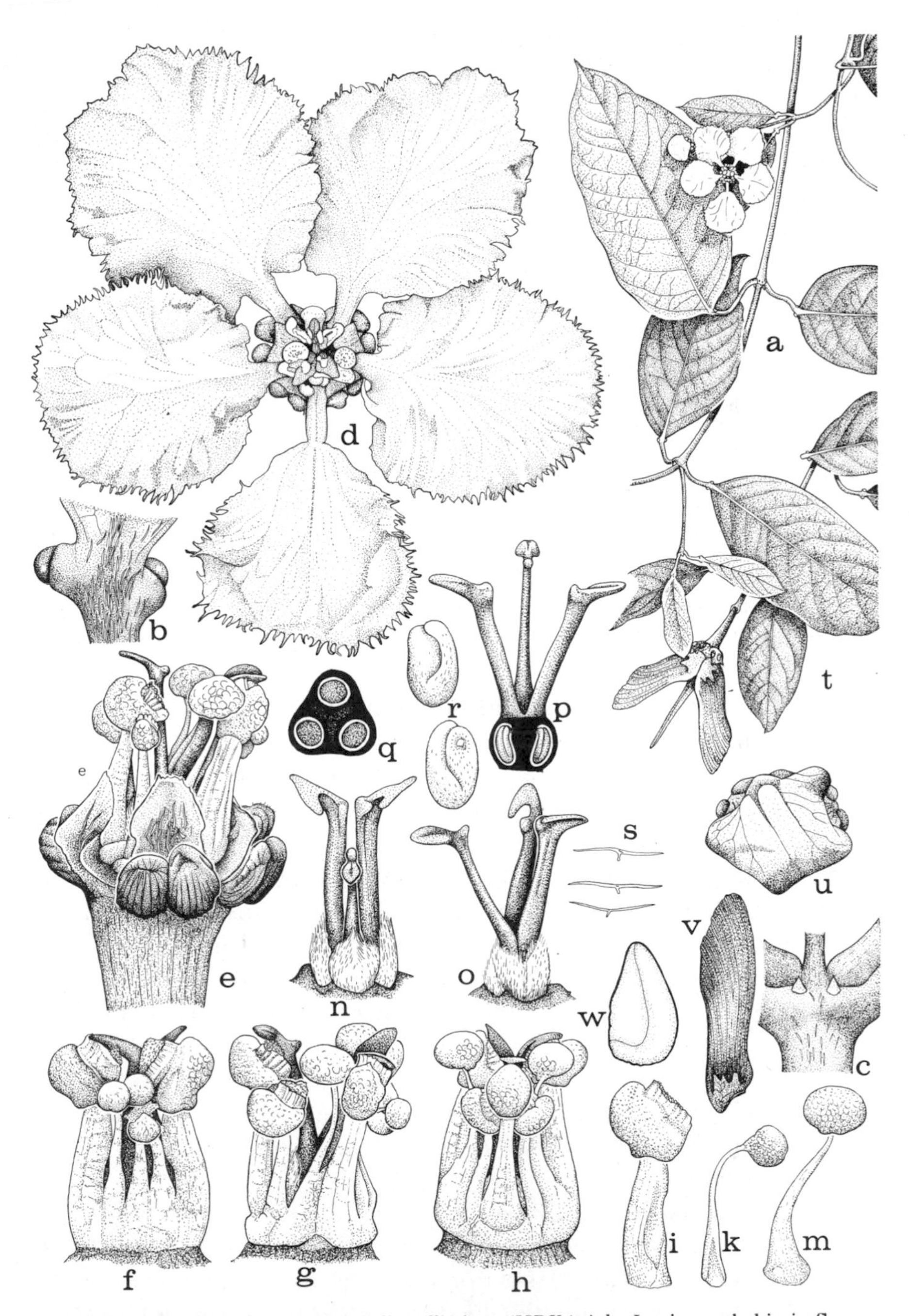

FIG. 5.48 Malpighiaceae. *Stigmaphyllon ellipticum* (HBK.) Adr. Jussieu. a, habit, in flower, ×1; b, petiolar glands, ×6; c, node, with minute stipules, ×6; d, flower, ×3; e, flower, with petals removed, showing sepalar glands, ×6; f, g, h, stamen-tube, ×6; i, fertile stamen, ×6; k, m, sterile stamens, ×6; n, o, pistil, ×6; p, pistil in partial long-section, ×6; q, schematic cross-section of ovary, ×9; r, ovules, ×12; s, malpighian hairs, ×25 t, habit, in fruit, ×½; u, flower bud, ×12; v, mericarp, ×1; w, long-section of seed, ×5. Note that figure d would better be rotated 180°, so that the odd petal, which serves as a flag to visiting bees, is uppermost.

with an enlarged connective; pollen-grains binucleate, 3–5-colporate or 4- to multiporate; nectary disk wanting; gynoecium of (2) 3 (–5) carpels set obliquely to the petals and generally united to form a compound, plurilocular ovary with axile placentas and distinct or only basally connate styles, these sometimes dissimilar; in spp. of *Bunchosia* the styles fully connate and the stigma barely lobed; stigmas terminal to often ventrally subterminal; carpels sometimes (as in *Banisteriopsis*) joined to what appears to be a pyramidal extension of the receptacle, but otherwise distinct; rarely (as in *Pterandra*) the carpels virtually distinct; ovules solitary in each locule, pendulous, epitropous, with ventral raphe, anatropous or hemitropous, bitegmic, crassinucellar; embryo-sac commonly tetrasporic and 16-nucleate, seldom monosporic or bisporic and 8-nucleate; endosperm-development nuclear. FRUIT commonly with winged to nutlike (seldom dehiscent) mericarps (often some of the carpels abortive), or sometimes a nut or drupe that does not separate into mericarps; seeds with large, oily, dicotyledonous, straight to sometimes curved or even circinate embryo, nearly or quite without endosperm. X = 6, 9–12+.

The family Malpighiaceae consists of about 60 genera and 1200 or more species, native to tropical and subtropical regions in both the Old and the New World, but especially well developed in South America. More than half of the species belong to only 8 genera, *Byrsonima* (150), *Heteropterys* (120), *Banisteriopsis* (100), *Tetrapterys* (90), *Stigmaphyllon* (80), *Bunchosia* (55), *Mascagnia* (55), and *Acridocarpus* (50). *Malpighia glabra* L. (including *M. punicifolia* L.), the Barbados cherry, is famous for the high concentration of vitamin C in its palatable drupes. Species of *Bunchosia* are cultivated in South America for their edible fruit. Several species of *Banisteriopsis* and related genera in the Amazon-Orinoco region of South America produce alkaloids that are used by aborigines as hallucinogens. The most famous of these is *B. caapi* (Spruce) Morton. The hallucinogenic species *Dipteropterys cabrerana* (Cuatrecasas) Gates is the plant that ethnobotanists have called *Banisteriopsis rusbyana,* which is not hallucinogenic.

The Malpighiaceae are morphologically transitional between the Linales and the Polygalales, and they might with some justification be included in either order. I opt for the Polygalales, because they (the Malpighiaceae) are apparently allied to the Vochysiaceae and Trigoniaceae, and because their inclusion in the Polygalales is less disturbing to the homogeneity of that order than their inclusion in the Linales would be to that group.

Most genera have multicellular glands variously disposed on the leaf. In a few species these are marginal and the leaf appears to be somewhat toothed (as in *Malpighia coccigera* L.) or ciliate (as in *Stigmaphyllon ciliatum* Adr. Juss.). Otherwise the leaves are never toothed, although they are lobed in *Stigmaphyllon*.

At one time I thought that the simple leaf of the Malpighiaceae was a unifoliolate compound leaf, but Dr. William Anderson has shown me that the glands which I took to mark a joint in the petiole are superficial and have no such significance. The Linales and Polygalales now appear to be fundamentally simple-leaved groups. If they have a compound-leaved ancestry, the direct structural evidence of it has been lost.

Pollen referred to the Malpighiaceae dates from the Middle Eocene.

The more archaic members of the family are all in the New World, and heavily concentrated in the Guayana Highlands and northern Amazonia. Oil-gathering anthophorid bees such as *Centris* are the principal pollinators of New World members of the Malpighiaceae. These pollinators are missing from the Old World, and the characteristic oil-glands of the sepals are correspondingly vestigial or wanting in Old World Malpighiaceae. (All teste W. R. Anderson.)

2. Family VOCHYSIACEAE A. St.-Hilaire 1820 nom. conserv., the Vochysia Family

Trees, shrubs, or sometimes woody vines, seldom herbs, with a resinous juice, provided with simple, unicellular, seldom malpighian hairs or sometimes with stellate hairs, commonly accumulating aluminum, often with tanniferous cells and mucilage-cells or -ducts scattered in the parenchymatous tissues, and with solitary or clustered crystals of calcium oxalate in some of the cells, often with traumatic vertical intercellular canals in the wood; nodes unilacunar; vessel-segments with simple perforations; imperforate tracheary elements with simple or narrowly bordered pits, sometimes septate; wood-rays heterocellular to less often homocellular, mixed uniseriate and pluriseriate, the latter up to 4 or 5 (–8) cells wide, or seldom all uniseriate; wood-parenchyma commonly in paratracheal bands, or sometimes merely vasicentric; internal phloem usually present, but sometimes wanting. LEAVES opposite or sometimes whorled, seldom alternate, simple, entire, leathery; epidermal cells often with a mucilaginous inner wall; stomates paracytic or anomocytic; stipules small, sometimes represented only by glands, or wanting. FLOWERS in racemes or seeming panicles with cymose ultimate components, bibracteolate, perfect, hypogynous or somewhat perigynous to sometimes eipgynous, obliquely irregular; sepals 5, imbricate, connate at the base, one of them commonly the largest and with gibbous or spurred base; petals (0) 1–3 or sometimes 5, convolute or imbricate, more or less unequal; stamens 1–5 (–7), usually only one antheriferous (this one antepetalous and across the flower from the spurred sepals) and the others (when present) staminodial, but in *Euphronia* several stamens antheriferous, and the one across from the spurred sepal staminodial; filaments distinct or connate into two groups; anthers dithecal, opening by longitudinal slits, sometimes with an expanded connective; pollen-grains tricolporate; nectary-disk or -glands wanting; gynoecium commonly of 3 carpels united to form a compound, trilocular ovary with axile placentas, or sometimes pseudomonomerous, with a single locule and 1 or 2 lateral or apical ovules; style solitary, with a capitate or lateral stigma, sometimes very short; ovules (1) 2-many in each locule, epitropous, anatropous or hemitropous, bitegmic, crassinucellar. FRUIT a loculicidal capsule or sometimes a samara winged by the persistent sepals; seeds often winged, sometimes hairy; embryo straight, with 2 cotyledons; endosperm wanting, or rarely well developed. N = 11.

The family Vochysiaceae as here defined consists of 7 genera and about 200 species, mostly native to tropical America, but with a single monotypic genus (*Erismadelphus*) in tropical western Africa. The two largest genera are *Vochysia* (100) and *Qualea* (65). None of the species is of much economic importance or especially well known to botanists generally.

3. Family TRIGONIACEAE Endlicher 1841 nom. conserv., the Trigonia Family

Trees, shrubs, or woody vines with simple, unicellular, basifixed hairs, or glabrous, often tanniferous, not accumulating aluminum; solitary or clustered crystals of calcium oxalate often present in some of the cells of the parenchymatous tissues; vessel-segments with simple perforations, or some of them with scalariform perforations; imperforate tracheary elements with bordered pits, sometimes septate; wood-rays heterocellular, mixed uniseriate and pluriseriate, the latter 2–4 cells wide; wood-parenchyma largely apotracheal, diffuse to diffuse-in aggregates or partly in bands; internal phloem wanting. LEAVES opposite (alternate in *Trigoniastrum*), simple, entire, pinnately veined; epidermis often with mucilaginous inner walls; stomates paracytic; stipules interpetiolar, often connate, deciduous. FLOWERS in various sorts of partly or wholly cymose inflorescences, bibracteolate or tribracteolate, perfect, hypogynous or slightly perigynous, strongly irregular,; sepals 5, imbricate, unequal, united into a cup below; corolla papilionaceous; petals 5, convolute in bud, the two anterior (lower, outer) ones forming an often saccate keel, the posterior (inner, upper) one forming a basally saccate or spurred standard, the 2 lateral petals or wings flat and spatulate; stamens 5–12 in all, 5–8 (often 6) of them with anthers, 0–4 staminodial, collectively unilateral on the anterior side of the flower, monadelphous (at least at the base) by their filaments; anthers dithecal, opening by longitudinal slits; pollen-grains 3–5-porate; nectary glands 2 or seldom only one, sometimes lobed or cleft, borne in front of the standard on the posterior side of the flower; gynoecium of 3 (4) carpels united to form a compound, pubescent, plurilocular ovary with axile placentas, or a unilocular ovary with more or less deeply intruded parietal placentas; style terminal, simple, with a capitate stigma; ovules 1–2 to rather numerous (and commonly biseriate) in each locule, epitropous, anatropous, bitegmic, crassinucellar. FRUIT a septicidal capsule in *Trigonia,* a 3-winged samara in *Humbertiodendron* and *Trigoniastrum*; seeds with a transverse or longitudinal, straight embryo, 2 thin, flat cotyledons, and a very short radicle, the seed-coat hairy in *Trigonia*; endosperm none. N = ca 10.

The family Trigoniaceae as here defined consists of 3 genera and 26 species, occurring in tropical and subtropical moist lowland forests, often along rivers. *Trigonia,* the largest genus, is widespread in tropical America. The monotypic genus *Humbertiodendron* is confined to Madagascar, and the likewise monotypic genus *Trigoniastrum* occurs in the Malay peninsula, Sumatra, and Borneo. The geographic pattern is discordant with current views on geohistory. The complex floral structure suggests an adaptation to insect-pollination similar to that found in the Fabaceae, but there is no direct evolutionary link between the two families. The rather different papilionaceous floral structure of the Polygalaceae is likewise evidently independent in evolutionary origin.

The genus *Euphronia* (*Lightia*), often included in the Trigoniaceae, is here referred to the Vochysiaceae, following Lleras.

The Trigoniaceae are generally admitted to be allied to the other families of the Polygalales. Relationships with the Sapindaceae (Sapindales) and Dichapetalaceae (Celastrales) have also been suggested. All three of these orders apparently have a common origin in the Rosales and show certain parallel as well as divergent evolutionary tendencies.

4. Family TREMANDRACEAE R. Brown ex A. P. de Candolle 1824 nom. conserv., the Tremandra Family

Xeromorphic shrubs, shrublets, or half-shrubs, often with winged stems, mostly of ericoid aspect, provided with simple or stellate, often glandular hairs, or seldom glabrous, often with solitary rhomboid crystals or sometimes clustered crystals of calcium oxalate in some of the cells of the parenchymatous tissues, and often with scattered tanniferous cells, frequently producing proanthocyanins and sometimes also ellagic acid, and sometimes cyanogenic; nodes unilacunar; vessel-segments small, mostly or all with simple perforations; imperforate tracheary elements usually of 2 types, most of them libriform fibers with simple pits, but some of them vascular tracheids with bordered pits; wood-rays heterocellular, most of them uniseriate, but some 2-several-seriate; wood-parenchyma diffuse to vasicentric or wanting. LEAVES alternate, opposite, or whorled, simple and entire or merely toothed, sometimes much-reduced; epidermis often mucilaginous; stomates anomocytic; petiole, if differentiated, with one large vascular bundle and sometimes also 2 smaller ones; stipules wanting. FLOWERS solitary in the axils, slender-pedicellate, perfect, regular, hypogynous, (3) 4–5-merous; sepals distinct or seldom connate at the base, valvate; petals distinct, induplicate-valvate; an annular extrastaminal nectary-disk with antepetalous lobes present in *Tremandra;* stamens twice as many as the petals; filaments short, distinct; anthers basifixed, more or less beaked except in *Tremandra,* 2–4-locular, opening by a single apical pore or pore-like slit; pollen-grains binucleate, tricolporate; gynoecium of 2 carpels united to form a compound, bilocular ovary with axile or commonly apical-axile placentas and a slender terminal style with a punctate stigma; ovules 1–2 (3) in each locule, pendulous, epitropous, with the micropyle directed upwards and outwards, anatropous, bitegmic, crassinucellar. FRUIT a flattened capsule, loculicidal and sometimes also septicidal; seeds glabrous or hairy, with an arilloid chalazal appendage except in *Platytheca;* embryo small or minute, straight, embedded in the copious endosperm.

The family Tremandraceae consists of 3 genera, native to Australia (especially western Australia) and Tasmania. The largest genus is

Tetratheca, with about 25 species. *Tremandra* has only 2 species, and *Platytheca* only one. None of the species is of much economic importance or well known to botanists generally.

5. Family POLYGALACEAE R. Brown in Flinders 1814 nom. conserv., the Milkwort Family

Herbs, shrubs, woody vines, or even small to medium-sized trees, very rarely achlorophyllous and mycoparasitic (*Epirixanthes*), glabrous or usually with simple, unicellular hairs, often with extrafloral nectaries, sometimes accumulating aluminum, often producing triterpenoid saponins and methyl salicylate, but not tanniferous, lacking both proanthocyanins and ellagic acid, and not cyanogenic, commonly with solitary or sometimes clustered crystals of calcium oxalate in some of the cells of the parenchymatous tissues, and in *Polygala* often with lysigenous secretory cavities or oil-ducts; nodes unilacunar; climbing forms often with anomalous secondary growth of concentric type, or with patches of included phloem; xylem of young stems usually forming a closed ring even in herbs, or if not so then the vascular bundles soon becoming connected by the activity of an interfascicular cambium; vessel-segments with simple perforations; imperforate tracheary elements with evidently bordered pits; wood-rays heterocellular or homocellular, all uniseriate, or some of them 2–3 (–16) cells wide and with elongate ends; wood parenchyma mainly paratracheal, sometimes wing-like, seldom (Moutabeae) mainly apotracheal. LEAVES alternate, rarely opposite or whorled, simple, entire, reduced to scales in *Epirixanthes* and some South American spp. of *Polygala*; stomates anomocytic or seldom paracytic; petiole generally with a single vascular bundle; stipules wanting, or represented by glands, these rarely spinescent. FLOWERS in terminal, axillary, or rarely extra-axillary spikes, racemes, or panicles, bracteate and almost always bibracteolate, perfect, hypogynous or sometimes (as in *Moutabea* and *Diclidanthera*) perigynous with a cupular hypanthium, usually strongly irregular (rarely only slightly so, as in *Diclidanthera*), commonly papilionaceous in aspect but not in structural detail, sometimes some of them cleistogamous; sepals mostly 5 and distinct, less frequently connate below (as in *Epirixanthes*), or the 2 lower ones connate (as in spp. of *Polygala* and *Monnina*); the 2 lateral (inner) sepals commonly larger than the others and petaloid, but sometimes (as in spp. of *Muraltia, Salomonia,* and *Moutabea*) all nearly alike; petals distinct from each other but typically adnate to the stamens to form a common tube, sometimes 5,

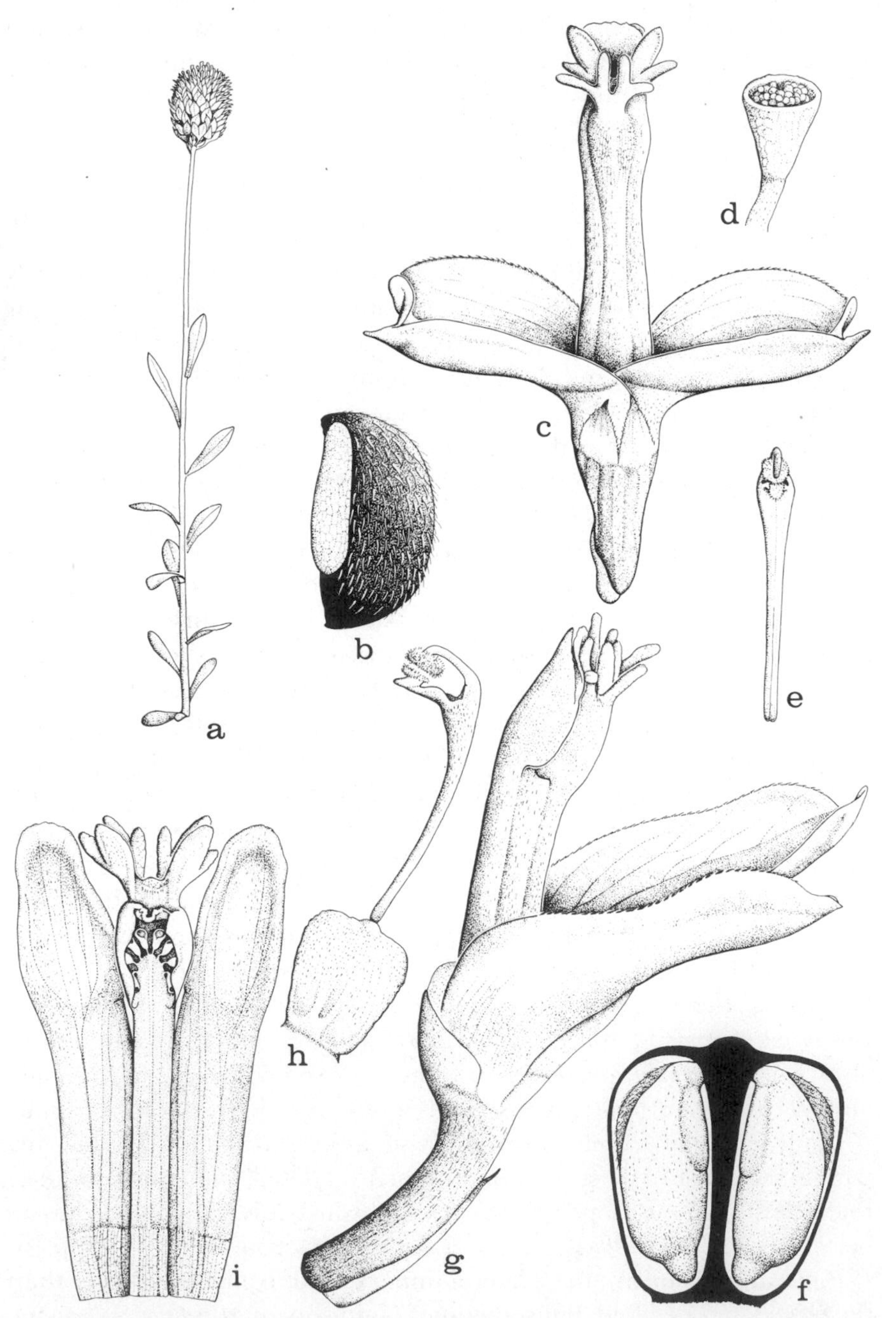

FIG. 5.49 Polygalaceae. *Polygala lutea* L. a, habit, ×½; b, seed, ×24; c, flower, ×8; d, anther, ×48; e, style, ×12; f, schematic long-section of ovary, ×24; g, side view of flower, ×12; h, pistil, ×12; i, corolla, opened out, showing adnate stamens, ×12.

as in *Carpolobia, Atroxima,* and *Moutabea,* but more often the 2 lateral ones much-reduced or completely suppressed, so that there are only 3 evident petals, the 2 upper ones and the lower median one, the latter boat-shaped and often fringed (or otherwise appendaged) at the tip; androecium commonly of 8 stamens in 2 cycles of 4 (one member of each ancestrally 5-merous cycle suppressed), or sometimes 10 or only 3–7; filaments generally connate (at least below the middle) into a cleft tube that is adnate at least below to the petals; anthers basifixed, tetrasporangiate with the ventral sporangia often reduced, or tri- or bisporangiate through suppression of the ventral sporangia, opening by 1 or 2 apical or subapical pores or very short slits, or the slits less often elongate, and sometimes running the full length of the anther; pollen-grains trinucleate or less often binucleate, of a distinctive polycolporate type; an intrastaminal annular nectary disk sometimes present, or the disk modified into an adaxial gland; gynoecium of 2 or up to 5 or even (*Eriandra*) 7 or 8 carpels united to form a plurilocular ovary with axile placentas, in *Securidaca* and most spp. of *Monnina* pseudomonomerous and unilocular; style terminal, simple, usually curved and apically bilobed, one lobe stigmatic, the other often ending in a tuft of hairs, but sometimes the style unlobed, with a capitate stigma; ovules solitary in each locule, pendulous, epitropous, with the micropyle directed upwards and outwards, anatropous to hemitropous, bitegmic, crassinucellar; endosperm-development nuclear. FRUIT a loculicidal capsule, or a nut, samara, or drupe, or capsule-like but indehiscent; seeds often hairy, and often with an evident micropylar aril (caruncle); embryo straight, with 2 cotyledons; endosperm well developed, oily and proteinaceous (sometimes also starchy), or not infrequently scanty or wanting. X = 5 – 11+. (Diclidantheraceae, Disantheraceae, Moutabeaceae)

The family Polygalaceae as here defined consists of about a dozen genera (or more, by some standards) and 750 species, nearly cosmopolitan in distribution, but wanting from New Zealand, Polynesia (except as introduced weeds) and the Arctic. About two-thirds of the species belong to the single large genus *Polygala* (500). Some species of *Polygala* have been used in folk medicine, and a few are cultivated for ornament.

Fossil samaras attributed to the genus *Securidaca* occur in Eocene deposits in Egypt and in the Rocky Mountains; leaves and pollen occur in Miocene deposits in Oregon. Pollen thought to represent the Polygalaceae occurs in Paleocene and more recent deposits.

The monotypic West-Australian genus *Emblingia* appears to be allied to the Polygalaceae but is distinctive in several respects: synsepalous calyx, deeply cleft on the abaxial side; flowers with an androgynophore; petals 2, alternating with the adaxial sepals; an adaxial unilateral nectary present between the 2 petals; pollen-grains tricolporate. Perhaps *Emblingia* should be treated as a monotypic family, but for the present I append it to the Polygalaceae.

6. Family XANTHOPHYLLACEAE Gagnep. in H. Lecomte 1909, the Xanthophyllum Family

Small trees, commonly accumulating aluminum, at least somtimes producing triterpenoid saponins, but not tanniferous, often with prismatic crystals of calcium oxalate in some of the cells of the parenchymatous tissues; nodes unilacunar; vessel-segments with simple perforations; imperforate tracheary elements with bordered pits; wood-rays heterocellular, mostly uniseriate; wood-parenchyma paratracheal, banded. LEAVES alternate, simple, coriaceous, nearly always with much-enlarged tracheary idioblasts terminal or subterminal to many of the veinlets; stomates anisocytic or sometimes paracytic; stipules wanting, or sometimes represented by crateriform glands. FLOWERS in axillary or terminal, raceme-like inflorescences, with sessile dichasial clusters along the axis, or in mixed panicles, perfect, hypogynous, strongly irregular; sepals 5, imbricate, the 2 inner a little longer than the 3 outer; petals 4–5, imbricate, unequal, sometimes clawed, the lowermost one internal, folded to form an often hairy keel; stamens 8; filaments more or less inflated and pubescent below, distinct or sometimes adnate to the claws of the petals, not forming a tube; anthers more or less dorsifixed, tetrasporangiate and dithecal, opening by longitudinal slits, often hairy, especially at the base; pollen-grains polycolporate; an annular intrastaminal nectary-disk present; gynoecium of 2 carpels united to form a stipitate, compound, unilocular ovary with 2 parietal placentas and 2–16 anatropous bitegmic, crassinucellar, ovules in all; a partial partition present at the base of the ovary; style terminal, simple, with a capitate, bilobed stigma. FRUIT globose, indehiscent, dry or fleshy-fibrous; seed solitary, without endosperm; embryo rich in stored fats.

The family Xanthophyllaceae consists of the single genus *Xanthophyllum,* with about 40 species native to the Indomalaysian region. *Xanthophyllum* is obviously allied to the Polygalaceae and has often been included in that family, but the differences seem as significant as those that separate other families in the order.

7. Family KRAMERIACEAE Dumortier 1829 nom. conserv., the Krameria Family

Hemiparasitic shrubs (sometimes arborescent) and perennial herbs, not saponiferous and not accumulating aluminum, but strongly tanniferous, with scattered tanniferous secretory cells in the parenchymatous tissues; clustered crystals of calcium oxalate commonly present in some of the cells of the parenchymatous tissues; vessel-segments with simple perforations; imperforate tracheary elements with bordered pits; wood-rays homocellular, narrow, commonly uniseriate; wood-parenchyma scanty-paractracheal. LEAVES alternate, simple and entire or rarely trifoliolate; stomates mainly paracytic or transitional to anomocytic; stipules wanting. FLOWERS solitary and axillary, or in terminal racemes, bibracteolate, perfect, hypogynous, strongly irregular, sepals (4) 5, petaloid, distinct, imbricate, unequal, the 3 outer ones commonly larger than the 2 inner and often nearly enclosing the rest of the flower, or seldom all the sepals reflexed; petals (4) 5, the 3 upper (adaxial) ones long-clawed, distinct or connate by their claws, the 2 lower (abaxial) ones very different, smaller, commonly broad, thick, sessile, often modified into lipid-secreting glands; nectary-disk wanting; stamens (3) 4, alternating with the upper petals but declined, a fifth sterile stamen rarely found on the anterior side of the flower; filaments thick, distinct or connate at the base, or adnate below to the claws of the upper petals; anthers tetrasporangiate and dithecal, opening by one or 2 terminal pores or short slits; pollen-grains 3 (4)-colporate, varying to apparently triporate, with horizontally elongate pores; gynoecium pseudomonomerous, the single pistil consisting of 2 fully united carpels, one well developed and fertile, the other reduced and empty; style obliquely terminal, with a discoid or punctiform stigma; ovules 2 in the single fertile locule, collateral and pendulous from near the top of the axial side of the locule, epitropous, the micropyle directed upwards and outwards, anatropous, bitegmic. FRUIT dry, indehiscent, 1-seeded, usually covered with bristles or spines, these often retrorsely barbed; seeds with a straight embryo and 2 thick cotyledons; endosperm wanting. X = 6.

The family Krameriaceae consists of the single genus *Krameria,* with about 15 species native from southwest U.S. to Argentina and Chile, mainly in dry regions. None of the species is of any great economic importance.

There has been a long-standing difference of opinion as to whether the Krameriaceae are more closely allied to the Polygalaceae or Caesalpiniaceae. The issue now appears to be resolved in favor of the polygalaceous relationship. Heimsch (1942) considered that the wood-anatomy of *Krameria* is similar to that of the Polygalaceae, but very different from that of legumes. The general floral structure, although it superficially suggests that of legumes, is actually more compatible with that of the Polygalaceae. More importantly, Leinfellner (1971) and Milby (1971) have shown that the pistil is basically bicarpellate, with one fertile and one sterile carpel. Milby also notes that the general floral anatomy of *Krameria* is very different from that of legumes, but he does not make any comparison with the Polygalaceae. Recent serological studies also support a relationship of *Krameria* to the Polygalaceae. One fly in the ointment of a polygalaceous affinity is the fact that *Krameria cytisoides* Cav. has trifoliolate leaves, whereas all other members of the Polygalales have simple leaves. The possible significance of the leaves of *K. cytisoides* to the ancestral leaf-form and broad-scale relationships of the genus remains to be investigated.

It is interesting and perhaps significant that *Krameria* is pollinated primarily by bees of the genus *Centris,* which collect oil from the flowers. The same species of *Centris* also collect oil from species of Malpighiaceae. Simpson and Neff (1978) consider that the association of *Centris* with Malpighiaceae is phyletically ancestral, and that with *Krameria* derived.

16. Order SAPINDALES Bentham & Hooker 1862

Autotrophic, terrestrial plants, woody or seldom herbaceous, chemically diverse, often producing resinous compounds or triterpenoid bitter substances, and frequently also bearing alkaloids, but lacking iridoid compounds, and only seldom with ellagic acid and seldom cyanogenic; vessel-segments generally with simple perforations (except mainly the Staphyleaceae). LEAVES variously alternate or opposite or whorled, usually compound or deeply cleft or lobed, seldom simple and unlobed, in most families lacking stipules or with only minute and caducous stipules. FLOWERS perfect or often unisexual, hypogynous, or seldom perigynous or epigynous, regular or somewhat irregular (strongly irregular only in Hippocastanaceae and Melianthaceae), generally well developed though often small (much-reduced only in Julianiaceae), commonly 4–5-merous, seldom trimerous or polymerous; petals distinct or seldom connate at the base, occasionally wanting; stamens distinct or sometimes connate by their filaments, most commonly 8 or 10 in 2 cycles (one cycle often incomplete), less often 4 or 5 and antesepalous, or sometimes (mainly in some Rutaceae and Zygophyllaceae) more than twice as many as the sepals or petals; pollen-grains triaperturate or of triaperturate-derived type, most commonly tricolporate; an extra- or intrastaminal nectary-disk commonly present, sometimes modified into a gynophore; gynoecium of 2–5 carpels united to form a compound ovary with usually axile (often basal-axile or apical-axile) placentas and distinct styles or a single common style, rarely the partitions not quite meeting in the center, or sometimes the carpels more than 5 or seemingly only one; ovules mostly only 1 or 2 in each locule (only one locule ovuliferous in many Anacardiaceae), seldom several (up to 12 in each locule), often seated on a placental obturator, epitropous or apotropous, anatropous to sometimes hemitropous or orthotropous or campylotropous, bitegmic or seldom unitegmic, crassinucellar; endosperm-development nuclear. FRUIT of various sorts; seeds with or without endosperm; embryo with 2 cotyledons.

The order Sapindales as here defined consists of 15 families and about 5400 species. More than half of these belong to only two families, the Sapindaceae and Rutaceae, each with about 1500 species. Another 2300 species belong to 6 well known but not large families, the Anacardiaceae (600), Burseraceae (600), Meliaceae (550), Zygophyllaceae (250), Aceraceae (110), and Simaroubaceae (150). The remaining

7 families have less than a hundred species in all, and most of these belong to the Staphyleaceae (50) or Hippocastanaceae (16).

The features common to most members of the Sapindales, which make it useful to distinguish them as a group from the Rosales, are the compound or cleft leaves, haplostemonous or diplostemonous androecium, well developed nectary-disk, and syncarpous ovary with a limited number (usually only 1 or 2) of ovules in each locule. All of these features can be found individually in the Rosales, but not in combination, except in some members of the Cunoniaceae.

Although a number of its families have in the past been referred to several different orders, the Sapindales are in my opinion a well characterized natural group. Only two families are really peripheral. The Staphyleaceae connect the Sapindales to the ancestral Rosales, in the vicinity of the Cunoniaceae, and the Zygophyllaceae are suggestive of the Geraniales. These two families also differ from the bulk of the order in usually having well developed stipules and in often having more than 2 ovules in each locule of the ovary. Collectively they constitute less than 6 percent of the order.

Ecologically the members of the Sapindales have little in common beyond their usually woody habit and limited number of seeds in each fruit. The nectary-disk is of course an adaptation to insect-pollination, but several related orders have a similar disk, and there is no reason to suppose that the pollinators of the Sapindales are in any way distinctive.

I here retain the same Sapindales because of its wide usage during the past century, although it is antedated by Lindley's names Acerales, Meliales, and Rutales.

SELECTED REFERENCES

Barkley, F. A. 1957. Generic key to the Sumac family (Anacardiaceae). Lloydia 20: 255–265.

Benseler, R. W. 1975. Floral biology of California buckeye. Madroño 23: 41–53.

Biesboer, D. D. 1976. Pollen morphology of the Aceraceae. Grana 15: 19–27.

Boesewinkel, F. D. 1977. Development of ovule and testa in Rutaceae I. *Ruta, Zanthoxylum* and *Skimmia*. Acta Bot. Neerl. 26: 193–211.

Boeswinkel, F. D., & F. Bouman. 1978. Development of ovule and testa in Rutaceae II. The unitegmic and pachychalazal seed of *Glycosmis* cf. *arborea* (Roxb.) DC. Acta Bot. Neerl. 27: 69–78.

Brizicky, G. K. 1962. The genera of Rutaceae in the southeastern United States. J. Arnold Arbor. 43: 1–22.

Brizicky, G. K. 1962. The genera of Simaroubaceae and Burseraceae in the southeastern United States. J. Arnold Arbor. 43: 173–186.

Brizicky, G. K. 1962. The genera of Anacardiaceae in the southeastern United States. J. Arnold Arbor. 43: 359–375.

Brizicky, G. K. 1963. The genera of Sapindales in the southeastern United States. J. Arnold Arbor. 44:462–501.
Copeland, H. F. 1955. The reproductive structures of *Pistacia chinensis* (Anacardiaceae). Phytomorphology 5: 440–449.
Copeland, H. F. 1959. The reproductive structures of *Schinus molle* (Anacardiaceae). Madroño 15: 14–25.
Copeland, H. F. 1961. Observations on the reproductive structures of *Anacardium occidentale*. Phytomorphology 11: 315–325.
Cronquist, A. 1944. Studies in Simaroubaceae. IV. Resume of the American genera. Brittonia 5: 128–147.
Cronquist, A. 1945. Additional notes on the Simaroubaceae. Brittonia 5: 469–470.
Desai, S. 1960. Cytology of Rutaceae and Simarubaceae. Cytologia 25: 28–35.
Fish, F., & P. G. Waterman. 1973. Chemosystematics in the Rutaceae. II. The chemosystematics of the *Zanthoxylum/Fagara* complex. Taxon 22: 177–203.
Gray, A. I., & P. G. Waterman. 1978. Coumarins in the Rutaceae. Phytochemistry 17: 845–864.
Grundwag, M. 1976. Embryology and fruit development in four species of *Pistacia* L. (Anacardiaceae). J. Linn. Soc., Bot. 73: 355–370.
Guédès, M. 1973. Carpel morphology and axis-sharing in syncarpy in some Rutaceae, with further comments on "New Morphology." J. Linn. Soc., Bot. 66: 55–74.
Guervin, C. 1961. Contribution à l'étude cytotaxinomique des Sapindacées et caryologique des Mélianthacées et des Didiéréacées. Rev. Cytol. Biol. Vég. 23: 49–86.
Gut, B. J. 1966. Beiträge zur Morphologie des Gynoeceums und der Blütenachse einiger Rutaceen. Bot. Jahrb. Syst. 85: 151–247.
el Hadidi, M. N. 1977. Tribulaceae as a distinct family. Publ. Cairo Univ. Herb. 7 & 8: 103–108.
Hall, B. A. 1951. The floral anatomy of the genus *Acer*. Amer. J. Bot. 38: 793–799.
Hall, B. A. 1954. The variability in the floral anatomy of *Acer Negundo*. Amer. J. Bot. 41: 529–532.
Hall, B. A. 1961. The floral anatomy of *Dipteronia*. Amer. J. Bot. 48: 918–924.
Hardin, J. W. 1956. Studies in the Hippocastanaceae. II. Inflorescence structure and distribution of perfect flowers. Amer. J. Bot. 43: 418–424.
Hardin, J. W. 1957. A revision of the American Hippocastanaceae. Brittonia 9: 145–195.
Hartl, D. 1957. Struktur und Herkunft des Endokarps der Rutaceen. Beitr. Biol. Pflanzen 34: 35–49.
Heimsch, C. 1940. Wood anatomy and pollen morphology of *Rhus* and allied genera. J. Arnold Arbor. 21: 279–291.
Hilger, H. H. 1978. Der multilakunäre Knoten einiger *Melianthus*- und *Greyia*-Arten im Vergleich mit anderen Knotentypen. Flora B 167: 165–176.
Inamdar, J. A. 1969. Epidermal structure, stomatal ontogeny and relationship of some Zygophyllaceae and Simarubaceae. Flora B 158: 360–368.
Inamdar, J. A., & R. C. Patel. 1970. Epidermal structure and development of stomata in vegetative and floral organs of *Fagonia cretica* Linn. Flora B 159: 63–70.
Jensen, U. 1974. Close relationships between Ranunculales and Rutales? Systematic considerations in the light of new results of comparative serological research. Serol. Mus. Bull. 50: 4–7.

Johri, B. M., & M. R. Ahuja. 1957. A contribution to the floral morphology and embryology of *Aegle marmelos* Correa. Phytomorphology 7: 10–24.

Kapil, R. N., & K. Ahluwalia. 1963. Embryology of *Peganum harmala* Linn. Phytomorphology 13: 127–140.

Kribs, D. A. 1930. Comparative anatomy of the woods of the Meliaceae. Amer. J. Bot. 17: 724–738.

Kryn, J. M. 1952 (1953). The anatomy of the wood of the Anacardiaceae and its bearing on the phylogeny and relationships of the family. Ph.D. diss., Univ. Michigan, Ann Arbor. Univ. Microfilms. Ann Arbor.

Lam, H. J. 1932 Beiträge zur Morphologie der Burseraceae insbesondere der Canarieae. Ann. Jard. Bot. Buitenzorg 42: 97–226.

Lam, H. J. 1938. Studies in phylogeny. II. On the phylogeny of the Malaysian Burseraceae-Canarieae in general and of *Haplolobus* in particular. Blumea 3: 126–158.

Leenhouts, P. W. 1978. The pollen morphology of Burseraceae. A taxonomic comment. Grana 17: 175–177.

Leroy, J.-F. 1959. Sur une petite famille de Sapindales propre à l'Afrique australe et à Madagascar: les Ptaeroxylaceae. Compt. Rend. Hebd. Séances Acad. Sci. 248: 1001–1003.

List, A., & F. C. Steward. 1965. The nucellus, embryo sac, endosperm, and embryo of *Aesculus* and their interdependence during growth. Ann. Bot. (London) II. 29: 1–15.

Lobreau-Callen, D., S. Nilsson, F. Albers, & H. Straka. 1978. Les Cneoraceae (Rutales): étude taxonomique, palynologique et systematique. Grana 17: 125–139.

Mauritzon, J. 1935. Über die Embryologie der Familie Rutaceae. Svensk Bot. Tidskr. 319–347.

Mauritzon, J. 1936. Zur Embryologie und systematischen Abgrenzung der Reihen Terebinthales und Celastrales. Bot. Not. 1936: 161–212.

Mitra, K., M. Mondal, & S. Saha. 1977. The pollen morphology of Burseraceae. Grana 16: 75–79.

Muller, J., & P. W. Leenhouts. 1976. A general survey of pollen types in Sapindaceae in relation to taxonomy. *In:* I. K. Ferguson & J. Muller, eds., The evolutionary significance of the exine, pp. 407–445. Linn. Soc. Symp. Ser. No. 1. Academic Press, London & New York.

Nair, N. C. 1959–1963. Studies on Meliaceae. I. Floral morphology and embryology of *Naregamia alata* W. & A. II. Floral morphology and embryology of *Melia azedarach* Linn.—A reinvestigation. V. Morphology and anatomy of the flower of the tribes Melieae, Trichileae and Swietenieae. VI. Morphology and anatomy of the flower of the tribe Cedrelieae and discussion on the floral anatomy of the family. J. Indian Bot. Soc. 38: 353–366; 367–378, 1959. 41: 226–242, 1962. 42: 177–189, 1963.

Nair, N. C., & T. C. Joseph. 1957. Floral morphology and embryology of *Samadera indica.* Bot. Gaz. 119: 104–115.

Nair, N. C., & R. K. Joshi. 1958. Floral morphology of some members of the Simaroubaceae. Bot. Gaz. 120: 88–99.

Nair, N. C., & K. S. Nathawat. 1958. Vascular anatomy of the flower of some species of Zygophyllaceae. J. Indian Bot. Soc. 37: 172–180.

Nair, N. C., & N. P. Sukumaran. 1960. Floral morphology and embryology of *Brucea amarissima.* Bot. Gaz. 121: 175–185.

Narayana, H. S., & C. G. Prakasa Rao. 1963. Floral morphology and embryology of *Seetzenia orientalis* Decne. Phytomorphology 13: 197–205.

Narayana, L. L. 1958, 1959. Floral anatomy of Meliaceae. I, II. J. Indian Bot. Soc. 37: 365–374, 1958. 38: 288–295, 1959.
Narayana, L. L. 1960. Embryology of the Staphyleaceae. Curr. Sci. 29: 403, 404.
Narayana, L. L. 1960. Studies in Burseraceae. I, II. J. Indian Bot. Soc. 39: 204–209; 402–409.
Narayana, L. L., & M. Sayeeduddin. 1958. Floral anatomy of Simarubaceae. I. J. Indian Bot. Soc. 37: 517–522.
Nene, P. M., & V. D. Tilak. 1977. Placentation in the Rutaceae. Proc. Indian Acad. Sci. 85B: 378–383.
Nooteboom, H. P. 1962. Generic delimitation in Simaroubaceae tribus Simaroubeae and a conspectus of the genus *Quassia* L. Blumea 11: 509–528.
Nooteboom, H. P. 1966. Flavonols, leuco-anthocyanins, cinnamic acids, and alkaloids in dried leaves of some Asiatic and Malesian Simaroubaceae. Blumea 14: 309–315.
Ogata, K. 1967. A systematic study of the genus *Acer*. Misc. Inform. Tokyo Univ. Forests 63: 89–206.
Pennington, T. D., & B. T. Styles. 1975. A generic monograph of the Meliaceae. Blumea 22: 419–540.
Pernet, R. 1972. Phytochimie des Burseracees. Lloydia 35: 280–287.
Phatak, V. G. 1971. Embryology of *Zygophyllum coccineum* L. and *Z. fabago* L. Proc. Kon. Nederl. Akad. Wetensch. C 74: 379–397.
Pijl, L. van der. 1957. On the arilloids of *Nephelium, Euphoria, Litchi* and *Aesculus*, and the seeds of Sapindaceae in general. Acta Bot. Neerl. 6: 618–641.
Polonsky, J. 1966. Les principes amers des Simarubacées. Pl. Med. 14: 107–116.
Porter, D. M. 1972. The genera of Zygophyllaceae in the southeastern United States. J. Arnold Arbor. 53: 531–552.
Record, S. J. 1941. American timbers of the Mahogany family. Trop. Woods 66: 7–33.
Record, S. J., & R. W. Hess. 1940. American woods of the family Rutaceae. Trop. Woods 64: 1–28.
Romero, E. J., & L. J. Hickey. 1976. A fossil leaf of Akaniaceae from Paleocene beds in Argentina. Bull. Torrey Bot. Club 103: 126–131.
Saleh, N. A. M., & M. N. El-Hadidi. 1977. An approach to the chemosystematics of the Zygophyllaceae. Biochem. Syst. Ecol. 5: 121–128.
Seigler, D. S., & W. Kawahara. 1976. New reports of cyanolipids from Sapindaceous plants. Bichem. Syst. Ecol. 4: 263–265.
Sharma, M. R. 1954. Studies in the family Anacardiaceae. I. Vascular anatomy of the flower of *Mangifera indica* L. Phytomorphology 4: 201–208.
Smith-White, S. 1954. Chromosome numbers in the Boronieae (Rutaceae) and their bearing on the evolutionary development of the tribe in the Australian flora. Austral. J. Bot. 2: 287–303.
Spongberg, S. 1971. The Staphyleaceae in the southeastern United States. J. Arnold Arbor. 52: 196–203.
Stern, W. L. 1952. The comparative anatomy of the xylem and the phylogeny of the Julianiaceae. Amer. J. Bot. 39: 220–229.
Stern, W. L., & G. K. Brizicky. 1960. The morphology and relationships of *Diomma*, gen. inc. sed. Mem. New York Bot. Gard. 10(2): 38–57.
Stöcklin, W., L. B. de Silva, & T. A. Geissman. 1969. Constituents of *Holacantha emoryi*. Phytochemistry 8: 1565–1569.
Straka, H., F. Albers, & A. Mondon. 1976. Die Stellung und Gliederung der Familie Cneoraceae (Rutales). Beitr. Biol. Pflanzen 52: 267–310.

Tanai, T. 1978. Taxonomical investigation of the living species of the genus *Acer* L. based on vein architecture of the leaves. J. Fac. Sci. Hokkaido Imp. Univ., Ser. 4, Geol. 18: 243–282.

Tang, Y. 1935. Notes on the systematic position of Bretschneideraceae as shown by its timber anatomy. Bull. Fan Mem. Inst. Biol. 6: 153–157.

Tiffney, B. H. 1979. Fruits and seeds of the Brandon Lignite III. *Turpinia* (Staphyleaceae). Brittonia 31: 39–51.

Tillson, A. H., & R. Bamford. 1938. The floral anatomy of the Aurantioideae. Amer. J. Bot. 25: 780–793.

Venning, F. D. 1948. The ontogeny of the laticiferous canals in the Anacardiaceae. Amer. J. Bot. 35: 637–644.

Waterman, P. G. 1975. Alkaloids of the Rutaceae: their distribution and systematic significance. Biochem. Syst. Ecol. 3: 149–180.

Webber, I. E. 1936. Systematic anatomy of the woods of the Simarubaceae. Amer. J. Bot. 23: 577–587.

Webber, I. E. 1941. Systematic anatomy of the woods of the "Burseraceae." Lilloa 6: 441–465.

Weberling, F. 1976. Die Pseudostipeln der Sapindaceae. Akad. Wiss. Abh. Math.-Naturwiss. Kl. 1976. Nr. 2: 1–27.

Weberling, F., & P. W. Leenhouts. 1965. Systematisch-morphologische Studien an Terebinthales-Familien (Burseraceae, Simaroubaceae, Meliaceae, Anacardiaceae, Sapindaceae). Akad. Wiss. Abh. Math.-Naturwiss. Kl. 1965. Nr. 10: 495–584.

Young, D. A. 1976. Flavonoid chemistry and the phylogenetic relationships of the Julianiaceae. Syst. Bot. 1: 149–162.

Агабабян, В. Ш. 1964. Морфологические типы пыльцы и систематика семейства Zygophyllaceae. Изв. АН Армянской ССР, Биол. Науки, 17(12): 39-45.

Анели, Н. А. 1960. Материалы к филогении семейства рутовых. Сборник Трудов Тбилисского Научно-иссл. Химико-Фармац. Инст. Министерства Здравоохранения Грузинской ССР, 9: 73-102.

Бобров, Е. Г. 1965. О происхождении флоры пустынь старого света в связи с обзором рода *Nitraria* L. Бот. Ж. 50: 1053-1067.

SYNOPTICAL ARRANGEMENT OF THE FAMILIES OF SAPINDALES

1 Leaves usually stipulate, often conspicuously so; ovules often more than 2 per carpel; seeds with endosperm, except some Zygophyllaceae.

2 Stamens mostly bicyclic or tricyclic, seldom unicyclic; ovules mostly epitropous, seldom apotropous; compound leaves usually without a terminal leaflet; plants generally not tanniferous .... 15. ZYGOPHYLLACEAE.

2 Stamens unicyclic; ovules apotropous; compound leaves with a terminal leaflet; plants usually tanniferous.

3 Leaves opposite, seldom alternate; disk annular, mostly intrastaminal (rarely poorly developed or wanting); flowers regular; carpels 2–3 (4); pollen-grains binucleate; vessel-segments with scalariform perforations; plants not producing ellagic acid .... 1. STAPHYLEACEAE.

3 Leaves alternate; disk unilateral, extrastaminal; flowers irregular; carpels 4 (5); pollen-grains trinucleate; vessel-segments with simple perforations; plants producing ellagic acid ....2. MELIANTHACEAE.

1 Leaves mostly exstipulate, the stipules when present small and caducous; ovules seldom more than 2 per carpel; seeds with or often without endosperm.

4 Disk mostly extrastaminal (and often unilateral) or wanting; ovules apotropous except in Akaniaceae; flowers regular or very often somewhat irregular; stamens very often 8; plants very often accumulating quebrachitol.

5 Leaves alternate (opposite in a few Sapindaceae).

6 Flowers perigynous; plants producing myrosin-cells and mustard-oils; most of the wood-rays multiseriate, 4–6 (–10) cells wide ....3. BRETSCHNEIDERACEAE.

6 Flowers hypogynous; myrosin-cells and mustard-oils wanting.

7 Seeds with endosperm and without an aril or sarcotesta; wood-rays mainly multiseriate; nectary-disk wanting; ovules epitropous, 2 in each locule .... 4. AKANIACEAE.

7 Seeds without endosperm, often with an aril or sarcotesta; wood-rays mostly uniseriate, seldom 2–3 (–6)-seriate; nectary-disk generally present; ovules apotropous, 1 or seldom 2 (rarely several) in each locule ....5. SAPINDACEAE.

5 Leaves opposite.

8 Flowers evidently irregular; fruit a (2) 3 (4)-carpellate, usually 1-seeded capsule; leaves palmately compound ....6. HIPPOCASTANACEAE.

8 Flowers regular; fruit a winged schizocarp, mostly of 2 carpels, typically a double samara; leaves simple and usually palmately lobed or veined, or in a few species either pinnately or palmately compound ....7. ACERACEAE.

4 Disk mostly intrastaminal, annular or sometimes modified into a gynophore, rarely extrastaminal or (Julianiaceae) wanting; ovules mostly epitropous except in Anacardiaceae and Julianiaceae; plants not known to accumulate quebrachitol, except some Anacardiaceae.

9 Plants strongly resinous, with vertical intercellular resin-canals in the bark and often also horizontal ones in the wood-rays, usually also with similar resin-ducts in the phloem of the larger veins of the leaves; ovules often attached to a short, broad placental obturator; plants only seldom producing alkaloids and only seldom with triterpenoid compounds.

10 Flowers of ordinary structure, perfect or unisexual, always

with a perianth and usually or always with a nectary-disk or gynophore; plants only seldom saponiferous.

11 Ovary with 2 (rarely only one) epitropous ovules in each of the (2) 3–5 locules; resin not notably allergenic or poisonous .... 8. BURSERACEAE.

11 Ovary with a single apotropous ovule in each of the (2) 3 locules, or more often only one locule ovuliferous, or the carpels rarely distinct or solitary and each (or one of them) with a single such ovule; resin often allergenic or poisonous to the touch .... 9. ANACARDIACEAE.

10 Flowers much-reduced, unisexual, the pistillate ones without perianth; disk and gynophore wanting; ovary tricarpellate, unilocular, with a single ovule; plants saponiferous .... 10. JULIANIACEAE.

9 Plants resinous or not, but without resin-ducts in the bark, wood-rays, and veins of the leaves; ovules with or more often without an obturator; plants very often producing triterpenoid substances and/or alkaloids, but only seldom saponiferous.

12 Leaves not glandular-punctate; plants often with scattered secretory cells, but generally without secretory cavities.

13 Stamens distinct; other features various, but not combined as in the Meliaceae.

14 Seeds without endosperm; leaves mostly compound, seldom simple; bark generally bitter; plants producing characteristic triterpenoid lactones called simaroubalides; flowers 3–8-merous, most commonly 5-merous; style solitary, or often the styles several and distinct; pollen-grains binucleate 11. SIMAROUBACEAE.

14 Seeds with copious endosperm; leaves simple; bark not bitter; plants not producing simaroubalides; flowers trimerous or sometimes tetramerous; style solitary; pollen-grains trinucleate .... 12. CNEORACEAE.

13 Stamens mostly connate by their filaments, seldom distinct; leaves usually compound; bark bitter; seeds usually with well developed endosperm, ovary with a single style; leaves commonly with resinous secretory cells at the boundary of the palisade and spongy mesophyll ....13. MELIACEAE.

12 Leaves glandular-punctate, the plants commonly with secretory cavities containing aromatic ethereal oils, these cavities scattered throughout the parenchymatous tissues and in the pericarp .... 14. RUTACEAE.

1. Family STAPHYLEACEAE Lindley 1829 nom. conserv., the Bladdernut Family

Shrubs or rather small trees, producing mucilage and tannin (often in mucilage-cells and tanniferous cells), probably at least sometimes with proanthocyanins, but without ellagic acid, not cyanogenic, not saponiferous, and without iridoid compounds; clustered and often also solitary crystals of calcium oxalate present in some of the cells of the parenchymatous tissues; nodes trilacunar; vessel-segments with oblique, scalariform perforation-plates that have 20–30 (–50) cross-bars, sometimes spirally thickened; imperforate tracheary elements with evidently bordered pits (often some of them true tracheids) in the Staphyleoideae, with simple pits in the Tapiscioideae; wood-rays heterocellular, mixed uniseriate and pluriseriate, the latter mostly 4–7 (–10) cells wide, with elongate or short ends; wood-parenchyma scanty-paratracheal, often limited to a few cells on the abaxial side of the vessels, sometimes also with a few diffusely scattered cells. LEAVES opposite (Staphyleoideae) or seldom alternate (Tapiscioideae), pinnately compound or often trifoliolate, occasionally unifoliolate (as in *Turpinia arguta* L.), generally with toothed leaflets; epidermis often mucilaginous; petiolar anatomy complex; stipules caducous, or wanting in some Tapiscioideae. FLOWERS small, in terminal or axillary, drooping panicles or sometimes racemes, regular, hypogynous, pentamerous, perfect or sometimes some or even all of them unisexual (the plants then dioecious); sepals imbricate, distinct (Staphyleoideae) or seldom connate for part or most of their length (Tapiscioideae), often petaloid; petals distinct, imbricate; stamens in a single series alternate with the petals, seated on or outside of the well developed, annular nectary-disk (disk poorly developed or wanting in (Tapiscioideae); anthers tetrasporangiate and dithecal, opening by longitudinal slits; pollen-grains binucleate, tricolporate; gynoecium of 2–3 (4) carpels united to form a compound, plurilocular ovary (distinct in *Euscaphis*); styles distinct, or connate only distally, or united into a common style; ovary superior, laterally and sometimes also apically lobed in Staphyleoideae, lobeless in Tapiscioideae, the base often embedded in the disk; ovules (1–2–) 6–12 in 2 rows in each locule on axile or basal-axile placentas (Staphyleoideae), or only 1 or 2 altogether in Tapiscioideae, apotropous, anatropous, bitegmic, crassinucellar, commonly ascending; endosperm-development nuclear. FRUIT an inflated capsule opening at the tip, the carpels separate (or separating) distally and opening along the ventral suture, or indehiscent and drupaceous

FIG. 5.50 Staphyleaceae. *Staphylea trifolia* L. a, habit, in fruit, $\times\frac{1}{4}$; b, habit, in flower, $\times\frac{1}{2}$; c, flower, from above, $\times 6$; d, flower, with perianth and 2 stamens removed, $\times 6$; e, flower, $\times 6$; f, anther, $\times 6$; g, schematic cross-section of ovary, $\times 12$; h, opened fruit, $\times 1$; i, internal surface of one valve of fruit, $\times 1$; k, node, with stipules, $\times 6$.

or baccate, or (*Euscaphis*) the fruit of distinct follicles; seeds often only 1 or 2 per locule, even when the ovules are more numerous; embryo straight, with 2 large, flat cotyledons; endosperm copious, fleshy, oily. X = 13.

The family Staphyleaceae as here defined consists of 5 genera and about 50 species, irregularly distributed in the Americas and in Eurasia, and extending also into the Malay Archipelago. *Staphylea* (10, North Temperate), *Turpinea* (35, C. and S. Amer. and tropical Asia) and *Euscaphis* (1, eastern Asia) make up the subfamily Staphyleoideae, on which the description of the family is primarily founded. *Tapiscia* (1, China) and *Huertea* (2, West Indies and northern S. Amer.) make up the well marked subfamily Tapiscioideae. Species of *Staphylea*, bladder-nut, are sometimes cultivated for ornament.

The Staphyleaceae have by different authors been included in the Sapindales, Celastrales, and Rosales (near Cunoniaceae). Inasmuch as the Celastrales and Sapindales are in my opinion derived in parallel fashion from the Rosales, the three different ordinal positions for the Staphyleaceae do not require any great conceptual difference as to their relationship. The Staphyleaceae would be anomalous in the Celastrales because of their compound leaves. In my opinion they form a connecting link between the Cunoniaceae (Rosales) and the Aceraceae and Sapindaceae (Sapindales), and could with almost equal propriety be put in either order.

On the whole, the Staphyleaceae have progressed further along the Sapindalean road than the Cunoniaceae. I find it useful to retain the Cunoniaceae in the more archaic order (Rosales) and refer the Staphyleaceae to the more advanced order (Sapindales). It is interesting and perhaps significant that the Staphyleaceae have the same basic chromosome number (13) as the Aceraceae. The characteristic inflated capsules of *Staphylea* are much like those of *Koelreuteria* in the Sapindaceae.

If the Staphyleaceae are included in the Sapindales, they must be regarded as the most archaic family in the order, because of their primitive wood-structure, their mostly stipulate leaves, usually several ovules in each locule,. well developed endosperm, and sometimes separate carpels. I will not argue with anyone who finds sufficient basis in these features to exclude the Staphyleaceae from the Sapindales and refer them to the Rosales near the Cunoniaceae.

Fossil leaves thought to represent *Turpinia* occur in Eocene deposits of the Wind River Basin, in Wyoming, and wood assigned to *Turpinia* is found in Eocene rocks in Yellowstone National Park, U.S.A. Seeds referrable to *Turpinia* first show up in the Oligocene. Seeds thought to represent *Staphylea* are common in Oligocene and more recent deposits.

2. Family MELIANTHACEAE Link 1831 nom. conserv., the Melianthus Family

Shrubs, rather small trees, or half-shrubs, tanniferous and producing ellagic acid, but without proanthocyanins, not saponiferous, not cyanogenic, and not producing mucilage; styloids (elongate, slender, prismatic crystals) commonly present in some of the cells of the parenchymatous tissues; stem at least sometimes with concentric medullary bundles that have central phloem; nodes multilacunar; vessel-segments with simple perforations; imperforate tracheary elements very short, with simple pits; wood-rays homocellular, multiseriate, mostly 3–9 cells wide, uniseriates wanting; wood-parenchyma paratracheal, rather scanty. LEAVES alternate, pinnately compound; styloids present parallel to the leaf-surface, often near the boundary of the palisade and spongy parenchyma; stomates anomocytic; petiole with a ring of vascular bundles and often also with some bundles included in the ring; stipules intrapetiolar, connate, often large. FLOWERS in racemes, often large, resupinate by twisting of the pedicel, perfect or less often unisexual, hypogynous; sepals 5, or 4 by fusion of 2, unequal, imbricate, distinct or connate below, one of them spurred or saccate-gibbous in *Melianthus;* petals distinct, clawed, imbricate, 5 and unequal, or one abortive; nectary-disk well developed, unilateral, extrastaminal; stamens 4 or 5, alternate with the petals; filaments distinct or basally connate; anthers tetrasporangiate and dithecal, opening by longitudinal slits; pollen-grains trinucleate tricolporate; gynoecium of 4 (5) carpels united to form a compound, plurilocular ovary with a simple, terminal, truncate or apically dentate style; ovules 2–5 in each locule (*Melianthus*), or solitary in each locule (*Bersama*), erect to pendulous, on axile-basal to axile-apical placentas, apotropous, anatropous, bitegmic, crassinucellar; endosperm-development nuclear. FRUIT capsular, loculicidal or opening only at the tip, often longitudinally sulcate, and sometimes apically lobed; seeds usually only 1 or 2 per locule, with small, straight, dicotyledonous embryo and copious, hard, oily and sometimes starchy endosperm, with (*Bersama*) or without (*Melianthus*) an aril. N = 18, 19.

The family Melianthaceae as here defined consists of only 2 genera, both African. *Melianthus* has about 6 species, native to South Africa.

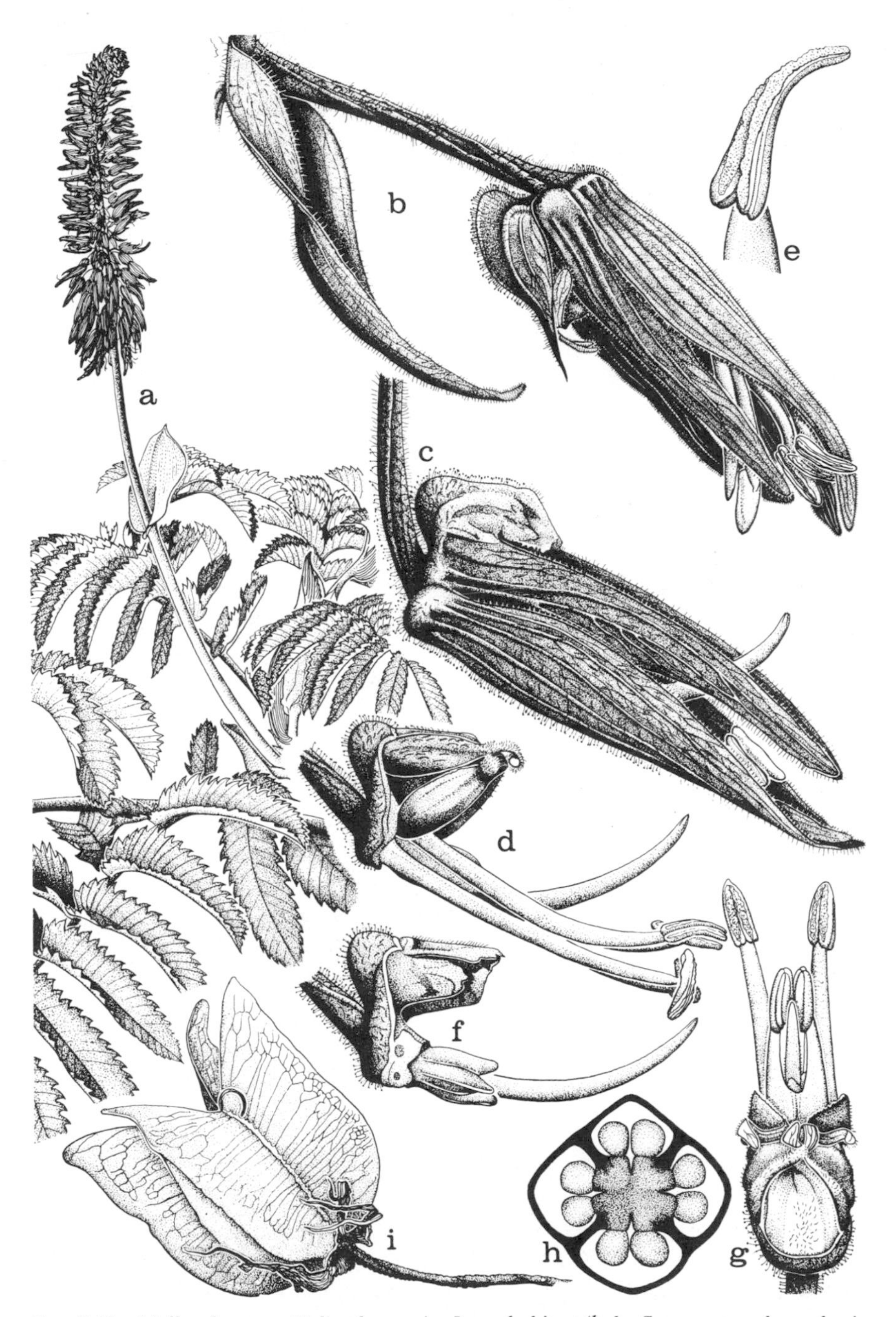

FIG. 5.51 Melianthaceae. *Melianthus major* L. a, habit, ×⅛; b, flower at early anthesis, ×2; c, flower, at late anthesis, ×2; d, flower, at late anthesis, with the 4 larger sepals removed, ×2; e, anther, ×8; f, flower, at later anthesis, with perianth (except one sepal) and stamens removed, showing pistil and nectary, ×2; g, flower, at early anthesis, with the sepals removed to show 4 petals surrounding the nectary, 4 stamens, and pistil, ×2; h, schematic cross-section of ovary, ×8; i, fruit, ×1.

Bersama has 2 highly polymorphic species (or up to 30 spp. in the opinion of some authors), native to tropical as well as southern Africa. *Melianthus major* L., sometimes cultivated for ornament, produces abundant nectar and is attractive to honey bees.

The Melianthaceae are generally conceded to be closely allied to the Sapindaceae, which lack endosperm, often have a curved embryo, and have only 1 or seldom 2 ovules per locule. The wood-structure of the Melianthaceae is considered to be more specialized than that of the Sapindaceae.

3. Family BRETSCHNEIDERACEAE Engler & Gilg 1924 nom. conserv., the Bretschneidera Family

Deciduous trees, producing mustard-oils and with myrosin-cells in the bark of branches and in the inflorescence; vessel-segments with simple and scalariform perforations, the latter with about 20 cross-bars; imperforate tracheary elements relatively thin-walled, with bordered pits; wood-rays heterocellular, some of them uniseriate, but most of them pluriseriate, 4–6 (–10) cells wide and up to 100 cells high, with short ends; wood-parenchyma terminal and scanty-paratracheal. LEAVES alternate, pinnately compound, without stipules. FLOWERS rather large, borne in terminal racemes, pinkish, perfect, perigynous, slightly irregular; calyx campanulate, 5-toothed; petals 5, distinct, unequal, clawed at the base, imbricate; stamens 8, the slender, hairy filaments attached to the rather thin, annular nectary-disk; anthers versatile; pollen-grains tricolpate or sometimes dicolpate; gynoecium of 3–5 carpels united to form a compound, plurilocular ovary with an elongate, curved style and capitate stigma; ovules 2 in each locule, apical-axile and pendulous. FRUIT capsular; seeds red, with a large, dicotyledonous embryo and no endosperm.

The family Bretschneideraceae consists of the single genus *Bretschneidera,* with the single species *B. sinensis* Hemsl., native to the mountains of western and southwestern China.

Bretschneidera has often in the past been included in the Hippocastanaceae or Sapindaceae. It would be anomalous in either family because of its perigynous flowers, mainly large and pluriseriate wood-rays, and myrosin-cells. Because of the myrosin-cells it has sometimes been referred to the Capparales, but otherwise it appears to have little in common with that group.

4. Family AKANIACEAE Stapf 1912 nom. conserv., the Akania Family

Small tree, producing alkaloids of unspecified nature, and also with proanthocyanins, but without ellagic acid and not saponiferous; secretory (tanniferous?) cells scattered in the parenchymatous tissues; clustered and large solitary crystals of calcium oxalate present in some of the cells of the parenchymatous tissues; young stems with a cycle of numerous small vascular bundles separated by medullary rays; vessel-segments with simple perforations, or a very few of them scalariform; imperforate tracheary elements with small, bordered pits, commonly septate; wood-rays heterocellular, large and multiseriate, uniseriates wanting or nearly so; wood-parenchyma scanty, paratracheal. LEAVES large, alternate, pinnately compound with offset, remotely spinulose-toothed leaflets; petiolules swollen distally as well as proximally; petiole with a ring of small vascular bundles, as in young stems; stipules wanting. FLOWERS in large mixed panicles, perfect, regular, hypogynous, pentamerous; sepals imbricate; petals distinct, convolute; stamens 8 (9), the 5 outer ones opposite the sepals; filaments distinct; anthers dithecal, opening by longitudinal slits; pollen-grains tricolporate; nectary-disk wanting; gynoecium of 3 carpels united to form a compound, trilocular ovary with a terminal style and a small, 3-lobed stigma; ovules 2 in each locule, axile, superposed, pendulous, epitropous, anatropous or hemitropous, bitegmic, crassinucellar. FRUIT a loculicidal capsule; seeds without an aril; embryo straight, with 2 cotyledons; endosperm copious, fleshy, smelling of bitter almonds.

The family Akaniaceae consists of the genus *Akania,* with a single species, *A. lucens* (F. Muell.) Airy Shaw (*A. hillii*), native to eastern Australia. *Akania* seems obviously to belong to the Sapindales as here defined, but it would be highly anomalous in any family other than its own. In the past it has often been referred to the Sapindaceae, or less commonly to the Staphyleaceae, but nearly all modern authors accept it as a distinct family. A fossil leaf considered to represent a species of *Akania* has been described from uppermost Paleocene deposits in Argentina.

5. Family SAPINDACEAE A. L. de Jussieu 1789 nom. conserv., the Soapberry Family

Trees, shrubs, or less often woody or even herbaceous vines, the vines with axillary tendrils representing modified inflorescences, and often

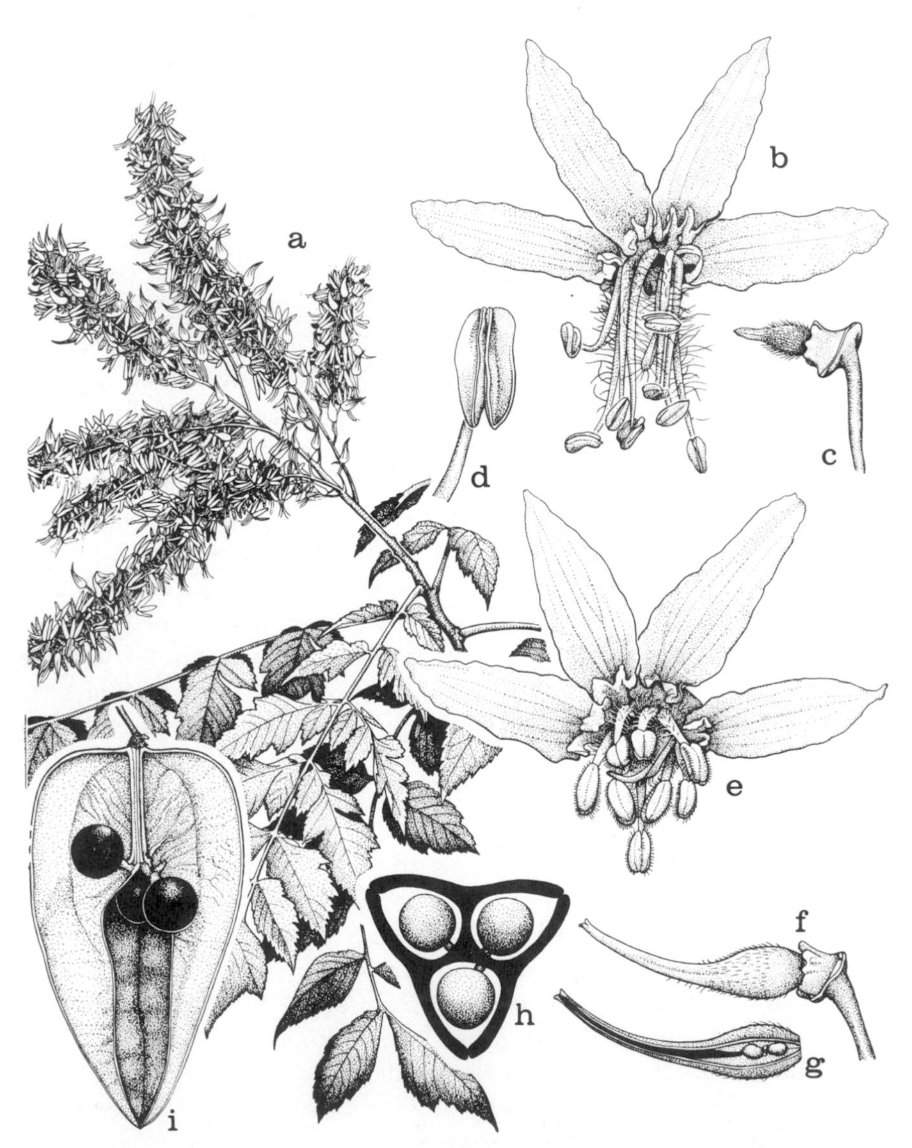

FIG. 5.52 Sapindaceae. *Koelreuteria paniculata* Laxm. a, habit, $\times\frac{1}{2}$; b, staminate flower, $\times 4$; c, receptacle, disk, and vestigial pistil of staminate flower, $\times 4$; d, stamen, $\times 12$; e, functionally pistillate flower, $\times 4$; f, receptacle, disk, and pistil of pistillate flower, $\times 4$; g, schematic long-section of pistil, $\times 4$; h, schematic cross-section of ovary, $\times 12$; i, fruit, in partial long-section, $\times 1$.

with anomalous stem-structure; plants generally tanniferous, commonly producing proanthocyanins but not ellagic acid, sometimes cyanogenic, seldom with silica grains in the wood, usually accumulating quebrachitol and triterpenoid saponins (these often highly toxic), the saponins in scattered secretory cells in some of the parenchymatous tissues; nodes trilacunar; vessel-segments with simple perforations; imperforate tracheary elements with simple pits, commonly septate; wood-rays commonly uniseriate and homocellular, seldom heterocellular or some of them 2–3 (–6) cells wide; wood-parenchyma mostly scanty and paratracheal. LEAVES alternate or very rarely opposite, pinnately (sometimes bipinnately) compound or trifoliolate, seldom simple; epidermis often with scattered mucilage-cells; petiolules often swollen proximally into a sort of pulvinus; stomates anomocytic or seldom paracytic; stipules wanting, except in some of the vines, which have small stipules. FLOWERS in terminal or axillary, mostly cymose or cymose-paniculate inflorescences, seldom solitary and axillary, individually small, hypogynous, regular or more often slightly irregular, perfect or more often functionally unisexual, with the androecium or gynoecium more or less reduced; sepals 4 or 5, distinct or sometimes connate below, imbricate or rarely valvate; petals distinct, often clawed, imbricate, mostly 4 or 5, seldom more than 5 or only 3, often with an internal, scale-like appendage toward the base that tends to conceal the nectary, or sometimes the petals wanting; extrastaminal nectary-disk commonly present, annular or often unilateral (or the stamens sometimes seated on the disk), rarely (*Dodonaea*) the disk minute and intrastaminal; stamens 4–10 (rarely more), often 8, apparently in a single cycle; filaments often hairy; anthers tetrasporangiate and dithecal, opening by longitudinal slits; pollen-grains binucleate, sometimes heteropolar, commonly tricolporate or syncolporate, seldom bicolporate or 3–4-porate, the structure of the exine highly variable; gynoecium of (2) 3 (–6) carpels united to form a compound, generally plurilocular ovary with a terminal, often lobed or cleft style, or the styles more or less distinct; typically a single ascending, apotropous ovule on the axile placenta in each locule, seldom 2 such ovules in each locule, or 1 or 2 apical-axile, pendulous ovules in each locule, or several spreading ovules in each locule, in *Koelreuteria* the partitions complete in the lower but not the upper part of the ovary, the ovules borne at the summit of the fully partitioned zone, one ovule ascending and one descending in each locule; ovules bitegmic, crassinucellar, anatropous to hemitropous or campylotropous, commonly without a defined funiculus, but broadly attached to a protruding portion of the placenta (i.e., the funiculus incorporated into the placental obturator, which is adnate to the ovule); endosperm-development nuclear. FRUIT of diverse types, fleshy or dry, dehiscent or indehiscent; seeds often with an aril or sarcotesta; embryo curved, oily and starchy, often with plicate or twisted cotyledons, the radicle often separated from the rest of the embryo by a deep fold in the testa that forms a radicular pocket; endosperm wanting. X = 10–16. (?Ptaeroxylaceae)

The family Sapindaceae as here defined consists of about 140 genera and 1500 species, widespread in tropical and subtropical regions, with relatively few members extending into temperate climates. Possibly the two largest genera are tendriliferous vines confined to the New World: *Serjania,* estimated at 220 species, and *Paullinia,* estimated at 150. Another possibly large genus is *Allophylus,* pantropical, considered by different authors to have as many as 190 species or only a single species. *Koelreuteria paniculata* Laxm., the golden rain tree, is well known in cultivation in temperate regions. The berries of *Sapindus saponaria* L., a tropical American Species, form a lather with water and can be used as soap, hence the name *Sapindus* (soap of the Indians). *Paullinia cupuna* H. B. K., the guarana, is an important crop in Amazonian Brasil, where the seeds are used in the preparation of a caffein-rich carbonated drink.

The small segregate family Ptaeroxylaceae Leroy is here included in the Sapindaceae with some hesitation. The Ptaeroxylaceae consist of 2 genera: *Ptaeroxylon,* with a single species in southern Africa; and *Cedrelopsis,* with several species in Madagascar. The group appears to differ from typical Sapindaceae chiefly in having an intrastaminal nectary disk, which is modified into a short gynophore. The taxonomic importance of the position of the nectary disk in the Sapindales has been overemphasized in the past, and needs to be re-evaluated. Its variability within the genus *Acer* (q.v.) seems to be well established.

The Sapindaceae can be traced back to the Eocene epoch on the basis of macrofossils, and to the Coniacian (below the middle of the Upper Cretaceous) on the basis of pollen.

6. Family HIPPOCASTANACEAE A. P. de Candolle 1824 nom. conserv., the Horse-Chestnut Family

Trees or shrubs, commonly (at least in *Aesculus*) storing carbohydrate as isokestose or the corresponding tetrasaccharide (but not as inulin), generally tanniferous, accumulating proanthocyanins but not ellagic

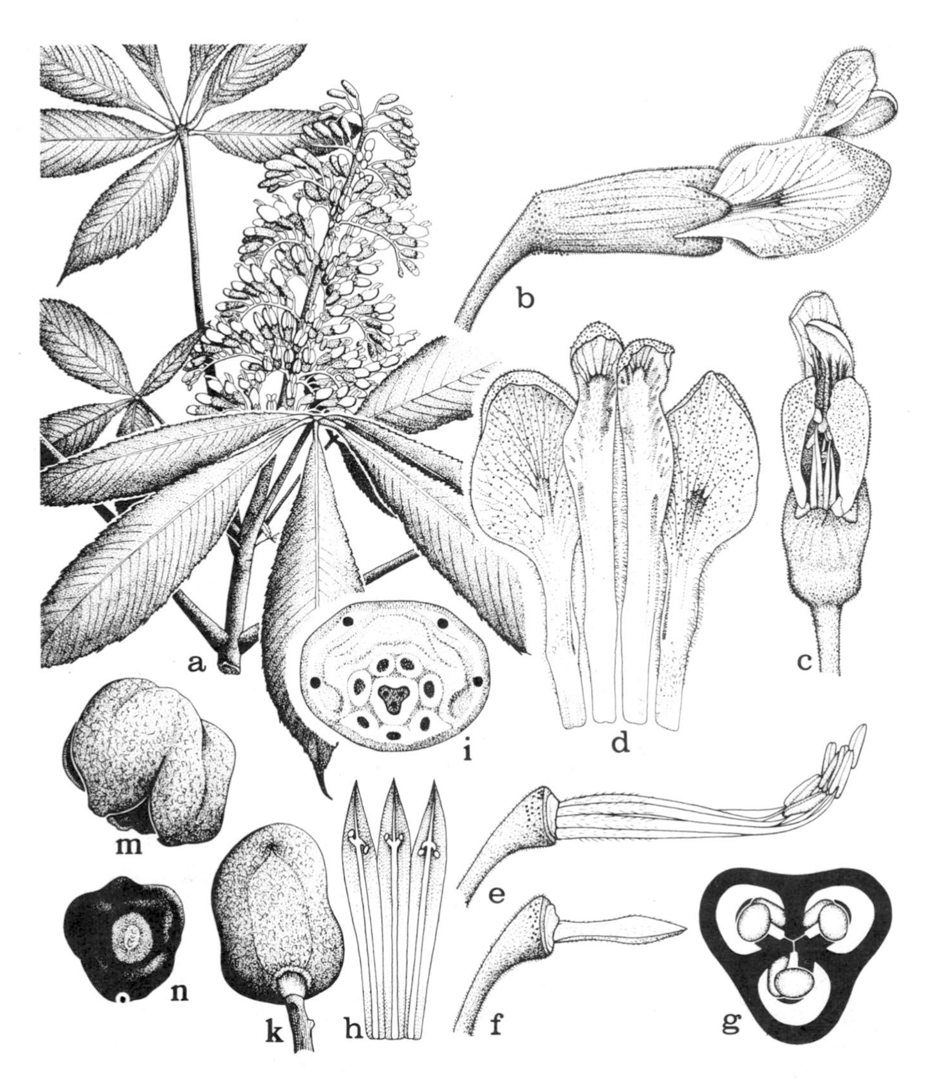

FIG. 5.53 Hippocastanaceae. *Aesculus octandra* Marsh. a, habit, $\times\frac{1}{4}$; b, c, flower, $\times 2$; d, four petals, $\times 2$; e, flower, with the perianth removed, $\times 2$; f, flower, with the perianth and stamens removed, $\times 2$; g, schematic cross-section of ovary, $\times 20$; h, internal view of dissected ovary, showing position of ovules, $\times 4$; i. schematic cross-section of base of flower, showing insertion of pistil, 7 stamens, 4 petals, and nectary disk; k, m, fruit, $\times\frac{1}{4}$; n, seed, $\times\frac{1}{4}$.

acid, and with scattered secretory cells in the parenchymatous tissues, these sometimes but not always saponiferous; not cyanogenic; clustered and sometimes also solitary crystals of calcium oxalate present in some of the cells of the parenchymatous tissues; nodes trilacunar or sometimes pentalacunar; vessel-segments with simple or mixed simple and scalariform perforations; imperforate tracheary elements with simple or slightly bordered pits; wood-rays uniseriate and homocellular or nearly so, or sometimes a few of them biseriate; wood-parenchyma terminal and scanty-paratracheal, or sometimes wanting. LEAVES opposite, palmately 3–11-foliolate, exstipulate. FLOWERS in terminal mixed panicles or racemes, rather large and showy, hypogynous, evidently irregular, perfect or usually many of them (especially the upper ones) functionally staminate; sepals 5, imbricate, nearly distinct (*Billia*) or connate to form a tube (*Aesculus*) petals 4–5, distinct, unequal, clawed, imbricate; nectary-disk small, extrastaminal and often unilateral; stamens (5) 6–8, distinct, the inner whorl of 5 complete, the outer whorl incomplete; anthers tetrasporangiate and dithecal, opening lengthwise; pollen-grains binucleate, tricolporate, with a striate exine; gynoecium of (2) 3 (4) carpels united to form a compound, plurilocular ovary with a terminal style and a dry, papillate, simple or obscurely lobed stigma; ovules 2 in each locule, superposed on the axile placenta, bitegmic and crassinucellar, anatropous to campylotropous or orthotropous; endosperm-development nuclear. FRUIT a loculicidal capsule, often unilocular and 1-seeded by abortion; seed large, with a hard testa and a very large hilum reflecting the incorporation of the funiculus into the placenta and the adnation of the placental obturator to the ovule; embryo large, curved, often starchy, one cotyledon generally much larger than the other, the radicle commonly separated from the rest of the embryo by a deep fold in the seed-coat that forms a radicular pocket; endosperm wanting. X = 20.

The family Hippocastanaceae consists of 2 genera, *Aesculus* and *Billia*. *Aesculus*, with deciduous, 5–11-foliolate leaves, has about 13 species, with an interrupted distribution in temperate North America, the Balkan Peninsula of Europe, and Asia as far south as Thailand and Vietnam. *Billia*, with evergreen, trifoliolate leaves, has 3 species, occurring from southern Mexico to northern South America. *Aesculus hippocastanum* L., the horse-chestnut, is well known in cultivation. *Aesculus octandra* Marsh., the yellow buckeye, is a valuable timber tree in the southern Appalachian Mts. of the U.S.A.

7. Family ACERACEAE A. L. de Jussieu 1789 nom. conserv., the Maple Family

Trees or less often shrubs, usually accumulating quebrachitol (but apparently not in *Dipteronia*), often saponiferous and usually tanniferous, commonly accumulating both ellagic acid and proanthocyanins, but not cyanogenic; various sorts of solitary or clustered crystals of calcium oxalate often present in some of the cells of the parenchymatous tissues; nodes trilacunar; vessel-segments with transverse or oblique, simple perforations, and in *Acer* with spiral terminal thickenings; imperforate tracheary elements with simple or seldom narrowly bordered pits, not septate; lumen of the vessels, imperforate tracheary elements, and cells of the pith often becoming more or less filled with deposits of calcium carbonate; wood-rays homocellular to slightly heterocellular, mixed uniseriate and pluriseriate, the latter mostly (2–) 4–7 (–18) cells wide, with short ends; wood parenchyma usually sparse, terminal and sometimes also paratracheal; secretory sacs containing latex or various other substances often present in the phloem of the leaf-veins and sometimes also of the stem, sometimes also scattered in the mesophyll. LEAVES opposite, mostly simple and palmately lobed or at least palmately veined, but in a few species pinnately veined and entire or merely toothed, or pinnately or palmately 3–5-foliolate, or (*Dipteronia*) pinnately 11–21-foliolate; epidermis commonly mucilaginous; stomates anomocytic; vascular bundles of the petiole commonly forming a ring, sometimes with included bundles; stipules usually wanting. FLOWERS small, entomophilous or seldom anemophilous, borne in corymbiform or umbelliform inflorescences, or sometimes in racemes or large panicles, regular, some or all of them functionally (or fully) unisexual, hypogynous or the staminate ones sometimes perigynous; sepals 5 or less often 4, rarely 6, imbricate, distinct or seldom connate below; petals 5 or less often 4, rarely 6, imbricate, distinct, often much like the sepals, or seldom wanting; stamens most commonly 8 (even when the perianth is 5-merous), less often 4, 5, 10, or about 12; filaments distinct, seated inside or outside the usually well developed nectary-disk, or sometimes engulfed by the disk or individually inset in it, or the disk seldom poorly developed or even wanting; anthers tetrasporangiate and dithecal, opening by longitudinal slits; pollen-grains binucleate, mostly tricolpate or tricolporate; gynoecium mostly of 2 carpels united to form a compound, bilocular ovary with distinct

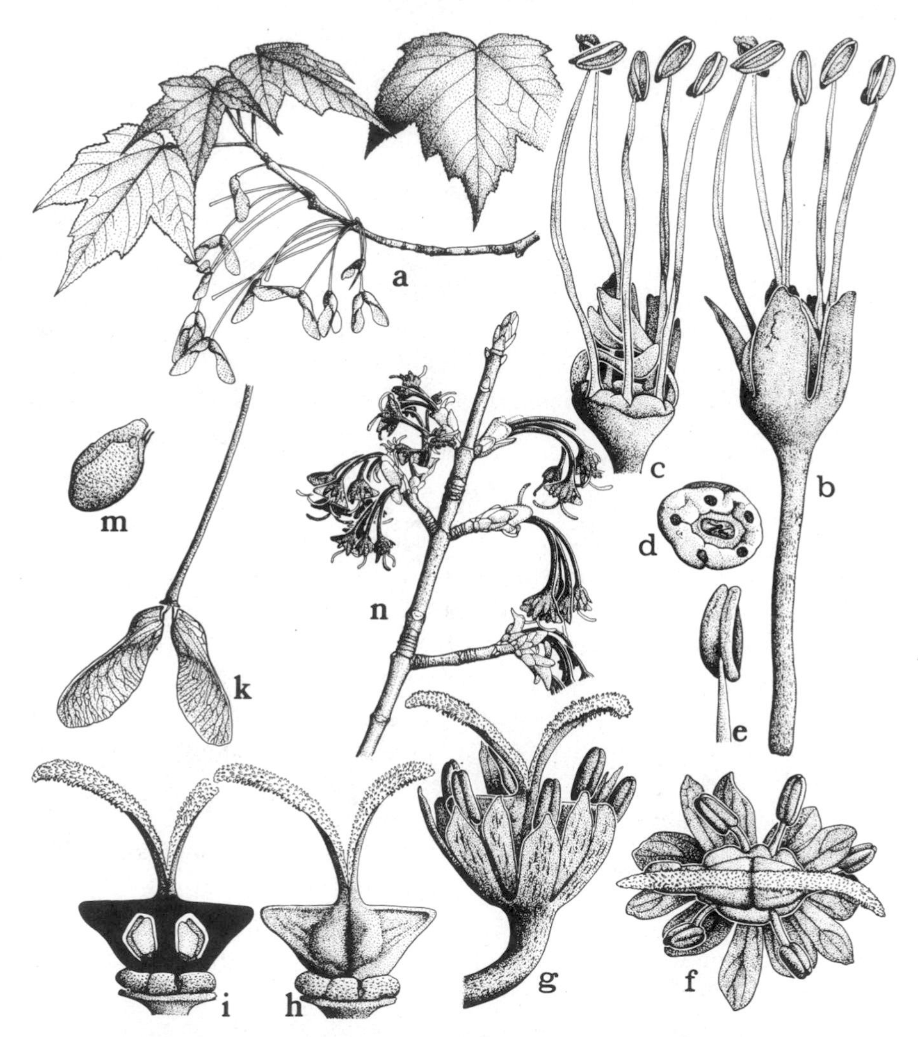

FIG. 5.54 Aceraceae. *Acer rubrum* L. a, habit, ×¼; b, staminate flower, ×6; c, staminate flower, with sepals and 2 petals removed, ×6; d, nectary disk and vestigial pistil of staminate flower, ×6; e, anther, ×12; f, g, functionally pistillate flower, ×6; h, pistil and nectary disk, ×6; i, pistil and nectary disk, in partial long-section, ×6; k, double samara, beginning to separate into two mericarps, ×1; m, seed, ×3; n, habit, with pistillate flowers, ×1.

or basally connate styles, or the number of carpels and locules seldom casually 3 or 4 or even up to 11, especially in the terminal flowers; ovary generally compressed at right angles to the septum, visibly so even at anthesis; ovules (1) 2 in each locule, collateral or superposed, apotropous, anatropous to nearly orthotropous, bitegmic, crassinucellar;

endosperm-development nuclear. FRUIT a winged schizocarp, commonly a double samara, the usually 1-seeded mericarps eventually separating from the persistent carpophore; embryo oily or starchy, with elongate radicle and 2 flat or plicate, green cotyledons; endosperm wanting. X = 13.

The family Aceraceae consists of the small genus *Dipteronia* (2 Chinese spp.), with the mericarps individually winged essentially all the way around, and the large (ca 110 ssp.) highly diversified genus *Acer* (maple), with the mericaps winged only on the abaxial side. *Acer* is widespread in both temperate and some tropical regions (notably Malesia), and is sometimes divided into 2 or more genera. The most distinctive segregate is *Negundo*, with pinnately 3–5-foliolate leaves and certain anatomical peculiarities, but some other species also have trifoliolate leaves. Maples are well known as forest trees and as the source of valuable lumber. Many species store sugar in the sap during the winter; *Acer saccharum* Marsh, is tapped in earliest spring to provide maple syrup.

Maples are well represented in the fossil record from the base of the Miocene to the present, and pollen considered to belong here dates from the Oligocene.

Authors are agreed that the Aceraceae are closely related to the Sapindaceae. The tribe Harpullieae of the Sapindaceae may be regarded as providing the closest approach of that family to the Aceraceae.

8. Family BURSERACEAE Kunth 1824 nom. conserv., the Frankincense Family

Trees or less often shrubs, with prominent vertical schizogenous resin-ducts (containing triterpenoid compounds and ethereal oils) in the bark, sometimes producing proanthocyanins, but not ellagic acid, sometimes cyanogenic, but only seldom saponiferous, often accumulating silica, especially in irregular silica-bodies in some of the cells; parenchymatous tissues often with scattered mucilage-cells, and commonly with solitary or clustered crystals of calcium oxalate in some of the other cells; nodes mostly pentalacunar; vessel-segments with simple perforations; imperforate tracheary elements mostly thin-walled, with small, simple or indistinctly bordered pits, usually septate; wood-rays mixed uniseriate and pluriseriate, or seldom wholly uniseriate, the uniseriate ones homocellular or less often heterocellular, the pluriseriate ones heterocellular and 2–6 cells wide, or wider especially when they contain schizogenous radial resin-ducts (intercellular canals), these sometimes connecting to vertical canals in the pith; wood-parenchyma mostly scanty-paratracheal. LEAVES alternate or rarely opposite, pinnately compound or trifoliolate, or seldom unifoliolate, sometimes pellucid-punctate; most genera with resin-ducts in the phloem of the larger veins; epidermis often with the cell-walls more or less silicified, and often containing some mucilaginous cells; stomates anomocytic; petiole commonly with a ring of vascular bundles with associated resin-ducts, and often with some included bundles as well; stipules only seldom present. FLOWERS borne in mixed panicles or less often in racemes or heads, regular, hypogynous or seldom perigynous, perfect or more often unisexual, the plants commonly dioecious; sepals and petals each (3) 4–5 and imbricate or sometimes valvate, the sepals usually connate below, but the petals usually distinct, or seldom the petals wanting; stamens typically bicyclic, but the antepetalous cycle often reduced, sometimes wholly suppressed; filaments distinct or rarely connate, borne outside or less often within the well developed, commonly annular nectary-disk; anthers tetrasporangiate and dithecal, opening by longitudinal slits; pollen-grains binucleate, commonly tricolporate, the exine often reticulate-striate; staminodia often present in the pistillate flowers; gynoecium of (2) 3–5 carpels united to form a compound, plurilocular ovary with a terminal style and lobed or capitate stigma; ovules pendulous on the axile placenta, (1) 2 in each locule, epitropous, with ventral raphe and upwardly and outwardly directed micropyle, anatropous, or hemitropous to campylotropous, bitegmic or seldom unitegmic, crassinucellar; endosperm-development nuclear. FRUIT a drupe with 1–5 1-seeded stones, or with a single plurilocular stone, or less often the fruit capsular; embryo oily, straight or curved, the 2 cotyledons usually lobed or cleft; endosperm virtually wanting. X = 11, 13, 23.

The family Burseraceae consists of some 16–20 genera and about 600 species, pantropical but especially well represented in tropical America and northeastern Africa. The largest genera are *Bursera* (100, tropical America), *Commiphora* (100, Africa to Arabia and India), *Protium* (80, tropics of both the Old and the New World), and *Canarium* (75, Old World tropics). Frankincense is obtained from *Boswellia carteri* Birdw. and related species, and myrrh from *Commiphora abyssinica* (Berg) Engl. and related species. Various sorts of gum, resin, and balsam are obtained from a number of other species. *Bursera simaruba* (L.) Sarg., called gumbo limbo or naked Indian, is a common and conspicuous species of the American tropics. *Aucoumea kleineana* Pierre, of tropical western Africa, is an important timber-tree, furnishing the gaboon mahogany of commerce.

The Burseraceae probably originated no later than the Eocene epoch. Paleogene fruits from England are attributed to the Burseraceae, as are some Eocene macrofossils from western U.S.A., and it has been suggested that the amber in the Eocene London Clay is of burseroid origin.

In the Englerian system the Burseraceae and Anacardiaceae are referred to different orders, because of the different orientation of the ovules (epitropous in Burseraceae, apotropous in Anacardiaceae). The two families are anatomically very similar, however, and many recent authors consider them to be closely allied.

9. Family ANACARDIACEAE Lindley 1830 nom. conserv., the Sumac Family

Trees, shrubs, or woody vines, rarely only half-shrubs, with well developed vertical schizogenous (or lysigenous) resin-ducts (or sometimes latex-channels) in the bark and in the phloem of the larger veins of the leaves, often also in the flowers and fruits and in the pith and other parenchymatous tissues, the resin often allergenic or poisonous to the touch; plants producing 5-deoxyflavonoids and biflavonyls, at least sometimes accumulating quebrachitol, and commonly tanniferous, with tanniferous cells or elongate sacs in the parenchymatous tissues, usually producing proanthocyanins and gallic acid but only seldom ellagic acid, only seldom saponiferous or cyanogenic; solitary or clustered crystals of calcium oxalate commonly present in some of the cells of the parenchymatous tissues, and silica grains sometimes present in some of the xylem cells, as in *Anacardium*; nodes mostly trilacunar; vessel-segments with simple perforations, or seldom some of the perforations scalariform or reticulate; imperforate tracheary elements with thick or thin walls and simple or indistinctly bordered pits, septate in about half of the genera; wood-rays commonly mixed uniseriate and pluriseriate, the uniseriate ones mostly homocellular, the others heterocellular, mostly 2–3 (–10) cells wide, in about two-thirds of the genera containing horizontal resin-ducts; wood-parenchyma mostly paratracheal, often scanty, sometimes wanting or diffuse. LEAVES alternate, or very rarely opposite or whorled, pinnately compound or trifoliolate, less often simple, epidermis often with scattered mucilaginous cells, and often

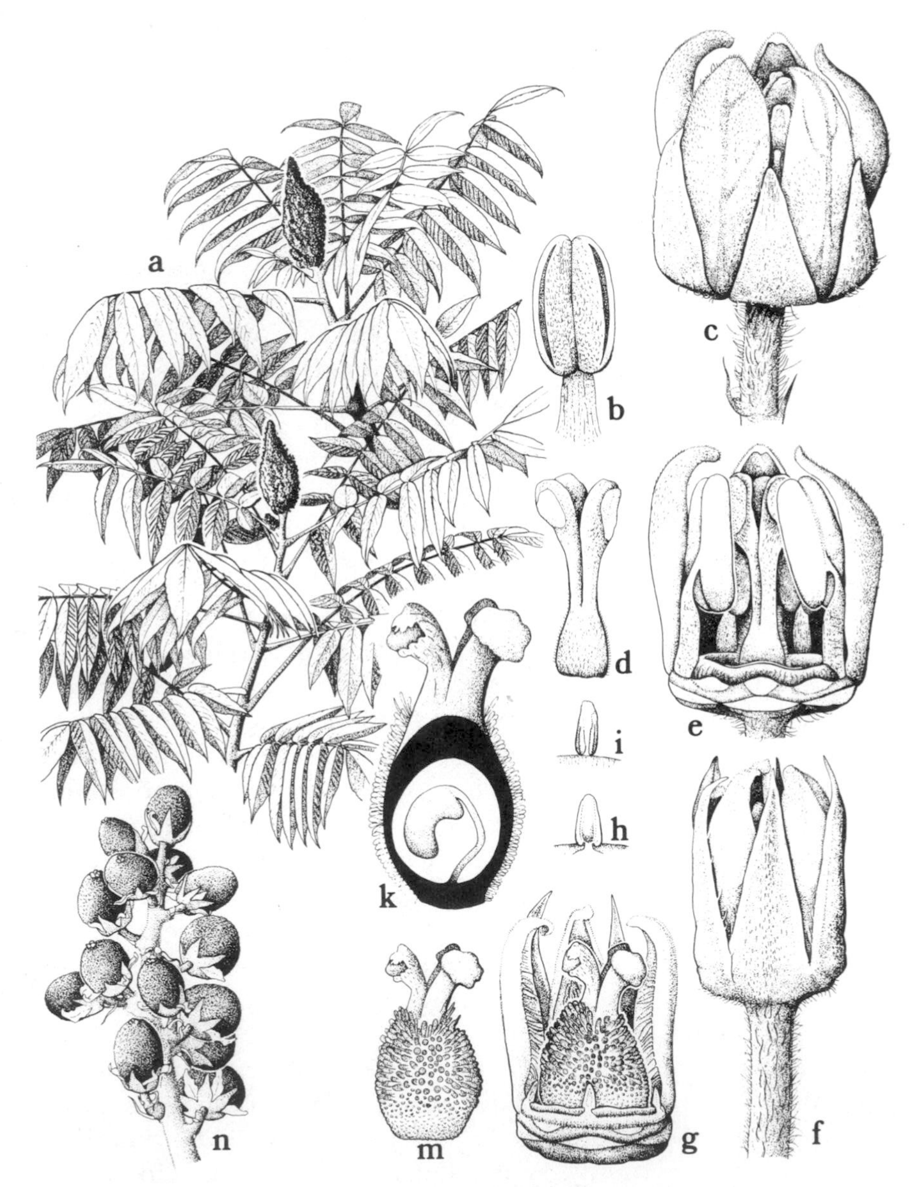

FIG. 5.55 Anacardiaceae. *Rhus glabra* L. a, habit, $\times\frac{1}{12}$; b, stamen, ×12; c, staminate flower, ×12; d, pistillode of staminate flower, ×12; e, staminate flower, in partial long-section, ×12; f, pistillate flower, ×12; g, pistillate flower, in partial long-section, ×12; h, i, staminodes of pistillate flower, ×24; k, pistil, in partial long-section, ×18; m, pistil, ×12; n, fruits, ×2.

with the cell-walls more or less silicified; stomates anomocytic; petiole with diverse sorts of complex architecture, but consistently with resin-canals in the phloem of the veins; stipules wanting, or rarely present but inconspicuous and vestigial. FLOWERS in terminal or axillary, often complex but eventually cymose inflorescences, small, regular, hypogynous or seldom perigynous or epigynous, perfect or more often unisexual, the unisexual flowers often with evident but nonfunctional parts of the other sex; sepals and petals each (3–) 5 (–7), valvate or imbricate, the sepals generally connate below, the petals distinct, or seldom the petals (or both the petals and the sepals) wanting; androecium diplostemonous, or often haplostemonous with the stamens antesepalous, or rarely of more numerous stamens or with only a single stamen fertile; filaments distinct or rarely connate at the base, borne outside or sometimes upon or seldom within the usually annular, often 5-lobed nectary-disk, or the disk sometimes modified into a short, stout gynophore; anthers tetrasporangiate and dithecal, opening by longitudinal slits; pollen-grains binucleate, tricolporate or triporate, less often 4–8-porate; gynoecium of (2) 3 (–5 or even 12) carpels united to form a compound, plurilocular ovary, or very often only one locule fully developed or the ovary pseudomonomerous, or seldom the gynoecium of several separate carpels (then often only one fully developed and fertile); styles distinct, or united to form a single common style; ovary with a single apical and pendulous to basal and erect, apotropous ovule in each carpel or in the single fertile carpel, often with a sort of placental obturator at the base of the funiculus; ovules anatropous, bitegmic or unitegmic, crassinucellar; endosperm-development nuclear. FRUIT commonly drupaceous, with more or less resinous and sometimes waxy or oily mesocarp; seeds with oily, curved or less often straight embryo and 2 expanded cotyledons; endosperm scanty or none. X = 7–16. (Blepharocaryaceae, Pistaciaceae, Podoaceae)

The family Anacardiaceae as here defined consists of some 60–80 genera and perhaps 600 species, chiefly pantropical in distribution, but with some species in temperate regions. The largest genus is *Rhus* (100, including *Schmaltzia*). *Toxicodendron* (poison ivy, poison oak, poison sumac) is notorious for causing dermatitis in susceptible individuals, but some other genera such as *Metopium* are even more potent. *Toxicodendron vernicifluum* (Stokes) Barkley, the traditional source of lacquer, is allergenic like the other species of the genus. The mango (*Mangifera indica* L.) and the pistachio nut (*Pistacia vera* L.) are innocuous to most people, and the cashew nut (*Anacardium occidentale* L) is rendered harmless by roasting.

Paleocene anacardiaceous wood is known from Patagonia, and the family is well represented in the Eocene in England, as well as in more recent strata. Wood attributed to *Rhus* occurs in Eocene deposits in Yellowstone National Park. The oldest fossil pollen of the family, similar to that of *Rhus* and some other genera, comes from the Paleocene of Europe.

10. Family JULIANIACEAE W. B. Hemsley 1906 nom. conserv., the Juliania Family

Very much like Anacardiaceae (such as *Rhus*) except for the more reduced flowers and the fruit; small trees or shrubs with well developed vertical schizogenous resin-ducts (bearing a thick, milky juice) in the bark and pith and horizontal resin-ducts in the wood-rays, saponiferous and tanniferous, producing proanthocyanins but probably not ellagic acid, not cyanogenic; clustered crystals of calcium oxalate present in some of the cells of the parenchymatous tissues; nodes trilacunar; vessel-segments with simple perforations, or a few of them reticulate; imperforate tracheary elements thin-walled, with simple or obscurely bordered pits, most or all of them septate; wood-rays nearly homocellular, the unseriates few, the others up to 6 cells wide; wood-parenchyma scanty-paratracheal. LEAVES alternate but commonly closely crowded toward the branch-tips, deciduous, pinnately compound, with serrate leaflets; veins with large resin-ducts in the phloem; stomates anomocytic; stipules wanting. FLOWERS very small, apetalous, unisexual (the plants dioecious), the staminate ones numerous in panicles, the pistillate ones in small, 3–4-flowered dichasia, each dichasium subtended by and semi-enclosed in an involucre; staminate flowers with 3–8 slender, basally connate sepals; stamens as many as and alternate with the sepals; anthers tetrasporangiate and dithecal, opening by longitudinal slits; pollen-grains (3–) 5–8-porate; nectary-disk and vestigial gynoecium wanting; pistillate flowers without perianth or staminodes, each consisting of a tricarpellate, unilocular ovary with a terminal, trifid style and flattened branches; ovule solitary, basal, erect, unitegmic, hemitropous, cupped at the base by a placental obturator. FRUIT a dry syncarp, consisting of the accrescent, thickened, subglobose involucre containing 1–2 compressed, hairy nuts more or less adnate to its wall; seeds without endosperm.

The family Julianiaceae consists of two small tropical American genera: *Orthopterygium*, with a single species in Peru, and *Amphipterygium* (*Juliania*), with 4 species in Central America. The Julianiaceae are much like the Anacardiaceae in aspect and in anatomical features, and also in some floral features such as the tricarpellate, unilocular ovary with the ovule seated on an obturator. The flavonoid chemistry is also interpreted to favor such a relationship. Although the affinities of the Julianiaceae have been much-debated in the past, modern opinion is virtually unanimous that they are florally reduced derivatives of the Anacardiaceae. Some authors would even include the Julianiaceae in the Anacardiaceae.

11. Family SIMAROUBACEAE A. P. de Candolle 1811 nom. conserv., the Quassia Family

Trees, shrubs, or seldom subshrubs, mostly with very bitter bark, wood, and seeds, producing characteristic triterpenoid lactones called simaroubalides (quassin, chaparin, glaucarubol, others), and commonly tanniferous, sometimes with proanthocyanins and sometimes with ellagic acid, but neither saponiferous nor cyanogenic, and not accumulating quebrachitol; scattered secretory cells of various sorts often present, less often secretory canals, these when present typically found near the protoxylem, whence they may extend out into the leaves; vertical canals sometimes also present elsewhere in the stem, but apparently mostly traumatic; solitary and clustered crystals of calcium oxalate often present in some of the cells of the parenchymatous tissues; nodes trilacunar or multilacunar; vessel-segments with simple perforations, or rarely some of them reticulate; imperforate tracheary elements with simple or bordered pits, sometimes septate; wood-rays homocellular to less often heterocellular, up to 5 (–10) cells wide, with short ends, or sometimes exclusively uniseriate or without uniseriates; wood-parenchyma of highly diverse types. LEAVES alternate or rarely opposite, pinnately compound to unifoliolate, or seldom simple and entire, never glandular-punctate; epidermis often mucilaginous and often silicified; stomates anomocytic or less often paracytic; mesophyll often containing sclereids; petiole with a complete ring of vascular tissue, often also with some medullary bundles; stipules mostly wanting, but present and intrapetiolar in *Irvingia* and its allies. FLOWERS in axillary or terminal racemes or mixed panicles or dichasial cymes, commonly small, regular, hypogynous, perfect or more often unisexual (then often with abortive parts

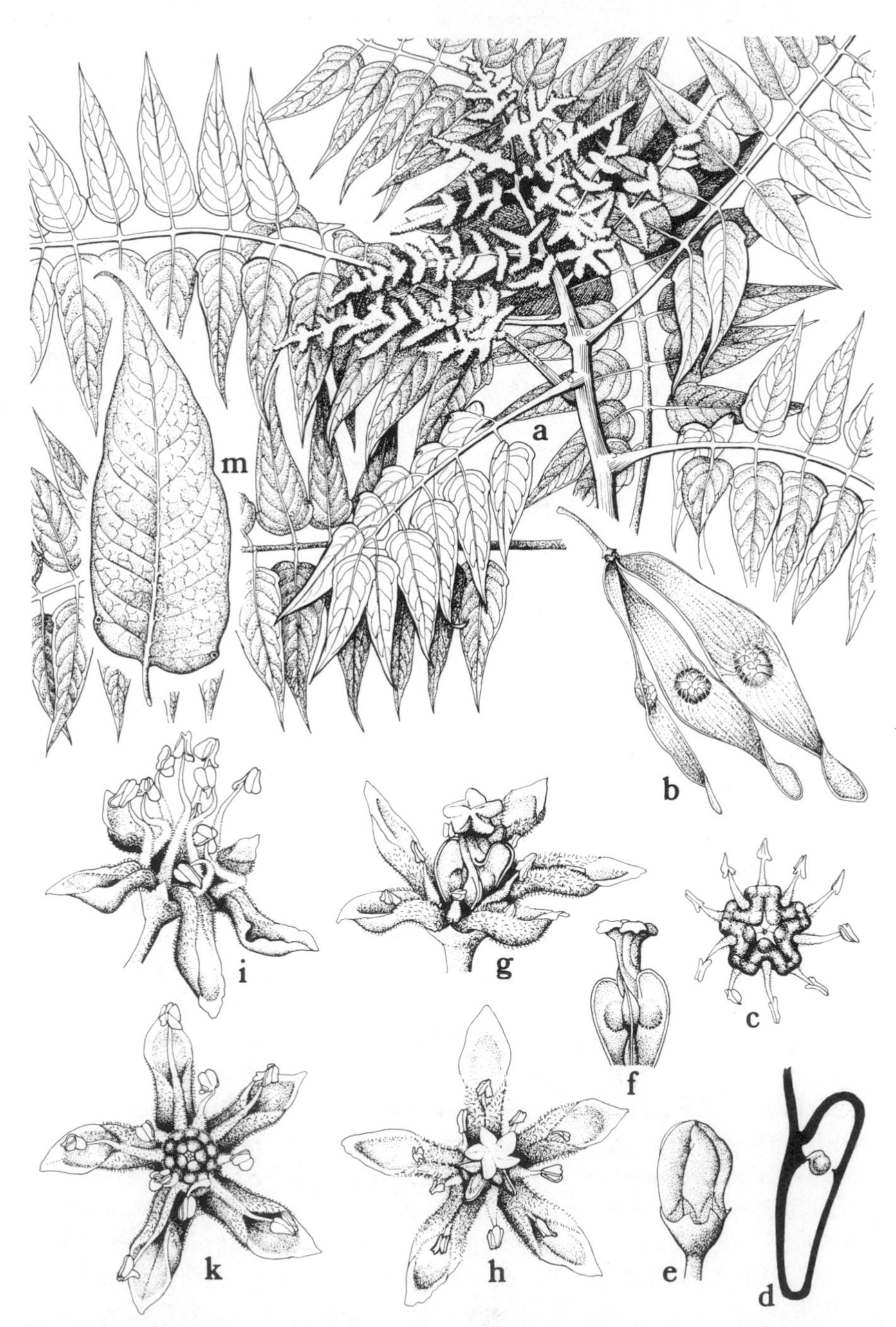

FIG. 5.56 Simaroubaceae. *Ailanthus altissima* (Miller) Swingle. a, habit, $\times\frac{1}{8}$; b, fruits, $\times 1$; c, stamens and disk of functionally pistillate flower, $\times 6$; d, schematic long-section of carpel, with a single ovule, $\times 12$; e, flower bud, $\times 6$; f, pistil, $\times 6$; g, h, functionally pistillate flower, $\times 6$; i, k, staminate flower, $\times 6$; m, leaflet, from beneath, $\times\frac{1}{2}$.

of the opposite sex) 3–8-merous; sepals most commonly 5, connate toward the base, or less often distinct, imbricate or valvate; petals distinct, imbricate or valvate, most commonly 5, or seldom wanting; stamens usually twice as many as the petals, or less often as many as and alternate with the petals, rarely more than twice as many as the petals; filaments distinct, often with a ventral appendage near the base; anthers tetrasporangiate and dithecal, opening by longitudinal slits; pollen-grains binucleate, tricolporate, the exine variously reticulate, striate, or spinulose; a well developed intrastaminal nectary-disk commonly present, sometimes modified into a short stout gynophore or androgynophore; gynoecium of 2–5 (–8) weakly to firmly united carpels, sometimes connate only by their styles, sometimes forming a plurilocular ovary with axile placentas and distinct styles or a single common style, or rarely the carpels wholly distinct (*Recchia, Picrolemma*, spp. of *Ailanthus*); ovules solitary or less often paired in the individual carpels or locules (seldom only one locule ovuliferous), apical and pendulous to basal and ascending-erect, epitropous or sometimes (as in *Alvaradoa*) apotropous, anatropous to hemitropous, bitegmic, crassinucellar; endosperm-development nuclear. FRUIT a capsule or samara, or seldom a drupe or berry, or often schizocarpic, separating into indehiscent, dry (sometimes samaroid) or fleshy mericarps; embryo oily, straight or curved, with 2 large, expanded cotyledons; endosperm very scanty or none. X = 8, 13+. (Irvingiaceae, Kirkiaceae)

The family Simaroubaceae as here defined consists of about 25 genera and 150 species, pantropical, with a few species extending into warm-temperate regions. The largest genus is *Picramnia*, with about 40 species native to the New-World tropics. Decoctions of the bark of *Quassia amara* L. and various species of *Simarouba* and other American genera have been used locally as antimalarials, but do not have a sufficient margin between the therapeutic and the toxic dose. Some species now have other specialized medical uses. *Ailanthus altissima* (Mill.) Swingle, a native of Siberia, is often cultivated as a street-tree in the U.S.A., under the name tree of heaven.

The genera of Simaroubaceae are mostly well defined, and the family is only loosely knit. Some authors would separate *Irvingia* and its allies as a family Irvingiaceae, and *Kirkia* as a family Kirkiaceae. I have not found a report of the chemistry of these genera. *Suriana* and several other small genera with simple leaves, often referred to the Simaroubaceae, are here treated as a separate family Surianaceae, in the order Rosales. *Suriana*, at least, lacks simaroubalides.

It is generally agreed that the Simaroubaceae are related to the Meliaceae, Rutaceae, and Burseraceae.

12. Family CNEORACEAE Link 1831 nom. conserv., the Cneorum Family

Shrubs, glabrous or with malpighian hairs, containing scattered secretory cells (with oily or resinous contents) in the parenchymatous tissues, notably in the cortex and mesophyll, producing coumarins and triterpenoid bitter substances, but not cyanogenic and not tanniferous, lacking both proanthocyanins and ellagic acid; crystals of calcium oxalate only seldom present in some of the cells of the parenchymatous tissues; vessel-segments with simple perforations; imperforate tracheary elements very short, with simple pits; wood-rays heterocellular to homocellular, mixed uniseriate and pluriseriate, the latter 2–3 cells wide. LEAVES alternate, small, simple, entire; stomates anomocytic; stipules wanting. FLOWERS solitary and axillary, or in small, few-flowered, axillary cymes (the peduncle sometimes adnate to the petiole), small, perfect, regular, hypogynous, trimerous in 2 spp., tetramerous in the third; sepals small, persistent, distinct or basally connate; petals distinct, elongate, imbricate; disk modified into a shortly columnar, nectariferous androgynophore; stamens as many as and alternate with the petals; filaments distinct, seated in pits in the androgynophore; anthers tetrasporangiate and dithecal, opening by longitudinal slits; pollen-grains globose, trinucleate, tricolporate or 4–6-colporate, with a striate-reticulate exine; gynoecium of 3 or 4 carpels united to form a compound, superior ovary with axile placentas and a terminal style with lobed stigmas; ovules (1) 2 in each locule, pendulous, collateral, epitropous, with ventral raphe, amphitropous, bitegmic and crassinucellar, the 2 ovules commonly more or less separated by a partial or nearly complete partition intruded from the carpellary midrib; endosperm-development nuclear. FRUIT of separating mericarpic drupelets, these with one or two seeds; seeds with strongly curved, dicotyledonous embryo and copious, fleshy, oily endosperm. X = 9.

The family Cneoraceae consists of the single genus *Cneorum*, with 3 species, one native to the western Mediterranean region, one to the Canary Islands, and one to Cuba. The species of the Canary Islands is tetramerous and has sometimes been treated as a distinct genus *Neochamaelea*.

Cneorum is generally admitted to be allied to families here referred to the Sapindales, notably the Rutaceae, Simaroubaceae, and Zygophyllaceae, but it would be a discordant element in any of these families.

13. Family MELIACEAE A. L. de Jussieu 1789 nom. conserv., the Mahogany Family

Trees (often pachycaulous) or shrubs, seldom half-shrubs or rarely herbs, with resinous secretory cells containing ethereal oils in the leaves, cortex and bark, and often also with scattered stone-cells, commonly producing triterpenoid bitter substances but only seldom saponiferous, not known to produce alkaloids or quebrachitol, but sometimes producing coumarins and sometimes tanniferous, often with proanthocyanins but without ellagic acid; nodes mostly pentalacunar; wood mostly hard and aromatic, sometimes silicified; vessel-segments with simple perforations; imperforate tracheary elements with simple or narrowly bordered pits, septate in many species; wood-rays most commonly uniseriate and heterocellular or less often homocellular, or sometimes 2–4 (–9) cells wide, but then usually with few or no uniseriates; traumatic intercellular canals sometimes present; bark bitter and astringent. LEAVES alternate, or rarely opposite (as in *Capuronianthus*, a monotypic genus of Madagascar), very often clustered at the branch-tips, pinnately or bipinnately compound, or less often trifoliolate, seldom unifoliolate or simple, not punctate; epidermis often silicified; stomates anomocytic; resinous secretory cells usually borne at the boundary between the palisade and spongy mesophyll; stipules wanting. FLOWERS in various sorts of mostly axillary inflorescences, less often terminal or cauliflorous or extra-axillary, mostly small, perfect or sometimes unisexual (the plants then mostly polygamous, but sometimes monoecious or dioecious), regular, hypogynous; sepals (2) 3–5 (–7), imbricate or open or rarely valvate, commonly connate below, sometimes the calyx-tube virtually entire, or closed and circumscissile; petals usually as many as and alternate with the sepals, or seldom up to 14 (the corolla biseriate in *Megaphyllaea*), distinct or seldom somewhat connate at the base (rarely to half-length), imbricate or convolute, or adnate to the filament-tube and valvate; androecium typically diplostemonous, less often haplostemonous with antesepalous stamens, rarely the stamens numerous (up to ca 25); stamens distinct (Cedreleae) or much more often the filaments connate into a tube that may have membranous teeth or appendages

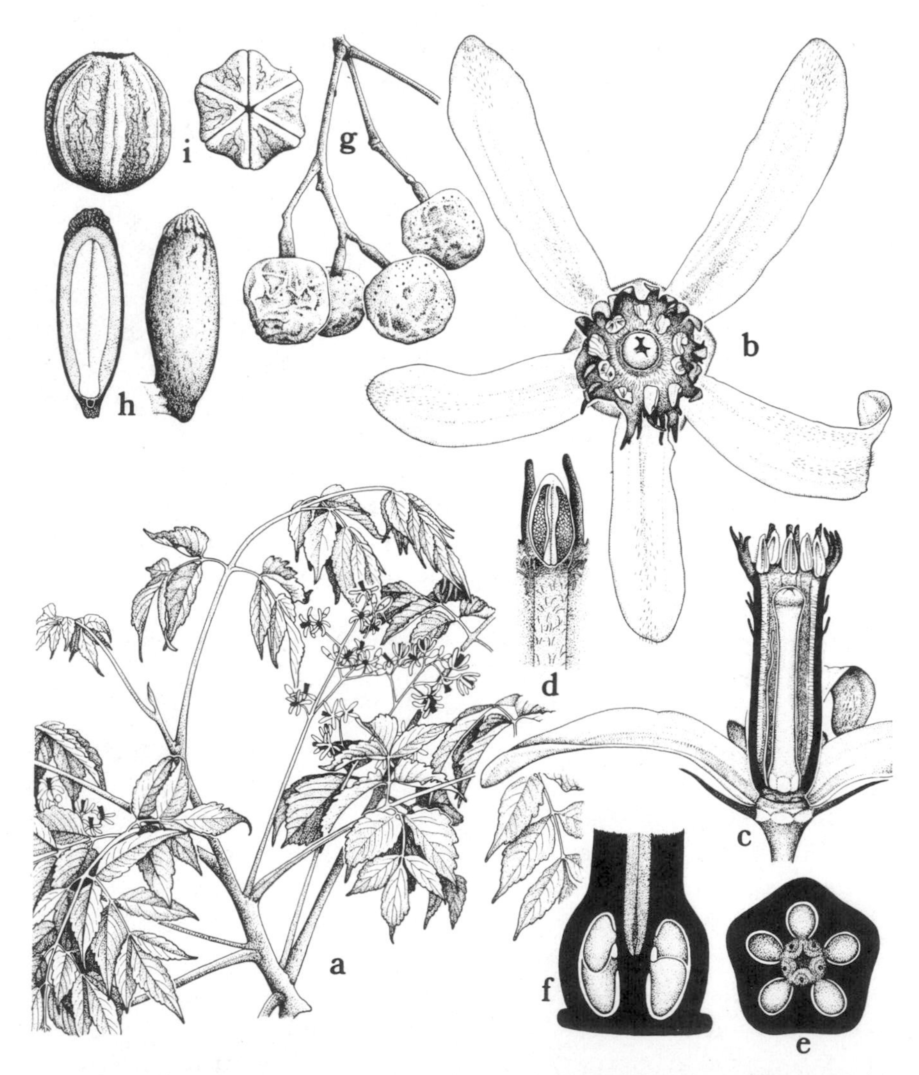

Fig. 5.57 Meliaceae. *Melia azedarach* L. a, habit, ×¼; b, flower, ×4; c, flower, in partial long-section, ×4; d, anther and part of filament-tube, ×8; e, schematic cross-section of ovary, ×16; f, schematic long-section of ovary, ×16; g, fruits, ×1; h, seed, in long-section and external view, ×4; i, stone, from the side and from beneath, ×2.

between or outside and opposite the anthers; anthers tetrasporangiate and dithecal; pollen-grains 2–3-nuclear, 2–5-colporate; nectary-disk annular, intrastaminal, sometimes adnate to the ovary, sometimes developed into an androgynophore; gynoecium of (1) 2–5 (–20) carpels united to form a compound, plurilocular ovary with typically axile placentation (the partitions often not reaching the summit), or rarely the ovary unilocular with intruded parietal placentas, as in *Heckeldora*, a monotypic African genus; ovary occasionally more or less sunken in the disk; style terminal, with a stigmatic head of varied form; ovules commonly 2 in each locule, less often one or several (–12), variously anatropous, campylotropous, or orthotropous, often provided with a placental obturator, generally pendulous and epitropous, with ventral raphe, bitegmic and crassinucellar; endosperm-development nuclear. FRUIT a septicidal or loculicidal capsule, less often a berry or drupe, very rarely a nut; seeds variously thin, dry, and winged, or wingless and then usually with an aril or sarcotesta; embryo spatulate, with 2 cotyledons; endosperm well developed, oily, fleshy, or sometimes wanting. X = 10–14+. (Aitoniaceae)

The family Meliaceae as here defined consists of some 51 genera and about 550 species, widespread in tropical and subtropical regions, with relatively few species in temperate climates. About half of the species belong to only 5 genera: *Aglaia* (100), *Trichilia* (65), *Dysoxylum* (60), *Chisocheton* (30), and *Turraea* (24). The family includes many important timber species. Mahogany is the wood of *Swietenia mahogani* (L.) Jacq., a native of the West Indies. *Cedrela odorata* L. and related species are used in making furniture. Pollen referred to the Meliaceae is known from Oligocene and more recent deposits.

Authors are agreed that the Meliaceae are related to the Rutaceae and Simaroubaceae.

14. Family RUTACEAE A. L. de Jussieu 1789 nom. conserv., the Rue Family

Aromatic trees or shrubs or seldom herbs, sometimes scandent, sometimes thorny, with calcium oxalate crystals of various form (sometimes raphides) in some of the cells of the parenchymatous tissues, commonly producing triterpenoid bitter substances, and usually with one or another of several alkaloids of diverse sorts (sometimes isoquinoline alkaloids, as in the Magnoliidae, sometimes acridine alkaloids, perhaps restricted to this family, or often various other kinds), typically producing coumarins (derived from cinnamic acid) and diverse sorts of phenolic compounds, but without quebrachitol, only seldom cyanogenic, seldom saponiferous, and seldom tanniferous, almost always lacking ellagic acid but sometimes with proanthocyanins, commonly with lysigenous secretory cavities containing aromatic ethereal oils scattered throughout the parenchymatous tissues and in the pericarp, sometimes with scattered ethereal oil cells in addition to (seldom instead of) the secretory cavities, and commonly also with scattered resin-cells; nodes trilacunar or sometimes unilacunar; vessel-segments usually with simple perforations; imperforate tracheary elements with simple or slightly bordered pits; wood-rays homocellular to somewhat heterocellular, mixed uniseriate and pluriseriate, the former usually few, the latter mostly 2–4 (–7) cells wide, with short ends, or seldom all the rays uniseriate; wood-parenchyma terminal and paratracheal (commonly vasicentric) or less often diffuse, rarely wanting; traumatic intercellular canals sometimes present; cambial region often yellowish; sieve-tubes with S-type plastids. LEAVES alternate or less commonly opposite, rarely whorled, usually pinnately compound or trifoliolate, seldom pinnately dissected or simple, pellucid-punctate because of the secretory cavities (sometimes only along the margin, as in *Evodea*); epidermis often silicified; stomates of various types; petiole commonly with a simple cylinder or an arc of vascular tissue; stipules wanting. FLOWERS in cymes or less often racemes or seldom solitary, rarely epiphyllous, perfect or seldom unisexual, hypogynous or seldom (as in *Adenandra*) perigynous, regular or rarely somewhat irregular; sepals 5 or less often 4, seldom only 2 or 3, distinct or connate toward the base, commonly imbricate; petals as many as and alternate with the sepals, distinct or sometimes connate below, imbricate or sometimes valvate, or rarely wanting; androecium most commonly diplostemonous (sometimes one set reduced to staminodes), less often of a single antepetalous cycle, or the stamens sometimes 3–4 times as many as the petals, or even up to 60 (the higher numbers commonly in taxa that have more or less numerous carpels), rarely only 2 or 3 stamens fertile and the others staminodial; filaments distinct or more or less connate toward the often dilated base; anthers tetrasporangiate and mostly dithecal, often gland-tipped, opening by longitudinal slits; pollen-grains binucleate or less often trinucleate, (2) 3–6 (–8)-colporate; nectary-disk intrastaminal, annular or sometimes unilateral, sometimes modified into a gynophore, or rarely obsolete; gynoecium of (2–) 4–5 (-many) carpels more or less clearly united to form a compound, plurilocular, often apically indented ovary with axile placentas and distinct styles or a single common style, or sometimes the carpel-bodies essentially distinct except for their coherent styles, rarely the partitions not joined in the center and the ovary thus unilocular with intruded parietal placentas, or rarely the gynoecium reduced to a single carpel; stigmas wet or dry; ovules (1) 2 (-several) in each locule, often superposed and pendulous, seldom biseriate, more or less distinctly epitropous, with ventral raphe and upwardly and outwardly directed micropyle, anatropous or hemitropous, bitegmic or rarely unitegmic (*Glycosmis*), crassinucellar; endosperm-development nuclear. FRUIT of diverse sorts; seeds with large, straight or curved, spatulate to sometimes linear, dicotyledonous embryo; endosperm more or less well developed and oily, or wanting. X = 7–11+, perhaps originally 9. (Flindersiaceae, Limoniaceae)

The family Rutaceae as here defined consists of about 150 genera and 1500 species, nearly cosmopolitan in distribution, but mostly tropical and subtropical, and especially well developed in South Africa and Australia. The most familiar members of the family are species of *Citrus*, producing juicy fruits prized for their flavor and their high content of vitamin C. The largest genus is *Zanthoxylum* (including *Fagara*), with 200 or more species, chiefly tropical. *Ruta graveolens* L., the common rue, was formerly much-used in medicine and as a spice-plant. It is unusual in the family in being an herb with twice-pinnatifid leaves.

Most authors are agreed that the Rutaceae are properly to be associated with the Simaroubaceae, Meliaceae, and other families here included in the Sapindales. On the other hand, the presence of isoquinoline alkaloids (which are otherwise largely restricted to the Magnoliidae) in some genera has led some authors to speculate that the Rutaceae may be more closely allied to the Magnoliidae than to the families of the Sapindales. The Rutaceae produce alkaloids of so many different groups, however, that it hardly seems warranted to seize upon one of these as indicating a relationship, especially when the proposed relationship is contrary to the rest of the evidence. Even on chemical evidence alone, the Rutaceae are linked to the Simaroubaceae, Meliaceae, Burseraceae, and Anacardiaceae by the presence in all of these families of triterpenoid bitter substances of essentially similar structure. It is noteworthy that Jensen (1974) found no evidence of relationship, on the basis of serological reactions, between *Phellodendron* (Rutaceae) and *Berberis*. The coumarins of the Rutaceae are very similar to those of the Apiaceae in some respects, but different in others.

15. Family ZYGOPHYLLACEAE R. Brown in Flinders 1814 nom. conserv., the Creosote-bush Family

Shrubs or half shrubs, less commonly small trees or annual or perennial herbs, usually (or very often) producing steroid or triterpenoid saponins,

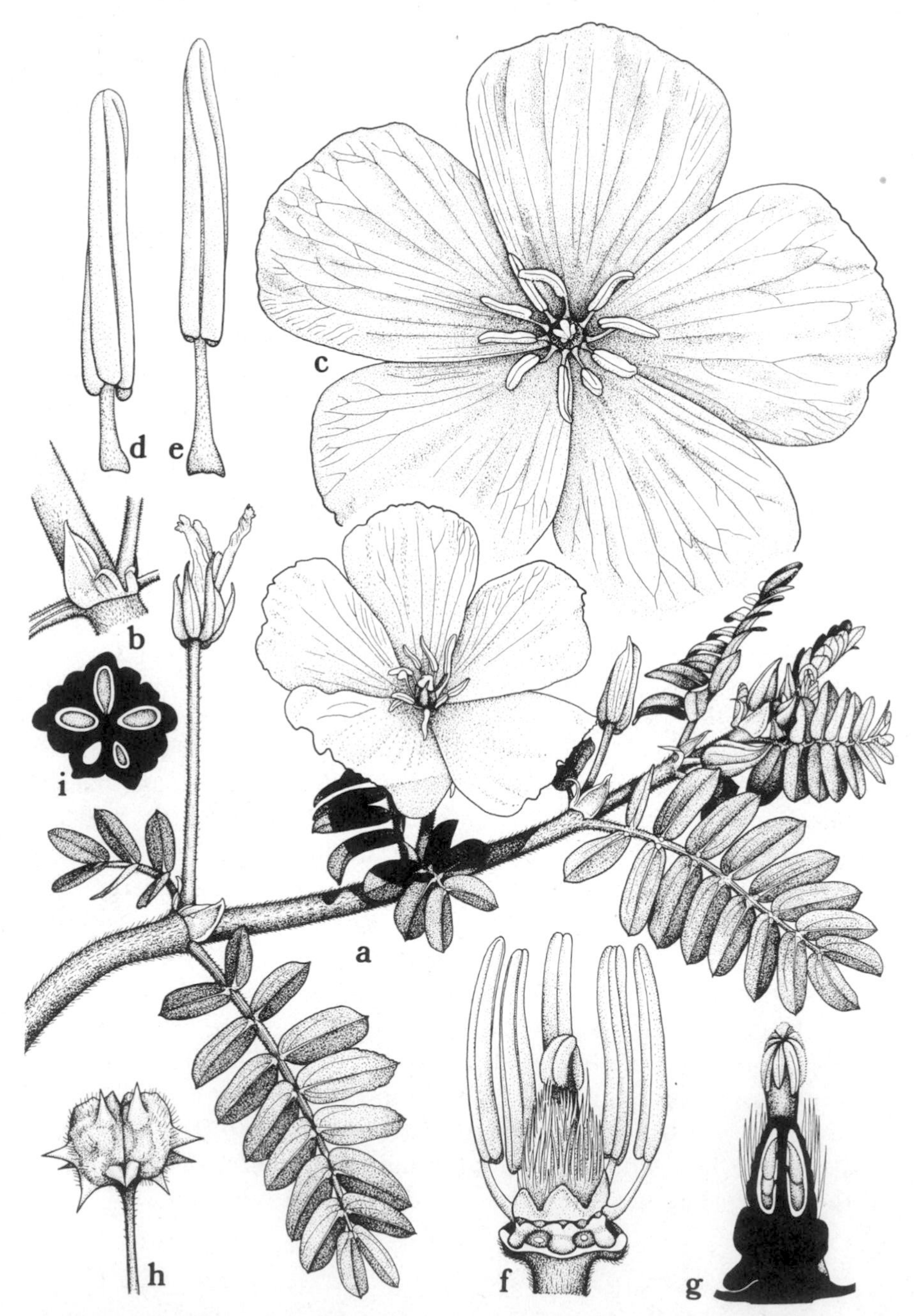

FIG. 5.58 Zygophyllaceae. *Tribulus cistoides* L. a, habit, ×1; b, node, with stipules, ×2; c, flower, ×2; d, e, stamens, ×6; f, side view of flower, with perianth and some stamens removed, ×4; g, pistil and disk, in partial long-section, ×6; h, fruit, ×1; i, schematic cross-section of ovary, ×8.

guaianolide sesquiterpenes, lignans, and sometimes indole-carboline or quinazoline alkaloids or mustard-oil, but only seldom with scattered mucilage-cells or mucilage cavities or tanniferous cells, generally not at all tanniferous, lacking both proanthocyanins and ellagic acid, not cyanogenic; clustered or solitary crystals of calcium oxalate (seldom raphides) generally present in some of the cells of the parenchymatous tissues and the epidermis, and stone-cells often present in the cortex; nodes trilacunar; stem often sympodial and swollen or jointed at the nodes, sometimes with anomalous secondary growth; vessel-segments short, with *simple* perforations; imperforate tracheary elements with bordered pits, consisting of libriform fibers and often also vasicentric tracheids; wood-rays commonly storied, all narrow and only 1–2 (3) cells wide and homocellular or slightly heterocellular, seldom (*Balanites*) much wider, up to 20 cells wide; wood-parenchyma mostly apotracheal, diffuse or in uniseriate bands, only seldom paratracheal. LEAVES opposite or less commonly alternate, most often pinnately compound but without a terminal leaflet, often bifoliolate, varying to sometimes trifoliolate or pinnately dissected, or even simple and entire, often strongly resinous, the mesophyll very often centric, with Kranz anatomy; stipules usually well developed, commonly slender, firm, and persistent, sometimes modified into spines, or seldom wanting. FLOWERS in cymose or seldom racemose inflorescences or solitary, perfect or rarely unisexual, regular or rarely irregular, hypogynous, 5-merous or less often 4-merous, rarely 6-merous; sepals distinct or sometimes connate below, variously imbricate or valvate; petals nearly always distinct, imbricate or convolute or rarely valvate, or sometimes wanting; androecium commonly of 2 cycles, or sometimes of 3 cycles (2 cycles then antepetalous), or seldom of only 1 cycle; filaments often glandular or appendiculate at the base; anthers tetrasporangiate and dithecal, opening by longitudinal slits; pollen-grains binucleate or seldom trinucleate, of diverse types, tricolporate or less often tricolpate or multiporate; nectary-disk usually well developed and intrastaminal (extrastaminal in *Kallstroemia*, filaments on the disk in *Tribulus*), seldom modified into a gynophore (or a gynophore sometimes formed by what appears to be the base of the ovary, as in *Guaiacum*); gynoecium of 5 or less often 4 carpels, seldom 6 or only 2, united to form a compound, plurilocular ovary with axile placentas and a slender (seldom very short) style with a capitate or sometimes lobed or cleft stigma or with several stigmas; ovules 1-several in each locule (rather numerous in *Peganum*), pendulous or seldom ascending, epitropous (apotropous in *Nitraria*), bitegmic, crassinucellar, sometimes with a funicular-placental obturator, anatropous to sometimes hemitropous or campylotropous or orthotropous, with an integumentary tapetum except in *Peganum*; endosperm development nuclear. FRUIT commonly a capsule or a schizocarp, seldom a berry or drupe; in *Kallstroemia* the 5 (6) carpels each divided by a septum between the two ovules, so that 10 (12) half-carpellary mericarps are eventually formed; seeds with straight or slightly curved, dicotyledonous embryo; endosperm hard and oily, or less often wanting. X = 6, 8–13+. (Balanitaceae, Nitrariaceae, Peganaceae, Tribulaceae)

The family Zygophyllaceae as here defined consists of about 30 genera and 250 species, mostly of arid tropical or subtropical regions, sometimes of saline habitats. The largest genus is *Zygophyllum*, with about 90 species occurring mainly from North Africa to central Asia. The very strong, hard, heavy wood of the tropical North America genus *Guaiacum*, called lignum vitae, has many commercial uses and was formerly thought to have miraculous medicinal properties. Some other familiar members of the family are: *Peganum harmala* L., of arid regions in Asia and the Mediterranean region, the source of a dye called turkey red; *Tribulus terrestria* L., puncture-vine, a weed of European origin, noted for its small, hard, spiny fruits; and species of *Larrea*, creosote-bush, which dominate some of the warm deserts of both North and South America.

The Zygophyllaceae form a loosely knit family, and several of the genera have sometimes been segregated as distinct families or referred to other families. For example, in 1966 Takhtajan put *Nitraria* (10) into a family Nitrariaceae, *Balanites* (20) into a family Balanitaceae, and *Peganum* (3) and *Malacocarpus* (1) into a family Peganaceae, and he transferred *Tetradiclis* (1) to the Rutaceae, even though it lacks the characteristic secretory cavities of that family. (In 1980 he returned the Peganaceae to the Zygophyllaceae.) The palynological data are also consistent with the exclusion of *Balanites* and *Nitraria* (separately) from the Zygophyllaceae. Such exclusions certainly produce a more homogeneous family Zygophyllaceae, but whether their net effect is an improvement in the scheme of classification is more debatable. The three segregate families, if recognized, remain as satellites of the Zygophyllaceae.

17. Order GERANIALES Lindley 1833

Mostly herbs, seldom shrubs, rarely even small trees, often tanniferous, with or without proanthocyanins and ellagic acid, only rarely cyanogenic, sometimes producing mustard-oil; vessel-segments mostly with simple perforations. LEAVES alternate, opposite, or whorled, usually compound or dissected or more or less deeply lobed, sometimes peltate and scarcely lobed, or (especially in the Balsaminaceae) simple and merely toothed or entire; stipules present or absent. FLOWERS hypogynous or only slightly perigynous, perfect or rarely unisexual, regular or slightly irregular, or strongly irregular, then with one of the sepals prolonged backwards into a conspicuous free spur; sepals 5 or less often 3 or 4, distinct, or connate only at the base; petals 5, or seldom 3, 4, or 8, distinct or in the Balsaminaceae one of them distinct and the others more or less connate into 2 lateral pairs, or rarely the petals wanting; stamens mostly 1–2 (3) times as many as the sepals or petals, distinct or often with more or less connate filaments, sometimes some of them staminodial; pollen-grains binucleate or trinucleate, mostly tricolpate or tricolporate, sometimes with 4 or more apertures; nectary-disk wanting or represented by 5 extrastaminal glands alternate with the petals, or the stamens of one set each with an adnate basal nectary; gynoecium of 3 or 5 (seldom 2, 4, or 8) carpels united to form a compound, plurilocular ovary with axile, apical-axile, or basal-axile placentation, or the carpels nearly distinct except for their common gynobasic style; style solitary, with as many branches or stigma-lobes as carpels, or the styles several and distinct, seldom the style solitary and with a capitate, unlobed stigma; ovules 1-several or rather numerous in each locule, epitropous or less often apotropous, anatropous to sometimes hemitropous or campylotropous, bitegmic or seldom unitegmic, crassinucellar (Geraniaceae) or tenuinucellar; endosperm-development nuclear or (Balsaminaceae) cellular. FRUIT a loculicidal capsule, or separating into 1-seeded, dehiscent or indehiscent mericarps, less often a septicidal capsule or a berry; seeds with copious endosperm (mainly in Oxalidaceae) or more often with scanty or no endosperm, sometimes containing unusual sorts of fatty acids.

The order Geraniales as here defined consists of 5 families and about 2600 species. The three largest families are the Oxalidaceae (900), Balsaminaceae (900), and Geraniaceae (700).

The close relationship of the Geraniaceae and Oxalidaceae has been evident to all. The tribe Geranieae and the large genus *Oxalis* are distinctive enough, but these two groups are connected by a series of smaller genera that are variously apportioned between the two families by different authors. I do not think it improves our understanding of relationships to recognize some of these small connecting groups as distinct families, as some authors have proposed. What we gain by sharpening the definition of the major groups we lose by increasing the number of families to be remembered.

Although the Tropaeolaceae are obviously distinctive, their relationship to the Geraniaceae is generally admitted. The characteristic spur on one of the sepals in the Tropaeolaceae is apparently homologous with a similar spur found in the large genus *Pelargonium* of the Geraniaceae, the chief difference being that the spur of *Pelargonium* is adnate to the pedicel and is therefore easily overlooked. Cytological studies also support the concept that the Oxalidaceae, Geraniaceae, and Tropaeolaceae are closely related.

The Limnanthaceae and Balsaminaceae are more controversial. Both are morphologically and cytologically somewhat removed from the core families of the order and from each other, without having any obvious affinities elsewhere. The Limnanthaceae have a unique type of pollen (according to Erdtman, 1952, general citations), which emphasizes their distinctness without providing any clue as to their relationships. According to the concepts that I presented in 1968 about the phyletic relationships and morphological divergence within the Sapindales-Geraniales-Linales-Polygalales complex, the Limnanthaceae fit into the Geraniales rather than into any of the related orders. It may be significant that the Limnanthaceae and Tropaeolaceae are alike in producing mustard-oil and erucic acid, which are otherwise largely restricted to the Capparales. Nothing else about these two families suggests the Capparales, however, and the cytochrome *c* of *Tropaeolum* is very unlike that of *Brassica*. (Cytochrome *c* of *Tropaeolum* differs in 9 amino acid positions from that of *Brassica*, which differs from that of *Triticum* in 10 positions.) Among recent students, Hutchinson, Takhtajan, Thorne, and Scholz (in the 12th edition of the Engler Syllabus) have agreed in placing the Limnanthaceae in the Geraniales.

On a purely morphological basis, the Balsaminaceae might be accommodated in either the Geraniales or the Polygalales as here conceived. They would be wholly isolated in the Polygalales, however, whereas they do have some similarity to the Tropaeolaceae in the Geraniales. The most obvious feature that these two families have in common is the conspicuous retrorse spur on one of the sepals. They also have rather similar pollen, and according to Huynh they have some parallel evolutionary tendencies. On the other hand, they differ in so many ways that one is tempted to consider the similarities as accidental. At our present level of knowledge it seems more appropriate to refer the Balsaminaceae to the Geraniales than to any other order.

The definition here adopted for the Geraniales is identical to that of Hutchinson. In this instance I believe his emphasis on growth-habit provides a better distinction between the Geraniales and Sapindales than does the emphasis on the orientation of the ovules in the traditional Englerian system. It may also be noted that none of the Geraniales as here defined has the annular or unilateral nectary-disk found in so many of the Sapindales.

There is nothing ecologically distinctive about the Geraniales. Nearly all of the species are herbs or soft shrubs, but this growth-form is well known in diverse other orders. Many of the Geraniales prefer moist or shady places, but *Erodium*, in the Geraniaceae, is a familiar vernal weed of dry regions. The curious, elastically dehiscent fruits of *Impatiens* (Balsaminaceae) are reminiscent of the technically very different but also elastically dehiscent fruits of *Geranium*. The Oxalidaceae have still another way of expelling the seeds mechanically, by means of a basal aril, but none of these methods of dispersal is highly effective. The long-awned, spirally twisting mericarps of *Erodium* are evidently well adapted to animal-transport and self-planting, but they help to illustrate the ecological diversity rather than the unity of seed-dispersal mechanisms within the order.

SELECTED REFERENCES

Behnke, H.-D., & T. J. Mabry. 1977. S-Type sieve-element plastids and anthocyanins in Vivianiaceae: Evidence against its inclusion into the Centrospermae. Pl. Syst. Evol. 126: 371–375.

Bortenschlager, S. 1967. Vorläufige Mitteilungen zur Pollenmorphologie in der Familie der Geraniaceen und ihre systematische Bedeutung. Grana Palynol. 7: 400–468.

Chadefaud, M. 1975. Sur la formule florale de la Capucine (*Tropaeolum majus* L.) Bull. Bot. Soc. France 121: 347–357.

Grey-Wilson, C. 1980. Studies in Balsaminaceae. V. *Hydrocera triflora*, its floral morphology and relationships with *Impatiens*. VI. Some observations on the floral vascular anatomy of *Impatiens*. Kew Bull. 35: 213–219; 221–227.

Herr, J. M. 1972. An extended investigation of the megagametophyte in *Oxalis corniculata* L. Adv. Pl. Morph. 1972: 92–101.

Herr, J. M., & M. L. Dowd. 1968. Development of the ovule and megagametophyte in *Oxalis corniculata* L. Phytomorphology 18: 43–53.

Huynh, K.-L. 1968, 1969. Morphologie du pollen des Tropaeolacées et des Balsaminacées. I, II, III. Grana Palyn. 8: 88–184; 277–516, 1968. 9: 34–49, 1969.

Huynh, K.-L. 1969. Etude du pollen des Oxalidaceae. Bot. Jahrb. Syst. 89: 272–303; 304–334.
Kjaer, A., J. Ø. Madsen, & Y. Maeda. 1978. Seed volatiles within the family Tropaeolaceae. Phytochemistry 17: 1285–1287.
Lefor, M. W. 1975. A taxonomic revision of the Vivianiaceae. Univ. Connecticut Occas. Papers 2: 225–255.
Leifertová, I., H. Bučková, & L. Natherová. 1965. K chemotaxonomii znaku tříslovin u rodu *Geranium*. Preslia 37: 413–418.
Léonard, J. 1950. *Lepidobotrys* Engl., type d'une famille nouvelle de spermatophytes: les Lepidobotryaceae. Bull. Jard. Bot. État. 20: 31–40.
Maheshwari, P., & B. M. Johri. 1956. The morphology and embryology of *Floerkea proserpinacoides* Willd. with a discussion on the systematic position of the family Limnanthaceae. Bot. Mag. (Tokyo) 69: 410–423.
Mason, C. T. 1951. Development of the embryo-sac in the genus *Limnanthes*. Amer. J. Bot. 38: 17–22.
Mason, C. T. 1952. A systematic study of the genus *Limnanthes* R. Br. Univ. Calif. Publ. Bot. 25: 455–512.
Mathur, N. 1956. The embryology of *Limnanthes*. Phytomorphology 6: 41–51.
Mosebach, G. 1934. Die Fruchtstielschwellung der Oxalidaceen und Geraniaceen. Jahrb. Wiss. Bot. 79: 353–384.
Narayana, L. L. 1963, 1965. Contributions to the embryology of Balsaminaceae. I. J. Indian Bot. Soc. 42: 102–109, 1963. 2. J. Jap. Bot. 40: 104–116, 1965.
Narayana, L. L. 1966. A contribution to the floral anatomy of Oxalidaceae. J. Jap. Bot. 41: 321–328.
Narayana, L. L. 1974. A contribution to the floral anatomy of Balsaminaceae. J. Jap. Bot. 49: 315–320.
Ornduff, R. 1971. Systematic studies of Limnanthaceae. Madroño 21: 103–111.
Ornduff, R., & T. J. Crovello. 1968. Numerical taxonomy of Limnanthaceae. Amer. J. Bot. 55: 173–182.
Parker, W. H., & B. A. Bohm. 1979. Flavonoids and taxonomy of the Limnanthaceae. Amer. J. Bot. 66: 191–197.
Robertson, K. R. 1972. The genera of Geraniaceae in the southeastern United States. J. Arnold Arbor. 53: 182–201.
Robertson, K. R. 1975. The Oxalidaceae in the southeastern United States. J. Arnold Arbor. 56: 223–239.
Tiwari, S. C., F. Bouman, & R. N. Kapil. 1977 (1978). Ovule ontogeny in *Tropaeolum majus*. Phytomorphology 27: 350–358.
Warburg, E. F. 1938. Taxonomy and relationship in the Geraniales in the light of their cytology. New Phytol. 37: 130–159; 189–210.
Wood, C. E. 1975. The Balsaminaceae in the southeastern United States. J. Arnold Arbor. 56: 413–426.

Каден, Н. Н. 1964. Морфология плодов гераниевых. Научн. Докл. Высшей Школы, Биол. Науки, 1964(2): 97-102.

SYNOPTICAL ARRANGEMENT OF THE FAMILIES OF GERANIALES

1 Flowers regular and not spurred, or in *Pelargonium* (Geraniaceae) somewhat irregular and with an inconspicuous spur adnate to the pedicel; stamens mostly 2 or 3 times as many as the sepals or petals, sometimes some of them staminodial; leaves mostly compound or deeply cleft, seldom simple and entire or merely toothed.

2 Annual or more often perennial herbs, generally with a sclerenchymatous sheath in the pericycle and sometimes with a functional cambium, or seldom woody plants, in any case not producing mustard-oil; ovules bitegmic, epitropous, usually pendulous when solitary or few, filaments connate at the base; style or styles terminal on the ovary, or (many Geraniaceae) the style gynobasic but strongly thickened to form a column.

3 Ovary generally with distinct, terminal styles, rarely (*Hypseocharis*) with a single terminal style; fruit a loculicidal capsule, or rarely a berry; endosperm usually copious; ovules tenuinucellar .... 1. OXALIDACEAE.

3 Ovary with a single (often thickened, columnar, and gynobasic) style, or rarely (*Biebersteinia*) with distinct styles; fruit typically of 5 1-seeded mericarps that separate elastically from the persistent central column (the mericarps often opening ventrally to release the seed), or in some of the smaller genera separating into mericarps but without a persistent column, or in some other small genera a loculicidal capsule; endosperm scanty or none, except in *Viviania*; ovules crassinucellar .... 2. GERANIACEAE.

2 Small annual herbs with a parenchymatous pericycle and without cambium, characteristically producing mustard-oil; ovules unitegemic, apotropous, solitary in each locule and erect or ascending; filaments distinct; style gynobasic, uniting the otherwise essentially distinct carpels .... 3. LIMNANTHACEAE.

1 Flowers more or less strongly irregular, one of the sepals with a conspicuous free spur (this poorly developed in one monotypic genus of Balsaminaceae); stamens 5 or 8, less than twice as many as the basic (unmodified) number of sepals or petals; leaves simple, not deeply lobed except in some Tropaeolaceae.

4 Leaves palmately veined, peltate or sometimes palmately lobed or cleft; stamens 8, with distinct filaments; carpels 3, each with a single epitropous ovule, the fruit schizocarpic; plants producing mustard-oil, but without raphides; endosperm-development nuclear; seeds with erucic acid, but without parinaric or acetic acid .... 4. TROPAEOLACEAE.

4 Leaves pinnately veined, not peltate; stamens 5, with more or less connate filaments; carpels (4) 5, each with 3-many apotropous ovules; fruit an elastically dehiscent capsule, or seldom a berrylike septicidal capsule; plants with raphides, but without mustard-oil; endosperm-development cellular; seeds with parinaric and acetic acid, but without erucic acid ....
.... 5. BALSAMINACEAE.

1. Family OXALIDACEAE R. Brown in Tuckey 1817 nom. conserv., the Wood-Sorrel Family

Herbs, often with tubers or bulbs, or sometimes half-shrubs or shrubs, rarely (*Averrhoa* and *Sarcotheca*) small trees, commonly accumulating both soluble and crystalline oxalates, generally with scattered tanniferous secretory cells and accumulating proanthocyanins, but without ellagic acid and only seldom cyanogenic; nodes trilacunar; stem in herbs with a ring of vascular bundles and usually a sclerenchymatous sheath in the pericycle, the bundles sometimes becoming connected by the activity of an interfascicular cambium; vessel-segments with simple perforations; in woody forms the imperforate tracheary elements septate and with simple pits, the rays uniseriate, heterocellular to almost homocellular, the wood-parenchyma mostly vasicentric and scanty. LEAVES alternate (sometimes all basal), palmately or pinnately compound or often trifoliolate, sometimes unifoliolate, often exhibiting sleep-movements, with the leaflets folded at night; stomates paracytic; mesophyll, at least in *Oxalis*, commonly with scattered secretory cavities containing red or brown substances; petiole commonly with a ring of vascular bundles; stipules tiny or wanting. FLOWERS in axillary, often pedunculate, cymose (sometimes umbelliform) inflorescences that may be reduced to a single flower, perfect, hypogynous, regular, pentamerous, mostly showing trimorphic heterostyly, sometimes some of them cleistogamous and apetalous; sepals 5, distinct, imbricate; petals 5, convolute or seldom imbricate, distinct or sometimes slightly connate at the base; stamens generally 10, more or less distinctly bicyclic, with the outer filaments shorter and antepetalous, the filaments all connate at the base, sometimes 5 of them without anthers, rarely (*Hypseocharis*) the stamens 15 in 3 series; nectary-disk wanting, but often the outer (antepetalous) filaments thickened and nectariferous below, or a nectary-gland borne at the outer base of the antepetalous stamens; anthers tetrasporangiate and dithecal, opening by longitudinal slits; pollen-grains binucleate or seldom trinucleate, commonly tricolpate or tricolporate; gynoecium of (3–) 5 carpels united to form a compound, plurilocular ovary with axile placentas and distinct, persistent styles, rarely (*Hypseocharis*) with a single style; stigmas commonly capitate or punctate, sometimes bilobed; ovules (1) 2-several in each locule, epitropous and more or less pendulous, with the micropyle directed outward and upward, anatropous or sometimes hemitropous, bitegmic, tenuinucellar; endosperm-development nuclear. FRUIT a loculicidal capsule,

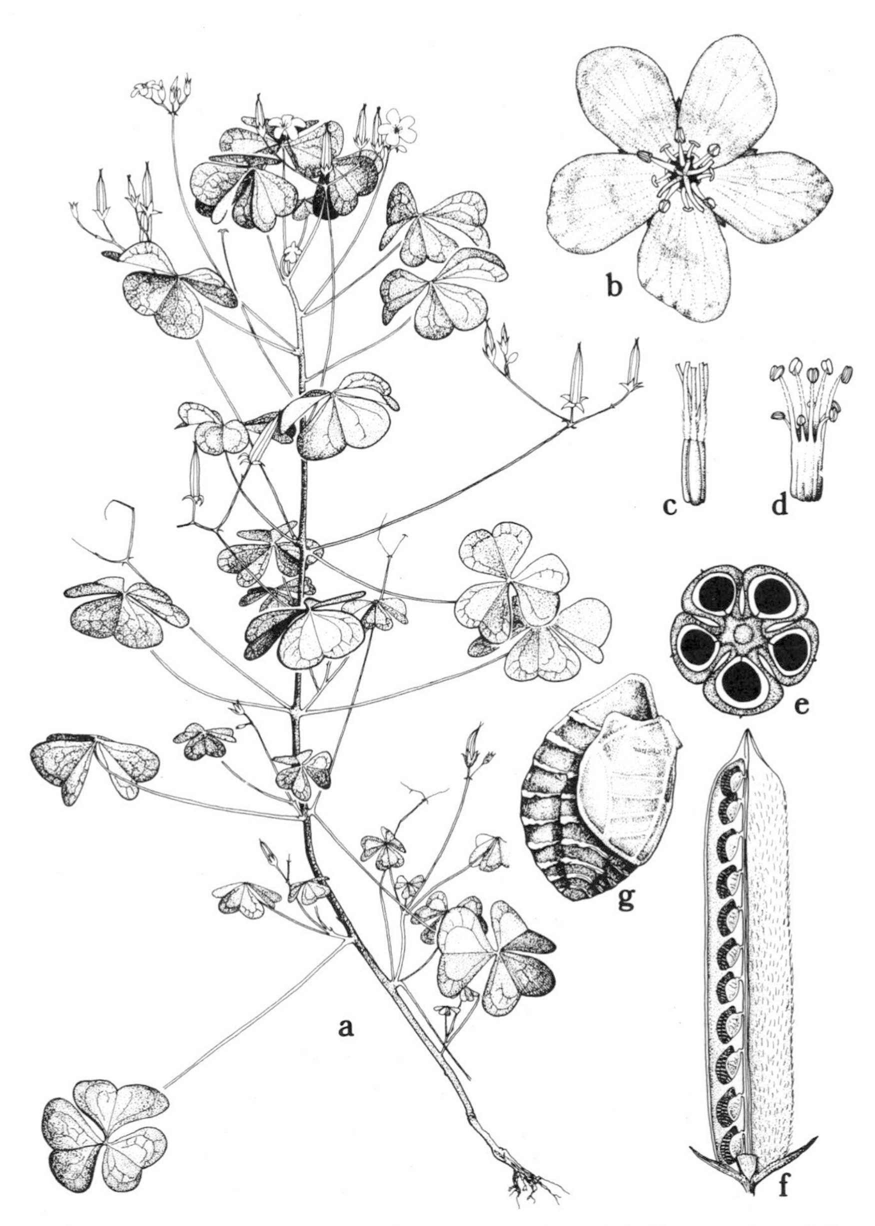

FIG. 5.59 Oxalidaceae. *Oxalis europaea* Jordan. a, habit, $\times\frac{1}{2}$; b, flower, $\times 4$; c, pistil, $\times 4$; d, stamen-tube, $\times 4$; e, schematic cross-section of ovary, $\times 20$; f, fruit, with one valve removed, $\times 4$; g, seed, with aril, $\times 20$.

or rarely (*Averrhoa*) a berry; seeds commonly with a basal aril that serves in their expulsion from the capsule; embryo large, straight, spatulate, with 2 cotyledons, embedded in the usually copious, oily fleshy endosperm. X = 5–12, most commonly 7. (Averrhoaceae, Hypseocharitaceae, Lepidobotryaceae)

The family Oxalidaceae consists of 7 or 8 genera and about 900 species, widespread in tropical and subtropical latitudes (but often at high altitudes), with relatively few species in temperate latitudes. By far the largest genus is *Oxalis*, with perhaps 800 species. Some species of *Oxalis* are familiar garden-weeds in North Temperate regions; a few others are cultivated as garden-ornamentals or as house-plants. Some other species of *Oxalis* are cultivated in the Andes for their edible tubers, and *Averrhoa carambola* L., the carambola, is widely cultivated in the tropics for its edible fruit.

2. Family GERANIACEAE A. L. de Jussieu 1789 nom. conserv., the Geranium Family

Herbs, or sometimes half-shrubs or shrubs, often tanniferous and producing ellagic acid or gallic acid or both, but only seldom with proanthocyanins, not saponiferous and not cyanogenic, very often producing aromatic oils in multicellular, capitate-glandular hairs; solitary or clustered crystals of calcium oxalate commonly present in some of the cells of the parenchymatous tissues; nodes trilacunar; stem commonly with a ring (or sometimes 2 rings) of vascular bundles and usually a sclerenchymatous sheath in the pericycle, the bundles sometimes becoming connected through the activity of an interfascicular cambium; vessel-segments with simple perforations; imperforate tracheary elements commonly with simple pits (bordered in *Viviania*), often septate, wood-rays in woody forms heterocellular, or very often wanting, the wood-parenchyma scanty-paratracheal; sieve-tubes with S-type plastids. LEAVES alternate or opposite, usually lobed or compound or dissected (either pinnately or palmately), seldom (as in *Viviania, Dirachma*) simple and entire or merely toothed; stomates mostly anomocytic (tending to be encyclocytic in *Dirachma*); petiole commonly with a ring (or sometimes 2 concentric rings) of vascular bundles; stipules usually present. FLOWERS mostly in cymose (often umbelliform) inflorescences, less often solitary and axillary, perfect or seldom unisexual, regular or (*Pelargonium*) somewhat irregular, hypogynous or nearly so; sepals 5 or seldom 4, imbricate or less often valvate, distinct

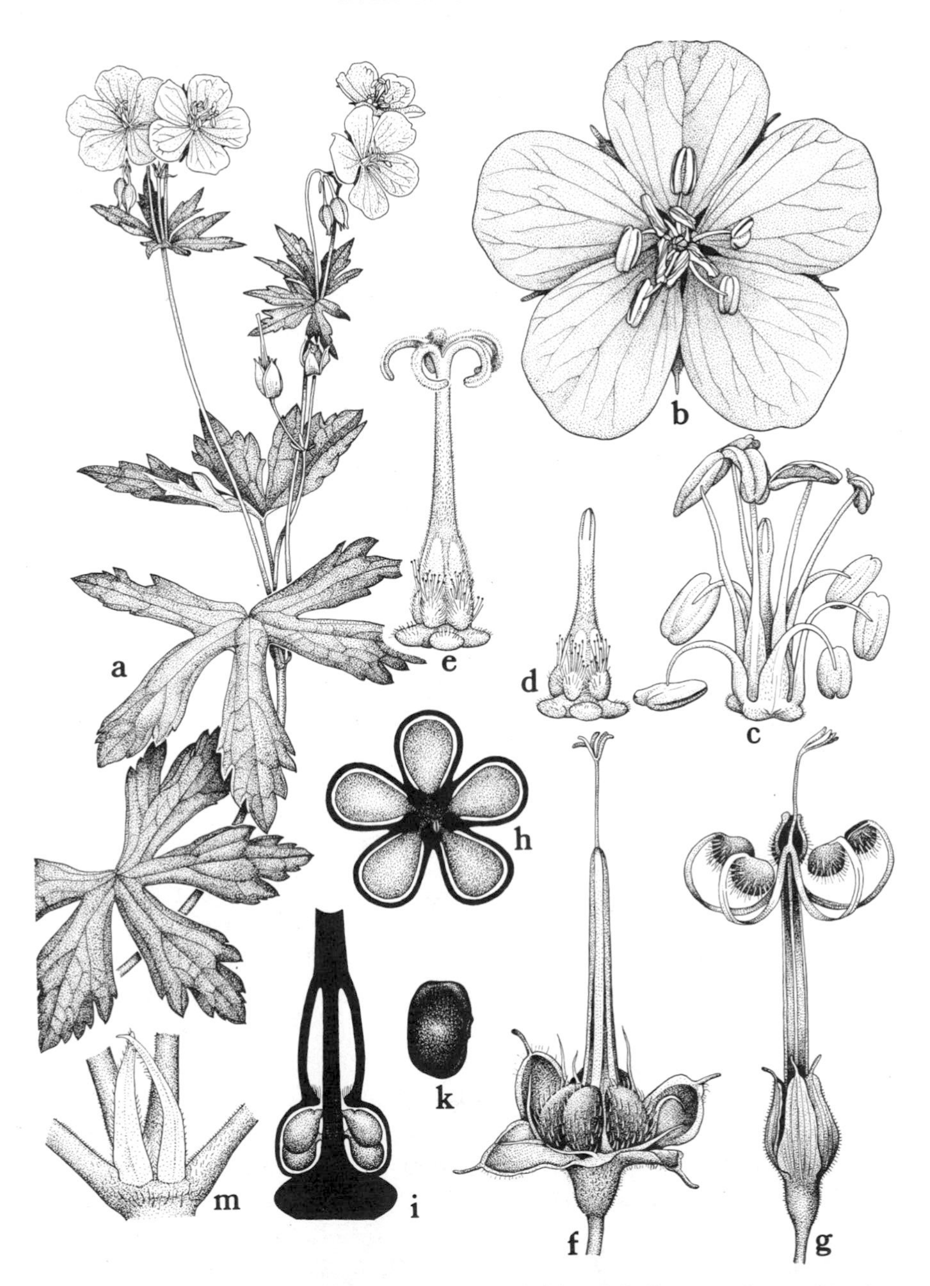

FIG. 5.60 Geraniaceae. *Geranium maculatum* L. a, habit, ×½; b, flower, ×2; c, androecium and gynoecium, ×4; d, pistil at stage of anther-dehiscence, ×4; e, pistil after anthers have fallen, ×4; f, mature, unopened fruit, with the calyx parted to expose the carpel-bases, ×2; g, mature fruit, after dehiscence, ×2; h, schematic cross-section of ovary, ×12; i, schematic long-section of ovary, ×8; k, seed, ×4; m, node, with stipules, ×4.

or connate below, seldom forming a shortly lobed tube, in *Pelargonium* the upper (adaxial) one commonly prolonged backwards into a spur-nectary that is adnate to the pedicel; an epicalyx present in *Balbisia*, *Dirachma*, and *Wendtia*; petals mostly 5, seldom 4 (8 in *Dirachma*), distinct, imbricate or seldom (but including *Balbisia* and *Dirachma*) convolute, or rarely wanting; extrastaminal nectary-glands usually alternate with the petals (wanting in *Pelargonium*); stamens most commonly 10 in 2 cycles, but some or all of the outer (antepetalous) cycle often staminodial (seldom only 2 of the stamens antheriferous), or rarely (*Monsonia, Sarcocaulon*) the stamens 15 in 3 cycles, in *Dirachma* the stamens 8 and antepetalous; filaments more or less connate at the base; anthers tetrasporangiate and dithecal, opening by longitudinal slits; pollen-grains trinucleate, tricolpate to more often tricolporate, or sometimes multiporate (as in *Viviania*) or inaperturate; gynoecium mostly of 5, less often 2 or 3 (8 in *Dirachma*) carpels united to form a compound, plurilocular, lobed or grooved ovary with axile placentas and a single style with distinct stigmas (styles distinct in *Biebersteinia*), in the Geranieae the gynoecium with a prominent, elongating, persistent central column (often called a stylar beak) to which the fertile, locular portion of the ovary appears to be attached in a lobed ring at the base; stigmas dry, papillate; ovules mostly 2 in each locule, superposed, at least the upper one pendulous, epitropous, with ventral raphe and upwardly directed micropyle (1 pendulous ovule per locule in *Biebersteinia* and sometimes also in *Wendtia* and *Rhynchotheca*, 1 erect ovule per locule in *Dirachma*, 2 rows of ovules in each locule in *Balbisia*), anatropous to campylotropous, bitegmic, crassinucellar; endosperm-development nuclear. FRUIT in the Geranieae commonly of 5 1-seeded mericarps that separate elastically and acropetally from the persistent central column, the mericarp often opening to discharge the seed, in the other genera a loculicidal capsule (*Viviania, Balbisia, Wendtia*) or separating into mericarps but without a persistent central column (*Biebersteinia, Rhynchotheca, Dirachma*); seeds with a straight or more often curved, dicotyledonous embryo; endosperm copious and oily in *Viviania*, otherwise scanty or none. X = 7–14. (Biebersteiniaceae, Dirachmaceae, Ledocarpaceae, Vivianiaceae)

The family Geraniaceae, as here broadly (and traditionally) defined, consists of some 11 genera and about 700 species, widespread in temperate and warm- temperate regions, with relatively few tropical species. The largest genera are *Geranium* (300, cosmopolitan), *Pelargonium* (250, chiefly South African, widely cultivated under the common name Geranium), and *Erodium* (75). These, together with *Monsonia* (30) and *Sarcocaulon* (6), make up the tribe Geranieae, a well marked natural group. The other 6 genera, encompassing scarcely 25 species in all, are aberrant in one way or another, and have sometimes been segregated into as many as 4 distinct families; Biebersteiniaceae (*Biebersteinia*, 5; *Rhynchotheca*, 1); Dirachmaceae (*Dirachma*, 1); Ledocarpaceae (*Balbisia*, 7; *Wendtia*, 3); and Vivianiaceae (*Viviania* sens. lat., 6). Furthermore, the genus *Hypseocharis* (8), here included in the Oxalidaceae, connects the Geraniaceae to the Oxalidaceae and has sometimes been treated as a distinct family Hypseocharitaceae.

It is possible that some of these genera, most notably *Viviania* and *Dirachma*, are not closely allied to the Geraniaceae, but I prefer to keep them here until their affinities are more clearly established. Some authors have sought to associate *Viviania* with the Caryophyllales, but the presence of anthocyanins and S-type sieve-tube plastids militates against the proposal. Lefor (1975) divides the traditional genus *Viviania* into 4 genera, 3 of them monotypic. For descriptive convenience the genus is here retained in the broad sense.

3. Family LIMNANTHACEAE R. Brown 1833 nom. conserv., the Meadow-Foam Family

Small, weak, subsucculent annual herbs, commonly of vernal pools or moist places, producing isokestose and sometimes also its corresponding fructose tetrasaccharide, but not inulin, characteristically with myrosin-cells and mustard oils, at least sometimes accumulating ellagic acid and proanthocyanins, neither saponiferous nor cyanogenic; stem with several distinct vascular bundles in a ring, without cambium, and with a parenchymatous pericycle. LEAVES alternate, pinnately parted or compound or dissected; stomates anomocytic; stipules wanting. FLOWERS solitary on long, axillary pedicels, perfect, regular, hypogynous or nearly so; calyx and corolla (4), 5-merous in *Limnanthes*, 3-merous in *Floerkea*; sepals distinct or nearly so, valvate; petals distinct, convolute; stamens twice as many as the petals, in 2 cycles, or in *Floerkea* sometimes as many as and alternate with the petals; filaments distinct, the antesepalous ones with an adnate nectary-gland internally at the base; anthers tetrasporangiate and dithecal, opening by longitudinal slits; pollen-grains binucleate, tricolpate or sometimes dicolpate; gynoecium in *Limnanthes* of (4) 5 carpels, in *Floerkea* of 2 or 3 carpels, these united by the gynobasic style, otherwise nearly distinct, the ovary deeply lobed into globular segments; style more or less deeply cleft, or undivided and then with lobed or capitate stigma; stigma dry, papillate; ovules solitary in each locule, basal, erect or ascending, apotropous (the micropyle directed downwards and outwards), anatropous, unitegmic, tenuinucellar; embryo-sac with an unusual ontogeny, tetrasporic, 4-nucleate in *Limnanthes*, 6-nucleate in *Floerkea*; endosperm-development nuclear. FRUIT separating into 1-seeded, indehiscent mericarps; seeds with straight embryo, 2 large cotyledons, and small radicle, containing unusual fats of the eicosenic and erucic acid types; endosperm wanting. X = 5.

The family Limnanthaceae consists of only 2 small genera, both native to temperate North America. *Limnanthes* has about 10 species, mostly in California. *Floerkea* has a single transcontinental species. *Limnanthes* has recently excited some interest as a possible new crop for the production of oil.

The affinities of the Limnanthaceae have been debated in the past, but recent authors are mostly in agreement that they are allied to the Geraniaceae.

4. Family TROPAEOLACEAE A. P. de Candolle 1824 nom. conserv., the Nasturtium Family

Subsucculent annual or perennial herbs, often scandent (with twining petioles), sometimes with tuberous roots, producing mustard-oil and with myrosin-cells at least in the cortex of both the stem and the root, sometimes tanniferous, but apparently lacking both proanthocyanins and ellagic acid, not cyanogenic; crystals of calcium oxalate only seldom present in some of the cells of the parenchymatous tissues; nodes trilacunar; vascular bundles in a single cycle surrounding the large pith, eventually becoming connected through the activity of an interfascicular cambium; vessel-segments mostly with simple perforations; pericycle without mechanical tissue. LEAVES alternate (or the lower opposite), palmately veined, peltate or palmately lobed or cleft; stomates anomocytic; myrosin present in the leaves, but myrosin-cells wanting from the leaves of at least some species; petiole with a ring of vascular bundles, often elongate; stipules present or absent, sometimes present only in the seedling stage. FLOWERS solitary and axillary, often long-pedunculate, showy, perfect, slightly perigynous, usually strongly irregular; sepals 5, distinct, imbricate, the adaxial one (or the 3 upper ones collectively) prolonged backwards into a nectariferous spur (spur

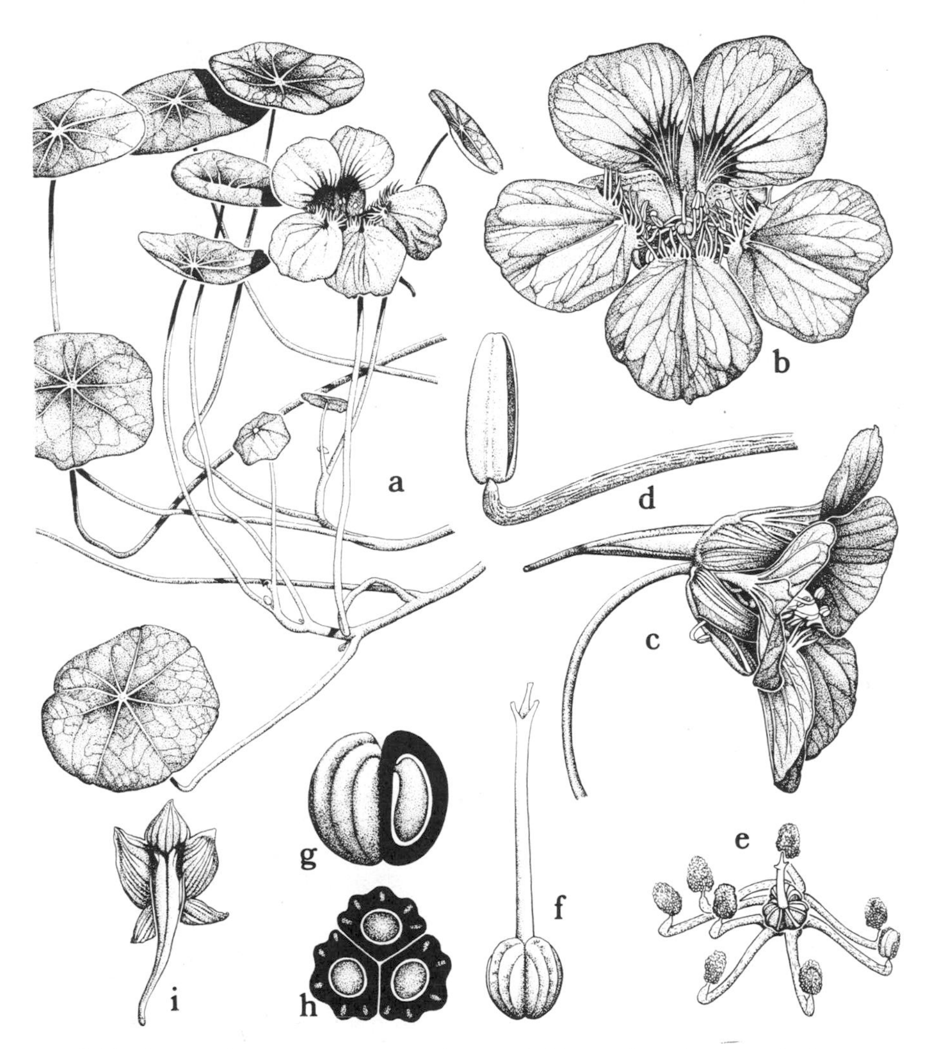

FIG. 5.61 Tropaeolaceae. *Tropaeolum majus* L. a, habit, ×½; b, c, flower, ×1; d, stamen, ×8; e, stamens and pistil, ×2; f, pistil, ×4; g, ovary, in partial long-section, ×8; h, schematic cross-section of ovary, ×8; i, calyx, ×1.

scarcely developed in *Trophaeastrum,* the flower nearly regular); petals 5, distinct, imbricate, clawed, the 3 lower (abaxial) ones generally different from the other 2, often with a hairy claw, or these 3 lower petals rarely wanting; stamens declined, 8 in 2 cycles of 4, reflecting the loss of one member of each of 2 ancestrally pentamerous cycles; filaments distinct; anthers small, basifixed, latrorse, tetrasporangiate, dithecal, opening by longitudinal slits; pollen-grains tricolporate or dicolporate, sometimes

2 colpi fully developed and one reduced; gynoecium of 3 carpels united to form a compound, trilocular ovary with a slender, terminal, apically trifid style; stigmas dry, not papillate; ovules solitary in each locule, apical-axile, pendulous, epitropous (the micropyle directed upwards and outwards), anatropous, bitegmic, tenuinucellar; endosperm-development nuclear. FRUIT separating into 1-seeded drupaceous or nutlike mericarps, or (*Magallana*) samaroid and only one mericarp maturing; seeds with a large, straight embryo and 2 thick cotyledons, containing erucic acid; endosperm wanting. X = 12–14.

The family Tropaeolaceae consists of 3 genera, all confined to the New World. *Tropaeolum* (90) occurs in the mountains from Mexico to Chile; *Magallana* (1) and *Trophaeastrum* (1) occur in Patagonia. *Tropaeolum majus* L. is the gardeners' nasturtium.

5. Family BALSAMINACEAE A. Richard 1822 nom. conserv., the Touch-me-not Family

Subsucculent annual or perennial herbs or seldom half-shrubs, nearly always glabrous, sometimes saponiferous, sometimes with tannin-sacs or otherwise tanniferous, often accumulating proanthocyanins but without ellagic acid, only seldom cyanogenic, with large mucilage-sacs in the parenchyma of the stem, and characteristically with raphide-sacs in the stem and leaves, these sometimes containing mucilage as well as raphides; stem more or less translucent; nodes unilacunar; vascular bundles in a ring, sometimes eventually becoming connected by the activity of a weakly developed interfascicular cambium; pericycle parenchymatous; vessel-segments with simple perforations. LEAVES alternate, opposite, or whorled, simple, entire or merely toothed, pinnately veined; stomates anomocytic, with some tendency toward the anisocytic type; petiole with an arc (not a ring) of vascular bundles; stipules represented only by a pair of small, petiolar glands, or wanting. FLOWERS solitary or in small, sometimes umbelliform cymes, perfect, hypogynous, strongly irregular, resupinate, the morphologically posterior (ventral, adaxial) sepal borne in the anterior (dorsal, abaxial) position; sepals 5 or more often 3, the 2 upper ones obsolete, the lowest one (as seen) somewhat petaloid and often saccate, produced backward into a slender spur-nectary; upper petal (as seen) distinct, external in bud, concave and often partly sepaloid, the 4 others distinct in *Hydrocera*, but in *Impatiens* connate into 2 lateral pairs, each pair suggesting a single equally or unequally bilobed petal; nectary-disk wanting; stamens 5, the

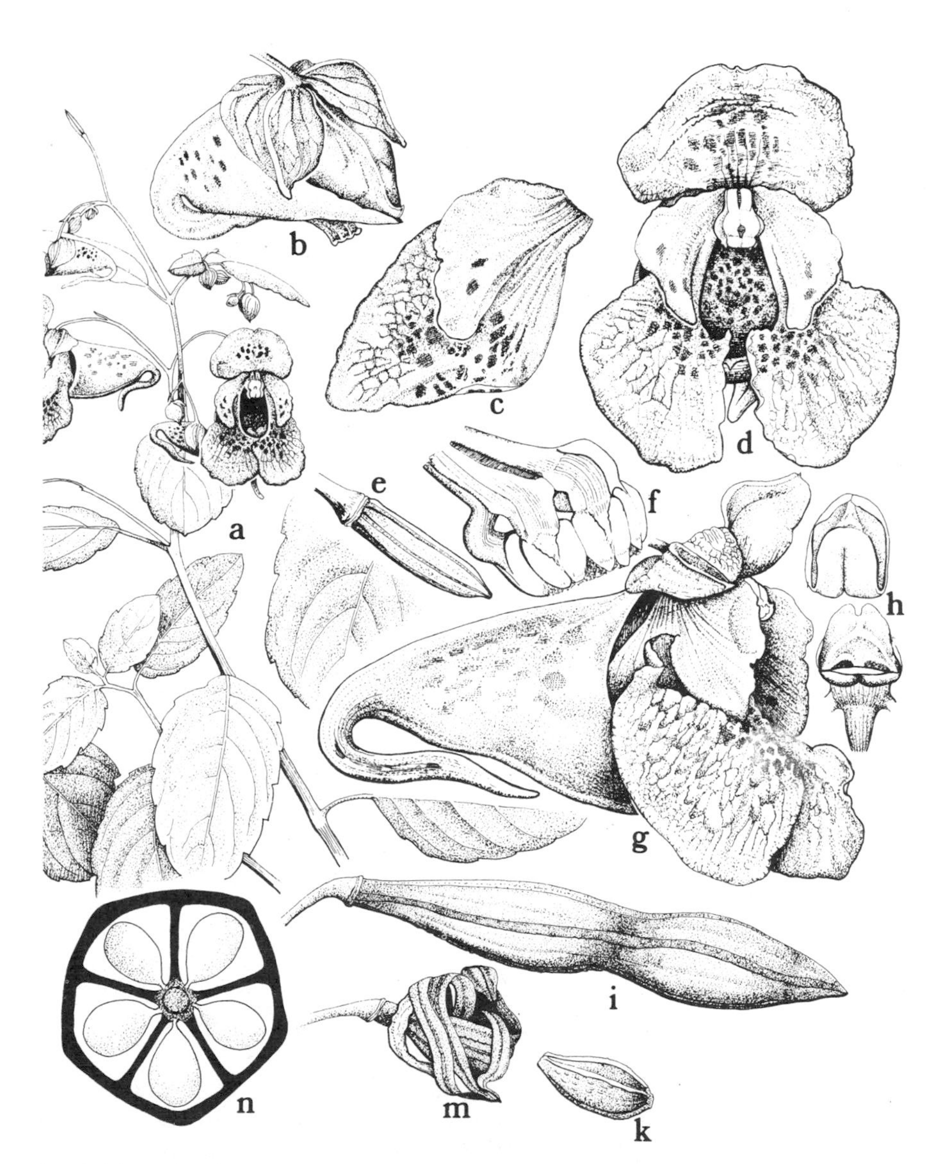

FIG. 5.62 Balsaminaceae. *Impatiens biflora* Walter. a, habit, ×1; b, flower bud, ×3; c, pair of lateral petals, ×3; d, front view of flower, ×3; e, pistil, ×6; f, androecium, ×6; g, side view of flower, ×3; h, two views of anther, ×6; i, fruit, ×3; k, seed, ×3; m, fruit, after dehiscence, ×3; n, schematic cross-section of ovary, ×20.

short, thick filaments connate at least above, and the short anthers connate or connivent, the androecium forming a deciduous calyptra over the ovary; anthers ordinarily tetrasporangiate and dithecal; pollen-sacs characteristically divided by trabeculae that separate the sporogenous tissue into islands; pollen-grains binucleate, tricolpate or sometimes 4–5-colpate; gynoecium of (4) 5 carpels united to form a compound, plurilocular ovary with axile placentas, a short style, and a single wet, nonpapillate stigma or 5 stigmas; carpels with 3 traces in *Impatiens*, 5 in *Hydrocera*; ovules several or many in each locule in *Impatiens*, only one per locule in *Hydrocera*, pendulous, apotropous (the micropyle directed upwards and inwards), anatropous, bitegmic or sometimes unitegmic by fusion, tenuinucellar, with an integumentary tapetum; endosperm-development cellular, with chalazal and often also micropylar haustoria. FRUIT in *Impatiens* an explosively dehiscent loculicidal capsule, the valves twisting in dehiscence, in *Hydrocera* a pentagonal, berry-like drupe, the stone eventually separating into 5 1-seeded pyrenes; embryo straight, with a short hypocotyl and 2 expanded cotyledons, containing large amounts of glycerides of parinaric and acetic acid; endosperm scanty in *Impatiens*, wanting in *Hydrocera*. X = 6–11.

The family Balsaminaceae consists of only 2 genera. *Impatiens* has about 450 species, native mainly to tropical Asia and Africa, but with a few species in temperate regions of both the Old and the New World. *Hydrocera* (*Tytonia*) has a single species native from India to Java. Several species of *Impatiens* are cultivated as garden ornamentals or as house-plants.

18. Order APIALES Nakai 1930

Plants woody (usually rather softly so) or herbaceous, commonly accumulating sesquiterpenes, triterpenoid compounds, and polyacetylenes, but only seldom alkaloids, hydrocyanic acid, or tannin, lacking ellagic acid and usually also lacking proanthocyanins, commonly storing carbohydrate as the trisaccharide umbelliferose, apparently restricted to this order: schizogenous secretory canals (or cavities) containing ethereal oils, resins, and gums usually well developed in the parenchymatous tissues; solitary secretory cells wanting; pith-bundles and cortical bundles often present; nodes mostly multilacunar, seldom trilacunar; vessel-segments with simple or seldom scalariform perforations; imperforate tracheary elements with small, simple pits, often septate; wood-rays heterocellular or seldom homocellular, mixed uniseriate and pluriseriate, the latter with short ends; wood-parenchyma mostly paratracheal and scanty. LEAVES alternate, rarely opposite or whorled, often very large, generally pinnately or ternately or palmately compound or dissected, or simple and palmately lobed, or occasionally simple and entire; stomates paracytic or of various other types; petiole commonly broad and sheathing, sometimes with stipular appendages, usually with a ring or arc of vascular bundles, and sometimes with medullary bundles as well. FLOWERS commonly small, mostly in umbels that are very often arranged into secondary umbels or into secondary inflorescences of various other sorts, but sometimes in heads or other sorts of inflorescences, perfect or seldom unisexual, epigynous or rarely secondarily hypogynous, very often but not always pentamerous; calyx commonly represented by small teeth around the top of the ovary, sometimes much-reduced or even obsolete; petals nearly always distinct, valvate or seldom imbricate, rarely wanting; stamens commonly as many as and alternate with the petals, seldom more numerous; anthers mostly tetrasporangiate and dithecal, opening by longitudinal slits; pollen-grains mostly trinucleate, commonly tricolporate; gynoecium of 2–5 (-many) carpels united to form a compound, inferior (very rarely superior or only half-inferior) ovary with as many locules as carpels (rarely pseudomonomerous and unilocular), and crowned by an epigynous nectary-disk; styles as many as the carpels, or seldom more or less connate into a single style; often enlarged at the base into a stylopodium that is confluent with the nectary-disk; ovules solitary in each locule (an additional abortive ovule also present), apical-axile, pendulous, epitropous, with ventral raphe and with the micropyle directed upwards and outwards, anatropous, unitegmic, crassinucellar to tenuinucellar, with an integumentary tapetum; endosperm-development nuclear. FRUIT a drupe (usually with as many stones as carpels) or a berry, or more often a schizocarp that typically has a persistent central carpophore to which the mericarps are apically attached; seeds with small, dicotyledonous embryo and abundant endosperm containing a considerable quantity of petroselinic acid, a compound otherwise known in *Garrya, Aucuba* (Cornaceae) and *Picrasma* (Simaroubaceae). (Araliales, Umbellales)

The order Apiales consists of 2 families and about 3700 species, of cosmopolitan distribution. The order has often been known under the irregular name Umbellales. The name Apiales, here used, is 16 years older than the name Araliales, which has recently gained some currency.

The families of Apiales have in the past often been submerged in the order Cornales. In my opinion this assignment reflects an overemphasis on the epigynous floral structure, a condition that has been independently attained in many different groups of angiosperms. In other respects the Apiales are much more like the Sapindales than like the Cornales. The Araliaceae would in fact fit comfortably into the Sapindales, near the Burseraceae, if the flowers were hypogynous instead of epigynous. Hegnauer has indicated in several papers that on chemical grounds the Apiales stand apart from the Cornales and might well be derived from the Sapindales (sensu mei).

It is widely agreed that the Apiaceae take their origin directly from the Araliaceae, though probably not from any modern genus. All of the features that have been used to distinguish the Apiaceae as a family can be found individually in the Araliaceae. In particular, the small (12 spp.) genus *Myodocarpus*, of New Caledonia, has bicarpellate, schizocarpic fruits with a central carpophore as in the Apiaceae, but has the inflorescence and vegetative structure of the Araliaceae. Indeed the wood of *Myodocarpus* is relatively primitive, with scalariform vessels that have 7–16 cross-bars. Eyde & Tseng (1971) speculate that *Myodocarpus* and its immediate allies belong to an evolutionary line that "diverged from the other Araliaceae perhaps as long ago as the Cretaceous Period."

A few of the Araliaceae have 10 or more petals, stamens, and carpels in a more or less regular, symmetrical arrangement, and a few others have numerous stamens and sometimes also carpels, but a more ordinary number of petals. These polymerous types have often been considered to be primitive within the family, but I challenged that concept in 1968. Subsequently Eyde & Tseng (1971) showed that very large numbers of stamens and carpels, as in *Tupidanthus*, are secondary, but they considered on the basis of floral anatomy that the moderately polymerous types are indeed primitive within the family in that respect. I find it difficult to believe that polymerous corollas are primitive in a group that is otherwise so advanced and so obviously allied to other advanced groups. Even the ancestral Rosales generally have a pentamerous perianth.

A moderately polymerous androecium might possibly be primitive in the Araliaceae, even if a polymerous corolla is not. If so, then the evolutionary connection of the Araliaceae to the Sapindales must be with the archaic rather than advanced members of that order, hardly short of the eventual connection to the Rosales. It may be significant that some members of the Rutaceae, a family to which the Araliaceae have been compared on chemical grounds, have a polymerous androecium and even a polymerous gynoecium, even though the perianth is pentamerous. In the Rutaceae, as in the Araliaceae, it is uncertain whether the forms with polymerous androecium and gynoecium have retained this polymery from a polymerous Rosalean ancestor, or whether the polymery reflects a reversion. We should not put ourselves in chains to the concept that evolution is always unidirectional.

The most nearly distinctive ecological feature of the Apiales is their means of chemical welfare. They forego such common weapons as tannins, HCN, iridoid compounds, and (in most genera) alkaloids, depending instead on monoterpenes, sesquiterpenes, triterpenoid saponins, polyacetylenes, coumarins, and phenyl-propanoid compounds. Their schizogenous secretory canals provide transport and delivery of at least some of these substances. These chemicals and canals are by no means unique to the Apiales, being shared in part with some members of the Sapindales as well as other groups (including notably the otherwise very dissimilar Asterales), but neither is the package common and widespread among angiosperms in general. Like other chemical defenses, those of the Apiales are only partly effective. Many species of Apiaceae are palatable to livestock and to herbarium-beetles.

In other respects the Apiales are not outstanding ecologically. The adaptive significance of epigyny is dubious at best, and there are of course many other epigynous groups of angiosperms. The fruit of the Apiaceae is morphologically unique, but ecologically undistinguished. In a few genera it is beset with hooks or barbs that adapt it to distribution by animals; in other genera it is dispersed by wind, and in others it has no very obvious means of dispersal.

SELECTED REFERENCES

Baumann, M. G. 1946. *Myodocarpus* und die Phylogenie der Umbelliferen-Frucht. Ber. Schweiz. Bot. Ges. 56: 13–112.

Burtt, B. L., & W. C. Dickison. 1975. The morphology and relationships of *Seemannaralia* (Araliaceae). Notes Roy. Bot. Gard. Edinburgh 33: 449–464.

Cerceau-Larrival, M. T. 1962. Plantules et pollens d'Ombelliferès. Leur intérêt systématique et phylogénique. Mém. Mus. Natl. Hist. Nat., Ser. B, Bot. 14: 1–166.

Constance, L., Tsan-Iang Chuang, & C. R. Bell. 1976. Chromosome numbers in Umbelliferae. V. Amer. J. Bot. 63: 608–625.

Crowden, R. K., J. B. Harborne, & V. H. Heywood. 1969. Chemosystematics of the Umbelliferae—a general survey. Phytochemistry 8: 1963–1984.

Dilcher, D. L., & G. E. Dolph. 1970. Fossil leaves of *Dendropanax* from Eocene sediments of southeastern North America. Amer. J. Bot. 57: 153–160.

Eyde, R. H., & C. C. Tseng. 1969. Flower of *Tetraplasandra gymnocarpa:* hypogyny with epigynous ancestry. Science 166: 506–508.

Eyde, R. H., & C. C. Tseng. 1971. What is the primitive floral structure of Araliaceae? J. Arnold Arbor. 52: 205–239.

Graham, S. A. 1966. The genera of Araliaceae in the southeastern United States. J. Arnold Arbor. 47: 126–136.

Gupta, S. C. 1964. The embryology of *Coriandrum sativum* L. and *Foeniculum vulgare* Mill. Phytomorphology 14: 530–547.

Guyot, M. 1966. Les types stomatiques des Ombellifères. Bull. Soc. Bot. France 113: 244–273.

Heywood, V. H., ed. 1971. The biology and chemistry of the Umbelliferae. J. Linn. Soc., Bot. 64 (Suppl. 1). Academic Press, London.

Hoar, C. S. 1915. A comparison of the stem anatomy of the cohort Umbelliflorae. Ann. Bot. (London) 29: 55–63.

Jackson, G. 1933. A study of the carpophore of the Umbelliferae. Amer. J. Bot. 20: 121–144.

Kaplan, D. R. 1970. Comparative development and morphological interpretation of 'rachis-leaves' in Umbelliferae. *In*: N. K. B. Robson, D. F. Cutler, & M. Gregory, eds., New research in plant anatomy, pp. 101–125. J. Linn. Soc., Bot. 63 (Suppl. 1). Academic Press. London.

Magin, N. 1977. Das Gynoecium der Apiaceae—Modell und Ontogenie. Ber. Deutsch. Bot. Ges. 90: 53–66.

Mittal, S. P. 1961. Studies in the Umbellales. II. The vegetative anatomy. J. Indian Bot. Soc. 40: 424–443.

Mohana Rao, P. R. 1972 (1973). Morphology and embryology of *Tieghemopanax sambucifolius* with comments on the affinities of the family Araliaceae. Phytomorphology 22: 75–87.

Philipson, W. R. 1970. Constant and variable features of the Araliaceae. *In*: N. K. B. Robson, D. F. Cutler, & M. Gregory, eds., New research in plant anatomy, pp. 87–100. J. Linn. Soc., Bot. 63 (Suppl. 1). Academic Press. London.

Pickering, J. L., & D. E. Fairbrothers. 1970. A serological comparison of Umbelliferae subfamilies. Amer. J. Bot. 57: 988–992.

Rodriguez, R. L. 1957. Systematic anatomical studies on *Myrrhidendron* and other woody Umbellales. Univ. Calif. Publ. Bot. 29: 145–318.

Roland-Heydacker, F., & M.-T. Cerceau-Larrival. 1978. Ultrastructure du tectum de pollens d'Ombellifères. Grana 17: 81–89.

Singh, D. 1954. Floral morphology and embryology of *Hedera nepalensis* K. Koch. Agra Univ. J. Res., Sci. 3: 289–299.

Thorne, R. F. 1973. Inclusion of the Apiaceae (Umbelliferae) in the Araliaceae. Notes Roy. Bot. Gard. Edinburgh 32: 161–165.

Tseng, C. C. 1967. Anatomical studies of flower and fruit in the Hydrocotyloideae (Umbelliferae). Univ. Calif. Publ. Bot. 42: 1–79.

Tseng, C. C. 1971. Light and scanning electron microscopic studies on pollen of *Tetraplasandra* (Araliaceae) and relatives. Amer. J. Bot. 58: 505–516.

Tseng, C. C., & J. R. Shoup. 1978. Pollen morphology of *Schefflera* (Araliaceae). Amer. J. Bot. 65: 384–394.

Грушвицкий, И. В., В. Н. Тихомиров, Е. С. Аксенов, Г. В. Шибакина. 1969. Сочный плод с карпофором у видов рода *Stilbocarpa* Decne et Planch. (Araliaceae.) Бюлл. Московск. Общ. Исп. Прир., Отд. Биол. 74(2): 64-76.

Козо-Полянский, Б. М. 1943. Морфологическое значение Dentes calycini у Apioideae. (К морфологии цветка зонтичных.) Советск. Бот. 4: 36-41.

Матюшенко, А. Н. 1949. Значение строения черешков для систематики расетений (на примере семейства зонтичных). Бюлл. Общ. естествоисп. Воронежск. Госуд. Унив., 1949: 35-45.

Первухина, Н. В. 1947. Материалы к изучению анатомии плодов зонтичных. Советск. Бот. 15: 27-31.

Первухина, Н. В. 1962. Природа нижней завязи зонтичных и некоторые вопросы "теории цветка." Труды Бот. Инст. АН СССР, сер. VII, 5: 31-45.

Тамамшян, С. Г. 1948. Вторичная гипогиния цветка зонтичных и принцип смены функций у растений. Докл. АН СССР, 61: 537-540.

Тихомиров, В. Н. 1958. Развитие завязи зонтичных в связи с вопросом о ее морфологической природе. Научн. Докл. Высшей Школы, 1958(1): 129-138.

Тихомиров, В. Н. 1961. Морфология завязи зонтичных. В сб.: Морфогенез растений, 2: 378-381. Изд. Московск. Унив.

Тихомиров, В. Н. 1961. К вопросу о происхождении плода зонтичных. В сб.: Морфогенез растений, 2: 478-480. Изд. Московск. Унив.

Тихомиров, В. Н. 1961. Морфогенез плода в семействе Umbelliferae. В сб.: Морфогенез растений, 2: 481-484. Изд. Московск. Унив.

Тихомиров, В. Н. 1961. О систематическом положении родов *Hydrocotyle* L. и *Centella* L. emend. Urban. Бот. Ж. 46: 584-586.

SYNOPTICAL ARRANGEMENT OF THE FAMILIES OF APIALES

1 Carpels 1-many, most often 5; fruit usually a drupe or berry, only rarely a schizocarp with a more or less well developed carpophore; flowers commonly in umbels or heads that are often grouped into various sorts of compound inflorescences, but only seldom forming regular compound umbels; trees, shrubs, or woody vines, only seldom perennial herbs .... 1. ARALIACEAE.

1 Carpels consistently 2; fruit a dry schizocarp, the mericarps usually attached apically to a persistent carpophore; flowers in the largest subfamily generally arranged into compound umbels, but in the 2 smaller subfamilies in heads or simple umbels or other sorts of inflorescences; herbs, rarely shrubs or trees .... 2. APIACEAE.

1. Family ARALIACEAE A. L. de Jussieu 1789 nom. conserv., the Ginseng Family

Trees (mostly soft-wooded and typically pachycaulous), shrubs, woody vines, or sometimes perennial herbs, often stellate-hairy and sometimes strongly prickly, accumulating triterpenoid saponins, other triterpenoid compounds, polyacetylenes (such as falcarinon), and sesquiterpenes, but only seldom producing alkaloids, lacking coumarins, generally not cyanogenic and generally not tanniferous, lacking ellagic acid and usually also lacking proanthocyanins, commonly storing carbohydrte as the trisaccharide umbelliferose; schizogenous secretory canals (or cavities) containing ethereal oils, resins, or gums generally well developed in the parenchymatous tissues of the stem and commonly also in the petioles and larger veins of the leaves; solitary secretory cells wanting; solitary or clustered crystals of calcium oxalate commonly present in some of the cells of the parenchymatous tissues; nodes multilacunar or seldom trilacunar; inverted vascular bundles commonly present in the pith; vessel-segments with simple perforations, or in a few genera some or all of them with scalariform perforations that have few to fairly numerous (up to ca 25) cross-bars; imperforate tracheary elements with small, simple pits, commonly septate; wood-rays heterocellular or rarely (as in *Hedera*) homocellular, mixed uniseriate and pluriseriate, the latter commonly 4–6 (–15) cells wide; wood-parenchyma paratracheal, often very scanty. LEAVES alternate, or rarely opposite or whorled, often very large, pinnately or palmately compound or dissected, or less often simple and palmately lobed (as in *Hedera*) or even entire (as in *Meryta* and spp. of *Myodocarpus*); stomates mostly paracytic or anisocytic, seldom anomocytic; petiole commonly broad and sheathing, sometimes with evident stipular appendages, usually with a ring (less often an arc) of vascular bundles, often with medullary bundles as well. FLOWERS commonly small, mostly in umbels that may be arranged into secondary inflorescences of various sorts, seldom in racemes or spikes or heads or more openly branched inflorescences, perfect or sometimes unisexual, regular or the outermost ones sometimes irregular, epigynous or rarely secondarily hypogynous (as in *Tetraplasandra*), very often but not always pentamerous; pedicel often jointed at the summit; calyx commonly represented by small teeth (isomerous with the petals) around the summit of the ovary, often much-reduced or even obsolete; petals (3–) 5 (–12), distinct (rarely

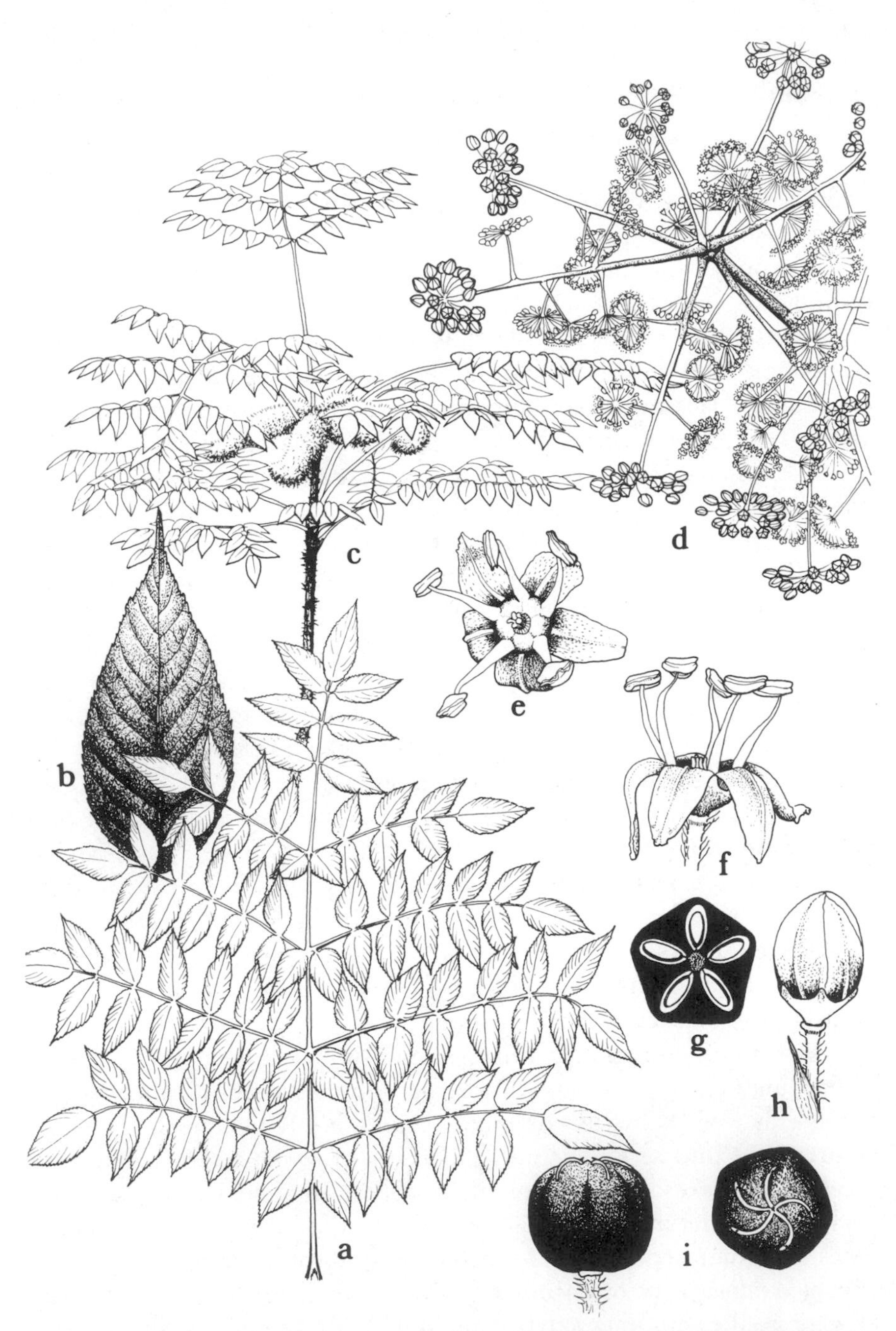

FIG. 5.63 Araliaceae. *Aralia spinosa* L. a, leaf, $\times\frac{1}{12}$; b, leaflet, $\times\frac{1}{2}$; c, habit, $\times\frac{1}{25}$; d, portion of inflorescence, $\times\frac{1}{2}$; e, f, flower, $\times 9$; g, schematic cross-section of ovary, $\times 15$; h, flower bud, $\times 9$; i, side (left) and top views of fruit, $\times 3$.

connate at the base, or even forming an undivided calyptra), deciduous, valvate or sometimes slightly imbricate; stamens commonly as many as and alternate with the petals, or seldom more or less numerous; anthers opening by longitudinal slits, mostly tetrasporangiate and dithecal, but sometimes with as many as 8 microsporangia; pollen-grains (2) 3-nucleate, most commonly tricolporate, the surface commonly reticulate, sometimes perforate; gynoecium of 2–5 (-many) carpels united to form a compound, inferior (rarely superior or only half-inferior) ovary with as many locules as carpels (or the ovary rarely pseudomonomerous and unilocular, as in *Arthrophyllum*); styles as many as the carpels, distinct or sometimes more or less connate into a common style, generally swollen at the base to form a more or less definite stylopodium that is confluent with the epigynous nectary-disk; stigmas wet or dry; intermediate carpellary bundles rejoining the ventral bundles at or above the level of attachment of the ovules; ovules solitary in each locule (an additional abortive one also present) apical-axile, pendulous, epitropous, with ventral raphe and with the micropyle directed upwards and outwards, anatropous, unitegmic, crassinucellar to tenuinucellar, with an integumentary tapetum, usually provided with a funicular obturator; in *Seemannaralia* the partition in the ovary complete only at the level of placentation, and only one mericarp maturing; embryo-sac monosporic or bisporic; endosperm-development nuclear. FRUIT a drupe (usually with as many stones as carpels) or a berry, rarely (as in *Myodocarpus*) a schizocarp with a persistent carpophore as in the Apiaceae, or a drupe with a carpophore (as in spp. of *Stilbocarpa*); seeds with a small, dicotyledonous embryo and abundant, oily endosperm containing large amounts of petroselenic acid. X = 11, 12+, most often 12.

The family Araliaceae consists of about 70 genera and 700 species, widespread in tropical and subtropical regions, with relatively few species in temperate climates. Nearly half of the species belong to only 3 genera, *Schefflera* (150, some cultivated as house-plants), *Oreopanax* (100), and *Polyscias* (75, some cultivated as house-plants). Some other familiar species of the family are *Hedera helix* L., English Ivy; *Oplopanax horridus* (Smith) Miq., devil's club; and *Panax ginseng* C. A. Mey., still prized in some cultures as an aphrodisiac and medical panacea.

The Araliaceae have an extensive fossil record. Megafossils considered to represent various genera of the family occur throughout the Tertiary and even into the Upper Cretaceous. Pollen referred to the Araliaceae dates from the Paleocene.

2. Family APIACEAE Lindley 1836 nom. conserv., the Carrot Family

Aromatic, often poisonous, perennial or less often annual herbs or seldom half-shrubs, rarely shrubs or soft-wooded trees, sometimes with anomalous secondary growth, glabrous or with various sorts of hairs, commonly producing polyacetylenes with a usually 17-carbon skeleton (these responsible for most of the poisonous effects, but the poisonous principle of *Conium* is an alkaloid), and often also with triterpenoid saponins and various sorts of coumarins (these derived from umbelliferone), and sometimes (notably in *Ferula*) producing sesquiterpene lactones, but only seldom cyanogenic and seldom tanniferous, lacking proanthocyanins, gallic acid, and ellagic acid, commonly storing carbohydrate as the trisaccharide umbelliferose; parenchymatous tissues permeated by schizogenous secretory canals containing monoterpenes (these ethereal and providing the characteristic odor of the plant), sesquiterpenes, and phenyl-propanoid compounds; solitary secretory cells wanting; crystals rarely formed; stem commonly with well developed peripheral collenchyma, sometimes with cortical bundles or pith-bundles (the latter sometimes inverted), the internodes generally becoming hollow; nodes mostly multilacunar, seldom trilacunar; vessel-segments mostly with simple perforations, seldom scalariform; imperforate tracheary elements with numerous small, simple pits; wood-rays (in woody forms) heterocellular, mixed uniseriate and pluriseriate, the pluriseriate ones up to 5 (–8) cells wide, with short ends; wood-parenchyma mostly paratracheal and scanty, sometimes vasicentric. LEAVES alternate or rarely opposite, often very large, pinnately or ternately (less often palmately) compound or dissected, or rarely simple (as for example in *Eryngium*, in which the parallel-veined simple leaf may be a modified petiole, and in *Lilaeopsis*, in which the unifacial, hollow leaf represents the rachis alone, with the pinnae suppressed); stomates of various types, most often paracytic or anomocytic or anisocytic; petiole commonly with a ring or arc of vascular bundles, sometimes with medullary bundles as well, usually with a broad, sheathing base, exstipulate or with mere stipular flanges. FLOWERS individually small, in the Apioideae generally borne in compound umbels, but in the Saniculoideae and Hydrocotyloideae usually borne in heads or simple umbels (which may even be reduced to single flowers) or rarely more or less dichasial, consistently epigynous, perfect or seldom unisexual, ordinarily pentamerous except for the dimerous

FIG. 5.64 Apiaceae. *Zizia trifoliata* (Michx.) Fern. a, habit, $\times \frac{1}{2}$; b, flower, from above, $\times 24$; c, flower, from the side, with 2 petals and 2 stamens removed; $\times 24$; d, e, anther, $\times 24$; f, pistil, in partial long-section, $\times 16$; g, fruit, with a bifid carpophore and 2 mericarps, $\times 8$; h, schematic cross-section of mericarp, $\times 8$.

gynoecium, regular or the marginal ones sometimes sterile or neutral and then often with an irregular, marginally expanded corolla, rarely individually unisexual and grouped into pseudanthia; umbels and umbellets often subtended by an involucre and involucel of distinct or connate bracts; calyx commonly represented by small teeth around the top of the ovary, often much-reduced or even obsolete; petals commonly white or yellow, less often purple or of other colors, distinct, typically inflexed at the tip, valvate, or rarely wanting; stamens alternate with the petals, borne on the epigynous nectary-disk; anthers tetrasporangiate and dithecal, opening by longitudinal slits; pollen-grains trinucleate, mostly tricolporate, the surface smooth or nearly so, the tectate-columellate exine-structure described as *simple* in some genera, variously *structured* in others; gynoecium of 2 carpels united to form a compound, inferior, bilocular ovary (rarely the ovary unilocular and pseudomonomerous); ovary superior in earliest ontogeny, becoming inferior by unequal interprimordial growth (sensu Sattler), the septum with an independent ontogenetic origin; styles distinct, often swollen at the base to form a stylopodium that is confluent with the nectary-disk, the stylopodium ontogenetically more nearly ovarian than stylar; stigmas wet, papillate; intermediate carpellary bundles rejoining the ventral bundles below the attachment of the ovules; ovules solitary in each locule (an additional abortive ovule also present), apical-axile, pendulous, epitropous, with ventral raphe and with the micropyle directed upwards and outwards, anatropous, with a massive single integument, crassinucellar to more often tenuinucellar, with an integumentary tapetum; embryo-sac of diverse types; endosperm-development nuclear. FRUIT a dry schizocarp, consisting of 2 mericarps united by their faces (the commissure); mericarps almost always separating at maturity, typically revealing a slender, central carpophore to which they are apically attached, the carpophore entire to deeply bifid, or sometimes obsolete by adnation of the separate halves to the commissural faces of the mericarps; carpophore wanting in most Hydrocotyloideae; seed-coat commonly adherent to the pericarp; embryo dicotyledonous, small or seldom more elongate and centric; endosperm copious, oily, containing large amounts of petroselenic and often also petroseledic acid. X = 4–12, most commonly 8 or 11. (Umbelliferae, nom. alt.; Hydrocotylaceae, Saniculaceae)

The family Apiaceae (often called Umbelliferae, the traditional name) consists of about 300 genera and 3000 species, nearly cosmopolitan in distribution, but best developed in North Temperate regions and to a lesser extent on tropical mountains. Many species are cultivated for food or spice. Among these are: *Anethum graveolens* L., dill; *Apium graveolens* L., celery; *Carum carvi* L., caraway; *Coriandrum sativum* L., coriander; *Daucus carota* L., carrot; *Petroselinum crispum* (Mill.) Nyman, parsley; and *Pastinaca sativa* L., parsnip. Many other members of the family, such as *Cicuta* and *Conium*, are highly poisonous. *Conium* is the hemlock of classical antiquity.

The Apiaceae consist of one large subfamily, the Apioideae, with some 250 genera, and two much smaller ones, the Saniculoideae, with about 9 genera, and the Hydrocotyloideae, with about 34. The latter two subfamilies mostly do not have the characteristic compound umbel that makes the Apioideae so obviously distinctive. The Hydrocotyloideae, in particular, stand apart because of their chiefly southern-hemisphere distribution and because of their fruit, which has a woody endocarp and usually lacks a free carpophore. The fruit is thus somewhat intermediate between that of typical Apiaceae and that of Araliaceae. It has been argued that the Hydrocotyloideae originated from the Araliaceae independently of the other Apiaceae, and that therefore one should either recognize an additional family Hydrocotylaceae or submerge the Apiaceae in the Araliaceae. The great majority of the Apiaceae form such an obvious group that I would not find it useful to put them into the same family with the Araliaceae, but the possibility that the Hydrocotyloideae should be recognized as a distinct family merits careful consideration by those with special knowledge of the order.

Pollen identified with the Apiaceae dates from the Eocene, one epoch later than the Araliaceae

VI. Subclass ASTERIDAE Takhtajan 1966

Woody or herbaceous dicotyledons with diverse sorts of repellents, these very often including iridoid compounds and/or various kinds of alkaloids (but not benzyl-isoquinoline alkaloids) or polyacetylenes or glycosides, but only seldom cyanogenic, seldom saponiferous, and seldom tanniferous, usually lacking both ellagic acid and proanthocyanins, and without betalains and mustard-oil; nodes in most families unilacunar; vessel-segments with simple or seldom scalariform or reticulate perforations; wood-rays generally with short ends; sieve-tubes with S-type plastids. LEAVES simple or less often variously compound or dissected, seldom much-reduced or highly modified. FLOWERS hypogynous to epigynous, but not strongly perigynous, usually well developed and often showy, seldom much-reduced; corolla usually sympetalous, rarely polypetalous and small, or wanting; stamens usually attached to the corolla-tube (sometimes only at the base of the tube, or even free), as many as and alternate with the lobes, or often some members of the cycle missing or staminodial, so that there are fewer stamens than corolla-lobes (paired half-stamens borne at the sinuses in *Adoxa*), in apetalous flowers the stamens never more than 5 except in the Theligonaceae, which may be pseudanthial; pollen-grains binucleate or trinucelate, triaperturate (typically tricolporate), or of triaperturate-derived type; flowers often with a nectary-disk that may represent a reduced cycle of stamens, but sometimes with other types of nectaries; carpels mostly 2–5 (very often 2), generally united to form a compound ovary with a single entire to cleft style, rarely with separate styles or sessile stigmas, rarely the carpels more than 5 or the ovary pseudomonomerous, or the carpels sometimes largely or wholly distinct except for the common style or stigma, the style then sometimes gynobasic, sometimes terminal; placentation variously axile, parietal, basal, apical, or even free-central; ovules 1-many in each locule (or some locules empty), typically tenuinucellar and with a massive single integument, very often with an integumentary tapetum, seldom bitegmic or crassinucellar or pseudocrassinucellar; endosperm-development nuclear or often cellular. Seeds with or without endosperm; embryo usually fairly large in relation to the size of the seed, embedded in the endosperm when endosperm is present, variously straight or curved, generally with

2 cotyledons, but in some parasitic or strongly mycotrophic groups not differentiated into parts.

The subclass Asteridae as here delimited consists of 11 orders, 49 families, and nearly 60,000 species. In terms of number of species it is about the same size as the Rosidae, but in terms of number of families it is surpassed by both the Rosidae and the Dilleniidae. About a third of the species of Asteridae belong to the family Asteraceae, which is the largest family of dicotyledons and one of the two largest families of plants.

The bulk of the Sympetalae of the traditional Englerian system belongs to the Asteridae, but the Diapensiales, Ericales, Ebenales, Primulales, Plumbaginales, and Cucurbitales are removed to other subclasses. Orders 6 to 11 of the Sympetalae in the current (1964) edition of the Engler Syllabus are collectively almost co-extensive with the Asteridae as here defined.

From the standpoint of practical recognition, the vast majority of the Asteridae can be distinguished from the vast majority of other dicotyledons by their sympetalous flowers, in which the stamens are isomerous and alternate with the corolla-lobes, or fewer than the corolla-lobes. Much less than 1 percent of the species of Asteridae fail this test, and probably no more than 1 percent of the species that meet the test do not belong to the Asteridae. The tenuinucellar ovule with a massive single integument is a further marker of the group, but there are more exceptions to this feature, both within and without the Asteridae.

Chemically, the Asteridae are noteworthy for the frequent occurrence of iridoid compounds, the usual absence of ellagic acid and proanthocyanins, and the apparently complete absence of betalains, mustard-oils, and benzyl-isoquinoline alkaloids. The absence of betalains and benzyl-isoquinoline alkaloids tends to set the Asteridae off from most of the Caryophyllidae and Magnoliidae, respectively, and the usual absence of ellagic acid and proanthocyanins tends to set them off from the Rosidae, Dilleniidae, and Hamamelidae. The distinction is far from absolute, however, because these substances are also missing from many members of the subclasses they are supposed to characterize. Iridoid compounds are certainly better represented in the Asteridae than in any other subclass, but they are lacking from the very large family Asteraceae, and will probably prove to occur in not more than about half the species of the subclass. Iridoids also occur in some members of the Rosidae and Dilleniidae, and even a few of the Hamamelidae.

The Asteridae are the most advanced subclass of dicotyledons, and possibly the most recently evolved (only the Caryophyllidae may be more recent). More than any other subclass, they exploit specialized pollinators and specialized means of presenting the pollen. It seems likely that the rise of the Asteridae is closely correlated with the evolution of insects capable of recognizing complex floral patterns.

It may be reasonable to surmise that in the continuing struggle between plants and their predators, the Asteridae abandoned the familiar weapons of tannins, saponins, and cyanide, in favor of the previously not much-exploited iridoid compounds and the alkaloids related to them. Continuing the speculation, one may suppose that, by the time the Asteraceae originated (at the end of the Oligocene), predators had already become adapted to tolerate iridoids and their allied alkaloids, so that the stage was set for another massive shift in defensive weapons. The tremendous success of the Asteraceae might well depend more on their chemical arsenal than on their specialized floral structure.

The ancestry of the Asteridae very probably lies in the order Rosales, sensu latissimo. Iridoid compounds, a sympetalous corolla, stamens isomerous and alternate with the petals, a compound pistil with numerous ovules on axile placentas, unitegmic ovules, and tenuinucellar ovules all occur in this order, but not in combination. The nectary-disk of many Asteridae also finds a ready precedent in the Rosales, as do the simple, stipulate, opposite leaves of the more archaic members of the

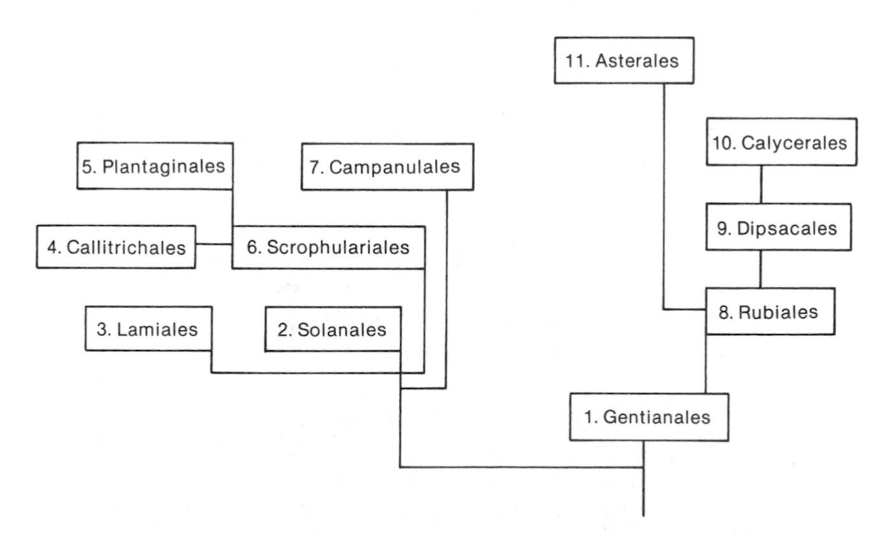

FIG. 6-1 Putative evolutionary relationships among the orders of Asteridae.

group. The combination of these features into a functional whole marks the transition from the ancestral Rosales to the first members of the Asteridae. All the other orders of Rosidae are already too advanced to serve as possible ancestors of the Asteridae.

The Asteridae make a relatively late entrance into the fossil record. There is no reason to suppose that the group originated before the beginning of the Tertiary period. Fossil pollen referred to the Apocynaceae dates from the Paleocene, and pollen or macro-fossils of families in several other orders turn up in the Eocene. The Asteridae began to play a prominent role in the flora of the world only during the Oligocene epoch.

SELECTED REFERENCES

Dahlgren, R. 1977. A note on the taxonomy of the "Sympetalae" and related groups. Cairo Univ. Herb. 7 & 8: 83–102.

Robyns, W. 1972. Outline of a new system of orders and families of Sympetalae. Bull. Jard. Bot. Nat. Belg. 42: 363–372.

Wagenitz, G. 1977. New aspects of the systematics of Asteridae. Pl. Syst. Evol., Suppl. 1: 375–395.

SYNOPTICAL ARRANGEMENT OF THE ORDERS OF ASTERIDAE

1 Flowers much-reduced, nearly or quite without perianth, either unisexual, or perfect and epigynous; stamen solitary, or seldom stamens 2–3; plants mostly aquatic .... 4. CALLITRICHALES.

1 Flowers generally with a more or less well developed perianth (typically a sympetalous corolla and a calyx or a pappus), but if essentially without perianth, then the stamens more than 3; plants terrestrial or seldom aquatic.

  2 Ovary with some exceptions (mainly some Gesneriaceae), superior.

    3 Plants nearly always with opposite or whorled leaves, and nearly always with internal phloem; flowers mostly regular or nearly so and with as many stamens as corolla-lobes; endosperm-development nuclear, rarely cellular; ovules usually without an integumentary tapetum, except in Apocynaceae; plants commonly producing alkaloids or iridoid compounds or both .... 1. GENTIANALES.

    3 Plants only rarely at once with opposite (or whorled) leaves and internal phloem, and then with irregular flowers that have fewer stamens than corolla-lobes; endosperm-development cellular or much less often nuclear; ovules often with an integumentary tapetum; plants with or without alkaloids and iridoid compounds.

4 Ovary generally consisting of 2 (–14) biovulate carpels with twice as many uniovulate lobes or locelli as carpels (a few exceptions in the Verbenaceae, but the carpels in any case with not more than 2 ovules each), very often (most Boraginaceae and Lamiaceae) the ovary consisting of 4 essentially distinct, half-carpellary lobes united mainly by their gynobasic style; fruit typically consisting of separate, closed half-carpels (nutlets) or of a drupe in which each seed has its own stone or its own locule of a compound stone; plants nearly always without internal phloem .... 3. LAMIALES.

4 Ovary consisting of 2–4 (–8) carpels with (1) 2-many ovules each (rarely the ovary pseudomonomerous), but the carpels only rarely divided into uniovulate segments; fruits diverse, most commonly capsular or baccate, not consisting of half-carpellary nutlets except in a few Convolvulaceae and Myoporaceae; plants sometimes (the Convolvulaceae consistently) with internal phloem.

5 Corolla scarious, persistent, generally regular; flowers mostly anemophilous, generally tetramerous as to the calyx, corolla, and androecium; leaves phyllodial, more or less parallel-veined, or sometimes much-reduced, very often all basal .... 5. PLANTAGINALES.

5 Corolla otherwise; flowers mostly entomophilous or ornithophilous, variously pentamerous or tetramerous or otherwise, with isomerous or anisomerous stamens; leaves various in form and structure, sometimes much-reduced, but not phyllodial and parallel-veined, and only seldom all basal.

6 Flowers mostly regular or nearly so and with as many functional stamens as corolla lobes, typically pentamerous; the principal exceptions are some Solanaceae with an irregular, 5-lobed corolla and only 4 stamens, but the Solanaceae have internal phloem, and the carpels, when (as usually) 2, are oblique to the median axis of the flower; plants often producing alkaloids, but only seldom iridoid compounds, and without orobanchin .... 2. SOLANALES.

6 Flowers mostly irregular and with fewer functional stamens than corolla-lobes, or sometimes with a regular, tetramerous corolla and 2 or 4 stamens, or the corolla rarely wanting; carpels, when (as usually) 2, median, not oblique; plants commonly producing orobanchin

and iridoid compounds, but only seldom alkaloids .....
....6. SCROPHULARIALES.

2 Ovary mostly inferior, seldom only half-inferior, superior only in a few Campanulales and Rubiales.

7 Flowers borne in various sorts of inflorescences, but if in involucrate heads then the heads generally basically cymose in structure; flowers with or without a specialized pollen-presentation mechanism similar to that of the next group; ovary with 1-several locules and 1-many ovules in each locule (or some locules empty).

8 Leaves, with few exceptions, alternate; stamens free from the corolla, or attached at the base of the tube (higher in the Pentaphragmataceae); flowers very often with a specialized pollen-presentation mechanism, the anthers connivent or connate into a tube around the style, which pushes out the pollen; plants usually herbaceous, sometimes secondarily woody, characteristically storing carbohydrates as inulin ...
....7. CAMPANULALES.

8 Leaves, with few exceptions, opposite or whorled; stamens attached to the corolla-tube, usually well above the base; flowers without the sort of pollen-presentation mechanism described above, except in the subfamily Ixoroideae of the Rubiaceae; plants variously woody or herbaceous, without inulin.

9 Stipules usually present and interpetiolar, bearing colleters on the inner surface (sometimes intrapetiolar, or reduced to mere interpetiolar lines, or enlarged into leaves); corolla typically regular and with isomerous stamens (much-reduced or wanting in Theligonaceae); endosperm-development nearly always nuclear; plants chiefly tropical and woody, but some members herbaceous, or of temperate regions, or both .... 8. RUBIALES.

9 Stipules typically none, when present usually small and adnate (at least basally) to the petiole, in any case without colleters; corolla regular or often irregular, the stamens as many as or often fewer or more numerous than the corrolla-lobes; endosperm-development cellular; plants woody or often herbaceous, of temperate or less often tropical regions .... 9. DIPSACALES.

7 Flowers borne in involucrate, centripetally flowering (rarely uniflorous) heads; anthers connate (or connivent) into a tube around the style, which pushes out the pollen; ovary unilo-

cuar, with a solitary ovule; plants characteristically storing carbohydrate as inulin.

10 Ovule apical and pendulous; pollen-grains binucleate; plants without a well developed secretory system, producing iridoid compounds but not polyacetylenes or sesquiterpene lactones; herbs with alternate leaves .... 10. CALYCERALES.

10 Ovule basal and erect; pollen-grains trinucleate; plants with a well developed secretory (either resiniferous or laticiferous) system, characteristically producing sesquiterpene lactones, usually also polyacetylenes, and often other chemically repellents, but without iridoid compounds; woody or herbaceous, with alternate, opposite, or whorled leaves .... 11. ASTERALES.

1. Order GENTIANALES Lindley 1833

Plants diverse in growth-habit, commonly producing iridoid compounds or one or another sort of alkaloid (including indole, oxindole, pyridine, steroid, and phenanthro-indolizidine groups) or both, and often producing cardiotonic glycosides, but not strongly tanniferous, without ellagic acid and usually without proanthocyanins, only seldom cyanogenic or saponiferous; calcium oxalate crystals of various forms commonly present in some of the cells of the parenchymatous tissues; vessel-segments with simple perforations, or seldom with scalariform perforations that mostly have only a few cross-bars; wood-rays mostly narrow and only 1–5 cells wide, or wanting, but sometimes up to 12 cells wide, or some even wider medullary rays sometimes present; internal phloem nearly always present, either as a continuous ring or as separate strands at the margin of the pith; cork usually superficial in origin. LEAVES opposite or sometimes whorled, only rarely alternate, simple and usually entire, with or more often without interpetiolar stipules, the stipules when present often with colleters. FLOWERS mostly regular or nearly so, commonly showy, sympetalous, with convolute or less often imbricate or valvate corolla-lobes, most commonly 5-merous (except as to the gynoecium), not infrequently 4-merous, seldom otherwise; stamens generally as many as and alternate with the corolla-lobes, the filaments attached to the corolla-tube; pollen-grains binucleate or often trinucleate, most often tricolporate; gynoecium most commonly bicarpellate, with one carpel anterior and the other posterior, but sometimes with up to 5 or even 8 carpels, the carpels united to form a compound ovary with axile or parietal (rarely free-central) placentas, or often the carpels distinct below and united only by their common style or style-head; ovary or ovaries wholly superior, or slightly sunken into the receptacle at the base; ovules (1–) numerous in each locule or on each placenta, anatropous to hemitropous or amphitropous, with a massive simple integument, tenuinucellar or (Asclepiadaceae) pseudocrassinucellar, without an integumentary tapetum except in the Apocynaceae; endosperm-development nuclear, rarely cellular. FRUIT most commonly a capsule or follicle, but sometimes of other types; seeds usually numerous, often comose or winged; embryo dicotyledonous (much-reduced in a few mycotrophic genera); endosperm well developed and oily to scanty or none.

The order Gentianales as here defined consists of 6 families and about 5500 species. The Apocynaceae and Asclepiadaceae have about

2000 species each, the Gentianaceae about 1000, the Loganiaceae about 500, and the Saccifoliaceae and Retziaceae only one each. The four larger families form a distinctive group whose members are obviously allied inter se. The Saccifoliaceae and Retziaceae are individually discussed.

The Buddlejaceae, Menyanthaceae, and Oleaceae, which have often been included in the Gentianales, are here excluded. They lack internal phloem, and they have an integumentary tapetum and cellular endosperm. The Menyanthaceae further differ in having alternate leaves, and the Oleaceae usually have only 2 stamens. Furthermore, there is good reason to suppose that the tetramerous, regular flower of most Buddlejaceae reflects reduction from a flower with pentamerous, irregular corolla and 4 stamens. The Menyanthaceae are here referred to the Solanales, and the Buddlejaceae and Oleaceae to the Scrophulariales.

Reasons for excluding the Rubiaceae from the Gentianales are discussed under the Rubiales.

The biological significance of the morphological and anatomical characters that distinguish the Gentianales as a group is doubtful. Diverse species of the order occur in a wide range of habitats that are also occupied by species of various other orders, and there is no obvious difference in the way that the Gentianales use these habitats, in contrast to other orders. The Gentianales heavily exploit iridoid compounds, cardiotonic glycosides, and certain groups of alkaloids as repellants of would-be predators, but repellents of these groups also occur in various other orders that are not closely allied to the Gentianales.

The principal characters marking the family Asclepiadaceae are clearly related to pollination, ensuring that many pollen-grains are transported in a group. The usually large number of ovules in an ovary is an obvious corollary of the method of pollination, but there is no evident advantage in having the carpels virtually distinct, so that one may remain barren even when the other is pollinated. The mechanism of pollination is so complex and subject to interruption that it frequently fails to function, and many flowers do not bear fruit. The large number of species of Asclepiadaceae bears witness to the viability of the mechanism of pollination, yet the Asclepiadaceae are no more successful than the equally large, closely related family Apocynaceae, in which pollination occurs in a more ordinary way. It seems reasonable to surmise that once the method of mass-pollination has been adopted there is a selective pressure toward perfecting the mechanism. Thus it should not be surprising that within the Asclepiadaceae the subfamily

Periplocoideae, with imperfectly developed pollinia, is a much smaller group than the Asclepiadoideae.

There is a fairly straight-line evolutionary series in floral morphology within the Apocynaceae and Asclepiadaceae collectively, from the Plumerioideae to the Apocynoideae to the Periplocoideae to the Secamoneae and thence to the other tribes of the Asclepiadoideae. A determined splitter could recognize 5 families here, and an equally determined lumper might see only one. It has long been customary, and I believe it is conceptually useful, to recognize just two families in the group. The distinction is drawn on the method of pollination, specifically on the presence of a translator in the Asclepiadaceae and its absence from the Apocynaceae. The several other differences between characteristic members of the two families present finely graded series or are subject to exception. The selective value of the progressive evolutionary separation of the carpels is wholly obscure. The evolutionary sequence of the separation, from the bottom of the gynoecium toward the top, is unique.

The fossil record as presently known and interpreted casts a modest amount of light on the evolutionary history of the Gentianales. The Apocynaceae can be traced back to the Paleocene on the basis of pollen, and to the Eocene on the basis of macrofossils. Pollen referred to the Asclepiadaceae dates from the Oligocene. Macrofossils of the Loganiaceae are recognized in the Eocene, of the Asclepiadaceae in the Oligocene, and of the Gentianaceae in the Miocene. Thus it appears that the group as a whole does not antedate the Tertiary, that the Apocynaceae and Loganiaceae may be of roughly comparable age, and that the Asclepiadaceae and Gentianaceae are more recently evolved. Such a pattern is, at least, compatible with the present-day morphology and chemistry of the order. The Gentianales are the only order of Asteridae that can be traced back beyond the Eocene on the basis of pollen.

SELECTED REFERENCES

Bendre, A. M. 1975. Studies in the family Loganiaceae. II. Embryology of *Buddleja* and *Strychnos*. J. Indian Bot. Soc. 54: 272–279.

Bissett, N. G. 1958, 1961. The occurrence of alkaloids in the Apocynaceae. Ann. Bogor. 3: 105–236, 1958. 4: 65–144, 1961.

Boke, N. 1948. Development of the perianth in *Vinca rosea* L. Amer. J. Bot. 35: 413–423.

Devi, H. M. 1962. Embryological studies in the Gentianaceae (Gentianoideae and Menyanthoideae). Proc. Indian Acad. Sci. B, 56: 195–216.

Devi, H. M. 1964. Embryological studies in Asclepiadaceae. Proc. Indian Acad. Sci. B, 60: 54–65.
Devi, H. M. 1971. Embryology of Apocynaceae. I. Plumiereae. J. Indian Bot. Soc. 50: 74–85.
Gopal Krishna, G., & V. Puri. 1962. Morphology of the flower of some Gentianaceae with special reference to placentation. Bot. Gaz. 124: 42–57.
Guédès, M. 1972. Stipules et ligules vraies chez quelques Apocynacées et Asclépiadacées. Compt. Rend. Hebd. Séances Acad. Sci. 274 Ser. D: 3218–3221.
Guédès, M., & J.-P. Gourret. 1973. Architecture des phyllomes végétatifs et périanthaires: cas de *Catharanthus roseus* (L.) G. Don (*Vinca rosea* L.) Flora 162: 309–334.
Hasselberg, G. B. E. 1937. Zur Morphologie des vegetativen Sprosses der Loganiaceen. Symb. Bot. Upsal. 2(3), viii–170 pp.
Joshi, A. C. 1936. Anatomy of the flowers of *Stellera chamaejasme*. J. Indian Bot. Soc. 15: 77–85.
Jovet, P. 1941. Aux confins des Rubiacées et des Loganiacées. Notul. Syst. (Paris) 10: 39–53.
Klett, W. 1924. Umfang und Inhalt der Familie der Loganiaceen. Bot. Arch. 5: 312–338.
Leeuwenberg, A. J. M. 1969. Notes on American Loganiaceae. IV. Revision of *Desfontainia* Ruiz et Pav. Acta Bot. Neerl. 18: 669–679.
Lindsey, A. A. 1940. Floral anatomy in the Gentianaceae. Amer. J. Bot. 27: 640–652.
Löve, D. 1953. Cytotaxonomical remarks on the Gentianaceae. Hereditas 39: 225–235.
Maguire, B., & J. M. Pires. 1978. Saccifoliaceae. In: The botany of the Guayana Highland—Part X. Mem. New York Bot. Gard. 29: 230–245.
Mohrbutter, C. 1936. Embryologische Studien an Loganiaceen. Planta 26: 64–80.
Moore, R. J. 1947. Cytotaxonomic studies in the Loganiaceae. I. Chromosome numbers and phylogeny in the Loganiaceae. Amer. J. Bot. 34: 527–538.
Nilsson, S. 1967. Pollen morphological studies in the Gentianaceae—Gentianinae. Grana Palynol. 7:46–143.
Nilsson, S., & J. J. Skvarla. 1969. Pollen morphology of saprophytic taxa in the Gentianaceae. Ann. Missouri Bot. Gard. 56: 420–438.
Ornduff, R. 1970. The systematics and breeding system of *Gelsemium* (Loganiaceae). J. Arnold Arbor. 51: 1–17.
Pichon, M. 1948–1950. Classification des Apocynacées I. Carissées et Ambélaniées. IX. Rauvolfiées, Alstoniées, Allamandées et Tabernémontanoidées. XXV. Échitoidées. Mém. Mus. Natl. Hist. Nat. N. Sér. 24: 111–181, 1948. 27: 153–251, 1948. Sér. B. 1: 1–174, 1950.
Post, D. M. 1958. Studies in Gentianaceae. I. Nodal anatomy of *Frasera* and *Swertia perennis*. Bot. Gaz. 120: 1–14.
Punt, W. 1978. Evolutionary trends in the Potalieae (Loganiaceae). Rev. Palaeobot. Palynol. 26: 313–335.
Punt, W., & P. W. Leenhouts. 1967. Pollen morphology and taxonomy in the Loganiaceae. Grana Palynol. 7: 469–516.
Rao, V. S., & A. Ganguli. 1963. The floral anatomy of some Asclepiadaceae. Proc. Indian Acad. Sci. 57: 15–44.
Safwat, F. M. 1962. The floral morphology of *Secamone* and the evolution of the pollinating apparatus in Asclepiadaceae. Ann. Missouri Bot. Gard. 49: 95–129.

Tournay, R., & A. Lawalrée. 1952. Une classification nouvelle des familles appartenant aux ordres des Ligustrales et des Contortées. Bull. Soc. Bot. France 99: 262–263.

Vijayaraghavan, M. R., & U. Padmanaban. 1969. Morphology and embryology of *Centaurium ramosissimum* Druce and affinities of the family Gentianaceae. Beitr. Biol. Pflanzen 46: 15–37.

Woodson, R. E. 1930. Studies in the Apocynaceae. I. A critical study of the Apocynoideae (With special reference to the genus *Apocynum*). Ann. Missouri Bot. Gard. 17: 1–212.

Woodson, R. E. 1941. The North American Asclepiadaceae. I. Perspective of the genera. Ann. Missouri Bot. Gard. 28: 193–244.

Woodson, R. E. 1954. The North American species of *Asclepias* L. Ann. Missouri Bot. Gard. 41: 1–211.

Woodson, R. E., & J. A. Moore. 1938. The vascular anatomy and comparative morphology of apocynaceous flowers. Bull. Torrey Bot. Club 65: 135–166.

Yamazaki, T. 1963. Embryology of *Mitrasacme alsinoides* var. *indica*. Sci. Rep. Tôhoku Imp. Univ., Ser. IV, Biol. 29: 201–205.

SYNOPTICAL ARRANGEMENT OF THE FAMILIES OF GENTIANALES

1 Plants without a latex-system, and without cardiotonic glycosides; style not especially thickened and modified distally; carpels fully united except for the often distinct or lobed stigmas, or seldom the style more deeply cleft.

2 Leaves not saccate-vaginate, nearly always opposite or whorled.

3 Leaves mostly with interpetiolar stipules, these sometimes reduced to mere connecting lines; ovary 2–3 (–5)-locular, with axile placentas (the partition sometimes imperfect in the upper part of the ovary); plants of diverse habit, often woody, often producing alkaloids as well as iridoid compounds .... 1. LOGANIACEAE.

3 Leaves exstipulate; plants commonly producing iridoid compounds but not alkaloids.

4 Ovary bilocular in the lower third or half, unilocular above, with 4 (–6) ovules attached in 2 pairs (or trios) at the slightly expanded summit of the partial partition, one ovule of each pair ascending into the unilocular part of the ovary, the other one (or 2) descending into one of the semilocules; low shrubs with whorled leaves and without internal phloem .... 2. RETZIACEAE.

4 Ovary generally unilocular, with parietal (often deeply intruded) or rarely free-central placentas, only seldom bilocular and with axile placentas; plants usually herbaceous, seldom woody, with opposite or seldom whorled (rarely alternate) leaves, and the stem with internal phloem .... 3. GENTIANACEAE.

2 Leaves saccate-vaginate, alternate, exstipulate; pulvinate subshrub; ovary bilocular, with axile placentas 4. SACCIFOLIACEAE.

1 Plants with a well developed latex system, and commonly producing cardiotonic glycosides as well as alkaloids; style, except in some Apocynaceae, thickened and modified at the tip; carpels often distinct toward the base and united only distally.

5 Androecium without translators, the pollen not forming pollinia; androecial corona wanting; carpels often united by part or all of the style below the thickened head, or even wholly united; plants commonly producing iridoid compounds .... 5. APOCYNACEAE.

5 Androecium provided with translators, the pollen coherent to form pollinia; androecial corona usually well developed, seldom wanting; carpels united only by the thickened style head; plants without iridoid compounds .... 6. ASCLEPIADACEAE.

1. Family LOGANIACEAE Martius 1827 nom. conserv., the Logania Family

Trees, shrubs, lianas, or herbs, glabrous or with unicellular or uniseriate hairs, sometimes accumulating aluminum, not tanniferous, lacking both proanthocyanins and ellagic acid, seldom cyanogenic and only rarely saponiferous, but commonly accumulating bitter substances, these consisting of iridoid compounds and alkaloids of the indole and oxindole groups, notably including the tryptophan alkaloids; calcium oxalate crystals of various forms commonly present in some of the cells of the parenchymatous tissues; cork usually superficial in origin; nodes unilacunar, with 1-several traces, but sometimes multilacunar, as in *Fagraea*; vessel-segments with simple or less often scalariform perforations, the latter with relatively few cross-bars; imperforate tracheary elements with simple or bordered pits; wood-rays heterocellular to homocellular, sometimes all uniseriate, sometimes many of them pluriseriate and up to 12 cells wide, with short ends; wood-parenchyma of diverse types; internal phloem characteristically present either as a continuous ring or as isolated bundles at the margin of the pith. LEAVES opposite, simple and entire or rarely (*Desfontainia*) spinulose-toothed; stomates of diverse types; stipules generally present, often provided with colleters, very often interpetiolar, sometimes reduced to mere interpetiolar lines. FLOWERS solitary or often in cymose inflorescences, commonly showy, perfect, regular or seldom (*Usteria*) with one of the calyx-lobes enlarged and petaloid; sepals 4–5, usually connate to form a toothed or more or less deeply cleft calyx with imbricate teeth or lobes, or the calyx seldom only 2 lobed; corolla sympetalous, the tube short to usually more or less elongate, with 4–5 (–15) imbricate, convolute, or valvate lobes; stamens as many as and alternate with the corolla-lobes (only 1 stamen in *Usteria*), attached to the corolla-tube or -throat; anthers tetrasporangiate and dithecal, opening by longitudinal slits; pollen-grains binucleate or trinucleate, tricolporate to less often tricolpate or triporate, seldom 2–6-porate; nectary disk generally wanting or only poorly developed; gynoecium of 2–3 carpels (5 in *Desfontainia*) united to form a compound, superior (rarely half-inferior) ovary with as many locules as carpels (the partitions sometimes imperfect in the upper part of the ovary); style terminal, capitate or shortly lobed, or seldom the styles more deeply cleft or twice two-lobed; ovules numerous to seldom few or solitary on axile (or distally parietal) placentas, anatropous to hemitropous, unitegmic, without an integumentary tapetum, tenuinucellar; endosperm-

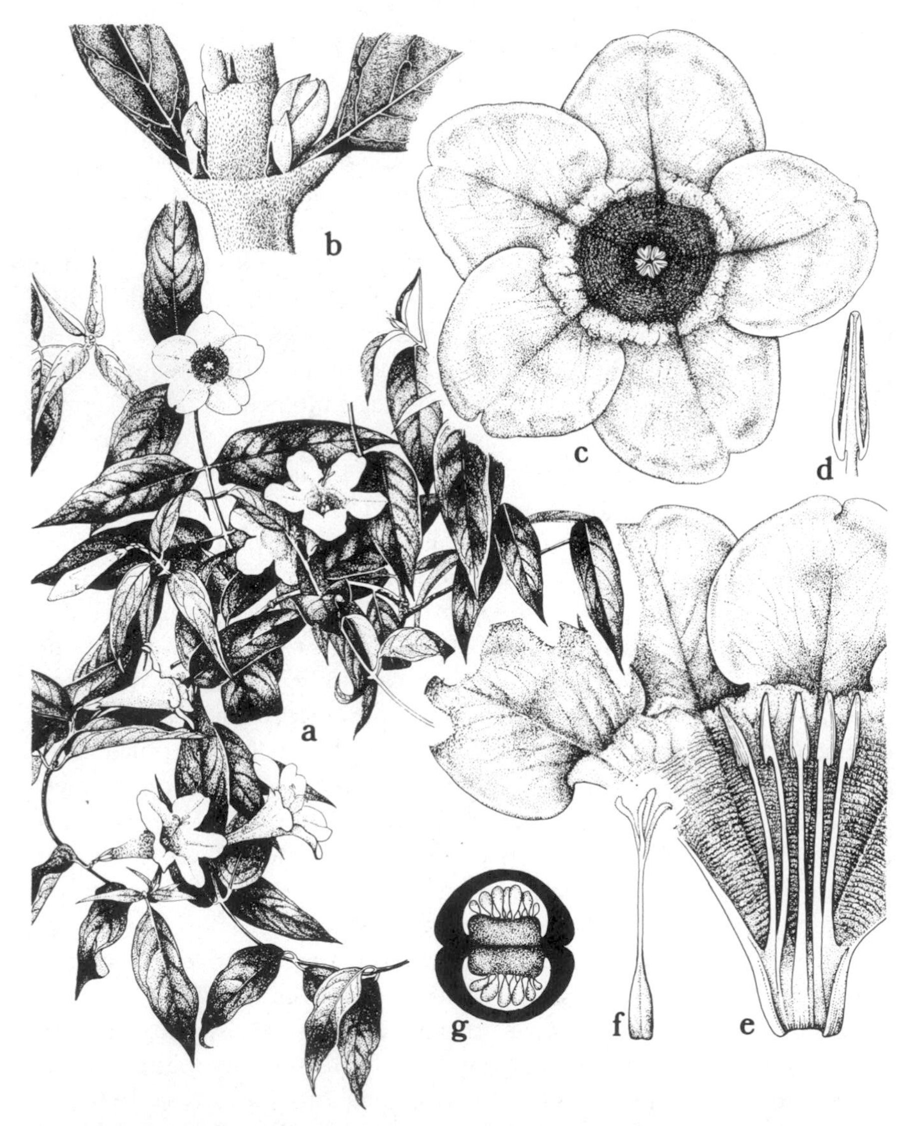

FIG. 6-2 Loganiaceae. *Gelsemium sempervirens* (L.) Aiton. a, habit, ×½; b, node, showing small stipules, ×8; c, top view of flower, ×2; d, anther, ×4; e, corolla, opened up, showing attached stamens, ×2; f, pistil, ×2; g, schematic cross-section of ovary, ×16.

development usually nuclear, but said to be helobial in *Mitrasacme*. FRUIT a capsule (this most often septicidal, sometimes loculicidal or circumscissile) or a berry, seldom a drupe; seeds sometimes winged at both ends or all around; endosperm with reserves of protein, oil, saccharose, and hemicellulose, but not starch; embryo dicotyledonous, straight or nearly so, often small, axially embedded in the endosperm. X = 6–12. (Antoniaceae, Desfontainiaceae, Potaliaceae, Spigeliaceae, Strychnaceae)

The family Loganiaceae as here defined consists of about 20 genera and 500 species, widespread in tropical and subtropical regions, with relatively few species in temperate climates. The largest genus by far is *Strychnos*, pantropical, with 150–200 species. Strychnine is obtained from *S. nux-vomica* L. Some of the alkaloids of curare are obtained from *Strychnos toxifera* Schomb. and related species.

The Loganiaceae form a loosely knit family. Some authors recognize as many as 5 additional segregate-families. The most distinctive of these is the Desfontainiaceae, consisting of the single genus *Desfontainia* (1–5 Andean and Costa Rican spp.), which lacks internal phloem and has spiny-toothed leaves and a 5-locular ovary. All of these segregates are generally considered to be allied to the Loganiaceae sens. strict. The question is the rank at which the groups should be received. I do not believe that their recognition as families facilitates understanding of the group.

The Buddlejaceae, often also included in the Loganiaceae, are here considered not only to form a distinct family, but also to belong to a different order, the Scrophulariales.

2. Family RETZIACEAE Bartling 1830, the Retzia Family

Simple or sparingly branched, evergreen, ericoid shrubs with several erect or closely ascending stout stems clustered on a large, persistent root-crown that regenerates new stems after fire; plants producing several iridoid compounds, including stilbericoside (known also in *Stilbe*, of the Verbenaceae) and unedoside (known also in *Stilbe* and in *Arbutus*, of the Ericaceae); most of the vessels of the stem with scalariform perforations; stem without internal phloem. LEAVES densely crowded, closely ascending, whorled in alternating sets of (3) 4, or 5, firm, simple, linear and entire, revolute-margined, silky when young, exstipulate; stomates anomocytic. FLOWERS adapted to pollination by birds, perfect, crowded on very short, compact, densely leafy axillary shoots along the upper part of the main stem, each flower closely subtended by a pair of leaf-like bracteoles about as long as the calyx; calyx tubular and somewhat unequally 5-lobed; corolla sympetalous, hairy outside, regular, with long, narrow, red or orange-red tube and 5 short, purplish-black, basally convolute lobes that are notably white-hairy especially

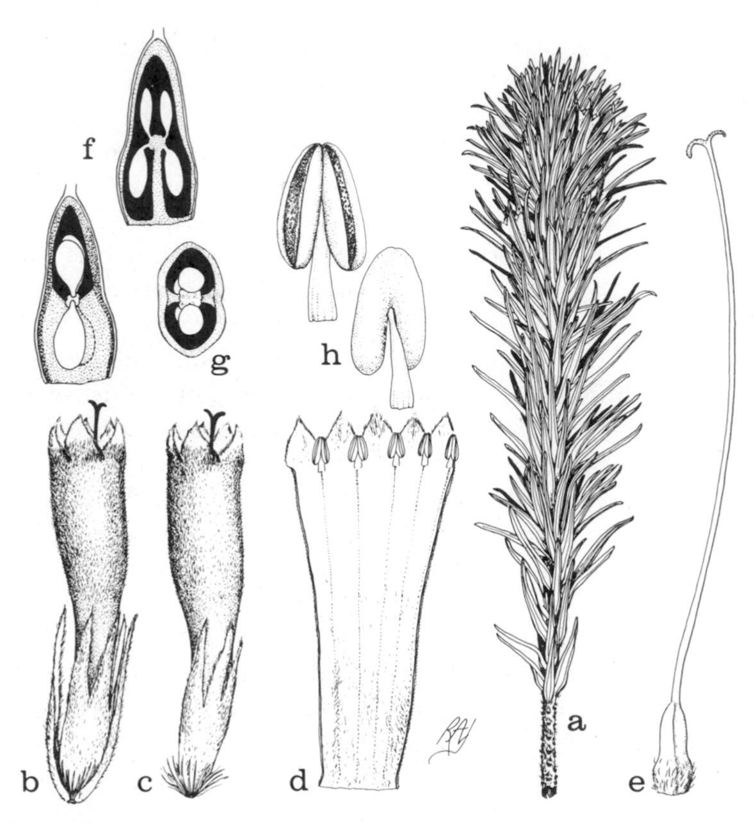

FIG. 6-3 Retziaceae. *Retzia capensis* Thunb. a, habit, $\times\frac{1}{6}$; b, c, flower, with and without subtending bracts, $\times 2$; d, corolla, opened out to show attachment of stamens, $\times 2$; e, pistil $\times 2$; f, two views of ovary, in schematic long-section, $\times 4$; g, ovary, in schematic cross-section, $\times 4$; h, ventral (left) and dorsal views of stamens, $\times 8$.

distally; stamens as many as and alternate with the corolla-lobes, attached at the summit of the corolla-tube, the filaments short, the anthers introrse, dorsifixed and deeply sagittate, dithecal, opening by longitudinal slits; pollen-grains tricolporate, finely reticulate; gynoecium of 2 carpels united to form a compound, superior ovary seated on a small nectary-disk; style terminal, elongate, exserted from the corolla, with a shortly 2-lobed stigma; ovary bilocular in the lower third or half, tending to be somewhat constricted at about the level of the top of the partial partition, unilocular above, the carpellary margins progressively less intruded distally; ovules anatropous, 4 (–6) in 2 pairs (or trios) at the slightly expanded placental summit of the partial partition; one ovule of each pair (or trio) ascending into the unilocular part of the ovary, the other one (or 2) descending into one of the 2 semilocules, each of the semilocules thus cupping one or 2 ovules. FRUIT a small capsule included in the persistent perianth, loculicidally dehiscent in the upper portion (opening from the top downward), each of the 2 valves sometimes becoming bifid; embryo straight, dicotyledonous, embedded in the copious, mealy endosperm. N = 12.

The family Retziaceae consists of the single genus and species *Retzia capensis* Thunb., native to the Cape Province of South Africa. The species is so distinctive that its affinities are uncertain. Traditionally it has been included in or associated with either the Solanaceae or the Loganiaceae. More recently, Dahlgren et al. (1979) have suggested that it should be associated (as a distinct family) with the Stilbaceae, a group here included in the Verbenaceae as a subfamily. Each of these positions has some advantages and some difficulties. [Dahlgren reference on p. 886.]

The principal advantages of a position near the Stilboideae (or Stilbaceae) are chemical, geographical, and habital. The Stilboideae are a South African group, and thus would have been physically available as a possible evolutionary source of *Retzia*. Several genera of Solanaceae (*Lycium, Withania, Solanum*) and Loganiaceae (*Anthocleista, Strychnos*) also occur in South Africa, but these are habitally quite different from *Retzia*, whereas the aspect of *Stilbe* is something like that of *Retzia*. The small South African genus *Gomphostigma*, which has been included in the Loganiaceae in the past, might also be loosely compared to *Retzia* in habit, but *Gomphostigma* belongs to a group that has more recently been removed from the Loganiaceae as a family Buddlejaceae and assigned to the Scrophulariales.

Dahlgren et al. lay heavy emphasis on the iridoid compounds of *Retzia* as indicators of affinity with the Stilboideae. Iridoid compounds are unknown in the Solanaceae, and indeed in the whole order Solanales except for the anomalous family Menyanthaceae. The Loganiaceae and other Gentianales do have iridoid compounds, but Dahlgren et al. maintain that these are different from those of *Retzia*. I find this argument less than compelling, since one of these characteristic iridoids of *Stilbe* and *Retzia* (unedoside) is also known in the Ericaceae, and all the iridoids of *Retzia* are said to belong to a biogenetically related group that is also known in *Mentzelia* (Loasaceae) and *Deutzia* (Hydrangeaceae). These genera are far removed from the Stilboideae in the present system.

The placentation and organization of the gynoecium in *Retzia* are distinctive, and unlike all the groups with which the genus might otherwise be compared. The Stilboideae typically have a bilocular ovary with only one ovule in each locule, or sometimes one of the locules is empty. In the present system the gynoecium of the Stilboideae is regarded as derived from that of typical Verbenaceae, in which (as in other characteristic Lamiales) a 4-ovulate, initially bilocular ovary becomes divided ontogentically into 4 uniovulate chambers by the intrusion of partitions from the carpellary midribs. Considerable evolutionary change would be required to derive the gynoecium of *Retzia* from either that of the Stilboideae or that of more typical Verbenaceae or other Lamiales. A bilocular ovary with axile placentation and several ovules in each locule would provide a more likely source.

The Verbenaceae typically have the stamens anisomerous with the corolla-lobes (most commonly 5 corolla-lobes and 4 stamens), and this pattern is standard for the subfamily Stilboideae. Most members of the Stilboideae and other Verbenaceae have a more or less zygomorphic corolla. Thus *Retzia*, with its regular corolla and isomerous stamens, is more primitive in these respects than the bulk of the Verbenaceae (including Stilboideae), and might better be compared with the Gentianales and Solanales. Furthermore, the aestivation of the corolla is convolute in *Retzia*, and at least typically imbricate in the Verbenaceae. The only families of Lamiales that characteristically have regular corolla with as many stamens as corolla-lobes are the Boraginaceae and the parasitic family Lennoaceae. These differ from *Retzia* in other important features, and the chemistry of the Boraginaceae, at least, is notably different. I am not enthusiastic about associating *Retzia* with any part of the Verbenaceae sens. lat., or with the Lamiales in general.

The Gentianales and Solanales offer better prospects as a possible home for the Retziaceae, but there are still problems. These two orders are at least hospitable in having a superior ovary, regular corolla, and stamens isomerous with the corolla-lobes. Convolute aestivation is very common in the Gentianales, and is also well known in the Solanales (Polemoniaceae, some Hydrophyllaceae). The whorled leaves of *Retzia* are fine for the Gentianales, but difficult for the Solanales. The absence of internal phloem is readily acceptable in the Solanales (although the Solanaceae do have it), but difficult for the Gentianales.

The ovular embryology of *Retzia* may eventually throw some light on the position of the genus. If *Retzia* belongs near the Loganiaceae in the Gentianales, it may be expected to have nuclear endosperm and lack an integumentary tapetum. If it belongs in the Solanales or Lamiales (or Scrophulariales) it may be expected to have an integumentary tapetum, and possible cellular endosperm. Quien viva lo verra.

At the present time *Retzia* seems best accommodated as a family in the Gentianales, near the Loganiaceae.

3. Family GENTIANACEAE A. L. de Jussieu 1789 nom. conserv., the Gentian Family

Herbs, often mycorhizal (seldom strongly mycotrophic, with reduced leaves and without chlorophyll), seldom half-shrubs, rarely shrubs or even small trees, glabrous or with various sorts of simple to uniseriate, sometimes glandular hairs, commonly accumulating xanthones and sometimes aluminum, not tanniferous, lacking both proanthocyanins and ellagic acid, seldom cyanogenic and only rarely saponiferous, but commonly accumulating bitter substances, these consisting of iridoid compounds, especially gentiopicroside and related substances; calcium oxalate crystals of various forms commonly present in some of the cells of the parenchymatous tissues; stem often winged; nodes unilacunar to trilacunar or sometimes multilacunar; xylem commonly forming a continuous cylinder, without rays or seldom with uniseriate rays; vessel-segments with simple perforations; internal phloem characteristically present either as a continuous ring or as isolated bundles at the margin of the pith; medullary bundles or isolated strands of phloem often present in the pith. LEAVES opposite or seldom whorled, only rarely alternate (as in spp. of *Swertia*), simple and usually entire, in *Bartonia* and the achlorophyllous genera (*Voyria, Voyriella, Leiphaimos*) reduced to scales; stomates anomocytic or anisocytic; scattered mucilage cells often present in the epidermis and mesophyll; petiole usually with an arc-shaped bicollateral vascular strand and small lateral bundles; stipules wanting. FLOWERS solitary or often in cymose, seldom racemose inflorescences, commonly showy, perfect or rarely unisexual, regular or nearly so; calyx usually with a short to well-developed tube and 4–5 (–12) imbricate or sometimes valvate or open lobes, seldom the lobes much-reduced or suppressed, and the tube then sometimes 2-cleft, or rarely the sepals 4–5 and distinct; sepals with 1–3 traces; corolla sympetalous, with short to usually more or less elongate tube (or saucer) and 4–5 (–12) usually convolute lobes, often with scales or nectary-pits within the tube, which is sometimes plicate toward the sinuses; petals with a single vascular trace; stamens as many as and alternate with the corolla-lobes, attached to the corolla-tube or -throat, rarely some of them staminodial or obsolete; anthers tetrasporangiate and dithecal, opening by longitudinal slits or seldom (as in *Exacum*) by terminal pores; pollen-grains binucleate or trinucleate, (2) 3 (4)-colporate or -porate, or even by reduction uniporate; a nectary-disk of one or another sort (or distinct nectary-glands) commonly surrounding the ovary; gynoecium

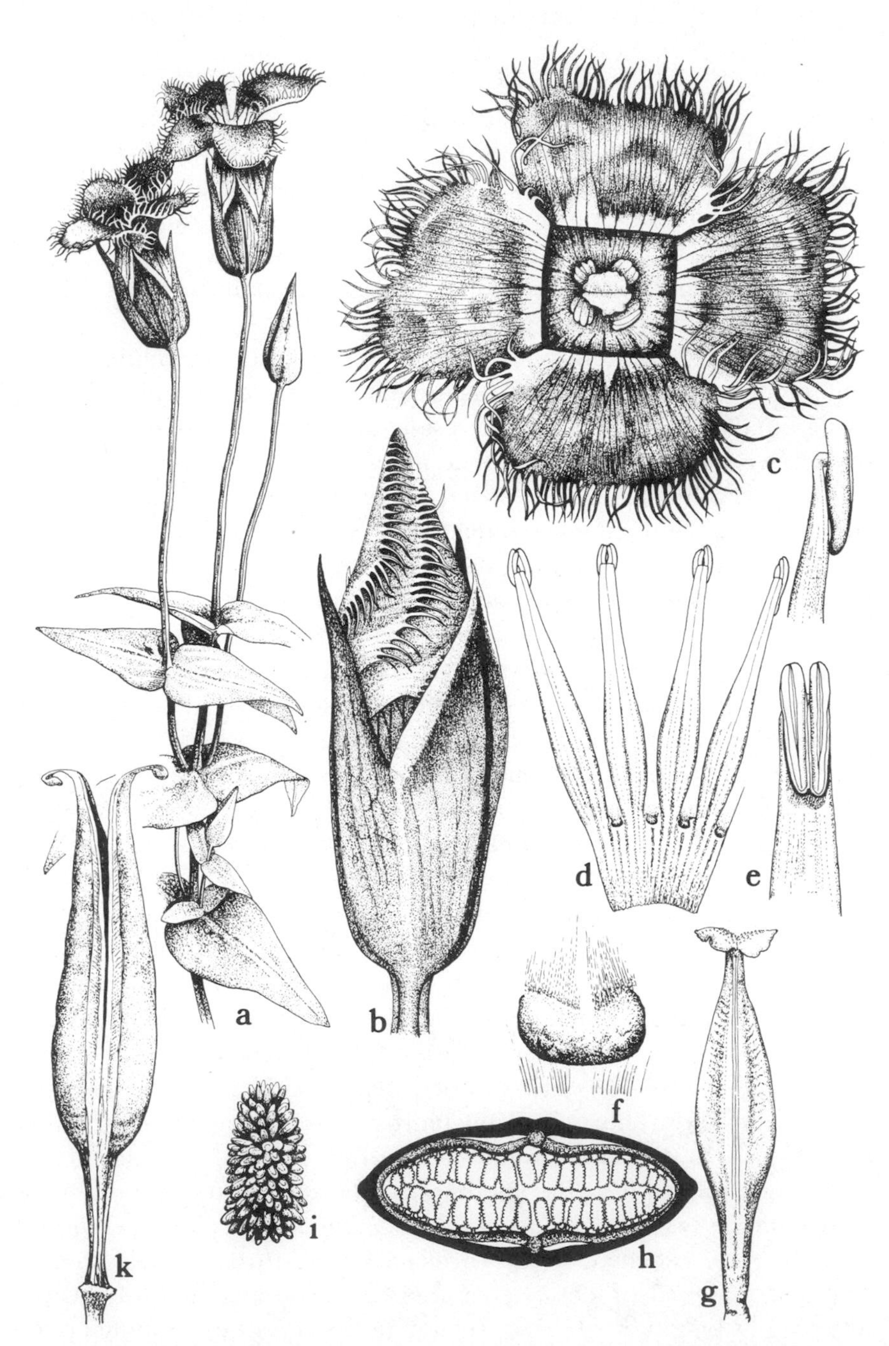

FIG. 6-4 Gentianaceae. *Gentiana crinita* Froel. a, habit, ×½; b, opening flower bud, ×2; c, top view of flower, ×2; d, stamens, attached to the corolla tube, ×2; e, two views of anther, ×4; f, nectary, ×16; g, pistil, ×2; h, schematic cross-section of ovary, ×8; i, seed, ×16; k, opened capsule, ×2.

of 2 carpels united to form a compound, superior, unilocular ovary with parietal placentas that may be more or less deeply intruded and bifid, or the ovary seldom bilocular with axile placentas, or rarely unilocular with a free-central placenta; style terminal, simple, with an entire or 2-lobed, wet, papillate stigma; in *Lomatogonium* the style wanting and the stigmas decurrent along the sides of the ovary; ovules more or less numerous, anatropous, tenuinucellar, unitegmic, only seldom (as in *Exacum*) with an integumentary tapetum; endosperm-development nuclear or rarely (in the achlorophyllous genera) cellular. FRUIT a septicidal capsule or rarely a berry; seeds typically with a small to straight and elongate, cylindric to spatulate, dicotyledonous embryo embedded in the abundant, oily endosperm, but in the achlorophyllous genera the seeds tiny, with undifferentiated embryo and scanty endosperm. X = 5–13+.

The family Gentianaceae consists of some 75 genera and about 1000 species, cosmopolitan in distribution, but most common in temperate and subtropical regions, and in tropical mountains. The largest genus by far is *Gentiana* (including *Gentianella*), with perhaps 400 species.

The Gentianaceae are obviously related to the Loganiaceae. All of the differences between the two families break down in one or another genus, but there is no controversy about which genera should be referred to which family.

4. Family SACCIFOLIACEAE Maguire & Pires 1978 the Saccifolium Family

Pulvinate subshrub, apparently producing iridoid compounds, without mucilage, and with well developed strands of internal phloem; vessel-segments with scalariform perforation-plates; wood-rays and wood-parenchyma none. LEAVES alternate, simple, small, sessile, closely crowded toward the branch-tips, with recurved margins, extrorsely saccate-vaginate distally; stomates anisocytic, borne on the inner surface of the sac; stipules wanting; several small glandular bodies present in the axil. FLOWERS solitary in the axils, fairly showy, perfect, regular; sepals (4) 5, connate at the base; corolla sympetalous, with a well developed tube and (4) 5 imbricate lobes; stamens as many as and alternate with the corolla-lobes, attached to the corolla-tube; anthers tetrasporangiate and dithecal, opening by longitudinal slits; connective prolonged as an evident point; pollen-grains tricolporate, with punctate, otherwise smooth exine; nectary-disk wanting; gynoecium of 2 carpels

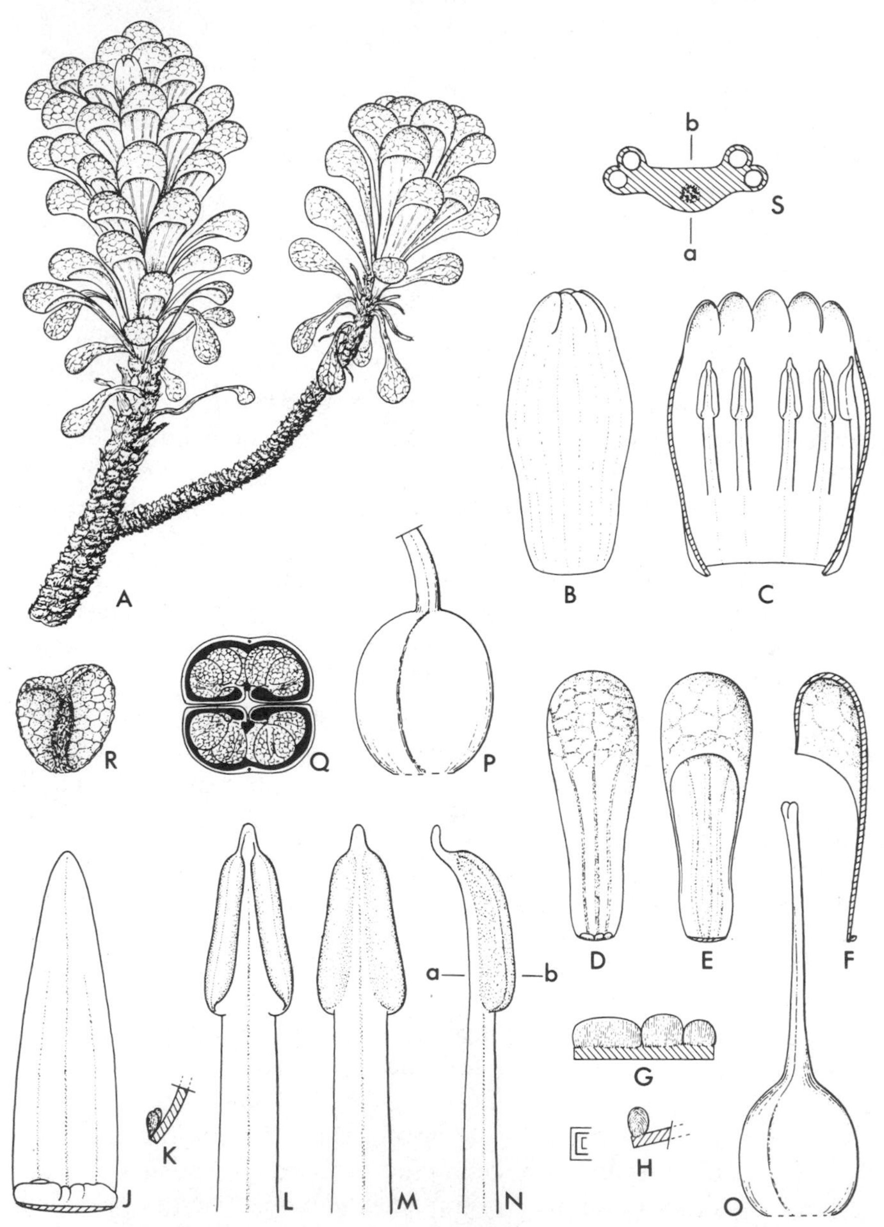

FIG. 6-5 Saccifoliaceae. *Saccifolium bandeirae* Maguire & Pires. A, habit, $\times\frac{9}{10}$; B, corolla, ready to open, with imbricate lobes, ×2.7; C, opened corolla, with attached stamens, ×2.7; D, E, F, adaxial, abaxial, and lateral views of leaf, ×1.8; G, glandular bodies at adaxial base of leaf, ×9; H, cross-section of one of the bodies in G, ×9; J, adaxial surface of sepal, with basal glands, ×9; K, cross-section showing basal gland of sepal, ×9; L, M, N, adaxial, abaxial, and lateral views of stamen, ×9: O, pistil, ×7; P, ovary, ×8; Q, cross-section of ovary, ×8; R, submature seed, ×18; S, cross-section of anther, a the dorsal side, b the ventral side, ×18. From Mem. New York Bot. Gard. 29: 243. 1978. Courtesy of Bassett Maguire and the New York Botanical Garden.

united to form a superior, bilocular ovary (the partitions imperfect toward the top of the ovary) with a terminal style and shortly bilobed stigma; ovules numerous on axile (distally intruded-parietal) placentas, anatropous, with a massive single integument, tenuinucellar. FRUIT unknown.

The family Saccifoliaceae consists of the single species *Saccifolium bandeirae* Maguire & Pires, native to an isolated "Lost World" mountain in southern Venezuela.

The affinities of *Saccifolium* are debatable. The sympetalous corolla, isomerous stamens alternate with the corolla-lobes, and tenuinucellar ovules with a single massive integument clearly mark it as a member of the Asteridae. The pentamerous, regular flowers and superior, bilocular ovary with a terminal style and numerous ovules restrict the choice of orders to the Gentianales and Solanales. These are also the only orders of Asteridae with internal phloem in a stem of otherwise normal anatomy. The alternate leaves are much more compatible with the Solanales, in which alternate leaves are common, than with the Gentianales, in which they are rare indeed.

In the Solanales *Saccifolium* would have to stand next to the Solanaceae, a family that is very poorly represented on the Guayana Highlands. The Solanaceae are rich in alkaloids, but like most other members of the Solanales they lack iridoid compounds. (Among the Solanales iridoid compounds are known only in the Menyanthaceae, which have often been included in the Gentianales). Preliminary chemical investigations suggest the presence of an iridoid compound in *Saccifolium*, whereas they do not disclose the presence of an alkaloid. Furthermore, the wood-structure of *Saccifolium*, without apparent medullary rays, resembles that of woody Gentianaceae but is unlike that of Solanaceae. Although there are only a few woody genera of Gentianaceae, these are well represented in the Guayana Highlands. Thus it seem advisable to consider a possible relationship of *Saccifolium* to the Gentianales, and particularly to the Gentianaceae.

Aside from the alternate leaves (which do occur in a few species of *Swertia*, in the Gentianaceae), *Saccifolium* appears to be perfectly at home in the Gentianales, on chemical as well as morphological grounds. No established family of the order can easily accommodate it, however. The Loganiaceae commonly have a bilocular ovary with axile placentation, as in *Saccifolium*, and woody forms are well represented in the family. On the other hand, the Loganiaceae not only have opposite leaves but also interpetiolar stipules, and their wood has well developed rays. The Apocynaceae and Asclepiadaceae can be excluded from serious consideration because of their latex-system and specialized floral features. That leaves the Gentianaceae. The vessels of *Saccifolium* have scalariform perforation-plates, which are so far unknown in the Gentianaceae. This feature needs to be evaluated with some caution, however, until the woody gentianoids from the Lost Worlds have been studied anatomically. *Saccifolium* differs from most of the Gentianaceae in its bilocular ovary with axile placentas, and in its imbricate corolla-lobes. These characters can be found individually in the Gentianaceae, but I believe not in combination. The unique leaves of course bespeak the isolated taxonomic position of *Saccifolium*.

5. Family APOCYNACEAE A. L. de Jussieu 1789 nom. conserv., the Dogbane Family

Trees, shrubs, herbs, or very often lianas, seldom succulent megaphytes (as in *Pachypodium*), glabrous or with various sorts of hairs, permeated by a well developed system of non-articulated, branched or unbranched laticifers, and producing diverse sorts of iridoid compounds, cardiotonic glycosides, and various kinds of alkaloids, notably of the indole (including tryptophan) and steroid groups, but usually not tanniferous, without ellagic acid and generally without proanthocyanins, sometimes cyanogenic, but only seldom with saponins; calcium oxalate crystals of various sorts often present in some of the cells of the parenchymatous tissues; nodes unilacunar; stem sometimes of anomalous structure; vessel-segments with simple perforations, or sometimes with short, scalariform perforation-plates that have only a few cross-bars; imperforate tracheary elements septate and with simple pits, or not septate and with bordered pits; wood-rays heterocellular to homocellular, 1–5 cells wide; wood-parenchyma of diverse types; internal phloem consistently present, either as a continuous ring or as isolated bundles at the margin of the pith, only very rarely wanting; pericycle generally with a continuous ring or separate strands of characteristic white cellulosic fibers; cork usually superficial in origin. LEAVES simple and entire, generally opposite or sometimes whorled, seldom (as in *Pachypodium*), the internodes so shortened that the leaves are not obviously opposite and appear to be in a close spiral, otherwise only rarely alternate; veins accompanied by laticifers; stomates paracytic or anomocytic, rarely actinocytic; petiole usually with an arc-shaped strand of xylem surrounded by phloem; stipules wanting, or rarely small and interpetiolar. FLOWERS in cymose

FIG. 6-6 Apocynaceae. *Allamanda cathartica* L. a, habit, ×½; b, flower ×1; c, d, portions of flower, the corolla opened up, ×1, 2; e, ovary and nectary-disk, ×2; f, opened fruit, ×½; g, ovary in schematic cross-section, ×4; h, seed, ×1.

or racemose inflorescences, or solitary, commonly showy, perfect, regular or nearly so, generally (4) 5-merous except as to the gynoecium; calyx synsepalous with imbricate lobes, the tube very often bearing glandular (or eglandular) scales within; corolla sympetalous, commonly funnelform or salverform, regular or nearly so, with convolute or rarely imbricate or valvate lobes, often bearing appendages within the tube, these sometimes connate; stamens as many as and alternate with the corolla-lobes, attached to the corolla-tube; filaments without coronal appendages (cf. Asclepiadaceae); anthers distinct or more or less closely connivent around the style-head, tetrasporangiate and usually dithecal, opening by longitudinal slits, often sagittate or sagittate-tailed at the base, the sporangia often restricted to the distal part of the anther; pollen-grains trinucleate, tricolporate or sometimes 2–3-porate, sometimes with granular instead of columellar infractectal structure, borne in monads or tetrads, sometimes loosely cohering, but not forming pollinia, the androecium without translators; 5 nectary-glands often present about the base of the ovary, alternating with the stamens, or the glands sometimes confluent into an annular disk, or reduced in number, or wanting; gynoecium of 2 (–8) carpels, these connate to varying degrees, primitively forming a superior, compound, bilocular ovary with a simple or only apically cleft style, in more advanced genera the ovary unilocular with intruded parietal placentas, or very often the carpels separate below for part of their length, sometimes forming 2 distinct ovaries united only by their common style and stigma, or even the styles largely distinct and united only by their thickened tip; style-head often thickened and specialized in structure, with variously restricted, wet or sometimes dry stigmatic surface, often subtended by a ring of hairs; ovules 2-many in each ovary or on each placenta of the compound ovary, commonly pendulous, amphitropous or anatropous or hemitropous, with a single massive integument, tenuinucellar, usually with an integumentary tapetum; endosperm-development nuclear. FRUIT of diverse types; seeds sometimes comose; embryo large, more or less spatulate, dicotyledonous; endosperm oily, copious to scanty. X = 8–12+. (Plocospermataceae, Plumeriaceae)

The family Apocynaceae as here (and customarily) defined consists of about 200 genera and 2000 species, widespread in tropical and subtropical regions, with relatively few genera and species in temperate climates. None of the genera is very large in relation to the family as a whole. *Tabernaemontana* (140), *Mandevilla* (115), *Rauvolfia* (100), *Parsonsia*) (100), and *Aspidosperma* (80) are among the largest. *Rauvolfia* is the source of tranquilizing drugs, and *Strophanthus* is the source of arrow-poisons and a cardiotonic glycoside, strophanthin, used in treating heart-ailments and as a precursor of cortisone. *Nerium oleander* L., *Vinca minor* L. (periwinkle), *Allamanda cathartica* L., and a number of other species are well known in cultivation.

The Apocynaceae consist of two well marked subfamilies, which might with some reason be treated as separate families. The more archaic subfamily, the Plumerioideae, has wholly fertile anthers that are distinct from each other and free from the style; the style is relatively unmodified, with a terminal stigma. The seeds are only seldom comose. The Apocynoideae have the anthers connivent to connate and more or less adnate to the style-head, which is enlarged and modified, and only the upper part of the anther bears pollen. The seeds are comose. Although well defined, the two subfamilies are obviously related, and collectively they form a natural group.

6. Family ASCLEPIADACEAE R. Brown 1810 nom. conserv., the Milkweed Family

Lianas, herbaceous vines, scrambling shrubs, or erect herbs, seldom erect shrubs or trees, sometimes succulent and with much-reduced leaves, often with fleshy underground parts, glabrous or with unicellular or uniseriate hairs or seldom with other kinds of hairs, permeated by a well developed system of non-articulated, branched or unbranched laticifers, commonly producing various sorts of cardiotonic glycosides and alkaloids of the indole, pyridine, and phenanthro-indolizidine groups, but apparently without ordinary iridoid compounds, only occasionally tanniferous, lacking ellagic acid and seldom with proanthocyanins, rarely cyanogenic or saponiferous; calcium oxalate crystals of various sorts often present in some of the cells of the parenchymatous tisuses; nodes unilacunar; stem often with anomalous structure; vessel-segments with simple perforations; imperforate tracheary elements with small, simple or bordered pits; wood-rays commonly 1–4 cells wide, a very few wide medullary rays sometimes also present; wood-parenchyma of diverse types; internal phloem consistently present, either as a continous ring or as isolated strands at the margin of the pith; pericycle generally containing some cellulosic fibers, these borne singly or in strands or even in concentric rings; cork usually superficial in origin. Leaves opposite or sometimes whorled, rarely alternate, simple and usually entire, rarely lobed or toothed; veins accompanied by laticifers;

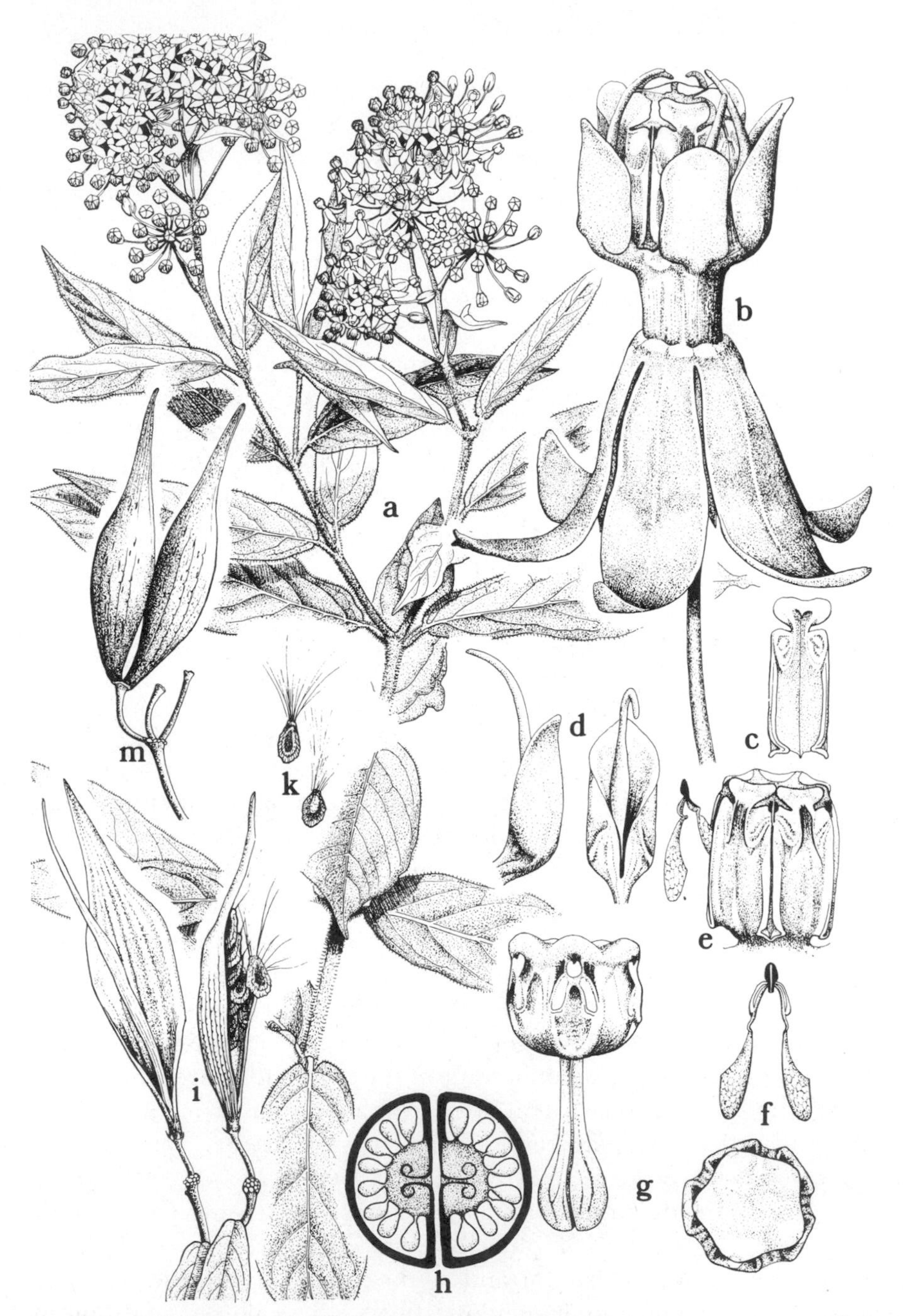

FIG. 6-7 Asclepiadaceae. *Asclepias incarnata* L. a, habit, $\times\frac{1}{2}$; b, flower $\times 10$; c, anther, $\times 10$; d, hood, in front and side views, $\times 10$; e, style-head and anthers, with a partially extracted pair of pollinia, $\times 10$; f, pollinia, with translator, $\times 10$; g, top and side views of pistil, the stamens removed, $\times 10$; h, schematic cross-section of gynoecium, $\times 20$; i, fruits, after dehiscence, $\times 1$; k, seeds, $\times 1$; m, unopened fruits, $\times 1$.

stomates paracytic, or less often anomocytic or anisocytic; petiole usually with an arc-shaped strand of xylem surrounded by phloem, sometimes with small peripheral bundles as well; stipules wanting or vestigial. FLOWERS in cymose (often umbelliform) or seldom racemose inflorescences, perfect or rarely functionally unisexual, regular or nearly so, pentamerous except for the gynoecium, specialized for mass-transfer of pollen by insects; sepals imbricate or valvate, shortly connate at the base, often reflexed; corolla sympetalous, with short to sometimes elongate tube and convolute or seldom imbricate or valvate, often spreading or reflexed lobes, sometimes with a thickened ring of distinct or connate scales in the throat; filaments inserted on the corolla-tube, usually toward its base, distinct in the Periplocoideae, connate into a usually short sheath around the style (and adherent to the style) in the Asclepiadoideae, the anthers in both subfamilies coherent or connate into a sheath and adherent to the thickened style-head, the combined filaments, anthers and style in the Asclepiadoideae forming a central column called a gynostegium; flowers usually with a well developed corona, consisting of diverse sorts of erect or more often incurved, commonly nectariferous, petaloid or often hood-shaped or partly horn-shaped, often structurally complex appendages arising from the external base of the filaments or from the region of union of the filaments and corolla, or rarely the filaments simple and unappendaged; an additional set of appendages sometimes arising from the backs or tips of the anthers; anthers basifixed, often ornamented along the margins, introrse, opening longitudinally or apically, dithecal, tetrasporangiate in the Periplocoideae and the genus *Secamone* of the Asclepiadoideae, bisporangiate in the remaining Asclepiadoideae; pollen-grains trinucleate, loosely to firmly coherent in masses called pollinia, 2 pollinia in each theca of the Periplocoideae and *Secamone*, a single pollinium in each theca of the remaining Asclepiadoideae; a special, acellular, yoke-shaped structure called a translator, derived from solidifed secretions of the anthers or style-head or both, serves to extract the pollinia from the anthers; each translator consists of arms (retinacula) joined at the middle by a 2-parted "gland" (corpusculum); one arm of the translator attaches to the pollinum (or pollinia) of one theca of one anther, and the other attaches to the pollinium (or pollinia) of the adjoining theca of an adjoining (not the same) anther; in the Periplocoideae the pollen-tetrads loosely coherent to form in each theca two distally open, cornucopia-shaped pollinia, each of which is entered by an expanded, sticky branch of the 2-cleft translator-arm; in the Asclepiadoideae the pollen-grains tightly coherent to form a solid pollinium (2 pollinia in *Secamone*) that is apically attached to the end of the translator-arm (the arm 2-cleft in *Secamone*); pollinia elongate and occupying most of the length of the anther, or often short and occupying only the distal part of the anther; nectary-disk wanting; gynoecium of 2 carpels forming separate, unilocular ovaries with distinct styles and united only by the common style-head, which is thickened and has restricted lateral stigmatic surfaces alternating with the anthers; ovaries wholly superior, or slightly sunken into the receptacle; ovules (1–) more or less numerous on marginal placentas, commonly pendulous, anatropous, unitegmic, pseudocrassinucellar (terminology of Davis), without an integumentary tapetum; endosperm-development nuclear. FRUIT of 2 distinct follicles (often only 1 developing); seeds usually flattened and with a terminal coma of long hairs; embryo straight, dicotyledonous; endosperm oily, rather scanty. X = 9–12. (Periplocaceae)

The family Asclepiadaceae as here (and customarily) defined consists of about 250 genera and 2000 species, widespread in tropical and subtropical regions, especially in Africa, with relatively few genera and species in temperate climates. None of the genera is very large in relation to the family as a whole. *Hoya* (150), *Ceropegia* (150), *Asclepias* (150), *Gonolobus* (100), *Secamone* (100), and *Stapelia* (75–100) are among the largest. *Cryptostegia* has excited some interest as a possible commercial source of rubber.

The Asclepiadaceae consist of 2 well marked subfamilies of very unequal size, which have sometimes been treated as distinct families. The smaller subfamily. Periplocoideae (40 genera, 200 species) has the translator-pollinium mechanism relatively poorly developed, in comparison to the Asclepiadoideae. The close relationship of the two groups is not in question.

2. Order SOLANALES Lindley 1833

Herbs, shrubs, vines, or seldom small trees, only rarely tall trees, often producing alkaloids, less often iridoid compounds, but only rarely tanniferous, generally lacking both ellagic acid and proanthocyanins, only rather seldom saponiferous or cyanogenic, and without cardiotonic glycosides; stem in the larger families with internal phloem; vessel-segments with simple perforations, or rarely some of the perforations with a few cross-bars. LEAVES most commonly alternate, sometimes (mainly in some genera of Polemoniaceae) opposite, simple and entire or toothed to variously cleft or compound or dissected; stipules wanting, unless the petiolar flanges of the Menyanthaceae are so interpreted. FLOWERS perfect, mostly regular or nearly so (distinctly irregular mainly in a few Solanaceae and Polemoniaceae), sympetalous, with diverse types of aestivation, most common 5-merous as to the calyx, corolla, and androecium, seldom 4-merous, rarely 3-merous or 6-merous; stamens generally as many as and alternate with the corolla-lobes (fewer in a few Solanaceae), attached to the corolla-tube; pollen-grains binucleate or much less often trinucleate, triaperturate or of various triaperturate-derived types; gynoecium most commonly bicarpellate, less often with up to 5 (rarely more) carpels, the carpels united to form a compound ovary with axile or basal-axile or parietal placentation, or seldom the carpels largely distinct and united mainly by their gynobasic style; rarely the ovary pseudomonomerous; ovary superior or rarely half-inferior, in most families surrounded by an annular nectary-disk; ovules (1) 2-many per carpel, anatropous to sometimes hemitropous or amphitropous, with a massive single integument, tenuinucellar or rarely (some Convolvulaceae) crassinucellar, very often with an integumentary tapetum; endosperm-development variously nuclear, cellular, or even helobial. FRUIT most commonly a capsule or berry, but sometimes of other types; seeds (1–) usually several or many; embryo with 2 cotyledons, or in the Cuscutaceae without differentiated cotyledons; endosperm copious to sometimes scanty or even wanting.

The order Solanales as here defined consists of 8 families and about 5000 species. More than four-fifths of the species belong to only two large families, the Solanaceae (2800) and Convolvulaceae (1500). At the other extreme, the family Duckeodendraceae has only a single species. The name Solanales is here adopted for the order because it is 5 years older than the more familiar name Polemoniales. It is also nomenclaturally convenient that the Solanaceae are the largest family in the order.

It has been widely agreed in the past that the Nolanaceae, Solanaceae and Convolvulaceae are closely related inter se, that the Polemoniaceae and Hydrophyllaceae form a pair, and that these two groups are related to each other. The relationship of the Cuscutaceae to the Convolvulaceae is also generally admitted; indeed the Cuscutaceae have often been submerged in the Convolvulaceae. Not all of these families remain associated in some of the newer systems, such as those of Dahlgren and Thorne. The system of Dahlgreen reflects what I consider an overemphasis on chemical features, and I simply do not understand the rationale for the system of Thorne. The relationships of the Duckeodendraceae and Menyanthaceae are discussed under the individual families.

The Solanales differ from the Gentianales in their mostly alternate leaves, in their often cellular (rather than nuclear) endosperm, in very often having an integumentary tapetum, and in the less consistent presence of internal phloem. Furthermore, the Gentianales generally produce iridoid compounds or cardiotonic glycosides or both, whereas the Solanales do not have cardiotonic glycosides, and only the small family Menyanthaceae has iridoids.

The Lamiales differ from the Solanales in their characteristic gynoecium and fruit, and the Scrophulariales differ in usually having an irregular corolla with fewer stamens than corolla-lobes. These two orders represent the realization of morphological tendencies that are evident but less fully developed in some of the Solanales. The Scrophulariales further differ from the Solanales in commonly having iridoid compounds and sometimes cardiotonic glycosides, but not alkaloids. These chemical features tend to link the Scrophulariales more directly to the Gentianales than to the Solanales.

The Solanales are evidently related to the Gentianales. Aside from the general similarities in floral structure, the two orders are linked by the common possession of internal phloem in most species of Solanales and nearly all of the Gentianales. Except for these two orders and the taxonomically remote order Myrtales, only a few widely scattered groups of dicots have internal phloem. Embryologically the Gentianales are more primitive than the Solanales; so far as these characters are concerned, the one order might well be directly ancestral to the other. Although in the angiosperms as a whole alternate leaves are more primitive than opposite leaves, I will hazard the guess that in this pair of orders, as in the Asterales, opposite leaves are primitive and alternate leaves derived. However, any common ancestor of the Solanales would probably be more primitive than the vast majority of both groups in having 5 carpels. The few 5-carpellate members of these orders do not appear to be especially primitive in other respects. The two orders should therefore be considered to have diverged from a common ancestor that combined the more primitive features of both.

The order Solanales as a whole is ecologically varied, and the adaptive significance of the characters that mark the group is at best debatable. Only a few members of the group are trees, but otherwise the order embraces a wide range of ecological types that also occur in other orders. In comparison with other Asteridae there is nothing unusual in the arrangement or structure of the leaves, the anatomy of the ovules, or the ontogeny of the endosperm of most members of the Solanales. They lack the specialized features of corolla, androecium, and gynoecium found in various combinations in more advanced orders of the Asteridae, but so do many of the Gentianales. Among the features that distinguish the Solanales from the Gentianales, only the nature of the chemical repellents is obviously important to the plants, but there is some overlap in the classes of alkaloids present in the Solanaceae and Loganiaceae.

Several of the individual families of Solanales do have some ecological significance. The Menyanthaceae are adapted to aquatic life, but most of the formal characters that mark the family are not obviously related to the habitat. The Convolvulaceae have exploited the twining vine habit more consistently and more successfully than most large families of plants, although they also show a wide range of other growth-forms. The Cuscutaceae have gone on from the twining vine habit of the Convolvulaceae to become parasitic. Both the Polemoniaceae and the Hydrophyllaceae have produced many annuals adapted to deserts and semideserts, but members of these families also occur in many other habitats, and the biological significance of the characters that distinguish the two families from each other is still unknown.

The fossil record of the Solanales is still scanty. Macrofossils attributed to the Solanaceae occur in Eocene and more recent deposits, but there is no significant record of fossil pollen for the family. D'Arcy (in Hawkes, Lester, & Skelding, eds.) doubts the identification of all purportedly solanaceous fossils. Pollen attributed to the Convolvulaceae goes back to the Eocene, but without accompanying macrofossils. Pollen attributed to the Polemoniaceae, likewise without accompanying macrofossils, dates from the Miocene epoch. Other members of the order appear only in post-Miocene deposits, if at all. Thus is appears that the order does not antedate the Tertiary period, and that most of its evolutionary diversification into families is post-Eocene.

SELECTED REFERENCES

Ahmad, K. J. 1964. Epidermal studies in *Solanum*. Lloydia 27: 243–250.

Alfaro, M. E., & A. Mesa. 1979. El origen morfologico del floema intraxilar en Nolanaceas y la posicion sistematica de esta familia. Bol. Soc. Argent. Bot. 18: 123–126.

Allard, H. A. 1947. The direction of twist of the corolla in the bud, and twining of the stems in Convolvulaceae and Dioscoreaceae. Castanea 12: 88–94.

Austin, D. F. 1973. The American Erycibeae (Convolvulaceae). *Maripa, Dicranostyles,* and *Lysiostyles*. II. Palynology. Pollen & Spores 15: 203–226.

Baehni, C. 1943. *Henoonia,* type d'une famille nouvelle? Boissiera 7: 346–358.

Baehni, C. 1946. L'ouverture du bouton chez les fleurs de Solanées. Candollea 10: 399–492.

Chuang, T.-I., W. C. Hsieh, & D. H. Wilken. 1978. Contribution of pollen morphology to systematics of *Collomia* (Polemoniaceae). Amer. J. Bot. 65: 450–458.

Constance, L. 1939. The genera of the tribe Hydrophylleae of the Hydrophyllaceae. Madroño 5: 28–33.

Crété, P. 1946, 1947. Embryogénie des Hydrophyllacées. Développement de l'embryon chez le *Phacelia tanacetifolia* Benth. Développement de l'embryon chez le *Nemophila insignis* Benth. Compt. Rend. Hebd. Séances Acad. Sci. 223; 459, 460, 1946. 224: 749–751, 1947.

Dahlgren, R., B. J. Nielsen, P. Goldblatt, & J. P. Rourke. 1979. Further notes on Retziaceae: its chemical contents and affinities. Ann. Missouri Bot. Gard. 66: 545–556.

D'Arcy, W. G., & R. C. Keating. 1973. The affinities of *Lithophytum*: A transfer from Solanaceae to Verbenaceae. Brittonia 25: 213–225.

Dawson, M. L. 1936. The floral morphology of the Polemoniaceae. Amer. J. Bot. 23: 501–511.

Finn, V. V. 1937. Vergleichende Embryologie und Karyologie einiger *Cuscuta*-Arten. (In Ukranian; German & Russian summary.) Žurn. Inst. Bot. Vseukrajins'k. Acad. Nauk 12(20): 83–99.

Goldblatt, P., & R. C. Keating. 1976. Chromosome cytology, pollen structure, and relationships of *Retzia capensis*. Ann. Missouri Bot. Gard. 63: 321–325.

Govil, C. M. 1972. Morphological studies in the family Convolvulaceae. IV. Vascular anatomy of the flower. Proc. Indian Acad. Sci. B, 75: 271–282.

Grant, V. 1959. Natural history of the Phlox family. Systematic Botany. Martinus Nijhoff. The Hague.

Grant, V., & K. A. Grant. 1965. Flower pollination in the Phlox family. Columbia Univ. Press. New York & London.

Hawkes, J. G., R. N. Lester, & A. D. Skelding, eds. 1979. The biology and taxonomy of the Solanaceae. Linn. Soc. Symposium 7. Academic Press. London, New York, San Francisco.

Herbst, E. E. 1972. 'N Morfologiese ondersoek van *Retzia capensis* Thunb. (English summary.) Thesis. University of Pretoria.

Inamdar, J. A., & G. S. R. Murthy. 1977. Vessels in some Solanaceae. Flora B 166: 441–447.

Inamdar, J. A., & R. C. Patel. 1969. Development of stomata in some Solanaceae. Flora B 158: 462–472.

Inamdar, J. A., & R. C. Patel. 1973. Structure, ontogeny and classification of trichomes in some Polemoniales. Feddes Repert. 83: 473–488.

Johnston, I. M. 1936. A study of the Nolanaceae. Proc. Amer. Acad. Arts 71: 1–87.

Johri, B. M. 1934. The development of the male and female gametophytes in *Cuscuta reflexa* Roxb. Proc. Indian Acad. Sci. B, 1: 283–289.

Johri, B. M., & B. Tiagi. 1952. Floral morphology and seed formation in *Cuscuta reflexa* Roxb. Phytomorphology 2: 162–180.

Kapil, R. N., P. N. Rustagi, & R. Venkataraman. 1968 (1969). A contribution to the embryology of the Polemoniaceae. Phytomorphology 18: 403–411.

Keeler, K. H. 1977. The extrafloral nectaries of *Ipomoea carnea* (Convolvulaceae). Amer. J. Bot. 64: 1182–1188.

Kennedy, P. B., & A. S. Crafts. 1931. The anatomy of *Convolvulus arvensis*, wild morning-glory or field bindweed. Hilgardia 5: 591–622.

Kuhlmann, J. G. 1947. Duckeodendraceae Kuhlmann (Nova familia). Arq. Serv. Florest. 3: 7–8.

Lindsey, A. A. 1938. Anatomical evidence for the Menyanthaceae. Amer. J. Bot. 25: 480–485.

Mason, H. L. 1945. The genus *Eriastrum* and the influence of Bentham and Gray upon the problem of generic confusion in Polemoniaceae. Madroño 8: 65–91.

Murray, M. A. 1945. Carpellary and placental structure in the Solanaceae. Bot. Gaz. 107: 243–260.

Nilsson, S. 1973. Pollen studies in the Menyanthaceae. *In*: Pollen and spores morphology of the recent plants, pp. 74–78. Proc. III, International Palynological Conference, Novosibirsk, USSR, 1971.

Nilsson, S., & R. Ornduff. 1973. Menyanthaceae Dum. World Pollen and Spore Flora 2: 1–20.

Patel, R. C., & J. A. Inamdar. 1971. Structure and ontogeny of stomata in some Polemoniales. Ann. Bot. (London) II. 35: 389–409.

Record, S. J. 1933. The woods of *Rhabdodendron* and *Duckeodendron*. Trop. Woods 33: 6–10.

Roberty, G. 1964. Les genres de Convolvulacées (esquisse). Boissiera 10: 129–156.

Robyns, W. 1930. L'organisation florale des Solanacées zygomorphes. Mém. Acad. Roy. Belgique, Cl. Sci. 11(8): 1–82.

Sayeedud-Din, M. 1953. Observations on the anatomy of some of the Convolvulaceae. Proc. Indian Acad. Sci. B, 37: 106–109.

Sengupta, S. 1972. On the pollen morphology of Convolvulaceae, with special reference to taxonomy. Rev. Palaeobot. Palynol. 13: 157–212.

Smith, D. M., C. W. Glennie, J. B. Harborne, & C. A. Williams. 1977. Flavonoid diversification in the Polemoniaceae. Biochem. Syst. Ecol. 5: 107–115.

Souèges, R. 1937. Embryogenie des Convolvulacées. Développement de l'embryon chez le *Convolvulus arvensis* L. Compt. Rend. Hebd. Séances Acad. Sci. 205: 813–815.

Souèges, R. 1939. Les lois du développement chez le *Polemonium caeruleum* L. Affinités des Polémoniaceées. Bull. Soc. Bot. France 86: 289–297.

Souèges, R. 1953. A propos de l'embryogénie des *Cuscuta*. Bull. Soc. Bot. France 100: 28–34.

Sripleng, A., & F. H. Smith. 1960. Anatomy of the seed of *Convolvulus arvensis*. Amer. J. Bot. 47: 386–392.

Stuchlik, L. 1967. Pollen morphology and taxonomy of the family Polemoniaceae. Rev. Palaeobot. Palynol. 4: 325–333.

Stuchlik, L. 1967. Pollen morphology in the Polemoniaceae. Grana Palynol. 7: 146–240.

Taylor, T. N., & D. A. Levin. 1975 (1976). Pollen morphology of Polemoniaceae in relation to systematics and pollination systems: scanning electron microscopy. Grana 15: 91–112.

Tiagi, B. 1951. A contribution to the morphology and embryology of *Cuscuta hyalina* Roth and *C. planiflora* Tenore. Phytomorphology 1: 9–21.

Tiagi, B. 1966. Floral morphology of *Cuscuta reflexa* Roxb. and *C. lupuliformis* Krocker with a brief review of the literature on the genus *Cuscuta*. Bot. Mag. (Tokyo) 79: 89–97.

Tucker, W. G. 1969. Serotaxonomy of the Solanaceae: a preliminary survey. Ann. Bot. (London) II. 33: 1–23.

Weberling, F. 1956. Weitere Untersuchungen zur Morphologie des Unterblattes bei den Dikotylen. III. Convolvulaceae; IV. Zygophyllaceae. Beitr. Biol. Pflanzen 33: 149–161.

Wherry, E. T. 1940. A provisional key to the Polemoniaceae. Bartonia 20: 14–17.

Wilson, K. A. 1960. The genera of Hydrophyllaceae and Polemoniaceae in the southeastern United States. J. Arnold Arbor. 41: 197–212.

Wilson, K. A. 1960. The genera of Convolvulaceae in the southeastern United States. J. Arnold Arbor. 41: 298–317.

Wojciechowska, B. 1972. Studia systematyczne nad nasionami rodziny Solanaceae Pers. (Subtitle: Systematic studies on the seeds of the Solanaceae family). Monogr. Bot. 36: 113–178.

Yuncker, T. G. 1932. The genus *Cuscuta*. Mem. Torrey Bot. Club 18: 113–331.

Данилова, М. Ф. 1952. О природе многокамерности плодов томатов (*Lycopersicon esculentum* Mill.). Труды Бот. Инст. АН СССР Сер. VII, 3: 87-146.

SYNOPTICAL ARRANGEMENT OF THE FAMILIES OF SOLANALES

1 Tall tree; fruit drupaceous, only one of the two locules fertile .... 1. DUCKEODENDRACEAE.

1 Herbs, shrubs, vines, or seldom small trees, only rarely (1 sp. of Convolvulaceae) tall trees; fruit of various types, most commonly capsular or baccate, only rarely drupaceous.

2 Stem with internal phloem; plants often producing alkaloids, but not iridoid compounds; always autotrophic.

3 Carpels (3) 5 (or in sets of 5); ovary with a terminal style and ripening into a schizocarp, or more often the gynoecium with a gynobasic style and ripening into 5 (or more) nutlets; endosperm-development cellular; Pacific slope of South America .... 2. NOLANACEAE.

3 Carpels 2 (rarely 3–5 or casually more); fruits diverse, but not schizocarpic; style not gynobasic except in a few bicarpellate Convolvulaceae; widely distributed.

4 Ovules and seeds (1–) more or less numerous; carpels, when 2, obliquely oriented; plants without latex; cotyledons not plicate; style simple, the stigma only shortly or scarcely lobed; endosperm-development most commonly cellular, but sometimes nuclear or helobial; herbs, shrubs, vines, or small trees .... 3. SOLANACEAE.

4 Ovules mostly 2 per carpel, basal and erect, only rarely more numerous; carpels, when 2, median (one anterior, the other posterior); plants generally with latex-canals or latex-cells; cotyledons plicate; style simple or often more or less deeply cleft, or the styles distinct; endosperm-development nuclear; plants most commonly twining herbaceous vines, but sometimes erect herbs or shrubs or even trees ....4. CONVOLVULACEAE.

2 Stem without internal phloem; plants sometimes producing iridoid compounds, but without alkaloids; autotrophic or (Cuscutaceae) parasitic.

5 Twining stem-parasites, not rooted in the ground at maturity, nearly or quite without chlorophyll; embryo without well differentiated cotyledons .... 5. CUSCUTACEAE.

5 Autotrophic plants, rooted in the ground; embryo with evident cotyledons.

6 Aquatic or semi-aquatic herbs, often with scattered vascular bundles in the stem; plants producing iridoid compounds; corolla-lobes valvate or induplicate-valvate or sometimes imbricate; carpels 2; ovules numerous on parietal placentas ....6. MENYANTHACEAE.

6 Terrestrial to semi-aquatic herbs or subshrubs; vascular bundles of the stem in a ring, or the xylem forming a continuous ring; plants without iridoid compounds; corolla-lobes imbricate or convolute.

7 Carpels (2) 3 (4); placentation axile; flowers generally

with a nectary disk around the ovary; corolla-lobes convolute; endosperm-development nuclear ....7. POLEMONIACEAE.

7 Carpels 2; placentation parietal or rarely axile; flowers without a nectary-disk; corolla-lobes imbricate or less often convolute; endosperm-development cellular or less often nuclear ....8. HYDROPHYLLACEAE.

1. Family DUCKEODENDRACEAE Kuhlmann 1947, the Duckeodendron Family

Tall tree; wood with large radial intercellular canals; vessel-segments with simple, slightly oblique perforations; imperforate tracheary elements with numerous small, bordered pits; wood-rays closely spaced, homocellular or nearly so, uniseriate or some of them partly biseriate; wood-parenchyma in tangential bands 1–2 cells thick. LEAVES alternate, simple, exstipulate. FLOWERS in small terminal cymes, perfect, regular, sympetalous, 5-merous except for the gynoecium; calyx persistent, tubular, with terminal teeth; corolla chloroleucous, funnelform, with elongate tube and short, imbricate lobes; stamens long-exserted, the filaments attached to the corolla-tube alternate with the lobes; anthers deeply sagittate, dithecal, opening by longitudinal slits; a well developed nectary-disk present around the base of the ovary; gynoecium of 2 carpels united to form a superior, compound, bilocular ovary with a terminal style and 2-lobed stigma; one anatropous ovule present in each locule, but only one locule fertile. FRUIT a large red, shining drupe with a bony, bilocular endocarp and fibrous mesocarp; fertile locule U-shaped, containing a U-shaped seed with a U-shaped embryo in scanty, oily endosperm; sterile locule straight or nearly so.

The family Duckeodendraceae consists of the single genus and species *Duckeodendron cestroides* Kuhlmann, native to the Amazon basin in Brasil. Proper systematic placement of *Duckeodendron* awaits more detailed morphological and chemical studies. Meanwhile it may stand as a distinct family of the Solanales.

2. Family NOLANACEAE Dumort. 1829 nom. conserv., the Nolana Family

Herbs or small shrubs, not tanniferous, lacking both ellagic acid and proanthocyanins, not cyanogenic, and without iridoid compounds; both glandular and eglandular hairs commonly present, the former with a 1–2-celled stalk and a unicellular or multicellular head, the latter commonly simple and uniseriate; secretory cells scattered in the parenchyma of the stem; some of the cells of the parenchymatous tissues containing crystal sand; vessel-segments with simple perforations; imperforate tracheary elements consisting of septate fibers; wood-rays uniseriate; internal phloem characteristically present, generally in small strands. LEAVES alternate (or those of the inflorescence in alternate pairs, the members of a pair unequal and both on the same side of the stem, as in many Solanaceae), simple, entire, commonly more or less succulent, sometimes small, ericoid, and unifacial; stomates tending to be anisocytic; petiole with a flattened, slightly arcuate strand of xylem surrounded by phloem, and with 2 small peripheral bundles; stipules wanting. FLOWERS mostly solitary in the axils, perfect, wholly pentamerous; calyx tubular-campanulate, with slightly imbricate (seldom valvate), often unequal lobes; corolla campanulate or funnelform, plicate between the lobes, regular or obscurely bilabiate; stamens unequal, 3 longer than the other 2, the filaments attached toward the base of the corolla-tube, alternate with the lobes; anthers dithecal, opening by longitudinal slits; pollen-grains binucleate, tricolporate; an annular, crenate or lobulate nectary-disk commonly present around the base of the ovary; gynoecium typically of (3) 5 carpels, these in *Alona* united to form a superior, pentalocular ovary with basal-axile placentation and a terminal style, but in *Nolana* largely distinct and united only by their gynobasic style, and in some spp. of *Nolana* the carpels in 2 or more concentric ranks of 5, these ranks originating phyletically by tangential division; stigma wet, peltate; ovules 1-several in each carpel, anatropous or hemitropous, unitegmic, tenuinucellar, with an integumentary tapetum; endosperm-development cellular. FRUIT of small, separate or separating mericarps with a stony endocarp (nutlets), each mericarp with 1-several seeds, these with curved or spiral, dicotyledonous embryo and copious, oily endosperm. X = 12.

The family Nolanaceae consists of 2 genera, *Nolana,* with about 60 species, and *Alona,* with only 6. Both are native to the Pacific slope of South America, in northern Chile and southern Peru, most commonly along the seashore. A single species occurs in the Galapagos Islands. Johnston's view that the family is derived from the Solanaceae is now widely accepted.

3. Family SOLANACEAE A. L. de Jussieu 1789 nom. conserv., the Potato Family

Herbs, shrubs, vines, or small trees, often sympodially branched, with diverse sorts of often stellate or otherwise branched hairs and often armed with prickles, generally with acylated anthocyanins, commonly producing various kinds of alkaloids (especially of the tropane, nicotine, and steroid groups), but without iridoid compounds and usually not tanniferous, lacking both ellagic acid and proanthocyanins, seldom cyanogenic; solitary or clustered crystals of calcium oxalate, various in

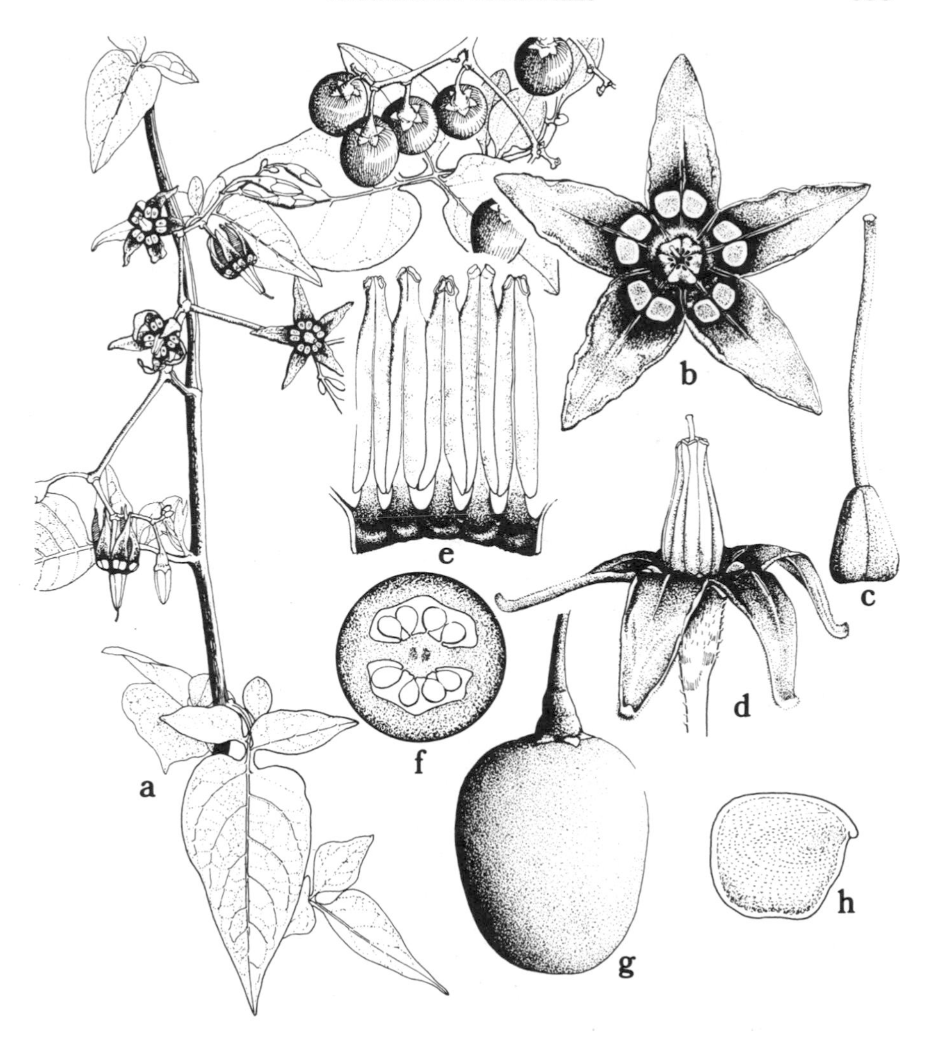

FIG. 6-8 Solanaceae. *Solanum dulcamara* L. a, habit, ×1; b, d, flower, ×3; c, pistil, ×6; e, androecium, opened out, ×6; f, schematic cross-section of ovary, ×15; g, fruit, ×3; h, seed, ×6.

form, often present in some of the cells of the parenchymatous tissues; stem sometimes with anomalous structure; nodes unilacunar, with a single broad trace or seldom 2 or 3 traces; vessel-segments with simple perforations; imperforate tracheary elements with simple or seldom bordered pits; wood-rays of diverse types, sometimes all uniseriate, sometimes some of them up to 8 cells wide, heterocellular to nearly homocellular; wood-parenchyma mostly either scanty-paratracheal or mainly apotracheal; internal phloem consistently present at the periphery of the pith, either as a complete cylinder or as separate strands, often as part of bicollateral bundles; cork variously superficial or deep-seated in origin. LEAVES alternate (those of the inflorescence often in alternate pairs, the members of a pair both on the same side of the stem and frequently unequal, the arrangement reflecting complex sympodial growth and lateral displacement of ontogenetically terminal inflorescences), simple and entire or variously toothed or cleft, sometimes pinnately compound or trifoliolate; stomates usually anomocytic, less often anisocytic or diacytic or paracytic; main veins with bicollateral bundles; petiole commonly with an arcuate bicollateral bundle (or the phloem wholly surrounding the xylem) and a few smaller lateral, also bicollateral bundles; stipules wanting. FLOWERS usually perfect, in various sorts of inflorescences that mostly appear to be ultimately cymose (determinate) in origin, or sometimes borne singly; corolla sympetalous, rotate to tubular, with (4) 5 (6) lobes, these commonly plicate (and sometimes also convolute) in bud, or seldom merely convolute, imbricate, or valvate, usually regular or nearly so, but sometimes (some Salpiglossoideae) distinctly irregular and even bilabiate; stamens attached to the corolla-tube, usually as many as and alternate with the lobes, but sometimes only 4 or even (*Schizanthus*) only 2 of them fertile, and the others staminodial or suppressed; anthers mostly tetrasporangiate and dithecal, often connivent, opening by longitudinal slits or by terminal pores or slits, sometimes one of the thecae more or less reduced, in *Browallia* and a few other genera monothecal; pollen-grains binucleate or seldom trinucleate, (2) 3–5 (6)-colpate or -colporate, or seldom inaperturate; nectary-disk usually present around the base of the ovary; gynoecium mostly of 2 carpels oriented obliquely to the median plane of the flower (neither collateral nor median, but diagonal) and united to form a compound, superior, bilocular ovary, occasionally 4-locular even though 2-carpellate, irregularly 3–5 locular in *Nicandra*, and casually pluricarpellate in some cultigens of several genera; in *Henoonia* the ovary pseudomonomerous, unilocular, with a single more or less parietal ovule; style terminal, with a mostly 2-lobed, generally wet stigma; ovules (1–) more or less numerous in each locule, on axile, often thickened placentas, anatropous to hemitropous or somewhat amphitropous, with a massive single integument, tenuinucellar, and with an integumentary tapetum; endosperm-development cellular, or less often nuclear or helobial. FRUIT a berry or a variously dehiscent, often septicidal capsule, or seldom a drupe; seeds with a dicotyledonous, generally linear, straight to more often curved (often subperipheral) embryo; endosperm mostly oily and proteinaceous, only seldom starchy, or the endosperm rarely wanting. X = 7–12+. (Goetziaceae, Salpiglossidaceae)

The family Solanaceae consists of about 85 genera and 2800 species, of nearly cosmopolitan distributions, but best developed in tropical South America. Many species are weedy or exploit disturbed habitats, even in tropical rain-forests. The woody habit is probably secondary within the family; it may relate in some way to the change in the inflorescence from strictly terminal to lateral-sympodial. The family is dominated by the very large genus *Solanum*, with perhaps 1400 species; *S. tuberosum* L. is the common potato. *Lycopersicon esculentum* Mill., perhaps equally well considered a species of *Solanum*, is the common tomato. Fruits of *Capsicum* (red pepper) are eaten or ground up as spice. Species of *Atropa* (bella-donna), *Datura* (Jimson-weed), *Duboisia, Hyoscyamus* (henbane), *Mandragora* (mandrake), and *Nicotiana* (tobacco) are familiar as the sources of certain alkaloids. Species of *Browallia, Lycium, Nicotiana, Petunia, Physalis, Salpiglossis, Schizanthus, Solanum* and other genera are cultivated for ornament.

Serological tests support the inclusion of *Salpiglossis* in the Solanaceae, but cast doubts on *Schizanthus*. Both of these genera are unusual in the family in having fewer functional stamens than corolla-lobes, and *Schizanthus* is further aberrant in having a bilabiate corolla, but they have the characteristic obliquely set carpels of the Solanaceae.

Tropane alkaloids have been found in 21 genera of the Solanaceae, including 3 (*Anthocercis, Anthotroche* and *Duboisia*) of the subfamily Salpiglossoideae. Outside the Solanaceae they are irregularly distributed in scattered genera of highly diverse families, including the Brassicaceae, Convolvulaceae, Dioscoreaceae, Erythroxylaceae, Euphorbiaceae, Orchidaceae, Proteaceae, and Rhizophoraceae.

4. Family CONVOLVULACEAE A. L. de Jussieu 1789 nom. conserv., the Morning-glory Family

Mostly herbs, commonly twining and climbing or prostrate, seldom erect or even arborescent shrubs, one monotypic genus (*Humbertia*) a tall tree, provided with diverse sorts of glandular and eglandular hairs, the latter often with 2 arms on a short stalk, or uniseriate and with an elongate terminal cell; plants often with acylated anthocyanins, often producing alkaloids of the indole or other groups, especially the ergoline

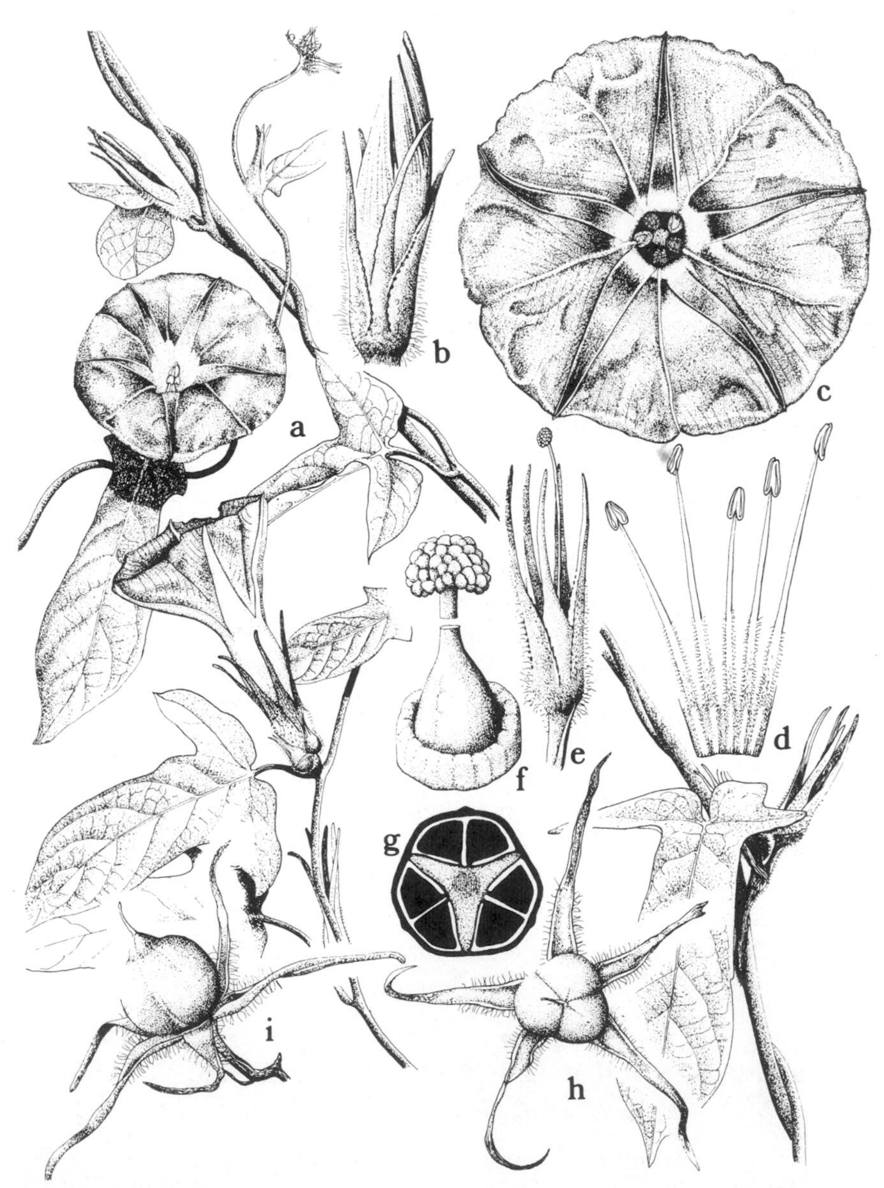

FIG. 6-9 Convolvulaceae. *Ipomoea hederacea* (L.) Jacq. a, habit, $\times\frac{1}{2}$; b, flower bud, $\times 1$; c, flower, from above, $\times 1$; d, stamens, $\times 1$; e, calyx and pistil, $\times 1$; f, pistil and nectary, $\times 5$; g, schematic cross-section of ovary, $\times 10$; h, i, 2 views of fruit, $\times 1$.

subgroup of the indole group, often cyanogenic, and commonly rich in caffeic acid and related compounds, but without iridoid compounds, only seldom saponiferous, and generally not tanniferous, lacking both proanthocyanins and ellagic acid; various sorts of solitary or clustered crystals of calcium oxalate commonly present in some of the cells of the parenchymatous tissues; stem often with anomalous secondary growth, and usually with articulated, non-anastomosing latex-canals or latex-cells, the latter scattered or in vertical rows, the latex often highly purgative; nodes unilacunar; xylem usually forming a continuous ring with only narrow rays; vessel-segments with simple perforations; imperforate tracheary elements with bordered pits; internal phloem consistently present in the stem, as separate strands, or in bicollateral bundles, or in inverted medullary bundles, or as a nearly continuous ring at the margin of the pith; interxylary phloem often also present; cork usually superficial in origin. LEAVES alternate, simple, entire or toothed to sometimes lobed; stomates paracytic or less often anisocytic, seldom anomocytic; veins bicollateral, the larger ones often with associated latex-cells; petiole commonly with an arc-shaped bicollateral vascular strand, and often with smaller accessory bundles; stipules wanting. FLOWERS often large and showy, borne in terminal or axillary dichasia, or often solitary in the axils, seldom in racemiform or paniculiform inflorescences, generally subtended by a pair of bracts, these sometimes enlarged and forming an involucre, perfect or rarely (*Hildebrandtia*) unisexual, 5-merous as to the calyx, corolla, and androecium (4-merous in *Hildebrandtia*); sepals imbricate, sometimes unequal, generally distinct or connate only at the base, seldom connate for much of their length; corolla sympetalous, commonly funnelform, evidently or scarcely lobed, regular (obliquely irregular in *Humbertia*), commonly induplicate-valvate and often also convolute in bud, or merely convolute when more strongly lobed, the convolution typically clockwise at least in *Convolvulus* and *Ipomoea*; stamens as many as and alternate with the lobes or connate members of the corolla, attached toward the base of the corolla-tube, the filaments often unequal; anthers tetrasporangiate or dithecal, opening by longitudinal slits; pollen-grains nearly always binucleate, tricolpate to pantoporate; an annular, commonly lobulate nectary-disk usually present around the base of the ovary; gynoecium of 2 (3–5) carpels united to form a superior, compound ovary with as many locules as carpels (rarely unilocular) and with a terminal, simple or often more or less deeply cleft style or with distinct styles, or seldom (*Dichondra* and *Falkia*) the bicarpellate ovary deeply 2- or 4-lobed, with the segments united mainly by the base of the deeply cleft gynobasic style; stigmas dry; carpels median (one anterior, the other posterior) when the ovary is bicarpellate; ovules mostly 2 per carpel (numerous in *Humbertia*), basal or basal-axile, erect, anatropous, apotropous, with the micropyle directed downward and outward, with a massive single integument, tenuinucellar or sometimes crassinucellar, with an integumentary tapetum; endosperm-development nuclear. FRUIT a loculicidal (or sometimes irregularly dehiscent) capsule, or less often indehiscent and baccate or nutlike; embryo large, straight or curved, with 2 plicate, often bifid cotyledons, embedded in a hard, often cartilaginous endosperm that has food reserves of oil, protein, and hemicellulose. X = 7–15+. (Dichondraceae, Humbertiaceae)

The family Convolvulaceae as here defined consists of about 50 genera and 1500 species, nearly cosmopolitan in distribution, but best developed in tropical and subtropical regions. The largest genera are *Ipomoea* (400, morning-glory) and *Convolvulus* (250, bindweed). *Ipomoea batatas* (L.) Lam. is the sweet potato.

The bulk of the family hangs together very well, but 3 small genera are aberrant, as noted in the description. *Humbertia* is sometimes taken as a separate family Humbertiaceae, and *Dichondra* and *Falkia* collectively are sometimes considered to form a family Dichondraceae. The taxonomic rank at which these groups should be received is debatable, but the relationships are not in dispute.

5. Family CUSCUTACEAE Dumortier 1829 nom. conserv., the Dodder Family

Twining, mostly annual herbs, with very little or usually no chlorophyll, parasitic, attaching to the host by intrusive haustoria, the small terrestrial root-system soon degenerating, so that the mature plant is not connected to the ground; hairs, when present, mostly unicellular or bicellular, not glandular; plants sometimes tanniferous and sometimes saponiferous, but lacking iridoid compounds and mostly not cyanogenic; stem thread-like, commonly orange or yellowish, with scattered secretory cells in the cortex, pericycle, and sometimes also the pith, the vascular system somewhat reduced, without internal phloem; vessel-segments with simple perforations. LEAVES much-reduced, scale-like but often with stomates, these anomocytic or surrounded by one or more circles of subsidiary cells. FLOWERS small, in dense, head-like or shortly spike-like, basically cymose inflorescences, with or without subtending bracts,

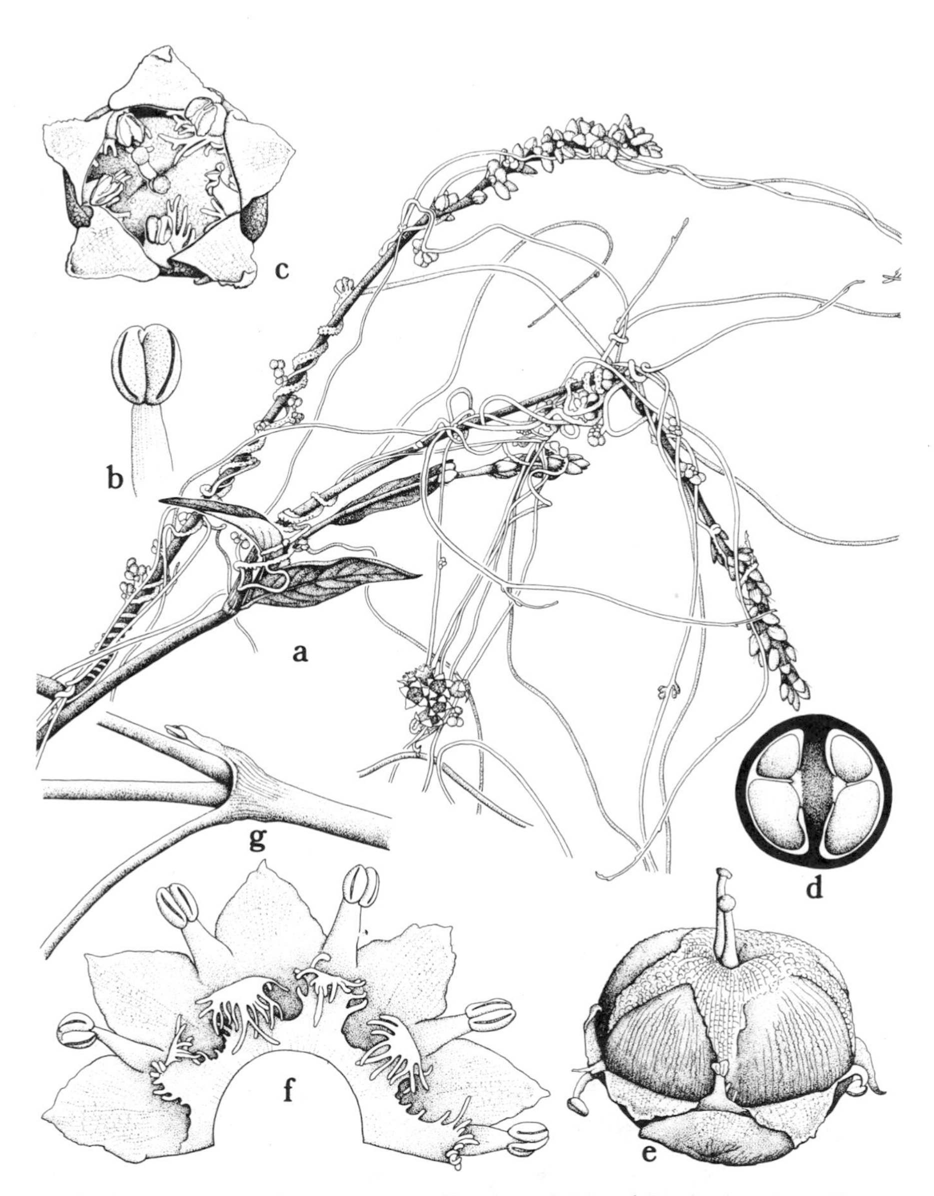

FIG. 6-10 Cuscutaceae. *Cuscuta pentagona* Engelm. a, habit, $\times\frac{1}{2}$; b, stamen, $\times 24$; c, flower, from above, $\times 12$; d, schematic cross-section of ovary, $\times 12$; e, fruit, $\times 12$; f, corolla, opened out to show stamens and staminodes, $\times 12$; g, nodal region, $\times 4$.

perfect, regular, 5-merous or less often 4-merous (rarely 3-merous) as to the calyx, corolla, and androecium; sepals each with a single vascular trace, persistent, distinct or often connate at the base or essentially throughout, so that the calyx-tube may be virtually entire; corolla mostly white or pink, sympetalous, with imbricate lobes, the tube usually bearing a whorl of variously fringed or cleft scales beneath (and aligned with) the stamens; filaments attached to the corolla-tube, alternate with the lobes; anthers short, tetrasporangiate and dithecal, opening by longitudinal slits; pollen-grains (2) 3-nucleate, smooth, 3–6-colpate; gynoecium of 2 (3) carpels united to form a compound, superior ovary with as many locules as carpels (the partitions sometimes incomplete); base of the ovary at least sometimes nectariferous; styles terminal and distinct, or united below into a more or less deeply cleft common style; stigmas dry, variously discoid, capitate, conic, or cylindric; ovules 2 in each locule, erect on basal-axile (or intruded-parietal) placentas, anatropous, with a massive single integument, tenuinucellar; a placental obturator commonly forming a canopy over the micropyle; endosperm-development nuclear; a several-celled suspensor-haustorium commonly developed. FRUIT a circumscissile or irregularly dehiscent capsule, or indehiscent and then often somewhat fleshy; embryo slender, filiform-cylindric, nearly or quite acotyledonous but sometimes with an enlargement at one end, peripheral and strongly curved or spirally wound around the starchy endosperm. X most commonly = 7, sometimes 15.

The family Cuscutaceae consists of the single genus *Cuscuta*, with about 150 species. The genus is nearly cosmopolitan, but best developed in the New World, especially in warmer regions.

It is generally agreed that *Cuscuta* is related to and derived from the Convolvulaceae, and many authors include the genus in that family. The differences seem as significant, however, as those separating other recognized families of the order.

6. Family MENYANTHACEAE Dumortier 1829 nom. conserv., the Buckbean Family

Aquatic or semi-aquatic herbs with a well-developed system of intercellular canals and spaces in the stem, producing various sorts of iridoid compounds but not gentiopicroside, occasionally saponiferous, but not cyanogenic and only seldom tanniferous, generally lacking both ellagic acid and proanthocyanins, and commonly storing carbohydrate as inulin; crystals of calcium oxalate wanting, but branching sclereids often

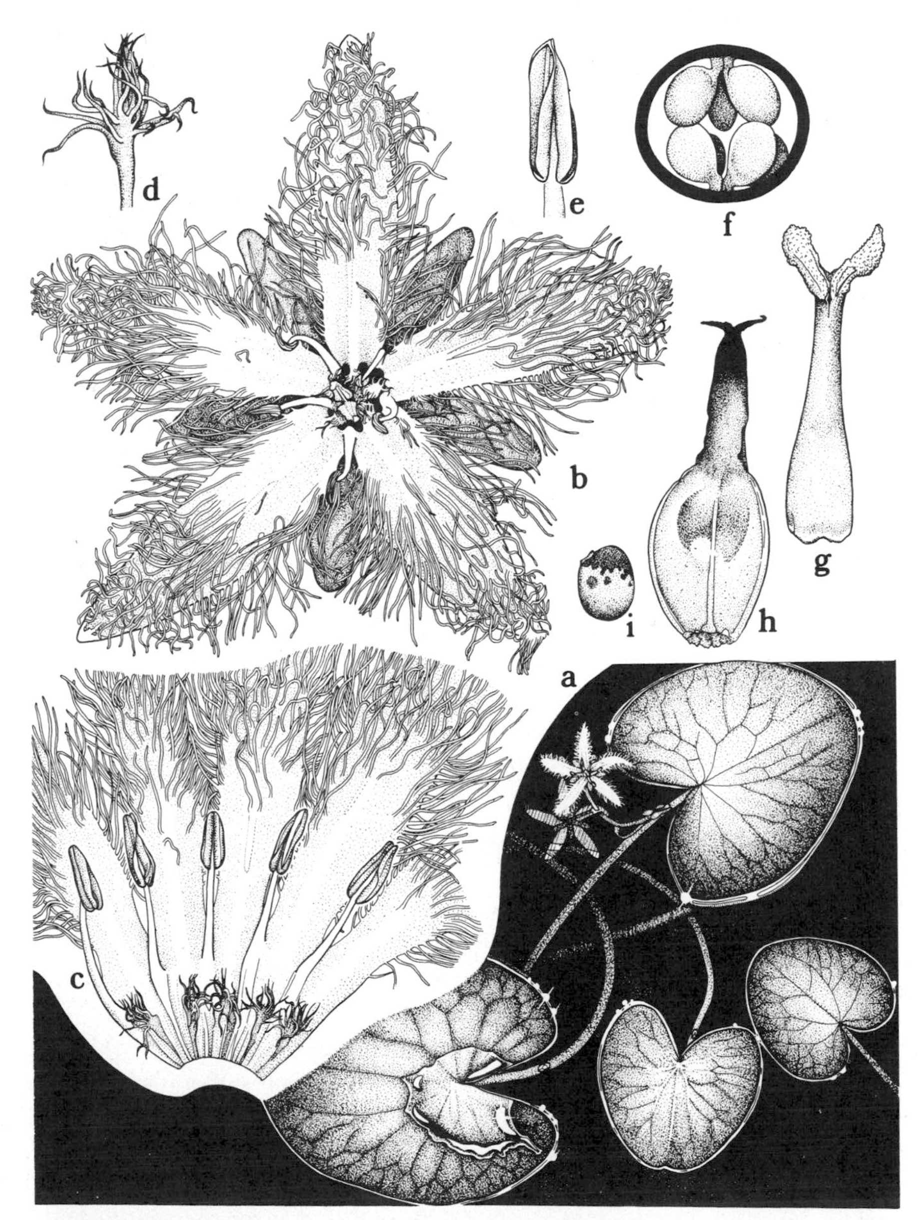

Fig. 6-11 Menyanthaceae. *Nymphoides indica* (Thwaites) Kuntze. a, habit, ×½; b, flower, from above, ×3; c, corolla, opened up, showing stamens and staminodes, ×3; d, staminode, ×6; e, anther, ×6; f, schematic cross-section of ovary, ×15; g, pistil, ×6, h, fruit, ×4; i, seed, ×6.

present in the parenchyma of the stem; vascular system of the stem more or less strongly dissected, without interfascicular cambium, the bundles often scattered to form an atactostele; nodes pentalacunar or trilacunar; vessel-segments usually or always with simple perforations; internal phloem wanting, but the inner portion of the vascular bundles generally composed of unlignified parenchyma. LEAVES alternate, simple and variously reniform, cordate, peltate, or linear, or (*Menyanthes*) trifoliolate; stomates anomocytic; petiole commonly with one or two arcs or a circle of collateral vascular bundles or with scattered bundles, sheathing at the base; stipules wanting, or represented only by the expanded wing-margins of the petiole. FLOWERS solitary or in diverse sorts of inflorescences, perfect, regular, typically pentamerous as to the calyx, corolla, and androecium; sepals evidently connate to essentially distinct; corolla sympetalous, with valvate or induplicate-valvate or imbricate lobes, the margins or the inner surface of the lobes often fimbriate or crested; petals said to have 3 vascular traces in *Nymphoides*; filaments attached to the corolla-tube alternate with the lobes; anthers mostly sagittate, tetrasporangiate and dithecal, opening by longitudinal slits; pollen-grains trinucleate, tricolporate; a set of fringed scales (staminodes?) sometimes present in the corolla-tube alternate with the stamens; a nectary-disk commonly present around the base of the ovary; gynoecium of 2 carpels united to form a compound, superior to half-inferior, unilocular ovary with a terminal style and 2-lobed, wet, papillate stigma; ovules numerous on the two parietal (often intruded) placentas, anatropous, tenuinucellar, with a massive single integument and an integumentary tapetum; endosperm-development cellular. FRUIT capsular, irregularly dehiscent or dehiscent by 2 or 4 valves, or sometimes indehiscent and baccate; seeds with a linear, axile, dicotyledonous embryo and copious, firm, oily endosperm. X = 9, 17.

The family Menyanthaceae consists of 5 genera and some 30–35 species, of cosmopolitan distribution. The largest genus is *Nymphoides* (20), with deeply cordate floating leaves, but the most familiar species is *Menyanthes trifoliata* L., the buckbean, a circumboreal plant of marshes, with emergent, trifoliolate leaves. The other genera are *Villarsia* (10), *Liparophyllum* (1), and *Nephrophyllidium* (1).

The Menyanthaceae have often been included in or associated with the Gentianaceae, from which they are amply distinguished by their alternate leaves, more dissected stele without internal phloem, cellular endosperm, integumentary tapetum, and absence of gentiopicroside. Nilsson (1973) considers that the pollen of Menyanthaceae is distinctive, but in *Menyanthes* and *Nephrophyllidium* (*Fauria*) is "reminiscent of certain Gentianaceous taxa." One may legitimately wonder what other families might have been called to mind had a wider net of comparison been cast. There is in any case enough diversity of pollen in both the Solanales and the Gentianales to accommodate the Menyanthaceae without difficulty.

The Menyanthaceae obviously belong to the Asteridae, but a long list of differences can be cited when they are compared to any other one family. The Solanales and the Gentianales are the only orders that cannot immediately be excluded from close affinity on the basis of gross floral and vegetative morphology.

Aside from the anatomical features associated with their aquatic habitat, the Menyanthaceae have no significant characters that are not well known in one or another of the two orders Solanales and Gentianales. Therefore it seems clear that they should be referred to one of these orders. The alternate leaves, integumentary tapetum, cellular endosperm, induplicate-valvate corolla, and absence of internal phloem would all be unusual in the Gentianales, but are common (or at least not unusual) in the Solanales. The one feature that points toward a close association with the Gentianales is the presence of iridoid compounds. These are very common in the Gentianales, but otherwise unknown in the Solanales. Iridoids are also common in several other orders of Asteridae, as well as in certain orders of both the Rosidae and the Dilleniidae. Their occasional presence in the Solanales should therefore not be surprising, especially inasmuch as the Asteridae form a closely knit group on the basis of both mega- and micromorphological features.

I conclude that the Menyanthaceae are better referred to the Solanales than to the Gentianales.

7. Family POLEMONIACEAE A. L. de Jussieu 1789 nom. conserv., the Phlox Family

Herbs, or occasionally shrubs, lianas (*Cobaea*) or even small trees (*Cantua*), often storing carbohydrate as inulin (unlike the Hydrophyllaceae and Convolvulaceae), often mephitic, the odor emanating from gland-tipped hairs, very often saponiferous (the saponins sometimes triterpenoid), but mostly not cyanogenic and only seldom tanniferous, lacking ellagic acid and usually lacking proanthocyanins, but otherwise with a diverse array of flavonoids, sometimes including acylated antho-

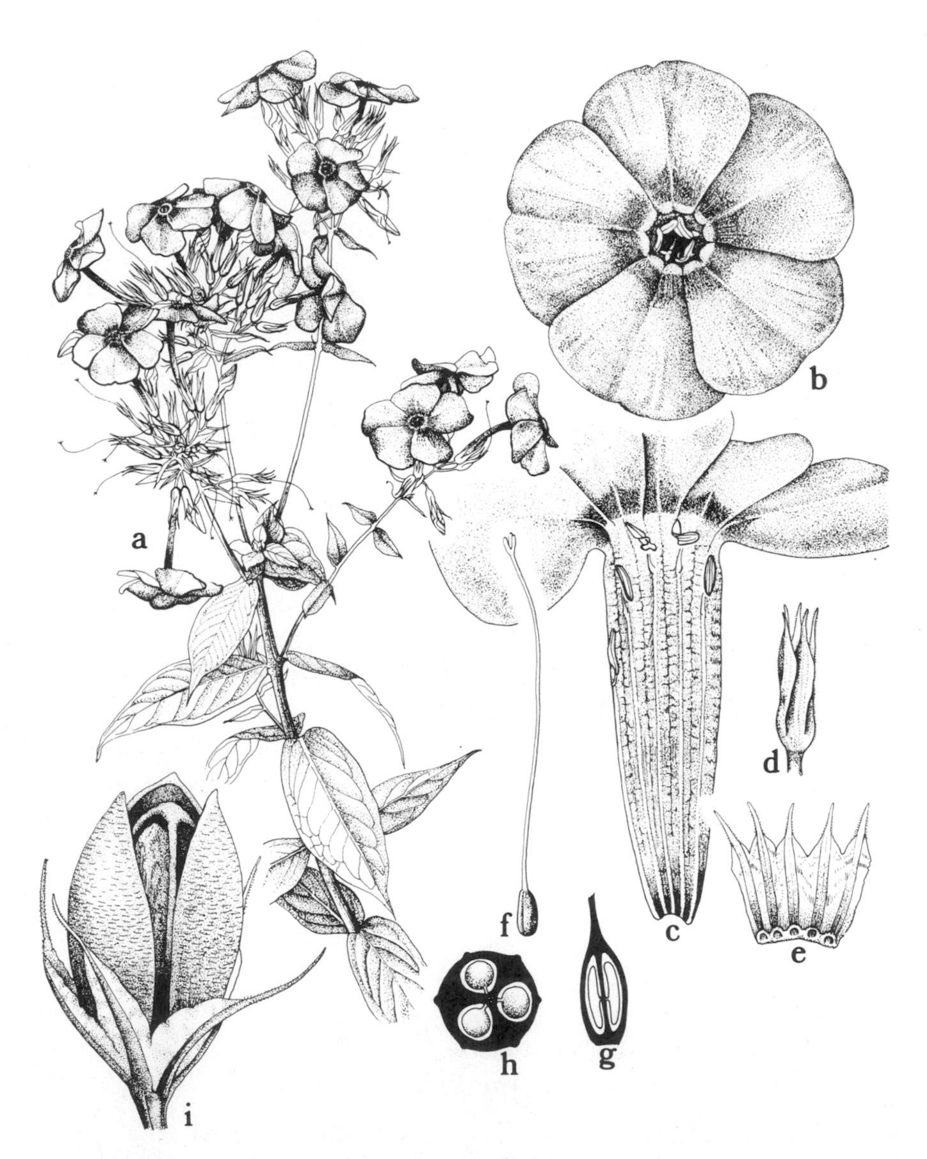

FIG. 6-12 Polemoniaceae. *Phlox paniculata* L. a, habit, $\times\frac{1}{2}$; b, flower, from above, $\times 2$; c, corolla, opened out, with attached stamens, $\times 2$; d, calyx, $\times 2$; e, calyx, opened out, from within, $\times 2$; f, pistil, $\times 2$; g, schematic long-section of ovary, $\times 6$; h, schematic cross-section of ovary, $\times 12$; i, opened fruit, $\times 6$.

cyanins; alkaloids and iridoid compounds wanting; solitary or clustered crystals of calcium oxalate sometimes present in some of the cells of the parenchymatous tissues, but more often wanting; nodes unilacunar; xylem commonly forming a continuous ring, even in herbaceous species, without rays or with only uniseriate rays; vessel-segments with simple perforations, or occasionally some of them scalariform with only a few cross-bars; imperforate tracheary elements with simple or inconspicuously bordered pits; phloem usually forming a continuous ring external to the xylem, or sometimes in small strands embedded in sclerenchyma. LEAVES opposite in *Phlox, Linanthus, Linanthastrum,* and sometimes *Leptodactylon,* and whorled in *Gymnosteris,* otherwise generally alternate (or only the lower opposite), simple and entire or merely toothed to pinnately compound or dissected, or sometimes palmatifid; stomates anomocytic or less often paracytic; petiole commonly with an arcuate collateral vascular strand, sometimes with some accessory lateral bundles; stipules wanting. FLOWERS solitary or more often in open or compact (often head-like), variously modified cymes, perfect, generally 5-merous as to the calyx, corolla, and androecium, seldom 4-merous, very rarely 6-merous; sepals essentially distinct in *Cobaea,* in the other genera connate to form a tube with equal or unequal lobes, the tube often with alternating green costae and hyaline intervals; corolla sympetalous, regular or occasionally somewhat bilabiate, its lobes convolute in bud; stamens alternate with the corolla-lobes, the filaments attached to the tube, sometimes at differing levels; anthers tetrasporangiate and dithecal, opening by longitudinal slits; pollen-grains binucleate, 4-many-colporate or -porate, often pantoporate; an annular nectary-disk nearly always present around the base of the ovary, the annulus sometimes lobed (and even vasculated) alternately with the stamens; gynoecium of 3 (seldom only 2, rarely 4) carpels united to form compound, superior ovary with axile placentas and as many locules as carpels; style terminal, simple, usually with separate stigmas (as many as the carpels); ovules 1-many in each locule, anatropous or hemitropous, with a massive single integument, tenuinucellar, provided with an integumentary tapetum; endosperm-development nuclear. FRUIT generally a capsule, loculicidally or seldom (*Cobaea*) septicidally dehiscent, or sometimes irregularly or scarcely dehiscent; seeds 1-many, often becoming mucilaginous when wetted; embryo straight or slightly curved, dicotyledonous, usually spatulate, surrounded by the copious and oily endosperm, or the endosperm seldom scanty or rarely essentially wanting. X primitively = 9, sometimes reduced to 8, 7, or 6. (Cobaeaceae)

The family Polemoniaceae consists of about 18 genera and nearly 300 species, best developed in temperate North America, especially in the Cordilleran region, but extending south to western South America, and north to Alaska and thence across most of temperate Eurasia. The largest genera are *Gilia* and *Phlox*, each with more than 50 species. Species of *Phlox* are common garden-ornamentals. Otherwise the family is of little economic importance.

The pollen of the Polemoniaceae is diverse in structure and surface-ornamentation, but Taylor and Levin (1975) could find no correlation with the equally diverse means of pollination!

Fossil pollen referred to the Polemoniaceae is known from Miocene and more recent deposits. The known megafossil record does not begin until the Pleistocene.

Most authors (Thorne excepted) agree that the Polemoniaceae are allied to both the Hydrophyllaceae and Convolvulaceae. The relationship to the Hydrophyllaceae is in my opinion the closer one. Metcalfe and Chalk have noted that, aside from lacking internal phloem, the Polemoniaceae are anatomically much like the Gentianaceae. This similarity reflects the close linkage among most of the families of Asteridae.

8. Family HYDROPHYLLACEAE R. Brown 1817 nom. conserv., the Waterleaf Family

Herbs or seldom (*Eriodictyon, Wigandia*) shrubs, often rough-hairy (the hairs often with calcified walls or containing a basal cystolith), or with gland-tipped hairs and then sometimes mephitic or otherwise odorous, only rarely saponiferous or tanniferous, lacking both ellagic acid and proanthocyanins as well as alkaloids and iridoid compounds, and not cyanogenic; clustered crystals of calcium oxalate only seldom present in some of the cells of the parenchymatous tissues; nodes unilacunar; vessel-segments with simple perforations; imperforate tracheary elements with bordered pits; woody forms with diffuse parenchyma and heterocellular to homocellular rays up to 3 or 4 cells wide. LEAVES alternate or sometimes partly or wholly opposite, simple and entire to pinnately cleft, compound, or dissected, rarely palmately compound; stomates anomocytic; petiole commonly with an arc-shaped collateral vascular strand and some accessory lateral bundles, seldom with a hollow, cylindrical vascular trace and an included small bundle; stipules wanting. FLOWERS solitary, or more often in variously modified (often

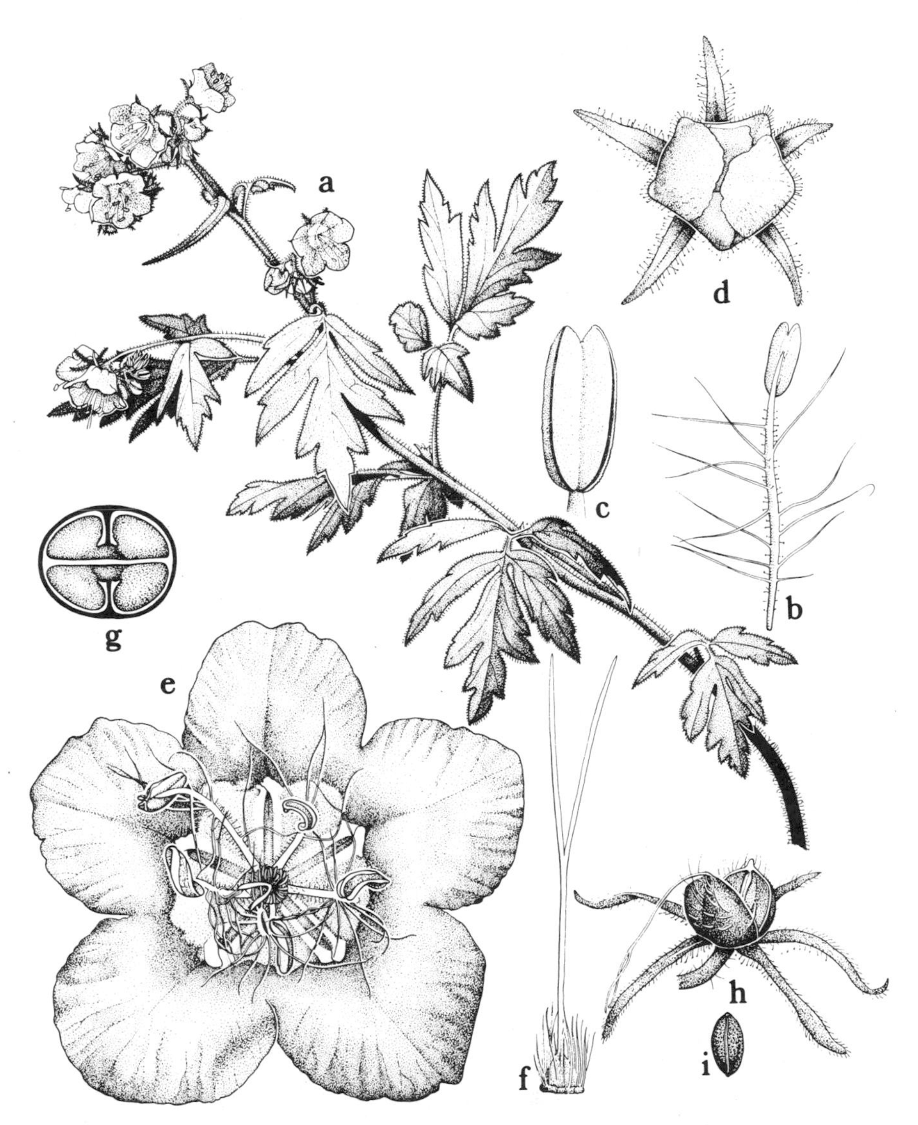

FIG. 6-13 Hydrophyllaceae. *Phacelia bipinnatifida* Michx. a, habit, ×½; b, stamen, ×4; c, anther, ×16; d, flower-bud, ×4; e, flower, from above, ×4; f, pistil, ×4; g, schematic cross-section of ovary, ×16; h, opened fruit, ×2; i, seed, ×2.

helicoid) cymes, perfect, sympetalous, generally 5-merous as to the calyx, corolla, and androecium, rarely 4-merous or (South African genus *Codon*) 10–12-merous; calyx cleft to the middle or more commonly to the base or nearly so, with imbricate segments, sometimes with appendages externally at the sinuses; corolla regular or nearly so, with imbricate or less commonly convolute lobes; stamens as many as and alternate with the corolla-lobes, attached toward the base or well up in the tube, very often flanked by a pair of small scales; anthers tetrasporangiate and dithecal, opening by longitudinal slits; pollen-grains tricolpate, tricolporate, or 5–6-colpate; nectary-disk nearly always wanting; gynoecium of 2 carpels united to form a compound, superior or rarely (as in *Nama stenocarpum*) half-inferior, mostly unilocular ovary; style terminal, usually more or less deeply bifid, sometimes cleft essentially to the base, as in spp. of *Nama*, or undivided (*Romanzoffia*); stigmas mostly capitate; placentas 2, usually parietal, but often more or less intruded, and sometimes meeting and joined, the ovary then bilocular and the placentation axile; ovules 2-many on each placenta, anatropous or amphitropous, with a massive single integument, tenuinucellar, provided with an integumentary tapetum; endosperm-development cellular or less often nuclear. FRUIT generally a loculicidal (sometimes also septicidal) capsule, or sometimes irregularly dehiscent or indehiscent; seeds with a small to fairly large, straight, spatulate or sometimes linear, dicotyledonous embryo surrounded by abundant or scanty, oily, more or less fleshy (seldom hard) endosperm. X = 5–13+.

The family Hydrophyllaceae consists of about 20 genera and 250 species, of fairly wide distribution, but most common in dry habitats in western United States. By far the largest genus is *Phacelia*, with perhaps 150 species. *Nemophila menziesii* Hooker & Arn., called baby blue-eyes, and a few other species are cultivated as garden ornamentals. Otherwise the family is of little economic interest.

The Hydrophyllaceae have usually been considered to be allied to the Polemoniaceae, on the one hand, and to the Boraginaceae, on the other. I agree. For purposes of formal classification, I find it useful to associate the Hydrophyllaceae with the Polemoniaceae, and to put the Boraginaceae in an allied order. This arrangement helps in the construction of orders that have a reasonable degree of morphological coherence. I prefer to avoid orders that can be characterized only by the list of families to be included.

The anomalous monotypic Mexican shrubby genus *Lithophytum* is here referred with some doubt to the Hydrophyllaceae. It is unique in the Asteridae in having the ovary open at the top, without style or stigma, and nearly unique in having the two carpels collateral instead of median or oblique, but in other respects it appears to be compatible with the Hydrophyllaceae. It has previously been associated with the Solanaceae or Verbenaceae, from both of which it appears to be amply distinguished. It differs from characteristic Solanaceae, inter alia, in its more or less distinctly opposite leaves, in its unilocular ovary with only one or two ovules on each parietal placenta, and in the absence of internal phloem and a nectary-disk. It differs from characteristic Verbenaceae in having 5 stamens and in its unilocular ovary with parietal placentas. I do not accept the interpretation of D'Arcy and Keating, based on vascular anatomy of the flower, that the gynoecium is 4-carpellate. *Lithophytum* is known from only two collections, neither of which bears fruit. It is possible that further study of more ample material will strengthen the case for treating it as a distinct family. A 4-carpellary interpretation of the gynoecium would of course support such a distinction.

3. Order LAMIALES Bromhead 1838

Herbs, shrubs, lianas, or trees, often producing iridoid compounds or alkaloids or aromatic oils, but only rarely cyanogenic or saponiferous, and generally not tanniferous, lacking both ellagic acid and proanthocyanins; vessel-segments mostly with simple perforations; internal phloem only very rarely present (a few Verbenaceae). LEAVES alternate or more often opposite, or occasionally whorled, simple and entire or toothed to occasionally compound, exstipulate. FLOWERS perfect or rarely unisexual, with a dichlamydeous, mostly pentamerous perianth, the corolla sympetalous, regular or more often irregular, with imbricate or sometimes convolute or even valvate lobes; stamens isomerous with the corolla-lobes and attached to the tube alternate with the lobes, or more often only 4 or 2; pollen-grains binucleate or trinucleate, triaperturate or of triaperturate-derived type; a partial or complete nectary-disk often present around the base of the ovary; gynoecium most commonly of 2 median carpels, but sometimes up to 5 (in Lennoaceae up to 14) carpels, these united by a gynobasic style, or more fully united to form a compound ovary with a terminal style; each carpel usually biovulate and generally divided into 2 uniovulate segments (either with an intrusive partition from the carpellary midrib or developing ab initio as separate lobes of the ovary), or seldom the ovary more or less reduced, the reduction reaching its extreme in the pseudomonomerous genus *Phryma* of the Verbenaceae, with a single ovule in the single locule; placentation ordinarily axile (often basal-axile or apical-axile); ovules anatropous to sometimes hemitropous or rarely (as in *Phryma*) orthotropous, with a massive single integument and commonly with an integumentary tapetum, tenuinucellar or seldom pseudocrassinucellar; endosperm-development cellular or less often (some Boraginaceae) nuclear, often with terminal haustoria. FRUIT most commonly of (1–) 4 half-carpellary separating nutlets, less often drupaceous, only rarely capsular; embryo dicotyledonous or (in the Lennoaceae) without differentiated cotyledons; endosperm copious to more often scanty or wanting.

The order Lamiales as here defined consists of 4 families and about 7800 species. The Lamiaceae (3200), Verbenaceae (2600) and Boraginaceae (2000) are all large families. The Lennoaceae (4–5) are a much smaller group.

The Verbenaceae and Lamiaceae are a closely related pair of families, which together include about three-fourths of the species in their order.

Most authors agree that the Lamiaceae represent the realization and culmination of trends that begin in the Verbenaceae. The boundary between the two families is arbitrary and in part merely conventional, but no other means of distinguishing between them seems more satisfactory, and the conceptual utility of recognizing two groups rather than only one has seldom been challenged.

Three subfamilies that are customarily referred to the Lamiaceae are transitional between the Lamiaceae and Verbenaceae in gynoecial structure. These are the Prostantheroideae, Ajugoideae, and Rosmarinoideae, collectively embracing less than 10 percent of the species of Lamiaceae. In these subfamilies the ovary is more or less strongly lobed, but the carpels are united below, so that the style is terminal and impressed rather than truly gynobasic. These subfamilies resemble the Lamiaceae rather than the Verbenaceae in odor and in the aspect of the flowers, although these features do not reliably separate the two families, even among genera whose position is clear on the basis of the gynoecium.

The Boraginaceae stand somewhat apart from the Verbenaceae-Lamiaceae in their alternate leaves and in some chemical features. The Boraginaceae frequently produce alkannin and pyrrolizidine alkaloids, and are not known to produce iridoid compounds, whereas the other two families frequently produce iridoid compounds, but not alkannin and only seldom alkaloids (these not of the same group as those of the Boraginaceae). Even so, some of the tropical, woody Boraginaceae with a terminal style are remarkably similar to some of the tropical, woody Verbenaceae. The gynoecium of the Boraginaceae covers the complete range from like that of typical Verbenaceae to like that of typical Lamiaceae, with the majority of species having a gynobasic style as in the Lamiaceae.

The gynoecium of the Lamiales provides another example of the evolutionary parallelism that permeates the angiosperms. If the more archaic members of the order were to die out, then the gynobasic style and half-carpellary nutlets of the bulk of the Boraginaceae and Lamiaceae would appear to be a unique, monophyletic feature.

The Lamiales are related to the Solanales and Gentianales. As noted in the discussion of the Solanales, the Lamiales reflect the realization of some morphological trends that are nascent in the former order. Many authors have noted a probable relationship between the Hydrophyllaceae (Solanales) and Boraginaceae. On purely morphological grounds, there is no obvious reason why the Lamiales might not be derived directly from the Solanales. The chemical picture is more complex. In the light of the distribution of alkaloids and iridoid compounds in the Gentianales, Solanales, and Lamiales, it might be safer to consider the relationship of the Lamiales to the Solanales to be more nearly fraternal than filial, with both orders coming from an ancestry near or in the Gentianales. If opposite leaves are primitive in this group of orders, as I have speculated, then it would be logical to suppose that the Verbenaceae and Lamiaceae have retained the ancestral decussate phyllotaxy, whereas the Boraginaceae and the bulk of the Solanales have reverted to the spiral phyllotaxy that is primitive for the angiosperms as a whole.

On the basis of the fossil pollen, it appears that the Boraginaceae can be traced back to the Oligocene, the Verbenaceae to the lower Miocene, and the Lamiaceae to the upper Miocene.

The biological importance of the characters that mark the Lamiales is doubtful. The structure of the mature gynoecium obviously influences the means of seed-dispersal, but we have yet to see a comprehensive explanation of the progressive competitive advantage to be obtained by the intrusion of false partitions in the ovary and the subsequent stepwise change from a drupaceous, unlobed fruit to separating, half-carpellary nutlets. The three major families of the order differ to some degree in the nature of their chemical repellents, but the functional significance of these differences remains to be elucidated. Judging by their abundance, diversity, and geographic and ecologic distribution, it does not appear that any one of the three major families has any obvious advantage over the others, although the Verbenaceae are more predominantly tropical than the other two.

The parasitic habit of the Lennoaceae is obviously a specialization. The possible advantage to the Lennoaceae in the increased number of carpels is wholly obscure. It is true enough that the increase in number of carpels increases the potential number of seeds, but this is a most unusual way to produce such an increase, if that is what is being selected for. The order is locked into a close relationship between the number of carpels and the number of ovules, but the three large families have obviously found it possible to govern the number of seeds by the number of flowers.

SELECTED REFERENCES

Carlquist, S. 1970. Wood anatomy of *Echium* (Boraginaceae). Aliso 7: 183–199.

Carrick, J. 1977. Studies in Australian Lamiaceae. *Eichlerago*, a new genus allied to *Prostanthera*. J. Adelaide Bot. Gard. 1: 115–122.

Copeland, H. F. 1935. The structure of the flower of *Pholisma arenarium*. Amer. J. Bot. 22: 366–383.
Drugg, W. S. 1962. Pollen morphology of the Lennoaceae. Amer. J. Bot. 49: 1027–1032.
El-Gazzar, A., & L. Watson. 1970. A taxonomic study of Labiatae and related genera. New Phytol. 69: 451–486.
Erdtman, G. 1945. Pollen morphology and plant taxonomy. IV. Labiatae, Verbenaceae and Avicenniaceae. Svensk Bot. Tidskr. 39: 279–285.
Hillson, C. J. 1959. Comparative studies of floral morphology of the Labiatae. Amer. J. Bot. 46: 451–459.
Inamdar, J. A. 1969. Epidermal structure and ontogeny of stomata in some Verbenaceae. Ann. Bot. (London) II. 33: 55–66.
Inamdar, J. A., & D. C. Bhatt. 1972. Structure and development of stomata in some Labiatae. Ann. Bot. (London) II. 36: 335–344.
Jaitly, S. C. 1968. Le développement de l'albumen chez les genres *Salvia, Hyptis, Ocimum, Pogostemon, Mentha* et *Leucas*. Bull. Soc. Bot. France 115: 373–378.
Johri, B. M., & I. K. Vasil. 1956. The embryology of *Ehretia laevis* Roxb. Phytomorphology 6: 134–143.
Junell, S. 1934. Zur Gynäceummorphologie und Systematik der Verbenaceen und Labiaten, nebst bemerkungen über ihre Samenentwicklung. Symb. Bot. Upsal. 1(4): 1–219.
Junell, S. 1937. Die Samenentwicklung bei einigen Labiaten. Svensk Bot. Tidskr. 31: 67–110.
Junell, S. 1938. Über den Fruchtknotenbau der Borraginazeen mit pseudomonomeren Gynäzeen. Svensk Bot. Tidskr. 32: 261–273.
Kapil, R. N., & R. S. Vani. 1966. *Nyctanthes arbor-tristis* Linn.: embryology and relationships. Phytomorphology 16: 553–563.
Kooiman, P. 1975. The occurrence of iridoid glycosides in the Verbenaceae. Acta Bot. Neerl. 24: 459–468.
Kundu, B. C., & Anima De. 1968. Taxonomic position of the genus *Nyctanthes*. Bull. Bot. Surv. India 10: 397–408.
Lawrence, J. R. 1937. A correlation of the taxonomy and the floral anatomy of certain of the Boraginaceae. Amer. J. Bot. 24: 433–444.
Maheshwari, J. K. 1954. Floral morphology and the embryology of *Lippia nodiflora* Rich. Phytomorphology 4: 217–230.
Nabli, M. A. 1976. Etude ultrastructurale comparée de l'exine chez quelques genres de Labiatae. *In:* I. K. Ferguson & J. Muller, eds., The evolutionary significance of the exine, pp. 499–525. Linn. Soc. Symp. Ser. No. 1. Academic Press. London & New York.
Rao, V. S. 1952. The floral anatomy of some Verbenaceae with special reference to the gynoecium. J. Indian Bot. Soc. 31: 297–315.
Record, S. J., & R. W. Hess. 1941. American woods of the family Verbenaceae. Trop. Woods 65: 4–21.
Record, S. J., & R. W. Hess. 1941. American woods of the family Boraginaceae Trop. Woods 67: 19–33.
Risch, C. 1956. Die Pollenkörner der Labiaten. Willdenowia 1: 617–641.
Saxena, M. R. 1975. Pollen morphology of the Nyctanthoideae (Verbenaceae). J. Indian Bot. Soc. 54: 71–74.
Shaw, H. K. Airy. 1952. Note on the taxonomic position of *Nyctanthes* L. and *Dimetra* Kerr. Kew Bull. 1952: 271–272.
Stant, M. Y. 1952. Anatomical evidence for including *Nyctanthes* and *Dimetra* in the Verbenaceae. Kew Bull. 1952: 273–276.

Suessenguth, K. 1927. Über die Gattung *Lennoa*. Flora 122:264–305.
Thieret, J. W. 1972. The Phrymaceae in the southeastern United States. J. Arnold Arbor. 53: 226–233.
Venkateswarlu, J., & B. Atchutaramamurti. 1955. Embryological studies in Boraginaceae. I. *Coldenia procumbens* Linn. J. Indian Bot. Soc. 34: 235–247.
Whipple, H. L. 1972. Structure and systematics of *Phryma leptostachya* L. J. Elisha Mitchell Sci. Soc. 88: 1–17.

Аветисян, Е. М. 1956. Морфология микроспор бурачниковых. Труды Бот. Инст. АН Армянской ССР, Биол. Науки, 10(1): 7-66.
Билимович, О. Ф. 1935. Значение анатоми и околоплодника Labiatae для их систематики. Труды Воронежск. Госуд. Унив. 7: 68-84.
Борзова, И. А. 1959. К вопросу о структуре экзины у сем. губоцветных. Докл. АН СССР 125: 1350-1352.
Борзова, И. А. 1960. К вопросу о происхождении шестибороздного типа пыльцы у губоцветных. Докл. АН СССР 133: 1465-1467.
Борзова, И. А. 1962. К вопросу о морфогении пыльцевых зерен губоцветных. В сб.: Докл. сов. палинологов к 1-й между-народн. палинол. конф., Москва: 33-37.
Борзова, И. А. 1962. Отдельные замечания по системе губо-цветных на основании данных палинологии. В сб.: Докл. сов. палинологов к 1-й международн. палинол. конф., Москва: 26-32.
Васильченко, И. Т. 1947. Морфология прорастания губоцветных (сем. Labiatae) в связи с их систематикой. Труды Бот. Инст. АН СССР, сер. 1, 6: 72-104.
Винская, С. С. 1949. Структура железистых органов губоцветных. Изв. Западно-Сибирского Филиала АН СССР, Сер. Биол.; Выпуск 1, Ботан. 3: 59-79.
Карташова, Н. Н. 1960. Некоторые данные по морфологии цветка губоцветных (Labiatae). Бот. Ж. 45: 109-114.
Черпакова, Н. В. 1956. О значении анатомии черешков губоцветных для их систематики и сырьеведения. Труды Воронежск. Унив. 36: 89-96.

SYNOPTICAL ARRANGEMENT OF THE FAMILIES OF LAMIALES

1 Leaves mostly alternate, usually entire; flowers mostly regular or nearly so and with as many stamens as corolla-lobes; stems not square; plants neither aromatic nor with iridoid compounds.

  2 Plants parasitic, without chlorophyll and with reduced, scale-like leaves; carpels 6–14; fruit a fleshy, eventually irregularly dehiscent capsule; seeds with well developed endosperm and globose, undifferentiated embryo .... 1. LENNOACEAE.

  2 Plants autotrophic, with chlorophyll and normal leaves; carpels 2 (–5); fruit usually either drupaceous or of separating, half-carpellary nutlets, rarely capsular; seeds with a dicotyledonous embryo and well developed to much more often scanty or no endosperm .... 2. BORAGINACEAE.

1 Leaves mostly opposite (or whorled), entire or often toothed or cleft, or sometimes compound; flowers mostly more or less irregular

and with 2–4 stamens, the exceptions (regular corolla and/or 5 stamens) being found chiefly among the Verbenaceae; young stems commonly square; iridoid and aromatic compounds common, though not universal.

3 Style terminal or nearly so, the ovary only shortly or not at all lobed at the top; plants seldom aromatic; fruit variously drupaceous, or capsular, or achenial, or often of half-carpellary nutlets as in the next family .... 3. VERBENACEAE.

3 Style commonly gynobasic, uniting the otherwise essentially distinct lobes of the ovary, or less commonly the ovary lobed only part way (one third or more) to the base; plants commonly aromatic; fruit of (1–) 4 half-carpellary nutlets or (rarely) drupelets ....4. LAMIACEAE.

1. Family LENNOACEAE Solms-Lauback 1870 nom. conserv., the Lennoa Family

Fleshy root-parasitic herbs, without chlorophyll, commonly provided with stalked, glandular hairs; vascular bundles collateral, the principal ones forming a ring, some smaller cortical bundles commonly also present; vessel-segments with simple perforations. LEAVES spirally arranged, reduced to mere scales. FLOWERS in dense, pyramidal to capitate or discoid inflorescences, perfect, regular or nearly so, 5–10-merous as to the calyx, corolla, and androecium; sepals narrow, distinct or nearly so; corolla sympetalous, with imbricate or induplicate-valvate lobes; stamens as many as and alternate with the corolla-lobes, the short filaments attached above the middle of the tube; anthers tetrasporangiate and dithecal, opening by longitudinal slits; pollen-grains binucleate, 3–4 (5)-colporate, and often with intervening, inaperturate colpi as well; nectary-disk wanting; gynoecium of 6–14 carpels united to form a compound, superior ovary with a stout, terminal style and a capitate or lobed stigma; each primary locule of the ovary divided by a median partition, so that the ovary has twice as many compartments (locelli) as carpels; ovules solitary in each compartment of the ovary, anatropous, epitropous, unitegmic, tenuinucellar; endosperm-development cellular. FRUIT a fleshy, eventually irregularly circumscissile capsule; seeds small, with an undifferentiated, globose embryo embedded in the copious, starchy endosperm. X = 9.

The family Lennoaceae consists of 3 genera and 4 or 5 species, native to the New World, from southwestern U.S.A. to Colombia and Venezuela. *Lennoa* has 2 species, *Pholisma* 1 or 2, and *Ammobroma* only one.

The Lennoaceae superficially resemble some of the Monotropaceae, and some authors have sought to ally these two groups. It should be noted, however, that in spite of the similarity in aspect, the Monotropaceae are mycotrophic, whereas the Lennoaceae are root-parasites. Most present-day authors consider that features of gross floral morphology, embryology, and palynology indicate a relationship of the Lennoaceae to the Hydrophyllaceae and Boraginaceae. Because of the morphology of the gynoecium it is more convenient to include the Lennoaceae in the Lamiales (with the Boraginaceae) than in the Solanales (with the Hydrophyllaceae).

2. Family BORAGINACEAE A. L. de Jussieu 1789 nom. conserv., the Borage Family

Herbs, less often shrubs or trees, seldom lianas, commonly provided with characteristic firm, unicellular hairs that have a basal cystolith and often calcified or silicified walls, sometimes with other sorts of hairs as well, often accumulating free silicic acid, often producing alkaloids of the pyrrolizidine group, and often bearing alkannin, a red naphthaquinone, in the roots, but without iridoid compounds, only seldom cyanogenic or saponiferous, and usually not tanniferous, generally lacking both ellagic acid and proanthocyanins, often accumulating fructosans (notably isoheptose and isokestose) as reserve carbohydrates, and allantoin (an amide) as a nitrogenous food-reserve; solitary or clustered crystals of calcium oxalate, varied in form, commonly present in some of the cells of the parenchymatous tissues; nodes unilacunar; xylem commonly forming a closed ring with only narrow rays, even in herbs; vessel-segments with simple (rarely reticulate) perforations; imperforate tracheary elements with simple or narrowly bordered pits, often septate; rays in woody forms heterocellular to homocellular, commonly most of them 4–6 cells wide, with only a few uniseriates; wood-parenchyma of diverse types. LEAVES alternate or seldom the lower ones (rarely all of them) opposite, simple and mostly entire; stomates mostly anomocytic; petiole with an arc of vascular bundles, or with a cylindrical vascular strand and some subsidiary bundles; stipules wanting. FLOWERS in diverse sorts of basically cymose inflorescences, the major branches commonly sympodial, helicoid cymes that elongate and straighten with maturity (forming false racemes or spikes), or seldom the flowers solitary and axillary; flowers mostly perfect (gynodioecious in *Echium*) sympetalous, regular or slightly irregular, (4) 5 (6)-merous as to the calyx, corolla, and androecium, sometimes some of them cleistogamous; sepals distinct or connate at the base or sometimes to the middle or above, imbricate or seldom valvate; corolla in most genera salverform, but sometimes tubular or funnelform, the lobes imbricate or convolute, rarely valvate; corolla-tube in the subfamily Boraginoideae generally with more or less evident, often hairy appendages (the fornices) at the summit opposite the lobes, the fornices being formed by invagination of the corolla-tube from the outside; stamens as many as and alternate with the corolla-lobes, attached to the tube; anthers tetrasporangiate and dithecal, opening by longitudinal slits; pollen-grains (2) 3-nucleate, from tricolporate or triporate to polycolpate

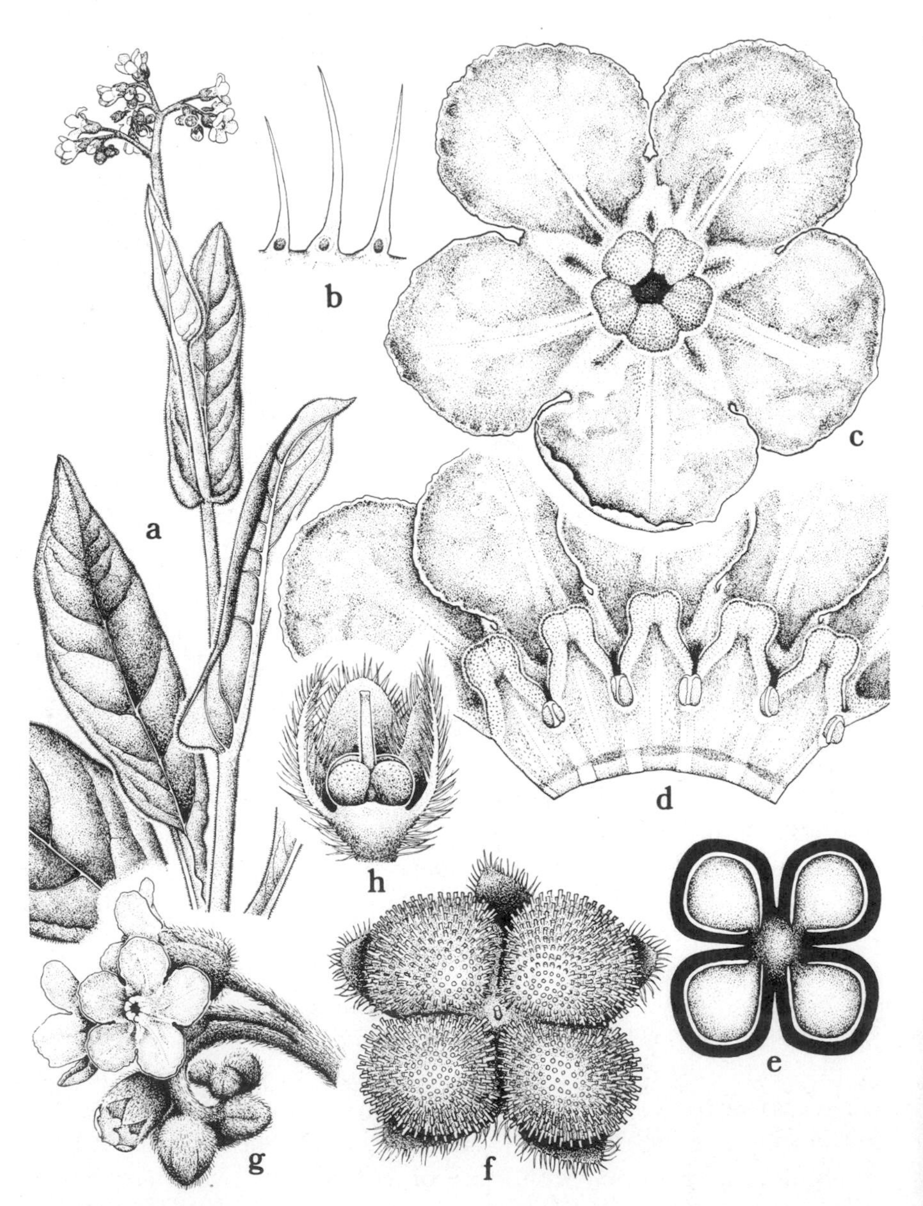

FIG. 6-14 Boraginaceae. *Cynoglossum virginianum* L. a, habit, $\times\frac{1}{2}$; b, trichomes, with basal cystolith, $\times 24$; c, flower, from above, $\times 6$; d, corolla, opened out, showing stamens and fornices, $\times 6$; e, schematic cross-section of ovary, $\times 12$; f, fruit of 4 nutlets, from above, $\times 4$; g, portion of inflorescence, $\times 2$; h, flower in partial long-section, with the corolla removed, $\times 6$.

or polycolporate, sometimes with 6 colpi, the alternate ones with and without a pore; an annular nectary-disk present around the base of the ovary, or wanting; gynoecium of 2 median carpels (4–5 in *Zoelleria*), united in varying degree to form a compound, superior ovary with usually twice as many segments or compartments as carpels, each such segment containing a single ovule; rarely the posterior carpel suppressed and the ovary with two 1-seeded chambers; in the subfamily Cordioideae the ovary entire, with a terminal, twice 2-cleft style, ripening into a dry and hard or fleshy drupe with a mostly 4-locular stone; in the subfamily Ehretioideae the ovary entire or 4-lobed, with a simple or bifid style or 2 separate styles, ripening into a drupe with two 2-seeded or four 1-seeded stones, or separating into 4 segments; in the subfamily Heliotropoideae the ovary entire or 4-lobed, the style arising from the summit or between the lobes, abruptly expanded at the top into a discoid stigma with a short, 2-lobed apical cone (or the stigma sessile), the ovary ripening into two (1) 2-seeded, separating nutlets or four 1-seeded separating nutlets; in the subfamily Boraginoideae (by far the largest subfamily) the ovary deeply 4-lobed, with a gynobasic, simple or 2-lobed style arising from between the otherwise essentially distinct lobes, ripening into (1-) 4 distinct, smooth or often variously ornamented nutlets individually attached to the gynobase; in the subfamily Wellstedioideae (a single genus *Wellstedia* with 2 spp.) the ovary compressed, bilocular, with a terminal style and a single pendulous ovule in each locule, ripening into a 1–2-seeded capsule; stigmas mostly dry and papillate; ovules anatropous to hemitropous, with a massive single integument and sometimes an integumentary tapetum, tenuinucellar, or in *Heliotropium* and *Ehretia* pseudocrassinucellar, sometimes with a placental obturator; endosperm-development nuclear or cellular. Embryo with 2 cotyledons (these in *Amsinckia* so deeply bifid that there appear to be 4), spathulate, straight and with the radicle pointed upwards, or sometimes more or less curved; endosperm well developed and oily in some of the more archaic genera, otherwise scanty or none. X = 4–12. (Ehretiaceae, Heliotropiaceae, Wellstediaceae)

The family Boraginaceae as here broadly (and customarily) defined consists of about a hundred genera and perhaps as many as 2000 species, of cosmopolitan distribution, but especially well developed in western North America and in the Mediterranean region, eastward into Asia. The largest genera are *Cordia*, *Tournefortia*, and *Heliotropium*, all chiefly tropical and each with about 200 or more species. *Cryptantha* (150, chiefly in Pacific North and South America) and *Onosma* (100,

Mediterranean region and eastward) are some other large genera. The common forget-me-not is *Myosotis sylvatica* Hoffm., a native of Europe. Decoctions of *Lithospermum ruderale* Dougl. have a pronounced contraceptive action, and their use by the Indians of western North America provided the inspiration for the development of commercial oral contraceptives.

The relationship of the Boraginaceae to the Hydrophyllaceae has long been evident to all. The similarity of the archaic, woody, tropical members of the Boraginaceae to archaic, woody, tropical members of the Verbenaceae does not so quickly come to the attention of botanists who live in North Temperate regions and are better acquainted with the more advanced herbaceous members of these families. In the pursuit of conceptually useful orders, I find it helpful to include the Boraginaceae in the Lamiales (thus emphasizing the relationship to the Verbenaceae).

The pollen of the Boraginaceae is said to be highly varied but still unified, most similar to that of the Hydrophyllaceae and Lennoaceae, but also comparable to that of the Lamiaceae, Convolvulaceae, and Polemoniaceae. Fossil pollen considered to represent the Boraginaceae occurs in lower Oligocene and more recent deposits.

3. Family VERBENACEAE Jaume St.-Hilaire 1805 nom. conserv., the Verbena Family

Herbs, shrubs, lianas, or small to large trees, sometimes (*Avicennia*) of mangrove habit, sometimes thorny, provided with diverse sorts of hairs (these sometimes with a basal cystolith and often with silicified or calcified walls), not notably aromatic, or sometimes with an odor approaching that of some Lamiaceae, commonly producing iridoid substances and often also the phenolic glycoside orobanchin, less often alkaloids, saponins or other triterpenoid compounds, and monoterpenoids, very often with highly methylated 6- or 8-hydroxy-flavonols, and at least sometimes with lapachol (a naphthaquinone allied to alkannin), but only seldom tanniferous (the bark tanniferous in *Avicennia*), lacking both ellagic acid and proanthocyanins; calcium oxalate crystals of various form commonly present in some of the cells of the parenchymatous tissues; young twigs very often quadrangular; stem sometimes with anomalous structure; nodes unilacunar, with 1-several traces; xylem even in herbs commonly forming a continuous cylinder with

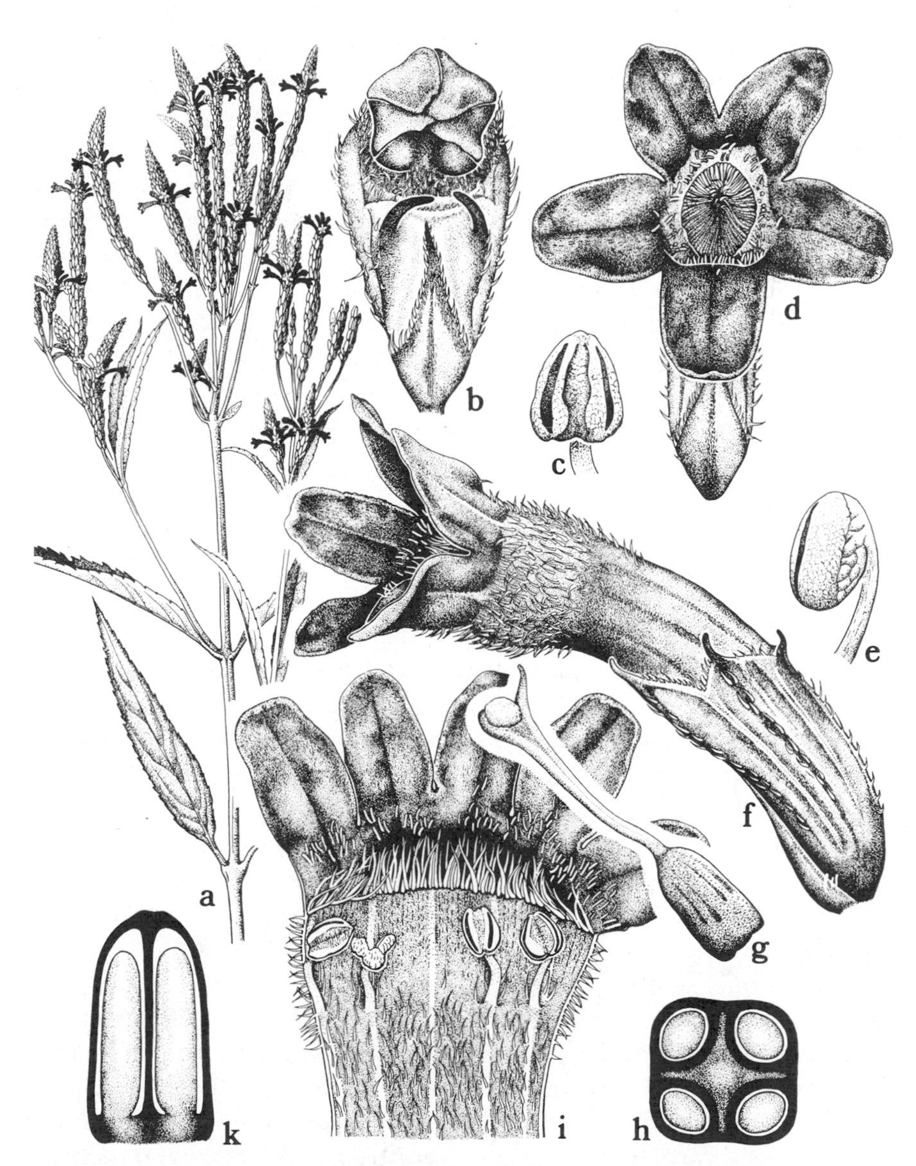

FIG. 6-15 Verbenaceae. *Verbena hastata* L. a, habit, $\times\frac{1}{2}$; b, flower bud, $\times 16$; c, e, anthers, $\times 32$; d, front view of flower, $\times 16$; f, side view of flower, $\times 16$; g, pistil, $\times 16$; h, schematic cross-section of ovary, $\times 20$; i, corolla, opened out, showing attached stamens, $\times 16$; k, schematic long-section of ovary, $\times 32$.

narrow rays; vessel-segments mostly with simple perforations; imperforate tracheary elements mostly septate, with simple or bordered pits; wood-rays heterocellular to homocellular, mostly mixed uniseriate and pluriseriate, the latter up to 4 or 5 (–12) cells wide, or rarely (*Pseudocarpidium*) all uniseriate; wood-parenchyma commonly wholly paratracheal; internal phloem only rarely present. LEAVES opposite or occasionally whorled, rarely alternate, simple or sometimes pinnately or palmately compound; stomates mostly anomocytic or diacytic, seldom paracytic; petiolar anatomy of diverse types; stipules wanting. Inflorescence of various types, essentially racemose (commonly a spike or head) in the Verbenoideae and Stilboideae, mainly cymose in the Viticoideae and some of the other subfamilies, often subtended by an involucre of colored bracts; FLOWERS perfect, or seldom some of them unisexual; sepals united to form a (4) 5 (–8)-toothed or -lobed, sometimes irregular calyx; corolla sympetalous, with (4) 5 (–8) imbricate lobes, often with a slender tube and spreading limb, less often campanulate, more or less irregular in most genera, sometimes bilabiate; stamens only rarely isomerous with the corolla-lobes, more often 4 (and then sometimes didynamous), or seldom only 2, some or all of the missing ones sometimes represented by staminodes; filaments attached to the corolla-tube alternate with the lobes (aligned with the sinuses); anthers tetrasporangiate and dithecal, opening by longitudinal slits; pollen-grains binucleate or seldom trinucleate, most commonly triaperturate; an annular nectary-disk weakly developed around the base of the ovary, or sometimes wanting, the corolla-tube nectariferous within in *Avicennia*; gynoecium most commonly of 2 median carpels united to form a compound, initially bilocular ovary that soon becomes divided into 4 uniovulate chambers by the intrusion of partitions from the carpellary midribs, the ovary sometimes shallowly 4-lobed laterally, its summit entire or only shortly depressed, the style accordingly terminal or arising from amongst the 4 short distal lobes; stigma at least sometimes bilobed, seldom punctate; variations on the gynoecial pattern include a 5-carpellary, 10-chambered ovary (spp. of *Geunsia*), a 4-carpellary, 8 chambered ovary (*Duranta*), reduction or suppression of one or the other of the 2 carpels (e.g., the anterior carpel suppressed and the fruit with two 1-seeded stones in *Lantana*, the posterior carpel reduced in spp. of *Cyanostegia*), absence of the secondary partition, so that the ovary is bilocular with 2 ovules in each locule (Chloanthoideae), a bilocular ovary with only one ovule in each locule (typical Stilboideae), or one of these locules empty (other Stilboideae), a pseudomonomerous ovary with a single locule and ovule (*Phryma*), and partial or complete failure or disintegration of the partitions so that the ovary is more or less clearly unilocular with 4 ovules on an apparently free-central placenta (Symphorematoideae, *Avicennia*); ovules anatropous to less often hemitropous or even virtually orthotropous (as in *Phryma*), erect or less often pendulous, apotropous or epitropous, the micropyle directed downward except in *Phryma*, tenuinucellar, with a single massive integument and usually with an integumentary tapetum (in *Avicennia* the tapetum lacking and the embryo-sac protruding from the ovule); endosperm-development cellular, with micropylar and chalazal haustoria. FRUIT drupaceous with 2 or 4 stones, or very often of 1-seeded, separating nutlets, or sometimes a 2-valved or 4-valved capsule; seeds with a straight, oily embryo that has 2 expanded, flat cotyledons; endosperm wanting in most genera, but well developed and more or less fleshy (oily) in the Avicennioideae, Chloanthoideae, and Stilboideae. X = 5–12. (Avicenniaceae, Chloanthaceae, Dicrastylidaceae, Nyctanthaceae, Phrymaceae, Stilbaceae, Symphoremataceae)

The family Verbenaceae as here defined consists of about 100 genera and 2600 species, of pantropical distribution; only a limited number of species occur in temperate regions. The largest genera are *Clerodendrum* (400), *Verbena* (250), *Vitex* (250), *Lippia* (200), *Premna* (200), and *Lantana* (150). *Tectona grandis* L.f., teak, originally of India and Burma, is well known for its hard, heavy, durable wood.

The Verbenaceae are diverse in habit and gynoecial structure, and several peripheral groups have often been extracted as distinct families. Notable among the segregates are the Avicenniaceae (*Avicennia*, ca. 12 spp.), Dicrastylidiaceae or Chloanthaceae (14 genera, ca. 90 spp.), Phrymaceae (1 sp.), Stilbaceae (5 genera, 12 spp.), and Symphoremataceae (3 genera, 35 spp.). These segregate families collectively include scarcely 6 percent of the species of the more broadly defined family Verbenaceae. Their relationships are not in dispute, and the taxonomic rank at which they should be recognized is purely a matter of taste.

It has been customary (though not universal) to treat the monotypic genus *Phryma* as a distinct family, mainly because of its specialized, pseudomonomerous gynoecium with a single ovule. The habital resemblance of *Phryma* to *Verbena* and its allies has been obvious to all, and a close relationship is generally admitted. The unusual gynoecium of *Phryma* is in pattern with other unusual types in the Verbenaceae. I see no value in keeping *Phryma* as a separate family if the other peripheral groups are to be retained in the Verbenaceae.

4. Family LAMIACEAE Lindley 1836 nom. conserv., the Mint Family

Herbs or shrubs, rarely small or middle-sized trees (up to 18 m tall in *Hyptis arborea* Benth.), with various sorts of hairs, commonly provided with short-stalked epidermal glands containing characteristic ethereal oils (these chemically diverse, but often monoterpenoid, sesquiterpenoid, or diterpenoid), the plants often also producing triterpenoid substances (but generally not saponins) and sometimes with iridoid compounds (but then mostly poor in essential oils) less often with alkaloids of the pyrrolidine or pyridine groups, commonly with acylated anthocyanins, only rarely cyanogenic, and generally without at least the ordinary kinds of tannins, lacking both ellagic acid and proanthocyanins, often accumulating potassium nitrate, and commonly storing carbohydrate as stachyose and/or oligogalactosides; calcium oxalate crystals of diverse sorts sometimes present in some of the cells of the parenchymatous tissues, but more often not; young stems commonly quadrangular, often with well developed collenchyma in the angles; nodes unilacunar, sometimes with 2 traces; xylem in young stems borne in distinct, collateral bundles, or sometimes forming a continuous ring with narrow rays; vessel-segments with simple perforations; imperforate tracheary elements with very small, simple pits, sometimes septate; wood-rays mostly heterocellular, mixed uniseriate and pluriseriate, the latter up to 4–12 cells wide; wood-parenchyma rather scanty, paratracheal. LEAVES opposite or sometimes whorled (alternate in *Icomum*), simple or occasionally pinnately compound; stomates commonly diacytic, less often some or all of them anomocytic; petiole commonly with a more or less arcuate vascular strand or with a ring of vascular bundles; stipules wanting. Inflorescence of various sorts, mostly of small, compact cymes axillary to leaves or bracts, forming a verticillaster at each node, collectively often forming a thyrse, or the axillary cymules sometimes reduced to single flowers, so that the inflorescence is essentially racemose; FLOWERS mostly bracteolate, perfect or some of them unisexual (the plants sometimes gynodioecious); calyx usually persistent (circumscissile at the base in fruit in *Aeolanthus* and *Icomum*, splitting longitudinally and only the ventral half persistent in fruit in many spp. of *Scutellaria*), more or less tubular with usually 5 teeth or lobes, sometimes

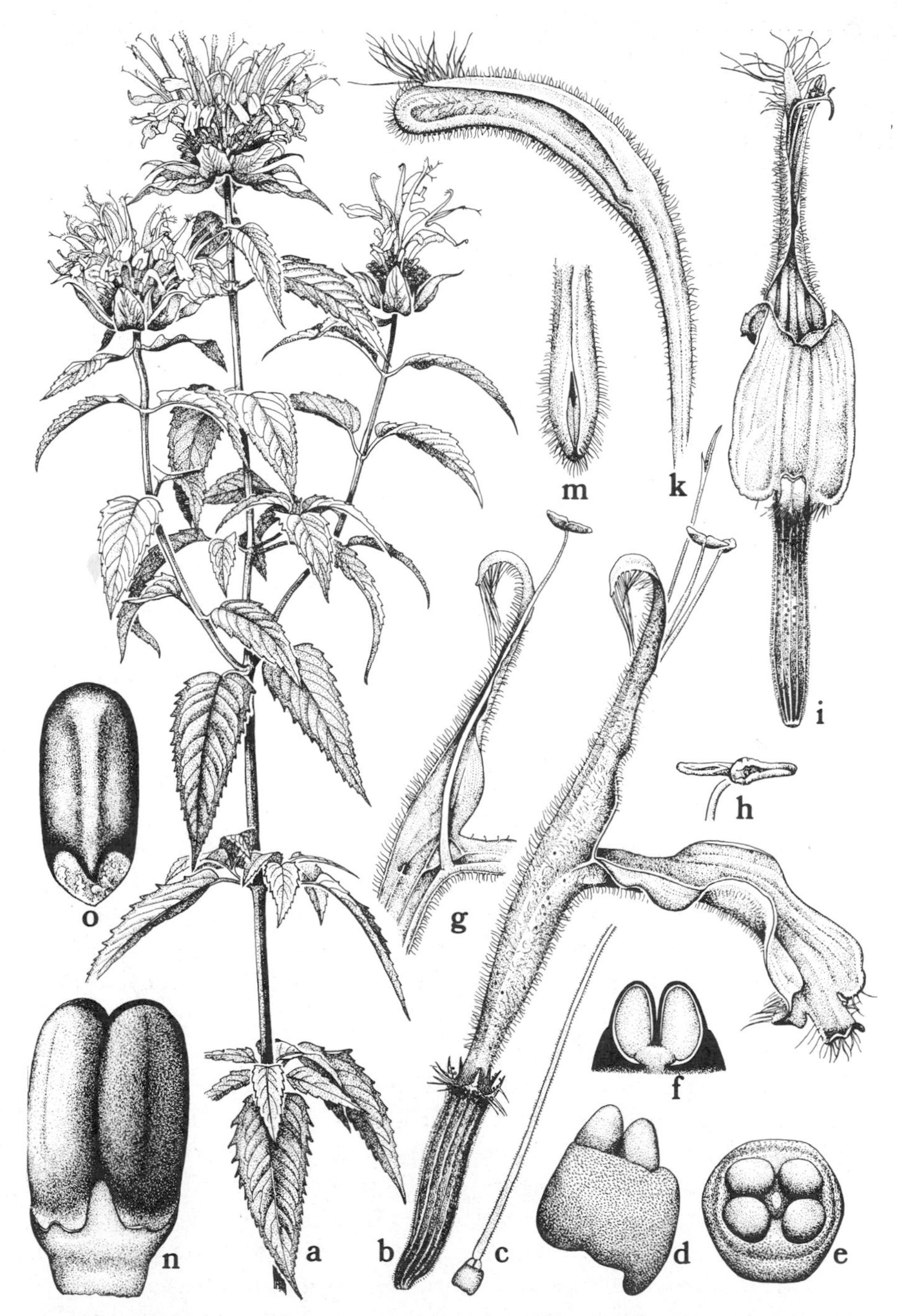

FIG. 6-16 Lamiaceae. *Monarda fistulosa* L. a, habit, ×½; b, flower, ×4; c, pistil, with basal nectary, ×4; d, e, ovary and basal nectary, the style removed, ×20; f, schematic long-section of ovary, without the style, ×20; g, portion of corolla in long-section, with one stamen, ×4; h, anther, ×8; i, front view of flower, ×4; k, corolla in bud, ×4; m, lower lip of corolla, in bud, ×4; n, fruit, ×20; o, ventral view of nutlet, ×20.

bilabiate, in *Hoslundia* enlarged, fleshy, and berry-like in fruit; corolla sympetalous and usually strongly irregular, with 5 imbricate lobes, mostly bilabiate, occasionally unilabiate, seldom nearly regular, then sometimes (as in *Mentha*) 4-lobed, one lobe slightly wider than the others and representing 2 connate lobes; stamens 4 and didynamous, or 2, then sometimes with an additional pair of staminodes; filaments attached to the corolla-tube, aligned with the sinuses; anthers mostly commonly tetrasporangiate and dithecal, opening by longitudinal slits, but one theca sometimes reduced or suppressed; connective sometimes transversely elongate so as to separate its pollen-sacs, most notably in *Salvia*, in which it articulates near its middle to the tip of the filament and bears a pollen-sac at each end or may be further modified so that the lower pollen-sac or the whole lower arm is more or less suppressed; pollen-sacs often attached to the connective only at the tip, becoming confluent and even explanate after dehiscence; pollen-grains mostly binucleate and tricolpate or trinucleate and 6-colpate, sometimes tetracolpate; an annular to unilateral (anterior) nectary-disk commonly present at the base of the ovary, which is superior and sometimes borne on a gynophore; gynoecium fundamentally bicarpellate, but each carpel longitudinally divided in half, so that the 4 essentially distinct segments of the ovary are united only by their gynobasic style, or the ovary less commonly merely 4-lobed for a third or more of its length, so that the style arises from between the lobes but is not gynobasic; style commonly cleft at the summit into 2 dry stigmas or stigma-lobes, one of the lobes often reduced or virtually suppressed; ovules solitary in each lobe of the ovary (4 in all), basal-axile, erect, anatropous to hemitropous, apotropous, with ventral raphe and downwardly directed micropyle, often with a funicular obturator, with a massive single integument and usually an integumentary tapetum, tenuinucellar, with terminal haustoria, the micropylar one usually the better-developed. FRUIT of (1–) 4 1-seeded nutlets with a hard pericarp (broadly winged in *Tinnea*), or rarely (as in *Prasium*) the nutlets drupaceous, with a fleshy exocarp; embryo dicotyledonous, straight, with the radicle directed downwards, or sometimes (as in *Scutellaria*) with the radicle curved and lying alongside one of the expanded, flat cotyledons; endosperm wanting or scanty (copious in *Tetrachondra*); germination mostly epigaeous. X = 5 – 11 +. (Labiatae, nom. altern.; Menthaceae; Tetrachondraceae)

The family Lamiaceae as here defined consists of about 200 genera and 3200 species, cosmopolitan in distribution, but especially abundant in the Mediterranean region and eastward into central Asia. More than half of the species belong to only 8 genera: *Salvia* (500), *Hyptis* (350), *Scutellaria* (200), *Coleus* (200), *Plectranthus* (200), *Stachys* (200), *Nepeta* (150, including catnip, *N. cataria* L.), and *Teucrium* (100). Some other familiar genera are *Lavandula* (lavender), *Marrubium* (horehound), *Mentha* (mint), and *Thymus* (thyme), cultivated for their aromatic ethereal oils.

4. Order CALLITRICHALES Lindley 1833

Submerged or emergent aquatic plants, or sometimes small, slender, terrestrial herbs of wet places, commonly producing iridoid substances, but not saponiferous, not cyanogenic, and not tanniferous, lacking both ellagic acid and proanthocyanins (chemistry of the Hydrostachyaceae unknown); vascular system much-reduced. LEAVES variously alternate, opposite, or whorled, simple to pinnately compound or dissected, without stipules, or with an inconspicuous intrapetiolar stipule (Hydrostachyaceae). FLOWERS solitary and sessile in the axils of leaves or bracts, small and inconspicuous, nearly or quite without perianth, perfect or unisexual, epigynous when perfect; calyx wanting, or (*Hippuris*) represented by a narrow rim around the top of the ovary; petals none; stamen solitary, or rarely 2–3, the anther tetrasporangiate and dithecal; gynoecium of 2 carpels united to form a compound, unilocular or 4-chambered ovary, or (*Hippuris*) pseudomonomerous and unilocular; styles terminal, elongate, filiform, stigmatic for most of their length, solitary in *Hippuris*, otherwise 2, distinct or occasionally united at the base; ovule(s) anatropous, unitegmic, tenuinucellar; endosperm-development cellular. FRUIT variously capsular, or of separating nutlets, or an achene or drupelet; embryo with 2 cotyledons; seeds with or without endosperm.

The order Callitrichales as here defined consists of 3 families and a little more than 50 species. Each family consists of only a single genus. The affinities of all three families have been debated in the past, and it has not been customary to associate them in a single order. Fairly recent embryological information for all three, and chemical information for the Callitrichaceae and Hippuridaceae, strongly suggest that each of the three groups belongs to the Asteridae, rather than with the Rosidan families to which they had been thought to be allied.

Once the three families have been transferred to the Asteridae, it becomes useful to think of them collectively as forming a group that has undergone vegetative and floral reduction in association with an aquatic habitat. The differences among them, although substantial, are not so great as to preclude such a classification. *Callitriche* resembles the Lamiales in the basic structure of its gynoecium, aside from the fact that the carpels are collateral rather than median, but it does not seem closely allied to any family of that order. Tendencies toward the lamialean gynoecium can be seen in some other orders of the Asteridae, including the Solanales (Nolanaceae and a few Convolvulaceae), Scrophulariales (some Globulariaceae), and Plantaginales (spp. of *Plantago*). Therefore the structure of the gynoecium does not require the inclusion of *Callitriche* in the Lamiales. The gynoecium of *Hydrostachys* is perfectly compatible with that of the Scrophulariales, and a number of recent authors have included the Hydrostachyaceae in the Scrophulariales. Aside from its aquatic habitat, *Hydrostachys* is habitally suggestive of *Plantago*, which is now generally conceded to be related to the Scrophulariaceae. The pseudomonomerous, inferior ovary of *Hippuris* is hard to reconcile with any petaliferous order of Asteridae, but on chemical grounds Hegnauer (1969) concludes that "Like *Callitriche*, *Hippuris* is most probably a wholly aquatic offshoot of the Tubiflorae which might have its nearest relatives in the Plantaginaceae (compare *Littorella*)." Thus all three genera can be associated in some way with the Plantaginaceae. The Plantaginaceae, in comparison with the more or less directly ancestral Scrophulariaceae, have taken the road of floral reduction, and *Littorella* has even become aquatic. The Callitrichales may be considered to have exploited more fully the tendencies toward floral reduction and aquatic habitat that are already evident in the Plantaginaceae. It is not required that the Plantaginaceae as we now know them be directly ancestral to the Callitrichales, but at least a collateral relationship seems to be indicated.

For purposes of a linear sequence of orders, it is convenient to put the Callitrichales and Plantaginales before the Scrophulariales, although from a strictly phylogenetic standpoint the Scrophulariales ought to come before the other two.

SELECTED REFERENCES

Cusset, C. 1973. Révision des Hydrostachyaceae. Adansonia sér. 2, 13: 75–119.

Hegnauer, R. 1969. Chemical evidence for the classification of some plant taxa. *In*: J. B. Harborne and T. Swain, eds., Perspectives in phytochemistry, pp. 121–138. Academic Press. London & New York.

Jäger-Zürn, I. 1965. Zur Frage der systematischen Stellung der Hydrostachyaceae auf Grund ihrer Embryologie, Blüten- und Infloreszenzmorphologie. Oesterr. Bot. Z. 112: 621–639.

Jørgensen, C. A. 1923. Studies on Callitrichaceae. Bot. Tidsskr. 38: 81–122.

McCully, M. E., & H. M. Dale. 1961. Heterophylly in *Hippuris*, a problem in identification. Canad. J. Bot. 39: 1099–1116.

Mauritzon, J. 1933. Über die systematische Stellung der Familien Hydrostachyaceae und Podostemonaceae. Bot. Not. 1933: 172–180.

Perol, C., & G. Cusset. 1968. Remarques sur l'*Hydrostachys maxima* Perr. Rev. Gén. Bot. 75: 403–438.

Rauh, W., & I. Jäger-Zürn. 1966. Zur Kenntnis der Hydrostachyaceae. I. Blütenmorphologische und embryologische Untersuchungen an Hydrostach-

yaceen unter besonderer Berücksichtung ihrer systematischen Stellung. Sitzungsber. Heidelberger Akad. Wiss., Math.-Naturwiss. Kl. 1966(1): 1–117.
Rauh, W., & I. Jäger-Zürn. 1967. Le problème de la position systématique des Hydrostachyacées. Adansonia sér. 2, 6: 515–523.

Трон, Е. Ж. 1959, 1960, 1962. Переходные анатомические структуры в листьях *Hippuris vulgaris* L. I, II, III. Бот. Ж. 44: 591-603, 1959. 45: 1271-1282, 1960. 47: 1100-1107, 1962.
Трон, Е. Ж. 1967. Анатомическое строение стебля *Hippuris vulgaris*. Бот. Ж. 52: 811-819.

SYNOPTICAL ARRANGEMENT OF THE FAMILIES OF CALLITRICHALES

1 Gynoecium seemingly of a single carpel, the ovary unilocular, with a single terminal ovule and a single style; flowers perfect and epigynous, or some or all of them unisexual; fruit an achene or drupelet ....1. HIPPURIDACEAE.

1 Gynoecium bicarpellate, the ovary with 4 or more ovules and 2 styles; flowers all unisexual.

 2 Ovary compartmented into 4 locelli, each locellus with a single pendulous, axile ovule; carpels collateral; fruit of 4 separating nutlets; pollen monadinous ....2. CALLITRICHACEAE.

 2 Ovary unilocular, with 2 parietal placentas and numerous ovules; ovules; carpels median; fruit capsular; pollen tetradinous ....3. HYDROSTACHYACEAE.

1. Family HIPPURIDACEAE Link 1821 nom. conserv., the Mare's-tail Family

Aquatic perennial herbs with a sympodial rhizome and erect, emergent stems, storing carbohydrate as stachyose, producing iridoid compounds, but not cyanogenic, not saponiferous, and not tanniferous, lacking both ellagic acid and proanthocyanins; erect stems with a well defined endodermis, and with the vascular system reduced to an axile strand of xylem surrounded by phloem; cortex permeated by an anastomosing system of vertically elongate intercellular spaces; calcium oxalate crystals wanting. LEAVES small, whorled in sets of (4–) 6–12 (–16), linear, entire, provided with minute, deciduous, peltate, glandular trichomes; stipules wanting. FLOWERS small and inconspicuous, anemophilous, solitary in

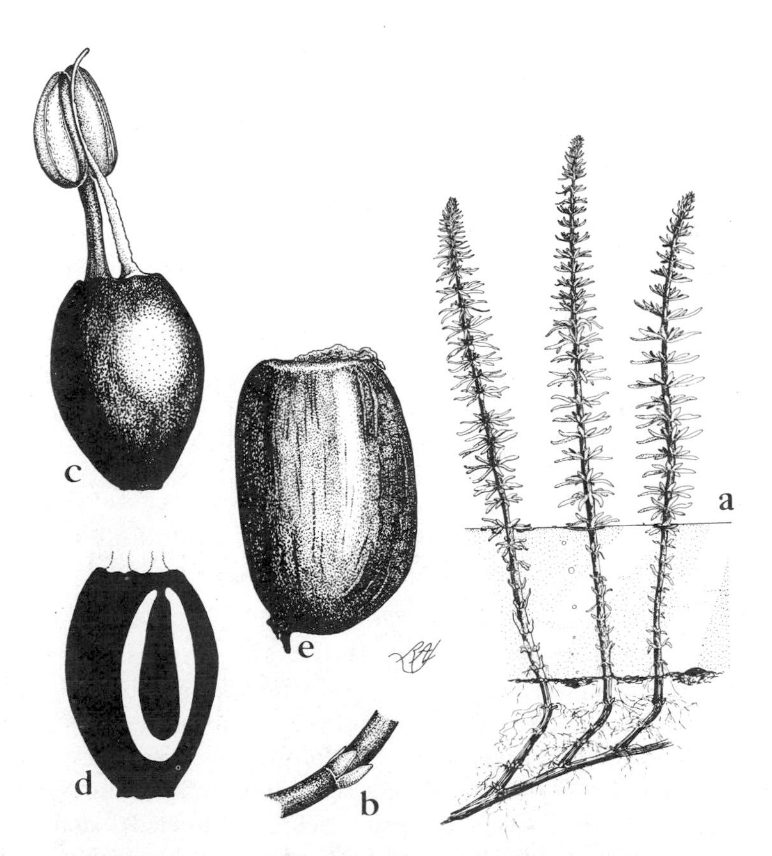

FIG. 6-17 Hippuridaceae. *Hippuris vulgaris* L. a, habit, $\times \frac{1}{4}$; b, nodal region of underground portion of stem, $\times 1\frac{1}{2}$; c, flower $\times 18$; d, schematic long-section of ovary, the ovule under the style, $\times 18$; e, fruit, $\times 18$.

the axils of the upper leaves, epigynous, perfect or sometimes some or all of them unisexual, the pistillate ones then above the staminate ones; calyx reduced to an inconspicuous, 2–4-lobed or subentire rim around the top of the ovary; petals none; stamen solitary, atop the ovary; filament slender; anther tetrasporangiate and dithecal, the style commonly lying in the groove between the pollen-sacs; pollen-grains trinucleate, 4–6-colporate; gynoecium pseudomonomerous, apparently consisting of a single carpel with an elongate, slender, terminal style that is dry and stigmatic its whole length; ovary inferior, unilocular; ovule solitary, apical and pendulous, anatropous, unitegmic, tenuinucellar, with closed micropyle, chalazogamous; endosperm-development cellular; embryo with a large haustorial suspensor. FRUIT an achene or drupelet with a very thin, fleshy exocarp; embryo elongate, straight, with 2 cotyledons; endosperm proteinaceous and very thin. X = 16 (or 8?).

The family Hippuridaceae consists of the single genus and species *Hippuris vulgaris* L. (if the species is broadly defined), widely distributed in temperate and boreal regions of the Northern Hemisphere, and in Australia and southern South America. Traditionally the Hippuridaceae have been associated with the Haloragaceae, a family here referred to the order Haloragales in the subclass Rosidae. More recently it has been pointed out that the ovular structure of *Hippuris* is more compatible with the Asteridae, and Hegnauer (1969, initial citations) believes on chemical grounds that *Hippuris* is related to the traditional order Tubiflorae, near the Plantaginaceae. The Tubiflorae are here considered to constitute several related orders of Asteridae.

2. Family CALLITRICHACEAE Link 1821 nom. conserv., the Water-starwort Family

Low, slender, commonly much-branched herbs, typically aquatic and submersed or emergent with distally floating stems, but sometimes terrestrial in wet places, commonly producing iridoid compounds, but not saponiferous, not cyanogenic, and not tanniferous, lacking both ellagic acid and proanthocyanins; calcium-oxalate crystals wanting; vascular system of the stem reduced to a slender, axial strand enclosing a very slender pith. LEAVES opposite (seldom whorled), small, often linear, entire, exstipulate. FLOWERS without perianth, unisexual, commonly subtended by a pair of horn-shaped bracteoles, borne singly in the axils, or seldom one flower of each sex in the same axil; staminate

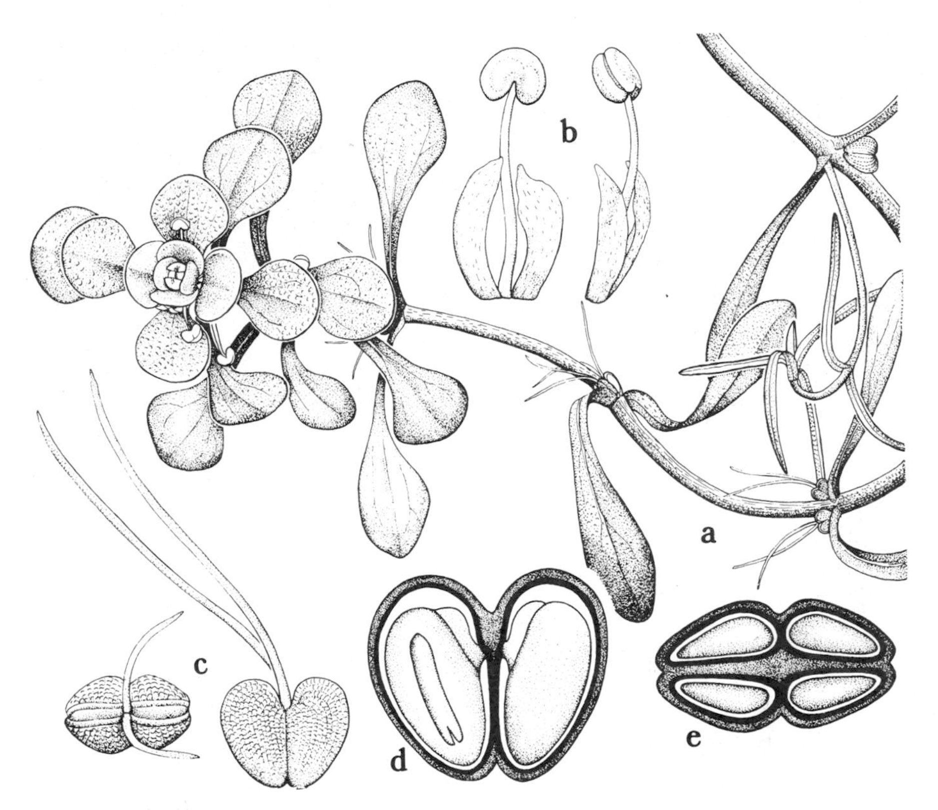

FIG. 6-18 Callitrichaceae. *Callitriche heterophylla* Pursh. a, habit, ×4; b, staminate flowers, ×16; c, pistillate flowers, ×16; d, schematic long-section of fruit, ×32; e, schematic cross-section of fruit, ×32.

flower consisting of a single stamen, or seldom 2–3 stamens; anthers tetrasporangiate and dithecal, opening by lateral, longitudinal slits that are apically confluent; pollen-grains binucleate or trinucleate, 3 (4)-colporate or inaperturate; pistillate flower consisting of a single bicarpellate, somewhat compressed, laterally 4-lobed (and sometimes winged), apically somewhat indented ovary, crowned by a pair of slender styles or a single deeply cleft style; carpels 2, collateral, the commissure edgewise to the axis of the stem; each primary locule of the ovary divided into 2 chambers by an intrusive partition from the carpellary midrib, so that the ovary is 4-chambered; each chamber with a single pendulous, axile, anatropous ovule with ventral raphe, the micropyle directed upward and outward; ovules with a massive single integument and an integumentary tapetum, tenuinucellar; endosperm-development cellular, with a micropylar haustorium and a smaller chalazal one. FRUIT dry, separating at maturity into 4 1-seeded units; seed-coat very thin; embryo straight or slightly curved, cylindrical, with 2 short cotyledons, embedded in the thin, oily endosperm. X = 3, 5+.

The family Callitrichaceae consists of the single genus *Callitriche*, with about 35 species, of nearly cosmopolitan distribution. The affinities of the family have been debated in the past. On the basis of gynoecial morphology, ovular anatomy, and the presence of iridoid compounds, most modern phylogenists believe that *Callitriche* belongs in the neighborhood of the Lamiales.

3. Family HYDROSTACHYACEAE Engler 1898 nom. conserv., the Hydrostachys Family

Submerged aquatic perennial herbs with a short, tuberous-thickened stem, a basal holdfast, fibrous roots, a cluster of basal leaves, and a central scape with a terminal spike; vascular system much-reduced, without vessels. LEAVES elongate, entire to 2–3 times pinnatifid, widened and ligular at the base, often covered with numerous small, scale-like or fringed appendages; stomates wanting; an inconspicuous, membranous, intrapetiolar stipule present. FLOWERS sessile in the axils of the bracts of the dense spike, small and inconspicuous, without a perianth, unisexual, the plants dioecious or seldom monoecious; staminate flowers consisting of a single extrorse stamen with a short filament and a tetrasporangiate, dithecal anther, the well separated pollen-sacs opening lengthwise; pollen tetradinous, binucleate, inaperturate; pistillate flowers consisting of a single bicarpellate, unilocular ovary with 2 parietal placentas and 2 elongate, persistent, filiform styles, these sometimes connate at the base; carpels median, one anterior, the other posterior; ovules numerous, anatropous, unitegmic, tenuinucellar; endosperm-development cellular, with a micropylar haustorium. FRUIT a septicidal capsule with numerous tiny seeds, lacking endosperm.

The family Hydrostachyaceae consists of the single genus *Hydrostachys*, with about 20 species native to Madagascar and tropical and southern Africa. Rauh and Jäger-Zürn think that the family is allied to the Scrophulariaceae and Plantaginaceae.

5. Order PLANTAGINALES Lindley 1833

The order consists of the single family Plantaginaceae.

SELECTED REFERENCES

Carlquist, S. 1970. Wood anatomy of insular species of *Plantago* and the problem of raylessness. Bull. Torrey Bot. Club 97: 353–361.

Cooper, G. O. 1942. Development of the ovule and formation of the seed in *Plantago lanceolata*. Amer. J. Bot. 29: 577–581.

Misra, R. C. 1966. Morphological studies in *Plantago*. III. Nodal anatomy. Proc. Indian Acad. Sci. 63B: 271–274.

Moncontie, C. 1969. Les stomates des Plantaginacées. Rev. Gén. Bot. 76: 491–529.

Stebbins, G. L., & A. Day. 1967. Cytogenetic evidence for long continued stability in the genus *Plantago*. Evolution 21: 409–428.

Василевская, В. К., М. П. Баранов, & Г. М. Борисовская. 1973. Строение розеточного растения *Plantago major* L. в первый год жизни. Бот. Ж. 58: 33-42.

1. Family PLANTAGINACEAE A. L. de Jussieu 1789 nom. conserv., the Plantain Family

Herbs or occasionally (some of the insular spp.) small shrubs or half-shrubs, commonly producing iridoid compounds and sometimes monoterpenoid alkaloids, sometimes saponiferous, but not cyanogenic, and not tanniferous, lacking both ellagic acid and proanthocyanins; crystals usually wanting; nodes variously unilacunar, trilacunar, or multilacunar; xylem in a ring of discrete bundles, or forming a continuous cylinder nearly or quite without rays; vestigial or well developed medullary bundles sometimes present; vessel-segments very small, with simple perforations; imperforate tracheary elements very short, with simple or narrowly bordered pits; wood-parenchyma wanting or nearly so. LEAVES commonly all basal and alternate, seldom cauline and alternate or opposite, simple, often sheathing at the base, the phyllodial, more or less parallel-veined blade apparently representing an expanded petiole, or the leaves sometimes much-reduced; stomates mostly diacytic, sometimes anomocytic; stipules wanting. FLOWERS mostly in pedunculate, bracteate spikes or heads, but without bracteoles (in *Littorella* the flowers basal in groups of 3, the central one staminate and long-pedicellate, the 2 lateral ones pistillate and sessile), typically chasmagamous, protogynous, and anemophilous, but sometimes cleistogamous in varying de-

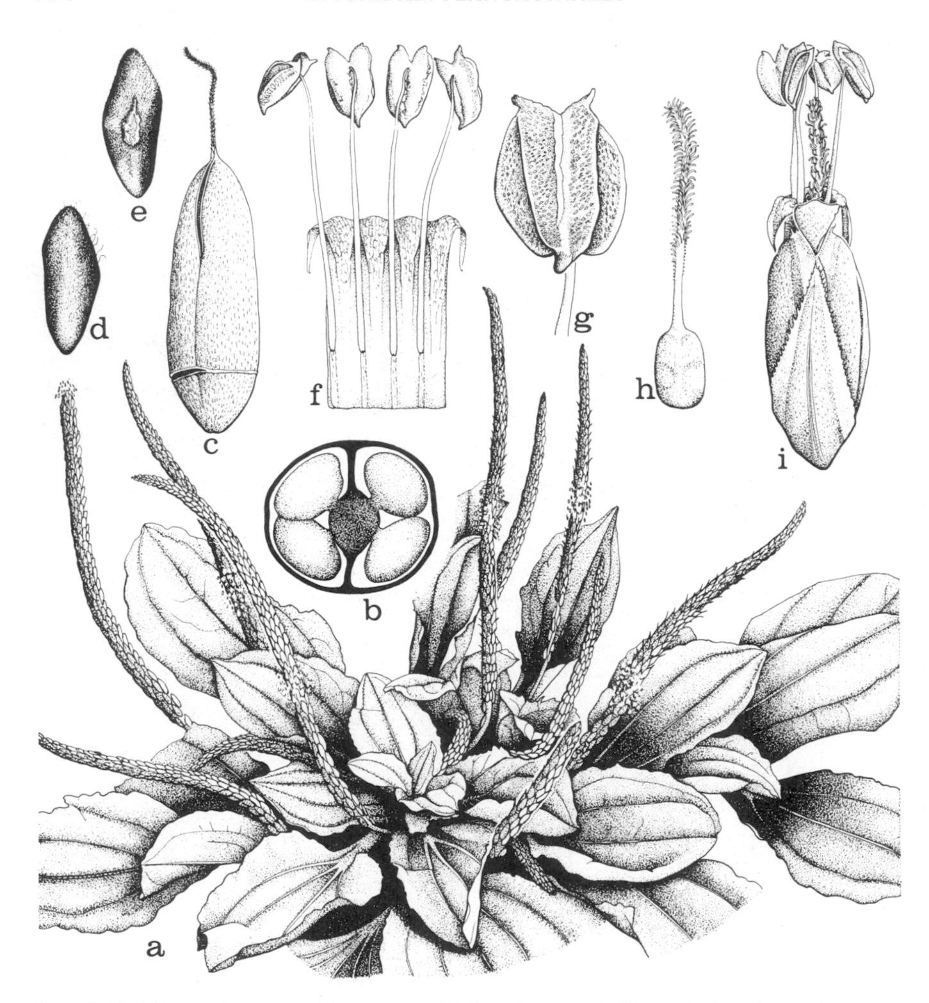

FIG. 6-19 Plantaginaceae. *Plantago rugelii* Decaisne. a, habit, ×½; b, schematic cross-section of ovary, ×24; c, capsule, ×8; d, e, seeds, ×8; f, corolla and stamens, ×8; g, anther, ×24; h, pistil, ×8; i, aspect of flower, ×8.

grees, individually small and not showy, mostly regular and perfect (plants monoecious in *Littorella*, gynomonoecious in *Bougueria*), commonly tetramerous (rarely trimerous) as to the calyx, corolla, and androecium; calyx lobed or cleft, the 2 abaxial segments sometimes more or less connate; corolla scarious, sympetalous, the teeth or lobes imbricate; stamens generally as many as and alternate with the corolla-lobes (only 1–2 in *Bougueria*), the filaments attached to the corolla-tube; anthers long-exserted, versatile, tetrasporangiate and dithecal, opening by longitudinal slits; pollen-grains binucleate or trinucleate, 4–14-porate; gynoecium of 2 median carpels united to form a compound, superior ovary, with a slender terminal style and a dry, usually 2-lobed stigma; ovary in *Plantago* basically bilocular with 1–40 ovules on an axile placenta in each locule, but occasionally appearing 4–locular because of intrusive partitions from the carpellary midribs, in *Bougueria* and *Littorella* the ovary unilocular and with a single basal ovule; ovules anatropous to hemitropous, with a massive single integument and an integumentary tapetum, tenuinucellar; endosperm-development cellular, with terminal haustoria. FRUIT a membranous, circumscissile capsule in *Plantago*, an achene or small nut enclosed in the persistent calyx in *Bougueria* and *Littorella*; embryo with 2 cotyledons, more or less spatulate, straight or seldom (*Bougueria*) curved; endosperm well developed, firm, translucent. X = 4–12+.

The family Plantaginaceae consists of 3 genera of very unequal size. The familiar cosmopolitan genus *Plantago* has about 250 species. *Littorella*, which is more or less aquatic, has only 3 species, and *Bougueria* is monotypic.

The affinities of the Plantaginaceae have been disputed in the past. Most present-day authors agree that they are related to the Scrophulariaceae, and Takhtajan includes them in the order Scrophulariales. Although I agree as to the relationships, I find it conceptually more useful to maintain the Plantaginales as a distinct order.

Derivation of the Plantaginaceae from the Scrophulariaceae or some similar ancestral group would imply the same sort of reduction from an irregular, 5-lobed corolla with 4 stamens to a regular, 4-lobed corolla with 4 stamens that appears to have taken place in the Buddlejaceae. The floral anatomy of some species suggests the possibility of a pentamerous ancestry.

The anemophilous habit of the Plantaginaceae must be of considerable importance to the plants, but the reasons for a change from entomophily are obscure. The Plantaginaceae grow in the same sorts of places as entomophilous plants, and there is nothing in their general structure, aside from the reduced corolla, that does not appear to be equally compatible with entomophily.

Pollen attributed to *Plantago* occurs in middle or late Miocene and more recent deposits.

6. Order SCROPHULARIALES Lindley 1833

Herbs or less often shrubs or even trees, sometimes vines, commonly producing iridoid compounds and the phenolic glycoside orobanchin, but only seldom with alkaloids, only rarely cyanogenic or saponiferous, and rarely tanniferous, lacking both ellagic acid and proanthocyanins; stems sometimes with anomalous structure, but only rarely with internal phloem; vessel-segments all or mostly with simple perforations. LEAVES opposite, alternate, or sometimes whorled, nearly always exstipulate, simple or sometimes (mainly Bignoniaceae and some Scrophulariaceae) compound, the dissected, leaf-like organs of many Lentibulariaceae possibley being modified stems. FLOWERS mostly perfect, generally sympetalous, most commonly with an irregular, basically pentamerous, often bilabiate corolla and only 4 or 2 functional stamens, less often the corolla tetramerous by reduction and regular or nearly so, rarely the corolla 5-lobed and essentially regular, or reduced and polypetalous, or even wanting; stamens attached to the corolla-tube, alternate with the lobes, in tetramerous corollas as many as the lobes, or only 2, in pentamerous corollas only rarely as many as the lobes, usually 4 functional, and the upper (adaxial, posterior) one wanting or staminodial, or only 2 functional, with either the anterior pair or the posterior-lateral pair also wanting or staminodial; pollen-grains binucleate or trinucleate, triaperturate or of various triaperturate-derived types; most families with an annular nectary-disk around the base of the ovary; gynoecium most commonly of 2 median carpels, rarely 3 or 4 carpels, the carpels united to form a compound, superior or sometimes partly or wholly inferior (mainly some Gesneriaceae, and *Trapella*, in the Pedaliaceae) ovary with axile or parietal or free-central placentation, seldom one of the 2 carpels more or less reduced, so that the ovary may be pseudomonomerous; style terminal, with usually 2 stigmas (one sometimes reduced or suppressed), or the style sometimes very short and the stigmas essentially sessile; ovules (1) 2-many per functional carpel, in the large majority of the species more or less numerous, only rarely the locules of the ovary divided into uniovulate locelli by partitions intrusive from the carpellary midribs; ovules anatropous to hemitropous or amphitropous, tenuinucellar, with a massive single integument and usually with an integumentary tapetum, at least toward the chalazal end; endosperm-development cellular (or seldom partly or wholly nuclear, as in some Acanthaceae), with terminal haustoria, or sometimes only the micropylar or the chalazal haustorium developed. FRUIT most commonly capsular, or sometimes indehiscent and then usually drupaceous or baccate, seldom of separating, drupe-like mericarps representing individual carpels (not half-carpels); seeds (1–) usually several or many; embryo dicotyledonous or (notably in the Orobanchaceae) without differentiated cotyledons; endosperm copious and oily to scanty or wanting.

The order Scrophulariales as here defined consists of 12 families and more than 11,000 species. About three-fourths of the species belong to only 3 large families, the Scrophulariaceae (4000), Acanthaceae (2500), and Gesneriaceae (2500). The Columelliaceae, often associated with the families of the Scrophulariales, are here referred to the Rosales.

The Scrophulariales differ from the closely related order Solanales in usually having an irregular (often bilabiate) corolla with fewer stamens than corolla-lobes, in the much less frequent presence of internal phloem, and in commonly producing iridoid compounds and orobanchin, but only seldom alkaloids. Exceptions can be found to all of these characters, but the clustering of the families is clear. It may reasonably be supposed that the floral differences between the two orders reflect increasing specialization by the Scrophulariales for pollination by specific insects or birds. The chemical differences presumably indicate a different choice of weapons in the necessary defense against predators. It may be noted that the Gentianales, which are more primitive in some respects than either the Scrophulariales or the Solanales, produce alkaloids as well as iridoid compounds, but are not known to have orobanchin. The ecological significance of internal phloem is obscure.

The Scrophulariaceae are not only the largest family of their order, but are also central to it. Four of the other families (Acanthaceae, Bignoniaceae, Globulariaceae, and Orobanchaceae) are connected to the Scrophulariaceae by transitional genera or groups of genera that have by different authors been referred to the central or the peripheral family. Although the other families of the order are more sharply limited, most of them may logically be considered to be specialized derivatives of the Scrophulariaceae. The Bignoniaceae may be more primitive than the Scrophulariaceae in being mainly woody, but in other respects such as the lack of endosperm, the often climbing habit, the often compound leaves, and the commonly winged seeds, the Bignoniaceae appear to be more advanced.

The Acanthaceae diverge from the Scrophulariaceae primarily in their explosively dehiscent fruit and specialized funiculus. The ecologic significance of the concurrent loss of endosperm and development of cystoliths is obscure. Perhaps the cystoliths serve a defensive function.

The Orobanchaceae diverge from the Scrophulariaceae in their parasitic habit and parietal placentation. The evolutionary journey toward parasitism obviously begins in the Scrophulariaceae; the Orobanchaceae merely occupy the house at the end of the road. The ecologic significance of the concurrent change to parietal placentation is obscure.

The Globulariaceae and Myoporaceae have taken the route of reduction in the number of ovules and formation of indehiscent, 1-seeded fruits or mericarps. The change obviously has ecological overtones, but it is not unusual in terms of the Asteridae as a whole, and it does not fit these families to any particular way of life not shared with other groups. These two families also share a special type of glandular hair, the ecological significance of which is unknown.

The Lentibulariaceae have become insectivorous and mostly aquatic. The ecologic significance of the concurrent change to free-central placentation is unknown.

The Gesneriaceae diverge from the Scrophulariaceae in their basically parietal placentation and frequently more or less inferior ovary. The selective significance of these changes is obscure.

The specialized, mucilaginous trichomes of the Pedaliaceae, which make the herbage somewhat slimy, may well serve a protective function. The ornamentation of the fruits in many Pedaliaceae is obviously an adaptation for zoochorous distribution of the seeds.

The Buddlejaceae and Oleaceae may have restored the ancestral regular corolla by reducing one of the lobes and becoming tetramerous. The selective forces that could drive such a change are not understood. The position of these two families at the beginning of the linear sequence in the Scrophulariales reflects an effort to cluster the members of the order around the Scrophulariales. The Buddlejaceae and Oleaceae are certainly not primitive (or archaic) within the order.

SELECTED REFERENCES

Ahmad, K. J. 1974. Cuticular studies in some Nelsonioideae (Acanthaceae). J. Linn. Soc., Bot. 68: 73–80.

Ahmad, K. J. 1974. Cuticular studies in some species of *Mendoncia* and *Thunbergia* (Acanthaceae). J. Linn. Soc., Bot. 69: 53–63.

Ahmad, K. J. 1978. Epidermal hairs of Acanthaceae. Blumea 24: 101–117.

Arekal, G. D. 1963. Contribution to the embryology of *Chelone glabra* L. Phytomorphology 13: 376–388.

Attawi, F. 1977. Morphologisch-anatomische Untersuchungen an den Haustorien einiger *Orobanche*-Arten. Ber. Deutsch. Bot. Ges. 90: 173–182.

Bhaduri, S. 1944. A contribution to the morphology of pollen grains of Acanthaceae and its bearing on taxonomy. Calcutta Univ. Jour. Dep. Sci. II. 1(4): 25–58.

Boeshore, I. 1920. The morphological continuity of Scrophulariaceae and Orobanchaceae. Contr. Bot. Lab. Morris Arbor. Univ. Pennsylvania 5: 139–177.

Bremekamp, C. E. B. 1953. The delimitation of Acanthaceae. Proc. Kon. Nederl. Akad. Wet. C 56: 533–546.

Bremekamp, C. E. B. 1965. Delimitation and subdivision of the Acanthaceae. Bull. Bot. Surv. India 7: 21–30.

Bunting, G. S., & J. A. Duke. 1961. *Sanango*: New Amazonian genus of Loganiaceae. Ann. Missouri Bot. Gard. 48: 269–274.

Burtt, B. L. 1963, 1970. Studies on the Gesneriaceae of the Old World. XXIV. Tentative keys to the tribes and genera. XXXI. Some aspects of functional evolution. Notes Roy. Bot. Gard. Edinburgh 24: 205–220, 1963. 30: 1–9, 1970.

Burtt, B. L. 1977. Classification above the genus, as exemplified by Gesneriaceae, with parallels from other groups. Pl. Syst. Evol., Suppl. 1: 97–109.

Buurman, J. 1977. Contribution to the pollenmorphology of the Bignoniaceae, with special reference to the tricolpate type. Pollen & Spores 19: 447–519.

Campbell, D. H. 1930. The relationships of *Paulownia*. Bull. Torrey Bot. Club 57: 47–50.

Casper, S. J. 1963. "Systematisch massgebende" Merkmale für die Einordnung der Lentibulariaceen in das System. Oesterr. Bot. Z. 110: 108–131.

Copeland, H. F. 1960. The reproductive structures of *Fraxinus velutina* (Oleaceae). Madroño 15: 161–172.

Crété, P. 1955. L'application de certaines données embryologiques à la systématique des Orobanchacées et de quelques familles voisins. Phytomorphology 5: 422–435.

De, A. 1966, 1967, 1968. Cytological, anatomical and palynological studies as an aid in tracing affinity and phylogeny in the family Acanthaceae. I. Cytological studies. II. Floral anatomy. III. General anatomy. IV. Palynology and final conclusion. Trans. Bose Research Inst. 29: 139–175, 1966. 30: 27–43: 51–65, 1967. 31: 17–29, 1968.

Farooq, M. 1964. Studies in the Lentibulariaceae. I. The embryology of *Utricularia stellaris* Linn. f. var. *inflexa* Clarke. Proc. Natl. Inst. Sci. India 30B: 263–299. III. The embryology of *Utricularia uliginosa* Vahl. Phytomorphology 15: 123–131.

Gentry, A. 1974. Coevolutionary patterns in Central American Bignoniaceae. Ann. Missouri Bot. Gard. 61: 728–759.

Gentry, A. H. 1980. Bignoniaceae—Part 1. (Crescentieae and Tourrettieae). Fl. Neotropica Monogr. 25.

Gentry, A. H., & A. S. Tomb. 1979 (1980). Taxonomic implications of Bignoniaceae palynology. Ann. Missouri Bot. Gard. 66: 756–777.

Goldblatt, P., & A. H. Gentry. 1979. Cytology of Bignoniaceae. Bot. Not. 132: 475–482.

Guédès, M. 1965. Remarques sur la placentation des Orobanchacées. Bull. Soc. Bot. France 111: 257–261.

Harborne, J. B. 1967. Comparative biochemistry of the flavonoids. VI. Flavonoid patterns in the Bignoniaceae and the Gesneriaceae. Phytochemistry 6: 1643–1651.

Hartl, D. 1956. Die Beziehungen zwischen den Plazenten der Lentibulariaceen und Scrophulariaceen nebst einem Exkurz über die Spezialisationsrichtungen der Plazentation. Beitr. Biol. Pflanzen 32: 471–490.
Heckard, L. R. 1962. Root parasitism in *Castilleja*. Bot. Gaz. 124: 21–29.
Huynh, K.-L. 1968. Étude de la morphologie du pollen du genre *Utricularia* L. Pollen & Spores 10: 11–55.
Ivanina, L. I. 1965 (1966). Application of the carpological method to the taxonomy of Gesneriaceae. Notes Roy. Bot. Gard. Edinburgh 26: 383–403.
Johnson, L. A. S. 1957. A review of the family Oleaceae. Contr. New South Wales Natl. Herb. 2: 395–418.
Johri, B. M., & H. Singh. 1959. The morphology, embryology and systematic position of *Elytraria acaulis* (Linn. f.) Lindau. Bot. Not. 112: 227–251.
Junell, S. 1961. Ovarian morphology and taxonomical position of Selagineae. Svensk Bot. Tidskr. 55: 168–192.
Karlström, P.-O. 1974. Embryological studies in the Acanthaceae. III. The genera *Barleria* L. and *Crabbea* Harv. Svensk Bot. Tidskr. 68: 121–135.
Khan, R. 1954. A contribution to the embryology of *Utricularia flexuosa* Vahl. Phytomorphology 4: 80–117.
Komiya, S. 1973. New subdivision of the Lentibulariaceae. J. Jap. Bot. 48: 147–153.
Kondo, K., M. Segawa, & K. Nehira. 1978. Anatomical studies on seeds and seedlings of some *Utricularia* (Lentibulariaceae). Brittonia 30: 76–88.
Kshetrapal, S., & Y. G. Tiagi. 1970. Structure, vascular anatomy and evolution of the gynoecium in family Oleaceae and their bearing on the systematic position of the genus *Nyctanthes* L. Acta Bot. Acad. Sci. Hung. 16: 143–151.
Li, H.-L. 1954. Trapellaceae, a familial segregate from the Asiatic flora. J. Wash. Acad. Sci. 44: 11–13.
Long, R. W. 1970. The genera of Acanthaceae in the southeastern United States. J. Arnold Arbor. 51: 257–309.
Lowry, J. B. 1973. *Rhabdothamnus solandri*: some phytochemical results. New Zealand J. Bot. 11: 555–560.
McIntyre, W. G., & M. A. Chrysler. 1943. The morphological nature of the photosynthetic organs of *Orchyllium endresii* as indicated by their vascular structure. Bull. Torrey Bot. Club 70: 252–260.
Maheshwari, J. K. 1961. The genus *Wightia* Wall. in India with a discussion on its systematic position. Bull. Bot. Surv. India 3: 31–35.
Mauritzon, J. 1934. Die Endosperm und Embryoentwicklung einiger Acanthaceen. Acta Univ. Lund. II. Sect. 2. 30(5): 1–41.
Mohan Ram, H. Y., & P. Masand. 1963. Embryology of *Nelsonia campestris* R. Br. Phytomorphology 13: 82–91.
Mohan Ram, H. Y., & M. Wadhi. 1964. Endosperm in Acanthaceae. Phytomorphology 14: 388–413.
Mohan Ram, H. Y., & M. Wadhi. 1965. Embryology and the delimitation of the Acanthaceae. Phytomorphology 15: 201–205.
Moore, R. J. 1948. Cytotaxonomic studies in the Loganiaceae. II. Embryology of *Polypremum procumbens* L. Amer. J. Bot. 35: 404–410.
Musselman, L. J., & W. C. Dickison. 1975. The structure and development of the haustorium in parasitic Scrophulariaceae. J. Linn. Soc., Bot. 70: 183–212.
Pennell, F. W. 1935. The Scrophulariaceae of eastern temperate North America. Acad. Nat. Sci. Philadelphia Monogr. 1.
Raj, B. 1961. Pollen morphological studies in the Acanthaceae. Grana Palynol. 3(1): 3–108.

Rao, V. S. 1953, 1954. The floral anatomy of some Bicarpellatae. I. Acanthaceae. II. Bignoniaceae. J. Univ. Bombay 21 (N.S.), part 5B: 1–34, 1953. 22 (N.S.), part 5B: 55–70, 1954.
Ratter, J. A. 1975. A survey of chromosome numbers in the Gesneriaceae of the Old World. Notes Roy. Bot. Gard. Edinburgh 33: 527–543.
Record, S. J., & K. W. Hess. 1940. American timbers of the family Bignoniaceae. Trop. Woods 63: 9–38.
Sax, K., & E. C. Abbe. 1932. Chromosome numbers and the anatomy of the secondary xylem in the Oleaceae. J. Arnold Arbor. 13: 37–48.
Schrock, G. F., & B. F. Palser. 1967. Floral development, anatomy, and embryology of *Collinsia heterophylla* with some notes on ten other species of *Collinsia* and on *Tonella tenella*. Bot. Gaz. 128: 83–104.
Singh, S. P. 1960. Morphological studies in some members of the family Pedaliaceae. I. *Sesamum indicum* D.C. Phytomorphology 10: 65–81.
Singh, S. P. 1960. Morphological studies in some members of the family Pedaliaceae. Agra Univ. J. Res., Sci. 9: 217–220.
Singh, V., & D. K. Jain. 1975. Floral development of *Justicia gendarussa* (Acanthaceae). J. Linn. Soc., Bot. 70: 243–253.
Singh, V., & D. K. Jain. 1975. Trichomes in Acanthaceae. I. General structure. J. Indian Bot. Soc. 54: 116–127.
Skog, L. E. 1976. A study of the tribe Gesnerieae, with a revision of *Gesneria* (Gesneriaceae: Gesnerioideae). Smithsonian Contr. Bot. 29: 1–182.
Souèges, R. 1940. Embryogénie des Loganiacées. Développement de l'embryon chez le *Buddleia variabilis* Hemsley. Compt. Rend. Hebd. Séances Acad. Sci. 211: 139–140.
Straka, H., & H.-D. Ihlenfeldt. 1965. Pollenmorpholigie und Systematik der Pedaliaceae R. Br. Beitr. Biol. Pflanzen 41: 175–207.
Straw, R. M. 1956. Adaptive morphology of the *Penstemon* flower. Phytomorphology 6: 112–119.
Straw, R. M. 1966. A redefinition of *Penstemon* (Scrophulariaceae). Brittonia 18: 80–95.
Suryakanta. [no initials] 1973. Pollen morphological studies in the Bignoniaceae. J. Palyn. (Lucknow) 9: 45–82.
Taylor, H. 1945. Cyto-taxonomy and phylogeny of the Oleaceae. Brittonia 5: 337–367.
Thieret, J. W. 1967. Supraspecific classification in the Scrophulariaceae: a review. Sida 3: 87–106.
Thieret, J. W. 1971. The genera of Orobanchaceae in the southeastern United States. J. Arnold Arbor. 52: 404–434.
Thieret, J. W. 1977. The Martyniaceae in the southeastern United States. J. Arnold Arbor. 58: 25–39.
Tiagi, B. 1956. A contribution to the embryology of *Striga orobanchoides* Benth. and *Striga euphrasioides* Benth. Bull. Torrey Bot. Club 83: 154–170.
Tiagi, B. 1963. Studies in the family Orobanchaceae. IV. Embryology of *Boschniackia himalaica* Hook. and *B. tuberosa* (Hook.) Jepson, with remarks on the evolution of the family. Bot. Not. 116: 81–93.
Trivedi, M. L., V. Khanna, & Shailja. 1976. Nodal anatomy of certain members of Bignoniaceae. Proc. Indian Acad. Sci. 84B: 31–36.
Wall, W. E., & R. W. Long. 1965. Megasporogenesis and embryo sac development in *Ruellia caroliniensis* (Acanthaceae). Bull. Torrey Bot. Club 92: 372–377.

Weber, A. 1973. Über die "Stipeln" von *Rhytidophyllum* Mart. (Gesneriaceae). Oesterr. Bot. Z. 121: 279–283.

Westfall, J. J. 1949. Cytological and embryological evidence for the reclassification of *Paulownia*. Amer. J. Bot. 36: 805.

Wilson, C. L. 1974. Floral anatomy in Gesneriaceae. I. Cyrtandroideae. II. Gesnerioideae. Bot. Gaz. 135: 247–256; 256–268.

Wilson, K. A., & C. E. Wood. 1959. The genera of Oleaceae in the southeastern United States. J. Arnold Arbor. 40: 369–384.

Афанасьева, Н. Г. 1959. Микроспорогенез и некоторые вопросы филогении норичниковых (Scrophulariaceae). Научн. Докл. Высш. Школы, Биол. Науки, 4: 114-122.

Афанасьева, Н. Г. 1960. К познанию филогении семейства норичниковых. Автореф. Дисс., Казань.

Каден, Н. Н., & С. А. Смирнова. 1964. К морфологии плодов норичниковых. Бюлл. Моск. Общ. Испыт. Прир., Отд. Биол., 69(3): 77-90.

Никитичева, З. И. 1968. Развитие пыльника и микроспорогенез у некоторых представителей Scrophulariaceae и Orobanchaceae. Бот. Ж. 53: 1704-1715.

Никитичева, З. И. 1971. Эмбриогенез некоторых паразитирующих видов из семейств Scrophulariaceae и Orobanchaceae. Бот. Ж. 56: 93-105.

Николаева, Э. В. 1969. Морфологическая характеристика цветков и пол у некоторых видов *Fraxinus* L. Бот. Ж. 54: 582-589.

Тиаги, Я. Д. 1962. Анатомическое изучение сосудистого оснащения цветка некоторых видов семейств Orobanchaceae и Scrophulariaceae. Вестн. Московск. Унив. сер. 6, Биол. 1962(2): 29-52.

SYNOPTICAL ARRANGEMENT OF THE FAMILIES OF SCROPHULARIALES

1 Plants not insectivorous, and only occasionally aquatic; placentation various, but not free-central.

  2 Corolla mostly 4-lobed and regular or nearly so, wanting in some Oleaceae; mostly woody plants with opposite or whorled leaves, rarely herbs.

    3 Stamens 4; ovules more or less numerous in each locule .... 1. BUDDLEJACEAE.

    3 Stamens 2, rarely 4; ovules most commonly 2 in each locule, sometimes 1–4, seldom numerous .... 2. OLEACEAE.

  2 Corolla mostly 5-lobed and/or more or less strongly irregular, rarely wanting; habit and leaves various.

    4 Seeds mostly with well developed endosperm (except many Gesneriaceae, these without the special features of any of the families in the next group).

      5 Placentation basically axile, the ovary typically bilocular, sometimes with one locule more or less reduced, or even suppressed so that the ovary is pseudomonomerous.

        6 Fruit usually a capsule, rarely a berry or a schizocarp; ovules (2–) more or less numerous in each locule .... 3. SCROPHULARIACEAE.

6 Fruit of 2 separating, often unequal nutlets or drupelets, or of a small nut or achene; ovules solitary (2) in each locule, or one locule empty or suppressed ....4. GLOBULARIACEAE.

5 Placentation basically parietal, seldom secondarily axile.

7 Plants parasitic, without chlorophyll; embryo minute, undifferentiated; leaves reduced and alternate; ovary superior ....6. OROBANCHACEAE.

7 Plants autotrophic; embryo well developed, with 2 cotyledons; leaves opposite, or rarely whorled or alternate, or the plants of anomalous vegetative structure; ovary superior to inferior ....7. GESNERIACEAE.

4 Seeds with scanty or no endosperm.

8 Fruit explosively dehiscent, the seeds with an enlarged and specialized funiculus that is typically developed into a jaculator: characteristic cystoliths usually present in some epidermal and parenchyma cells .... 8. ACANTHACEAE.

8 Fruit indehiscent or dehiscent, but not explosively so, the funiculus of ordinary type; cystoliths wanting.

9 Herbs (rarely shrubs) with specialized mucilaginous hairs, the herbage commonly slimy; fruits very often with hooks or horns or prickles, or sometimes winged ....9. PEDALIACEAE.

9 Trees, shrubs, or woody vines, only rarely herbs, without specialized mucilaginous hairs; fruit otherwise.

10 Ovules more or less numerous in each locule; fruit usually a capsule, or seldom fleshy and indehiscent but not a drupe; plants erect or often climbing, with simple or more often compound, opposite or whorled or rarely alternate leaves .... 10. BIGNONIACEAE.

10 Ovules (1) 2–8 in each locule, or one locule empty; fruit a drupe, or sometimes separating into 1-seeded, drupe-like segments; leaves simple.

11 Twiners, without specialized secretory cavities; leaves opposite ....11. MENDONCIACEAE.

11 Erect shrubs or small trees, most genera with scattered secretory cavities; leaves alternate, seldom opposite .... 5. MYOPORACEAE.

1 Insectivorous herbs, aquatic or of wet places; placentation free-central .... 12. LENTIBULARIACEAE.

1. Family BUDDLEJACEAE Wilhelm 1910 nom. conserv., the Butterfly-bush Family

Shrubs, or trees, seldom herbs (as in *Polypremum*), often with lepidote or stellate or branching hairs and/or glandular hairs, but without unicellular or uniseriate simple hairs (except sometimes on the inner surface of the corolla), commonly producing orobanchin and iridoid

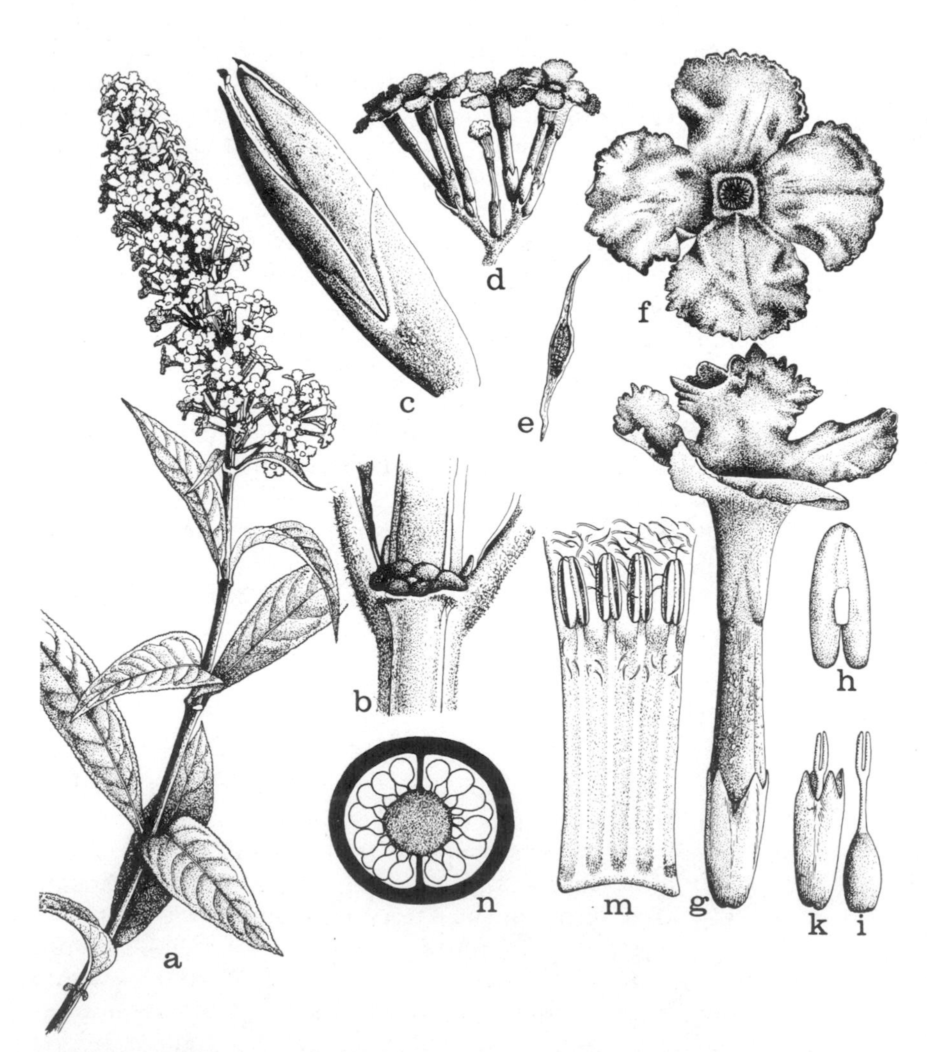

Fig. 6-20 Buddlejaceae. *Buddleja davidii* Franchet; a, habit, ×½; b, node, with stipules, ×2; c, opening fruit, ×6; d, cymule, ×2; e, seed, ×6; f, g, flower, ×6; h, anther, ×12; i, pistil, ×6; k, calyx, with included pistil, ×6; m, corolla-tube, opened out to show anthers, ×6; n, schematic cross-section of ovary, ×24.

compounds and often saponiferous, but without alkaloids and only seldom tanniferous, lacking both ellagic acid and proanthocyanins; vessel-segments with simple perforations; vasicentric tracheids sometimes present, the imperforate tracheary elements otherwise septate and with simple or obscurely bordered pits; wood-rays heterocellular, 1–3 cells wide; wood-parenchyma sparse, paratracheal; internal phloem wanting. LEAVES opposite or less often whorled, rarely (a few spp. of *Buddleja*) alternate, simple, entire to more often toothed or lobed; stomates mostly anomocytic; interpetiolar stipules commonly represented by a mere line, seldom (*Adenoplusia*) well developed. FLOWERS in various sorts of inflorescences, perfect or often functionally dioecious (at least in the New World species), mostly 4-merous as to the calyx, corolla and androecium (5-merous in *Peltanthera* and casually in *Buddleja*; 5-merous as to the calyx and corolla only in *Sanango*); calyx synsepalous and lobed, or sometimes very deeply cleft; corolla sympetalous, essentially regular (ventricose in *Sanango*), the lobes imbricate or rarely (*Peltanthera*) valvate; stamens attached to the corolla-tube alternate with its lobes (the posterior one absent or represented by a small staminode in *Sanango*); anthers tetrasporangiate and dithecal, opening by longitudinal slits; pollen grains binucleate, 3 (4)-colporate; gynoecium of 2 median carpels united to form a compound, bilocular, superior or (*Polypremum*) half-inferior ovary; style solitary, terminal, with a capitate or 2-lobed stigma; ovules numerous on thickened, axile placentas, hemitropous or amphitropous, tenuinucellar, with a massive single integument and usually an integumentary tapetum; endosperm-development cellular, usually with terminal haustoria. FRUIT commonly a septicidal capsule, rarely fleshy and indehiscent (*Adenoplea*, *Adenoplusia*, *Nicodemia*); seeds often winged, with small to large, dicotyledonous embryo axially embedded in the oily, copious to scanty endosperm. X = 19 (11 in *Polypremum*).

The family Buddlejaceae consists of about 10 genera and 150 species, mainly tropical and subtropical. The largest genus is *Buddleja*, the butterfly-bush, with about a hundred species of both the Old and the New World, some of them cultivated for ornament. The next largest is *Nuxia*, with about 30 spp. of Africa and Madagascar.

Unpublished studies by Piechura show strong serological affinities between the Buddlejaceae and several tested families of Scrophulariales (personal communication, 1979).

The position of the monotypic American genus *Polypremum* is debatable. It clearly belongs to the Loganiaceae when that family is defined in the traditional broad sense, but its position when the Buddlejaceae are held as a distinct family is not yet clear. On embryological features it goes with the Buddlejaceae (Moore, 1948). On the other hand, Punt & Leenhouts (1967) consider that its pollen resembles that of *Spigelia* (Loganiaceae), whereas the pollen of other genera of Buddlejaceae is said to be much like that of the Scrophulariaceae, and unlike that of Loganiacea. *Polypremum* has no internal phloem. Its other anatomical and chemical features do not appear to have been studied.

Sanango provides an interesting approach to the typical scrophulariaceous flower in having a somewhat irregular, 5 lobed corolla with only 4 functional stamens, the 5th stamen being small and staminodial or vestigial.

2. Family OLEACEAE Hoffmannsegg & Link 1813–1820 nom. conserv., the Olive Family

Trees or shrubs, sometimes climbing, mostly with unilacunar, 1-trace nodes, very often with peltate secretory trichomes, and sometimes with extrafloral nectaries composed of groups of secretory hairs; plants commonly producing mannitol, iridoid compounds, and the phenolic glycosides orobanchin and syringin, but generally without alkaloids and not cyanogenic, sometimes with triterpenoid saponins or other triterpenoid compounds, and sometimes somewhat tanniferous, but lacking both ellagic acid and proanthocyanins; small crystals of calcium oxalate commonly found in some of the cells of the parenchymatous tissues; vessel-segments with simple perforations, or seldom some of them with a few cross-bars; imperforate tracheary elements with simple to sometimes evidently bordered pits, occasionally septate; wood-rays heterocellular to homocellular, mixed uniseriate and pluriseriate, the latter 2–4 (5) cells wide, with short ends; wood-parenchyma in most genera mainly paratracheal, but sometimes (as in *Forsythia*) mainly apotracheal and diffuse, or wanting; internal phloem wanting; cork arising superficially. LEAVES opposite or rarely (spp. of *Jasminum*) alternate, simple to pinnately compound or trifoliolate or unifoliolate, often containing sclereids; stomates commonly anomocytic; petiole generally with an arcuate vascular strand, sometimes with smaller accessory lateral bundles; stipules wanting. Inflorescence fundamentally cymose, but often racemiform or paniculiform, or the flowers sometimes solitary. FLOWERS mostly rather small, regular, perfect or sometimes some or all of them unisexual (the plants sometimes dioecious); calyx mostly small, 4 (–15)-

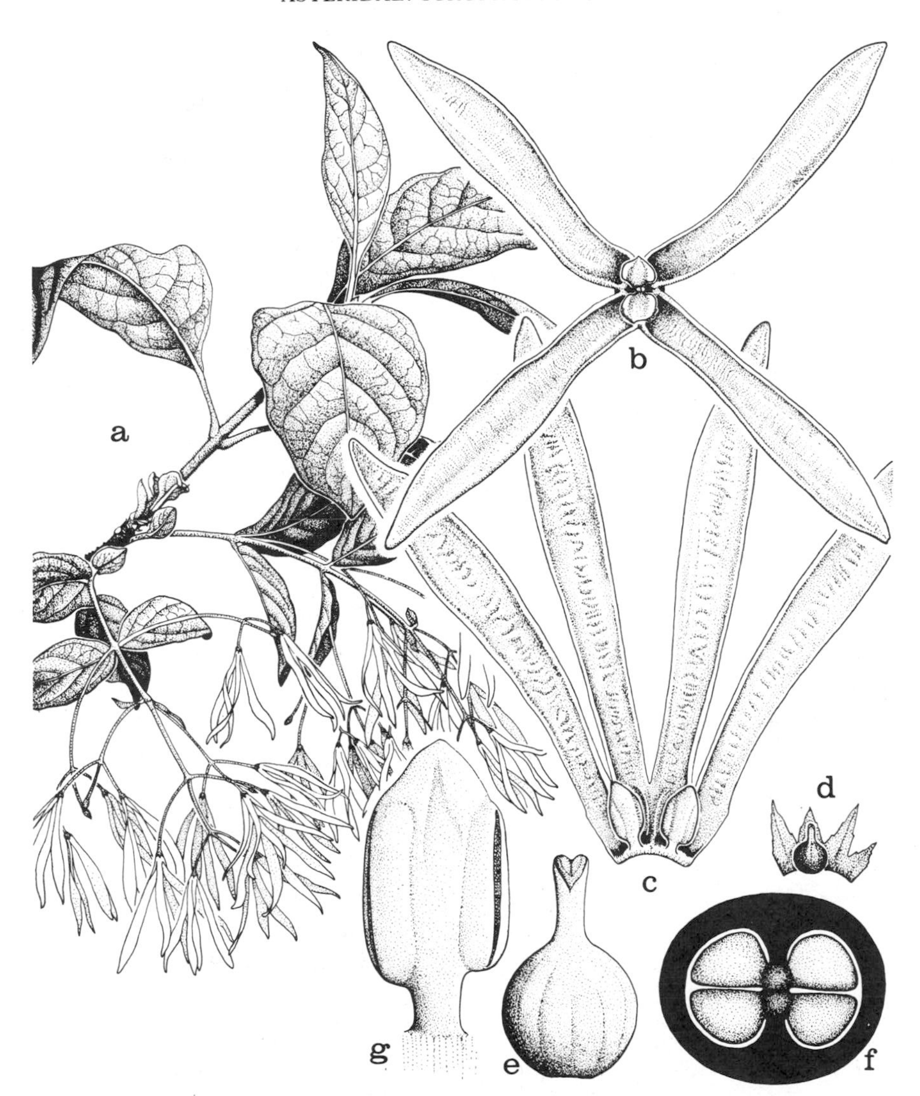

FIG. 6-21 Oleaceae. *Chionanthus virginicus* L. a, habit, ×½; b, flower, from above, ×3; c, corolla, opened out, ×3; d, calyx and pistil, ×3; e, pistil, ×15; f, schematic cross-section of ovary, ×30; g, stamen, ×15.

lobed, the lobes valvate, or rarely (spp. of *Fraxinus*) the calyx wanting; corolla typically sympetalous and 4-lobed, the lobes imbricate or induplicate-valvate or convolute, rarely the corolla up to 12-lobed, or sometimes (as in spp. of *Fraxinus*) the petals essentially distinct, or even (as in *Forestiera* and other spp. of *Fraxinus*) obsolete; stamens commonly 2, attached to the corolla-tube when the corolla is sympetalous, aligned with the sinuses, transversely or less often medially placed, rarely 4 (as in *Hesperelaea* and *Tessarandra*); anthers dithecal, the thecae placed back to back, opening by longitudinal slits; pollen-grains binucleate, tricolporate or tricolpate; a nectary-disk sometimes present around the base of the ovary, but more often wanting, or the entire ovary-wall nectariferous; gynoecium of 2 carpels (mostly median) united to form a compound, superior, bilocular ovary with a terminal style and a dry, 2-lobed stigma, or the stigma sessile; ovules most commonly 2 in each locule, sometimes 1–4, seldom numerous, borne on axile placentas, anatropous or amphitropous, tenuinucellar, with a massive single integument and an intergumentary tapetum; endosperm-development cellular, with a micropylar but not a chalazal haustorium. FRUIT variously a loculicidal or circumscissile capsule, or a samara, berry, or drupe, often 1-seeded; seeds with a straight, spathulate, dicotyledonous embryo embedded in the oily, thick-walled endosperm, or without endosperm. X = 10, 11, 13, 14, 23, 24. (Fraxinaceae, Syringaceae)

The family Oleaceae consists of some 30 genera and 600 species, nearly cosmopolitan in distribution, but best developed in Asia and Malesia. Fully half of the species belong to only two genera, *Jasminum* (jasmine, 200) and *Chionanthus* (125, including *Linociera*); and more than a quarter belong to only 5 more, *Fraxinus* (ash, 60), *Ligustrum* (privet, 40), *Noronhia* (40), *Syringa* (lilac, 30), and *Olea* (olive, 20).

The affinities of the Oleaceae are in dispute. Traditionally they have been included in the Gentianales, where they are anomalous in embryological features, in having only 2 stamens, and in not having internal phloem. Tournay and Lawalrée propose to associate the Oleaceae, Buddlejaceae, and Menyanthaceae in an order Ligustrales, but I do not believe that the Menyanthaceae are intimately related to the other two families.

The only obvious problem in putting the Oleaceae into the Scrophulariales is that the 4-lobed corolla is essentially regular. In this feature they resemble the bulk of the Buddlejaceae, which are now often (as here) considered to be allied to the Scrophulariaceae.

Fossil pollen of Oleaceae is known from Upper Miocene and more recent deposits.

Unpublished serological studies by Piechura (personal communication) would be consistent with the placement of the Oleaceae in either the Gentianales or the Scrophulariales. Serologically the Oleaceae appear to form a link between the two orders.

3. Family SCROPHULARIACEAE A L. de Jussieu 1789 nom. conserv., the Figwort Family

Mostly herbs, seldom shrubs or even small trees (*Halleria*), autotrophic or not infrequently hemiparasitic on the roots of other plants, rarely (as in *Harveya*) wholly parasitic, often provided with hairs that have a basal cystolith, as in the Boraginaceae, often also with various other sorts of hairs, these frequently glandular; plants commonly (but not always) producing orobanchin and iridoid compounds (which may cause the leaves to turn blackish in drying), not infrequently with triterpenoid saponins, and sometimes with cardiotonic glycosides (as in *Digitalis*), but only seldom with alkaloids, rarely cyanogenic, and usually not tanniferous, lacking both ellagic acid and proanthocyanins; calcium oxalate crystals only seldom present in the parenchymatous tissues; nodes unilacunar; xylem in young stems commonly forming a closed cylinder without medullary rays, but sometimes in bundles separated by evident rays; vessel-segments with simple perforations; imperforate tracheary elements commonly with simple pits; wood-rays of various types, heterocellular to homocellular; wood-parenchyma very scanty or wanting; internal phloem wanting. LEAVES alternate or opposite, seldom whorled, simple to sometimes pinnately dissected; stomates commonly anomocytic, rarely anisocytic; petiole commonly with an arc-shaped vascular strand, or with an arc of vascular bundles; stipules wanting. FLOWERS in various sorts of determinate or indeterminate inflorescences, often a thryse, raceme, or spike, or sometimes solitary, perfect, usually evidently irregular; calyx commonly deeply (2) 4–5-lobed or cleft, with imbricate or valvate segments; corolla sympetalous, slightly to usually evidently irregular, sometimes spurred or saccate at the base, often bilabiate, the 5 or less commonly 4 (–8) lobes imbricate or valvate (corolla wanting in spp. of *Besseya*); stamens attached to the corolla-tube, alternate with the lobes, sometimes 5 and all functional, as in *Verbascum*, but usually only 4, or only 4 functional, the uppermost (adaxial) one staminodial (as in *Penstemon* and *Scrophularia*) or wanting, or sometimes only 2 (3), the abaxial (lower) pair also reduced or wanting; anthers tetrasporangiate and dithecal, sometimes (as in the Manuleae) becoming unithecal during ontogeny, the pollen-sacs sometimes unequal or unequally placed, opening by longitudinal slits, or the pollen-sacs confluent distally and opening by one continuous slit; pollen-grains binucleate or sometimes (as in *Euphrasia*) trinucleate, 2–7-aperturate, most commonly

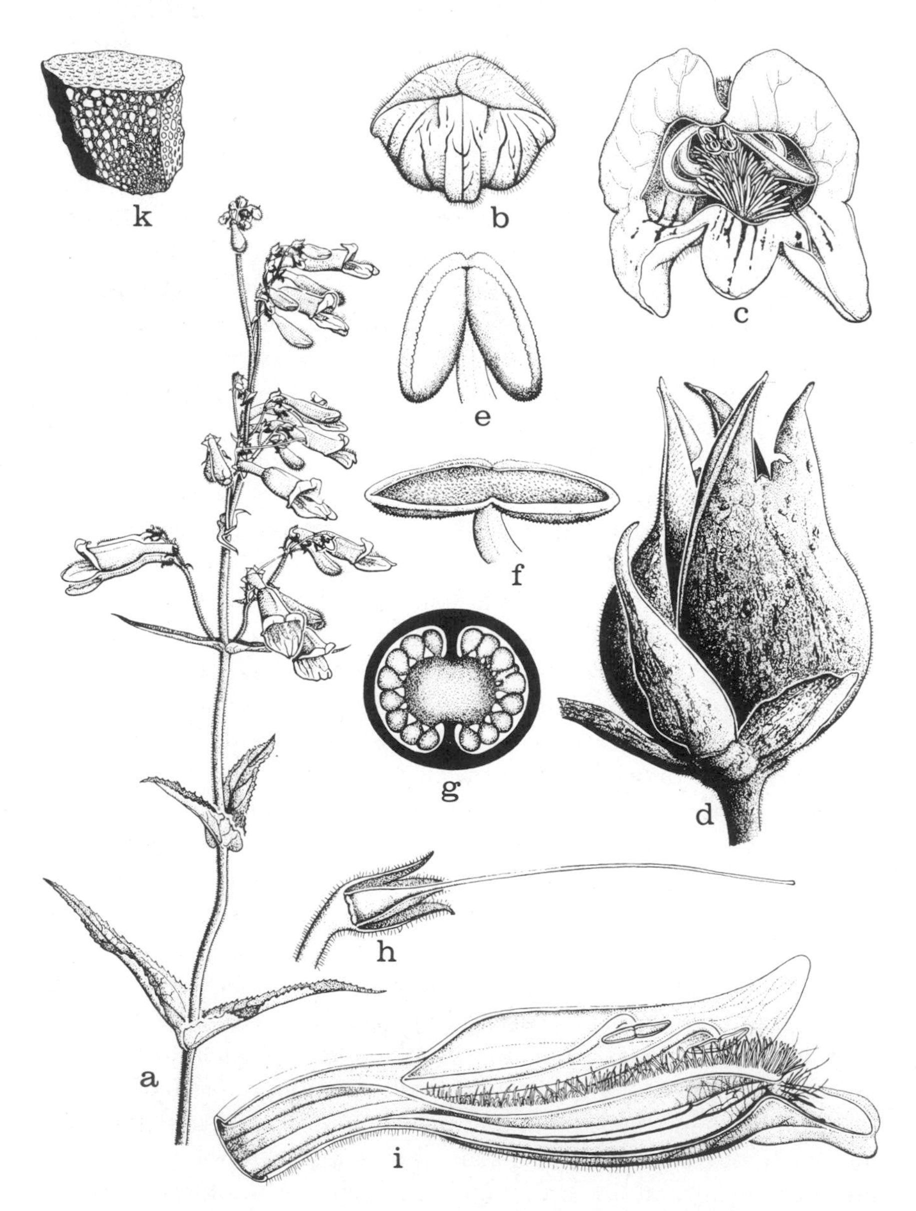

FIG. 6-22 Scrophulariaceae. *Penstemon canescens* (Britton) Britton. a, habit, $\times\frac{1}{2}$; b, flower bud, $\times 3$; c, front view of flower, $\times 3$; d, opened fruit, $\times 6$; e, anther, before dehiscence, $\times 12$; f, anther, after dehiscence, $\times 12$; g, schematic cross-section of ovary, $\times 12$; h, calyx and pistil, $\times 3$; i, corolla, in long-section, $\times 3$; k, seed, $\times 18$.

tricolporate; a unilateral or annular nectary-disk commonly present at the base of the ovary; gynoecium of 2 median carpels united to form a compound, superior, bilocular ovary with a terminal style and a simple or 2-lobed, usually wet stigma (very rarely tricarpellate and trilocular, as in spp. of *Bowkeria*); ovules (2–) more or less numerous in each locule, on axile placentas, anatropous or hemitropous to seldom amphitrophous or campylotropous, tenuinucellar, with a massive single integument and an integumentary tapetum that is open at the micropylar end, the embryo-sac sometimes protruding from the micropyle; endosperm-development cellular, with terminal haustoria. FRUIT commonly a septicidal capsule, less often loculicidal or opening by pores, rarely a berry, very rarely (*Lagotis*) a schizocarp; seeds angular or winged, with a straight or slightly curved, short to linear or spatulate, dicotyledonous embryo (cotyledons reduced but distinguishable in the nongreen parasite *Lathraea*); endosperm oily. X = 6+ (Ellisiophyllaceae, Rhinanthaceae)

The family Scrophulariaceae as here defined consists of about 190 genera and 4000 species, of cosmopolitan distribution, but most abundant in temperate regions and in tropical mountains. About half of the species belong to only 8 genera, *Pedicularis* (500), *Calceolaria* (300), *Verbascum* (300), *Veronica* (300), *Penstemon* (280), *Castilleja* (150), *Linaria* (150), and *Scrophularia* (150). Some other familiar genera are *Antirrhinum* (40, *A. majus* L. is the common snapdragon), *Digitalis* (20, *D. purpurea* L. is foxglove), and *Mimulus* (120). *Paulownia*, often included in the Scrophulariaceae, is here referred to the Bignoniaceae.

4. Family GLOBULARIACEAE A. P. de Candolle in Lamarck & de Candolle 1805 nom. conserv., the Globularia Family

Heath-like herbs or small shrubs with short, uniseriate hairs and small, glandular hairs, the latter with a short stalk-cell and a head of 2 (4) cells separated by vertical partitions; plants producing iridoid compounds, and (at least in *Globularia*) with a characteristic bitter principle, globularin, that may be chemically related to cinnamic acid, but not cyanogenic and not tanniferous, lacking both ellagic acid and proanthocyanins, only seldom bearing crystals in the parenchymatous tissues; xylem commonly forming a closed cylinder with rays only 1–2 cells wide; vessel-segments with simple perforations; imperforate tracheary elements relatively thin-walled; wood-parenchyma scanty. LEAVES alternate or rarely (*Globulariopsis*) opposite, simple and small, entire; stomates anomocytic, or seldom diacytic or paracytic; stipules wanting. FLOWERS individually rather small, in spikes or involucrate heads or compound corymbs, perfect, more or less strongly irregular; calyx (2) 3–5-toothed or -lobed, or spathaceous and adnate to the bract, or rarely of 2 essentially distinct members; corolla sympetalous, slightly to strongly irregular, often bilabiate, with 4 or 5 imbricate lobes, sometimes, as in *Hebenstretia*, unilabiate and 4-lobed, split in front; stamens attached to the corolla-tube alternate with its lobes, mostly 4, the fifth (upper, posterior) one wanting or rarely represented by a small staminode; sometimes the stamens only 2, or rarely all 5 fully developed; anthers initially dithecal, but the thecae apically confluent at maturity and opening by a single distal slit; pollen-grains binucleate (2) 3-colporate; an annular or unilateral nectary-disk or a single nectary-gland often present at the base of the ovary; gynoecium basically bicarpellate, the carpels united to form a compound, superior ovary with a terminal style and capitate or shortly 2-lobed stigma; ovary sometimes symmetrically bilocular, sometimes with one locule fully developed and the other more or less reduced, or even unilocular and pseudomonomerous, the posterior carpel suppressed; ovules solitary (2) in each functional locule, pendulous from near the summit of the partition, variously apotropous or epitropous or pleurotropous, or seldom 2 in each locule and superposed, the lower one pendulous, the upper one erect, in any case anatropous, tenuinucellar, with a massive single integument and an integumentary tapetum; endosperm-development cellular, with terminal haustoria. FRUIT of 2 separating, often unequal nutlets, or of a single small nut or achene that may be enclosed in the persistent calyx; seeds with straight, dicotyledonous embryo embedded in the oily endosperm. X = 8. (Selaginaceae)

The family Globulariaceae as here defined consists of about 10 genera and 300 species, native to Africa, Madagascar, Europe, and western Asia. The largest genus is *Selago*, with about 180 species, followed by *Walafrida* and *Hebenstretia* with about 40 each, and *Globularia*, with about 25.

It is customary to restrict the Globulariaceae to *Globularia* and *Cockburnia*, with a pseudomonomerous gynoecium, and to refer the other genera to the Scrophulariaceae as a separate subfamily Selaginoideae or tribe Selagineae. The Selaginoideae, however, show the complete range from a fully dimerous to a pseudomonomerous gynoecium, so that no clear line can be drawn on this basis. I believe the arrangement here presented is more nearly natural than the traditional one, and provides a better distinction between the families.

The Selaginoideae and the Globulariaceae sens. strict. are anatomically so much alike that Metcalfe & Chalk (1950, general citations) found it useful to treat them collectively as a group, separately from the Scrophulariaceae. One of the unifying anatomical features of the Globulariaceae (sensu mei) is the structure of the glandular trichomes, with a single stalk-cell and a head of 2 or seldom 4 cells separated by vertical partitions. The somewhat similar glandular trichomes of the Myoporaceae (another small family related to the Scrophulariaceae) differ in having 2 or more stalk-cells.

The connection of the Globulariaceae to the Scrophulariaceae is through the tribe Manuleae of the latter family. *Glumicalyx*, in the Manuleae, has only 8–12 ovules per locule.

5. Family MYOPORACEAE R. Brown 1810 nom. conserv., the Myoporum Family

Shrubs or sometimes small trees, often with specialized, lepidote or plumose indument, or with gland-tipped hairs, the gland with only vertical partitions; plants producing iridoid compounds and sometimes sesquiterpenes and acetylenic compounds, only seldom cyanogenic or saponiferous, sometimes tanniferous, but without ellagic acid and proanthocyanins, some species producing manna, with mannitol as a principal constituent; plants commonly with calcium oxalate crystals in some of the cells of the parenchymatous tissues, and, except in the small genus *Oftia*, with scattered small secretory cavities, lined by an epithelium, in the stem and usually also in the leaves, the leaves therefore often appearing pellucid-dotted; vessels with simple perforations; imperforate tracheary elements with small, simple pits; wood-rays heterocellular, mixed uniseriate and pluriseriate, the latter mostly 2–3 cells wide; wood-parenchyma paratracheal; internal phloem present only in *Oftia*. LEAVES alternate or seldom opposite, simple, entire or toothed, often reduced; stipules wanting; petiole with an arc-shaped vascular strand and some smaller accessory bundles. FLOWERS in small axillary cymes, or solitary in the axils, perfect; calyx synsepalous, the 5 lobes imbricate or open; corolla sympetalous, from essentially regular (as in *Myoporum*) to strongly irregular (as in *Pholidia*), often bilabiate, the lobes imbricate; stamens attached to the corolla-tube alternate with the lobes, mostly 4, the 5th (upper, posterior) one wanting or represented by a staminode, or seldom all 5 stamens fully developed; pollen-sacs contiguous only distally, where they become confluent, opening by

FIG. 6-23 Myoporaceae. *Myoporum laetum* G. Forster. a, habit, ×½; b, f, flower, ×4; c, flower bud, ×4; d, anther, before dehiscence, ×8; e, anther, after dehiscence, ×8; g, flower in long-section, ×4; h, i, schematic cross-section of bilocular and trilocular ovaries, ×12; k, fruit, ×2; m, lower surface of leaf, showing abundant pellucid secretory cavities, ×32.

longitudinal slits; pollen-grains binucleate, 2–4-colporate, often with 2 pores in each furrow; gynoecium of 2 carpels united to form a compound, superior ovary with a terminal (sometimes impressed) style and a simple stigma; ovary bilocular, with (1) 2 ovules pendulous from near the summit of the partition in each locule, or with 4–8 ovules in each locule, superposed in pairs, or the locules divided by supernumerary partitions, so that the ovary has 4–10 uniovulate compartments; ovules anatropous, tenuinucellar, with a massive single integument and with an integumentary tapetum around the chalazal two-thirds of the embryo-sac; endosperm-development cellular, with terminal haustoria. FRUIT drupaceous, the stone at least sometimes with 2 or 4 locules, or the fruit sometimes separating into 1-seeded, drupelike segments; seeds with a straight or slightly curved, dicotyledonous embryo and scanty or no endosperm. X = 27 (*Myoporum*).

The family Myoporaceae consists of 3 or 4 genera and about 125 species. *Pholidia* (including *Eremophila*), with about 90 species, and *Myoporum*, with about 30, make up the core of the family. Both of these genera are basically Australian, but *Myoporum* also extends to eastern Asia and the Pacific isalnds. The monotypic genus *Bontia*, occurring in the West Indies and northern South America, is the only representative of the family in the New World. *Oftia* (2 South African species) is anomalous in having internal phloem and in lacking the characteristic secretory cavities of the family. Thorne has suggested (personal communication) that *Oftia* is better referred to the Scrophulariaceae. This may well be correct, but its internal phloem would be equally anomalous in that family.

6. Family OROBANCHACEAE Vent. 1799 nom. conserv., the Broom-rape Family

Annual or perennial herbaceous root-parasites, often fleshy, without chlorophyll, provided with glandular hairs that have a 2-several-celled stalk and a several-celled head provided with only vertical partitions, often with simple, nonglandular hairs as well; root-system scarcely developed, the radicle becoming a haustorium that penetrates the root of the host, subsequent haustoria developing internally or externally from the hypocotyledonary region; plants producing mannitol and commonly iridoid compounds, accumulating the phenolic glycoside orobanchin and sometimes also monoterpenoid alkaloids, often tanniferous, but lacking both ellagic acid and proanthocyanins, neither

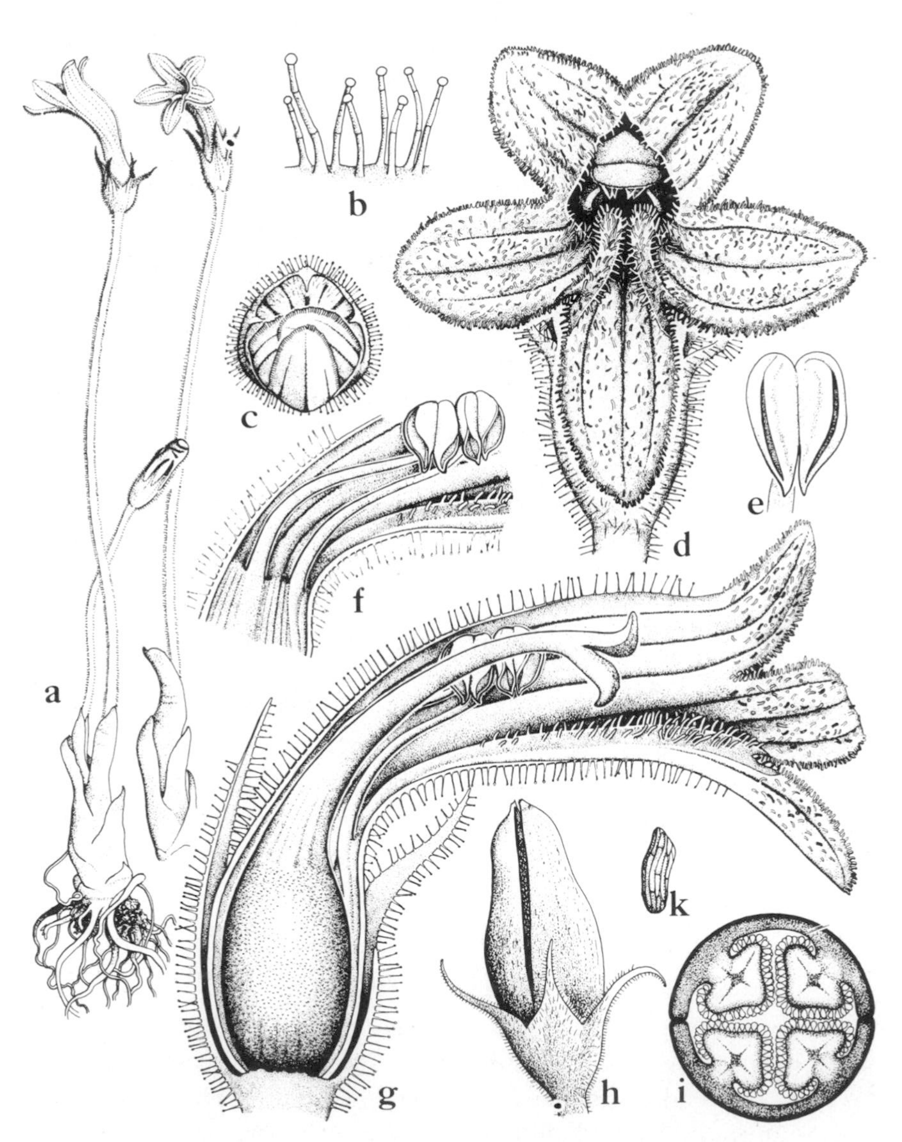

FIG. 6-24 Orobanchaceae. *Orobanche uniflora* L. a, habit, ×1; b, gland-tipped trichomes, ×18; c, flower bud, with imbricate corolla-lobes, ×6; d, front view of flower, ×3; e, anther, ×12; f, portion of interior of corolla, with a pair of stamens, ×6; g, flower, in partial long-section, ×6; h, fruit, ×3; i, schematic cross-section of ovary, ×9; k, seed, ×36.

cyanogenic nor saponiferous; crystals of calcium oxalate seldom produced; vascular bundles in one or sometimes 2 cycles in the stem, the phloem better-developed than the xylem; vessel-segments with simple perforations. LEAVES reduced to mere alternate (often crowded) scales, the stomates commonly present but often disorganized, the mesophyll composed largely or wholly of spherical cells; stipules wanting. FLOWERS perfect, borne in bracteate terminal racemes or spikes, or seldom solitary; calyx regular or irregular, synsepalous and (1–) 4–5-lobed or cleft, or divided to the base above and below, the segments open or valvate; corolla sympetalous, irregular and commonly more or less bilabiate (seldom nearly regular), the tube often curved, the 5 lobes imbricate, the adaxial (posterior) ones internal; stamens attached to the corolla-tube alternate with the lobes, 4 fully developed and generally paired, the fifth (adaxial, posterior) one staminodial or wanting; anthers dorsifixed, tetrasporangiate and dithecal, opening by longitudinal slits; sometimes one theca reduced or modified; pollen-grains binucleate, tricolporate or rarely inaperturate; gynoecium of 2 median carpels (rarely 3 carpels) united to form a compound, superior ovary with a slender style and a capitate or crateriform to often 2–4-lobed stigma; ovary unilocular (rarely bilocular at the base), with (2) 4 (6) intruded parietal placentas, these simple or often branched and T-shaped in cross-section; ovules numerous, anatropous, tenuinucellar, with a massive single integument and an integumentary tapetum; endosperm-development cellular, with terminal haustoria. FRUIT a loculicidal capsule, each of the 2 (3) valves commonly bearing 2 placentas; seeds numerous, small, with minute, ellipsoid or globose, undifferentiated embryo embedded in the oily endosperm. X = 12, 18–21.

The family Orobanchaceae consists of about 17 genera and 150 species, wide-spread in the Northern Hemisphere, best-developed in subtropical and temperate parts of the Old World. About two-thirds of the species belong to the single genus *Orobanche* (100).

It is generally believed that the Orobanchaceae are related to and derived from the Scrophulariaceae. Many of the Scrophulariaceae are hemiparasitic. A few, such as *Harveya*, lack chlorophyll but still have a bilocular ovary and an embryo with more or less differentiated cotyledons. *Lathraea*, with 2 intruded, bifid placentas and reduced but distinguishable cotyledons, has been put in one family or the other by different authors.

The morphology of the ovary of the Orobanchaceae has been debated in the literature. Some authors have maintained that because of the 4 parietal placentas it must be 4-carpellary, even though the fruit splits into only 2 valves. The 2-carpellary interpretation here adopted is more in harmony with the generally admitted relationship of the Orobanchaceae to the Scrophulariaceae. As Bentham and Hooker point out, the 4 placentas in *Orobanche* are sometimes equally spaced, but sometimes paired. Each pair might well be comparable to a single bifid placenta such as those of the Gesneriaceae.

7. Family GESNERIACEAE Dumortier 1822 nom. conserv., the Gesneriad Family

Herbs or half-shrubs, rarely shrubs or small trees, sometimes lianas or epiphytes, the herbs often of anomalous structure, the cotyledons becoming unequal after germination in the subfamily Cyrtandroideae, and sometimes (as in spp. of *Streptocarpus*) the larger cotyledon becoming foliar, serving as the essential photosynthetic organ, enlarging by intercalary growth, and bearing the flowers; plants commonly with uniseriate hairs that have a thickened, often calcified or silicified terminal cell, or with stalked glands, or both, commonly accumulating orobanchin but apparently without iridoid compounds, not cyanogenic, not saponiferous, and only seldom tanniferous, lacking both ellagic acid and proanthocyanins; calcium oxalate crystals of various forms commonly present in some of the cells of the parenchymatous tissues; nodes unilacunar or seldom trilacunar; vascular bundles in herbs commonly becoming joined by secondary growth; vessel-segments with simple perforations; imperforate tracheary elements septate, with simple pits; wood-rays homocellular or nearly so, mostly pluriseriate, up to as much as 17 cells wide, with few or no uniseriates; wood-parenchyma wanting, or sometimes paratracheal and then usually scanty. Leaves opposite or rarely whorled or alternate, sometimes all alike, sometimes one member of a pair more or less reduced, sometimes all basal, generally simple and entire or toothed, rarely pinnatifid; stomates commonly anisocytic; petiolar anatomy diverse; stipules wanting, or seldom irregularly developed as basal auricles. Flowers solitary in the axils of the leaves, or in various kinds of cymose or seldom racemose inflorescences, sometimes even epiphyllous, perfect, often large and showy; sepals 5, distinct or more often united into a lobed tube, valvate or rarely imbricate; corolla sympetalous, evidently irregular and usually bilabiate, often also saccate-spurred at the base, varying to sometimes nearly or quite regular, its 5 lobes imbricate, the posterior (adaxial) ones generally

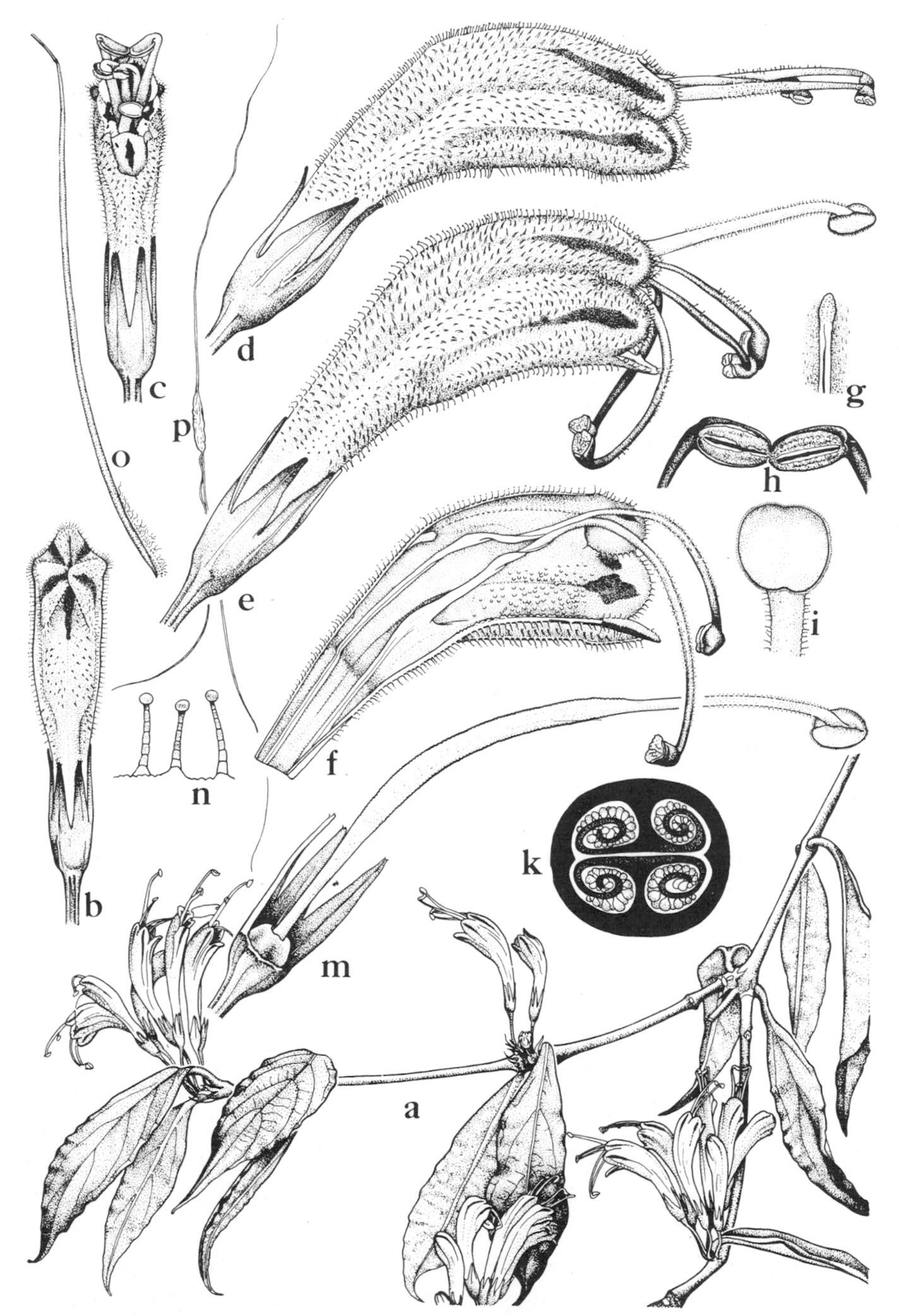

FIG. 6-25 Gesneriaceae. *Aeschynanthus sikkimensis* (C. B. Clarke) Stapf. a, habit, $\times\frac{1}{2}$; b, flower, in late bud, $\times 2$; c, flower, front view, beginning to open, $\times 2$; d, e, flower in early and late anthesis, $\times 2$; f, corolla in long-section, with 2 attached stamens, $\times 2$; g, staminode, $\times 4$; h, anthers, $\times 4$; i, stigma, $\times 4$; k, schematic cross-section of ovary, $\times 16$; m, pistil and part of calyx, $\times 2$; n, gland-tipped trichomes, $\times 20$; o, fruit, $\times\frac{1}{4}$; p, seed, $\times 8$.

internal; stamens attached to the corolla-tube alternate with the lobes, commonly 4, with the anthers connivent in pairs or all together, the posterior one reduced or wanting, less often only a single pair of stamens developed, rarely (as in *Ramonda* and spp. of *Sinningia*) all 5 stamens developed; 1–3 staminodes often present in place of the missing stamens; anthers dithecal, opening by longitudinal slits; pollen-grains (2) 3 (4)-colporate; nectary-disk commonly present at the base of the ovary, sometimes annular or cupular, sometimes unilateral, or sometimes of discrete glands, rarely wanting; gynoecium of 2 median carpels united to form a compound, superior or more or less inferior ovary with a terminal, slender style and a usually 2-lobed, wet or dry stigma; ovary typically unilocular and with 2 parietal placentas, these commonly more or less intruded and bifurcate, sometimes (as in *Monophyllaea*) more or less joined in the center and dividing the ovary into 2 chambers; ovules numerous, anatropous, tenuinucellar, with a massive single integument (reputedly sometimes 2 integuments) and an integumentary tapetum; endosperm-development cellular, with terminal haustoria. FRUIT a loculicidal or seldom septicidal capsule, or less often a berry; seeds numerous and small, with a straight, dicotyledonous embryo embedded in the oily endosperm (Gesnerioideae), or the endosperm wanting (Cyrtandroideae). X = 4–17+; 8 or 9 may be the ancestral number for the Cyrtandroideae.

The family Gesneriaceae consists of about 120 genera and 2500 species, pantropical in distribution, with only a few species in temperate regions, as in the Pyrenees and the mts. of the Balkan Peninsula. The largest genus by far is *Cyrtandra*, variously estimated at 200 to 600 species, followed by *Columnea*, with 150 to 200. Species of *Saintpaulia* (African violet) and *Sinningia* (gloxinia, the name misapplied) are familiar ornamental house-plants, and many others are grown by a devoted coterie of gesneriad-fanciers.

The Gesneriaceae are generally considered to consist of two well marked subfamilies, the Gesnerioideae and Cyrtandroideae. The Gesnerioideae, almost entirely confined to the New World, have equal cotyledons, and the ovary is often partly or wholly inferior. The Cyrtandroideae, almost entirely confined to the Old World, have cotyledons that become unequal after germination, and the ovary is always superior. It is possible that several small genera in the Old World should be treated as a third, archaic subfamily, lacking the specialized features of both of the other groups.

The Gesneriaceae are generally admitted to be closely allied to the

Scrophulariaceae, from which they differ most notably in their mainly parietal placentation and frequently more or less inferior ovary. The axile placentation of some Gesneriaceae appears to be derived from parietal placentation.

8. Family ACANTHACEAE A L. de Jussieu 1789 nom. conserv., the Acanthus Family

Herbs or less often shrubs, often twining, rarely trees, very diverse in habit, even including some mangrove spp. (such as *Acanthus ilicifolius* L.), provided with various types of glandular and eglandular hairs, often with anomalous secondary growth, commonly accumulating orobanchin and often also iridoid compounds, quinazoline or quinoline alkaloids, and diterpenoid bitter substances, but only rarely cyanogenic, rarely saponiferous, and rarely tanniferous, lacking both ellagic acid and proanthocyanins; various types of silicified cystoliths commonly present in some of the parenchyma and/or epidermal cells of both stem and leaf (these wanting in the subfamilies Nelsonioideae and Thunbergioideae); calcium oxalate crystals of various form often also present in some of the cells of the parenchymatous tissues; bundles of acicular fibrils, resembling large raphides, often present in some of the cells of the phloem of the stem and leaves; nodes unilacunar; xylem commonly (not always) forming a closed cylinder with narrow rays, even in herbs; vessel-segments small, with simple perforations; imperforate tracheary elements with simple pits, usually septate; wood-rays commonly homocellular, seldom heterocellular, 1–6 cells wide, sometimes mixed uniseriate and pluriseriate; wood-parenchyma scanty, paratracheal; internal phloem sometimes present. LEAVES opposite (alternate in Nelsonioideae), simple, sometimes spiny and thistle-like; stomates almost always diacytic; petiole commonly with an arcuate vascular strand, or with an arc of discrete bundles, but sometimes with other types of anatomy; stipules wanting. FLOWERS in various sorts of cymose or less often racemose inflorescences, or sometimes solitary, perfect, bracteate and commonly also bracteolate, the bracts and bracteoles often petaloid and showy; calyx synsepalous, more or less deeply (4) 5 (–16)-lobed, the lobes imbricate or valvate, or the lobes sometimes suppressed; corolla sympetalous, from essentially regular to more often irregular, commonly bilabiate and 5-lobed, with imbricate or convolute lobes but the upper lip sometimes suppressed; stamens attached to the corolla-tube, alternate with the lobes, commonly 4 or 2 and paired, the missing (upper)

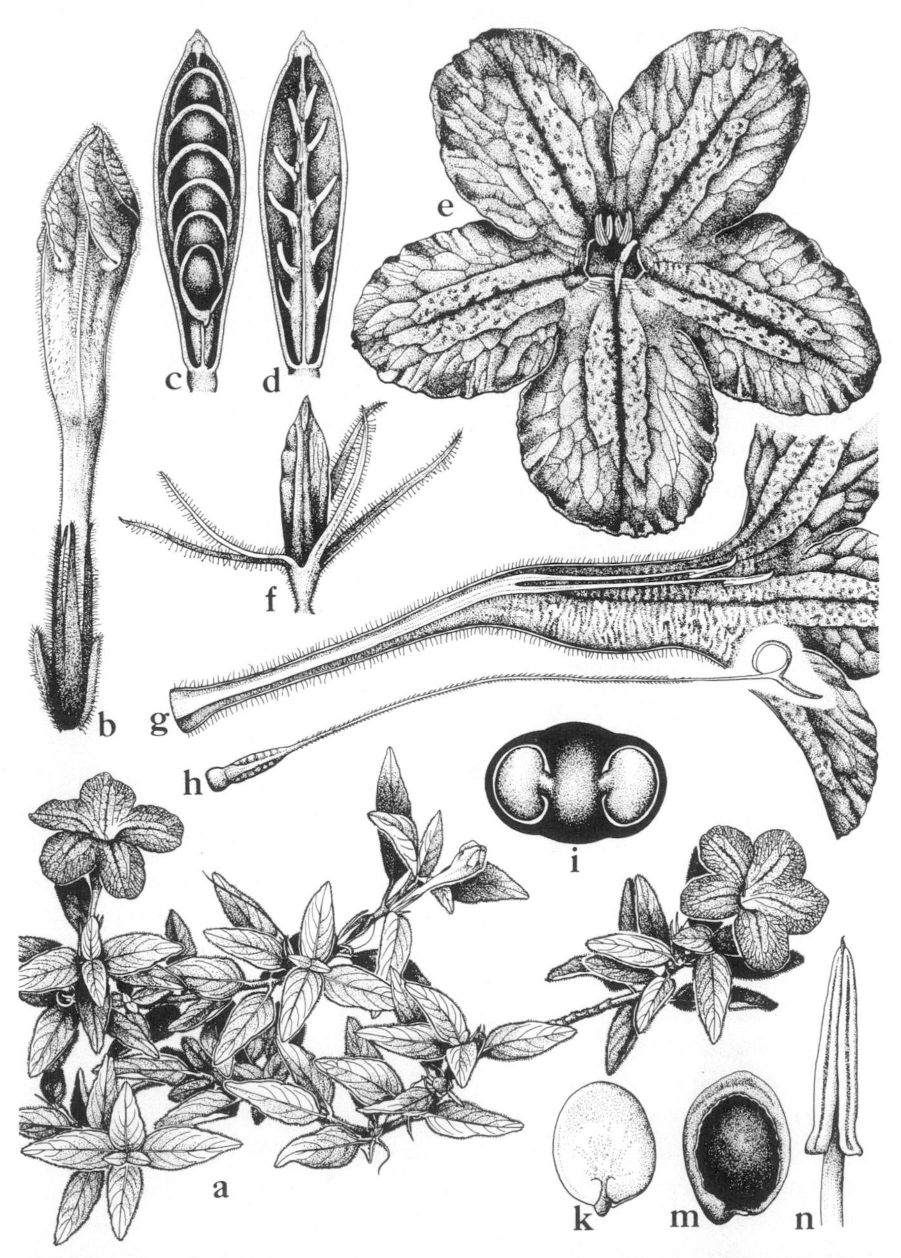

FIG. 6-26 Acanthaceae. *Ruellia humilis* Nutt. a, habit, ×½; b, flower, in late bud, ×2; c, fruit in long-section, showing seeds, ×4; d, interior surface of valve of fruit, after the seeds have been dispersed, showing jaculators, ×4; e, front view of flower, ×2; f, fruit and persistent calyx, ×2; g, corolla in long-section, with 2 attached stamens, ×2; h, pistil, ×2; i, schematic cross-section of ovary, ×12; k, embryo, ×8; m, seed, ×8; n, anther, ×8.

member(s) sometimes represented by staminodes; rarely all 5 stamens fully developed (*Pentstemonacanthus*); anthers typically tetrasporangiate and dithecal, opening by longitudinal slits, the pollen-sacs sometimes parallel and juxtaposed, or sometimes widely separated on a modified connective, or one of them sometimes reduced or suppressed; pollen-grains binucleate or trinucleate, most commonly tricolporate, but diverse in architecture, including tricolpate, triporate, diporate, pantoporate, and inaperturate types; an annular nectary-disk commonly present around the base of the ovary; gynoecium of 2 median carpels united to form a compound, superior, generally bilocular ovary (ovary unilocular, with intruded parietal placentas in *Elytraria*, of the Nelsonioideae); style terminal, slender, with a dry, funnelform or more often 2-lobed stigma, or the upper stigma-lobe sometimes reduced or suppressed; ovules most commonly 2 in each locule, superposed or sometimes collateral, less often (as in *Ruellia*) up to 10 in each locule, in the Nelsonioideae more or less numerous in 2 rows in each locule (or on each intruded parietal placenta), each with a more or less strongly modified funiculus which is typically developed into a hook-shaped jaculator that functions in flinging out the seeds (funiculus more cushion-shaped and forming a sort of obturator in the Nelsonioideae and Thunbergioideae), anatropous to amphitropous or campylotropous, tenuinucellar, only rarely (as in some Nelsonioideae) with an integumentary tapetum; endosperm-development cellular, or sometimes partly or wholly nuclear, with terminal haustoria, or sometimes only the micropylar haustorium developed. FRUIT a loculicidal, explosively dehiscent (at least in the Acanthoideae) capsule; seed-coat very thin, often becoming mucilaginous when wetted; embryo large, dicotyledonous, straight and spathulate or more often more or less curved or bent; endosperm usually wanting, but in the Nelsonioideae more or less well developed, oily, and ruminate. X = 7–21. (Thunbergiaceae)

The family Acanthaceae as here defined consists of about 250 genera and 2500 species, widespread in tropical regions, with only a few species in temperate climates. More than half of the species belong to only 7 genera, *Justicia* (including *Beloperone*, 300), *Ruellia* (250), *Barleria* (250), *Strobilanthes* (200), *Thunbergia* (200), *Dicliptera* (180), and *Aphelandra* (150). The sculpture of the pollen is much-used in the delimitation of genera, and nearly half of the recognized genera are monotypic. Species of *Acanthus*, *Aphelandra*, *Justicia*, *Ruellia*, *Strobilanthes*, and some other genera are sometimes cultivated for ornament.

Nearly nine-tenths of the species and all but 9 of the genera of

Acanthaceae belong to the well marked subfamily Acanthoideae, which is agreed by all to be a natural group. The other two subfamilies (Nelsonioideae, Thunbergioideae) have the funiculus less strongly modified, not forming a jaculator, and they lack the characteristic cystoliths of the Acanthoideae. The Thunbergioideae have sometimes been considered to form a distinct family, and the Nelsonioideae have sometimes been referred to the Scrophulariaceae. The Nelsonioideae and Thunbergioideae represent way-stations along the route from the Scrophulariaceae to the Acanthoideae. Detailed embryological and palynological studies emphasize their relationship with the Acanthoideae, but they are anomalous in either family, without being sufficiently distinctive to warrant separate status. The traditional (and here accepted) line between the Scrophulariaceae and Acanthaceae is purely arbitrary, but it seems as good as any other, and the connecting forms are not numerous enough to warrant uniting the two families.

The Mendonciaceae, often treated as another subfamily of Acanthaceae, are more distinctive than the Nelsonioideae and Thunbergioideae, and are here treated as a separate family.

Fossil pollen resembling that of various genera of Acanthaceae is known from Lower Miocene and more recent deposits.

9. Family PEDALIACEAE R. Brown 1810 nom. conserv., the Sesame Family

Annual or perennial herbs, terrestrial or sometimes (*Trapella*) aquatic, rarely half-shrubs or shrubs, beset on all herbaceous parts (at least when young) with characteristic trichomes that have a short stalk and a broad head composed of 4 or more cells filled with mucilage, sometimes also with some uniseriate hairs, the surface often appearing slimy; plants generally storing carbohydrate as stachyose rather than starch, and commonly producing iridoid compounds, but not cyanogenic, not saponiferous, and not tanniferous, lacking both ellagic acid and proanthocyanins; crystals small or none; primary xylem tending to form a continuous ring with mostly narrow rays and a few broader ones, vessel-segments with simple perforations; imperforate tracheary elements with simple pits. LEAVES opposite, or the upper ones sometimes alternate, simple and entire to toothed or lobed; stomates mostly anomocytic; petiole with an arcuate vascular strand and some smaller accessory lateral bundles; stipules wanting. FLOWERS perfect, borne in terminal racemes, or solitary and axillary, or in axillary simple dichasia, often with 1 or 2 characteristic glands (extrafloral nectaries) representing abortive flowers axillary to as many bracts at the base of the pedicel, and often with a pair of bracteoles just beneath the calyx; sepals persistent, (4) 5 and nearly distinct, or often united to form a lobed calyx; corolla sympetalous, irregular, sometimes spurred or saccate at the base, the limb often oblique or more or less bilabiate, the 5 lobes imbricate; stamens attached to the corolla-tube alternate with the lobes, 4 and generally paired, the fifth (posterior) one represented by a small staminode, or sometimes the stamens only 2 (the posterior-lateral ones) and then sometimes accompanied by 2 staminodes representing the anterior-stamens; anthers tetrasporangiate and dithecal, opening by longitudinal slits; pollen-grains binucleate or trinucleate, from tricolporate or triporate to more often polycolpate or pantoporate, or sometimes with numerous anastomosing apertures; an annular or adaxially expanded nectary-disk commonly present around the base of the ovary; gynoecium of 2 median carpels united to form a compound, superior (inferior in *Trapella*) ovary with a terminal style and a 2-lobed stigma; ovary bilocular, with axile placentas, sometimes divided into 4 locelli by the intrusion of partitions from the carpellary midribs, or unilocular with intruded, forked, parietal placentas, these sometimes joined in the center to produce a secondarily bilocular or 4-locellar ovary; rarely, as in *Josephinia*, the ovary with 8 uniovulate locelli, presumably representing 4 carpels; ovules 1-many in each locule or locellus (the anterior locule reduced and empty in *Trapella*), anatropous, tenuinucellar, with a single massive integument and an integumentary tapetum that may be incomplete at the micropylar end; endosperm-development cellular, with terminal haustoria. FRUIT variously a loculicidal capsule, or loculicidal after the separation of a soft, deciduous exocarp, or a drupe or nut, the capsule or the hardened endocarp commonly provided with terminal or lateral horns or hooks or prickles serving in zoochorous distribution, or sometimes winged; seeds with a straight, spathulate, dicotyledonous embryo; endosperm scanty and oily, or none. X = 8, 13, 14, 15, 18. (Martyniaceae, Trapellaceae)

The family Pedaliaceae as here defined consists of about 20 genera and 80 species, occurring chiefly in the tropics, especially along the seacoast or in arid regions, with only a few species in temperate climates. The largest genus is *Sesamum*, with about 20 species. *Sesamum indicum* L., sesame, is cultivated for its aromatic, oily seeds. Species of *Proboscidea* (unicorn-plant) and *Martynia* are also well known in cultivation.

Many authors distinguish the Martyniaceae as a New World family with terminal inflorescences and parietal placentation, in contrast to the Pedaliaceae proper, an Old World group with axillary flowers and axile placentation. The gynoecial difference is more theoretical than actual, however, inasmuch as the fruit in Martyniaceae tends to become 4-locellar by intrusion and union of the forked parietal placentas. The Old-World and New-World groups are so obviously related, and the total number of species is so small, that I think it is conceptually more useful to include them all in the same family. Those who find it helpful may recognize 3 subfamilies, the Pedalioideae, Martynioideae, and Trapelloideae.

The aquatic, central and eastern Asiatic genus *Trapella,* with only 1 or 2 species, is ecologically, morphologically, and geographically distinctive. On the other hand, its relationship to the Pedaliaceae is plain enough, and its inclusion in the family causes no other problems. Monotypic families should be tolerated only when there is no reasonable alternative.

10. Family BIGNONIACEAE A. L. de Jussieu 1789 nom. conserv., the Trumpet-creeper Family

Trees, shrubs, or very often woody vines with a twining stem or climbing by means of adventitious roots or various sorts of tendrils, only rarely herbaceous vines (*Tourrettia*) or ordinary herbs (*Argylia, Incarvillea*); the woody vines generally with a characteristic sort of anomalous vascular structure; plants provided with various sorts of hairs, and commonly producing iridoid compounds, orobanchin (a phenolic glycoside), and often lapachol (a quinone), sometimes producing indole or monoterpenoid alkaloids, but only seldom saponiferous and seldom tanniferous, lacking both ellagic acid and proanthocyanins; small crystals of calcium oxalate often present in some of the cells of the parenchymatous tissues; nodes unilacunar, with 3-several traces; secondary growth very often anomalous, especially in the vines; vessel-segments commonly with simple perforations, in some genera some of them with reticulate plates or scalariform plates with numerous cross-bars; imperforate tracheary elements with small, simple or very narrowly bordered pits; wood-rays homocellular or sometimes heterocellular, in the erect, woody species commonly 2–4 cells wide, with few or no uniseriates, or seldom all uniseriate, but in the climbing species commonly 5–13 cells wide; wood-parenchyma mostly paratracheal. LEAVES opposite or sometimes whorled, rarely alternate, simple or more often pinnately 1–3 times compound,

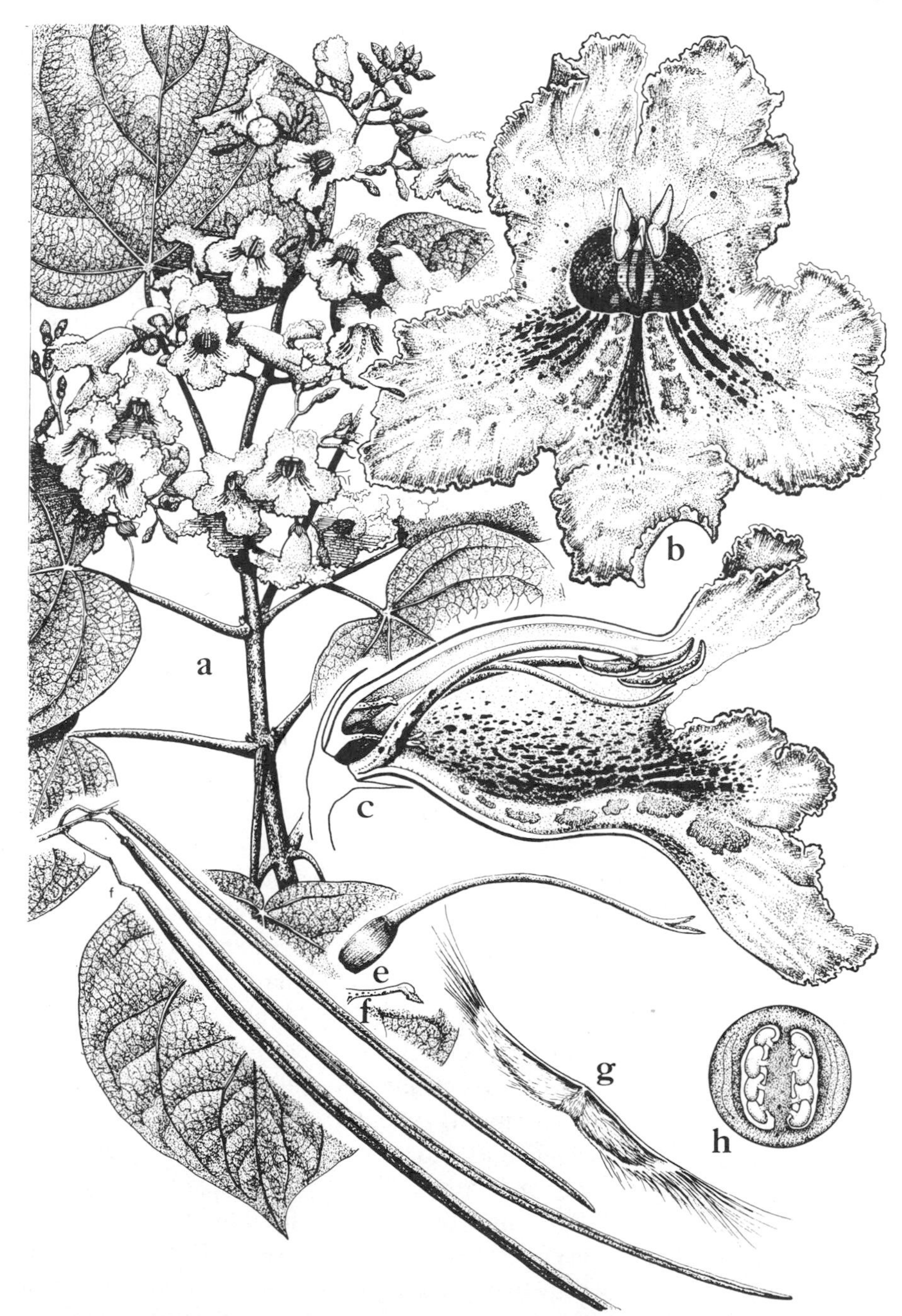

FIG. 6-27 Bignoniaceae. *Catalpa bignonioides* Walter. a, habit, $\times \frac{1}{4}$; b, front view of flower, $\times 2$; c, flower in long-section, $\times 2$; d, pistil, $\times 2$; e, staminode, $\times 2$; f, fruits, $\times \frac{1}{4}$; g, seed, $\times 2$; h, schematic cross-section of ovary, $\times 20$.

or trifoliolate or palmately compound, the terminal leaflet and sometimes also some of the adjacent lateral ones sometimes modified into tendrils; stomates commonly anomocytic, surrounded by a rather large number of ordinary epidermal cells, or sometimes paracytic, rarely diacytic; petiolar anatomy of diverse types; stipules wanting. FLOWERS mostly large and showy, in various sorts of cymose to racemose inflorescences or solitary, bracteate and bracteolate, perfect; calyx synsepalous, with mostly 5 teeth or lobes, sometimes bilabiate, or sometimes the lobes suppressed, calyptrate in spp. of *Lundia*; corolla sympetalous, more or less strongly irregular, often bilabiate, seldom nearly regular, the 5 lobes imbricate or rarely valvate; stamens attached to the corolla-tube, alternate with the lobes, commonly 4 and paired, the fifth (upper, adaxial) one staminodial or wanting, or sometimes (as in *Catalpa*) only 2 stamens fertile and the other 3 staminodial, rarely all 5 stamens fully developed (as in *Oroxylum*); anthers tetrasporangiate or bisporangiate, dithecal, the pollen-sacs often unequally set, opening by longitudinal slits; pollen-grains binucleate, variously borne in monads, tetrads, or rarely polyads, highly diverse in architecture, variously tricolporate or 2–12-colpate, less often inaperturate or of diverse other types; an annular or sometimes cupular nectary-disk usually present around the base of the ovary; gynoecium of 2 median carpels united to form a compound, superior ovary with a terminal style and 2-lobed stigma; ovary bilocular, with 2 axile placentas in each locule, or unilocular with 2 or 4 more or less intruded parietal placentas, or (*Tourrettia*) 4-locular with the ovules uniseriate in each locule; ovules numerous, anatropous or hemitropous, commonly erect, with the micropyle directed downward, tenuinucellar, with a massive single integument and an integumentary tapetum; endosperm-development cellular, with a chalazal and sometimes also a micropylar haustorium. FRUIT a bivalved, septicidal or loculicidal capsule, or sometimes septifragal, very often with a replum, seldom (Crescentieae and Coleae) fleshy and indehiscent; seeds mostly flat, in capsular fruits commonly winged; embryo straight, the 2 cotyledons usually enlarged and more or less foliaceous; endosperm wanting, or seldom (*Paulownia*) present and oily but rather scanty. X most commonly = 20, but 7 may be basic for the family.

The family Bignoniaceae consists of more than 100 genera and perhaps 800 species, mainly tropical, best developed in tropical America. *Tabebuia*, with about 100 species, may be the largest genus. *Spathodea* (flame-tree, African tulip-tree) and *Jacaranda* are familiar street-trees in tropical cities; *Catalpa* and *Paulownia* (empress-tree) are often planted as street-trees or specimen-trees in temperate and warm-temperate regions. Species of *Campsis* and *Bignonia*, called trumpet-creepers, are often cultivated for ornament in tropical and warm-temperate climates. *Crescentia cujete* L., of tropical America, is the calabash-tree.

Fossil pollen attributed to the Bignoniaceae occurs in deposits of middle Eocene and more recent age. Well preserved flowers and fruits attributed to *Catalpa* have been found in the Eocene London Clay.

Gentry (1980) recognizes 8 tribes, as indicated below. The small tribe Schlegelieae (3 genera) is regarded as transitional to the Scrophulariaceae. *Paulownia*, here referred to the Tecomeae, forms another link between the two families. It has endospermous seeds, as in the Scrophulariaceae, but it is a tree with leaves very much like those of *Catalpa*. It shows stronger serological reactions with tested members of the Scrophulariaceae than with tested members of the Bignoniaceae (Piechura, personal communication 1979).

SYNOPTICAL ARRANGEMENT OF THE TRIBES OF BIGNONIACEAE

a Fruit dehiscent; trees, shrubs, herbs, or compound-leaved vines.
  b Placentation parietal; friut without a septum; wiry vines of the Andes .... ECCREMOCARPEAE.
  b Placentation axile; fruit with a septum.
    c Herbaceous or rather succulent vines; inflorescence subspicate, the upper flowers mostly sterile; ovary 4-locular; capsule densely uncinate-spiny, bur-like, the 4 valves not splitting to the base; New World .... TOURRETTIEAE.
    c Woody vines, trees, shrubs, or erect herbs; inflorescence never spicate, without sterile flowers; ovary bilocular; fruit various, but not as above.
      d Lianas with the terminal leaflet frequently modified into a tendril (or derived suffrutescent or treelet forms without tendrils); stems mostly with anomalous phloem arms; fruits usually dehiscent parallel to the septum (or occasionally 4-valved); New World .... BIGNONIEAE.
      d Trees, shrubs, herbs, or non-tendrillate lianas; stems without anomalous phloem arms.
        e Fruit dehiscent parallel to the septum, or 4-valved; tropical Asia .... OROXYLEAE.
        e Fruit mostly dehiscent perpendicular to the septum; widespread .... TECOMEAE.
a Fruit indehiscent; trees, shrubs, or simple-leaved hemi-epiphytic lianas.

f Leaves mostly pinnately compound; terrestrial plants, the flowers in most genera pollinated by bees; Madagascar and Africa .... COLEAE.
f Leaves simple to palmately compound; New World.
 g Flowers pollinated by insects or humming birds; leaves simple and opposite; usually epiphytic shrubs or hemi-epiphytic lianas; seeds with nonfoliaceous cotyledons ... SCHLEGELIEAE.
 g Flowers pollinated by bats; leaves alternate, or palmately compound, or both; terrestrial trees or shrubs; seeds with foliaceous cotyledons .... CRESCENTIEAE.

11. Family MENDONCIACEAE Bremekamp 1953, the Mendoncia Family

Twining shrubs with jointed twigs and anomalous secondary growth, tanniferous but not saponiferous; hairs glandular and eglandular, much as in the Acanthaceae; cystoliths wanting; vessel-segments with simple perforations. LEAVES opposite, simple, entire; stomates diacytic; stipules wanting. FLOWERS axillary or sometimes in terminal racemes, subtended by 2 large, spathe-like bracteoles, perfect; calyx reduced, annular or truncate or shortly lobed; corolla sympetalous, regular or somewhat irregular, the 5 lobes convolute; stamens 4, attached to the corolla-tube alternate with the lobes, paired, the fifth (upper) one wanting or represented by a staminode; anthers dithecal; pollen-grains with 4–6 very short colpi; a prominent, cupular nectary-disk present around the base of the ovary; gynoecium of 2 carpels united to form a compound, superior, bilocular ovary, or one locule more or less reduced or even suppressed; style terminal, with a small, bilobed stigma, the lobes often unequal; ovules 2 in each locule, or 2 in the one fertile locule, collateral, presumably unitegmic and tenuinucellar, the funiculus of ordinary type, neither thickened nor otherwise much-modified. FRUIT a drupe with a unilocular (rarely bilocular) stone and 1 or 2 seeds; embryo dicotyledonous, the cotyledons twice folded; endosperm wanting.

The family Mendonciaceae consists of 2 to 4 genera, depending on the definitions, and about 60 species, native to South America, tropical Africa, and Madagascar. Nearly all of the species belong to the genus *Mendoncia*.

Traditionally the Mendonciaceae have been included in the Acanthaceae, but they lack both the cystoliths and the specialized mechanism of seed-dispersal which characterize that family. No other family is any more hospitable than the Acanthaceae, and therefore it seems necessary to recognize the Mendonciaceae as a distinct family.

12. Family LENTIBULARIACEAE L. C. Richard in Poiteau & Turpin 1808 nom. conserv., the Bladderwort Family

Insectivorous herbs, aquatic or of wet places, rooted in the substrate, or rootless and often free-floating with the photosynthetic organs sub-

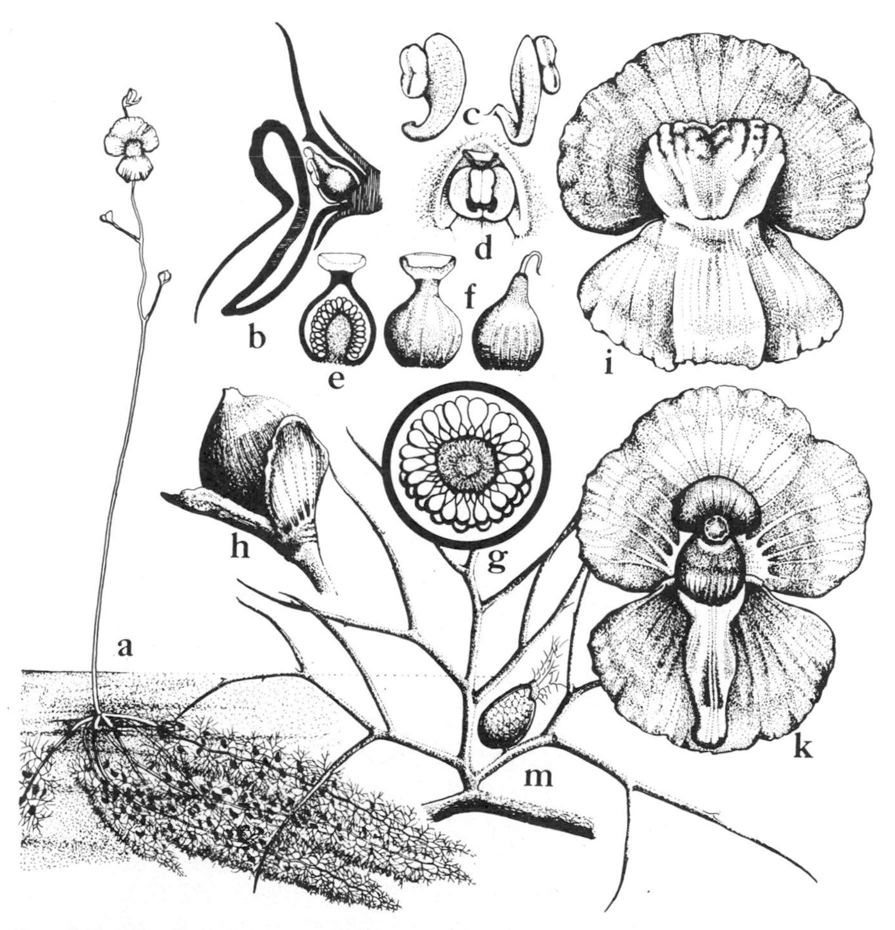

Fig. 6-28 Lentibulariaceae. *Utricularia geminiscapa* Benj. a, habit, $\times\frac{1}{2}$; b, diagram of flower, in long-section, ×3; c, stamens, ×8; d, orientation of stamens in corolla-throat, ×3; e, schematic long-section of ovary, ×8; f, two views of pistil, ×8; g, schematic cross-section of ovary, ×15; h, fruit, ×3; i, front view of flower, ×3; k, back view of flower, with the calyx evident, ×3; m, portion of "leaf," with a bladder, ×8.

merged, provided with stalked and/or sessile glands, these most abundant and complex in *Pinguicula*, in which the leaf-surface is viscid or slimy; plant commonly producing iridoid compounds and sometimes also tropane alkaloids, but not cyanogenic, not saponiferous, and only seldom tanniferous, lacking both ellagic acid and proanthocyanins; small crystals of calcium oxalate only seldom present in some of the cells of the parenchymatous tissues; stem with well developed vascular bundles (*Pinguicula*) or more often with the vascular system more or less strongly reduced or modified, the phloem commonly detached from the xylem. Foliage LEAVES simple, alternate, and crowded into basal rosettes in *Pinguicula* and *Genlisea*, the latter genus also with complex, tubular trap-leaves arising separately from the rhizome; stomates diacytic or seldom anomocytic, sometimes wanting; stipules wanting; stem in *Utricularia*, *Polypompholyx*, and *Biovularia* commonly bearing dissected, alternate or sometimes whorled branches or appendages, these photosynthetic and sometimes considered to be leaves, but not leaflike in anatomy, sometimes anatomically resembling the stem, sometimes more simple, and sometimes poorly developed or wanting, so that the stem appears to be merely bracteate or even naked; submerged branches ("leaves"), when present, bearing characteristic small bladders with a trap-door entrance, ingesting small insects and crustaceans. FLOWERS perfect, solitary and terminal on a bractless scape in *Pinguicula*, otherwise in a bracteate raceme (rarely reduced to a single flower) terminating the naked or subnaked stem or peduncle, the bracts clearly representing reduced leaves; calyx persistent, equally 4–5-lobed or cleft, or more often deeply cleft laterally to form an upper and a lower lip that may or may not give external evidence of consisting of more than one sepal each; corolla sympetalous, commonly showy, irregular, bilabiate and more or less distinctly 5-lobed, the lobes imbricate; lower lip evidently spurred or saccate at the base; anthers 2 (the anterior pair), tetrasporangiate and unithecal or with 2 confluent thecae, sometimes with an external form suggesting the ancestral dithecal condition, sometimes not; pollen-grains binucleate or trinucleate, tricolporate to multicolporate; nectary-disk wanting; gynoecium of 2 median carpels united to form a compound, unilocular ovary with a free-central (or basal) placenta; style wanting or very short; stigma wet, papillate, unequally 2-lobed, the upper lobe reduced or even suppressed; ovules numerous and somewhat sunken into the placenta (only 2 in *Biovularia*), anatropous, tenuinucellar, with a single massive integument and a saccate integumentary tapetum that is open at the micropylar end, the embryo-sac sometimes protruding from the micropyle; endosperm-development cellular, with terminal haustoria, or at least with a micropylar haustorium. FRUIT mostly capsular, circumscissile or opening by 2–4 valves or irregularly, in *Biovularia* indehiscent and 1-seeded; seeds small; embryo scarcely or not at all differentiated into parts; endosperm wanting. X = 6, 8, 9, 11, 21. (Pinguiculaceae, Utriculariaceae)

The family Lentibulariaceae consists of 5 genera and a little more than 200 species, of cosmopolitan distribution. The largest genus by far is *Utricularia*, the bladderwort, with about 150 species. Next comes *Pinguicula*, butterwort, with about 35. The other genera are *Genlisea* (15), *Polypompholyx* (2), and *Biovularia* (4). The last two are perhaps better included in *Utricularia*.

It is generally agreed that the Lentibulariaceae are derived from the Scrophulariaceae. The taxonomic limits of the Lentibulariaceae are not in question.

7. Order CAMPANULALES Lindley 1833

Herbs, or sometimes secondarily woody and shrubby or rarely even arborescent plants, commonly storing carbohydrate as inulin, sometimes producing pyridine alkaloids (Lobelioideae), sometimes iridoid compounds (some Stylidiaceae and Goodeniaceae), and sometimes polyacetylenes (Campanulaceae), but only rarely cyanogenic and rarely saponiferous, and not tanniferous or at least not strongly so, usually lacking both ellagic acid and proanthocyanins; stem sometimes with anomalous structure, but without internal phloem; nodes unilacunar or seldom 3–5-lacunar; vessel-segments with simple or seldom scalariform perforations. LEAVES mostly alternate, seldom opposite or whorled, simple, exstipulate. FLOWERS perfect or rarely unisexual, very often with a specialized pollen-presentation mechanism involving connate or connivent, introrsely dehiscent anthers and a modified style that pushes through the anther-tube and collects or expels the pollen; corolla nearly always sympetalous (polypetalous in the small family Donatiaceae and in a few Pentaphragmataceae), regular to strongly irregular, usually 5-lobed, the lobes valvate (sometimes induplicate) or less often imbricate; stamens usually as many as and alternate with the corolla-lobes, free from the corolla or attached to the base of the corolla-tube (attached just below the sinuses of the corolla in *Pentaphragma*), but in the Stylidiaceae and Donatiaceae the stamens only 2 or 3 and in the Stylidiaceae adnate to the style; pollen-grains of various triaperturate or triaperturate-derived types; gynoecium of 2–3 (–5) carpels united to form a compound, inferior or seldom superior ovary with as many locules as carpels, or the ovary sometimes unilocular either by failure of the partitions or by reduction of one carpel and locule; style terminal, very often with an indusium or a fringe of collecting hairs just beneath the stigma(s); ovules (1) 2-many per carpel (in the Brunoniaceae the ovule solitary in the single locule of a bicarpellate ovary), characteristically anatropous and tenuinucellar, with a massive single integument and an integumentary tapetum that is often incomplete at the micropylar end; endosperm-development cellular, with or without terminal haustoria. FRUIT most commonly a capsule, but sometimes a drupe, berry, achene, or small nut; seeds with dicotyledonous embryo and usually with copious, oily endosperm, but nearly or quite without endosperm in the Brunoniaceae and Sphenocleaceae.

The order Campanulales as here defined consists of 7 families and about 2500 species. About four-fifths of the species belong to the single family Campanulaceae, which is here broadly interpreted to include the Lobelioideae, and most of the remainder belong to the Goodeniaceae (300). The Stylidiaceae have about 150 species, the Pentaphragmataceae about 30, the Sphenocleaceae and Donatiaceae only 2 each, and the Brunoniaceae only one. The last four families have only a single genus each.

The Campanulales are Asteridae with a typically inferior ovary, mostly alternate leaves, and stamens mostly either free from the corolla or attached at the base of the corolla-tube. They characteristically store carbohydrate as inulin. They do not heavily exploit iridoids or tannins as repellents, and except for the pyridine alkaloids of some Lobelioideae, alkaloids are not important constituents. The vast majority of the species are herbaceous, and the woody species appear to have an herbaceous ancestry.

Three of the families and more than nine-tenths of the species of Campanulales have a specialized pollen-presentation mechanism of a type that appears to have been independently evolved also in the Asteraceae, Calyceraceae, and the subfamily Ixoroideae of the Rubiaceae. Even within the Campanulales, the mechanism may well have arisen separately in the Campanulaceae and Goodeniaceae, inasmuch as it is claimed that the collecting hairs of the first family are not homologous with the indusium of the second. The same evolutionary possibility has evidently been exploited several times within the Asteridae.

The pollen-presentation mechanism of the Campanulales would seem a priori to be of great importance to the plant. Certainly differences in its nature furnish important taxonomic characters in the delimitation of families within the order. On the other hand, it is not easy to see how this complex mechanism is a real improvement over more ordinary arrangements. A study of the kinds of pollinators in the order as a whole might be instructive. The basically herbaceous habit of the order is of course significant ecologically, but it is also far from unique. The other characters that collectively distinguish the Campanulales from the other orders of Asteridae are also difficult to interpret in terms of survival value.

The ancestry of the Campanulales is to be sought in or near the Solanales. Members of the Campanulales that have regular flowers would fit fairly comfortably into the Solanales if they were deprived of the advanced features of epigyny and specialized pollen-presentation.

SELECTED REFERENCES

Brizicky, G. K. 1966. The Goodeniaceae in the southeastern United States. J. Arnold Arbor. 47: 293–300.

Bronckers, F., & F. Stainier. Contribution à l'étude morphologique du pollen de la famille des Stylidiaceae. Grana 12: 1–22.

Carlquist, S. 1969. Studies in Stylidiaceae: New taxa, field observations, evolutionary tendencies. Aliso 7: 13–64.

Carlquist, S. 1969. Wood anatomy in Goodeniaceae and the problem of insular woodiness. Ann. Missouri Bot. Gard. 56: 358–390.

Carlquist, S. 1969. Wood anatomy of Lobelioideae (Campanulaceae). Biotropica 1: 47–72.

Carlquist, S. 1976. New species of *Stylidium*, and notes on Stylidiaceae from southwestern Australia. Aliso 8: 447–463.

Carolin, R. C. 1959. Floral structure and anatomy in the family Goodeniaceae Dumort. Proc. Linn. Soc. New South Wales 84: 242–255.

Carolin, R. C. 1960. Floral structure and anatomy in the family Stylidiaceae Swartz. Proc. Linn. Soc. New South Wales 85: 189–196.

Carolin, R. C. 1960. The structures involved in the presentation of pollen to visiting insects in the order Campanales. Proc. Linn. Soc. New South Wales 85: 197–207.

Carolin, R. C. 1966. Seeds and fruit of the Goodeniaceae. Proc. Linn. Soc. New South Wales 91: 58–83.

Carolin, R. C. 1967. The concept of the inflorescence in the order Campanulales. Proc. Linn. Soc. New South Wales 92: 7–26.

Carolin, R. C. 1971. The trichomes of the Goodeniaceae. Proc. Linn. Soc. New South Wales 96: 8–22.

Carolin, R. C. 1978. The systematic relationships of *Brunonia.* Brunonia 1: 9–29.

Chapman, J. L. 1966. Comparative palynology in Campanulaceae. Trans. Kansas Acad. Sci. 69: 197–204.

Crété, R. 1956. Contribution à l'étude de l'albumen et de l'embryon chez les Campanulacées et les Lobéliacées. Bull. Soc. Bot. France 103: 446–454.

Dunbar, A. 1975. On pollen of Campanulaceae and related families with special reference to the surface ultrastructure. I. Campanulaceae subfam. Campanuloidae. II. Campanulaceae subfam. Cyphioidae and subfam. Lobelioidae; Goodeniaceae; Sphenocleaceae. Bot. Not. 128: 73–101; 102–118.

Dunbar, A. 1978. Pollen morphology and taxonomic position of the genus *Pentaphragma* Wall. (Pentaphragmataceae). Grana 17: 141–147.

Erdtman, G., & C. R. Metcalfe. 1963. Affinities of certain genera *incertae sedis* suggested by pollen morphology and vegetative anatomy. III. The Campanulaceous affinity of *Berenice arguta* Tulasne. Kew Bull. 17: 253–256.

Gupta, D. P. 1959. Vascular anatomy of the flower of *Sphenoclea zeylanica* Gaertn. and some other related species. Proc. Natl. Inst. Sci. India 25B: 55–64.

Kapil, R. N., & M. R. Vijayaraghavan. 1965. Embryology of *Pentaphragma horsfieldii* (Miq.) Airy Shaw with a discussion on the systematic position. Phytomorphology 15: 93–102.

Kaplan, D. R. 1967. Floral morphology, organogenesis and interpretation of the inferior ovary in *Downingia bacigalupii.* Amer. J. Bot. 54: 1274–1290.

Maheshwari, P. 1956. The embryology of angiosperms, a retrospect and prospect. Curr. Sci. 25: 106–110.

Peacock, W. J. 1963. Chromosome numbers and cytoevolution in the Goodeniaceae. Proc. Linn. Soc. New South Wales 88: 8–27.

Rapson, L. J. 1953. Vegetative anatomy in *Donatia*, *Phyllacne*, *Forstera* and *Oreostylidium* and its taxonomic significance. Trans. & Proc. Roy. Soc. New Zealand 80: 399–402.

Rosén, W. 1935. Beitrag zur Embryologie der Stylidiaceen. Bot. Not. 1935: 273–278.

Rosén, W. 1938. Beiträge zur Kenntnis der Embryologie der Goodeniaceen. Acta Horti Gothob. 12: 1–10.

Rosén, W. 1946. Further notes on the embryology of the Goodeniaceae. Acta Horti Gothob. 16: 235–249.

Rosén, W. 1949. Endosperm development in Campanulaceae and closely related families. Bot. Not. 1949: 137–147.

Shetler, S. G. 1979. Pollen-collecting hairs of *Campanula* (Campanulaceae). Taxon 28: 205–215.

Subramanyam, K. 1950. A contribution to our knowledge of the systematic position of the Sphenocleaceae. Proc. Indian Acad. Sci. B 31: 60–65.

Subramanyam, K. 1951. A morphological study of *Stylidium graminifolium*. Lloydia 14: 65–81.

Tjon Sie Fat, L. 1978. Contribution to the knowledge of cyanogenesis in angiosperms. 2. Communication. Cyanogenesis in Campanulaceae. Proc. Kon. Nederl. Akad. Wetensch. ser. C. 81: 126–131.

Vijayaraghavan, M. R. & U. Malik. 1972. Morphology and embryology of *Scaevola frutescens* K. and affinities of the family Goodeniaceae. Bot. Not. 125: 241–254.

Аветисян, Е. М. 1967. Морфология пыльцы сем. Campanulaceae и близких к нему семейств (Sphenocleaceae, Lobeliaceae, Cyphiaceae) в связи с вопросами их систематики и филогении. Труды Бот. Инст. Армянской ССР, 16: 5-41.

Аветисян, Е. М. 1973. Палинология порядка Campanulales s. l. В сб.: Морфология и происхождение современных растений. Труды III Международной Палинологической Конф. СССР, Новосибирск, 1971: 90-93. Изд. Наука, Ленинградск. Отд. Ленинград.

Шулькина, Т. В. 1978. Жизненные формы в семействе Campanulaceae Juss., их географическое распространение и связь с таксономией. Бот. Ж. 63: 153-169.

SYNOPTICAL ARRANGEMENT OF THE FAMILIES OF CAMPANULALES

1 Style without an indusium, but often with collecting hairs.

2 Stamens as many as the corolla-lobes, typically 5, free from the style; anthers introrse.

3 Style glabrous, and with a single stigma (or the stigma sessile); plants without a latex-system.

4 Corolla-lobes valvate; fruit a berry; flowers borne in dense, sympodial, helicoid cymes that have conspicuous bracts .... 1. PENTAPHRAGMATACEAE.

4 Corolla-lobes imbricate; fruit a circumscissile capsule; flowers

borne in dense, terminal, inconspicuously bracteate spikes .... 2. SPHENOCLEACEAE.

3 Style with well developed collecting hairs just below the 2–3 (–5) stigmas; plants with a well developed latex-system ....3. CAMPANULACEAE.

2 Stamens 2 or 3, fewer than the corolla-lobes or petals; anthers extrorse; plants without a latex-system.

5 Petals united to form a 5-lobed corolla; stamens wholly adnate to the style, together with which they form a column .... 4. STYLIDIACEAE.

5 Petals 5–10, distinct; stamens free from the style; filaments distinct .... 5. DONATIACEAE.

1 Style with a more or less cupulate indusium just beneath the stigma, but without collecting hairs.

6 Flowers regular, borne in involucrate, cymose heads; endosperm wanting; ovary superior, unilocular, with a single basal ovule .... 6. BRUNONIACEAE.

6 Flowers irregular, borne in various sorts of inflorescences, but not in involucrate heads; endosperm well developed; ovary mostly inferior, seldom only half-inferior, rarely essentially superior, (1) 2 (4)-locular, with 2 or more ovules .... 7. GOODENIACEAE.

1. Family PENTAPHRAGMATACEAE J. G. Agardh 1858 nom. conserv., the Pentaphragma Family

Perennial, coarse, somewhat succulent herbs, often with branching hairs, without a latex-system and without alkaloids; vessel-segments with scalariform perforations. LEAVES alternate, simple, often relatively large, usually asymmetrical at the base; stomates surrounded by 3 or 4 cells somewhat differentiated from the ordinary epidermal cells; stipules wanting. FLOWERS in dense, sympodial, axillary or extra-axillary, helicoid cymes with conspicuous, membranous bracts, perfect or rarely unisexual; sepals 5, unequal, imbricate, persistent; corolla usually fleshy or cartilaginous, sympetalous or seldom polypetalous, with (4) 5 valvate lobes (or petals); stamens as many as and alternate with the corolla-lobes, attached to the corolla-tube shortly below the sinuses (or to the margins of the top of the ovary in polypetalous spp.) filaments distinct, persistent, or virtually wanting; anthers basifixed, tetrasporangiate and dithecal, opening by longitudinal slits; pollen-grains binucleate, trilobate, smooth, the exine nearly solid, with a thin, baculate hollow, not evidently tectate-columellate; gynoecium of 2 or 3 carpels united to form a compound, bilocular or trilocular, inferior ovary, the ovary joined to the hypanthium only by longitudinal septa usually more or less aligned with the filaments, leaving 5 (or 6) intervening nectariferous pits; style short and thick, with a massive, terminal, glabrous stigma, lacking collecting hairs; ovules numerous on the axile placentas, anatropous, with a thin single integument, tenuinucellar; embryo-sac protruding from the micropyle; endosperm-development cellular, with a unicellular micropylar haustorium and without a chalazal haustorium. FRUIT a berry with the perianth persistent at the tip; seeds minute, with dicotyledonous embryo and copious, starchy endosperm.

The family Pentaphragmataceae consists of the single genus *Pentaphragma*, with about 30 species native to southeastern Asia and some of the nearby Pacific islands. The genus has customarily been included in the Campanulaceae, but more recently it has been noted to differ from typical members of that family in several ways, most notably in having the stamens attached well up in the corolla-tube and in lacking the characteristic pollen-presentation mechanism. Some authors not only remove *Pentaphragma* from the Campanulaceae, but also exclude it from the order. On embryological grounds Kapil and Vijayaraghavan affirm its affinity with the Campanulaceae. Avetisian (1967, 1973) considers the pollen to be rather similar to that of some genera of Campanulaceae such as *Platycodon* and *Canarina*, but Dunbar (1978) thinks it is very different from that of Campanulaceae, especially in the internal structure of the exine.

2. Family SPHENOCLEACEAE A. P. de Candolle 1839 nom. conserv., the Sphenoclea Family

Annual herbs of wet places, habitally resembling some spp. of *Phytolacca*; stem somewhat succulent, with large vertical air-passages in the cortex, but without typical laticifers, these perhaps represented by occasional elongate, rather broad cells in the phloem with granular contents; clustered crystals of calcium oxalate present in some of the cells of the parenchymatous tissues; xylem and phloem forming a continuous cylinder traversed by medullary rays 1–2 cells wide; vessel-segments with simple perforations. LEAVES alternate, simple, entire; stomates tetracytic, with 2 supporting cells parallel to the guard-cells, and 2 at right angles to them; petiole with an arcuate vascular strand; stipules wanting. FLOWERS perfect, regular, borne in the axils of small bracts in dense, terminal spikes, bibracteolate; calyx synsepalous, persistent, the 5 lobes imbricate, connivent over the top of the ovary; corolla small, sympetalous, caducous, urceolate-campanulate, the 5 lobes imbricate; stamens attached to the base of the corolla-tube, as many as and alternate with the lobes; filaments short; anthers tetrasporangiate and dithecal, opening by longitudinal slits; pollen-grains trinucleate, tricolporate, the surface of the exine reticulate; gynoecium of 2 carpels united to form a compound, inferior or half-inferior, bilocular ovary; stigma capitate, sessile or borne on a very short style, without collecting hairs; ovules numerous on the large, spongy, axile placentas, anatropous, tenuinucellar, with a massive single integument; endosperm-development cellular, with terminal haustoria. FRUIT a membranous, circumscissile capsule; seeds numerous, small, with a straight, dicotyledonous embryo and very scanty or no endosperm. N = 12.

The family Sphenocleaceae consists of the single genus *Sphenoclea*, with only two species, one pantropical, one West African. *Sphenoclea* has traditionally been associated with or included in the Campanulaceae. This view has been challenged by Airy Shaw in favor of an alliance with the Phytolaccaceae, but the embryological and other differences between *Phytolacca* and *Sphenoclea* are so overwhelming that the habital similarity must be purely coincidental. *Sphenoclea* is embryologically and palynologically much like the Campanulaceae, and its relationship to that family seems well established. Hutchinson (1973 in general citations) vigorously presents some of the essential differences between *Sphenoclea* and *Phytolacca*.

3. Family CAMPANULACEAE A. L. de Jussieu 1789 nom. conserv., the Bellflower Family

Plants mostly herbaceous, but sometimes secondarily woody and forming shrubs or even small trees (notably some tree-Lobelias); hairs when present mostly unicellular, sometimes becoming silicified or calcified; cystoliths often present at the base of the hairs or in surrounding epidermal cells; plants storing carbohydrate as inulin, commonly accumulating polyacetylenes and sometimes (some Lobelioideae) pyridine alkaloids, but lacking iridoid compounds, only rarely cyanogenic (and then containing triglochinin, a tyrosine derivative, unlike the cyanogenic members of the Asteraceae), rarely saponiferous, and usually not strongly tanniferous, lacking proanthocyanins and only seldom with ellagic acid; nodes unilacunar; small, needle-like crystals of calcium oxalate often present in some of the cells of the parenchymatous tissues; articulated, anastomosing laticifers forming a well developed system in the phloem of the stems and leaves, and commonly extending into the surrounding tissues as well; xylem and phloem nearly always forming a continuous cylinder; vessel-segments mostly with simple perforations, seldom some or even all of the perforations scalariform; imperforate tracheary elements with simple or indistinctly bordered pits; wood-rays commonly 4–8 cells wide; wood-parenchyma scanty-paratracheal, or often wanting; medullary bundles or medullary strands of phloem often present. LEAVES simple, alternate or seldom opposite, rarely (as in *Ostrowskia*) whorled; epidermal cells often with silicified or calcified walls; stomates commonly anomocytic; petiole mostly with an arc-shaped vascular strand; stipules wanting. Inflorescence of diverse racemose or mixed or sometimes strictly cymose types; FLOWERS perfect or rarely unisexual, mostly (3–) 5 (–10)-merous, epigynous (the hypanthium sometimes shortly prolonged beyond the ovary) or seldom merely half-epigynous or rarely (*Cyananthus*, with 5 carpels) merely perigynous, almost hypogynous; calyx with imbricate or valvate lobes or segments; odd sepal posterior (its back to the axis of the inflorescence) in Campanuloideae, morphologically anterior in Lobelioideae but then the flower resupinate by twisting of the pedicel, so that the odd sepal appears to be posterior as in the Campanuloideae; corolla sympetalous,

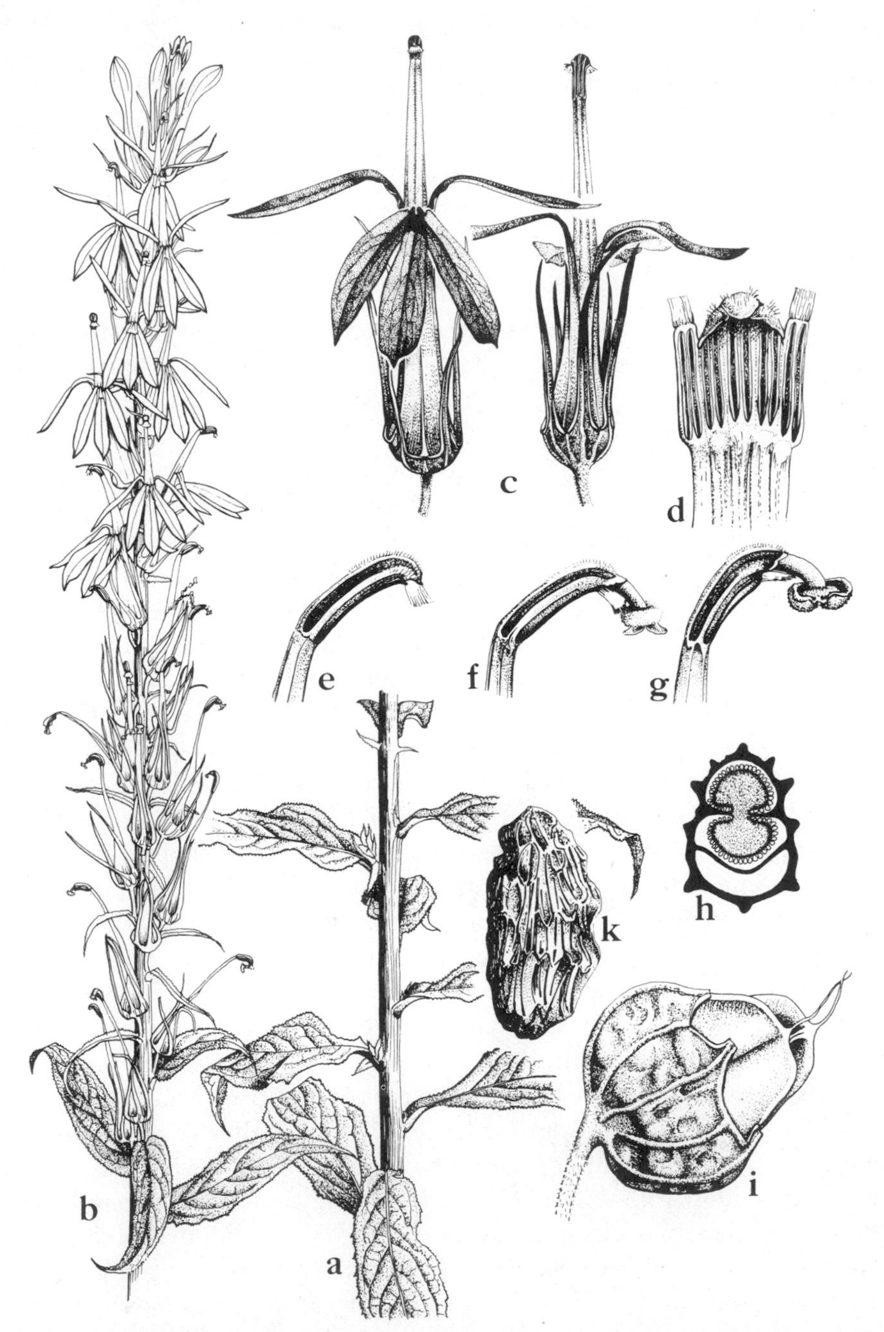

FIG. 6-29 Campanulaceae. *Lobelia cardinalis* L. a, portion of stem and leaves $\times\frac{1}{2}$; b, inflorescence, $\times\frac{1}{2}$; c, two views of flower, $\times 1\frac{1}{2}$; d, stamen-tube, opened out, $\times 3$; e, f, g, successive stages of emergence of the style from the anther-tube, $\times 3$; h, schematic cross-section of ovary, $\times 3$; i, mature fruit, after dehiscence, with the calyx-lobes removed, $\times 3$; k, seed, $\times 30$.

with valvate lobes, regular or nearly so (and often campanulate) in the Campanuloideae, highly irregular in the Lobelioideae, in the latter group resupinate, so that the 3-lobed morphologically upper lip appears as the lower lip; morphologically lower (visually upper) lip in the Lobelioideae 2-lobed or 2-cleft, sometimes so deeply cleft as to lose its identity, the corolla then appearing unilabiate; corolla-tube in the Lobelioideae often fenestrate; some of the flowers sometimes cleistogamous and apetalous; pollen-presentation mechanism complex, the anthers connivent (Campanuloideae) or connate (Lobelioideae) to form a tube into which the pollen is shed, the style provided with a fringe of collecting hairs below the initially appressed stigmas, growing up through the anther tube and pushing out the pollen, after which the stigmas spread apart; collecting hairs, at least in the Campanuloideae, eventually retracted into the style by invagination; stamens as many as and alternate with the lobes of the corolla; filaments attached to the annular, epigynous nectary-disk or to the base of the corolla (to a hypanthium in *Cyananthus*), distinct at the base, but in the Lobelioideae often connate above; anthers tetrasporangiate and dithecal, introrse, opening by longitudinal slits, in the Campanuloideae separating after anthesis; pollen-grains 2–3-nucleate, variously (2) 3–12-aperturate, the exine usually spinulose in the Campanuloideae, usually reticulate in the Lobelioideae; gynoecium of 2–5 carpels (mostly 3 in Campanuloideae, but 2 and median in Lobelioideae) united to form a compound, inferior or seldom only half-inferior, rarely superior (*Cyananthus*) ovary, commonly with as many locules as carpels, but in some Lobelioideae unilocular with 2 parietal placentas, and in some Campanuloideae the primary locules divided into locelli by the intrusion of partitions from the carpellary midribs; stigmas wet or dry; ovary commonly crowned by a nectary-disk; ovules numerous on axile (rarely parietal) placentas, anatropous, tenuinucellar, with a massive single integument and an integumentary tapetum that is often incomplete at the micropylar end; endosperm-development cellular, with terminal haustoria. FRUIT commonly a capsule, often poricidal or opening by longitudinal slits that may be more numerous than the carpels, less often the fruit a berry; seeds numerous, small, with a straight, short to spatulate, dicotyledonous embryo embedded in the oily endosperm, or seldom (as in *Cephalostigma*) the endosperm starchy. X = 6–17. (Lobeliaceae, Cyphiaceae)

The family Campanulaceae as here defined consists of about 70 genera and perhaps as many as 2000 species, of cosmopolitan distribution. There are two well marked subfamilies, the Campanuloideae and Lobelioideae, connected by a small group of transitional genera that are sometimes treated as a third subfamily Cyphioideae. The existence of the two major groups, and the close relationship between them, are not in dispute. The Lobelioideae are the more advanced group, marked by their highly irregular, resupinate flowers and connate anthers. Whether to treat the lobeliads as a subfamily of the Campanulaceae or as a distinct family is purely a matter of taste.

About half of the species in the Campanulaceae belong to only 4 genera, *Campanula* (300), *Lobelia* (300), *Centropogon* (200), and *Siphocampylus* (200). A number of species of *Campanula* (bellflower) and *Lobelia* are familiar garden-ornamentals. *Campanula medium* L. is the Canterbury bell, and *C. rotundifolia* L. is the harebell, or bluebell of Scotland.

4. Family STYLIDIACEAE R. Brown 1810 nom. conserv., the Trigger-plant Family

Small, often xerophytic herbs, or seldom undershrubs, often bearing stalked glands with a bicellular or multicellular head, most commonly scapose and with a basal rosette of grass-like leaves, but sometimes with a creeping or shortly erect stem bearing crowded small leaves, or an erect stem with spaced pseudowhorls of leaves; plants storing carbohydrate as inulin, sometimes producing iridoid compounds, sometimes with scattered tanniferous cells but not cyanogenic, and only seldom saponiferous; stem anatomically complex, varying in different species; laticifers wanting. LEAVES simple, basically alternate; stomates commonly anomocytic; stipules wanting. FLOWERS in terminal, bracteate racemes or branching cymes, or solitary in the upper axils, perfect or sometimes unisexual, epigynous; calyx (2–) 5 (–7)-lobed; corolla sympetalous, with 5 imbricate lobes, irregular, with the morphologically anterior (median) lobe differentiated (and often turned back) as a labellum, but the flower commonly half-resupinate, so that the labellum is lateral, or seldom the corolla essentially regular; stamens 2, free from the corolla but wholly adnate to the style, together with which they form a stylar column; anthers extrorse, tetrasporangiate and dithecal (the thecae set end to end and sometimes apically confluent), opening by longitudinal slits; pollen-grains binucleate or trinucleate, 3–8-colpate; an epigynous nectary-disk (or a pair of nectary-glands) often present; gynoecium of 2 median carpels united to form a compound, more or less completely bilocular, inferior ovary with axile to free-central placentation, or the posterior locule reduced or suppressed, so that the ovary is pseudomonomerous; stylar column often irritable, bent to the labellar side, snapping to an oppositely bent position when touched; stigmas dry, papillate, diverging above the anthers; ovules numerous, anatropous, tenuinucellar, with a massive single integument and an integumentary tapetum; endosperm-development cellular, with terminal haustoria. FRUIT capsular, or seldom indehiscent; seeds few to more often numerous, with small, dicotyledonous (frequently monocotyledonous) embryo embedded in the oily endosperm. N = 15, 18.

The family Stylidiaceae consists of 5 genera and about 155 species, native to Australasia, southern and especially southeastern Asia, and southernmost South America. The largest genus by far is *Stylidium*, with about 140 species.

5. Family DONATIACEAE Takhtajan ex Dostál 1957 nom. conserv., the Donatia Family

Dwarf, subalpine, pulvinate-caespitose herbs, storing carbohydrate as inulin, and with scattered tanniferous cells in the stems, roots and leaves; crystals wanting; stem with a ring of vascular bundles; vessel-segments with oblique scalariform perforation plates. LEAVES densely crowded, spirally arranged, linear, coriaceous, hairy in the axils; stomates paracytic; mesophyll with scattered mucilage-cells; stipules wanting. FLOWERS solitary and terminal, sessile, perfect, essentially regular, epigynous; calyx-lobes 5–7, equal or unequal; petals 5–10, distinct, imbricate; stamens 2 or 3, inserted on top of the ovary within an epigynous nectary-disk, free from the style; filaments distinct; anthers extrorse, dithecal, opening by longitudinal slits; pollen-grains 3 (4)-colporate; gynoecium of 2 or 3 carpels united to form a compound, bilocular or trilocular ovary; styles distinct, recurved, with a capitate stigma; ovules numerous, on axile placentas in the upper part of the locules. FRUIT turbinate, dry, indehiscent; seeds few, with minute, dicotyledonous embryo and abundant, oily endosperm.

The family Donatiaceae consists of the single genus *Donatia*, with 2 species, one in southern South America, the other in New Zealand and Tasmania.

Donatia has often been included in the Stylidiaceae, presumably because it produces inulin, because it has epigynous flowers with only 2 or 3 stamens, because it is habitally similar to some of the Stylidiaceae, and because its geographic distribution is concordant with that of the Stylidiaceae. Some other authors, among recent ones notably Hutchinson, have treated *Donatia* as a family allied to the Saxifragaceae. I can readily agree that *Donatia* is a discordant element that needs to be extracted from the Stylidiaceae. Its proper position in the system is more doubtful. It is certainly anomalous in the Campanulales, and in the Asteridae as a whole, in its distinct petals. In other respects, however, it is readily compatible with the Campanulales, and Avetisian (1973) considers that the pollen is harmonious with that of other members of the order. I prefer to leave *Donatia* in the Campanulales until more definitive evidence for its removal is forthcoming.

6. Family BRUNONIACEAE Dumortier 1829 nom. conserv., the Brunonia Family

Scapose perennial herbs with trichomes of a special type, these with a short, thin-walled, erect basal cell and a 2-several-celled arm attached at right angles to the basal cell and appressed to the surface; plants storing carbohydrate as inulin, not cyanogenic, and without iridoid compounds. LEAVES simple, entire, all basal; stomates with 4 subsidiary cells, 2 of them parallel to the guard-cells; stipules wanting. FLOWERS perfect, regular, borne in terminal, involucrate, cymose heads; calyx 5-lobed; corolla blue, sympetalous, with 5 valvate lobes; stamens 5, attached to the base of the corolla-tube alternate with the lobes; filaments distinct; pollen-presentation mechanism similar to that of the Goodeniaceae (but without hairs on the indusium), the dithecal anthers connate into a tube and opening introrsely; pollen-grains binucleate, tricolporate; gynoecium of 2 median carpels united to form a compound, superior, unilocular ovary; style with a subterminal cupulate indusium around the small stigma; ovary with a septum at the base only, the solitary ovule inserted on one side of the septum; ovule erect, anatropous, tenuinucellar, with a massive single integument and an integumentary tapetum; endosperm-development not known. FRUIT an achene or small nut surrounded by the indurated tube of the persistent calyx; seeds without endosperm; embryo straight, with a minute radicle and 2 thickened cotyledons. X = 9.

The family Brunoniaceae consists of the single genus and species *Brunonia australis* Smith, native to Australia. It is generally agreed that *Brunonia* is allied to the Goodeniaceae, but so distinctive that it is best treated in its own family.

7. Family GOODENIACEAE R. Brown 1810 nom. conserv., the Goodenia Family

Mostly perennial herbs, but some species secondarily woody and shrubby or even arborescent (mainly in some insular spp. of *Scaevola*); hairs variously simple or branched or stellate, but unlike those of *Brunonia*; plants commonly storing carbohydrate as inulin, often poisonous, sometimes saponiferous and sometimes producing iridoid compounds, but only seldom cyanogenic and not tanniferous, lacking proanthocyanins; crystals of calcium oxalate commonly present in some of the cells of the parenchymatous tissues; laticifers wanting; nodes unilacunar or sometimes 3–5-lacunar; vessel-segments with simple or sometimes scalariform perforations; imperforate tracheary elements with bordered pits; wood-rays heterocellular to homocellular, mixed uniseriate and pluriseriate, the latter 2–7 cells wide, with elongate ends; wood-parenchyma vasicentric or diffuse or both, or scanty or wanting. LEAVES simple, alternate, rarely opposite or whorled; stomates anomocytic; sclereids commonly present in the mesophyll and sometimes in the cortex of the stem as well; stipules wanting. FLOWERS perfect, proterandrous, variously in cymes, racemes, or heads, or solitary in the axils; calyx tubular, mostly (3–) 5-lobed; corolla sympetalous, irregular, bilabiate or sometimes unilabiate (i.e., the upper lip bifid to the base), the 5 lobes valvate, often induplicate; pollen-presentation mechanism complex, the anthers connivent or connate to form a tube into which the pollen is shed, the style with a cupular indusium below the small stigma, growing up through the anther-tube, and collecting the pollen, which is subsequently dusted onto visiting insects; indusium with a fringe of very short hairs around the top; stamens 5, alternate with the corolla-lobes, free from the corolla or attached to the base of the corolla-tube; anthers tetrasporangiate and dithecal, introrse, opening by longitudinal slits; pollen-grains binucleate, commonly tricolporate or triporate; 1 or 2 intrastaminal nectary-glands sometimes present; gynoecium typically of 2 median carpels, these united to form a compound, inferior to sometimes only half-inferior or rarely (*Velleia*) essentially superior ovary with 2 locules or seldom only one locule; in *Scaevola porocarya* F. Muell. the ovary and fruit 4 locular, with one ovule and seed per locule, the 2 lateral locules set a little below the 2 median ones; in some other spp. of *Scaevola* 2 vestigial lateral locules present in addition to the 2 well developed median ones; ovules 1-many in each

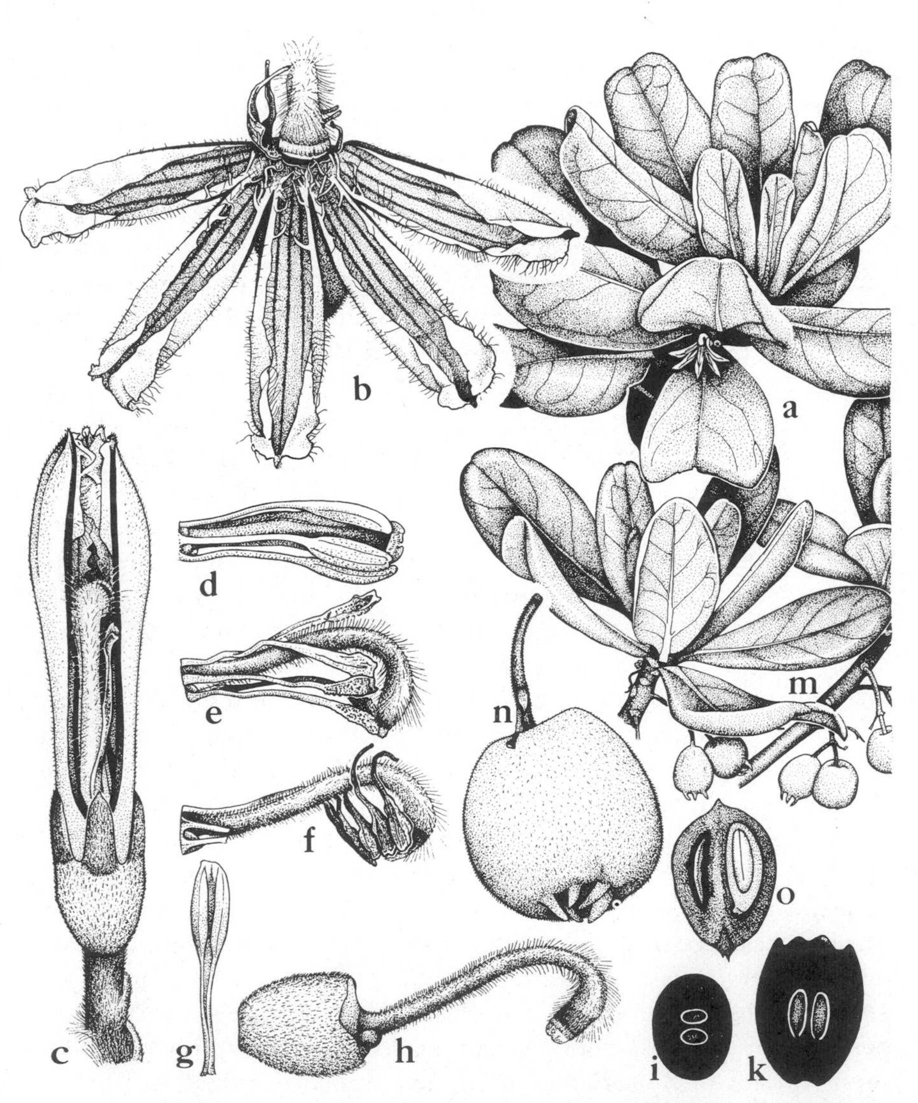

FIG. 6-30 Goodeniaceae. *Scaevola taccada* (Gaertn.) Roxb. a, habit, in flower, ×½; b, front view of flower, ×4; c, flower bud, ready to open, from above, ×4; d, e, f, successive stages in the growth of the style through the anther-tube, ×4; g, ventral view of stamen before dehiscence, ×4; h, pistil, ×4; i, k, schematic cross- and long-sections of ovary, ×4; m, habit, in fruit, ×½; n, fruit, ×2; o, schematic long-section of stone, ×4.

locule, mostly erect or ascending on axile or basal-axile placentas, anatropous, tenuinucellar, with a massive single integument and an integumentary tapetum; endosperm-development cellular, without haustoria. FRUIT commonly a capsule, less often a drupe or a small nut; seeds with a straight, dicotyledonous embryo embedded in the copious, oily endosperm. X = 7–9.

The family Goodeniaceae consists of about 14 genera and 300 species, primarily Australian, but extending also to New Zealand, Japan, and various tropical and subtropical parts of both the Old and the New World. The largest genera are *Goodenia* (100) and *Scaevola* (80). Carolin (1978) suggests that the Campanulaceae evolved primarily in western Gondwanaland, and the Goodeniaceae in eastern Gondwanaland.

Pollen considered to represent the Goodeniaceae occurs in Oligocene and more recent deposits.

8. Order RUBIALES Bentham & Hooker 1873

Woody or less often herbaceous plants, without internal phloem, often with raphides or other sorts of crystals, commonly producing iridoid compounds and often other types of repellents as well, but without ellagic acid and usually without proanthocyanins; vessel-segments all or nearly all with simple perforations. LEAVES simple and mostly entire, commonly opposite and with interpetiolar stipules, less often with intrapetiolar stipules, or whorled and without stipules (i.e., the interpetiolar stipules transformed into leaves), rarely the upper or all leaves alternate by suppression of one member of each opposite pair; stipules commonly bearing colleters on the inner surface; stomates mostly paracytic. FLOWERS perfect or less often unisexual, epigynous, borne in various sorts of basically cymose inflorescences, or seldom solitary; calyx mostly 4–5-lobed, with open aestivation, sometimes much-reduced or obsolete; corolla sympetalous, regular or rarely somewhat irregular, wanting in the staminate flowers of Theligonaceae; stamens in the Rubiaceae attached to the corolla-tube, as many as and alternate with the lobes, in the Theligonaceae 6–30 and free from the tiny apparent perianth, the seeming staminate flower perhaps a pseudanthium; anthers tetrasporangiate and dithecal, opening by longitudinal slits; pollen-grains of various triaperturate-derived types, rarely inaperturate; gynoecium in the Rubiaceae of 2 (seldom 3–5 or more) carpels united to form a compound, nearly always inferior ovary with a terminal style, in the Theligonaceae the ovary inferior and pseudomonomerous, with a basilateral style; ovules 1-many in each locule, anatropous to hemitropous or campylotropous, with a massive single integument, tenuinucellar (or the nucellus obsolete), without an integumentary tapetum; endosperm-development nearly always nuclear. FRUIT of diverse types; embryo with 2 cotyledons; endosperm generally well developed and oily, seldom scanty or wanting.

The order Rubiales consists of the very large family Rubiaceae and the very small satellite-family Theligonaceae. The two families are discussed separately. The name Rubiales is here preferred over the older name Cinchonales Lindley (1833).

SELECTED REFERENCES

Blaser, J. L. 1954. The morphology of the flower and inflorescence of *Mitchella repens*. Amer. J. Bot. 41: 533–539.

Bremekamp, C. E. B. 1957. On the position of *Platycarpum* Humb. et Bonpl., *Henriquezia* Spruce ex Bth., and *Gleasonia* Standl. Acta Bot. Neerl. 6: 351–377.
Bremekamp, C. E. B. 1966. Remarks on the position, the delimitation and the subdivision of the Rubiaceae. Acta Bot. Neerl. 15: 1–33.
Darwin, S P. 1976. The subfamilial, tribal, and subtribal nomenclature of the Rubiaceae. Taxon 25: 595–610.
Fagerlind, F. 1937. Embryologische, zytologische und bestäubungsexperimentelle Studien in der Familie Rubiaceae nebst Bemerkungen über einige Polyploiditätsprobleme. Acta Horti Berg. 11: 195–470.
Fukuoka, N. 1978. Studies in the floral anatomy and morphology of Rubiaceae I. Hedyotideae (*Anotis*, *Argostemma* and *Clarkella*). Acta Phytotax. Geobot. 29: 85–94.
Koek-Noorman, J. 1970. A contribution to the wood anatomy of the Cinchoneae, Coptosapelteae, and Naucleeae (Rubiaceae). Acta Bot. Neerl. 19: 154–164.
Koek-Noorman, J. 1972. The wood anatomy of Gardenieae, Ixoreae and Mussaendeae (Rubiaceae). Acta Bot. Neerl. 21: 301–320.
Koek-Noorman, J. 1977. Systematische Holzanatomie einiger Rubiaceen. Ber. Deutsch. Bot. Ges. 90: 183–190.
Koek-Noorman, J., & P. Hogeweg. 1974. Wood anatomy of Vanguerieae, Cinchoneae, Condamineae, and Rondeletieae (Rubiaceae). Acta Bot. Neerl. 23: 627–653.
Kooiman, P. 1971. Ein phytochemischer Beitrag zur Lösung des Verwandtschaftsproblems der Theligonaceae. Oesterr. Bot. Z. 119: 395–398.
Lee, Y. S. & D. E. Fairbrothers. 1978. Serological approaches to the systematics of the Rubiaceae and related families. Taxon 27: 159–185.
Lersten, N. R. 1975. Colleter types in Rubiaceae, especially in relation to the bacterial leaf nodule symbiosis. J. Linn. Soc., Bot. 71: 311–319.
Mabry, T. J., I. J. Eifert, C. Chang, H. Mabry, & C. Kidd. 1975. Theligonaceae: Pigment and ultrastructural evidence which excludes it from the order Centrospermae. Biochem. Syst. Ecol. 3: 53–55.
Melhem, T. S. A., C. L. B. Rossi, & M. S. F. Silvestre. 1974. Pollen morphological studies in Rubiaceae. Hoehnea 4: 49–70.
Pant, D. D., & Bharati Mehra. 1965. Ontogeny of stomata in some Rubiaceae. Phytomorphology 15: 300–310.
Poucques, M.-L. de. 1949. Recherches caryologiques sur les Rubiales. Rev. Gén. Bot. 56: 5–27; 74–138; 172–188.
Praglowski, J. 1973. The pollen morphology of the Theligonaceae with reference to taxonomy. Pollen & Spores 15: 385–396.
Siddiqui, S. A., & S. B. Siddiqui. 1968. Studies in the Rubiaceae I. A contribution to the embryology of *Oldenlandia dichotoma* Hook.f. II. A contribution to the embryology of *Borreria stricta* Linn. Beitr. Biol. Pflanzen 44: 343–351; 353–360.
Verdcourt, B. 1958. Remarks on the classification of the Rubiaceae. Bull. Jard. Bot. État 28: 209–290.
Wagenitz, G. 1959. Die systematische Stellung der Rubiaceae. Ein Beitrag zum System der Sympetalen. Bot. Jahrb. Syst. 79: 17–35.
Weberling, F. 1977. Beiträge zur Morphologie der Rubiaceen-Infloreszenzen. Ber. Deutsch. Bot. Ges. 90: 191–209.
Wunderlich, R. 1971. Die systematische Stellung von *Theligonum*. Oesterr. Bot. Z. 119: 329–394.

Зубкова, И. Г. 1971. Модусы образования устьичного аппарата в сем. Rubiaceae. Бот. Ж. 56: 1816–1819.

SYNOPTICAL ARRANGEMENT OF THE FAMILIES OF RUBIALES

1 Flowers perfect or rarely unisexual, zoophilous (mainly entomophilous) or rarely anemophilous, usually with both a calyx and a sympetalous corolla, or the calyx sometimes obsolete; stamens as many as and alternate with the corolla-lobes, attached to the corolla-tube; ovary composed of 2 (or more) carpels, with a terminal style, and with 1-many anatropous to hemitropous ovules in each locule; embryo straight, or seldom curved 1. RUBIACEAE.

1 Flowers unisexual, anemophilous, the staminate ones (perhaps pseudanthia) with 6–30 stamens free from the small, inconspicuous perianth (or involucre) of 2–5 segments, the pistillate ones pseudomonomerous, with a basilateral style surrounded by a tubular, 2–4-toothed perianth; ovule solitary, basal, campylotropous; embryo strongly curved .... 2. THELIGONACEAE.

1. Family RUBIACEAE A. L. de Jussieu 1789 nom. conserv., the Madder Family

Trees, shrubs, woody vines, or less often herbs, sometimes myrmecophilous, sometimes with anomalous secondary growth, but without internal phloem, often accumulating aluminum, and producing a wide range of chemical repellents, commonly including iridoid compounds, anthraquinones, and various sorts of alkaloids (especially indole alkaloids, but also quinoline, isoquinoline, and purine alkaloids), sometimes tanniferous and producing proanthocyanins, but apparently without ellagic acid, often with triterpenes and occasionally saponiferous, but only rather seldom cyanogenic; secretory cells or cavities of various sorts often present in the parenchymatous tissues, as also cells containing crystals of calcium oxalate, these varied in form, often as raphides, which occur in the epidermis of the leaves as well as in the parenchymatous tissues; nodes unilacunar or less often trilacunar; vessel-segments all or nearly all with simple perforations; imperforate tracheary elements in about one-fourth of the genera septate and with simple or slightly bordered pits, in the others non-septate and with more or less evidently bordered pits; wood-rays heterocellular, mixed uniseriate and pluriseriate, the latter mostly 2–3 (–10) cells wide; wood-parenchyma variously apotracheal, paratracheal, or wanting; sieve-tubes with S-type plastids. LEAVES simple and mostly entire (lobed in *Pentagonia*), commonly decussately opposite and with connate (less often distinct) interpetiolar stipules, or sometimes apparently whorled (the interpetiolar stipules transformed into leaves and sometimes increased in number), seldom the stipules intrapetiolar, or reduced to a mere line connecting the opposite leaves; rarely (*Didymochlamys*) the leaves alternate by suppression of one member of a pair at each node; stipules commonly bearing colleters on the inner surface, these producing mucilage and serving to protect the growing tip of the shoot; stomates commonly paracytic; petiole with an arc-shaped or cylindrical vascular strand, sometimes with smaller medullary bundles. FLOWERS in various sorts of basically cymose inflorescences, or rarely solitary, perfect or rarely unisexual, nearly always epigynous, often heterostylic, zoophilous (mainly entomophilous) or rarely anemophilous; calyx mostly 4–5-lobed, with open aestivation, the lobes often small, sometimes even obsolete, or one or more of them sometimes enlarged and brightly colored (as in *Mussaenda*); corolla sympetalous, (3) 4–5-lobed, or rarely 8–10-lobed, with valvate, imbricate, or convolute lobes, regular or rarely somewhat irregular and even bilabiate; stamens as many as and alternate

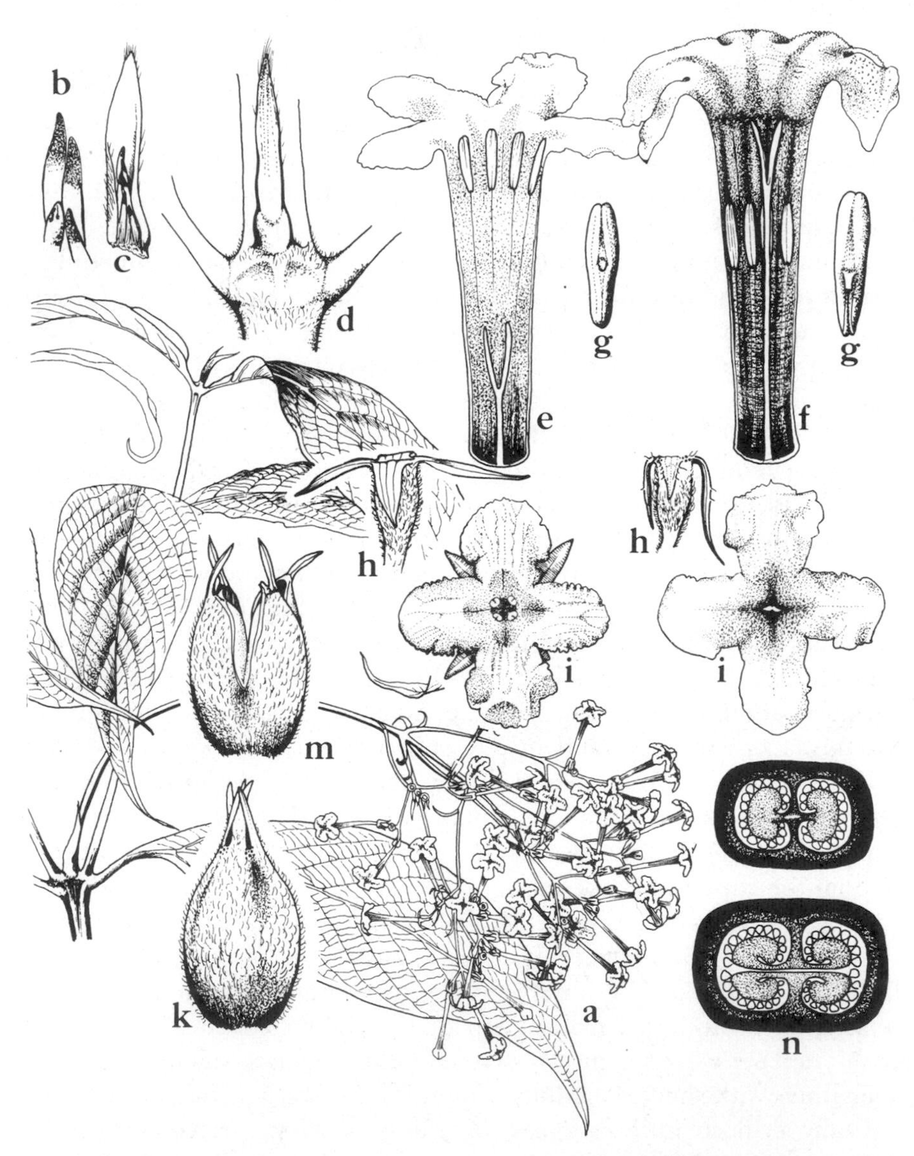

FIG. 6-31 Rubiaceae. *Rondeletia ovata* Rusby. a, habit, $\times\frac{1}{2}$; b, colleters, $\times 12$; c, ventral side of stipule, with colleters, $\times 3$; d, node, with interpetiolar stipule, $\times 3$, e, f, internal view of thrum and pin flowers, respectively, $\times 3$; g, anthers from thrum and pin flowers, $\times 6$; h, ovary and calyx of thrum (left) and pin (right) flowers, $\times 3$; i, view of thrum (left) and pin (right) flowers, from above, $\times 3$; k, maturing fruit, $\times 3$; m, opened fruit, $\times 3$; n, schematic cross-section of ovary of thrum (above) and pin (below) flowers, $\times 10$.

with the lobes of the corolla, attached within the corolla-tube or at its mouth; anthers tetrasporangiate and dithecal, opening by longitudinal slits; pollen-grains binucleate or trinucleate, of various triaperturate or triaperturate-derived types (most often tricolporate), seldom inaperturate; an epigynous nectary-disk commonly present; gynoecium of 2 (seldom 3–5 or even more) carpels united to form a compound, inferior (seldom only half-inferior, rarely essentially superior as in *Gaertnera* and *Pagamea*) ovary with axile (varying to essentially basal or essentially apical) placentas and as many locules as carpels, or seldom (as in *Gardenia*) unilocular and with parietal placentas; style slender, terminal, with a lobed or capitate, dry or sometimes wet stigma, or the styles distinct (as in *Galium*); ovules 1-many in each locule, anatropous to hemitropous, often with a funicular obturator, with a massive single integument, tenuinucellar (or the nucellus obsolete, as in *Houstonia*), without an integumentary tapetum; endosperm-development nuclear, or rarely cellular; a suspensor-haustorium often produced. FRUIT a capsule, berry, or drupe, or dry and indehiscent or schizocarpic, sometimes dicoccous (as in *Galium*); seeds with a dicotyledonous, spathulate or sometimes linear, straight or seldom somewhat curved embryo axially embedded in the well developed, oily endosperm (which may also have reserves of starch or hemicellulose), or sometimes the endosperm scanty or wanting (as in the Guettardoideae). X = 6–17, most commonly 11, less often 9. (Henriqueziaceae, Naucleaceae)

The family Rubiaceae includes about 450 genera and 6500 species. The vast majority of the species occur in tropical and subtropical regions, but the well known, cosmopolitan genus *Galium* (300) is common in North Temperate regions as well as elsewhere, and there are also some other extratropical groups. By far the largest genus is *Psychotria* (700). *Cinchona* is well known as the source of quinine and related alkaloids. *Coffea*, chiefly *C. arabica* L., is the source of coffee. *Rubia tinctorum* L., madder, is the traditional source of a red dye, alizarin, now prepared artificially.

Pollen considered to represent the Rubiaceae is known from Eocene and more recent deposits. Stipulate leaves that have been referred to the Rubiaceae are also known from Middle Eocene strata in southeastern United States. Even if these identifications are correct, however, the evolutionary radiation of the group appears to be largely post-Eocene.

The Rubiaceae are Asteridae with opposite leaves and interpetiolar stipules, or with whorled leaves and no stipules. The flowers have an inferior ovary and a regular corolla with isomerous stamens attached to the tube. These features distinguish the bulk of the Rubiaceae from all other families. The occasional exceptions to some of them need not becloud our recognition of the basic pattern. As additional characters it may be noted that the stipules commonly bear colleters, that the vast majority of the species have well developed endosperm, and that the endosperm nearly always follows the nuclear pattern of development.

The small genera *Henriquezia* and *Platycarpum*, which have fairly recently been segregated as a family Henriqueziaceae, are here retained in the Rubiaceae. At the present stage of knowledge, I do not believe that the admitted peculiarities (irregular corolla, no endosperm, no colleters, filaments bent at the base) of these genera require familial segregation, and I am not convinced by comparisons with families of the "Tubiflorae" as possible allies. It may be noted that *Gleasonia*, which Bremekamp keeps in the Rubiaceae, also lacks endosperm, and that there are some other resemblances among *Gleasonia*, *Henriquezia*, and *Platycarpum* which have caused these genera to be associated in the past.

The Rubiaceae form a connecting link between the Gentianales and the Dipsacales, and would be an aberrant element in either order. Each of these orders is relatively homogeneous and well defined without the Rubiaceae. The characters of the Rubiaceae, Dipsacales, and Gentianales have been discussed at some length by Wagenitz. He concludes that the Rubiaceae should be included in the Gentianales, and that the resemblance of the Rubiaceae to the Caprifoliaceae (Dipsacales) is due to convergence rather than close relationship. I see no need to deny the one relationship in affirming the other. In my opinion the Loganiaceae (Gentianales) stand near to the ancestry of the Rubiaceae, which in turn stand near to the ancestry of the Caprifoliaceae. The other families of the Dipsacales appear to be derived from the Caprifoliaceae.

The characters that link the Rubiaceae to the Gentianales as a whole or specifically to the Loganiaceae are: (1) nuclear endosperm; (2) well developed stipules; (3) specialized glandular appendages, called colleters, on the inner surface of the stipules; (4) the wall-structure of the pollen; and (5) the frequent presence of complex indole (tryptophane) alkaloids. There are also two genera, *Gaertnera* and *Pagamea*, which have usually been included in the Rubiaceae but have sometimes been referred to the Loganiaceae instead because of their superior ovary. The pollen of these two genera is typically rubiaceous, and differs from that of the Loganiaceae (Walter Lewis, personal communication).

The Rubiaceae differ sharply from the Gentianales in having an inferior ovary (with the exceptions noted above), and in lacking internal phloem. These characters tend to ally the Rubiaceae with the Dipsacales. Within the Dipsacales, the Rubiaceae are most nearly like the Caprifoliaceae, which as we have noted are the most archaic family of the order. Botanists have repeatedly expressed doubt that the Caprifoliaceae should be maintained as a family separate from the Rubiaceae.

The Caprifoliaceae and other Dipsacales differ from the Rubiaceae, however, in having a cellular pattern of endosperm-development, and in lacking colleters. The Dipsacales are sometimes said to lack stipules, but that is an overstatement. The Caprifoliaceae often have small stipules, and occasionally the stipules are well developed and conspicuous, as in *Viburnum ellipticum* Hook. Such stipules as there are in the Caprifoliaceae are usually petiolar (basally adnate to the petiole), however, rather than interpetiolar as in the Rubiaceae. I am not convinced by efforts to interpret all stipules in the Caprifoliaceae as mere pseudostipules, of different nature from true stipules. Such interpretations smack of preconceived conclusions that the Caprifoliaceae do not have stipules.

It should be noted that although the classical taxonomic characters have been fairly well observed, the more recondite characters such as structure of the pollen, arrangement of the vascular tissue, development of the endosperm, and distribution of the various secondary metabolites are still very inadequately known in the Rubiaceae and many other families. Examination of more species is likely to bring surprises. If it required specialized equipment and hours or days of time to determine the position of the ovary, we might well be unaware that the ovary in *Gaertnera* and *Pagamea* is superior, instead of inferior as in other members of the family.

According to how one weighs the evidence, the Rubiaceae could be referred to the Gentianales, or to the Dipsacales, or put into a separate order. If the family consisted of only a handful of species, it could easily be included in either order as a peripheral group, transitional to the other order. In actual fact, there are about as many species of Rubiaceae as of Gentianales and Dipsacales combined, and the Rubiaceae would dominate either order. It therefore seems useful to maintain the Rubiaceae in their own order.

A recent serological study by Lee & Fairbrothers (1978) gives a somewhat mixed bag of results, depending on which of the tested genera within a given family are emphasized. On the whole the results suggest a mutual interrelationship among the Rubiaceae, Dipsacales (especially Caprifoliaceae), Gentianales, and Cornales, to the exclusion of the Apiales, Magnoliales, and Caryophyllales. (The last two of these orders were represented by a single species each). I find the association of the Cornales with the several orders of Asteridae a bit difficult, because on morphological grounds the nearest common ancestor of these groups ought to be something in or allied to the Rosales. (No taxa of Rosales were included in the study.) Otherwise the results are wholly compatible with my position that the Rubiales form a link between the Gentianales and Dipsacales.

The recent association of the Theligonaceae with the Rubiaceae may complicate the phraseology of the argument about the proper position of the Rubiaceae, but it does not significantly affect the consideration of the evidence.

2. Family THELIGONACEAE Dumortier 1829 nom. conserv., the Theligonum Family

Annual or perennial herbs, with raphides, tanniferous and producing iridoid compounds; xylem forming a continuous cylinder, including radial rows of very numerous, small vessels, the stem otherwise lacking mechanical tissue; sieve-tubes with S-type plastids. LEAVES simple, entire, opposite below, alternate above by suppression of one member of each pair; stomates paracytic; petiole with 3 vascular bundles in an arc; stipules interpetiolar, bearing colleters on the inner side near the tip. FLOWERS anemophilous, unisexual, the staminate ones borne singly or paired opposite the leaves at the upper nodes, the pistillate ones mostly in simple, axillary dichasia at the lower nodes; staminate flowers with a small, inconspicuous, simple apparent perianth that is closed in bud but soon becomes cleft to the base into 2–5 recurved segments; stamens 6–30, sometimes basally united into groups of 2, 4, or 6, whence it is supposed that the seeming flower may be a pseudanthium of several naked flowers, and the seeming perianth an involucre; anthers tetrasporangiate and dithecal, opening by longitudinal slits; pollen-grains trinucleate, 4–8-zonoporate (unlike all investigated Rubiaceae); pistillate flowers epigynous, pseudomonomerous, with a basilateral style surrounded by a small, tubular perianth (interpreted as a corolla) with 2–4 valvate teeth; ovule solitary, basal in the single locule, campylotropous, without a funiculus, tenuinucellar, with a massive single integument, without an integumentary tapetum; endosperm-development nuclear; developing embryo without a suspensor-haustorium. FRUIT a small drupe; seed with a strongly curved, dicotyledonous embryo embedded in the copious, oily and proteinaceous endosperm. N = 10, 11.

The family Theligonaceae consists of the single genus *Theligonum* (*Cynocrambe*), with 3 species, occurring from temperate eastern Asia to the Mediterranean region and the Canary Islands.

The taxonomic affinity of *Theligonum*, long uncertain and controversial, was clarified by the recent careful study of Wunderlich (1971). Accepting her view of the relationships, I still find it useful to continue to treat the Theligonaceae as a distinct family. Wunderlich's conclusion was anticipated by S. S. Nenyukov in an unpublished manuscript, dated 1939, Systematic position of the family Theligonaceae (my translation from the Russian), which Academician Takhtajan has shown me in Leningrad. Nenyukov lost his life in the Second World War, and his manuscript has only recently come to light.

9. Order DIPSACALES Lindley 1833

Herbs, shrubs, or more or less woody vines, seldom (some Caprifoliaceae) small trees, producing iridoid substances and sometimes various sorts of alkaloids, and sometimes somewhat tanniferous, with or more often without proanthocyanins and ellagic acid; stem without internal phloem; vessel-segments with simple or less often scalariform perforations. LEAVES opposite or sometimes whorled (basal leaves and rhizome-scales in *Adoxa* alternate), simple or less often compound; stomates mostly anomocytic, seldom paracytic or anisocytic, stipules wanting except in a few Caprifoliaceae, and then lacking colleters. FLOWERS perfect or seldom unisexual, epigynous or (Adoxaceae) half-epigynous, borne in various sorts of mostly basically cymose or mixed inflorescences, sometimes in cymose heads; corolla sympetalous, regular to evidently irregular, sometimes bilabiate, usually 5-lobed, or 4-lobed by fusion of 2, the lobes imbricate or sometimes valvate; stamens attached well up in the corolla-tube or at the throat, as many as and alternate with the lobes, or fewer than the lobes, or (Adoxaceae) twice as many as the lobes and paired at the sinuses; pollen-grains trinucleate or seldom binucleate, mostly tricolporate; gynoecium of 2–5 carpels united to form a compound, inferior or (Adoxaceae) half-inferior ovary with as many locules as carpels, or the ovary with only one fully developed and functional locule, the others reduced or obsolete; style terminal, with a capitate or lobed stigma, or (Adoxaceae) the styles distinct; ovules in the Caprifoliaceae 1-many in each locule or in the one fertile locule, in the other families solitary and pendulous from near the apex in each fertile locule (Adoxaceae) or in the single fertile locule, anatropous and usually tenuinucellar, with a single massive integument and sometimes an integumentary tapetum; endosperm-development cellular, without haustoria. FRUIT of diverse types, dehiscent or indehiscent, dry or fleshy; seeds with a dicotyledonous embryo, with or without endosperm.

The order Dipsacales as here defined consists of 4 families and nearly a thousand species. The Caprifoliaceae (400), Valerianaceae (300) and Dipsacaceae (270) are all in the same size-range and make up the generally recognized core of the order. The monotypic family Adoxaceae is more controversial and has often been associated with the Saxifragaceae.

The Caprifoliaceae are obviously the most archaic family in the Dipsacales. The Valerianaceae appear to take their origin directly from the Caprifoliaceae. The Dipsacaceae are a little more removed, but evidently of the same ancestry. The Caprifoliaceae can be traced back to the middle Eocene, on the basis of the fossil pollen, and seeds and the Dipsacaceae dates from the middle Miocene.

The Dipsacales are usually considered to be allied to the Rubiales, from which they differ in their more often herbaceous habit, usually anomocytic stomates, usual absence of stipules and complete absence of colleters, often irregular corolla and often anisomerous stamens, and cellular endosperm-development. All of these features represent evolutionary advances over the condition in the Rubiaceae.

I take the absence of colleters in the Dipsacales to reflect the reduction of stipules in the group. In those few Caprifoliaceae that have well developed stipules the stipules may be secondarily enlarged. Indeed Weberling considers that the stipules of the Caprifoliaceae are not homologous with ordinary stipules and should be considered only as pseudostipules. Taking a more relaxed view of the bearing of phylogeny on morphological terminology, I find his position too extreme. Even so, his interpretation helps to reconcile the absence of colleters in stipulate Caprifoliaceae with a putative origin of the Caprifoliaceae from ancient, 5-carpellate Rubiaceae.

The principal fly in the ointment of a Rubialean ancestry for the Caprifoliaceae (and thus for the Dipsacales) is *Adoxa. Adoxa* differs from the vast majority of Asteridae in the presumably primitive feature of separate styles. Nonetheless it appears to be allied to the Caprifoliaceae, especially to *Sambucus.* I am not prepared to resolve this anomaly. Any attempt to divorce *Sambucus* from the Caprifoliaceae, or to use *Adoxa* to establish a direct link between the Caprifoliaceae and the Rosales, raises even more intractable problems.

The adaptive significance of most of the characters that distinguish the Dipsacales from other orders of Asteridae is doubtful. If the Dipsacales are indeed derived from the Rubiales, then they have given up the stipular colleters that may reasonably be supposed to help protect the terminal bud in the latter group. Within the Dipsacales there is a progression from woody to herbaceous habit, from regular to irregular corolla, from isomerous to anisomerous stamens, from several fertile locules to only one, and from endospermous to non-endospermous seeds. Each of these changes has also taken place in other orders of angiosperms, and there is nothing ecologically distinctive about the order or any of its families. The involucel or epicalyx of the Dipsacaceae is morphologically distinctive, but its functional significance, especially in the dense heads of most genera of the family, is obscure.

SELECTED REFERENCES

Bremekamp, C. E. B. 1939. On the position of the genera *Carlemannia* Benth. and *Sylviantus* Hook. f. Recueil Trav. Bot. Néerl. 36: 372.

Clarke, G. 1978. Pollen morphology and generic relationships in the Valerianaceae. Grana 17: 61–75.

Ehrendorfer, F. 1964. Über stammesgeschichtliche Differenzierungsmuster bei den Dipsacaceen. Ber. Deutsch. Bot. Ges. 77: (83)–(94).

Ehrendorfer, F. 1964. Evolution and karyotype differentiation in a family of flowering plants: Dipsacaceae. Genetics Today; Proc. XI Int. Cong. of Genetics, The Hague. 1963. pp. 399–407.

Ferguson, I. K. 1965. The genera of Valerianaceae and Dipsacaceae in the southeastern United States. J. Arnold Arbor. 46: 218–231.

Ferguson, I. K. 1966. The genera of Caprifoliaceae in the southeastern United States. J. Arnold Arbor. 47: 33–59.

Fukuoka, N. 1968. Phylogeny of the tribe Linnaeeae. Acta Phytotax. Geobot. 23: 82–94. (In Japanese, with English summary.)

Fukuoka, N. 1974. Floral morphology of *Adoxa moschatellina*. Acta Phytotax. Geobot. 26: 65–76.

Fukuoka, N. 1975. Studies in the systematics of Caprifoliaceae 2. Acta Phytotax. Geobot. 26: 133–139.

Hillebrand, G. R. 1966. Phytoserological systematic investigation of the genus *Viburnum*. Ph.D. Thesis. Rutgers University.

Hillebrand, G. R., & D. E. Fairbrothers. 1970. Serological investigation of the systematic position of the Caprifoliaceae. I. Correspondence with selected Rubiaceae and Cornaceae. Amer. J. Bot. 57: 810–815.

Hillebrand, G. R., & D. E. Fairbrothers. 1970. Phytoserological systematic survey of the Caprifoliaceae. Brittonia 22: 125–133.

Katina, Z. F. 1953. Anatomical data on the localization of essential oil in some species of Valerian. (In Ukrainian; Russian summary.) Bot. Zhur. Kiev 10: 81–86.

Moissl, E. 1941. Vergleichende embryologische Studien über die Familie der Caprifoliaceae. Oesterr. Bot. Z. 90: 153–212.

Patel, V. C., & J. J. Skvarla. 1979. Valerianaceae pollen morphology. Pollen & Spores 21: 81–104.

Persidsky, D. 1939. Gynoeceum evolution in the family Caprifoliaceae. (In Ukrainian; summaries in Russian and English.) J. Inst. Bot. Acad. Sci. Rss. Ukr. 21–22 (29–30): 45–75.

Shaw, H. K. Airy. 1965. On a new species of the genus *Silvianthus* Hook. f., and on the family Carlemanniaceae. Kew Bull. 19: 507–512.

Sprague, T. A. 1927. The morphology and taxonomic position of the Adoxaceae. J. Linn. Soc., Bot. 47: 471–487.

Troll, W., & F. Weberling. 1966. Die Infloreszenzen der Caprifoliaceen und ihre systematische Bedeutung. Akad. Wiss. Abh. Math.-Naturwiss. Kl. 1966. Nr. 4: 459–605.

Vijayaraghavan, M. R., & G. S. Sarveshwari. 1968. Embryology and systematic position of *Morina longifolia* Wall. Bot. Not. 121: 383–402.

Weberling, F. 1961. Die Infloreszenzen der Valerianaceen und ihre systematische Bedeutung. Akad. Wiss. Abh. Math.-Naturwiss. Kl., Nr. 5: 151–281.

Weberling, F. 1977. Vergleichende und entwicklungsgeschichtliche Untersuchungen über die Haarformen der Dipsacales. Beitr. Biol. Pflanzen 53: 61–89.

Wilkinson, A. M. 1948. Floral anatomy and morphology of some species of the tribe Lonicerae of the Caprifoliaceae. Amer. J. Bot. 35: 261–271.
Wilkinson, A. M. 1948. Floral anatomy and morphology of some species of the tribes Linnaeeae and Sambuceae of the Caprifoliaceae. Amer. J. Bot. 35: 365–371.
Wilkinson, A. M. 1948. Floral anatomy and morphology of the genus *Viburnum* of the Caprifoliaceae. Amer. J. Bot. 35: 455–465.
Wilkinson, A. M. 1949. Floral anatomy and morphology of *Triosteum* and of the Caprifoliaceae in general. Amer. J. Bot. 36: 481–489.

Артюшенко, З, Т. 1948. О морфологической сущности нижней завязи некоторых представителей сем. Caprifoliaceae. Бот. Ж. 33: 202-212.
Артюшенко, З. Т. 1951. Развитие цветка и плода жимолостных. Труды Бот. Инст. АН СССР, сер. VII, 2: 131-169.
Винокурова, Л. В. 1959. Палинологические данные к систематике семейств Dispsacaceae и Morinaceae. Пробл. Бот. 4: 51-67.
Камелина, О. П., & М. С. Яковлев. 1974. Развитие зародышевого мешка в роде *Morina* L. Бот. Ж. 59: 1609-1617.
Камелина, О. П., & М. С. Яковлев. 1976. Развитие пыльника и микрогаметогенез у представителей сем. Dipsacaceae и Morinaceae. Бот. Ж. 61: 932-945.

SYNOPTICAL ARRANGEMENT OF THE FAMILIES OF DIPSACALES

1 Plants mostly woody, seldom herbaceous; staments mostly as many as the corolla-lobes, seldom fewer; ovules often more than 1 per locule; seeds with well developed endosperm 1. CAPRIFOLIACEAE.

1 Plants herbaceous, or rarely shrubby; stamens seldom (except some Dipsacaceae) of the same number as the corolla-lobes; ovules not more than 1 per locule.

 2 Stamens 8 or 10, twice as many as the corolla-lobes, paired at the sinuses of the corolla, each with only a single pollen-sac; fruit a small, dry drupe with several stones; endosperm copious ....2. ADOXACEAE.

 2 Stamens 1–4, as many as or usually fewer than the corolla-lobes, each with 2 pollen-sacs; fruit dry, 1-seeded; endosperm mostly scanty or none, more copious in some Dipsacaceae.

  3 Flowers without an epicalyx or involucel (except in *Triplostegia*), borne in various sorts of inflorescences, but not in involucrate heads, ovary basically tricarpellate, often more or less evidently trilocular, with 1 fertile and 2 sterile locules, or pseudomonomerous, with a single locule 3. VALERIANACEAE.

  3 Flowers nearly always individually enclosed or subtended by a more or less cupulate epicalyx or involucel, mostly borne in compact, involucrate, basically cymose heads (in axillary verticillasters in *Morina*); ovary basically bicarpellate, but pseudomonomerous and strictly unilocular ....4. DIPSACACEAE.

1. Family CAPRIFOLIACEAE A. L. de Jussieu 1789 nom. conserv. the Honeysuckle Family

Shrubs or more or less woody vines, or sometimes small trees, seldom herbs, with diverse kinds of trichomes (very often glandular hairs), often accumulating one or another sort of phenolic heteroside (especially glycosides), commonly producing iridoid compounds, often saponiferous, and often more or less tanniferous, frequently with proanthocyanins and sometimes with ellagic acid, but only rarely cyanogenic and rarely producing alkaloids; crystals of calcium oxalate, of varied form, often present in some of the cells of the parenchymatous tissues; scattered secretory cells often present; nodes mostly trilacunar, but sometimes unilacunar or pentalacunar; vessel-segments with simple or more often scalariform perforations, sometimes with numerous cross-bars; imperforate tracheary elements with bordered or seldom simple pits; wood-rays heterocellular, mixed uniseriate and pluriseriate, the latter up to about 5 cells wide; wood-parenchyma commonly diffuse, sometimes scanty-paratracheal. LEAVES opposite, simple (pinnately compound in *Sambucus*); stomates anomocytic, or sometimes some of them paracytic; petiole with diverse sorts of vascular anatomy; stipules mostly wanting or vestigial, when present usually small and adnate (at least basally) to the petiole, in species of *Sambucus* and *Viburnum* taking the form of extrafloral nectaries at the base of the petiole, in any case lacking colleters; winter-buds commonly with well developed bud-scales (in contrast to the Rubiaceae), but sometimes naked. FLOWERS in various sorts of mostly cymose or mixed inflorescences, commonly bracteolate, perfect, epigynous, with a distinct constriction just beneath the calyx-limb, mostly (4) 5-merous as to the calyx, corolla, and androecium (the calyx-lobes seldom much-reduced); calyx commonly small, its lobes or teeth imbricate or open in bud, more or less accrescent in fruit; corolla sympetalous, with imbricate or sometimes valvate lobes, regular or often more or less irregular, sometimes bilabiate, the lower part of the corolla-tube often nectariferous, sometimes gibbous or spurred (and nectariferous) at the base; stamens attached to the corolla-tube alternate with the lobes, or sometimes only 4 (*Linnaea*) or 2 (*Carlemannia, Silvianthus*) even when the perianth is pentamerous; anthers dorsifixed and versatile, tetrasporangiate and dithecal, opening by longitudinal slits; pollen-grains trinucleate, commonly tricolporate or triporate; gynoecium of 2–5 (–8) carpels united to form a compound, inferior or seldom only half-inferior ovary with typically as many locules as carpels and with

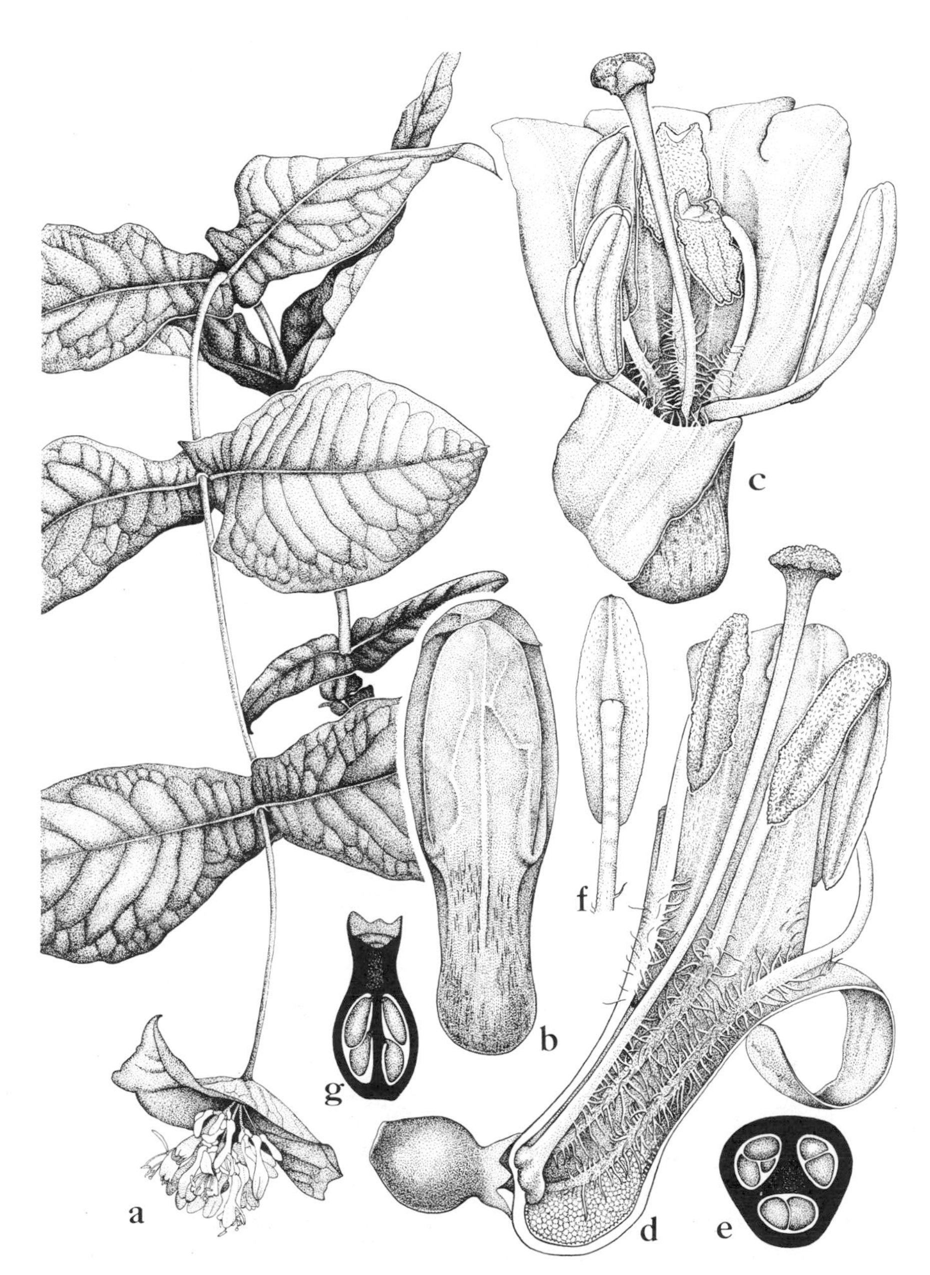

FIG. 6-32 Caprifoliaceae. *Lonicera dioica* L. a, habit, ×1; b, flower, in late bud, from above, ×6; c, flower, ×6; d, flower in partial long-section, ×6; e, schematic cross-section of ovary, ×12; f, anther, ×6; g, schematic long-section of ovary, ×6.

axile placentation, or the partitions sometimes failing to meet in the upper part of the ovary, sometimes only one locule fertile, the other locules with abortive ovules or empty and more or less strongly reduced, in spp. of *Viburnum* the gynoecium pseudomonomerous; a nectary sometimes present atop the ovary; style terminal, with a wet, capitate or lobed stigma, or the stigma(s) subsessile; ovules 1-many in each locule, pendulous, anatropous, with a massive single integument, tenuinucellar (crassinucellar in *Viburnum*); endosperm-development cellular or reputedly sometimes nuclear. FRUIT a capsule, or a berry, or a drupe with as many seedlike stones as fertile locules, or dry and indehiscent; seeds commonly with a small, basal to linear and axile, straight, dicotyledonous embryo and oily, fleshy endosperm. X = 8–12+, most often 8 or 9. (Carlemanniaceae, Sambucaceae)

The family Caprifoliaceae as here defined consists of about 15 genera and 400 species, mostly of North Temperate and boreal regions, and of tropical mountains. Three-fourths of the species belong to the two large genera *Lonicera* and *Viburnum*, with about 150 species each. *Abelia*, *Linnaea*, *Sambucus* (elderberry), *Symphoricarpos* (snowberry, coralberry) and *Weigela* are some other familiar genera. The genera *Alseuosmia*, *Memecylanthus*, and *Periomphale* are here excluded from the Caprifoliaceae and recognized as a family Alseuosmiaceae in the order Rosales.

Sambucus and *Viburnum* stand somewhat apart from the rest of the Caprifoliaceae and from each other, and have been the subject of much comment and phylogenetic speculation. *Sambucus* is the more distinctive of the two, but no other family presents itself as a likely close relative. Both the Hydrangeaceae and the Cornaceae have been suggested as possible ancestors for *Viburnum*, and the similarity of some species of *Hydrangea* to species of *Viburnum* in both vegetative and floral aspect is striking. More detailed comparisons reveal so many differences that any close relationship seems most unlikely. *Hydrangea* differs from *Viburnum* in having raphide-sacs, and in its separate petals, diplostemonous flowers, binucleate pollen, bilocular ovary, separate styles, capsular fruits with numerous seeds and other features. *Cornus* differs from *Viburnum* in its tetramerous flowers, separate petals, binucleate pollen, bilocular ovary with an ovule in each locule, bilocular endocarp, and other features. There is no more reason to remove *Viburnum* from the Caprifoliaceae and associate it with *Hydrangea* or *Cornus* than to associate a part of *Euphorbia* with *Cereus* (Cactaceae).

Serological studies by Hillebrand (1966), reported in part, with additions, by Hillebrand and Fairbrothers in 1970, suggest a close relationship of the Caprifoliaceae to the Cornaceae, and a much less close relationship to the Rubiaceae. Within the Caprifoliaceae, the several genera tested form an interlocking group, with *Viburnum* being the most distinctive, followed by *Sambucus.* Thus these serological results are reasonably compatible with widely held views about the relationship among the genera of Caprifoliaceae, but at variance with the usual view of the position of the family in the system.

If Hillebrand's results were taken at face value, the Caprifoliaceae would have to be associated with the Cornales. Here we face the problem that the Asteridae appear to be a highly natural group, in which the Caprifoliaceae are not archaic. Even if we try to extricate the Caprifoliaceae or the Dipsacales as a whole from the rest of the Asteridae, we must still seek an ancestor with capsular fruits and numerous seeds. Such an ancestor would be more archaic than anything in the Cornales, and the two groups could scarcely have a common ancestor short of the Rosales. Inasmuch as the Asteridae as a whole are here considered to be of Rosalean origin, we could thus eventually effect a partial reconciliation of Hillebrand's results with my own views on relationships among the dicotyledons.

A more recent study by Lee and Fairbrothers, discussed and cited here under the Rubiales, indicates mutual interrelationships among the Caprifoliaceae, Rubiaceae, and some families of the Gentianales and Cornales. In this study the serological results take on a somewhat different flavor, at least, and do not argue against a reasonably close relationship of the Caprifoliaceae to the Rubiaceae.

Admitting as always the possibility of error, I am reluctant to place decisive weight on the serological evidence, especially inasmuch as we are still wholly unaware of what seeming affinities might be uncovered by more extensive application of the same methods. How would members of the Caprifoliaceae react with members of the Asteraceae, or Hydrangeaceae, or Grossulariaceae, or Ranunculaceae? We simply do not know. Strong serological reactions between otherwise very different groups are so frequent that Moritz and his associates sought to devise methods to deal with the problems created by these "antisystematic reactions" (a term rejected by the current generation of serologists). The wide differences in reactions by different species in the same genus also indicate a need for caution. Thus on a scale of 100 for reaction with itself, *Cornus florida* gives an 84 percent reaction with *C. kousa*, 58 percent with *C. stolonifera*, and only 20 percent with *C. racemosa.* I understand from the late Marion Johnson that a rabbit can be so strongly sensitized that its blood-serum will react to almost anything. At the higher levels of sensitivity, which are required to show more distant relationships, it would seem that the possibility of an "antisystematic reaction" might be substantially increased. Serological data, like other taxonomic data, need to be evaluated in the context of the rest of the information about a group.

2. Family ADOXACEAE R. E. Trautvetter 1853 nom. conserv., the Moschatel Family

Delicate perennial herb with a musky odor, arising from a short, scaly rhizome, with 1-several alternate radical leaves and a single pair of opposite cauline leaves; plants producing iridoid compounds, but not cyanogenic, poor in polyphenols, without ellagic acid and without crystals; rhizome with a dorsal and a ventral flattened collateral vascular bundle, separated by a pith; erect stems with 2 well developed vascular bundles, each supplying a leaf, and 2 smaller bundles leading to the inflorescence. LEAVES trifoliolate, with the leaflets again cleft, bearing multicellular, tanniferous secretory hairs when young; stomates anomycytic; stipules wanting; scales of the rhizome alternate. FLOWERS mostly 5 in a compact, headlike terminal cyme, perfect, half-epigynous, with a shortly prolonged hypanthium, sympetalous; sepals of the central flower commonly 2, of the lateral ones 3; corolla rotate, that of the central flower regular and commonly 4-lobed, that of the lateral ones slightly irregular and commonly 5-lobed, the lobes imbricate, each with a nectary at the base of the upper side, the nectaries composed of cushion-like groups of multicellular, sessile or short-stalked, clavate glands; stamens attached to the corolla-tube, paired at the sinuses, each pair evidently reflecting the division of an original stamen into two halves; anthers bisporangiate and unithecal, opening by a longitudinal slit; pollen-grains trinucleate, tricolporate, somewhat similar to those of *Sambucus*; gynoecium of (2) 3–5 carpels (most often 5 carpels in the lateral flowers and 4 in the central one) united to form a compound, half-inferior ovary with distinct styles and capitate, dry, papillate stigmas; locules as many as the carpels; ovules solitary in each locule, apical-axile, pendulous, anatropous, tenuinucellar, with a massive single integument and an integumentary tapetum; endosperm-development cellular. FRUIT a small, dry drupe with distinct stones; seeds with a small, dicotyledonous embryo and copious, oily endosperm. X = 9.

The family Adoxaceae consists of the single circumboreal species

Adoxa moschatellina L. The affinities of *Adoxa* have been debated, some authors favoring an alliance with the Saxifragaceae, others an alliance with the Caprifoliaceae through *Sambucus*. Features of classical morphology might permit either interpretation, with the sympetalous corolla favoring the Caprifoliaceae, and the distinct styles favoring the Saxifragaceae. In my opinion the embryological, palynological, and chemical evidence strongly supports the traditional position near the Caprifoliaceae. [Two new monotypic genera have recently been described from China.]

3. Family VALERIANACEAE Batsch 1802 nom. conserv., the Valerian Family

Herbs, or rarely shrubs, often with multicellular, clavate, glandular hairs, with a characteristic rank odor caused by the monoterpenoid and sesquiterpenoid ethereal oils borne in idioblastic cells especially in the cortex and cork of the rhizomes and roots, commonly also producing iridoid compounds and sometimes monoterpene alkaloids and triterpenoid saponins, but not cyanogenic and not tanniferous, lacking both ellagic acid and proanthocyanins; crystals recorded only in *Patrinia*; nodes trilacunar; vascular bundles of the stem at first distinct, but soon becoming connected by prosenchymatous elements and a narrow zone of phloem; vessel-segments very small and short, with simple perforations; imperforate tracheary elements with simple or narrowly bordered pits; wood-rays nearly or quite homocellular, mostly 4–6 cells wide, sometimes wanting; wood-parenchyma wanting. LEAVES opposite (sometimes all basal), simple and entire to pinnatifid or pinnately compound; stomates anomocytic; petiole with 3 vascular bundles near the base, these distally confluent into an arc; stipules wanting. FLOWERS in a compound dichasium, or a monochasium, or a thryse, or in other sorts of open or congested cymose inflorescences, but not in an involucrate head, bracteate and often also bracteolate, but not with an epicalyx except in *Triplostegia*, perfect or sometimes unisexual, epigynous; calyx in *Nardostachys* of 5 fairly well developed segments, in the other genera variously reduced or modified or obsolete, sometimes represented by inconspicuous teeth, often (as in *Valeriana*) of up to 20 segments that are inrolled at anthesis to form a ring around the base of the corolla, unrolling and expanding in fruit, becoming setaceous, plumose, and pappus-like; corolla sympetalous, from essentially regular to more often more or less strongly irregular and sometimes bilabiate, the tube often gibbous or spurred (and nectariferous) near the base; corolla-lobes (3–) 5, imbricate; stamens fewer than the corolla-lobes, sometimes 4 (*Nar-*

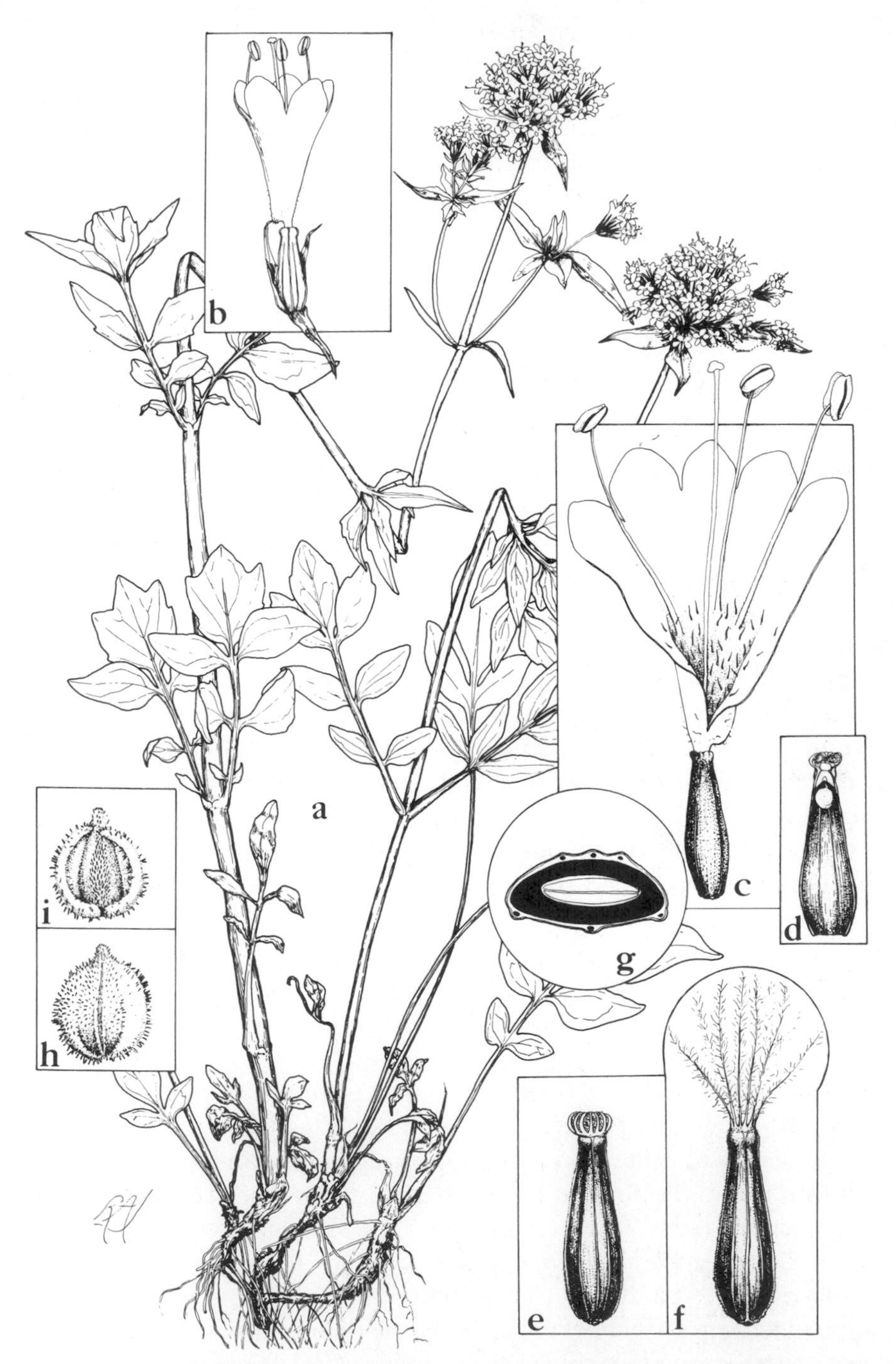

Fig. 6-33 Valerianaceae. a–g, *Valeriana sitchensis* Bong. a, habit, $\times\frac{1}{2}$; b, flower, $\times 3$; c, flower, with the corolla opened to show internal structure, $\times 6$; d, ovary in long-section, $\times 12$; e, submature fruit, the calyx not yet expanded, $\times 7$; f, mature fruit, with pappus-like calyx, $\times 7$; g, schematic cross-section of fruit, $\times 20$; h, i, *Plectritis macrocera* T. & G. two views of fruit.

dostachys, *Triplostegia*, and spp. of *Patrinia*), more often 3, sometimes only 1 or 2, attached to the corolla-tube, aligned with the sinuses; anthers versatile, tetrasporangiate and dithecal, opening by longitudinal slits; pollen-grains trinucleate or reputedly sometimes binucleate, 3 (4–)-colporate, some genera with simple columellae, others with elongate, branched columellae at the poles, each aperture surrounded by a "halo" from which the columellae are missing; gynoecium basically of 3 carpels united to form a compound, inferior ovary with a terminal style and entire or lobed stigma, but two of the carpels (and their locules) more or less strongly reduced and empty, or even obsolete, the single fertile locule with a single pendulous apical ovule, this anatropous, tenuinucellar, with a massive single integument and an integumentary tapetum enclosing the embryo-sac; endosperm-development cellular. FRUIT dry, one-seeded, achene-like, sometimes winged, often crowned by an accrescent, plumose calyx; seeds with a large, straight, oily, dicotyledonous embryo, and without endosperm. X = 7–12, most commonly 9. (Triplostegiaceae)

The family Valerianaceae consists of about 13 genera and 300 species, nearly cosmopolitan in distribution, but best-developed in North Temperate regions and in the Andes. About two-thirds of the species belong to the single large genus *Valeriana*, and half (or more) of the remainder belong to *Valerianella*.

It is generally agreed that the Valerianaceae are allied on the one hand to the Caprifoliaceae, and on the other to the Dipsacaceae, being more advanced than the former group and more archaic than the latter. *Nardostachys* (2 spp. of the Himalayan region), with 4 stamens and a well developed calyx, is regarded as the closest approach to the Caprifoliaceae. *Triplostegia* (2 spp. of se. Asia and Malaysia) approaches the Dipsacaceae, and has sometimes been referred to that family. It has a thyrsoid inflorescence, 4 stamens, a minute calyx, 3 carpels (2 much-reduced), and a double epicalyx, the outer of 4 conspicuous, persistent, basally connate members, the inner 8-ribbed and urceolate.

4. Family DIPSACACEAE A. L. de Jussieu 1789 nom. conserv., the Teasel Family

Herbs, or rarely shrubs, often with multicellular, clavate, glandular hairs, or capitate-glandular (*Knautia*) or tufted hairs (*Cephalaria*), commonly producing iridoid compounds and sometimes also monoterpenoid alkaloids or α-naphthaphenanthridine alkaloids, but not cyanogenic,

FIG. 6-34 Dipsacaceae. *Scabiosa graminifolia* L. a, habit, $\times\frac{1}{2}$; b, front view of flower from central part of head, $\times 4$; c, front view of marginal flower, $\times 4$; d, schematic long-section of ovary, $\times 8$; e, side view of flower from central part of head, $\times 4$; f, bract, $\times 4$; g, flower from central part of head, in partial long-section, $\times 4$; h, fruit, $\times 4$.

only seldom saponiferous, and not tanniferous, lacking both ellagic acid and proanthocyanins; crystals of calcium oxalate often present in some of the cells of the parenchymatous tissues, usually clustered; nodes trilacunar; vascular bundles at first distinct, but soon becoming connected by phloem and prosenchymatous elements so as to form a continuous ring; vessel-segments with simple perforations, or sometimes some of them scalariform; medullary rays mostly 1–2 cells wide. LEAVES opposite or sometimes whorled, simple and entire or toothed to deeply pinnatifid or pinnately dissected, sometimes prickly on the margins;

stomates mostly anomocytic, seldom anisocytic; petiole with several vascular bundles arranged in an arc; stipules wanting. FLOWERS characteristically grouped into dense, cymose, involucrate heads with a conspicuously bracteate to merely hairy or naked receptacle (in *Cephalaria* the heads often with a mixed or even virtually racemose sequence), but in *Morina* borne in axillary verticillasters, in any case perfect and epigynous, and nearly always individually subtended by a more or less cupulate, apically toothed or subentire epicalyx or involucel (possibly representing a pair of bracteoles) that may be adnate to the ovary toward the base; epicalyx wanting in a few spp. of *Cephalaria* and *Succisa*; calyx mostly small, cupulate or more or less deeply cut into 4 or 5 segments or up to 10 teeth or pappus-like bristles, or seldom obsolete; corolla sympetalous, slightly to evidently irregular, 5-lobed, or 4-lobed by fusion of 2, the lobes imbricate; stamens 4, or seldom only 2 or 3, attached toward the summit of the corolla-tube, alternate with the lobes; anthers tetrasporangiate and dithecal, opening by longitudinal slits; pollen-grains trinucleate (binucleate in *Morina*), most often tricolporate, but sometimes triporate (as in *Morina*) or tetracolporate, sometimes with a halo rather similar to that of the Valerianaceae; gynoecium basically of 2 carpels united to form a compound ovary, but one carpel obsolete, so that the ovary is unilocular and pseudomonomerous; style terminal, slender, with a dry, nonpapillate, entire or 2-lobed stigma; a nectary borne around the style on the summit of the ovary; ovule solitary, pendulous from near the top of the locule, anatropous, tenuinucellar, with a massive single integument and an integumentary tapetum; endosperm-development cellular. FRUIT an achene, enclosed (except at the top) by the epicalyx and commonly crowned by the persistent calyx; seeds with a large, straight, spathulate, dicotyledonous embryo and rather scanty, fleshy, oily endosperm. X = 5–10, perhaps originally 9, soon becoming 10, and thereafter variously reduced; 17 in *Morina*. (Morinaceae)

The family Dipsacaceae as here defined consists of 10 genera and about 270 species, native to Eurasia and Africa, especially in and directly east of the Mediterranean region. About three-fourths of the species belong to only 3 genera, *Scabiosa* (80), *Cephalaria* (65), and *Knautia* (60). The most familiar species in North America is *Dipsacus sylvestris* Hudson, which has become a widespread weed.

Morina (ca 17 Eurasian spp.) stands somewhat apart from the other genera, and has sometimes been treated as a distinct family. The following comparison may be made:

Morina	*typical Dipsacaceae*
Flowers in axillary verticillasters	Flowers in heads
Functional stamens 2	Functional stamens 4
Pollen-grains binucleate, triporate, the wall composed of 3 layers	Pollen-grains trinucleate, most often tricolporate, the wall composed of 2 layers
Fibrous thickenings present in all layers of the anther and connective	Fibrous thickenings in the anthers present only in the endothecium, or sometimes also in the epidermis
Sperms with a little cytoplasm	Sperms naked
Alkaloids wanting	Alkaloids often present
X = 17	X = 5–10

In spite of these differences, the relationship of *Morina* to the Dipsacaceae is not in dispute. It is only the rank of the group that is debated. This is one of the few places where Thorne splits and Hutchinson and I lump.

Botanists are agreed that the Dipsacaceae are allied to the Valerianaceae. The Dipsacaceae are the more advanced of the two families in the structure of the inflorescence and gynoecium and in the development of an epicalyx, but the Valerianaceae are the more advanced in lacking endosperm. Therefore, as Takhtajan has pointed out, the relationship between the two families is probably fraternal rather than paternal-filial.

10. Order CALYCERALES Takhtajan 1966

The order consists of the single family Calyceraceae.

SELECTED REFERENCES

Skvarla, J. J., B. L. Turner, V. C. Patel, & A. S. Tomb. 1977 (1978). Pollen morphology in the Compositae and in morphologically related families. *In*: V. H. Heywood, J. B. Harborne, & B. L. Turner, eds., The biology and chemistry of the Compositae, pp. 141–248. Academic Press. London, New York. & San Francisco.

Аветисян, Е. М. 1980. Палиноморфология семейства Calyceraceae. В кн.: Систематика и зволюция высших растений: 57-64. Наука, Ленигpад.

1. Family CALYCERACEAE L. C. Richard 1820 nom. conserv., the Calycera Family

Herbs, storing carbohydrate as inulin, commonly producing iridoid compounds, but not tanniferous, lacking proanthocyanins and presumably also ellagic acid, and also without latex or secretory cavities; clustered crystals of calcium oxalate sometimes present in some of the cells of the parenchymatous tissues; vascular bundles separated by broad medullary rays, or embedded in a sheath of mechanical tissue; vessel-segments with simple perforations; imperforate tracheary elements with simple pits. Leaves alternate, simple and entire or pinnately lobed; stomates anomocytic; stipules wanting. FLOWERS borne in involucrate, centripetally flowering heads, epigynous, perfect or seldom functionally unisexual, with a specialized pollen-presentation mechanism, the anthers connate into a tube, the pollen released into the interior of the anther-tube and pushed out by the growth of the style; calyx with (4) 5 (6) small lobes or teeth; corolla sympetalous, regular or somewhat irregular, with (4) 5 (6) valvate lobes; stamens as many as and alternate with the corolla-lobes, attached near the summit of the tube; filaments more or less connate, at least toward the base; anthers tetrasporangiate and dithecal, introrse, opening by longitudinal slits; pollen-grains binucleate, tricolporate; gynoecium basically of 2 carpels, these united to form a compound, inferior, unilocular, pseudomonomerous ovary; style terminal, growing up through the anther-tube and pushing out the pollen; stigma solitary, capitate; ovule solitary, pendulous from the top of the locule, anatropous, tenuinucellar, with a massive single integument; endosperm-development cellular, without haustoria. FRUIT an achene, crowned by the persistent calyx; seed with a straight, dicotyledonous embryo and oily, abundant or scanty endosperm. X = 8, 15, 18, 21.

The family Calyceraceae consists of 6 genera and perhaps 60 species, native to Central and South America. The largest genera are *Calycera* and *Boöpis,* with about 15 species each. None of the species is economically important or familiar to botanists generally. The family has attracted botanical attention mainly because of its role in speculation about the ancestry and relationships of the Asteraceae.

The Calyceraceae have variously been included in the Campanulales, the Dipsacales, or a separate order Calycerales. They resemble the Campanulales, and differ from the Dipsacales, in their alternate leaves, in producing inulin, and in their specialized pollen-presentation mechanism. (A somewhat similar pollen-presentation mechanism also exists in the subfamily Ixoroideae of the Rubiaceae, and a very similar one in the Asteraceae). They resemble the Dipsacales, and differ from characteristic members of the Campanulales, in having the filaments attached near the summit of the corolla-tube, and in a series of embryological features that have been commented on by Poddubnaja-Arnoldi (p. 376 of 1964, general citations, in Russian). The unilocular ovary with a single apical, pendulous ovule is fully in harmony with the Dipsacales, but would be unique in the Campanulales.

The involucrate, centripetally flowering heads and the pollen-presentation mechanism of the Calyceraceae strongly suggest the Asteraceae, and the pollen is also very much like that of many Asteraceae in the structure and ornamentation of the wall (Skvarla et al., 1977). On the other hand, the two families differ in ways which suggest that the relationship between them is only collateral. The solitary ovule is apical in the Calyceraceae, and basal in the Asteraceae. Conceivably the ovule might have moved (phyletically), from the apex to the base of the ovary, or vice versa, after the uniovulate condition had been achieved, but parallel reductions from an ancestor with a bilocular, pluriovulate ovary seem more likely. The Asteraceae appear to be primitively woody and opposite-leaved. Herbaceous Asteraceae with alternate leaves, comparable in habit to the Calyceraceae, are advanced within their family in these respects. On the other hand, the Calyceraceae are more primitive than the Asteraceae in their binucleate pollen and fairly well developed endosperm. The two families have very different chemical defenses. The Calyceraceae produce iridoid compounds but do not have an evident secretory system, and they apparently lack the polyacetylenes and sesquiterpene lactones that are so common in the Asteraceae. The Asteraceae, in contrast, do not produce iridoid compounds, and they have a well developed secretory system (either resiniferous or laticiferous). Thus it seems likely that the admitted similarities between the Calyceraceae and Asteraceae reflect parallelism from a fairly remote common ancestry.

The difficulties in placing the Calyceraceae properly in the system provide another example of the pervasive parallelism that besets our efforts to decipher relationships among the angiosperms. The whole subclass Asteridae forms a closely knit group, which resists ill-conceived efforts to dismember it on the basis of a limited set of data.

In my view, the Calyceraceae are related to and probably derived from the Dipsacales. Their nearest common ancestry with the Asteraceae probably lies in or near the Rubiales. In this connection we should note that Poddubnaja-Arnoldi (cited above) considers that inclusion of the Dipsacaceae and Calyceraceae in the Rubiales would be in accord with the embryological data, and that from an embryological point of view the Calyceraceae, Dipsacaceae, and Asteraceae are closely related. In 1968 I included the Calyceraceae in the Dipsacales as a somewhat aberrant family. Without any change in my views on the relationships of the group, I now find it more useful to follow Takhtajan in treating the Calyceraceae as a monotypic order.

The very modest evolutionary success of the Calyceraceae, as compared to the tremendous success of the Asteraceae, is a legitimate subject for inquiry and speculation. I suggest that the explanation lies not in the morphology but in the chemistry. According to the principles of chemical evolution that I have expounded elsewhere (1977, in general citations), the chemical defenses of the Asteraceae probably evolved after the iridoids found in many other members of the subclass had begun to lose their effectiveness. (See further discussion under Asteraceae.) The Calyceraceae are presumably latecomers to the evolutionary scene. At least they probably do not antedate the Asteraceae. Their floral and vegetative morphology should be just as effective as that of herbaceous Asteraceae, but with a limited and outmoded chemical arsenal they do not have a strong competitive position.

11. Order ASTERALES Lindley 1833

The order consists of the single family Asteraceae.

SELECTED REFERENCES

Baagøe, J. 1978. Taxonomical application of ligule microcharacters in Compositae. Bot. Tidsskr. 72: 125–147.

Bolick, M. R., 1978. Taxonomic, evolutionary, and functional considerations of Compositae pollen ultrastructure and sculpture. Pl. Syst. Evol. 130: 209–218.

Boulter, D., J. T. Gleaves, B. G. Haslett, D. Peacock, & U. Jensen. 1978. The relationships of 8 tribes of the Compositae as suggested by plastocyanin amino acid sequence data. Phytochemistry 17: 1585–1589.

Burnett, W. C., S. B. Jones, & T. J. Mabry. 1977. Evolutionary implications of herbivory on *Vernonia* (Compositae). Pl. Syst. Evol. 128: 277–286.

Burtt, B. L. 1961. Compositae and the study of functional evolution. Trans. & Proc. Bot. Soc. Edinburgh 39: 216–232.

Carlquist, S. 1966. Wood anatomy of Compositae: a summary, with comments on factors controlling wood evolution. Aliso 6: 25–44.

Carlquist, S. 1976. Tribal interrelationships and phylogeny of the Asteraceae. Aliso 8: 465–492.

Crepet, W. L., & T. F. Steussy. 1978. A reinvestigation of the fossil *Viguera cronquistii* (Compositae). Brittonia 30: 483–491.

Crété, R. 1956. Contribution à l'étude de l'albumen et de l'embryon chez les Campanulacées et les Lobéliacées. Bull. Soc. Bot. France 103: 446–454.

Cronquist, A. 1955. Phylogeny and taxonomy of the Compositae. Amer. Midl. Naturalist 53: 478–511.

Cronquist, A. 1977. The Compositae revisited. Brittonia 29: 137–153.

Harling, G. 1950, 1951. Embryological studies in the Compositae. I. Anthemideae—Anthemidinae. II. Anthemideae—Chrysantheminae. III. Astereae. Acta Horti Berg. 15: 135–168, 1950. 16: 1–56; 73–120, 1951.

Heywood, V. H., J. B. Harborne, & B. L. Turner, eds. 1977 (1978). The biology and chemistry of the Compositae. Two volumes. Academic Press. London, New York, San Francisco.

Manilal, K. S. 1971. Vascularization of corolla of the Compositae. J. Indian Bot. Soc. 50: 189–196.

Philipson, W. R. 1953. The relationships of the Compositae, particularly as illustrated by the morphology of the inflorescence in the Rubiales and the Campanulatae. Phytomorphology 3: 391–404.

Richard, A. 1964. Étude du sac embryonnaire et de la fécondation chez quelques Composées. Rev. Cytol. Biol. Vég. 27: 1–44.

Rodriguez, E., G. H. N. Towers, & J. C. Mitchell. 1976. Biological activities of sesquiterpene lactones. Phytochemistry 15: 1573–1580.

Skvarla, J. J., & B. L. Turner. 1966. Systematic implications from electron microscopic studies of Compositae pollen—a review. Ann. Missouri Bot. Gard. 53: 220–256.

Stix, H. 1960. Pollenmorphologische Untersuchungen an Compositen. Grana Palynol. 2(2): 41–114.

Wagenitz, G. 1975 (1976). Systematics and phylogeny of the Compositae (Asteraceae). Pl. Syst. Evol. 125: 29–46.

Zohary, M. 1950. Evolutionary trends in the fruiting head of Compositae. Evolution 4: 103–109.

Александров, В. Г., & М. И. Савченко. 1951. Об особенностях истории развития плода и семени в семействе сложноцветных. Труды Бот. Инст. АН СССР, сер, VII, 2: 5–98.

Сенянинова-Корчагина, М. В. 1952. О природе нижней завязи у сложноцветных. Бюлл. Московск. Общ. Исп. Прир., Отд. Биол., нов. сер. 57(4): 63–75.

Тамамшян, С. Г. 1965. "Сверхэволюционные" формы чашечки и их значение для филогенетической проблемы Asteraceae Link. В кн.: Проблемы филогении растений. Труды Московск. Общ. Исп. Прир. 13: 161–174.

1. Family ASTERACEAE Dumortier 1822 nom. conserv., the Aster Family

Herbs or less often shrubs, seldom small to medium-sized trees, glabrous or often with various sorts of glandular or eglandular hairs, commonly storing carbohydrate as polyfructosans, notably inulin, and commonly producing polyacetylenes (these borne in the resin-canals, but largely absent from the Lactuceae and Senecioneae); polyacetylenes characterized by the presence of cyclic, aromatic or heterocyclic end-groups, as contrasted to the mainly aliphatic polyacetylenes of the Campanulaceae and Araliales), bitter sesquiterpenes (especially sesquiterpene lactones), terpenoid volatile essential oils, and often one or another sort of alkaloid (notably the pyrrolizidine alkaloids, often called *Senecio* alkaloids, of the Senecioneae and some Eupatorieae), occasionally cyanogenic (valine-derived or phenylalanine-derived), but without iridoid compounds, and not tanniferous, or at least not strongly so, generally lacking both ellagic acid and proanthocyanins; the Lactuceae and a few genera of other tribes with a system of articulated laticifers in the phloem, containing a characteristic triterpene-rich latex, other tribes with a more or less well developed system of schizogenous resin-ducts that are often lined with an epithelium, the two secretory systems largely alternative rather than co-extensive, but not homologous; scattered latex-cells occasionally present in addition to the resin-ducts; crystals of calcium oxalate only seldom present in some of the cells of the parenchymatous tissues; nodes trilacunar to multilacunar; stem often showing well developed secondary growth, even when herbaceous and with initially separate vascular bundles, sometimes with anomalous secondary growth, and sometimes with medullary or cortical bundles; vessel-segments com-

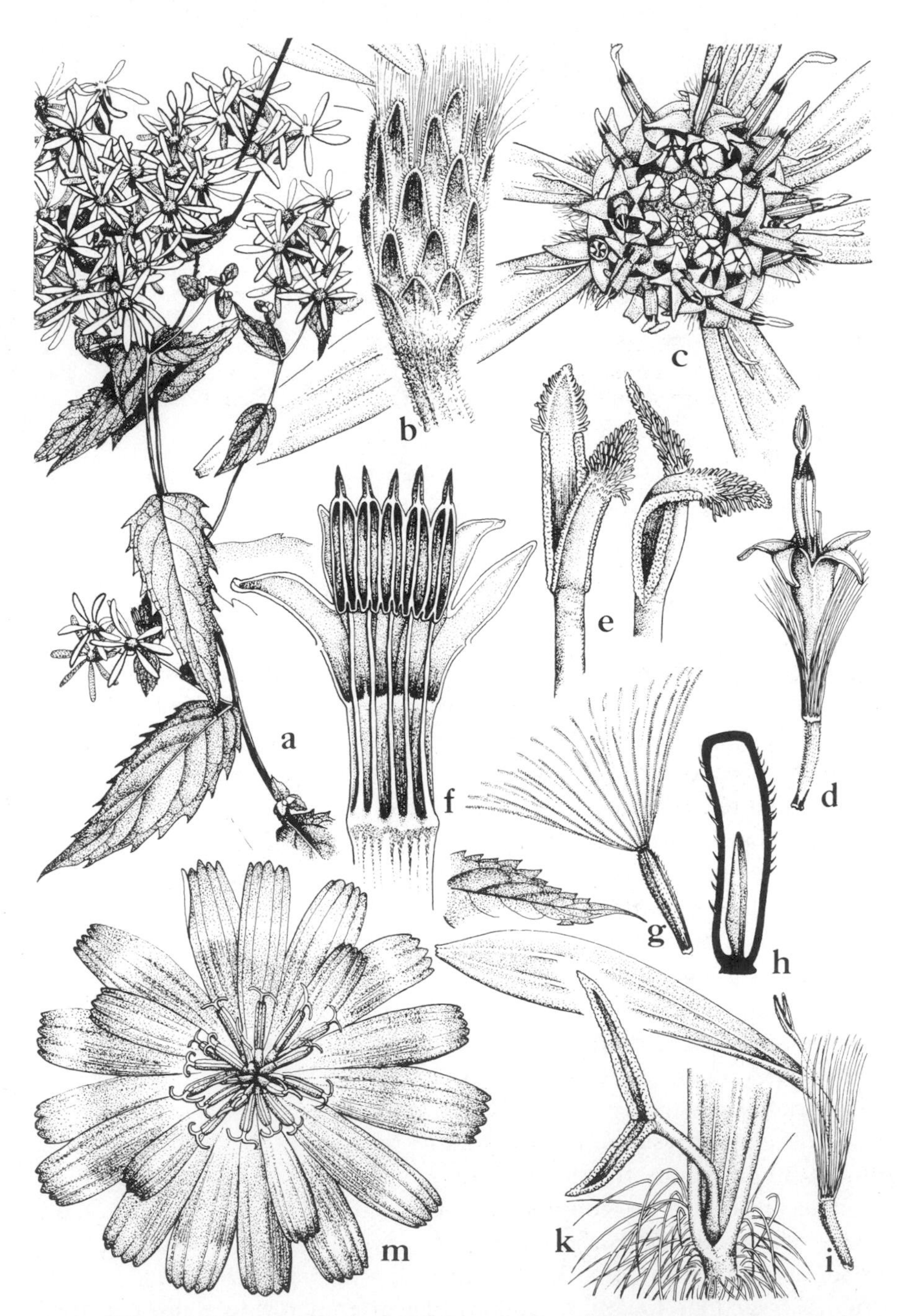

FIG. 6-35 Asteraceae. a–k, *Aster divaricatus* L. a, habit, ×½; b, involucre, ×5; c, top view of head, ×5; d, disk-flower, ×5; e, style-branches, ×20, f, disk-flower, opened out to show stamens, ×10; g, achene with pappus, ×5; h, schematic long-section of ovary, ×10; i, ray-flower, ×5; k, portion of ray-flower, with style, ×20. m, *Cichorium intybus* L., top view of head, ×2.

monly with simple perforations, or sometimes some of them scalariform or reticulate; imperforate tracheary elements small, with simple or sometimes very narrowly bordered pits; wood-rays mostly heterocellular, mixed uniseriate and pluriseriate, seldom all one or the other, the pluriseriate ones usually the more abundant, mostly 4–10 (–18) cells wide; wood-parenchyma scanty and paratracheal. LEAVES alternate or less commonly opposite; seldom whorled, simple and entire or toothed to sometimes compound or variously dissected; stomates anomocytic or occasionally anisocytic; petiole commonly with an arc of vascular bundles, sometimes with other arrangements; stipules wanting. Inflorescence of 1-many dense heads with (1–) several to very numerous sessile flowers on a common receptacle, the head nearly always subtended by an involucre of 1-several series of bracts (involucre wanting in *Psilocarphus*), and flowering in strictly racemose sequence (mixed sequence in *Espeletia*), but the sequence of flowering *among* the heads nearly always cymose or mixed; heads occasionally aggregated into cymose secondary heads that may have a secondary involucre; common receptacle of the head flat to convex, conic, or cylindric, sometimes (mainly in the Heliantheae) chaffy, i.e., with a bract subtending each flower, or sometimes (mainly in the Cynareae) densely bristly without any clear positional relationship of bristles to flowers, but most often naked, i.e., without bracts or bristles; FLOWERS epigynous, sympetalous, perfect, or some of them pistillate or neutral or functionally staminate, the perfect and functionally staminate ones with a specialized pollen-presentation mechanism, the anthers connate (or connivent) into a tube, the pollen released into the interior of the anther-tube and pushed out by growth of the style; heads variously radiate, discoid, disciform, or ligulate, according to the kind and arrangement of the flowers, or in many Mutisieae of still another special type described below; *radiate* heads with marginal pistillate (or neutral) flowers and central perfect (or functionally staminate) flowers, the former called *ray-flowers*, the latter called *disk-flowers*; corolla of the ray-flowers tubular at the base, prolonged on the outer side into a flat, more or less strap-shaped ligule or ray representing 3 lobes of an ancestrally pentamerous corolla (these lobes sometimes evident as apical teeth), the other 2 (adaxial) lobes much-reduced or commonly obsolete; corolla of the disk-flowers tubular, regular and (4) 5-toothed or -lobed, with valvate teeth or lobes, the veins of the corolla running from the base to the sinuses, where they fork and follow the margin to the tip of the lobe, joining the vein from the other margin; *discoid* heads with only disk-flowers, as described above; *disciform* heads with central disk-flowers and marginal pistillate flowers with a reduced, inconspicuous slender, tubular corolla lacking a ligule, or with only such reduced pistillate flowers, or the pistillate corolla rarely obsolete; *ligulate* heads (characteristic of and almost entirely restricted to the Lactuceae) wholly of perfect flowers, the corolla superficially resembling that of ray-flowers of radiate heads, but comprising all 5 corolla-lobes, as if the originally tubular corolla had been slit between two lobes on the inner (upper) side; in most members of the Mutisieae some or all of the flowers with a bilabiate corolla, the outer lip generally the larger, but the marginal flowers sometimes ligulate as in the Lactuceae, and the central ones (or all of them) sometimes tubular and regular as in the other tribes, then usually with notably long and slender corolla-lobes; calyx in flowers of all sorts strongly modified, forming a characteristic *pappus* atop the ovary and fruit, or sometimes obsolete; pappus variously of (1) 2-many scales, awns, or capillary or plumose bristles, or sometimes the scales connate to form a crown, the pappus-members sometimes in more than one series and/or of more than one type; stamens as many as the corolla-lobes; filaments attached to the corolla-tube, alternate with the lobes, distinct or rarely connate; anthers tetrasporangiate (rarely bisporangiate) and dithecal, nearly always provided with a short, apical appendage, obtuse to often sagittate at the base, and in three tribes (Cynareae, Inuleae, Mutisieae) commonly prolonged into basal tails, the pollen-sacs more or less elongate, connate (seldom only connivent) into a tube, introrse, opening by longitudinal slits and releasing the pollen into the anther-tube; pollen-grains trinucleate, mostly tricolporate, spinulose or (especially in the Lactuceae) lophate, often caveate (i.e., some of the columellae detached from the foot-layer of the exine); gynoecium of 2 median carpels united to form a compound, inferior, unilocular ovary with a terminal style that has 2 short or elongate branches; style in perfect and functionally staminate flowers growing up through the anther-tube and pushing out the pollen, the branches commonly separating thereafter; style branches variously hairy or papillate externally and sometimes also internally, sometimes stigmatic across the whole inner surface, but more often with ventromarginal stigmatic lines that often stop short of the summit, so that there is a sterile (often shortly hairy) appendage; stigmas dry; in functionally staminate flowers the style-branches commonly not fully developed and often not separating; nectary commonly a thickened scale or cup atop the ovary around or alongside the base of the style; ovule solitary, basal, erect, anatropous, tenuinucellar, with a massive single integument and an integumentary tapetum; embryo-sac variously monosporic, bisporic, or tetrasporic; endosperm-development nuclear or more often cellular, without haustoria. FRUIT an achene, crowned by the persistent pappus, or the pappus sometimes deciduous or wanting; embryo oily, straight, with 2 expanded cotyledons and a short to long hypocotyl; endosperm usually said to be wanting; but sometimes present as a very thin peripheral layer, demonstrable in microscopic section. X = 2–19+, perhaps originally 9. (Compositae, nom. altern., Ambrosiaceae, Carduaceae, Cichoriaceae)

The Asteraceae consist of more than 1100 genera and perhaps as many as 20,000 species, cosmopolitan in distribution, but best represented in temperate or subtropical regions that are not densely forested. Very many of the genera are ill-defined. The family may be organized into about 13 tribes. Among these, the Lactuceae are often considered to represent a distinct subfamily Cichorioideae (Liguliflorae) in contrast to the subfamily Asteroideae (Tubuliflorae) for the remaining tribes, which are largely confluent. Species of many genera are cultivated for ornament, and a few for food or oil. The largest genera are *Senecio* (1500, Senecioneae), *Vernonia* (900, Vernonieae), *Hieracium* (reputedly 800 or more, Lactuceae), *Eupatorium* (600, Eupatorieae), *Centaurea* (600, Cynareae), *Cousinia* (600, Cynareae), *Helichrysum* (500, Inuleae), *Baccharis* (400, Astereae), and *Artemisia* (400, Anthemideae). The following genera are also noteworthy for their abundance, or their large number of species, or their familiarity in cultivation or as weeds: *Ambrosia, Bidens, Coreopsis, Cosmos, Dahlia, Helianthus, Rudbeckia, Tagetes, Verbesina, Viguiera*, and *Xanthium*, all in the Heliantheae; *Aster, Bellis, Callistephus, Erigeron, Haplopappus*, and *Solidago*, in the Astereae; *Arctium, Carduus, Carthamus, Cirsium, Echinops*, and *Saussurea*, in the Cynareae; *Crepis, Lactuca, Sonchus*, and *Taraxacum*, in the Lactuceae; *Ageratum, Mikania*, and *Stevia*, in the Eupatorieae; *Anthemis, Chrysanthemum*, and *Tanacetum*, in the Anthemideae; *Calendula*, in the Calenduleae; *Gerbera*, in the Mutisieae; and *Arctotis*, in the Arctotideae. *Liabum* and *Munnozia*, with about 50 species each, are the largest genera in the small tribe Liabeae, which is otherwise unrepresented in the foregoing lists.

The Asteraceae are one of the most successful families of flowering plants, represented by numerous genera, species, and individuals. Not many of them are forest trees, and only a few are aquatic, but otherwise they are highly diversified not only in habitat and life-form but also in the methods of pollination and seed-dispersal. Some of the Cynareae and Mutisieae with elongate corollas appear to be designed for special pollinators such as hummingbirds and long-tongued insects, but otherwise the nectar as well as the pollen in most species is available to all comers, and a single species may be visited by numerous kinds of insects. (I well remember the role of *Chrysothamnus* in augmenting the collection of insects that I made for a high-school class in biology). Wind-pollinated, self-pollinated, and apomictic forms also occur. The fruits are commonly dispersed by wind or externally by mammals, less often by ants or other agents. Some have no obvious means of dispersal, but like the bedbug in the doggerel verse, they get there just the same.

The ecologically distinctive features of the Asteraceae are the pseudanthial heads with a specialized pollen-presentation mechanism, and their particular array of chemical weapons. None of these features is unique, but collectively they distinguish the family from all others.

I believe that the chemistry is more important than the morphology in fostering the evolutionary success of the Asteraceae. The pseudanthial head and the complex pollen-presentation mechanism are duplicated in the small and unsuccessful family Calyceraceae, which has an undistinguished set of chemical repellents. The monotypic family Brunoniaceae has flowers with a similar pollen-presentation mechanism crowded into cymose heads. A similar pollen-presentation mechanism, but without the pseudanthial heads, is developed to varying degrees in the Campanulaceae, Goodeniaceae, and the subfamily Ixoroideae of the Rubiaceae. The largest of these groups, other than the Asteraceae, is the Campanulaceae, which are also notable for having their own characteristic set of repellents.

I suggest that the initial success of the Asteraceae grew out of their discovery of the effective defensive combination of polyacetylenes and sesquiterpene lactones, before these had been exploited by any other group. The continued expansion of the family has been fostered by its chemical evolutionary lability, which has permitted it to develop and exploit other new repellents. It is probably no accident that the largest and most diversified genus in the family is *Senecio*. *Senecio* and its immediate allies have a special set of alkaloids, shared with only few other families (most notably the Boraginaceae, in which they probably also play a major role in the success of the group). The Lactuceae, which do not enter the fossil record until the middle of the Miocene, have retained the sesquiterpene lactones, but introduced a latex-system in place of the polyacetylene-bearing resin-system of other tribes. The Anthemideae have their own characteristic odor (properly, a set of similar odors), as does the subtribe Tagetinae of the Heliantheae, indicating an intensive exploitation of various possibilities among volatile monoterpenes and terpenoids. *Tagetes* has an apparently well justified reputation for killing nematodes in the surrounding soil. The future role of the Asteraceae in the vegetation of the world will probably depend more on their continuing chemical evolution than on morphological changes within the group.

The affinities of the Asteraceae have been vigorously but inconclusively debated. Traditionally they have been thought to be allied to the Campanulaceae, and often even included in the same order. The specialized pollen-presentation mechanism in the two families has often been used to bolster such an association. Unfortunately for this argument, a similar mechanism also exists in the Calyceraceae and in the subfamily Ixoroideae of the Rubiaceae. This mechanism is not so well developed in the Ixoroideae as it is in the Asteraceae, Calyceraceae, and the subfamily Lobelioideae of the Campanulaceae, being more nearly comparable to that in the Campanuloideae, but it indicates the evolutionary possibilities within the Rubiaceae.

Patterns of relationship within the Asteraceae strongly suggest that their immediate ancestors must have been shrubs or small trees with opposite leaves and a cymose inflorescence. The basically herbaceous, alternate-leaved Campanulaceae and their immediate allies therefore do not provide a suitable starting point. Furthermore, the two families differ in a set of embryological details which tend to associate the Asteraceae with the Calyceraceae, Dipsacales, and Rubiales (Crété, 1956; p. 376 in Poddubnaja-Arnoldi, 1964, general citations, in Russian). As I have pointed out under that family, the Calyceraceae can be no more than collateral allies of the Asteraceae, in spite of the many similarities. The Dipsacaceae are likewise removed from serious consideration as possible ancestors of the Asteraceae because of their herbaceous habit and apical (rather than basal) ovule. Nothing in the Dipsacales more advanced than the Caprifoliaceae is morphologically suitable to be ancestral to the Asteraceae. The frequently nuclear endosperm-development of the Asteraceae suggests that even the Caprifoliaceae may be too advanced, and that we must look further back, to the vicinity of the Rubiaceae, for the origin of the Asteraceae.

Before complacently accepting a rubialean affinity for the Asteraceae, we should emphasize again that the Asteraceae are chemically as well as morphologically distinctive. On chemical grounds nothing in the Asteridae seems to be a very likely ancestor for them, although some similarities can be pointed out to the Campanulales (inulin, polyacetylenes) and Boraginaceae (Senecio alkaloids). The family that may be chemically most nearly like the Asteraceae is the Apiaceae, in the subclass Rosidae, but the morphological differences make any attempt to link these two families patently absurd.

All considered, I am convinced that the Asteraceae properly belong to the large subclass to which they give the name. Their distinctive chemical features can easily be interpreted as a new and successful set of weapons, developed at a time when predators were becoming adapted to the iridoid compounds of more ancient members of the subclass. I believe that the ancestry of the Asteraceae probably lies in or near the Rubiaceae, along a line parallel in some respects to the line leading to the Dipsacales and Calycerales.

The fossils tell us nothing about the ancestry of the Asteraceae except that their origin is probably relatively recent. Pollen representing the group enters the fossil record at many places in the world at about the end of the Oligocene epoch. A macrofossil of about this same age from Montana appears to be a composite head and has been described as *Viguiera cronquistii* H. Becker, but its identity as a composite has recently been challenged. Earlier fossils ascribed to the Asteraceae are probably incorrectly identified. The Lactuceae may be the most recently evolved tribe in the family. Their distinctive pollen does not enter the fossil record until about the middle of the Miocene epoch, some 18 million years ago.

LILIOPSIDA

Xanthorrhoea australis R. Br., Grass-tree, in the Gibraltar Range of eastern Australia.

Class LILIOPSIDA Cronquist, Takhtajan & Zimmermann 1966, the Monocotyledons

Plants herbaceous or less often woody, never with typical secondary growth, but sometimes with a special kind of secondary growth that results in the formation of complete vascular bundles and associated ground-tissue in the stem; vascular bundles of the stem closed (i.e., without a cambium), usually scattered or in 2 or more rings; vessels often confined to the roots, or wholly wanting, less commonly present also in the shoot, or in the stem but not in the leaves; vessels in the roots more likely to have simple (instead of scalariform) perforations than those in the shoot; plastids of the sieve-tubes with cuneate proteinaceous inclusions; root-system wholly adventitious; root-hairs typically arising only from certain specialized epidermal cells; plants only rather seldom notably tanniferous, sometimes producing proanthocyanins, but not ellagic acid; cyanogenesis, when present, typically following a pathway based on tyrosine. LEAVES typically with parallel or pinnate-parallel venation, varying to sometimes more or less distinctly net-veined (most notably in many Araceae), very often with a well defined basal sheath, the blade often slender and without a petiole, but sometimes broader, or distinctly petiolate, or both; blade typically developing from a portion of the leaf-primordium somewhat behind the tip and maturing basipetally, the primordial tip inactive or producing only a terminal point or small appendage (Vorläuferspitze) on the blade. FLOWERS very often with septal or septal-derived nectaries, but also often with other kinds of nectaries or without nectar; floral parts, when of definite number, typically borne in sets of 3, seldom 4 or 2 (carpels often fewer than 3) never 5 (except for the stamens in some Zingiberales); pollen-grains typically uniaperturate or of uniaperturate-derived type, only very rarely triaperturate. Cotyledon one (the embryo often modified so that the cotyledon appears to be terminal and the

FIG. 7.1 *Cypripedium acaule* Aiton, Moccasin flower, an orchid native to eastern United States.

the plumule lateral), almost never 2, or not infrequently the embryo not differentiated into parts. (Liliatae, a variant spelling; Monocotyledoneae)

The Liliopsida consist of 5 subclasses, 19 orders, 65 families, and about 50,000 species. No one of the subclasses can be considered basal to the others. The Alismatidae, or some of them, retain a number primitive floral features, as noted in the discussion under that subclass, but they are not ancestral to the other groups.

It is now widely agreed that the monocots were derived from primitive dicots, and that the monocots must therefore follow rather than precede the dicots in any proper linear sequence. The solitary cotyledon, parallel-veined leaves, absence of a cambium, the dissected stele, and the adventitious root-system of the monocots are all regarded as secondary rather than primitive characters in the angiosperms as a whole, and any plant that was more primitive than the monocots in these several respects would certainly be a dicot. The monocots are more primitive than the bulk of the dicots in mostly having uniaperturate pollen, but several of the archaic families of dicots also have uniaperturate pollen, so there is no problem here.

The fossil pollen record suggests that the origin of monocotyledons from primitive dicotyledons in Aptian-Albian time was the first significant dichotomy in the evolutionary diversification of the angiosperms. The wide variety of monocotyledonous leaves found throughout the Upper Cretaceous attests to the continuing diversification of the monocots during this time, but most of these leaves cannot be referred with any certainty to modern groups. The first modern family of monocots to be clearly represented in the fossil record is the Arecaceae (subclass Arecidae), near the base of the Santonian (or perhaps in the Coniacian) epoch, but palms are surely not primitive monocots. Their large, distinctive, readily fossilized leaves merely make the group easy to recognize from its inception. Pollen that probably represents the Cyperales or Restionales (subclass Commelinidae) appears in the Upper Cretaceous, probably before the Maestrichtian, and the distinctive leaves of the Zingiberales show up in the Maestrichtian. Pollen thought to represent *Pandanus* occurs in Maestrichtian deposits in North America. The Alismatidae and Liliidae are not certainly recognizable before the Tertiary, but some of the miscellaneous Upper Cretaceous monocot fossils might well belong to one or the other of these groups. We can be reasonably sure that the palms and the Zingiberales did not arise long before the first appearance of their characteristic leaves in the fossil record, but we cannot be so confident that other large groups did not long antedate their first identifiable fossils. Thus the fossil record as presently understood is compatible with any of several different views about the Cretaceous diversification of the monocots. The principal constraint is the recognition of a very early dichotomy between monocots and dicots.

The dicots that gave rise to the monocots must have had apocarpous flowers with a fairly ordinary (not highly specialized) perianth, and with uniaperturate pollen. They must have been herbs without a very active cambium, and they presumably had laminar placentation. The only modern order of dicots that fits these specifications is the Nymphaeales.

It is not here suggested that the Nymphaeales are directly ancestral to the monocots as a whole, but rather that the premonocotyledonous dicots were probably something like the modern Nymphaeales. It is noteworthy that an aquatic group of angiosperms with leaves much like those of the modern Nymphaeales was already proliferating in the Albian epoch of the Lower Cretaceous. The modern Nymphaeales are aquatic, they mostly lack vessels, and they show tendencies toward the fusion of two cotyledons into one. Although the idea has never gained wide acceptance, some botanists still insist that the Nymphaeales are better grouped with the monocotyledons than with the dicotyledons.

The monocots are here considered to be of aquatic ancestry, and the typical parallel-veined leaf is considered to be a modified, bladeless petiole. This morphological interpretation of the monocot leaf was proposed a century and a half ago by de Candolle, and was further elaborated in evolutionary terms by Arber in 1925. It is the only hypothesis known to me that permits all the information about monocots to fall into place and make sense. Even the ontogeny of the typical monocot leaf (as noted in the description of the class) is highly compatible with the petiolar hypothesis, although this fact seems to have escaped those who have elucidated the ontogeny.

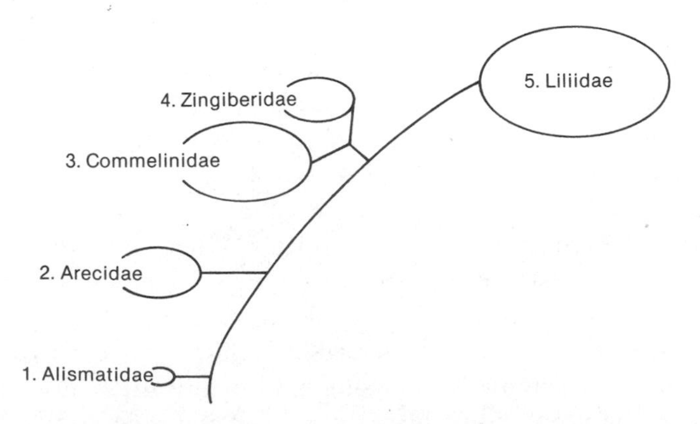

Fig. 7.2 Putative relationships among the subclasses of Liliopsida. The size of the balloons is roughly proportional to the number of species.

As aquatic herbs, the early monocots were preadapted to evolve terrestrial herbaceous forms, filling a niche not then effectively occupied by dicots. Nevertheless, the evolution of an efficient water-conducting system, of expanded, net-veined leaves, woody stems, arborescent habit, and a means of secondary thickening have evidently not been easy for them. Aroids, palms, and the Zingiberales have taken three essentially different routes in the evolutionary expansion of leaves, and the difference is reflected in the mature morphology.

Recognition of major groups of monocots by their aspect is not so consistently difficult and frustrating as a comparable effort among the dicots. Such recognition therefore becomes a feasible addition to the other objectives of the system of monocots. A taxonomic system should always be as simple and easy to use as is consistent with naturalness. The scheme presented here reflects my attention to these matters.

SELECTED REFERENCES

Boudet, Al. M., An. Boudet, & H. Bouyssou. 1977. Taxonomic distribution of isoenzymes of dehydroquinate hydrolase in the angiosperms. Phytochemistry 16: 919–922.

Boyd, L. 1932. Monocotylous seedlings. Morphological studies in the post-seminal development of the embryo. Trans. & Proc. Bot. Soc. Edinburgh 31: 1–224.

Burger, W. C. 1977. The Piperales and the monocots—alternate hypotheses for the origin of monocotyledonous flowers. Bot. Rev. 43: 345–393.

Cheadle, V. I. 1942. The occurrence and types of vessels in the various organs of the plant in the Monocotyledoneae. Amer. J. Bot. 29: 441–450.

Cheadle, V. I. 1943. Vessel specialization in the late metazylem of the various organs in the Monocotyledoneae. Amer. J. Bot. 30: 484–490.

Cheadle, V. I. 1944. Specialization of vessels within the xylem of each organ in the Monocotyledoneae. Amer. J. Bot. 31: 81–92.

Cheadle, V. I. 1948. Observations on the phloem in the Monocotyledoneae. II. Additional data on the occurrence and phylogenetic specialization in structure of the sieve tubes in the metaphloem. Amer. J. Bot. 35: 129–131.

Cheadle, V. I. 1953. Independent origin of vessels in the monocotyledons and dicotyledons. Phytomorphology 3: 23–44.

Cheadle, V. I., & N. B. Whitford. 1941. Observations on the phloem in the Monocotyledoneae. I. The occurrence and phylogenetic specialization in structure of the sieve tubes in the metaphloem. Amer. J. Bot. 28: 623–627.

Clifford, H. T. 1977. Quantitative studies of inter-relationships amongst the Liliatae. Pl. Syst. Evol. Suppl. 1: 77–95.

Cronquist, A. 1974. Thoughts on the origin of the monocotyledons. Birbal Sahni Inst. Palaeobot. Lucknow, Special Publ. 1: 19–24.

Daumann, E. 1970. Das Blütennektarium der Monocotyledonen unter besonderer Berücksichtigung seiner systematischen and phylogenetischen Bedeutung. Feddes Repert. 80: 463–590.

Deyl, M. 1955. The evolution of the plants and the taxonomy of the monocotyledons. Acta Mus. Natl. Prag. 11B(6), Botanica 3: 1–143.

Doyle, J. A. 1973. Fossil evidence on early evolution of the monocotyledons. Quart. Rev. Biol. 48: 399–413.

Erdtman, G. 1944. Pollen morphology and plant taxonomy. II. Notes on some monocotyledonous pollen types. Svensk Bot. Tidskr. 38: 163–168.

Fahn, A. 1954. Metaxylem elements in some families of the Monocotyledoneae. New Phytol. 53: 530–540.

Goldblatt, P. 1980. Polyploidy in angiosperms: monocotyledons. In: W. H. Lewis, ed., Polyploidy: Biological relevance. Plenum. New York.

Haines, R. W., & K. A. Lye. 1979. Monocotylar seedlings: a review of evidence supporting an origin by fusion. J. Linn. Soc., Bot. 78: 123–140.

Holttum, R. E. 1955. Growth-habits of monocotyledons—variations on a theme. Phytomorphology 5: 399–413.

Huber, H. 1977. The treatment of the monocotyledons in an evolutionary system of classification. Pl. Syst. Evol., Suppl. 1: 285–298.

Kaplan, D. R. 1973. The problem of leaf morphology and evolution in the monocotyledons. Quart. Rev. Biol. 48: 437–457.

Kimura, Y. 1956. Système et phylogénie des Monocotylédones. Notul. Syst. (Paris)15: 137–159.

Kuschel, G. 1971. Entomology of the Aucklands and other islands south of New Zealand: Coleoptera: Curculionidae. Pacific Insects Monogr. 27: 225–259.

Lowe, J. 1961. The Phylogeny of monocotyledons. New Phytol. 60: 355–387.

Slater, J. A. 1976. Monocots and chinch bugs: A study of host plant relationships in the Lygaeid subfamily Blissinae (Hemiptera: Lygaeidae).Biotropica 8: 143–165.

Stebbins, G.L., & G. S. Khush. 1961. Variations in the organization of the stomatal complex in the leaf epidermis of Monocotyledons and its bearing on their phylogeny. Amer. J. Bot. 48: 51–59.

Tomlinson, P. B. 1974. Development of the stomatal complex as a taxonomic character in the monocotyledons. Taxon 23: 109–128.

Tomlinson, P. B., & M. H. Zimmermann. 1969. Vascular anatomy of monocotyledons with secondary growth—an introduction. J. Arnold Arbor. 50: 159–179.

Wagner, P. 1977. Vessel types of the monocotyledons: a survey. Bot. Not. 130: 383–402.

Williams, C. A., J. B. Harborne, & H. T. Clifford. 1971. Flavonoid patterns in the monocotyledons. Flavonols and flavones in some families associated with the Poaceae. Phytochemistry 10: 1059–1063.

Zimmerman, M. H., & P. B. Tomlinson. 1972. The vascular system of monocotyledonous stems. Bot. Gaz. 133: 141–155.

Кудряшов, Л. В. 1964. Происхождение односемядольности (на примере Helobiae). Бот. Ж. 49: 473-486.

Куприянова, Л. А. 1948. Морфология пыльцы однодольных растений. Труды Бот. Инст. АН СССР, сер. 1, 7: 163-262.

Куприянова, Л. А. 1954. Палинологические данные о филогении класса однодольных растений. Вопр. Бот. 1: 113-119. (Also in French: pp. 120-127.) Изд. АН СССР, Москва-Ленинград.

Хохряков, А. Р. 1973. Новая система однодольных на эколого-географической основе. Заметки по сист. и геогр. раст., Тбилисск. Бот. Инст. 30: 67-73.

SYNOPTICAL ARRANGEMENT OF THE SUBCLASSES OF LILIOPSIDA

1 Plants either with apocarpous (sometimes monocarpous) flowers, or more or less aquatic, or very often both, always herbaceous, but never thalloid; vascular system generally not strongly lignified, often much reduced, the vessels confined to the roots, or wanting; endosperm mostly wanting, not starchy when present; subsidiary cells mostly 2; pollen trinucleate ....I. ALISMATIDAE.

1 Plants with syncarpous (or seldom pseudomonomerous) flowers except in some arborescent taxa, usually terrestrial (or epiphytic),much less often more or less aquatic, although sometimes not only aquatic but also thalloid and free-floating; vessels, endosperm, stomates, and pollen various, but not combined as in the Alismatidae.

2 Flowers usually numerous, usually small, and subtended by a prominent spathe (or several spathes), often aggregated into a spadix (but the inflorescence much reduced in the Lemnaceae); plants very often either arborescent, or with relatively broad leaves that do not have typical parallel venation, or both (but sometimes lacking both of these features, and in the Lemnaceae even thalloid and free-floating); septal nectaries wanting except in many Arecaceae; subsidiary cells typically 4, less often 2 or more than 4; vessels generally present in all vegetative organs, except among the Arales; endosperm not starchy except in some Arales ....II. ARECIDAE.

2 Flowers few to numerous and small to large, but never aggregated into a spadix, and usually without a distinct spathe (inflorescence with 1 or more spathe-like bracts in many Zingiberidae, but the flowers then larger and showy); plants herbaceous or much less often arborescent; leaves narrow and parallel-veined, or less often broader and more or less net-veined, or with a special type of pinnate-parallel venation; nectaries, stomates, vessels, and endosperm various, but not combined as in typical Arecidae.

3 Nectar and nectaries mostly wanting; perianth in the more archaic families trimerous and with well differentiated sepals and petals, in the more advanced families reduced and chaffy and often not obviously trimerous, or wanting, the families with reduced perianth typically adapted to wind pollination; ovary always superior (or nude); vessels generally present in all vegetative organs; endosperm wholly or in large part starchy, commonly mealy and with compound starch grains, without significant reserves of hemicellulose and usually also without significant reserves of oil, or rarely the endosperm wanting; stomates mostly with 2 subsidiary cells, seldom

without subsidiary cells or with more than 2 ....III. COMMELINIDAE.

3 Nectar and nectaries of one or another sort (often septal nectaries) generally present; perianth generally well developed, not reduced and chaffy, the flowers typically adapted to pollination by insects or other kinds of animals; ovary superior or very often inferior; vessels most often confined to the roots, but sometimes occurring also in the stem or in all vegetative organs; endosperm and stomates various.

4 Sepals usually well differentiated from the petals, often green and herbaceous, sometimes petaloid in texture but still unlike the petals; endosperm typically starchy and mealy, with compound starch grains, only seldom hard and presumably with reserves of hemicellulose; subsidiary cells (2) 4 or more; plants not obviously mycotrophic; leaves narrow and parallel-veined, or equally often with a broad, expanded, petiolate blade and a characteristic pinnate-parallel venation .... IV. ZINGIBERIDAE.

4 Sepals usually petaloid in form and texture, sometimes differentiated from the petals but still petaloid in appearance, only rarely green and herbaceous; endosperm when present typically very hard, with reserves of hemicellulose, protein, and oil, less commonly starchy, but not mealy in texture, the starch grains typically simple rather than compound, or very often the endosperm wanting; subsidiary cells 2 or often none, only seldom 4; plants often strongly mycotrophic; leaves typically narrow and parallel-veined, seldom broader and even more or less net-veined, but without the characteristic pinnate-parallel venation of many Zingiberidae .... V. LILIIDAE.

I. Subclass ALISMATIDAE Takhtajan 1966

Herbs, either aquatic, or of wet places, or mycotrophic and without chlorophyll; vascular system generally not strongly lignified, often much-reduced; vessels confined to the roots, or wanting; plastids of the sieve-tubes, so far as known, with cuneate proteinaceous inclusions. LEAVES simple, alternate or less often opposite or whorled, basically parallel-veined, with or without an expanded blade, commonly sheathing at the base; stem commonly (except in the Triuridales) with some small intravaginal scales at the nodes; leaves in Triuridales reduced and scale-like. FLOWERS from fairly large and showy to often small and inconspicuous, variously pollinated by insects, wind, or water, regular or irregular, hypogynous or sometimes epigynous, perfect or unisexual, borne in various sorts of inflorescences, most commonly a raceme or spike, with or without subtending bracts, the inflorescence sometimes subtended by a spathe; perianth in the more archaic groups trimerous, with 3 sepals and 3 petals, but in the more modified groups variously reduced or even wanting, seldom (some Triuridaceae) of up to 10 basally connate tepals in a single cycle; stamens 1-many; pollen nearly always trinucleate, monosulcate or inaperturate, or less often 2-many-porate; gynoecium of 1-many nearly or quite distinct carpels, or of 2–3 carpels united to form a unilocular, pseudomonomerous ovary, or of several carpels united to form a compound ovary with parietal or intruded-laminar placentation, or of several carpels weakly united to a central axis but otherwise distinct, never forming a distinctly plurilocular ovary with axile placentation; ovules 1-many in each carpel or locule, variously marginal, laminar, basal, apical, or parietal, apparently always bitegmic, crassinucellar or pseudocrassinucellar, in the Triuridaceae tenuinucellar; endosperm-development helobial or less commonly nuclear or even cellular. FRUIT most commonly of follicles or follicular mericarps or achenes or drupelets, sometimes of other types; seeds without endosperm, except in the Triuridales; embryo with a single cotyledon, or (Triuridales) not differentiated into parts.

The subclass Alismatidae as here defined consists of 4 orders, 16 families, and scarcely 500 species. The Alismatales, Hydrocharitales, and Najadales are closely related; these have often been treated as parts of a single order, usually under the name Helobiae. The Triuridales are a more isolated group.

The Alismatidae have often been considered to be the most archaic group of Liliopsida. They can scarcely be on the main line of evolution of the class, however, because a primitive monocot should have binucleate pollen and endospermous seeds. They are here considered to be a near-basal side-branch, a relictual group that has retained a number of primitive characters. The apocarpous gynoecium of most members, combined with the chiefly uniaperturate pollen of the Liliopsida as a whole, indicates that any connection of the Liliopsida to the Magnoliopsida must be to the archaic subclass Magnoliidae. It should be noted, however, that the ontogeny of the pleiomerous androecium in the Alismatidae is quite different from that in the Magnoliidae, so that the evolutionary homology can be questioned.

Within the Magnoliidae the aquatic order Nymphaeales presents the closest approach to the Alismatidae. The concept that the Liliopsida as a whole are of aquatic origin has already been discussed.

The trinucleate pollen of the Alismatidae may well relate to their aquatic habitat. Trinucleate pollen typically requires specialized conditions for germination, and a disproportionately high number of aquatic angiosperms have trinucleate pollen. However, judged by the presently extant members of the subclass, it appears that trinucleate pollen in the group may have antedated subsurface pollination. The trinucleate condition should probably therefore be regarded as preadapted to subsurface pollination, rather than as evolved in response to it.

The biological significance of the helobial pattern of endosperm-development is wholly obscure, as is also the usual disappearance of the endosperm before maturity of the seed.

The fossil record clearly carries the Alismatidae back to the Paleocene epoch, but records from the Upper Cretaceous are doubtful or incorrect.

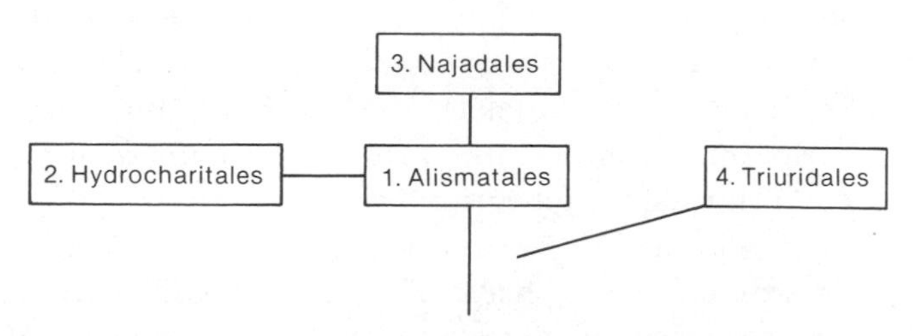

FIG. 7.3 Putative relationships among the orders of Alismatidae.

SELECTED REFERENCES

Eber, E. 1934. Karpellbau und Plazentationverhältnisse in der Reihe der Helobiae. Mit einem Anhang über die verwandschaftlichen Beziehungen zwischen Ranales und Helobiae. Flora 127: 273–330.

Harborne, J. B., & C. A. Williams. 1976. Occurrence of sulphated flavones and caffeic acid esters in members of the Fluviales. Biochem. Syst. Ecol. 4: 37–41.

Kaul, R. B. 1978. Morphology of germination and establishment of aquatic seedlings in Alismataceae and Hydrocharitaceae. Aquatic Bot. 5: 139–147.

Wilder, G. J. 1975. Phylogenetic trends in the Alismatidae (Monocotyledoneae). Bot. Gaz. 136: 159–170.

Мейер, Н. Р. 1966. Изучение морфологии пыльцы представителей Nymphaeaceae и группы Helobiae для систематики и филогении. В сб.: Значение палинологического анализа для стратиграфии и палеофлористики: 30–41. Изд. Наука, Москва.

SYNOPTICAL ARRANGEMENT OF THE ORDERS OF ALISMATIDAE

1 Seeds without endosperm; plants aquatic or semi-aquatic, green, not mycotrophic.

  2 Perianth generally differentiated into evident sepals and petals; flowers often bracteate.

    3 Flowers hypogynous; carpels distinct or only basally connate .... 1. ALISMATALES.

    3 Flowers epigynous, with a compound ovary and parietal or often a modified sort of laminar placentation .... 2. HYDROCHARITALES.

  2 Perianth, when present, not differentiated into sepals and petals; bracts wanting or small and inconspicuous, except in Scheuchzeriaceae, but the perianth sometimes consisting of only a single, bract-like tepal .... 3. NAJADALES.

1 Seeds with well developed endosperm; plants terrestrial, mycotrophic, without chlorophyll .... 4. TRIURIDALES.

1. Order ALISMATALES Lindley 1833

Glabrous or subglabrous, perennial, aquatic or semi-aquatic herbs, strongly aerenchymatous, the xylem only slightly or scarcely lignified; vessels confined to the roots, but tracheids present in the stem and leaves as well as in the roots; vessel-segments with simple or scalariform perforations; vascular bundles of the stem scattered, or sometimes in two concentric rings; well developed schizogenous laticifers, lined with an epithelium, present in the stem and leaves except in Butomaceae; sieve-tube plastids with cuneate proteinaceous inclusions. LEAVES alternate, closely crowded near the tip of the rhizome or stolon, thus commonly appearing basal, usually somewhat expanded and sheathing (but not closed) at the base, sometimes differentiated into blade and petiole, less often not, the blade when present commonly with an acrodromous or campylodromous variant of parallel venation; stomates paracytic or tetracytic; stem with intravaginal scales at the nodes. Scape or peduncle terminating in an involucrate, cymose umbel, or in a raceme or panicle, or sometimes in a solitary flower; FLOWERS generally each subtended by a bract, regular, hypogynous, trimerous or in part polymerous, perfect or seldom some or all of them unisexual; perianth of 3 green or seldom anthocyanic, usually persistent sepals and 3 white or less often anthocyanic, deciduous or less often persistent petals; stamens 3-many, when numerous associated with trunk-bundles and originating in either centripetal or centrifugal sequence; typically the first-formed stamens arranged in antepetalous pairs to form a cycle of 6, each such pair of stamens arising together with a petal from a common primordium; anthers tetrasporangiate and dithecal, opening by longitudinal slits; pollen-grains trinucleate, monosulcate in the Butomaceae, otherwise with 2-many pores or seldom inaperturate; gynoecium of 3–20 distinct (or basally connate) carpels in a single whorl, or of more or less numerous, distinct carpels; carpels when numerous appearing on first inspection to be spirally arranged, but actually in successive whorls, the later (centripetal) whorls with fewer members; nectaries receptacular, between or amongst the carpels, or at the base of and between some of the various floral appendages, often at the basal margins of the carpels; ovules more or less numerous and scattered over the inner surface of the carpel, or solitary (-several) and basal or basal-marginal, anatropous to amphitropous or campylotropous, bitegmic, pseudocrassinucellar; endosperm wanting; embryo straight or horseshoe-shaped, with a single terminal cotyledon and lateral plumule, germination epigeal.

The order Alismatales as here defined consists of 3 families and fewer than a hundred species, of cosmopolitan distribution. The Alismataceae are by far the largest family.

The three families of Alismatales exploit essentially similar habitats in essentially similar ways. It is reasonable to suppose that the laticifers of the Limnocharitaceae and Alismataceae serve some sort of protective function, and one could further speculate that they may be an important factor in the greater success of these two families, as compared to the Butomaceae. It is also possible to surmise that there is some survival-value in having pollen-grains with several apertures instead of only one, although the survival value may well be only for the individual pollen grain rather than for the species (like the red throats of nestling birds). It might also be possible to see a difference in adaptive strategies between follicular and achaenial fruits, although in this instance there is no obvious difference in the results. A more determined imagination is required to see survival-value or ecologic differentiation in the disposition of the ovules, the shape of the embryo, and the number of megaspores involved in the development of the embryo-sac.

SELECTED REFERENCES

Argue, C. L. 1971. Pollen of the Butomaceae and Alismataceae. I. Development of the pollen wall in *Butomus umbellatus* L. Grana 11: 131–144.

Argue, C. L. 1976. Pollen studies in the Alismataceae with special reference to taxonomy. Pollen & Spores 18: 161–201.

Charlton, W. A. 1973. Studies in the Alismataceae. I. Developmental morphology of *Echinodorus tenellus*. II. Inflorescences of Alismataceae. Canad. J. Bot. 46: 1345–1360, 1968. 51: 775–789, 1973.

Charlton, W. A., & A. Ahmed. 1973. Studies in the Alismataceae. III. Floral anatomy of *Ranalisma humile*. IV. Developmental morphology of *Ranalisma humile* and comparisons with two members of the Butomaceae, *Hydrocleis nymphoides* and *Butomus umbellatus*. Canad. J. Bot. 51: 891–897; 899–910.

Daumann, E. 1964. Zur Morphologie der Blüte von *Alisma plantago-aquatica* L. Preslia 36: 226–239.

Johri, B. M. 1936. The life-history of *Butomopsis lanceolata* Kunth. Proc. Indian Acad. Sci. B 4: 139–162.

Johri, B. M. 1938. The embryo-sac of *Hydrocleis nymphoides* Buchen. Beih. Bot. Centralbl. 58(A): 165–172.

Johri, B. M. 1938. The embryo-sac of *Limnocharis emarginata* L. New Phytol. 37: 279–285.

Kaul, R. B. 1965. Development and vasculature of the androecium in the Butomaceae. Amer. J. Bot. 52: 624. (Abstract).

Kaul, R. B. 1967. Development and vasculature of the flowers of *Lophotocarpus calycinus* and *Sagittaria latifolia* (Alismaceae). Amer. J. Bot. 54: 914–920.
Kaul, R. B. 1967. Ontogeny and anatomy of the flower of *Limnocharis flava* (Butomaceae).Amer. J. Bot. 54: 1223–1230.
Kaul, R. B. 1968. Floral development and vasculature in *Hydrocleis nymphoides* (Butomaceae). Amer. J. Bot. 55: 236–242.
Kaul, R. B. 1976. Conduplicate and specialized carpels in the Alismatales. Amer. J. Bot. 63: 175–182.
Leins, P., & P. Stadler. 1973. Entwicklungsgeschichtliche Untersuchungen am Androeceum der Alismatales. Oesterr. Bot. Z. 121: 51–63.
Markgraf, F. 1936. Blütenbau und Verwandschaft bei den einfachsten Helobiae. Ber. Deutsch. Bot. Ges. 54: 191–229.
Mayr, F. 1943. Beiträge zur Anatomie der Alismataceen. Die Blattanatomie von *Caldesia parnassifolia* (Bassi) Parl. Beih. Bot. Centralbl. 62: 61–77.
Pichon, M. 1946. Sur les Alismatacées et les Butomacées. Notul. Syst. (Paris)12: 170–183.
Rao, Y. S. 1953. Karyo-systematic studies in Helobiales. I. Butomaceae. Proc. Natl. Inst. Sci. India 19: 563–581.
Roper, R. B. 1952. The embryo-sac of *Butomus umbellatus* L. Phytomorphology 2: 61–74.
Sattler, R., & V. Singh. 1973. Floral development of *Hydrocleis nymphoides*. Canad. J. Bot. 51: 2455–2458.
Sattler, R., & V. Singh. 1977. Floral organogenesis of *Limnocharis flava*. Canad. J. Bot. 55: 1076–1086.
Sattler, R., & V. Singh. 1978. Floral organogenesis of *Echinodorus amazonicus* Rataj and floral construction of the Alismatales. J. Linn. Soc., Bot. 77: 141–156.
Singh, V., & R. Sattler. 1972. Floral development of *Alismà triviale*. Canad. J. Bot. 50: 619–627.
Singh, V., & R. Sattler. 1973. Nonspiral androecium and gynoecium of *Sagittaria latifolia*. Canad. J. Bot. 51: 1093–1095.
Singh, V., & R. Sattler. 1974. Floral development of *Butomus umbellatus*. Canad. J. Bot. 52: 223–230.
Singh, V., & R. Sattler. 1977. Development of the inflorescence and flower of *Sagittaria cuneata*. Canad. J. Bot. 55: 1087–1105.
Stant, M. Y. 1964. Anatomy of the Alismataceae. J. Linn. Soc., Bot 59: 1–42.
Stant, M. Y. 1967. Anatomy of the Butomaceae. J. Linn. Soc., Bot. 60: 31–60.
Troll, W. 1932. Beiträge zur Morphologie des Gynaeceums. II. Über das Gynaeceum von *Limnocharis* Humb. & Bonpl. Planta 17: 453–460.
Wodehouse, R. P. 1936. Pollen grains in the identification and classification of plants. VIII. The Alismataceae. Amer. J. Bot. 23: 535–539.

Мейер, Н. Р. 1966. О развитии пыльцевых зерен Helobiae **и их связи с нимфейными. Бот. Ж. 51: 1736-1740.**

SYNOPTICAL ARRANGEMENT OF THE FAMILIES OF ALISMATALES

1 Pollen-grains monosculcate; embryo sac monosporic; embryo straight; plants without secretory canals; leaves linear, not differentiated into blade and petiole; placentation laminar ...... 1. BUTOMACEAE.

1 Pollen-grains (2–)4-many porate, or sometimes inaperturate; embryo sac bisporic; embryo horseshoe-shaped; plants usually with schizogenous secretory canals that are lined with an epithelium; leaves typically but not always with a petiole and an expanded blade; placentation various.
   2 Ovules several or many, scattered over the inner surface of the carpel; fruits dehiscent ....2. LIMNOCHARITACEAE.
   2 Ovules solitary, seldom 2-several, generally ventral-basal; fruits indehiscent, or seldom dehiscent at the base ....3. ALISMATACEAE.

1. Family BUTOMACEAE L. C. Richard 1815 nom. conserv., the Flowering Rush Family

Glabrous, perennial, emergent aquatic herbs from creeping, dorsiventral, starchy, edible rhizomes, without secretory canals, strongly aerenchymatous, the xylem not much lignified; vessels confined to the roots, with transverse simple perforations; tracheids present in the stem and leaves as well as in the roots; rhizome with scattered tanniferous cells (containing proanthocyanins) and with small crystals in a few of the cells; sieve-tube plastids with cuneate proteinaceous inclusions. LEAVES distichous in origin at the tip of the rhizome, parallel-veined, linear, erect and more or less triquetrous, not differentiated into blade and petiole, but the base somewhat expanded and sheathing; stomates mostly paracytic; stem with intravaginal scales at the nodes. Scape axillary, erect, terminating in a complex, cymose umbel subtended by 3 bracts; FLOWERS numerous, perfect, regular, hypogynous, trimerous; perianth of 6 persistent, distinct members in 2 cycles, the 3 sepals a little smaller and more greenish that the 3 pink petals; stamens 9, the outer cycle consisting of 3 pairs of obliquely antesepalous members, the inner (and ontogenetically subsequent) cycle of 3 directly antepetalous members, so that all 9 are at equal angles from the center; anthers basifixed, tetrasporangiate and dithecal, opening by longitudinal slits; pollen-grains trinucleate, monosulcate; gynoecium of 6 conduplicate, distally unsealed carpels, these connate at the very base into a ring, otherwise distinct, each with a short, terminal style ending in a dry, papillate, shortly bilobed and shortly decurrent stigma (i.e., the margins of the pistils stigmatic in the uppermost portion) carpels nectariferous on the lower lateral surfaces; ovules numerous, scattered over the inner surface of the carpel (laminar placentation), anatropous, bitegmic, pseudocrassinucellar; embryo-sac monosporic; endosperm-development helobial. FRUIT of separate follicles; endosperm wanting; embryo straight, with a single terminal cotyledon and lateral plumule. X = 13, the chromosomes small.

The family Butomaceae as here defined consists of the single species *Butomus umbellatus* L., native to temperate Eurasia.

Seeds thought to represent *Butomus* occur in Oligocene and younger rocks. Pollen similar to that of *Butomus* (and some other monocotyledonous genera) dates back at least to the Miocene.

2. Family LIMNOCHARITACEAE Takhtajan 1954,[1] the Water-poppy Family

Glabrous or subglabrous, perennial, aquatic herbs, free-floating or rooted in the substrate and emergent, strongly aerenchymatous, the xylem scarcely lignified; vessels confined to the roots, with more or less oblique, scalariform perforations; stem and leaves with thin-walled tracheids (these also in the roots) and well developed schizogenous laticifers, these lined with an epithelium; no tannin; no crystals; sieve-tube plastids with cuneate proteinaceous inclusions. LEAVES in a close spiral (almost distichous in origin) at the tip of the rhizome or floating stolon, petiolate, with a more or less expanded blade showing an acrodromous, perfect, basal or suprabasal variant of parallel venation; stomates mostly paracytic; stem with intravaginal scales at the nodes. Scape or peduncle terminating in an involucrate, cymose umbel or in a solitary flower; FLOWERS perfect, regular, hypogynous, trimerous; perianth of 3 green, persistent sepals and 3 larger, deciduous petals; stamens 3-many, in the latter case associated with trunk-bundles and originating in centrifugal sequence, the outer ones often staminodial; anthers tetrasporangiate and dithecal, opening by longitudinal slits; pollen-grains globose, inaperturate or with 4-many pores; gynoecium of 3, or 5–9, or 12–20 carpels in a single whorl, these distinct except sometimes at the very base, conduplicate and distally unsealed, each with a short, terminal style ending in a shortly decurrent stigma, or the stigma sessile; lateral surface of the carpels nectariferous toward the base; ovules numerous, scattered over the inner surface of the carpel, anatropous to more or less campylotropous, bitegmic, pseudocrassinucellar; embryo-sac bisporic; endosperm-development helobial. FRUIT of distinct follicles; endosperm wanting; embryo horseshoe-shaped, with a single terminal cotyledon and lateral plumule; germination epigeal. X = 7, 8, 10, the chromosomes large.

The family Limnocharitaceae consists of 3 genera (*Limnocharis, Hydrocleys,* and *Tenagocharis*) and 7–12 species, native to tropical and subtropical regions in both the Old and the New World. *Hydrocleys*

[1]A Latin diagnosis, not given by Takhtajan, is provided here:
Herbae aquaticae laticiferae; flores hermaphroditi hypogyni, sepalis 3, petalis 3, staminibus 3-numerosis; pollinis grana inaperturata vel 3-∞-porata; carpella 3–20 unicycla, distincta vel tantum basi connata, placentatione laminali, ovulis numerosis; fructus ex folliculis constituti; semina exalbuminosa, embryone hippocrepidiformi, cotyledone solitario. Type: *Limnocharis* Humb. & Bonpl. Pl. Aequin. 1: 116. 1808.

nymphoides (Willd.) Buchenau, the water-poppy, is occasionally grown as an aquarium-plant.

In *Limnocharis* the carpels arise initially as 3 groups of 3 in antesepalous position, followed by additional carpels so as to produce a whorl of 12–20. In *Hydrocleys* the carpels all arise simultaneously in a single whorl.

The Limnocharitaceae have usually been included in the Butomaceae, because of the laminar placentation and follicular fruits of the two groups. In a series of other features the Limnocharitaceae differ from *Butomus* and resemble the Alismataceae. It therefore seems appropriate to follow Takhtajan in treating the Limnocharitaceae as a distinct family.

3. Family ALISMATACEAE Vent. 1799 nom. conserv., the Water-plantain Family

Glabrous, perennial, usually aquatic (or marsh) herbs from creeping rhizomes, producing C-glycosyl flavones and possible steroidal saponins, but typically without proanthocyanins, strongly aerenchymatous, the xylem scarcely lignified; vessels confined to the roots, with simple or scalariform perforations; stem and leaves with thin-walled tracheids (these also in the roots) and well developed schizogenous laticifers, these lined with an epithelium (*Sagittaria* said to lack laticifers and have tanniferous cells); various sorts of single and clustered crystals present in some of the cells of the parenchymatous tissues; sieve-tube plastids with cuneate proteinaceous inclusions. LEAVES basal, fundamentally alternate, commonly with a well developed, basally sheathing but open petiole and an elliptic to sagittate or hastate blade showing an acrodromous or campylodromous variant of parallel venation, but sometimes the blade reduced or suppressed (especially in deep-water forms) and the petiole flattened to form a somewhat grass-like, parallel-veined leaf; stomates tetracytic; stem with intravaginal scales at the nodes. Scape ending in a terminal inflorescence with the pedicels or primary branches commonly in whorls of 3; FLOWERS axillary to bracts, hypogynous, regular, perfect or seldom some or all of them unisexual; perianth of 3 imbricate green sepals and 3 imbricate, deciduous, usually white petals; stamens (3) 6-many, when 6 borne in a single cycle composed of a lateral pair for each petal, when numerous seemingly spiralled, but actually whorled, associated with trunk-bundles, and originating in centripetal sequence, the upper cycles with successively fewer members; anthers tetrasporangiate and dithecal, extrorse, opening by longitudinal

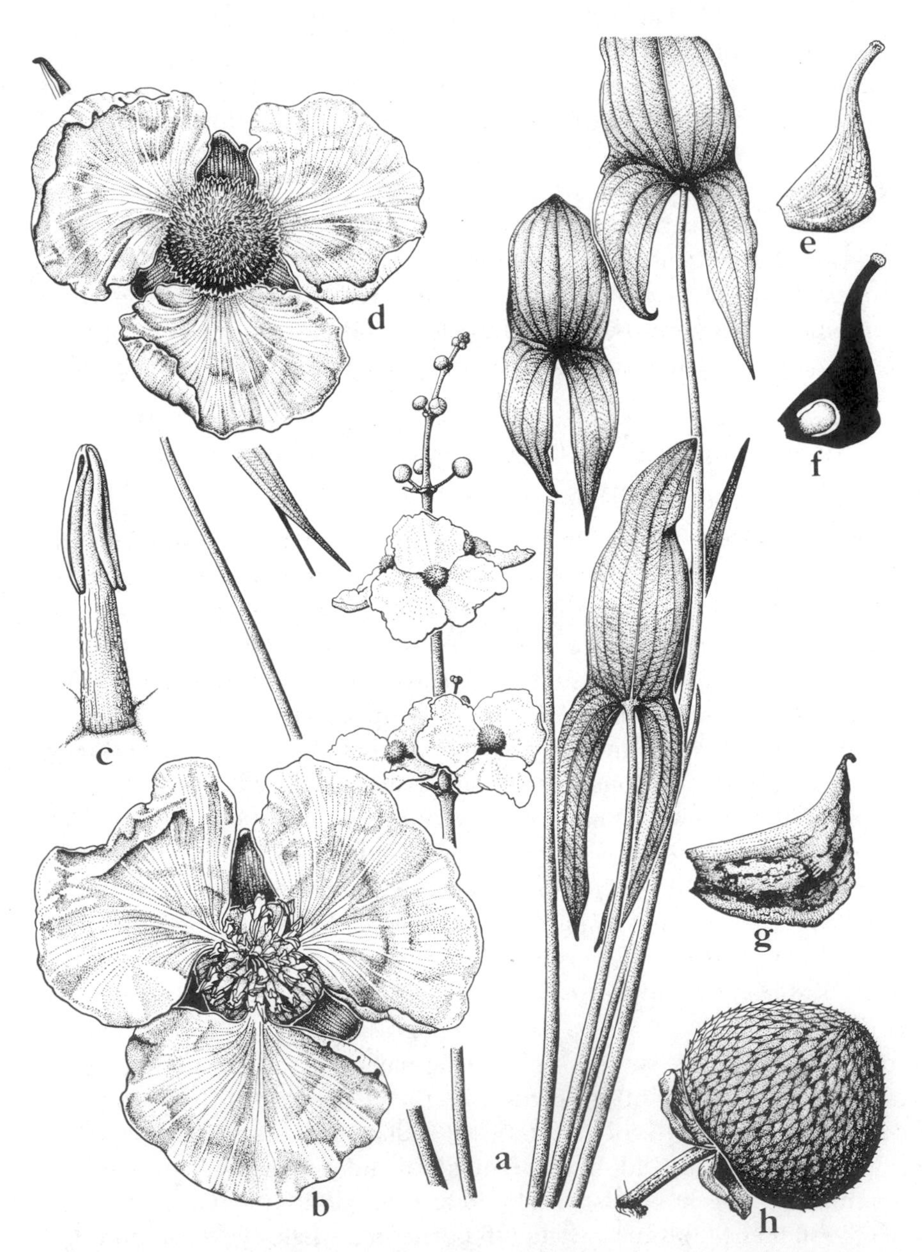

FIG. 7.4 Alismataceae. *Sagittaria latifolia* Willd. a, habit, $\times\frac{1}{2}$; b, staminate flower, $\times 2$; c, stamen, $\times 12$; d, pistillate flower, $\times 2$; e, pistil, $\times 12$; f, pistil, the ovary in long-section, $\times 12$; g, achene, $\times 6$; h, fruiting head of achenes, $\times 2$.

slits; pollen-grains globose, trinucleate, mostly pantoporate, with 9–29 pores, seldom (*Caldesia*) 2 (3)-porate; gynoecium of 3, 6, or more or less numerous, distinct, usually closed carpels in a single whorl or in an apparent spiral, in the latter case actually in successive cycles with fewer members in the upper cycles; each carpel with a terminal or basilateral style and often with a decurrent stigma; nectaries borne on the receptacle at the base of and between some of the various floral appendages, or often at the basal margins of the carpels; ovules solitary (-several) in each carpel, ventral-basal, in *Damasonium* reported to be transitional to laminar, variously anatropous to amphitropous, bitegmic, weakly pseudocrassinucellar, sometimes with a weakly developed integumentary tapetum; embryo-sac bisporic; endosperm-development helobial or seldom nuclear. FRUIT of distinct achenes or seldom of basally dehiscent follicles; endosperm wanting; embryo horse-shoe shaped, with a single terminal cotyledon and lateral plumule; germination epigeal. X = 5–13.

The family Alismataceae consists of about a dozen genera and 75 species, cosmopolitan in distribution, but best developed in the Northern Hemisphere. The largest genera are *Echinodorus* (25), and *Sagittaria* (20, arrowhead). *Alisma* (5), the water-plantain, is also common and widespread.

A single leaf that looks very much like *Sagittaria* has been found in Cenomanian deposits (Dakota sandstone) in Kansas (personal communication from David Dilcher), but in the absence of other material it would be premature to make a positive identification. Leaves and associated fruits that may well represent the Alismataceae occur in Paleocene deposits, and more securely identified fruits and seeds date from the Oligocene.

2. Order HYDROCHARITALES Lindley 1833

The order consists of the single family Hydrocharitaceae.

SELECTED REFERENCES

Ancibor, E. 1979. Systematic anatomy of vegetative organs of the Hydrocharitaceae. J. Linn. Soc., Bot. 78: 237–266.

Baude, E. 1956. Die Embryoentwicklung von *Stratiotes aloides* L. Planta 46: 649–671.

Brunaud, A. 1976, 1977. Ramification chez les Hydrocharitaceae. I. Ontogénie du système des pousses. II. Organisation des rameaux latéraux. Rev. Gén. Bot. 83: 397–413, 1976. 84: 137–157, 1977.

Govindappa, D. A., & T. R. B. Naidu. 1956. The embryo-sac and endosperm of *Blyxa oryzetorum* Hook. f. J. Indian Bot. Soc. 35: 417–422.

Islam, A. S. 1950. A contribution to the life history of *Ottelia alismoides* Pers. J. Indian Bot. Soc. 29: 79–91.

Kaul, R. B. 1968. Floral morphology and phylogeny in the Hydrocharitaceae. Phytomorphology 18: 13–35.

Kaul, R. B. 1969. Morphology and development of the flowers of *Boottia cordata, Ottelia alismoides*, and their synthetic hybrid (Hydrocharitaceae). Amer. J. Bot. 58: 951–959.

Kaul, R. 1970. Evolution and adaptation of inflorescences in the Hydrocharitaceae. Amer. J. Bot. 57: 708–715.

Lakshmanan, K. K. 1961, 1963. Embryological studies in the Hydrocharitaceae. I. *Blyxa octandra* Planch. J. Madras Univ. 31(B): 133–142, 1961. II. *Halophila ovata* Gaudich. J. Indian Bot. Soc. 42: 15–18, 1963. III. *Nechamandra alternifolia*. Phyton (Buenos Aires) 20: 49–58.

Rangasamy, K. 1941. A morphological study of the flower of *Blyxa echinosperma* Hook. f. J. Indian Bot. Soc. 20: 123–133.

Troll, W. 1931. Beiträge zur Morphologie des Gynaeceums. I. Über das Gynaeceum der Hydrocharitaceen. Planta 14: 1–18.

1. Family HYDROCHARITACEAE A. L. de Jussieu 1789 nom. conserv. the Tape-grass Family

Perennial (rarely annual) herbs of fresh water or coastal marine water, wholly submerged or partly emergent, rooted to the substrate or sometimes free-floating, commonly with scattered tanniferous cells containing proanthocyanins, but without ellagic acid, without alkaloids, and only seldom cyanogenic; anthocyanins produced by some spp.; crystals wanting; vessels confined to the roots, elongate, with oblique end-walls and scalariform perforation-plates, or often wanting; plastids of the sieve-tubes with cuneate proteinaceous inclusions; stem aerechymatous, with a much-reduced conducting system, the mechanical tissues consisting mostly of collenchyma. LEAVES basal or cauline, variously alternate, opposite, or whorled, often more or less sheathing or stipular-expanded at the base, consistently subtending intravaginal (axillary) scales, sometimes differentiated into a blade and petiole, the blade with an acrodromous or campylodromous variant or parallel venation, sometimes narrow, elongate and sessile, with parallel veins or only a midrib; leaf-margins very often beset with unicellular, thick-walled prickle-hairs (these seldom present also along the main veins), the plants otherwise usually glabrous, seldom with long unicellular hairs; epidermis chlorophyllous; stomates, when present, mostly paracytic, but more often wanting. FLOWERS small and inconspicuous to fairly large and showy, regular or (*Vallisneria*) slightly irregular, epigynous, perfect or more often unisexual, the plants then dioecious or sometimes monoecious; inflorescence a compact, usually few-flowered cyme (especially the staminate flowers) or a single flower (especially the pistillate flowers), collectively subtended by (1) 2 distinct or more often more or less connate bracts that form a sessile to long-pedunculate spathe; perianth commonly of 3 distinct, green sepals and 3 distinct, white or colored petals, these attached directly to the summit of the ovary or elevated on a long, slender hypanthium, or the petals sometimes wanting, rarely the whole perianth wanting; stamens 2 or 3 to numerous in 1-several trimerous or ancestrally trimerous cycles, sometimes paired in front of the sepals, when numerous developing in centripetal sequence, the inner or the outer ones sometimes staminodial and nectariferous; anthers tetrasporangiate or seldom bisporangiate, dithecal, opening by longitudinal slits; pollen grains globose, monosulcate or more often inaperturate, monadinous, or in 2 marine genera (*Thalassia* and *Halophila*) united into threadlike moniliform chains; male flowers often breaking loose from the submerged inflorescence and floating on the surface, as commonly in *Elodea, Vallisneria, Enhalus, Hydrilla, Lagarosiphon,* and *Nechamandra*, but in some spp. of *Elodea* the pollen grains are shed and float to the surface; gynoecium of (2) 3–6 (–20) carpels weakly united to form a compound, inferior, unilocular ovary, often with more or less deeply intruded partial partitions (the slightly or scarcely connate carpellary margins); styles as many as the carpels, often bilobed or bifid, sometimes shortly connate at the base into a common style; stigmas dry, papillate; placentation laminar, with the ovules scattered over the surfaces of the partial partitions, or merely parietal when the partitions are not intruded; ovules anatropous or seldom orthotropous, bitegmic, crassinucellar; embryo-sac monosporic; endosperm-development helobial. FRUIT submerged, globose to linear, dry or fleshy, usually opening irregularly; seeds several to numerous, without endosperm, or seldom (as in *Ottelia*) with scanty endosperm; embryo straight, with a stout radicle, a small, lateral plumule, and a single obliquely terminal cotyledon; germination epigeal. X = 7–12.

The family Hydrocharitaceae consists of about 15 genera and scarcely a hundred species, of cosmopolitan distribution. The most familiar genera are *Elodea* (*Anacharis*, 15) and *Vallisneria* (ca 8), both frequently grown in aquaria. The three marine genera (*Enhalus, Halophila,* and *Thalassia*) are of particular interest. The chains of pollen-grains in *Thalassia* and *Halophila* appear to be functionally comparable to the filamentous grains of the Zosteraceae. *Thalassia testudinum* Banks ex König, the Caribbean turtle-grass, grows at depths of as much as 30 m.

Seeds confidently assigned by paleobotanists to the modern genus *Stratiotes* occur in deposits of Eocene and more recent age. Pollen assigned to *Stratiotes* dates from the Oligocene. Remains of *Ottelia* have also been recorded from the Eocene. Upper Cretaceous rhizomes described as *Thalassiocharis* have also been attributed to the Hydrocharitaceae by some authors.

3. Order NAJADALES Nakai, 1930

Perennial (seldom annual) herbs, aquatic or semi-aquatic, rooted to the substrate, submerged or emergent; xylem only slightly or scarcely lignified; vessels confined to the roots, or wanting; plastids of the sieve-tubes with cuneate proteinaceous inclusions (at least in the Potamogetonaceae). LEAVES basal or cauline, alternate or less often opposite or whorled, commonly more or less sheathing at the base, and often with a ligule at the juncture of the sheath with the blade or petiole, or the sheath and blade sometimes separately attached at the node, so that the sheath appears as an axillary stipule; portion of the leaf above the sheath sometimes narrow and not differentiated into blade and petiole, then generally with only a midvein or a few parallel veins, sometimes with a slender petiole and an expanded blade with an acrodromous variant of parallel venation; stomates when present commonly paracytic, but sometimes tetracytic; stem generally with intravaginal scales (or long hairs) at the nodes. FLOWERS mostly small and inconspicuous, mostly anemophilous or hydrophilous, regular or irregular, hypogynous, perfect or unisexual, borne in terminal or axillary spikes or racemes, or in small axillary sympodial cymes, or solitary in the axils, not obviously bracteate except in *Scheuchzeria* (inconspicuously so in Cymodoceaceae); perianth of 6 distinct tepals in 2 sets of 3 (or 2 sets of 2), but not differentiated into calyx and corolla, or of 1–4 tepals in a single series, or wanting; stamens 1–6, or in the Aponogetonaceae sometimes more numerous, distinct, or sometimes connate, typically tetrasporangiate and dithecal, but sometimes unisporangiate or bisporangiate and monothecal, opening by longitudinal slits; pollen-grains trinucleate or rarely binucleate, monosulcate or more often inaperturate, often without an exine; gynoecium of 1-several separate carpels, or less often the carpels united to form a unilocular compound ovary, or the carpels adnate to a central axis from which they later separate; ovules solitary in each locule or separate carpel, or less often 2-several, variously anatropous, campylotropous, or orthotropous, bitegmic, crassinucellar or pseudocrassinucellar; embryo-sac monosporic or seldom bisporic; endosperm-development helobial or less commonly nuclear, rarely cellular. FRUIT of distinct follicles, or follicular mericarps, or an achene or drupelet; seeds without endosperm, embryo with a single cotyledon.

The order Najadales as here defined consists of 10 families and a little more than 200 species. The largest family is the Potamogetonaceae, with perhaps as many as a hundred species in the single genus *Potamogeton*.

The Najadales are Alismatidae in which the perianth, when present, is not differentiated into evident sepals and petals. Except in *Scheuchzeria* the flowers are not individually subtended by bracts, but the perianth sometimes consists of a single bract-like tepal.

Scheuchzeria forms a sort of connecting link between the Alismatales and Najadales. It resembles the Alismatales in its biseriate perianth and in having each flower subtended by a bract, but otherwise it appears to be a relatively archaic member of the Najadales. The Aponogetonaceae stand somewhat apart from the rest of the order and may perhaps represent a separate reduction from the Alismatales. The remaining families of Najadales all evidently hang together, although the differences among them are sharp enough.

The evolutionary history of the Najadales is in large part a story of floral reduction associated with progressive adaptation to aquatic and eventually marine habitats. The Scheuchzeriaceae and Juncaginaceae are typically emergent plants of marshy places. The Aponogetonaceae and Potamogetonaceae are fresh-water aquatics with submerged or floating leaves, but often with emergent inflorescences. The Najadaceae, Ruppiaceae, and Zannichelliaceae are submerged aquatics of fresh or brackish water, and the Posidoniaceae, Zosteraceae and Cymodoceaceae are submerged marine plants.

The reduction of the perianth in the Najadales doubtless reflects abandonment of insect-pollination, but the concomitant reduction in number of stamens, carpels, and ovules is more difficult to interpret in ecological terms. The thread-like pollen-grains of the Posidoniaceae, Zosteraceae and Cymodoceaceae may well be related to the submersed, marine habitat of these three families. It is not hard to believe that in moving water a long thread has more chance of brushing against a stigma than does a little ball. It is interesting to note that two submersed marine genera of the Hydrocharitaceae have the pollen-grains united into thread-like tetrads—apparently a different means of achieving a similar result.

SELECTED REFERENCES

Agrawal, J. S. 1952. The embryology of *Lilaea subulata* H. B. K. with a discussion on its systematic position. Phytomorphology 2: 15–29.

Arber, A. 1940. Studies in flower structure. VI. On the residual vascular tissue in the apices of reproductive shoots, with special reference to *Lilaea* and *Amherstia*. Ann. Bot. (London) II 4: 617–627.

Dahlgren, K. V. O. 1939. Endosperm- und Embryobildung bei *Zostera marina*. Bot. Not. 1939: 607–615.

Daumann, E. 1963. Zur Frage nach dem Ursprung der Hydrogamie. Zugleich ein Beitrag zur Blütenökologie von *Potamogeton*. Preslia 35: 23–30.

Gardner, R. O. 1976. Binucleate pollen in *Triglochin* L. New Zealand J. Bot. 14: 115, 116.

Hartog, C. den. 1970. Sea-Grasses of the World. Verh. Kon. Ned. Akad. Wetensch. Afd. Natuurk. Tweede Sect. 59(1): 1–275.

Haynes, R. R. 1977. The Najadaceae in the southeastern United States. J. Arnold Arbor. 58: 161–170.

Haynes, R. R. 1978. The Potamogetonaceae in the southeastern United States. J. Arnold Arbor. 59: 170–191.

Isaac, F. M. 1969 (1970). Floral structure and germination in *Cymodocea ciliata*. Phytomorphology 19: 44–51.

Kay, Q. O. N. 1971. Floral structure in the marine angiosperms *Cymodocea serrulata* and *Thalassodendron ciliatum* (*Cymodocea ciliata*). J. Linn. Soc., Bot. 64: 423–429.

Kirkman, H. 1975. Male floral structure in the marine angiosperm *Cymodocea serrulata* (R. Br.) Ascherson & Magnus (Zannichelliaceae). J. Linn. Soc., Bot. 70: 267–268.

Kuo, J. 1978. Morphology, anatomy and histochemistry of the Australian seagrasses of the genus *Posidonia* König (Posidoniaceae). I. Leaf blade and leaf sheath of *Posidonia australis* Hook. f. Aquatic Bot. 5: 163–170.

Kuo, J., & M. L. Cambridge. 1978. Morphology, anatomy and histochemistry of the Australian seagrasses of the genus *Posidonia* König (Posidoniaceae).II. Rhizome and root of *Posidonia australis* Hook. f. Aquatic Bot. 5: 191–206.

Posluszny, U., & R. Sattler. 1973. Floral development of *Potamogeton densus*. Canad. J. Bot. 51: 647–656.

Posluszny, U., & R. Sattler. 1974. Floral development of *Potamogeton richardsonii*. Amer. J. Bot. 61: 209–216.

Posluszny, U., & R. Sattler. 1974. Floral development of *Ruppia maritima* var. *maritima*. Canad. J. Bot. 52: 1607–1612.

Posluszny, U., & R. Sattler. 1976. Floral development of *Zannichellia palustris*. Canad. J. Bot. 54: 651–662.

Posluszny, U., & R. Sattler. 1976. Floral development of *Najas flexilis*. Canad. J. Bot. 54: 1140–1151.

Posluszny, U., & P. B. Tomlinson. 1977. Morphology and development of floral shoots and organs in certain Zannichelliaceae. J. Linn. Soc., Bot. 75: 21–46.

Reinecke, P. 1964. A contribution to the morphology of *Zannichellia aschersoniana* Graebn. J. S. African Bot. 30: 93–101.

Roth, I. 1961. Histogenese der Laubblätter von *Zostera nana*. Bot. Jahrb. Syst. 80: 500–507.

Ruijgrok, H. W. L. 1974. Cyanogenese bei *Scheuchzeria palustris*. Phytochemistry 13: 161–162.

Sane, Y. K. 1939. A contribution to the embryology of the Aponogetonaceae. J. Indian Bot. Soc. 18: 79–91.

Shah, C. K. 1970. Embryogenesis in Monocotyledons: III. *Aponogeton natans* (L.) Engl. et Krause. Proc. Natl. Acad. Sci., India 40B: 1–5.

Singh, V. 1964–1966. Morphological and anatomical studies in Helobiae. I.

Vegetative anatomy of some members of Potamogetonaceae. III. Vascular anatomy of the node and flower of Najadaceae. IV. Vegetative and floral anatomy of Aponogetonaceae. V. Vascular anatomy of the flower of *Lilaea scilloides* (Poir) Hamm. VII. Vascular anatomy of the flower of *Butomus umbellatus* Linn. Proc. Indian Acad. Sci. B, 60: 214–231, 1964. 61: 98–108; 147–159; 316–325, 1965. 63: 313–320, 1966. II. Vascular anatomy of the flower of Potamogetonaceae. Bot. Gaz. 126: 137–144, 1965.

Singh, V., & R. Sattler. 1977. Floral development of *Aponogeton natans* and *A. undulatus*. Canad. J. Bot. 55: 1106–1120.

Souèges, R. 1943. Embryogénie des Scheuchzériacées. Développement de l'embryon chez le *Triglochin maritimum* L. Compt. Rend. Hebd. Séances Acad. Sci. 216: 746–748.

Stenar, H. 1935. Embryologische Beobachtungen über *Scheuchzeria palustris* L. Bot. Not. 1935: 78–86.

Swamy, B. G. L., & K. K. Lakshmanan. 1962. Contributions to the embryology of the Najadaceae. J. Indian Bot. Soc. 41: 247–267.

Tomlinson, P. B., & U. Posluszny. 1976. Generic limits in the Zannichelliaceae (sensu Dumortier). Taxon 25: 273–279.

Tomlinson, P. B., & U. Posluszny. 1978. Aspects of floral morphology and development in the seagrass *Syringodium filiforme* (Cymodoceaceae).Bot. Gaz. 139: 333–345.

van Bruggen, H. W. E. 1968–1973. Revision of the genus *Aponogeton* (Aponogetonaceae). Blumea 16: 243–265, 1968. 17: 121–137, 1969. 18: 457–487, 1970. Bull. Jard. Bot. Natl. Belg. 43: 193–233, 1973.

Vijayraghavan, M. R., & A. Vidya Kumari. 1974. Embryology and systematic position of *Zannichellia palustris* L. J. Indian Bot. Soc. 53: 292–302.

Wilde, W. J. J. O. de. 1961. The morphological evolution and taxonomic value of the spathe in *Najas*, with descriptions of the new Asiatic-Malaysian taxa. Acta Bot. Neerl. 10: 164–170.

Yamashita, T. 1976. Über die Pollenbildung bei *Halodule pinifolia* und *H. uninervis*. Beitr. Biol. Pflanzen 52: 217–226.

Жилин, С. Г. 1974. Первый третичный вид рола *Aponogeton* (Aponogetonaceae). Бот. Ж. 59: 1203-1206.

Савич, Е. И. 1968. Формирование археспория и происхождение тапетума у Helobiae. Бот. Ж. 53: 514-523.

SYNOPTICAL ARRANGEMENT OF THE FAMILIES OF NAJADALES

1 Ovules (1) 2-several in each of the (2) 3-several distinct or proximally connate carpels; fruits follicular; stamens 6 or more.

2 Aquatic plants with floating leaf-blades or with wholly submersed leaves; tepals 1–3 (–6), sometimes petaloid; inflorescence typically a simple or basally forking spike, bractless, although the perianth sometimes consists of a single bract-like tepal; pollen-grains monosulcate, borne in monads ... 1. APONOGETONACEAE.

2 Emergent marsh-plants; tepals 6, never petaloid; inflorescence a raceme, each pedicel subtended by a bract (unique in the order); pollen-grains inaperturate, borne in dyads ....2. SCHEUCHZERIACEAE.

1 Ovules solitary in each distinct carpel or in each locule of a compound ovary; fruits mostly indehiscent or only tardily and irregularly dehiscent, except in the Juncaginaceae; stamens fewer than 6, except commonly in the Juncaginaceae.

3 Leaves all basal, the plants scapose and with a terminal spike or raceme, commonly largely emergent .... 3. JUNCAGINACEAE.

3 Leaves, or many of them, cauline, sometimes with a floating blade, but not emergent (sometimes exposed at low tide); inflorescence various.

4 Pollen-grains globose or isobilateral, not thread-like; plants of fresh or alkaline or brackish water (*Ruppia* seldom marine).

5 Flowers perfect.

6 Tepals 4; stamens 4; pollen-grains globose; fruiting carpels sessile; ovule ventromarginal, generally near the base; plants of fresh water .... 4. POTAMOGETONACEAE.

6 Tepals 0; stamens 2; pollen-grains of a unique, isobilateral type; fruiting carpels long stipitate; ovule pendulous from the apex; plants chiefly of brackish or alkaline water .... 5. RUPPIACEAE.

5 Flowers unisexual.

7 Pistil solitary, forming a unilocular ovary surmounted by 2–4 elongate stigmas; ovule basal, erect 6. NAJADACEAE.

7 Pistils (1)3–4 (–9), each with a short or elongate style and stigma; ovule ventral-apical, pendulous .... 7. ZANNICHELLIACEAE.

4 Pollen-grains thread-like; plants marine.

8 Flowers perfect; stamens 3, distinct; gynoecium of a single unicarpellate pistil .... 8. POSIDONIACEAE.

8 Flowers unisexual; stamen solitary, or stamens 2 and connate; carpels 2.

9 Gynoecium of 2 separate pistils; stamens 2, united back to back; tanniferous cells present .... 9. CYMODOCEACEAE.

9 Gynoecium of a single unilocular, bicarpellate pistil; stamen solitary; tanniferous cells wanting ...... 10. ZOSTERACEAE.

1. Family APONOGETONACEAE J. G. Agardh 1858 nom. conserv., the Cape-pondweed Family

Perennial, glabrous fresh-water herbs with a short, sympodial, tuberous-thickened rhizome or corm, rooted in the substrate, provided with well developed secretory canals containing oil or latex or tannin (proathocyanins); anthocyanins produced at least by some species; vessels with spiral thickenings, confined to the roots, or vessels wanting. LEAVES all basal, usually with a long petiole and a floating blade that has a few main parallel veins and numerous transverse secondary veins, or sometimes the leaves wholly submersed and with a scarcely expanded blade; small crystals of calcium oxalate present in the mesophyll and to some extent in other parenchymatous tissues; petiole subtending some intravaginal scales at the somewhat sheathing base. Inflorescence exserted from the water on a leafless scape that has a caducous (rarely persistent) spathe just beneath the simple or basally 2–10 forked spike; in *Aponogeton ranunculiflorus* the inflorescence contracted to form a pseudanthium, the sterile flowers reduced to single tepals so as to form a single seemingly polypetalous perianth around the perfect flower; FLOWERS small, hypogynous, perfect or rarely unisexual (the plants then dioecious); perianth of (1) 2 (–6) distinct, commonly persistent, often petaloid tepals, the tepal when solitary sometimes sessile by a broad base and simulating a petaloid bract, or the perianth sometimes wanting; stamens free and distinct, mostly 6 in 2 cycles, or sometimes more numerous (up to about 25) in 3–4 cycles; anthers tetrasporangiate and dithecal, opening by longitudinal slits; pollen-grains ellipsoid, trinucleate or seldom binucleate, monosulcate, borne in monads; gynoecium of (2) 3–6 (–9 or more) carpels, these basally and adaxially coherent for about half or two-thirds of their length, separating at maturity; each carpel narrowed to a short style with a ventral stigmatic groove; nectaries septal; ovules (1) 2–8, basal-marginal, anatropous, bitegmic, crassinucellar; embryo-sac monosporic; endosperm-development helobial. FRUIT of distinct follicles; seeds without endosperm; embryo straight, with a single terminal cotyledon, the plumule often fitting into a lateral groove (or undeveloped). X = 8.

The family Aponogetonaceae consists of the single genus *Aponogeton*, with about 40 species native to the tropics of the Old World and to South Africa. *Aponogeton distachyon* L., the cape pondweed, is sometimes cultivated in aquaria. Fossil leaves considered to represent *Aponogeton* have recently been described from Tertiary deposits near the Aral Sea in the USSR.

2. Family SCHEUCHZERIACEAE Rudolfi 1830 nom. conserv., the Scheuchzeria Family

Rhizomatous perennial herbs of bogs and other very wet places, cyanogenic (through a pathway based on tyrosine) and tanniferous, containing proanthocyanins; crystals of calcium oxalate present in some of the cells of the parenchymatous tissues; vessels with scalariform perforation-plates, confined to the roots. LEAVES alternate, cauline as well as basal, with a prominent ligule at the juncture of the blade with the open sheath; blade elongate, slender, semiterete, with an evident pore at the tip; stomates tetracytic; stem with numerous long, intravaginal hairs at the nodes. FLOWERS in terminal, bracteate racemes, perfect,

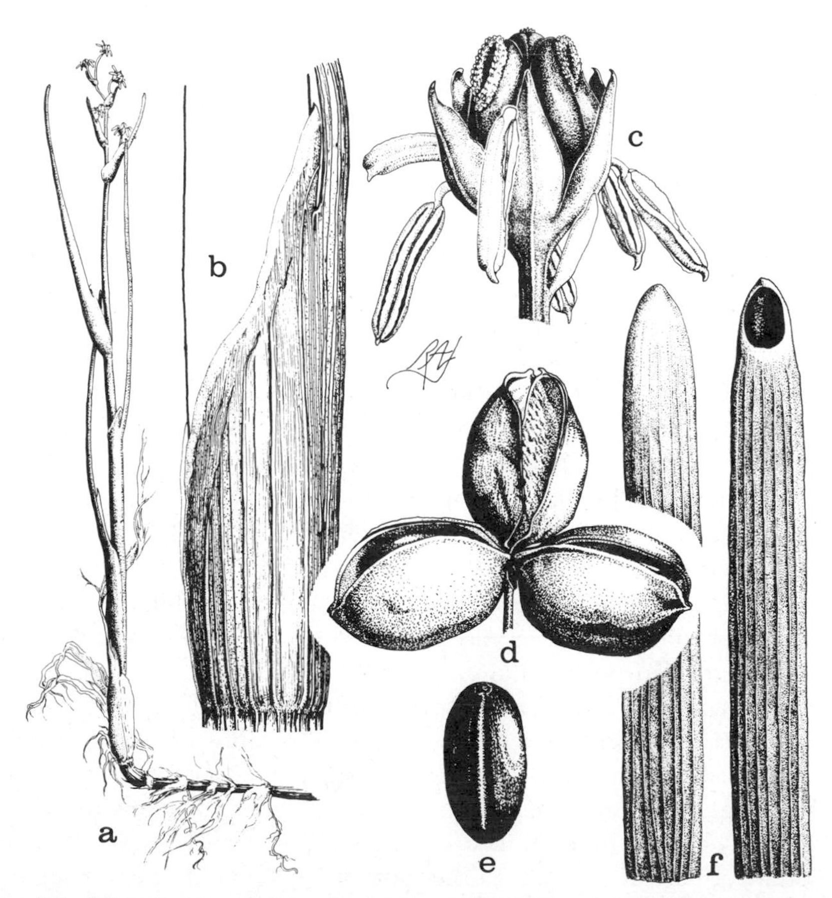

FIG. 7.5 Scheuchzeriaceae. *Scheuchzeria palustris* L. a, habit, $\times \frac{1}{2}$; b, leaf sheath and auricle, $\times 5$; c, flower, $\times 12$; d, fruits, $\times 7$; e, seed, $\times 7$; f, leaf tip, abaxial (left) and adaxial (right) views, showing adaxially subterminal pore, $\times 7$.

anemophilous, hypogynous, trimerous, without nectaries; perianth of 2 similar cycles of 3 distinct, yellow-green tepals; stamens 6, free and distinct; filaments short, anthers elongate, tetrasporangiate and dithecal, extrorse, opening by longitudinal slits; pollen-grains borne in dyads, trinucleate, inaperturate; gynoecium of 3 (–6) carpels, these slightly connate at the base, otherwise distinct, each with a dry, papillate, extrorse, sessile stigma; ovules 2 (-several), basal-marginal, erect, anatropous, bitegmic, crassinucellar; embryo-sac monosporic; endosperm-development helobial. FRUIT of recurved-spreading follicles; seeds without endosperm; embryo with a small plumule and a single rounded cotyledon. N = 11.

The family Scheuchzeriaceae consists of the single species *Scheuchzeria palustris* L., native to the cooler parts of the Northern Hemisphere.

3. Family JUNCAGINACEAE L. C. Richard 1808 nom. conserv., the Arrow-grass Family

Perennial or seldom annual, commonly rhizomatous herbs of bogs and other very wet, often saline places, seldom more distinctly aquatic but with at least the inflorescence emergent or floating; plants with elongate secretory canals, mostly cyanogenic (through a pathway based on tyrosine) and producing C-glycosyl flavones, but not tanniferous, lacking proanthocyanins; crystals of calcium oxalate at least sometimes present in some of the cells of the parenchymatous tissues; vessels usually or always with scalariform perforations, confined to the roots. LEAVES mainly or wholly basal, alternate, with an evident ligule at the juncture of the slender blade with the open sheath (the blade suppressed in *Maundia*), or the elongate leaf not divided into blade and sheath; stomates tetracytic; intravaginal scales present at the nodes. FLOWERS small and inconspicuous, perfect or unisexual, anemophilous, borne in terminal, bractless spikes or racemes (but in *Lilaea* the single tepal of the staminate and perfect flowers bract-like, and some of the pistillate flowers borne in the basal axils); perianth most commonly of 6 similar and distinct tepals in 2 cycles, but sometimes of only 4 biseriate tepals, or 3 uniseriate tepals, and in *Lilaea* the staminate and perfect flowers with a single tepal and the pistillate flowers naked; stamens mostly 6, seldom 3, 4 or 8, and in *Lilaea* only one, the very short filament generally adnate at the base to the base of the tepal; anthers elongate, tetrasporangiate and dithecal, extrorse, opening by longitudinal slits; pollen-grains globose, inaperturate, binucleate or trinucleate; gynoe-

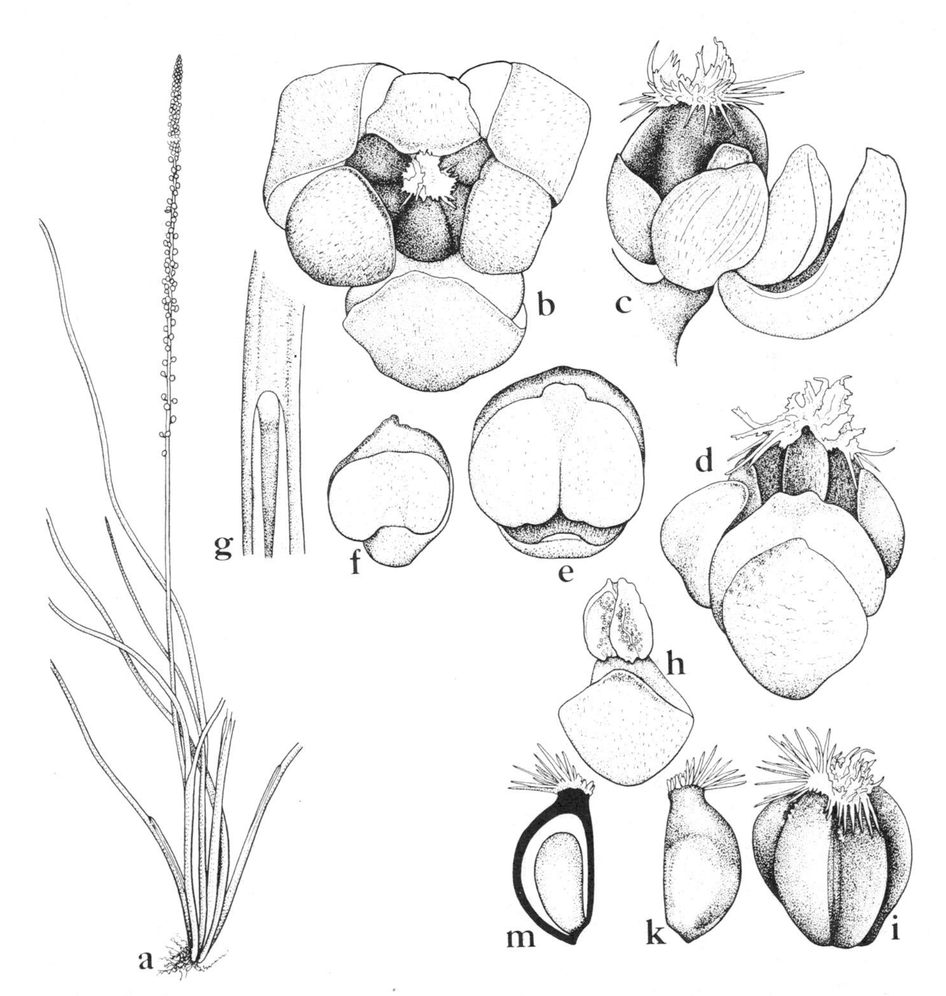

FIG. 7.6 Juncaginaceae. *Triglochin maritimum* L. a, habit, ×½; b, perfect flower from lower part of inflorescence, ×20; c, side view of flower from upper part of inflorescence, with one stamen fully developed, ×20; d, dorsal view of flower from upper part of inflorescence, ×20; e, ventral view of stamen and its outer-cycle tepal, ×20; f, ventral view of stamen and its inner-cycle tepal, ×20; g, ventral view of portion of leaf, showing ligule; h, stamen, after dehiscence, the tepal reflexed, ×20; i, pistil, ×20; k, single carpel, detached from the central axis, ×20; m, single carpel, in long section, ×20.

cium mostly of 6 carpels (but alternate ones sometimes sterile or even obsolete), less often (*Tetroncium, Maundia*) of 4 carpels, these generally adnate to a central axis from which they later separate (distinct in *Cycnogeton*), in *Lilaea* the single ovary unilocular but trigonous and probably tricarpellate; stigma mostly sessile or on a short style, but the basal pistillate flowers of *Lilaea* with an elongate, filiform style; ovules solitary in each locule or carpel, bitegmic and crassinucellar, mostly ventral-basal, erect, and anatropous, but in *Maundia* apical, pendulous, and orthotropous; embryo-sac monosporic; endosperm-development nuclear. FRUIT in *Lilaea* dry and indehiscent, otherwise basically follicular, the mature carpels in *Triglochin, Tetroncium,* and *Maundia* separating from the persistent central axis and opening ventrally; seed solitary, without endosperm; embryo straight, with a single cotyledon. X = 6, 8, 9. (Lilaeaceae)

The family Juncaginaceae as here defined consists of 5 genera and scarcely 20 species, widespread in temperate and cold regions of both the Northern and the Southern Hemisphere. *Triglochin* has about 15 species; the other genera are monotypic. *Triglochin* is an occasional cause of poisoning of livestock, because of its HCN. Otherwise the family is of no economic significance.

4 Family POTAMOGETONACEAE Dumortier 1829 nom. conserv., the Pondweed Family

Glabrous perennial herbs of fresh water, rooted in the substrate, with creeping sympodial rhizomes and erect, leafy branches, producing C-glycosyl flavones and often somewhat tanniferous (with proanthocyanins), lacking anthocyanins, the red pigment of some spp. being rhodoxanthin; crystals and saponins wanting; vessels with scalariform perforation-plates, confined to the roots; xylem in the stem represented mainly by cavities, some spp. with a ring of 8–12 vascular bundles and many small cortical bundles, others with a complete vascular cylinder in which the phloem surrounds the xylem-cavity; nodes trilacunar; plastids of the sieve-tubes with cuneate proteinaceous inclusions. LEAVES alternate or sometimes the uppermost ones (rarely all of them) subopposite, served by 3 traces, parallel-veined (or the blade with a single midvein), with a well developed basal open sheath, the blade or petiole sometimes attached near the top of the sheath, which projects shortly beyond the juncture as a ventral ligule, sometimes attached farther down the sheath or directly to the node, so that the sheath appears to form a large, adaxial or intrapetiolar sheathing stipule; sheath subtending some small scales at the node; sometimes all leaves submersed and with filiform or flattened and grasslike blade, but often the upper leaves with a floating, expanded blade and a slender petiole; stomates paracytic. FLOWERS small, anemophilous, or seldom hydro-

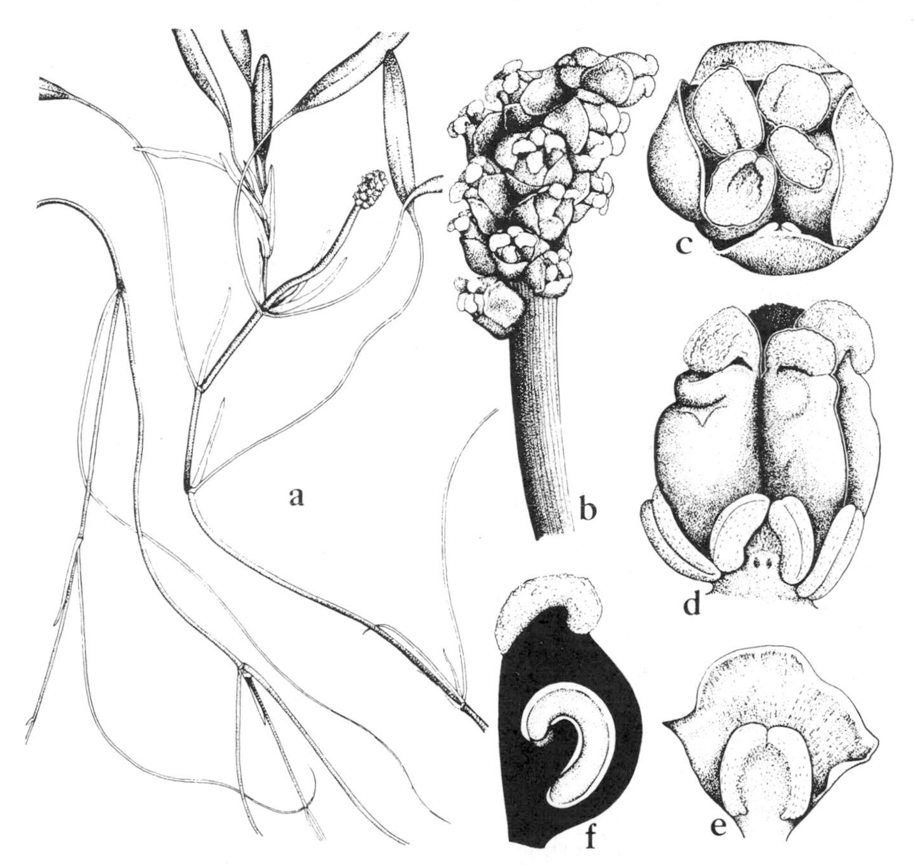

FIG. 7.7 Potamogetonaceae. *Potamogeton oakesianus* Robbins. a, habit, $\times\frac{1}{2}$; b, inflorescence, $\times 3$; c, flower, from above, $\times 12$; d, androecium and gynoecium of flower, the perianth removed, $\times 12$; e, ventral view of tepal and stamen, $\times 12$; f, carpel in partial long-section, $\times 12$.

philous perfect, regular, hypogynous, borne in axillary and terminal, bractless, often fleshy spikes that are elevated a little above the water; perianth a single whorl of 4 distinct, fleshly-firm, valvate, shortly clawed tepals; stamens 4, opposite the tepals and basally adnate to the claw; tepal-primordia initiated before and separately from the stamen-primordia, the two kinds of organs becoming united only during subsequent growth; anthers virtually sessile, dithecal, extrorse, opening by longitudinal slits; pollen-grains ellipsoid, trinucleate, inaperturate but with an elongate vestige of a sulcus; gynoecium of 4 distinct carpels, these alternate with the stamens, each with a short terminal style or a sessile stigma; stigma dry, not papillate; ovule solitary, attached to the ventral margin of the carpel (generally toward the base), bitegmic and crassinucellar, orthotropous at least at first, often becoming campylotropous or anatropous when the embryo-sac is mature; embryo-sac monosporic; endosperm-development helobial. FRUIT of distinct achenes or drupelets, generally buoyant in the water, the pericarp partly aerenchymatous; seed without endosperm; embryo with a large hypocotyl and a single obliquely terminal cotyledon that encloses the plumule. X = 7?, 13–15.

The family Potamogetonaceae consists of the single cosmopolitan genus *Potamogeton*, with perhaps a hundred species. Aside from their ecological role as food-producers in fresh-water habitats, they are of no economic importance.

Fruits attributed to the Potamogetonaceae occur in Paleocene and more recent deposits, and *Potamogeton* is represented by leaves, fruits, seeds, and pollen of Oligocene age in both England and the USSR.

The tepals of *Potamogeton* have sometimes been interpreted as appendages of the stamens. I see no need for such a complicated interpretation, and the separate ontogenetic origin of tepals and stamens argues against it. The interpretation of the flower as pseudanthial is equally unnecessary.

Dr. Rolf Sattler tells me that in some species of *Potamogeton*, such as *P. richardsonii* (Benn.) Rydb., floral bracts are initiated as primordia, but are not visible at maturity. In others such as *P. densus* L. the bracts are completely wanting, even as primordia.

5. Family RUPPIACEAE Hutchinson 1934 nom. conserv., the Ditch-grass Family

Glabrous, submerged aquatic herbs, chiefly of alkaline or brackish water, but sometimes in sea-water, rooted in the substrate; crystals and saponins wanting; vessels wanting; stem monopodial below, sympodial above; flowering stem-tip very compact and congested, ending in a tiny, terminal inflorescence; stem elongating by development of an axillary bud just beneath the inflorescence, the axis again terminating in an inflorescence, etc.; typically 4 leaves present between successive inflorescences. LEAVES alternate or opposite, linear or setaceous, with a single midvein, expanded at the base into a well developed, distally open sheath that is sometimes interpreted as a pair of elongate, membranous stipules adnate to the blade for nearly their whole length; 2 intravaginal scales present in the axil. FLOWERS perfect, very small and inconspicuous, borne in very short, mostly 2-flowered, terminal spikes, each such spike initially concealed in the sheath of the uppermost vegetative leaf, and immediately subtended by a small, inconspicuous, hyaline, sheathing, nonvasculated prophyll that in later growth may come to be associated with the upper flower only; peduncle eventually much-elongate and commonly becoming spirally twisted; tepals wanting, or (according to interpretation) represented by the tiny, abaxial, unvasculated appendage near the tip of the anther-connective; stamens 2, opposite; anthers nearly sessile, tetrasporangiate and dithecal, extrorse, the 2 relatively large pollen-sacs well separated on the expanded connective, opening by longitudinal slits; pollen-grains trinucleate, of a unique, isobilateral type; gynoecium of (2–) 4 (–8) distinct carpels, each forming a pistil with an expanded, ventro-apical stigma, each pistil becoming elevated on a slender stipe in fruit, so that the several pistils of a flower form an umbelliform cluster; ovule solitary, ventral-apical, pendulous, campylotropous, bitegmic, crassinucellar; embryo-sac monosporic; endosperm-development helobial. FRUIT of ovoid, commonly asymmetrical drupelets; seed without endosperm; embryo with a single obliquely terminal cotyledon and a lateral plumule. X = 8, 10.

The family Ruppiaceae consists of a single highly variable, nearly cosmopolitan species, *Ruppia maritima* L. It is generally agreed that *Ruppia* is related to but more advanced than *Potamogeton*, and many authors include the two in the same family. In the context of the order as a whole, the two genera are sufficiently different so that in my opinion they are better considered to represent distinct families.

6. Family NAJADACEAE A. L. de Jussieu 1789 nom. conserv., the Water-nymph Family

Submerged herbs of fresh or sometimes brackish water, rooted in the substrate, at least sometimes tanniferous, but not cyanogenic, and without crystals; stems slender, freely branched, with a much-reduced conducting system; vessels wanting from all organs. LEAVES subopposite or apparently whorled (whorled leaves at least sometimes consist of 2 subopposite leaves, plus a leaf on a very short shoot in one of the axils), linear or subulate, 1-nerved, toothed or entire, somewhat expanded and sheathing at the base, and commonly subtending 2 small axillary scales, not ligulate; stomates wanting. FLOWERS very small and inconspicuous, hydrophilous, solitary (-few) in the axils, unisexual, the plants monoecious or dioecious; generally 2 leaves present on the stem between successive inflorescences, staminate flower nearly always subtended by an outer involucre of scales that are distinct or more often more or less connate into a cup or tube (often called a spathe), and by a very thin and membranous, flask-shaped inner involucre (sometimes considered to be a perianth) with a narrow, more or less bilobed mouth; stamen solitary, basal; anther virtually sessile (but the stalk beneath the inner involucre lengthening in anthesis), tetrasporangiate and dithecal, or (as in *Najas flexilis*) with a single microsporangium, opening irregularly; pollen-grains spheroidal, trinucleate, without an exine; pistillate flower naked, without a perianth or involucre, or with an inconspicuous, membranous involucre more or less adherent to the ovary; pistil solitary, with a single unilocular ovary surmounted by 2–4 elongate stigmas; ovule solitary, basal, erect, anatropous, bitegmic, crassinucellar; embryo-sac monosporic; endosperm-development variously nuclear, helobial, or cellular. FRUIT indehiscent, with a very thin pericarp closely surrounding the seed; seed without endosperm; embryo straight, with a single obliquely terminal cotyledon and a lateral plumule. X = 6, 7.

The family Najadaceae consists of the single genus *Najas*, with about 35 species, of cosmopolitan distribution.

Seeds thought to represent *Najas* occur sparsely in Oligocene and more recent deposits in both England and the USSR.

7. Family ZANNICHELLIACEAE Dumortier 1829 nom. conserv., the Horned pondweed Family

Glabrous, submerged herbs of fresh or alkaline or brackish water, with much-branched, thread-like stems, more or less well differentiated creeping rhizomes, and unbranched roots anchored in the substrate, not saponiferous, and without crystals; flavonoid constituents at least sometimes including flavonoid sulphates; vascular system much-reduced, the stem with a single axial bundle, this with a protoxylem lacuna surrounded by phloem; vessels wanting from all organs. LEAVES alternate and distichous, varying to subopposite, or (in *Zannichellia*) the distal ones in seeming whorls of 3 or 4, all (except the reduced ones of the rhizome) with a linear, 1-nerved or incompletely 3-nerved blade, this typically expanded below into an open basal sheath that is wrapped completely around the stem and is ligulate at the summit, but in *Zannichellia* the blade attached directly to the node or to the lowest part of the sheath, the sheath then well developed but free from the blade for most or all of its length, or obsolete; intravaginal scales filiform, inconspicuous, usually a lateral pair at each node. FLOWERS small and inconspicuous, hydrophilous, in small, axillary, usually complex and sympodial inflorescences, unisexual, the plant monoecious or in *Lepilaena* possibly dioecious; 1–3 leaves present between successive inflorescences; perianth wanting, or of 3 tiny, scale-like, unvasculated, basally connate tepals; androecium originating from a single primordium; stamen solitary, or the stamens sometimes (by interpretation) 2 or 3 and fully connate so that there is only a single apparent anther which then has 8 or 12 microsporangia and 4–6 thecae; anther when simple either tetrasporangiate and dithecal or (*Althenia*) bisporangiate and monothecal; anther opening by one or more longitudinal slits; connective sometimes shortly prolonged; pollen-grains spheroidal, trinucleate, lacking an exine, or sometimes with an exine and monosulcate; gynoecium of (1–) 3–4 (–9) distinct carpels, each with a short or elongate style and a variously capitate or peltate to funnelform, ligular, or feathery, dry stigma; ovule solitary, ventral-apical, pendulous, anatropous by the time the embryo-sac is mature, bitegmic, crassinucellar; embryo-sac bisporic; endosperm-development helobial. FRUIT of achenes or drupelets; seed without endosperm; embryo with a single circinately coiled cotyledon above the swollen hypocotyl; germination epigaeous. X = 6–8.

The family Zannichelliaceae as here defined consists of 4 genera and 7 or 8 species. The familiar genus *Zannichellia* has only a single cosmopolitan species, *Z. palustris* L. *Althenia* (1 or 2) is Mediterranean, *Lepilaena* (4) is Australian, and *Vleisia* (1) is South African.

8. Family POSIDONIACEAE Lotsy 1911 nom. conserv., the Posidonia Family

Glabrous, perennial marine herbs with laterally flattened creeping rhizomes, growing wholly submerged or on rocks that may be exposed at low tide; plants without crystals, but with scattered tanniferous cells containing proanthocyanins; erect stems ending in bundles of distichously arranged leaves; xylem poorly developed; vessels wanting from all organs; vascular bundles closed. Rhizome with a central stele (delimited by an endodermis) that has a central fluted core of xylem with strands of phloem in the hollows (as in typical roots), but also with scattered vascular bundles and fiber-bundles in the cortex; or (in the Mediterranean species) the rhizome with a central stele, some smaller lateral steles all in one plane, and scattered vascular bundles in the cortex. LEAVES alternate, distichous on the rhizomes, linear, parallel-veined, ligulate at the juncture of sheath and blade, the sheath open, persistent after the blade has been shed; intravaginal scales present at the nodes. FLOWERS small, hydrophilous, perfect, proterandrous, sessile, borne in a modified cymose inflorescence that appears to be a terminal, pedunculate, compound spike, each spike subtended by 2 or 4 bracts, but the flowers not individually bracteate; perianth of 3 vestigial scales, or wanting; stamens 3 (4); anthers sessile, tetrasporangiate and dithecal, the 2 well separated pollen-sacs borne toward the base of the outer side of the much-expanded, more or less laminar (and apicallyappendiculate) connective, opening by longitudinal slits; pollen-grains trinucleate, filamentous, without an exine; gynoecium apparently of a single closed carpel, with an oblique, sessile, discoid, irregularly lobed stigma; ovule solitary, orthotropous, pendulous from the carpellary margin. FRUIT small, fleshy, buoyant in water; seed without endosperm; embryo straight, with a prominent hypocotyl, a single small, lateral cotyledon, and a terminal plumule.

The family Posidoniaceae consists of the single genus *Posidonia*, with 3 species, one Mediterranean, the other two Australian. Hartog considers that *Posidonia* is accurately identified from Eocene deposits in Europe, but Daghlian (in ms.) challenges the identification of all fossils assigned to the Posidoniaceae.

Posidonia has sometimes been included in the Zosteraceae or in a broadly defined family Potamogetonaceae (along with *Zostera*). I am not enthusiastic about monotypic families in general, but the inclusion of *Posidonia* in any other family could only be justified as part of a wholesale slaughter of other small families in the Najadales.

9. Family CYMODOCEACEAE N. Taylor 1909 nom. conserv., the Manatee-grass Family

Glabrous, submerged perennial marine herbs from creeping rhizomes, with scattered tanniferous cells (containing proanthocyanins) in leaves and other parts, but without crystals; vessels wanting from all organs; stem in at least some spp. sympodial, the apparently axillary inflorescences being morphologically terminal. LEAVES alternate or apparently opposite, distichous, linear, 3-several-nerved, with an open basal sheath and a ligule at the juncture of sheath and blade; intravaginal scales present at the nodes. FLOWERS small, unisexual (the plants dioecious), hydrophilous, naked, borne in small, bracteate, axillary cymes or racemes of small cymes, or solitary or paired in the axils; anthers paired on a common filament, or virtually sessile, united back to back, tetrasporangiate and dithecal (a total of 8 microsporangia per flower), opening by longitudinal slits, provided with an apical appendage except in *Syringodium*; pollen-grains trinucleate, filamentous, up to about 1 mm long, without exine; gynoecinm of 2 distinct carpels, each forming a pistil with a terminal style that may be bifid or trifid; ovule solitary, pendulous from the apex of the locule, orthotropous, bitegmic, crassinucellar; endosperm-development nuclear. FRUIT small, hard, indehiscent; seed without endosperm; embryo with a single cotyledon. X = 7 in *Cymodocea*.

The family Cymodoceaceae consists of some 5 genera and 18 species, native to tropical and subtropical seacoasts. *Halodule* (*Diplanthera*) has about 7 species, *Cymodocea* has about 5, and *Amphibolis, Syringodium,* and *Thalassodendron* have 2 each. None of the genera or species is well known to many botanists.

In *Syringodium* the staminate flower is initially subtended by a girdling, bract-like scale which may originate from 4 primordia that become confluent. The scale is no longer visible at anthesis.

The Cymodoceaceae have often been included in the Zannichelliaceae, from which they differ in their marine habitat and filamentous pollen. They have also been included in the Zosteraceae, from which they differ in the nature of the inflorescence and the structure of the staminate flowers. In an order in which all the recognized families are small and sharply delimited, it seems better to retain the Cymodoceaceae as a distinct family than to force them into some other group.

According to the den Hartog, some Eocene fossils from Europe are reliably indentified as *Cymodocea*, but Daghlian (in ms.) considers that the family is not certainly represented in the fossil record.

10. Family ZOSTERACEAE Dumortier 1829 nom conserv., the Eel-grass Family

Glabrous, perennial marine herbs with creeping, sometimes tuberous-thickened rhizomes, growing wholly submerged at depths of sometimes as much as 50 m, or on rocks that may be exposed at low tide; plants lacking crystals and tanniferous cells; flavonoid constituents commonly including flavonoid sulphates; stem simple or branched, often flattened; vessels wanting from all organs. LEAVES alternate, commonly distichous, parallel-veined or with only a midvein, linear or filiform, generally ligulate at the juncture of sheath and blade, the sheath open or closed, commonly with stipule-like flanges; intravaginal scales present at the nodes. FLOWERS small, hydrophilous, unisexual, sessile, arranged in two rows on one side of a flattened spadix enclosed in a spathe (staminate and pistillate flowers alternating in each row in *Zostera* and *Heterozostera*, plants dioecious in *Phyllospadix*), the axis of the spadix (in *Heterozostera* and most spp. of *Zostera*) or of the staminate spadix (in *Phyllospadix*) often with a series of more or less well developed marginal lobes or appendages (called retinacula) that fold over and more or less cover the flowers (or some of them), the flowers otherwise bractless and without a perianth; stamen solitary, sessile; anther tetrasporangiate and dithecal, opening by longitudinal slits; pollen-grains binucleate or trinucleate, filamentous, up to about 2 mm long, without exine; gynoecium of 2 carpels united to form a compound, unilocular ovary with basally connate styles; ovule solitary, pendulous from the apex of the locule, orthotropous, bitegmic, pseudocrassinucellar; endosperm-development nuclear. FRUIT small, drupaceous, or firm and eventually opening irregularly; seed without endosperm; embryo with a single basally closed and sheathing cotyledon lying in a groove of the enlarged hypocotyl. X = 6, 10.

The family Zosteraceae as here defined consists of 3 genera, *Zostera* (12), *Phyllospadix* (5), and *Heterozostera* (1), found along the seacoast in temperate to subarctic or subtropical regions. *Zostera marina* L., the common eel-grass, is an important winter-food for waterfowl.

The morphological nature of the retinacula is debatable. They have been interpreted as representing tepals (cf. *Aponogeton*), or as enlarged appendages of the connective of the stamens. A case might be made for interpreting them as bracts. Finally, they may be just what they look like—marginal lobes of a flattened axis. Dr. Rolf Sattler tells me that the retinacula are initiated separately on the axis and may most reasonably be interpreted as bracts.

Old reports of fossils of Zosteraceae from Eocene and even Upper Cretaceous deposits have been challenged by more recent workers.

4. Order TRIURIDALES Engler 1897

Small, mycotrophic herbs without chlorophyll; stem slender, simple, erect, its vascular bundles in a single ring, attached to the inner surface of a continuous sclerenchymatous sheath, tending to be irregular in form and position of the xylem and phloem; some scalariform vessels present in the roots only. LEAVES reduced to alternate scales. FLOWERS small, perfect or unisexual, regular, hypogynous or partly (less than half) epigynous, borne in a terminal, bracteate raceme or corymb, or seldom in a sort of cyme; perianth of (3–)6 (–10) tepals in a single series or in 2 series of each, the tepals distinct or connate toward the base, not petaloid; stamens (2) 3–6, sometimes some of them staminodial; anthers 2–4-thecal, the thecae sometimes distally confluent, opening by longitudinal or transverse slits, the connective often more or less prolonged; pollen-grains monosulcate or inaperturate; gynoecium of 3-many carpels, these distinct or connate only at the base; ovules 1-many in each carpel, on basal and marginal placentas. FRUIT follicular, with 1-many seeds; seeds with well developed endosperm.

The order Triuridales as here defined consists of only the Triuridaceae and Petrosaviaceae. The possible affinities of the families are separately discussed.

SELECTED REFERENCES

Green, P. S., & O. Solbrig. 1966. *Sciaphila dolichostyla* (Triuridaceae). J. Arnold Arbor. 47: 266–269.

Stant, M. Y. 1970. Anatomy of *Petrosavia stellaris* Becc., a saprophytic monocotyledon. *In*: N. K. B. Robson, D. F. Cutler, & M. Gregory, eds., New research in plant anatomy, pp. 147–161. J. Linn. Soc., Bot. 63, Supplement 1. Academic Press. London & New York.

Sterling, C. 1978. Comparative morphology of the carpel in the Liliaceae: Hewardieae, Petrosavieae, and Tricyrteae. J. Linn. Soc., Bot. 77: 95–106.

SYNOPTICAL ARRANGEMENT OF THE FAMILIES OF TRIURIDALES

1 Carpels 3, each with numerous marginal ovules; seeds numerous; flowers perfect; tepals 6, in 2 cycles of 3 each .... 1. PETROSAVIACEAE.

1 Carpels 6–50, each with a single basal ovule; seed solitary; flowers unisexual or seldom perfect; tepals 3–10, in a single cycle .... 2. TRIURIDACEAE.

1. Family PETROSAVIACEAE Hutchinson 1934 nom. conserv., the Petrosavia Family

Small, mycotrophic herbs without chlorophyll; rhizome well developed, covered with sheathing scale-leaves; stem slender, simple, erect; vascular bundles of the stem in a single ring, attached to the inner surface of a continuous sclerenchymatous sheath, tending to be irregular in form and position of the xylem and phloem; some scalariform vessels present in the roots only. LEAVES reduced to alternate scales; stomates anomocytic. FLOWERS small, borne in a terminal corymb or corymbiform raceme, each subtended by a small bract, perfect, regular, trimerous, appearing hypogynous and apocarpous, but actually partly epigynous and basally syncarpous, the tepals attached about at the level of separation of the carpels; perianth of 6 persistent, colorless tepals in 2 cycles of 3 each, those of the inner series broader than the outer but not petaloid; stamens 6; filaments well developed, adnate to the base of the tepals; anthers dithecal, introrse, opening by longitudinal slits; connective very shortly prolonged; pollen-grains monosulcate; gynoecium of 3 carpels, these connate toward the base, otherwise distinct, each with a short style and a small, subcapitate stigma; septal nectaries present; ovules numerous on marginal (basally axile) placentas, bitegmic, anatropous to campylotropous. FRUIT of recurved-spreading, basally united follicles; seeds with endosperm.

The family Petrosaviaceae consists of only the genus *Petrosavia*, with 2 species native to forests from southern China and southern Japan to the Malay Peninsula and Borneo. *Petrosavia* has often been included in the Liliaceae, subfamily Melanthoideae, where it is aberrant in its nearly distinct carpels as well as in its mycotrophic habit. Stant (1970) has commented that "My recent investigations . . . . have revealed a striking identity between the anatomical structure of the stem, root and leaf of the Triuridaceae and Petrosaviaceae. The resemblance is so complete that I would have no hesitation in placing *Petrosavia* in the family Triuridaceae on the basis of anatomical evidence."

2. Family TRIURIDACEAE Gardner 1843 nom. conserv., the Triuris Family

Small, mycotrophic herbs without chlorophyll, sometimes anthocyanic; stem slender, simple, erect, its vascular bundles in a single ring, attached to the inner surface of a continuous sclerenchymatous sheath, tending

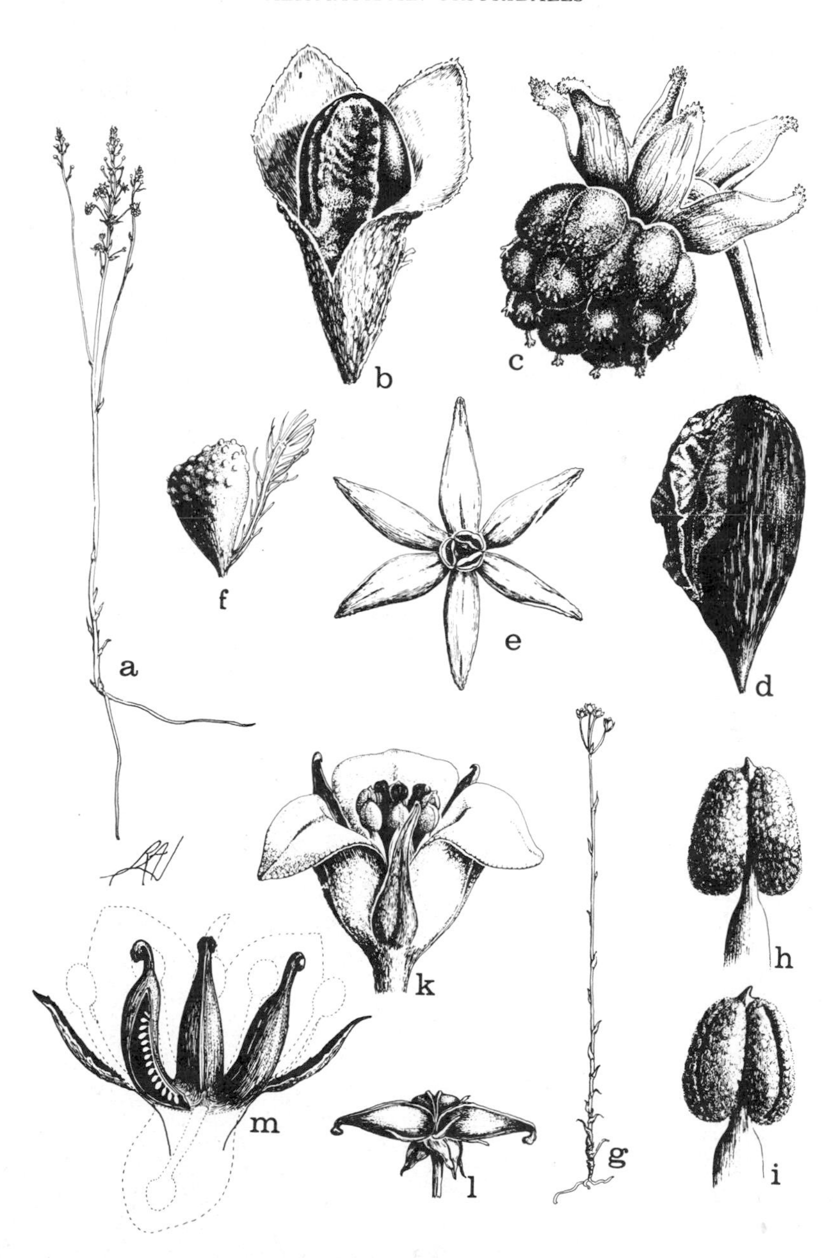

FIG. 7.8 Above, Triuridaceae. *Sciaphila albescens* Benth. a, habit, $\times\frac{1}{2}$; b, dorsal view of opened fruit and seed, $\times 40$; c, pistillate flower, $\times 15$; d, seed, $\times 40$; e, staminate flower, from above, $\times 15$; f, pistil, with basilateral style, $\times 30$. Below, Petrosaviaceae. *Petrosavia stellaris* Becc. g, habit, $\times\frac{1}{2}$; h, i, dorsal and ventral view of anther, $\times 50$; k, flower, $\times 10$; l, opened fruit, $\times 5$; m, schematic view of flower in partial section, showing carpels united at base, the ovary inferior up to the level of separation of the carpels, $\times 10$.

to be irregular in form and position of the xylem and phloem; some scalariform vessels present in the roots only. LEAVES reduced to alternate scales, sometimes very few. FLOWERS small, unisexual (the plants monoecious or dioecious), or seldom (spp. of *Sciaphila*) perfect, regular, hypogynous, borne in a terminal raceme or seldom a sort of cyme, each subtended by a small bract; perianth of 3–10 (most commonly 6) valvate tepals in a single cycle; these often connate at the base, and often appendiculate-prolonged at the tip, commonly reflexed after anthesis; stamens (2) 3 or 6, sometimes some of them staminodial; filaments short, or the anthers sessile and sometimes even partly sunken in the receptacle; anthers 2–4-thecal, extrorse, the thecae often apically confluent, opening by longitudinal or transverse slits; connective often with an elongate, slender, terminal appendage; pollen-grains trinucleate, monosulcate in *Sciaphila*, otherwise inaperturate; gynoecium of 6–50 distinct carpels, each with a terminal, lateral, or basilateral style; ovules solitary in each carpel. basal, erect, anatropous, bitegmic, tenuinucellar; endosperm-development nuclear. FRUIT of small, thick-walled follicles; seeds with copious, proteinaceous and oily endosperm; embryo small, undifferentiated. X = 11, 12, 14.

The family Triuridaceae consists of 7 genera and about 70 species, widespread in tropical and subtropical regions, but seldom seen or collected. The largest genus is *Sciaphila*, with about 50 species.

The Triuridaceae are taxonomically isolated. Aside from the probable relationship with the small family Petrosaviaceae, they stand off sharply from all other groups. The more or less numerous, separate carpels suggest that the Triuridaceae should be associated with the Alismatidae, and the trinucleate pollen is of course quite compatible with such an association. On the other hand, the terrestrial, mycotrophic habit and the well developed endosperm of the seeds are out of harmony with the other orders of the subclass. It may well eventually become necessary to establish the Triuridales as a distinct subclass, but at the present state of knowledge I do not think it is useful.

II. Subclass ARECIDAE Takhtajan 1966

Herbs (sometimes thalloid), shrubs, vines, or trees, the woody species sometimes with some limited secondary growth that does not lead to the formation of new vascular tissue; plants usually producing raphides; vessels present in all vegetative organs, or restricted to the roots and stems, or to the roots, or none; plastids of the sieve-tubes with cuneate proteinaceous inclusions; vascular bundles of the stem closed, more or less scattered, or the vascular system vestigial or wanting. LEAVES alternate, sometimes all basal or all in a terminal crown, variously narrow and parallel-veined to broader and plicate or net-veined, or the leaves not differentiated; stomates mostly paracytic or tetracytic, or sometimes with more than 4 subsidiary cells. FLOWERS crowded and usually rather small, often borne in a spadix (the inflorescence usually subtended by one or more spathes), variously adapted to pollination by wind, insects, birds, or even bats, perfect or often unisexual, sometimes with septal nectaries, hypogynous or nude, or with the ovary sunken into the axis of the spadix, never distinctly epigynous; perianth often more or less well developed (though usually rather small), and then usually in 2 series of (2) 3 tepals, but also often reduced and vestigial, or wanting; stamens 1-many; pollen-grains binucleate or sometimes trinucleate, variously uniaperturate or with more than one aperture; gynoecium of (1–) 3 (-numerous) carpels, these united to form a compound ovary, or less often the carpels distinct; ovules 1-many in each locule or on each placenta, various in form, bitegmic, crassinucellar or pseudocrassinucellar, rarely tenuinucellar; endosperm-development variously helobial, nuclear, or cellular. FRUIT nearly always indehiscent, most commonly a berry or a fleshy or dry drupe, sometimes multiple; endosperm oily (and often also proteinaceous) or seldom somewhat starchy, and often with reserves of hemicellulose in the thickened cell-walls, or the seeds sometimes without endosperm; embryo monocotyledonous.

The subclass Arecidae consists of 4 orders and only 5 families, about 5600 species in all. More than half of the species belong to the single order Arecales, which includes only the family Arecaceae (palms). A loose affinity among the 4 orders is widely recognized, but the group is morphologically and ecologically diverse. The palms have the longest

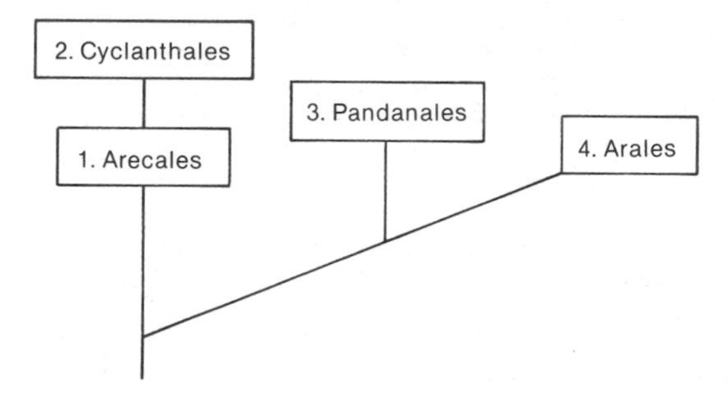

Fig. 8.1 Putative relationships among the orders of Arecidae.

fossil record in the subclass, but it is difficult to visualize them as being ancestral to any group other than the Cyclanthales.

SYNOPTICAL ARRANGEMENT OF THE ORDERS OF ARECIDAE

1 Leaves characteristically with an expanded, plicate blade that has a unique ontogeny; vessels typically present in all vegetative organs.
  2 Inflorescence usually large and branched, although the branches are sometimes somewhat spadix-like; perianth usually well developed (though typically rather small), only seldom vestigial or wanting; ovules solitary in each of the (1–) 3 (–10) locules; endosperm-development nuclear; plants very often arborescent and with a terminal crown of large leaves, less often climbing or acaulescent; leaves seldom merely bifid, except in the seeding stage .... 1. ARECALES.
  2 Inflorescence a spadix; perianth minute or none; ovules more or less numerous in the single locule; endosperm-development helobial; plants typically herbaceous and acaulescent or with a short, erect stem, less often erect half-shrubs or lianas; leaves often bifid, but sometimes flabellately lobed or entire .... 2. CYCLANTHALES.
1 Leaves with or without an expanded blade, but in any case not plicate; inflorescence a spadix, except in some Pandanales.
  3 Leaves appearing to be in 3 or 4 spirals because of the unique, spiral growth of the stem, parallel-veined, narrow and usually elongate, very often firm and xeromorphic; vessels generally present in all vegetative organs; endosperm-development nuclear; woody, very often arborescent or coarsely shrubby plants, or sometimes woody climbers .... 3. PANDANALES.
  3 Leaves not appearing to be spiralled, sometimes narrow and parallel-veined, but very often with an expanded and more or

less distinctly net-veined blade, or the plant thalloid and without leaves; vessels mostly confined to the roots, seldom present also in the stems, or sometimes wanting; endosperm-development cellular or sometimes helobial; herbs or scrambling slender shrubs or climbing vines, never arborescent ....4. ARALES.

1. Order ARECALES Nakai 1930

The order consists of the single family Arecaceae (Palmae). (Principes)

SELECTED REFERENCES

Arber, A. 1922. On the development and morphology of the leaves of palms. Proc. Roy. Soc. London Ser. B. Biol. Sci. 93: 249–261.

Bosch, E. 1947. Blütenmorphologische und zytologische Untersuchungen an Palmen. Ber. Schweiz. Bot. Ges. 57: 37–100.

Burret, M., & E. Potztal. 1956. Systematische Übersicht über die Palmen. Willdenowia 1: 59–74; 350–385.

Corner, E. J. H. 1966. The natural history of Palms. Univ. California Press. Berkeley.

Daghlian, C. P. 1978. Coryphoid palms from the Lower and Middle Eocene of southeastern North America. Palaeontographica Abt. B, 166: 44–82.

Eames, A. J. 1953. Neglected morphology of the palm leaf. Phytomorphology 3: 172–189.

Mahabalé, T. S. 1967. Pollen grains in Palmae. Rev. Palaeobot. Palynol. 4: 299–304.

Moore, H. E. 1963. An annotated checklist of cultivated palms. Principes 7: 119–182.

Moore, H. E. 1973. The major groups of palms and their distribution. Gentes Herb. 11: 27–141.

Moore, H. E., & N. W. Uhl. 1973. Palms and the origin and evolution of monocotyledons. Quart. Rev. Biol. 48: 414–436.

Periasamy, K. 1962. Morphological and ontogenetic studies in palms—I. Development of the plicate condition in the palm-leaf. Phytomorphology 12: 54–64.

Read, R. W., & L. J. Hickey. 1972. A revised classification of fossil palm and palm-like leaves. Taxon 21: 129–137.

Scott, R. A., P. L. Williams, L. C. Craig, E. S. Barghoorn, L. J. Hickey, & H. D. MacGinitie. 1972. "Pre-Cretaceous" angiosperms from Utah: Evidence for Tertiary age of the palm woods and roots. Amer. J. Bot. 59: 886–896.

Sowunmi, M. A. 1972. Pollen morphology of the Palmae and its bearing on taxonomy. Rev. Palaeobot. Palynol. 13: 1–80.

Thanikaimoni, G. 1970 (1971). Les palmiers: Palynologie et systématique. Trav. Sect. Sci. Techn. Inst. Franç. Pondichéry 11: 1–286.

Thanikaimoni, G. 1970. Pollen morphology, classification and phylogeny of Palmae. Adansonia sér. 2, 10: 347–365.

Tomlinson, P. B. 1960. Seedling leaves in palms and their morphological significance. J. Arnold Arbor. 41: 414–428.

Tomlinson, P. B. 1960–1962. Essays on the morphology of palms. Principes 4: 56–61; 140–143, 1960. 5: 8–12; 46–53; 83–89; 117–124, 1961. 6: 44–52; 122–124, 1962.

Tomlinson, P. B. 1961. Vol. 2. Palmae. *In:* C. R. Metcalfe, ed. Anatomy of the monocotyledons. Clarendon Press. Oxford.

Tomlinson, P. B. 1962. The leaf base in palms, its morphology and mechanical biology. J. Arnold Arbor. 43: 23–50.

Uhl, N. W., & H. E. Moore. 1971. The palm gynoecium. Amer. J. Bot. 58: 945–992.
Uhl, N. W., & H. E. Moore. 1973. The protection of pollen and ovules in palms. Principes 17: 111–149.
Uhl, N. W., & H. E. Moore. 1977. Centrifugal stamen initiation in phytelephantoid palms. Amer. J. Bot. 64: 1152–1161.
Uhl, N. W., & H. E. Moore. 1977. Correlations of inflorescence, flower structure, and floral anatomy with pollination in some palms. Biotropica 9: 170–190.
Williams, C. A., & J. B. Harborne. 1973. Negatively charged flavones and tricin as chemosystematic markers in the Palmae. Phytochemistry 12: 2417–2430.

1. Family ARECACEAE C. H. Schultz-Schultzenstein 1832 nom. conserv., the Palm Family

Plants most commonly slender trees (to 60 m) or stout shrubs, with an unbranched trunk (this smooth or spiny or sometimes covered by persistent leaf-bases) and a terminal crown of large leaves, but the trunk sometimes short or virtually suppressed, or the main stem creeping or subterranean and with a crown of leaves at the surface of the ground, or sometimes the stem more slender and with well spaced leaves, and then often climbing (even to more than 150 m in spp. of *Calamus*), only rarely the stem sparingly branched; stem sometimes with a small amount of diffuse secondary thickening (especially near the base) that does not result in the formation of any new vascular tissue; plants commonly with raphides (sometimes in specialized raphide-cells), and accumulating silica (notably in some of the cell walls) and various sorts of polyphenols, very often with tanniferous cells or canals and producing proanthocyanins, sometimes producing pyridine alkaloids, and occasionally saponiferous, but only seldom cyanogenic; flavonoid constituents very often including flavone sulphates (such as tricin) and C-glycosyl flavones; vascular bundles closed, very numerous, scattered in a ground tissue that often contains many very hard, silicified fibers; vessels commonly present in all vegetative organs, those in the roots most commonly with simple perforations, those in the shoot with simple or often scalariform perforations; plastids of the sieve-tubes with cuneate proteinaceous inclusions; roots mycorhizal, without root-hairs. LEAVES evergreen, alternate (but often densely crowded at the stem-tip), mostly large or very large, with a basal sheath (this at first tubular but often splitting during growth and thus open at maturity), usually a petiole, and an expanded, pinnately or less often palmately cleft or compound or twice compound or rarely entire blade; axillary buds mostly wanting; blade simple in early ontogeny, commonly splitting during growth and

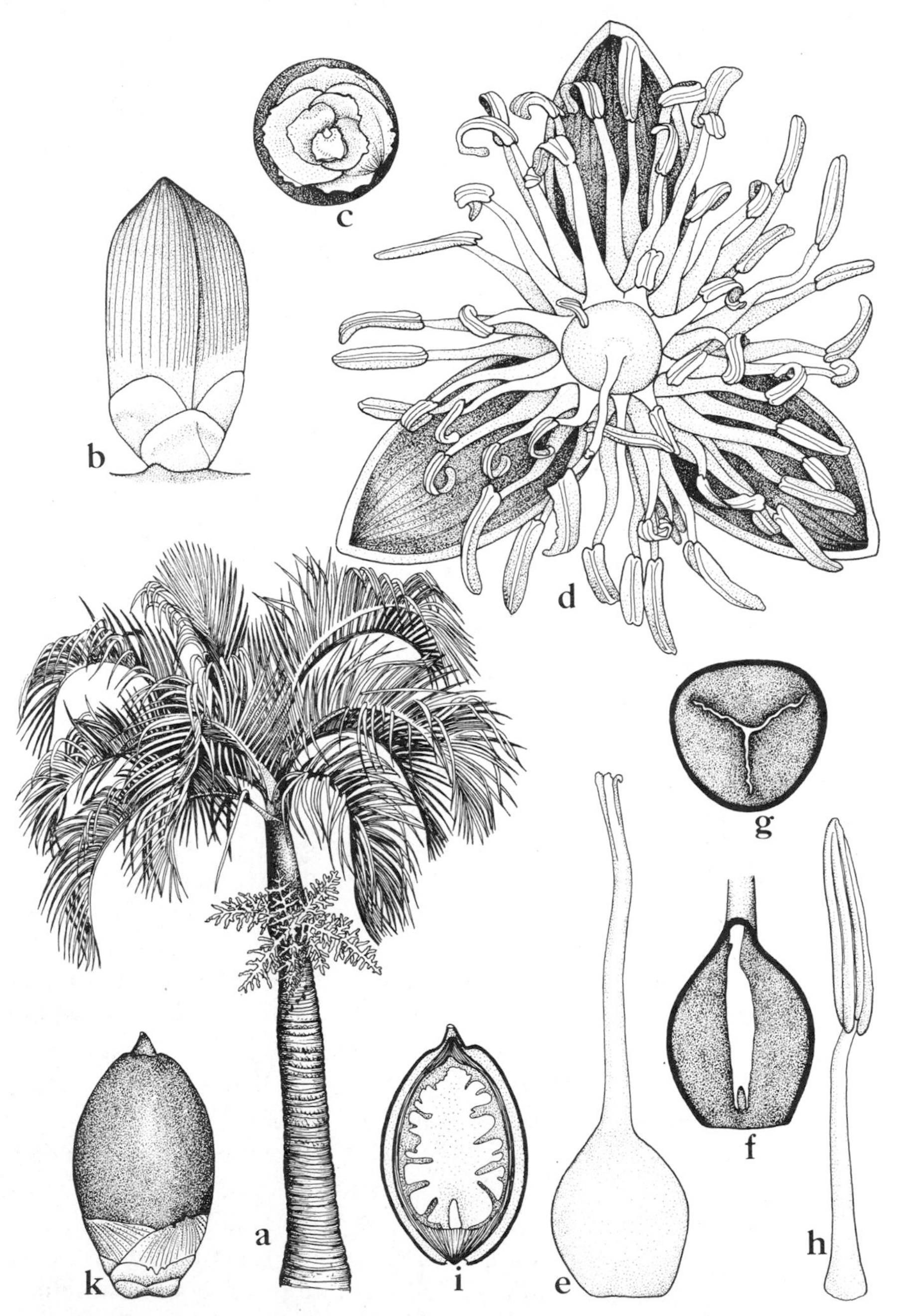

Fig. 8.2 Arecaceae. *Veitchia merrillii* (Becc.) H. E. Moore. a, habit, $\times \frac{1}{24}$; b, flower bud, $\times 3$; c, base of fruit, showing persistent, spirally arranged tepals and bracteoles, $\times 1$; d, flower, from above, $\times 3$; e, pistillode of staminate flower, $\times 6$; f, long-section of pistillode, with basal sterile ovule, $\times 6$; g, cross-section of pistillode, above the level of the sterile ovule, $\times 6$; h, stamen, $\times 6$; i, long-section of fruit, $\times 1$; k, fruit, with persistent bracteoles and tepals $\times 1$.

becoming strongly plicate (the numerous more or less parallel ribs alternately elevated and depressed), reflecting a unique ontogeny shared only with the Cyclanthaceae; a ligule sometimes present as a prolongation of the sheath in front of the petiole or on both sides of the petiole or surrounding the stem; sheath, blade, and/or petiole sometimes armed with spines; stomates tetracytic, or sometimes with 2 subsidiary cells at each side as well as one at each end; seedling-leaves variously pinnate or palmate or undivided, or often simple and broadly notched at the tip, with pinnate-parallel veins (as in mature leaves of the Cyclanthaceae), rarely this latter type of leaf retained throughout the life of the plant. Inflorescences most commonly axillary, sometimes terminal or supra- or infra-axillary, generally large, paniculately branched or seldom simple, the ultimate axes sometimes thickened and spadix-like; peduncle bearing a prophyll and 1-several spathes; FLOWERS numerous, usually individually small (but sometimes large or very large, as in the pistillate flowers of *Lodoicea* and *Phytelephas*, and even showy, as in the staminate flowers of *Arenga*), short-pedicellate or much more often sessile, sometimes sunken into the axis, inserted on the axis individually or in complexly organized groups, entomophilous or less often anemophilous, perfect or more often unisexual (the plants then monoecious or dioecious), basically trimerous, regular or sometimes slightly irregular; tepals typically 6 in 2 cycles of 3, leathery or fleshy, variously white, green, yellow, orange, or red, the outer ones (sepals) sometimes similar to the inner (petals), but much more often smaller and somewhat differentiated from the petals, the sepals distinct or sometimes connate below, mostly imbricate, the petals distinct or connate below, usually valvate in staminate flowers and imbricate in pistillate ones; seldom the tepals 2 + 2, or spirally arranged and of unstable number (up to 10), or the perianth vestigial or wanting; stamens (3) 6–900+, most commonly 6 in 2 cycles of 3; highly multistaminate flowers sometimes with the stamens initiated centrifugally and associated with trunk-bundles, sometimes centripetal and without trunk-bundles; filaments free and distinct, or variously connate, or adnate to the tepals, or both connate and adnate; anthers tetrasporangiate and dithecal, latrorse, opening by longitudinal slits; pollen-grains binucleate, most often monosulcate, but varying to uniulcerate or trichotomosulcate or tetrachotomosulcate or virtually tricolpate, or dicolpate or diporate or bisulcate, and in *Sclerosperma mannii* G. Mann & Wendl. triporate (fide Erdtman & Singh, 1957); staminodes often present, distinct or sometimes connate into a cup or tube; gynoecium apocarpous (in 13 genera) or more often syncarpous with 3 (–10) carpels united to form a compound, superior, plurilocular ovary, or sometimes pseudomonomerous, with only one locule or with one fertile and 2 abortive locules; septal nectaries often present, or the nectaries sometimes borne at the base of the stamens, or often wanting; styles distinct or connate, or short and thickened so that the stigmas appear to be sessile; stigmas dry; ovules solitary in each locule (or in each fertile locule), on an axile to apical-axile or basal-axile or basal placenta, sometimes with a funicular obturator, anatropous or less often hemitropous, campylotropous, or orthotropous, bitegmic, weakly crassinucellar; endosperm-development nuclear. FRUIT typically indehiscent, most commonly fleshy, sometimes dry and fibrous, often drupaceous with the endocarp adnate to or free from the testa, seldom more or less dehiscent, the pericarp splitting (as in spp. of *Astrocaryum* and *Socratea*); seed solitary (–10); endosperm commonly with reserves of oil (typically including large amounts of lauric acid), hemicellulose (in the thickened cell-walls), and some protein, but not starch, sometimes ruminate; embryo typically with a minute radicle and plumule lying in a small slit on the side of the large, solitary cotyledon. X = 13–18, perhaps originally 18. (Palmae, nom altern.; Nypaceae, Phytelephasiaceae)

The well marked family Arecaceae has about 200 (or a few more) genera and nearly 3000 species, nearly all of them restricted to tropical or warm-temperate regions. The largest genera are *Calamus* (300+) and *Bactris* (200+). *Cocos nucifera* L. (coconut palm), *Phoenix dactylifera* L. (date palm), *Arenga pinnata* (Wurmb.) Merrill (sugar palm), and *Areca catechu* L. (betel-nut palm) are some familiar, economically important palms. *Lodoicea maldivica* (J. F. Gmel.) Pers., the Seychelles palm or double coconut, has the largest seeds in the world.

The organization of the genera of palms into tribes and subfamilies is still fluid. Moore (1973) considers that there are 15 major groups, but Potztal has only 9 in the current Engler Syllabus.

The palms have a continuous fossil record extending back to about the middle of the Upper Cretaceous period (near the base of the Campanian, some 80 million years ago). The distinctive pollen of the modern genus *Nypa* is found in rocks of Maestrichtian and more recent age. Reports of pre-Cretaceous palms are based on misidentifications (e.g., *Sanmiguelia*, from Triassic deposits in Colorado), or on stratigraphic confusion (e.g., *Palmoxylon*, from purportedly Jurassic but actually Tertiary deposits in Utah). Doyle (1973), cited under Liliopsida) considers that "The pollen and megafossil record of palms beginning with putatively primitive costapalmate leaves in the late Upper Cretaceous (Senonian), suggests that the palms, like the grasses, originated relatively late in the history of monocotyledons."

Palms are the only monocotyledons to combine an arborescent habit, a broad leaf-blade, and a well developed vascular system that has vessels in all vegetative organs. This obviously functional syndrome approaches that of woody dicotyledons, but palms lack an adequate means of secondary growth and a means of expanding the coverage of the crown. Furthermore, palms have never developed the deciduous habit, and with minor exceptions they have not adapted to temperate or cold climates. Thus their ecological amplitude is limited, as compared with that of woody dicots. They do well in tropical regions that are moist enough to support evergreen tree-growth, but not moist enough to support a dense forest, and they are also common components of the understory in tropical rain-forests. It is not difficult to suppose that the Cyclanthaceae may share a common ancestry with some of these smaller palms of the forests.

2. Order CYCLANTHALES J. H. Schaffner 1929

The order consists of the single family Cyclanthaceae. (Synanthae)

SELECTED REFERENCES

Harling, G. 1946. Studien über den Blütenbau und die Embryologie der Familie Cyclanthaceae. Svensk Bot. Tidskr. 40: 257–272.

Harling, G. 1958. Monograph of the Cyclanthaceae. Acta Horti Berg. 18: 1–428.

Wilder, G. J. 1976. Structure and development of leaves in *Carludovicia palmata* (Cyclanthaceae) with reference to other Cyclanthaceae and Palmae. Amer. J. Bot. 63: 1237–1256.

1. Family CYCLANTHACEAE Dumortier 1829 nom. conserv., the Panama Hat Family

Perennial, mostly herbaceous plants without secondary growth, rhizomatous and acaulescent or with a short, erect stem, or sometimes slender, erect half-shrubs, or not infrequently lianas climbing by means of adventitious roots, sometimes virtually epiphytic; mucilage-canals generally present in the roots, stems, and petioles, but in *Cyclanthus* only in the inflorescence; tanniferous cells present in most organs, except in *Cyclanthus*; plants commonly with raphide-sacs and sometimes also with other types of calcium oxalate crystals; silica-cells wanting; vessels with scalariform perforations commonly present in all vegetative organs, or at least in the roots and leaves; vascular bundles closed, scattered, but mostly in the cortex; plastids of the sieve-tubes with cuneate proteinaceous inclusions. LEAVES alternate, either distichous or spirally arranged, with a sheath, petiole, and expanded blade, commonly more or less palm-like in aspect and even plicate in the manner of palm leaves (and with a similar ontogeny), the blade with parallel or parallel-pinnate venation and with evident cross-veins, most commonly bilobed or bifid (the primary segments sometimes again cleft), sometimes flabellately lobed as in fan-palms, seldom simple and entire; stomates tetracytic. Inflorescences axillary, with 2-several large, deciduous spathes subtending the usually coarse spadix; FLOWERS pollinated by beetles, very numerous and tiny, much-reduced in structure, sessile (the pistillate ones often more or less embedded in the axis of the spadix), unisexual, both types borne in the same spadix, but differently arranged in

Fig. 8.3 Cyclanthaceae. *Sphaeradenia* sp. a, habit at late anthesis, $\times \frac{1}{16}$; b, inflorescence at early anthesis, ×8; c, staminate flower, from above, showing absence of tepals from the ventral side (the upper side as drawn), ×8; d, dorsal view of staminate flower, ×8; e, stamen, ×8; f, infertile anther of staminode, ×8; g, pistillate flower, with very long staminodes, ×8; h, pistillate flower, from above, with the staminodes removed (their position shown by scars), ×8; i, schematic view of pistil, in partial long-section, with placentas descending from near the summit, ×4.

different genera; staminate flowers with 4–24 tiny, distinct or connate tepals in 1 or 2 series, sometimes all disposed on one side of the flower, or the perianth reduced to an entire or toothed cup or ring, or wanting and the staminate flowers then sometimes confluent; stamens 6 to more often more or less numerous; filaments connate at the base; anthers tetrasporangiate and dithecal, opening by longitudinal slits; pollen-grains binucleate, monosulcate to uniporate; nectaries wanting; pistillate flowers commonly with 4 tiny, distinct or connate tepals and as many staminodes opposite and partly adnate to them, the staminodes often much-elongate and thread-like; gynoecium of 4 carpels united to form a compound, unilocular ovary with 4 parietal or nearly apical placentas and as many stigmas, these sessile or on a short common style, or seldom the pistil pseudomonomerous, with a single placenta and a single stigma; in *Cyclanthus* the pistillate flowers confluent into rings around the spadix, each ring with a common ovular chamber and numerous placentas and ovules; ovules more or less numerous, anatropous, bitegmic, weakly crassinucellar; endosperm-development helobial. FRUIT indehiscent, fleshy, juicy, berry-like, the individual pistils often coalescent to form a multiple fruit; seeds several to numerous, embryo small to medium-sized, straight, cylindric, with a single cotyledon; endosperm copious and soft to hard, with reserves of oil and protein and often hemicellulose, seldom any starch. X = 9–16.

The well marked family Cyclanthaceae consists of some 11 genera and about 180 species, confined to tropical America. Most of the species occur in moist or wet forests or along streams. Nearly half of the species belong to the genus *Asplundia*. The monotypic genus *Cyclanthus* is so distinctive within the family that many authors put it into its own subfamily, Cyclanthoideae. The remaining genera are then considered to form a subfamily Carludovicoideae. Panama hats are traditionally made from the leaves of *Carludovica palmata* Ruiz & Pavon.

Authors are agreed that the Cyclanthaceae are allied to the palms, aroids, and screw-pines, but sufficiently distinct from each of these groups to warrant being placed in a separate, unifamilial order. The family is perhaps most closely allied to the palms, and it may eventually become useful to include both groups in the same order.

Macrofossils referred to the Cyclanthaceae data from the Eocene epoch, but fossil pollen of the family has yet to be recognized.

3. Order PANDANALES Lindley 1833

The order consists of the single family Pandanaceae.

SELECTED REFERENCES

Fagerlind, F. 1940. Stempelbau und Embryosackentwicklung bei einigen Pandanazeen. Ann. Jard. Bot. Buitenzorg 49: 55–78.

Kam, Y. K., & B. C. Stone. 1970. Morphological studies in Pandanaceae. IV. Stomate structure in some Mascarene and Madagascar *Pandanus* and its meaning for infrageneric taxonomy. Adansonia Sér. 2, 10: 219–246.

Lee, L. L., & B. C. Stone. 1971. Notes on systematic foliar anatomy of the genus *Freycinetia* (Pandanaceae). J. Jap. Bot. 46: 207–220.

Nambudiri, E. M. V., & W. D. Tidwell. 1978. On probable affinities of *Viracarpon* Sahni from the Deccan Intertrappean flora of India. Palaeontographica 166: 30–43.

North, C. A., & A. J. Willis. 1971. Contributions to the anatomy of *Sararanga* (Pandanaceae). J. Linn. Soc., Bot. 64: 411–421.

Pijl, L. van der. 1956. Remarks on pollination by bats in *Freycinetia, Duabanga* and *Haplophragma,* and on chiropterophily in general. Acta Bot. Neerl. 5: 135–144.

Stone, B. C. 1968. Morphological studies in Pandanaceae. I. Staminodia and pistillodia of *Pandanus* and their hypothetical significance. Phytomorphology 18: 498–509.

Stone, B. C. 1970. Observations on the genus *Pandanus* in Madagascar. J. Linn. Soc., Bot. 63: 97–131.

Stone, B. C. 1972. A reconsideration of the evolutionary status of the family Pandanaceae and its significance in monocotyledon phylogeny. Quart. Rev. Biol. 47: 34–45.

Stone, B. C. 1974. Towards an improved infrageneric classification in *Pandanus* (Pandanaceae). Bot. Jahrb. Syst. 94: 459–540.

Strömberg, B. 1956. The embryo-sac development of the genus *Freycinetia.* Svensk Bot. Tidskr. 50: 129–134.

Tomlinson, P. B. 1965. A study of stomatal structure in Pandanaceae. Pacific Sci. 19: 38–54.

Zimmermann, M. H., P. B. Tomlinson, & J. LeClaire. 1974. Vascular construction and development in the stems of certain Pandanaceae. J. Linn. Soc., Bot. 68: 21–41.

1. Family PANDANACEAE R. Brown 1810 nom. conserv., the Screw-Pine Family

Trees, shrubs, or (*Freycinetia*) lianas, the stem without secondary growth except sometimes for some nonvascular thickening associated with the origin of the adventitious roots, often appearing to be dichotomously branched as a result of sympodial growth of the axis after flowering, in *Pandanus* commonly with stout, adventitious prop-roots descending at an angle from the lower part of the stem to the ground; plants not accumulating silica, but with well developed raphides, these commonly in bundles in mucilaginous cells or channels throughout the shoot; other sorts of calcium oxalate crystals commonly also present in scattered idioblasts; vascular bundles closed, scattered but tending to be more crowded toward the periphery of the stem, some of the bundles in the crowded zone commonly compound (2 or 3 within a common sheath); vessels with scalariform perforations commonly present in all vegetative organs; plastids of the sieve-tubes with cuneate proteinaceous inclusions. LEAVES alternate, in 3 or (*Sararanga*) 4 ranks, appearing to be in spirals because of the spiral growth of the stem-tip, simple, with an open, sheathing base and a narrow, usually elongate (even to 5 m), often very firm and xeromorphic blade, glabrous, pallelel-veined, usually spiny along the margins and midrib; epidermal cells in longitudinal files; stomates more or less distinctly tetracytic, restricted to the intercostal files, mostly on the lower epidermis. FLOWERS unisexual (the plants dioecious), very small and numerous, pollinated by wind, or less often by insects, or sometimes by birds or even bats, without bracts or bracteoles except in *Sararanga*, so much reduced that the structure is subject to diverse interpretations, borne in a terminal spadix subtended by (1–) several large spathes, or in several racemosely arranged spadices, or (*Sararanga*) in a large terminal panicle; in *Sararanga* each flower subtended by a small, saucer-like, often lobed cup, the perianth otherwise wanting; staminate flowers not only closely crowded but often (especially in *Pandanus*) not sharply limited from one another, each with 1-many stamens; filaments distinct, or often connate into phalanges for most of their length (each such phalanx considered to represent a flower); anthers tetrasporangiate and dithecal; pollen-grains binucleate, uniaperturate; pistillate flowers with 1-many carpels (up to 70 or 80 in *Sararanga*), these distinct or often connate into phalanges, in *Freycinetia* 2–12 carpels united to form a compound, unilocular ovary with parietal placentation; carpels in *Pandanus* often incompletely sealed, without a style but with an open terminal stigmatic portion; ovules solitary in each carpel in *Pandanus* and *Sararanga*, several to numerous in *Freycinetia*, anatropous, bitegmic, and crassinucellar; endosperm-development nuclear. FRUIT of crowded but distinct, unicarpellate drupes, or more often of polydrupes corresponding to the phalanges of carpels, or sometimes the fruit baccate (*Freycinetia*) or woody; seeds rather small, with oily endosperm and small, monocotyledonous embryo. X = 30.

The family Pandanaceae consists of 3 genera: *Pandanus*, the screw-pine, with numerous, perhaps 500–600 species; *Freycinetia*, with about 180, and *Sararanga*, with only 2. The family is entirely confined to the Old World, and is best developed in tropical regions, especially Malesia, Melanesia, and Madagascar. Relatively few species occur in temperate climates, as in New Zealand and southern China and Japan. Many species are maritime, but others occur well inland. Some species are used locally for fiber.

The Pandanaceae can be traced back to the Maestrichtian on the basis of fossil pollen, and to the lower Eocene on the basis of fossil fruits.

4. Order ARALES Lindley 1833

Herbs (sometimes thalloid and free-floating) or sometimes scrambling slender shrubs or climbing vines or virtually epiphytes, usually producing raphides. LEAVES (when differentiated) alternate, often all basal, usually with a distinct, basally sheathing petiole and an expanded, simple and entire to variously cleft or perforate or even compound blade, or the leaf rarely narrow and ensiform, variously parallel-veined or pinnately to palmately net-veined or of some transitional type; stomates with 2 or more subsidiary cells. Inflorescence unbranched, a spadix subtended by a spathe (much-reduced and often without a spathe in the Lemnaceae), with tiny, perfect or unisexual flowers; floral bracts wanting; perianth wanting, or of 4 or 6 (8) minute, distinct or connate tepals in 2 cycles; stamens 1–6 (8), opposite the tepals when these are present; anthers mostly tetrasporangiate and dithecal, sometimes bisporangiate and monothecal; pollen-grains 2–3-nucleate, variously uniaperturate or 2–4-sulcate or -porate, or inaperturate; gynoecium of (2) 3(–15) carpels united to form a compound, plurilocular ovary with axile (to axile-basal or axile-apical) placentation, or sometimes the ovary unilocular with parietal placentation, or apparently monomerous; ovules variously orthotropous to more often anatropous, bitegmic, crassinucellar or pseudocrassinucellar; endosperm-development usually cellular, but sometimes helobial. FRUIT a berry or utricle, or seldom opening irregularly, or the whole spadix sometimes ripening as a multiple fruit; seeds 1-many; embryo monocotyledonous, relatively large, embedded in the copious to scanty, oily or partly starchy endosperm, or the seed sometimes without endosperm. X = 7–17+. (Lemnales)

The order Arales consists of 2 well defined families of very unequal size. The probable error in estimating the number of species of Araceae is greater than the estimated number of species of Lemnaceae.

SELECTED REFERENCES

Blanc, P. 1977. Contribution à l'étude des Aracées. I. Remarques sur la croissance monopodiale. II. Remarques sur la croissance sympodiale chez l'*Anthurium scandens* Engl., le *Philodendron fenzlii* Engl. et le *Philodendron speciosum* Schott. Rev. Gén. Bot. 84: 115–126; 319–331.

Crepet, W. L. 1978. Investigations of angiosperms from the Eocene of North America: an aroid inflorescence. Rev. Palaeobot. Palynol. 25: 241–252.

Dilcher, D. L., & C. P. Daghlian. 1977. Investigations of angiosperms from the

Eocene of southeastern North America: *Philodendron* leaf remains. Amer. J. Bot. 64: 526–534.

Eyde, R. H., D. H. Nicolson, & P. Sherwin. 1967. A survey of floral anatomy in Araceae. Amer. J. Bot. 54: 478–497.

Hamashima, S. 1978. Seed germination of three *Lemna* species. J. Jap. Bot. 53: 28–31. (In Japanese, with English summary.)

Hartog, C. den, & F. von der Plas. 1970. A synopsis of the Lemnaceae. Blumea 18: 355–368.

Hillman, W. S. 1961. The Lemnaceae, or duckweeds. A review of the descriptive and experimental literature. Bot. Rev. 27: 221–287.

Hotta, M. 1971. Study of the family Araceae. General remarks. J. Jap. Bot. 20: 269–310.

Kaplan, D. R. 1970. Comparative foliar histogenesis in *Acorus calamus* and its bearing on the phyllode theory of monocotyledonous leaves. Amer. J. Bot. 57: 331–361.

Landolt, E. 1957. Physiologische und ökologische Untersuchungen an Lemnaceen. Ber. Schweiz. Bot. Ges. 67: 271–410.

Lawalrée, A. 1945. La position systématique des Lemnaceae et leur classification. Bull. Soc. Roy. Bot. Belgique 77: 27–38.

Lawalrée, A. 1952. L'émbryologie des Lemnaceae. Observations sur *Lemna minor*. Cellule 54: 305–326.

McClure, J. W., & R. E. Alston. 1966. A chemotaxonomic study of Lemnaceae. Amer. J. Bot. 53: 849–860.

Maheshwari, S. C. 1954. The embryology of *Wolffia*. Phytomorphology 4: 355–365.

Maheshwari, S. C. 1956. The endosperm and embryo of *Lemna* and systematic position of Lemnaceae. Phytomorphology 6: 51–55.

Maheshwari, S. C. 1958. *Spirodela polyrrhiza*: the link between the aroids and the duckweeds. Nature 181: 1745–1746.

Maheshwari, S. C., & R. N. Kapil. 1963. Morphological and embryological studies on the Lemnaceae. I. The floral structure and gametophytes of *Lemna paucicostata*. Amer. J. Bot. 50: 677–686.

Maheshwari, S. C., & P. P. Khanna. 1956. The embryology of *Arisaema wallichianum* Hook. f. and the systematic position of the Araceae. Phytomorphology 6: 379–388.

Nahrstedt, A. 1975. Triglochinin in Araceen. Phytochemistry 14: 2627–2628.

Parmelee, J. A., & B. D. O. Savile. 1954. Life history and relationships of rusts of *Sparganium* and *Acorus*. Mycologia 46: 823–836.

Souèges, R. 1959. Embryogénie des Lemnacées. Développement de l'embryon chez le *Lemna minor* L. Compt. Rend. Hebd. Séances Acad. Sci. 248: 1896–1900.

Thanikaimoni, G. 1969. Esquisse palynologique des Aracées. Trav. Sect. Sci. Techn. Inst. Franç. Pondichéry 5(5): 1–31.

Webber, E. E. 1960. Observations on the epidermal structure and stomatal apparatus of some members of the Araceae. Rhodora 62: 251–258.

Wilson, K. A. 1960. The genera of the Arales in the southeastern United States. J. Arnold Arbor. 41: 47–72.

Иванова, И. Е. 1970. Некоторые особенности цветения и опыления рясок (Lemnaceae S. Gray). **Бот. Ж. 55: 649-659.**

Иванова, И. Е. 1973. К систематике семейства Lemnaceae S. Gray. **Бот. Ж. 58: 1413-1423.**

Ростовцева, З. П. 1967. Органогенез и особенности анатомического строения *Wolffia arrhiza* (L.) Wimm. в вегетативном периоде её жизни. Бот. Ж. 52: 1177–1184.

Савченко, М. И., & Е. Н. Маня. 1970. Сравитедьно-анатомические исследования покрывала и оси соцветия некоторых ароидых (Araceae). Бот. Ж. 55: 406–421.

SYNOPTICAL ARRANGEMENT OF THE FAMILIES OF ARALES

1 Plants with roots, stems, and leaves, terrestrial (or epiphytic) or sometimes more or less aquatic, but only rarely free-floating; flowers very numerous in a spadix; vascular system well developed, the plants commonly with vessels in the roots and tracheids in all vegetative organs; ovary usually with more than one carpel, only rarely apparently monomerous .... 1. ARACEAE.

1 Plants thalloid, free-floating, with or without 1-several short, slender roots; flowers (rarely produced) only 2–3 (4) in an infloresence; vascular system much reduced, the plants lacking both vessels and tracheids, or sometimes with tracheids in the roots; ovary pseudomonomerous ....2. LEMNACEAE.

1. Family ARACEAE A. L. de Jussieu 1789 nom. conserv., the Arum Family

Small to very large terrestrial herbs from creeping or tuberous rhizomes or corms, or sometimes scrambling slender shrubs or climbing vines with aerial roots (some anchoring the plant to the host, others descending to the ground), or virtually epiphytes, rarely (*Pistia*) free-floating aquatics, commonly with bundles of raphides in idioblasts scattered throughout the shoot (not in *Acorus*, which has ethereal oil cells instead), often cyanogenic (tyrosine-derived), sometimes saponiferous, often with proanthocyanins, and sometimes with alkaloids (especially of the indole group) or other sorts of poisons; tracheids present throughout the vegetative body, but vessels (with scalariform perforations) mostly confined to the roots, seldom well developed also in the stem; stem sympodial or rarely monopodial, without secondary growth; vascular bundles closed, scattered, very often containing articulated laticifers or mucilage-ducts (these sometimes tanniferous, bearing proanthocyanins and other substances) alongside the phloem; laticifers also well scattered in the parenchymatous tissues in the Colocasioideae; plastids of the sieve-tubes with cuneate proteinaceous inclusions; roots mycorhizal, without root-hairs. LEAVES alternate, often all basal, usually with a distinct, basally sheathing petiole and an expanded, simple and entire to variously cleft or perforate or even compound blade (narrow and ensiform in *Acorus*), variously parallel-veined or pinnately or palmately essentially net-veined; leaf ontogeny often or usually acropetal, as in typical dicotyledons, instead of basipetal, as in most monocotyledons; stomates paracytic or tetracytic, or with more than 4 subsidiary cells. Infloresence unbranched, a spadix, often malodorous, usually terminating a unit of the sympodium (stem) and nearly always subtended by a more or less prominent, often brightly colored spathe; FLOWERS very numerous and tiny, without bracts, pollinated by insects (especially flies), or sometimes by wind; perfect or unisexual (the plants then monoecious, with the staminate flowers uppermost in the spadix, or rarely dioecious); perfect flowers naked (Monsteroideae) or with a small but definite perianth of 4 or 6 (8) distinct or connate tepals in 2 cycles, but in unisexual flowers the perianth generally much-reduced or wanting; stamens (1–) 4 or 6 (8), opposite the tepals when these are present, distinct or sometimes partly or wholly connate; filaments mostly short and broad; anthers tetrasporangiate and dithecal or rarely monothecal, or sometimes bisporangiate and monothecal, opening by terminal pores

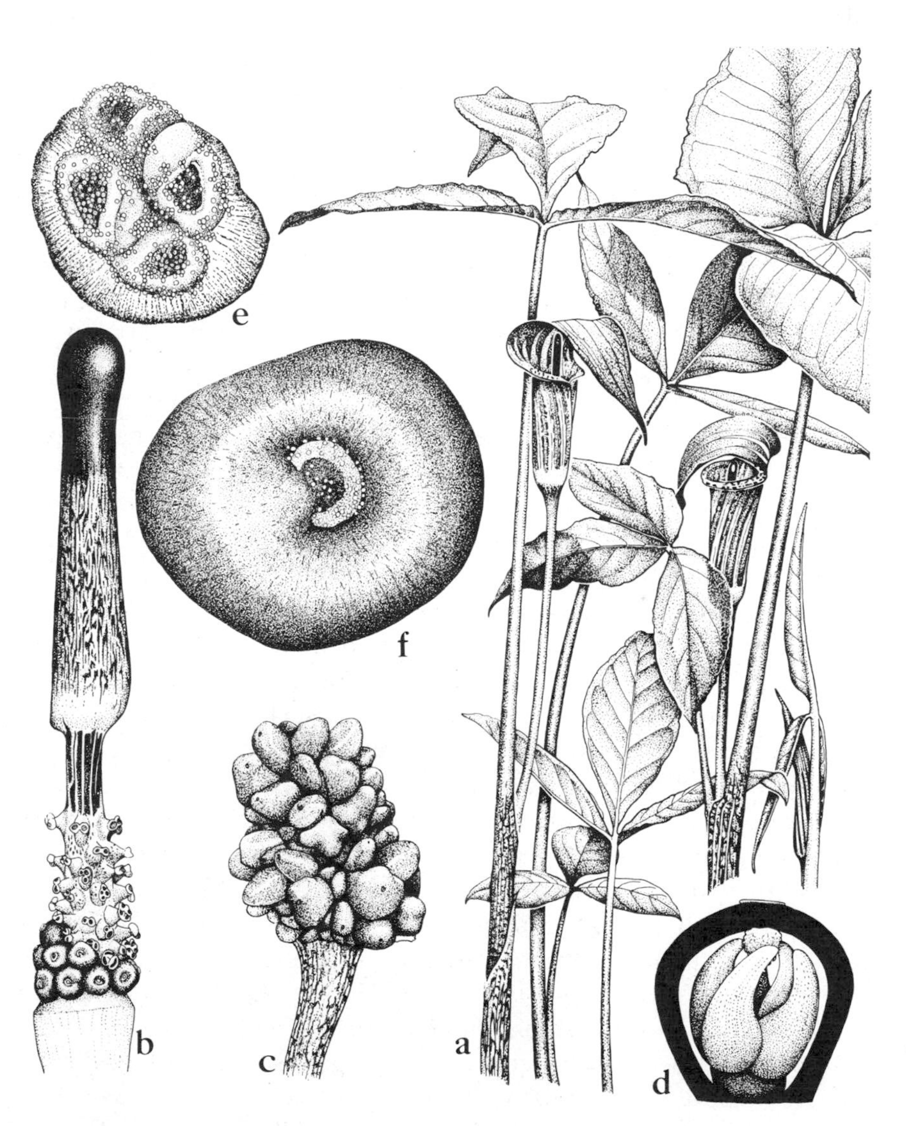

Fig. 8.4 Araceae. *Arisaema triphyllum* (L.) Schott. a, habit, $\times\frac{1}{2}$; b, spadix, the upper portion sterile, $\times 2$; c, fruiting spadix, $\times\frac{1}{2}$; d, schematic long-section of ovary, $\times 12$; e, staminate flower, from above, $\times 24$; f, pistillate flower (consisting only of the pistil), from above, $\times 24$.

or slits or longitudinal slits; pollen-grains 2–3-nucleate, variously monosulcate or with 2–4 sulci or pores, or often inaperturate; gynoecium of (2) 3 (–15) carpels united to form a compound, plurilocular ovary with axile (to axile-basal or axile-apical) placentation, or sometimes the ovary unilocular with parietal placentation, or seldom pseudomonomerous; style terminal and short, or the stigmas sessile; ovary superior; ovules 1-many in each locule, variously orthotropous or hemitropous to more often anatropous, bitegmic, pseudocrassinucellar or rarely (*Pistia*) tenuinucellar, at least sometimes with an integumentary tapetum; endosperm-development usually cellular, but sometimes helobial. FRUIT usually a berry, seldom dry or leathery and opening irregularly, or the whole spadix ripening as a multiple fruit; seeds 1-many; embryo rather large, monocotyledonous or sometimes with a vestigial second cotyledon opposite the principal one, linear, axially embedded in the usually copious, oily and sometimes also starchy endosperm, or the seed sometimes without endosperm. X = 7–17+. (Acoraceae, Pistiaceae)

The family Araceae consists of about 110 genera and 1800 species, the vast majority of them tropical or subtropical. Probably at least half of the species belong to only 4 genera, *Anthurium* (500), *Philodendron* (250), *Arisaema* (100+), and *Amorphophallus* (nearly 100). Most aroids are herbs of the forest floor, or vines climbing on forest trees. Some early-flowering genera of temperate regions, such as *Symplocarpus* and *Lysichiton*, develop relatively high temperatures, in the neighborhood of 40 or 45 degrees C, in the early growth of the spadix.

Philodendron and *Dieffenbachia* (dumb cane) are well known as house plants, popular not only because of their decorative foliage, but also because they require relatively little light. *Dieffenbachia* is notoriously dangerous when eaten. It contains an obscure proteinaceous poison as well as highly irritating raphides that cause the mouth and tongue to swell, sometimes to the point of strangulation. The poisonous quality of *Philodendron* is also well established, but less widely known. Babies and household pets eat the leaves of either genus at the risk of death.

Leaves plausibly referred to *Philodendron* and *Peltandra* have been found in Eocene deposits in the United States (*Philodendron* in Tennessee, *Peltandra* in North Dakota), and aroid spadices also occur in Middle Eocene strata in the United States. Fragmentary leaves from the Paleocene of Kazakhstan are probably araceous. Pollen that may represent *Spathiphyllum* has been found in Paleocene deposits in Colombia, and undoubted aroid pollen occurs in Miocene and more recent deposits elsewhere in the world.

Acorus stands somewhat apart from the rest of the family, and some botanists have thought it might be related to *Typha* instead of to the Araceae. Serological studies (Lee and Fairbrothers 1972, cited under Typhales) disclose no affinity between the two genera, however.

2. Family LEMNACEAE S. F. Gray 1821 nom. conserv., the Duckweed Family

Small or very small, free-floating, glabrous, thalloid plants of quiet water, rootless or with 1-several simple, unbranched roots, wholly without xylem, or (*Spirodela*) with tracheids in the roots; plants with or without raphides, sometimes accumulating manganese, and sometimes with scattered tanniferous cells, but not cyanogenic; frond (thalloid shoot) fleshy or membranous, flat and subrotund to linear, or thickened and nearly globular, sometimes with 1-several visible nerves, but without tracheids or vessels, provided with 2 reproductive pouches, one on each margin near the base, or with a single pouch on the upper surface near the base; reproduction almost wholly vegetative, by budding in the pouch or pouches, the new plants sometimes remaining attached to the parent plant for some time, but eventually breaking away. Inflorescence (rarely produced) borne in one of the 2 marginal pouches, or in the pouch on the upper surface, in the former case consisting of 2 (3) staminate flowers and one pistillate flower on a short axis, collectively surrounded by a small, vestigal, membranous spathe, in the latter case consisting of one staminate and one pistillate flower, without a spathe; staminate flower consisting of a single stamen, without a perianth; anther tetrasporangiate and dithecal, opening by longitudinal or transverse slits, or reputedly sometimes bisporangiate and monothecal; pollen-grains spheroidal, trinucleate, uniporate; pistillate flower consisting of a single pseudomonomerous, unilocular gynoecium with a short terminal style; ovules 1–7, basal, bitegmic, crassinucellar, orthotropous at first, but often becoming more or less hemitropous, campylotropous, or anatropous after fertilization; inner integument forming a persistent operculum at the micropylar end of the seed; embryo-sac bisporic, except in *Spirodela*; endosperm-development variously reported as cellular or helobial. FRUIT a small utricle with 1–4 seeds; embryo relatively large, straight, with a single large cotyledon, sometimes lacking a radicle; endosperm wanting, or rather scanty, sheathing the embryo, and containing starch as well as other food-reserves. X = 8, 10, 11, 21.

The family Lemnaceae consists of 6 genera, of cosmopolitan distribution. The most familiar genera are *Lemna*, with about 9 species, and *Spirodela*, with 4. *Wolffia* (the smallest angiosperm) has 7 species, *Wolffiella* 5, *Pseudowolffia* 3, and *Wolffiopsis* only one.

The Lemnaceae occupy a distinctive ecological niche that has scarcely been entered by other angiosperms. Their structure and reproduction reflect the habitat.

It is widely agreed that the Lemnaceae are related to and derived from the Araceae. *Pistia*, a free-floating aquatic aroid with a relatively small and few-flowered spadix, is seen as pointing the way toward *Spirodela*, the least-reduced genus of Lemnaceae.

III. Subclass COMMELINIDAE Takhtajan 1966

Herbs, or seldom somewhat woody plants, without secondary growth and not obviously mycotrophic; vessels usually present in all vegetative organs; plastids of the sieve-tubes with cuneate proteinaceous inclusions; vascular bundles of the stem closed, scattered (often concentrated toward the periphery) or in 2 cycles, seldom in only 1 cycle. LEAVES alternate (sometimes all basal), simple and entire or nearly so, parallel-veined, usually with an open or closed basal sheath and a narrow and elongate to sometimes shorter and broader blade, but sometimes without a sheath, or the blade reduced or wanting; stomates mostly paracytic, with 2 subsidiary cells, seldom anomocytic or with more than 2 subsidiary cells. FLOWERS strictly hypogynous, perfect or unisexual, without nectaries or nectar (except for the nectaries just within the tip of the petals in some Eriocaulaceae), pollinated by pollen-gathering insects (nectar-gathering in spp. of Eriocaulaceae) or more often wind-pollinated or self-pollinated (or apomictic); perianth in the more archaic families trimerous and with well differentiated sepals and petals, in the more advanced families reduced and chaffy or bristly and often not obviously trimerous, or wanting; stamens mostly 3 or 6, seldom 1, 2, or many; pollen-grains variously binucleate or trinucleate, uniaperturate (often uniporate) or less commonly inaperturate or with more than one aperture; gynoecium of 2 or 3 (4) carpels united to form a compound, superior ovary, or seldom pseudomonomerous, with variously axile, parietal, free-central, basal, or apical placentation; ovules 1-many, anatropous to hemitropous, orthotropous, or rarely campylotropous, bitegmic or less often unitegmic, variously crassinucellar, pseudocrassinucellar, or tenuinucellar; endosperm-development most commonly nuclear, sometimes helobial. FRUIT usually dry, seldom fleshy, variously dehiscent or indehiscent; endosperm wholly or in large part starchy, often mealy and with compound starch-grains, frequently with a proteinaceous layer or segment as well, but usually without significant reserves of hemicellulose or oil, rarely wanting; embryo with a single cotyledon, or not differentiated into parts.

The subclass Commelinidae as here defined consists of 7 orders, 16 families, and about 15,000 species. More than half of the species belong to the single family Poaceae, and the Poaceae and Cyperaceae together account for about four-fifths of the species. The most archaic order of

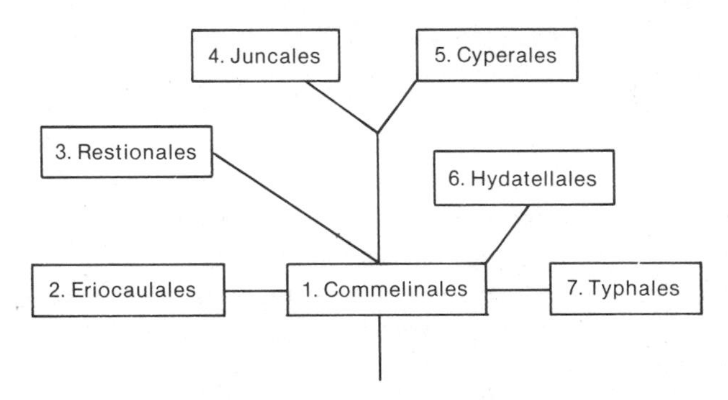

Fig. 9.1 Putative relationships among the orders of Commelinidae.

the group, the Commelinales, has flowers pollinated by pollen-gathering insects. The other orders show varying degrees of floral reduction associated with adaptation to wind-pollination. Some of the Eriocaulales have reverted to insect-pollination and developed nectaries just within the tip of the tiny petals, quite unlike the septal nectaries that are so common in other groups of monocotyledons.

The fossil record shows that the Commelinidae are a fairly ancient group, but it does not yet show the connections among the several orders. The grasses, one of the most advanced families, can be traced back to the Paleocene, and possibly to the Maestrichtian segment of the Upper Cretaceous. Smooth, monoporate pollen that presumably represents the Commelinidae first appears in the fossil record in the Senonian, above the middle of the Upper Cretaceous. Presumably the subclass as a whole antedates the Maestrichtian, but there is no reason to suppose that it antedates the Upper Cretaceous.

SYNOPTICAL ARRANGEMENT OF THE ORDERS OF COMMELINIDAE

1 Flowers commonly perfect, and with more or less showy petals that are well differentiated from the sepals, adapted to pollination by insects .... 1. COMMELINALES.

1 Flowers perfect or unisexual, without showy petals, the perianth (when present) sometimes in 2 series, but dry and chaffy, the flowers (except some Eriocaulaceae) adapted to wind-pollination or self-pollination or pollination by water.

2 Ovules various, but never at once solitary, pendulous from near the summit of the ovary, and anatropous; plants terrestrial or less often aquatic.

3 Ovary with 1–3 fertile locules and as many stigmas, each fertile

locule with a single pendulous, orthotropous ovule; embryo peripheral to the endosperm.

4 Flowers aggregated into dense, pseudanthial, involucrate heads (usually both sexes in the same head), pollinated by wind or often by insects; ovules tenuinucellar; anthers usually tetrasporangiate and dithecal; leaves all basal, without a well differentiated sheath ....2. ERIOCAULALES.

4 Flowers seldom aggregated into dense heads, usually or always wind-pollinated or self-pollinated; ovules crassinucellar to sometimes tenuinucellar; anthers bisporangiate and monothecal, or less often tetrasporangiate and dithecal; leaves cauline or basal, and with a more or less well differentiated basal sheath ....3. RESTIONALES.

3 Ovary otherwise, either with more stigmas than locules, or with more than one ovule per locule, the ovules of various form and position; embryo peripheral (Poaceae) or embedded (Cyperaceae, Juncales) in the endosperm; flowers perfect or unisexual.

5 Ovary with 1–3 locules and 3-many ovules; fruit capsular; pollen-grains in tetrads; flowers with an evident (though small) biseriate, chaffy perianth, borne in open or compact inflorescences unlike those of the Cyperales ....4. JUNCALES.

5 Ovary with a single locule and ovule; fruit indehiscent; pollen-grains in monads or pseudomonads; flowers borne in characteristic spikes or spikelets, without a clearly biseriate, chaffy perianth .... 5. CYPERALES.

2 Ovules solitary (or solitary in each locule), pendulous from near the summit of the ovary, anatropous; plants aquatic or semiaquatic.

6 Seeds with well developed perisperm and virtually no endosperm (unique in the subclass); vascular system much-reduced, the vessels confined to the roots, the stem with only 2 or 3 vascular bundles (unique in the subclass); flowers hydrophilous (unique in the subclass) or autogamous ....6. HYDATELLALES.

6 Seeds with well developed endosperm and no perisperm; vascular system well developed, with vessels in all vegetative organs, and with numerous vascular bundles in the stem; flowers anemophilous (or autogamous) ....7. TYPHALES.

1. Order COMMELINALES Lindley 1833

Perennial or less often annual herbs, without secondary growth; vessels present in all vegetative organs, or at least in the stems and roots; vascular bundles of the stem closed, generally scattered; plastids of the sieve-tubes with cuneate proteinaceous inclusions. LEAVES alternate, often all basal, simple, with or rarely (Mayacaceae) without a basal sheath, and with a narrow to fairly broad, parallel-veined blade that is sometimes separated from the sheath by a slender petiole; stomates variously paracytic or anomocytic or with several subsidiary cells. FLOWERS entomophilous but without nectaries or nectar, trimerous, hypogynous, commonly perfect, regular or irregular, with well differentiated calyx and corolla, the corolla very often ephemeral; sepals 3, green or chaffy, or seldom petaloid but unlike the petals, distinct or connate below, rarely only 2 by reduction and loss of one; petals 3, distinct or less often connate at the base or forming a slender, 3-lobed tube; stamens (1–) 3 or 6, or 3 polliniferous and 3 staminodial; anthers mostly tetrasporangiate and dithecal, or sometimes tetrathecal, or seldom bisporangiate and dithecal, opening by longitudinal slits or by apical or basal pores; pollen-grains binucleate or seldom trinucleate, monosulcate or inaperturate, seldom bisulcate or trichotomosulcate or with a pair of short sulci in addition to the principal one; gynoecium of 3 carpels united to form a compound, trilocular or unilocular ovary topped by a simple or trifid style, seldom 1 or 2 carpels more or less reduced or even suppressed; ovules 1-several on each of the axile, parietal, or free-central placentas, anatropous to hemitropous, orthotropous, or campylotropous, bitegmic, crassinucellar or tenuinucellar; endosperm-development nuclear. FRUIT a loculicidal capsule, or seldom indehiscent and then sometimes fleshy; seeds with abundant, mealy endosperm containing reserve starch and often also protein, seldom also some oil, the starch-grains compound or (Rapateaceae) simple; seed-coat in 2 families (Commelinaceae and Mayacaceae) with a discoid thickening marking the position of the embryo; embryo mostly small and capping the endosperm at the micropylar end, variously lenticular or thick-cylindric or conic, or with a conic radicle and expanded, discoid cap, the cotyledon generally not clearly differentiated.

The order Commelinales as here defined consists of 4 families and about 1000 species. The Commelinaceae, with about 700 species, are by far the largest family, followed by the Xyridaceae (200+), Rapateaceae (100), and Mayacaceae (4). Within the order, the Rapateaceae and

Xyridaceae are a closely related pair, the Xyridaceae perhaps derived from the Rapateaceae. Embryological features (Stevenson, unpublished) suggest that the Mayacaceae are more closely related to the Xyridaceae than to the Commelinaceae.

Ecologically, the Commelinales are characterized by having relatively simple adaptations favoring pollination by pollen-gathering insects. The other orders of the subclass have more or less reduced flowers that are in most families wind-pollinated or self-pollinated. Members of the Zingiberidae and Liliidae, on the other hand, commonly have nectariferous flowers, often with much more complex adaptations for pollination.

Aside from pollination, the Commelinales are ecologically rather diverse. They are all essentially herbaceous, but they occur in a wide range of habitats. The Mayacaceae are essentially aquatic. The Xyridaceae mostly occur in wet or marshy places but are rarely truly aquatic. The Rapateaceae and Commelinaceae range from ordinary mesophytes to marsh-plants or occasional xerophytes.

SELECTED REFERENCES

Brenan, J. P. M. 1966. The classification of Commelinaceae. J. Linn. Soc., Bot. 59: 340–370.

Brückner, G. 1926. Beiträge zur Anatomie, Morphologie und Systematik der Commelinaceae. Bot. Jahrb. Syst. 61 (Beiblatt 137): 1–70.

Carlquist, S. 1960. Anatomy of Guayana Xyridaceae: *Abolboda, Orectanthe,* and *Achlyphila*. Mem. New York Bot. Gard. 10(2): 65–117.

Carlquist, S. 1961. Pollen morphology of Rapateaceae. Aliso 5: 39–66.

Carlquist, S. 1966. Anatomy of Rapateaceae—roots and stems. Phytomorphology 16: 17–38.

Chikkannaiah, P. S. 1963. Embryology of some members of the family Commelinaceae: *Commelina subulata* Roth. Phytomorphology 13: 174–184.

Forman, L. L. 1962. *Aëtheolirion*, a new genus of Commelinaceae from Thailand, with notes on allied genera. Kew Bull. 16: 211–221.

Guervin, C., & C. Le Coq. 1966. Caryologie des Commélinacées: Application à quelques problèmes relatifs à leur évolution. Rev. Cytol. Biol. Vég. 29: 267–334.

Hamann, U. 1961. Merkmalbestand und Verwandschaftbeziehungen der Farinosae. Willdenowia 2: 639–768.

Hamann, U. 1962. Über Bau und Entwicklung der Philydraceae und über die Begriffe "mehliges Nährgewebe" und "Farinosae." Bot. Jahrb. Syst. 81: 397–407.

Hamann, U. 1962. Weiteres über Merkmalsbestand und Verwandtschaftsbeziehungen der "Farinosae." Wildenowia 3: 169–207.

Jones, K., & C. Jopling. 1972. Chromosomes and the classification of the Commelinaceae. J. Linn. Soc., Bot. 65: 129–162.

Kral, R. 1966. *Xyris* (Xyridaceae) of the continental United States and Canada. Sida 2: 177–260.

Lourteig, A. 1952. Mayacaceae. Notul. Syst. (Paris) 14: 234–248.
Maguire, B., J. J. Wurdack, & collaborators. 1958. The botany of the Guayana Highland—Part III. Mem. New York Bot. Gard. 10(1): 1–156.
Maguire, B., J. J. Wurdack, & collaborators. 1960. The botany of the Guayana Highland—Part IV. Mem. New York Bot. Gard. 10(2): 1–37.
Maheshwari, S. C., & B. Baldev. 1958. A contribution to the morphology and embryology of *Commelina forskalaei* Vahl. Phytomorphology 8: 277–298.
Murty, Y. S., N. P. Saxena, & V. Singh. 1974. Floral morphology of the Commelinaceae. J. Indian Bot. Soc. 53: 127–136.
Pichon, M. 1946. Sur les Commélinacées. Notul. Syst. (Paris) 12: 217–242.
Rohweder, O. 1963. Anatomische und histogenetische Untersuchungen an Laubsprossen und Blüten der Commelinaceen. Bot. Jahrb. Syst. 82: 1–99.
Rohweder, O. 1969. Beiträge zur Blütenmorphologie und -anatomie der Commelinaceen mit Anmerkungen zur Begrenzung und Gliederung der Familie. Ber. Schweiz. Bot. Ges. 79: 199–220.
Rowley, J. R., & A. O. Dahl. 1962. The aperture of the pollen grain in *Commelinantia*. Pollen & Spores 4: 221–232.
Thieret, J. W. 1975. The Mayacaceae in the southeastern United States. J. Arnold Arbor. 56: 248–255.
Tomlinson, P. B. 1966. Anatomical data in the classification of Commelinaceae. J. Linn. Soc., Bot. 59: 371–395.
Tomlinson, P. B. 1969. Vol. 3. Commelinales-Zingiberales. *In*: C. R. Metcalfe, ed. Anatomy of the monocotyledons. Clarendon Press. Oxford.
Woodson, R. E. 1942. Commentary on the North American genera of Commelinaceae. Ann. Missouri Bot. Gard. 29: 141–154.

SYNOPTICAL ARRANGEMENT OF THE FAMILIES OF COMMELINALES

1 Leaf-sheath open, often not well differentiated from the blade, or (Mayacaceae) the leaf without a sheath.

  2 Leaves mostly or all clustered at the base, with a basal, open sheath and narrow blade; flowers usually in a compact inflorescence terminating a long scape or peduncle.

    3 Stamens 6, opening by apical or subapical pores or short, pore-like slits; inflorescence generally a compact cluster (often a head) of spikelets, each spikelet with several bracts subtending the single flower .... 1. RAPATEACEAE.

    3 Stamens generally 3, often accompanied by 3 staminodes, rarely 6 and all polliniferous; anthers opening by longitudinal slits; inflorescence usually a simple, racemose head or short, stout spike, rarely open ....2. XYRIDACEAE.

  2 Leaves well distributed along the stem, numerous and linear or filiform, without a sheath; flowers pedicellate in the axils of the crowded, bract-like leaves at the stem-tips; stamens 3, anthers opening by apical pores or short, pore-like slits .... 3. MAYACACEAE.

1 Leaf with a closed sheath and a well defined (narrow or broad), commonly somewhat succulent blade ....4. COMMELINACEAE.

1. Family RAPATEACEAE Dumortier 1829 nom. conserv., the Rapatea Family

Perennial, often coarse herbs from a stout rhizome, often accumulating aluminum (a most unusual feature among monocotyledons), commonly with scattered tanniferous cells in the parenchymatous tissues, without calcium oxalate crystals, but some of the epidermal cells containing several or many silica-bodies; vessels present in all vegetative organs (or at least in the roots and stems), with scalariform or sometimes simple perforations; vascular bundles of the stem closed, scattered. LEAVES all basal, distichous, with a folded, open sheath and a narrow, often gladiate, parallel-veined blade; stomates paracytic or anomocytic or tetracytic. Inflorescence borne at the summit of a scape (scape suppressed and the inflorescence surrounded by the leaves in the monotypic West-African genus *Maschalocephalus*), very often with an involucre of (1) 2-several large bracts subtending the capitate or unilaterally racemose cluster of spikelets; each spikelet with a single terminal flower and several imbricate, spirally arranged sterile bracts beneath it; FLOWERS trimerous, perfect, regular, entomophilous, hypogynous with strongly differentiated calyx and corolla; nectar and nectaries wanting; sepals 3, chaffy-translucent, firm, imbricate, often more or less connate at the base; petals 3, ephemeral, yellow or seldom red, usually connate below to form a tube with imbricate lobes; stamens 6, in 2 cycles of 3; filaments short, generally attached to the corolla-tube and often connate at the base; anthers basifixed, usually surpassed by the corolla, generally tetrasporangiate and dithecal, seldom bisporangiate and monothecal, the thecae often confluent distally, the anther opening by 1, 2, or 4 apical pores or short slits; pollen-grains much like those of *Xyris*, monosulcate or seldom trichotomosulcate or bisulcate; gynoecium of 3 carpels united to form a compound, trilocular ovary, the partitions sometimes imperfect near the top, in *Spathanthus* only one locule fertile; style solitary and terminal, with a simple stigma; ovules anatropous, several or numerous in each locule on axile placentas, or only 1–2 in each locule (or in the one fertile locule) and axile-basal. FRUIT a loculicidal capsule; seeds with a small, lenticular embryo lying near the hilum alongside the copious, starchy, mealy endosperm, in which the starch-grains are mostly simple rather than compound. 2n = 22 (*Maschalocephalus*).

The family Rapateaceae consists of 16 genera and nearly a hundred species, native to tropical South America, with a single genus and species in tropical West Africa. The family is best developed in the Guayana Highland region of northern South America. The largest genera are *Rapatea* and *Stegolepis*, with about 20 species each. Botanists are agreed that the Rapateaceae are allied to the Xyridaceae.

2. Family XYRIDACEAE C. A. Agardh 1823 nom. conserv., the Yellow-eyed Grass Family

Perennial or less often annual herbs, mostly of wet places, fibrous-rooted from rosette-stems or sometimes from creeping rhizomes, sometimes tanniferous, sometimes saponiferous, and sometimes producing crystals of calcium oxalate, but without raphides, and not accumulating aluminum; vessels with simple perforations, present in all vegetative organs; vascular bundles of the stem closed, scattered, but sometimes confined to the outer portion; roots producing root-hairs. LEAVES mostly or all basal, distichous or less often spirally arranged, with an open sheath and a narrow, flat to cylindric or filiform, very often equitant and unifacial, parallel-veined (or univeined) blade; stomates paracytic or anomocytic. Scapes or peduncles usually numerous, each terminating in a globose to cylindric head or dense spike (inflorescence open and few-flowered in *Achlyphila*); FLOWERS sessile (long-pedicellate in *Achlyphila*) in the axils of the firm, tough, spirally arranged, closely imbricate bracts, without nectaries or nectar, odorless, pollinated by pollen-gathering bees, perfect, hypogynous, trimerous, with strongly differentiated calyx and corolla; sepals typically 3, the outer (anterior) one thin and membranous, more or less enclosing the rest of the flower, but thrown back at anthesis, or sometimes more or less reduced or even obsolete; lateral sepals 2, boat-shaped, keeled, chaffy-scarious; corolla regular or slightly irregular, yellow or less often blue or white, commonly ephemeral, consisting of 3 distinct, clawed petals, or the petals sometimes connate at the base or connate for much of their length to form a long, slender tube; stamens typically 3, opposite the petals or corolla-lobes, sometimes alternating with as many staminodes that are usually bifid at the summit, or rarely the stamens 6 and all fertile; filaments short, adnate to the petals or the corolla-tube; anthers tetrasporangiate and dithecal, extrorse or introrse, opening by longitudinal slits; pollen-grains binucleate or sometimes trinucleate, mostly monosulcate or inaperturate; gynoecium of 3 carpels united to form a compound, unilocular or basally trilocular to seldom fully trilocular ovary, the placentation accordingly parietal, free-central, or axile; style terminal,

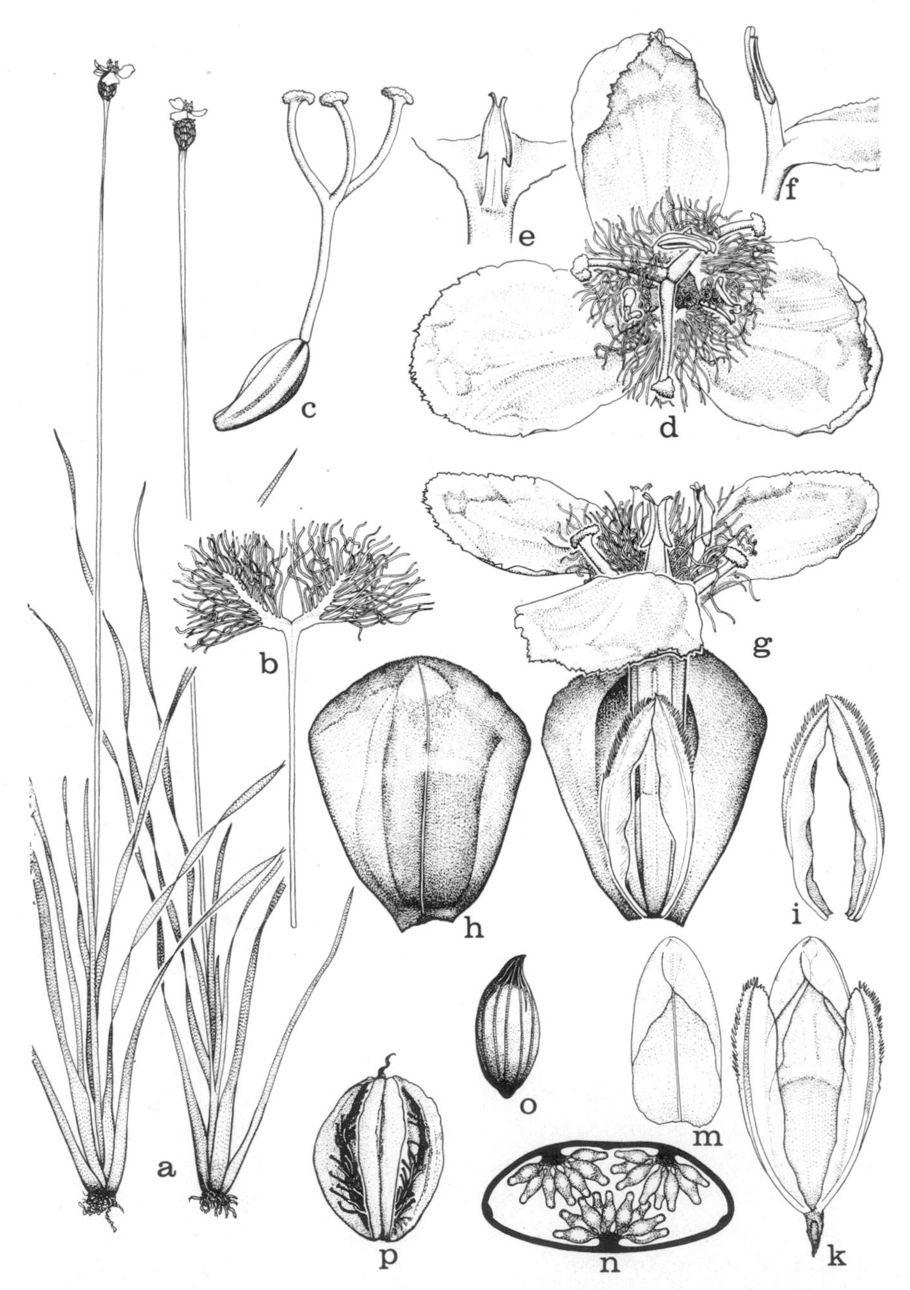

Fig. 9.2 Xyridaceae. *Xyris torta* J. E. Smith. a, habit, ×½; b, staminode, ×6; c, pistil, ×6; d, flower, from above, ×6; e, f, two views of stamen, attached to corolla, ×6; g, ventral view of flower, ×6; h, bract, ×6; i, lateral sepals, ×6; k, ventral view of flower bud, ×6; m, ventral view of dorsal sepal, ×6; n, schematic cross-section of ovary, ×18; o, seed, ×36; p, opening fruit, ×6.

simple or trifid; ovules (1–) several or numerous on each placenta, orthotropous to anatropous or weakly campylotropous, bitegmic, crassinucellar to tenuinucellar; endosperm-development nuclear, sometimes with haustoria. FRUIT capsular, loculicidal or sometimes opening irregularly, sometimes enclosed by the persistent corolla-tube; seeds small, the copious, more or less mealy, starchy and proteinaceous endosperm with compound starch-grains, and sometimes also with some oil; embryo lying alongside the endosperm, small to large, lenticular or shield-shaped, scarcely differentiated into parts. X = 8, 9, 13, 17. (Abolbodaceae)

The family Xyridaceae as here defined consists of 4 genera and more than 200 species, widespread in tropical and subtropical regions, with relatively few species extending into temperate climates. The vast bulk of the species belongs to the single genus *Xyris*. The other three genera *Abolboda, Achlyphila*, and *Orectanthe*) have only about 20 species altogether.

3. Family MAYACACEAE Kunth 1842 nom. conserv., the Mayaca Family

Small, soft, creeping herbs of shallow water or of very wet places, rooted in the substrate and often submerged (at least seasonally), without secretory cells and lacking both silica and crystals of calcium oxalate; vessels with reticulate (or scalariform) perforations, present in the roots and stems, perhaps also in the leaves; vascular bundles of the stem typically 3 in number, the slender central cylinder delimited from the aerenchymatous cortex by an endodermis. LEAVES numerous, spirally arranged, well distributed along the stem, sessile, without a basal sheath, rather small, lanceolate to linear or filiform, commonly bidentate at the tip, 1-nerved, with a longitudinal air-canal on each side of the midstrip; stomates paracytic; short, uniseriate, filamentous, ephemeral hairs initially present in the axils, protecting the shoot-apex, the plants otherwise glabrous. FLOWERS pedicellate, ontogenetically terminal, but seemingly borne in the axils of the crowded, bract-like leaves at the stem-tips, aerial, perfect, regular, hypogynous, trimerous, with strongly differentiated calyx and corolla, without nectaries or nectar; sepals 3, distinct, green, valvate or nearly so; petals 3, distinct, white or pale, imbricate; stamens 3, alternate with the petals; filaments short, slender; anthers basifixed, tetrasporangiate and tetrathecal, or reputedly seldom bisporangiate and dithecal, opening by apical or subapical pores or short, pore-like slits; pollen-grains binucleate, monosulcate; gynoecium of 3 carpels united to form a compound, unilocular ovary with a slender

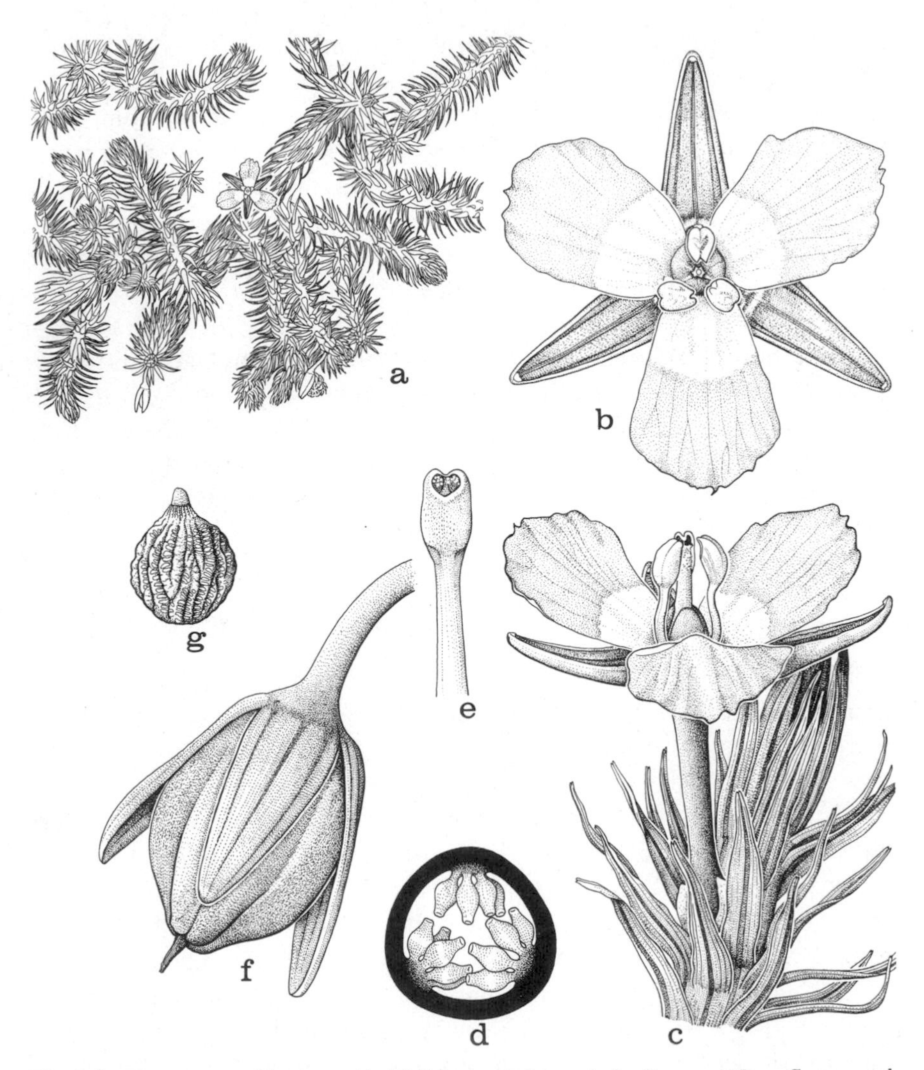

Fig. 9.3 Mayacaceae. *Mayaca aubletii* Michx. a, habit, ×1; b, flower, ×6; c, flower and portion of leafy branch, ×6; d, schematic cross-section of ovary, ×24; e, stamen, ×12; f, developing fruit, ×6; g, seed, ×12.

terminal style and 1–3 short stigmas; ovules several or numerous on each of the 3 parietal placentas, orthotropous, bitegmic, almost tenuinucellate. FRUIT a loculicidal capsule, often immersed on a deflexed pedicel; seeds with copious, mealy, starchy and proteinaceous endosperm that has compound starch-grains; embryo small, shaped like a centrally unipapillate disk, forming an apical cap on the endosperm just beneath an opercular "embryostega" at the micropylar end of the seed-coat.

The family Mayacaceae consists of the single genus *Mayaca*, with 4 or more species, one native to tropical West Africa, the others to tropical and warm-temperate America.

4. Family COMMELINACEAE R. Brown 1810 nom. conserv., the Spiderwort Family

Perennial or less commonly annual herbs, without secondary growth, often somewhat succulent, rarely twining or epiphytic, almost invariably with 3-celled, glandular microhairs and often also with unicellular or uniseriate macrohairs, commonly (but not in *Cartonema*) with mucilage-cells or -canals each containing a bundle of raphides, and with idioblasts containing silica-bodies in the epidermis of the stem and leaf in some spp. of *Tradescantia* and other genera, only seldom saponiferous, not accumulating aluminum, occasionally bearing proanthocyanins that may or may not be concentrated in tanniferous idioblasts; principal flavonoids consisting of flavone C-glycosides; stem swollen at the nodes; large, simple starch-grains stored in the rhizome; vessels present in all vegetative organs, with simple or in part scalariform perforations (vessels confined to roots in *Cartonema*); vascular bundles of the stem closed, scattered, but the stem commonly with an endodermis-like sheath separating the vascular region from an outer cortex; plastids of the sieve-tubes with cuneate proteinaceous inclusions; roots producing root-hairs, not mycorhizal. LEAVES alternate, simple, with closed sheath and narrow or somewhat expanded, parallel-veined blade that may be separated from the sheath by a slender petiole, the several main veins very often connected by a dense system of regular transverse veins; leaf-blade in bud with the opposite halves rolled separately against the midrib, the abaxial surface outermost; stomates commonly with 4 or 6 subsidiary cells, but sometimes with only 2. FLOWERS hypogynous, commonly perfect, regular or more or less irregular, trimerous, usually entomophilous but without nectaries or nectar, borne in terminal, axillary, or leaf-opposed, basically cymose inflorescences, the inflorescence or its shoot sometimes breaking through the subtending sheath, a folded spathaceous leafy bract often subtending each inflorescence, or the flowers seldom solitary or in apparent spikes or racemes; sepals 3, generally green, but sometimes petaloid and colored (as in *Dichorisandra*), with open or imbricate aestivation, distinct or rarely connate

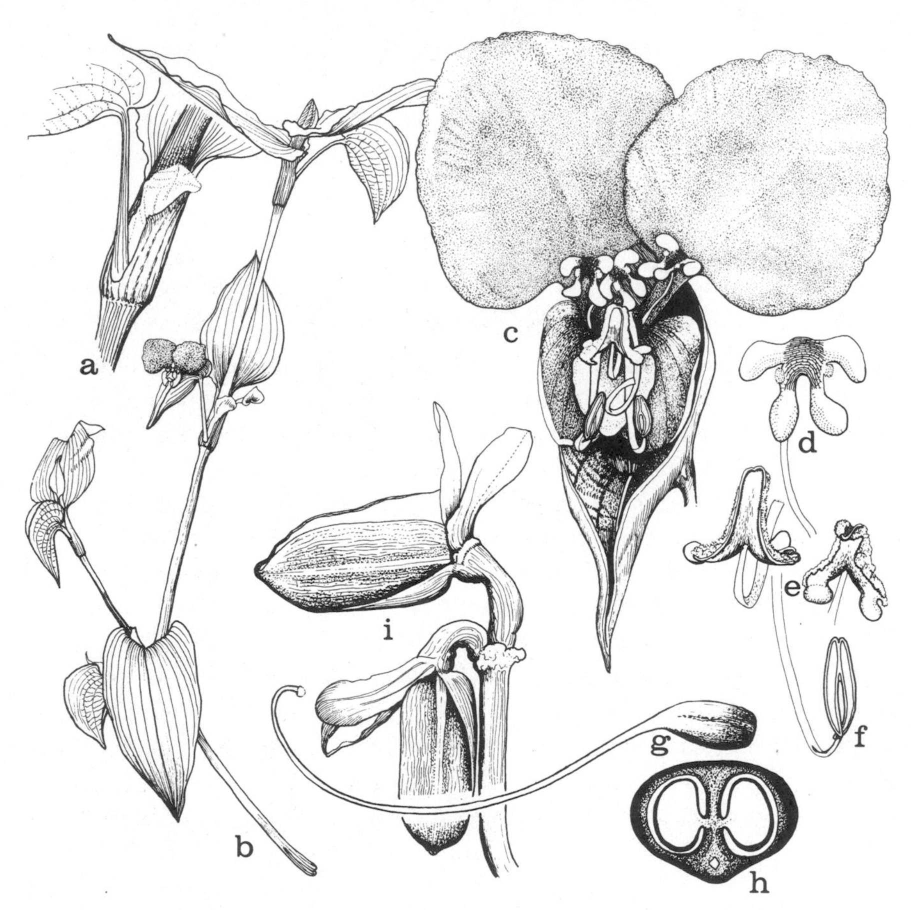

Fig. 9.4 Commelinaceae. *Commelina communis* L. a, detail of nodal region, $\times\frac{3}{4}$; b, habit, $\times\frac{1}{2}$; c, flower, from above, $\times 3$; d, e, f, upper, lateral, and lower stamens $\times 6$; g, pistil, $\times 6$; h, schematic cross-section of ovary, $\times 12$; i, fruits, $\times 3$.

below; petals 3, ephemeral, mostly blue or white, distinct and sometimes clawed (as in *Commelina* and *Aneilema*), or sometimes connate below to form a short or elongate tube, all alike, or one of them differently colored and/or more or less strongly reduced; stamens most commonly 6 in 2 cycles of 3, but sometimes only 3 functional and the others staminodial or suppressed, rarely the stamen solitary; filaments usually slender, often conspicuously long-hairy; anthers basifixed or versatile, often with an expanded connective, sometimes dissimilar inter se, tetrasporangiate and dithecal, opening by longitudinal slits or seldom by apical or basal pores (or a single apical pore); pollen-grains binucleate or seldom trinucleate, monosulcate or rarely (*Commelinantia*) with 2 additional short sulci; gynoecium of 3 carpels united to form a compound, superior, trilocular ovary, or the ovary often unilocular in the upper quarter, or one or two of the locules often imperfectly developed or even suppressed; style terminal, hollow, with a capitate or penicillate or trilobed, wet or dry stigma; ovules 1-several in each locule (or in each functional locule) on the axile placentas, orthotropous to hemitropous or anatropous, bitegmic, crassinucellar; endosperm-development nuclear. FRUIT commonly a loculicidal capsule, seldom indehiscent (and then sometimes fleshy); seeds with abundant, mealy endosperm and compound starch-grains; embryo small, capping the endosperm at one end, thick-cylindric to conic or capitate, with a single terminal cotyledon and lateral plumule, or sometimes with a vestigial second cotyledon opposite the principal one, so that the ancestrally terminal position of the plumule is shown; position of the embryo more or less clearly marked externally by a disk-like or conic opercular swelling (the embryostega) of the seed-coat. X = 4–19. (Cartonemataceae)

The family Commelinaceae as here defined consists of about 50 genera and 700 species, widespread in tropical and subtropical regions. The largest genus is *Commelina*, with 150–200 species. *Zebrina pendula* Schnizl., the Wandering Jew, is a common house-plant; *Rhoeo spathacea* (Swartz) W. T. Stearn is often cultivated in greenhouses; and species of *Tradescantia*, spiderwort (especially *T. virginiana* L.) are used as garden ornamentals.

The status of *Cartonema*, recently taken as a separate family by some authors, has been adequately dealt with by Brenan and by Rohweder. *Cartonema* and *Triceratella* individually stand somewhat apart from the bulk of the family, but their affinity to the Commelinaceae is clear, and there is no need to fragment the family.

2. Order ERIOCAULALES Nakai 1930

The order consists of the single family Eriocaulaceae.

SELECTED REFERENCES

Hare, C. L. 1950. The structure and development of *Eriocaulon septangulare* With. J. Linn. Soc., Bot. 53: 422–448.

Kral, R. 1966. Eriocaulaceae of continental North America north of Mexico. Sida 2: 285–332.

Thanikaimoni, G. 1965. Contribution to the pollen morphology of Eriocaulaceae. Pollen & Spores 7: 181–191.

1. Family ERIOCAULACEAE Desvaux 1828 nom. conserv., the Pipewort Family

Perennial or seldom annual, mostly rather small herbs, without secondary thickening, growing in wet places or emergent from shallow water, or seldom fully submerged; various sorts of calcium oxalate crystals, but not raphides, often present in some of the cells of the parenchymatous tissues; vessels with simple or scalariform perforations, present in all vegetative organs; vascular bundles of the stem closed, mostly in 2 cycles; plastids of the sieve-tubes with cuneate proteinaceous inclusions; roots producing root-hairs. LEAVES alternate, all closely crowded at the base, parallel-veined, narrow and grass-like, but without a well differentiated basal sheath; stomates paracytic. Inflorescence a dense, centripetally flowering, commonly whitish or grayish or lead-colored head terminating the scape, subtended by an involucre of chaffy bracts; common receptacle of the head provided with chaffy bracts subtending the flowers, or merely hairy, or naked; FLOWERS individually very small, anemophilous or often entomophilous, without nectaries or nectar except for the nectariferous glands just within the tip of the petals in *Eriocaulon*, often stipitate, trimerous or less often dimerous, regular or irregular, hypogynous, unisexual, both sexes intermingled in the same head, or the pistillate flowers marginal, or rarely the heads unisexual and the plants dioecious; sepals 2 or 3, distinct, or connate below to form a lobed tube, or connate to form a spathe-like scale; petals 2 or 3, distinct or variously connate, or wanting; staminate flowers with the filaments adnate to the corolla-tube, or often with a slender, stipe-like androphore (above the calyx) at the top of which the filaments and petals (when these are present) diverge; stamens in dimerous flowers

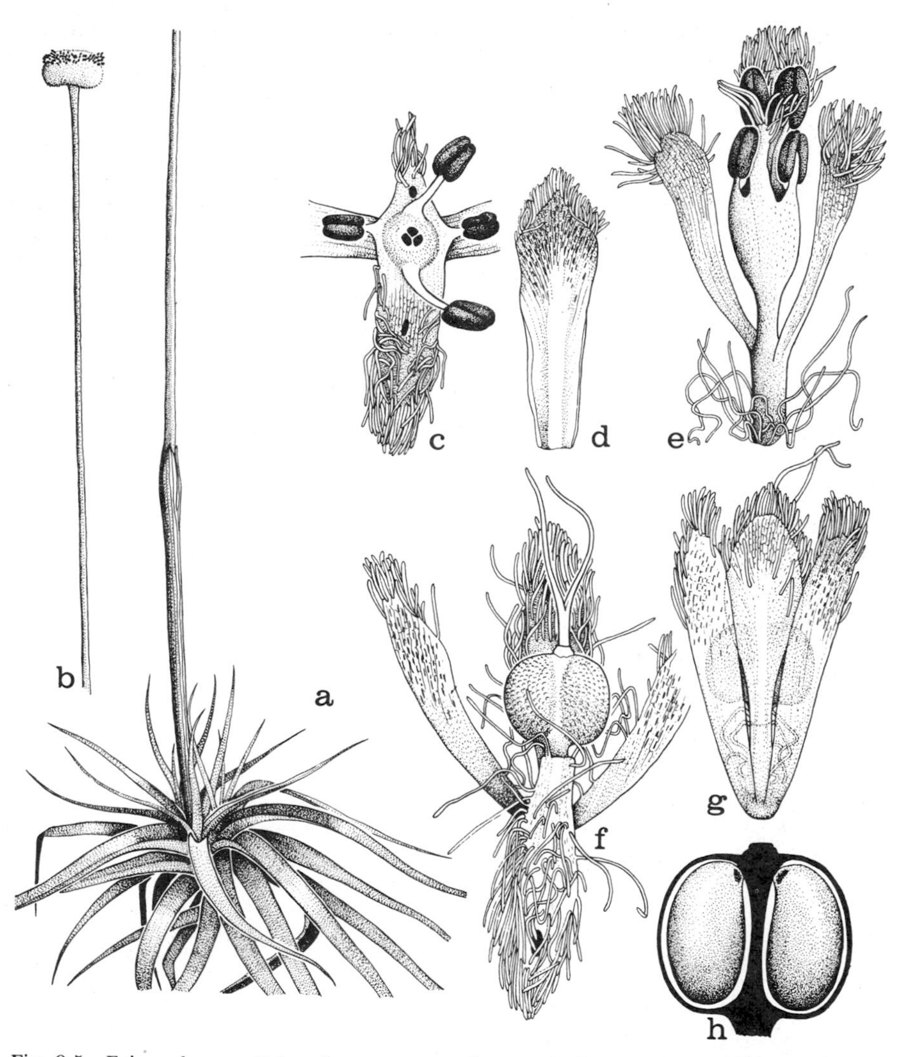

Fig. 9.5 Eriocaulaceae. *Eriocaulon compressum* Lam. a, b, habit, ×½; c, staminate flower, from above, ×12; d, bract, ×12; e, staminate flower, from the side, ×12; f, pistillate flower opened out to show pistil, ×12; g, dorsal view of pistillate flower, ×12; h, schematic long-section of ovary, ×24.

2 or 4, in trimerous flowers 3 or 6 (rarely only 1), always opposite the petals (or alternate with the sepals) when only 2 or 3; anthers small, introrse, mostly tetrasporangiate and dithecal, seldom bisporangiate and unithecal, opening by longitudinal slits; pollen-grains trinucleate, with a single elongate, spiral or convoluted sulcus; gynoecium of 2 or 3 carpels united to form a compound, superior, often stipitate ovary

with as many locules as carpels; style terminal, with as many primary branches as carpels, sometimes again branched, often with prominent appendages below the primary branches; ovules solitary in each locule, ventral-apical, pendulous, orthotropous, bitegmic, tenuinucellar; endosperm-development probably nuclear. FRUIT a loculicidal capsule; seeds with abundant mealy endosperm containing reserve starch in compound grains; embryo small, lenticular, undifferentiated, forming a cap atop the endosperm at the micropylar end of the seed.

The family Eriocaulaceae consists of about 13 genera and 1200 species, widespread in tropical and subtropical regions, with only a few species in temperate climates. The vast majority of the species belong to only 3 genera, *Paepalanthus* (500), *Eriocaulon* (400), and *Syngonanthus* (200). Inflorescences of species of *Syngonanthus* are sometimes sold as "everlastings."

The Eriocaulaceae are much like the Xyridaceae in habit, with clustered basal leaves and a terminal, racemose head on a long peduncle or scape. There is no obvious reason why the Eriocaulaceae might not have been derived directly from the Xyridaceae or from some similar common ancestor with 6 functional stamens.

The Eriocaulaceae are not clearly represented in the fossil record.

3. Order RESTIONALES J. H. Schaffner 1929

Perennial or less often annual herbs (sometimes lianas) without secondary growth; vessels generally present in all vegetative organs; vascular bundles of the stem closed, commonly scattered or bicyclic; plastids of the sieve-tubes with cuneate proteinaceous inclusions. LEAVES alternate, sometimes all basal, simple, with an open or closed basal sheath and a more or less grass-like or rush-like blade with parallel veins or a single central vein, or the blade often reduced or even suppressed; stomates paracytic and grass-like. FLOWERS small and inconspicuous, in various sorts of inflorescences, often grouped into spikelets, sometimes into pseudanthia, anemophilous or sometimes autogamous, hypogynous, most commonly trimerous but sometimes more reduced, perfect or more often unisexual; perianth small and chaffy, or somewhat petaloid (but still small) in Flagellariaceae, uniseriate or biseriate, or wanting; stamens 6 in 2 cycles of 3, or often 3 (4) in a single cycle, or sometimes only 1 or 2; anthers tetrasporangiate or bisporangiate, accordingly dithecal or monothecal, opening by longitudinal slits; pollen-grains binucleate or trinucleate, uniporate and more or less graminoid; gynoecium most commonly of 3 carpels united to form a compound, trilocular ovary, less often of 2 carpels forming a bilocular ovary, or sometimes the ovary unilocular and apparently composed of a single carpel; ovules solitary and pendulous in each locule, orthotropous, bitegmic, crassinucellar to sometimes tenuinucellar; endosperm-development nuclear. FRUIT variously indehiscent and dry or drupaceous, or dehiscent and more or less capsular; seeds with abundant, mealy, starch-bearing endosperm; embryo lenticular or conic, scarcely differentiated into parts, small and capping or lying alongside the endosperm at the micropylar end.

The order Restionales as here defined consists of 4 families and about 450 species, best developed in tropical and south-temperate parts of the Old World. About nine-tenths of the species belong to the single family Restionaceae.

SELECTED REFERENCES

Chanda, S. 1966. On the pollen morphology of the Centrolepidaceae, Restionaceae and Flagellariaceae, with special reference to taxonomy. Grana Palynol. 6: 355–415.

Cutler, D. F. 1966. Anatomy and taxonomy of the Restionaceae. Jodrell Lab. Notes 4: 1–25.

Cutler, D. F. 1968. Anatomy and taxonomy of certain monocotyledonous families. Proc. Linn. Soc. London 179: 261–267.
Cutler, D. F., & H. K. Airy Shaw. 1965. Anarthriaceae and Ecdeiocoleaceae: two new monocotyledonous families, separated from the Restionaceae. Kew Bull. 19: 489–499.
Hamann, U. 1963. Über die Entwicklung und den Bau des Spaltöffnungsapparats der Centrolepidaceae. Bot. Jahrb. Syst. 82: 316–320.
Hamann, U. 1975. Neue Untersuchungen zur Embryologie und Systematik der Centrolepidaceae. Bot. Jahrb. Syst. 96: 154–191.
Krupko, S. 1962. Embryological and cytological investigations in *Hypodiscus aristatus* Nees (Restionaceae). J. S. African Bot. 28: 21–44.
Lee, D. W., Yap Kim Pin, & Liew Foo Yew. 1975. Serological evidence on the distinctness of the monocotyledonous families Flagellariaceae, Hanguanaceae and Joinvilleaceae. J. Linn. Soc., Bot. 70: 77–81.
Newell, T. K. 1969. A study of the genus *Joinvillea* (Flagellariaceae). J. Arnold Arbor. 50: 527–555.
Prakash, N. 1969 (1970). The floral development and embryology of *Centrolepis fascicularis*. Phytomorphology 19: 285–291.
Smithson, E. 1957. The comparative anatomy of the Flagellariaceae. Kew Bull. 1956: 491–501.
Tomlinson, P. B., & A. C. Smith. 1970. Joinvilleaceae, a new family of monocotyledons. Taxon 19: 887–889.
Tomlinson, P. B., & U. Posluszny. 1977. Features of dichotomizing apices in *Flagellaria indica* (Monocotyledones). Amer. J. Bot. 64: 1057–1065.

Баранова, М. А. 1975. Стоматографическое исследование сем. Flagellariaceae. Бот. Ж. 60: 1690-1697.

SYNOPTICAL ARRANGEMENT OF THE FAMILIES OF RESTIONALES

1 Stamens 6; anthers tetrasporangiate and dithecal; flowers perfect; perianth present; leaves chiefly or wholly cauline, with a well developed blade.
  2 Perianth somewhat petaloid; solid-stemmed lianas; leaf-blade circinately inrolled in bud, cirrhose at the tip; sheath closed .... 1. FLAGELLARIACEAE.
  2 Perianth chaffy; hollow stemmed, erect, coarse herbs; leaves plicate in bud, not cirrhose; sheath open ... 2. JOINVILLEACEAE.
1 Stamens 1–3 (4); anthers, except in a few Restionaceae, bisporangiate and monothecal; flowers unisexual or rarely perfect.
  3 Perianth usually present; stamens (1–) 3 (4); leaves in most genera largely cauline, with an open sheath and much-reduced or no blade .... 3. RESTIONACEAE.
  3 Perianth wanting; stamen 1 (2); leaves clustered at the base, with a more or less well defined open sheath and an elongate, slender blade .... 4. CENTROLEPIDACEAE.

1. Family FLAGELLARIACEAE Dumortier 1829 nom. conserv., the Flagellaria Family

Glabrous, high-climbing, cyanogenic lianas without secondary growth, arising from sympodial rhizomes that accumulate sucrose but not starch, the stem lacking axillary buds, but the apical meristem forking at intervals to produce pseudo-axillary or nearly dichotomous branching; cell walls, especially of the epidermis, tending to be silicified, and some of the cells of the sclerenchyma containing each a round silica-body; scattered secretory cells present in the parenchymatous tissues; stem with solid internodes and scattered, closed vascular bundles; vessels in all vegetative organs, with simple or scalariform perforations; plastids of the sieve-tubes with cuneate proteinaceous inclusions. LEAVES alternate, well distributed along the stem, with a closed sheath and a somewhat grass-like blade that may be separated from the sheath by a short petiole, but without a ligule; blade parallel-veined, circinately inrolled in bud, with a cirrhose tip at maturity; epidermal cells somewhat irregular, but with some tendency to be arranged in longitudinal files; stomates paracytic, grass-like; mesophyll with secretory canals and with small, clustered crystals of calcium oxalate in some of the cells. Inflorescence a terminal, branched, bracteate panicle; FLOWERS relatively small and inconspicuous, sessile, perfect, regular, hypogynous, trimerous throughout, apparently anemophilous; perianth of 6 persistent, distinct, somewhat corolloid white tepals in 2 series; stamens 6, distinct; anthers basifixed, sagittate, tetrasporangiate and dithecal, slightly introrse, opening by longitudinal slits; pollen-grains binucleate, uniporate, graminoid; gynoecium of 3 carpels united to form a compound, superior, trilocular ovary; styles terminal, distinct or shortly connate at the base, stigmatic for most of their length; stigmas dry, papillate; ovules solitary in each locule on the axile placenta, orthotropous or reputedly anatropous. FRUIT drupaceous, with a single stone and mostly only 1 or 2 seeds; seed with copious, mealy endosperm containing reserve starch in simple grains; embryo small, lenticular, undifferentiated, capping the endosperm, its position marked externally by a well developed embryostega. $2n = 38$.

The family Flagellariaceae consists of the single genus *Flagellaria*, with 3 species native to tropical regions in the Old World. Fossil pollen assigned to *Flagellaria* occurs in rocks of Miocene and more recent age.

2. Family JOINVILLEACEAE A. C. Smith & Tomlinson 1970, the Joinvillea Family

Coarse, erect herbs from a short, sympodial rhizome, hairy or bristly, without secondary growth, the stem unbranched, hollow except at the nodes; vessels with simple or scalariform perforations present in all vegetative organs; cell walls, especially of the epidermis, more or less strongly silicified, as in grasses; secretory cells absent, but the shoot with abundant silica-bodies, each of these occupying the lumen of a cell next to the fibers sheathing the vascular bundles, these closed. LEAVES alternate, well distributed along the stem, with an open sheath and a somewhat grass-like blade, provided with a small ligule and a pair of auricles at the juncture of sheath and blade; blade parallel-veined, plicate in bud, not cirrhose; most of the veins entering the blade from the sheath, but a few originating from the proximal part of the midrib; veins converging distally, but the peripheral ones running into the margin at a narrow angle; epidermal cells arranged in longitudinal files; stomates paracytic, grass-like. Inflorescence a terminal, much-branched, bracteate panicle; bracteoles caducous; FLOWERS relatively small and inconspicuous, sessile, perfect, regular, hypogynous, trimerous throughout; perianth of 6 persistent, distinct or only basally connate, imbricate, chaffy tepals in 2 series; stamens 6, distinct; anthers basifixed, sagittate, tetrasporangiate and dithecal, latrorse, opening by longitudinal slits; pollen-grains uniporate, graminoid; gynoecium of 3 carpels united to form a compound, trilocular, superior ovary; styles terminal, distinct or shortly connate at the base, persistent, plumose-stigmatic for most of their length; ovules solitary in each locule on the axile placenta, pendulous, orthotropous. FRUIT drupaceous, more or less triquetrous, with a single stone and 1–3 seeds; seeds with copious, mealy, starchy endosperm capped by the small, lenticular, undifferentiated embryo.

The family Joinvilleaceae as here defined consists of the single genus *Joinvillea*, with 2 species native to the Pacific Islands. *Joinvillea* has often been associated with *Flagellaria* in the family Flagellariaceae, but the two groups seem as different as other families of the order and subclass. Pollen of *Joinvillea* is known from Miocene and more recent deposits.

3. Family RESTIONACEAE R. Brown 1810 nom. conserv., the Restio Family

Xeromorphic, rhizomatous perennial herbs without secondary growth, glabrous or with various sorts of unicellular or multicellular (sometimes plate-like) hairs, very often tanniferous and commonly producing proanthocyanins, but not cyanogenic, generally without crystals, but commonly (not in all genera) some of the cells in the parenchyma-sheath containing one or more large silica-bodies, the epidermis often with several silica-bodies in long cells and a solitary one in the short cells; epidermal cells of the stem and leaves arranged in longitudinal files; stomates paracytic, grasslike, longitudinally oriented in alignment with the epidermal cells, often sunken; vessels generally present in all vegetative organs (or wanting from the leaves), with variously oblique or transverse, and simple to scalariform or reticulate perforations; stems simple or branched, with solid nodes and solid or hollow internodes, photosynthetic, with a subsurface layer of chlorenchyma and usually a sclerenchymatous sheath internal to the chlorenchyma; vascular bundles closed, more or less scattered, or often all embedded in the sclerenchymatous sheath but still in more than one cycle, or seldom in a single cycle. LEAVES alternate, mostly scattered along the stem, rarely (as in *Anarthria*) all basal, usually reduced to an open sheath with scarcely any blade, but sometimes with a well developed, terete to flattened and equitant blade, as in spp. in *Anarthria*; ligule usually wanting. FLOWERS small and inconspicuous, anemophilous, unisexual (the plants mostly dioecious) or rarely perfect, regular or nearly so, essentially hypogynous, borne singly in the axils of chaffy bracts in (1–) several- to many-flowered spikelets (seldom with a pair of glume-like bracteoles as well, as in *Anarthria*), or in much-branched inflorescences that sometimes have leafy bracts or bracteoles subtending the flowers; spikelets simple or less often compound, variously arranged, generally each with a sheath-like spathe at the base (not in *Ecdeiocolea*): pistillate and staminate inflorescences often dissimilar; staminate flowers sometimes with a pistillode, and pistillate flowers sometimes with staminodes; perianth chaffy, generally in 2 cycles of 3 scale-like tepals, seldom one or both cycles reduced to 1 or 2 members, or one or even both cycles suppressed; stamens (1–) 3 (4), opposite the inner tepals; filaments distinct or rarely connate, anthers mostly bisporangiate and monothecal, seldom tetrasporangiate and dithecal, introrse or seldom latrorse, opening by longitudinal slits; pollen-grains binucleate or trinucleate, uniporate, more or less graminoid (commonly with a circular pore bordered by a distinct annulus), sometimes with a large ragged pore, as in the Centrolepidaceae, sometimes with a smaller one; gynoecium most commonly of 3 carpels united to form a compound, trilocular ovary, less often of 2 carpels forming a bilocular ovary, or seldom the ovary unilocular and apparently composed of a single carpel; styles distinct or connate below, with elongate, often plumose stigma; a single pendulous, apical or apical-axile, orthotropous, bitegmic, crassinucellar (tenuinucellar, fide Dahlgren) ovule present in each locule of the ovary; endosperm-development nuclear. FRUIT an achene or small nut or a loculicidal capsule; seeds 1–3, often fewer than the ovules of the flower, pendulous, with copious, mealy endosperm containing compound starch-grains; embryo small, lenticular, capping the endosperm. X = 6–13+. (Anarthriaceae, Ecdeiocoleaceae)

The family Restionaceae as here defined consists of about 30 genera and 400 species, widely distributed in the Southern Hemisphere, best developed in Australia and South Africa. They characteristically occur in nutrient-poor soils. The largest genus is *Restio*, with 100+ species. On anatomical and palynological grounds the name *Restio* is here restricted to the South African and Malgache species, and the Australian species often referred to *Restio* are considered to form several other genera.

The small Australian genera *Anarthria* (5) and *Ecdeiocolea* (1) stand apart from the bulk of the family, especially in having tetrasporangiate, dithecal anthers and in lacking the characteristic sclerenchymatous sheath of the stem. *Lyginia* (1) also has tetrasporangiate, dithecal anthers, but anatomically it is concordant with the rest of the family. Although at an earlier time I followed Cutler and Airy Shaw in recognizing the Anarthriaceae and Ecdeiocoleaceae as distinct families, I now consider it more useful to include them in the Restionaceae, to which their affinity is plain enough.

Fossil pollen thought to represent the Restionaceae occurs in rocks of Maestrichtian and more recent age. Another type of pollen thought to represent either the Restionaceae or the Centrolepidaceae dates from the Paleocene.

4. Family CENTROLEPIDACEAE Endlicher 1836 nom. conserv., the Centrolepis Family

Small, tufted, annual or seldom perennial, grass-like or moss-like herbs without secondary growth, lacking crystals and silica-bodies, and only seldom tanniferous; vessels with scalariform to reticulate perforations, generally present in all vegetative organs, or seldom wanting from the leaves; roots without a differentiated pericycle; root-hairs arising near the margin rather than the center of the surface of their respective epidermal cells; stem solid, the basal part (or the short, more or less erect rhizome) commonly with an interrupted amphivasal vascular sheath, the remainder generally with 2 cycles of vascular bundles (the outer smaller) and with well developed mechanical tissues and a subsurface chlorenchymatous layer. LEAVES generally all tufted at the base, alternate, with a more or less well defined open basal sheath and a linear or filiform blade, often with an adaxial ligule at the juncture of sheath and blade, usually with a single central vascular bundle, sometimes with a pair of smaller lateral bundles as well; stomates paracytic; somewhat grass-like. Inflorescence a terminal spike or head subtended by 1-several leafy bracts, with 2-many distichous bracts subtending individual flowers or pseudanthia or spikelets, or the bracts subtending the spikelets vestigial; FLOWERS anemophilous or autogamous, basically naked and unisexual, with a single stamen or a single carpel, but often aggregated into bisexual or pistillate pseudanthia with 1 or 2 stamens and/or 1-several carpels which may be collaterally or spirally united (or superposed) to form an apparent compound pistil; anther versatile on a slender filament, dorsifixed, bisporangiate, monothecal; pollen-grains trinucleate, uniporate, resembling those of the Poaceae and Restionaceae, but with a mostly larger and more ragged aperture; each carpel with a single style and an elongate, decurrent stigma; ovule solitary, pendulous, orthotropous, bitegmic, weakly crassinucellar or tenuinucellar; endosperm-development nuclear. FRUIT small, dry, membranous, usually opening on the back to release the seed; seed with copious, starchy, mealy endosperm containing compound starch-grains, without perisperm; embryo small, conic, not fully differentiated, with a single shield-like cotyledon, capping the endosperm or lying alongside it at the tip of the seed. X = 10–13.

The family Centrolepidaceae as here defined consists of 4 genera and about 35 species, native to Australia, southeastern Asia, the Pacific islands, and southernmost South America, mostly in nutrient-poor soils. Most of the species belong to the genus *Centrolepis* (25).

4. Order JUNCALES Lindley 1833

Perennial or less often annual herbs, or seldom large shrubs, without secondary growth, sometimes tanniferous (with proanthocyanins); vessels commonly present in all vegetative organs, with scalariform or sometimes simple perforations; vascular bundles of the stem closed, in 1–2 cycles or scattered; plastids of the sieve-tubes with cuneate proteinaceous inclusions (Thurniaceae not investigated). LEAVES alternate, sometimes all basal, simple, with an open or closed sheath and a more or less elongate, narrow, parallel-veined blade, or the blade sometimes more or less reduced or suppressed; stomates paracytic, with 2 subsidiary cells, or seldom some of them tetracytic. FLOWERS individually small and inconspicuous, in various sorts of inflorescences, perfect or rarely unisexual, anemophilous or sometimes autogamous, rarely secondarily entomophilous, hypogynous, most commonly trimerous; perianth generally small and chaffy, biseriate or rarely uniseriate, the tepals of the inner and outer series similar, free and distinct; stamens generally 6 in 2 cycles of 3, less commonly 3 in a single cycle, rarely only 2 ; anthers basifixed, tetrasporangiate and dithecal, opening by longitudinal slits; pollen-grains borne in tetrads, trinucleate, uniaperturate; gynoecium of 3 carpels united to form a compound, superior, trilocular or unilocular ovary; style terminal, often short, with 3 stigmatic branches, or sometimes the styles 3 and distinct; ovules 1-many per carpel, anatropous, bitegmic, crassinucellar (Thurniaceae not studied); endosperm-development helobial (Thurniaceae not studied). FRUIT nearly always a loculicidal capsule, with 3-many seeds; embryo small, monocotyledonous, embedded in the copious, starchy endosperm. Chromosomes with diffuse centromere (Thurniaceae not studied).

The order Juncales as here defined consists of 2 families, the Juncaceae, with about 300 species, and the Thurniaceae, with only 3. The probable error in estimating the number of species of Juncaceae is greater than the number of species in the Thurniaceae, and the order Juncales may be said to have about 300 (rather than 303) species.

Subsequent to (and consequent upon) the general recognition of the importance of floral reduction in the evolution of the angiosperms, it has become customary to consider the Juncaceae as florally reduced descendants of the Liliaceae or some lily-like ancestor. There is indeed a superficial resemblance between the two groups, but they characteristically differ in their vascular architecture (vessels in all vegetative organs in Juncaceae, confined to the roots in Liliaceae), stomatal structure (paracytic in Juncaceae, usually anomocytic in Liliaceae), and food-reserves of the seed (starch in Juncaceae, protein or oil or hemicellulose in Liliaceae). In all of these respects the Juncaceae resemble typical Commelinidae rather than typical Liliidae. Furthermore, if the Thurniaceae are related to the Juncaceae, as is now generally believed, we must take note of the fact that the silica-bodies of *Thurnia* are similar to those of the Rapateaceae, Restionaceae and some Cyperaceae. Somewhat different silica-bodies occur in the Poaceae. Silica-bodies are virtually unknown in the Liliidae. The Juncales are fully at home in the Commelinidae, where they are here placed, but would be highly aberrant in the Liliidae.

Placement of the Juncales in the Liliidae would furthermore present us with a taxonomic and phyletic dilemma. If the Juncales are derived from the Commelinales, parallel to the Restionaceae, then there is no internal contradiction in the interlocking set of resemblances among the Juncaceae, Cyperaceae, Poaceae, and Restionales. If, on the other hand, the Juncales are derived from the Liliales, then it is difficult to make a coherent scheme. The Cyperaceae resemble both the Juncaceae and the Poaceae, and the Poaceae also appear to be allied to the Restionales, which in turn are allied to the Commelinales.

It should be noted that the connection between the Restionales on the one hand, and the Juncales, Cyperaceae, and Poaceae on the other appears to be indirect, through a common ancestry in or near the Commelinales. A more direct connection seems unlikely, especially inasmuch as the orthotropous ovules of the Restionales are not likely to be phyletically antecedent to the anatropous ovules found in the Juncales, Cyperaceae, and some of the Poaceae. Neither can the Juncales, with tetradinous pollen, be directly ancestral to the Poaceae and Restionales.

If there is anything ecologically distinctive about the Juncales, as compared to the other orders of the Commelinidae with reduced, wind-pollinated flowers, the fact has escaped detection. The adaptive significance of the known differences between the Juncaceae and Cyperaceae, for example, is obscure. The reasons why the Juncaceae are less successful than the more abundant and varied Cyperaceae remain to be discovered.

According to Savile (1979, initial citations) the array of fungal parasites strongly supports the unity of the Commelinidae (exclusive of the Typhales), and emphasizes the distinction of this group from the Liliaceae and other Liliidae. These data further indicate a closer relationship between the Juncaceae and Cyperaceae than between either of these families and the Poaceae. I find nothing disturbing here, even though I find it useful to maintain the Juncales as an order distinct from the Cyperales.

What is more surprising is that all rusts of the Juncaceae are advanced members of the 3 (and only) lineages that appear to have originated and diversified as parasites on members of the Cyperaceae, subfamily Cyperoideae. Savile has accordingly proposed that the Juncaceae evolved from the Cyperaceae through the establishment of mutations restoring an ancestral, trimerous, bicyclic perianth. I find this interpretation difficult, the more so because it would also be necessary to restore capsular fruits with numerous seeds.

SELECTED REFERENCES

Barnard, C. 1958. Floral histogenesis in monocotyledons. III. The Juncaceae. Austral. J. Bot. 6: 285–298.

Boudet, Al. M., An. Boudet, & H. Bouyssou. 1977. Taxonomic distribution of isoenzymes of dehydroquinate hydrolase in the angiosperms. Phytochemistry 16: 919–922.

Cutler, D. F. 1965. Vegetative anatomy of Thurniaceae. Kew Bull. 19: 431–441.

Cutler, D. F. 1969. Vol. 4. Juncales. *In:* C. R. Metcalfe, ed. Anatomy of the monocotyledons. Clarendon Press. Oxford.

Williams, C. A., & J. B. Harborne, 1975. Luteolin and daphnetin derivatives in the Juncaceae and their systematic significance. Biochem. Syst. Ecol. 3: 181–190.

Wulff, H. D. 1939. Die Pollenentwicklung der Juncaceen nebst einer Auswertung der embryologischen Befunde hinsichtlich einer Verwandtschaft zwischen den Juncaceen und Cyperaceen. Jahrb. Wiss. Bot. 87: 533–556.

Zandee, M. 1976. Beobachtungen über Cyanogenese in der Gattung *Juncus*. Proc. Kon. Nederl. Akad. Wetensch. ser. C. 79: 529–543.

SYNOPTICAL ARRANGEMENT OF THE FAMILIES OF JUNCALES

1 Inflorescence of diverse sorts, but not as in the Thurniaceae; vascular bundles of the leaf of ordinary structure, with abaxial phloem, not as in the Thurniaceae; cells without silica-bodies .... 1. JUNCACEAE.

1 Inflorescence of one or more dense heads subtended by spreading, leafy bracts; vascular bundles of the leaf in vertical pairs, the lower (and smaller) bundle of a pair with the phloem on top (adaxial), facing the phloem of the upper bundle; silica-bodies present in some of the cells of the parenchyma and epidermis .... 2. THURNIACEAE.

1. Family JUNCACEAE A. L. de Jussieu 1789 nom. conserv., the Rush Family

Herbs or seldom (*Prionium*) rather large shrubs, usually perennial, often from a creeping, sympodial, starchy rhizome, or seldom annual, in any case lacking secondary growth, at least sometimes containing the free form of dehydroquinate hydroxylase, in addition to the bound form that is common to angiosperms in general, sometimes tanniferous and producing proanthocyanins, and sometimes cyanogenic (tyrosine-derived), but without crystals of calcium oxalate and not accumulating silica; flavonoid pigments characteristically (but not in *Prionium*) including large amounts of free flavones, among them luteolin and/or its 5-methyl ester, and also often including flavonoid sulfates; C-glycosylflavones and tricin usually wanting; vessels generally present in all vegetative organs, with simple or mixed simple and scalariform perforations; stem commonly photosynthetic, solid or often hollow, its vascular bundles closed, commonly in 1 or 2 rings, less often (but notably in *Prionium*) scattered; plastids of the sieve-tubes with cuneate proteinaceous inclusions; roots said not to be mycorhizal. LEAVES alternate, often all basal or confined to the lower part of the stem, simple, parallel-veined, with an open or closed sheath and usually a flat to channeled, terete, ensiform-folded, or centric and cylindric and hollow blade, sometimes with cross-partitions, or the blade sometimes much-reduced or wholly suppressed; leaf-sheath often shortly prolonged on both sides into a pair of auricles at the juncture with the blade, the auricles sometimes more or less confluent to form a short, adaxial ligule; epidermal cells usually more or less elongate longitudinally; stomates also longitudinally oriented, paracytic, with 2 subsidiary cells. FLOWERS individually small and inconspicuous, variously in open and branching, basically cymose inflorescences, or in compact, head-like or spike-like clusters, or seldom solitary, generally perfect, or rarely unisexual and the plants then dioceious, mostly anemophilous, sometimes autogamous, seldom secondarily entomophilous but without nectaries or nectar; perianth small, chaffy or leathery, commonly greenish to brownish or blackish, rarely white or yellowish or anthocyanic and somewhat corolloid, ordinarily in 2 more or less similar sets of 3 distinct tepals, rarely in sets of 2, or one set suppressed; stamens free and distinct, most commonly in 2 cycles of 3 each, less often only 3 (the inner cycle suppressed), rarely only 2; anthers basifixed, tetrasporangiate and dithecal, opening by longitudinal slits; pollen-grains borne in tetrads,

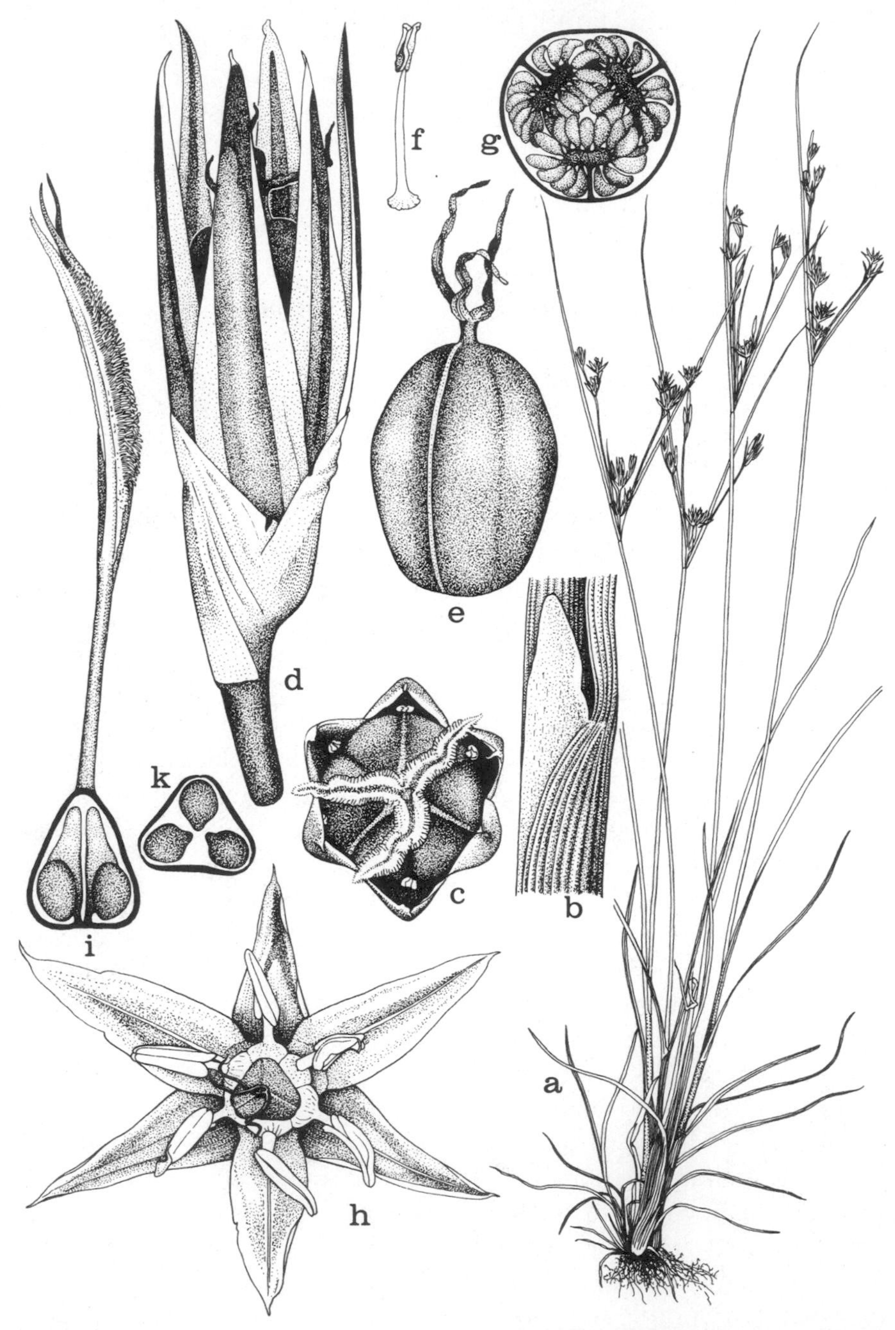

Fig. 9.6 Juncaceae. a–g, *Juncus tenuis* Willd. a, habit, ×½; b, nodal region, with auricle, ×8; c, flower, from above, after anthesis, ×16; d, side view of flower after anthesis, with persistent bracteoles and tepals, ×16; e, pistil, ×16; f, stamen, ×16; g, schematic cross-section of ovary, ×16. h–k, *Luzula acuminata* Raf. h, flower, from above, ×8; i, pistil, in partial long-section, ×16; k, schematic cross-section of ovary, ×16.

trinucleate, monoporate (ulceroidate); gynoecium of 3 carpels united to form a compound superior, trilocular or unilocular ovary with accordingly axile or parietal placentation, or (*Luzula*) with 3 basal ovules in a tricarpellate, unilocular ovary; style terminal, often short, with 3 stigmatic branches, or seldom the styles 3 and distinct; ovules 3-many, always more numerous than the locules, anatropous, bitegmic, crassinucellar; endosperm-development helobial. FRUIT a small, loculicidal capsule, or very rarely indehiscent; embryo small, straight, broad, embedded in the starchy endosperm, with a terminal cotyledon and lateral plumule. Chromosomes with diffuse centromere, at least in *Luzula*; x = 3–36.

The family Juncaceae consists of about 8 genera and 300 species, mostly of temperate or cold regions, or of tropical mountains. The two large genera, *Juncus* and *Luzula*, are widespread but best developed in the Northern Hemisphere. The six small genera are wholly confined to the Southern Hemisphere.

Bate-Smith (1967) considers that chemically the Juncaceae have much in common with the Poaceae, and Barnard (1958) considers that the pattern of floral development in the investigated members of the Juncaceae is similar to that of investigated members of the Cyperaceae and Poaceae. Williams and Harborne (1975) emphasize the chemical distinctness of the Juncaceae, but see some similarities to the Poaceae and Cyperaceae.

The fossil record of the Juncaceae is scanty and does not certainly carry the family back beyond the Miocene.

2. Family THURNIACEAE Engler 1907, nom. conserv., the Thurnia Family

Coarse perennial herbs from an upright rhizome, without secondary growth, with scattered tanniferous cells, and with spheroidal silica-bodies (several per cell) in some of the parenchyma and epidermal cells of the shoot, but without crystals of calcium oxalate; luteolin and its 5-methyl ester wanting; vessels present in all vegetative organs, with slanting end-walls and scalariform perforations; stem bluntly triangular or quadrangular; vascular bundles closed, scattered. LEAVES alternate, simple, parallel-veined, all basal, with sheathing base and elongate, leathery, flat or (in cross-section) v-shaped, sometimes prickly-margined blade; vascular bundles in vertical pairs, the lower (and smaller) bundle of the pair with the phloem on top (adaxial), facing the phloem of the upper bundle; epidermal cells from slightly longer than wide (oriented with the blade) to more often somewhat wider than long (oriented crosswise to the blade); stomates paracytic, with 2 subsidiary cells, or sometimes some of them tetracytic. FLOWERS small and individually inconspicuous, anemophilous, perfect, borne in one or more large, dense, terminal, racemose heads subtended by spreading, leafy bracts; perianth small, chaffy, in 2 similar sets of 3 distinct tepals; stamens 6 in 2 cycles of 3, distinct, the filament somewhat adnate to its tepal at the base; anthers basifixed, tetrasporangiate and dithecal, opening by longitudinal slits; pollen borne in tetrads, monoporate, similar to that of the Juncaceae; gynoecium of 3 carpels united to form a compound, superior, trilocular ovary with 3 or more erect, anatropous ovules toward the base; ovary tapering distally, scarcely produced into a style, with 3 elongate terminal stigmas. FRUIT a loculicidal capsule, commonly with 3 spindle-shaped seeds, 1 in each locule; embryo small, more or less cylindric, monocotyledonous, embedded in the lower part of the copious, mealy, starchy endosperm.

The family Thurniaceae consists of the single genus *Thurnia*, with 3 species native to the Amazon basin and Guayana. *Thurnia* has generally been considered to be allied to the Rapateaceae or to the Juncaceae. It resembles the Rapateaceae in habit and in having silica-bodies in some of the cells, but it is more like the Juncaceae in the structure of the perianth, stamens, and pollen. Recent opinion mostly favors the juncaceous alliance. As noted above, the Thurniaceae help to tie the Juncales to the Commelinidae rather than to the Liliidae.

5. Order CYPERALES G. T. Burnett 1835

Perennial or less often annual herbs, or seldom more or less woody or even arborescent plants, but in any case without secondary growth; plants generally containing the free form of dehydroquinate hydroxylase; flavonoid constituents commonly including C-glycosylflavones and tricin; cell-walls tending to be silicified; vessels with simple or scalariform perforations, commonly present in all vegetative organs; vascular bundles of the stem closed, more or less scattered, but often concentrated toward the periphery; plastids of the sieve-tubes with cuneate proteinaceous inclusions. LEAVES mostly in 2 or 3 ranks, with an open or closed sheath and a parallel-veined, usually narrow and more or less elongate blade, or the blade sometimes broader or more or less reduced or even suppressed; an adaxial ligule often present at the juncture of sheath and blade; some of the epidermal cells containing 1 or more distinctive silica-bodies; stomates paracytic or seldom tetracytic, typically arranged in straight, longitudinal, 1–2-seriate rows; guard-cells dumbbell-shaped. FLOWERS individually small and inconspicuous, arranged in simple or often complex spikes or spikelets that are usually organized into secondary inflorescences, perfect or unisexual, anemophilous or sometimes autogamous or apomictic, generally each subtended by a chaffy bract (scale) or enclosed between a pair of scales; perianth of 1–3 (seldom more) short scales or 1-many short or elongate bristles, or none; stamens most often 3, less commonly 1, 2, or 6, rarely numerous; anthers tetrasporangiate and dithecal, opening by longitudinal slits; basifixed but in the Poaceae very deeply sagittate, pollen-grains borne singly, in seeming or actual monads, trinucleate, mostly uniporate; gynoecium of 2 or 3 (4) carpels united to form a compound, superior, unilocular ovary with a terminal style and as many stigmatic branches as carpels, or the stigmas sometimes virtually sessile; ovule solitary, bitegmic or seldom unitegmic, crassinucellar or pseudocrassinucellar to seldom tenuinucellar; endosperm-development nuclear. FRUIT indehiscent, usually dry, with a single seed; endosperm copious, partly or wholly starchy and often mealy, or rarely wanting; embryo with a single cotyledon, embedded in or peripheral to the endosperm. (Glumiflorae, Graminales, Poales)

The order Cyperales as here defined consists of 2 large and widely distributed families, the Cyperaceae (nearly 4000) and Poaceae (8000). The Poaceae are even more successful, as compared to the Cyperaceae, than the number of species would suggest. Grasses form the dominant vegetation over considerable portions of the earth.

If the formal morphological characters usually cited as distinguishing the Poaceae from the Cyperaceae have any great adaptive significance, the fact remains to be established. One may reasonably speculate instead that it is the intercalary meristem of the leaf that plays a major role in the greater success of the grasses. The leaves of monocots in general mature from the tip downwards; in grasses the base of the leaf-blade remains more or less permanently immature and meristematic. Some of the Cyperaceae, notably species of *Carex*, share this feature to some extent, but in general it is much less developed in the Cyperaceae than in the Poaceae. Anyone who has pushed a lawn-mower should recognize the significance of the intercalary meristem in permitting a plant to withstand grazing.

Although the Cyperaceae and Poaceae have traditionally been associated in older systems of classification, in recent years many authors have taken each family to form a separate, unifamilial order. This latter view has been conditioned in large part by the obvious relationship of the Poaceae to the Restionaceae, and the Cyperaceae to the Juncaceae. When the Juncaceae were considered to be reduced derivatives of the Liliaceae, as in early versions of the Besseyan approach to relationships among angiosperms, it seemed necessary to divorce the Cyperaceae from the Poaceae. The two families are indeed clearly distinct and well defined, and it is not difficult to point out a series of differences between them. On the other hand, now that the Juncaceae as well as the Cyperaceae and Poaceae have found their proper home along with the Restionaceae in the subclass Commelinidae, it is possible to reassess the ordinal classification.

The Commelinidae almost completely lack nectaries and nectar, and the dominant evolutionary trend within the group has been toward floral reduction associated with wind-pollination. The Typhales stand somewhat apart from the other wind-pollinated orders, but the Restionales, Juncales, and Cyperales as here defined form a group with overlapping similarities and differences.

In the Restionales-Juncales-Cyperales group, as in some others, a consideration of the overlapping sets of similarities among related taxa leads phylogenists into a thicket with no established exit. If the similarity between A and B in character *x* is considered to be inherited from a common ancestry, then the similarity between B and C in character *y* must be due to close parallelism, and vice versa. This sort of parallelism, however, reflects a basic genetic similarity of the ancestors, which conditioned them to undergo similar evolutionary changes.

At this point we must back out of the thicket, or cut our way through it, and put together the things that are now most alike. For strictly phylogenetic purposes it is important to distinguish similarities due to close parallelism from similarities inherited from a common ancestry, but for taxonomic purposes this distinction is less significant. Many similarities between groups that on first inspection seem to be inherited from a common ancestry turn out on more careful consideration to reflect close parallelism instead, but that is not sufficient reason to dissolve the groups. A consistent attempt to do so would in fact be destructive to the taxonomic system. We should of course refrain from using a few similarities of convergence to group fundamentally different things into the same taxon, but when the similarities are pervasive it is not taxonomically necessary that they result strictly from monophylesis. Taxonomy can present no more than a muddy reflection of phylogeny.

On the basis of all the available information, including chemical and anatomical as well as gross morphological features, I believe that the Cyperaceae and Poaceae are closely related lines diverging at an angle from a common source. They have many characters in common, and they have both reached the stage of floral reduction at which the ovary is unilocular and uniovulate, but still obviously composed of more than one carpel. It may be significant that Bate-Smith (1968, in general citations) considers that the Poaceae and Cyperaceae "are not inconsistent in their chemistry." The two families are, all considered, too similar to be dissociated; they must stand side by side in the system. Putting them in separate but adjacent orders is pointless and confronts the purposes of classification. In my opinion the two families are most usefully considered to form a single order.

SELECTED REFERENCES

Barnard, C. 1957. Floral histogenesis is monocotyledons. I. The Gramineae. II. The Cyperaceae. Austral J. Bot. 5: 1–20; 115–128.

Blaser, H. W. 1941, 1944. Studies in the morphology of the Cyperaceae. I. Morphology of flowers. A. Scirpoid genera. B. Rynchosporoid genera. II. The prophyll. Amer. J. Bot. 28: 542–551; 832–838, 1941. 31: 53–64, 1944.

Brown, W. V. 1965. The grass embryo—A rebuttal. Phytomorphology 15: 274–284.

Brown, W. V. 1977. The Kranz syndrome and its subtypes in grass systematics. Mem. Torrey Bot. Club 23(3): 1–97.

Butzin, F. 1965. Neue Untersuchungen über die Blüte der Gramineae. Inaugural-Dissertation. Berlin.

Butzin, F. 1977. Evolution der Infloreszenzen in der Borstenhirsen-Verwandtschaft. Willdenowia 8: 67–79.
Calderón, C. E., & T. R. Soderstrom. 1980. The genera of Bambusoideae (Poaceae) of the American continent: Keys and comments. Smithsonian Contr. Bot. 44: 1–27.
Celakovsky, L. 1889. Über den Ahrchenbau der brasilianische Grasgattung *Streptochaeta* Schrad. Sitzungsber. Konigl. Böhm. Ges. Wiss. Prag. Math.-Naturwiss. Cl. 1: 14–42.
Cheadle, V. I. 1955. The taxonomic use of specialization of vessels in the metaxylem of Gramineae, Cyperaceae, Juncaceae, and Restionaceae. J. Arnold Arbor. 36: 141–157.
Cheadle, V. I., & H. Kosakai. 1972. Vessels in the Cyperaceae. Bot. Gaz. 133: 214–223.
Clifford, H. T. 1961. Floral evolution in the family Gramineae. Evolution 15: 455–460.
Clifford, H. T. 1970. Monocotyledon classification with special reference to the origin of grasses (Poaceae). *In:* N. K. B. Robson, D. F. Cutler, & M. Gregory, eds., New research in plant anatomy, pp. 25–34. J. Linn. Soc., Bot. 63, Suppl. 1. Academic Press. London.
Cocucci, A. E., & M. E. Astegiano. 1978. Interpretacion del embryon de las Poaceas. Kurtziana 11: 41–54.
Dunbar, A. 1973. Pollen development in the *Eleocharis palustria* group (Cyperaceae). I. Ultrastructure and ontogeny. Bot. Not. 126: 197–254.
Eiten, L. T. 1976. Inflorescence units in the Cyperaceae. Ann. Missouri Bot. Gard. 63: 81–112.
Goller, H. 1977. Beiträge zur Anatomie adulter Gramineenwurzeln in Hinblick auf taxonomische Verwendbarkeit. Beitr. Biol. Pflanzen 53: 217–307.
Gregory, M., & C. R. Metcalfe. 1967. Bibliography for the anatomy of the Cyperaceae. Notes Jodrell Lab. 5: 1–17.
Guédès, M., & P. Dupuy. 1976. Comparative morphology of lodicules in grasses. J. Linn. Soc., Bot. 73: 317–331.
Guignard, J. L. 1961. Recherches sur l'embrogénie des Graminées; rapports des Graminées avec les autres Monocotylédones. Ann. Sci. Nat. Bot. sér. 12, 2: 491–610.
Harborne, J. B. 1971. Distribution and taxonomic significance of flavonoids in the leaves of the Cyperaceae. Phytochemistry 10: 1569–1574.
Harborne, J. B., & C. A. Williams. 1976. Flavonoid patterns in leaves of Gramineae. Biochem. Syst. Ecol. 4: 267–280.
Hartley, W. 1958, 1973. Studies on the origin, evolution, and distribution of the Gramineae. II. The tribe Paniceae. V. The subfamily Festucoideae. Austral. J. Bot. 6: 343–357, 1958. 21: 201–234, 1973.
Holttum, R. E. 1948. The spikelet in Cyperaceae. Bot. Rev. 14: 525–541.
Johnson, C., & W. V. Brown. 1973. Grass leaf ultrastructural variations. Amer. J. Bot. 60: 727–735.
Kinges, H. 1961. Merkmale des Gramineenembryos. Ein Beitrag zur Systematik der Gräser. Bot. Jahrb. Syst. 81: 50–93.
Koyama, T. 1961. Classification of the family Cyperaceae (1). J. Fac. Sci. Univ. Tokyo, Sect. 3, Bot. 8: 37–148.
Metcalfe, C. R. 1960. Vol. 1. Gramineae. *In:* C. R. Metcalfe, ed. Anatomy of the monocotyledons. Clarendon Press. Oxford.
Metcalfe, C. R. 1969. Anatomy as an aid to classifying the Cyperaceae. Amer. J. Bot. 56: 782–790.

Metcalfe, C. R. 1971. Vol. 5. Cyperaceae. *In:* C. R. Metcalfe, ed. Anatomy of the monocotyledons. Clarendon Press. Oxford.
Mora, L. E. 1960. Beiträge zur Entwicklungsgeschichte und vergleichenden Morphologie der Cyperaceen. Beitr. Biol. Pflanzen 35: 253–341.
Page, V. M. 1951. Morphology of the spikelet of *Streptochaeta*. Bull. Torrey Bot. Club 78: 22–37.
Prat, H. 1936. La systematique des Graminées. Ann. Sci. Nat. Bot. Sér. 10. 18: 165–258.
Prat, H. 1960. Vers une classification naturelle des Graminées. Bull. Soc. Bot. France 107: 32–79.
Rath, S. P., & S. N. Patnaik. 1974. Cytological studies in Cyperaceae with special reference to its taxonomy. Cytologia 39: 341–352.
Raynal, J. 1973. Notes Cypérologiques: 19: Contribution á la classification de la sous-famille des Cyperoideae. Adansonia sér. 2, 13: 145–171.
Reeder, J. R. 1953. The embryo of *Streptochaeta* and its bearing on the homology of the coleoptile. Amer. J. Bot. 40: 77–80.
Reeder, J. R. 1957. The embryo in grass systematics. Amer. J. Bot. 44: 756–768.
Reeder, J. R. 1962. The bambusoid embryo: a reappraisal. Amer. J. Bot. 49: 639–641.
Roth, I. 1955. Zur morphologischen Deutung des Grasembryos und verwandter Embryotypen. Flora 142: 564–600.
Schultze-Motel W. 1959. Entwicklungsgeschichtliche und vergleichend-morphologische Untersuchungen im Blütenbereich der Cyperaceae. Bot. Jahrb. Syst. 78: 129–170.
Shah, C. K. 1965. Embryogeny in some Cyperaceae. Phytomorphology 15: 1–9.
Shah, C. K., & N. Neelakandan. 1971. Embryogeny in some Cyperaceae II. Beitr. Biol. Pflanzen 47: 215–227.
Sharma, A. K., & A. K. Bal, 1956. A cytological investigation of some members of the Cyperaceae. Phyton (Buenos Aires) 6: 7–22.
Stebbins, G. L. 1956. Cytogenetics and evolution of the grass family. Amer. J. Bot. 43: 890–905.
Tjon Sie Fat, L. 1977. Contribution to the knowledge of cyanogenesis in angiosperms. 1. Communication. Cyanogenesis in some grasses (Poaceae [= Gramineae]). Proc. Kon. Nederl. Akad. Wetensch. ser. C. 80: 227–237.
Tjon Sie Fat, L., & F. van Valen. 1977. Contribution to the knowledge of cyanogenesis in angiosperms. 5. Communication. Cyanogenesis in some grasses. II. Proc. Kon. Nederl. Akad. Wetensch. ser. C. 81: 204–210.
Williams, C. A., & J. B. Harborne. 1977. Flavonoid chemistry and plant geography in the Cyperaceae. Biochem. Syst. Ecol. 5: 45–51.
Williams, C. A., J. B. Harborne, & H. T. Clifford. 1971. Flavonoid patterns in the monocotyledons. Flavonols and flavones in some families associated with the Poaceae. Phytochemistry 10: 1059–1063.

Авдулов, Н. П. 1931. Карио-систематическое исследование семейства злаков. Приложение 44 к Трудам по прикладной ботанике, генетике и селекции.
Добротворская, А. В. 1962. Морфологические особенности лодикул у некоторых представителей семейства злаковых. Труды Бот. Инст. АН СССР, сер. VII, 5: 148-165.
Пашков, Г. Д. 1951. О морфологической природе корневого влагалища злаков. Бот. Ж. 36: 597-606.
Петрова, Л. Р. 1973. Морфология репродуктивных органов некоторых видов

Bambusoideae (к филогении подсемейства). Бюлл. Московск. Общ. Исп. Прир. Отд. Биол. 78(4): 113–123.

Петрова, Л. Р., & Н. Н. Цвелев. 1974. Об эволюции соцветия злаков Poaceae, природе и функциях лодикул. Бот. Ж. 59: 1713–1720.

Рожевиц, Р. Ю. 1945. Система злаков в связи с их эволюцией. Сб. Научн. Работ., Бот. Инст. АН СССР 1: 25–40.

Цвелев, Н. Н. 1969. Некоторые вопросы эволюции злаков (Poaceae). Бот. Ж. 54: 361–373.

Цвелев, Н. Н. 1974. О направлениях эволюции вегетативных органов злаков (Poaceae). Бот. Ж. 59: 1241–1253.

Цвелев, Н. Н. 1975. О природе частей зародыша злаков (Poaceae) в связи с происхождением односемядольности. Бюлл. Общ. Исп. Прир., Отд. Биол. 80(3): 68–75.

Цвелев, Н. Н., & П. Г. Жукова. 1974. О наименьшем основном числе хромосом в сем. Poaceae. Бот. Ж. 59: 265–269.

Яковлев, М. С. 1948. Морфологические типы зародыша и филогения злаков. Докл. АН Армянской ССР 8: 127–134.

Яковлев, М. С. 1950. Структура эндосперма и зародыша злаков как систематический признак. Труды Бот. Инст. АН СССР, сер, VII, 1: 121–218.

SYNOPTICAL ARRANGEMENT OF THE FAMILIES OF CYPERALES

1 Flowers spirally or less often distichously arranged on the axis of the spike or spikelet, usually each flower seemingly or actually subtended by only a single scale, without an evident scale between the flower and the axis; seed-coat generally free from the pericarp; leaf-sheath usually closed; stem usually solid, very often triangular; carpels 3 or less often 2; embryo embedded in the endosperm; pollen-grains borne in pseudomonads; chromosomes often with a diffuse centromere .... 1. CYPERACEAE.

1 Flowers distichously arranged on the axis of the spikelet (or only one per spikelet), each flower ordinarily subtended by a pair of scales (lemma and palea), the palea inserted between the flower and the axis; seed-coat usually adnate to the pericarp; leaf-sheath usually open; stem usually hollow, never triangular; carpels 2, seldom 3; embryo peripheral to the endosperm; pollen-grains borne in true monads; chromosomes monocentric ... 2. POACEAE.

1. Family CYPERACEAE A. L. de Jussieu 1789 nom. conserv., the Sedge Family

Mostly perennial, very often rhizomatous herbs (the rhizomes often starchy), less often annual, very rarely shrubby, frequently of moist or wet habitats, without secondary growth, often with proanthocyanins in scattered tanniferous cells, sometimes producing simple indole alkaloids, and sometimes producing ethereal oils, but only seldom cyanogenic or saponiferous; plants generally containing the free form of dehydroquinate hydroxylase, as well as the bound form that is common to angiosperms in general; flavonoid constituents commonly including C-glycosylflavones and tricin (a flavone sulfate), and often 5-oxy-methyl flavones; cell-walls often more or less silicified, the epidermal cells overlying the vascular bundles of the stem and leaves generally with 1-several silica-bodies on the inner wall, these typically conical, sometimes of more complex shape; vessels with simple or scalariform perforations, commonly present in all vegetative organs; stems triangular or less often terete, solid or seldom hollow, with scattered, closed vascular bundles; plastids of the sieve-tubes with cuneate proteinaceous inclusions; roots usually with root hairs (but not in *Eleocharis*) and not mycorhizal. LEAVES alternate, very often in 3 ranks, with a closed (rarely open) sheath and a usually narrow and more or less elongate, grass-like, parallel-veined blade, but varying to terete and unifacial, or even flattened and unifacial, or the blade sometimes more or less reduced or suppressed; an adaxial ligule sometimes present at the juncture of sheath and blade, but usually not so well developed as in the Poaceae; stomates paracytic or rarely tetracytic, typically aligned in straight, longitudinal rows; guard-cells dumbbell-shaped. FLOWERS individually small and inconspicuous, commonly anemophilous, or rarely entomophilous (as in *Dichromena*), perfect or often unisexual (the plant then monoecious or rarely dioecious), sessile in the axils of spirally or distichously arranged bracts (scales), forming spikes or spikelets that are sometimes solitary and terminal, but more often arranged in various sorts of secondary inflorescences; a small bract only very rarely present between the flower and the axis of the spike or spikelet (aside from the perigynium of *Carex* and *Kobresia*); perianth of 1-many (often 6) short or elongate bristles or short scales, or none; stamens most often 3, sometimes only 2 or 1, rarely 6; anthers basifixed, tetrasporangiate and dithecal, opening by longitudinal slits; pollen-grains trinucleate, uniporate or seldom 2–4-porate (rather graminoid in *Mapania*), borne in pseudomonads (cryptotetrads), 3 of the 4 nuclei formed by meiosis in

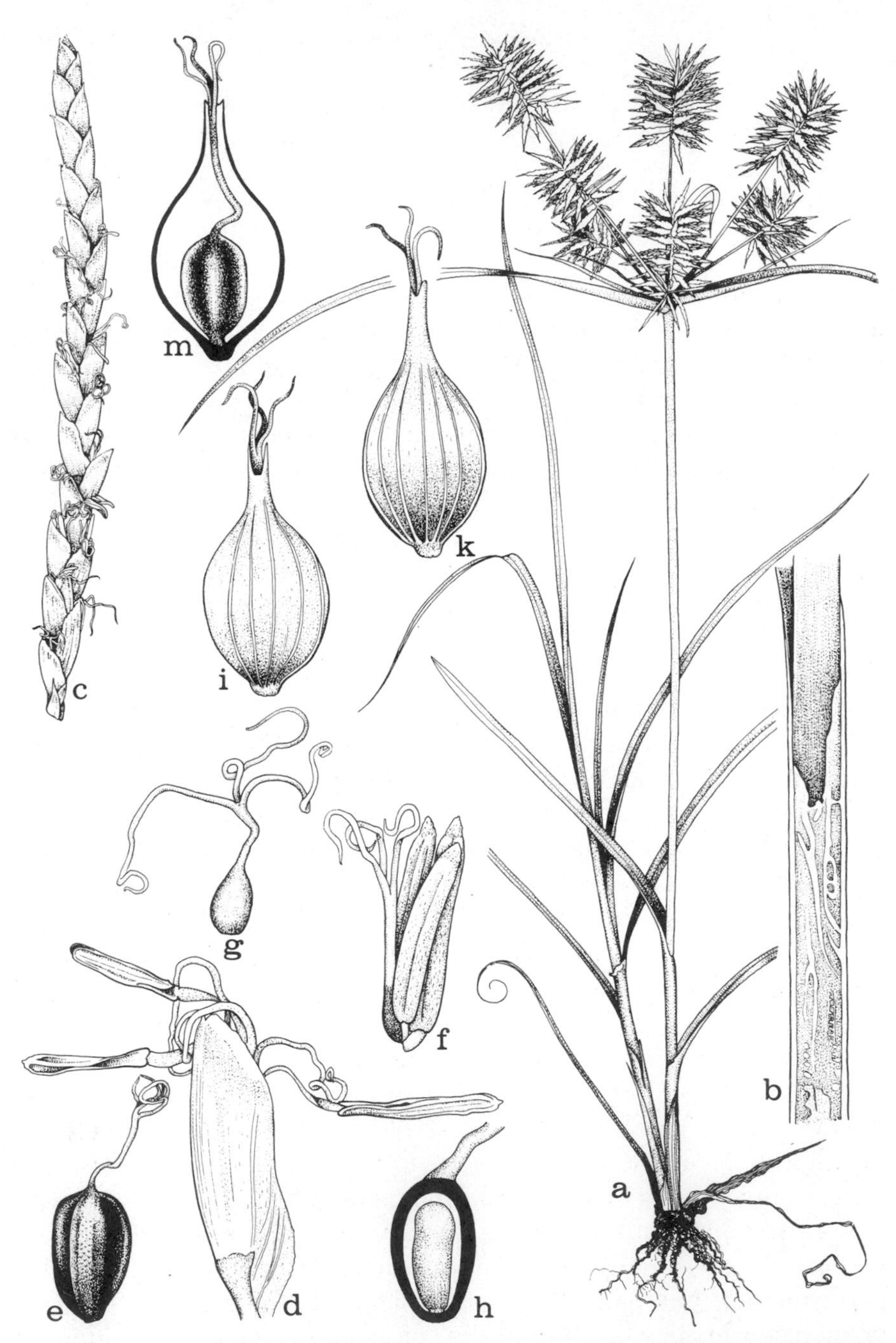

Fig. 9.7 Cyperaceae. a–h, *Cyperus esculentus* L. a, habit, $\times\frac{1}{2}$; b, portion of stem and leaf-sheath, $\times 3$; c, spikelet, $\times 4$; d, scale and flower, $\times 16$; e, achene, $\times 12$; f, flower, from the side, with the scale removed, $\times 16$; g, pistil, $\times 16$; h, schematic long-section of ovary, $\times 32$. i–m, *Carex rostrata* Stokes. i, k, two views of perigynium, $\times 6$; m, schematic long-section of perigynium, with achene enclosed, $\times 6$.

the microspore mother cells soon degenerating; gynoecium of 3 or less often 2 (rarely 4) carpels united to form a compound, superior, unilocular ovary with a terminal style and as many dry stigmatic branches as carpels; ovules solitary, centrally basal, anatropous, bitegmic, crassinucellar, sometimes provided with an obturator; endosperm-development nuclear. FRUIT a trigonous or lenticular achene, the seed free from the pericarp; embryo small to middle-sized, usually with a terminal cotyledon and lateral or nearly basal plumule, embedded in the copious endosperm, which is more or less starchy as well as oily and has an outer proteinaceous layer. Chromosomes often with diffuse centromere; x = 5–60+. (Kobresiaceae)

The well defined family Cyperaceae has about 70 genera and nearly 4000 species, of cosmopolitan distribution, but most abundant in temperate regions. About two-thirds of the species belong to only 6 genera: *Carex* (1100), *Cyperus* (600), *Scirpus* (250), *Rhynchospora* (250), *Fimbristylis* (200), and *Scleria* (200).

The Cyperaceae are of relatively little economic importance. Some of them furnish useful forage for livestock. Stems of some species of *Cyperus* are used in basketry, and those of *Cyperus papyrus* L. were used to produce papyrus, the writing-paper of ancient Egypt. The rhizomes of *Cyperus esculentus* L. and other species are edible.

The statement in the description that there are as many styles as carpels is a convenient oversimplification. Carpellary reduction proceeds by grades, and the number of styles may not match the number of discernible carpels in the ovary.

The pistillate spikes of *Carex* clearly represent compound inflorescences. Each pistillate flower is subtended by a bract, and enclosed by another bract, open at the top, which has its back toward the axis of the spike and often shows an evident suture (the connate margins) toward the summit of the abaxial side. The style protrudes through the apical orifice of the enclosing bract, which is called the perigynium. In a few species of *Carex*, the solitary pistillate flower within the perigynium is evidently lateral on a small rachilla that is prolonged beyond the ovary. The rachilla is considered to represent a vestigial ultimate branch of a compound inflorescence.

Some authors maintain that the apparently perfect flowers of *Scirpus* and some other genera are actually pseudanthia, partial inflorescences in which the apparent perianth represents ancestral bracts, each stamen represents a separate flower, and the pistil represents still another flower. Under this interpretation, the true flowers of all Cyperaceae are unisexual and without perianth. Even if this pseudanthial interpretation were correct (which Occam's Razor makes me doubt), it would be more useful from a purely descriptive standpoint to follow the classical interpretation that the flowers and flower-parts are just what they seem. Descriptive morphology loses its function if it is intelligible only to a limited coterie of specialists.

Good fossils of the Cyperaceae (both fruits and associated probable pollen) date from the Eocene of several areas.

2. Family POACEAE Barnhart 1895 nom. conserv., the Grass Family

Mostly perennial, often rhizomatous herbs, less commonly annual, only rather seldom (as in most bamboos) more or less woody and even arborescent, but in any case without secondary thickening; plants generally containing the free form of dehydroquinate hydroxylase, as well as the bound form that is common to angiosperms in general, very often accumulating ferulic acid, often producing fructosans (notably of a type called phlein), sometimes producing isoquinoline, simple indole, or pyrrolizidine alkaloids, and occasionally cyanogenic (tyrosine-derived), but only rarely tanniferous (from proanthocyanins) and without crystals of calcium oxalate; flavonoid constituents commonly including large amount of C-glycosylflavones and tricin (a flavone sulfate); cell-walls, especially of the epidermis of the shoot, more or less strongly silicified, and the epidermis also with specialized short cells that contain each a silica-body (these of various forms, but unlike those of the Cyperaceae); vessels commonly present in all vegetative organs, with simple perforations, or sometimes some of them scalariform; stems terete (seldom flattened), usually with hollow internodes (especially in genera of temperate regions), vascular bundles closed, more or less scattered, but concentrated toward the periphery of the stem; plastids of the sieve-tubes with cuneate proteinaceous inclusions; roots often producing root-hairs, but often forming endomycorhizae as well. LEAVES distichous or rarely spirally arranged, but not 3-ranked, with an open (rarely closed) sheath and a parallel-veined, usually narrow and elongate blade that commonly has a well developed intercalary meristem at the base, the blade often with a pair of small marginal auricles at the base, seldom (many bamboos) constricted into a petiolar base above the sheath; a membranous ligule generally present at the juncture of the sheath and blade on the adaxial side, the ligule sometimes well developed

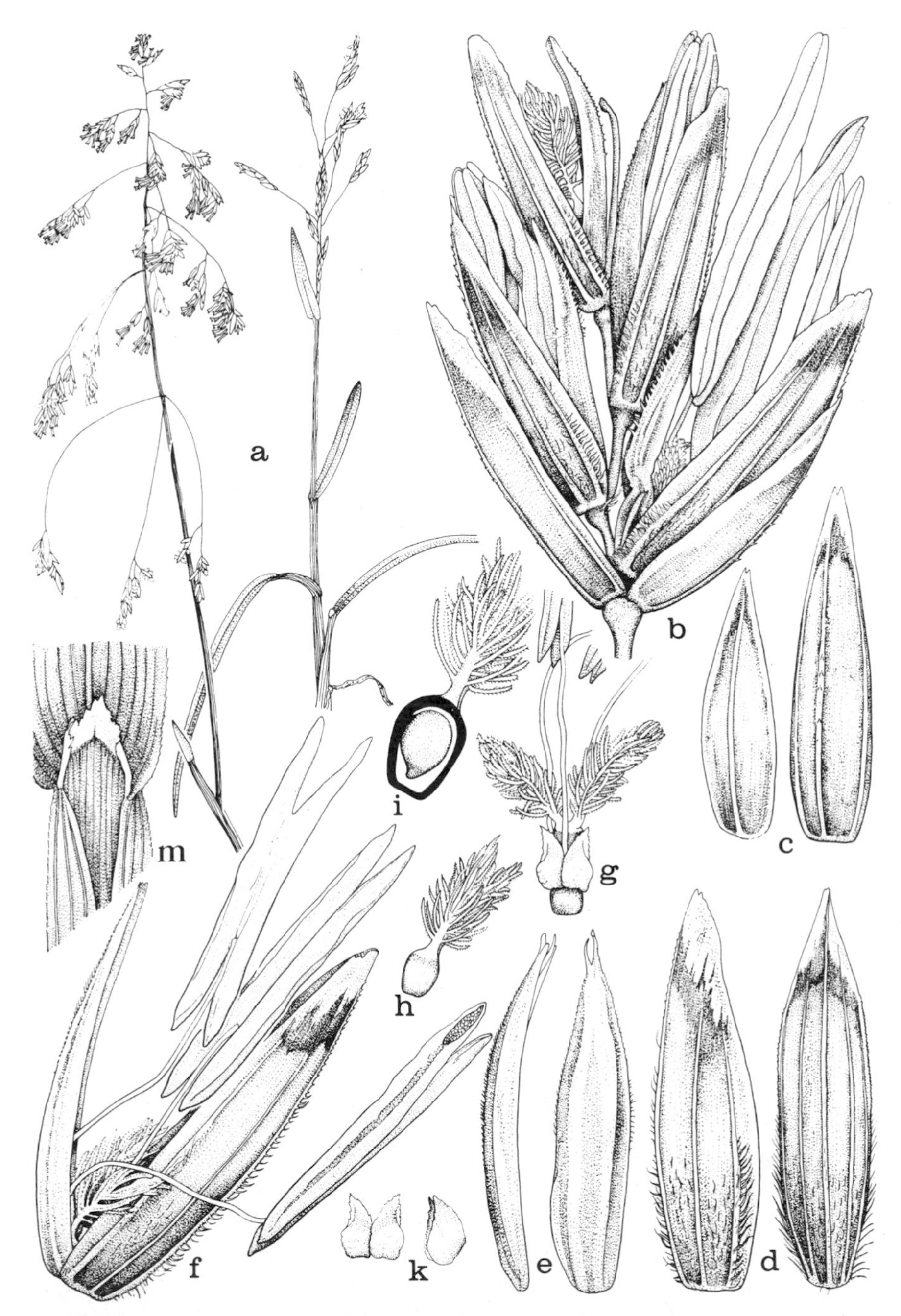

Fig. 9.8. Poaceae. *Poa autumnalis* Muhl. a, habit, $\times\frac{1}{2}$; b, spikelet, $\times 12$; c, outer (left) and inner glumes, $\times 12$; d, lemma, in side (left) and dorsal view, $\times 12$; e, palea, in side (left) and external view, $\times 12$; f, a single floret, $\times 12$; g, portion of flower, the lemma and palea removed, $\times 12$; h, pistil, $\times 12$; i, pistil, the ovary in schematic long-section, $\times 24$; k, lodicules, $\times 12$; m, ventral view of leaf at juncture of sheath and blade, showing ligule, $\times 6$.

and more than 1 mm (or even more than 1 cm) long, sometimes reduced or represented by a row of hairs, seldom wholly suppressed; epidermis typically of alternating files of long and short cells, some of the short cells containing each a silica-body, the shape of which differs in different groups of genera; stomates paracytic, aligned in straight, longitudinal, uniseriate or biseriate rows; guard-cells dumbbell-shaped. FLOWERS individually small and inconspicuous, commonly anemophilous but sometimes autogamous or apomictic or even (*Pariana*) entomophilous, perfect or sometimes unisexual, arranged in 1- to many-flowered spikelets that are themselves arranged in a determinate or mixed secondary inflorescence that most commonly has the form of a panicle but is sometimes spike-like or raceme-like; each spikelet distichously organized, typically with a pair of subopposite small bracts (*glumes*) at the base and 1 to several or many florets alternating on opposite sides of an often zigzag axis (the *rachilla*) above the glumes; first (lower) glume sometimes more or less reduced or even wholly suppressed, rarely both glumes suppressed; each floret typically consisting of a pair of subopposite enclosing or subtending scales (the *lemma* and *palea*), two or three much smaller scales (the *lodicules*) above these, and the androecium and gynoecium; outer (lower) scale (the lemma) of the floret, interpreted as a subtending bract, generally with a midvein and one or more pairs of lateral veins, the midvein often excurrent as a short or elongate, dorsal or terminal awn; inner (upper, adaxial) scale (the palea) of the floret placed with its back to the rachilla, typically bearing 2 main veins and interpreted as two connate members of an ancestral trimerous outer cycle of tepals, the third member suppressed; palea generally enfolded by and shorter than the lemma, sometimes much shorter, or sometimes (as in some bamboos) longer than the lemma, rarely wholly suppressed; lower floret of the spikelet sometimes represented by an empty, glume-like sterile lemma, especially when the first glume is reduced or suppressed, or the sterile lemma sometimes subtending a staminate flower with vestigial or no palea; lodicules, apparently representing the inner cycle of tepals, mostly tiny and inconspicuous, (1) 2 or (*Stipa* and Bambusoideae) 3, or seldom up to 6 or even more in some Bambusoideae; stamens most often 3, sometimes 6 (especially in the Bambusoideae), seldom 2 or only 1, rarely (*Ochlandra*, of the Bambusoideae) more than 100; anthers elongate, basifixed but so deeply sagittate as to appear versatile, tetrasporangiate and dithecal, opening by longitudinal slits; pollen-grains borne in monads, trinucleate, uniporate, typically with a small pore that has a raised or thickened

. im, nearly smooth; gynoecium of 2 or (Bambusoideae) 3 carpels united to form a compound, superior, unilocular ovary with accordingly 2 or 3 stigmas, these often large and feathery (stigmas connate in *Zea*); ovule solitary, subapical to nearly basal on the lateral wall of the ovary, orthotropous to hemitropous or rarely campylotropous to almost anatropous, bitegmic or rarely unitegmic, pseudocrassinucellar or sometimes tenuinucellar; endosperm-development nuclear. FRUIT, called a *caryopsis*, usually tightly enclosed by the persistent lemma and palea, indehiscent, usually dry (seldom fleshy), the coat of the single seed adnate to the pericarp; rarely (as in *Sporobolus*) the fruit falling free from the lemma and palea, and the seed-coat free from the pericarp, the inner portion of the pericarp then becoming mucilaginous when wet and expelling the seed on drying; endosperm copious, in large part starchy and usually mealy, with compound or simple starch-grains, usually with a proteinaceous segment or outer layer as well, and sometimes also oily, rarely (*Melocanna*) the endosperm wanting; embryo basilateral, peripheral to the endosperm, straight, complex in structure, with a well developed plumule, radicle, and enlarged lateral cotyledon (the *scutellum*), the plumule ensheathed by a cylindric, apically closed *coleoptile*, and the radicle similarly ensheathed by an apically closed *coleorhiza*, the nature of the coleoptile and coleorhiza variously interpreted, the coleoptile perhaps the first true leaf, perhaps a highly modified segment of the cotyledon. Chromosomes monocentric; x = 2–23+. (Gramineae, nom. altern.; Anomochloaceae, Bambusaceae, Streptochaetaceae)

The well defined family Poaceae has about 500 genera and 8000 species, of cosmopolitan distribution, but most abundant in tropical and north-temperate semi-arid regions with seasonal rainfall, where they form extensive grasslands. The largest genus is *Panicum*, with about 400 species, followed by *Poa* (300), *Eragrostis* (300), *Stipa* (250), and *Paspalum* (200).

The grasses are by far the most important family of plants to human society. They provide staple foods for man, and the principal forage for countless grazing animals. *Triticum* (wheat), *Oryza* (rice), *Zea* (maize), *Avena* (oats), *Saccharum* (sugar cane), *Sorghum, Secale* (rye), *Hordeum* (barley), and *Poa* (bluegrass) are among the more important genera.

The morphology of the grass spikelet has been variously interpreted, partly because of differing views as to the general course of evolution in flowering plants. The interpretation here adopted reflects the view that grasses are derived from ancestors with trimerous flowers and two cycles of tepals, and draws on the work of a series of botanists from Celakovsky (1889) onwards.

The small (3 spp.) South American genus *Streptochaeta* is especially significant to the morphological interpretation of the grass spikelet. The spikelet of *Streptochaeta* has only a single flower. There are 4 or 5 small, irregularly arranged glumes. The single lemma is evidently longer than the glumes. The palea is shorter than the lemma and bifid almost to the base, so that it is readily interpreted as consisting of 2 basally connate members. A vestige of a third member, completing the whorl of 3, can be seen in early ontogeny. The 3 lodicules are well developed, leathery, and longer than the palea. There are 6 stamens, and the long style has 3 short stigmas. The elongate caryopsis is free from the lemma and palea. *Streptochaeta* resembles many of the bamboos in having 6 stamens, 3 lodicules, and 3 stigmas; it has a relatively broad leaf-blade with a petiolar constriction at the base, and its leaf-anatomy is bambusoid. Its affinity with the Bambusoideae seems to be well established, but authors who wish to emphasize its distinctive qualities separate it as a subfamily Streptochaetoideae (Butzin, 1965).

The taxonomic organization of the grasses into subfamilies and tribes is in a state of flux. The classic work of Avdulov (1931) and of Prat (1936) represents a turning point in grass taxonomy. Many microscopic anatomical characters, as well as chromosome-number, have proved to be of critical importance in assessing relationships, and some of the traditional tribes now appear to be heterogeneous and unnatural. A minimum of 3 subfamilies must be recognized. These are the Bambusoideae, Pooideae, and Panicoideae.

The Bambusoideae (bamboos) are mostly woody, sometimes even arborescent, and have perennial leaves. The leaf-blade is often relatively broad, with evident cross-veins between the longitudinal veins, and it is often attached to the sheath by a petiole. Bamboos usually have 3 lodicules, and they often have 3 stigmas. Many of them have 6 stamens. The floral structure of bamboos is now generally regarded as relatively primitive within the family, but the vegetative structure is not.

The Panicoideae are a specialized group with 2 dissimilar florets in each spikelet. The upper floret is fertile and of ordinary structure, but the lower one is more or less reduced. The lower floret is typically represented only by an empty lemma, but sometimes there is a vestigial palea, and there may even be 3 stamens, but (with rare exceptions) there is no gynoecium. The first glume is commonly more or less reduced or even vestigial. Sometimes it is wholly suppressed, so that the sterile lemma has all the features of another glume, and the basic structure of the spikelet can be determined only indirectly.

The Pooideae, if broadly defined, are by far the largest subfamily. They lack the syndromes of features that set the Bambusoideae and Panicoideae apart, but they are otherwise diverse inter se. Further study will doubtless lead to a narrower definition of the Pooideae and the recognition of several additional subfamilies, but there is at present no agreement as to the number and definition of these other groups. One view is that the Eragrostoideae, Oryzoideae, and Arundinoideae should be recognized as additional subfamilies.

Smooth, monoporate pollen, characteristic of the grasses and most Restionales, first appears in the fossil record in Senonian deposits, above the middle of the Upper Cretaceous. More distinctly graminoid pollen dates from the Paleocene, and becomes abundant in the Lower Eocene. Fruits thought to represent grasses also date from the Eocene. Grasslands appear to have been an important feature of the vegetation of the earth from the Eocene to the present.

6. Order HYDATELLALES Cronquist 1980

The order consists of the single family Hydatellaceae.

SELECTED REFERENCES

Edgar, E. 1966. The male flowers of *Hydatella inconspicua* (Cheesem.) Cheesem. (Centrolepidaceae). New Zealand J. Bot. 4: 153–158.

Hamann, U. 1976. Hydatellaceae—a new family of Monocotyledoneae. New Zealand J. Bot. 14: 193–196.

Hamann, U., K. Kaplan, & T. Rübsamen. 1979. Über die Samenschalenstruktur der Hydatellaceae (Monocotyledoneae) und die systematische Stellung von *Hydatella filamentosa*. Bot. Jahrb. Syst. 100: 555–563.

1. Family HYDATELLACEAE Hamann 1976, the Hydatella Family

Small, tufted, submerged or partly emergent aquatic annual herbs without secondary growth, strongly aerenchymatous, without mechanical tissues and with much-reduced xylem, the vessels confined to the roots; plants lacking crystals and silica-bodies, but sometimes with tannin in some of the floral tissues; stem commonly with 2 or 3 vascular bundles that may surround a central pith and with chlorenchymatous subepidermal tissue. LEAVES all tufted at the base, alternate, without a well defined sheath, slender, subterete or somewhat compressed, centric in structure, with a single vascular bundle; stomates anomocytic or wanting. Inflorescence a terminal unisexual (*Hydatella*) or bisexual (*Trithuria*) head with 2-several bracts on a short axis, each subtending a single flower or several flowers, in *Hydatella* both types of unisexual head borne on the same plant; FLOWERS hydrophilous or autogamous, naked and unisexual, each composed of a single stamen or a single pseudomonomerous pistil, not separated by bracts other than the primary bracts of the head; anther basifixed on a rather stout filament, tetrasporangiate, dithecal; pollen-grains trinucleate, of an unusual, monocolpate type; pistil composed of a unilocular ovary bearing at the summit 2–3 (*Trithuria*) or 5–10 (*Hydatella*) uniseriate filaments which may represent styles; ovule solitary, pendulous from the apex of the locule, anatropous, bitegmic, crassinucellar; endosperm-development probably cellular. FRUIT small, dry, opening by 3 valves (*Trithuria*) or indehiscent (*Hydatella*); seed solitary, with a well developed, starchy perisperm, but virtually without endosperm; embryo minute, lenticular, scarcely differentiated, capping the perisperm at the micropylar end, its position marked by an opercular swelling of the seed-coat.

The family Hydatellaceae consists of 2 small genera, *Hydatella* (4) and *Trithuria* (3), native to Australia, New Zealand, and Tasmania. These two genera have customarily been assigned to the Centrolepidaceae, which they resemble in habit, but from which they differ in numerous other ways. The affinities of the Hydatellaceae are uncertain, but they seem to be best accommodated in the Commelinidae. The opercular swelling of the seed-coat suggests a possible eventual relationship with the Commelinales.

7. Order TYPHALES Lindley 1833

Rhizomatous perennial herbs of shallow water and wet places, emergent or with distally floating stem and leaves, producing raphides and at least sometimes proanthocyanins, and sometimes cyanogenic, but without saponins; vessels (with scalariform perforations) and tanniferous cells present in all vegetative organs; stem without secondary growth; vascular bundles closed, numerous and more or less scattered; plastids of the sieve-tubes with cuneate proteinaceous inclusions. LEAVES alternate, distichous, with strongly sheathing base and elongate, parallel-veined, linear blade of somewhat spongy texture; stomates paracytic. FLOWERS anemophilous (or autogamous), strictly unisexual (the plants monoecious), numerous, sessile or nearly so in dense spikes or globose heads; perianth of 1-several inconspicuous tepals or more or less numerous slender bristles; stamens 1–8, often 3, the filaments distinct or often connate below; anthers basifixed, tetrasporangiate and dithecal; pollen-grains uniporate; gynoecium most commonly pseudomonomerous, with a single fertile carpel, less often with 2–3 fully developed carpels connate to form a compound ovary, the stigmas or branches of the common style accordingly 1–3; ovule solitary (or solitary in each locule), pendulous from near the top of the locule, apotropous, anatropous, bitegmic, crassinucellar; endosperm-development helobial. FRUIT small, dry; seed with a straight embryo that has a single terminal cotyledon and a lateral plumule; embryo surrounded by a copious, mealy endosperm and a thin perisperm, the endosperm containing starch, protein, and oil. X = 15.

The order Typhales consists of 2 closely related small families, the Typhaceae and Sparganiaceae, each with only a single genus. There are only about two dozen species in all. A good case can be made for including *Sparganium* in the Typhaceae, but it is sufficiently different from *Typha* so that there is no serious harm in continuing the customary practice of putting each of the two genera into its own family.

The Typhales are taxonomically isolated. Serological tests emphasize the remoteness of the order from others, without providing a clear guide to its affinities. *Acorus*, mentioned below, was among the genera that did not react significantly with *Typha*.

A relationship of *Sparganium eurycarpum* Engelm. to *Acorus* (Araceae) has been suggested on the basis of their common susceptibility to the rust fungus *Uromyces sparganii* (Savile, 1979, initial citations, and earlier), and there is some habital similarity between *Typha* and *Acorus*. On the other hand, the Araceae as a group differ notably from the Typhaceae in their mostly nonstarchy endosperm and in having vessels only in the roots.

The Pandanales, in which the Typhales have often been included, differ in their arborescent habit, spiral growth-pattern, mostly tetracytic stomates, and nonstarchy endosperm. Furthermore, the evolutionary development of the spadix of the Pandanaceae can be traced within the family itself. Whatever similarity there is between the inflorescence of the Pandanaceae and that of the Typhales reflects parallelism or convergence, rather than inheritance from a common ancestry. On the basis of embryological features, Asplund (1972) considers the Typhales well removed from the Pandanales. On the other hand, the Typhales are cytologically rather similar to the Pandanales; in both groups the chromosomes are all small, with a probable base number of 15. The admittedly incomplete fossil record suggests that the Pandanales are the more recently evolved group, dating back only to the Miocene, whereas the Typhales go back well beyond the Miocene, apparently all the way to the Upper Cretaceous. On the basis of present evidence, a close relationship between the Typhales and Pandanales does not seem likely.

The Typhales are here included in the Commelinidae because of their floral reduction associated with wind-pollination, their paracytic stomates, the presence of vessels in all vegetative organs, and their mealy, starch-bearing endosperm. All of these features are perfectly fine for the Commelinidae, but collectively they are difficult to reconcile with any other subclass of monocotyledons. The Commelinidae provide a more hospitable home for the Typhales than do the Arecidae, the only other subclass with which they might possibly be associated.

Pollen considered to represent the Typhales dates from the Paleocene. Fruits showing a clear differentiation of the two families date from the Oligocene. Some late Upper Cretaceous fruits may represent *Sparganium*; some of these have up to 5 carpels.

SELECTED REFERENCES

Asplund, I. 1972. Embryological studies in the genus *Typha*. Svensk Bot. Tidskr. 66: 1–17.

Graef, P. E. 1955. Ovule and embryo sac development in *Typha latifolia* and *Typha angustifolia*. Amer. J. Bot. 42: 806–809.

Lee, D. W., & D. E. Fairbrothers. 1972. Taxonomic placement of the Typhales within the monocotyledons; preliminary serological investigation. Taxon 21: 39–44.

Müller-Doblies, U. 1969. Über die Blütenstände und Blüten sowie zur Embryologie von *Sparganium*. Bot. Jahrb. Syst. 89: 359–450.
Müller-Doblies, D. 1970. Über die Verwandtschaft von *Typha* und *Sparganium* im Infloreszens- und Blütenbau. Bot. Jahrb. Syst. 89: 451–562.

SYNOPTICAL ARRANGEMENT OF THE FAMILIES OF TYPHALES

1 Inflorescence of dense, globose heads; perianth of the pistillate flowers of 2–6 small tepals; fruits sessile or nearly so, hydrochorous or endozoochorous .... 1. SPARGANIACEAE.

1 Inflorescence a dense, elongate, cylindric spike; perianth of the pistillate flowers mostly of numerous capillary bristles; fruits long-stipitate, anemochorous .... 2. TYPHACEAE.

1. Family SPARGANIACEAE Rudolphi 1830 nom. conserv., the Bur-reed Family

Perennial herbs from starchy rhizomes, emergent from shallow water or with the stem and leaves distally floating, commonly producing raphides and other sorts of calcium oxalate crystals, with scattered tanniferous cells containing proanthocyanins in all organs, but not cyanogenic and not saponiferous; vessels with scalariform perforations, present in all vegetative organs; stems simple or branched; vascular bundles numerous, closed, scattered; plastids of the sieve-tubes with cuneate proteinaceous inclusions; root-hairs well developed. LEAVES alternate, distichous, with a sheathing base and an elongate, parallel-veined, linear blade, this commonly spongy in texture, with intersecting partitions; stomates paracytic. FLOWERS anemophilous and proterogynous, or autogamous, strictly unisexual (the plants monoecious), sessile or nearly so, grouped into dense, unisexual, globular, complex heads, these sessile or short-pedunculate along the main stem or its primary branches, the staminate ones uppermost; pistillate heads axillary to leafy bracts (i.e., to the upper leaves), or supra-axillary (the peduncle adnate to the internode for some distance), the staminate ones often without subtending bracts; staminate flowers with an apparent perianth of 1–6 small, scale-like tepals, one or more of which may actually represent subtending bracts, but the positional relationship of these to each other and to the androecium often becoming obscure during ontogeny, so that the staminate head may come to appear as a mass of irregularly disposed stamens and scales; stamens 1–8, opposite the tepals when equal in number, distinct or with the filaments sometimes connate below; anthers basifixed, tetrasporangiate and dithecal; pollen-grains borne in monads, uniporate, trinucleate; pistillate flowers usually each axillary to a bract within the head, but this bract sometimes difficult to differentiate from the (2) 3–4 (–6) small, greenish, often spatulate, hypogynous tepals in a single cycle or more or less irregularly in 2 cycles; gynoecium most commonly pseudomonomerous, with a single more or less adaxial fertile carpel and a strongly reduced, abaxial, sterile one, but sometimes of 2 or 3 fully developed carpels united to form a compound ovary with as many locules; style short, tipped by the oblong or more often linear-elongate, dry, papillate stigma(s); ovule solitary in the single locule (or in each of the 2–3 locules), pendulous from near the summit, apotropous, anatropous, bitegmic, crassinucellar; endosperm-development helobial. FRUIT small, dry, indehiscent, with spongy exocarp and stony endocarp, usually subtended by the persistent

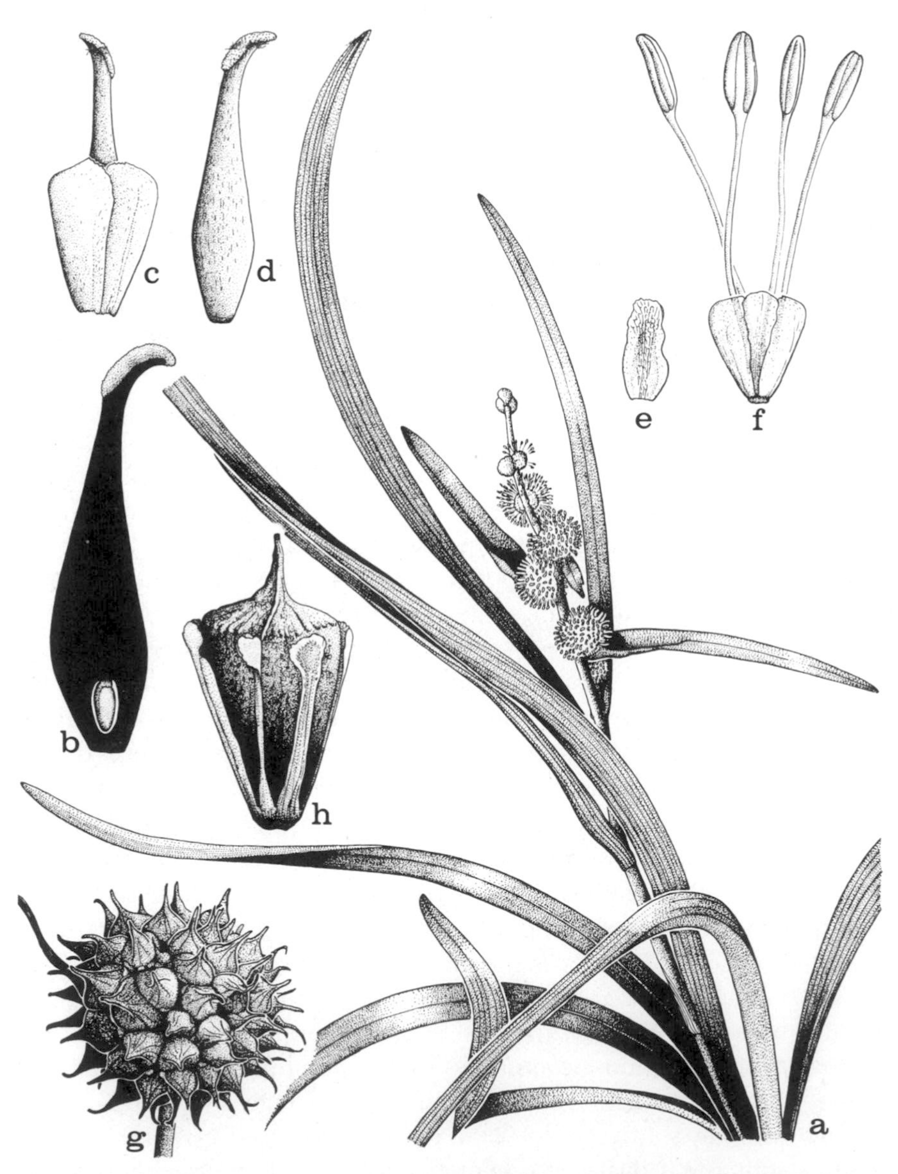

Fig. 9.9 Sparganiaceae. a–f, *Sparganium americanum* Nutt. a, habit, ×½; b, pistil, in partial long-section, showing solitary, pendulous ovule, ×9; c, pistillate flower, ×6, d, pistil, ×6; e, tepal of staminate flower, ×6; f, staminate flower, ×6. g, h, *Sparganium eurycarpum* Engelm. g, fruiting head, ×1; h, achene with persistent perianth, ×3.

tepals as well as beaked by the persistent style, distributed by water or endozootically; seed with a straight, linear, axile embryo that has a terminal cotyledon and a lateral plumule, the embryo surrounded by a copious endosperm and thin perisperm, the endosperm mealy, containing starch as well as protein and oil. X = 15.

The family Sparganiaceae consists of the single genus *Sparganium*, with about 13 species, chiefly of North Temperate regions, but extending south to Australia and New Zealand in the Old World.

2. Family TYPHACEAE A. L. de Jussieu 1789 nom. conserv., the Cat-tail Family

Glabrous perennial herbs from starchy rhizomes, often densely colonial, emergent from shallow water or growing in very wet soil, commonly producing raphides and other sorts of calcium oxalate crystals, and with scattered tanniferous cells (containing proanthocyanins) in all organs, sometimes cyanogenic, but not saponiferous; vessels with scalariform perforations, present in all vegetative organs; stems erect, simple, terminating in an inflorescence; vascular bundles closed, numerous and scattered but concentrated toward the periphery of the stem, the interior of the stem soft and pithy; plastids of the sieve-tubes with cuneate proteinaceous inclusions; root-hairs more or less well developed. LEAVES alternate, distichous, most of them basal or near-basal, with a strongly sheathing base and an elongate, parallel-veined, flattened blade, this commonly spongy in texture, with longitudinal partitions connecting the upper and lower surface, and with numerous cross-partitions connecting the longitudinal ones; stomates paracytic. FLOWERS anemophilous and proterogynous, strictly unisexual (the plants monoecious), very numerous, in a very dense, elongate-cylindric, complex spike, many or all of them axillary to short, bristle-like bracts, the staminate ones in the upper part of the spike, the pistillate ones in the lower part, the two parts of the spike sharply marked, separated or contiguous; staminate flowers with a modified perianth of 0–3 (–8) slender, capillary bristles or slender scales; stamens (1–) 3 (–8), the short filaments distinct or connate for much of their length; anthers basifixed, tetrasporangiate and dithecal, with a broad connective that is prolonged beyond the pollen-sacs; pollen-grains borne in monads or tetrads, binucleate, uniporate; pistillate flowers with a hypogynous modified perianth of more or less numerous slender, capillary bristles (these sometimes distally thickened) or narrow scales in 1–4 irregular cycles, some of the members sometimes connate into small groups, and often some of them adnate below to the lower part of the gynophore; gynoecium of a single carpel (the ovary considered to be pseudomonomerous) elevated on a gynophore that elongates in fruit to form a slender stipe; style terminal on the carpel, persistent and elongating in fruit, with a shortly decurrent, linear to spatulate, dry, nonpapillate stigma; ovule solitary, pendulous from near the summit of the locule, apotropous, anatropous, bitegmic, crassinucellar; endosperm-development helobial; sterile pistillate flowers with abortive ovary intermingled with the fertile ones. FRUIT distributed by wind, small, dry, 1-seeded, eventually opening; seed with a slender, cylindric, straight embryo that has a single terminal cotyledon and a lateral plumule; embryo surrounded by a fairly copious endosperm and thin perisperm, the endosperm mealy, containing starch as well as protein and oil. X = 15.

The family Typhaceae consists of the single genus *Typha*, with about 10 species, of cosmopolitan distribution. Cat-tails are familiar plants of marshes throughout most of the world.

IV. Subclass ZINGIBERIDAE Cronquist 1978

Terrestrial or epiphytic herbs (sometimes emergent from shallow water), or seldom small trees with a simple, unbranched trunk, in any case without secondary growth and not obviously mycotrophic; sieve tube plastids with cuneate proteinaceous crystalloids and also starch grains; vessels restricted to the roots, or sometimes present also in the stems or in all vegetative organs; vascular bundles of the stem closed, scattered; plants commonly with scattered silica cells, and often producing raphides. LEAVES alternate (sometimes all basal), with sheathing base, the blade narrow and parallel-veined, or broader, petiolate, and with pinnate-parallel venation; stomates with (2) 4 or more subsidiary cells. Inflorescence commonly (not always) with a series of large, conspicuous, often brightly colored bracts, each subtending a small partial inflorescence or a single flower; FLOWERS perfect or sometimes functionally unisexual, regular to strongly irregular, basically trimerous, hypogynous to much more often epigynous, distinctly heterochlamydeous, adapted to pollination by insects, birds, or bats, only rarely by wind; sepals 3, distinct or connate into a lobed tube (or into a spathe), green and more or less herbaceous to somewhat petaloid; but unlike the petals; petals basically 3, distinct or connate, or joined together with petaloid staminodes; stamens 6 in 2 cycles of 3, or only one or 5 stamens functional, the others wanting or represented by more or less petaloid staminodes; pollen binucleate or trinucleate, monosulcate to multiporate or often inaperturate; gynoecium typically of 3 carpels united to form a compound, superior to much more often inferior, trilocular ovary with a terminal style, but sometimes unilocular, or with one or two of the carpels reduced and empty or even obsolete; septal nectaries commonly present, opening at the top of the ovary, ovules 1 to much more often several or numerous in each locule, anatropous to sometimes campylotropous, bitegmic, crassinucellar; endosperm-development helobial or nuclear. FRUIT capsular or baccate, or seldom schizocarpic or multiple; seeds in capsular fruits often arillate or winged or plumose; embryo monocotyledonous; endosperm copious, or partly replaced by a well developed perisperm, in either case the food storage tissue of the seed starchy and very often mealy, typically with compound starch grains (as also in the rhizome).

The subclass Zingiberidae consists of 2 orders, 9 families, and about

3800 species. The two orders, Bromeliales and Zingiberales, are of nearly equal size in terms of number of species, but the Bromeliales all belong to the single family Bromeliaceae, whereas the Zingiberales are organized into 8 families.

The two orders of Zingiberidae have in the past usually been associated with the orders of the Commelinidae or Liliidae, sometimes the Bromeliales with one group (Commelinidae) and the Zingiberales with the other (Liliidae). They are discordant elements in either group. They resemble the Liliidae (and differ from the Commelinidae) in commonly having septal nectaries and in usually having the vessels confined to the roots. They resemble the Commelinidae (and differ from most Liliidae) in their starchy endosperm with compound starch grains, and they further differ from typical Liliidae (and resemble the Commelinales) in having the sepals well differentiated from the petals, often green and herbaceous in texture. They differ from both the Liliidae and the Commelinidae in that the number of subsidiary cells around the stomates is usually 4 or more.

The definition of both the Commelinidae and the Liliidae can be considerably sharpened by excluding the Bromeliales and Zingiberales. Although these two orders differ in general appearance and ecological adaptation, they agree in the several features that separate them collectively from the Liliidae and the Commelinidae. The inflorescence in characteristic members of the Bromeliales and Zingiberales also has a certain similarity of aspect (relating to the large, showy bracts), and differs from that of other monocotyledons. Williams & Harborne (1977, under Zingiberales) see a possible association of the Zingiberales and Bromeliales on the basis of the flavonoids. Thus it appears that the purposes of taxonomy are best served by grouping the two orders into a separate subclass.

A putative common ancestor of the Bromeliales and Zingiberales might have been much like some of the more archaic, terrestrial Bromeliaceae, but less xerophytic.

SELECTED REFERENCES

Cronquist, A. 1978. The Zingiberidae, a new subclass of Liliopsida (Monocotyledons). Brittonia 30: 505.

SYNOPTICAL ARRANGEMENT OF THE ORDERS OF ZINGIBERIDAE

1 Functional stamens 6; flowers regular or sometimes somewhat irregular; xerophytes and epiphytes with narrow, parallel-veined,

often firm and spiny-margined leaves, the blade continuous with the sheath, not petiolate; subsidiary cells most commonly 4 .... 1. BROMELIALES.

1 Functional stamens 1 or 5, very rarely 6; flowers more or less strongly irregular; mesophytes (or emergent aquatics), very often growing on the forest floor; leaves with a sheath, petiole, and expanded, entire blade that has a prominent midrib and numerous primary lateral veins in a characteristic pinnate-parallel arrangement; subsidiary cells most commonly more than 4 .... 2. ZINGIBERALES.

1. Order BROMELIALES Lindley 1833

The order consists of the single family Bromeliaceae.

SELECTED REFERENCES

Benzing, D. H., & A. Renfrow. 1971. Significance of the patterns of CO_2 exchange to the ecology and phylogeny of the Tillandsioideae (Bromeliaceae). Bull. Torrey Bot. Club 98: 322–327.

Benzing, D. H., J. Seemann, & A. Renfrow. 1978. The foliar epidermis in Tillandsioideae (Bromeliaceae) and its role in habitat selection. Amer. J. Bot. 65: 359–365.

Cheadle, V. I. 1955. Conducting elements in the xylem of the Bromeliaceae. Bull. Bromeliad Soc. 5: 3–7.

Ehler, N., & R. Schill. 1973. Die Pollenmorphologie der Bromeliaceae. Pollen & Spores 15: 13–49.

Marchant, C. J. 1967. Chromosome evolution in the Bromeliaceae. Kew Bull. 21: 161–168.

Sharma, A. K., & I. Ghosh. 1971. Cytotaxonomy of the family Bromeliaceae. Cytologia 36: 237–247.

Smith, L. B. 1934. Geographical evidence on the lines of evolution in the Bromeliaceae. Bot. Jahrb. Syst. 66: 446–468.

Smith, L. B., & R. J. Downs. 1974, 1977. Bromeliaceae. Part 1. Pitcairnioideae. Part 2. Tillandsioideae. Fl. Neotropica Monogr. 14, 1974. Monogr. 14, part 2, 1977.

Smith, L. B., & C. E. Wood. 1975. The genera of Bromeliaceae in the southeastern United States. J. Arnold Arbor. 56: 375–397.

Williams, C. A. 1978. The systematic implications of the complexity of the leaf flavonoids in the Bromeliaceae. Phytochemistry 17: 729–734.

1. Family BROMELIACEAE A. L. de Jussieu 1789 nom. conserv., the Bromeliad Family

Mostly short-stemmed epiphytic herbs, but sometimes terrestrial xerophytes, in *Puya* with a stout, erect, simple, rather woody stem; plants without alkaloids, but usually or always accumulating papain-like proteolytic enzymes in one or another organ or tissue, at least sometimes producing steroid saponins, commonly with mucilage-canals and raphide-sacs in all organs, and characteristically with small, round silica-bodies more or less embedded in the thick inner periclinal walls of the epidermal cells (one per cell), commonly with refractive, often yellow or yellow-green tannin-granules in many of the parenchyma-cells of the shoot, these at least sometimes containing proanthocyanins; vessels restricted to the roots, or present in the roots and stems but not in the leaves and inflorescence-axes, or well developed in all vegetative organs,

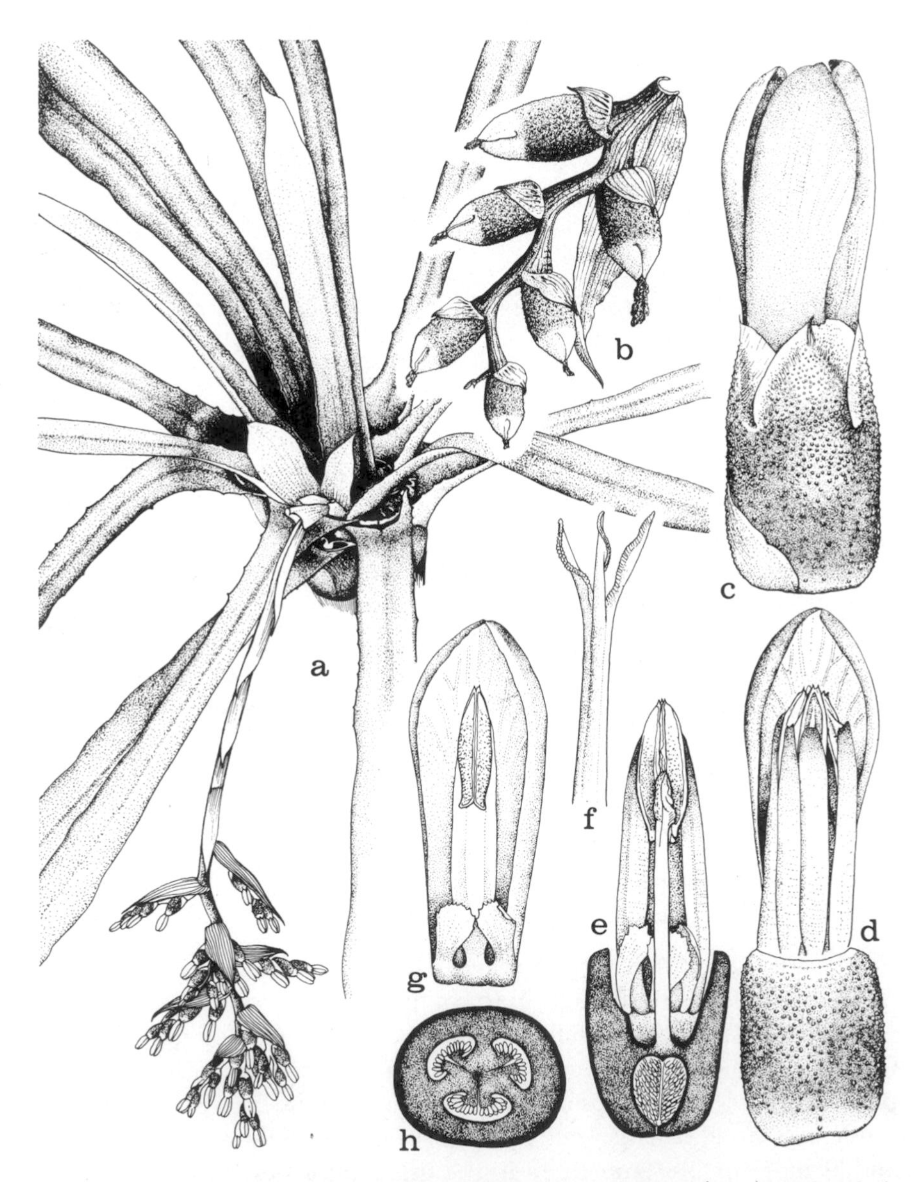

FIG. 10.1 Bromeliaceae. *Aechmea weilbachii* Didrichsen. a, habit, $\times\frac{1}{4}$; b, maturing fruit, $\times 1$; c, flower, $\times 3$; d, flower, with sepals and 2 petals removed, $\times 3$; e, flower in long-section, $\times 3$; f, style and stigmas, $\times 6$; g, single petal, with attached stamen and basal scales, $\times 3$; h, schematic cross-section of ovary, $\times 3$.

generally with scalariform perforation-plates, or those in the roots sometimes with simple perforations; vascular bundles closed, those of the stem scattered within an endodermis-like sclerenchymatous sheath, but more crowded toward the periphery than in the middle. LEAVES alternate, most commonly rosulate on the short stem, but sometimes dispersed on a more elongate stem, narrow, parallel-veined, simple, entire or more often spinose-serrate, often very firm and strongly xeromorphic, in most genera provided with stalked, peltate, water-absorbing scales at least when young, usually suberect or arcuate-spreading and concave on the upper surface, so that rainwater is channeled into a reservoir formed collectively by the broadly sheathing, often red or otherwise brightly colored bases, the reservoir often serving as a habitat for various insects or even for species of *Utricularia* that are restricted thereto; cuticle typically very thick; stomates tetracytic, or sometimes with 6 subsidiary cells; a prominent layer of water-storage tissue often present between the epidermis and the interior, chlorenchymatous portion of the leaf. FLOWERS perfect or sometimes functionally unisexual, regular or slightly irregular, trimerous, hypogynous to more often partly or wholly epigynous, borne in simple or compound spikes or racemes or heads that are usually provided with conspicuous, brightly colored bracts, or seldom solitary, adapted to pollination by insects or sometimes birds or bats, or seldom (*Navia*) by wind, or even cleistogamous; sepals and petals each 3, unlike, the sepals green and herbaceous to frequently more or less petaloid in texture, distinct or connate below; petals distinct or shortly connate below, often brightly colored, commonly provided along the basal margins with a pair of scale-like appendages that sometimes function as nectaries, and the ovary regularly with septal nectaries; stamens 6, in 2 sets of 3, all free and distinct, or often connate or collectively or individually adnate to the separate tepals; anthers tetrasporangiate and dithecal, opening by longitudinal slits; pollen-grains binucleate, monosulcate or disulcate or biporate, or with several pores; gynoecium of 3 carpels united to form a compound, trilocular, superior or very often partly or wholly inferior ovary with a terminal, often trifid style; stigmas papillate, wet or seldom dry; ovules few to usually more or less numerous on axile placentas, anatropous or seldom campylotropous, bitegmic, crassinucellar; endosperm-development helobial. FRUIT a berry or less often a usually septicidal capsule, or seldom multiple and fleshy (as in *Ananas*); seeds in capsular fruits usually winged or plumose; embryo small to fairly large, monocotyledonous, usually peripheral, seldom axile, situated at the base of the copious, mealy endosperm, which has starch in compound grains. X = 8–28, very often 25. (Tillandsiaceae)

The family Bromeliaceae consists of about 45 genera and 2000 species, wholly American except for a single species of *Pitcairnia* (*P. feliciana*) of West Tropical Africa). More than two-thirds of the species belong to only 6 genera, *Tillandsia* (400), *Pitcairnia* (250),*Vriesia* (200), *Aechmea* (150), *Puya* (140), and *Guzmania* (120). Most of the species are tropical, but *Tillandsia usneoides* L., the Spanish moss, extends north on the Atlantic coast of North America to Maryland, and few other bromeliads occur as far south as central Argentina. Spanish moss is unusual in the family in its elongate, slender, lax, branching stems. Many members of the Bromeliaceae are grown indoors for ornament, especially by a devoted coterie of bromeliad-fanciers. The cultivated pineapple is *Ananas comosus* (L.) Merrill.

The terrestrial, xerophytic bromeliads such as *Puya* are regarded as the most archaic members of the family. The xerophytic habit may be regarded as a pre-adaptation for growth as an epiphyte, as also in the Cactaceae. Smith considers *Puya* to be phyletically basal within the family. He also sees an eventual connection of the Bromeliaceae to the Rapateaceae, with *Navia* (Pitcairnioideae) as the most nearly connecting genus.

It is customary to recognize 3 subfamilies, but some authors recognize *Navia* as an additional subfamily. The Bromelioideae have about as many species as the other subfamilies collectively.

1 Ovary superior or seldom half-inferior; fruit a capsule.
  2 Plants mostly terrestrial; seeds mostly winged or with other sorts of appendages, but not with a plumose crown (unappendaged in *Navia*); leaves entire or often spiny-toothed .... PITCAIRNIOIDEAE.
  2 Plants mostly epiphytic; seeds with a plumose crown of hairs; leaves entire .... TILLANDSIOIDEAE.
1 Ovary inferior, or rarely only half-inferior; fruit a berry, or seldom multiple and fleshy; seeds without wings or other appendages; leaves mostly spiny-toothed; epiphytes or less often terrestrial plants. .... BROMELIOIDEAE.

2. Order ZINGIBERALES Nakai 1930

Perennial herbs from sympodial, very short to elongate, often tuberous-thickened and starchy rhizomes (the starch commonly in large, compound grains), or small trees with a simple, unbranched trunk lacking secondary growth; plants commonly producing calcium oxalate, sometimes in the form of raphides, and generally tanniferous, with proanthocyanins; ethereal oil cells present in the parenchymatous tissues of the largest family (Zingiberaceae); plastids of the sieve-tubes with cuneate proteinaceous inclusions; vessels with simple or scalariform perforations, confined to the roots, or sometimes present also in the shoot; vascular bundles closed, those of the stem scattered, but the ones toward the periphery sometimes more numerous and more crowded than the internal ones; silica-cells of one or another sort present in some of the tissues, often next to the veins of the leaves. LEAVES distichous or spirally arranged, small to often large or very large, with an open or closed sheath, usually a distinct petiole, and an expanded, simple, entire blade that is rolled up from one side to the other in bud; blade with a prominent midrib and numerous primary lateral veins in a characteristic pinnate-parallel arrangement, the lateral veins often curved distally and joining the marginal vein, the blade usually easily frayed between the primary lateral veins, even though slender cross-veins are commonly present; stomates paracytic or more often surrounded by 4 or more subsidiary cells, these sometimes only weakly differentiated from the other epidermal cells; petiole usually with prominent longitudinal air-cavities; leaf-sheaths typically firm, elongate, and overlapping, concealing the relatively slender stem. Inflorescence terminating the main stem, or sometimes on a separate short shoot from the rhizome, typically of short, few-flowered cymes axillary to (1–) several or many prominent, racemosely arranged, firm, folded or boat-shaped or spathe-like bracts, but sometimes with only one flower in the axil of each bract, or the bracts wanting; FLOWERS mostly short-lived, perfect or sometimes unisexual, epigynous, more or less strongly irregular, basically trimerous but with a more or less modified symmetry (or asymmetric); perianth heterochlamydeous; sepals 3, distinct or connate below to form a tube, rarely connate into an open, spathe-like structure, green or sometimes colored or even petaloid, but in any case unlike the petals; corolla basically of 3 petals, these usually dissimilar, distinct or more often connate inter se or some of them joined to the sepals; functional stamens 1 or 5, rarely 6, but one to several staminodes of various (often petaloid) form often present when there is only one functional stamen; pollen-grains monosulcate to multiporate, or often inaperturate, often with thickened intine and much-reduced exine; gynoecium typically of 3 carpels united to form a compound, inferior ovary with axile placentation, but sometimes the ovary unilocular, with parietal placentation, or one or two of the carpels more or less reduced or even obsolete, and the placentation sometimes virtually basal; stigmas wet, in most families papillate; septal nectaries commonly present, opening to the summit of the ovary, the orifice sometimes elaborated into a more or less prominent scale; ovules solitary to more often numerous in each locule, anatropous to sometimes campylotropous, bitegmic, crassinucellar; endosperm-development nuclear or helobial. FRUIT most commonly a capsule, but sometimes a schizocarp, or fleshy and indehiscent; seeds often operculate next to the radicle, and often arillate; embryo straight or curved, monocotyledonous, surrounded by a more or less well developed endosperm and a copious perisperm, these starchy, with large, often compound starch-grains, often mealy in texture, but sometimes very hard. X perhaps originally = 11. (Scitamineae, Musales)

The order Zingiberales is well characterized and sharply defined; its limits have occasioned no controversy. The order is here considered (following Tomlinson) to consist of 8 families, with a total of about 1800 species. The largest family is the Zingiberaceae, with about 1000 species, followed by the Marantaceae, with about 400. The Strelitziaceae and Heliconiaceae have often been included in the Musaceae, and the Costaceae in the Zingiberaceae, but these segregate families are as distinctive as the families that have traditionally been recognized. One could wish that the families in all orders were as well marked and sharply defined as those of the Zingiberales.

The Zingiberales are almost exclusively tropical. Most members of the family occur in moist forests or in wet open places, but some are adapted to seasonally dry habitats. The broad but fragile leaf-blade of the group may reflect adaptation to a forest habitat, with low light-intensity and but little wind.

Leaves that represent the Zingiberales date from the Maestrichtian; some of these Maestrichtian leaves have characters of the Zingiberaceae. Fruits considered to be clearly of the Zingiberaceae date from the Eocene, and fruits believed to represent the Musaceae also date from the Eocene.

SELECTED REFERENCES

Berger, F. 1958. Zur Samenanatomie der Zingiberazeen-Gattungen *Elettaria, Amomum* and *Aframomum*. Sci. Pharmaceutica 26: 224–258.

Boehm, K. 1931. Embryologische Untersuchungen an Zingiberaceen. Planta 14: 411–440.

Hickey, L. J., & R. K. Peterson. 1978. *Zingiberopsis*, a fossil genus of the ginger family from Late Cretaceous to Early Eocene sediments of western interior North America. Canad. J. Bot. 56: 1136–1152.

Holttum, R. E. 1950. The Zingiberaceae of the Malay Peninsula. Gard. Bull. Singapore 13: 1–249.

Lane, I. E. 1955. Genera and generic relationships in Musaceae. Mitt. Bot. Staatssamml. Munchen 2 (Heft 13): 114–131.

Maas, P. J. M. 1972. Costoideae (Zingiberaceae). Fl. Neotropica Monogr. 8.

Mahanty, H. K. 1970. A cytological study of the Zingiberales with special reference to their taxonomy. Cytologia 35: 13–49.

Mauritzon, J. 1936. Samenbau und Embryologie einiger Scitamineen. Acta Univ. Lund. 31(9): 1–31.

Nakai, T. 1941. Notulae ad plantas Asiae orientalis (XVI). J. Jap. Bot. 17: 189–210.

Pai, R. M. 1966. The floral anatomy of *Kaempferia rosea* Schweinf. ex Benth. with special reference to the glands in the Zingiberaceae. Proc. Indian Acad. Sci. 64B: 83–90.

Panchaksharappa, M. G. 1962. Embryological studies in the family Zingiberaceae I. *Costus speciosus* Smith. Phytomorphology 12: 418–430.

Punt, W. 1968. Pollen morphology of the American species of the subfamily Costoideae (Zingiberaceae). Rev. Palaeobot. Palynol. 7: 31–43.

Raghavan, T. S., & K. R. Venkatasubban. 1943. Cytological studies in the family Zingiberaceae with special reference to chromosome number and cytotaxonomy. Proc. Indian Acad. Sci. B. 17:118–132.

Rao, V. S. 1963. The epigynous glands of Zingiberaceae. New Phytologist 62: 342–349.

Rao, V. S., & others. 1954–1961. The floral anatomy of some Scitamineae—Part I. J. Indian Bot. Soc. 33: 118–147, 1954. II, III, IV. J. Univ. Bombay 28 (New Series): 82–114, 1959. 1–19, 1960. 29 (New Series): 134–150, 1961.

Saad, S. I., & R. K. Ibrahim. 1965. Palynological and biochemical studies of Scitamineae. J. Palyn. (Lucknow) 1: 62–66.

Skutch, A. F. 1932. Anatomy of the axis of the banana. Bot. Gaz. 93: 233–258.

Skvarla, J. J., & J. R. Rowley. 1970. The pollen wall of *Canna* and its similarity to the germinal apertures of other pollen. Amer. J. Bot. 57: 519–529.

Spearing, J. K. 1977. A note on closed leaf-sheaths in Zingiberaceae-Zingiberoideae. Notes Roy. Bot. Gard. Edinburgh 35: 217–220.

Stone, D. E., S. C. Sellers, & W. J. Kress. 1979 (1980). Ontogeny of exineless pollen in *Heliconia*, a banana relative. Ann. Missouri Bot. Gard. 66: 731–755.

Tilak, V. D., & R. M. Pai. 1966. Studies in the floral morphology of the Marantaceae. I. Vascular anatomy of the flower of *Schumannianthus virgatus* Rolfe, with special reference to the labellum. Canad. J. Bot. 44: 1365–1370.

Tomlinson, P. B. 1956. Studies in the systematic anatomy of the Zingiberaceae. J. Linn. Soc. Bot. 55: 547–592.

Tomlinson, P. B. 1959. An anatomical approach to the classification of the Musaceae. J. Linn. Soc., Bot. 55: 779–809.
Tomlinson, P. B. 1960. The anatomy of *Phenakospermum* (Musaceae). J. Arnold Arbor. 41: 287–297.
Tomlinson, P. B. 1961. The anatomy of *Canna*. J. Linn. Soc. Bot. 56: 467–473.
Tomlinson, P. B. 1961. Morphological and anatomical characteristics of the Marantaceae. J. Linn. Soc., Bot. 58: 55–78.
Tomlinson, P. B. 1962. Phylogeny of the Scitamineae—morphological and anatomical considerations. Evolution 16: 192–213.
Tomlinson, P. B. 1969. Vol. 3. Commelinales-Zingiberales. *In*: C. R. Metcalfe, ed. Anatomy of the monocotyledons. Clarendon Press. Oxford.
Tran Van Nam. 1975. Costaceae et Zingiberaceae: Leurs appareils ligulaires. Adansonia sér. 2, 14: 561–570.
van de Venter, H. A. 1976. Notes on the morphology of the embryo and seedling of *Strelitzia reginae* Ait. J. S. African Bot. 42: 63–69.
Williams, C. A., & J. B. Harborne. 1977. The leaf flavonoids of the Zingiberales. Biochem. Syst. Ecol. 5: 221–229.

SYNOPTICAL ARRANGEMENT OF THE FAMILIES OF ZINGIBERALES

1 Functional stamens 5 or seldom 6, each with 2 pollen-sacs; plants with raphide-sacs; guard-cells symmetrical except in Lowiaceae.

2 Ovary not prolonged into a hypanthium-like neck; inflorescence with 1-many folded or boat-shaped or spathe-like main bracts, each subtending or enfolding a compact, few-flowered, monochasial cyme; flowers nectariferous, sweet-smelling, adapted to pollination by birds, bats, or insects; some of the tepals often more or less connate.

3 Flowers perfect; leaves distichous; plants without laticifers; fruit capsular or schizocarpic.

4 Ovules numerous in each locule; fruit capsular; seeds arillate; median sepal anterior (abaxial) .... 1. STRELITZIACEAE.

4 Ovules solitary in each locule; fruit schizocarpic; seeds not arillate; median sepal posterior (adaxial) .... 2. HELICONIACEAE.

3 Flowers functionally unisexual; leaves spirally arranged; plants with laticifers; fruit fleshy, indehiscent; seeds not arillate ....3. MUSACEAE.

2 Ovary conspicuously prolonged into a slender, hypanthium-like neck; inflorescence of irregularly branched axillary cymes, these without specialized subtending bracts; flowers malodorous, presumably adapted to pollination by flies; tepals all distinct .... 4. LOWIACEAE.

1 Functional stamen 1, with 1 or 2 pollen-sacs; plants without raphide-sacs; guard-cells asymmetrical except in most Cannaceae.

5 Stamen with 2 pollen-sacs, often not strongly petaloid; flowers

bilaterally symmetrical; endosperm-development helobial; sepals connate below.

6 Leaves distichous; sheaths mostly open; plants aromatic, with abundant ethereal oil cells; labellum formed from 2 connate staminodes of the inner staminal cycle; two stamens of the outer cycle often developed as small or petaloid staminodes flanking the fertile stamen or adnate to the labellum .... 5. ZINGIBERACEAE.

6 Leaves spirally arranged; sheaths initially closed; plants not aromatic, without ethereal oil cells; labellum formed from 5 connate staminodes (2 from the inner cycle, all 3 of the outer cycle) .... 6. COSTACEAE.

5 Stamen with a single pollen-sac, the blade strongly petaloid; flowers asymmetrical; endosperm-development nuclear; sepals distinct.

7 Ovules more or less numerous in each of the 3 locules of the ovary; stem with mucilage-canals; leaves spirally arranged; embryo straight; seeds not arillate; flowers not paired .... 7. CANNACEAE.

7 Ovule solitary in the single locule or in each of the 3 locules of the ovary; stem without mucilage-canals; leaves more or less distichous; embryo usually strongly curved or plicate; seeds mostly arillate; flowers borne in mirror-image pairs .... 8. MARANTACEAE.

1. Family STRELITZIACEAE Hutchinson 1934 nom. conserv., the Bird-of-Paradise Flower Family

Large, glabrous perennial herbs from short or elongate sympodial rhizomes, or shrubs or small, banana-like trees with a simple, unbranched trunk lacking secondary growth, with raphide-sacs in all parts of the shoot, poor in flavonoids, but sometimes somewhat tanniferous (from proanthocyanins); root with an unusual stelar structure, including scattered vessels and strands of phloem through the ground tissue, each phloem strand containing a single sieve-tube; plastids of the sieve-tubes containing cuneate, proteinaceous crystalloids; vessels with scalariform or partly simple perforations, confined to the roots, or present also in the stem; vascular bundles closed, those in the stem scattered, but the ones toward the periphery crowded and each provided with a fibrous sheath; silica-cells of two types, one type thick-walled and superifical, containing a spherical silica-body, the other type adjacent to the vascular bundles on the inner side, thin-walled, and containing a druse-like silica-body. LEAVES medium-sized to very large, distichous, with a short basal sheath, long petiole, and expanded, simple, entire blade (but the blade sometimes fragmenting and appearing to be pinnatifid) that is rolled up from one side to the other in bud; blade with a prominent midrib and numerous lateral veins in a pinnate-parallel arrangement, the lateral veins extending to the margin and curved up into the marginal vein; stomates (at least in *Strelitzia*) weakly encyclocytic, the numerous subsidiary cells not sharply differentiated from the other epidermal cells; guard-cells symmetrical; petiole with numerous air-canals both above and below the principal arc of vascular bundles. Inflorescence terminal or axillary, with 1-several large, firm, folded, boat-shaped or spathe-like, distichous, racemosely arranged bracts, each subtending and more or less enveloping a compact, few-flowered, monochasial cyme; FLOWERS individually bracteate, perfect, epigynous, slightly to strongly irregular, adapted to pollination by birds or insects; tepals 6, in 2 cycles, all more or less petaloid but not all alike; sepals 3, free and distinct, or more or less adnate to the petals, the median one abaxial; petals 3, slightly to more often strongly dissimilar, the lateral petals larger than the median (adaxial) one, in *Strelitzia* the 2 lateral petals strongly asymmetrical and connivent to form a large, bilaterally symmetrical, arrow-shaped organ that enfolds the style and filaments; stamens in *Ravenala* 6, in the other genera 5 functional and one (adjacent to the adaxial petal) small and staminodial or wanting; filaments rigid and elongate; anthers linear, tetrasporangiate and

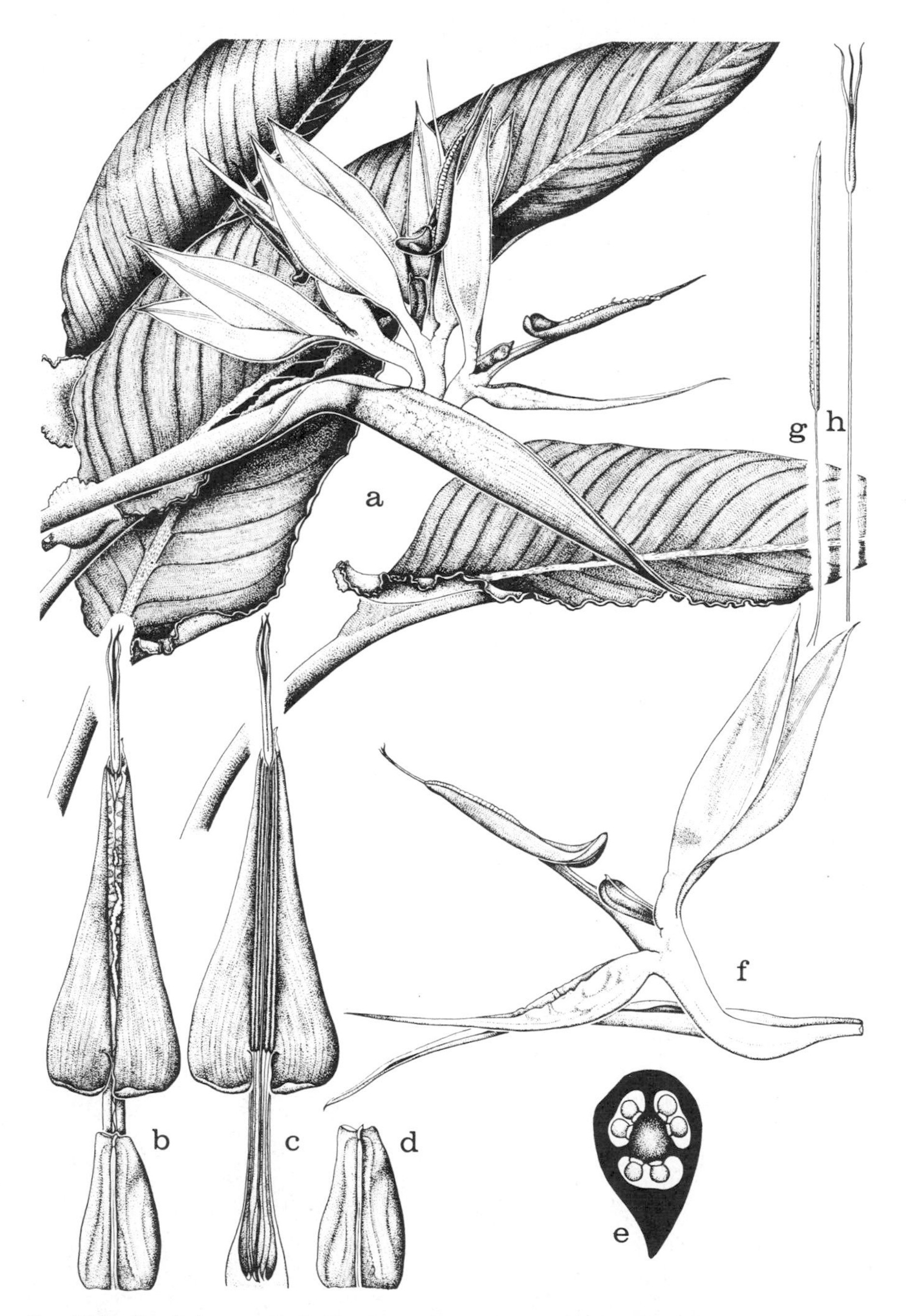

FIG. 10.2 Strelitziaceae. *Strelitzia reginae* Aiton. a, habit, $\times\frac{1}{2}$; b, portion of flower, showing 3 petals and exserted style, $\times 1$; c, portion of flower, as in b but the median petal removed and the 2 connivent petals opened out to show the stamens, $\times 1$; d, median petal, $\times 1$; e, schematic cross-section of ovary, $\times 2$; f, single flower, with its subtending bract, $\times\frac{1}{2}$; g, stamen, $\times 1$; h, style and stigmas, $\times 1$.

dithecal, opening by longitudinal slits; pollen-grains inaperturate, the intine thickened, but the exine very thin; gynoecium of 3 carpels united to form a compound, trilocular, inferior ovary with septal nectaries, a terminal style, and 3 wet, papillate stigmas; ovules numerous in each locule on the axile placentas, anatropous, bitegmic, crassinucellar; endosperm-development nuclear. FRUIT a loculicidal capsule; seeds numerous, arillate and with an operculum next to the radicle; embryo straight, subcylindric, with a massive terminal cotyledon, the plumule lateral and close to the radicle; endosperm and perisperm copious, starchy and mealy. X = 7, 8, 11.

The family Strelitziaceae consists of only 3 genera and 7 species, all tropical. *Strelitzia* (5), the bird-of-paradise flower, is native to South Africa. *Phenakospermum* (1) occurs in tropical South America, and *Ravenala* (1), the traveler's tree, is confined to Madagascar. *Strelitzia reginae* Ait. is familiar in cultivation in moist, tropical countries and under glass farther north. *Ravenala madagascariensis* Sonn., with large, banana-like leaves in a conspicuously distichous arrangement, is frequently grown as a spectacular specimen-plant in the moist tropics.

2. Family HELICONIACEAE Nakai 1941, the Heliconia Family

Large perennial herbs from sympodial rhizomes, acaulescent or with slender, unbranched aerial stems, glabrous or sometimes with uniseriate, branching hairs, relatively poor in flavonoids, though tanniferous, and with raphide-sacs in all parts; vessels confined to the roots, with elongate, scalariform perforation-plates; vascular bundles closed, those in the stem scattered, but the ones toward the periphery crowded and each provided with a fibrous sheath; silica-cells present next to the vascular bundles toward the inner side, each with irregularly thickened walls and containing a trough-shaped silica-body. LEAVES distichous, with a long basal sheath, long petiole, and expanded, simple, entire blade that is rolled up from one side to the other in bud; ligule wanting; blade with a prominent midrib and numerous lateral veins in a pinnate-parallel arrangement, the lateral veins extending to the margin and curved up into the marginal vein; stomates paracytic; guard-cells symmetrical; petiole with a single row of rather large air-canals on each side of the principal arc of vascular bundles. Inflorescence terminal, with large, colored, showy, distichous, keeled or boat-shaped, usually well spaced bracts, each bract subtending and more or less enveloping a compact few-flowered, monochasial cyme; FLOWERS perfect, epigynous, strongly irregular, adapted to pollination by insects; tepals 6, in 2 cycles, all petaloid but unlike, the median posterior (adaxial) one of the outer series distinct from the others, the remaining 5 mostly connate into a 5-toothed, boat-shaped structure; functional stamens 5, tetrasporangiate and dithecal, opening by longitudinal slits; pollen-grains more or less similar to those of *Canna*, inaperturate, with a thickened intine, the exine very thin, with scattered small spinules; distal hemisphere of the intine permeated by radially arranged channels; 6th stamen represented by a small staminode adnate to the odd tepal; gynoecium of 3 carpels united to form a compound, trilocular, inferior ovary with a slender style and capitate or 3-lobed, wet, papillate stigma; ovules solitary in each locule, basal-axile, anatropous, presumably bitegmic and crassinucellar; endosperm-development nuclear. FRUIT a schizocarp, separating into (2) 3 1-seeded, usually blue, fleshy mericarps; seed-coat with an operculum opposite the radicle, not arillate; embryo straight, monocotyledonous, poorly differentiated when the seed is first mature; endosperm and perisperm copious. X = 8–13, most commonly 12.

The family Heliconiaceae consists of the single rather large genus *Heliconia*, with a hundred or more species. *Heliconia* is native mainly to tropical and subtropical South and Central America, but one species is widespread in the islands of the southwestern Pacific. Several species of *Heliconia* are cultivated under glass in temperate regions for ornament.

One of the distinctive features of *Heliconia* is the inverted symmetry of the flower, as compared to that of the Strelitziaceae, Musaceae, Zingiberaceae, and Costaceae. The median sepal in *Heliconia* is adaxial, whereas in the other families it is abaxial.

3. Family MUSACEAE A. L. de Jussieu 1789 nom. conserv., the Banana Family

Large or very large, coarse, often tree-like, perennial herbs from a massive sympodial corm, dying back to the ground after flowering, in *Ensete* monocarpic, but in *Musa* suckering and pleiocarpic; plants glabrous throughout, sometimes producing indole alkaloids, tanniferous (sometimes with proanthocyanins) and producing 3-deoxyanthocyanins (otherwise rare among monocotyledons), but otherwise poor in flavonoids, and with raphide-sacs in all parts; roots with an unusual stelar structure, including scattered vessels and strands of phloem throughout the ground tissue, each phloem-strand containing several sieve-tubes;

FIG. 10.3 Musaceae. *Musa* X *paradisiaca* L. a, habit, × $\frac{1}{20}$; b, side view of flower, ×1; c, front view of flower, ×2; d, stamen, ×2; e, adaxial petal, ×2; f, style and stigma, ×2.

plastids of the sieve-tubes with cuneate proteinaceous inclusions; vessels mostly confined to the roots, with simple or scalariform perforations; aerial stem soft, with little mechanical tissue, supported mainly by the sheathing bases of the leaves; vascular bundles closed, scattered, but more or less crowded in a zone shortly within the epidermis, so that the stem has a thin, nearly unvasculated cortex; articulated laticifers well scattered throughout the shoot, some of them associated with the leaf-trace bundles; silica-cells present next to the vascular bundles on the inner side, each with a trough-shaped silica-body and usually with unevenly thickened walls. LEAVES large to very large, spirally arranged, with a coarse basal sheath, long petiole, and expanded, simple, entire (but easily frayed) blade that is rolled up from one side to the other in bud, the sheathing bases closely over-lapping and appressed to each other, forming a hollow false trunk (pseudostem) from which the petioles depart at the summit; blade with a prominent midrib and numerous lateral veins in a pinnate-parallel arrangement, the lateral veins extending to the margin and curved up into the marginal vein; stomates surrounded by several subsidiary cells that are only weakly differentiated from the other epidermal cells; guard-cells symmetrical; mesophyll with transverse, parenchymatous, vasculated or unvasculated septa interrupting the large air-spaces parallel to the main veins; petiole with a single row of large air-spaces beneath the principal arc of vascular bundles. Axis of the inflorescence arising from the corm, growing up through the tube formed by the leaf-sheaths, and exserted from the summit, commonly turned toward one side or drooping; exserted portion of the axis bearing a close succession of large, leathery-firm, keeled or boat-shaped, spirally arranged bracts, each of which subtends and initially more or less envelops a compact, few-flowered, bractless, monochasial cyme, the bracts becoming recurved, exposing the flowers at anthesis; FLOWERS epigynous, irregular, strongly nectariferous (from septal nectaries) and adapted to pollination by birds or bats, functionally unisexual, those subtended by the lower bracts pistillate, with reduced, nonfunctional stamens, those subtended by the upper bracts staminate, with a reduced, nonfunctional ovary; tepals 6, all petaloid but unlike, basically in 2 series, but the 3 sepals and 2 of the petals joined together into a usually 5-toothed or -lobed, initially tubular structure that soon splits along one side, the third (adaxial) petal free; stamens mostly 5, the blank opposite the free petal, or sometimes the 6th stamen represented by a small staminode, rarely the stamens 6 and all functional; filaments slender, distinct; anthers linear, tetrasporangiate and dithecal, opening by longitudinal slits; pollen-grains inaperturate, the intine thickened, but the exine very thin; gynoecium of 3 carpels united to form a compound, trilocular, inferior ovary with a slender, terminal style and 3-lobed, wet, papillate stigma; ovules borne on axile placentas, numerous in each locule, anatropous, bitegmic, crassinucellar; endosperm-development nuclear. FRUIT fleshy, with a firmer, separable exocarp; seeds few to numerous, with a thick, hard testa that has an operculum next to the radicle, not arillate, or the aril funicular embryo monocotyledonous, straight or (*Ensete*) curved; endosperm and perisperm copious, starchy and mealy, with large, excentric, compound starch-grains. X = 9, 10, 11, 16, 17.

The family Musaceae consists of 2 genera, confined to tropical and subtropical regions in the Old World. *Musa* has about 35 species, and *Ensete* about 7. Several species are cultivated for food and fiber. Manila hemp is obtained from the petioles of *Musa textilis* Née. The common cultivated banana, *Musa* × *paradisiaca* L., is a sterile triploid that produces no seeds.

4. Family LOWIACEAE Ridley 1924 nom. conserv., the Orchidantha Family

Rather small, perennial, acaulescent, wholly glabrous herbs from erect sympodial rhizomes, tanniferous, but otherwise poor in flavonoids, and bearing raphide-sacs in all parts; vessels confined to the roots, with elongate, scalariform perforation-plates and rather numerous (ca 20–40) cross-bars; vascular bundles closed, those in the rhizome scattered; silica-cells present next to the vascular bundles toward the inner side, each with unevenly thickened walls and a hat-shaped silica-body. LEAVES distichous, with an expanded, sheathing base, well developed petiole, and expanded, simple, entire blade that is rolled up from one side to the other in bud; venation pinnate-parallel, the primary lateral veins diverging from the midrib at intervals, directed forward at a rather narrow angle, and converging again toward the tip, but many of the veins not reaching the tip, either anastomosing with an adjacent vein or ending at a cross-vein; palisade-tissue wanting; stomates paracytic; guard-cells asymmetrical; petiole with numerous air-canals both above and below the principal arc of vascular bundles. Inflorescence of irregularly branched, axillary cymes, without subtending bracts; FLOWERS individually bracteate, malodorous, presumably pollinated by flies, individually bracteate, perfect, strongly irregular, somewhat orchid-like in aspect, epigynous, the ovary conspicuously prolonged above into a slender, hypanthium-like (but solid) neck; tepals 6, all petaloid, but unlike; sepals 3, narrow, distinct; petals 3, very unequal, distinct, the 2 lateral ones small, the anterior (median abaxial) one enlarged to form an elliptic or spatulate labellum; stamens 5 (the blank opposite the labellum), unilaterally disposed and adnate to the base of the petals; anthers tetrasporangiate and dithecal, opening by longitudinal slits; pollen-grains inaperturate; gynoecium of 3 carpels united to form a compound, trilocular, inferior ovary with a slender, terminal style and 3 large, laciniate, wet, nonpapillate stigmas; ovules numerous on the axile placentas, anatropous, presumably bitegmic and crassinucellar. FRUIT a capsule; seeds numerous, surrounded by a 3-lobed aril, and with an operculum opposite the radicle; embryo monocotyledonous; starch-grains in the food-reserves of the seed more or less isodiametric, with faceted surface. X = 9. (Orchidanthaceae)

The family Lowiaceae consists of the single genus *Orchidantha* (including *Lowia*), with about 6 species native to southern China, the Malay peninsula, and some of the Pacific islands. None of the species is economically important or familiar in cultivation.

5. Family ZINGIBERACEAE Lindley 1835 nom. conserv., the Ginger Family

Small to large, caulescent, aromatic perennial herbs with short or elongate, sympodial, starchy and often tuberous-thickened rhizomes (the starch-grains large and simple), glabrous or often with unicellular (rarely multicellular) hairs, producing calcium oxalate but without raphides, with abundant flavonoids, and often somewhat tanniferous (from proanthocyanins), and bearing in all parts scattered secretory cells containing ethereal oils and often other characteristic substances, these commonly including monoterpenoids, sesquiterpenoids, ketones, and/or phenyl-propanoid compounds; vessels with scalariform or less often simple perforations, mostly confined to the roots, but sometimes present also in the stem or throughout the shoot; plastids of the sieve-tubes with cuneate proteinaceous inclusions; aerial stem short, unbranched, usually divided into cortex and pith by a well developed fibrous cylinder; vascular bundles closed, those of the cortex each with a fibrous sheath, those of the pith without; silica-cells mostly restricted to epidermal cells above and below the veins of the leaves, each containing a spherical silica-body, or sometimes many of the leaf-

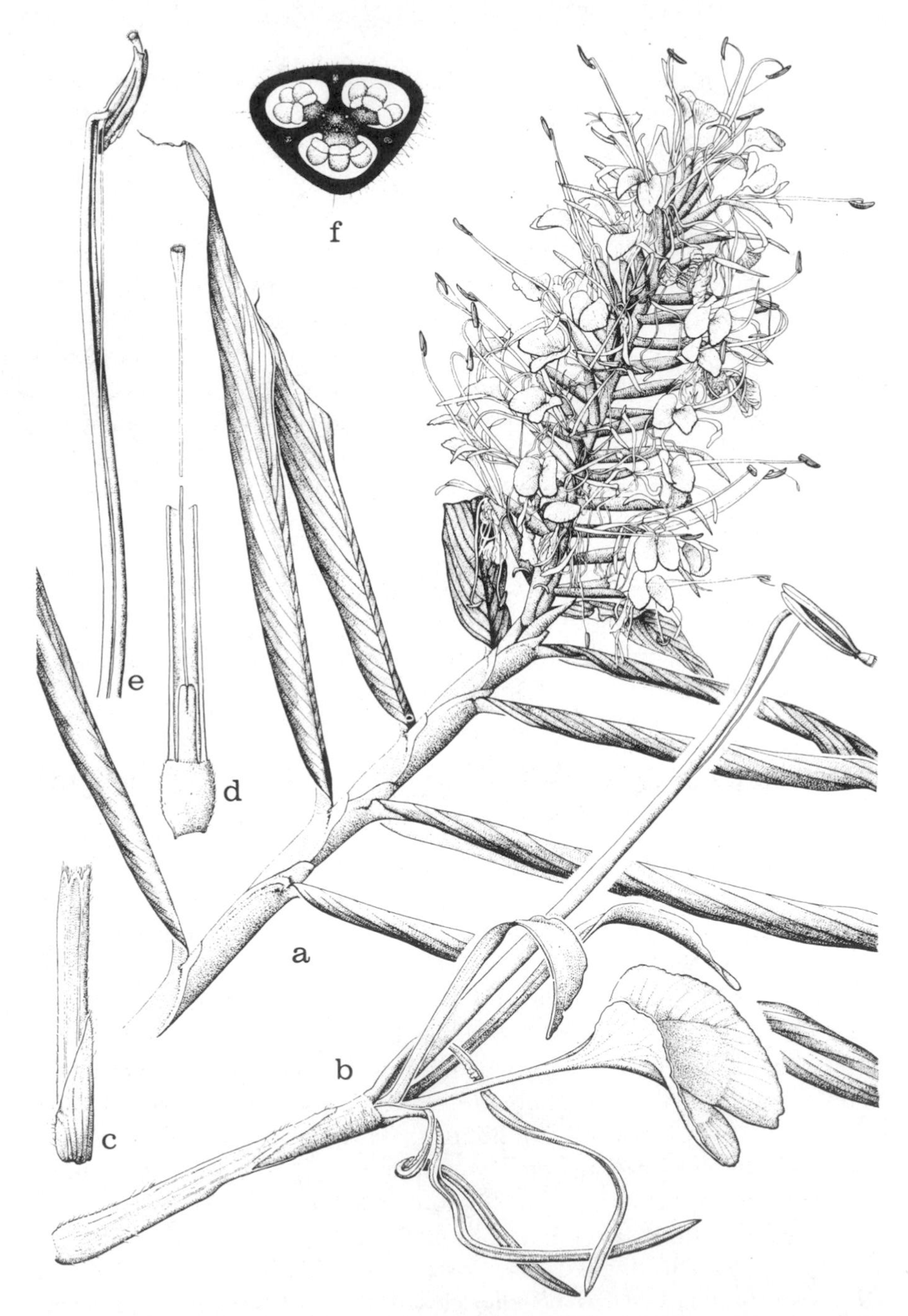

FIG. 10.4 Zingiberaceae. *Hedychium* hybrid. a, habit, $\times \frac{1}{2}$; b, flower, $\times 2$; c, bract and calyx, $\times 2$; d, section showing inferior ovary, portion of filament-tube, and style, $\times 2$; e, androecium, with partially enfolded and apically exserted style, $\times 2$; f, schematic cross-section of ovary, $\times 16$.

epidermal cells containing silica sand. LEAVES distichous, with an elongate, open (seldom closed) sheath (the sheaths connivent and forming a pseudostem), usually a long petiole (petiole sometimes short or wanting), and an expanded, simple, entire blade that is rolled up from one side to the other in bud; blade with a prominent midrib and usually numerous lateral veins in a pinnate-parallel arrangement, the lateral veins extending to the margin and curved into the marginal vein; stomates most commonly with 4 subsidiary cells, but sometimes with only 2, or more than 4; guard-cells asymmetrical; petiole with a single series of air-cavities alternating with (and more or less distinctly adaxial to) the vascular bundles of the principal arc; an adaxial ligule present at the juncture of petiole and sheath. Inflorescence terminal on the principal leafy stem, or on a short, separate, sheath-covered stem arising directly from the rhizome, its main bracts spirally arranged (or only one or even wanting); cymes axillary to the principal bracts of the inflorescence, short or sometimes elongate, or occasionally (as in *Zingiber*) only one flower present in the axil of each bract; FLOWERS perfect, epigynous, irregular but bilaterally symmetrical (zygomorphic), mostly ephemeral, fundamentally trimerous, but with a modified androecium, only one stamen polliniferous; sepals 3, green or greenish, or virtually colorless, not petaloid, united below to form a tube with short, similar or dissimilar lobes, or sometimes the calyx spathe-like and split to the base on one side; petals 3, united below to form a tube with short lobes, the posterior (median adaxial) lobe often larger than the others; androecium ancestrally of 6 stamens in 2 cycles of 3, but only the median (adaxial) stamen of the inner cycle functional, the other 2 stamens of the inner cycle sterile and connate to form a petaloid staminode called a labellum, this of highly diverse forms in different genera, commonly the most conspicuous part of the flower, but sometimes much-reduced; two stamens of the outer cycle often developed as small or petaloid staminodes flanking the fertile stamen or (*Zingiber*) adnate to the labellum, but the insertion of the lateral staminodes sometimes (as in *Hedychium*) modified so that they appear to be internal to the labellum but external to the functional stamen, the anterior (median abaxial) member of the outer staminal cycle uniformly wanting; filament of the fertile stamen variously long and slender, or short and broad, or virtually obsolete; anther tetrasporangiate and dithecal, the two pollen-sacs generally well separated on the adaxial side of an often broad and laminar connective; pollen-grains binucleate, monosulcate or inaperturate, the intine thickened, but the exine very thin; gynoecium of 3 carpels united to form a compound, inferior, trilocular or much less often unilocular ovary, the placentation accordingly axile or parietal (or essentially basal), rarely (as reported in *Scaphochlamys burkillii*) the partitions developed only at the base, so that the placentation is virtually free-central; style terminal, elongate, slender, commonly lying between the pollen-sacs of the anther; stigma wet, papillate, protruding beyond the anther, often funnelform; 2 variously developed nectaries, perhaps representing the emergent, modified tips of ancestral septal nectaries, commonly developed atop the ovary, internal to the androecium; ovules more or less numerous, anatropous, bitegmic, crassinucellar; endosperm-development helobial. FRUIT usually capsular, but sometimes baccate or dry and indehiscent; seeds with an operculum next to the radicle, usually with a prominent, commonly lobed aril; embryo straight, linear, monocotyledonous, centrally embedded in the endosperm; endosperm and especially perisperm abundant, hard or more often mealy, starchy, with large, compound starch-grains; germination hypogaeal. X = 9–18+.

The family Zingiberaceae consists of about 47 genera and a thousand species, native to tropical regions, especially in southern and southeastern Asia. The largest genus is *Alpinia*, with more than 200 species. Three tribes may be recognized. The Globbeae (4 genera, 100 spp.) have a unilocular ovary. The other two tribes are trilocular. The Zingibereae (18 genera, 300 spp.) have petaloid lateral staminodes (adnate to the labellum in *Zingiber*) and a dense inflorescence with closely overlapping main bracts. The Alpineae (25 genera, 600 spp.) have small or no lateral staminodes and a usually more open inflorescence.

Several members of the Zingiberaceae are used as spices or as a source of starch. The rhizomes of *Zingiber officinalis* Roscoe are ground up to produce ginger, and those of *Curcuma domestica* Val. provide the spice and yellow dye called turmeric, which is a common ingredient of curry powder. Cardamon is obtained from the seeds of various spp. of *Amomum* and *Elettaria*, especially *E. cardamomum* (L.) Maton. East Indian arrow-root starch is made from the rhizomes of several species of *Curcuma*.

6. Family COSTACEAE Nakai 1941, the Costus Family

Caulescent (rarely acaulescent) perennial herbs from sympodial, usually tuberous-thickened (sometimes very short) rhizomes, glabrous or often with multicellular, uniseriate (seldom unicellular) hairs, producing

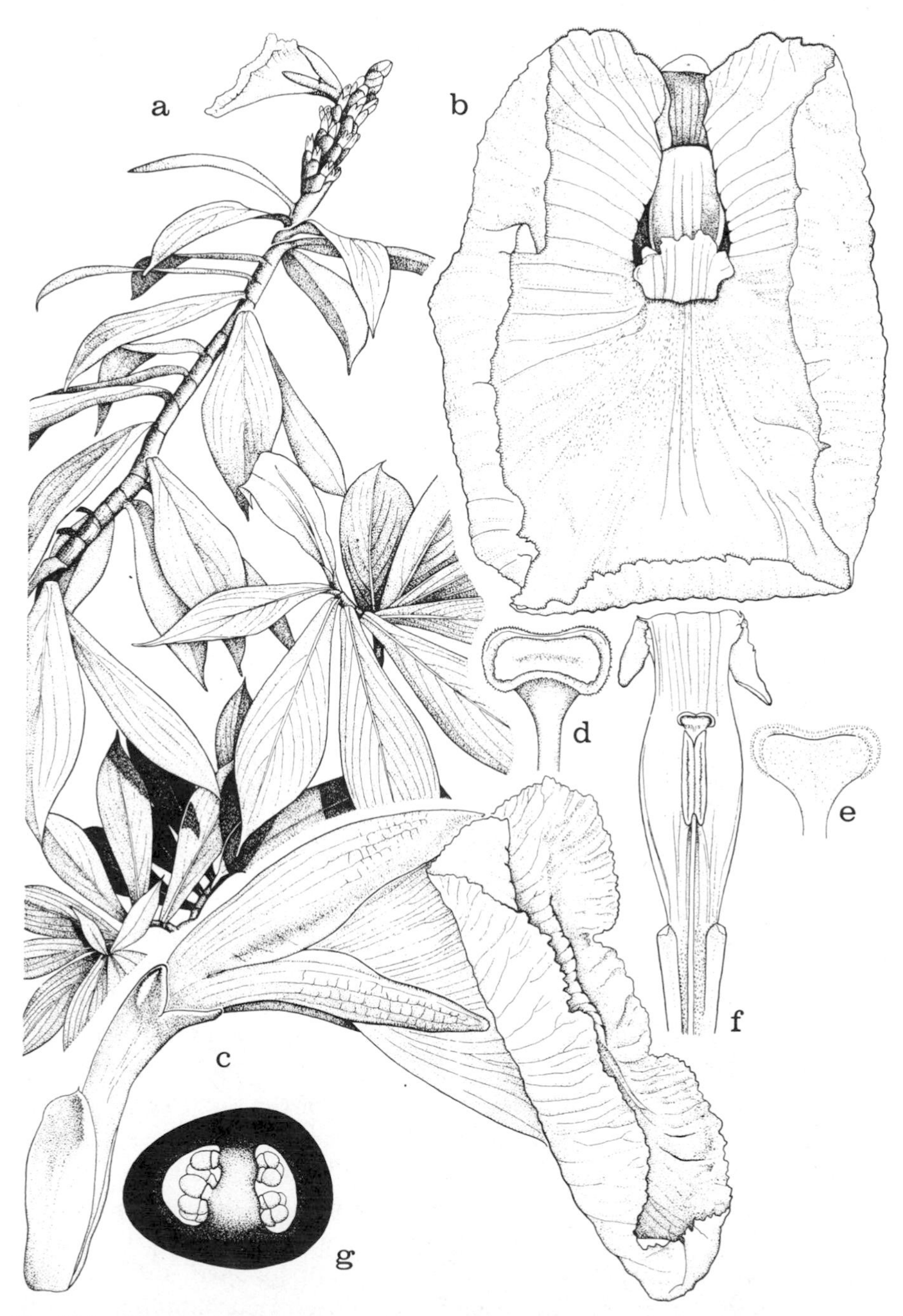

FIG. 10.5 Costaceae. *Dimerocostus strobilaceus* Kuntze. a, habit, ×½; b, front view of flower, ×1; c, side view of flower, ×1; d, e, two views of stigma, ×4; f, ventral surface of anther, the style lying between the two pollen-sacs; g, schematic cross-section of ovary, ×4.

calcium oxalate but not raphides, poor in flavonoids, and not aromatic, lacking ethereal-oil cells; aerial stems simple or sometimes branched, often spirally twisted, divided internally into a thin, nonvascular cortex and a large pith by a well developed, fluted, fibrous cylinder; vascular bundles closed, those of the aerial stem all internal to the fluted cylinder, the ones nearest to it each with a fibrous sheath, the others without; vessels with scalariform perforations, confined to the roots, or sometimes also in the stem; silica-cells thin-walled, next to the vascular bundles on the inner side, each provided with a druse-like silica-body. LEAVES spirally arranged, with a ventral ligule, a short, initially closed (but later sometimes ruptured) sheath, a very short but definite petiole, and an expanded, simple, entire blade that is rolled up from one side to the other in bud; blade with a prominent midrib and more or less numerous lateral veins in a pinnate-parallel arrangement, or seldom (as in *Monocostus*) the venation more distinctly parallel, with all the main veins arising at or near the base; stomates surrounded by several subsidiary cells; petiole without air-cavities, or with a single series of rather poorly developed air-cavities more or less alternating with and adaxial to the vascular bundles of the principal arc. Inflorescence a terminal, bracteate, dense head or spike, borne either on the principal leafy stem or less often on a separate, leafy shoot, or rarely the flowers solitary in the axils of the upper leaves (*Monocostus*); bracts with a linear, nectariferous callus just below the tip; FLOWERS pollinated by insects or bats, perfect, epigynous, irregular but bilaterally symmetrical (zygomorphic), fundamentally trimerous but with a modified androecium, only one stamen polliniferous; sepals 3, green, not petaloid, united below to form a tube with short lobes; petals 3, somewhat fleshy, unequal, connate below to form a lobed tube, the lobes convolute in aestivation, the ventral (median adaxial) lobe larger than the others and more or less hooded; androecium ancestrally of 6 stamens, in 2 cycles of 3, but only the median (adaxial) stamen of the inner cycle functional, the other 2 members of the inner cycle joined with the 3 of the outer cycle to form a large, petaloid, often more or less 3-lobed or 5-lobed labellum, the flower without additional staminodes; fertile stamen attached to the corolla-tube, expanded and petaloid, bearing the 2 pollen-sacs on its inner surface; pollen-sacs generally well separated, individually bisporangiate, opening by a longitudinal slit, often appendaged at the summit, pollen-grains large, smooth, dicolpate to more often pantoporate with 5–16 pores, the intine thickened but the exine very thin; gynoecium of 3 carpels united to form a compound, inferior, trilocular ovary with axile placentation, or sometimes one of the carpels virtually obsolete, so that the ovary is bilocular; style terminal, elongate, slender, commonly lying between the pollen-sacs of the anther, beyond which the wet, papillate stigma protrudes; septal nectaries present, with a depressed orifice atop the ovary, often 2 well developed and one reduced; ovules anatropous, bitegmic, crassinucellar; endosperm-development helobial. FRUIT crowned by the persistent calyx, capsular or rarely indehiscent; seeds with a well developed aril, and with an operculum next to the prominent radicle; seed coat formed from the outer integument only; embryo straight, monocotyledonous; endosperm and perisperm generally abundant, starchy, mealy, with large, compound starch-grains; germination epigaeal. X most commonly = 9, rarely 7 or 8.

The family Costaceae consists of 4 genera and about 150 species, pantropical in distribution, but best developed in the New World. The plants nearly always grow in wet, shady habitats, often in rain-forests. The largest genus is *Costus*, with about 130 species. *Tapeinocheilos* has about 20 species, *Dimerocostus* 2, and *Monocostus* only one.

The Costaceae have often been included as a subfamily of the Zingiberaceae, but they form a group as distinctive as the other families of the order.

7. Family CANNACEAE A. L. de Jussieu 1789 nom. conserv., the Canna Family

Glabrous, erect, perennial herbs, often coarse and robust, from sympodial, often tuberous-thickened, starchy rhizomes (these with large, compound, asymmetrical starch-grains), producing various sorts of crystals of calcium oxalate, but not raphides, with scattered tanniferous cells (containing proanthocyanins) in both the root and shoot, but otherwise poor in flavonoids; mucilage-canals present in both the rhizome and the aerial stem; vessels confined to the roots, their perforations simple or sometimes with a few cross-bars; vascular bundles closed, those in the stem scattered, but the ones toward the periphery of the aerial stem crowded and each provided with a fibrous sheath, the others sheathless; silica-cells present next to the inner side of the vascular bundles, thin-walled, each containing a druse-like silica-body, LEAVES large, spirally arranged, with a short, open sheath gradually passing into the petiole, and an expanded, simple, entire blade that is rolled up from one side to the other in bud; blade with a prominent midrib and numerous lateral veins in a pinnate-parallel arrangement;

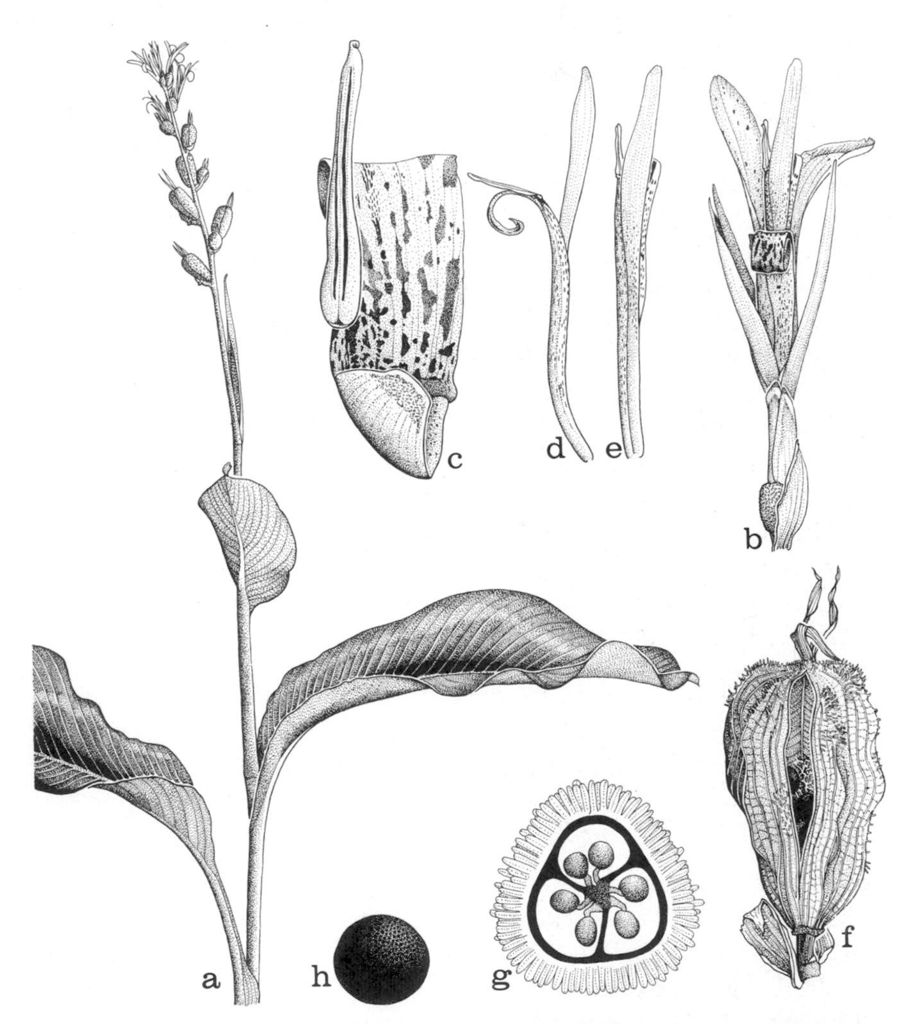

FIG. 10.6 Cannaceae. *Canna indica* L. a, habit, ×⅛; b, flower, ×1; c, stamen, ×4, d, e, two views of stamen, the lower part wrapped around the style, ×1; f, opening fruit, ×1; g, schematic cross-section of ovary, ×4; h, seed, ×2.

stomates accompanied by 2, 4, or several subsidiary cells; guard-cells mostly symmetrical; petiole lacking both a ligule and a pulvinus, containing a single row of air-canals abaxial to the principal arc of vascular bundles. Inflorescence terminal, usually openly branched, commonly with short, 2-flowered cymules axillary to the principal bracts; FLOWERS showy, perfect, epigynous, highly irregular, obliquely oriented to the axis so that no organ is clearly median, fundamentally trimerous but with a modified androecium, only one stamen bearing pollen; sepals 3, green or purplish, not petaloid, distinct, spirally arranged, similar or somewhat dissimilar, persistent in fruit; petals, stamens, and staminodes all joined into a short or elongate tube at the base; petals 3, one smaller than the other 2; functional stamen the middle member of the inner staminal cycle, nearly median-adaxial, petaloid, bearing a single bisporangiate pollen-sac along the more nearly median edge, the other pollen-sac suppressed; pollen-grains trinucleate, inaperturate but functionally holo-aperturate, the intine thickened, the exine very thin, with scattered small spinules; at least one petaloid staminode (the labellum, regarded as the outermost member of the inner staminal cycle) always present, this outcurved or outrolled; 1–4 (most commonly 2) additional petaloid staminodes of varying form and size, called wings, usually also present; gynoecium of 3 carpels united to form a compound, inferior, trilocular ovary with axile placentation; style petaloid; stigmas wet, papillate; septal nectaries present; ovules more or less numerous in each locule, anatropous, bitegmic, crassinucellar; endosperm-development nuclear. FRUIT a usually warty capsule, reputedly sometimes indehiscent; seed with an operculum next to the radicle, not arillate; embryo straight, linear, monocotyledonous, surrounded by a thin, starchy endosperm and a copious, very hard, starchy perisperm, the starch-grains simple, more or less spindle-shaped. X = 9.

The family Cannaceae consists of the single genus *Canna*, with about 50 species native to tropical and subtropical parts of the New World. The commonly cultivated garden cannas are mostly of hybrid origin, with *C. indica* L. as a principal parent. *C. edulis* Kew-Gawl., is cultivated in tropical regions for its starchy, tuberous, edible rhizomes.

The labellum of *Canna* is not homologous with the labellum of the Zingiberaceae and Costaceae.

Fossil leaves thought to represent the Cannaceae occur in Eocene deposits both in North America and in the Soviet Union.

8. Family MARANTACEAE Petersen in Engler & Prantl 1888 nom. conserv., the Prayer-Plant Family

Few-leaved perennial herbs from sympodial, mostly starchy, often tuberous-thickened rhizomes, glabrous or often with basally embedded unicellular hairs, rich in flavonoids (sometimes including proanthocyanins), commonly producing crystals of calcium oxalate but not raphides,

FIG. 10.7 Marantaceae. *Maranta leuconeura* E. Morren. a, habit, $\times\frac{1}{3}$; b, pair of flowers, $\times 5$; c, cross-section of ovary, one locule fertile, $\times 25$; d, sepal, e, petal; f, g, antero-lateral staminodes of outer cycle; h, style-holding staminode; i, labellum; k, stamen; m, style, after release.

accumulating silica in the form of stegmata and in special silica-cells; starch-grains of vegetative organs large and simple; plastids of the sieve-tubes with cuneate proteinaceous inclusions; vessels with simple or scalariform perforations, mostly restricted to the roots, but sometimes also in the stem, rarely even in the leaves; vascular bundles closed, those in the stem scattered, but the ones toward the periphery crowded (mainly in a single cycle) and each provided with a fibrous sheath, the others sheathless. LEAVES more or less distinctly distichous, in most species most of them borne at or near the ground-level, with an open sheath, a distinct, sometimes winged petiole, and an expanded, simple, entire, often color-patterned blade that is rolled up from one side to the other in bud; blade with a prominent midrib and numerous spreading-ascending lateral veins in a pinnate-parallel arrangement; stomates accompanied by 2, 4, or more subsidiary cells; guard-cells asymmetrical; silica-cells mostly of 2 types, some adjacent to the veins, with unevenly thickened walls, and each containing a hat-shaped silica-body, others scattered in the mesophyll, thin-walled and spherical, each with a central, druse-like silica body; petiole with a single arc of air-canals abaxial to the principal arc of vascular bundles, and modified at or near the summit into a swollen pulvinus that controls the orientation of the blade. Inflorescence a terminal spike or panicle or head, sometimes on a separate stem arising directly from the rhizome, each main bract subtending one or more pairs of asymmetrical, mirror-image flowers (the flower-pair thus bilaterally symmetrical); FLOWERS perfect, epigynous, fundamentally trimerous but with a modified androecium (only one stamen bearing pollen) and often with a reduced, pseudo-monomerous gynoecium; sepals 3, distinct, not petaloid; petals 3, usually white, connate below into a tube, one often hood-like and larger than the other 2; functional stamen the median (posterior) member of the inner cycle, petaloid, bearing a single bisporangiate pollen-sac along one edge, the other pollen-sac suppressed; staminodes (2) 3 or 4, petaloid but rather small, all different, two of them from the inner staminal cycle, the other 1 or 2 (when present) being postero-lateral members of the outer cycle; median (anterior) member of the outer staminal cycle always suppressed; staminodes of the inner cycle both antero-lateral, one forming a hood (the labellum) over the style and stigma before anthesis, the other generally leathery or callous-thickened and often serving as a landing platform for insects; corolla, androecium, and style forming a complex mechanism for insect-pollination, the style before anthesis held straight by the hooded staminode, but developing an internal tension, and elastically escaping from the hood when the latter is touched, bending abruptly so that the stigma faces downward; pollen deposited from the stamen onto the style (below the stigma) before the flower opens; pollen-grains inaperturate, with thin exine; gynoecium of 3 carpels united to form a compound, inferior, trilocular ovary, or often 2 of the carpels reduced and empty or even obsolete; style terminal, often expanded and lobed at the summit, with the wet, nonpapillate stigma in the depression between the lobes; septal nectaries well developed, opening at the summit of the ovary; ovules solitary in each locule or in the single fertile locule, nearly basal, anatropous to campylotropous, bitegmic, crassinucellar; endosperm-development nuclear. FRUIT a capsule or a berry; seeds with a basal operculum (next to the radicle) and usually with a basal aril as well; embryo monocotyledonous, linear, usually strongly curved or plicate (seldom straight), surrounded by an abundant, starchy, mealy perisperm with compound starch-grains. X = 4–14+.

The family Marantaceae consists of about 30 genera and 400 species, pantropical in distribution, but most abundant in the New World. By far the largest genus is *Calathea,* with about 150 species. West Indian arrowroot starch is obtained from the rhizomes of *Maranta arundinacea* L. Various other species of *Maranta* are grown indoors for their decorative, patterned leaves.

Fossils thought to represent the Marantaceae occur in Eocene depostis of southern England.

Of the potentially 6 stamens in the androecium of the Marantaceae, the anterior member of the outer cycle is always suppressed. One or both postero-lateral members of the outer cycle are usually developed as more or less petaloid staminodia. The functional stamen is the posterior member of the inner cycle. The two antero-lateral members of the inner cycle differ inter se: one forms the hood (labellum), the other the landing platform.

V. Subclass LILIIDAE Takhtajan 1966

Terestrial or epiphytic or seldom aquatic herbs (often geophytes from rhizomes, bulbs, or corms), often strongly mycotrophic and sometimes without chlorophyll, or sometimes shrubs or vines or even trees, the woody (and a few herbaceous) forms often with secondary growth of monocotyledonous type; plants often producing alkaloids or steroidal saponins; vessels usually confined to the roots, but in a few families and scattered genera present also in the shoot; plastids of the sieve-tubes with cuneate proteinaceous inclusions; vascular bundles of the stem closed, mostly scattered, less often in 2-several cycles. LEAVES alternate, seldom opposite or whorled (sometimes all basal), simple and entire or merely toothed to rarely somewhat dissected, narrow and parallel-veined to sometimes broader and then often more or less distinctly net-veined (but the net-veined types showing evidence of a parallel-veined ancestry), with or sometimes without a basal sheath, and with or much more often without a distinct petiole, sometimes reduced to mere scales; stomates most commonly anomocytic or paracytic, seldom tetracytic. FLOWERS hypogynous to very often epigynous, perfect or less often unisexual, regular or often highly irregular, usually with nectaries of one or another sort (often septal nectaries) and adapted to pollination by insects (or birds or bats), only seldom reduced and anemophilous or autogamous; tepals generally in 2 series of 3 and all petaloid, all alike or the two series or the members of a single series more or less differentiated inter se, but the outer series only seldom green and herbaceous, or rarely the perianth reduced and chaffy; stamens mostly 1, 3, or 6, seldom 2, 4, or more than 6; pollen-grains binucleate or seldom trinucleate, mostly monosulcate or inaperturate or with scarcely developed exine, only seldom with more than one aperture; gynoecium usually of 3 carpels united to form a compound, superior or very often inferior ovary with typically axile or parietal placentation, rarely the ovary pseudomonomerous or with 3 only basally united or virtually distinct carpels; ovules (1-) several or many in each locule, mostly bitegmic, crassinucellar or tenuinucellar; endosperm-development variously helobial or nuclear or cellular, often not proceeding to completion. FRUIT most commonly capsular, but occasionally of other types; endosperm often wanting, when present typically very hard, with reserves of hemicellulose (in the thickened cell walls), protein, and oil, less commonly starchy, but the starch-grains then often simple rather than compound, and the endosperm fleshy rather than mealy; embryo with a single cotyledon, or not differentiated into parts, often very small.

The subclass Liliidae as here narrowly defined consists of 2 orders, 19 families, and about 25,000 species. More than four-fifths of the species belong to only two families, the Liliaceae and Orchidaceae.

The Liliidae characteristically have showy flowers, with the tepals all petaloid, and they have intensively exploited insect-pollination. With few exceptions, they have not taken the path of floral reduction, and none of them has a spadix. Although some few Liliidae are arborescent, some have broad, net-veined leaves, and some have vessels throughout the shoot as well as the root, none of them has coordinated these features into a working system that could present a competitive challenge to woody dicotyledons.

SYNOPTICAL ARRANGEMENT OF THE ORDERS OF LILIIDAE

1 Plants not obviously mycotrophic; seeds of ordinary number and structure, usually with well developed embryo and endosperm; most families and genera with septal nectaries, sometimes with other kinds of nectaries in addition or instead; ovary superior or inferior .... 1. LILIALES.

1 Plants strongly mycotrophic, sometimes without chlorophyll; seeds very numerous and tiny, with minute, usually undifferentiated embryo and very little or no endosperm; nectaries diverse, but only seldom septal; ovary inferior .... 2. ORCHIDALES.

1. Order LILIALES Lindley 1833

Autotrophic (and not obviously mycotrophic) perennial (rarely annual) herbs, often geophytes from rhizomes, bulbs, or corms, less often herbaceous or woody vines, or shrubs or even trees, with or much more often without secondary growth of monocotyledonous type; vessels confined to the roots, or less often also present in the shoot; plastids of the sieve-tubes with cuneate proteinaceous inclusions; plants often with chelidonic acid and often producing steroid saponins or alkaloids (these mainly derived from phenylalanine and/or tyrosine); vascular bundles of the stem closed and mostly scattered or in several irregular cycles, seldom in only two cycles or even in only one cycle. LEAVES alternate, or seldom opposite or whorled, sometimes all basal, simple and entire or merely toothed to rarely dissected, narrow and parallel-veined to broader and sometimes more or less distinctly net-veined (but the net-veined types showing evidence of a parallel-veined ancestry), with or sometimes without a basal sheath, and with or more often without a distinct petiole; stomates most commonly anomocytic, but sometimes paracytic or tetracytic. FLOWERS in various sorts of inflorescences, perfect or less often unisexual, most commonly entomophilous, less often ornithophilous, or sometimes autogamous, seldom reduced and apparently anemophilous, mostly trimerous, seldom dimerous or tetramerous; tepals generally in 2 series and all petaloid, distinct or united below to form a perianth-tube, all alike, or the two series differentiated, or the members of a series differentiated inter se, but the outer series only seldom green and sepaloid, or seldom all the tepals small and chaffy; stamens most commonly as many as the tepals, sometimes only half as many, rarely more numerous or only one; anthers tetrasporangiate and dithecal, opening longitudinally or seldom by terminal pores or short slits; pollen-grains binucleate or seldom trinucleate, most commonly monosulcate, but sometimes inaperturate or with more than one aperture; gynoecium most commonly of 3 carpels united to form a compound, superior or inferior ovary with axile placentation, or pseudomonomerous, or the carpels united only toward the base; septal nectaries commonly present, but the flowers sometimes with other kinds of nectaries in addition or instead, or without nectaries; ovules (1-) several or numerous in each locule, anatropous to sometimes hemitropous, campylotropous or orthotropous, bitegmic and more or less distinctly crassinucellar, only rarely tenuinucellar; endosperm-development helobial or nuclear. FRUIT most commonly a capsule, less often a berry or drupe, or dry and indehiscent; seeds commonly with abundant endosperm, this most often very hard and with reserves of hemicellulose (in the thickened cell-walls), protein, and oil, less commonly starchy, but the starch-grains simple rather than compound, so that the endosperm is fleshy in texture rather than mealy; in the Cyanastraceae a starchy chalazosperm instead of endosperm; embryo well developed, usually with a terminal cotyledon and lateral plumule, less often with a lateral cotyledon and subterminal plumule.

The order Liliales as here defined consists of 15 families and nearly 8000 species, of cosmopolitan distribution. About half of the species belong to the single family Liliaceae, which is often considered to be the most "typical" family of monocotyledons. The next largest family is the Iridaceae, with about 1500 species. The Agavaceae, Aloeaceae, Dioscoreaceae, Smilacaceae, and Velloziaceae are middle-sized families, with 250 to 600 species each. The small families Cyanastraceae, Haemodoraceae, Hanguanaceae, Philydraceae, Pontederiaceae, Stemonaceae, Taccaceae, and Xanthorrhoeaceae have fewer than 250 species collectively.

I take the narrow, parallel-veined leaf to be primitive within the Liliales. Broader, more or less net-veined types have evolved repeatedly, but still show traces of the ancestral condition. Often they have several main veins that arch out from the base and converge toward the tip. It is not difficult to envisage the evolution of this type of venation by increase in the width of the blade, without any increase in the number of main veins. Concomitantly, the connecting cross-veins are elaborated to vasculate the expanded space between the main veins.

I take the primitive type of endosperm in the Liliales to be fleshy or cartilaginous, with food-reserves of protein, oil, and possibly some starch, but not hemicellulose. This type is readily compatible with the endosperm of archaic dicotyledons. Both strongly starchy endosperm and very hard endosperm with reserves of hemicellulose in the thickened cell-walls appear to be advanced features.

Although the Dioscoreaceae and Taccaceae do not seem to be primitive in other respects, they have a more primitive embryo than the other families of the order. The embryo is slipper-shaped or obliquely ovate, the cotyledon is evidently lateral, and the plumule is more or less distinctly terminal. In the other families the embryo is mostly barrel-shaped or ellipsoid to ovoid or cylindric, with a terminal cotyledon and a tiny, lateral, often scarcely distinguishable plumule that may be sunken into a small pocket.

Except possibly for the Hanguanaceae, all the families here referred to the Liliales evidently belong together, but I have not been able to resolve the evidence on their interrelationships into a coherent phylogenetic scheme. The Haemodoraceae, Pontederiaceae, and Philydraceae have relatively primitive endosperm, but each of these three families is specialized in its own way and cannot be regarded as basal to the order. The Liliaceae are central to most or all of the rest of the group. The Iridaceae, Aloeaceae, Agavaceae, and Velloziaceae appear to be derived directly from the Liliaceae. Although the last three of these families are habitally somewhat similar, each probably originated independently of the others. The Xanthorrhoeaceae may have originated from the Agavaceae, or perhaps more directly from the Liliaceae. The Cyanastraceae are linked to the Liliaceae through such genera as *Tecophilaea* and *Cyanella,* which have similar poricidal anthers. The Smilacaceae seem to originate directly from the Liliaceae, but they also resemble the Stemonaceae and Dioscoreaceae. The Dioscoreaceae, although advanced in other respects, have a more primitive embryo than either the Liliaceae or the Smilacaceae. The embryo, the broad, net-veined leaves, and the inferior ovary link the Taccaceae to the Dioscoreaceae, but the endosperm of the Taccaceae lacks the specialized hemicellulose reserves that characterize the Dioscoreaceae and most other families of Liliales. The Hanguanaceae may be related to the Xanthorrhoeaceae, or they may be misplaced in the Liliales.

Pollen thought to represent the Liliales enters the fossil record at about the base of the Maestrichtian epoch of the Upper Cretaceous, some 70 million years before present.

SELECTED REFERENCES

Ayensu, E. S. 1966. Taxonomic status of *Trichopus*: anatomical evidence. J. Linn. Soc., Bot. 59: 425–430.

Ayensu, E. S. 1968. The anatomy of *Barbaceniopsis,* a new genus recently described in the Velloziaceae. Amer. J. Bot. 55: 399–405.

Ayensu, E. S. 1968. Comparative vegetative anatomy of the Stemonaceae (Roxburghiaceae). Bot. Gaz. 129: 160–165.

Ayensu, E. S. 1969. Aspects of the complex nodal anatomy of the Dioscoreaceae. J. Arnold Arbor. 50: 124–137.

Ayensu, E. S. 1969. Leaf-anatomy and systematics of Old World Velloziaceae. Kew Bull. 23: 315–335.

Ayensu, E. S. 1970. Analysis of the complex vascularity in stems of *Dioscorea composita*. J. Arnold Arbor. 51: 228–240.

Ayensu, E. S. 1972. Vol. 6. Dioscoreales. *In*: C. R. Metcalfe, ed. Anatomy of the monocotyledons. Clarendon Press. Oxford.

Ayensu, E. S. 1973. Biological and morphological aspects of Velloziaceae. Biotropica 5: 135–149.
Ayensu, E. S. 1974. Leaf anatomy and systematics of New World Velloziaceae. Smithsonian Contr. Bot. No. 15: 1–125.
Ayensu, E. S., & D. G. Coursey, 1972. Guinea yams. The botany, ethnobotany, use and possible future of yams in West Africa. Economic Botany 26: 301–318.
Ayensu, E. S., & J. J. Skvarla. 1974. Fine structure of Velloziaceae pollen. Bull. Torrey Bot. Club 101: 250–266.
Barnard, C. 1960. Floral histogenesis in the Monocotyledones. IV. The Liliaceae. Austral. J. Bot. 8: 213–225.
Bayer, M. B. 1972. Reinstatement of the genera *Astroloba* & *Poellnitzia* Uitew. (Liliaceae-Aloineae). Natl. Cact. Succ. J. 27: 77–79.
Berg, R. Y. 1962. Contribution to the comparative embryology of the Liliaceae: *Scoliopus, Trillium, Paris* and *Medeola*. Skr. Norske Vidensk.-Akad. Oslo, Mat.-Naturvidensk. Kl. Ny Ser., 4: 1–64.
Björnstad, I. N. 1970. Comparative embryology of Asparagoideae-Polygonateae, Liliaceae. Nytt Mag. Bot. 17: 169–207.
Blunden, G., & K. Jewers. 1973. The comparative leaf anatomy of *Agave, Beschorneria, Doryanthes* and *Furcraea* species (Agavaceae: Agaveae). J. Linn. Soc., Bot. 66: 157–179.
Brandham, P. E. 1971. The chromosomes of the Liliaceae: II. Polyploidy and karyotype variation in the Aloineae. Kew Bull. 25: 381–389.
Burkill, I. H. 1960. The organography and evolution of Dioscoreaceae, the family of the yams. J. Linn. Soc., Bot. 56: 319–412.
Buxbaum, F. 1954. Morphologie der Blüte und Frucht von Alstroemeria und Anschluss der Alstroemerioideae bei den echten Liliaceae. Oesterr. Bot. Z. 101: 337–352.
Buxbaum F. 1958. Der morphologische Typus und die systematische Stellung der Gattung *Calochortus*. Beitr. Biol. Pflanzen 34: 405–452.
Cave, M. S. 1948. Sporogenesis and embryo sac development of *Hesperocallis* and *Leucocrinum* in relation to their systematic position. Amer. J. Bot. 35: 343–349.
Cave, M. S. 1955. Sporogenesis and the female gametophyte of *Phormium tenax*. Phytomorphology 5: 247–253.
Cave, M. S. 1964. Cytological observations on some genera of the Agavaceae. Madroño 17: 163–170.
Chanda, S., & K. Ghash. 1976. Pollen morphology and its evolutionary significance in Xanthorrhoeaceae. *In*: I. K. Ferguson & J. Muller, eds., The evolutionary significance of the exine, pp. 527–559. Linn. Soc. Symp. Ser. No. 1. Academic Press. London & New York.
Cheadle, V. I. 1937. Secondary growth by means of a thickening ring in certain monocotyledons. Bot. Gaz. 98: 535–555.
Cheadle, V. I. 1963. Vessels in Iridaceae. Phytomorphology 13: 245–248.
Cheadle, V. I. 1968 (1969). Vessels in Haemodorales. Phytomorphology 18: 412–420.
Cheadle, V. I. 1969 (1970). Vessels in Amaryllidaceae and Tecophilaeaceae. Phytomorphology 19: 8–16.
Cheadle, V. I. 1970. Vessels in Pontederiaceae, Ruscaceae, Smilacaceae and Trilliaceae. *In*: N. K. B. Robson, D. F. Cutler, & M. Gregory, eds., New research in plant anatomy, pp. 45–50. J. Linn. Soc., Bot. 63, Suppl. 1. Academic Press. London & New York.

Cheadle, V. I., & H. Kosakai. 1971 (1972). Vessels in Liliaceae. Phytomorphology 21: 320–333.

Clausen, R. T. 1940. A review of the Cyanastraceae. Gentes Herb. 4: 293–304.

Daumann, E. 1965. Das Blütennektarium bei den Pontederiaceen und die systematische Stellung dieser Familie. Preslia 37: 407–412.

Drenth, E. 1972. A revision of the family Taccaceae. Blumea 20: 367–406.

Fahn, A. 1954. The anatomical structure of the Xanthorrhoeaceae Dumort. J. Linn. Soc., Bot. 55: 158–184.

Goldblatt, P. 1971. Cytological and morphological studies in the southern African Iridaceae. J. S. African Bot. 37: 317–460.

Gómez-Pompa, A., R. Villalobos-Pietrini, & A. Chimal. 1971. Studies in the Agavaceae. I. Chromosome morphology and number of seven species. Madroño 21: 208–221.

Granick, E. B. 1944. A karyosystematic study of the genus *Agave*. Amer. J. Bot. 31: 283–298.

Hamann, U. 1963. Die Embryologie von *Philydrum lanuginosum* (Monocotyledoneae—Philydraceae). Ber. Deutsch. Bot. Ges. 76: 203–208.

Hamann, U. 1966. Embryologische, morphologisch-anatomische und systematische Untersuchungen an Philydraceen. Willdenowia Beih. 4: 1–178.

Hirsch, A. M. 1977. A developmental study of the phylloclades of *Ruscus aculeatus* L. J. Linn. Soc., Bot. 74: 355–365.

Huber, H. 1969. Die Samenmerkmale und Verwandtschaftsverhältnisse der Liliifloren. Mitt. Bot. Staatssamml. München 8: 219–538.

Kapil, R. N., & K. Walia. 1965. The embryology of *Philydrum lanuginosum* Banks, ex Gaertn. and the systematic position of the Philydraceae. Beitr. Biol. Pflanzen 41: 381–404.

Kupchan, S. M., J. H. Zimmerman, & A. Afonso. 1961. The alkaloids and taxonomy of *Veratrum* and related genera. Lloydia 24: 1–26.

Lenz, L. W. 1976. The nature of the floral appendages in four species of *Dichelostemma* (Liliaceae). Aliso 4: 379–381.

Lewis, G. J. 1954. Some aspects of the morphology, phylogeny and taxonomy of the South African Iridaceae. Ann. S. African Mus. 40: 15–113.

McKelvey, S. D., & K. Sax. 1933. Taxonomic and cytological relationships of *Yucca* and *Agave*. J. Arnold Arbor. 14: 76–81.

Nagaraja Rao, C. 1955. Embryology of *Trichopus zeylanicus* Gaertn. J. Indian Bot. Soc. 34: 213–221.

Nair, P. K. K., & M. Sharma. 1965. Pollen morphology of Liliaceae. J. Palyn. (Lucknow) 1: 39–61.

Nietsch, H. 1941. Zur systematischen Stellung von *Cyanastrum*. Oesterr. Bot. Z. 90: 31–52.

Rao, T. S., & R. R. Rao. 1961. Pollen morphology of Pontederiaceae. Pollen & Spores 3: 45–46.

Rao, V. S. 1969. The vascular anatomy of the flowers of *Tacca pinnatifida*. J. Univ. Bombay 38 (New Series), Part 5B: 18–24.

Satô, D. 1935. Analysis of the karyotypes in *Yucca, Agave* and the related genera with special reference to the phylogenetic significance. Jap. J. Genet. 11: 272–278.

Satô, D. 1938. Karyotype alteration and phylogeny. IV. Karyotypes in Amaryllidaceae with special reference to the SAT-chromosome. Cytologia 9: 203–242.

Satô, D. 1942. Karyotype alteration and phylogeny in Liliaceae and allied families. Jap. J. Bot. 12: 57–161.

Savile, D. B. O. 1962. Taxonomic disposition of *Allium*. Nature 196: 792.
Schaffer, W. M., & M. V. Schaffer. 1977. The reproductive biology of the Agavaceae. I. Pollen and nectar production in four Arizona agaves. Southw. Naturalist 22: 157–167.
Schlittler, J. 1949. Die Systematische Stellung der *Petermannia* F. v. Muell. und ihrer phylogenetischen Beziehungen zu den Luzuriagoideae Engl. und den Dioscoreaceae Lindl. Vierteljahrsschr. Naturf. Ges. Zurich 94, Beih. 1: 1–28.
Schnarf, K. 1948. Der Umfang der Lilioideae im natürlichen System. Oesterr. Bot. Z. 95: 257–269.
Schulze, W. 1971. Beiträge zur Pollenmorphologie der Iridaceae und ihre Bedeutung für die Taxonomie. Feddes Repert. 82: 101–124.
Schwartz, O. 1926. Anatomische, morphologische und systematische Untersuchungen über die Pontederiaceen. Beih. Bot. Centralbl. 42(1): 263–320.
Sen, S. 1975. Cytotaxonomy of Liliales. Feddes Repert. 86: 255–305.
Shah, G. L. & B. V. Gopal. 1970. Structure and development of stomata on the vegetative and floral organs of some Amaryllidaceae. Ann. Bot. (London) II. 34: 737–749.
Singh, V. 1962. Vascular anatomy of the flower of some species of the Pontederiaceae. Proc. Indian Acad. Sci. B, 56: 339–353.
Singh, V. 1972. Floral morphology of the Amaryllidaceae. I.Subfamily Amaryllidioideae. Canad. J. Bot. 50: 1555–1565.
Smith, L. B. 1962. A synopsis of the American Velloziaceae. Contr. U. S. Natl. Herb. 35: 251–292.
Smith, L. B., & E. S. Ayensu. 1974. Classification of Old World Velloziaceae. Kew Bull. 29: 181–205.
Smith, L. B., & E. S. Ayensu. 1976. A revision of American Velloziaceae. Smithsonian Contr. Bot. 30: 1–172.
Smithson, E. 1957. The comparative anatomy of the Flagellariaceae. Kew Bull. 1956: 491–501.
Souèges, R. 1956. Embryogénie des Pontédériacées. Développement de l'embryon chez le *Pontederia cordata* L. Compt. Rend. Hebd. Séances Acad. Sci. 242: 2080–2083.
Stenar, A. H. S. 1951. Zur Embryologie von *Haemanthus Katherinae* Bak., nebst Erörterungen über das helobiale Endosperm in den Amaryllidaceae und Liliaceae. Acta Horti Berg. 16: 57–72.
Stenar, A. H. S. 1953. The embryo sac type in *Smilacina, Polygonatum* and *Theropogon* (Liliaceae). Phytomorphology 3: 326–338.
Sterling, C. 1972–1977. Comparative morphology of the carpel in the Liliaceae. J. Linn. Soc., Bot. 65: 163–171, 1971. 66: 75–82; 213–221, 1972, 1973. 67: 149–156, 1973. 68: 115–125, 1974. 70: 341–349, 1975. 74: 345–354, 1977.
Stevens, P. F. 1978. Generic limits in the Xeroteae (Liliaceae sensu lato). J. Arnold Arbor. 59: 129–155.
Suto, T. 1936. List of chromosome number and idiogram types in Liliaceae and Amaryllidaceae. Jap. J. Genet. 12: 107–112; 157–162; 221–231.
Swamy, B. G. L. 1964. Observations on the floral morphology and embryology of *Stemona tuberosa* Lour. Phytomorphology 14: 458–468.
Tomlinson, P. B. 1965. Notes on the anatomy of *Aphyllanthes* (Liliaceae) and comparison with Eriocaulaceae. J. Linn. Soc., Bot. 59: 163–173.
Tomlinson, P. B., & E. S. Ayensu. 1968. Morphology and anatomy of *Croomia pauciflora* (Stemonaceae). J. Arnold Arbor. 49: 260–275.
Tomlinson, P. B., & E. S. Ayensu. 1969. Notes on the vegetative morphology

and anatomy of Petermanniaceae (Monocotyledones). J. Linn. Soc., Bot. 62: 17–26.

Tomlinson, P. B. & J. B. Fisher. 1971. Morphological studies in *Cordyline* (Agavaceae). I. Introduction and general morphology. J. Arnold Arbor. 52: 459–478.

Traub, H. P. 1957. Classification of the Amaryllidaceae—subfamilies, tribes and genera. Plant Life 13: 76–83.

Vaikos, N. P., S. K. Markandeya, & R. M. Pai. 1978. The floral anatomy of the Liliaceae. Tribe Aloineae. Indian J. Bot. 1: 61–68.

Valen, F. van. 1978. Contribution to the knowledge of cyanogenesis in angiosperms. 3. Communication. Cyanogenesis in Liliaceae. Proc. Kon. Nederl. Akad. Wetensch. ser. C. 81: 132–140.

Whitaker, T. 1934. Chromosome constitution in certain monocotyledons. J. Arnold Arbor. 15: 135–143.

Williams, C. A. 1975. Biosystematics of the Monocotyledoneae—Flavonoid patterns in leaves of the Liliaceae. Biochem. Syst. Ecol. 3: 229–244.

Wunderlich, R. 1936. Vergleichende Untersuchungen von Pollenkörnern einiger Liliaceen und Amaryllidaceen. Oesterr. Bot. Z. 85: 30–55.

Wunderlich, R. 1950. Die Agavaceen Hutchinsons im Lichte ihrer Embryologie, ihres Gynözeum-, Staubblatt-und Blattbaues. Oesterr. Bot. Z. 97: 437–502.

Ильинская, И. А., & А. Г. Штефырца. 1971. Ископаемые Smilacaceae и Dioscoreaceae из миоцена Молдавии. Бот. Ж. 56: 175-184.

Комар, Г. А. 1978. Ариллусы и ариллусоподобные образования у некоторых Liliales. Бот. Ж. 63: 937-955.

Хохряков, А. П. 1971. Соматическая эволюция и систематика лилейных, 4-8. Вопросы интродукции растений. Главный Ботанический Сад АН СССР. (Сборник научных работ молодых учених ГБС АН СССР.)

Чупов, В. С., & Н. Г. Кутявина. 1978. Сравительное иммуноэлектрофоретическое исследование белков семян Лилейных. Бот. Ж. 63: 473-493.

Чупов, В. С., Н. Г. Кутявина, & Н. С. Морозова. 1978. Систематические связи между родами *Hosta, Yucca* (Liliaceae) и *Agave* (Amaryllidaceae). "Тезисы VI делегатского сьезда ВБО" Кишинёв, Сентябрь 1978. Изд. Наука. Ленинград 1978.

SYNOPTICAL ARRANGEMENT OF THE FAMILIES OF LILIALES

1 Food-reserves of the seed consisting mainly or wholly of starch, sometimes with lesser amounts of protein, oil, and hemicellulose; endosperm never very hard; stomates most commonly paracytic, sometimes tetracytic or anomocytic; herbs.

  2 Stamen solitary; staminodes none; tepals apparently 4 and distinct; flowers without septal nectaries; principal leaves with an ensiform or subulate, unifacial blade .... 1. PHILYDRACEAE.

  2 Stamens (including staminodes) at least 3, more often 6; tepals usually 6 (very rarely 4), distinct or often connate below to form a perianth-tube; flowers mostly with septal nectaries.

3 Seeds with endosperm, not chalazosperm; plants usually with raphides; anthers usually opening by longitudinal slits, seldom by terminal pores; fruit with (1–) several or many seeds.

4 Aquatic or semi-aquatic plants; leaves mostly with a distinct (sometimes inflated) petiole and an expanded, bifacial blade with curved-convergent main veins and evident cross-veins; inflorescence glabrous, or at least not conspicuously long-hairy; vessels confined to the roots .... 2. PONTEDERIACEAE.

4 Terrestrial geophytes; leaves mostly linear and parallel-veined, with an equitant, unifacial blade; inflorescence usually conspicuously long-hairy, seldom glabrous; vessels present in all vegetative organs ....3. HAEMODORACEAE.

3 Seeds with chalazosperm, not endosperm; plant without raphides; anthers opening by terminal pores or short slits; fruit mostly 1-seeded; leaves with an expanded, bifacial blade, curved-convergent main veins, and evident cross-veins .... 4. CYANASTRACEAE.

1 Food-reserves of the seed consisting mainly or wholly of protein, oil, and usually also hemicellulose (the latter in the thickened cell-walls), only occasionally with a significant amount of starch; endosperm typically very hard; stomates most commonly anomocytic, but sometimes paracytic or tetracytic; herbs, woody or herbaceous vines, shrubs, or even trees.

5 Leaves mostly narrow, parallel-veined, and without a distinct petiole, the blade sessile or with a basal sheath, sometimes broader and even net-veined (as in *Trillium*), but only seldom (as in *Hosta*) with a broad, net-veined blade on a distinct petiole.

6 Habit mostly lilioid, i.e, the plants geophytes with soft, annual (rarely perennial) leaves, the stem herbaceous or nearly so and usually without secondary growth; only seldom approaching the next group in habit.

7 Stamens mostly as many as the tepals (typically 6), seldom more numerous or only 3; ovary superior or less often inferior; plants very often with raphides in some of the cells .... 5. LILIACEAE.

7 Stamens 3, opposite the outer tepals; ovary inferior except in the monotypic Tasmanian genus *Isophysis*; plants commonly with prismatic crystals of calcium oxalate in some of the cells, but without raphides .... 6. IRIDACEAE.

6 Habit agavoid or yuccoid, i.e, the plants coarse, often shrubby or arborescent xerophytes, with firm or succulent, mostly perennial leaves, the stem very often with secondary growth.

8 Plants without secondary growth, and without raphides; lower

part of the stem covered by persistent adventitious roots; ovary inferior ....7. VELLOZIACEAE.

8 Plants generally with secondary growth (not in *Kniphofia*, of the Aloaceae) and with raphides; aerial stem without persistent adventitious roots; ovary superior except in many Agavaceae.

9 Perianth evidently corolloid, often showy.

10 Ovules orthotropous to sometimes hemitropous; plants commonly producing anthraquinones but not steroid saponins; vascular bundles of the leaves at least typically with a large cap of wide, thin-walled cells at the phloem pole (the colored sap, when present, borne in these cells); plants mostly of Africa, Madagascar, and nearby places .... 8. ALOEACEAE.

10 Ovules anatropous to sometimes campylotropous; plants commonly producing steroid saponins but not anthraquinones; vascular bundles of the leaves with a well developed fibrous cap at the phloem pole; plants widespread in warm, dry regions ....9. AGAVACEAE.

9 Perianth dry and chaffy, usually small, only the inner tepals seldom somewhat petaloid; plants of Australia and some of the islands of the southwestern Pacific ....10. XANTHORRHOEACEAE.

5 Leaves mostly with a well defined petiole and a broad, more or less net-veined blade.

11 Leaves chiefly or all basal; stem or scape terminating in an inflorescence; flowers without nectar.

12 Ovary superior; flowers small, unisexual, borne in a large, terminal panicle; perianth more or less chaffy, scarcely corolloid; ovules orthotropous, solitary in each locule; fruit drupaceous; stomates tetracytic 11. HANGUANACEAE.

12 Ovary inferior; flowers larger, perfect, borne in an involucrate cymose umbel; perianth more or less distinctly corolloid; ovules anatropous to campylotropous, more or less numerous; fruit a berry to seldom a capsule; stomates anomocytic .... 12. TACCACEAE.

11 Leaves chiefly or all cauline; flowers in various sorts of inflorescences, often producing nectar; plants very often climbing.

13 Flowers dimerous, with 4 tepals, 4 stamens, and 2 carpels united to form a unilocular ovary with basal or apical placentation; vessels mostly confined to the roots .... 13. STEMONACEAE.

13 Flowers trimerous, with 6 tepals, 6 stamens, and 3 carpels united to form a trilocular (seldom unilocular) ovary with axile (parietal) placentation; vessels usually present in all vegetative organs.

14 Ovary superior (inferior in the monotypic Australian genus *Petermannia*); plants climbing, herbaceous or slenderly woody vines, or less often erect herbs or branching shrubs, usually with petiolar tendrils (tendrils leaf-opposed in *Petermannia*, wanting in some other genera); rhizome often tuberous, but plants without the characteristic basal "tuber" of the next family; plants without alkaloids; septal nectaries wanting .... 14. SMILACACEAE.

14 Ovary inferior; plants twining-climbing herbaceous vines without tendrils, or sometimes erect herbs; plants commonly with a large basal "tuber" derived from the lowest internodes and/or the hypocotyl; plants often producing alkaloids; both septal and tepalar nectaries commonly present .... 15. DIOSCOREACEAE.

1. Family PHILYDRACEAE Link 1821 nom. conserv., the Philydrum Family

Erect perennial herbs from a stout rhizome or tuber or thickened stem-base, with scattered tanniferous cells and abundant proanthocyanins, but not saponiferous, sometimes with solitary styloid crystals in some of the parenchyma-cells, but without raphides; vessels with scalariform perforation-plates, confined to the roots; vascular bundles of the stem closed, scattered; no secondary growth. LEAVES parallel-veined, the basal and lower cauline ones distichous, with a well developed sheath and an ensiform or subulate, unifacial blade, the others smaller and spirally arranged; stomates paracytic or tetracytic. FLOWERS sessile in the axils of the bracts in a simple or often paniculately branched spike, perfect, hypogynous, irregular, modified from a trimerous ancestry; perianth of 4 members, the 2 outer relatively large and median (one adaxial, the other abaxial), the 2 inner much smaller and antero-lateral (considered to be petals), all petaloid; adaxial (upper) perianth-member considered to represent the adaxial petal and 2 adjacent sepals, and sometimes showing 3 terminal teeth; lower tepal considered to represent a single sepal; stamen solitary, opposite the lower sepal, its filament basally adnate to the lower sepal, the 2 adjacent petals, and the base of the ovary to form a short or sometimes more or less elongate column; anther basifixed, tetrasporangiate and dithecal, opening by longitudinal slits; the pollen-sacs sometimes helically coiled; pollen-grains binucleate, sometimes tetradinous, the exine with diverse sorts of sculpture, sometimes more or less trisulcate; gynoecium of 3 carpels united to form a compound, superior ovary that is trilocular with axile placentas, or unilocular with deeply intruded parietal placentas; no septal nectaries; style terminal, with a dry, capitate or punctate or obscurely trilobed stigma; ovules numerous, anatropous, bitegmic, barely crassinucellar; endosperm-development helobial. FRUIT usually a loculicidal (sometimes also septicidal) capsule, seldom only tardily and irregularly dehiscent; seeds more or less numerous and small; embryo straight, cylindrical, with a terminal cotyledon and lateral plumule, embedded in the fleshy endosperm, which contains food-reserves of starch (as simple grains), protein, and generally also oil. X = 8, 17.

The family Philydraceae consists of only 4 genera and 5 species: *Philydrum* (1), *Philydrella* (1), *Orthothylax* (1), and *Helmholtzia* (2). All of the genera occur in Australia, and some of them extend onto the islands of the western Pacific (as far north as southern Japan) and the mainland of southeast Asia.

2. Family PONTEDERIACEAE Kunth 1816 nom. conserv., the Water-Hyacinth Family

Glabrous, aquatic or semi-aquatic herbs, free-floating or rooted to the substrate, perennial or seldom annual, with scattered tanniferous cells containing proanthocyanins, commonly producing raphides and sometimes other sorts of calcium oxalate crystals, but only seldom cyanogenic, and not saponiferous; vessels with simple or scalariform perforations, confined to the roots, or seldom also present in the stem; vascular bundles of the stem closed, scattered; stem sympodial, the successive axes ending in inflorescences. LEAVES in a basal rosette or distributed along the stem, distichous to sometimes spirally arranged, mostly with an open or proximally closed basal sheath, a distinct (sometimes inflated) petiole, and an expanded, floating or emersed blade with parallel, curved-convergent veins, seldom (as in *Eurystemon*) the petiole hardly differentiated and the blade scarcely expanded, or the leaf linear and virtually sheathless (*Zosterella*), or even filiform (*Hydrothrix*); stomates paracytic. FLOWERS in terminal racemes or spikes or panicles, or solitary and terminal, perfect, regular or irregular, the inflorescence subtended by a generally bladeless sheath; perianth of 6 (in *Schollerupsis* 4) petaloid tepals in 2 series, these usually connate at the base to form a perianth-tube, but virtually distinct in *Monochoria*; stamens most commonly 6 in 2 cycles, seldom only 3 (with or without additional staminodes), in *Hydrothrix* only one, but then accompanied by 2 staminodes; filaments slender, distinct, adnate to the perianth-tube; anthers tetrasporangiate and dithecal, opening by longitudinal slits or rarely by terminal pores; pollen grains 2 (3)-nucleate, with 1 or 2 (3) distal or subequatorial colpi; gynoecium mostly of 3 carpels united to form a compound, superior, trilocular ovary with axile placentation, or a unilocular ovary with intruded parietal placentas, but pseudomonomerous in *Pontederia* and *Reussia*; septal nectaries present except in *Heteranthera*; style slender, with a dry, terminal stigma or 3 (6) short stigmas; ovules numerous (solitary, terminal, and pendulous in *Pontederia* and *Reussia*), anatropous, bitegmic, crassinucellar; endosperm-development helobial. FRUIT a many-seeded, loculicidal capsule, or (in *Pontederia* and *Reussia*) dry, 1-seeded, and indehiscent; seeds small, longitudinally ribbed; embryo axile, cylindric, with a terminal cotyledon and lateral plumule, surrounded by the copious, starchy, mealy endosperm with simple (not aggregated) starch-grains; endosperm with an outer aleurone layer. X = 8, 14, 15.

The family Pontederiaceae consists of about 9 genera and 30 species,

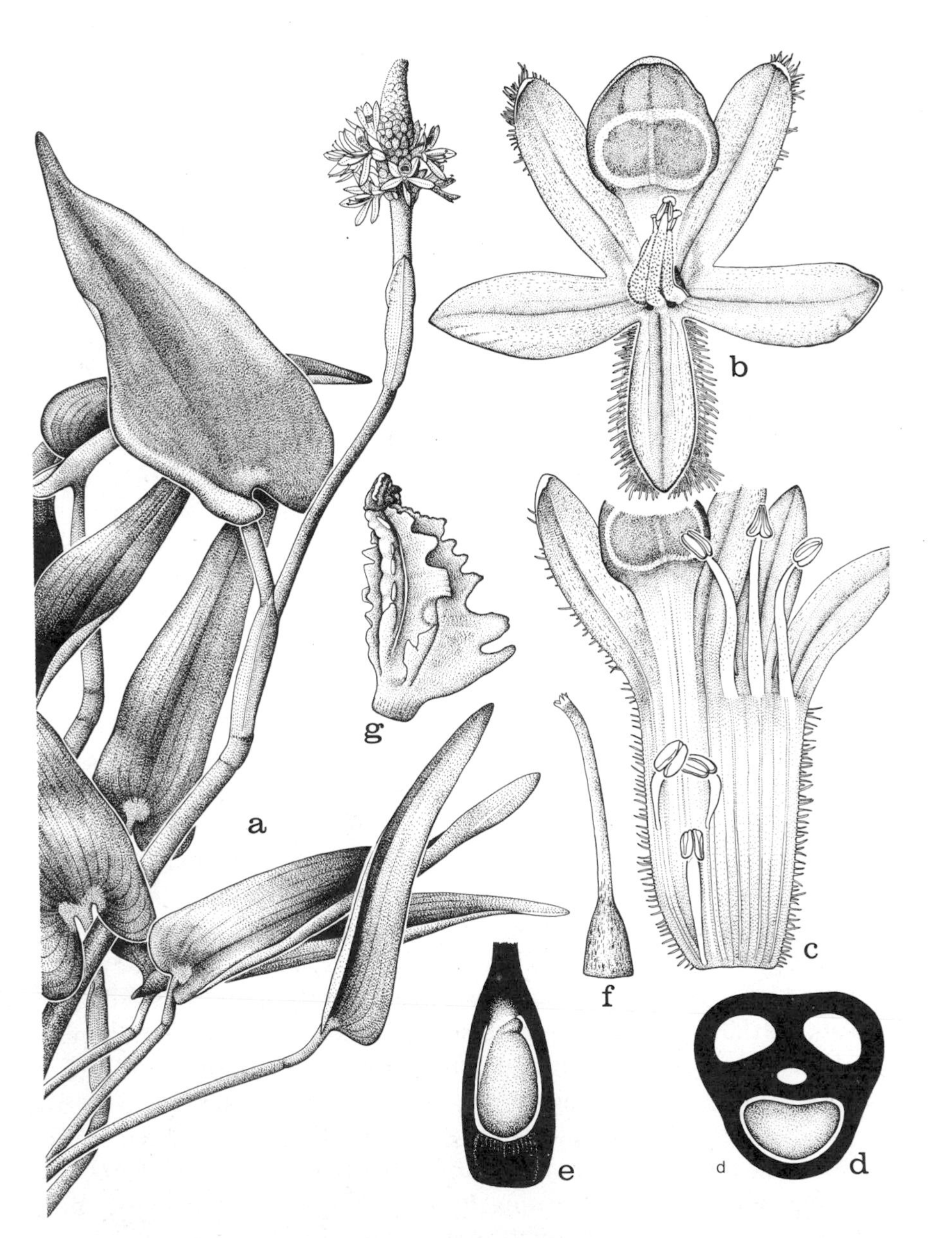

Fig. 11.1 Pontederiaceae. *Pontederia cordata* L. a, habit, $\times\frac{1}{2}$; b, flower, $\times 6$; c, corolla, the tube slit and laid open to show stamens, $\times 6$; d, schematic cross-section of ovary, showing one fertile and two empty locules, $\times 24$; e, schematic long-section of ovary, with a single pendulous ovule, $\times 12$; f, pistil, $\times 6$; g, fruit, $\times 3$.

widespread in tropical and subtropical regions, with a few species extending well into the North Temperate zone. The largest genus is *Heteranthera*, with about 8 species, followed by *Eichhornia*, with 6 or 7. The remaining genera are *Pontederia* (3–4), *Reussia* (3–4), *Monochoria* (3), Zosterella (2), *Hydrothrix* (1), *Scholleropis* (1), and *Eurystemon* (1). The water-hyacinth is *Eichhornia crassipes* (Mart.) Solms-Laub.

Eichhornia, Reussia, and *Pontederia* are unusual among monocots (and among angiosperms with irregular flowers) in having a number of heterostylic species. Heterostyly here is furthermore of an unusual, tristylic type.

3. Family HAEMODORACEAE R. Brown 1810 nom. conserv., the Bloodwort Family

Perennial geophytic herbs of lilioid habit, fibrous-rooted from a tuber or short rhizome, usually with raphides and chelidonic acid, and often with a characteristic polyphenolic red pigment (several kinds of phenalenone) in the roots and rhizomes, but seldom saponiferous, only slightly or not at all tanniferous (proanthocyanins), and not cyanogenic; vessels with simple or less often scalariform perforations, usually confined to the roots, but sometimes present also in the stem, seldom wholly wanting; vascular bundles of the stem closed, scattered except in *Lophiola*, which has 2 cycles in the aerial stem and only one cycle in the rhizome; roots typically producing root-hairs. Foliage LEAVES all radical, with sheathing base and linear, parallel-veined equitant, unifacial blade; cauline leaves small or wanting; stomates paracytic, varying to almost anomocytic (*Lophiola*). FLOWERS borne in panicles, racemes, cymes, or cymose umbels, often in a number of racemosely arranged cymes, perfect, regular or sometimes slightly irregular, trimerous, hypogynous to epigynous; inflorescence usually conspicuously long-hairy, seldom glabrous, perianth of 6 persistent, distinct or more or less connate tepals in 1 (Conostyloideae) or 2 (Haemodoroideae) cycles, the perianth-tube when present short to elongate, straight or curved; stamens 3 or 6, opposite the petals (inner perianth-members) when 3, the filaments free or adnate to the perianth-tube; anthers basifixed or versatile, tetrasporangiate and dithecal, opening by longitudinal slits; pollen-grains binucleate, monosulcate (Haemodoroideae) or with 2–8 apertures (Conostyloideae); gynoecium of 3 carpels united to form a compound, trilocular, superior to more often inferior ovary with axile placentation; septal nectaries present, but often not well developed; style slender, with a capitate stigma; ovules 1-many in each locule, anatropous to

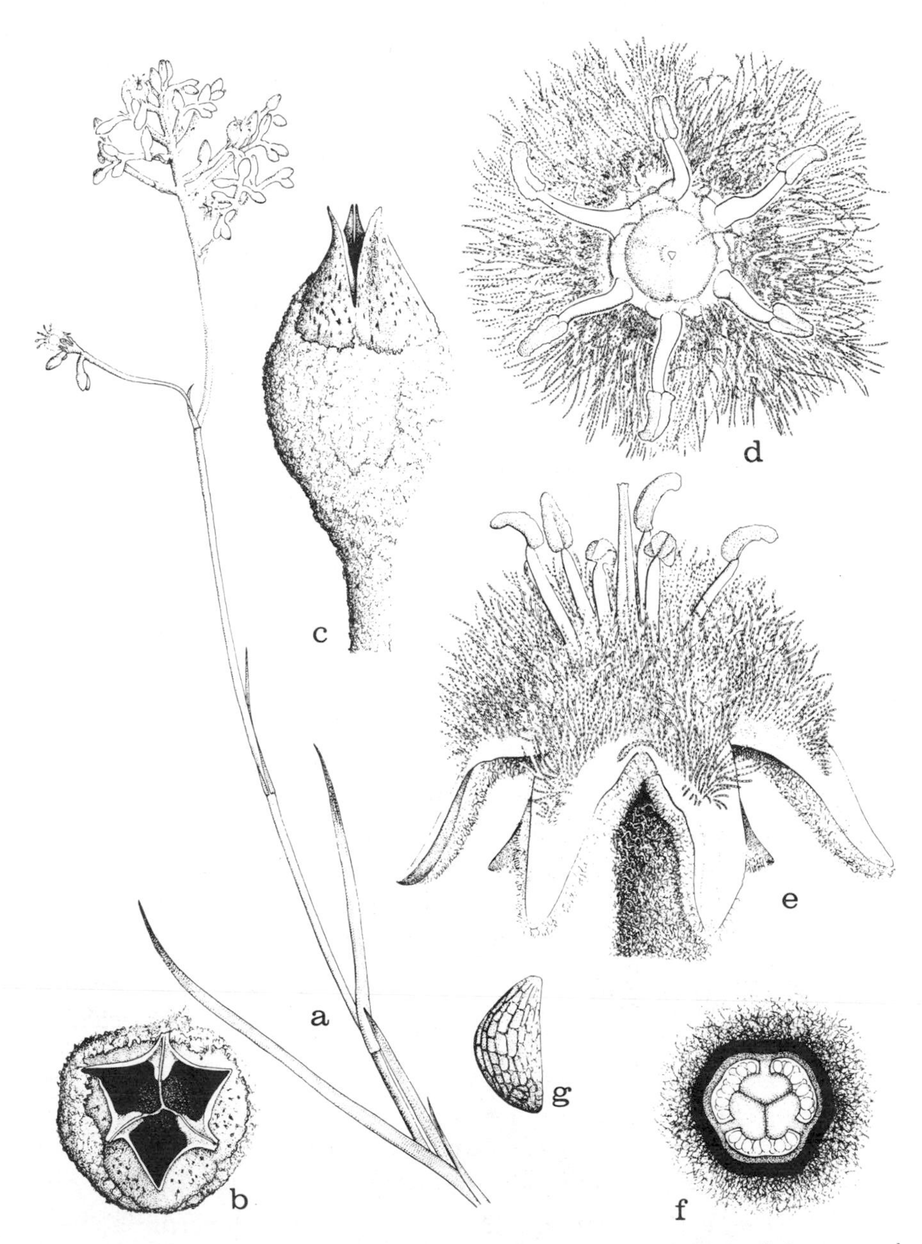

Fig. 11.2 Haemodoraceae. *Lophiola americana* (Pursh) A. Wood. a, habit $\times \frac{1}{2}$; b, c, opened fruit, $\times 8$; d, e, flower, $\times 8$; f, schematic cross-section of ovary, $\times 8$; g, seed, $\times 16$.

sometimes hemitropous or more often orthotropous, bitegmic, crassinucellar; endosperm-development helobial. FRUIT a loculicidal capsule; seeds (1–) several or many; embryo small, with a terminal cotyledon and lateral plumule, embedded in the copious endosperm; endosperm fleshy, starchy (with compound starch-grains) but also with some stored protein, oil, and hemicellulose. X = 4–8, 15+.

The family Haemodoraceae as here defined consists of some 16 genera and probably fewer than a hundred species, mostly of the Southern Hemisphere, but reaching as far north as Massachusetts in the eastern United States. The largest genus is *Conostylis*, with about 25 species, followed by *Haemodorum*, with about 20; both genera are principally Australian.

4. Family CYANASTRACEAE Engler 1900 nom. conserv., the Cyanastrum Family

Perennial herbs from a tuberous-thickened base, lacking raphides and tannins, and not cyanogenic; vessels with scalariform perforations, confined to the roots; stem with scattered, closed vascular bundles, without secondary growth. LEAVES all essentially basal, with a closed sheath, a more or less evident petiole, and a relatively broad, lance-oblong to often broadly cordate blade, the principal veins connected by evident cross-veins; stomates paracytic; blade and petiole with schizogenous channels containing an oily, refractive substance. FLOWERS borne in a raceme or panicle terminating a short to elongate scape, perfect, regular, trimerous; perianth of 6 evidently biseriate, petaloid, blue tepals that are connate into a short tube at the base; stamens 6, in 2 cycles; filaments short, basally adnate to the perianth; anthers basifixed, tetrasporangiate and dithecal, opening by terminal pores or short slits; pollen-grains binucleate, monosulcate or seldom trichotomosulcate; gynoecium of 3 carpels united to form a compound, half-inferior, trilocular and strongly lobed ovary; style filiform, deeply indented between the lobes of the ovary, with 3 short stigmas; septal nectaries present; ovules 2 in each locule, basal-axile, anatropous, bitegmic and crassinucellar, with a funicular obturator; endosperm-development variously reported as nuclear or helobial. FRUIT capsular, commonly 1-seeded; embryo with a large, terminal cotyledon and sunken, lateral plumule, filling about half the seed, the other half filled by a large chalazosperm containing starch (in compound grains) and oil; true endosperm wanting. X = 11, 12.

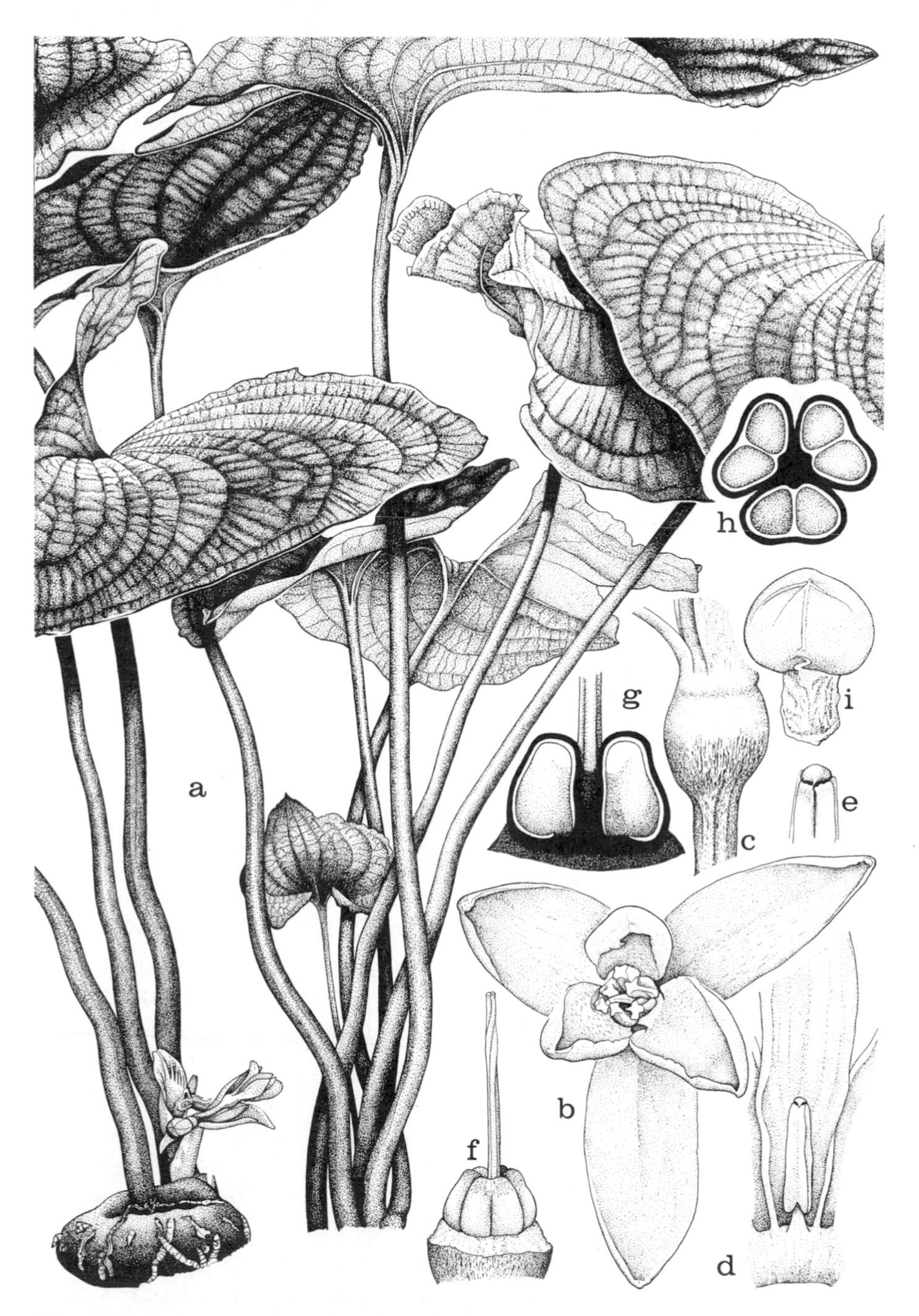

Fig. 11.3 Cyanastraceae. *Cyanastrum cordifolium* Oliver. a, habit, $\times\frac{1}{2}$; b, flower, from above, $\times 4$; c, base of flower, $\times 4$; d, stamen, the filament adnate to the base of a tepal, $\times 4$; e, apical pores of anther, $\times 8$, f, pistil, $\times 4$; g, schematic long-section of ovary, $\times 8$; h, schematic cross-section of ovary, $\times 8$; i, fruit, $\times 2$.

The family Cyanastraceae consists of the single genus *Cyanastrum*, with about 7 species native to forests of tropical Africa. It is generally thought to be related to the *Tecophilaea*-group of Liliaceae, especially *Cyanella* and *Odontostomum*, but it does not in fact look much like these genera.

5. Family LILIACEAE A. L. de Jussieu 1789 nom. conserv., the Lily Family

Perennial, mostly geophytic, often poisonous herbs (rarely wiry shrubs) from a starchy rhizome, bulb, or corm, usually dying back to the ground each year, but in some tropical genera such as *Crinum* and *Astelia* with a persistent but soft aerial stem, only rarely with some secondary growth of monocotyledonous type (as in *Veratrum*), usually producing either steroid saponins, or steroid alkaloids, or alkaloids derived from phenylalanine and/or tyrosine (but both saponins and alkaloids apparently wanting from the Alstroemerioideae), very often accumulating chelidonic acid and sometimes anthraquinones, often with raphides or other sorts of calcium oxalate crystals, often with mucilage cells or canals, and in *Allium* with articulated laticifers, seldom (as in *Aletris*) accumulating aluminum, only rarely with proanthyanins, and very rarely (*Chlorophytum*) cyanogenic, probably (unlike other monocotyledons) through a phenylalanine-based pathway; starch present in the vegetative organs of some genera, but not others; vessels with scalariform or simple perforations, usually confined to the roots, but sometimes also present in the stems; vascular bundles of the stem closed, scattered or in several irregular cycles; plastids of the sieve-tubes with cuneate proteinaceous inclusions; roots sometimes producing root-hairs, but more often not. LEAVES simple, alternate or less often opposite or whorled, often all basal, neither very firm, nor very succulent, nor (with few exceptions, such as *Crinum, Dianella* and *Xerophyllum*) persisting from year to year, usually narrow and parallel-veined, with or without a sheathing base, sometimes equitant or terete and unifacial, sometimes relatively broad and sessile (as in *Veratrum*) or petiolate (as in *Convallaria*) but still essentially parallel-veined, only seldom broad, sessile, and more or less distinctly net-veined (as in *Trillium*), or broad, petiolate, and approaching the zingiberalean type of pinnate-parallel venation (as in *Hosta*), sometimes much-reduced (as in *Asparagus*; stomates mostly anomocytic, only rarely paracytic (as in *Astelia* and *Hosta*) or tetracytic (as in *Curculigo*). FLOWERS borne in various sorts of inflorescences, often a raceme, spike,

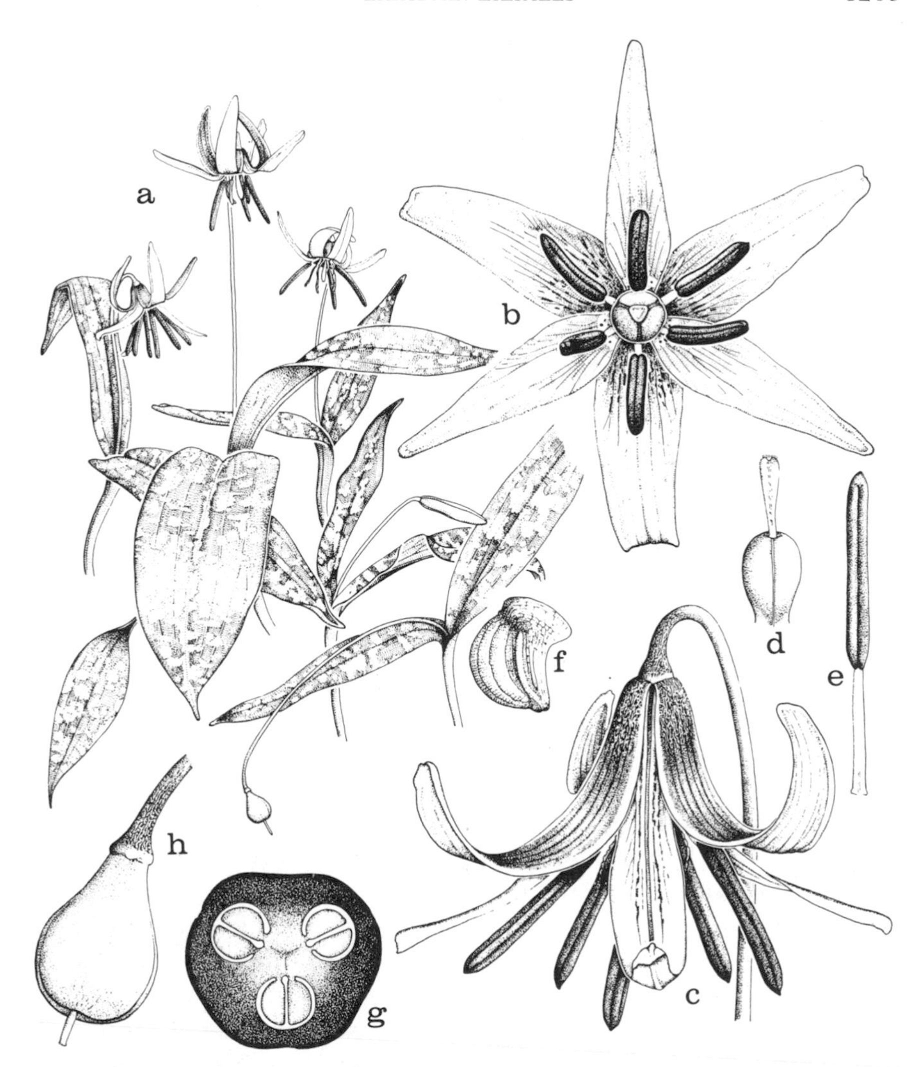

Fig. 11.4 Liliaceae. *Erythronium americanum* Ker-Gawl. a, habit, ×½; b, c, flower, ×2; d, pistil, ×2; e, stamen, ×2; f, seed, ×4; g, schematic cross-section of ovary, ×8; h, capsule, ×2.

panicle, or involucrate, cymose umbel, or sometimes solitary or paired in the axils of the leaves, mostly perfect, regular or less often slightly irregular, mostly entomophilous, trimerous or rarely dimerous or tetramerous; tepals in 2 usually similar and petaloid cycles, often very showy, occasionally (as in *Trillium* and *Calochortus*) the outer set narrower, greener, and more sepaloid; all the tepals distinct, or less often all of them joined toward the base to form a perianth-tube, this (especially in many members of the Amaryllidoideae) often bearing a corona in addition to the perianth-segments, the corona considered on ontogenetic and vascular evidence to be derived from the androecium in some genera, and from the perianth in others; nectaries most often septal, but sometimes borne at the base of the tepals or stamens instead, or seldom wanting; stamens usually as many as the tepals, seldom only 3, rarely up to 12, all free and distinct, or sometimes borne on the perianth-tube or connate by their filaments; 3 petaloid staminodes rarely alternating with 3 normal stamens; anthers tetrasporangiate and dithecal (the thecae sometimes confluent), opening by longitudinal slits or seldom by terminal pores; pollen-grains binucleate or seldom trinucleate, monosulcate, or seldom trichotomosulcate, bisulcate, tetraporate, spiraperturate, or inaperturate; gynoecium of (2) 3 (4) carpels united to form a compound, superior to less often inferior ovary with axile or seldom deeply intruded parietal placentation, capped by a solitary style with as many stigmas as carpels, or the stigma sometimes solitary and capitate, seldom (as in *Tofieldia*) with 3 distinct styles; stigmas papillate, dry or less often wet; ovules (1–) several or numerous in each locule, anatropous to sometimes campylotropous or rarely essentially orthotropous, bitegmic or rarely unitegmic, crassinucellar or seldom tenuinucellar, sometimes provided with an integumentary obturator; embryo-sac of various types; endosperm-development helobial or nuclear. FRUIT usually a loculicidal or septicidal capsule, less often a berry, seldom nutlike; seeds often flat and wind-distributed, sometimes thicker and with an aril or sarcotesta or elaisomic strophiole (of diverse ontogenetic origins in different genera); embryo more or less linear, with a terminal cotyledon and lateral plumule, centrally embedded in the copious, usually very hard endosperm, which has food-reserves of protein, oil and hemicellulose (in the thickened cell-walls), but only rarely a significant amount of starch. X = 3–27+. (Agapanthaceae, Alliaceae, Alstroemeriaceae, Amaryllidaceae, Asparagaceae, Asteliaceae, Anthericaceae, Aphyllanthaceae, Calochortaceae, Colchicaceae, Convallariaceae, Dianellaceae, Hemerocallidaceae, Hyacinthaceae, Hypoxidaceae, Herreriaceae, Melanthiaceae, Ruscaceae, Tecophilaeaceae, Tricyrtidaceae, Trilliaceae)

The family Liliaceae as here broadly defined consists of about 280 genera and nearly 4000 species, widespread throughout the world, but most abundant and varied in fairly dry, temperate to subtropical regions. The long list of segregate families bears witness to the diversity of the group, but the mutual affinity among all the members is also widely recognized. The Asparagaceae (with much-reduced leaves and evident phylloclades), Trilliaceae (with whorled, more or less net-veined leaves and green, herbaceous sepals), Amaryllidaceae and Alstroemeriaceae (both with inferior ovary, and the former with a distinctive group of alkaloids), and Alliaceae (with amaryllis-like, involucrate, umbelloid inflorescence, articulated laticifers, and superior ovary) are among the more distinctive segregate families, but precise morphological definition even of these groups is not easy. It is possible that studies now underway (such as by Clifford & Dahlgren, and by Takhtajan) will permit the reorganization of the Liliaceae into a series of smaller and more homogeneous families, but I do not feel able to undertake such a reorganization at the present time.

Many members of the Liliaceae are familiar garden or indoor ornamentals, including species of *Amaryllis, Aspidistra, Chionodoxa, Colchicum, Convallaria, Galanthus, Gloriosa, Hemerocallis, Hippeastrum, Hosta, Hyacinthus, Lilium, Muscari, Narcissus, Ornithogalum, Scilla, Trillium,* and *Tulipa. Asparagus* and *Allium* (onion, garlic, chives) are cultivated for food, and the bulbs of *Camassia* were a favorite food of western Amerindians. Many lilies, on the other hand, are highly poisonous. *Zigadenus* has the appropriate common name of death-camas, and *Veratrum* has been identified as the cause of a congenital deformity in lambs.

Pollen considered to represent the Liliaceae occurs in upper Eocene and more recent deposits, and pollen that probably represents either the Liliaceae or the Agavaceae dates from the Maestrichtian.

6. Family IRIDACEAE A. L. de Jussieu 1789 nom. conserv., the Iris Family

Perennial, mostly geophytic herbs from sympodial rhizomes, corms, or bulbs, or seldom evergreen herbs (South African genera *Aristea, Bobartia, Dietes, Pillansia,* and *Dierama*—ca 100 spp. in all) or even low half-shrubs with some secondary growth of monocotyledonous type (South African genera *Klattia, Nivenia,* and *Witsenia*—11 spp in all), rarely (3 spp. of *Sisyrinchium*) annual, very often with long, prismatic crystals of calcium oxalate in some of the cells (consistently so in the Ixioideae), but without raphides, sometimes poisonous, often tanniferous (from proanthocyanins), often saponiferous, often with one or another sort of terpenoid, often with mangiferin (a glycoxanthone), and sometimes producing

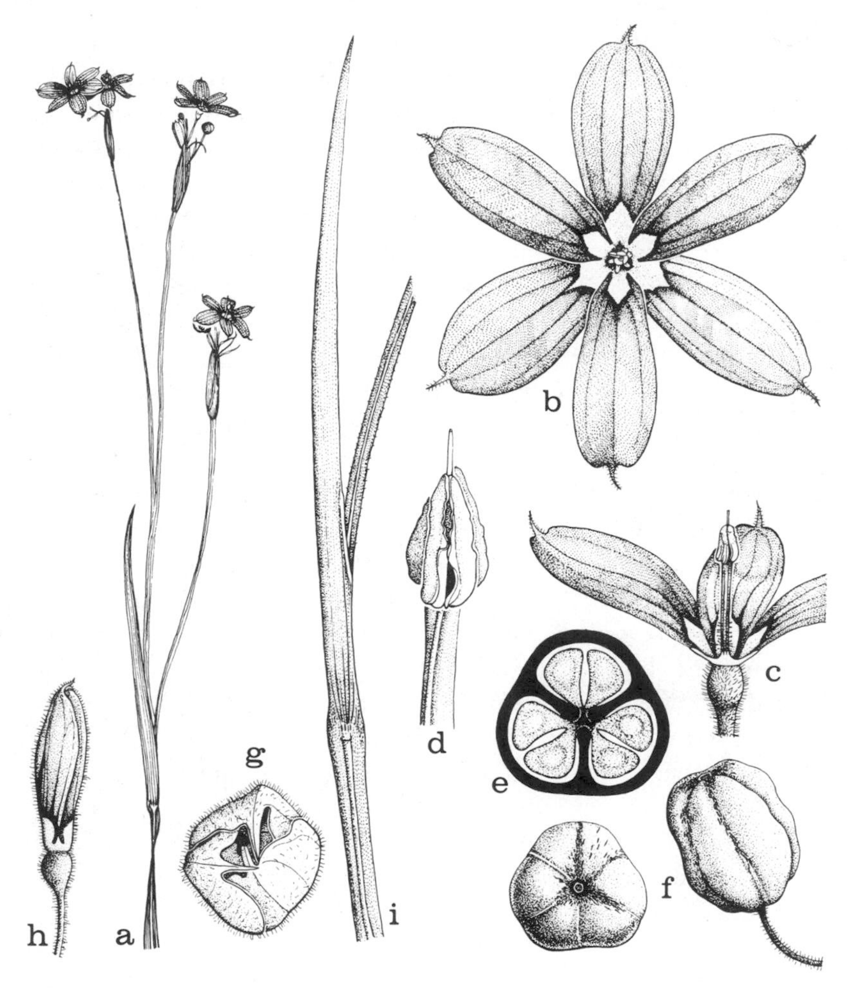

Fig. 11.5 Iridaceae. *Sisyrinchium angustifolium* Miller. a, habit, $\times\frac{1}{2}$; b, flower, from above, $\times 3$; c, cut-away view of flower, $\times 3$; d, stamens and exserted style, $\times 12$; e, schematic cross-section of ovary, $\times 18$; f, two views of fruit, $\times 3$; g, flower bud, from above, $\times 9$; h, flower in late bud, ready to open, $\times 3$; i, leaf and portion of stem, $\times 1$.

naphthaquinones or anthraquinones, but without alkaloids and only rarely cyanogenic; vegetative organs often storing starch, but sometimes fructosans instead; vessels with simple or scalariform perforations, confined to the roots, or seldom (as in *Sisyrinchium*) present in the aerial organs as well; stem with scattered, closed vascular bundles; plastids of the sieve-tubes with cuneate proteinaceous inclusions; roots commonly mycorhizal, without root hairs. LEAVES parallel-veined, narrow, mostly distichous, with an open, sheathing base and an often equitant and gladiate, unifacial blade, very rarely with a petiole and a more or less expanded blade. FLOWERS in various sorts of terminal inflorescences, typically large and showy, commonly individually subtended by a bract, and often collectively subtended by one or two expanded, bladeless sheaths forming a spathe, or sometimes terminal and solitary, in any case perfect, regular or often irregular; perianth of 6 petaloid tepals in 2 cycles, all alike, or the two cycles or the members of a single cycle more or less strongly differentiated, often all connate below to form a perianth-tube; nectaries septal, or at the base of the tepals, or on the inner tepals (as in *Tigridia*), or (*Iris*) at the base of the stamens, or (*Sisyrinchium*) wanting; stamens 3 (2 in *Diplarrhena*), opposite the outer tepals; filaments often connate below to form a tube; anthers tetrasporangiate and dithecal, extrorse, opening by longitudinal slits (by apical pores in *Cobana*) pollen-grains 2 (3)-nucleate, commonly monosulcate, seldom dicolpate or spiraperturate or inaperturate; gynoecium of 3 carpels united to form a compound, inferior, trilocular ovary with axile placentation (ovary superior in the monotypic Tasmanian genus *Isophysis*, unilocular and with parietal placentation in the monotypic European Mediterranean genus *Hermodactylus*); style terminal, 3-lobed, the branches sometimes again divided or often expanded and petaloid with the stigma on the outer side of the branch rather than at the tip; stigmas dry, papillate; ovules (1–) more or less numerous in each locule, anatropous, bitegmic and crassinucellar (tending toward a tenuinucellar condition in *Sisyrinchium*); endosperm-development nuclear. FRUIT a loculicidal capsule; seeds sometimes arillate or with a sarcotesta; embryo rather small, linear, mostly straight, with a terminal cotyledon and a lateral plumule; endosperm fleshy, with reserves of hemicellulose (in the thickened, pitted cell-walls), protein, and oil, but generally without starch. X = 3–19+. (Gladiolaceae, Hewardiaceae, Isophysidaceae, Ixiaceae)

The family Iridaceae consists of some 80 genera and about 1500 species, cosmopolitan in distribution, but most abundant and diversified in Africa; at least 900 species occur in southern Africa, south of the tropic of Capricorn. About half of the species belong to only 8 genera: *Iris* (200), *Gladiolus* (150), *Moraea* (100), *Romulea* (80), *Crocus* (75), *Watsonia* (70), *Babiana* (60), and *Sisyrinchium* (60). *Ixia* (45), *Tigridia* (10–15), and *Freesia* are also well known as ornamentals.

The Iridaceae are obviously related to but more advanced than the Liliaceae.

7. Family VELLOZIACEAE Endlicher 1841 nom. conserv., the Vellozia Family

Stout, simple or often subdichotomously branched, xerophytic shrubs, the relatively slender stem appearing very thick because of the persistent leafsheaths on the upper part and the persistent, appressed, adventitious roots on the lower part; plants without raphides, without saponins, and not cyanogenic, but commonly secreting resin or gum, and often somewhat tanniferous; vessels with simple perforations regularly present in the roots, and scalariform vessels sometimes present in the shoot (at least in the leaves); vascular bundles scattered, closed, the stem without secondary growth. LEAVES clustered at the branch-tips, firm, narrow, parallel-veined, perennial, the blade eventually deciduous from the persistent sheath; stomates paracytic or seldom tetracytic. FLOWERS solitary at the branch-tips in the axils of the leaves, on short or elongate peduncles, perfect or seldom (*Barbaceniopsis*) functionally unisexual, regular, epigynous, trimerous, small to more often large and showy; perianth of 6 petaloid tepals (in 2 cycles) that are distinct or often united into a short or seldom more elongate tube, the perianth-tube often with 6 separated or united corona-members external to and usually adnate to the stamens; stamens 6 in 2 cycles, or (*Vellozia*) in 6 bundles of 2-several; filaments free and distinct, or attached to the perianth-tube or to the tepals, sometimes with ventral basal appendages; anthers basifixed to medifixed, linear, introrse, opening by longitudinal slits; pollen-grains monosulcate or inaperturate, often in tetrads; gynoecium of 3 carpels united to form a compound, inferior, trilocular ovary with axile placentation; style slender, with a capitate or shortly trilobed stigma; placentas lamellar, extending out into the locules, distally thickened or forked; ovules very numerous, anatropous, bitegmic, pseudocrassinucellar, sometimes with a funicular obturator. FRUIT a loculicidal, often sticky capsule; seeds numerous, compressed, hard; embryo small, with a terminal cotyledon and a lateral plumule, embedded in the copious, usually very hard endosperm, this usually with food-reserves of hemicellulose (in the thickened cell-walls), protein, and oil, but sometimes the endosperm more starchy, with only limited amounts of hemicellulose, oil, and protein.

The family Velloziaceae consists of 6 genera and about 250 species, confined to South America, Africa, Madagascar, and southern Arabia. The largest genus is *Vellozia*, with about 120 species, followed by *Barbacenia*, with about 100, and *Xerophyta*, with about 30. *Barbaceniopsis* has only 3 species, and *Talbotia* and *Nanuza* are monotypic.

8. Family ALOACEAE* Batsch 1802, the Aloe Family

Stout, simple or sparingly branched shrubs or branching stout trees up to several m high, with leaves crowning the branches, or sometimes the plants with an erect, slender, woody stem and a terminal crown of leaves in the manner of palms (but the leaves not palm-like), or often coarse, short-stemmed, more or less herbaceous plants arising from a short rhizome or erect caudex, plants presumably with crassulacean acid metabolism, commonly with corky, raphide-containing cells in at least some of the tissues, and commonly producing anthraquinones and chelidonic acid, but not known to produce tannins, steroid saponins, or alkaloids; starch wanting, even from vegetative organs; vessels confined to the roots; often some of the roots tuberous-thickened; stem with secondary growth of the monocotyledonous type, the vascular bundles closed and scattered. LEAVES simple, alternate, sessile, crowded in dense rosettes at the ends of the stem and branches or at the ground-level on a short main stem, perennial and more or less strongly fleshy-succulent, often prickly along the margins, but not strongly (or not at all) spine-pointed, parallel-veined, but the veins commonly obscure; stomates mostly sunken, more or less distinctly tetracytic; vascular bundles at least typically with a large cap of wide, thin-walled cells at the phloem pole (in the same position as a cap of fibers in typical Agavaceae), the colored sap, when present, borne in these large, thin-walled cells. FLOWERS borne in a terminal spike, raceme, or panicle on a relatively slender, scape-like axillary shoot, perfect, mostly ornithophilous or entomophilous, trimerous throughout; tepals evidently in 2 cycles of 3, all petaloid and often fleshy, connivent or connate into a straight or sometimes ventricose tube (sometimes the 3 outer ones connate into a tube, but the 3 inner free and distinct), the limb regular or sometimes bilabiate; stamens 6, hypogynous, tetrasporangiate and dithecal, opening by longitudinal slits; pollen-grains monosulcate; gynoecium of 3 carpels united to form a compound, superior, trilocular ovary with axile placentation and at least usually with septal nectaries; style terminal, with a punctate or discoid to somewhat trilobed stigma; ovules numerous in each locule, orthotropous to hemitropous, bitegmic, crassinucellar. FRUIT a loculicidal capsule, rarely fleshy and berry-like (*Lomatophyllum*) but still dehiscent; seeds commonly winged or flattened; embryo straight, with a terminal cotyledon and lateral plumule, centrally embedded in the copious, very hard endosperm, this with thick, pitted cell-walls composed in large part of hemicellulose food-reserves, and

*The name might better be spelled Aloaceae. A. C. 1983.

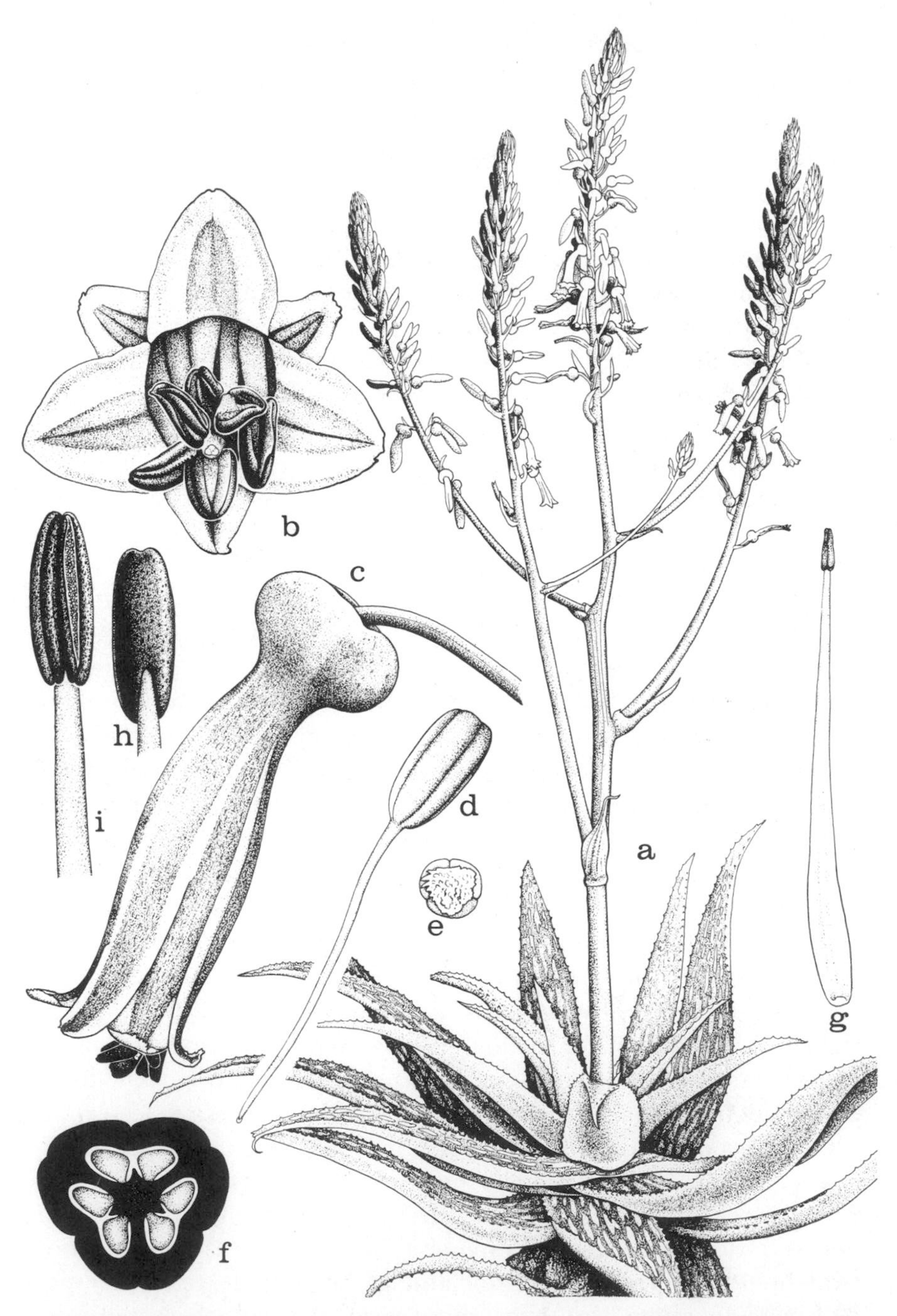

Fig. 11.6. Aloeaceae. *Aloe ammophila* Reynolds. a, habit, $\times\frac{1}{6}$; b, flower, from above, ×4; c, side-view of flower, ×2; d, pistil, ×2; e, stigma, ×24; f, schematic cross-section of ovary, ×6; g, stamen, ×2; h, i, two views of anther, ×8.

commonly also with considerable amounts of stored oil and protein. X mostly = 7, with 4 long and 3 short chromosomes.

The family Aloaceae as here defined consists of 5 genera and perhaps as many as 700 species, native to Africa, Madagascar, Arabia, and some nearby islands, but best developed in South Africa. The genera here recognized are *Aloe* (350, including *Aloinella, Chamaealoe, Guillauminia*, and *Leptaloe*), *Haworthia* (200, including *Astroloba, Chortolirion*, and *Poellnitzia*), *Gasteria* (70), *Kniphofia* (70, including *Notosceptrum*), and *Lomatophyllum* (15). These estimates of the number of species may all be unduly high, reflecting excessive splitting by enthusiastic horticulturists. A purgative called bitter aloes is made from the expressed juice of species of *Aloe*; the active ingredients are aloin and related compounds, which are anthraquinones.

I have some qualms about including *Kniphofia* in the Aloaceae. Its flowers and inflorescence are very aloeoid, but the stem lacks secondary growth (at least in the investigated species), and the leaves are not strongly succulent and hardly if at all perennial. Thus a good case could be made for referring *Kniphofia* to the Liliaceae. Regardless of which way it is thrown, *Kniphofia* is transitional between the Aloaceae and Liliaceae. The existence of such a transitional genus could be used to support the retention of the Aloaceae in the Liliaceae, but that would create problems regarding the characterization of the Agavaceae, Velloziaceae, and Xanthorrhoeaceae.

Bulbine, which sometimes approaches the Aloaceae in aspect, is here retained in the Liliaceae.

The Aloaceae and Agavaceae are parallel derivatives from the Liliaceae. The differences between the two groups are rather slight, but they do suggest that the two originated from different parts of the ancestral Liliaceae. Thus it seems better to treat the two derived groups as separate families than to combine them into one. Reasons for keeping the Aloaceae and Agavaceae out of the Liliaceae are discussed under the Agavaceae.

9. Family AGAVACEAE Endlicher 1841 nom. conserv., the Century-Plant Family

Stout, simple or sparingly branched, often arborescent shrubs or even trees, seldom scandent, or often coarse, short-stemmed, more or less herbaceous plants arising from a short rhizome or erect caudex, at least sometimes with crassulacean acid metabolism, commonly with raphides

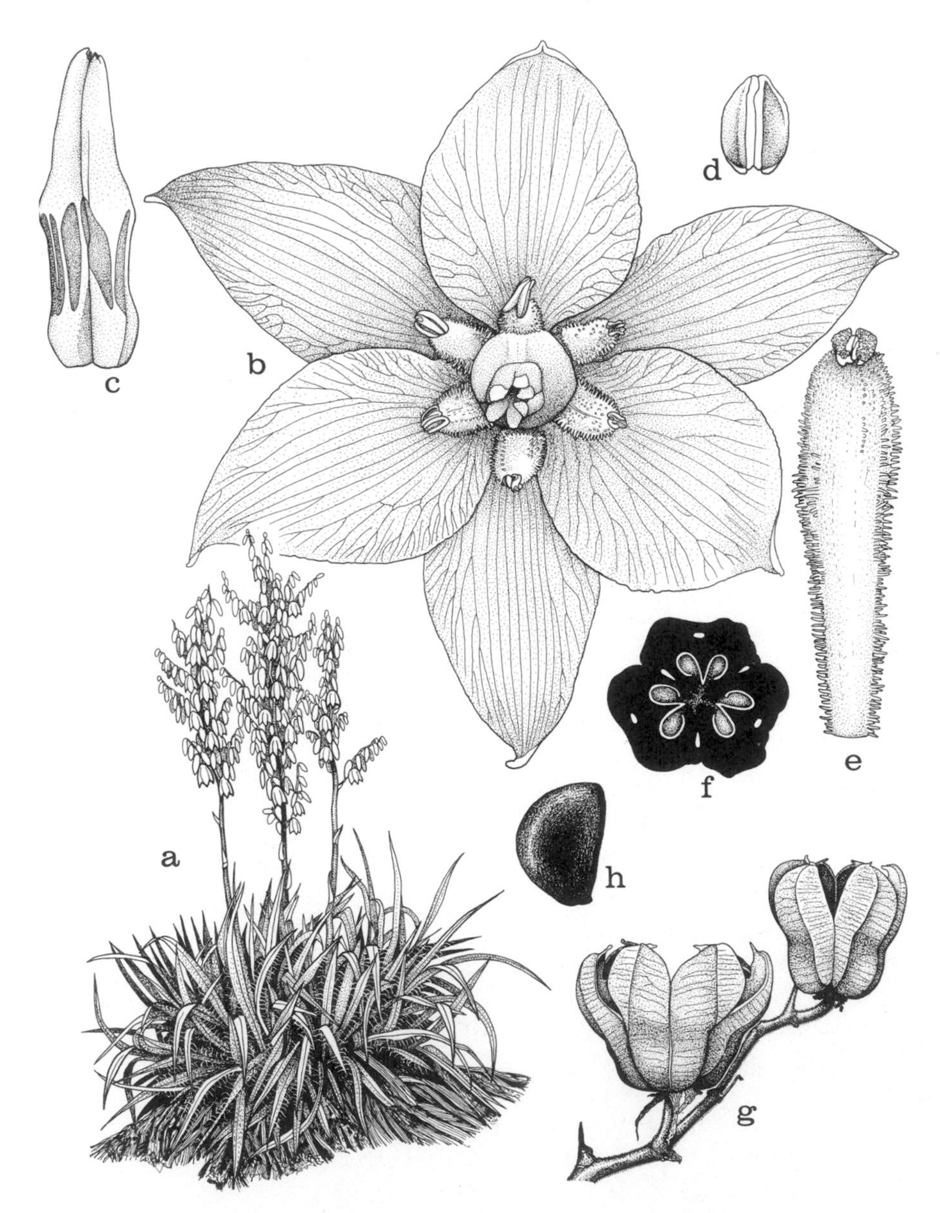

Fig. 11.7 Agavaceae. *Yucca filamentosa* L. a, habit, $\times \frac{1}{16}$; b, flower, from above, $\times 2$; c, pistil, $\times 2$; d, anther, $\times 4$; e, stamen, $\times 4$; f, schematic cross-section of ovary, $\times 4$; g, fruits, $\times 1$; h, seed, $\times 3$.

and/or pseudoraphides and chelidonic acid, and often accumulating steroid saponins, but only seldom tanniferous (from proanthocyanins), rarely if ever cyanogenic, and not known to produce alkaloids; food-reserves in vegetative organs consisting largely of polyfructosans, without starch; vessels with simple or scalariform perforations, mostly confined to the roots, or also present in the leaves but not in the stems; stem commonly with secondary growth of monocotyledonous type, the vascular bundles closed and scattered. LEAVES simple, alternate, sessile, tending to be crowded in dense rosettes at the ends of the stem and branches, or at the ground-level on a short main stem, perennial and more or less strongly thickened, but usually leathery or firm-succulent rather than softly succulent, often prickly on the margins and often firmly spine-pointed, parallel-veined, but the veins commonly obscure; stomates anomocytic or often tetracytic, or (*Doryanthes*) paracytic; vascular bundles with a well developed fibrous cap at the phloem pole. FLOWERS borne in rather massive, often dense racemes or panicles or heads terminating the stems, or axillary and subterminal, perfect or sometimes unisexual (the plants then polygamo-dioecious to dioecious), variously pollinated by insects, birds, or bats, trimerous throughout; tepals in 2 cycles of 3, all petaloid, often thick and fleshy, distinct or connate below to form a short or more or less elongate tube; stamens 6, the filaments distinct, adnate to the base of the tepals or to the perianth-tube; anthers mostly dorsifixed (basifixed in *Doryanthes*), tetrasporangiate and dithecal, opening by longitudinal slits; pollen-grains binucleate, monosulcate, or sometimes trichotomosulcate or with 2 distal pores; gynoecium of 3 carpels united to form a compound, superior or inferior, trilocular ovary (the primary locules sometimes, as in spp. of *Yucca*, partly or completely divided in half by an ingrowth from the carpellary midvein) with axile placentation (unilocular and with parietal placentation in *Dasylirion*), generally with septal nectaries, or seldom the flower without nectaries; style terminal, with 3 stigmas, or the stigmas sometimes sessile; ovules (1–) several to numerous in each locule, anatropous to sometimes campylotropous, bitegmic, crassinucellar; endosperm-development helobial. FRUIT a loculicidal capsule or a berry; seeds flattened; embryo straight, linear, with a terminal cotyledon and lateral plumule, centrally embedded in the copious, very hard endosperm, this with thick, pitted cell-walls composed in large part of hemicellulose food-reserves, and commonly also with considerable amounts of stored oil and protein, but without starch; perisperm sometimes (*Yucca*) present around the endosperm. X = 16–30+, with a few large and more numerous small chromosomes. (Dracaenaceae, Nolinaceae)

The family Agavaceae as here defined consists of about 18 genera and nearly 600 species, native to warm, mostly arid regions in both the Old World and the New. Only a few species occur in distinctly temperate climates. The largest genus by far is *Agave*, with about 300 species, followed by *Dracaena* (80), *Sansevieria* (50), *Yucca* (35), and *Nolina* (30). The remaining genera have fewer than a hundred species in all.

The leaves of *Agave* and some other genera provide fiber for cordage, and the thickened stem-base in *Agave* is used in Mexico to make the alcoholic beverages pulque and tequila. *Agave* is famous for its monocarpic habit. The plants live for several or many years with a cluster of fleshy-firm leaves at the ground-level and without an obvious stem. Then a nearly naked, coarse, flowering stem grows up very rapidly and produces a massive terminal inflorescence. The plant dies as the seeds mature. *Yucca* is noteworthy for its dependence on the moth *Tegeticula* (*Pronuba*) for pollination. The female moth lays eggs in the ovary, and then carefully pollinates the flower.

Pollen attributed to the Agavaceae is known from Eocene and more recent deposits.

Yucca, with a superior ovary, and *Agave*, with an inferior ovary, share an unusual karyotype with 5 large and 25 small chromosomes. This similarity, noted by McKelvey and Sax in 1933, led to the revival and redefinition of the Agavaceae as a widely accepted family. *Yucca* had previously been referred to the Liliaceae, and *Agave* to the Amaryllidaceae. The karyotype now seems less important, because some species of *Hosta*, an otherwise very different genus of the Liliaceae, resemble *Yucca* and *Agave* in this respect, whereas some of the other genera referred to the Agavaceae on habital grounds do not (e.g., n = 19 in *Nolina*, 24 in *Doryanthes*). The combination of a few large and more numerous small chromosomes is in fact not uncommon in the lilioid monocotyledons, but it does not correlate well enough with other features to be of critical taxonomic importance.

The only real basis for maintaining the Agavaceae as a family distinct from the Liliaceae is the specialized growth-habit. Removal of the Agavaceae from the Liliaceae implies also the removal of the Aloeaceae, since these two groups differ from the ancestral Liliaceae in essentially the same way. On the other hand, inclusion of these two groups in the Liliaceae would eliminate most of the justification for keeping the

Velloziaceae as a family, and if all these are included in the Liliaceae, then the familial status of the Xanthorrhoeaceae also becomes shaky.

Acceptance of these several segregate families does not solve all the taxonomic problems, however. It is not at all certain that either the Agavaceae or the Xanthorrhoeaceae as now constituted form a natural group. *Phormium* and *Doryanthes*, for example, are now usually included in the Agavaceae, but they may not properly belong with the other genera. On serological grounds *Yucca* and *Agave* seem to stand apart from a group consisting of *Dracaena, Nolina, Sansevieria,* and probably *Cordyline,* but the serological results do not suggest any distinctly better arrangement of these genera with regard to each other or to the Liliaceae (data from Chupov & Kutjavina, 1978, in Russian). In another very brief paper, these authors and N. S. Morozova report a close serological linkage of *Hosta, Agave,* and *Yucca,* without presenting the supporting data.

The temperate North American genus *Xerophyllum* is a monocarpic perennial with firm, perennial basal leaves, and on that basis it might be a candidate for inclusion in the Agavaceae, but it is almost certainly more closely allied to *Tofieldia,* which it resembles cytologically (X = 15) and in having separate styles. *Tofieldia* is considered to be one of the more archaic genera of the Liliaceae, having nothing to do with the Agavaceae.

The difference in leaf-anatomy between the Agavaceae and Aloaceae, as noted in the descriptions, holds true at least for large numbers of species in the larger genera. Its consistency as a distinguishing character for the families can only be determined by further study.

10. Family XANTHORRHOEACEAE Dumortier 1829 nom. conserv., the Blackboy Family

Stout, simple or sparingly branched, often somewhat arborescent shrubs, or coarse, short-stemmed or acaulescent herbs, arising from a thick, woody caudex or a short rhizome, commonly with scattered raphide-cells, commonly producing chelidonic acid, and at least sometimes producing naphthaquinones, but not cyanogenic, and not known to produce saponins or alkaloids; vessels mostly confined to the roots, but present also in the leaves of *Xanthorrhoea* and *Acanthocarpus* (these scalariform), mostly with simple perforations, seldom scalariform; stem commonly (not always) with secondary growth of monocotyledonous type, the vascular bundles closed and scattered, mostly amphivasal (U-shaped in *Kingia*). LEAVES simple, alternate, often sheathing, linear or narrowly oblong, xeromorphic, firm and often pungent, parallel-veined; old leaf-bases commonly persistent on the stem; stomates anomocytic or less often paracytic. FLOWERS mostly small and borne in spikes, heads, or mixed panicles (large, solitary, and surrounded by radical leaves in *Baxteria*, small and terminating leafy shoots in *Calectasia*), perfect or sometimes unisexual (and in *Lomandra* mostly dioecious), trimerous, individually not showy; tepals in 2 cycles of 3, all distinct, or the inner ones seldom connate at the base, all dry and chaffy, or the inner ones rarely colored; stamens 6, the outer set free, the inner commonly adnate to the base of the inner tepals; anthers basifixed, or dorsifixed and versatile, dithecal, opening by longitudinal slits; pollen-grains monosulcate or bisulcate, or sometimes zonosulcate or spiraperturate or even pantocolpate; gynoecium of 3 carpels united to form a compound, superior, trilocular ovary with axile or axile-basal placentation, or seldom the ovary unilocular with 3 basal ovules; style wth 3 stigmas, or the styles 3; ovules 1-many in each locule, hemitropous to almost campylotropus or reputedly sometimes anatropous, bitegmic, crassinucellar. FRUIT a loculicidal capsule, or sometimes a 1-seeded nut; embryo straight (rarely curved), with a terminal cotyledon and lateral plumule, centrally embedded in the copious, horny endosperm, this with food-reserves of protein, oil, and often also hemicellulose, the latter in the thick, pitted cell-walls. X = 7, 8, 11, 17, 24.

The family Xanthorrhoeaceae is restricted to Australia, Tasmania, New Guinea, and New Caledonia. There are 9 genera: *Lomandra* (30), *Xanthorrhoea* (15), *Chamaexeros* (3), *Dasypogon* (2), *Acanthocarpus* (1), *Baxteria* (1), *Calectasia* (1), *Kingia* (1), and *Romnalda* (1). The group may be derived from the Agavaceae, from which it differs most notably in its small, chaffy flowers. It is possible that the several genera do not all properly belong together. *Lomandra, Chamaexeros, Acanthocarpus*, and *Romnalda*, at least, do appear to be mutually related.

11. Family HANGUANACEAE Airy Shaw 1965, the Hanguana Family

Robust, erect, perennial herbs of moist or wet places, arising from a short, stout rhizome, and sometimes with elongate, creeping or floating stolons; aerial parts beset with short, multicellular, branching hairs set in pits, and with many tanniferous cells (containing proanthocyanins);

vessels wanting; rhizome with a well developed endodermis separating the cortex from the central cylinder, which has scattered, closed fibrovascular bundles. LEAVES mostly basal, with a long basally expanded and sheathing petiolar base, the blade rolled up from one side to the other in bud (as in the Zingiberales), and with pinnate-parallel venation (almost exactly as in *Orchidantha*, of the Lowiaceae), the rather numerous, more or less parallel primary lateral veins diverging at intervals from the strong midrib at a narrow angle, converging again distally, but not all reaching the tip, some of them joining with adjacent veins, and some ending at a cross-vein; stomates tetracytic. FLOWERS small, unisexual (the plant dioecious), regular, trimerous, borne in a large, conspicuously bracteate, terminal panicle that contains scattered silica-cells; perianth of 6 persistent tepals in 2 cycles of 3, these small, somewhat chaffy, greenish or yellowish, the inner often a little larger than the others and sometimes red-dotted; stamens 6; filaments adnate to the base of the tepals; anthers small, basifixed, dithecal, opening by longitudinal slits; pollen-grains inaperturate, with thin exine; pistillate flowers with 6 antherless staminodes, 3 smaller and 3 larger; gynoecium of 3 carpels united to form a compound, trilocular, superior ovary with axile placentation; stigma sessile, broad, triangular-discoid; a single pendulous, nearly orthotropous ovule with a funicular obturator present in each locule. FRUIT drupaceous, with 1–3 seeds; seeds with a thick testa, copious endosperm, and small embryo.

The family Hanguanaceae consists of the single genus *Hanguana*, with 1 or 2 species, native to Malesia and Ceylon. The systematic position of *Hanguana* is debatable. In recent systems it has most often been associated with or included in the Flagellariaceae. On palynological and anatomical features it is aberrant in the Flagellariaceae and resembles *Lomandra*, in the Xanthorrhoeaceae. Further study is needed to clarify the position of both genera.

12. Family TACCACEAE Dumortier 1829 nom. conserv., the Tacca Family

Perennial herbs from solid, starchy, tuberous-thickened, globose or elongate, vertical or horizontal rhizomes, with well developed raphide-cells in the stem and leaves, but not tanniferous and not cyanogenic; vessels restricted to the roots, their perforations scalariform, with numerous crossbars; stem with scattered, closed vascular bundles, without secondary growth. LEAVES all basal, large, long-petiolate, with entire and broadly elliptic to much-dissected blade, the venation parallel or palmate and with a secondary network; stomates anomocytic; mesophyll with scattered mucilage cells as well as raphide sacs. FLOWERS borne in an involucrate, cymose umbel atop a scape, perfect, regular, epigynous, trimerous, without nectar, or with more or less well developed septal nectaries; tepals 6, all dark-colored and more or less petaloid, often connate toward the base into a short tube with lobes in 2 alternating whorls; stamens 6, in 2 cycles, attached to the perianth-tube or to the base of the tepals; filaments short, flat and somewhat petaloid, forming, together with a broad connective, a sort of hood over the inflexed anther; anthers tetrasporangiate and dithecal, opening by longitudinal slits; pollen-grains binucleate, monosulcate, with thin exine; gynoecium of 3 carpels united to form a compound, unilocular, 6-ribbed, inferior ovary with more or less intruded parietal placentas; style very short, the 3 stigmatic branches often petaloid and incurved; stigmas dry, papillate, ovules numerous, anatropous to campylotropous, bitegmic, crassinucellar; endosperm-development nuclear. FRUIT a berry or seldom a loculicidal capsule; seeds 10-numerous; embryo small but well differentiated, with a lateral cotyledon and terminal plumule; endosperm copious, more or less cartilaginous, with reserves of protein and fat, but without starch and with little or no hemicellulose. N = 15.

The family Taccaceae consists of the genus *Tacca* (including *Schizocapsa*), with about 10 species, pantropical in distribution, but best developed in southeast Asia and Polynesia. The tubers of *Tacca leontopetaloides* (L.) Kuntze are the source of an arrowroot starch.

13. Family STEMONACEAE Engler in Engler & Prantl 1887 nom. conserv., the Stemona Family

Erect herbs or herbaceous vines or low shrubs from tuberous roots or creeping rhizomes, commonly producing lactone alkaloids of a unique type, at least sometimes saponiferous, and often (not in *Croomia*) with raphides or other sorts of calcium oxalate crystals, but not cyanogenic and not tanniferous; vessels with scalariform perforations, confined to the roots or sometimes extending also into the stem; vascular bundles of the stem closed, borne in one ring or 2 adjacent rings; no secondary growth. LEAVES variously alternate, opposite, or whorled, petiolate, the broad blade with 5–11 (–15) curved-convergent main veins and numerous transverse secondary veins; stomates anomocytic. FLOWERS in few- to many-flowered axillary cymes, or solitary and axillary, sometimes ill-scented, without nectaries (or at least without septal nectaries), perfect or seldom unisexual, regular, dimerous; perianth of 4 petaloid to sepaloid tepals in 2 series, distinct or seldom connate at the base; stamens 4, in 2 series or all in one series; filaments distinct or nearly so; anthers tetrasporangiate and dithecal, the connective conspicuously prolonged in *Stemona*; pollen-grains monosulcate, binucleate; gynoecium of 2 carpels united to form a compound, unilocular, superior or seldom (*Stichoneuron*) half-inferior ovary (by another interpretation the gynoecium has only a single carpel); stigmas sessile or subsessile; ovules 2-many, basal or apical, anatropous, bitegmic, crassinucellar; endosperm-development nuclear. FRUIT a bivalved capsule; seeds elongate, with an appendage on the hilum and sometimes also toward the chalazal end; embryo small, with a terminal cotyledon and lateral plumule; endosperm copious, with food-reserves of protein and fat and sometimes hemicellulose, but the cell-walls not pitted even when thickened; some starch commonly present around the cotyledon. X = 7. (Croomiaceae, Roxburghiaceae)

The family Stemonaceae consists of 3 genera and about 30 species. *Stemona* (*Roxburghia*) has about 25 species, *Croomia* 3, and *Stichoneuron* 2. Except for one species of *Croomia* that is native to southeastern United States, all species of Stemonaceae are confined to eastern Asia, Malesia, and northern Australia.

14. Family SMILACACEAE Ventenat 1799 nom. conserv., the Catbrier Family

Climbing, herbaceous or slenderly woody vines, or less often erect herbs or branching shrubs, arising from creeping, often tuber-bearing or tuberous-thickened, starchy rhizomes, generally with raphides, commonly producing steroid saponins, and sometimes somewhat tanniferous (from proanthocyanins), but not known to be cyanogenic or to produce alkaloids; vessels with simple or scalariform perforations, commonly present in all aerial vegetative organs as well as in the roots, seldom in the roots and stems only, rarely (as in *Eustrephus* and *Petermannia*) confined to the roots; stems leafy, commonly branched, often with recurved prickles, sometimes twining, espcially in the genera without tendrils; vascular bundles scattered, closed; no secondary growth; roots mycorhizal, without root-hairs. LEAVES alternate or sometimes (as in *Rhipogonum*) opposite, generally petiolate, and commonly (though not in some of the smaller genera) with a pair of tendrils arising

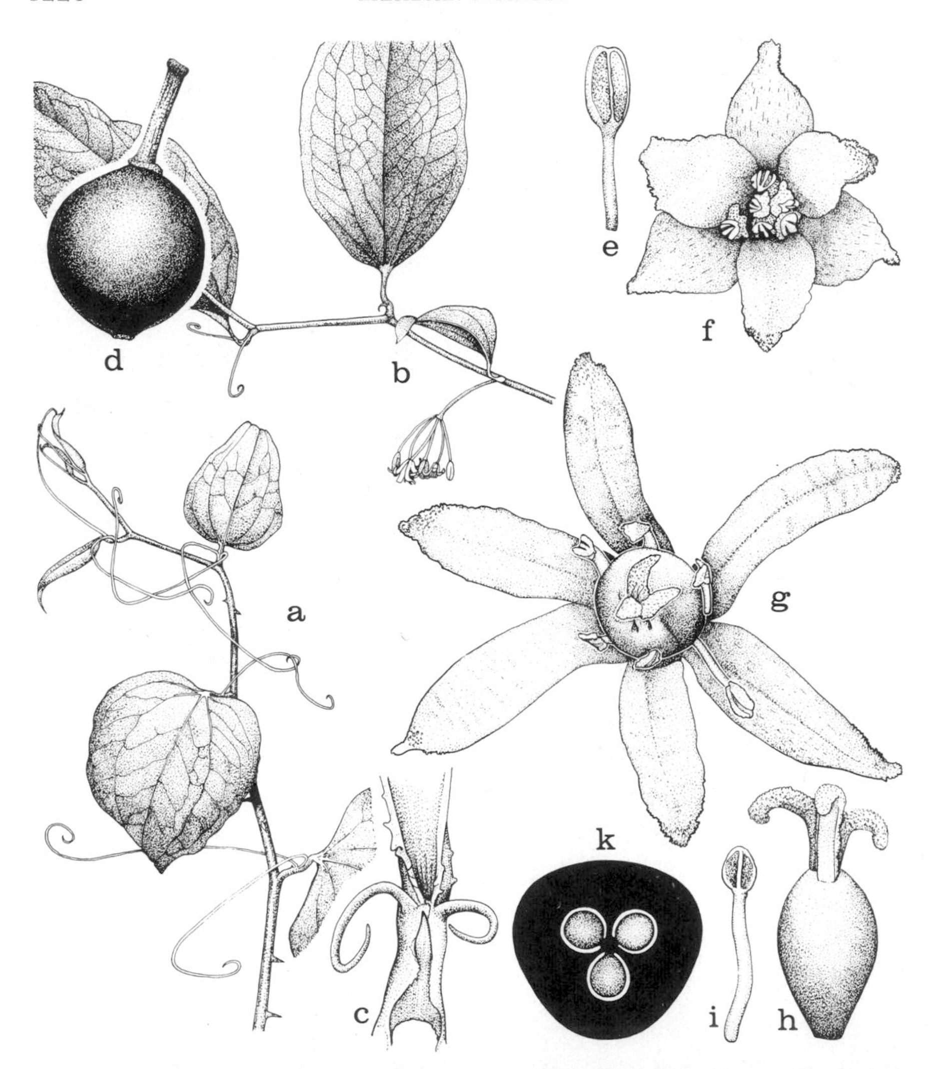

Fig. 11.8 Smilacaceae. *Smilax hispida* Muhl. a, b, habit, ×½; c, base of leaf, with stipular tendrils, ×4; d, fruit, ×4; e, stamen, ×8; f, staminate flower, ×8; g, pistillate flower, ×8; h, pistil, ×8; i, stamen from pistillate flower, ×8; k, schematic cross-section of ovary, ×16.

from the petiole near its junction with the short stipular flange or open sheath; blade well developed and expanded, commonly with 3–7 "parallel," curved-convergent main veins connected by an evident network of smaller veins, in *Petermannia* with several pinnate-parallel primary lateral veins arising at intervals from the midrib, but still connected by an evident network of smaller veins; stomates anomocytic;

leaf-opposed, branching tendrils present in *Petermannia*, which lacks the petiolar tendrils; petioles firm and somewhat hook-curved in some of the nontendriliferous genera, such as *Rhipogonum*. FLOWERS most commonly in axillary umbels, or racemes or spikes of umbels, or in some of the smaller genera in various other sorts of axillary or terminal inflorescences (these leaf-opposed in *Petermannia*), or solitary and axillary, individually small in *Smilax*, but larger and more showy in some of the other genera, regular, trimerous, sometimes foetid, unisexual (the plants dioecious) in *Smilax*, but perfect in some of the other genera; pistillate flowers commonly with staminodia, but staminate flowers without a vestigial pistil; tepals basically 6 in 2 series, generally all alike and more or less petaloid, distinct or shortly connate below, in *Heterosmilax* united to form a tubular or urceolate, 6-toothed or 3-toothed perianth; stamens most commonly 6, seldom more numerous (up to 18) or only 3; filaments free and distinct, or sometimes adnate to the tepals or perianth-tube, or in *Heterosmilax* and some spp. of *Smilax* more or less connate into a column, nectaries commonly present at the inner base of the tepals, or at the base of the stamens or staminodes, but septal nectaries wanting; anthers tetrasporangiate and dithecal (the thecae eventually confluent in *Smilax*), opening by longitudinal slits; pollen-grains binucleate, monosulcate or sometimes trichotomosulcate or inaperturate; gynoecium generally of 3 carpels united to form a compound, superior (inferior in *Petermannia*), trilocular or sometimes unilocular ovary, the placentation accordingly axile or parietal; styles 3 and distinct, or connate only at the base, or sometimes the style solitary and more elongate, with a capitate or 3-lobed stigma; in *Smilax pumila* the gynoecium apparently monomerous, with a single carpel and locule and a single style; ovules 1-many in each locule or on each placenta, bitegmic, crassinucellar, anatropous or hemitropous to campylotropous or more often orthotropous; endosperm-development nuclear. FRUIT a berry with 1–3 (less often numerous) seeds, the pericarp rarely eventually becoming dry and dehiscent; embryo mostly small, with a terminal cotyledon and lateral plumule, axially embedded in the very hard endosperm, which has food-reserves of lipid, protein, and hemicellulose (the cell-walls commonly thickened and often pitted), and often also some starch. X = 10, 13–16+. (Lapageriaceae, Luzuriagaceae, Petermanniaceae, Philesiaceae, Rhipogonaceae)

The family Smilacaceae, as here broadly defined, consists of about a dozen genera, widespread in tropical and subtropical regions, most diversified in the Southern Hemisphere, but also well represented in parts of the North Temperate zone. The family is dominated by the large genus *Smilax*, with about 300 species. *Heterosmilax* is variously estimated at 6 to 15 species, *Rhipogonum* at 7, *Luzuriaga* at 3, and *Pseudosmilax* at 2. The genera *Behnia, Elachanthera, Eustrephus, Geitonoplesium, Lapageria, Petermannia*, and *Philesia* are monotypic. The starchy tubers of *Smilax* are edible, and have been used in primitive cultures. The flavoring agent sarsaparilla, which is chemically related to the steroid saponins, is obtained from the tubers of several tropical species of *Smilax*.

Iljinskaja and Stefyrtsa (1971) consider that fossil leaves of *Smilax* and *Dioscorea* occur in Miocene deposits in Moldavia (USSR). *Smilax* pollen is recorded from the upper Miocene of Mexico.

The Smilacaceae have traditionally been included in the Liliaceae, where they are aberrant in their mostly vesseliferous shoot, climbing, mostly tendriliferous, often woody habit, and petiolate leaves with broad, net-veined blade. The habit, leaves, and vasculature of the Smilacaceae suggest the Dioscoreaceae, and thus the inclusion of the Smilacaceae in the Liliaceae tends to undercut the distinctions between the Liliaceae and Dioscoreaceae. Although the Smilacaceae are not so sharply set off from the Liliaceae as are the Dioscoreaceae, I think it is conceptually more useful to associate them with the Dioscoreaceae (as a distinct family) than to force them into the Liliaceae.

Petermannia, with an inferior ovary as in the Dioscoreaceae, is perhaps the most distinctive of the genera here assigned to the Smilacaceae. It is clearly a part of the smilacaceous alliance, however, and I see nothing to be gained by establishing a separate family for a single species, the affinities of which are not in serious dispute.

15. Family DIOSCOREACEAE R. Brown 1810 nom. conserv., the Yam Family

Twining-climbing or seldom erect, sometimes spiny herbs from a fleshy-thickened, starchy rhizome, or much more often from a large, basal "tuber" (derived from the lowest internode(s) of the stem and/or the hypocotyl) that may continue to thicken year after year; plants generally with raphides (in mucilaginous idioblasts), commonly producing steroid saponins and often accumulating lactone alkaloids and chelidonic acid, often tanniferous (from proanthocyanins), but only seldom cyanogenic; vessels present in all vegetative organs (or sometimes wanting from the leaf-blades and rhizomes), at least sometimes with scalariform perfo-

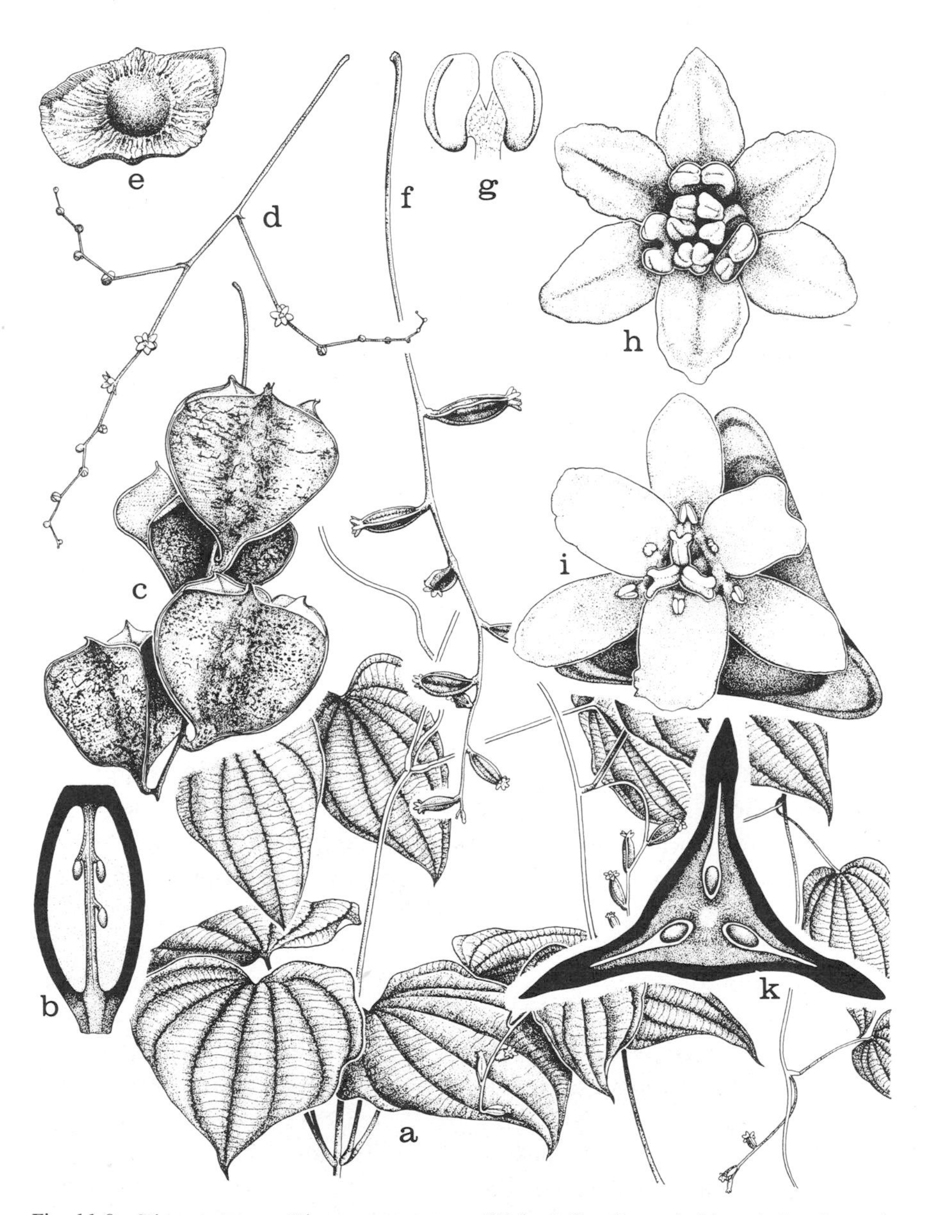

Fig. 11.9 Dioscoreaceae. *Dioscorea quaternata* (Walter) Gmelin. a, habit, $\times\frac{1}{2}$; b, schematic long-section of ovary, $\times 12$; c, fruits, $\times 1$; d, staminate inflorescence, $\times 1$; e, seed, $\times 2$; f, pistillate inflorescence, $\times 1$; g, stamen, $\times 24$; h, staminate flower, $\times 12$; i, pistillate flower, $\times 12$; k, schematic cross-section of ovary, $\times 12$.

rations; vascular bundles of the stem closed, relatively few, in 2 dissimilar cycles, or in a single cycle of 2 alternating types; no secondary growth except in the basal tuber; plastids of the sieve-tubes with cuneate proteinaceous inclusions; nodal anatomy complex; roots commonly mycorhizal, without root-hairs. LEAVES alternate or rarely opposite, generally with a distinct blade and petiole, the petiole commonly twisted or jointed at the base and often with a stipule-like flange; blade broad, often cordate at the base, entire to less often palmately lobed or cleft or even compound, commonly with 3–13 "parallel," curved-convergent main veins and a network of smaller veins, often with embedded nectaries or mucilaginous pits, some of which harbor nitrogen-fixing bacteria; stomates mostly anomocytic, seldom of other types. FLOWERS small, variously in racemes, spikes, or panicles, unisexual (the plants dioecious) or much less often perfect, consistently regular, epigynous, and trimerous; tepals 6, all about alike, petaloid or somewhat bract-like, mostly connate at the base into a short tube; both septal and tepalar nectaries commonly present; stamens usually 6 in 2 cycles, but the inner set sometimes staminodial or obsolete; filaments distinct or shortly connate at the base, attached to the base of the perianth-tube; anthers tetrasporangiate and dithecal, opening by longitudinal slits, often well separated on the broad connective; pollen-grains binucleate, monosulcate to less often disulcate or trichotomosulcate, or (*Avetra*) 4-porate or even 5-porate; gynoecium of 3 carpels united to form a compound, inferior, trilocular ovary with axile placentation; styles distinct, or connate below, with separate stigmas; ovules 2 (-many) in each locule, anatropous, bitegmic, crassinucellar; endosperm-development nuclear. FRUIT capsular, often triangular and 3-winged, or rarely (*Tamus*) a berry, or indehiscent and samaroid (*Trichopus* and *Rajania*); seeds mostly winged; embryo small but well differentiated, with a subterminal plumule and a broad, lateral cotyledon (sometimes with a vestigial second cotyledon), axially embedded in the copious, very hard endosperm, which has food-reserves of protein, oil, and hemicellulose, the latter in the thickened cell-walls. X = 9(?), 10, 12, 14+. (Cladophyllaceae, Stenomeridaceae, Tamaceae, Trichopodaceae)

The family Dioscoreaceae as here defined consists of about 6 genera, widespread in tropical and subtropical regions, with relatively few species extending into the North Temperate zone. The family is dominated by the large genus *Dioscorea*, with perhaps 600 species. The other genera, including *Avetra* (1), *Rajania* (20), *Stenomeris* (2), *Tamus* (4–5), and *Trichopus* (1), have scarcely 30 species in all. Several species of *Dioscorea* have long been cultivated in tropical regions especially in Africa, for their edible, starchy "tubers." More recently *Dioscorea* has excited medical interest as a source of raw materials for the manufacture of cortisone, a steroid.

Stenomeris and *Trichopus* stand somewhat apart from the remainder of the family and from each other. Both have perfect flowers, as does also *Avetra*, in contrast to the unisexual flowers of the remaining genera, and *Trichopus* has a short stem with only a single leaf. The relationship of these genera to *Dioscorea* is not in dispute, however, and I see nothing to be gained by establishing trivial satellite families for them.

Fossils of Eocene and later age have been credibly assigned to the Dioscoreaceae. The genus *Dioscorea* may well antedate the wide separation of South America from Africa.

2. Order ORCHIDALES Bromhead 1838

Strongly mycotrophic, sometimes achlorophyllous herbs, without secondary growth; plants very often with raphides in some of the cells; vessels usually confined to the roots, seldom also present in the stem; vascular bundles of the stem scattered, closed; plastids of the sieve-tubes with cuneate proteinaceous inclusions. LEAVES alternate, or seldom opposite or whorled, sometimes all basal, parallel-veined, simple, entire, with a usually closed sheath at the base, or sometimes reduced to mere scales; stomates paracytic or less often anomocytic, seldom tetracytic. FLOWERS regular to highly irregular, epigynous, basically trimerous (but often with only one or two stamens), perfect or seldom unisexual, most commonly entomophilous, seldom ornithophilous, rarely autogamous; tepals generally in 2 series and all petaloid, distinct or united below to form a perianth-tube, or the outer series seldom greener and more sepaloid, sometimes 2 or all 3 of the sepals connate; stamens in nearly all of the Orchidaceae solitary or only 2, and united to the style, but in the other families 3 or 6 and free from the style; anthers tetrasporangiate and dithecal; pollen-grains binucleate or seldom trinucleate, uniaperturate or inaperturate, or without an exine; gynoecium of 3 carpels united to form a compound, inferior ovary, most commonly unilocular and with parietal placentation, but sometimes trilocular and with axile placentation; nectaries of diverse sorts; ovules very numerous and tiny, anatropous, bitegmic (seldom unitegmic), tenuinucellar; endosperm-development variously nuclear, cellular, or helobial, or the endosperm-nucleus not formed or not dividing. FRUIT capsular; seeds very numerous and tiny, without endosperm; embryo minute, often composed of only a few cells, undifferentiated or with a weakly differentiated cotyledon.

The order Orchidales as here defined consists of 4 families of very unequal size. The total number of species in the 3 smaller families (less than 150) is much less than the probable error in estimating the number of species of Orchidaceae (15,000 or more).

The Orchidales differ from the Liliales essentially in their strongly mycotrophic habit, and in their very numerous, tiny seeds with a minute, mostly undifferentiated embryo and no endosperm. The reduction of the embryo is at least in part a consequence of mycotrophy; these features are also associated in other groups of angiosperms. The large number of seeds, their small size, and the lack of endosperm are presumably due to other factors. At least there are many other mycotrophic angiosperms with seeds of ordinary size and a well developed endosperm. The ovary in the Orchidales is always inferior, and only seldom has typical septal nectaries, although the ovarian nectaries of some Burmanniaceae and Orchidaceae may well be derived from septal nectaries.

The combination of mycotrophy and numerous, tiny seeds offers certain evolutionary opportunities as well as imposing some limitations. The plants are physiologically dependent on their fungal symbionts, sometimes even for food, sometimes only for other factors as yet not fully understood, but in any case they can grow only where the fungal symbiont finds suitable conditions. The dust-like seeds of the Orchidaceae are admirably adapted to being carried by wind and lodging in the bark of trees, and many orchids are epiphytes. The production of many ovules is of course of no value if the ovules do not get fertilized. One way to increase the likelihood of fertilization is to offer special attractions to a limited set of potential pollinators, and to have the pollen-grains stick together in masses so that many are transported at once. This strategy puts the plants in thrall to their pollinators, but it opens the door to explosive speciation.

Only the Orchidaceae have efficiently exploited the evolutionary opportunities of the order. The floral characters that distinguish the Orchidaceae from their immediate allies clearly reflect progressive specialization for massive transfer of pollen by specific pollinators.

The Orchidales are evidently derived from the Liliales. All the characters in which the Orchidales differ from the Liliales represent evolutionary advances, i.e., they are phyletically acropetal. Within the Liliales, only the epigynous segment of the Liliaceae has the characters from which those of the Orchidaceae, Burmanniaceae, and Corsiaceae might well have arisen. Although the Orchidaceae never have more than 3 stamens, 2 of these are considered to come from the ancestral inner cycle, and one from the outer. Thus a hexandrous ancestry seems likely. The Geosiridaceae, with 3 antesepalous stamens, might have arisen from the Iridaceae rather than directly from the Liliaceae. I have expounded elsewhere the concept that absolute monophylesis is unattainable in a reasonable taxonomic system.

SELECTED REFERENCES

Abe, K. 1972. Contributions to the embryology of the family Orchidaceae. VII. A comparative study of the orchid embryo sac. Sci. Rep. Tôhoku Imp. Univ., ser. 4, Biol. 36: 179–201.

Ames, O. 1946. The evolution of the orchid flower. Amer. Orchid Soc. Bull. 14: 355–360.
Ames, O. 1948. Orchids in retrospect; a collection of essays on the Orchidaceae. Botanical Museum of Harvard Univ. Cambridge, Mass.
Barber, H. N. 1942. The pollen-grain division in the Orchidaceae. J. Genetics 43: 97–103.
Dodson, C. H. 1962. The importance of pollination in the evolution of the orchids of tropical America. Amer. Orchid Soc. Bull. 31: 525–534; 641–649; 731–735.
Dressler, R. L. 1961. The structure of the orchid flower. Missouri Bot. Gard. Bull. 49: 60–69.
Dressler, R. L., & C. H. Dodson. 1960. Classification and phylogeny in the Orchidaceae. Ann. Missouri Bot. Gard. 47: 25–68.
Garay, L. A. 1960, 1972. On the origin of the Orchidaceae. Bot. Mus. Leafl. 19: 57–96, 1960. II. J. Arnold Arbor. 53: 202–215, 1972.
Jonker, F. P. 1938. A monograph of the Burmanniaceae. Meded. Bot. Mus. Herb. Rijks Univ. Utrecht 51: 1–279.
Jonker, F. P. 1939. Les Géosiridacées, une nouvelle famille de Madagascar. Recueil Trav. Bot. Néerl. 36: 473–479.
Kores, P., D. A. White, & L. B. Thien. 1978. Chromosomes of *Corsia* (Corsiaceae). Amer. J. Bot. 65: 584–585.
Pai, R. M. 1966. Studies in the floral morphology and anatomy of the Burmanniaceae. I. Vascular anatomy of the flower of *Burmannia pusilla* (Wall. ex Miers) Thw. Proc. Indian Acad. Sci. B63: 301–308.
Poddubnaya-Arnoldi, V. A. 1967 (1968). Comparative embryology of the Orchidaceae. Phytomorphology 17: 312–320.
Rao, V. S. 1969. Certain salient features in the floral anatomy of *Burmannia, Gymnosiphon* and *Thismia*. J. Indian Bot. Soc. 48(1–2): 22–29.
Rao, V. S. 1974. The relationships of the Apostasiaceae on the basis of floral anatomy. J. Linn. Soc., Bot. 68: 319–327.
Schmid, R., & M. J. Schmid. 1973. Fossils attributed to the Orchidaceae. Amer. Orchid Soc. Bull. 42: 17–27.
Swamy, B. G. L. 1948. Vascular anatomy of orchid flowers. Bot. Mus. Leafl. 13: 61–95.
Swamy, B. G. L. 1949. Embryological studies in the Orchidaceae. I. Gametophytes. II. Embryogeny. Amer. Midl. Naturalist 41: 184–201; 202–232.
Vermeulen, P. 1966. The system of the Orchidales. Acta Bot. Neerl. 15: 224–253.
Williams, N. H. 1979. Subsidiary cells in the Orchidaceae: their general distribution with special reference to development in the Oncidieae. J. Linn. Soc., Bot. 78: 41–66.
Withner, C. L., ed. 1959. The Orchids: A scientific survey. New York, The Ronald Press. (Vol. 32 of Chronica Botanica.)

Терёхин, Э. С., & О. П. Камелина. 1969. Эндосперм Orchidaceae (к вопросу о редукции). Бот. Ж. 54: 657-666.

SYNOPTICAL ARRANGEMENT OF THE FAMILIES OF ORCHIDALES

1 Stamens 3 or 6, symmetrically arranged and free from the style; pollen-grains not cohering in pollinia; terrestrial

plants, most species with reduced leaves and without chlorophyll.

2 Flowers regular or only moderately irregular; stamens 3 or 6.

3 Stamens 3, opposite the sepals .... 1. GEOSIRIDACEAE.

3 Stamens 6, or more often 3 and opposite the petals ....2. BURMANNIACEAE.

2 Flowers highly irregular, the upper sepal large, cordate-ovate, and enclosing the other 5 tepals in bud; stamens 6 .... 3. CORSIACEAE.

1 Stamen usually solitary, seldom the stamens 2, or rarely 3, in any case all on the same side of the flower; pollen-grains usually but not always cohering in pollinia; terrestrial or very often epiphytic plants, with or less often without chlorophyll; flowers with few exceptions strongly irregular .... 4. ORCHIDACEAE.

1. Family GEOSIRIDACEAE Jonker 1939 nom. conserv., the Earth-Iris Family

Small, mycotrophic herbs without chlorophyll, with a slender, scaly rhizome and simple or branched aerial stems. LEAVES alternate, reduced and scale-like. FLOWERS solitary or in small, open, bracteate cymes, perfect, regular, epigynous, trimerous; tepals 6, biseriate, all petaloid, connate at the base into a short perianth-tube, the outer series (sepals) imbricate, the inner series (petals) convolute; stamens 3, attached toward the base of the perianth-tube opposite the sepals; filaments short, distinct; anthers basifixed, extrorse, tetrasporangiate and dithecal; pollen-grains uniaperturate; gynoecium of 3 carpels united to form a compound, inferior, trilocular ovary; style terminal, with 3 short branches, each ending in a flat stigma; ovules very numerous on strongly branched, axile placentas. FRUIT a triangular-obconic capsule with an annulus at the truncate summit; seeds numerous, minute.

The family Geosiridaceae consists of the single genus and species *Geosiris aphylla* Baillon, native to Madagascar and some other islands of the Indian Ocean. Botanists are agreed that it is allied to the Iridaceae, on the one hand, and the Burmanniaceae, on the other. Jonker's proposal that *Geosiris* should form a monotypic family is now widely accepted.

2. Family BURMANNIACEAE Blume 1827 nom. conserv., the Burmannia Family

Small, mycotrophic, annual or perennial herbs, often from rhizomes or tubers, with or much more often without chlorophyll, sometimes with raphides in some of the cells, the chlorophyllous spp. sometimes with scalariform vessels in the roots, stems, and leaves. LEAVES alternate, commonly colorless to yellowish or reddish and reduced to scales, but sometimes green and better developed, simple, entire, linear or lanceolate and clustered around the base of the stem; stomates mostly anomocytic. FLOWERS in terminal cymes or racemes, or solitary and terminal, perfect, regular or sometimes somewhat irregular, epigynous; perianth corolloid, tubular or campanulate, 6-lobed or rarely 3-lobed, the tube sometimes 3-angled or 3-winged; sepals (outer perianth-lobes) valvate, the petals mostly smaller and induplicate-valvate, or sometimes wanting; 3 or all 6 of the perianth-lobes often with an elongate, slender, terminal appendage; stamens 6, or often only 3 and then opposite the petals, borne on the perianth-tube, sessile or on short filaments, sometimes connate by their anthers to form a tube around the style; anthers tetrasporangiate and dithecal, opening by transverse lateral slits or introrse longitudinal slits, the pollen-sacs often well separated on an expanded and frequently ornate or appendiculate connective; pollen-grains monosulcate to uniporate, or inaperturate, binucleate or trinucleate; gynoecium of 3 carpels united to form a compound, inferior, trilocular or unilocular ovary (sometimes trilocular below and unilocular above) with accordingly axile or parietal placentation, the parietal placentas sometimes deeply intruded; nectary-glands apical on the ovary, or internal near the summit as septal nectaries or wanting; ovules very numerous and small, anatropous, bitegmic, tenuinucellar; endosperm-development cellular or helobial or nuclear. FRUIT capsular (seldom fleshy), often winged, circumscissile, or opening irregularly, or sometimes dehiscent by valves; seeds very numerous and tiny, with scanty or virtually no endosperm; embryo tiny and undifferentiated, often consisting of only 4–10 cells when the seed is shed. X = 6, 8. (Tripterellaceae, Thismiaceae)

The family Burmanniaceae as here defined consists of about 20 genera and 130 species, of pantropical distribution, with only a few species in temperate regions. Nearly half of the species belong to the single genus *Burmannia* (60), and the genera *Ptychomeria* (20) and *Thismia* (25) together make up another third of the family.

3. Family CORSIACEAE Beccari 1878, the Corsia Family

Mycotrophic perennial herbs from rhizomes or tubers, without chlorophyll, at least sometimes saponiferous. LEAVES alternate, reduced to fairly large scales. FLOWERS solitary and terminal, perfect or unisexual, strongly irregular, epigynous; perianth corolloid, tubular below, with 6 lobes in 2 cycles; posterior (upper, adaxial) perianth-segment large, colored, sometimes with a large gland toward the base within, more or less cordate-ovate, enclosing the other 5 segments in bud, these elongate, mostly linear-spatulate, eventually pendulous; stamens 6; filaments short, distinct, borne on the perianth-tube; anthers tetrasporangiate and dithecal, extrorse, opening by longitudinal slits; pollen-grains monosulcate; gynoecium of 3 carpels united to form a compound, inferior, unilocular ovary with more or less strongly-intruded, bifurcate placentas; style short, with 3 thick stigmas; ovules numerous, tiny. FRUIT capsular, dehiscent by 3 valves; seeds very numerous and small, with tiny, undifferentiated embryo and very scanty endosperm. X = 9. (Arachnitidaceae, Achratinitaceae)

The family Corsiaceae consists of 2 genera and about 9 species. *Arachnitis* (*Achratinis*) is a monotypic Chilean genus with unisexual flowers. The remaining species of the family belong to the genus *Corsia*, which has perfect flowers and is restricted to New Guinea. Botanists are agreed that the Corsiaceae are allied to the Burmanniaceae.

4. Family ORCHIDACEAE A. L. de Jussieu 1789 nom. conserv., the Orchid Family

Strongly mycotrophic, terrestrial or often epiphytic perennial herbs, sometimes (as in *Vanilla*) climbing, usually green and photosynthetic, but some terrestrial spp. without chlorophyll, and 3 Australian spp. wholly subterranean; plants apparently always with raphides in some of the cells, and often with mucilage-cells, commonly with crassulacean acid metabolism, and often producing one or another sort of alkaloid, but only occasionally saponiferous or somewhat tanniferous, and not cyanogenic; vegetative organs commonly storing starch, the grains sometimes compound; vessels with scalariform or simple perforations, often present in the roots, seldom also in the stems, rarely even in the leaves; plastids of the sieve-tubes with cuneate proteinaceous inclusions; epiphytic spp. with modified aerial roots, the epidermis proliferated into a spongy, water-absorbing *velamen* usually several cells thick (the cells dead, air-filled when dry); stem-base (or the whole stem) especially in epiphytic spp. often thickened to form a pseudobulb, but terrestrial spp. commonly rhizomatous or with corms or tubers; vascular bundles of the stem scattered, closed, without secondary growth. LEAVES alternate (sometimes distichous) or seldom opposite or whorled, sometimes all basal, or reduced to mere scales (notably in the spp. without chlorophyll), simple and entire, variously convolute or conduplicate in aestivation, parallel-veined, often somewhat fleshy, sheathing at the base, the sheath nearly always closed; stomates paracytic or less often anomocytic, seldom tetracytic. FLOWERS borne in racemes, spikes, or panicles, generally individually subtended by a bract, or sometimes solitary in the leaf-axils or on a scape, perfect or seldom unisexual (the plants then monoecious or dioecious), epigynous, commonly but not always resupinate (twisted in ontogeny so that the morphologically adaxial side appears to be abaxial), usually very strongly irregular and bilaterally symmetrical, but virtually regular in the Apostasioideae; perianth typically of 6 tepals in 2 series, all petaloid, or the sepals sometimes greener and more foliaceous in texture; sepals all alike, or

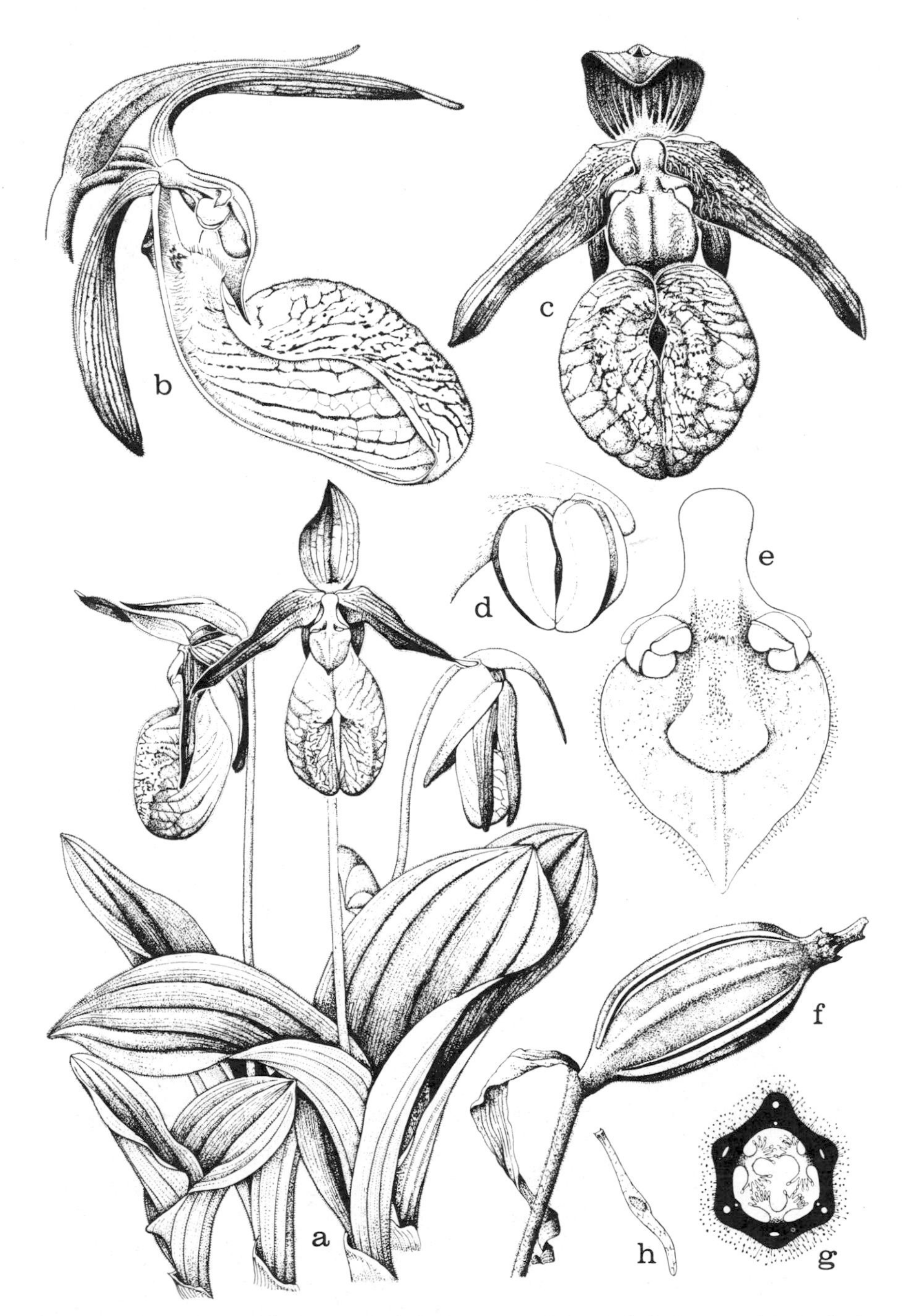

Fig. 11.10 Orchidaceae. *Cypripedium acaule* Aiton. a, habit, $\times\frac{1}{2}$; b, flower, with the lip in long-section, $\times 1$; c, front view of flower, $\times 1$; d, fertile stamen, $\times 4$; e, column, as seen from beneath, showing 2 fertile stamens, stigma, and dilated staminode, $\times 2$; f, fruit, $\times 1$; g, schematic cross-section of ovary, after fertilization, with developing ovules, $\times 4$; h, seed, $\times 12$.

the median one (morphologically abaxial, but seemingly adaxial) somewhat unlike the other two in form or color, or two or all three sepals connate; petals 3, the median one (morphologically adaxial, but seemingly abaxial) strongly differentiated from the others (except in the Apostasioideae), commonly larger and differing in form and also in color, forming a lip (*labellum*), the two lateral petals commonly but not always more or less similar to the sepals; nectaries of various sorts, sometimes forming a hollow spur produced backward or downward from the base of the lip, sometimes at the sepal-tips, sometimes a cup within the perianth atop the ovary, sometimes embedded in the ovary wall and opening to the summit (modified septal nectaries), or sometimes of still other forms; extrafloral nectaries sometimes present on the pedicels, bracts, or leaf-sheaths; stamen usually one, seldom two, rarely three, all on the morphologically abaxial (but usually seemingly adaxial) side of the flower, on the opposite side from the lip; stamen when solitary fully adnate to the style, forming a *column* (*gynostemium*), this solitary stamen being the morphologically abaxial member of the ancestral outer cycle, all the other stamens being suppressed (the two adjacent members of the inner cycle commonly forming a part of the column, as shown by the vascular anatomy, or seldom represented by staminodes); stamens when three (*Neuwiedia*) with the filaments adnate to the style only at the base, when two sometimes also free from the style for most of their length (*Apostasia*), but sometimes (Cypripedioideae) largely adnate to the style as in the unistaminate species, the third stamen then commonly represented by an expanded staminode; anther(s) tetrasporangiate and dithecal, the two thecae sometimes widely separated, opening by longitudinal slits; pollen-grains binucleate, uniaperturate or sometimes biaperturate (reported to be triporate in *Vanilla*!), monadinous and with a thin exine in the Apostasioideae and Cypripedioideae, but tetradinous, virtually without exine, and organized into pollinia in the Orchidoideae, the members of a tetrad connected by plasmodesmata, and the tetrads united by acellular, elastic (viscin) threads derived from the tapetum; pollinia 1–6 in each pollen-sac, sometimes subdivided into smaller massulae; one or the other end of a pollinium often prolonged into a slender tip (caudicle); gynoecium of 3 carpels united to form a compound, inferior, unilocular (trilocular in Apostasioideae) ovary with expanded parietal placentas (axile in Apostasioideae); style column (gynostemium) typically stout and terminated by the solitary anther, which is subtended by an enlarged rostellum derived from part or all of the adjacent stigma-lobe; rostellum separating the anther from the other two (often connate) stigma-lobes and commonly preventing self-pollination; caudicles of the pollinia often attached to the rostellum, a portion of which comes off as a sticky pad (*viscidium*) with the pollinia when the pollen is transferred; in the most advanced spp. the pollinia attached to a caudicle with a gland at the tip; in diandrous and triandrous genera the stigma wet, papillate, terminal on the style or style-column and more or less evidently 3-lobed; column often with a basal outgrowth, the column-foot, to which the lip (and often the lateral sepals) are jointed; ovules very numerous and tiny, commonly not developing until after pollination, anatropous, bitegmic (seldom unitegmic) and tenuinucellar; endosperm-development nuclear, but terminating after the production of only 2–4 (–10) nuclei, after which the endosperm degenerates, or the endosperm-nucleus degenerating without dividing, or often an endosperm-nucleus not formed. FRUIT mostly capsular, opening by 3 (6) longitudinal slits, but remaining closed at the top and bottom; seeds very numerous (from more than a thousand to several million) and tiny; embryo minute, sometimes consisting of only a few cells, morphologically undifferentiated or seldom with a poorly differentiated cotyledon; endosperm wanting; association with an appropriate fungus ordinarily necessary for normal germination. X = 6–29+. (Apostasiaceae, Cypripediaceae, Limodoraceae, Neottiaceae, Thyridiaceae, Vanillaceae)

The family Orchidaceae as here broadly (and traditionally) defined, consists of up to 1000 genera and 15–20,000 species; some estimates run as high as 30,000 species. The only other family of comparable size is the Asteraceae. In terms of total biomass, on the other hand, the Orchidaceae are probably not one of the largest families. They must surely be outweighed by the grasses, and probably even by such relatively small families as the Fagaceae.

When the Orchidaceae are (as here) taken in the broad sense, the family is absolutely sharply limited. As with the Asteraceae, there is not one genus or species whose status as a member of the family is in any doubt.

Orchids are cosmopolitan in distribution, but most abundant and diversified in tropical forests, where many of them are epiphytes. The family is characterized by numerous diverse, often bizarre specializations for pollination by particular species of insects (or sometimes birds or bats), but pollination can occasionally be effected by a "wrong" pollinator. The adaptations to specific pollinators are among the principal barriers to interbreeding, and vigorous, fertile hydrids between different genera are often readily produced if the pollen is artificially transferred. Many genera of the Orchidaceae are cultivated for their spectacular flowers, and *Vanilla* is the source of a well known flavoring material.

The largest genera of Orchidaceae are *Dendrobium* and *Bulbophyllum*, reputedly with about 1500 species each, followed by *Pleurothallis*, with about 1000, and *Epidendrum*, with about 800. *Cattleya*, prized for corsages, is one of the smaller genera, with only about 60 species. These several genera are all tropical, epiphytic members of the subfamily Orchidoideae.

The Orchidaceae have no significant fossil record. Neither the pollen nor the flowers are well adapted to fossilization, and fossil leaves and stems of orchids might be hard to distinguish from those of various other monocots.

The Orchidaceae consist of 3 well marked subfamilies of very unequal size. The Apostasioideae have only 2 genera, *Apostasia* (including *Adactylus*, 10), and *Neuwiedia* (10). The Cypripedioideae have 4 genera, *Cypripedium* (50), *Paphiopedilum* (50), *Phragmipedium* (10), and *Selenipedium* (3). All of the remaining genera belong to the Orchidoideae as here defined, but some authors divide the Orchidoideae into several smaller and less well characterized subfamilies such as the Neottioideae and Kerosphaeroideae. The three subfamilies here recognized may be characterized as follows:

1 Stamens 2 or 3; all 3 stigma-lobes equally developed; no rostellum; pollen-grains in monads, and not aggregated into pollinia.

  2 Perianth only slightly or scarcely irregular, without a saccate lip; placentation axile; style-column short, the free part of the style more elongate; pollen-grains not cohering .... APOSTASIOIDEAE.

  2 Perianth strongly irregular, with a deeply saccate lip; placentation parietal; style-column more elongate, the free part of the style short; pollen-grains sticky, tending to cohere irregularly .... CYPRIPEDIOIDEAE.

1 Stamen 1; one of the stigma-lobes, or part of it, modified into a rostellum; pollen-grains in tetrads, and aggregated into pollinia; perianth strongly irregular; placentation parietal .... ORCHIDOIDEAE.

INDEX

Names of accepted taxa at the rank of family and higher are in boldface type. Subfamilies, tribes, and taxonomic or nomenclatural synonyms of higher taxa are in roman type, as are names of authors mentioned in the text. All generic names are in italic type, whether accepted or not. Page numbers in italics indicate illustrations.